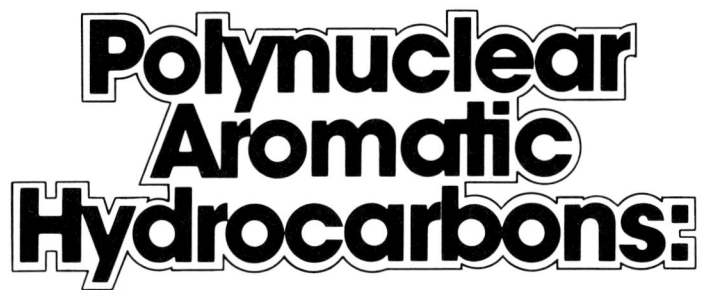

Chemistry and Biological Effects

# Polynuclear Aromatic Hydrocarbons:
## Chemistry and Biological Effects

Alf Bjørseth
Analytical and Environmental Chemistry
Battelle's Columbus Laboratories

and

Anthony J. Dennis
Biomedical Sciences
Battelle's Columbus Laboratories

**Fourth International Symposium**
**Sponsored by**

U.S. Environmental Protection Agency
Battelle Memorial Institute
Battelle's Columbus Laboratories
Electric Power Research Institute

Columbus, Ohio

**Library of Congress Cataloging in Publication Data**

**International Symposium on Polynuclear Aromatic Hydrocarbons, 4th, Battelle Columbus Laboratories, 1979.**

Polynuclear aromatic hydrocarbons.
"Fourth International Symposium sponsored by U.S. Environmental Protection Agency . . . (et al.)"
Includes bibliographies and index.
1. Hydrocarbons—Toxicology—Congresses.
2. Carcinogens—Congresses.
3. Aromatic compounds—Toxicology—Congresses.
4. Polycyclic compounds—Toxicology—Congresses.
5. Hydrocarbons—Analysis—Congresses.
6. Environmentally induced diseases—Congresses.
I.   Bjorseth, Alf.
II.  Dennis, Anthony J., 1948—
III. United States. Environmental Protection Agency.
IV.  Title.

RC 268. 7.H9I57 1979    616.99'4071    80-17877
ISBN 0-935470-05-0

**Copyright © 1980 Battelle Memorial Institute**

No part of this book may be reproduced in any form or by any electronic or mechanical means, including information storage and retrieval devices or systems, without prior written permission from the publisher, except that brief passages may be quoted for reviews.

**Preface**

More than any other single class of compound, the polynuclear aromatic hydrocarbons (PAH) have become synonymous with environmentally related hazardous products. The formation of PAH results from the incomplete combustion of hydrocarbons and are naturally present in many forms of vegetation and fossil fuel. Thus, many species of PAH are produced as a result of coal and oil combustion for commercial energy generation, use of various types of gasoline and diesel fuels, refuse burning power plants, wood burning stoves, and many other forms of effluents resulting from personal and commercial energy production.

The PAH are among the earliest classes of compounds to be demonstrated as carcinogens in man in classic studies describing tumors in chimney sweeps and coke-oven workers. Subsequently, many PAH species have been shown to be carcinogenic, mutagenic, and/or co-carcinogenic in animals, apparently as a result of oxidation of the PAH, aromatic ring.

Thus controlling emissions for such a ubiquitous and hazardous class of materials is of great importance to government regulatory agencies, combustion-related industries, and the general populace of individual energy consumers. The prerequisites for this control are a complete knowledge of the chemistry of the formation of PAH, the ability to detect PAH in as sensitive and cost-effective fashion as possible and finally, a complete grasp of the fate and activity of PAH in biological systems including man.

It is the intent of this continuing series of International Symposia to provide a forum for the presentation of the latest efforts in these areas. The chemistry papers include important new and sensitive methods of detecting PAH in real world effluents and new methods of separation of closely related species. The biologically oriented papers provide a large volume of information on PAH metabolite formation, the carcinogenic activity of substituted PAH and the application of several new approaches for the detection of the biological effects of PAH at the cellular level.

The Fourth International Symposium on Polynuclear Aromatic Hydrocarbons, on which this volume is based, was held at Battelle's Columbus Laboratories in October 1979. This series will continue in 1980 with the annual symposium and as in the past, we intend that these meetings will again provide a valuable forum and resource for leading scientists working in the area.

**The Editors**

## Contributors

**S. Ali**
Department for Environmental
 Studies
University of Illinois
Urbana, Illinois 61801

**A. Amin**
Naylor Dana Institute for
 Disease Prevention
American Health Foundation
Valhalla, New York 10595

**A. W. Andren**
Water Chemistry Program
University of Wisconsin
Madison, Wisconsin 53706

**M.S. Arnott**
The University of Texas
 System Cancer Center
M. D. Anderson Hospital
 and Tumor Institute
Houston, Texas 77030

**S. W. Ashurst**
Department of Biochemistry
University of Surrey
Guildford GU2 5XH
Surrey, England

**H. A. Autrup**
Human Tissue Studies
Laboratory of Experimental
 Pathology
National Cancer Institute
Bethesda, Maryland 20205
 and
Institute of Cancer Research
Columbia University
College of Physicans and
 Surgeons
New York, New York 10032

**W. M. Baird**
The Wistar Institute of
 Anatomy and Biology
36th Street at Spruce
Philadelphia, Pennsylvania 19104

**W. L. Banwart**
Department of Agronomy
Univerity of Illinois
Urbana, Illinois 61801

**D. O. Bassett**
Calgon Analytical Laboratories
Calgon Corporation
Pittsburgh, Pennsylvania 15230

**D. Basu**
Center for Chemical Hazard
 Assessment
Syracuse Research Corporation
Syracuse, New York 13210

**U. Bayer**
Institut de Recherches
Scientifiques sur le Cancer
F-94800 Ville juif, France

**V. Bedenko**
Naylor Dana Institute for
 Disease Prevention
American Health Foundation
Valhalla, New York 10595

**E. P Benditt**
Department of Pharmacology
 and Pathology
School of Medicine
University of Washington
Seattle, Washington 98195

**D. R. Bevan**
Gray Freshwater Biological
  Institute and Department
  of Biochemistry
University of Minnesota
P. O. Box 100
Navarre, Minnesota 55392

**C.A.H. Bigger**
Chemical Carcinogenesis Program
National Cancer Institute Frederick
  Cancer Research Center
Frederick, Maryland 21701

**A. Bjørseth**
Battelle Columbus Laboratories
505 King Avenue
Columbus, Ohio 43201
  and
Central Institute for Industrial
  Research
Blindern, Oslo 3
Norway

**J. A. Bond**
Department of Pharmacology
  and Pathology
School of Medicine
University of Washington
Seattle, Washington 98195

**W. J. Bonnett**
Organic Analytical Research Division
Center for Analytical Chemistry
National Bureau of Standards
Washington, DC 20234

**R. K. Boutwell**
McArdle Labortory for Cancer
  Research
University of Wisconsin
Madison, Wisconsin 53706

**D. Brash**
School of Medicine
The Ohio State University
Columbus, Ohio 43210

**C. Bruce**
Ministry of Agriculture, Food,
  and Fish
Toory Research Station
135 Abbey Road
Aberdeen, Scotland

**M. K. Buening**
Department of Biochemistry
  and Drug Metabolism
Hoffman-LaRoche, Incorporated
Nutley, New Jersey 07110

**D. Busbee**
Department of Biology/Genetics
  Center
North Texas State University
Denton, Texas 76203
  and
Veterans Administration Hospital
Houston, Texas 77030

**L. M. Calle**
Department of Chemistry
Ohio University
Athens, Ohio 45701

**F. Cannova**
Department of Chemical Engineering
  and Chemistry
New Jersey Institute of Technology
Newark, New Jersey 07102

**E. Cantrell**
Pharmacology Department
Texas College of Osteopathic
  Medicine
Fort Worth, Texas 76107

**R. L. Carpenter**
Inhalation Toxicology Research
  Institute
Lovelace Biomedical and
  Environmental Research Institute
P.O. Box 5890
Albuquerque, New Mexico 87115

**J. H. Carver**
Biomedical Sciences Division
Lawrence Livermore Laboratory
Livermore, California 94550

**R. N. Castle**
Department of Chemistry
Brigham Young University
Provo, Utah 84602

**J. E. Caton**
Analytical Chemistry Division
Oak Ridge National Laboratory
Oak Ridge, Tennessee 37830

**E. Cavalieri**
Eppley Institute for Research
 in Cancer
University of Nebraska Medical
 Center
42nd and Dewey Avenue
Omaha, Nebraska 68105

**F. D. Cazer**
Radiochemistry Laboratory
Comprehensive Cancer Center
The Ohio State University
Columbus, Ohio 43210

**J. M. Charpentier**
Institut Francais Du Pétrole
1 et 4 Avenue de Bois Préau
92506 Rueil Malnaison, France

**R. L. Chang**
Department of Biochemistry and
 Drug Metabolism
Hoffman-LaRoche, Incorporated
Nutley, New Jersey 07110

**O. Chaude**
Institut Francais De Pétrole
1 et 4 Avenue de Bois Préau
92506 Rueil Malnaison, France

**M. W. Chou**
Department of Pharmacology
School of Medicine
Uniformed Services
University of the Health
 Services
Bethesda, Maryland 20014

**G. M. Cohen**
Department of Biochemistry
University of Surrey
Guildford GU2 5XH
Surrey, England

**A. H. Conney**
Department of Biochemistry
 and Drug Metabolism
Hoffman-LaRoche, Incorporated
Nutley, New Jersey 07110

**M. C. Cronen**
University of Tennessee
Oak Ridge Graduate School
 of Biomedical Sciences
 and Biology Division
Oak Ridge National Laboratory
Oak Ridge, Tennessee 37830

**J. P. Crowley**
Battelle Columbus Laboratories
505 King Avenue
Columbus, Ohio 43201

**S. M. D'Ambrosio**
School of Medicine
The Ohio State University
Columbus, Ohio 43210

**F. B. Daniel**
Division of Medicinal Chemistry
College of Pharmacy
The Ohio State University
Columbus, Ohio 43210

and

Environmental Protection Agency
Cincinnati, Ohio 45268

**G. H. Daub**
Department of Chemistry
University of New Mexico
Albuquerque, New Mexico 87131

**A. J. Dennis**
Battelle Columbus Laboratories
505 King Avenue
Columbus, Ohio 43201

**J. DiGiovanni**
McArdle Laboratory for Cancer
 Research
University of Wisconsin
Madison, Wisconsin 53706

**A. Dipple**
Chemical Carcinogenesis Program
National Cancer Institute Frederick
 Cancer Research Center
Frederick, Maryland 21701

**M. A. Drum**
Battelle Columbus Laboratories
505 King Avenue
Columbus, Ohio 43201

**C. S. Dudney**
Health and Safety Research Division
Oak Ridge National Laboratory
Oak Ridge, Tennessee 37830

**R. U. Dumaswala**
The Wistar Institute of Anatomy
 and Biology
36th Street at Spruce
Philadelphia, Pennsylvania 19104

**B. P. Dunn**
Environmental Carcinogenesis Unit
British Columbia Cancer Research
 Center
601 W. 10th Avenue
Vancouver, British Columbia
Canada V5Z 1L3

**J. P. Durand**
Institut Francais De Pétrole
1 et 4 Avenue de Bois Préau
92506 Rueil Malmaison, France

**U. Edlund**
Department of Organic Chemistry
Chemical Institute
University of Umeå
S-901 87 Umeå Sweden

**D. Elam**
Chemistry Department
Auburn University
Auburn, Ohio 36830

**L. E. Ellis**
Department of Chemistry
Ohio University
Athens, Ohio 45701

**J. Ezike**
Scientific Research Laboratory
Ford Motor Company
Dearborn, Michigan 48121

**T. J. Facklam**
Battelle Columbus Laboratories
505 King Avenue
Columbus, Ohio 43201

**J. S. Felton**
Biomedical Sciences Division
Lawrence Livermore Laboratory
Livermore, California 94550

**P. P. Fu**
National Center for Toxicological
 Research
Jefferson, Arkansas 72079

**P. S. Furcinitti**
Health and Safety Research
 Division
Oak Ridge National Laboratory
Oak Ridge, Tennessee 37830

**E. Fujimori**
Department of Fine Structure
Boston Biomedical Research
 Institute
20 Staniford Street
Boston, Massachusetts 02114

**R. B. Gammage**
Health and Safety Research
 Division
Oak Ridge National Laboratory
Oak Ridge, Tennessee 37830

**E. S. Gibson**
DOFASCO Limited
Hamilton, Ontario
Canada L8N 3J5

**R. E. Gibson**
Department of Radiology
School of Medicine
The Ohio State University
Columbus, Ohio 43210

**P. Giorgio**
Department of Chemical Engineering
 and Chemistry
New Jersey Institute of Technology
Newark, New Jersey 07102

**D. J. Gmur**
National Oceanic and Atmospheric
 Administration
2725 Montlake Boulevard East
Seattle, Washington 98112

**A. Greenberg**
Department of Chemical Engineering
 and Chemistry
New Jersey Institute of Technology
Newark, New Jersey 07102

**W. H. Griest**
Analytical Chemistry Division
Oak Ridge National Laboratory
Oak Ridge, Tennessee 37830

**A. C. Griffin**
The University of Texas
System Cancer Center
M. D. Anderson Hospital
 and Tumor Institute
Houston, Texas 77030

**G. D. Griffin**
Health and Safety Research
 Division
Oak Ridge National Laboratory
Oak Ridge, Tennessee 37830

**G. Grimmer**
Biomedical Institute for Environmental
 Carcinogens
Sieker Landstraβe 19
D-2070 Ahrensburg
Federal Republic of Germany

**M. R. Guerin**
Analytical Chemistry Division
Oak Ridge National Laboratory
Oak Ridge, Tennessee 37830

**H. Güsten**
Kernforschungszentrum Karlsruhe
 Institut für Radiochemie
7500 Karlsruhe Postfach 3640
Federal Republic of Germany

**R. L. Hanson**
Inhalation Toxicology Research
 Institute
Lovelace Biomedical and Environmental
 Research Institute
P..O. Box 5890
Albuquerque, New Mexico 87112

**R. Hardy**
Ministry of Agriculture, Food,
 and Fish
Toory Research Station
135 Abbey Road
Aberdeen, Scotland

**C. C. Harris**
Human Tissue Studies
Laboratory of Experimental
 Pathology
National Cancer Institute
Bethesda, Maryland 20205

and

Institute of Cancer Research
Columbia University
College of Physicians and
 Surgeons
New York, New York, 10032

**R. W. Hart**
Department of Radiology
College of Medicine
The Ohio State University
Columbus, Ohio 43210

**R. G. Harvey**
Ben May Laboratory for Cancer Research
University of Chicago
Chicago, Illinois 60637

**J. J. Hassett**
Department of Agronomy
University of Illinois
Urbana, Illinois 61801

**S. S. Hecht**
Naylor Dana Institute for
 Disease Prevention
American Health Foundation
Valhalla, New York, 10595

**G. Heinrich**
Kernforschungszentrum Karlsruhe
 Institut für Radiochemie
7500 Karlsruhe Postfach 3640
Federal Republic of Germany

**D. M. Hembree**
Department of Chemistry
University of Tennessee
Knoxville, Tennessee 37916

**M. Hermann**
Institut Francais Du Pétrole
1 et 4 Avenue de Bois Préau
92506 Rueil Malmaison, France

**C. F. Higgins**
Analytical Chemistry Division
Oak Ridge National Laboratory
Oak Ridge, Tennessee 37830

**D. Hoffmann**
Naylor Dana Institute for Disease
  Prevention
American Health Foundation
Valhalla, New York 10595

**M. Hofnung**
Institut Pasteur
28, Rue du Di Roux
75724 Paris Cedex 15 France

**M. Homola**
Institut für Landespflege
Universitat Freiburg
D-7800 Freiburg FRG

**P. H. Howard**
Center for Chemical Hazard Assessment
Syracuse Research Corporation
Syracuse, New York 13210

**W. R. Hudgins**
Chemical Carcinogenesis Program
NCI Frederick Cancer Research Center
Frederick, Maryland 21701

**M. M. Hughes**
Department of Chemistry
Colorado State University
Fort Collins, Colorado 80523

**G. T. Hunt**
GCA/Technology Division
Burlington Road
Bedford, Massachusetts 01730

**M. N. Inbasekaren**
Division of Medicinal Chemistry
College of Pharmacy
The Ohio State University
Columbus, Ohio 43210

**R. S. Isaac**
Division of Pharmacodynamics and
  Toxicology
College of Pharmacy
University of Kentucky
Lexington, Kentucky 40506

**R. P. Iyer**
Biology Division
Oak Ridge National Laboratories
Oak Ridge, Tennessee 37830

**J. Jacob**
Biochemisches Institut fur
  Umweltcorcinogene
D-2070 Ahrensburg 1 Holst.
Sieker Landstraβe 19
West Germany

**M. R. Jachau**
Department of Pharmacology
  and Pathology
School of Medicine
University of Washington
Seattle, Washington 98195

**A. M. Jeffrey**
Human Tissue Studies
Laboartories of Experimental Pathology
National Cancer Institute
Bethesda, Maryland 20205

   and

Institute of Cancer Research
Columbia University
College of Physicians and Surgeons
New York, New York 10032

**D. M. Jerina**
Laboratory of Bioorganic Chemistry
Section on Oxidation Mechanisms
National Institute of Arthritis
  Metabolism and Digestive Diseases
National Institutes of Health
Bethesda, Maryland 20205

   and

Laboratory of Bioorganic Chemistry
National Institute of Arthritis,
  Metabolism and Digestive Diseases
National Institutes of Health
Bethesda, Maryland 20014

**P. W. Jones**
Electric Power Research Institute
Palo Alto, California 94303

**T. D. Jones**
Health and Safety Research Division
Oak Ridge National Laboratory
Oak Ridge, Tennessee 37830

**N. J. Joyce**
Department of Radiology
School of Medicine
The Ohio State University
Columbus, Ohio 43210

**A. Kahn**
Department of Agronomy
University of Illinois
Urbana, Illinois 61801

**C. Kaiser**
Department of Biochemistry
McMaster University
Hamilton, Ontario
Canada L8S 459

**A. Kerr**
DOFASCO Ltd.
Hamilton, Ontario
Canada L8N 3J5

**C. A. Kieda**
Calgon Analytical Laboratories
Calgon Corporation
Pittsburgh, Pennsylvania 15230

**M. Kilgore**
Chemistry Department
Auburn University
Auburn, Alabama 36830

**W. Kim**
Department of Medicinal Chemistry
The Ohio State University
Columbus, Ohio 43210

**G. S. Kishore**
McArdle Laboratory for Cancer
  Research
University of Wisconsin
Madison, Wisconsin 53706

**J. D. Kittle, Jr.**
Department of Chemistry
Ohio University
Athens, Ohio 45701

**M. G. Knize**
Biomedical Sciences Division
Lawrence Livermore Laboratory
Livermore, California 94550

**R. M. Kocan**
Department of Pharmacology
  and Pathology
School of Medicine
University of Washington
Seattle, Washington 98195

**A. Kootstra**
University of Tennessee
Oak Ridge Graduate School of
  Biomedical Sciences and
  Biology Division
Oak Ridge National Laboratory
Oak Ridge, Tennessee 37830

**M. M. Krahn**
National Oceanic and Atmospheric
  Administration
2725 Montlake Boulevard East
Seattle, Washington 98112

**S. Kumar**
Department of Chemistry
University of Oklahoma
Norman, Oklahoma 73019

**J. R. Lakowicz**
Gray Freshwater Biological Institute
  and Department of Biochemistry
University of Minnesota
P.O. Box 100
Navarre, Minnesota 55392

**D. A. Lane**
SCIEX Incorporated
55 Glencameron Road #202
Thornhill, Ontario
Canada L3T 1P2

**R. C. Lao**
Chemistry Division
Air Pollution Control Directorate
Environmental Canada
Ottawa, Canada K1A 1C8

**P. R. Mackie**
Ministry of Agriculture, Food,
  and Fish
Toory Research Station
135 Abbey Road
Aberdeen, Scotland

**M. C. MacLeod**
Biology Division
Oak Ridge National Laboratory
The University of Tennessee
Oak Ridge Graduate School of
  Biomedical Sciences
Oak Ridge, Tennessee 37830

**G. Mamantov**
Department of Chemistry
University of Tennessee
Knoxville, Tennessee 37916

**J. R. Maple**
Department of Chemistry
University of Tennessee
Knoxville, Tennessee 37916

**M. Marshall**
Biochemistry Department
M. D. Anderson Hospital
Houston, Texas 77030

**M. H. Marshall**
The University of Texas System Cancer
  Center
M. D. Anderson Hospital and Tumor
  Institute
Houston, Texas 77030

**M. V. Marshall**
North Texas State University
Denton, Texas 76203
  and
The University of Texas System Cancer
  Center
M. D. Anderson Hospital and Tumor
  Institute
Houston, Texas 77030

**R. R. Martin**
Baylor College of Medicine
Houston, Texas 77030
  and
Department of Medicine
Baylor College of Medicine
Houston, Texas 77030

**W. E. May**
Organic Analytical Research Division
Center for Analytical Chemistry
National Bureau of Standards
Washington, DC 20234

**P. R. McCalla**
Department of Biochemistry
McMaster University
Hamilton, Ontario
Canada L58 4J9

**M. W. McElroy**
Electric Power Research Institute
Palo Alto, California 94303

**A. S. McGill**
Ministry of Agriculture, Food,
  and Fish
Toory Research Station
135 Abbey Road
Aberdeen, Scotland

**T. McLemore**
Department of Medicine
Baylor College of Medicine
Houston, Texas 77030
  and
Veterans Administration Hospital
Houston, Texas 77030

**J. C. Means**
Chesapeake Biological Laboratory
University of Maryland
Solomons, Maryland 20688

**P. Melius**
Chemistry Department
Auburn University
Auburn, Alabama 36830

**J. M. Meuser**
Battelle Columbus Laboratories
505 King Avenue
Columbus, Ohio 43201

**D. Mhaskar**
Department of Radiology
School of Medicine
The Ohio State University
Columbus, Ohio 43210

**A. H. Miguel**
Instituto de Quimica
Universidade Federal de Rio de Janeiro
21941 Ilha do Funao
Rio de Janeiro
Brazil

**C. J. Moore**
Biology Division
Oak Ridge National Laboratory
The University of Tennessee
Oak Ridge Graduate School of
 Biomedical Sciences
Oak Ridge, Tennessee 37830

**F. R. Moore**
Battelle Columbus Laboratories
505 King Avenue
Columbus, Ohio 43201

**R. C. Moschel**
Chemical Carcinogenesis Program
NCI Frederick Cancer Research Center
Frederick, Maryland 21701

**D.F.S. Natusch**
Department of Chemistry
Colorado State University
Fort Collins, Colorado 80523

**K. W. Naujack**
Biochemical Institute of Environmental
 Carcinogens
Sieker Landstrabe 19
D-2070 Ahrensburg, F.R.G.

**G. J. Newton**
Inhalation Toxicology Research Institute
Lovelace Biomedical and Environmental
 Research Institute
P.O. Box 5890
Albuquerque, New Mexico 87112

**B. Norden**
Department of Organic Chemistry
Chemical Institution
Univesity of Umeå
S-90187 Umeå
Sweden

**H. G. Nowicki**
Calgon Analytical Laboratories
Calgon Corporation
Pittsburgh, Pennsylvania 15230

**I. J. Ocasio**
Department of Chemistry
Ohio University
Athens, Ohio 45701

**B. Olufsen**
Central Institute for
 Industrial Research
Blindern, Oslo 3
Norway

**C. J. Omiecinski**
Department of Pharmacology
School of Medicine
University of Washington
Seattle, Washington 98195

**J. Ortman**
Department of Radiology
School of Medicine
The Ohio State University
Columbus, Ohio 43210

**D. S. Orwig**
Biomedical Sciences Division
Lawrence Livermore Laboratory
Livermore, California 94550

**N. Pangaro**
BCA/Tecnology Division
Burlington Road
Bedford, Massachusetts 01730

**K. Parker**
Biology Department
University of Calgary
Calgary, Alberta
Canada T2N 1N4

**E. Peake**
Kananaskis Centre for Environmental
 Research
University of Calgary
Calgary, Alberta
Canada T2N 1N4

**N. Pétroff**
Institut Francais De Pétrole
1 et 4 Avenue de Bois Préau
92506 Rueil Malmaison, France

**W. R. Pierson**
Scientific Research Lab
Ford Motor Company
Dearborn, Michigan 48121

**S. K. Quan**
SCIEX Incorporation
55 Glencameron Road #202
Thornhill, Ontario
Canada L3T 1P2

**W. K. Robbins**
Exxon Research and Engineering Co.
Florham Park, New Jersey 07932

**W. L. Roberts**
Department of Chemistry
The Ohio State University
Columbus, Ohio 43210

**E. Rogan**
Roswell Park Memorial Institute
Buffalo, New York 14263

**R. Roth**
Midwest Research Institute
425 Volker Blvd.
Kansas City, Missouri 64110

**L.M.S. Rübenich**
Instituto de Quimica
Universidade Federal de Rio de Janeiro
21941 Ilha do Funao
Rio de Janeiro
Brazile

**T. Sakuma**
SCIEX Incorporation
55 Glencameron Road #202
Thornill, Ontario
Canada L3T 1P2

**E. P. Salazar**
Biomedical Sciences Division
Lawrence Livermore Laboratory
Livermore, California 94550

**J. Santodonato**
Center for Chemical Hazard Assessment
Syracuse Research Corporation
Syracuse, New York 13210

**K. Schmidt**
Department of Radiology
School of Medicine
The Ohio State University
Columbus, Ohio 43210

**A. Schmoldt**
Pharmakologisches Institut
  d. Universitat Hamburg
D-2000 Hamburg 20,
Martinstraβe 52
West Germany

**D. Schneider**
Biochemical Institute of
  Environmental Carcinogens
Sieker Landstrabe 19
D-2070 Ahrensburg, F.R.G.

**W. P. Schoor**
United States Environmental
  Protection Agency
Environmental Research Labortory
Sabine Island
Gulf Breeze, Florida 22561

**A. Secrist, III**
Department of Chemistry
The Ohio State University
Columbus, Ohio 43210

**J. K. Selkirk**
Biology Division
Oak Ridge National Laboratory
The University of Tennessee
Oak Ridge Graduate School
  of Biomedical Sciences
Oak Ridge, Tennessee 37830

**Y. M. Sheikh**
Division of Medicinal Chemistry
College of Pharmacy
The Ohio State University
Columbus, Ohio 43210

**D. Siebert**
Institut fur Tandespflege
Universitat Freiburg
D-7800 Freiburg, FRG

**D. Sinha**
Eppley Institute for Research
  in Cancer
University of Nebraska
  Medical Center
Omaha, Nebraska 68105

**T. J. Slaga**
Biology Division
Oak Ridge National Laboratory
Oak Ridge, Tennessee 37830
and
University of Tennessee
Oak Ridge Graduate School of Biomedical
 Sciences and Biology Division
Oak Ridge National Laboratory
Oak Ridge, Tennessee 37830

**R. D. Smillie**
Ontario Ministry of the Environment
Laboratory Services Branch
P.O. Box 213
Rexdale, Ontario
Canada M92 5L1

**E. M. Smith**
Arthur D. Little, Incorporated
Acorn Park
Cambridge, Massachusetts 02140

**T. W. Sonnichsen**
KVB, Inc.
A Research-Cottrell Co.
Tustin, California 92680

**B. N. Srinivasan**
Department of Fine Structure
Boston Biomedical Research Institute
20 Staniford Street
Boston, Massachusetts 02114

**J. W. Strand**
Water Chemistry Program
University of Wisconsin
Madison, Wisconsin 53706

**P. E. Strup**
Battelle Columbus Laboratories
505 King Avenue
Columbus, Ohio 43201

**P. D. Sullivan**
Department of Chemistry
Ohio University
Athens, Ohio 45701

**S. J. Swarin**
Analytical Chemical Department
General Motors Research Laboratories
Warren, Michigan 48090

**M. Tada**
Laboratory of Bio-organic Chemistry
Section on Oxidation Mechanisms
national Institute of Arthritis
 Metabolism and Digestive Diseases
National Institutes of Health
Bethesda, Maryland 20205

**B. Tan**
Chemistry Department
Auburn University
Auburn, Alabama 36830

**D. R. Taylor**
Department of Chemistry
Colorado State University
Fort Collins, Colorado 80523

**D. R. Thakker**
Laboratory of Bio-organic Chemistry
Section on Oxidation Mechanisms
National Institute of Arthritis,
 Metabolism and Digestive Diseases
National Institutes of Health
Bethesda, Maryland 20205

**R. S. Thomas**
Chemistry Division
Air Pollution Control Directorate
Environment Canada
Ottawa, Canada K1A 1C8

**L. Tulley**
Naylor Dana Institute for Disease
 Prevention
American Health Foundation
Valhalla, New York 10595

**J. P. Vandecasteele**
Institut Francais De Pétrole
1 et 4 Avenue de Bois Préau
92506 Rueil Malmaison, France

**U. Varanasi**
National Oceanic and Atmospheric
 Administration
2725 Montlake Boulevard East
Seattle, Washington 98112

**T. Vo-Dinh**
Health and Safety Research Division
Oak Ridge National Laboratory
Oak Ridge, Tennessee 37830

**P. J. Walsh**
Health and Safety Research Division
Oak Ridge National Laboratory
Oak Ridge, Tennessee 37830

**D. T. Wang**
Ontario Ministry of the Environment
Laboratory Services Branch
P.O. Box 213
Rexdale, Ontario
Canada M9W 5L1

**A. Wani**
Department of Radiology
School of Medicine
The Ohio State University
Columbus, Ohio 43210

**E. L. Wehry**
Department of Chemistry
University of Tennessee
Knoxville, Tennessee 37916

**N. Weill**
Institut Pasteur
28, Rue du Di Roux
75724 Paris Cedex 15
Paris

**D. L. Whalen**
Laboratory of Chemical Dynamics
University of Maryland
Baltimore County
Baltimore, Maryland 21228

**C. M. White**
U.S. Department of Energy
Pittsburgh Energy Technical Center
Pittsburgh, Pennsylvania 15213

**K. J. Whittle**
Ministry of Agriculture, Food,
  and Fish
Toory Research Station
135 Abbey Road
Aberdeen, Scotland

**J. E. Wilkinson**
Battelle Columbus Laboratories
505 King Avenue
Columbus, Ohio 43201

**C. Willey**
Department of Chemistry
Brigham Young University
Provo, Utah 84602

**R. L. Williams**
Environmental Science Department
General Motors Research Laboratories
Warren, Michigan 48090

**S. A. Wise**
Organic Analytical Research Division
Center for Analytical Chemistry
National Bureau of Standards
Washington, D.C. 20234

**P. G. Wislocki**
Department of Animal Drug Metabolism
Merck, Sharp, and Dome Research
  Laboratories
Rahway, New Jersey 07065

**D. T. Witiak**
Division of Medicinal Chemistry
The Ohio State University
Columbus, Ohio 43210

**S. Wold**
Department of Organic Chemistry
Chemical Institute
University of Umea°
S-90187 Umea
Sweden

**A. W. Wood**
Department of Biochemistry and
  Drug Metabolism
Hoffman-LaRoche, Inc.
Nutley, New Jersey 07110

**S. G. Wood**
Department of Environmental Studies
University of Illinois
Urbana, Illinois 61801

**N. Wray**
Department of Medicine
Baylor College of Medicine
Houston, Texas 77030
    and
Veterans Administration Hospital
Houston, Texas 77030

**H. Yagel**
Laboratory of Bio-organic Chemistry
National Intitute of Arthritis,
  Metabolism, Digestive Diseases
National Institute of Health
Bethesda, Maryland 20014

**H. Yagi**
Laboratory of Bio-organic Chemistry
Section on Oxidation Mechanisms
National Institute of Arthritis,
  Metabolism, Digestive Diseases
National Institute of Health
Bethesda, Maryland 20205

**S. K. Yang**
Department of Pharmacology
School of Medicine
Uniformed Services
University of the Health Sciences
Bethesda, Maryland 20014

**L. B. Yeatts, Jr.**
Analytical Chemistry Division
Oak Ridge National Laboratory
Oak Ridge, Tennessee 37830

**R. Yokoyama**
Department of Chemical Engineering
  and Chemistry
New Jersey Institute of Technology
Newark, New Jersey 07102

**S. G. Zelenski**
GCA/Technology Division
Burlington Road
Bedford, Massachusetts 07130

**M. V. Zeller**
Physical Electronic Division
Perkin Elmer Corporation
Eden Prairie, Minnesota 55344

**Acknowledgments**

This symposium was co-sponsored by the Environmental Protection Agency (Industrial Environmental Research Laboratory and Health Effects Research Laboratory, Research Triangle Park, N.C.), The Electric Power Research Institute (EPRI Environmental Analysis Division, Palo Alto, California) and Battelle Memorial Institute (Columbus, Ohio). Further, many Battelle staff members, including the staff of the Communications and Public Relations Department especially the Conference Coordination Group and the Report and Library Services continue to be instrumental in the smooth functioning of the symposium with a special acknowledgement for the primary support provided by Debbi Bay.

**Contents**

Chemical Transformations of Particulate Polycyclic
Organic Matter
M. M. Hughes, D.F.S., Natusch, D. R. Taylor,
and M. V. Zeller .................................................. 1

Analysis of Water-Soluble Conjugates Produced by Hamster
Embryo Cells Exposed to Polynuclear Aromatic Hydrocarbons
M. C. MacLeod, C. J. Moore, and J. K. Selkirk .................. 9

Use of Mixed Phases for Enhanced Gas-Chromatographic
Separation of Polycyclic Aromatic Hydrocarbons: Preliminary
Studies with Liquid Crystals
R. J. Laub and W. L. Roberts .................................. 25

Separation and Identification of Sulfur Heterocycles in
Coal-Derived Products
M. L. Lee, C. Willey, R. N. Castle, and C. M. White ........... 59

Quantitative Evaluation of Priority Pollutant Polycyclic Aromatic
Hydrocarbons at One Part per Billion Using EPA Recommended
Priority Pollutant Protocol
H. G. Nowicki, C. A. Kieda, and D. O. Bassett ................. 75

Metabolism of Benzo(a)Pyrene in Cultured Human Bronchus,
Trachea, Colon, and Esophagus
H. Autrup, A. M. Jeffrey, and C. C. Harris .................... 89

Changes in PAH-Profiles in Different Areas of a City
During the Year
G. Grimmer, K.-W. Naujack, and D. Schneider .................. 107

Polyaromatic Hydrocarbons in Aerosols over Lake Michigan,
Fluxes to the Lake
J. W. Strand and A. W. Andren ................................ 127

Room Temperature Phosphorimetry for the Analysis of
Environmental Systems
T. Vo-Dinh and R. B. Gammage ................................. 139

The Influence of Epoxide Hydratase Inhibition on the Genetic
Activity of Benzo(a)Pyrene in the Liver Microsome Test Using Mitotic
Gene Conversion in Yeasts as a Mutagenicity Test System
U. Bayer, M. Homola, and D. Siebert ........................ 153

The Effect of Antioxidants on the Mutagenicity of
Benzo(a)Pyrene and Derivatives
P. D. Sullivan, L. M. Calle, I. J. Ocasio,
J. D. Kittle, Jr., and L. E. Ellis ............................. 163

Mutation Induction at Multiple Gene Loci in Chinese Hamster
Ovary Cells: Comparisons of Benzo(a)Pyrene Metabolism by
Organ Homogenates and Intact Cells
J. H. Carver, E. P. Salazar, M. G. Knize,
D. S. Orwig, and J. S. Felton ............................... 177

Analysis of Polynuclear Aromatic Hydrocarbons on the
Airborne Particulates of Urban New Jersey
A. Greenberg, R. Yokoyama, P. Giorgio, and F. Cannova ........ 193

Real-Time Analysis of Gas Phase Polycyclic Aromatic
Hydrocarbons Using a Mobile Atmospheric Pressure
Chemical Ionization Mass Spectrometer System
D. A. Lane, T. Sakuma, and E.S.K. Quan ..................... 199

Rat Mammary Gland Versus Mouse Skin: Different Mechanisms of
Activation of Aromatic Hydrocarbons
E. Cavalieri, D. Sinha, and E. Rogan ........................ 215

Hematin-Mediated Increases in the Monooxygenation of
Polynuclear Aromatic Hydrocarbons
C. J. Omiecinski and M. R. Juchau .......................... 233

Inhibition of Pulmonary Metabolism of Benzo(a)Pyrene
Produced by Acute-Tobacco Smoke Exposure
W. C. Lubawy and R. S. Isaac .............................. 243

Manganic Acetate and Horseradish Peroxidase/Hydrogen Peroxide:
In Vitro Models of Activation of Aromatic Hydrocarbons by
One-Electron Oxidation
E. G. Rogan, R. Roth, and E. Cavalieri ...................... 259

Comparative Metabolism of Dihydrodiols of Polycyclic
Aromatic Hydrocarbons to Bay-Region Diol Epoxides
D. R. Thakker, W. Levin, H. Yagi, M. Tada,
A. H. Conney, and D. M. Jerina ............................. 267

Reproductive Survival and Macromolecular Synthesis in Cultured
Mammalian Cells Exposed to Single Metabolites and Mixtures
of Metabolites of Benzo(a)Pyrene
G. D. Griffin, C. S. Dudney, P. S. Furcinitti,
T. D. Jones, and P. J. Walsh .................................. 287

Benzo(a)Pyrene Activation and Detoxification by Human
Pulmonary Alveolar Macrophages and Lymphocytes
M. V. Marshall, T. L. McLemore, R. R. Martin,
M. H. Marshall, N. P. Wray, D. L. Busbee,
E. T. Cantrell, M. S. Arnott, and A. C. Griffin ................ 299

Benzo(a)Pyrene Radicals and Other Derivatives Formed in
Interactions with Cysteine/Serum Albumin
B. N. Srinivasan and E. Fujimori ............................. 319

Polynuclear Aromatic Hydrocarbons in Norwegian Drinking
Water Resources
B. Olufsen ................................................. 333

Structure-Carcinogenicity Studies of Polycyclic Aromatic
Hydrocarbons: A Pattern Recognition Approach
B. Nordén, U. Edlund, and S. Wold .......................... 345

Polycyclic Aromatic Hydrocarbons in Marine Sediments, Bivalves,
and Seaweeds: Analysis by High-Pressure Liquid Chromatography
B. P. Dunn ................................................. 367

The Tissue Hydrocarbon Burden of Mussels from Various Sites
Around the Scottish Coast
P. R. Mackie, R. Hardy, K. J. Whittle, C. Bruce,
and A. S. McGill ........................................... 379

Sorption Properties of Polynuclear Aromatic Hydrocarbons and
Sediments: Heterocyclic and Substituted Compounds
J. C. Means, J. J. Hassett, S. G. Wood, W. L. Banwart, S. Ali,
and A. Khan ............................................... 395

Analysis of PAH in Environmental Samples by High Temperature
Stable Glass Capillary Columns
J. M. Meuser, F. R. Moore, P. E. Strup, J. E. Wilkinson, and
A. Bjørseth ................................................ 405

On the Metabolic Activation of the Benzofluoranthenes
S. S. Hecht, E. LaVoie, S. Amin, V. Bedenko, and
D. Hoffmann ............................................... 417

Multimedia Human Exposure and Carcinogenic Risk
Assessment for Environmental PAH
*J. Santodonato, D. Basu, and P. H. Howard* .................. 435

Metabolism and Subsequent Binding of Benzo(a)Pyrene to
DNA in Pleuronectid and Salmonid Fish
*U. Varanasi, D. J. Gmur and M. M. Krahn* ................... 455

Benzo(a)Pryene-DNA Adduct Formation in Cells: Time-Dependent
Differences in the Benzo(a)Pryene-DNA Adducts Present
*W. M. Baird and R. U. Dumaswala* .......................... 471

Oxidative and Nonoxidative Metabolism of Polycyclic Aromatic
Hydrocarbons in Rabbit and Chicken Aortas and in Human
Fetal Smooth-Muscle Cells
*J. A. Bond, R. M. Kocan, E. P. Benditt, and M. R. Juchau* ....... 489

Hydrocarbon-Deoxyribonucleoside Adducts In Vivo and In Vitro
and Their Relationship to Carcinogenicity
*G. M. Cohen, S. W. Ashurst, J. K. Selkirk,, and T. J. Slaga* ...... 503

Recognition of DNA Damage In Vitro and In Vivo
*S. M. D'Ambrosio, F. B. Daniel, D. Brash, R. E. Gibson,
N. J. Joyce, W. Kim, D. Mhaskar, J. Ortman, K. Schmidt,
A. Wani, D. Witiak, and R. W. Hart* ......................... 523

The Problem of PAH Degradation During Filter Collection
of Airborne Particulates—An Evaluation of Several
Commonly Used Filter Media
*F. S.-C. Lee, W. R. Pierson, and J. Ezike* ..................... 543

Proxy Methods and Compounds for Workplace Monitoring of
Polynuclear Aromatic Hydrocarbons
*R. B. Gammage and A. Bjørseth* ............................ 565

Mutagenic Material in Air Particles in a Steel Foundry
*C. Kaiser, A. Kerr, D. R. McCalla, J. N. Lockington,
and E. S. Gibson* ......................................... 579

Comparison of SIM GC/MS and HPLC for the Detection of
Polynuclear Aromatic Hydrocarbons in Fly Ash Collected
from Stationary Combustion Sources
*S. G. Zelenski, G. T. Hunt, and N. Pangaro* ................... 589

Chemical Characterization of Polynuclear Aromatic Hydrocarbons in
Airborne Effluents from an Experimental Fluidized Bed Combustor
*R. L. Hanson, R. L. Carpenter, and G. J. Newton* .............. 599

Use of PAH Tracers During Sampling of Coal Fired Boilers
T. W. Sonnichsen, M. W. McElroy, and A. Bjørseth ............. 617

Binding of BaP Diol Epoxide to Chromosomal Histone Proteins
A. Kootstra, M. C. Cronen, and T. J. Slaga .................... 633

Metabolism of 6-, 7-, 8-, and 12-Methylbenz(a)—Anthracenes and Hydroxymethylbenz(a)Anthracenes
S. K. Yang, M. W. Chou, and P. P. Fu ....................... 645

DNA Binding of 7,12-Dimethylbenz(a)Anthracene (DMBA) and Related Compounds
R. C. Moschel, C.A.H. Bigger, W. R. Hudgins, and A. Dipple .... 663

Benzo(e)Pyrene Dihydrodiols and Diol Epoxides: Chemistry, Mutagenicity and Tumorigenicity
R. E. Lehr, S. Kumar, W. Levin, A. W. Wood, R. L. Chang, M. K. Buening, A. H. Conney, D. L. Whalen, D. R. Thakker, H. Yagi, and D. M. Jerina ................................ 675

A Study of the 7,12-Dimethylbenz(a)Anthracene (DMBA) Bay Region Involvement in the Production of Carcinogen and Mutagen Metabolites
Y. M. Sheikh, M. N. Inbasekaran, F. B. Daniel, F. D. Cazer, R. W. Hart, and D. T. Witiak ................................ 689

Metabolism of 7,12-Dimethylbenz[a]-Anthracene: Quantitation of Metabolite Formations in Rat Liver Microsomes and a Reconstituted Enzyme System Containing Highly Purified Cytochrome P-450 and P-448
S. K. Yang, M. W. Chou, P. G. Wislocki, and A.Y.H. Lu ........ 733

Comparison of the Skin Tumor-Initiating Activities of Dihydrodiols, Diol-Epoxides, and Methylated Derivatives of Various Polycyclic Aromatic Hydrocarbons
T. J. Slaga, R. P. Iyer, W. Lyga, A. Secrist, III, G. H. Daub, and R. G. Harvey ............................................. 753

Liquid Chromatographic Determination of Benzo[a]Pyrene in Diesel Exhaust Particulate: Verification of the Collection and Analytical Methods
S. J. Swarin and R. L. Williams ............................. 771

Normal- and Reverse-Phase Liquid Chromatographic Separations of Polycyclic Aromatic Hydrocarbons
S. A. Wise, W. J. Bonnett, and W. E. May .................... 791

Gas-Chromatographic Profile-Analysis of PAH Metabolites
From Rat Liver Microsomes and Cells in Culture
*J. Jacob, G. Grimmer, and A. Schmoldt* ..................... 807

Extraction and Recovery of Polycyclic Aromatic Hydrocarbons
from Highly Sorptive Matrices Such as Fly Ash
*W. H. Griest, J. E. Caton, M. R. Guerin, L. B. Yeatts, Jr., and
C. E. Higgins* ............................................. 819

The Volatility of PAH and Possible Losses in Ambient Sampling
*R. C. Lao and R. S. Thomas* ................................ 829

Solvent Extraction of Polynuclear Aromatic Hydrocarbons
*W. K. Robbins* ............................................ 841

A Comparison of Extraction Techniques for Polynuclear Aromatic
Hydrocarbon Analysis of Industrial Effluents and Natural Waters
*R. D. Smillie and D. T. Wang* .............................. 863

Microsomal Uptake of Benzo(a)Pyrene: Effect of Adsorption to
Asbestos, Hematite, Silica, and Carbon Black
*J. R. Lakowicz and D. R. Bevan* ............................ 879

Correlations of Mutagenic Activity with Polynuclear Aromatic
Hydrocarbon Content of Various Mineral Oils
*M. Hermann, J. P. Durand, J. M. Charpentier, O. Chaudé,
M. Hofnung, N. Pétroff, J.-P. Vandecasteele, and N. Weill* ....... 899

High Aryl Hydrocarbon Hydroxylase Inducibility is Positively
Correlated with Occurrence of Lung Cancer
*D. Busbee, T. McLemore, R. R. Martin, N. Wray, M. Marshall,
and E. Cantrell* ............................................ 917

2,3,7,8-Tetrachlorodibenzo-p-dioxin (TCDD)-Induced
Alterations in Oxidative and Nonoxidative Biotransformation of PAH
in Mouse Skin: Role in Anticarcinogenesis by TCDD
*J. DiGiovanni, G. S. Kishore, T. J. Slaga, and
R. K. Boutwell* ............................................ 935

Membrane Changes Associated with Aqueous Extracts of
Fossil-Fuel Generated Respirable Particulates
*T. J. Facklam, J. P. Crowley, M. A. Drum, and A. J. Dennis* ..... 955

Sensitized Fluorescence Detection of PAH
*E. M. Smith and P. L. Levins* .............................. 973

Fluorescence Spectroscopic Properties of Carcinogenic and
Airborne Polynuclear Aromatic Hydrocarbons
*G. Heinrich and H. Güsten* .................................. 983

Recent Developments in Matrix Isolation Spectroscopic Analysis
of Polynuclear Aromatic Hydrocarbons
*E. L. Wehry, G. Mamantov, D. M. Hembree, and J. R. Maple* ... 1005

Polynuclear Aromatic Hydrocarbons and the Mutagenicity
of Used Crankcase Oils
*E. Peake and K. Parker* .................................. 1025

Mutagenicity, Tumor Initiating Activity, and Metabolism of
Tricyclic Polynuclear Aromatic Hydrocarbons
*E. LaVoie, L. Tulley, V. Bedenko, and D. Hoffmann* ........... 1041

Mixed Function Oxidase Inducibility and Polyaromatic
Hydrocarbon Metabolism in the Mullet, Sea Catfish, and
Gulf Killifish
*P. Melius, D. Elam, M. Kilgore, B. Tan, and W. P. Schoor* ...... 1059

Submicron Size Distributions of Particulate Polycyclic Aromatic
Hydrocarbons in Combustion Source Emissions
*A. H. Miguel and L.M.S. Rubenich* ......................... 1077

Measurement and Environmental Impact of PAH—
Some Closing Remarks
*P. W. Jones* ............................................ 1085

# CHEMICAL TRANSFORMATIONS OF PARTICULATE POLYCYCLIC ORGANIC MATTER

**M. M. Hughes\*, D.F.S. Natusch\*, D. R. Taylor\*, and Mary V. Zeller\*\***

\*Department of Chemistry
Colorado State University
Fort Collins, Colorado 80523
\*\*Physical Electronics Division
Perkin Elmer Corporation
Eden Prairie, MN 55344

## INTRODUCTION

Although the formation of polynuclear aromatic hydrocarbons (PAH) as the result of fossil fuel combustion is well known (2), the fate of these compounds in the atmosphere is not well understood. The fact that most PAH are associated with particulate surfaces is clearly established (2,7) but the reactions of the surface-associated molecules are only now being investigated. Two general types of reactions appear to be possible. The first type of reaction involves the photochemical and nonphotochemical oxidation of individual PAH. The second involves studies of the chemical reactions undergone by PAH adsorbed on particle surfaces and exposed to gases such as NO, $NO_2$, $SO_2$, and $SO_3$, which are likely to be encountered in plumes from most fossil fuel combustion and conversion sources.

Summarizing the oxidative behavior of adsorbed polycyclic organic matter (POM) (including both PAH as well as several nitrogen-containing heterocyclic compounds), it can be stated that the actual process of adsorption significantly alters the chemical characteristics of many members of this class of compounds (3,4). Thus, adsorption of pyrene, phenanthrene, fluoranthene, anthracene, and benzo(a)pyrene onto coal fly ash, activated carbon, or graphite particles results in effective stabilization of these compounds *against* photochemical decomposition. On the other hand, adsorption of fluorene, benzo(a)fluorene, benzo(b)fluorene, 9,10-dimethylanthracene, and 4-azafluorene onto these same types of particles

## 2 CHEMICAL TRANSFORMATIONS OF PARTICULATE POM

results in spontaneous dark oxidation to the corresponding quinone or ketone. However, neither type of behavior is observed when these compounds are adsorbed onto glass, alumina, or silica surfaces. One possibly significant, albeit tentative, finding of these studies was that apparently only those compounds containing a benzylic carbon atom underwent oxidation.

The second possible type of reaction of particle associated PAH molecules is that which can occur with other atmospheric gases. A report indicating that such reactions can produce compounds with increased or previously unobserved mutagenic activity (as indicated by the Ames test) has appeared recently in the literature (8). In these studies, it was found that both benzo(a)pyrene and perylene dispersed on glass fiber filters reacted to form nitro derivatives when exposed to an atmosphere containing 1 ppm $NO_2$, and that the products were mutagenic (8) even though one of the materials (perylene) did not previously exhibit such behavior.

We have studied the reactions of two PAH adsorbed onto different substrates with a number of gases known to exist in the plumes of fossil fuel combustion and conversion sources, and have found that both the gas and the particulate surface significantly affect the reactions observed.

## EXPERIMENTAL

Pyrene and benzo(a)pyrene (50 to 150 $\mu g/g$) were adsorbed onto the surface of several different particulate substrates (alumina, coal fly ash, and silica gel) by using techniques that are described elsewhere (5,6). Samples were then exposed to the individual gases NO, $NO_2$, $SO_2$, or $SO_3$ at room temperature for a period of 6 (NO) or 12 ($NO_2$, $SO_2$, $SO_3$) hours. Concentrations of approximately 100 ppm were used in all cases except that of $SO_3$, where precise quantification proved unfeasible because of experimental difficulties. Exposure was conducted both in the presence and absence of ultraviolet irradiation as provided by a vitalite, quartzline lamp and natural daylight in order to observe whether the reactions might be photochemically induced. Following exposure, the samples were soxhlet extracted for 12 hours with benzene and the extract was subjected to high-performance liquid chromatography using a Bondapak C18 reverse-phase column and a water-methanol gradient elution program for the mobile phase. The chromatographic peaks were identified by high-resolution mass spectrometry following collection of fractions containing each separated compound. Electron Spectroscopy for Chemical Analysis (ESCA) was employed to examine particle surfaces both before and after exposure to the reactant gases. The results of these experiments are presented below.

## RESULTS AND DISCUSSION

In general, only two of the gases studied ($NO_2$ and $SO_3$) reacted with the two PAH used in this study. Neither NO nor $SO_2$ reacted significantly under the experimental conditions employed. None of the PAH-gas reactions were influenced by exposure to UV irradiation. Chromatograms of the extract resulting from the exposure of pyrene and benzo(a)pyrene adsorbed on fly ash to $NO_2$ and $SO_3$ are shown in Figures 1 and 2, respectively, together with control chromatograms of extracted but unexposed material. Prior to gas exposure only one major peak, the parent compound peak, A, is evident in both cases. Following exposure to $NO_2$, however, a second peak, labeled B in both figures, is observable. Mass spectrometric analyses indicate a mononitro derivative in both cases. The smaller peaks, labeled C-F in Figure 2, appear to be either a benzene

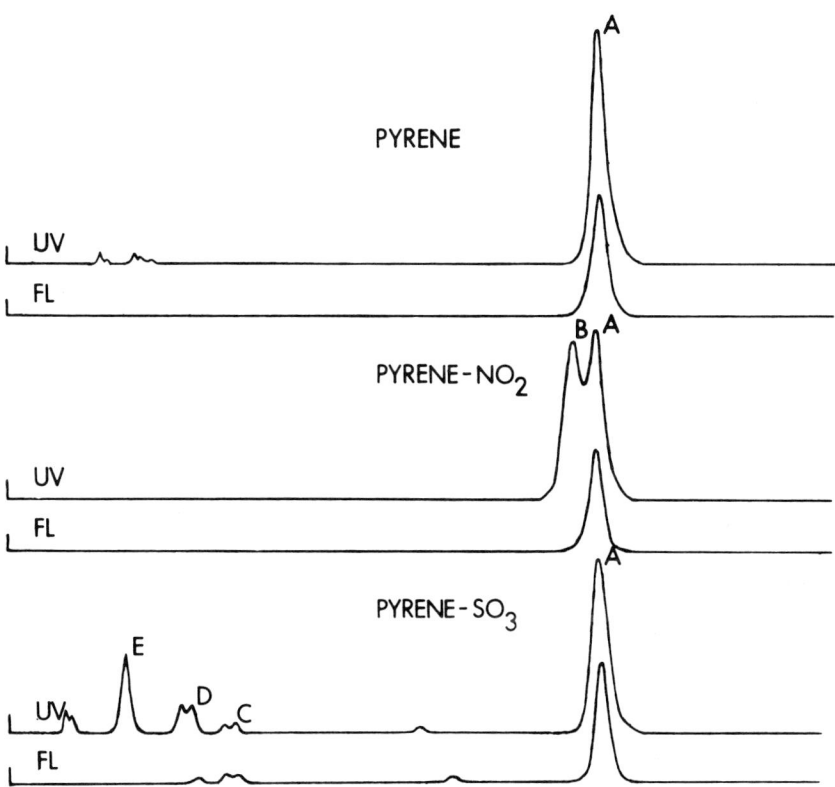

**FIGURE 1.** Chromatograms of extract from pyrene adsorbed on coal fly ash before and after exposure to $NO_2$ and $SO_3$. Both UV and fluorescence traces are shown. For peak identification, see text.

## 4 CHEMICAL TRANSFORMATIONS OF PARTICULATE POM

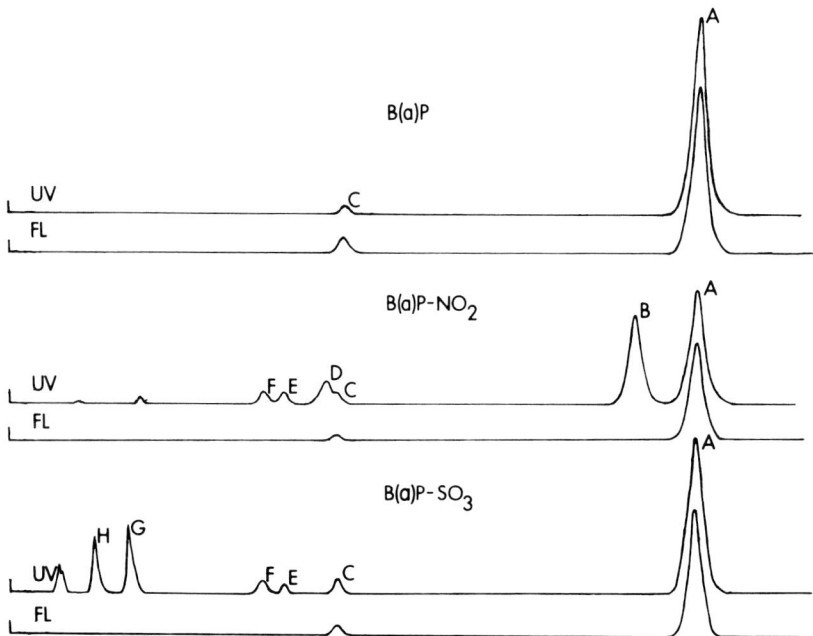

**FIGURE 2.** Chromatograms of extract from benzo(a)pyrene adsorbed on coal fly ash before and after exposure to $NO_2$ and $SO_3$. Both UV and fluorescence traces are shown. For peak identification see text.

impurity (peak C) or else small amounts of oxidized BaP. Insufficient material was available to permit mass spectrometric indentification of these smaller peaks.

The results of the PAH-$SO_3$ reactions were not as clear. Although several peaks (C-E in Figure 1 for pyrene, and G-I in Figure 2 for benzo(a)-pyrene) were not present following $SO_3$ exposure, it was unclear whether these peaks were actually products of the reaction or were decomposition/ oxidation products. It does appear that these substances are not simple substitution products analogous to the previously observed mononitro compounds. If reactions with $SO_3$ are occurring, it appears that the PAH itself must be decomposing.

The observed reactions appear to be highly surface dependent. Three different surfaces were investigated in this regard, viz., alumina, fly ash, and silica gel. Only the reactions of adsorbed pyrene and $NO_2$ were examined in these experiments. The results indicate that pyrene reacts similarly when absorbed on alumina and fly ash, but that the reactions on the silica gel surface are significantly different. Figure 3 shows the chromatogram of pyrene-silica gel extract obtained immediately after exposure to $NO_2$ and following aging for 3 weeks. Initially, three strong

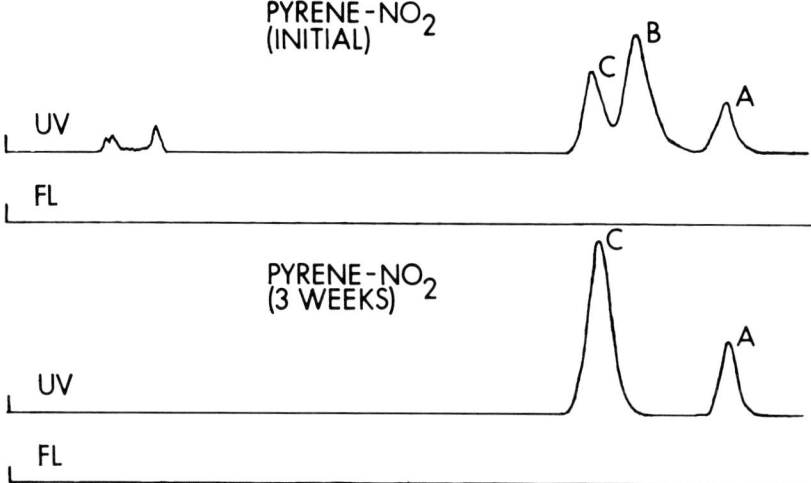

**FIGURE 3.** Chromatogram of extract from pyrene adsorbed on silica exposed to $NO_2$, showing changes with aging.

peaks are evident, none of which correspond to pyrene. Peaks A and C have been identified as different dinitro pyrene products, and peak B appears to be the mononitro product previously observed in the fly ash and alumina systems. After 3 weeks, however, the mononitro pyrene has disappeared but the two products remain and it appears that at least one of the dinitro products has increased by a comparable amount, suggesting that conversion, rather than decomposition, of the mononitro pyrene has occurred. Since the acidic surface of silica gel and expected existence of $HNO_3$ on the surface would be expected to promote nitration, it is probable that nitration is at least partially responsible for the reactions, as suggested by Pitts et al (8).

Further information concerning PAH-surface interactions were obtained from surface analytical studies by using ESCA. A low-resolution scan of a fly ash surface after exposure to pyrene is shown in Figure 4A, and a scan following exposure of this same material to $NO_2$ is shown in Figure 4B. As can be seen in Figure 4, the principal difference between the two scans is the appearance of a nitrogen peak following $NO_2$ exposure. A high-resolution scan of the nitrogen region at liquid nitrogen temperature is presented in Figure 5. Three distinct regions are evident in this spectrum. The largest peak at ~408 eV, corresponds to an inorganic nitrate (probably nitric acid); the peak at ~404 eV corresponds to that expected for an organic nitrate; and the peak at ~399 eV is probably that of a reduced nitrogen species such as a nitrile, amino, or pyridino compound. Observation of the behavior of these three regions at both room temperature and 80 C shows a gradual depletion of the high binding energy

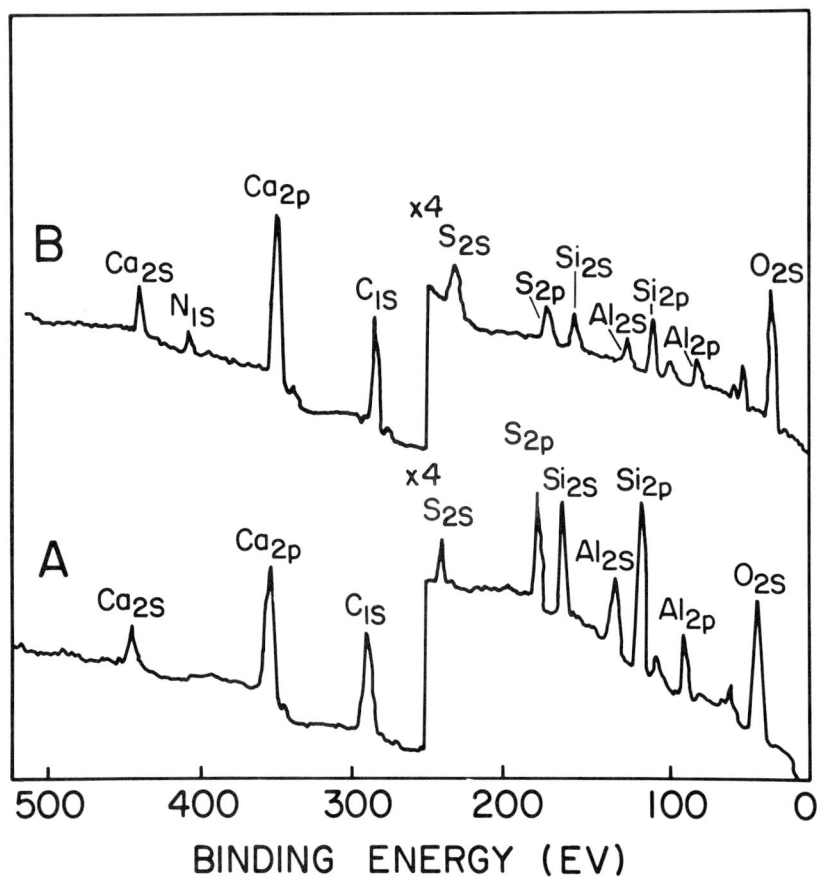

**FIGURE 4.** Low-resolution ESCA spectra of a (A) coal fly ash surface following pyrene adsorption, and (B) a coal fly ash surface following adsorption of pyrene and exposure to $NO_2$.

nitrogen species and a slight increase in the reduced nitrogen species with increasing temperature.

These results suggest the existence of nitric acid on the surface of the fly ash and are consistent with the idea that nitric acid is either a nitrating agent or a catalyst in these systems, as in those studied by Pitts et al (8). No reduced nitrogen species were extracted or identified but the formation of such species has been reported on soot by Chang and Novakov (1). No useful information about PAH reaction products was obtained from examination of the high-resolution carbon and sulfur peaks.

In summary, the data indicate that particulate PAH can undergo a variety of reactions when exposed to normal plume gases. The PAH studied appear to react most readily with $NO_2$, forming both mononitro

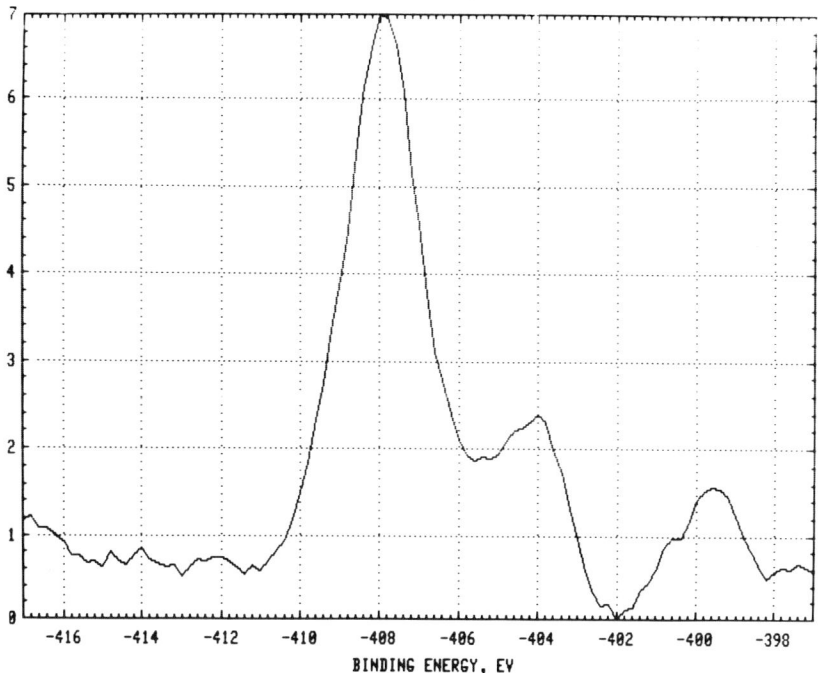

**FIGURE 5.** High-resolution ESCA spectrum of a coal fly ash surface following adsorption of pyrene and exposure to $NO_2$.

and dinitro products and possibly reduced species as well. Some reaction occurs with $SO_3$ but none were observed with either NO or $SO_2$ under similar experimental conditions. The nature of the reactions appears to differ depending on the surface involved, with reactions proceeding most readily on silica gel. It is believed that the acidity of the surface, particularly the presence of nitric acid, facilitates the reactions between the PAH and $NO_2$ gas.

## ACKNOWLEDGMENT

The authors gratefully wish to acknowledge the support of the U. S. Department of Energy for its support through contracts EP-78-S-02-4960.A000 and EE-77-S-02-4347.

## REFERENCES

1. Chang, S. G. and Novakov, T. (1975): Formation of pollution particulate nitrogen compounds by NO-soot and $NH_3$-soot gas-particle surface reactions. *Atmos. Environ.* 9:495-504.

2. Committee on Biologic Effects of Atmospheric Pollutants (1972): *Particulate polycylic organic matter.* National Academy of Sciences, Washington, D. C.
3. Natusch, D.F.S. (1978): Potentially carcinogenic species emitted from fossil fueled power plants. *Environ. Health Perspectives* 22: 79-90.
4. Korfmacher, W. A., Natusch, D.F.S., Taylor, D. R., Mamantov, G., and Wehry, E. L. (1979): Oxidative transformations of polycyclic aromatic hydrocarbons adsorbed on coal fly ash. *Science (in press).*
5. Korfmacher, W. A., Natusch, D.F.S., Taylor, D. R., Wehry, E. L. and Mamantov, G. (1979): Thermal and photochemical decomposition of particulate PAH. In *Polynuclear Aromatic Hydrocarbons,* P. W. Jones and P. Leber, Eds. pp. 165–169, Ann Arbor Science, Ann Arbor, Michigan.
6. Miguel, A. H., Korfmacher, W. A., Wehry, E. L., Mamantov, G. and Natusch, D.F.S. (1979): Apparatus for vapor-phase absorption of polycyclic organic matter onto particulate surfaces. *Environ. Sci. Technol.* 13:1229-1232.
7. Miguel, A. H. and Natusch, D.F.S. (1975): Diffusion cell for the preparation of dilute vapor concentrations *Environ. Sci. Technol.* 47:1705-1707.
8. Pitts, J. N., Jr., Van Cauwenberghe, K. A., Grosjean, D., Schmid, J. P., Fitz, D., Belser, W. L., Jr., Knudson, G. B. and Hynds, P. M. (1978): Atomspheric reactions of polycylic aromatic hydrocarbons: facile formation of mutagenic nitro derivatives. *Science* 202:515-519.

# ANALYSIS OF WATER-SOLUBLE CONJUGATES PRODUCED BY HAMSTER EMBRYO CELLS EXPOSED TO POLYNUCLEAR AROMATIC HYDROCARBONS

M. C. MacLeod, C. J. Moore, and J. K. Selkirk

Biology Division
Oak Ridge National Laboratory

University of Tennessee-Oak Ridge Graduate
School of Biomedical Sciences
Oak Ridge, Tennessee 37830

## INTRODUCTION

Polynuclear aromatic hydrocarbons (PAH) are metabolized by a complex enzyme system comprising the microsomal monooxygenase aryl hydrocarbon hydroxylase (AHH) and a number of cytoplasmic transferases (13-15, 37), which has evolved as the major detoxification system for xenobiotics. While the liver is the primary organ for drug detoxification, AHH activity appears to be ubiquitous in mammalian tissues. The enzyme has been extensively studied in microsomal and reconstituted systems (15, 26, 33, 34, 37) where the absence of competing and/or subsequent reactions allows for the determination of the initial rates of reaction in the various regions of the substrate molecule. Highly reactive epoxides are formed as intermediates in the detoxification process, and the reaction of these epoxides with critical cellular macromolecular targets is thought to play a key role in the cytotoxic, mutagenic, and carcinogenic effects of PAH (9, 15, 30, 37).

Metabolism of PAH has also been studied in whole animals (7, 8, 36) and in cell culture (1, 2, 10-12, 16, 17, 24, 27, 28) and it has been noted

(33, 35) that the pattern of metabolites formed in isolated microsomal systems is often different from that formed by the intact cells or tissues from which the microsomes were derived, which suggests the need to study the intact system more closely.

Such studies have revealed that conjugation of the metabolites following microsomal metabolism is important in the generation of the overall pattern. For example, Baird et al (2) reported a 72 percent conversion to water-soluble metabolites in hamster embryo cell (HEC) cultures incubated 24 hours with a low concentration of benzo(a)pyrene (BaP). Even with an eightfold higher concentration, we have shown that conversion was complete within 48 hours (unpublished observations). The critical importance of conjugation as a competing reaction was emphasized in a study of the fate of BaP-7,8-dihydrodiol formed metabolically in BALB/c mouse embryo cell cultures (27). As shown in Figure 1, BaP-7,8-dihydrodiol is eliminated by excretion as the free diol during the first 24 hours of incubation with $^3$H-BaP. Subsequent to this, conjugation to glucuronic acid gains in importance and by 72 hours essentially all of the 7,8-dihydrodiol is found as the glucuronide

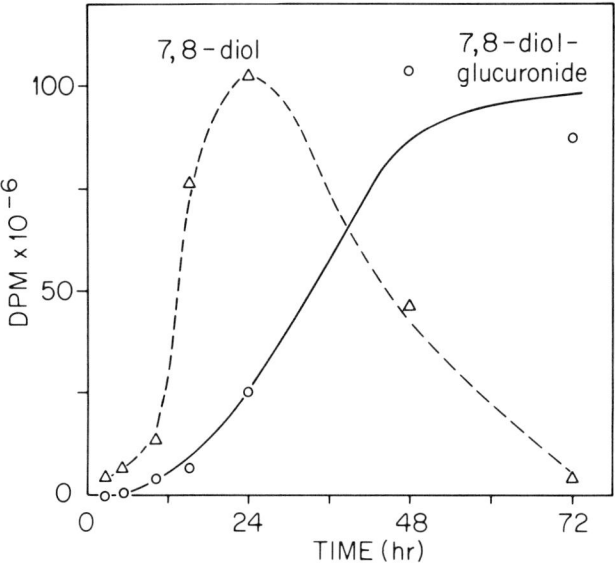

**FIGURE 1.** Conjugation of BaP-7, 8-dihydrodiol by mouse embryo cells. BALB/c mouse embryo cell cultures were treated with 4 μM BaP for periods up to 3 days. The amount of BaP-7, 8-dihydrodiol free in the extracellular medium or present as the glucuronide conjugate (after treatment with β-glucuronidase) was determined by high-pressure liquid chromatography.

conjugate. Furthermore, in some systems (12) glucuronidation appears to be the exclusive pathway for eliminating BaP-7, 8-dihydrodiol. Thus it is likely that in some situations the formation of potentially important metabolites will be completely overlooked unless both organic solvent-soluble and $H_2O$-soluble metabolites are analyzed in detail. As can be seen in Table 1, a number of conjugation reactions are possible. Early workers (4-9, 36) attempted, with the limited analytical capability available at the time, to assess the importance of all branches of the metabolic pattern in whole animals, tissue slices, and homogenates, whereas more recent work in cell culture has concentrated on the glucuronide and sulfate conjugation pathways (2, 10-12, 27, 28). In our own work with HEC cultures, we have found that at all times during the incubations with a variety of PAH substrates glucuronides constitute from 40 to 60 percent of the total water-soluble metabolites. Thus it has become of great interest to analyze the remaining water-soluble derivatives since they constitute a relatively large fraction of the total metabolites. In this paper we present evidence suggesting that the major conjugation pathway in HEC cultures other than glucuronidation is conjugation to a cysteine-containing moiety, presumably glutathione (GSH).

TABLE 1. Possible Pathways for PAH Detoxification

| Process | Substrate |
|---|---|
| Epoxide hydration | Epoxide |
| Glucuronic acid conjugation | Phenol, dihydrodiol, quinone |
| Macromolecular interaction | Epoxide |
| Sulfate conjugation | Phenol |
| Glutathione conjugation | Epoxide |

## MATERIALS AND METHODS

Sources for most of the materials were as described previously (28). Other sources were: Raney Nickel Catalyst, Grace Chemical, South Pittsburgh, TN; 3-benzo(a)pyrene-$\beta$-D-glucopyranosiduronic acid (3-GA-BaP), Midwest Research, Kansas City, MO; Sephadex LH-20, Pharmacia, Piscataway, NJ; ($^{35}$S)-L-cysteine hydrochloride, Amersham, Arlington Heights, IL; aryl sulfatase, Catalog #S9754, Sigma, St. Louis, MO; SepPak-C18, Waters, Milford, MA.

The basic experimental plan, outlined in Table 2, has been described in previous communications (28). The ethyl acetate extraction at pH 4.5 (step 3) removes the same fraction of $^3$H-BaP-derived radioactivity from the aqueous phase as does treatment with $\beta$-glucuronidase. The final aqueous phase is greater than 90 percent resistant to $\beta$-glucuronidase, while the material extracted into ethyl acetate at pH 4.5 is susceptible to $\beta$-glucuronidase [88 percent cleavage as monitored by high-pressure liquid chromatography (HPLC)].

Sephadex LH-20 (9.2 g dry weight) was swelled in distilled water, degassed under reduced pressure and packed in a 16-mm column. All eluants were degassed before application to the column. Samples of the final aqueous phase were applied to the column after clarification by centrifugation at 12,000 x g for 10 min. Elution was with 80 to 100 ml H$_2$O followed by 80 to 100 ml MeOH.

To label cellular glutathione pools biosynthetically, HEC cultures were incubated 18 hours in 5.0 ml of medium containing 71.5 $\mu$Ci of $^{35}$S-L cysteine (83.92 mCi/mmol). The $^{35}$S-containing medium was then replaced with 10 ml of fresh medium to which was added 40 nmol of the indicated hydrocarbon in 0.044 ml dimethylsulfoxide (DMSO) or DMSO alone, and incubation was continued for 24 hours.

**TABLE 2. Experimental Design**

1. Cells + $^3$H-PAH, 24 hours
2. Extract 4 times with 2 vol. ethyl acetate
3. Acidify, extract 4 times with 2 vol. ethyl acetate
4. (a) Clarify by centrifugation, absorb to SepPak C18 cartridge; and elute with methanol or (b) clarify by centrifugation, chromatograph on Sephadex LH-20, and elute with methanol.

## Macromolecular Binding and Sulfate Conjugation

Covalent binding of PAH metabolites to cellular macromolecules was first demonstrated in mouse skin (9, 20, 30) and has since been well documented in cultured cells (18, 25), including HEC (3, 21, 28, 29). It is generally thought that the cytotoxic, mutagenic, and carcinogenic properties of PAH result from such interactions, presumably through modulation of the function of critical target sites. It is possible that binding of reactive intermediates to noncritical, cellular macromolecules represents a detoxification mechanism, decreasing the probability of reaction with target sites which lead to carcinogenesis. We found that the total amount of covalent binding of $^3$H-BaP to cellular macromolecules is less than 1 percent of the overall metabolism, making this a quantitatively minor pathway. An additional 5 percent of the metabolic products are associated with proteins in the medium (assayed by methanol precipitation). This material can be separated from other water-soluble metabolites by LH-20 chromatography (see below).

Sulfate conjugation of BaP phenols has been shown to be an important pathway in both respiratory (10, 31) and hepatic tissue (24). The conjugates are soluble in ethyl acetate and are cleaved by aryl sulfatase. In experiments utilizing inorganic-($^{35}$S) sulfate labeling (G. M. Cohen, personal communication) or aryl sulfatase treatment (M. C. MacLeod and B. K. Mansfield, unpublished data) of the organic solvent-soluble metabolites produced by HEC cultures during a 24-hour incubation with BaP, no sulfate conjugates could be detected by HPLC. Sulfatase treatment of the water-soluble metabolites followed by ethyl acetate extraction yielded 0.75 percent as an upper limit to the fraction of the total metabolites which could be sulfate conjugates. These data do not, however, rule out the possibility of a more elaborate sulfate conjugate which is resistant to cleavage by aryl sulfatase.

## Chromatography of Water-Soluble Metabolites

In preliminary attempts to fractionate the nonglucuronide water-soluble metabolites, we filtered the aqueous medium through cartridges containing an octadecylsilane sorbent (SepPak-C18). After washing with distilled $H_2O$, bound material amounting to 65 percent of the radioactivity was eluted with MeOH. This preparation had chromatographic and fluorescent properties identical to metabolite AIII described below (data not shown). However, due to the very low capacity of the SepPak cartridges, we developed chromatography on Sephadex LH-20 as an alternative preparative method.

Three peaks of radioactivity are obtained when the water-soluble nonglucuronide metabolites of BaP are chromatographed on Sephadex LH-20 (Figure 2). The first peak to elute from the column (labeled AI, Figure 2) is co-chromatographic with bovine serum albumin and contains all of the trichloroacetic acid (TCA)-precipitable radioactivity in the preparation. When fractions corresponding to AI are pooled, the radioactivity (18.5 percent of the total) is 38 percent TCA-precipitable suggesting that a major component of AI is PAH bound to protein.

Partially overlapping AI is a broad peak of material which gives a fluorescent product when reacted with fluorescamine, a reagent specific

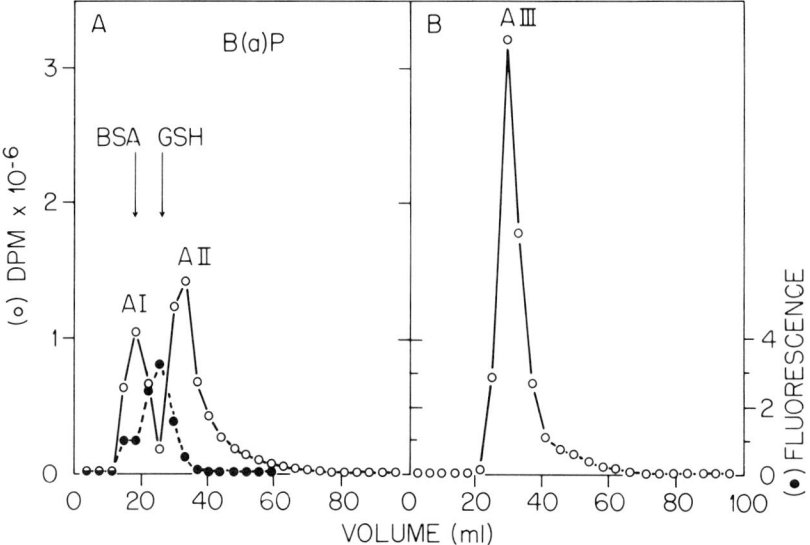

FIGURE 2. Sephadex LH-20 chromatography of water-soluble metabolites of BaP. The extensively extracted and acid-treated medium was injected (2.0 ml) onto a column of Sephadex LH-20 equilibrated with $H_2O$ and chromatographed at a flow rate of about 1 ml/min with $H_2O$ as the eluant. A. Aliquots (100 µl) of the first 20 fractions were reacted with fluorescamine as previously described, and 1.0 ml aliquots of all fractions were mixed with 10 ml Aquasol for liquid scintillation counting (LSC). B. After the radioactivity in the eluant had reached background levels, material bound to the column was eluted with 100 percent methanol. We have recently found that with some batches of Sephadex LH-20, superior resolution can be obtained by substituting phosphate-buffered saline (Dulbecco's 'A', KC Biological Lenexa KS) for $H_2O$.

for primary amines (38). In separate runs, samples of glutathione, glycine, and phenylalanine chromatographed in this region, and we presume that the peak is due primarily to the amino acid components of the growth medium.

A second radioactive peak (AII, 29.9 percent of the total) also elutes with $H_2O$, on the trailing edge of the peak of fluorescamine-positive material. The nature of the radioactivity in this peak is at present unknown.

The major peak (AIII, 52.5 percent of the total) is eluted with methanol. When fractions containing AIII are pooled, concentrated, and analyzed by HPLC (Figure 3) a single peak results which has a retention time approximately 0.6 min longer than that for 3-GA-BaP on this reverse-phase system. In Figure 4 we compare the fluorescence excitation (panel A) and emission (panel B) spectra of peak AIII with spectra of the glucuronide conjugate of 3-OH-BaP (3-GA-BaP). The spectra are very

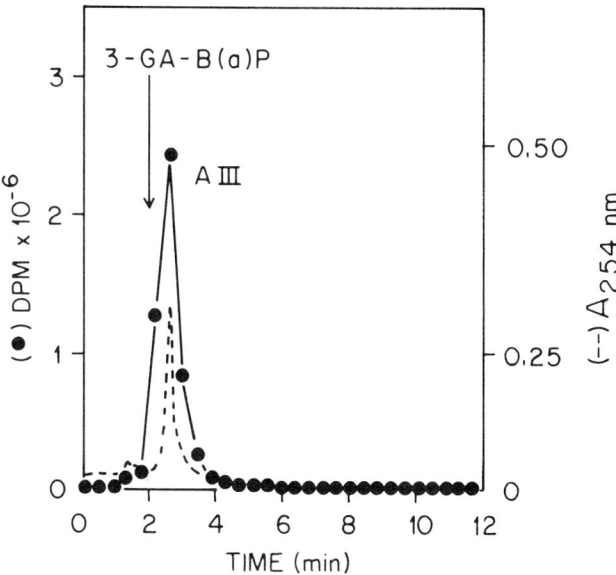

**FIGURE 3.** HPLC analysis of peak AIII. Sephadex LH-20 peak AIII was concentrated, adjusted to 50 percent methanol, 50 percent $H_2O$, and injected at 45°C onto a Zorbax C8 column (0.46 x 25 cm; Du Pont Instruments, Wilmington, DE) running with 50 percent MeOH at a flow rate of 1.5 ml/min. Absorbance was monitored at 254 nm, and 0.3 min fractions were collected for LSC.

similar but the main peaks exhibit a 4 to 5 nm shift. Spectra with similar patterns and exhibiting small shifts in wavelength are also found for 3-OH-BaP and 9-OH-BaP (data not shown). We therefore take this spectral data to suggest that peak AIII is a conjugate of BaP in which the BaP moiety retains full aromaticity. Furthermore, the fluorescent spectra of AIII and 3-GA-BaP do not change when the apparent pH of the methanolic solution is raised above 12 with NaOH. Under these conditions, all 12 monohydroxy BaPs ionize to the corresponding phenolate ions with a concomitant red shift of C. 100 nm in the emission spectra (23). We conclude that AIII does not contain an ionizable phenolic group.

## AIII is not Derived from a Phenol

The similarity in retention times and spectral properties between AIII and 3-GA-BaP suggested that AIII might be derived by conjugation of a small molecule to a BaP phenol. To test this hypothesis we incubated HEC cultures with tritium-labeled-3-OH-BaP or 9-OH-BaP, the major metabolically formed phenols (34). The major metabolites found in such

## 16 BaP CONJUGATES FORMED IN CELL CULTURE

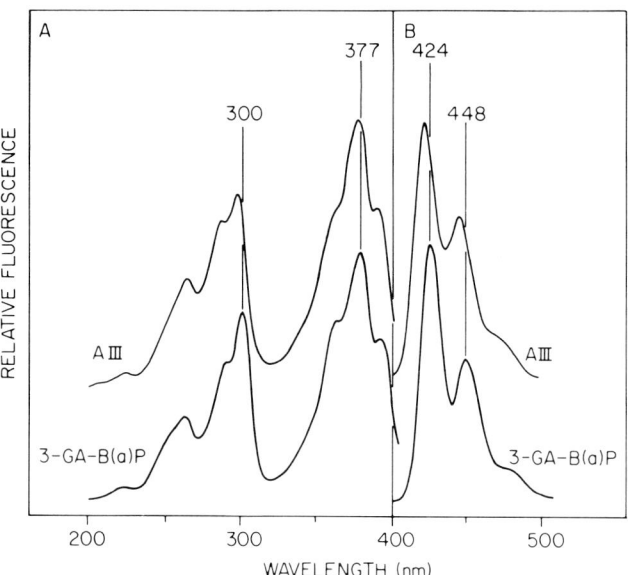

**FIGURE 4.** Fluorescence spectra of AIII and 3-GA-BaP. Relative fluorescence was determined on samples of peak AIII and synthetic 3-GA-BaP (both in 100 percent MeOH) on a Farrand Mark II spectrofluorometer. A. Uncorrected excitation spectra were collected by analyzing at 425 nm (3-GA-BaP) or at 420 nm (AIII). B. Emission spectra were obtained using 377 nm (3-GA-BaP) or 374 nm (AIII) as the excitation wavelength. For both compounds, the spectra were unaffected by the addition of NaOH to about 1 N (apparent pH >12).

incubations are the corresponding glucuronide conjugates (11). LH-20-chromatography of the remaining water-soluble metabolites derived from 3-OH- and 9-OH-BaP is shown in Figure 5. It can be seen that in both cases essentially all of the radioactivity elutes with $H_2O$ and only background levels are found in the position where AIII elutes from the column. Thus, 3-OH- and 9-OH-BaP are metabolized but do not give rise to a peak with the chromatographic properties of AIII. Since these are the major phenolic metabolites of BaP in HEC, this implies that AIII is not derived from a phenol.

### AIII is Probably a Glutathione Conjugate

The remaining possibility is that AIII represents a conjugation of glutathione (GSH) to a carbonium-ion formed from an epoxide. Taking BaP-9,10-oxide as an example (Figure 6, **1** and **2**), the result is the formation of -OH and -SG groups on adjacent carbons which is inconsistant with our finding that AIII does not contain an ionizable -OH

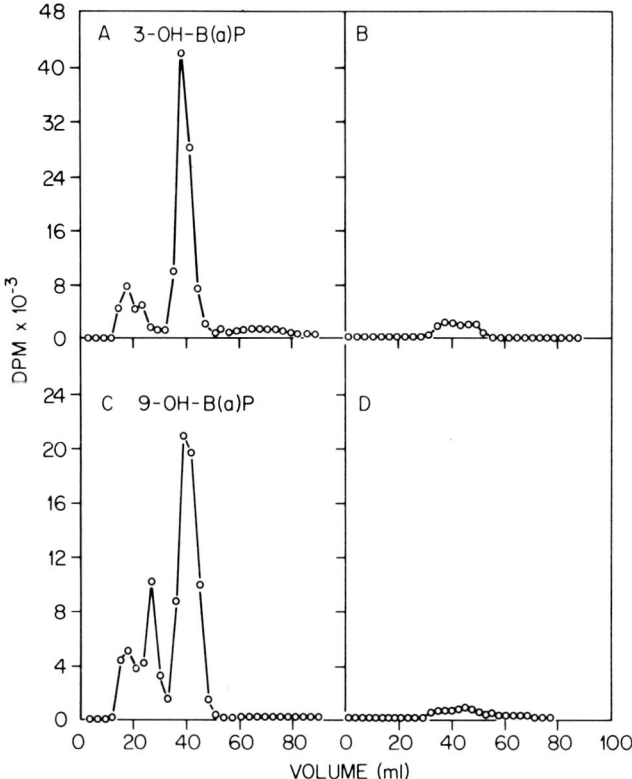

**FIGURE 5.** Water-soluble metabolites of 3-OH- and 9-OH-BaP. Chromatography was performed as in Figure 2 but with samples of medium from cultures treated with 3-OH-BaP (A and B) or with 9-OH-BaP (C and D). The medium from the 9-OH-BaP incubation had been previously treated with β-glucuronidase. Similar profiles were obtained in untreated samples (data not presented).

(see above). However, it has been shown (22) that the GSH conjugate of naphthalene-1,2-oxide is easily dehydrated with acid, accompanied by a migration of the glutathionyl moiety (for the example in Figure 6 this results in structure **3**). Since acid treatment is used to prepare AIII, a structure analogous to **3** had to be considered. To test this, we treated AIII with Raney Nickel which has been shown to specifically catalyze the hydrolysis of C-S bonds (6, 32) and should therefore yield BaP (Figure 6, **4**). Analysis of AIII by HPLC before and after treatment with Raney Nickel is shown in Figure 7. Metabolite AIII chromatographed as a single peak under these conditions with a retention time of 2 min (panel A). After Raney Nickel treatment (panel B), the major product (86 percent) co-chromatographed with BaP (retention time 11 min) and

## 18  BaP CONJUGATES FORMED IN CELL CULTURE

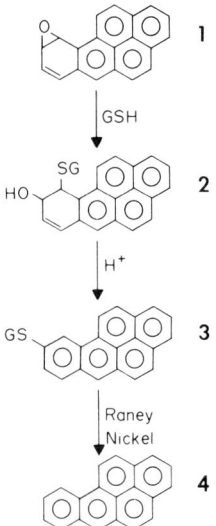

FIGURE 6. Formation, dehydration and hydrolysis of a conjugate of BaP-9,10-oxide with glutathione.

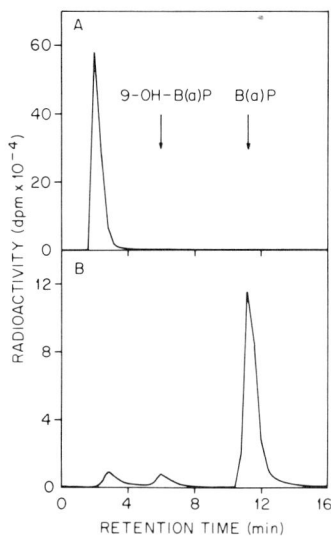

FIGURE 7. HPLC analysis of AIII ± Raney Nickel treatment. A sample of AIII in 1 ml of MeOH was treated with about 1 g of an aqueous slurry of Raney Nickel No. 28 for 2 hours at 42 C, then filtered and concentrated. HPLC analysis was on Zorbax C8 in 80 percent methanol, 45°C, 1.5 ml/min. A. Untreated AIII. B. Raney Nickel-treated AIII.

demonstrated fluorescence excitation and emission maxima characteristic of BaP (data not shown). An additional 7 percent of the radioactivity was found in the phenolic region of the chromatogram and may represent molecules which escaped acid dehydration.

## SUMMARY

We have presented evidence which suggests that a major metabolite of BaP formed by HEC cultures, AIII, is a glutathione conjugate. Since arene oxides are known to form adducts with glutathione (19, 37), we presume that metabolite AIII derives from an epoxide intermediate and has been acid dehydrated during preparation. Because of the loss of the hydroxyl function we have not identified the original site of epoxidation. If AIII derives from BaP-2, 3-oxide, as suggested by the similarities between the fluorescence spectra of AIII and 3-GA-BaP, it would account for the fact that isolated HEC microsomes make reasonably large amounts of 3-OH-BaP (35), whereas only small amounts of 3-OH-BaP have been detected in the ethyl acetate-soluble and glucuronide fractions of intact cell incubations (28). The methodology presented here should be useful in preparing large amounts of AIII for more rigorous structural analysis.

The general distribution of PAH in a typical 24-hour incubation of BaP with HEC is given in Table 3. Glucuronide conjugates, organic solvent-soluble metabolites, and the putative GSH conjugate are formed in approximately equal proportions, and together constitute over 75 percent of the total metabolic products. Clearly, an incomplete and probably inaccurate picture of the overall metabolic pattern will result if any of these fractions are ignored. Given the apparent selectivity of the conjugation pathways for specific metabolites, the demonstrated time dependency, and the system-to-system variation (1, 10, 12, 27, 28, 33, 35),

**TABLE 3. Overall Distribution of Metabolites**

| Fraction | % of Total |
| --- | --- |
| Glucuronides | 23 |
| Organic solvent-soluble metabolites | 21 |
| AIII | 20 |
| Unmetabolized BaP | 17 |
| Other water-soluble metabolites (AII) | 11 |
| Macromolecular bound metabolites | 8 |

Note: HEC cultures were incubated 24 hours with 4 $\mu$M BaP and the various fractions prepared as described above. The last entry represents the sum of intracellular water-soluble radioactivity and fraction AI from the LH-20 chromatography. Percentages were reconstructed from several representative experiments.

it becomes especially important to construct a complete description of the metabolism at more than one time point in the process. Only in this way can metabolism be related to biological endpoints in a meaningful way.

## ACKNOWLEDGMENTS

This research was supported jointly by the National Cancer Institute under Interagency Agreement (4-5-63) and the Office of Health and Environmental Research, U. S. Department of Energy, under Contract W-7405-eng-26 with the Union Carbide Corporation. C. J. Moore is a postdoctoral investigator supported by Carcinogenesis Training Grant CA 09336 from the National Institutes of Health to The University of Tennessee—Oak Ridge Graduate School of Biomedical Sciences. Expert technical help was provided by B. K. Mansfield and K. Dearstone and secretarial assistance was from N. Trent.

## REFERENCES

1. Autrup, H. (1979): Separation of water-soluble metabolites of benzo(a)pyrene formed by cultured human cells. *Biochem. Pharmacol.* 28: 1727-1730.
2. Baird, W. M., Chern, C. J., and Diamond, L. (1977): Formation of benzo(a)pyrene-glucuronic acid conjugates in hamster embryo cell cultures. *Cancer Res.* 37: 3190-3197.
3. Baird, W. M., and Diamond, L. (1977): The nature of benzo(a)pyrene-DNA adducts formed in hamster embryo cells depends on the length of time of exposure to benzo(a)pyrene. *Biochem. Biophys. Res. Comm.* 77: 162-167.
4. Booth, J., Boyland, E., and Sims, P. (1960): Metabolism of polycyclic compounds 15. The conversion of naphthalene into a derivative of glutathione by rat liver slices. *Biochem. J.* 74: 117-122.
5. Booth, J., Keysell, G. R., and Sims, P. (1973): Formation of glutathione conjugates as metabolites of 7,12-dimethylbenz(a)-anthracene by rat liver homogenates. *Biochem. Pharmacol.* 22: 1781-1791.
6. Boyland, E., and Sims, P. (1960): Metabolism of polycyclic compounds 16. The metabolism of 1, 2-dihydronaphthalene and 1, 2-epoxy-1, 2, 3, 4 tetrahydronaphthalene. *Biochem. J.* 77: 175-181.
7. Boyland, E., and Sims, P. (1962): Metabolism of polycyclic compounds 20. The metabolism of phenanthrene in rabbits and rats: mercapturic acids and related compounds. *Biochem. J.* 84: 564-570.

8. Boyland, E., and Sims, P. (1962): Metabolism of polycyclic compounds 21. The metabolism of phenanthrene in rabbits and rats: dihydrodihydroxy compounds and related glucosiduronic acids. *Biochem. J.* 84: 571-582.
9. Brookes, P., and Lawley, P. D. (1964): Evidence of the binding of polynuclear aromatic hydrocarbons to the nucleic acids of mouse skin: Relation between carcinogenic power of hydrocarbons and their binding to DNA. *Nature* 202: 781-782.
10. Cohen, G. M., Haws, S. M., Moore, B. P., and Bridges, J. W. (1976): Benzo(a)pyren-3-yl hydrogen sulfate, a major ethyl acetate-extractable metabolite of benzo(a)pyrene in human, hamster, and rat lung cultures. *Biochem. Pharmacol.* 25: 2561-2570.
11. Cohen, G. M., MacLeod, M. C., Moore, C. J., and Selkirk, J. K. The metabolism and macromolecular binding of carcinogenic and non-carcinogenic metabolites of BaP by hamster embryo cells. *Cancer Res.* (in press).
12. Cohen, G. M., Marchok, A. C., Nettesheim, P., Steele, V. E., Nelson, F., Huang, S., and Selkirk, J. K. (1979): Comparative metabolism of benzo(a)pyrene in organ and cell cultures derived from rat tracheas. *Cancer Res.* 39: 1980-1984.
13. Conney, A. H. (1967): Pharmacological implications of microsomal enzyme induction. *Pharmacol. Rev.* 19: 317-366.
14. Conney, A. H., Miller, E. C., and Miller, J. A. (1957): Substrate-induced synthesis and other properties of benzpyrene hydroxylase in rat liver. *J. Biol. Chem.* 228: 753-766.
15. DePierre, J. W., and Ernster, L. (1978): The metabolism of polycyclic hydrocarbons and its relationship to cancer. *Biochem. Biophys. Acta.* 473: 149-186.
16. Diamond, L. (1971): Metabolism of polycyclic hydrocarbons in mammalian cell cultures. *Int. J. Cancer* 8: 451-462.
17. Diamond, L., Sardet, C., and Rothblat, G. H. (1968): The metabolism of 7, 12-dimethylbenz(a)anthracene in cell cultures. *Int. J. Cancer* 3: 838-849.
18. Grover, P. L., Forrester, J. A., and Sims, P. (1971): Reactivity of the K-region epoxides of some polycyclic hydrocarbons toward the nucleic acids and proteins of BHK21 cells. *Biochem. Pharmacol.* 20: 1297-1302.
19. Hayakawa, T., Udenfriend, S., Yagi, H., and Jerina, D. M. (1975): Substrate and inhibitors of hepatic glutathione-S-epoxide transferase. *Arch. Biochem. Biophys.* 170: 438-451.
20. Heidelberger, C., and Moldenhaur, M. G. (1956): The interaction of carcinogenic hydrocarbons with tissue constituents. IV. A quantitative study of the binding to skin proteins of several $^{14}$C-labeled hydrocarbons. *Cancer Res.* 16: 442-449.

21. Ivanovic, V., Geacintov, N. E., Yamasaki, H., and Weinstein, I. B. (1978): DNA and RNA adducts formed in hamster embryo cell cultures exposed to benzo(a)pyrene. *Biochemistry* 17: 1597-1603.
22. Jeffery, A. M., and Jerina, D. M. (1975): Novel rearrangements during dehydration of nucleophile adducts of arene oxides. A reappraisal of premercapturic acid structures. *J. Amer. Chem. Soc.* 97: 4427-4428.
23. Jerina, D. M., Yagi, H., Hernandez, O., Dansette, P. M., Wood, A. W., Levin, W., Chang, R. L., Wislocki, P. G., and Conney, A. H. (1976): Synthesis and biologic activity of potential benzo(a)pyrene metabolites. In: *Polynuclear Aromatic Hydrocarbons: Chemistry, Metabolism and Carcinogenesis,* edited by R. Freudenthal and P. W. Jones, pp. 91-113, Raven Press, New York.
24. Jones, C. A., Moore, B. P., Cohen, G. M., Fry, J. R., and Bridges, J. W. (1978): Studies on the metabolism and excretion of benzo(a)pyrene in isolated adult rat hepatocytes. *Biochem. Pharmacol.* 27: 693-702.
25. Kuroki, T., and Heidelberger, C. (1971): The binding of polycyclic aromatic hydrocarbons to the DNA, RNA, and proteins of transformable cells in culture. *Cancer Res.* 31: 2168-2176.
26. Lu, A.Y.H., Levin, W., Vore, M., Conney, A. H., Thakker, D. R., Holder, G., and Jerina, D. M. (1976): Metabolism of BaP by purified liver microsomal cytochrome P448 and epoxide hydrase. In: *Polynuclear Aromatic Hydrocarbons: Chemistry, Metabolism, and Carcinogenesis,* R. Freudenthal and P. W. Jones, Eds., pp. 115-126, Raven Press, New York.
27. MacLeod, M. C., Cohen, G. M., and Selkirk, J. K. (1979): Differences in detoxification pathways for benzo(a)pyrene and benzo(e)pyrene in rodent fibroblasts and epithelial cells. *Proc. Amer. Assoc. Cancer Res.* 20: 268-273.
28. MacLeod, M. C., Cohen, G. M., and Selkirk, J. K. (1979): Metabolism and macromolecular binding of the carcinogen benzo(a)pyrene and its relatively inert isomer benzo(e)pyrene by hamster embryo cells. *Cancer Res.* 39: 3463-3470.
29. MacLeod, M. C., Mansfield, B. K., Huff, A., and Selkirk, J. K. (1979): Simultaneous preparation of nuclear DNA, RNA, and protein from carcinogen-treated hamster embryo fibroblasts. *Anal. Biochem.* 97: 410-417.
30. Miller, E. C. (1951): Studies on the formation of protein-bound derivatives of 3, 4-benzpyrene in the epidermal fraction of mouse skin. *Cancer Res.* 11: 100-108.

31. Moore, B. P., and Cohen, G. M. (1978): Metabolism of benzo(a)pyrene and its major metabolites to ethyl acetate-soluble and water-soluble metabolites by cultured rodent trachea. *Cancer Res.* 38: 3066-3075.
32. Pettit, G. R., and Van Tamelen, E. E. (1962): Desulfurization with Raney nickel. *Organic Reactions* 12: 356-529.
33. Selkirk, J. K. (1977): Divergence of metabolic activation systems for short-term mutagenesis assays. *Nature (London)* 270: 604-607.
34. Selkirk, J. K., Croy, R. G., and Gelboin, H. V. (1976): High-pressure liquid chromatographic separation of 10 benzo(a)pyrene phenols and the identification of 1-phenol and 7-phenol as new metabolites. *Cancer Res.* 36: 922-926.
35. Selkirk, J. K., Croy, R. G., Wiebel, F. J., and Gelboin, H. V. (1976): Differences in benzo(a)pyrene metabolism between rodent liver microsomes and embryonic cells. *Cancer Res.* 36: 4476-4479.
36. Sims, P. (1962): Metabolism of polycyclic compounds 19. The metabolism of phenanthrene in rabbits and rats: phenols and sulphuric esters. *Biochem. J.* 84: 558-563.
37. Sims, P., and Grover, P. L. (1974): Epoxides in polycyclic aromatic hydrocarbon metabolism and carcinogenesis. *Adv. Cancer Res.* 20: 165-274.
38. Udenfriend, S., Stein, S., Bihler, P., Dairman, W., Leimgruber, W., and Weigele, M. (1972): Fluorescamine: A reagent for assay of amino acids, peptides, proteins and primary amines in the picomole range. *Science* 178: 871-872.

# USE OF MIXED PHASES FOR ENHANCED GAS-CHROMATOGRAPHIC SEPARATION OF POLYCYCLIC AROMATIC HYDROCARBONS: PRELIMINARY STUDIES WITH LIQUID CRYSTALS

R. J. Laub and W. L. Roberts
Department of Chemistry
The Ohio State University
Columbus, Ohio 43210

## INTRODUCTION

Although the ubiquitous presence of polycyclic aromatic hydrocarbons (PAH) is today taken for granted, recognition of the distribution of PAH throughout the environment has come to predominate slowly and only in the past 30 years, commensurate with developments in analytical instrumentation, techniques, and procedures. For example, although the carcinogenicity of tobacco resins and tars (1) was recognized as early as 1936 as being due to benzpyrenes (2) and although by 1940 teas, coffees, and other foodstuffs were also known to contain carcinogenic hydrocarbons (3), emphasis was placed upon the development of analytical procedures for PAH only in the 1950s largely because of attempts at linking the occurrence of lung cancer with the use of cigarettes (4-15). Simultaneously, PAH were shown to be present in, for example, sewage effluents (16,17) and of greater concern, in urban and rural air (18-35) and water (36). Current controversy regarding energy sources (and the lack thereof) has, however, caused a shift in interest in PAH analysis toward petroleum and its constituents (37-40), a trend which is likely to continue in the foreseeable future.

A variety of methods have been developed over the years for PAH analysis. Column chromatography followed by UV spectroscopy of the resultant fractions was pioneered by Wedgewood and Cooper (16) in 1953 and later applied with some success by Cooper and co-workers (5-8) to a variety of sample types. Wynder and Wright (41) subsequently developed a comprehensive fractionation scheme based upon solute partitioning between pairs of solvents, which was later simplified by Hoffman and Wynder (20), who also reported partition coefficients of PAH with cyclohexane/water + methanol and with cyclohexane/nitromethane. The method was later (42) expanded in 1960 to include paper and column chromatography, and UV examination of the resultant PAH-containing fractions. Sawicki and Miller (18) devised an interesting color test for PAH in 1958 which eventually (21) led to their use of a combination of techniques (including column chromatograpy) in 1960. Short and Young (43) provided an innovative modification to column chromatographic fractionation in 1969 with the dispersion of pyromellitic dianhydride (a charge-transfer Lewis acid) throughout the packing. Sample cleanup was considered in detail by Dubois and co-workers (31), who also employed what amounts to UV and fluorescence background subtraction spectroscopy. Hoffman and Rathkamp (44) and Popl, Stejskal, and Mostecky (45) reported improved column chromatographic techniques for PAH analysis in 1972-1975; the former group utilized subsequent gas-chromatographic (GC) and mass-spectrometric (MS) separation and identification methods, whereas the latter workers employed fluorescence and phosphorescence spectroscopy as advocated in a number of earlier studies (46-51). Various forms of paper (52-54), thin-layer (25,27,55-59), and gel (51) chromatography have also been employed for sample fractionation, and electroluminescence measurement of submicrogram amounts of PAH has been described (60,61), the latter involving a technique that appears to have been overlooked in the search for selective detectors for HPLC.

Gas chromatography has, of course, been used extensively as a separations tool for PAH: Carugno (62), Sawicki, Fox, and Elbert (63), Lijinsky et al (64), and many others (65-73) studied the use of the technique for polycyclic and polynuclear aromatic hydrocarbons prior to 1964. In 1965, Wilmhurst (74) published the retentions of 50 PAH with packed columns containing SE-30 and SE-52, while Cantuti and co-workers (28) reported improved separations with glass capillary columns with SE-30, SE-52, and XE-60 phases. Vidal-Madjar and Guiochon (75) used copper phthalocyanine as an adsorbent for PAH analysis by gas-solid chromatography, while Zane (76) employed graphitized carbon black. Gump (77), on the other hand, utilized inorganic salt adsorbents to take advantage of selectivity due to charge-transfer complexation for the GSC separation of benzpyrenes. Sato, Matsui, and Ikekawa (78) also found that lithium nitrate dispersed in SE-30 proved to be a selective stationary phase for 15 aromatic hydrocarbons.

The separation of PAH seems at present most easily carried out by glass-capillary GC, largely because of the pioneering work of Carugno and Rossi (28), Cartoni (79), Bhatia (32), Lao et al (33), Zafiriou (34), Uden et al (39) and, most recently, Bjorseth and Eklund (80), Borwitzky and Schomburg (81), and Bernaert (82), in addition to the many others who have popularized and made practicable the related methodologies and procedures. Thus, although initially appearing to hold much promise, the techniques of, e.g., supercritical fluid chromatography (83) and liquid chromatography (84-90) are by comparison today little used for PAH analysis (see, however, the recent work of Scott and Kucera (91) and Figures 10 through 16 therein which may mark the reversal of this trend).

As is the case in many areas of GC analysis, considerable effort has been expended in order to identify stationary phases which are "selective" for PAH, that is, phases which provide maximum resolution and hence minimum time of analysis. Seemingly tailor-made for this task are mesogenic compounds (liquid crystals). First employed in GC by Kelker (92) and by Dewar and Schroeder (93), the use of such phases is becoming increasingly popular for a number of types of separations, as indicated in the reviews by Schroeder (94), Kelker and Von-Schivizhoffen (95), Kelker (96), and Kraus and Winterfeld (97). A number of workers have also described and discussed the use of a variety of types of mesogens (98-116), and the list continues to grow. For example, Zielinski and his colleagues (117-124) have developed and employed N,N'-bis(p-phenylbenzylidene)-$\alpha,\alpha'$-bi-p-toluidine (BPhBT), N,N'-bis(p-hexyloxybenzylidene)-$\alpha,\alpha'$-bi-p-toluidine (BHxBT), N,N'-bis(p-methoxybenzylidene)-$\alpha,\alpha'$-bi-p-toluidine (BMBT), and N,N'-bis(p-butoxybenzylidene)-$\alpha,\alpha'$-bi-p-toluidine (BBBT) phases, the structures and properties of which are shown in Table 1. Of these, BMBT has been found to give excessive column bleed with packed columns (117,118,120) while retentions with BMBT and with BBBT were said to be excessive (120), although the latter drawback is reduced somewhat when open-tubular capillary (as opposed to packed) columns (124) are employed. BHxBT has, however, a relatively small nematic range (48°C). The two most useful phases in Table 1 therefore appear to be BBBT and BPhBT, the former being chosen for this initial study.

In general terms, however, the use of any single stationary phase for gas-chromatographic separations limits the type or types of solute-solvent interactions one can employ for the analysis at hand. For example, SE-30 provides by and large a "boiling-point" separation, i.e., one based virtually upon solute vapor pressure. Thus, benzo(a)pyrene elutes together with benzo(e)pyrene with such stationary phases, and no separation is achieved. Liquid crystals, on the other hand, provide separations based predominantly upon molecular geometry. For example (120), benzo(d)-fluoranthene elutes just prior to benzo(e)pyrene, which is then followed by perylene and benzo(a)pyrene with a stationary phase composed of BPhBT. Clearly, it would be of considerable utility to be able to take

## Table 1. Structure and Properties of Liquid Crystal Stationary Phases

| Abbreviation | Structure | Molecular Weight (g mol$^{-1}$) | Transition Temperature (°C) | | |
|---|---|---|---|---|---|
| | | | Smectic | Nematic | Isotropic |
| BMBT | $H_3CO$–⟨O⟩–CH = N–⟨O⟩–$(CH_2)_2$–⟨O⟩–N = CH–⟨O⟩–$OCH_3$ | 448 | — | 181 | 320 |
| BBBT | $H_9C_4O$–⟨O⟩–CH = N–⟨O⟩–$(CH_2)_2$–⟨O⟩–N = CH–⟨O⟩–$OC_4H_9$ | 532 | 159 | 188 | 303 |
| BPhBT | ⟨O⟩–⟨O⟩–CH = N–⟨O⟩–$(CH_2)_2$–⟨O⟩–N = CH–⟨O⟩–⟨O⟩ | 540 | — | 257 | 403 |
| BHxBT | $H_{13}C_6O$–⟨O⟩–CH = N–⟨O⟩–$(CH_2)_2$–⟨O⟩–N = CH–⟨O⟩–$OC_6H_{13}$ | 588 | 127(I), 203(II) | 229 | 276 |

advantage simultaneously of a number of different types of solute/solvent interactions in order to enhance particular separations.

Several approaches for achieving separations on the above basis have, in fact, been devised over the years, Now (regrettably) called "two-dimensional" chromatography, a variety of multiple stationary-phase systems have been described (125-135) including those employing multiple-column switching (136-146), multiple-column multiple-temperature switching (147-148), and the use of multiple mobile phases (149-154), the subject having recently been reviewed briefly by Bertsch (155). Unfortunately and with few exceptions (156) prior to 1975, the use of any one or more of the above-mentioned methodologies was as hindered and as imprecise as the selection of a pure stationary phase for a given separation; that is, optimization of various components of the system (e.g., column length and temperature) proved largely to be what can most charitably be characterized as a matter of trial and error, the usual procedure (e.g., ref. 157) amounted to trying initially a solvent and set of conditions thought to be selective for the type of sample at hand if guidelines for the separation could not be located in the literature. A column of length often dictated by "standard" sizes available to the analyst was then packed with this stationary phase/support material and used at some operating temperature. If the subsequent analysis was not (for whatever reason) considered satisfactory, the column temperature was lowered and the run repeated. If still less than adequate, the procedure might well have been repeated several times with different stationary phases. Finally, the analyst may have resorted to ever-increasing lengths of capillary columns or multiple-column systems, which under ideal conditions offer the advantage of separation efficiency (however this is defined) which is greater than that possible with packed columns.

In contrast, Laub, Purnell, and their colleagues have recently formulated and established extremely simple and versatile predictive techniques for *quantitative* preselection of *all* conditions (including column length) required to effect a given separation by gas chromatography. The method, called the plenary optimization strategy, is based upon the use of multicomponent stationary phases and has thus far been applied with complete success to the GLC separation of hydrocarbon mixtures with squalane/dinonyl phthalute (DNP) binary phases (158-160) and with squalane/DNP/di-n-propyl tetrachlorophthalate (DPTC) ternary mixed phases (160), to the analysis of underivatized steroids with methylsilicone gum/trifluoropropylsilicone gum (160), to aromatic amines and hydrocarbons with methylsilicone gum/2,4,7-trinitrofluorenone (161), to vinyl chloride industrial stills residues with squalane/DNP (162), to chlorinated phenol and cresol pesticide metabolite residues with methylphenylsilicone/polyethylene glycol (163), and to the GSC and

GLSC separation of all commercially available $C_1$ to $C_4$ hydrocarbons (164). It has further been shown to offer the means of quantitative optimization of column temperature in GSC, GLSC, and HPLC (165), of mobile-phase composition in HPLC (166,167), and of solvent systems in NMR spectroscopic (168) and electrochemical analysis (169). Rarely, however, does an analyst have prior knowledge of the number of solutes in a given mixture nor in many cases the origin, let alone history, of the sample to be resolved. The optimization strategy was easily adapted to samples of initially unknown content and complexity by Laub and Purnell (162,163). Its completely successful application to analysis of such solute mixtures provides further illustration of the estimable productive power of the technique [which has since been computerized (160,170) for manipulation of large numbers of solutes].

The preliminary studies reported here are designed to capitalize on and to expand these developments by inclusion of mixed GLC phases comprising, in part, liquid crystals, for PAH analysis. The catholic nature of the methodologies is such, however, that virtually any sample type is amenable to analysis by the described techniques and procedures.

## THEORY

Many workers have attempted the prediction of relative as well as absolute retentions in efforts to enhance GC separations. For example, Zielinski and Martire (171) evaluated various solutes with monofunctional stationary phases and related the data through changes in retention indices (172). Functional-group contributions to solute retention volumes have also been tabulated in similar studies (173,174). Although these efforts have provided some success in limited cases, a theory or method with predictive capabilities for solving a broader range of analytical separations is clearly desirable. This is particularly true insofar as heretofore profferred methodologies leave much to be desired with respect to practical operating conditions. Part of the difficulty stems from the use of equivocal data. For example, McReynolds (175) determined the specific retention volumes of the solutes listed in Table 2 with a column containing squalane to which had been added 2 percent w/w of a surfactant. This seemingly trivial amount is sufficient to alter retention times and even retention order, as pointed out as early as in 1955 by Purnell and Spencer (176). Thus, McReynolds found that chloroform eluted after n-hexane with his (mixed-phase) solvent system, whereas chloroform elutes prior to n-hexane with pure squalane. Risby and colleagues (174) unfortunately failed to take these inconsistencies into account when they formulated functional-group contributions to retentions from McReynold's data. Their method of

TABLE 2. Specific Retention Volumes (ml g$^{-1}$) for Listed Solutes with Squalane

| Solute | Ref. 175 | | Refs. 178, 181 | |
|---|---|---|---|---|
| | 80°C | 100°C | 80°C | 100°C |
| n-Hexane | 50.2 | 30.7 | 60.33 | 34.07 |
| Benzene | 88.7 | 50.2 | 87.39 | 49.68 |
| Toluene | 226 | 118 | 213.8 | 110.2 |
| Chloroform | 55.1 | 31.7 | 51.71 | 30.75 |
| Carbon tetrachloride | 89.0 | 49.8 | 92.65 | 52.37 |
| Furan | 17.6 | 11.1 | 18.36 | 12.04 |

predicting specific retention volumes at temperatures other than those that McReynolds employed thus fails badly. For example, the specific retention volume of n-hexane is forecast, according to their relations, to be 186.5 ml g$^{-1}$ at 30°C, whereas the average of a number of independent GLC (177,178) and static (179,180) determinations of high accuracy is 343.3 ml g$^{-1}$.

It is hardly surprising in view of the foregoing that multicomponent sorbents have been little used in in analytical GC, since prediction of retention behavior with pure stationary phases, let alone their mixtures, would seem to be intractable. Nevertheless, there have been several noteworthy efforts made in this direction. The strategies developed have without exception been founded upon some or other function which relates solute retentions with pure solvents to those obtained with their mixtures. The simplest function which has received the most attention is that originally proposed by Primavesi (128), which has since been verified and discussed by numerous others (182,183):

$$V^\circ_{g(A,S)} = w_A V^\circ_{g(A)} + w_S V^\circ_{g(S)} \qquad (1)$$

where $V^\circ_{g(A)}$ and $V^\circ_{g(S)}$ are the solute specific retention volumes with pure stationary phase A and with pure stationary phase S, and where $w_A$ and $w_S$ are the weight fractions of A and S, respectively, in the binary solvent. Since

$$\phi_i = \frac{w_i\, \rho_{i,s}}{\rho^\circ_i} \qquad (2)$$

and

$$K^\circ_{R(i)} = \frac{(V^\circ_{g(i)})\, T\, \rho^\circ_i}{(273)} \qquad (3)$$

where $\phi$ is volume fraction, $\rho$ is a density, and $K^\circ_R$ a solute partition coefficient, Equation (1) is precisely equivalent to

$$K^\circ_{R(A,S)} = \phi_A K^\circ_{R(A)} + \phi_S K^\circ_{R(S)} \qquad (4)$$

a relation derived originally by Purnell and Vargas de Andrade (184) which was later christened the diachoric solutions hypothesis by Laub and Purnell (185). Equation (1) is more readily applied in analytical applications, however, since the density and molecular weight of the stationary phase are not required for measurement of $V^\circ_g$ data. Thus, it may be employed with polydisperse solvents (186,187) (such as silicone gums) which, generally speaking, require protracted (188) techniques for density measurements at elevated temperatures.

The ability to predict solute retentions with mixed phases (or serial columns) (130) from data relating to pure solvents is an important advance in gas chromatography. Consider, for example, the separation of a three-component solute mixture in which two of the components overlap with one phase while another pair overlaps with a second phase. Even though neither of the two pure solvents will provide resolution, there is every likelihood that some mixture of the two phases will effect the desired separation.

Figure 1 illustrates the situation for the hypothetical solutes labelled 1, 2, and 3 with phases A and S, where straight lines have been drawn between

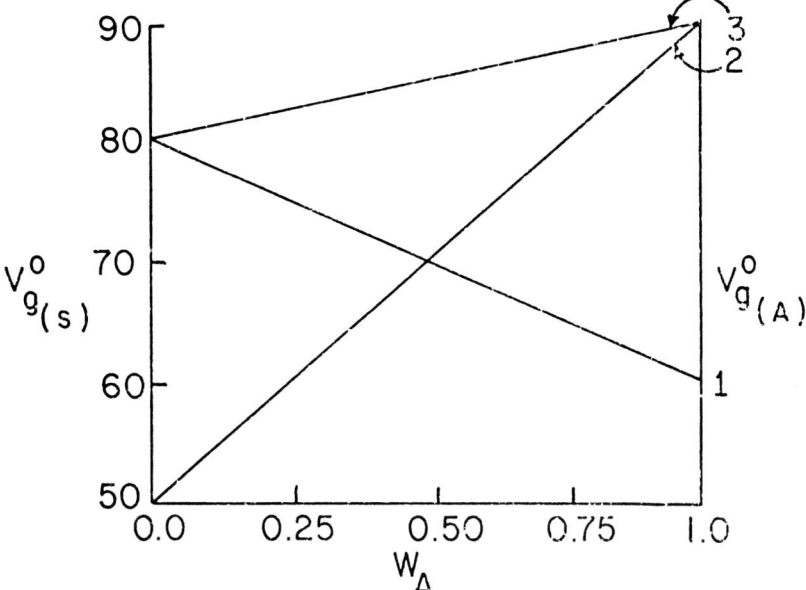

**FIGURE 1.** "Straight-line" plots of $V_g^\circ$ versus $w_A$ for three hypothetical solutes in accordance with Equation (1).

the two end-points $V^o_{g(s)}$ and $V^o_{g(a)}$ for each solute in accordance with Equation 1.

Inspection of the figure shows that overlap will occur at $w_A = 0.5$ in addition to those at $w_A = 0$ and $w_A = 1$, but that separation will be achieved at roughly 0.25 and at 0.75. More precise values of $w^{opt}_A$ could be determined by calculation [Equation (1)] and tabulation of $V^o_{g(s)}$ for each solute at intervals of $w_A$, the weight fraction which gives the largest difference (separation) between retention volumes being that which offers the highest resolution. Manual data reduction with this procedure becomes tedious for more than a few solutes but it can be computerized for up to 50 solutes with mixtures of three or more stationary phases (135).

Laub and Purnell (158) developed a method of graphical representation of retention data which simplifies considerably the procedure described above (partition coefficients and volume fractions were used in their initital studies but since Equation (3) is equivalent to Equation (1), their method is compatible with either $K^o_R$ or $V^o_g$ data). First, these workers noted that the relative retention of two solutes $\alpha$ may, according to Equation (1), be cast in the form:

$$\alpha_{i/j} = \frac{V^o_{g(A,S)_i}}{V^o_{g(A,S)_j}} = \frac{w_A V^o_{g(A)_i} + w_S V^o_{g(S)_i}}{w_A V^o_{g(A)_j} + w_S V^o_{g(S)_j}} \quad (5)$$

Separation of the two solutes i and j may therefore be calculated as a function of $w_A$ from data pertaining only to the pure phases provided that, for each solute, $V^o_g$ is a linear function of $w_A$. When this condition is met, Equation (5) represents an *exact* description of solute separation over the *entire* range, $W_A = 0$ to 1.

Plots of $\alpha$ versus column composition offer further insight into the separation at hand (158). Such a plot is illustrated in Figure 2 where the end-point data of Figure 1 have been used. When, as shown, $\alpha$ is maintained $\geq 1.00$ by inverting the i/j assignment of identity of the solutes, these plots resemble a set of inverted and partially overlapped triangles. Regions in which no overlaps occur amount to "windows" of separation which provide, upon inspection, quantitative information concerning resolution of the mixture at hand. First, the optimum A + S stationary-phase composition is found exactly from location of the highest window. In the present instance, the highest window is, marginally, that at $w_A = 0.348 = w^{opt}_A$. Second, the alpha lines which form the sides of the window identify the two pairs of solutes which are the most difficult to resolve at that column composition. Here the pairs: 3/2 and 2/1 offer the greatest difficulty (lowest alpha) at $w^{opt}_A$. Next, the height of the window yields the

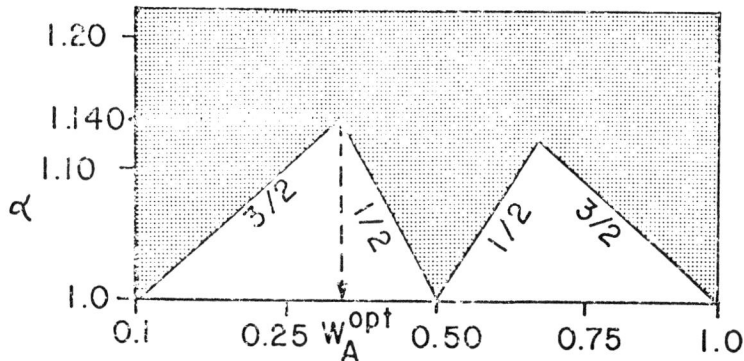

FIGURE 2. "Window" diagram for solutes of Figure 1 (cf. text).

precise value of alpha of the most difficult pairs, which in the present instance is 1.140. Since the number of plates required to effect resolution of a given mixture may be calculated from the relation (189):

$$N_{req} = 36 \left(\frac{\alpha}{\alpha - 1}\right)^2 \left(\frac{k' + 1}{k'}\right)^2 \qquad (6)$$

this $\alpha$ value gives, for $k' > 10$, immediate assessment of the column efficiency necessary to perform the separation. Since, further, the analyst will have determined from earlier $V_g^o$ measurements and in the course of practical experience the number of plates obtained per foot of column, the column length $L_{req}$ which will baseline resolve the mixture is specified. Finally, reference back to the "straight-line" diagram (Figure 1) provides the order of elution at the optimum A+S composition (here: 2, 1, and 3 at $w_A = 0.348$). Thus, the window-diagram procedure provides upon inspection: the optimum stationary phase composition, the most difficult pairs to separate at $w_A^{opt}$, their $\alpha$ value, the number of plates and hence the column length required to effect the separation, and the elution order. The only data required initially are the solute specific retention volumes with the two pure phases.

It may well be argued at this point that measurement of specific retention volumes or partition coefficients is not often undertaken and, indeed, is usually considered to be an onerous task in analytical GLC. This is due primarily to the supposition that considerable time and equipment are required for such measurements. In contrast to this somewhat specious view, Laub and co-workers (177,178) have shown that inexpensive and simple equipment and techniques can yield partition and activity coefficient data with an accuracy exceeding ± 1 percent. Nevertheless, the vast majority of retention data has been reported as relative to some standard available to the analyst, since $\alpha$ values with a given phase are calculable directly from raw retention times and to a first approximation are independent of flow rate, liquid loading, and type or length of column. Laub and co-workers (161,163) therefore recast their relations and procedures entirely in terms of alpha values and capacity factors, and showed that the window-diagram technique could be implemented solely with these data. Most recently, Pecsok and Apffel (190) and Laub (191) have also demonstrated that window diagrams may be constructed from data from various forms of retention index schemes.

Finally, although a mixture of most-difficult alpha corresponding to 1.03 can in favorable cases (160,192) be resolved with packed columns, a lower limit of 1.08 ($N_{req}$ = 6500 plates) represents a more realistic figure for most analytical laboratories. Pairs of stationary phases, say A and S, for which all windows for a given separation are less than this limit would accordingly be rejected by an analyst and a new set of stationary phases would be sought. This amounts only to chromatographing the sample with additional pure phases, say B, C, ..., and programming the data through previously described (159,169) algorithms. All combinations of A, B, C, ...., and S may, however, yield alpha values which are still too small for practical consideration. Open-tubular capillary columns, on the other hand, are today commonly fabricated and employed which routinely provide 50,000 to 100,000 plates, which is sufficient to baseline resolve solutes (for $k' > 10$) of $\alpha$ of 1.02. Use of capillary columns therefore permits greater flexibility in the choice of appropriate stationary-phase pairs; combination of such columns with the window-diagram technique forms the basis of this work.

## EXPERIMENTAL

Solid support (120/140 mesh Chromosorb G, AW-DMCS), column tubing (nickel) and stationary phases were obtained either from Alltech Associates (Arlington Heights, Illinois) or from Applied Science (State College, Pennsylvania) and were used as received. Packed columns (10 ft by 1/8-in. OD; 1 percent w/w; 250°C; 25 psig He) were fabricated as previously described(160). Capillary columns (220°C; 10 psig $H_2$) were drawn from Pyrex glass with a Shimadzu GDM-1 glass drawing machine to 1-mm OD by 0.25-mm ID in 50-m lengths. The columns were subsequently leached (HCl; 160°C for 5 hours) and silylated (HMDS; 400°C for 12 hours), following which coating was carried out in the static mode, either with pentane or with methylene chloride solvent.

The GC instrument used throughout was a Varian Model 3700, the injector and detector ports of which were modified in our laboratory in order to accommodate SGE 1/16-inch unions and a dropping-needle (Chrompack, Los Angeles, California) injector.

## RESULTS AND DISCUSSION

Application of the optimization strategy advocated here requires that a stationary-phase pair (or pairs) be located such that retention inversions and/or altered retention behavior obtain from one phase to the other. Thus, a preliminary search for such pairs need not involve columns of high efficiency nor, in fact, complete resolution of all solutes. Figures 3 to 5 illustrate such a search for suitable stationary phases for the solutes listed in Table 3. Figure 3 shows, for example, the separation of these solutes with OV-101, where solutes 1 and 2 and numbers 5 to 7 are fully overlapped. Figure 4 illustrates the separation under identical conditions with PS-176, a polysulfone phase claimed (193) to be as "selective" as cyano stationary phases. However, solutes 1 and 2 are still fully overlapped, while solutes 5 to 7 are only partially resolved. Retentions, on the other hand, have increased markedly over those seen in Figure 3. Figure 5 shows the separation achieved with BBBT: all solutes are clearly resolved and, in addition, are eluted in less that 10 minutes. Efficiency of the column, however, in contrast with the previous two chromatograms, is poor. Hence, while the column performs the separation at hand, it would not be suitable for the more commonly encountered types of PAH samples which are routinely found to contain in excess of 50 aromatic hydrocarbons. However, the initial task, namely, location of suitable stationary-phase pairs for subsequent use with the window diagram procedure, has been accomplished insofar as BBBT gives retentions which differ considerably from those obtained with either OV-101 or with PS-176. (Note, on the other hand, that OV-101 with PS-176 is not a suitable pair for application of the optimization strategy since the retention order and indeed, the number and kind of peak overlaps are very nearly identical with both phases.)

To increase the separation efficiency (hence, practicability) of the phases at hand for PAH analysis, 20-m capillary columns were fabricated from SE-30, SE-52, and BBBT. Figure 6 shows the separation of solutes 1 to 8 with SE-30 where, not unexpectedly, the same overlaps as those in Figure 3 (OV-101) are found. Thus, even though the column provides enhanced efficiency over packed columns, anthracene (number 2) is not resolved from phenanthrene (number 1), and chrysene (number 5), triphenylene (number 6), and benz(a)anthracene (number 7) coelute. Figure 7 illustrates the separation obtained with SE-52, a phase which has

TABLE 3. Elution Order for Eight Polycyclic Aromatic Hydrocarbons with Methylsilicone (OV-101), N,N'-Bis[p-butoxybenzylidene]-$\alpha,\alpha$-bi-p-toluidine (BBBT), and Polyphenylsulfone (PS-176) Stationary Phases at 250°C

| Solute No. | Name | Structure | Stationary Phase | | |
|---|---|---|---|---|---|
| | | | OV-101 | BBBT | PS-176 |
| 1 | Phenanthrene | | 1a | 1 | 1a |
| 2 | Anthracene | | 1b | 2 | 1b |
| 3 | Fluoranthene | | 2 | 3 | 2 |
| 4 | Pyrene | | 3 | 4 | 3 |
| 5 | Chrysene | | 4a | 7 | 6 |
| 6 | Triphenylene | | 4b | 5 | 5 |
| 7 | Benz(a)anthracene | | 4c | 6 | 4 |
| 8 | Naphthacene | | 5 | 8 | 7 |

often been recommended for PAH analysis. Anthracene and phenanthrene are overlapped still, while benz(a)anthracene has been resolved only partially from chrysene and triphenylene which, however, remain unresolved. The column efficiency obtained is comparable to that found with SE-30. Figure 8 illustrates baseline resolution of all solutes with a 20-m capillary column containing BBBT. Once again, however, the column efficiency leaves much to be desired: all peaks tail noticeably and the elution time of naphthacene exceeds 38 minutes

Plots of relative retention versus $w_{BBBT}$ for the stationary-phase pair SE-30 + BBBT are shown in Figure 9, where the relation (161,163)

$$\alpha \cdot \beta = \alpha^\circ_{(SE-30)} \, w_{SE-30} + \alpha^\circ_{(BBBT)} \, w_{BBBT} \, \beta^\circ_{(BBBT)} \qquad (7)$$

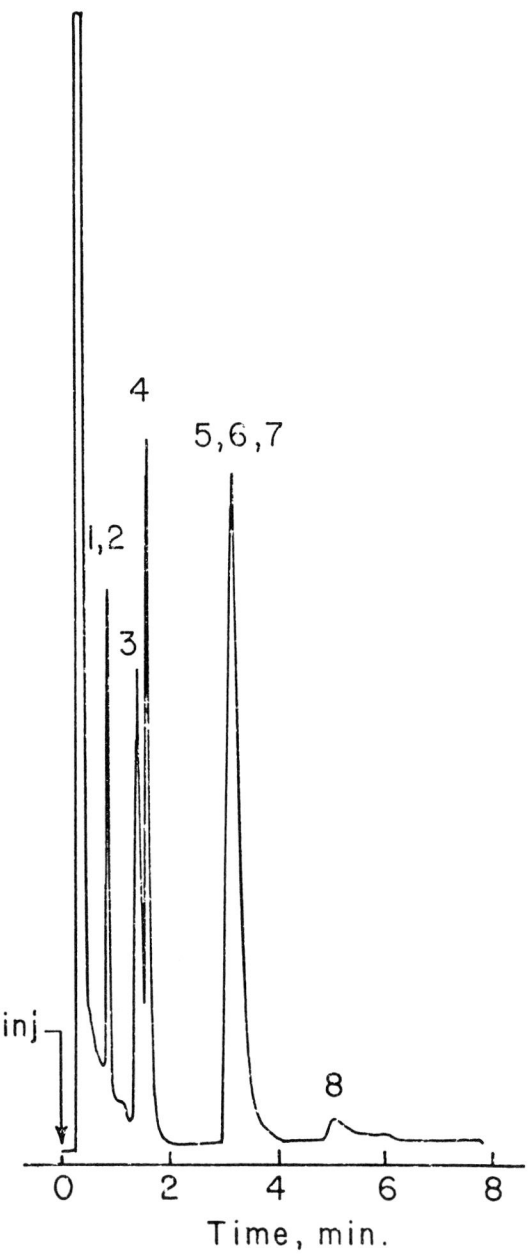

FIGURE 3. Chromatogram of solutes of Table 3 with OV-101 packed column (1 percent w/w on Chromosorb G AW-DMCS 120/140 mesh; 10 ft 1/8-in. OD nickel tubing; 25 psig He inlet pressure; 250°C column Temperature).

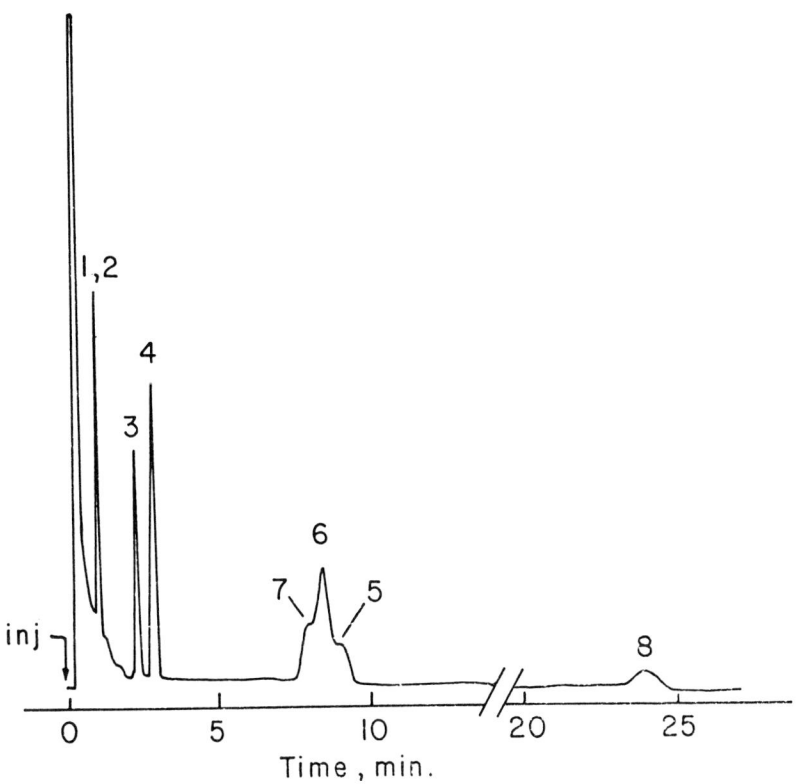

**FIGURE 4.** Chromatogram of solutes of Table 3 with PS-176 packed column; conditions as in Figure 3.

benz(a)anthracene (number 7), $\beta$ is the capacity factor ratio of the standard (number 7) ($=k'_{(SE-30)} + _{BBBT}{}^{\infty}k'_{(SE-30)}$) with the mixed phase, $\alpha^\circ_{c(SE=30)}$ and $\alpha^\circ_{(BBBT)}$ are the solute retentions relative to number 7 with each of the pure phases, $\beta^\circ_{(BBBT)}$ is the capacity factor ratio of the standard (number 7) ($=k'_{(BBBT)}/k'_{(SE-30)}$) with the pure phases, and $w_{SE-30}$ and $w_{BBBT}$ are the weight fractions of SE-30 and BBBT in the SE-30 + BBBT (mixed) phase respectively. The plots as shown were constructed solely from the pure-phase data, i.e., $\alpha^\circ_{(SE-30)}$ ($w_{BBBT} = 0$) and $\alpha^\circ_{(BBBT)} \cdot \beta^\circ_{(BBBT)}$ ($w_{BBBT} = 1.0$) as listed in Table 4, and illustrate graphically that poor separation of the PAH is found with SE-30, whereas excellent separation is obtained with BBBT. Nevertheless, and as was true for the packed-column chromatograms, the column efficiency observed for BBBT leaves much to be desired, and thus the 20-m capillary column is of little use for solute mixtures other than the simple example utilized here.

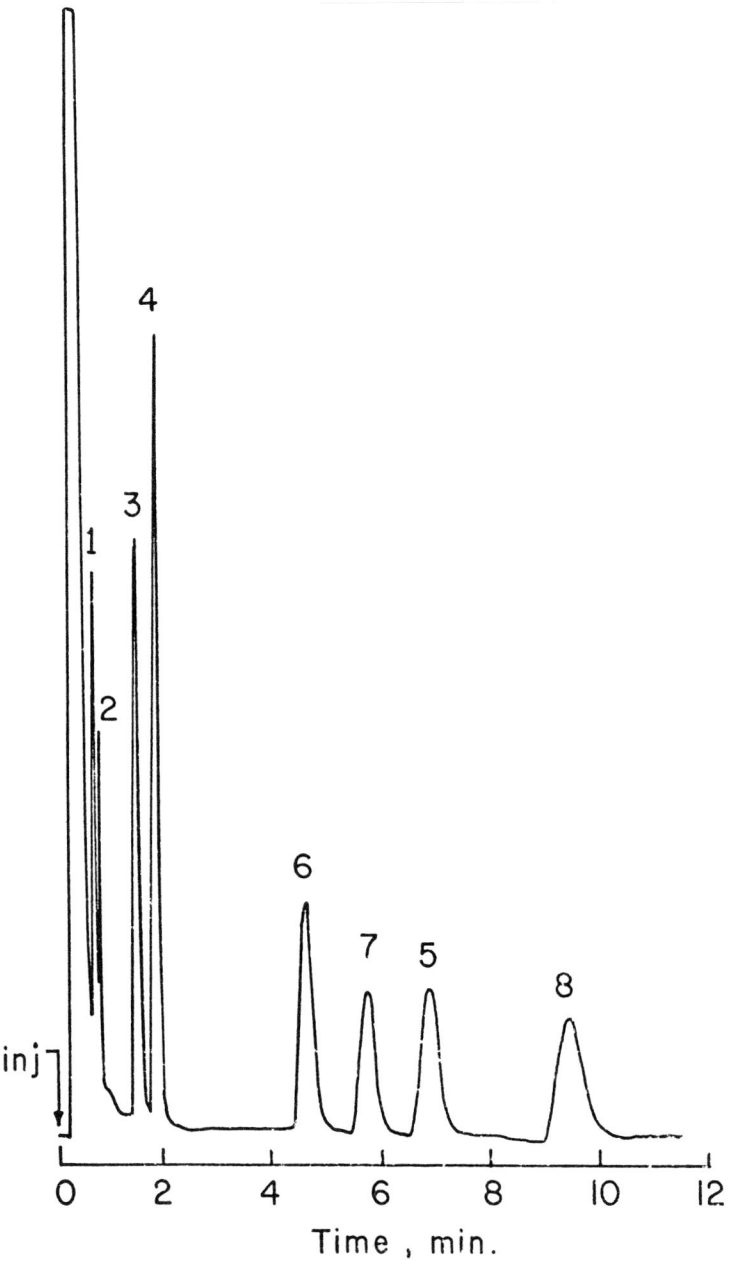

FIGURE 5. Chromatogram of solutes of Table 3 with BBBT packed column; conditions as in Figure 3.

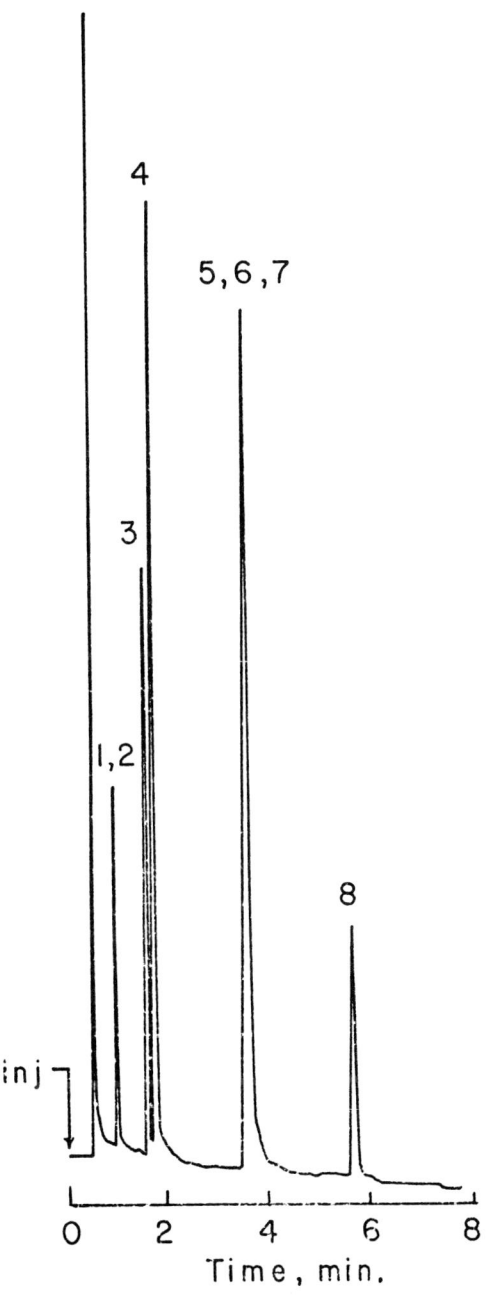

FIGURE 6. Chromatogram of solutes of Table 3 with SE-30 glass capillary column (20 m, 0.25-mm ID; 10 psig $H_2$ inlet pressure; 220°C column temperature).

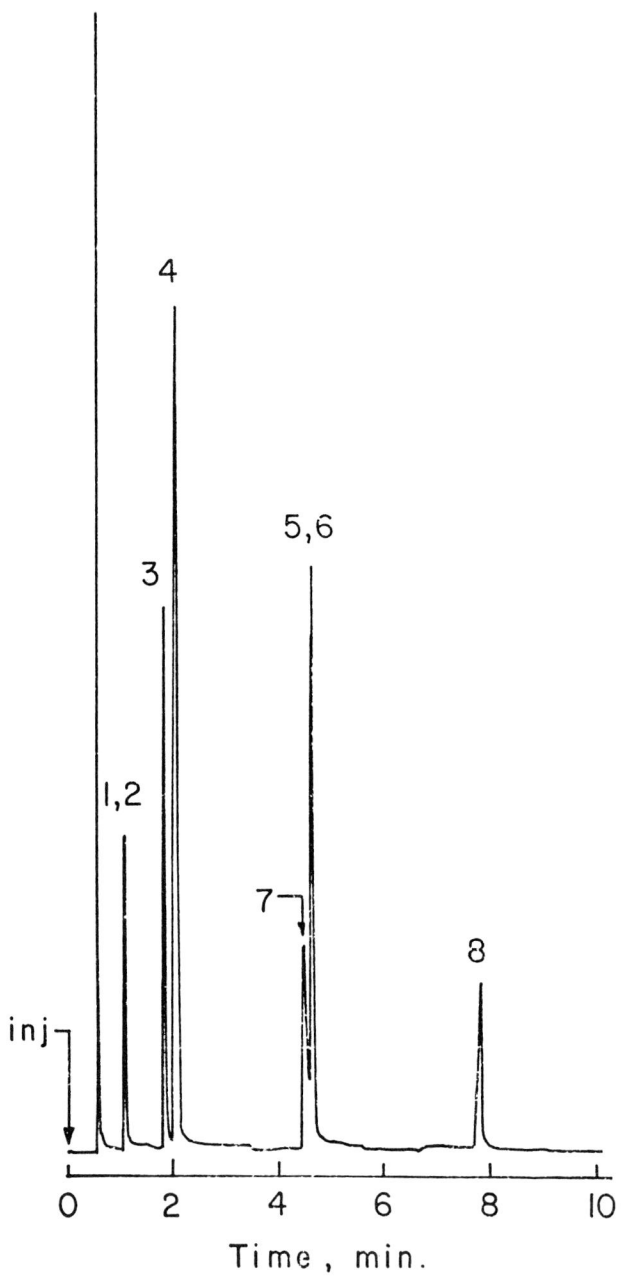

FIGURE 7. Chromatogram of solutes of Table 3 with SE-52 glass capillary column; conditions as in Figure 6.

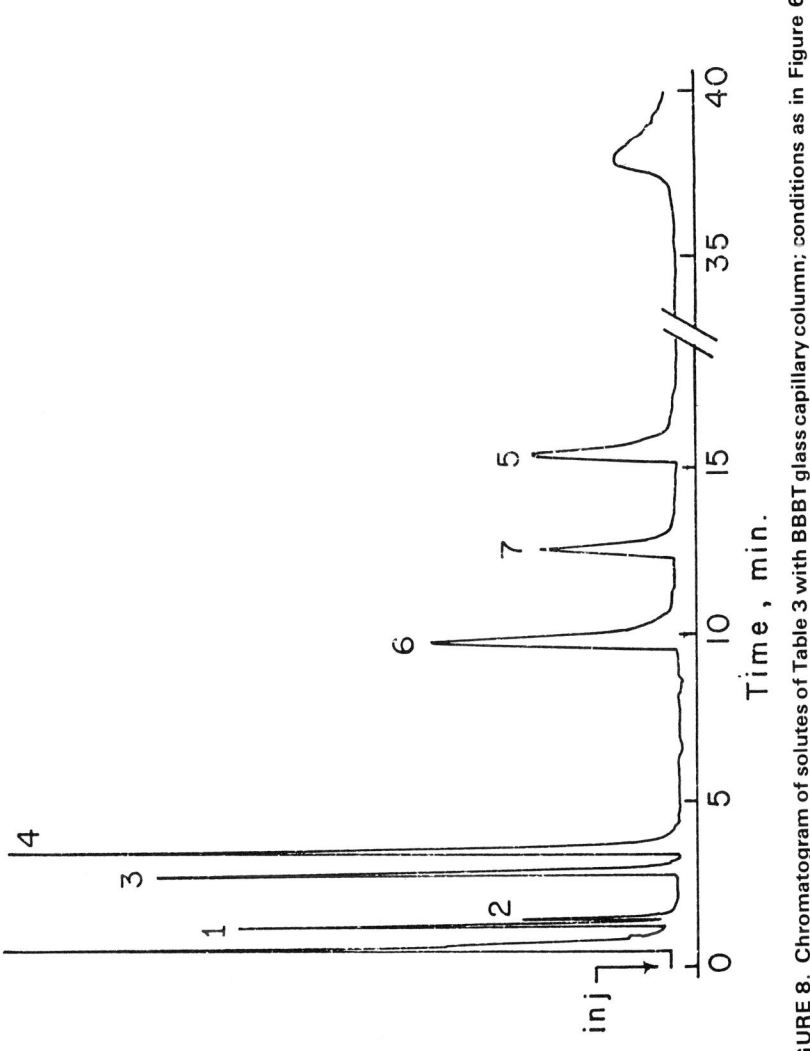

FIGURE 8. Chromatogram of solutes of Table 3 with BBBT glass capillary column; conditions as in Figure 6.

## 44 MIXED PHASE GC OF PAH

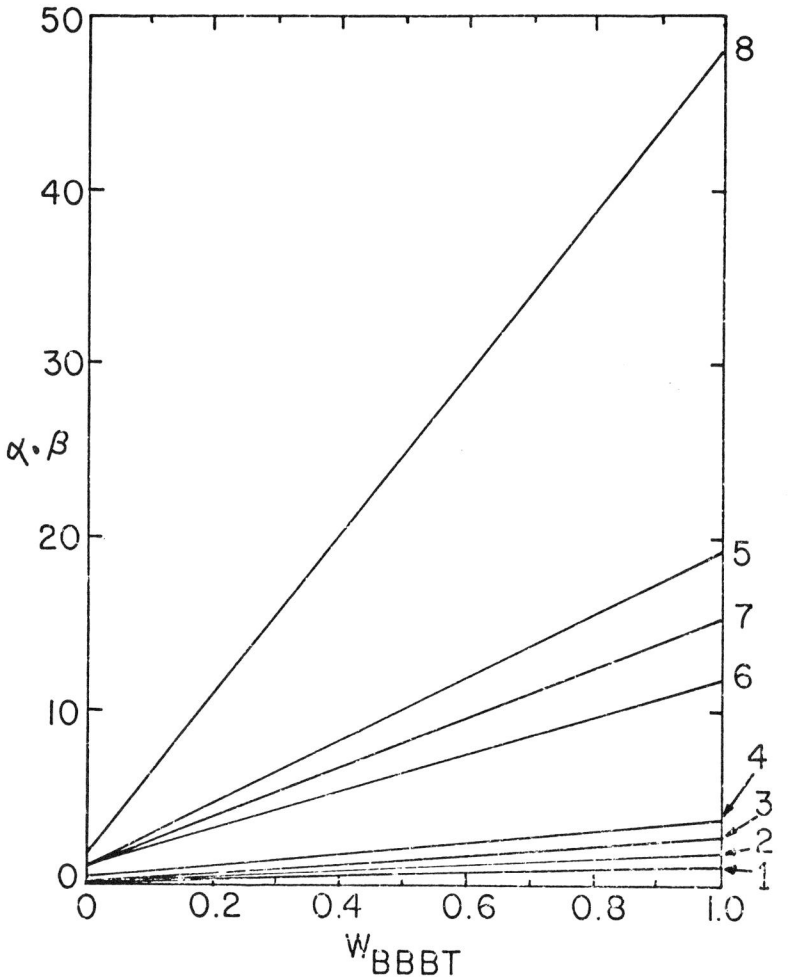

**FIGURE 9.** Plots of $\alpha \cdot \beta$ [cf. Equation (7)] versus weight fraction of BBBT in BBBT + SE-30 mixed phase at 220°C. See Table 3 for solute key to numbers.

A number of approaches have, of course, been advocated for the enhancement of capillary column efficiency, the latest (and what should prove to be a major) innovation involving the use of fused silica materials (194,195). With regard to liquid crystal stationary phases, Rogers and coworkers (196-199) have shown that peak symmetry can be improved by application of an electric field across the radius of glass capillary columns. Alternatively, Witkiewicz, Suprynowicz, and Dabrowski (200) employed packed microbore columns and obtained approximately 6800 plates per meter with cyanoazoxybenzene alkyl carbonate phases.

In contrast to the above, the window diagram method may be employed (utilizing only existing and simple procedures) for producing capillary columns of acceptable efficiency with mesogenic phases via optimization of secondary criteria, e.g. analysis time, temperature, column length, etc. for subsequent analytical tasks once the primary goal, namely, separation of the mixture at hand via identification of appropriate pairs of stationary phases, has been achieved. In the present instance, for example, all eight of the solutes of Table 3 were resolved with a pure-phase column at a single temperature and the only task left is enhancement of the column efficiency to render it useful for more complex solute mixtures.

Figure 10 shows plots of relative retention ($\alpha$) as a function of the weight fraction of BBBT in SE-30 for the four most difficult solute pairs to resolve, as calculated from Equations 5 and 7 and the data of Table 4. The pair 7/5 (benz(a)anthracene with chrysene) is seen to be the most difficult at all column compositions, followed by 7/6 (benz(a)anthracene with triphenylene), 2/1 (anthracene with phenanthrene) and 4/3 (pyrene with fluoranthene), the latter two pairs appearing to offer roughly the same degree of difficulty. The plots are somewhat deceiving, however, since no account has yet been taken of the solute capacity factors which affect the number of plates required to achieve baseline resolution in accordance with equation 6.

Average capacity factors, $\alpha$ values, and $N_{req}$ for the most difficult pairs are listed in Table 5 and plots of $N_{req}$ versus $w_{BBBT}$ are shown in Figure 11, the latter providing illustration that, in the present instance, the pair 2/1 offers the greatest difficulty of separation. Indeed, reference to the plots indicates that a neat BBBT column yielding only 1750 plates would be sufficient to resolve all pairs. Alternatively, 6000 plates would be required of a column containing $w_{BBBT}$ = 0.20. A reasonable compromise between $N_{req}$ and $w_{BBBT}$ appeared in this case to be a weight fraction of 1/3. Figure 12 illustrates the resultant chromatogram where all solutes are well resolved and, more importantly, the column efficiency clearly approximates that obtained with SE-30 (Figure 6). Thus, the use of mixed phases enables the extraordinary selectivity of BBBT to be preserved, while at the same time retaining the efficiency of SE-30. The situation could hardly be more advantageous. Finally, Figure 13 shows the separation of 13 PAH with the mixed-phase column of Figure 12 at 265°C. Benzo(e)pyrene (number 9), perylene (number 10), and benzo(a)pyrene (number 11) are baseline-resolved, while dibenz(a,c)anthracene (number 12) elutes a full 2 minutes prior to dibenz(a,h)anthracene (number 13). In addition, the analysis requires less than 8 minutes overall. Thus, the mixed-phase column appears to be useful for the more realistic situations in which larger numbers of solutes are commonplace, especially in view of the fact the chromatogram of Figure 13 was produced under isothermal conditions.

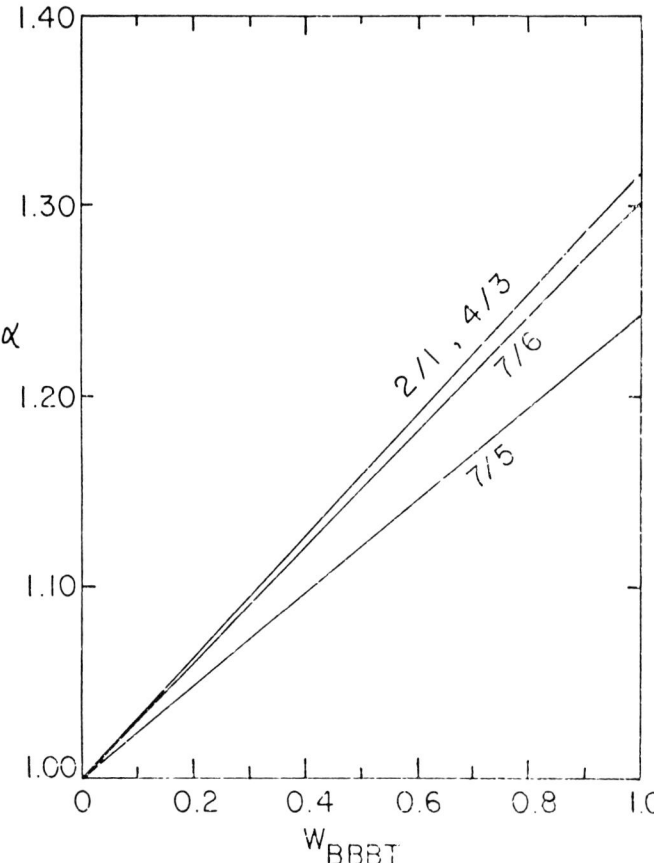

**FIGURE 10.** Window diagram for most difficult solute pairs of Figure 9. See Table 3 for key to solute numbers.

**Table 4. Relative Retention Data for Solutes of Table 3 with SE-30 (Figure 6) and BBBT (Figure 8) Stationary Phases at 220°C**

| Solute No. | $\alpha^\circ_{(SE-30)}$ | $\alpha^\circ_{(BBBT)} \cdot \beta^\circ_{(BBBT)}$ [a] |
|---|---|---|
| 1 | 0.138 | 0.888 |
| 2 | 0.138 | 1.171 |
| 3 | 0.338 | 2.857 |
| 4 | 0.393 | 3.770 |
| 5 | 1.000 | 19.12 |
| 6 | 1.000 | 11.83 |
| 7 | 1.000 | 15.40 |
| 8 | 1.682 | 48.03 |

[a] $\beta^\circ_{(BBBT)} = 3.925$.

TABLE 5. Average Capacity Factor[a] (k'), Alpha (α), and N_req[b] Data for Indicated Solute Pairs of Table 3 with SE-30 + BBBT Mixed Stationary Phases at 220°C

| Solute Pair | $W_{BBBT}$ | | | | | | | | | | | | | | |
|---|---|---|---|---|---|---|---|---|---|---|---|---|---|---|---|
| | 0.0 | | | 0.2 | | | 0.4 | | | 0.6 | | | 0.8 | | | 1.0 | | |
| | α | k' | $N_{req}$ | α | k' | $N_{req}$ | α | k' | $N_{req}$ | α | k' | $N_{req}$ | α | k' | $N_{req}$ | α | k' | $N_{req}$ |
| 7/5 | 1.000 | 5.35 | ∞ | 1.192 | 8.99 | 1715 | 1.220 | 12.63 | 1290 | 1.232 | 16.26 | 1145 | 1.238 | 19.90 | 1075 | 1.242 | 23.54 | 1030 |
| 7/6 | 1.000 | 5.35 | ∞ | 1.226 | 7.99 | 1340 | 1.268 | 10.64 | 965 | 1.286 | 13.28 | 840 | 1.296 | 15.92 | 780 | 1.302 | 18.56 | 745 |
| 2/1 | 1.000 | 0.737 | ∞ | 1.197 | 0.871 | 6135 | 1.259 | 1.00 | 3390 | 1.289 | 1.14 | 2530 | 1.307 | 1.27 | 2085 | 1.319 | 1.40 | 1805 |
| 4/3 | 1.163 | 1.96 | 4185 | 1.269 | 2.47 | 1580 | 1.296 | 2.98 | 1230 | 1.308 | 3.49 | 1075 | 1.315 | 4.01 | 980 | 1.320 | 4.52 | 915 |

[a] $k' = w_{(SE-30)} k'_{(SE-30)} + w_{BBBT} k'_{(BBBT)}$.
[b] Cf. Equation 6.

## 48 MIXED PHASE GC OF PAH

In summary, the optimization strategy has been shown to be an effective means of enhancing the separation of PAH with mixed phases. Of particular importance is the finding that such phases may comprise, in part, nematogenic compounds without adversely affecting the inherent efficiency of cosolvent phases. Indeed, this feature of the work should find considerable utility in a number of areas in glass-capillary gas chromatography insofar as it is a matter of experience that the most

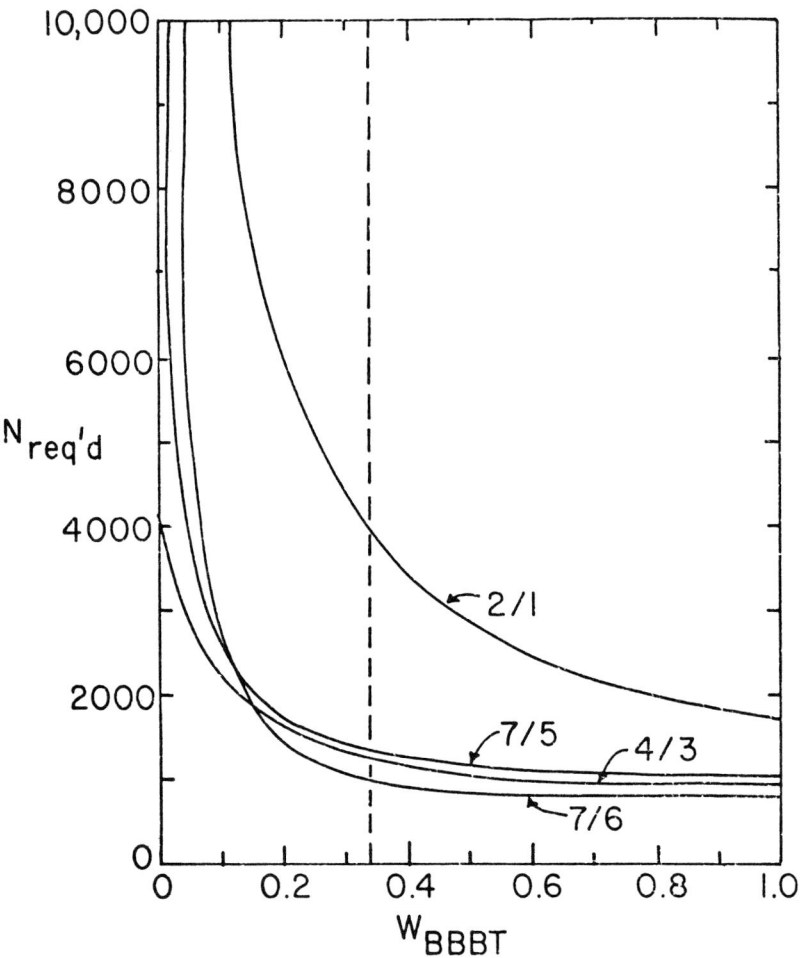

FIGURE 11. Plots of $N_{req}$ [cf. Equation (6) and Table 5] versus weight fraction of BBBT in BBBT + SE-30 mixed phase at 220°C. See Table 3 for key to solute numbers.

selective stationary phases often prove in practice to yield the poorest column efficiencies. The practicability of the mixed-solvent approach is currently under further investigation in our laboratory, and we hope to report the results of these studies shortly.

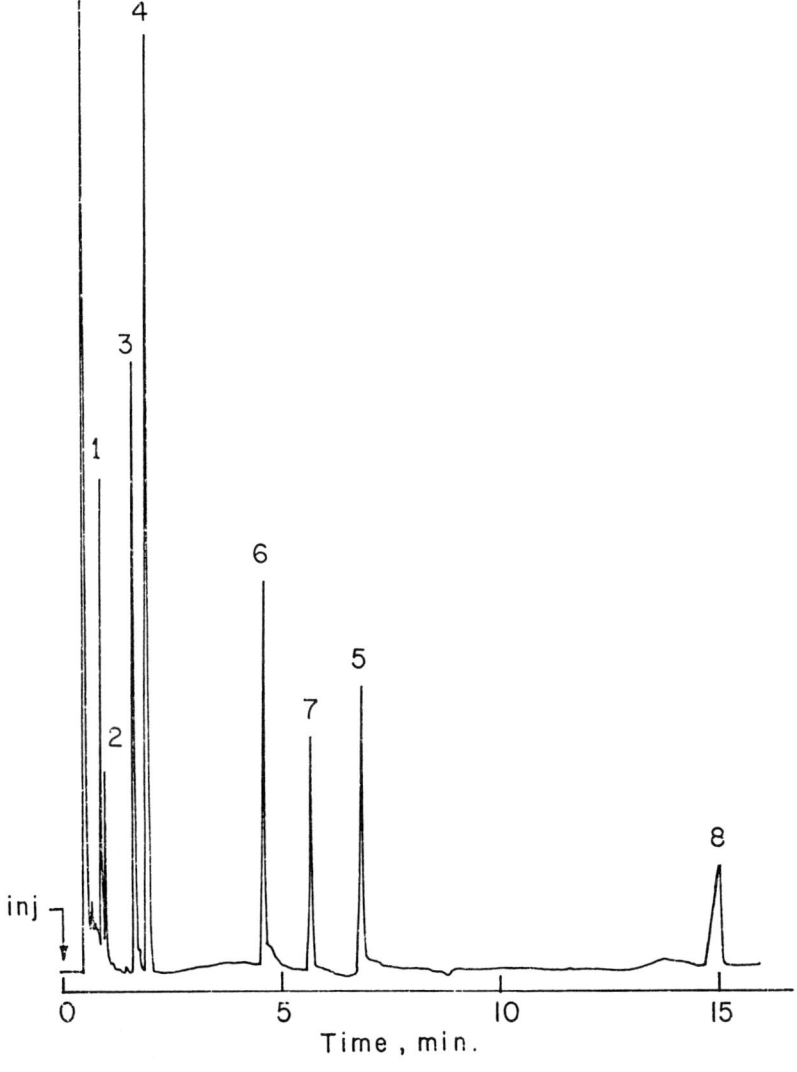

**FIGURE 12.** Chromatogram of solutes of Table 3 with a glass capillary column containing one-third BBBT + two-thirds SE-30; conditions as in Figure 6.

## 50 MIXED PHASE GC OF PAH

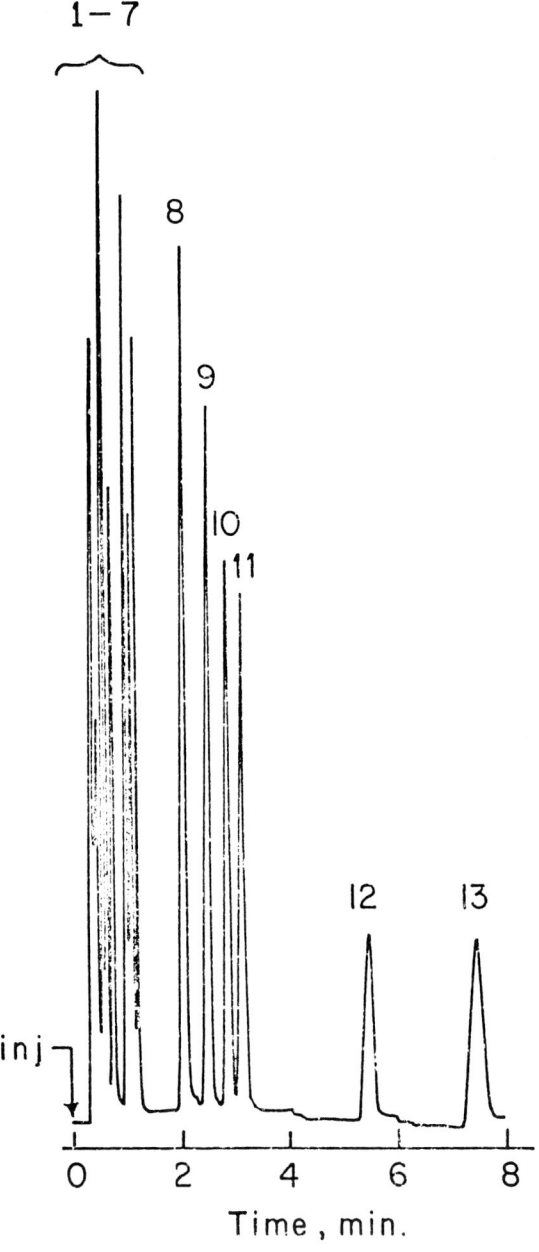

FIGURE 13. Chromatogram of solutes of Table 3 in addition to benzo(e)pyrene (number 9), perylene (number 10), benzo(a)pyrene (number 11), dibenz(a,c) anthracene (number 12, and dibenz(a,h)anthracene (number 13) at 265°C with the column of Figure 12.

## ACKNOWLEDGMENT

We gratefully acknowledge the support of the National Science Foundation, Grant No. CHE-7820477.

## REFERENCES

1. Brosch, A. (1900): *Virchoms Arch. Pathol. Anat. u. Physiol.* 162:32.
2. Roffo, A. H. (1936a, 1936b, 1939a, 1939b, and 1941): *Bol. Inst. Med. Exptl.* No. 42; *Rev. Sud-American Endocrinol. Immunol. Quimoterap.* 20:550; *Bol. Inst. Med. Exptl. Estudio Cancer* 15:349; *Prensa Med. Argentina* 26:721; *Z. Krebsforsch.* 49:588; and *Schweiz. Med. Wochschr.* 71:549.
3. Roffo, A. H.,(1940): *Bol. Inst. Med. Exptl. Estudio Cancer* 17:661.
4. Wynder, E. L., Graham, E. A., and Croninger, A. B., (1953): *Cancer Res.* 13:855.
5. Cooper, R. L., and Lindsey, A. J.,(1953): *Chem. Ind.* 1205
6. Commins, B. T., Cooper, R. L., and Lindsey, A. J., (1954): *Brit. J. Cancer* 8:296.
7. Cooper, R. L., Lindsey, A. J., and Waller, R. E., (1954): *Chem. Ind.* 1418.
8. Cooper, R. L., and Lindsey, A. J., (1955): *Brit. J. Cancer* 9:304.
9. Rand, H. J.,Cardon, S. Z., Alvord, E. T., and Burhan, A., (1957): *Amer. J. Surg.* 94:438.
10. Lyons, M. J., (1958): *Nature* 182:178.
11. Van Duuren, B. L., and Kosak, A. I., (1958): *J. Org. Chem.* 23:473.
12. Bentley, H. R., and Burgan, J. G., (1958): *Analyst* 83:442.
13. Van Duuren, B. L., (1958): *J. Nat. Cancer Inst.* 21:1.
14. Johnstone, R.A.W., and Plimmer, J. R., (1959): *Chem. Rev.* 59:885.
15. Stedman, R. L., (1968): *Chem. Rev.* 68:153.
16. Wedgewood, P., and Cooper, R. L., (1951,1953): *Chem. Ind.* 1066; *Analyst* 78:170.
17. Borneff, J., and Kunte, H., (1967): *Arch. Hyg. u. Bakter.* 151:202.
18. Sawicki, E., and Miller, R. R., (1958): *Anal. Chem.* 30:109.
19. Kennaway, E., and Lindsey, A. J., (1958): *Brit. Med. Bull.* 14:124.
20. Hoffmann, D., and Wynder, E. L., (1960): *Anal. Chem.* 32:295.
21. Sawicki, E., Elbert, W., Stanley, T. W., Hanser, T. R., and Fox, F. T., (1960a, 1960b): *Int. J. Air Poll.* 2:273; *Anal. Chem.* 32:810.
22. Sawicki, E., Hanser, T. R., Elbert, W. C., Fox, F. T., and Meeker, J. E., (1962): *Amer. Ind. Hyg. Assoc. J.* 23:137.
23. Lyons, M. J., (1962): *Nat. Cancer Inst. Monograph.* 9:193.
24. Liberti, A., Cartoni, G. P., and Cantuti, V., (1964): *J. Chromatogr.* 15:141.

25. Sawicki, E., Stanley, T. W., Elbert, W. C., and Pfaff, J. D., (1964): *Anal. Chem.* 36:497.
26. Cantuti, V., Cartoni, G. P., Liberti, A., and Torri, A. G., (1965): *J. Chromatogr.* 17:60.
27. Kunte, H., (1967): *Arch. Hyg. u. Bakter.* 151:193.
28. Carugno, N., and Rossi, S., (1967): *J. Gas Chromatogr.* 5:103.
29. Wallcane, L., (1969): *J. Environ. Sci. Technol.* 3:948.
30. Strömberg, L. E., and Widmark, G., (1970): *J. Chromatogr.* 49:334.
31. Dubois, L., Zdrojewski, A., Jennawar, P., and Monkman, J. L., (1970): *Atmospher. Environ.* 4:199.
32. Bhatia, K., (1971): *Anal. Chem.* 43:609.
33. Lao, R. C., Thomas, R. S., Oja, H., and Dubois, L., (1973): *Anal. Chem.* 45:908.
34. Zafiriou, O. C., (1973): *Anal. Chem.* 45:952.
35. Lane, D. A., Moe, H. K., and Katz, M., *Anal. Chem.* 45:1776.
36. Zobell, C. E., (1971): *Proceedings of Joint Conference on Prevention and Control of Oil Spills,* Washington, D.C., p. 441.
37. Uden, P. C., Carpenter, A., Jr., DiSanzo, F. P., Hackett, H. F., Siggia, S., and Henderson, D. E., (1978): In: *Advances in Chemistry Series No. 170: Analytical Chemistry of Liquid Fuel Sources - Tar Sands, Oil Shale, Coal, and Petroleum*, P. C. Uden, S. Siggia, and H. B. Jensen, Eds., Chapter 15, American Chemical Society, Washington, D.C.
38. D'Alonzo, R. P., Carpenter, A. P., Jr., Siggia, S., and Uden, P. C., (1978): *Anal. Chem.* 50:326.
39. Uden, P. C., Carpenter, A. P., Jr., Hackett, H. M., Henderson, D. E., and Siggia, S., (1979): *Anal. Chem.* 51:38.
40. DiSanzo, F. P., Uden, P. C., and Siggia, S., (1979): Paper no. 665 presented at the Pittsburgh Conference on Analytical Chemistry and Applied Spectroscopy, Cleveland, Ohio, March.
41. Wynder, E. L., and Wright, G., (1957): *Cancer* 10:255.
42. Hoffmann, D., and Wynder, E. L., (1960): *Cancer* 13:1062.
43. Short, G. D., and Young, R., (1969): *Analyst* 94:259.
44. Hoffmann, D., and Rathkamp, G., (1972): *Anal. Chem.* 44:899.
45. Popl, M., Stejskal, M., and Mostecky, J., (1975): *Anal. Chem.* 47:1947.
46. Van Duuren, B. L., (1960): *Anal. Chem.* 32:1436.
47. Thomas, J. F., Mukai, M., and Tebbens, B. D., (1962): *Nat. Cancer Inst. Monograph* 9:127.
48. Davis, H. J., Lee, L. A., and Davidson, T. R., *Anal. Chem.* 38:1752.
49. Jäger, J., (1968): *Atmospher, Environ.* 2:293.
50. Freed, D. J., and Faulkner, L. R., *Anal. Chem.* 44:1194.
51. McKay, J. F., and Latham, D. R., (1973): *Anal. Chem.* 45:1050.

52. Tarbell, D. S., Brooker, E. G., Vanterpool, A., Conway, W., Claus, C. J., and Hall, T. J., (1955): *J. Amer. Chem. Soc.* 77:767.
53. Bergmann, E. D., and Gruenwald, T., (1957): *J. Appl. Chem.* 7:15.
54. Spotswood, T. M., (1959): *J. Chromatogr.* 2:90.
55. Matsushita, H., Suzuki, Y., and Sakabe, H., (1963): *Bull. Chem. Soc. Japan* 36:1371.
56. Köhler, M., Golder, H., and Schiesser, R., *Z. Anal. Chem.* 206:430.
57. Hood, L.V.S., and Winefordner, J. D., (1968): *Anal. Chim. Acta* 42:199.
58. Libickova, V., Stuchlik, M., and Krasnec, L., (1969): *J. Chromatogr.* 45:278.
59. Strömberg, L. E., and Widmark, G., (1970a, 1970b): *J. Chromatogr.* 47:27, 49:334.
60. Fleet, B., Kirkbright, G. F., and Pickford, C. J., (1968): *Talanta* 15:556.
61. Fleet, B., Keliher, P. N., Kirkbright, G. F., and Pickford, C. J., (1969): *Analyst* 94:847.
62. Carugno, N., (1962): *Natl. Cancer Inst. Monograph* 9:171.
63. Sawicki, E., Fox, F. T., and Elbert, W., (1962): *Amer. Ind. Hyg. Assoc. J.* 23:482.
64. Lijinsky, W., Domsky, I., Mason, G., Ramahi, H. Y., and Safavi, T., (1963): *Anal. Chem.* 35:952.
65. Dupire, F., and Botquin, G., (1958): *Anal. Chim. Acta* 18:282.
66. Dupire, F., (1959a, 1959b, 1962): *Z. Anal. Chem.* 170:317; *Ind. Chim. Belge Suppl.* 1:159; *Natl. Cancer Inst. Monograph* 9:183.
67. Carugno, N., and Grovannozzi-Sermanni, G., (1959): *Tobacco*, 63:285.
68. Gudzinowicz, B. J., and Smith, W. R., (1960): *Anal. Chem.* 32:1767.
69. Ferrero, P., (1960, 1961): *Ind. Chim. Belge* 25:237; *Chimia (Aarau)* 15:333.
70. Wood, L. J., (1961): *J. Appl. Chem.* 11:130.
71. Sauerland, H. D., (1963): *Brennstoff.-Chem.* 44:37.
72. Farrand, R., (1963): *Chim. Anal.* 45:133.
73. Lijinsky, W., and Mason, G., (1963): *J. Gas Chromatogr.* 1(9):12.
74. Wilmhurst, J. R., (1965): *J. Chromatogr.* 17:50.
75. Vidal-Madjar, C., and Guiochon, G., (1967): *Nature* 215:1372.
76. Zane, A., (1968): *J. Chromatogr.* 38:130.
77. Gump, B. H., *J. Chromatogr. Sci.* 7:755.
78. Sato, K., Matsui, M., and Ikekawa, N., (1968): *Bunseki Kagaku* 17:639.
79. Cartoni, G. P., (1968): *Estrat. Cron. Chim.*, No. 22 December, 1968.
80. Bjorseth, A., and Eklund, G., (1979): *J. High Resolut. Chromatogr. Chromatogr. Commun.* 2:22.

81. Borwitzky, H., and Schomburg, G., (1979): *J. Chromatogr.* 170:99.
82. Bernaert, H., (1979): *J. Chromatogr.* 173:109.
83. Sie, S. T., and Rijnders, G.W.A., (1967): *Separ. Sci.* 2:729, 755.
84. Jentoft, R. E., and Gouw, T. H., (1968): *Anal. Chem.* 40:1787.
85. Lankmayr, E. P., and Mueller, K., (1979): *J. Chromatogr.* 170:139.
86. Nielsen, T., (1979): *J. Chromatogr.* 170:147.
87. Toussaint, G., and Walker, E. A., (1979): *J. Chromatogr.* 171:448.
88. Robertson, D. J., Groth, R. H., Gardner, D. G., and Glastris, E. G. (1979): *J. Air Pollut. Control Assoc.* 29:143.
89. Takami, K., Ishitani, H., Kuge, Y., and Asada, S. (1979): *Nippon Kagaku Kaishi* 233.
90. Dutkiewicz, T., Masny, N., Ryborz, S., Maslowski, J., and Grabka, A. (1979): *Chem. Anal.* 24:191.
91. Scott, R.P.W., and Kucera, P. (1979): *J. Chromatogr.* 169:51.
92. Kelker, H. (1963a, 1963b): *Ber. Bunsenges. Phys. Chem.* 67:698; *Z. Anal. Chem.* 198:254.
93. Dewar, M.J.S., and Schroeder, J. P. (1964): *J. Amer. Chem. Soc.* 86:5235.
94. Schroeder, J. P. (1974): In: *Liquid Crystals and Plastic Crystals*, Vol. 1, G. W. Gray and P. A. Winsor, Eds., p. 356, Ellis Horwood, Chichester, England.
95. Kelker, H., and Von-Schivizhoffen, E. (1968): In: *Advances in Chromatography*, Vol. 6, J. C. Giddings and R. A. Keller, Eds., p. 247, Marcel Dekker, New York.
96. Kelker, H. (1978): *Adv. Liq. Cryst.* 3:237.
97. Kraus, G., and Winterfeld, A. (1978): *Wiss. Z. Martin Luther Univ. Halle Wittenberg, Math.-Naturwiss, Reihe* 27(5):83.
98. De Leone, M. E., Ramos, A., and Manjarrez, A. (1978): *Rev. Soc. Quim. Mex.* 22:126.
99. Egrets, V. A., Garusov, A. V., and Vigdergauz, M. S. (1978): *Zh. Fiz. Khim.* 52:2702.
100. Miyata, I., and Kishimoto, H. (1978): *Yakugaku Zasshi* 98:1629.
101. Pham, H. V., Nguyen, B. H., Nguyen, X. D., Dang, V. L., and Ta, P. H. (1978): *Tap San Hao Hoc* 16:29.
102. Strand, J. W., and Andren, A. W. (1978): *Anal. Chem.* 50:1508.
103. Tesarik, K., Frycka, J., and Ghyczy, S. (1978): *J. Chromatogr.* 148:223.
104. Witkiewicz, Z., Suprynowicz, Z., Wojcik, J., and Dabrowski, R. (1978a, 1978b): *Pol J. Chem.* 52:1495; *J. Chromatogr.* 152:323.
105. Popiel, S., and Witkiewicz, Z. (1978): *Biul. Wojsk. Akad. Tech.* 27(4):29,39.
106. Witkiemicz, Z., and Popiel, S. (1978): *J. Chromatogr.* 154:60.
107. Dabrowski, R., and Witkiewicz, Z., *Biul. Wojsk. Akad. Techn.* 28(2):101.

108. Lester, R. (1978): *J. Chromatogr.* 156:55.
109. Vernon, F., and Khakoo, A. N. (1978): *J. Chromatogr.* 157:412.
110. Bocquet, J. F., and Pommier, C. (1978): *J. Chromatogr.* 166:357.
111. Lochmueller, C. H., and Hinshaw, J. V., Jr. (1979): *J. Chromatogr.* 171:407.
112. Witkiewicz, Z., and Waclawczyk, A. (1979): *J. Chromatogr.* 173:43.
113. Santoro, A., Modica, R., Pazlialunga, S., and Bartosek, I. (1979): *Toxicol. Lett.* 3:85.
114. Koizumi, H., and Suzuki, Y. (1979): *Bunseki Kagaku* 28:92.
115. Deutsch, L. J. (1979): *Diss. Abstr. Int. B* 39:4848.
116. Solsky, J. F. (1978): *Diss. Abstr. Int. B* 39:1315.
117. Janini, G. M., Johnston, K., and Zielinski, W. L., Jr. (1975): *Anal. Chem.* 47:670.
118. Janini, G. M., Muschik, G. M., and Zielinski, W. L., Jr. (1976): *Anal. Chem.* 48:809.
119. Zielinski, W. L., Jr., Johnston, K., and Muschik, G. M. (1976): *Anal. Chem.* 48:907.
120. Janini, G. M., Muschik, G. M., Schroer, J. A., and Zielinski, W. L., Jr. (1976): *Anal. Chem.* 48:1879, 1974.
121. Janini, G. M., Muschik, G. M., Manning, W., and Zielinski, W. L., Jr. (1977): Paper No. 360, presented at the Pittsburgh Conference on Analytical Chemistry and Applied Spectroscopy, Cleveland, Ohio.
122. Janini, G. M., Shaikh, B., and Zielinski, W. L., Jr. (1977): *J. Chromatogr.* 132:136.
123. Zielinski, W. L., Jr., Janini, G. M., Muschik, G. M., Sato, R. I., Miller, M. M., Young, R. M., Jerina, D. M., Thakker, D., Yagi, H., and Levin, W. (1978): Paper No. 210 presented at the Pittsburgh Conference on Analytical Chemistry and Applied Spectroscopy, Cleveland, Ohio.
124. Zielinski, W. L., Jr., Scanlan, R. A., and Miller, M. M. (1979): Paper No. 663 presented at the Pittsburgh Conference on Analytical Chemistry and Applied Spectroscopy, Cleveland, Ohio.
125. Tenney, H. M. (1958): *Anal. Chem.* 30:2.
126. McFadden, W. H. (1958): *Anal. Chem.* 30:479.
127. Zlatkis, A., O'Brien, L., and Scholly, P. R. (1958): *Nature* 181:1794.
128. Primavesi, G. R. (1959): *Nature* 184:2010.
129. Maier, H. J., and Karpathy, O. C. (1962): *J. Chromatogr.* 8:308.
130. Hildebrand, G. P., and Reilley, C. N. (1964): *Anal. Chem.* 36:47.
131. Porter, R. S., Hinkins, R. L., Tornheim, L., and Johnson, J. F. (1964): *Anal. Chem.* 36:260.
132. Littlewood, A. B., and Willmott, F. W. (1967): *J. Gas Chromatogr.* 5:543.

133. Touchstone, J. C., Wu, C. H., Nikolski, A., and Murawec, T. (1967): *J. Chromatogr.* 29:235.
134. Touchstone, J. C., Murawec, T., and Nikolski, A. (1970): *J. Chromatogr. Sci.* 8:221.
135. Molera, M. J., Dominguez, J.A.G., and Biarge, J. F. (1969, 1973): *Sci.* 7:305; 11:538.
136. Fredericks, E. M., and Brooks, F. R. (1956): *Anal. Chem.* 28:297.
137. Dietz, W. A. (1958): In *Gas Chromatography*, V. J. Coates, H. J. Noebels, and I. S. Fagerson, Eds., p. 87, Academic Press, New York.
138. Mayer, R. A. (1958): *ibid*, p. 93.
139. Villalobos, R., Brace, R. O., and Johns, T. (1961): In *Gas Chromatography* H. J. Noebels, R. F. Wall, and N. Brenner, Eds., p. 39, Academic Press, New York.
140. Purnell, J. H. (1962): *Gas Chromatography*, pp. 360-366, Wiley, New York.
141. Villalobos, R., and Turner, G. S. (1963): In *Gas Chromatography*, L. Fomber, Ed., p. 105, Academic Press, New York.
142. Nuss, G. R., *ibid*, p. 119.
143. Deans, D. R. (1969, 1968): In *Gas Chromatography 1968*, C.L.A. Harbourn, Ed., p. 447, Institute of Petroleum, London; *Chromatographia* 1:18.
144. Deans, D. R., and Scott, I. (1973): *Anal. Chem.* 45:1137.
145. Schomburg, G., Husmann, H., and Weeke, F. (1975): *J. Chromatogr.* 112:205.
146. Hoevermann, W., Harke, L., Krüger, E., Schuster, R., and Strubert, W. (1977): *Siemens Analytical Application Note* No. 179, Siemens A. G., Karlsruhe, W. Germany.
147. Pretorius, V., Smuts, T. W., and Moncrieff, J. (1978): *J. High Resolut. Chromatogr. Chromatogr. Commun.* 1:200.
148. Kaiser, R. E., and Rieder, R. I. (1978): *J. High Resolut. Chromatogr. Chromatogr. Commun.* 1:201.
149. Miller, R. J., Stearns, S. D., and Freeman, R. R. (1979): *J. High Resolut. Chromatogr. Chromatogr. Commun.* 2:55.
150. Kwantes, A., and Rijnders, G.W.A. (1958): In *Gas Chromatography 1958*, D. H. Desty, Ed., p. 125, Butterworths, London.
151. Desty, D. H., and Goldup, A. (1960): In *Gas Chromatography 1960*, R.P.W. Scott, Ed., p. 162, Butterworths, London.
152. Goldup, A., Luckhurst, G. R., and Swanton, W. T. (1962): *Nature* 193:
153. Everett, D. H., and Stoddart, C.T.H. (1961):*Trans. Faraday Soc.* 57:746.
154. Pretorius, V. (1978): *J. High Resolut. Chromatogr. Chromatogr. Commun.* 1:199.

155. Bertsch, W. (1978, 1979): *J. High Resolut. Chromatogr. Chromatogr. Commun.* 1:187, 289; 2:85.
156. Massart, D. L., Dijkstra, A., and Kaufman, L. (1978): *Evaluation and Optimization of Laboratory Methods and Analytical Procedures: A Survey of Statistical and Mathematical Techniques*, Elsevier, Amsterdam.
157. Supina, W. R. (1977): In *Modern Practice of Gas Chromatography*, R. L. Grob, Ed., pp. 137-138, Wiley-Interscience, New York.
158. Laub, R. J., and Purnell, J. H. (1975): *J. Chromatogr.* 112:71.
159. Laub, R. J., and Purnell, J. H. (1976): *Anal. Chem.* 48:799.
160. Laub, R. J., Purnell, J. H., and Williams, P. S. (1977): *J. Chromatogr.* 134:249.
161. Laub, R. J., Purnell, J. H., Summers, D. M., and Williams, P. S. (1978): *J. Chromatogr.* 155:1.
162. Laub, R. J., and Purnell, J. H. (1976): *Anal. Chem.* 48:1720.
163. Laub, R. J., and Purnell, J. H. (1978): *J. Chromatogr.* 161:59.
164. Al-Thamir, W. K., Laub, R. J., and Purnell, J. H. (1977): *J. Chromatogr.* 142:3.
165. Laub, R. J., and Purnell, J. H. (1978): *J. Chromatogr.* 161:49.
166. Scott, R.P.W. (1976): *J. Chromatogr.* 122:35.
167. Deming, S. N., and Turoff, M.L.H. (1978): *Anal. Chem.* 50:546.
168. Laub, R. J., Pelter, A., and Purnell, J. H. (1979): *Anal. Chem.* 51:1878.
169. Anderson, L. B., and Laub, R. J. (1979): Paper No. 10, presented at the Eleventh Central Regional Meeting of the American Chemical Society, Columbus, Ohio, 1979; *Anal. Chim. Acta* (in preparation).
170. Laub, R. J., Purnell, J. H., and Williams, P. S. (1977): *Anal. Chim. Acta* 95:135.
171. Zielinski, W. L., Jr., and Martire, D. E. (1976): *Anal. Chem.* 48:1111.
172. Kovats, E. sz. (1958): *Helv. Chim. Acta* 41:1915.
173. Novak, J., Ruzickova, J., Wicar, S., and Janak, J. (1973): *Anal. Chem.* 45:1365.
174. Figgins, C. E., Risby, T. H., and Jurs, P. C. (1976): *J. Chromatogr. Sci.* 14:453.
175. McReynolds, W. O. (1966): *Gas Chromatographic Retention Data*, Preston Technical Abstracts, Niles, Ill.
176. Purnell, J. H., and Spencer, M. S. (1955): *Nature* 175:988.
177. Laub, R. J., Purnell, J. H., Williams, P. S., Harbison, M.W.P., and Martire, D. E. (1978): *J. Chromatogr.* 155:233.
178. Harbison, M.W.P., Laub, R. J., Martire, D. E., Purnell, J. H., and Williams, P. S. (1979): *J. Phys. Chem.* 83:1262.

179. Ashworth, A. J. (1973): *J. Chem. Soc. Faraday Trans. I* 69:459.
180. Ashworth, A. J., and Hooker, D. M. (1977, 1979, 1976): *J. Chromatogr.* 131:399; 174:307; *J. Chem. Soc. Faraday Trans. I* 72:2240.
181. Harbison, M.W.P., Laub, R. J., Martire, D. E., Purnell, J. H., and Williams, P. S. (in preparation): *J. Chromatogr.*
182. Laub, R. J., and Pecsok, R. L. (1978): In: *Physicochemical Applications of Gas Chromatography*, Chapter 6, Wiley-Interscience, New York.
183. Laub, R. J., and Wellington, C. A. (1979): In *Molecular Association*, Vol. 2, R. Foster, Ed., Chapter 3, Academic Press, London.
184. Purnell, J. H., and Vargas de Andrade, J. M. (1975): *J. Amer. Chem. Soc.* 97:3585, 3590.
185. Laub, R. J., and Purnell, J. H. (1976): *J. Amer. Chem. Soc.* 98:30, 35.
186. Martynyuk, R. N., and Vigdergauz, M. S. (1976): *Chromatographia* 9:454.
187. Klein, J., and Widdecke, H. (1978): *J. Chromatogr.* 147:384.
188. Kanchencko, Y. A., Berezkin, V. G., Mysak, A. V., and Paskal, L. P. (1974): *Zavod. Lab.* 40:1450.
189. Purnell, J. H. (1960): *J. Chem. Soc.* 1268.
190. Pecsok, R. L., and Apffel, J. (1979): *Anal. Chem.* 51:594.
191. Laub, R. J. submitted for publication in *Anal. Chem.*
192. Laub, R. J., and Purnell, J. H., *J. High Resolut. Chromatogr. Chromatogr. Commun.* (in press).
193. Ballantine, J. A., and Williams, K. (1978): *J. Chromatogr.* 148:504.
194. Dandeneau, R., and Zerenner, E. H. (1979): *J. High Resolut. Chromatogr. Chromatogr. Commun.* 2:351.
195. Dandeneau, R., Bente, P., Rooney, T., and Hiskes, R. (1979): *Amer. Lab.* 11(9):61.
196. Taylor, P. J., Culp, R. A., Lochmüller, C. H., Rogers, L. B., and Barrall, E. M., II (1971): *Separ. Sci.* 6:841.
197. Taylor, P. J., Ntukogu, A. O., Metcalf, S. S., and Rogers, L. B. (1973): *Separ. Sci.* 8:245.
198. Westerberg, R. B., Van Lenten, F. J., and Rogers, L. B. (1975): *Separ. Sci.* 10:593.
199. Conaway, J. E., and Rogers, L. B. (1978): *Separ. Sci. Techol.* 13:303.
200. Witkiewicz, Z., Suprynowicz, Z., and Dabrowski, R. (1979): *J. Chromatogr.* 175:37.

# SEPARATION AND IDENTIFICATION OF SULFUR HETEROCYCLES IN COAL-DERIVED PRODUCTS

**M. L. Lee\*, C. Willey\*, R. N. Castle\*, and C. M. White\*\***
  \*Department of Chemistry
  Brigham Young University
  Provo, Utah 84602
  \*\*U.S. Department of Energy
  Pittsburgh Energy Technology Center
  Pittsburgh, Pennsylvania 15213

## INTRODUCTION

The carcinogenic and mutagenic activities demonstrated by complex mixtures of polycyclic aromatic compounds (PAC) have stimulated increased efforts toward the identification of individual mixture components. Considerable work has previously been done on the analysis of the polycylic aromatic hydrocarbon (PAH) fraction of various samples, and considerable work is presently being done on the nitrogen heterocycles. One area that has been somewhat neglected is the sulfur heterocycle fraction. This is surprising in light of the relatively high concentrations of sulfur found in fossil fuels. Elemental analysis of petroleum has shown that it contains up to 6 percent sulfur (1). The total sulfur in coal ranges from 0.2 to 11 percent, although in most cases it is between 1 and 3 percent (2). The amount of organic sulfur in coal is usually one-half to one-third of the total sulfur.

Under reducing conditions during the combustion of petroleum, coal, and other organic fuels, PAH are among the principal combustion by-products (3). Coal-conversion products and by-products contain high levels of these compounds (4). Under the same conditions, fuel-bound sulfur is incorporated into relatively stable aromatic systems with structures similar to the hydrocarbons. A number of sulfur heterocycles have been tentatively identified in air particulate matter (5), in coal-combustion products (3), and in carbon black (6).

The detailed characterization of the sulfur-containing compounds in coal-derived products is important for two reasons. First, the human health effects of these products are a direct result of the specific structures of the individual components in the mixtures, as has been found within the class of PAH. Different isomers exhibit varying degrees of carcinogenic activity. For example, benzo(b)naphtho(1,2-d)thiophene (I) and benzo(b)naphtho(2,1-d)thiophene (II) are noncarcinogenic (7); benzo(b)-phenanthro(2,3-d)thiophene (III) and benzo(b)phenanthro(3,2-d)thiophene (IV) are moderately carcinogenic (8); and benzo(b)phenanthro(3,4-d)thiophene (V) is a very potent carcinogen (8).

Second, the removal of sulfur from fuels is desirable to prevent the formation of noxious sulfur gases during combustion and to prevent catalyst poisoning in coal conversion reactors. The identification of specific sulfur-containing species in coal-derived fuels can provide insights into better methods of sulfur removal.

In this paper, the analytical methodology for the separation and identification of sulfur heterocycles in complex aromatic mixtures is described. The isolation of the sulfur heterocycle fraction from a composite aromatic fraction was accomplished by oxidation of the aromatic sulfur compounds to the corresponding sulfones, followed by column adsorption chromatography and subsequent reduction back to the sulfides. Capillary-column gas chromatography in conjunction with sulfur-selective detection and mass spectrometry was used for resolution and structural identification of individual mixture constituents.

These methods are described as applied to the analysis of a coal gasification tar. This sample was chosen for preliminary studies because of its relatively high organosulfur content and because of the possible use of gasification tars as feedstocks for coal liquefaction processes.

## EXPERIMENTAL

A 1.318 kg sample of tar from the condensate produced from the gasification of an Illinois No. 6 coal in the Synthane gasifier (9) was distilled* first at atmospheric pressure and at temperatures from ambient to 200°C. This fraction (fraction 1) represented 15.6 wt % of the original tar and was slightly less than one-half (6.8 wt %) water. The material boiling between 200°C and 290°C** (fraction 2) was distilled under reduced pressure (1 torr) and represented 22.1 wt % of the original tar. The remaining fraction (fraction 3) amounted to 60.6 wt % of the original tar. The unaccountable material (1.7 percent) is a result of losses of the more volatile components during distillation. Elemental analysis of these three fractions gave sulfur concentrations of 4.1, 2.3, and 2.9 percent, respectively.

Sulfur-rich fractions were isolated from distillate fractions 2 and 3 according to the general scheme illustrated in Figure 1. Preliminary cleanup of these fractions was accomplished by first adsorbing

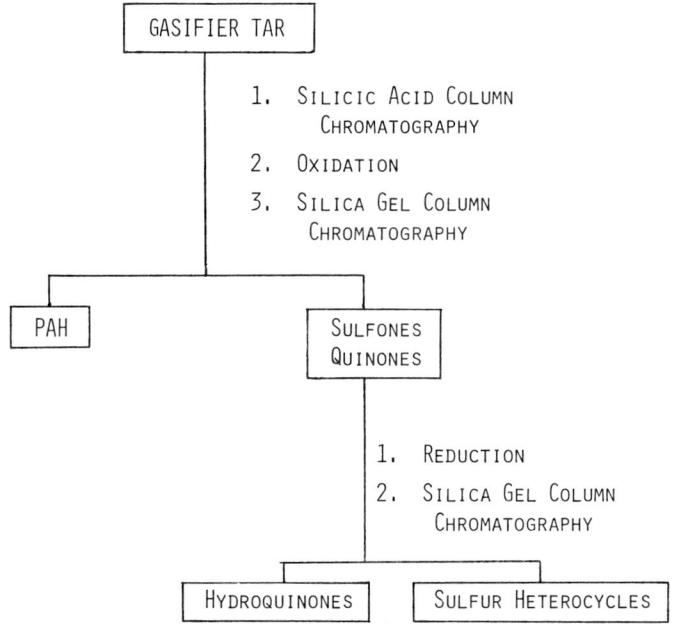

**FIGURE 1.** Sulfur heterocycle separation scheme.

---

*The distillation of the gasification tar was performed by Conoco Coal Development Company, Library, Pennsylvania, under the direction of Edward Obermiller.
**This temperature (290°C) has been corrected to atmospheric pressure.

approximately 2 g of material on 0.5 g of activated silicic acid (Mallinckrodt 100 mesh). This was done by dissolving the tar in a silicic acid/methylene chloride slurry and then evaporating the methylene chloride to dryness. The sample was then transferred to a column containing 2 g of activated silicic acid and eluted with approximately 300 ml of hexane. The hexane eluate was taken to dryness in a rotary evaporator at 45°C.

Samples were then treated according to a modified procedure of Drushel and Sommers (10) in order to isolate the sulfur-rich fraction. The residue from silicic acid chromatography was taken up in 50 ml of benzene; 50 ml of glacial acetic acid were added and the mixture was then refluxed. Approximately 20 ml of 30 percent $H_2O_2$ were added to the refluxing mixture over a period of 1 hour. Refluxing was continued for 16 hours. This procedure quantitatively oxidized the sulfur heterocycles to their corresponding sulfones.

The cooled mixture was washed five times with 50-ml portions of distilled water in a separatory funnel to remove any excess $H_2O_2$ and acetic acid. The $H_2O$ washings were then back extracted twice with 50-ml portions of benzene.

The oxidized material was separated from the unoxidized material by evaporating the combined benzene solutions to 1 to 2 ml, adding 0.5 g of silica gel (Baker Analyzed 60/200 mesh), evaporating the remaining benzene solvent, introducing the adsorbed material on a 20-g silica gel column, and eluting with benzene. The first 60 to 80 ml contained unoxidized compounds, most of which were PAH. The oxidized fraction was subsequently collected by elution with 200 ml of 1:1 benzene/methanol. This fraction contained sulfones and some oxidized PAH.

The sulfones were reduced back to the sulfides according to the procedure of Bordwell and McKellin (11). After the benzene/methanol solution was evaporated to dryness, a suspension was formed by adding 25 ml of anhydrous ethyl ether to the residue. This suspension was reduced by adding it dropwise to a refluxing suspension of 50 ml of ethyl ether containing 1 g of finely powdered $LiAlH_4$. Addition was continued for a period of 1 hour with stirring. The mixture was refluxed for an additional 2 hours, after which $H_2O$ was added dropwise to decompose the excess $LiAlH_4$. The ether portion was separated from the inorganic precipitate and excess water. The precipitate was then washed twice with 10-ml portions of ethyl ether and twice with 10-ml portions of methylene chloride. All ethyl ether and methylene chloride solutions were combined and evaporated to a volume of 1 to 2 ml.

The sulfides were separated from the hydroquinones (formed by reduction of oxidized PAH) by column adsorption chromatography on 10 g of silica gel using hexane as eluent. The first 250 ml contained the sulfides. The hexane solution was evaporated in a rotary evaporator to the appropriate volume for gas chromatography.

The fractions were chromatographed on glass capillary columns coated with SE-52 methylphenylsilicone stationary phase. A Tracor* Model 550 gas chromatograph equipped with a dual FID/FPD was used to obtain simultaneous chromatograms of the total organic fraction. A Varian Model 3700 gas chromatograph was used to obtain the chromatograms shown in Figures 2 through 4, and a Varian Model 1400 gas chromatograph was used to obtain the chromatograms shown in Figures 8 and 9. A Hewlett-Packard 5985A gas chromatograph/mass spectrometer/data system equipped with a glass capillary column was used to obtain mass spectral information on resolved components.

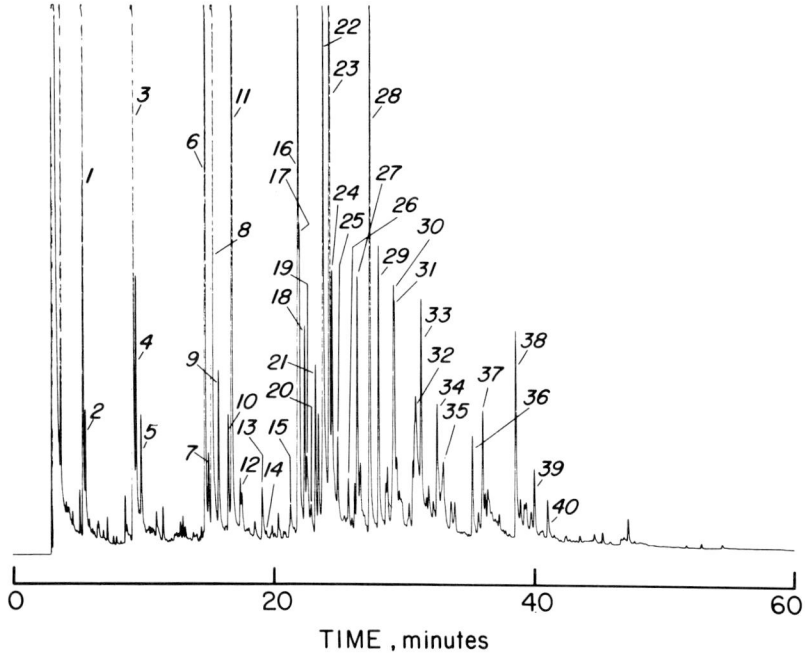

FIGURE 2. Capillary column gas chromatogram of distillate fraction 1. Peak numbers refer to compounds identified in Table 1. Conditions: 30 m x 0.28 mm glass capillary column coated with SE-52, temperature programmed from 50°C to 250°C at 2°C/min.

---

*The use of brand names facilitates understanding and does not necessarily imply endorsement by the U.S. Department of Energy.

## 64 SULFUR HETEROCYCLES IN COAL-DERIVED PRODUCTS

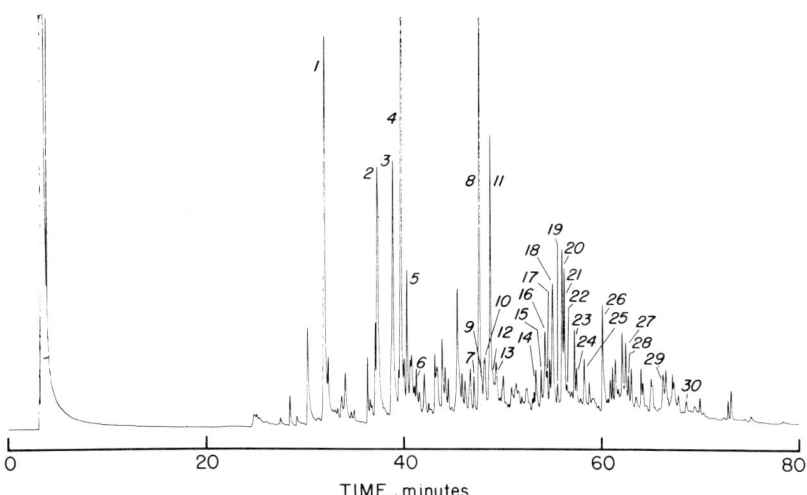

FIGURE 3. Capillary column gas chromatogram of distillate fraction 2. Peak numbers refer to compounds identified in Table 2. Conditions: 30 m x 0.28 mm glass capillary column coated with SE-52, temperature programmed from 50°C to 250°C at 2°C/min.

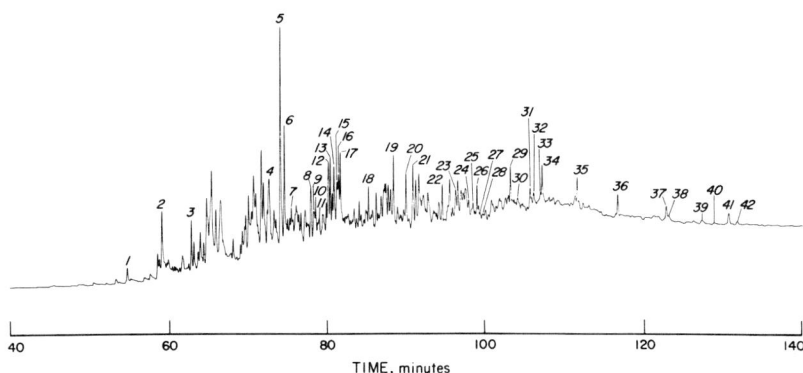

FIGURE 4. Capillary column gas chromatogram of distillate fraction 3. Peak numbers refer to compounds identified in Table 3. Conditions: 30 m x 0.28 mm glass capillary column coated with SE-52, temperature programmed from 50°C to 250°C at 2°C/min.

## RESULTS AND DISCUSSION

Capillary-column gas chromatograms of the three distillation cuts of the gasifier tar are shown in Figures 2 through 4. The compounds identified by gas chromatography/mass spectrometry (GC/MS) are given in Tables 1 through 3, respectively. These fractions are extremely complex with the PAH being the major class of compounds present. The FID and sulfur-selective FPD chromatograms that were simultaneously obtained for each

**TABLE 1.** Compounds Identified in Distillate Fraction 1

| Peak[a] | Molecular Weight | Compound |
|---|---|---|
| 1 | 78 | Benzene |
| 2 | 84 | Thiophene |
| 3 | 92 | Toluene |
| 4 | 98 | 2-methylthiophene |
| 5 | 98 | 3-methylthiophene |
| 6 | 106 | Ethylbenzene |
| 7 | 112 | 2-Ethylthiophene |
| 8 | 106 and 112 | m- and/or p-Xylene and 2,5-dimethylthiophene |
| 9 | 112 | 3-Ethylthiophene |
| 10 | 112 | 2,4-Dimethylthiophono |
| 11 | 106 | o-Xylene |
| 12 | 112 | 2,3-Dimethylthiophene |
| 13 | 120 | $C_3$-benzene |
| 14 | 112 | 3,4-Dimethylthiophene |
| 15 | 120 | $C_3$-benzene |
| 16 | 120 | $C_3$-benzene |
| 17 | 126 | $C_3$-thiophene |
| 18 | 120 | $C_3$-benzene |
| 19 | 126 | $C_3$-thiophene |
| 20 | 126 | $C_3$-thiophene |
| 21 | 126 | $C_3$-thiophene |
| 22 | 120 | $C_3$-benzene |
| 23 | 120 | $C_3$-benzene |
| 24 | 118 | Benzofuran |
| 25 | 142 | n-Decane |
| 26 | 126 | $C_3$-thiophene |
| 27 | 120 | $C_3$-benzene |
| 28 | 118 | Indan |
| 29 | 116 | Indene |
| 30 | 134 | $C_4$-benzene |
| 31 | 134 | $C_4$-benzene |
| 32 | 134 | $C_4$-benzene |
| 33 | 132 | Methylindan |
| 34 | 132 | Methylbenzofuran |
| 35 | 156 | n-Undecane |
| 36 | 132 | Methylindan |
| 37 | 132 | Methylindan |
| 38 | 128 | Naphthalene |
| 39 | 134 | Benzo(b)thiophene |
| 40 | 170 | n-Dodecane |

[a]Numbers refer to peaks in Figure 2.

fraction are compared in Figures 5 through 7. These indicate the presence of a large number of sulfur-containing compounds. Many of the sulfur compounds were sufficiently abundant that mass spectra could be obtained, but many others were buried under the more abundant PAH, and mass spectral data were ambiguous.

## 66 SULFUR HETEROCYCLES IN COAL-DERIVED PRODUCTS

TABLE 2. Compounds Identified in Distillate Fraction 2

| Peak[a] | Molecular Weight | Compound |
|---|---|---|
| 1 | 132 | Methylindan |
| 2 | 132 | Methylindan |
| 3 | 132 | Methylindan |
| 4 | 128 | Naphthalene |
| 5 | 134 | Benzo(b)thiophene |
| 6 | 170 | n-Dodecane |
| 7 | 148 | $C_1$-benzothiophene |
| 8 | 142 | 2-Methylnaphthalene |
| 9 | 148 | 2-Methylbenzo(b)thiophene |
| 10 | 148 | $C_1$-benzothiophene |
| 11 | 142 | 1-Methylnaphthalene |
| 12 | 148 | 3-Methylbenzo(b)thiophene |
| 13 | 184 | n-Tridecane |
| 14 | 154 | Biphenyl |
| 15 | 162 | $C_2$-benzothiophene |
| 16 | 156 | 2-ethylnaphthalene |
| 17 | 162 | $C_2$-benzothiophene |
| 18 | 156 and 162 | 2,6- and/or 2,7-Dimethylnaphthalene and $C_2$-benzothiophene |
| 19 | 162 | 5-Ethylbenzo(b)thiophene |
| 20 | 156 and 162 | 1,3- and/or 1,7-Dimethylnaphthalene and $C_2$-benzothiophene |
| 21 | 156 | 1,6-Dimethylnaphthalene |
| 22 | 162 | $C_2$-benzothiophene |
| 23 | 156 | 2,3- and/or 1,4-Dimethylnaphthalene |
| 24 | 156 | 1,5-Dimethylnaphthalene |
| 25 | 156 | 1,2-Dimethylnaphthalene |
| 26 | 154 | Acenaphthene |
| 27 | 168 | Dibenzofuran |
| 28 | 212 | n-Pentadecane |
| 29 | 166 | Fluorene |
| 30 | 226 | n-Hexadecane |

[a]Numbers refer to peaks in Figure 3.

The FPD is extremely useful for preliminary screening for sulfur compounds in aromatic fractions but has several drawbacks. The FPD seldom gives linear response, and response quenching due to other co-eluting compounds has been observed (12). Quenching is a serious problem when the sulfur compounds are present as trace components. Furthermore, even though the FPD is a good selective indicator for sulfur compounds, it is a poor identification tool, and positive identification must be confirmed by mass spectrometry.

The separation of the sulfur heterocycle fraction from the PAH is the first step that must be accomplished if detailed structural information on the organosulfur compounds is desired. The isolation of this fraction is not an easy task because of the similarities in chemical properties of these

TABLE 3. Compounds Identified in Distillate Fraction 3

| Peak[a] | Molecular Weight | Compound |
|---|---|---|
| 1 | 154 | Acenaphthene |
| 2 | 168 | Dibenzofuran |
| 3 | 166 | Fluorene |
| 4 | 184 | Dibenzothiophene |
| 5 | 178 and 184 | Phenanthrene and naphtho(2,1-b)thiophene |
| 6 | 178 | Anthracene |
| 7 | 184 | Naphthothiophene |
| 8 | 167 and 198 | Carbazole and $C_1$-dibenzothiophene |
| 9 | 198 | $C_1$-dibenzothiophene or $C_1$-naphthothiophene |
| 10 | 204 | 1-Phenylnaphthalene |
| 11 | 198 | $C_1$-dibenzothiophene or $C_1$-napthothiophene |
| 12 | 192 | 3-Methylphenanthrene |
| 13 | 192 | 2-Methylphenanthrene |
| 14 | 192 | 2-Methylanthracene |
| 15 | 192 | 9- and/or 4-Methylphenanthrene |
| 16 | 192 | 1-Methylanthracene |
| 17 | 192 | 1-Methylphenanthrene |
| 18 | 204 | 2-Phenylnaphthalene |
| 19 | 202 | Fluoranthene |
| 20 | 208 | Phenanthro(4,5-bcd)thiophene |
| 21 | 202 | Pyrene |
| 22 | 296 | n-Uncosane |
| 23 | 216 | Benzo(a)fluorene |
| 24 | 216 | Benzo(b)fluorene |
| 25 | 216 | 4-Methylpyrene |
| 26 | 310 | n-Docosane |
| 27 | | Unidentified |
| 28 | 216 | 1-Methylpyrene |
| 29 | 324 | n-Tricosane |
| 30 | 234 | Benzo(b)naphtho(2,1-d)thiophene |
| 31 | 234 | Isomers of benzo(b)naphtho(2,1-d)thiophene |
| 32 | 234 | Isomers of benzo(b)naphtho(2,1-d)thiophene |
| 33 | 228 | Benz(a)anthracene |
| 34 | 228 | Chrysene and/or triphenylene |
| 35 | 352 | n-Pentacosane |
| 36 | 366 | n-Hexacosane |
| 37 | 380 and 252 | n-Heptacosane and benzo(j)fluoranthene |
| 38 | 252 | benzo(k)fluoranthene |
| 39 | 252 | Benzo(e)pyrene |
| 40 | 252 | Benzo(a)pyrene |
| 41 | 394 | n-Octacosane |
| 42 | 252 | Perylene |

[a]Numbers refer to peaks in Figure 4.

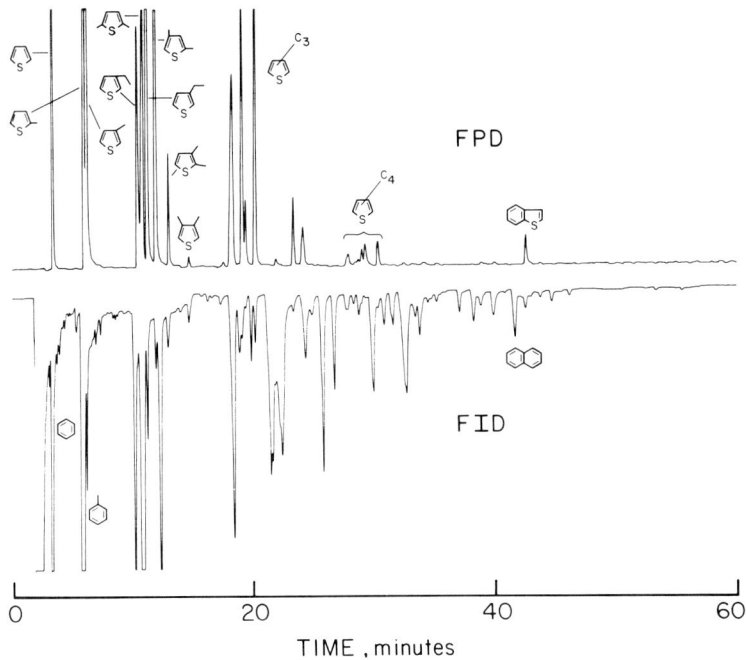

**FIGURE 5.** FID/FPD dual trace capillary column gas chromatograms of distillate fraction 1. Conditions: 12 m x 0.28 mm glass capillary column coated with SE-52, temperature programmed from 50°C to 230°C at 2°C/min.

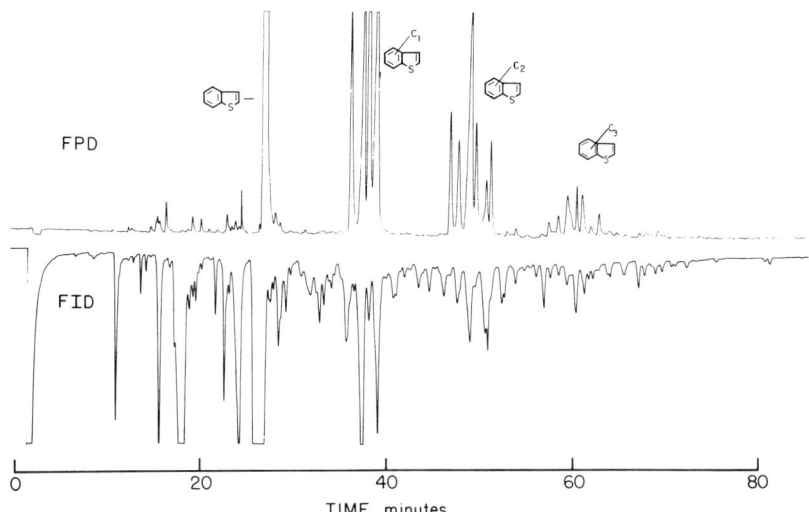

**FIGURE 6.** FID/FPD dual trace capillary column gas chromatograms of distillate fraction 2. Conditions: 12 m x 0.28 mm glass capillary column coated with SE-52, temperature programmed from 90°C to 230°C at 2°C/min.

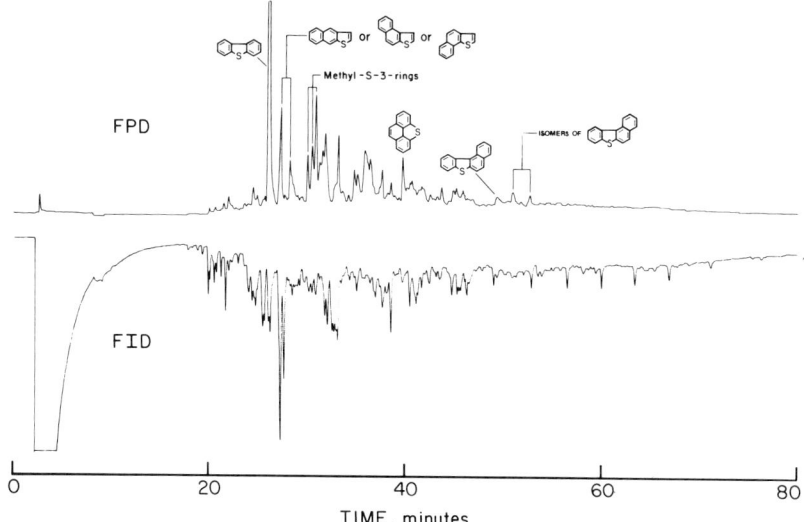

FIGURE 7. FID/FPD dual trace capillary column gas chromatograms of distillate fraction 3. Conditions: 12 m x 0.28 mm glass capillary column coated with SE-52, temperature programmed from 120°C to 230°C at 4°C/min.

compounds to the PAH. In previous studies (3,5), the sulfur heterocycles have always been associated with the PAH fraction.

The isolation of a sulfur-rich fraction is not a new problem. Most previous studies have been aimed at oxidation of the sulfur compounds to sulfoxides or sulfones followed by column-adsorption chromatography (10). The problems that have been encountered include:

(a) The oxidation of the sulfides to sulfoxides or sulfones increases their polarity and decreases their volatility, thus rendering them more difficult to analyze by gas chromatography.
(b) The reduction of sulfones back to the sulfides requires a very strong reducing agent which often reduces the unsaturated hydrocarbon portion of the compounds.
(c) In order to oxidize the sulfides only to the sulfoxides requires an equimolar ratio of oxidizing agent to sulfur content which is difficult to determine in complex mixtures containing relatively low concentrations of sulfur compounds.

It was found in this study that selective oxidation to only the sulfoxide was extremely difficult, and that further oxidation to the sulfone followed by reduction back to the sulfide gave satisfactory results. Figures 8 and 9 show the chromatograms obtained with the FID for the sulfur heterocycle fractions isolated from distillate fractions 2 and 3 by this procedure. Tables 4 and 5 list the compounds identified by GC/MS. As is seen in these figures, this isolation procedure is very effective in isolating the sulfur-rich

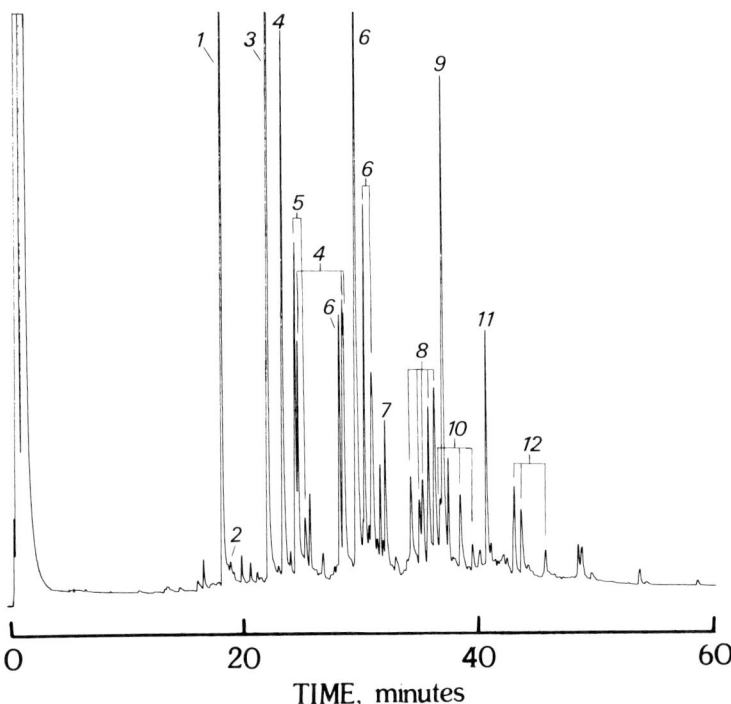

FIGURE 8. Capillary column gas chromatogram of the sulfur heterocycle fraction isolated from distillate fraction 2. Peak numbers refer to compounds identified in Table 4. Conditions: 20 m x 0.28 mm glass capillary column coated with SE-52, temperature programmed from 40°C to 250°C at 2°C/min.

fractions from aromatic mixtures. The presence of several PAH in these fractions (especially in Figure 8) results from some overlap in the PAH and sulfone bands during separation by column-adsorption chromatography.

In evaluating the integrity of this method for isolating sulfur-rich fractions, two questions must be answered: First, is the reduction step in this procedure selective for the sulfur atom only, or does reduction of the aromatic ring occur? As is seen in Figure 9 and Table 5, no hydrogenated compounds are detected in fraction 3. On the other hand, the benzothiophenes in fraction 2 have been reduced to some extent as can be seen by comparison of Figures 6 and 8 and Table 4. The double bond between the 2 and 3 positions on the benzothiophene ring is more susceptible to reduction than other positions (11). To determine whether this tendency for reduction is unique for the benzothiophenes alone, or is also expected for higher molecular weight homologs, phenanthro(2,1-b) thiophene was subjected to similar reducing conditions. It was found that this compound was not reduced, but remained totally aromatic.

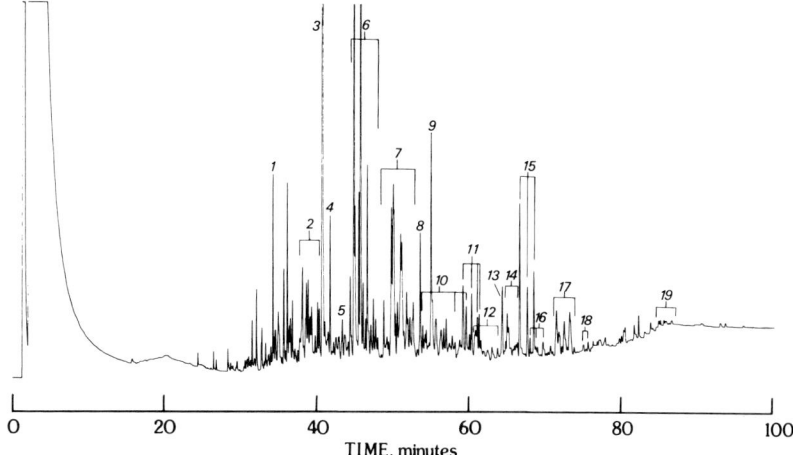

**FIGURE 9.** Capillary column gas chromatogram of the sulfur heterocycle fraction isolated from distillate fraction 3. Peak numbers refer to compounds identified in Table 5. Conditions: 30 m x 0.28 mm glass capillary column coated with SE-52, temperature programmed from 40°C to 125°C at 10°C/min and then 125°C to 250°C at 2°C/min.

The second question involves the possible oxidation of the sulfur heterocycle at a position other than the sulfur atom. The result would be the subsequent reduction to a hydroquinone and removal from the sulfur-rich fraction in the column-adsorption chromatographic step. This question is presently under study.

**TABLE 4.** Compounds Identified in Sulfur Heterocycle Fraction Isolated from Fraction 2

| Peak[a] | Molecular Weight | Compound |
|---|---|---|
| 1 | 128 | Naphthalene |
| 2 | 134 | Benzo(b)thiophene |
| 3 | 136 | 2,3-Dihydrobenzo(b)thiophene |
| 4 | 150 | $C_1$-dihydrobenzothiophene |
| 5 | 148 | $C_1$-benzothiophene |
| 6 | 164 | $C_2$-dihydrobenzothiophene |
| 7 | 162 | $C_2$-benzothiophene |
| 8 | 178 | $C_3$-dihydrobenzothiophene |
| 9 | 168 | Dibenzofuran |
| 10 | 176 | $C_3$-benzothiophene |
| 11 | 166 | Fluorene |
| 12 | 182 | $C_1$-dibenzofuran |

[a]Numbers refer to peaks in Figure 8.

TABLE 5. Compounds Identified in Sulfur Heterocycle Fraction
Isolated from Fraction 3

| Peak | Molecular Weight | Compound |
|---|---|---|
| 1 | 166 | Fluorene |
| 2 | 190 and 204 | $C_4$- and $C_5$-Benzothiophene |
| 3 | 184 | Dibenzothiophene |
| 4 | 178 and 184 | Phenanthrene and naphtho(2,1-b)-thiophene |
| 5 | 184 | Naphthothiophene |
| 6 | 198 | $C_1$-dibenzothiophene |
| 7 | 212 | $C_2$-dibenzothiophene |
| 8 | 202 | Fluoranthene |
| 9 | 208 | Phenanthro(4,5-bcd)thiophene |
| 10 | 226 | $C_3$-dibenzothiophene |
| 11 | 222 | $C_1$-phenanthro(4,5-bcd)thiophene |
| 12 | 240 | $C_4$-dibenzothiophene |
| 13 | 254 | $C_5$-dibenzothiophene |
| 14 | 236 | $C_2$-phenanthro(4,5-bcd)thiophene |
| 15 | 234 | Isomers of benzo(b)naphtho(2,1-d)-thiophene |
| 16 | 250 | $C_3$-phenanthro(4,5-bcd)thiophene |
| 17 | 248 | Isomers of $C_1$-benzo(b)naphtho(2,1-d)-thiophene |
| 18 | 262 | Isomers of $C_2$-benzo(b)naphtho(2,1-d)-thiophene |
| 19 | 258 | Isomers of chryseno(4,5-bcd)thiophene |

[a]Numbers refer to peaks in Figure 9.

The positive identification of most of the isomeric components in these complex mixtures is presently impossible because of the lack of standard reference compounds. Comparison of chromatographic retention data and mass spectral information about mixture constituents with data from synthesized reference compounds is necessary. New mass spectrometric ionization techniques, which can differentiate between many PAH isomers (13), may also provide helpful information.

## ACKNOWLEDGMENT

This study was supported by the U.S. Department of Energy, Division of Biomedical and Environmental Research, Contract No. DE-ACO2-79EV10237. This work was partially accomplished at the Pittsburgh Energy Technology Center at which time M.L. Lee was a Fossil Energy Travel Program Participant under the administration of the Oak Ridge Associated Universities, University Programs.

## REFERENCES

1. Sweeney, W.J. (1950): *Petroleum and Its Products,* W.J. Sweeney, New York.
2. Attar, A. (1978): Chemistry, thermodynamics and kinetics of reactions of sulfur in coal-gas reactions: A review. *Fuel* 57:201-211.
3. Lee, M.L., Prado, G.P., Howard, J.B., and Hites, R.A. (1977): Source identification of urban airborne polycyclic aromatic hydrocarbons by gas chromatographic mass spectrometry and high resolution mass spectrometry. *Biomed. Mass Spectrom.* 4:182-186.
4. Guerin, M.R. (1978): In: *Polycyclic Hydrocarbons and Cancer,* H.V. Gelboin and P. O. P. TS'O, Eds., pp. 3-42, Academic Press, New York.
5. Lee, M.L., Novotny, M., and Bartle, K.D. (1976). Gas chromatography/mass spectrometric and nuclear magnetic resonance determination of polynuclear aromatic hydrocarbons in airborne particulates. *Anal. Chem.* 48:1566-1572.
6. Lee, M.L., and Hites, R.A. (1976): Characterization of sulfur-containing polycyclic aromatic compounds in carbon blacks. *Anal. Chem.* 48:1890-1893.
7. Tilak, B.D. (1960): Carcinogenesis by thiophene isosters of polycyclic hydrocarbons. Synthesis of condensed thiophenes. *Tetrahedron* 9:76-95.
8. Croisy, A., personal communication.
9. McMichael, W.J., Forney, A.J., Haynes, W.P., Strakey, J.P., Gasior, S.J., and Kornosky, R.M. (1977): Synthane gasifier effluent streams. Pittsburgh Energy Research Center Progress Report PERC/RI-77/4, 37 pp.
10. Drushel, H.V., and Sommers, A.L. (1967): Isolation and characterization of sulfur compounds in high-boiling petroleum fractions. *Anal. Chem.* 39:1819-1829.
11. Bordwell, F.G., and McKellin, W.H. (1951): The reduction of sulfones to sulfides, *J. Amer. Chem. Soc.* 73:2251-2253.
12. Grice, H.W., Yates, M.L., and David, D.J. (1970): Response characteristics of the melpar flame photometric detector. *J. Chromatog. Sci.* 8:90-94.
13. Lee, M.L., Vassilaros, D.L., Pipkin, W.S., and Sorensen, W.L. (1979): Combined glass capillary-column gas chromatography and mixed charge exchange-chemical ionization mass spectrometry of isomeric polycyclic aromatic hydrocarbons. *NBS Special Publication* 519:731-738.

# QUANTITATIVE EVALUATION OF PRIORITY POLLUTANT POLYCYCLIC AROMATIC HYDROCARBONS AT ONE PART PER BILLION USING EPA RECOMMENDED PRIORITY POLLUTANT PROTOCOL

**H. G. Nowicki, C. A. Kieda, and D. O. Bassett**

Calgon Analytical Laboratories, Calgon Corporation
Subsidiary of Merck and Co., Inc.
Pittsburgh, Pennsylvania 15230

In the past, water quality has been judged primarily by utilizing test methods such as biochemical oxygen demand (BOD), chemical oxygen demand (COD), suspended solids, conductivity, and other gross pollution measurements. The present trend, as a result of the development of water quality criteria for a large number of target compounds which are potentially hazardous, emphasizes the critical concentrations of specific contaminants in waters as indicators of water quality. The U. S. Environmental Protection Agency's (EPA) Quality Criteria for Water (1) establishes standards for approximately 50 chemical parameters or characteristics of water. The EPA has been required by the Consent Decree, (arising from the Environmental Defense Fund's lawsuit concerning the control of hazardous chemicals) to promulgate water quality criteria for approximately 50 additional compounds. Presently, the critical concentrations for 129 compounds, 114 organic compounds and 15 inorganic chemicals, will be used as the basis for judging water quality.

Of the 114 specific organic compounds, 16 belong to the polynuclear aromatic hydrocarbon (PAH) class (2). Several problems are involved in analysis of these PAH. Adequate cleanup is required for a measurement sensitivity of part per billion (ppb), and the PAH frequently must be separated from other organic substances in complex mixtures often encountered in environmental samples. These mixtures often contain potential interfences at relatively high part-per-million (ppm) concentrations. The specific PAH have a very broad volatility range from acenaphthylene through benzo(a)pyrene (BaP). A high degree of separation efficiency is necessary to measure phenanthrene, anthracene, and other geometric isomers.

An important aspect in the interpretation of chemical information with respect to water quality, and often the most frequently overlooked, is the evaluation of the quality of the data that serve as the basis for comparison with the water quality criteria. This report summarizes experimental findings from the evaluation of the EPA recommended protocol (2) to measure 13 of the priority pollutant PAH. Segments of the evaluation were: (1) gas chromatography-mass spectrometry (GC/MS) analysis of standard solutions, (2) Kuderna-Danish (K-D) concentration of PAH-spiked methylene chloride, and (3) total method (extraction, concentration, and GC/MS analysis). It must be emphasized that the purpose of this study was to determine the application of this technique at the 1-ppb level rather than at the 10-ppb level as stated in the protocol.

The experiment reported here was intended to serve as a preliminary study to be followed by a more extensive investigation of analytical variation in low-level PAH measurements.

## ANALYTICAL METHODS

### Preparation of Standards

The PAH compounds studied contained two to five aromatic rings. In addition to the priority pollutant PAH, several alkyl-substituted PAH and nitrogen- and sulfur-containing analogs were included. All compounds studied are listed in Table 1. These compounds range in molecular weight from 128 for naphthalene to 278 for dibenzo(a,h)anthracene.

The highest purity standards commerically available or solutions were purchased. Several of the compounds had to be obtained as dilute solutions (1000 $\mu$g/ml or 100 $\mu$g/ml) in hydrocarbon solvents because of their inherently low solubilities in methylene chloride, the solvent of choice. Purchase of dilute solutions of carcinogenic PAH was also necessary for safe laboratory handling of carcinogens. Solid compounds were weighed on a microbalance, dissolved in a high-purity solvent (methylene chloride, cyclohexane, or toluene*), and diluted to working standard concentration levels of approximately 1000 $\mu$g/ml. Composite PAH standard solutions were prepared by dilution. The three deuterated PAH compounds—naphthalene-$d_8$, anthracene-$d_{10}$, and chrysene-$d_{12}$ were obtained from the Merck Isotope Division for use as GC/MS internal standards and for extraction efficiency measurements.

---

*Solvents other than methylene chloride were sometimes necessary to dissolve PAH at suitable standard concentration.

## Chromatography

Because of the isomeric nature of many of the PAH compounds of interest in this study, a high-resolution (capillary column) separation technique was required for qualitative indentification. GC/MS analysis using packed GC columns for separation could not provide the chromatographic resolution requirements for measuring the PAH of interest, especially if other organic interferences were present. Glass capillary column GC/MS has a high resolving power and is capable of measuring low levels of PAH even in the presence of other organic interferences.

The most efficient analysis was obtained with an 11 m x 0.25-mm-ID glass column coated with SE-52. This stationary phase shows superior isomer resolution efficiency compared with that of a 30 m-SE-30 column. The SE-52 column was generously donated by Curt M. White of Pittsburgh Energy Technology Center in Pittsburgh, Pennsylvania, at a time before the commercialization of this column by the supply vendors. A new retention index system for PAH has been developed using this stationary phase (3).

Figure 1 shows an electron impact ionization (EI) GC/MS chromatogram of a PAH composite standard mixture. Each peak represents approximately 10 ng of PAH injected onto the column. The peak identities are listed in Table 1. The peak labelled DFTPP (decafluorotriphenylphosphine) is a mass calibration reference compound.

## Gas Chromatography - Mass Spectrometer Parameters

The instrument used in this evaluation was the Finnigan 4023 GC/MS system. Helium was employed as the carrier gas at a flow of 2 ml/min, linear gas velocity of 65 cm/sec. All injections were in the splitless mode (Grob) with delay vent of 40 seconds with injector temperature indicated at 260°C. The column was held as 50°C for 2 minutes, temperature programmed to 250°C at 5°C/min and held for 4650 scans (46.5 minutes).

The temperature in the GC/MS transfer-line region was approximately 270°C. Parameters for the electron impact ionization mode of mass spectral analysis were: electron energy, 70 eV; emission current, 0.5 mA; cycle time, 0.6 sec/scan (100 scans/min) from 45 to 400 atomic mass units; electron multiplier voltage, 1.7 kV; and preamp sensitivity, $10^{-7}$ amp/V. The GC/MS conditions were selected to optimize separation of the specific PAH. Hence, these conditions may not provide the "best" separation of other organics.

Mass spectrometer performance was evaluated with decafluorotriphenylphosphine (DFTPP) as initially proposed by Eichelberger et al (4). DFTPP is employed to document mass assignment, mass peak resolution, and the ion abundance scale. Priority pollutant PAH which have been

TABLE 1. Compounds in PAH Composite Standard Mixture

| Compound No. | Scan No. (Ret. Time)[a] | Relative Ret. Time[a] | Compound | Formula | MW | Comments |
|---|---|---|---|---|---|---|
| 1 | 641 | 0.310 | Naphthalene-$d_8$ | $C_{10}D_8$ | 136 | Deuterated standard |
| 2 | 646 | 0.313 | Naphthalene | $C_{10}H_8$ | 128 | Priority pollutant |
| 3 | 941 | 0.455 | 2-methylnaphthalene | $C_{11}H_{10}$ | 142 | Impurity of 1-methylnaphthalene |
| 4 | 978 | 0.473 | 1-methylnaphthalene | $C_{11}H_{10}$ | 142 | |
| 5 | 1191 | 0.576 | 2-ethylnaphthalene | $C_{12}H_{12}$ | 156 | |
| 6 | 1316 | 0.637 | Acenaphthylene | $C_{12}H_8$ | 152 | Priority pollutant |
| 7 | 1383 | 0.669 | 1,8-dimethylnaphthalene | $C_{12}H_{12}$ | 156 | |
| 8 | 1401 | 0.678 | Acenaphthene | $C_{12}H_{10}$ | 154 | Priority pollutant |
| 9 | 1631 | 0.789 | Fluorene | $C_{13}H_{10}$ | 166 | Priority pollutant |
| 10 | 1997 | 0.966 | Dibenzothiophene | $C_{12}H_8S$ | 184 | Heteroatom PAH |
| 11 | 2054 | 0.994 | Phenanthrene | $C_{14}H_{10}$ | 178 | Priority pollutant |
| 12 | 2067 | 1.000 | Anthracene-$d_{10}$ | $C_{14}D_{10}$ | 188 | Deuterated standard |
| 13 | 2073 | 1.003 | Anthracene | $C_{14}H_{10}$ | 178 | Priority pollutant |
| 14 | 2168 | 1.049 | 3,4-benzoquinoline | $C_{13}H_9N$ | 179 | Heteroatom PAH |
| 15 | 2318 | 1.121 | 2-methylanthracene | $C_{15}H_{12}$ | 192 | |
| 16 | 2414 | 1.168 | 9-methylanthracene | $C_{15}H_{12}$ | 192 | |
| 17 | 2598 | 1.257 | Fluoranthene | $C_{16}H_{10}$ | 202 | Priority pollutant |
| 18 | 2690 | 1.301 | Pyrene | $C_{16}H_{10}$ | 202 | Priority pollutant |
| 19 | 2741 | 1.326 | 9,10-dimethylanthracene | $C_{16}H_{14}$ | 206 | |
| 20 | 2908 | 1.407 | 2,3-benzofluorene | $C_{17}H_{12}$ | 216 | |
| 21 | 3274 | 1.584 | Benzo(a)anthracene | $C_{18}H_{12}$ | 228 | Priority pollutant |
| 22 | 3280 | 1.587 | Chrysene-$d_{12}$ | $C_{18}D_{12}$ | 240 | Deuterated standard |
| 23 | 3291 | 1.592 | Chrysene | $C_{18}H_{12}$ | 228 | Priority pollutant |
| 24 | 3749 | 1.814 | Benzo(b)fluoranthene | $C_{20}H_{12}$ | 252 | Priority pollutant |
| 25 | 3768 | 1.823 | Benzo(k)fluoranthene | $C_{20}H_{12}$ | 252 | Priority pollutant |
| 26 | 3846 | 1.861 | Benzo(e)pyrene | $C_{20}H_{12}$ | 252 | |
| 27 | 3873 | 1.874 | Benzo(a)pyrene | $C_{20}H_{12}$ | 252 | Priority pollutant |
| 28 | 4274 | 2.068 | 1,2,5,6-dibenzoacridine | $C_{21}H_{13}N$ | 279 | Heteroatom PAH |
| 29 | 4358 | 2.108 | Indeno(1,2,3-c,d)pyrene | $C_{22}H_{12}$ | 276 | Priority pollutant |
| 30 | 4392 | 2.125 | Dibenzo(a,h)anthracene | $C_{22}H_{14}$ | 278 | Priority pollutant |
| 31 | 4463 | 2.159 | Benzo(g,h,i)perylene | $C_{22}H_{12}$ | 276 | Priority pollutant |

[a]The scan number (retention time) and relative retention time are the means of seven replicates.

evaluated in this study are indicated by the peak numbers circled in Figure 1.

## PRECISION STUDIES

The method used for sample preparation prior to GC/MS analysis was the EPA's protocol for priority pollutant measurements. The method

EVALUATION OF EPA PRIORITY POLLUTANT PAH METHOD 79

consists of three major parts: (1) pH-adjusted methylene chloride liquid-liquid extraction, (2) Kuderna-Danish concentration, and (3) GC/MS analysis. Since there are no published statistical data available on this

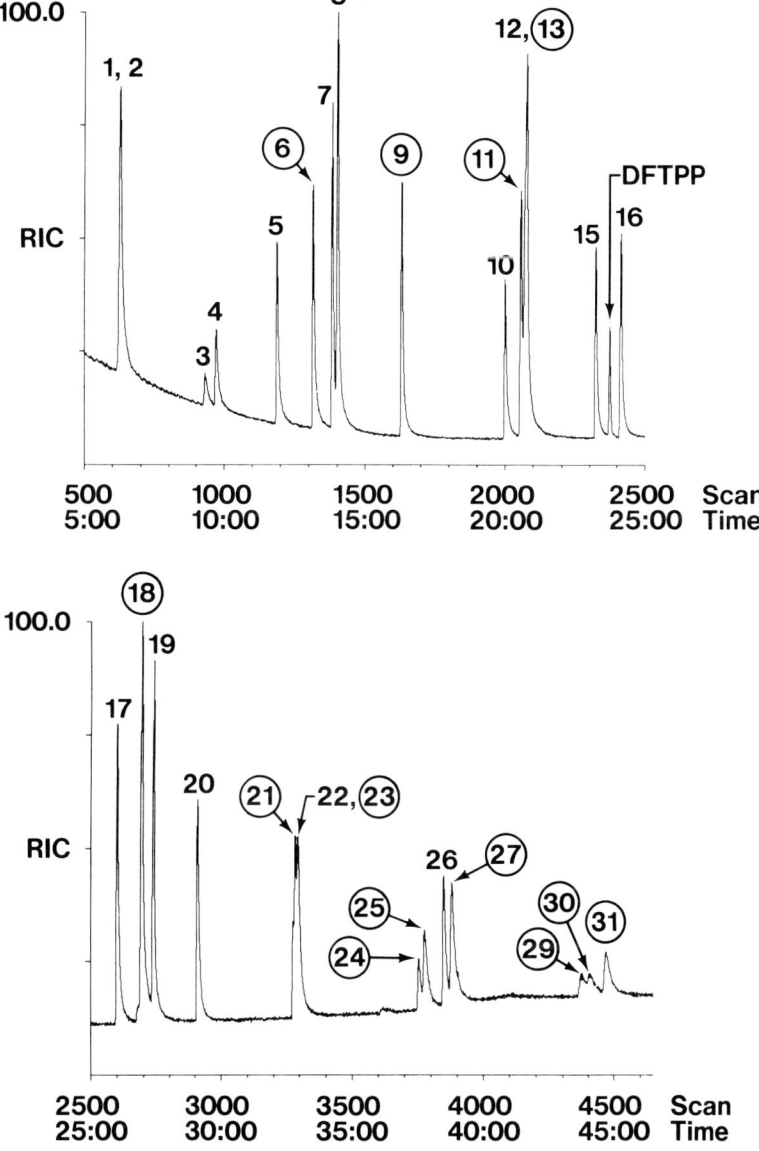

FIGURE 1. Regenerated total ion chromatogram of composite standard PAH ($\approx$10 ng each) mixture. See Table 1 for the compound identification and the text for experimetal parameters.

technique, we decided to determine statistical variations associated with various steps in the procedure. Because of the large volume of data to be analyzed by considering all 31 PAH compounds we studied, it was decided to analyze statistically only the 13 PAH designated by EPA as priority pollutants.

## Instrumental Precision

To evaluate short-term GC/MS system variations under specific analysis conditions, seven replicate analyses of a standard PAH mixture containing 5.4 to 6.6 ng/$\mu$l of each compound (equivalent to water concentrations of 0.54 to 0.66 ppb) were performed utilizing the proposed sample preparation technique with the assumption of 100 percent recovery of each compound. Data acquisition and instrumental parameters used in this phase of the work are presented in the previous section. Replicate analyses were performed by a single operator over a 2-day period with a daily mass spectrometer calibration using perfluorotributylamine (FC-43) and DFTPP.

## Kuderna-Danish Concentration Precision

Replicate analyses of a standard PAH mixture after Kuderna-Danish evaporative concentration were run over a 2-day period to determine the effect of concentration losses, if any, on the statistical variation of PAH analyses.

A 125-$\mu$l volume of standard PAH mixture containing between 6.7 ng/$\mu$l and 8.3 ng/$\mu$l of each of the PAH compounds of interest was added to 300 ml of methylene chloride in each of seven Kuderna-Danish evaporative concentrators. The solutions were concentrated to approximately 2 ml on a boiling water bath and cooled to ambient temperature. The ground glass portion of the Snyder column and concentrator tube was rinsed with methylene chloride. Using a thin-layer chromatography ultraviolet lamp (San Gabriel, California), this area of the K-D glassware has been observed to trap fluorescent residue. Final concentration to 0.1 ml was performed with a gentle stream of nitrogen. The final concentrates were each spiked with anthracene-$d_{10}$ immediately prior to GC/MS analysis so that the anthracene-$d_{10}$ internal standard response would be independent of Kuderna-Danish concentration variations. The amount of each PAH expected to be present in the final concentrates (assuming 100 percent recovery) was 840 ng to 1040 ng—approximately equal to a concentration equivalent to 1 ppb in 1 liter of water.

# EVALUATION OF EPA PRIORITY POLLUTANT PAH METHOD 81

## Total Method Precision

Replicate analyses of prepared aqueous PAH solutions were run over a 2-day period to determine the precision of the total method including extraction, concentration, and GC/MS analysis.

A 125-$\mu$l volume of standard PAH mixture containing between 6.7 ng/$\mu$l and 8.3 ng/$\mu$l of each of the PAH compounds of interest was injected (subsurface) into each of seven-1L volumes of "high purity" water contained in 2L glass separatory funnels. This gave approximately 1 ppb concentrations of each PAH (840 ng/l to 1040 ng/l). Chrysene-$d_{12}$ was added to each separatory funnel at the 1 ppb level as a recovery standard. The aqueous samples were adjusted to pH 11 with NaOH solution (prepared with "high purity" water and extracted with methylene chloride prior to use) and extracted with methylene chloride as specified in the EPA's priority pollutant protocol (2). The methylene chloride extracts were passed through purified anhydrous sodium sulfate to remove water and collected in Kuderna-Danish flask-evaporative concentrators. The extracts were concentrated as described above. Prior to GC/MS analysis, the concentrated extracts were spiked with anthracene-$d_{10}$ at a concentration of 10 ng/$\mu$l again to ensure that a precisely known amount of reference standard would be present.

Blanks consisting of (1) 300 ml of methylene chloride K-D concentrated to 0.1 ml and (2) 1 liter of high purity water (Hydroservice—resin and granular activated carbon treated) extracted with methylene chloride and the extract concentrated to 0.1 ml were analyzed respectively with each series of replicate standards for these precision studies. No PAH were detected in any of the blanks.

## ANALYSIS OF EXPERIMENTAL DATA

Statistical data (5) concerned with repeatability of gas chromatography absolute retention time for seven replicate analyses were calculated for the 13 priority pollutant PAH. The mean relative standard deviation at the 95 percent confidence level, range of observations and sequential sign deviation from the mean were determined.

Table 2 contains the data for the acenaphthylene evaluation; experimental data for the other 12 PAH on the priority pollutant list as well as anthracene-$d_{10}$ and chrysene-$d_{12}$, the "extraction efficiency" monitor have also been generated. Statistical data contained in Table 2 include: column 1—the scan number, a measure of GC retention time; columns 2 and 3 — limited mass (LM) area measurements of the two most intense characteristic ions in the electron impact (EI) mass spectrum of each PAH

TABLE 2. Statistical Data for Acenaphthylene

|  | Scan No. | LM 152 Area Injected | LM 76 Area Injected | LM 152 Area/ LM 188 Area | LM 76 Area/ LM 94 Area |
|---|---|---|---|---|---|
|  |  | Instrumental | (7 replicates) |  |  |
| Mean | 1316 | 9272 | 2093 | 1.69 | 1.53 |
| Std. Dev. | 2.93 | 973 | 209 | 0.18 | 0.20 |
| RSD % | 0.22 | 10.5 | 10.0 | 10.6 | 13.1 |
| RSD %, 95% CL | 0.44 | 21.0 | 20.0 | 21.2 | 26.2 |
| Range | 8 | 3084 | 619 | 0.43 | 0.56 |
| Seq. Sign Dev. | --+0+++ | --++-++ | --++-++ | +---+++ | +---++-- |
| ng Injected | 11.6 |  |  |  |  |
|  | Kuderna-Danish Concentration (6 replicates) | | | | |
| Mean | 1312 | 1937 | 236 | 0.75 | 0.80 |
| Std. Dev. | 2.82 | 592 | 79 | 0.22 | 0.27 |
| RSD % | 0.21 | 30.6 | 33.5 | 29.3 | 33.8 |
| RSD%, 95% CL | 0.42 | 61.2 | 67 | 58.6 | 67.6 |
| Range | 8 | 1724 | 226 | 0.58 | 0.74 |
| Seq. Sign Dev. | -0+0+- | -+-++- | -+-++- | -+-++- | -++++- |
| ng Injected | 18.9 |  |  |  |  |
|  | Total Method (7 replicates) | | | | |
| Mean | 1316 | 4981 | 795 | 0.82 | 0.53 |
| Std. Dev. | 2.83 | 849 | 135 | 0.16 | 0.15 |
| RSD % | 0.21 | 17.1 | 16.9 | 19.5 | 28.8 |
| RSD %, 95% CL | 0.42 | 34.2 | 33.8 | 39 | 57.6 |
| Range | 9 | 2347 | 397 | 0.48 | 0.34 |
| Seq. Sign Dev. | +++-0-- | ---+++- | -+-+++- | +++---- | +++---- |
| ng Injected | 18.2 |  |  |  |  |

on a response (area counts/nanogram injected) basis; and columns 4 and 5 — ratios of the LM area measurements of the two most intense characteristic ions in the PAH EI mass spectrum to the LM area measurements of the two most intense characteristic ions in the mass spectrum of anthracene-$d_{10}$, the internal standard, whose two most intense characteristic ions are at m/e 188 (the molecular ion) and m/e 94. The ions selected for PAH measurements always include the molecular ion and frequently an ion at one-half the molecular weight. As stated previously, measurements are on the basis of area counts/nanogram injected for relevant comparisons. Ideally, the internal standard (anthracene-$d_{10}$) being a PAH should compensate for any instrument or injection variations from run to run.

## EXPERIMENTAL RESULTS AND DISCUSSION

The most important step in any utilization of chemical data for judging the significance of an analytical result to water quality is an evaluation of the reliability of the data. Often this phase of the study is ignored or not adequately studied.

The data contained in Table 2 indicate that the repeatibility of the gas chromatographic absolute retention time is a strength of the priority pollutant PAH method; this is also true for the majority of other PAH studied. The individual method segment data [instrumental analysis of standard solution, Kuderna-Danish (K-D) concentration of PAH-spiked methylene chloride, and total method of extraction, K-D concentration, and GC/MS analysis] were obtained over a 2-day analysis period by a single operator. The instrumental precision data were obtained approximately 1 month before the total method data. Thus, short-term and relatively long-term precision of absolute retention time was demonstrated. The absolute retention time RSD at the 95 percent confidence level ranges from 0.07 to 0.50 for the 15 compounds reported. There is a relationship between elution order and variation; the first eight compounds vary from 0.14 to 0.59 ($\bar{x}$ = 0.28 with S.D. 0.15), whereas the last seven compounds vary from 0.07 to 0.14 ($\bar{x}$ = 0.09 with S.D. 0.040). This was expected as the early eluters are expected to be more affected by minor changes in initial column temperature, while late eluters are unaffected by early temperature variations. The relative-retention-time data in Table 1 are the means of seven replicates. This form of retention-time presentation is considered less variable than absolute retention, especially when runs are compared over extended periods of time. However, we have found very little variation in absolute retention on a short-term basis.

The ratios of LM PAH area to LM area anthracene-$d_{10}$ should show less variation than those of LM of PAH uncompensated for by internal standard (IS). In the majority of instances (22 of 26), comparison of Column 2 with Column 4 (LM of molecular ion of compound of interest and LM of molecular to molecular ion of internal standard) and Column 3 with Column 5 (LM of half-ion and LM of half-ion relative to half-ion of anthracene-$d_{10}$) demonstrates less varation at the 95 percent confidence level. LM area counts represent the signal that the GC/MS data system receives at a specific mass-to-charge (m/e) ratio. Limited mass implies that only the signal due to a specific m/e is being examined. Selectivity is high enough that coeluting compounds having different fragment ions in their mass spectra can be quantified without interference by measuring LM areas of noninterfering fragments.

Table 3 was constructed to simplify overall comparison of the statistical data for the priority pollutant PAH. The scatter in the data is seen to be dependent upon elution order and molecular weight. Table 3 considers only the most intense fragment ion LM area/nanogram injected for further simplification.

Table 3 shows that the instrumental RSD's of the LM areas for the most intense ions for all PAH examined are below 50 percent with a majority of the compounds showing RSD's below 30 percent. As noted before, the compounds showing a higher percentage of RSD are generally the PAH compounds of higher molecular weights. Table 3 shows the difference in precision between the instrumental precision and the total method precision. The effect is approximately a doubling of the percentage of RSD's by going through the entire procedure. At a concentration level of 1 ppb, the percent RSD statistical inference at the 95 percent confidence level indicates $1 \pm 1$ ppb for three of the 13 compounds evaluated. Four compounds indicate $1 \pm 0.8$ ppb, seven indicate $1 \pm 0.6$ ppb, and one has been determined to vary $1 \pm 0.4$ ppb.

The LM area response data for each PAH relative to anthracene-$d_{10}$ after extraction and concentration are in relatively good agreement with the response data after Kuderna-Danish (K-D) concentration for most of the PAH studied, confirming that the extraction of low ppb concentrations of PAH from water is not a serious source of PAH loss. The K-D concentration is again indicated as the critical step. It must be pointed out that the K-D evaluation was performed with dry methylene chloride. Thus, there are subtle differences in the K-D component in the total method evaluation compared with the direct spiking of methylene chloride for the K-D evaluation.

The LM area for the most intense ion precision data after K-D concentration is given in the second column of Table 3. The data show that a majority of the PAH studied have over 50 percent RSD for the LM area of the most intense ion/nanogram injected. This indicates that the K-D concentration step is a major source of statistical variation in quantitative measurements. The area response data for each PAH ion measured relative to the reference anthracene-$d_{10}$ LM area data confirm the variability of results obtained after concentration. The PAH compounds with molecular weights above 200 generally show the poorest repeatability.

Although the statistics indicate a problem with this stage of the analytical procedure, it must be remembered that these data approximate the data expected for aqueous PAH concentrations of 1 ppb. Thus, dibenzo(a,h)anthracene, the PAH with the greatest percentage of RSD (164 percent) in this stage of the study, would have a RSD of 328 percent at the confidence level which would correspond to $1 \pm 3$ ppb. The PAH with the smallest percentage of RSD (16 percent), anthracene, would have a RSD of 32 percent at the 95 percent confidence level which would correspond to $1 \pm 0.3$ ppb. This observation was expected (6) because anthracene and anthracene-$d_{10}$ would be expected to be the most similar in physical properties; other PAH would tend to vary in physical

TABLE 3. Grouping of PAH Priority Pollutants According to the Percentage of Standard Deviation of the Measured Limited Mass Most Intense Ion Area/Nanogram Injected at Selected Precision Study Stages

| 10 to 20% RSD[a] | 20 to 30% RSD | 30 to 40% RSD | 40 to 50% RSD | >50% RSD |
|---|---|---|---|---|
| **Instrument Precision (Standard)** | | | | |
| Acenaphthylene | Benzo(a)anthracene | Dibenzo(a,h)anthracene | | |
| Fluorene | Chrysene-d | | | |
| Phenanthrene | Benzo(b)fluoranthene | | | |
| Anthracene-$d_{10}$ | Indeno(1,2,3-c,d)pyrene | | | |
| Anthracene | | | | |
| Pyrene | | | | |
| Chrysene | | | | |
| Benzo(k)fluoranthene | | | | |
| Benzo(a)pyrene | | | | |
| Benzo(g,h,i)perylene | | | | |
| **Kuderna-Danish Concentration Precision** | | | | |
| Anthracene | | Acenaphthylene | Fluorene | Benzo(a)anthracene |
| | | Phenanthrene | Chrysene | Benzo(b)fluoranthene |
| | | Anthracene-$d_{10}$ | | Benzo(k)fluoranthene |
| | | Pyrene | | Benzo(a)pyrene |
| | | | | Indeno (1,2,3-c,d)pyrene |
| | | | | Dibenzo(a)anthracene |
| | | | | Benzo(g,h,i)perylene |
| **Total Method Precision** | | | | |
| Acenaphthylene | Fluorene | Benzo(a)anthracene | Phenanthrene | |
| | Anthracene-$d_{10}$ | Chrysene-$d_{12}$ | Indeno(1,2,3-c,d)pyrene | |
| | Anthracene | Chrysene | Dibenzo(a,h)anthracene | |
| | Pyrene | Benzo(b)fluoranthene | | |
| | Benzo(k)fluoranthene | | | |
| | Benzo(a)pyrene | | | |
| | Benzo(g,h,i)perylene | | | |

[a] RSD percentage is one sigma confidence level.

properties from the internal standard. Sauter (7) has demonstrated that the chromatographic peak shape of the priority pollutant PAH varies significantly from the internal standard, which elutes with the most Gaussian peak shape. It was concluded (7) that anthracene-$d_{10}$ does not exactly trace the elution of either napthalene or dibenzo(a,h)anthrance.

## SUMMARY OF RESULTS

Aqueous samples containing approximately 1 ppb of 29 polynuclear aromatic hydrocarbons (PAH) and heteroatom analogs have been extracted, concentrated, and analyzed by glass capillary column GC/MS. Thirteen of these compounds, designated priority pollutants by the EPA, were observed to give total method recoveries ranging from 42 to 169 percent, based upon statistical inference of the mean values from seven replicate analyses. The total method recovery seems to be dependent upon the molecular weight of the PAH being measured, e.g., acenaphthylene (MW = 152), the lowest molecular weight priority pollutant PAH, was recovered at 54 percent, while chrysene (MW = 228) was recovered at 80 percent and compounds with higher molecular weights had lower recoveries. Dibenzo(a,h)anthracene was atypical with 169 percent recovery; this compound was found to show the largest variation in all respects studied relative to the other 12 members of the PAH class of organic compounds on the priority pollutant list. Absolute retention times of all compounds analyzed showed RSD's of 0.25 percent or less.

Precision studies of the individual analytical method components, i.e., liquid-liquid extraction, K-D concentration, and capillary column GC/MS analysis, indicate that repeatability variation is greatest after K-D concentration. Instrumental precision studies showed 9 PAH with quantitative RSD's between 10 and 20 percent, 3 PAH with RSD's in the range 20 to 30 percent, and 1 PAH with an RSD between 30 and 40 percent. K-D concentration of a PAH standard of similar concentration showed seven components with RSD's greater than 50 percent, 2 PAH with RSD between 40 and 50 percent, 3 PAH with RSD in the 30 to 40 percent range, and 1 PAH with an RSD between 10 and 20 percent.

## ACKNOWLEDGEMENTS

We wish to thank R. F. Devine for providing assistance in statistical preparation and analysis of the experimental data. V. Current performed the sample preparations in close collaboration with C. A. Kieda, the project senior chemist. We also thank S. Wagoner for her efforts in behalf of this work.

## REFERENCES

1. U. S. Environmental Protection Agency (1976): *Quality Criteria for Water,* U. S. Government Printing Office, Washington, D. C.
2. U. S. Environmetal Protection Agency, (1978): "Sampling and analysis procedures for survey of industrial effluents for priority pollutants" Environmental Monitoring and Support Laboratory (EMSL), Cincinnati, Ohio.
3. Lee, M. L., Vassilaros, D. L., White, C. M., and Novotny, M. (1979): Retention indices for programmed-temperature capillary-column gas chromatography of polycyclic aromatic hydrocarbons. *Anal. Chem.* 51(6): 768-744.
4. Eichelberger, J. W., Harris, L. F., and Budde, W. L. (1975): Reference compound to calibrate ion abundance measurements in gas chromatography-mass spectrometry systems. *Anal. Chem.* 47:995.
5. Meyer, Stuart L. (1975): *Data Analysis for Scientist and Engineers,* John Wiley and Sons Inc., New York.
6. Millard, B. J. (1978): *Quantitative Mass Spectrometry,* Heyden.
7. Sauter, A. D., Kieda, C. A., Devine, R. F., and Nowicki, H. G. (1979): Quantitative determination of priority pollutants—gas chromatography-mass spectrometry response factor variation. *Measurement of Organic Pollutants in Water and Waste Water.* ASTM STP 686:221-233.

# METABOLISM OF BENZO(a)PYRENE IN CULTURED HUMAN BRONCHUS, TRACHEA, COLON, AND ESOPHAGUS

**H. Autrup, A. M. Jeffrey, and C. C. Harris**

Human Tissue Studies Section
Laboratory of Experimental Pathology
National Cancer Institute
Bethesda, Maryland 20205
and
Institute of Cancer Research
Columbia University
College of Physicians and Surgeons
New York, New York 10032

## INTRODUCTION

Polynuclear aromatic hydrocarbons (PAH), such as benzo(a)pyrene (BaP), are widespread environmental pollutants and likely human health hazards. However, all these compounds are biologically inactive per se, and require metabolic activation to exert their cytotoxic, mutagenic, and carcinogenic effects. The activation can take place within the target cells themselves, or activated metabolites can be transferred from one cell type to another. The metabolism of BaP has been extensively studied in microsomal preparations, cells, and intact tissues from humans (5,6,8,11,18,19,22,25,26,32,39,42,44,45,48,52). The main emphasis of these studies has been to determine the pathways of activation into the ultimate carcinogenic forms and their reaction with DNA. The covalent binding of carcinogens to nuclear DNA is currently considered an important event in the initiation of the carcinogenic process, although reaction with other cellular macromolecules also may be important. A positive correlation between DNA binding and the carcinogenic activity of different PAH has been found (29). Further, in V-79 cells the mutation frequencies are related to the extent of modification of DNA rather than to the chemical structure of PAH (51).

The cellular toxicity of a chemical depends in part on the cell's ability to activate/deactivate the chemical. The ratio of metabolic activation to deactivation of a chemical procarcinogen is an important determinant of carcinogenicity in experimental animal models and is likely to play a similar role in determining carcinogenic risk in the exposed individual. This ratio also may explain the relative organ specificity of most chemical carcinogens, as the specific target tissues may vary in ability to detoxify the active metabolites.

Extensive work on BaP metabolism in human tissues has been done using microsomal preparations. However, marked quantitative and qualitative differences in primary metabolites formed by microsomal preparation and intact cells have been found (13,46). Furthermore, cell-free systems lack the enzymatic activity to conjugate the primary metabolites into water-soluble products. This emphasizes the need for using intact cell systems for such metabolic studies. Using explant culture systems for human tissues (2,10,23,24), we and our co-workers have studies the metabolism of BaP by cultured human bronchus, colon, esophagus, peripheral lung, and trachea (6,8,22,24,48). The advantage of using the explant culture system for the study of carcinogen metabolism is manifold: (1) an intact cellular system is being studied; (2) the effects of carcinogens can be studied in the different cell types which form the epithelium; (3) interaction between different cell types can be investigated; and (4) the explant culture systems provide a link between experimental animals and humans, as the metabolism of the carcinogen can be studied on the same level of biological organization.

## MATERIALS AND METHODS

Specimens of bronchus, colon, and esophagus showing grossly normal morphology were collected at the time of surgery for either cancerous or benign lesions or at the time of autopsy immediately following death resulting from head trauma (49). The tissues were cultured under chemically defined conditions (2,10,24). Bronchial and esophageal specimens were dissected into pieces measuring 1 cm$^2$ and placed in 60 mm plastic dishes with the epithelial layer facing the gas interface. Two ml of CMRL-1066 medium, supplemented with 2 mM L-glutamine, insulin (1.0 $\mu$g/ml), hydrocortisone (0.1 $\mu$g/ml), $\beta$-retinyl acetate (0.1 $\mu$g/ml), penicillin G (100 units/ml), and streptomycin (100 $\mu$g/ml), was added to each dish. The dishes were placed in a controlled-atmosphere chamber, flushed with 50% $O_2$-45% $N_2$-5% $CO_2$, and incubated at 36.5°C (under rocking; 10 cycles/min). Human colon was cultured under slightly modified conditions (2). Bronchus and esophagus were cultured for 7 days prior to incubation with radioactively labeled BaP to reduce the effect of exogenous inducers of the mixed-function oxidase system, while colon was cultured for 24 hours. The viability of the tissues was monitored by high-resolution light microscopy (35). After incubation with BaP, the mucosal layers were separated from the stroma and DNA isolated by the phenol extraction procedure and purified on a CsCl gradient (26). The tissue culture medium was saved, and metabolites released into the media, both organo- and water-soluble, were determined by chromatographic techniques (3,53). The BaP-DNA adducts were identified through separation of enzyme-digested DNA (31) by high-pressure liquid chromatography.

## RESULTS AND DISCUSSION

### Binding of BaP to Cellular DNA

The metabolism of BaP is mediated, in part, by aryl hydrocarbon hydroxylases which are mixed-function oxidases. The pathway leading to the predominant BaP-DNA adducts is through the formation of *trans*-7,8-diol, which is thought to be the proximate carcinogenic form of BaP. Further activation of 7,8-diol to a diol epoxide is required for reaction with DNA. Recent experiments with reconstituted mixed-function oxidases indicate that different "cytochrome P-450s" are involved in these two steps of metabolism of BaP (18).

BaP was significantly metabolized into those intermediates that reacted with cellular DNA in all investigated human tissues. The highest mean binding level was found in bronchus (22 ± 24 pmoles per 10 mg DNA; mean ± SD range 1 to 151; 129 cases), the intermediate level in trachea (10.8 ± 5.9; 3 to 17; 5 cases) and esophagus (10.2 ± 9.3; 0.3 to 29.3; 13 cases), and the lowest level in colon (5.9 ± 5.8; 0.2 to 25.0; 103 cases) (Figure 1), with the mean binding level in bronchus being equivalent to 1 modification per $10^6$ nucleosides. The binding level varied among the individual cases—from 75-fold in bronchus (21) to 100-fold in esophagus (24) and 135-fold in colon (4), with the binding-level values having a unimodal distribution. A similar variation of the activity of aryl hydrocarbon hydroxylase (AHH) in human placenta (42), monocytes (11,40), and liver (32) has previously been reported. Studies of the activity and inducibility of aryl hydrocarbon hydroxylase (AHH) in both monozygotic and dizygotic twin pairs indicate that both are heritable

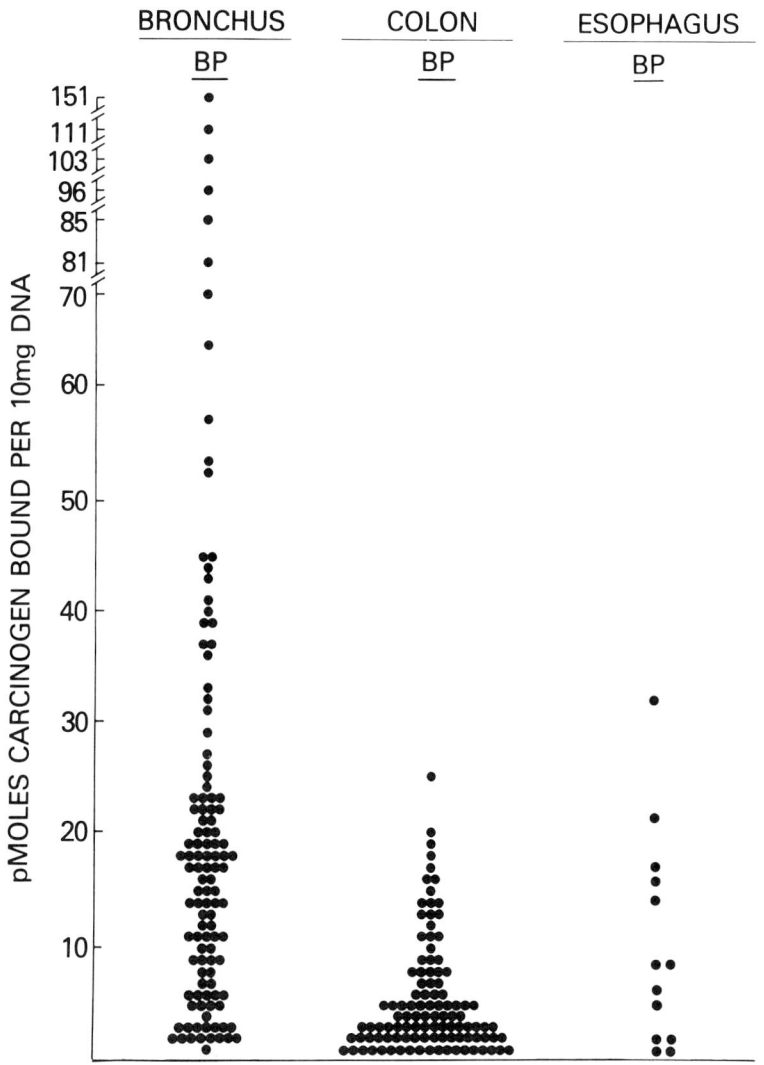

FIGURE 1. Binding level of benzo(a)pyrene to DNA in cultured human bronchus (115 cases), colon (106 cases), and esophagus (13 cases).

traits, and that inducibility is controlled by a single or a very few polymorphic genes (1,33,34,41). Several factors for the wide interindividual variation in our cultured tissues were considered. The interindividual variation attributable to the methodology was minimal, and the viability of the tissue, as monitored by high-resolution light microscopy was good in all of the reported cases; however, subtle changes in cellular physiology could, in part, account for some of the observed differences. Some of the interindividual variation observed in colon could, in part, be explained by the different anatomical origin of the colonic tissues. The highest binding level was observed in the transverse and the ascending colon for tissues from different subjects (Figure 2). When the different anatomical segments of the colon were obtained from the same patient, the highest binding level also was found in the ascending and the transverse colon. A higher level (2- to 3-fold) of binding also was

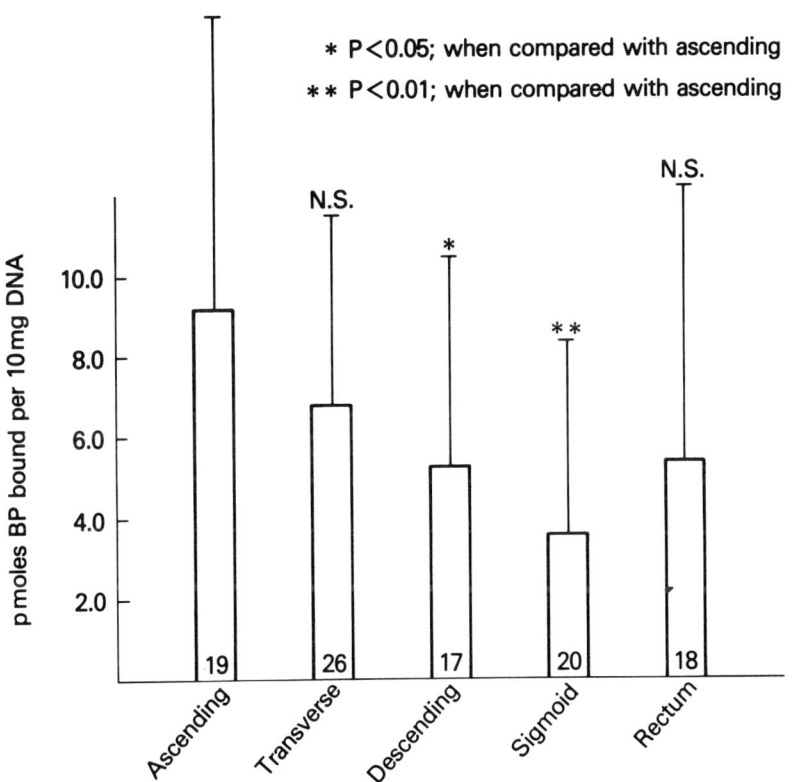

FIGURE 2. The mean binding level of benzo(a)pyrene to DNA in different anatomical segment of cultured human colon. Numbers in the bars represent the number of patients.

found in the duodenum than in the transverse colon from the same patient. No major difference in PAH binding was observed in the tracheobronchial tract (8) or esophagus (24) because of the anatomical site of the tissues. When human colon was obtained from the same patient within a 1-month interval, no difference in the level of binding of BaP to DNA was observed even though day-to-day variation in AHH has been reported in both monocytes and lymphocytes (40). Although our tissues appear normal when checked by high-resolution light microscopy, we have previously shown that the binding level was higher in colon obtained from noncancerous patients than in that from normal tissues from colon cancer patients; however, this also could reflect an age difference in the patients, as the mean age of the noncancer patients was 20 years less than that of the cancer patients.

A positive correlation was found among the binding levels of BaP to bronchial, colonic, and esophageal tissues from the same patients (Figure 3) (r = 0.579; P > 0.05; 7 cases). These results indicate that the metabolism by one cell type could be indicative of an overall ability to activate BaP into DNA binding metabolites. This basic premise could be useful to identify individuals having high risk for chemically induced cancers, assuming that a positive correlation exists between the binding of BaP to DNA in the target cells and in an easily obtainable cell type. Binding to DNA may be a better end point for this determination than the formation of fluorescent products (e.g., BaP phenols) which are generally a result of detoxification of the carcinogen. A positive correlation between the binding of BaP and 1,2-dimethylhydrazine to colonic DNA (2) and between BaP and N-nitrosodimethylamine in esophageal DNA has previously been reported (24).

## BaP-DNA Adducts

Isolation of the BaP-DNA adducts from the cultured tissues by high-pressure liquid chromatography (Table 1) indicates that the major adduct is formed by trans-addition of ($\pm$) ($7\beta,8\alpha$)-dihydroxy-($9\alpha,10\alpha$)-epoxy-7,8,9,10-tetrahydrobenzo(a)pyrene (BPDE I) to the exocyclic 2-amino group of guanine (50). Both stereoisomeric forms of BPDE I reacted with guanine, the 7R form being the most predominant in cultured human tissues. The relative distribution between adducts formed by reaction of guanine with either 7S-BPDE I or 7R-BPDE I appears to vary among animal species, the 7S-BPDE I being the major active metabolite in trachea from rats (8). Experiments in a cell-free system indicated that the ratio of the two enantiomers depended on the secondary structure of DNA: the 7S form reacted preferentially with guanine in double-stranded

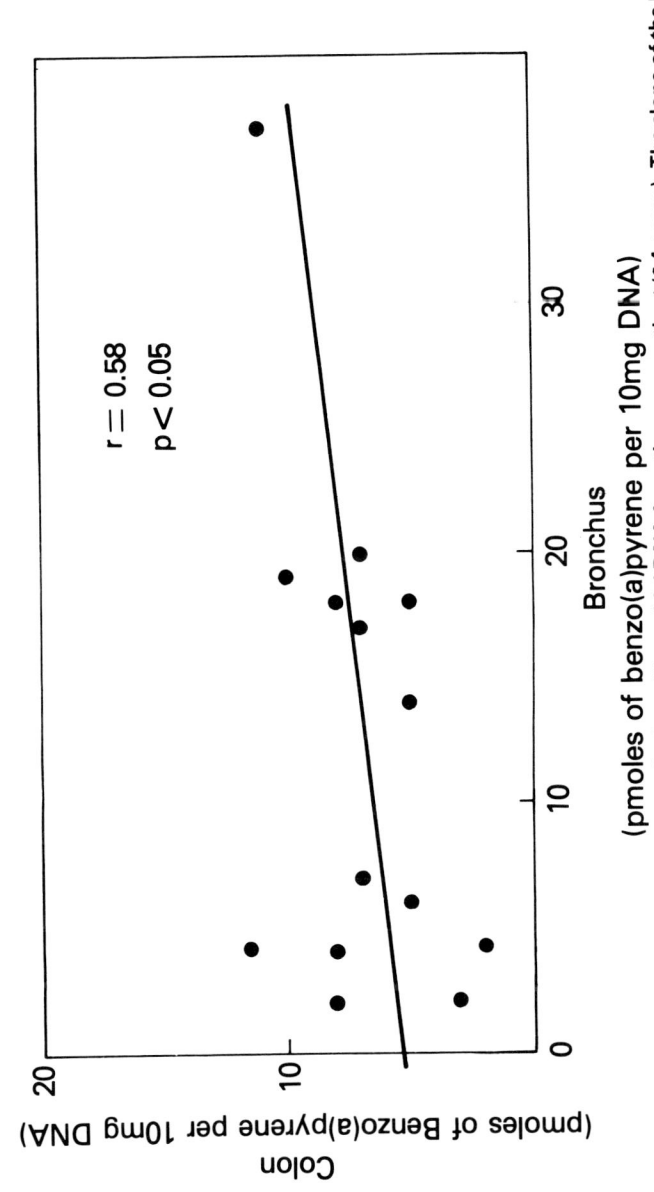

FIGURE 3. Binding level of benzo(a)pyrene to colonic and bronchial DNA from the same patient (14 cases). The slope of the line is calculated using the least-squares method.

TABLE 1. Benzo(a)pyrene—DNA Adducts Formed by
Cultured Human Tissues[a]

| Tissue | BaP-Cytosine | 7S-BPDE I-Guanine | 7R-BPDE I-Guanine | BPDE II-Guanine |
|---|---|---|---|---|
| Bronchus[b] | 17 | 21 | 56 | 7 |
| Colon[c] | — | — | 92 ± 12 | 8 ± 12 |
| Esophagus[d] | 28 | 7 | 59 | 5.5 |

[a]The results are expressed as percentage of total identified adducts.
[b]Reference 8.
[c]Mean ± S.D. of DNA isolated from five individual patients (Reference 6).
[d]Reference 24.

DNA, while equal reactivity with guanine was found in single-stranded DNA (36). Minor amounts of adducts also were formed between the 2-amino group of guanine and $(7\beta,8\alpha)$-dihydroxy-$(9\beta,10\beta)$-epoxy-7,8,9,10-tetrahydrobenzo(a)pyrene (BPDE II). The relative distribution of these two adducts differed in colonic DNA from different patients. The biological significance of these two modifications is unknown, but a different rate of removal of the two adducts has been suggested. Minor amounts of unidentified BaP-DNA adducts with retention times similar to those of BaP-cytosine adducts also were observed in bronchus, esophagus, and trachea (8,24,30). Adducts formed between the BPDEs and the 2-amino group of guanine also have been observed in human peripheral lung (15,47), in human diploid fibroblasts (9), and when human liver microsomes were incubated with BaP and DNA (14). A more complex adduct pattern is generally observed when the activation is performed in a cell-free system (14,43). The biological consequences of the formation of BaP-DNA adducts seen in human tissues is presently unknown. However, the formation of the BaP-DNA adduct in humans appears to resemble that found in tissues from experimental animals in which BaP proved to be carcinogenic.

## Organo- and Water-Soluble Metabolites

The total metabolism of BaP by cultured human tissues was higher in bronchus and trachea than in colon (Table 2). The interindividual variation noted here was not as great as seen in the level of binding to DNA. A ratio of organo-soluble to water-soluble metabolites greater than 1 (mean value) was found in all four tissues. Although a ratio greater than one was observed in cultured tracheobronchial tissues from animal species susceptible to the carcinogenic action of BaP, a ratio of less than

**TABLE 2. Total Metabolism of BaP by Cultured Human Tissues[a]**

|  | Bronchus | Trachea | Colon | Esophagus |
|---|---|---|---|---|
| Total metabolism[a] | 1,335 ± 551 | 1,092 ± 256 | 260 ± 186 | 381 ± 133 |
| Organic extractable[a] | 834 ± 459 | 783 ± 227 | 222 ± 155 | 360 ± 212 |
| Water-soluble[a] | 501 ± .276 | 309 ± 115 | 36 ± 16 | 164 ± 280 |
| Sulfates[b] | 44 ± 8 | 56 ± 5 | 45 ± 5 | 34 ± 12 |
| Glucuronides[b] | 7 ± 2 | 12 ± 5 | 7 ± 5 | 18 ± 9 |
| Glutathione[b] | 51 ± 8 | 32 ± 9 | 48 ± 7 | 48 ± 10 |

[a] In pmoles BaP metabolites per $\mu$g DNA. The results are expressed as the mean ± S.D. of 5 (bronchus), 8 (colon), 7 (esophagus), and 4 (trachea) patients.
[b] Percentage of total water-soluble metabolites.

one was observed in tracheobronchial tissues from species in which BaP was either weakly carcinogenic or has not yet proved to be carcinogenic (8). These results could indicate a low level of detoxifying enzymes in cultured human target tissues. The water-soluble metabolites accounted for about 20 to 60 percent of the total metabolism in bronchus and trachea and for about 10 to 40 percent in cultured colon. Determination of both the activation and deactivation of a potential carcinogen is important, because it could, in part, explain the species and tissue specificity of the carcinogen.

Separation of the water-soluble metabolites by liquid chromatography (3) into the three major fractions of conjugates—sulfate esters, glucuronides, and glutathione—indicates that sulfate esters and glutathione conjugates are of major importance in cultured human tissues. Glucuronides, which are the major detoxification products in intact animals and in animal cell cultures, accounted for only 6 to 18 percent of the total water-soluble metabolites in cultured human tissues. Some of this discrepancy could be caused by our experimental system. However, the level of precursors for the different conjugation pathways in the media was sufficient, so they did not represent a limiting factor. Similarly, glucuronides were of major importance in the presumed detoxification of BaP by cultured mouse trachea and bovine bronchus, using the same culture technique (8). Using an indirect method, Mehta and Cohen showed that sulfate conjugation was predominant in human lung tissues, while glucuronide was the major detoxification product in rodent lung tissue (37).

No method is presently available to separate the individual conjugated metabolites. Enzymatic hydrolysis of the glucuronides and sulfate esters and subsequent separation of the organo-soluble metabolites by HPLC showed that all the primary metabolites were substrate for both conjugation sytems. BaP diols and BaP tetrols, or perhaps their epoxide intermediates, are major substrates for UDP-glucuronyltransferase (3). BaP-7,8 epoxide was found to be a good substrate for UDP-glucuronyltransferase in a cell-free system, whereas the tetrols and the trans-7,8-diol were poor substrates (38). The phenols were mostly conjugated by PAPS-sulfotransferase in cultured human colon.

The pattern of organo-soluble metabolites of BaP in cultured human tissues differs from that of the metabolites formed by microsomal preparations (see Table 3). Tetrols and diols were the major metabolites, while phenols were seen only in minor amounts. The relative amount of 9,10-diol cannot be calculated, as these compounds have the same chromatographic properties as (7/8,9)-triol and (7,9,10/8)-tetrol under our HPLC conditions. Nearly equal amounts of 4,5-diol and 7,8-diol were formed in trachea and bronchus, while more 7,8-diol was formed by cultured colon. A higher ratio of tetrols to diols was found in bronchus and trachea than in colon and esophagus; this could indicate that the enzyme systems responsible for the further oxidation of the diols are saturated in colon, while further metabolism of the diols could take place in tracheobronchial tissues. The trans-7,8-diol is more readily metabolized and bound to DNA than trans-9,10-diol, trans-4,5-diol, or the parent BaP (52). Phenols and diols were identified as the major metabolites by short-term incubation of lymphocytes and monocytes with BaP (39), and in human lung and liver microsomal preparations. Quinones accounted for 13 to 30 percent of the metabolites in cultured human tissues, but this high level could possibly be due either to nonenzymatic formation from phenols during the storage of the sample prior to analysis or to the long incubation time with BaP. A wide interindividual variation also was seen in the relative distribution of the organo-soluble metabolites of BaP, similar to the observations in monocytes (40). The presence of 7,10/8,9-tetrol and 7/8,9,10-tetrol is an indication of the formation of BPDE I, inasmuch as the diol epoxides are unstable in aqueous media. The relative amount of BPDE II cannot be deduced because (7,9,10/8)-tetrol co-chromatographed with 9,10-diol.

Indirect proof for the release of BPDEs into the media was found by using either human bronchial tissue or pulmonary alveolar macrophages (27,28) as an activating system in a tissue or cell mediated mutagenesis assay using Chinese hamster V-79 cells as indicator cells.

Some difference in the metabolite profile of BaP also was observed in bronchial tissues from lung cancer and from normal patients. A higher percentage of the peak containing 9,10-diol (7/8,9)-triol and (7,9,10/8)-tetrol was formed by explants from noncancerous patients (25).

TABLE 3. Pattern of Organo-Soluble BaP Metabolites Formed by Cultured Human Tissues[a]

| Metabolites | Retention Time (seconds) | Trachea | Bronchus | Colon | Esophagus |
|---|---|---|---|---|---|
| (7,10/8,9)-Tetrol | 450 | 10.2 ± 3.1 | 5.0 ± 2.4 | 16.4 ± 6.2 | 2.7 ± 2.8 |
| (7,9/8,10)-Tetrol | 660 | 10.4 ± 1.9 | 11.3 ± 3.9 | 3.5 ± 2.4 | 1.7 ± 1.5 |
| (7,9/10/8)-Tetrol | 930 | 0.8 ± 1.3 | 19.9 ± 5.9 | ND[c] | 5.0 ± 3.3 |
| (7,9,10/8)-Tetrol (7/8,9)-Triol and trans-9,10-Diol | 1140 | 28.4 ± 11.3 | 21.2 ± 8.2 | 12.3 ± 10.1 | 19.3 ± 10.7 |
| trans-4,5-Diol | 1740 | 3.9 ± 1.1 | 3.9 ± 3.4 | 7.8 ± 4.2 | 5.7 ± 3.8 |
| trans-7,8-Diol | 1920 | 4.9 ± 0.4 | 4.4 ± 1.4 | 12.9 ± 5.0 | 8.2 ± 4.9 |
| 3-Hydroxy | 2310 | 2.4 ± 0.7 | 2.4 ± 0.4 | 1.8 ± 1.0 | 3.8 ± 2.2 |
| 9-Hydroxy | 2550 | 4.1 ± 0.2 | 4.1 ± 3.0 | ND[c] | 6.8 ± 3.3 |
| Quinones | | 26.4 ± 11.9[3] | 11.5 ± 7.9 | 13.1 ± 6.2 | 30.8 ± 8.5 |
| Unidentified[b] | 2640–2970 | 3.1 ± 1.8 | 16.5 ± 12.2 | 32.2 ± 16.8 | 17.9 ± 4.6 |

[a] Mean values ± S.D. for bronchus (6 cases, Reference 8); colon (5 cases, Reference 6); esophagus (8 cases, Reference 24); and trachea (4 cases, Reference 8).
[b] Unidentified metabolites were eluted prior to (7,10/8,9)-tetrol, presumably sulfate esters (Reference 16), and in the region of phenols and quinones.
[c] N.D. = not detected.

## Effect of Exogenous Chemicals on BaP Metabolism

Several exogenous compounds have been found to alter the metabolism of BaP and to modify the carcinogenic response. Butylated hydroxyanisole (BHT) and disulfiram (DS), two antioxidants, inhibit the neoplastic effect of many chemical carcinogens, including BaP. Coincubation of bronchial and colonic explants with DS and BHT significantly decreased the level of binding to DNA (6,22). The inhibiting effect of BHT is probably mediated by an enhanced activity of glutathione-transferase (12) which would compete for the reactive species of BaP (epoxides and diol epoxide), thereby leaving less to react with the cellular DNA. An opposite effect was seen when bronchial explants were incubated with BaP and aminophylline, a drug previously used clinically to treat chronic bronchitis. An enhanced level of BaP binding to bronchial DNA was found to coincide with a significant decrease in the amount of glutathione conjugates (20) and a simultaneous increase in the level of tetrols, whereas the total metabolism was unaffected. Similarly, the active ingredient of a laxative, Bisacodyl, enhanced the binding level of BaP to colonic DNA. This effect is probably mediated through partial inhibition of the detoxifying enzymes. This last result indicates the importance of testing a drug not only for its own carcinogenicity but also for examining its effect on other known carcinogens.

Bile acids, which may play a role in colon carcinogenesis, increased the level of binding of BaP to colonic DNA. When colonic explants were coincubated with either lithocholic acid or taurodeoxycholic acid, an enhanced level of binding and a decrease in the total amount of water-soluble metabolites was observed (6). However, there was some indication that the bile acids also might have an effect on the uptake of BaP into mucosal cells by decreasing the viscosity of the protective layer of mucus.

## CONCLUSION

Cultured human bronchus, colon, esophagus, and trachea metabolize BaP into water-soluble and organo-soluble metabolites, and to intermediates that bind to cellular DNA. The metabolism was qualitatively similar in all four tissues, and the major BaP-DNA adduct was formed by trans-addition of 7R-BPDE I to the 2-amino group of guanine. This pathway is similar to the pathway in tissues from animals susceptible to the carcinogenic action of BaP. The binding levels of BaP to DNA were lower in colon than in the esophagus and the tracheobronchial tissues, and a wide interindividual variation was found in binding levels.

By using an explant culture system, we have shown that human target tissues can activate BaP and that the activation pathway is similar to that found in tissues and cell cultures from experimental animal species in which BaP is a known carcinogen. These results provide further evidence of the validity of the explant culture system for comparing human and animal model data in chemical carcinogenesis.

## ACKNOWLEDGMENTS

This paper reviews and summarizes data collected with the aid of collaborators. We thank Dr. B. F. Trump, Department of Pathology, University of Maryland; Dr. L. Smith, National Naval Medical Center; and Dr. P. Schafer, VA Hospital, Washington, D. C., for supplying the human tissues. The technical assistance of Ms. R. Schwartz and Mr. H. Tate and the secretarial assistance of Ms. S. Dorfman are greatly appreciated.

## REFERENCES

1. Atlas, S. A., Vesell, E. S., and Nebert, D. W. (1976): Genetic control of interindividual variations in the inducibility of aryl hydrocarbons by hydroxylase in cultured human lymphocytes. *Cancer Res.* 36:4619-4630.
2. Autrup, H. (1980): Explant culture of human colon. In: *Methods and Perspectives in Cell Biology,* C. C. Harris, B. F. Trump, and G. D. Stoner, Eds., Academic Press, New York (in press).
3. Autrup, H. (1979): Separation of water-soluble metabolites from cultured human colon. *Biochem. Pharmacol.* 28:1727-1730.
4. Autrup, H. A., Barrett, L. A., Jackson, F. E., Jesudason, M. L., Stoner, G. D., Phelps, P., Trump, B. F., and Harris, C. C. (1978): Explant culture of human colon. *Gastroenterology* 74:1248-1257.
5. Autrup, H., Harris, C. C., Stoner, G., Selkirk, J. K., Schafer, P. W., and Trump, B. F. (1978): Metabolism of $^3$H-benzo(a)pyrene by cultured human bronchus and pulmonary alveolar macrophages. *Lab. Invest.* 38:217-224.
6. Autrup, H., Harris, C. C., Trump, B. F., and Jeffrey, A. M. (1978): Metabolism of benzo(a)pyrene and identification of the major benzo(a)pyrene-DNA adducts in cultured human colon. *Cancer Res.* 38:3689-3696.
7. Autrup, H., Harris, C. C., Trump, B. F., Schafer, P. W., Stoner, G. D., and Hsu, I-C. (1979): Uptake of benzo(a)pyrene-ferric oxide particulates by human pulmonary alveolar macrophages and the

release of benzo(a)pyrene and its metabolites. *Proc. Soc. Biol. Med.* 161:280-284.
8. Autrup, H., Wefald, F. C., Jeffrey, A. M., Tate, H., Schwartz, R. D., Trump, B. F., and Harris, C. C. (1980): Metabolism of benzo(a)pyrene by cultured tracheobronchial tissues from mice, rats, hamsters, bovine, and humans. *Int. J. Cancer* (in press).
9. Baird, W. M., and Diamond, L. (1978): Metabolism and DNA binding of polycyclic aromatic hydrocarbons by human diploid fibroblast. *Int. J. Cancer,* 22:189-195.
10. Barrett, L. A., McDowell, E. M., Frank, A. L., Harris, C. C., and Trump, B. F. (1976): Long-term organ culture of human bronchial epithelium. *Cancer Res.* 36:1003-1010.
11. Bast, R. C., Whitlock, J. P., Miller, H., Rapp, H. J., and Gelboin, H. V. (1974): Aryl hydrocarbon [benzo(a)pyrene hydroxylase] in human peripheral blood monocytes. *Nature* 250:664-665.
12. Benson, A. M., Batzinger, R. P., Ou, S.-Y., Bueding, E., Cha, Y.-N., Talalay, P. (1978): Elevation of hepatic glutathione S-transferase activities and protection against mutagenic metabolites of benzo(a)pyrene by dietary antioxidants. *Cancer Res.* 38:4486-4495.
13. Bigger, C.A.H., Tomaszewski, J. E., and Dipple, A. (1978): Differences between products of binding of 7,12-dimethylbenz(a)anthracene to DNA in mouse skin and in a rat liver microsomal system. *Biochem. Biophys. Res. Commun.* 80:229-235.
14. Boobis, A. R., Atlas, S. A., and Nebert, D. W. (1978): Carcinogenic benzo(a)pyrene metabolites bound to DNA: metabolic formation by human cultured lymphocytes and by human liver microsomes. *Pharmacology* 17:241-248.
15. Cerutti, P., Shinohara, K., Ide, M.-L, and Remsen, J. (1978): Formation and repair of benzo(a)pyrene-induced DNA damage in mammalian cells. In: *Polycyclic Hydrocarbons and Cancer,* Vol. 2. H. V. Gelboin and P.O.P. T'so, Eds., pp. 203-212, Academic Press, New York.
16. Cohen, G. M., Haws, S. M., Moore, B. P., and Bridges, J. W. (1976): Benzo(a)pyrene-3-yl hydrogen sulphate, a major ethyl acetate-extractable metabolite in human, hamster and rat lung cultures. *Biochem. Pharmacol.* 25:2561-2570.
17. Cohen, G. M., Mehta, R., and Meredith-Brown, M. (1979): Large interindividual variations in metabolism of benzo(a)pyrene by peripheral lung tissues from lung cancer patients. *Int. J. Cancer* 24:129-133.
18. Deutsch, J., Leutz, J. C., Yang, S. K., Gelboin, H. V., Chiang, Y. L., Vatsis, K. P., and Coon, M. J. (1978): Regio- and stereoselectivity of various forms of purified cytochrome P-450 in the metabolism of

benzo(a)pyrene and (-)trans-7,8-dihydroxy-7,8-dihydrobenzo(a)pyrene as shown by product formation and binding to DNA. *Proc. Natl. Acad. Sci. U.S.A.* 75:3123-3127.
19. Grover, P. L., Pal, K., Hewer, A., and Sims, P. (1976): The involvement of a diol-epoxide in the metabolic activation of benzo(a)pyrene in human bronchial mucosa and in mouse skin. *Int. J. Cancer* 18:1-6.
20. Harris, C. C., Ashworth, H., Trump, B. F., and Autrup, H. (1980): Aminophylline enhances the metabolic activation of benzo(a)pyrene in cultured human bronchus (submitted for publication).
21. Harris, C. C., Autrup, H., Connor, R., Barrett, L. A., McDowell, E. M., and Trump, B. F. (1976): Interindividual variation in binding of benzo(a)pyrene to DNA in cultured human bronchi. *Science* 194:1067-1069.
22. Harris, C. C., Autrup, H., and Stoner, G. (1978): Metabolism of benzo(a)pyrene in cultured human tissues and cells. In: *Polycyclic Hydrocarbons and Cancer,* Vol. 2, H. V. Gelboin and P.O.P. T'so, Eds., pp. 331-342, Academic Press, New York.
23. Harris, C. C., Autrup, H., Stoner, G., and Trump, B. F. (1978): Model systems using human lung for carcinogenesis studies. In: *Pathogenesis and Therapy of Lung Cancer,* C. C. Harris, Ed., pp. 559-597, Marcel Dekker, Inc., New York.
24. Harris, C. C., Autrup, H., Stoner, G. D., Trump, B. F., Hillman, E., Schafer, P. W., Weinstein, I. B., and Jeffrey, A. M. (1979): Metabolism of benzo(a)pyrene, nitrosomethylamine and N-nitrosopyrrolidine and identification of the major carcinogen-DNA adducts formed in cultured human esophagus. *Cancer Res.* 39:4401-4406.
25. Harris, C. C., Autrup, H., Stoner, G., Yang, S. K., Leutz, J. C., Gelboin, H. V., Selkirk, J. K., Connor, R. J., Barrett, L. A., Jones, R. T., McDowell, E. M., and Trump, B. F. (1977): Metabolism of benzo(a)pyrene and 7,12-dimethylbenz(a)anthracene in cultured human tissues: bronchus and pancreatic duct. *Cancer Res.* 37:3349-3355.
26. Harris, C. C., Frank, A., van Haaften, C., Kaufman, D. G., Connor, R., Jackson, F. E., Barrett, L. A., McDowell, E. M., and Trump, B. F. (1976): Binding of [$^3$H]benzo(a)pyrene to DNA in cultured human bronchus. *Cancer Res.* 36:1011-1018.
27. Harris, C. C., Hsu, I.-C., Trump, B. F., and Selkirk, J. K. (1978): Human pulmonary alveolar macrophages metabolise benzo(a)pyrene to proximate and ultimate mutagens. *Nature* 272:633-634.
28. Hsu, I.-C., Stone, G. D., Autrup, H., Trump, B. F., Selkirk, J. K., and Harris, C. C. (1978): Human bronchus-mediated mutagenesis of mammalian cells by carcinogenic polynuclear aromatic hydrocarbons. *Proc. Natl. Acad. Sci. U.S.A.* 75:2003-2007.

29. Huberman, E., and Sachs, L. (1977): DNA binding and its relationship to carcinogenesis by different polycyclic hydrocarbons. *Int. J. Cancer* 19:122-127.
30. Jeffrey, A. M., Weinstein, I. B., Jennette, K. W., Grzeskowiak, K., Nakanishi, K., Harvey, R. G., Autrup, H., and Harris, C. (1977): Structures of benzo(a)pyrene-nucleic acid adducts formed in human and bovine bronchial explants. *Nature* 269:348-350.
31. Jennette, K., Jeffrey, A. M., Blobstein, S. H., Beland, F. A., Harvey, R. G., and Weinstein, I. B. (1976): Nucleoside adducts from the in vitro reaction of benzo(a)pyrene-7,8-dihydrodial-9,10-oxide or benzo-(a)pyrene 4,5-oxide with nucleic acids. *Biochemistry* 16:932-938.
32. Kapitulnik, J., Poppers, P. J., and Conney, A. H., (1977): Comparative metabolism of benzo(a)pyrene and drugs in human liver. *Clin. Pharmacol* 21:166-176.
33. Kellermann, G., Luyten-Kellermann, M., and Shaw, C. R. (1973): Genetic variation of aryl hydrocarbon hydroxylase in human lymphocytes. *Am. J. Hum. Genet.* 25:327-331.
34. Kellermann, G., Shaw, C. R., and Luyten-Kellermann, M. (1973): Aryl hydrocarbon hydroxylase inducibility and bronchogenic carcinoma. *N. Engl. J. Med.* 289:934-935.
35. McDowell, E. M., and Trump, B. F. (1976): Histological fixatives suitable for diagnostic light and electron microscopy. *Arch. Pathol. Lab. Med.* 100:405-414.
36. Meehan, T., and Straub, K. (1979): Double-stranded DNA stereoselectivity inds benzo(a)pyrene diol epoxides. *Nature* 277:410-412.
37. Mehta, R., and Cohen, G. M. (1979): Major differences in the extent of conjugation with glucuronic acid and sulphate in human peripheral lung. *Biochem. Pharmacol.* 28:2479-2484.
38. Nemoto, N., and Gelboin, H. V. (1976): Enzymatic conjugation of benzo(a)pyrene oxides, phenols and dihydrodiols with UDP-glucuronic acid. *Biochem. Pharmacol.* 25:1221-1226.
39. Okano, P., Miller, H. N., Robinson, R. C., and Gelboin, H. V. (1979): Comparison of benzo(a)pyrene and (-)-trans-7,8-dihydroxy-7,8-dihydrobenzo(a)pyrene metabolism in human blood monocytes and lymphocytes. *Cancer Res.* 39:3184-3193.
40. Okuda, T., Vessel, E. S., Plotkin, E., Tarone, R., Bast, R. C., and Gelboin, H. V. (1977): Interindividual and intraindividual variations in aryl hydrocarbons hydroxylase in monocytes from monozygatic and dizygotic twins. *Cancer Res.* 37:3904-3911.
41. Paigen, B., Ward, E., Steenland, K., Houten, L., Gurtoo, H. L., and Minowada, J. (1978): Aryl hydrocarbon hydroxylase in cultured lymphocytes of twins. *Am. J. Hum. Genet.* 30:561-571.

42. Pelkonen, O. (1976): Metabolism of benzo(a)pyrene in adult and fetal tissues. In: *Polynuclear Aromatic Hydrocarbons: Chemistry, Metabolism and Carcinogenesis,* by R. I. Freudental and P. W. Jones, Eds., pp. 9-21, Raven Press, New York.
43. Pelkonen, O., Boobis, A. R., Yagi, H., Jerina, D. M., and Nebert, D. W. (1978): Tentative identification of benzo(a)pyrene metabolite-nucleoside complexes produced in vitro by mouse liver microsomes. *Mol. Pharmacol.* 14:306-322.
44. Prough, R. A., Patrizi, V. W., Okita, R. T., Masters, B.S.S., and Jakobsson, S. W. (1979): Characteristics of benzo(a)pyrene metabolism by kidney, liver and lung microsomal fractions from rodents and humans. *Cancer Res.* 39:1199-1206.
45. Prough, R. A., Sipal, Z., and Jakobsson, S. W. (1977): Metabolism of benzo(a)pyrene by human lung microsomal fractions. *Life Sciences* 21:1629-1636.
46. Selkirk, J. K. (1977): Divergence of metabolic activation systems for short-term mutagenesis assays. *Nature* 270:604-607.
47. Shinohara, K., and Cerutti, P. A. (1977): Formation of benzo(a)pyrene DNA adducts in peripheral human lung tissue. *Cancer Letters* 3:303-310.
48. Stoner, G. D., Harris, C. C., Autrup, H., Trump, B. F., Kingsbury, E. W., and Myers, G. A. (1978): Explant culture of human peripheral lung tissues. I. Metabolism of benzo(a)pyrene. *Lab. Invest.* 38:685-692.
49. Trump, B., McDowell, E., Barrett, L., Frank, A., and Harris, C. (1974): Studies of ultrastructure, cytochemistry and organ culture of human bronchial epithelium. In: *Experimental Lung Cancer,* E. Karbe and J. J. Park, Eds., pp. 548-558, Springer-Verlag, New York.
50. Weinstein, I. B., Jeffrey, A. M., Jennette, K. W., Blobstein, S. H., Harvey, R. G., Harris, C., Autrup, H., Kasai, H., and Nakanishi, K. (1976): Benzo(a)pyrene diol epoxides as intermediates in nucleic acid binding in vitro and in vivo. *Science* 193:592-595.
51. Wigley, C. B., Newbold, R. F., Amos, J., and Brookes, P. (1979): Cell-mediated mutagenesis in cultured Chinese hamster cells by polycyclic hydrocarbons: Mutagenicity and DNA reaction related to carcinogenicity in a series of compounds. *Int. J. Cancer* 23:691-696.
52. Yang, S. K., Gelboin, H. V., Trump, B. F., Autrup, H., and Harris, C. C. (1977): Metabolic activation and DNA binding of benzo(a)pyrene in cultured human bronchus. *Cancer Res.* 37:1207-1212.
53. Yang, S. K., Roller, P. P., and Gelboin, H. V. (1977): Enzymatic mechanism of benzo(a)pyrene conversion to phenols and diols and an improved high pressure liquid chromatographic separation of benzo(a)pyrene derivatives. *Biochemistry* 16:3680-3686.

# CHANGES IN PAH-PROFILES IN DIFFERENT AREAS OF A CITY DURING THE YEAR

G. Grimmer, K.-W. Naujack, and D. Schneider

Biochemical Institute for Environmental Carcinogens
Sieker Landstraβe 19,
D-2070 Ahrensburg, F.R.G.

Among the polycyclic aromatic hydrocarbons (PAH), not only benzo-(a)pyrene (BaP) produces carcinomas when applied to the skin of mice. In the case of automobile exhaust gas condensate, the fraction of PAH accounts for more than 90 percent of the carcinogenic effect of the total condensate in this bioassay, but only 10 percent of the total carcinogenicity can be explained by BaP content in this exhaust condensate (1). In tobacco smoke condensate, the content of BaP accounts for only about 1 percent of the total carcinogenicity in the above-mentioned test (2,3).

There is no doubt that this is true for air pollution also. Therefore, analysis of the most widely spread PAH gives more information than the determination of BaP. Presently, the most efficient separation method for complex PAH mixtures is glass capillary gas chromatography (GCGC), a method which is used for the PAH-profile analysis of air pollutants (4-10).

## PAH-Profile

The term "profile" represents the relative composition of the mixture of PAH present in a specimen and recorded by a mass-dependent detector (e.g., a glass capillary gas chromatogram recorded by a flame-ionisation detector represents a "PAH-profile").

## Purpose of the Study

The purpose of this investigation was to compare local concentrations of several PAH in some selected areas of an industrial city. To record, furthermore, the temporary changes in the concentration at these sampling-stations, a 1-hour-sample was collected each month. This was performed by the Landesanstalt für Immissionsschutz in Essen.

PAH-profiles in five areas, surrounded by four stations each, were investigated weekly during this year and the preceding one.

## TECHNIQUE OF MEASURMENT

### The Sampling System and the Analytical Method

The sampling system consists of a low-volume air sampler, equipped with a flowmeter, which collects 10 m$^3$/hr (SARTORIUS, Göttingen, type Portikon). The air sampler is connected to a filter container. The glass fiber filter, with a collecting area of 490 cm$^2$, is not impregnated. We have found that by using control filters, this is not necessary if low flow rates are used. The commercially available filters are installed in a plastic casing material. It fits in a container which is connected to the air sampler (Figure 1).

The PAH collected on the filter are extracted with toluene (120 ml). Before heating, 300 ng of benzo(a)chrysene are added to the solution. For the enrichment of PAH, a chromatography on Sephadex LH-20 (10 g) with isopropanol is necessary. The fraction from 50 to 180 ml contains only PAH with more than 3 rings. This fraction is sufficiently pure for gas chromatography (11).

To separate the mixture of PAH, glass capillary gas chromatography is used. The capillaries are coated with Silicone OV 17. In contrast to polydimethylsiloxanes, such as SE 30 or OV 101, a separation of benz(a)-antracene, cyclopenta(cd)pyrene, and chrysene can be achieved with this coating. This is true for the triplet benzo(ghi)fluoranthene, benzo(b)-naphtho(2,1-d)thiophene, and benzo(c)phenanthrene also as shown in Figure 2. The PAH mixture made soluble in 5 µl toluene, is injected at 100°C without splitting, and after a few minutes the column is heated to 260°C.

### Arrangement of the Sampling Stations in the City

In Essen, an industrial city of 700,000 inhabitants in the F.R.G., four areas were selected inside the city and one outside. These areas are polluted by typical emittants:
  (I) An area with hand-stoked residential coal heatings
 (II) An area with oil heating only
(III) Stations in a tunnel with car traffic
(IV) An area with coke ovens.

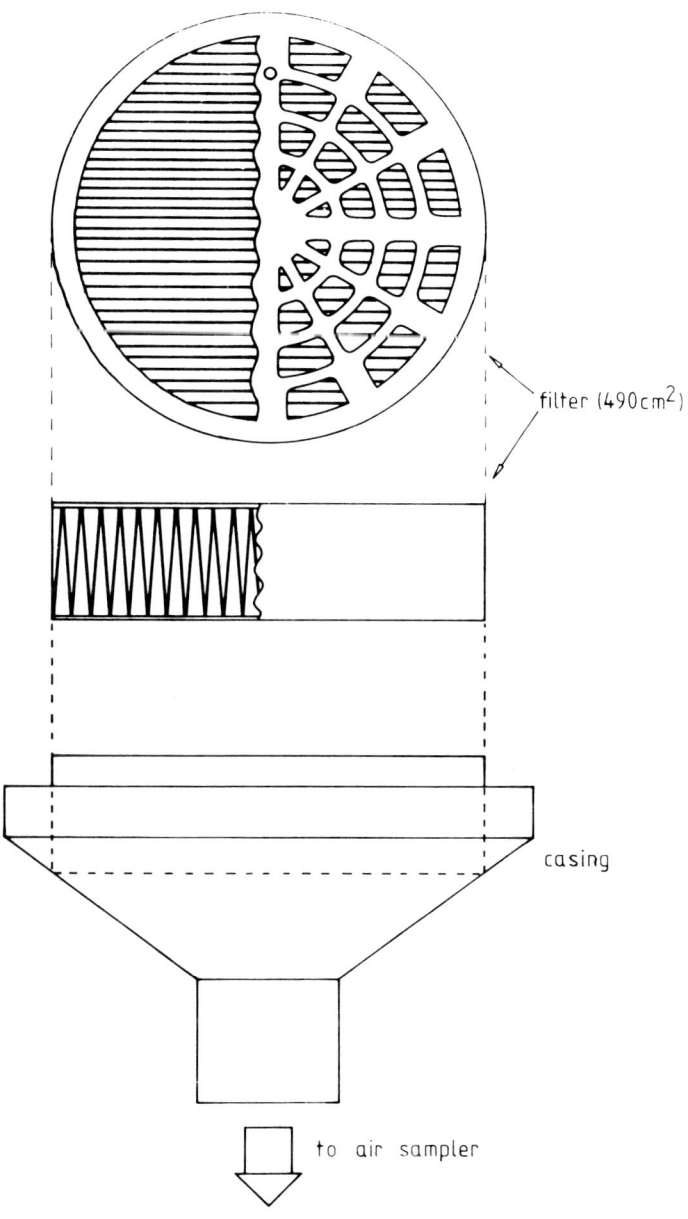

**FIGURE 1.** Collecting arrangement for air suspended matter (air pollution).

## 110 PAH—PROFILES OF DIFFERENT AREAS IN A CITY

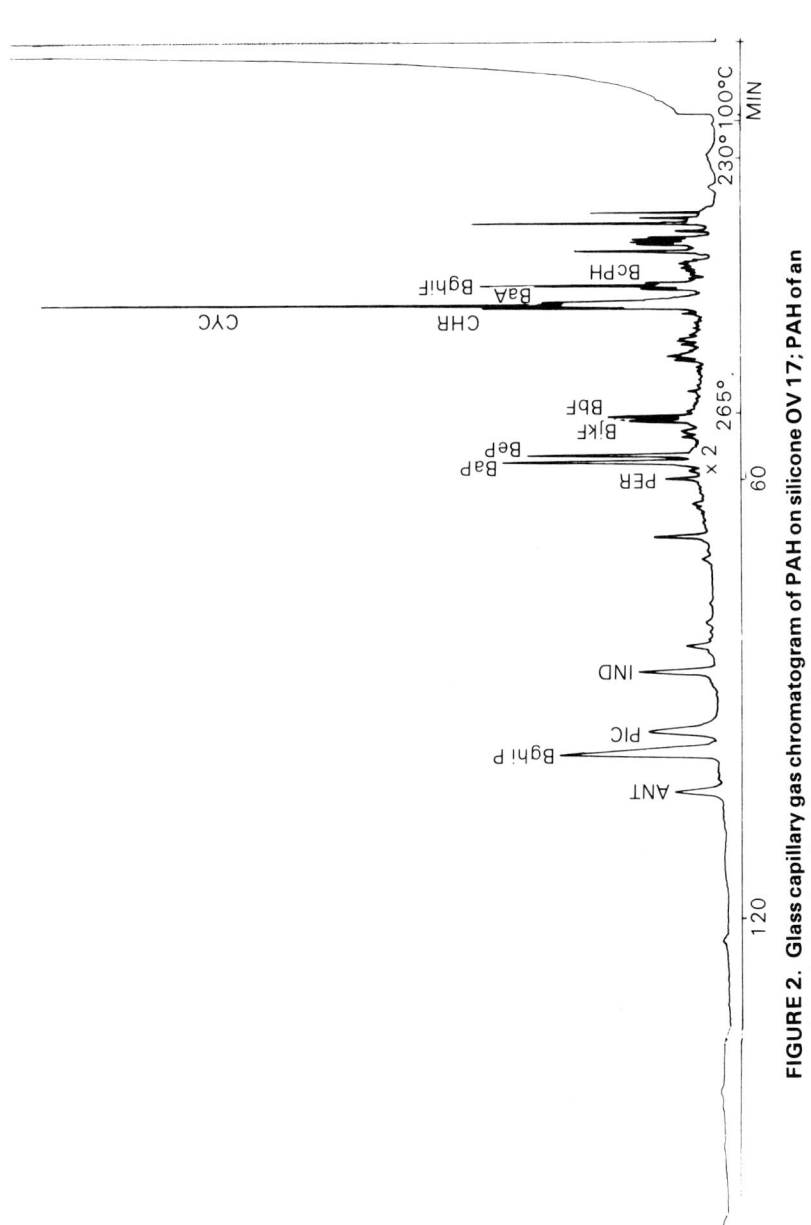

FIGURE 2. Glass capillary gas chromatogram of PAH on silicone OV 17; PAH of an area in a city burdened with automobile exhaust (tunnel).

# PAH—PROFILES OF DIFFERENT AREAS IN A CITY

The last one (V) was a rural area outside of the city. Each area was surrounded by four sampling stations. The distances between the different areas was 3.5 to 8 km (details in Figure 3). Note that the fifth (rural) area is located outside of the city limits shown in this figure.

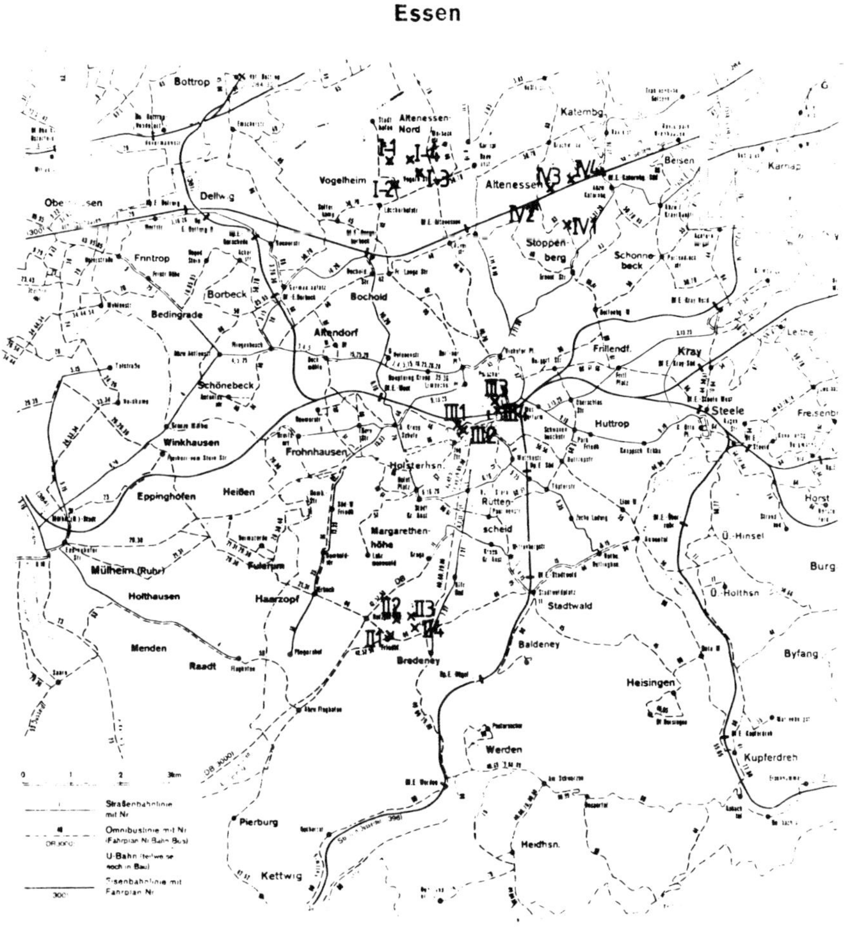

**Essen**

**FIGURE 3.** Map of the investigated industrial city (Essen). Localization of the four areas.

## RESULTS

### Comparison of the PAH-Profiles from Different Sampling Stations

Figure 4 demonstrates a glass capillary gas chromatogram (GCGC) of the PAH-profile from the coal-heating area. From this GCGC, 15 PAH

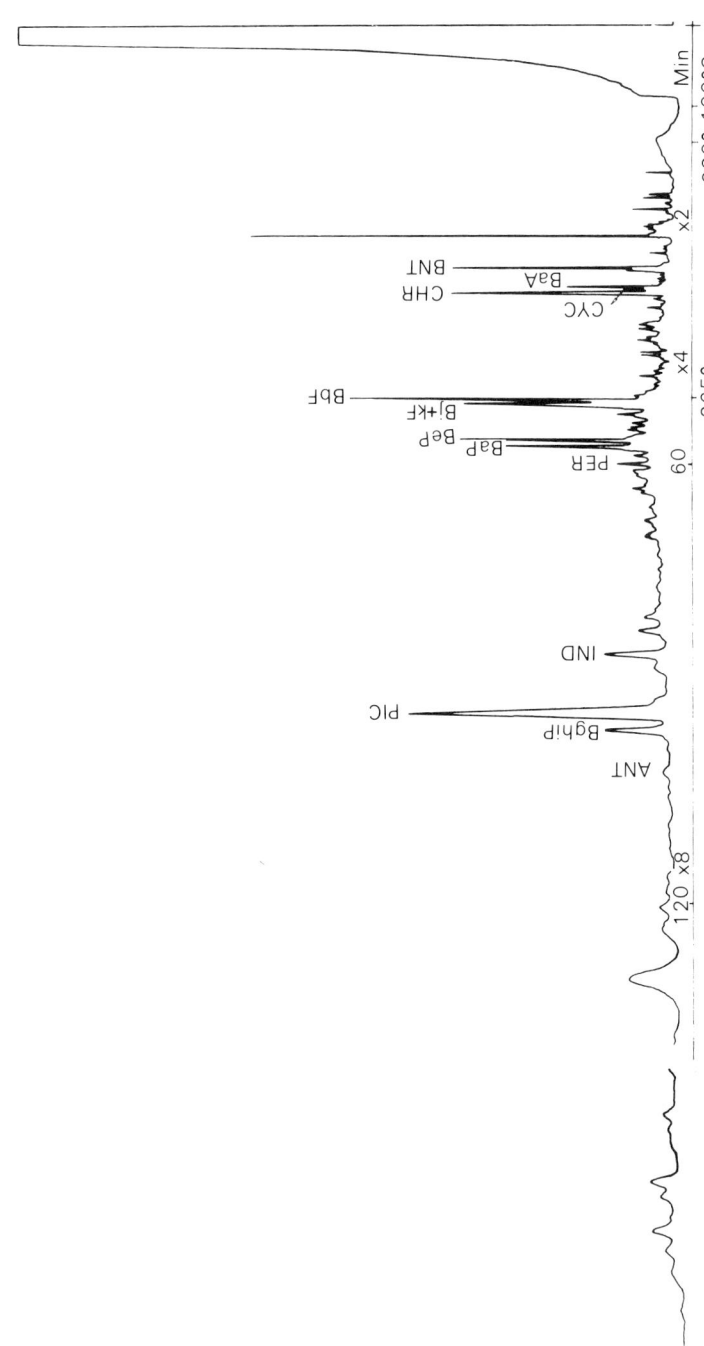

FIGURE 4. Glass capillary gas chromatogram of PAH on silicone OV 17; PAH of an area in the city burdened with residental coal heatings.

were selected: benzo(b)naphtho(2,1-d)thiophene, benzo(c)phenanthrene, benzo(ghi)fluoranthene, benz(a)anthracene, cyclopenta(cd)pyrene, chrysene, benzo(b)fluoranthene, benzo(j + k)fluoranthene, benzo(e)pyrene, benzo(a)pyrene, perylene, indeno(1,2,3-cd)pyrene, benzo(ghi)perylene, anthanthrene, and coronene.

To compare the PAH-profiles, it was helpful to standardize them. For the representation of PAH concentrations for a selected PAH, the PAH was compared with the benzo(e)pyrene-concentration. In this case the BeP-concentration is defined as 1.0, and all other PAH are given in ratio to BeP (Figure 5). This figure shows 15 PAH in ratio to BeP. The doublet of each PAH stands for:

(a) The Sampling Station I-1 of the coal-heating area, 10 m$^3$ collected from 8:48 to 10:14 (February 21, 1979)
(b) The same station and the same volume but collected from 10:24 to 11:50.

The concentrations of the PAH in the first and the second hour are within the margin of errors also [e.g., BeP/m$^3$ for (a) is 95.6 ng/m$^3$, and for (b) it is 96.1 ng/m$^3$ (Figure 6)]. Therefore, the repeatability of the total sampling procedure and analysis is satisfactory. Figure 7 shows the PAH-profiles of the same station in January, February, March, and April 1979, which are very similar. In contrast to this, the absolute concentrations of the PAH are not constant. This is demonstrated in Figure 8 for benzo(a)pyrene (BaP). These are the concentrations of BaP at Sampling Stations I-1, I-2, I-3, and I-4 (coal-heating area). During the period of heating from October to March, the concentrations at I-1 varied from 6 ng to 60 ng/m$^3$. This was observed for other sampling stations in this area of coal heating also. The distances between the four stations surrounding this area were between 100 and 400 meters.

Results from another area, the tunnel with automobile traffic, are demonstrated in Figure 9. The second profile is the PAH emission from a passenger car on a chassis dynamometer driven during the Europa-drive-cycle (ECE-reglement 15), which simulates city traffic (12,13). In this case, cyclopenta(cd)pyrene is the PAH with the highest concentration. This is in contrast to the emission of coal heating, which produces only a very small amount of cyclopenta(cd)pyrene. Furthermore, Figure 9 demonstrates that the profile inside the tunnel and that of the bench test are very similar. A difference can be observed in the content of benzo(b)naphtho(2,1-d)thiophene (BNT); this PAH is absent in automobile exhaust, but present in the air of the tunnel. BNT could originate from Diesel engines, which blow out thiophenes. Also, at this sampling station the PAH-profiles are very similar in January, February, March, and April (Figure 10).

The levels of PAH concentrations depend on the density of the traffic within the tunnel. This is demonstrated in Figure 11, representing the BaP

# 114 PAH—PROFILES OF DIFFERENT AREAS IN A CITY

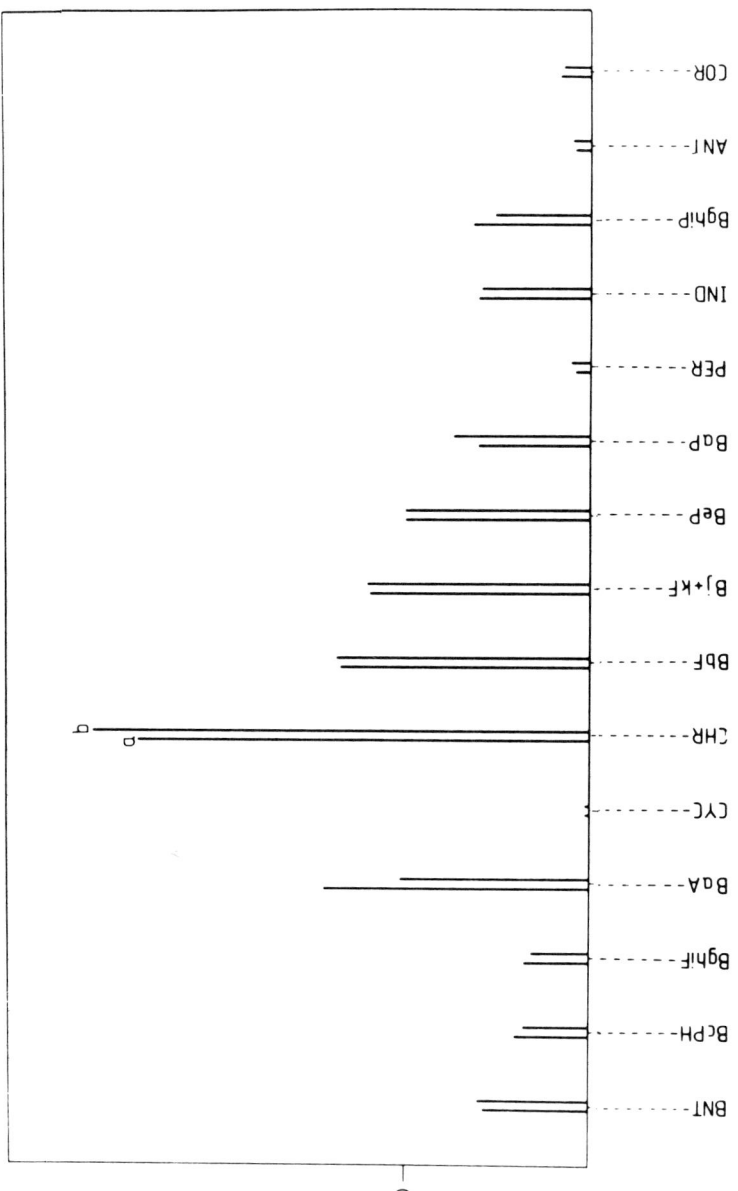

FIGURE 5. Schematic of PAH profiles (relative to BeP = 1.0). (a) Collecting station I-1 of the coal heating area, February 21, 1979 -collecting times: (a) 8.48 - 10.14 and (b) dto - collecting time: 10.24 - 11.50.

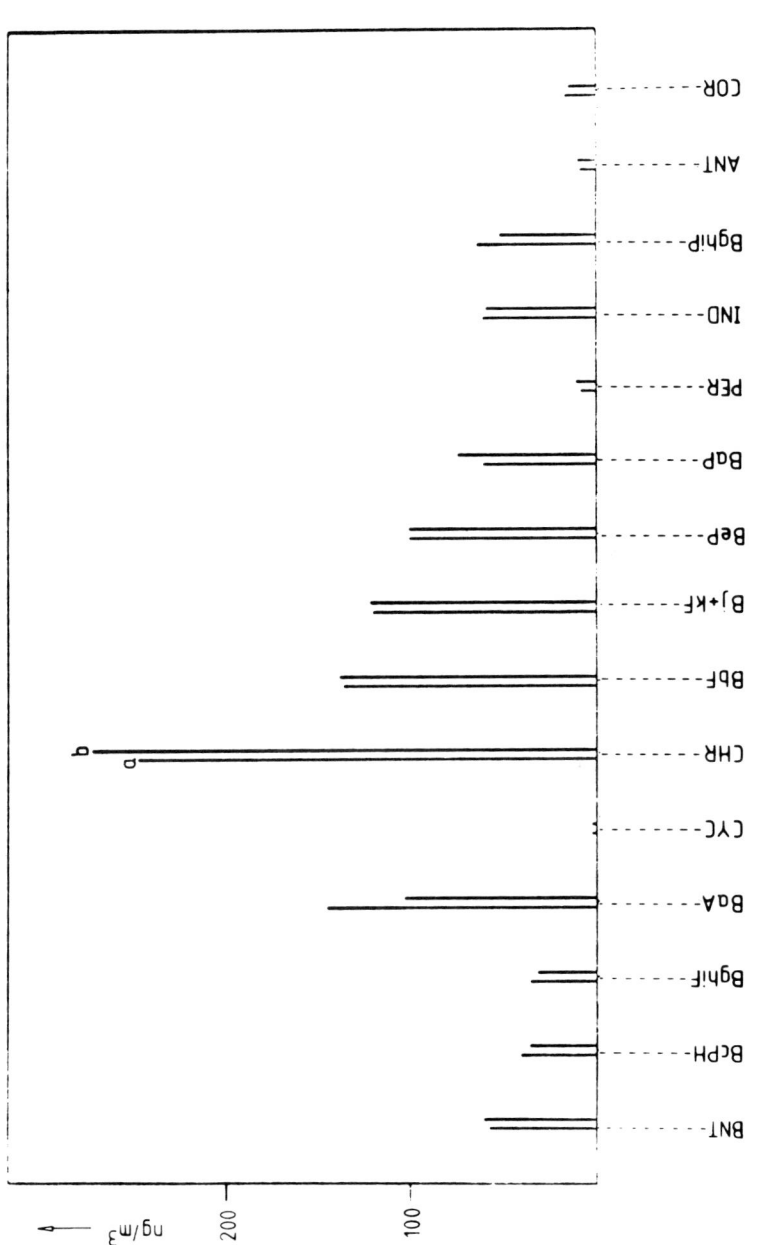

FIGURE 6. Schematic of PAH profiles (absolute PAH concentrations) in the collecting station of the coal heating area, February 21, 1979.: (a) from 8.48 - 10.14 and (b) from 10.24 - 11.50.

# 116  PAH—PROFILES OF DIFFERENT AREAS IN A CITY

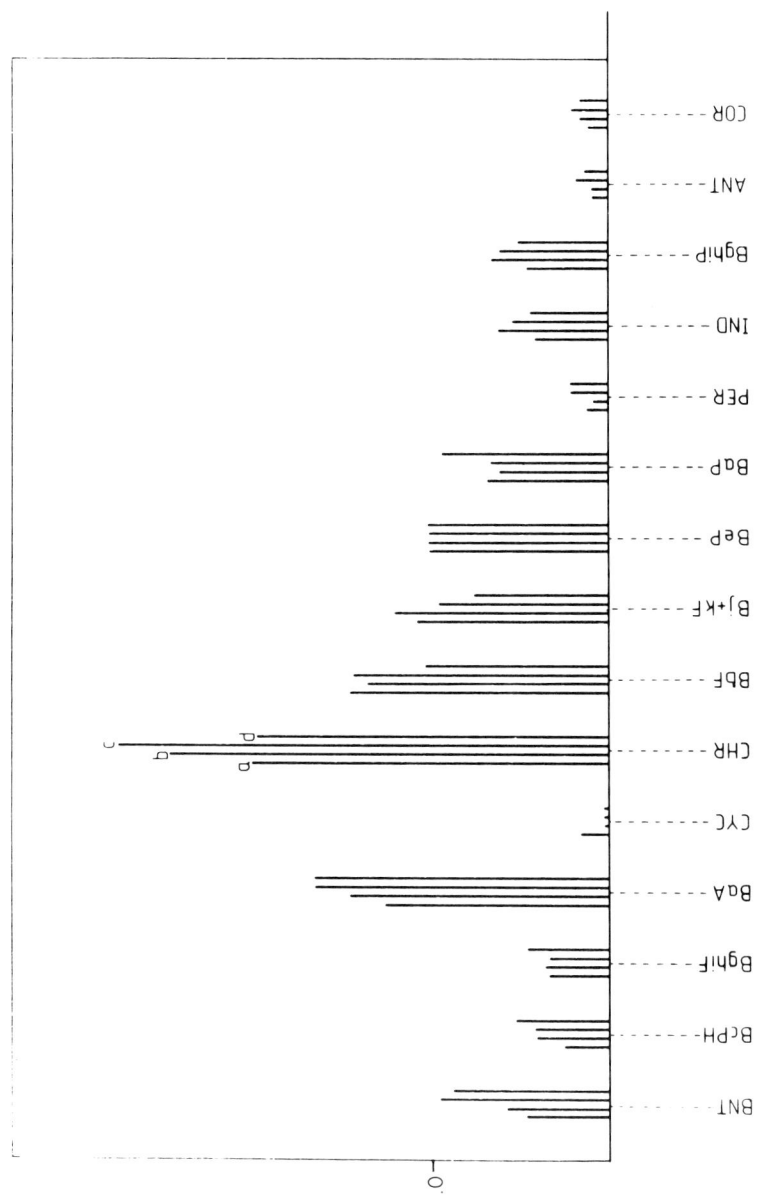

FIGURE 7. Schematic of PAH profiles (relative to BeP = 1.0) in the collecting station I-1 of the coal heating area in (a) January, (b) Februray, (c) March, and (d) April.

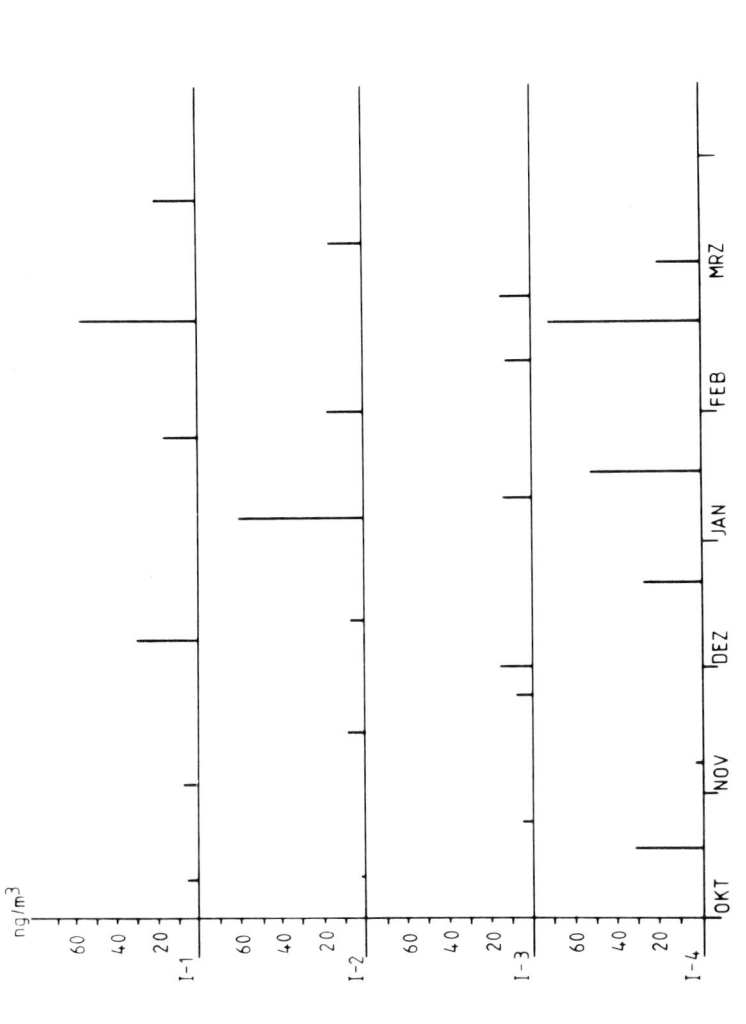

FIGURE 8. Schematic of BaP concentrations in collecting stations I-1, I-2, I-3, and I-4 surrounding the coal heating area; collected from October 1978 to March 1979.

# 118 PAH—PROFILES OF DIFFERENT AREAS IN A CITY

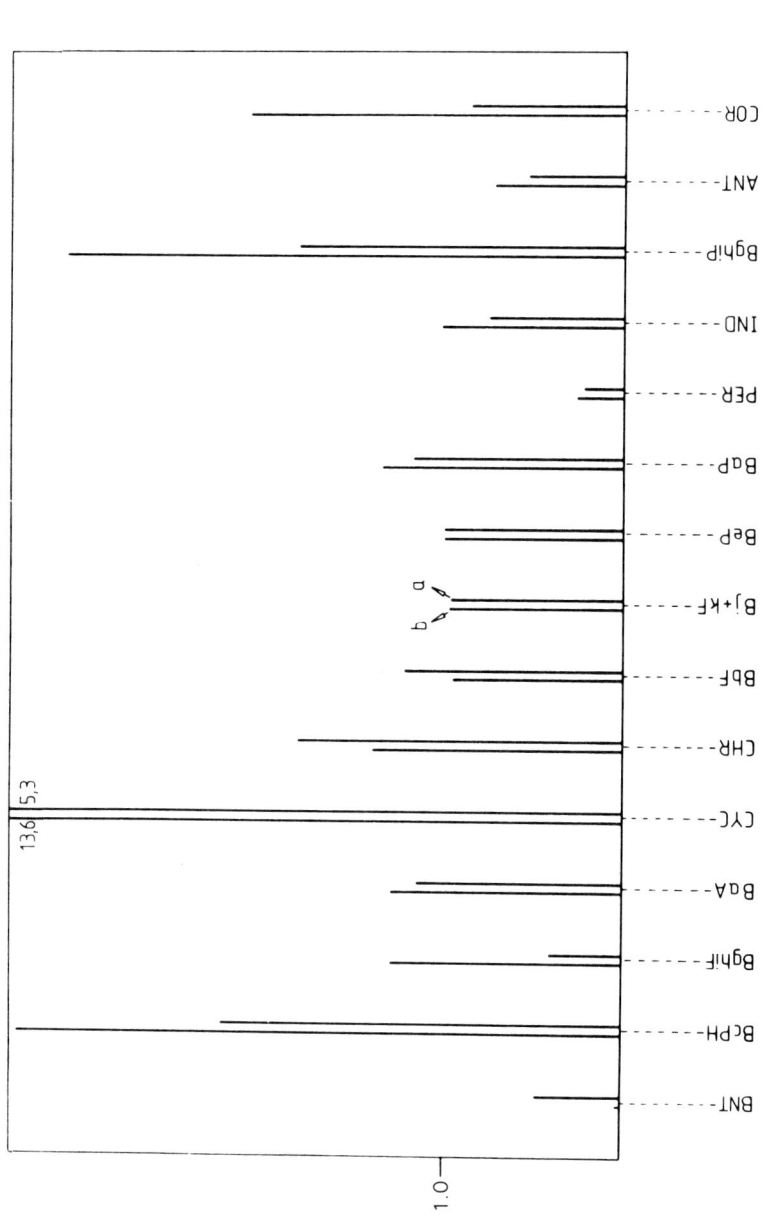

FIGURE 9. Schematic of PAH profiles (relative to BeP = 1.0) comparing (a) the collecting station III-4 data in the automobile traffic area (tunnel) and (b) PAH emission of a passenger car on a chassis dynamometer, simulating city traffic (Europa-Test).

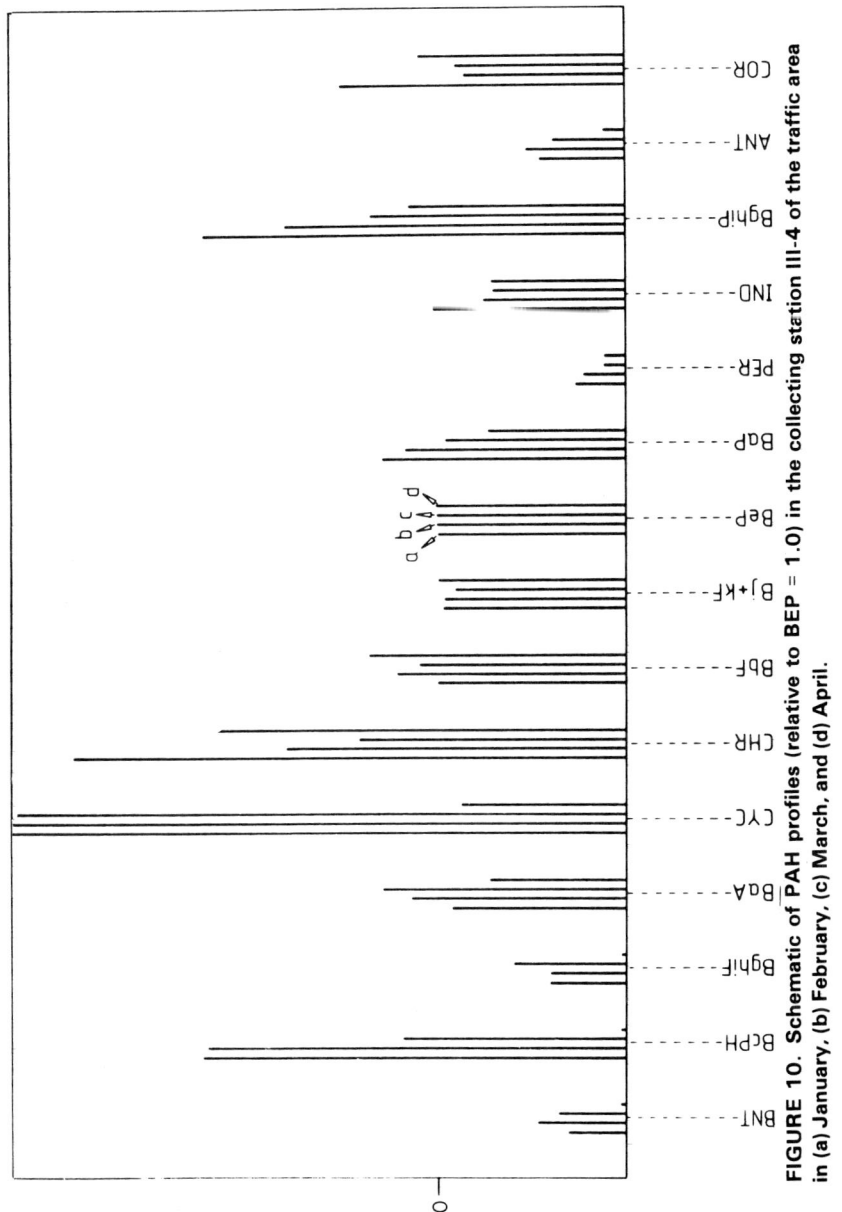

FIGURE 10. Schematic of PAH profiles (relative to BEP = 1.0) in the collecting station III-4 of the traffic area in (a) January, (b) February, (c) March, and (d) April.

## 120 PAH—PROFILES OF DIFFERENT AREAS IN A CITY

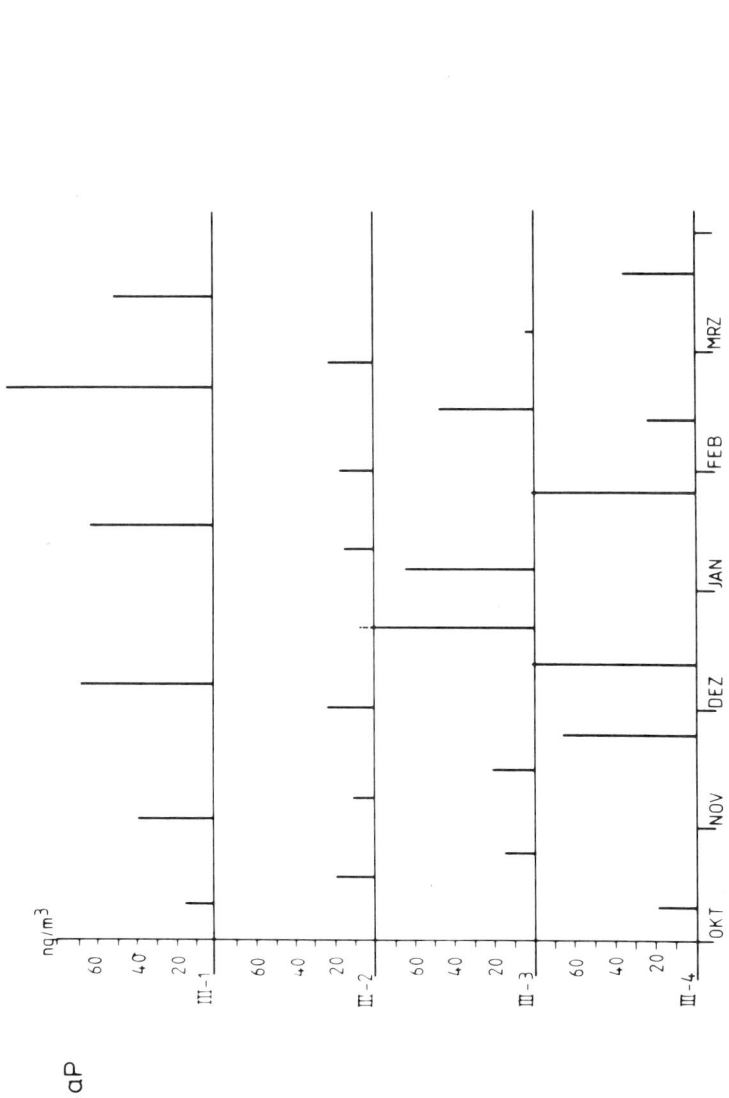

FIGURE 11. Schematic of BaP concentrations in III-1, III-2, III-4 (all inside the tunnel), and III-3 (entrance of the tunnel) from October 1978 to March 1979.

concentration during the period of October 78 to April 79. The figure plots the four sampling stations. The range of the BaP concentration extends from a few nanograms to more than 100 ng/m$^3$. This oscillation in the BaP concentration in air can be observed in the coal-heating area too. The large differences in the BaP concentration at a distinct point from one day to another seem to be normal and not exceptional.

Figure 12 shows a similar oscillation from October 78 to March 79, which varies from 1.3 to 66 ng BaP/m$^3$ for the oil heating area. This is similar in the coke-oven area. In Sampling Station IV-3, the lowest BaP concentration is 11.2 and the highest is 208 ng BaP/m$^3$ (Figure 13). In the rural area the same oscillation can be observed (Figure 14) but on a very low level. The BaP concentration varies from 0.5 ng/m$^3$ to 8 ng BaP/m$^3$.

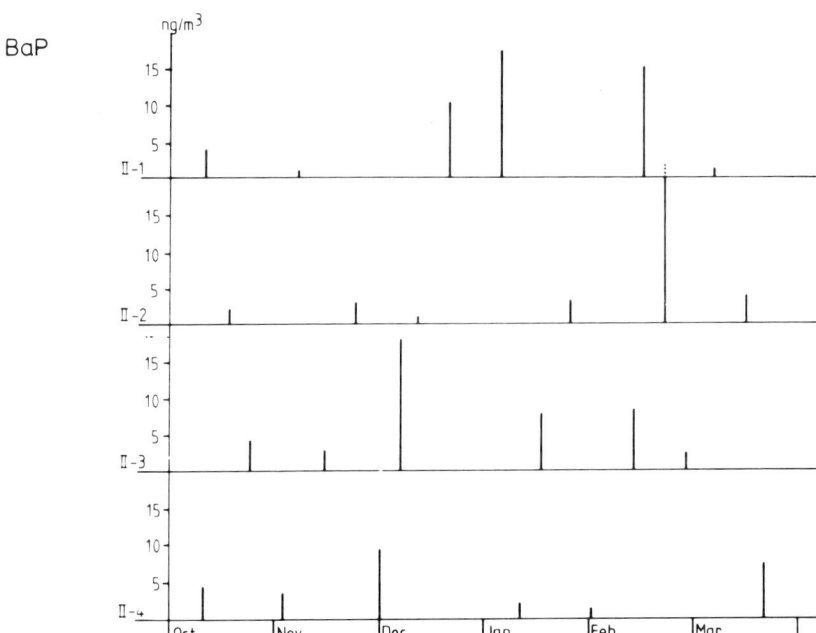

**FIGURE 12.** Schematic of BaP concentrations in II-1, II-2, II-3, and II-4, surrounding the oil heating area, collected from October 1978 to March 1979.

## CONCLUSIONS

The concentration of BaP/m$^3$ in ambient air varies strongly at each sampling station during the heating period for all 20 sampling stations investigated monthly from October 78 to March 79. The concentrations vary more than tenfold. In the same range, the BaP/m$^3$ varies from one

## 122 PAH—PROFILES OF DIFFERENT AREAS IN A CITY

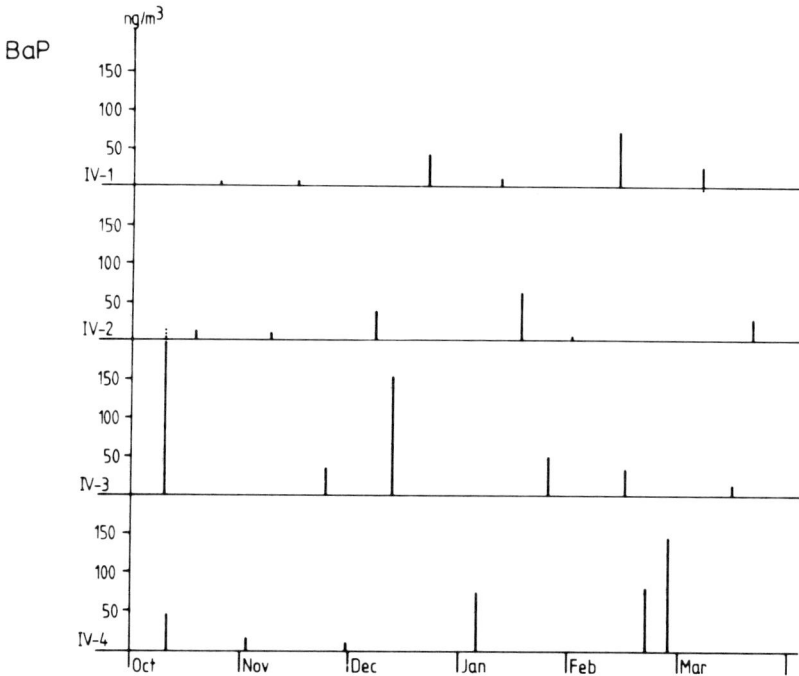

FIGURE 13. Schematic of BaP concentrations in IV-1, IV-2, IV-3, and IV-4, surrounding the coke ovens area, collected from October 1978 to March 1979.

station to another if measured on the same day. Therefore, it is very difficult to give an average value for a city which is representative of such a large area and over a long period.

The PAH mixture (PAH-profile) differs in many cases from area to area, which is demonstrated in Figure 15. In the traffic area (tunnel) cyclopenta(cd)pyrene predominates. This is not the case in the other areas. Area I (coal heating) and Area IV (coke ovens) seem to be similar as far as the PAH-profiles are concerned. Perhaps the different PAH-profiles are helpful to recognize the mean sources of air pollution.

## ACKNOWLEDGMENTS

We wish to thank Dr. M. Buck and Dr. H. Ixfeld, Landesanstalt für Immissionsschutz, D-4300 Essen, for operations planning of the air sampling stations and for the technical organization thereof.

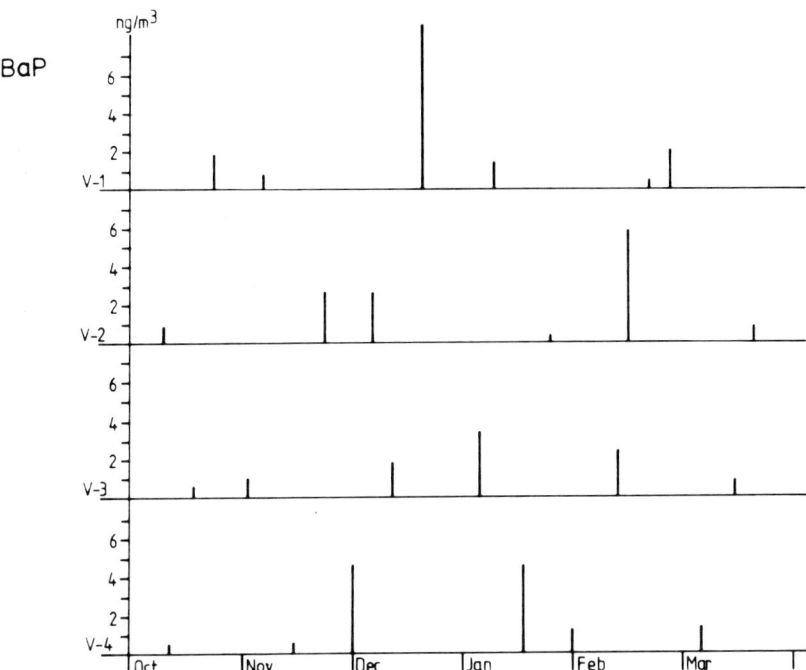

**FIGURE 14.** Schematic of BaP concentrations in V-1, V-2, V-3, and V-4, surrounding a rural area outside of the city, collected from October 1978 to March 1979.

## REFERENCES

1. Misfeld, J., and Timm, J. (1978): The tumor-producing effect of automobile exhaust condensate and fractions thereof. Part III: Mathematical-statistical evaluation of the test results. *J. Environ. Path. Toxicol.* 1:747-772
2. Dontenwill, W., Chevalier, H.-J., Harke, H.-P., Klimisch, H.-J., Brune, H, Fleischmann, B., und Keller, W. (1976): Experimentelle Untersuchungen über die tumorerzeugende Wirkung von Zigarettenrauch-Kondensaten an der Mäusehaut. *Z. Krebsforsch.* 85:155-167.
3. Misfeld, J., und Weber, K. H. (1972): Tierexperimente mit Tabakrauch-Kondensaten und ihre statistische Beurteilung. *Planta Medica* 22:282-292.
4. Cantuti, V., Cartoni, G. P., Liberti, A., and Torri, A. G. (1965): Improved evaluation of polynuclear hydrocarbons in atmospheric dust by gas chromatography. *J. Chromatogr.* 17:60-65.
5. Grimmer, G. (1972): Die quantitative Bestimmung von polycyclischen Aromaten mit der Kapillargaschromatographie. *Erdöl und Kohle, Erdgas* 25:339-343.

# 124 PAH—PROFILES OF DIFFERENT AREAS IN A CITY

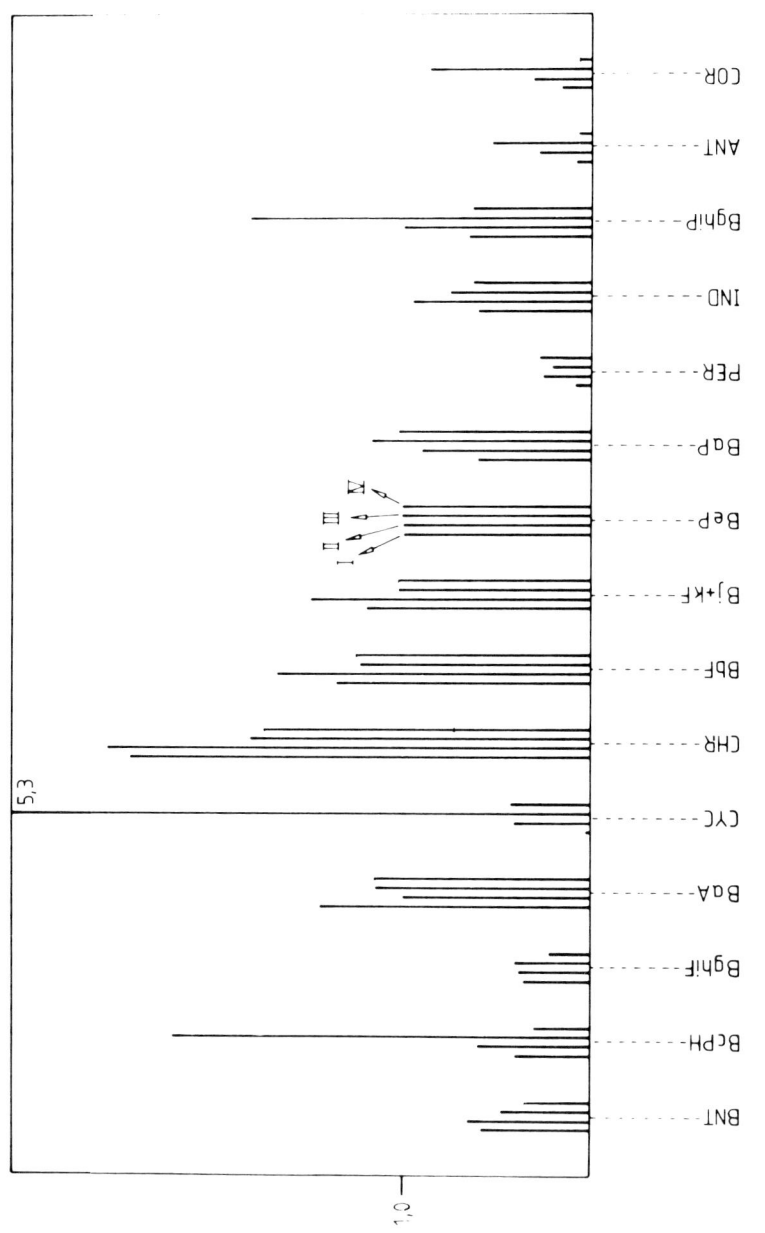

FIGURE 15. Schematic of PAH profiles (relative to BeP = 1.0) of the following areas: I, coal heating area, station I; II, oil heating area, station II; III, automobile traffic (tunnel), station III; IV, coke ovens area, station IV.

6. Novotny, M., Lee, M. L., and Bartle, K. D. (1974): The methods for fractionation, analytical separation and identification of polynuclear aromatic hydrocarbons in complex mixtures. *J. Chromatogr. Sci.* 12: 606-612.
7. Lee, M. L., and Novotny, M. (1976): Gas chromatography/mass spectrometric and nuclear magnetic resonance determination of polynuclear aromatic hydrocarbons in airborne particulates. *Anal. Chem.* 48:1566-1572.
8. Bjørseth, A., and Lunde, G. (1977): Analysis of the polycyclic aromatic hydrocarbons content of airborne particulate pollutants in a Søderberg paste plant. *Am. Ind. Hyg. Assoc. J.* 38:224-228.
9. Bjørseth, A. (1977): Analysis of polycyclic aromatic hydrocarbons in particulate matter by glass capillary gas chromatography *Anal. Chim. Acta* 94:21-27.
10. Lunde, G., and Bjørseth, A. (1977): Polycyclic aromatic hydrocarbons in long-range transported aerosols. *Nature* 268:518-519.
11. Grimmer, G., and Böhnke, H. (1975): Polycyclic aromatic hydrocarbon profile analysis in high-protein foods, oils, and fats by gas chromatography. *J. Ass. Off. Anal. Chem.* 58:725-733.
12. Grimmer, G., Hildebrandt, A., and Böhnke, H. (1973): Investigation on the carcinogenic burden by air pollution in man. III. Sampling and analysis of PAH in automobile exhaust gas. *Zbl. Bakt. Hyg., I. Abt. Orig. B* 158:35-49.
13. Grimmer, G., Böhnke, H., and Glaser, A. (1977): Investigation on carcinogenic burden by air pollution in man. XV. PAH in automobile exhaust gas—an inventory. *Zbl. Bakt. Hyg., I. Abt Orig. B* 164:218-234.

# POLYAROMATIC HYDROCARBONS IN AEROSOLS OVER LAKE MICHIGAN, FLUXES TO THE LAKE

**J. W. Strand and A. W. Andren**
Water Chemistry Program
University of Wisconsin
Madison, Wisconsin 53706

## INTRODUCTION

The increased use of fossil fuels as a source of energy has resulted in an increase in organic combustion by-products in the air. Although this pollution has been recognized, little information exists on the composition and fate of the organic fraction of aerosols. Once organic aerosols are formed those physical, chemical, and meteorological factors that can influence their deposition from air to water, as well as other pathways, have not been adequately documented and real values to work with are meager or nonexistent.

The total composition of organic matter in aerosol particles has not been fully investigated and virtually no information exists on the organic fraction of aerosols directly over Lake Michigan. This becomes more important in view of work done on aerosol trace element input and the small ratio of drainage basin to surface area of Lake Michigan. Atmospheric deposition of certain elements (18) and nutrients (11) to Lake Michigan exceed all other source inputs, and atmospheric input to the Lake basin goes, in most instances, directly to the water in an undiluted and unmodified form. For these reasons, it was hypothesized that polynuclear hydrocarbons (PAH) deposition to the Lake was of importance and posed a potential source of water pollution.

## AEROSOL SAMPLE COLLECTION

Atmospheric particle samples were collected aboard the University of Michigan's Laurention, the University of Wisconsin's System the Neeskay, and the Environmental Protection Agency's the Roger R. Simons. Sampling was continous while the ships were under way in both the southern and central basins of Lake Michigan, and was conducted during the spring, summer, and fall seasons from June 1975 to August 1977.

High-volume aerosol samplers were run in duplicate and located on the bow of each ship ahead of smoke stacks, ventilation systems, and from 2 to 7.6 meters above the water line. These samplers were positioned such that they were as far apart as possible and precautions were taken to secure all samplers with rope, tape, and foul-weather proofing.

To avoid contamination of the filters from the sampling pumps, exhaust tubing was connected and channeled overboard (5). To avoid contamination from the ship itself, the high-volume samplers were turned off whenever the wind blew within the arc coming from the ship and not the Lake.

The high-volume samplers collected bulk particulates on 8 x 10-inch glass-fiber filters ranging from 100 $\mu$m down to the rated limit of the glass-fiber filters of >99.9 percent for 0.3 $\mu$m particles. Flow rate was controlled by a constant-temperature thermal anemometer, which corrected errors caused by filter loading and line voltage changes as well as giving a value at standard temperature and pressure (9). Flow rates for the high-volume samplers were calibrated in the laboratory, before each cruise, with a top loading orifice and water manometer.

Glass-fiber filters were soaked in 6N $H_2SO_4$, extracted with ether and combusted for 3 hours at 450°C. In order to minimize weight changes with relative humidity, the filters were also weighted before and after collection in a constant temperature-relative humidity chamber (15). Once the samples were collected they were stored in clean aluminum pouches, in the dark at 4°C.

To minimize loses of PAH by volatilization from high-volume samplers flows were maintained at 1.4$m^3$/min and collection periods of from 24 to 48 hours (16). This allowed nearly quantitative recoveries of tetracyclic and larger PAH ring systems.

## ISOLATION OF POLYAROMATIC HYDROCARBONS

The glass-fiber filters were then soxhlet extracted with methanol (7), water was added to a 4:1 ratio, and this solution was again extracted with cyclohexane. The cyclohexane was subsequently washed with acid and base, then extracted with nitromethane for a concentrated fraction of PAH. This method of liquid-liquid partition was evaluated using a radiolabeled tracer of $^3$H-benzo (a) pyrene and followed through each step by liquid scintillation. The flow diagram and recovery values (Figure 1)

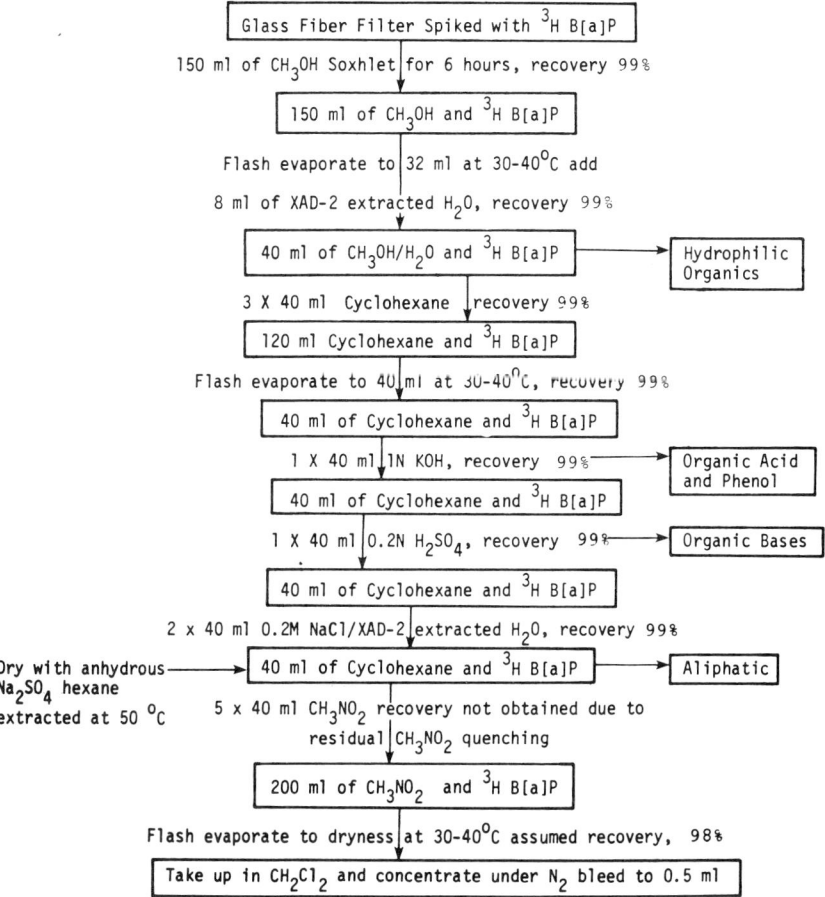

FIGURE 1. Flow Diagram for low concentration acid and base partition of $^3$H BaP.

reveal no loss with a 1 M KOH wash, 0.2 N $H_2SO_4$ wash, and a 0.2 N NaCl/$H_2O$ wash. The NaCl salted out the water soluble benzo(a)pyrene. Minimal losses of $^3$H BaP in these low concentration acids and bases indicated the lack of protolysis and provided a complete extraction and enrichment of aromatics with partition coefficients equal to or greater than BaP, as well as providing other important organic fractions.

The PAH fraction was analyzed by gas chromatography using a Hewlett-Packard 5730A equipped with dual flame ionization and a linear temperature programmer. Chromatograms were generated at 1 mv full scale on a Linear Electronics strip chart recorder equipped with a peak area integrator. Nitrogen carrier gas flow was regulated with calibrated Brooks mass flow controllers. Air flow was maintained at 240 ml/min and

## 130 AIRBORNE PAH OVER LAKE MICHIGAN

hydrogen at 39 ml/min to optimize the detector response. Particle screens and molecular sieve traps were placed on all gas lines, and an Oxisorb cartridge was placed on the carrier gas line. The purity of the carrier and hydrogen was greater than 99.9 percent, which minimized noise and prolonged column life (14).

Liquid nematic crystals were synthesized and used as liquid phases (14), as well as Dexsil 300 and Ultra Bond Carbowax-20M. The resolution of PAH on nematic crystals was exceptional; however, the thermal stability was not satisfactory. The ultrathin chemically bonded Carbowax-20M provided the necessary resolution and stability sought in the analysis of environmental PAH samples.

PAH absorb light between 350 and 450 nm and are known to photooxidize at these wavelengths (12). To minimize sample or standard losses, yellow fluorescent lights were installed with a wavelength longer than 450 nm.

## DESCRIPTION OF SHIP SAMPLING AND METEOROLOGICAL DATA

Cruise tracts are listed in Table 1, and Figure 2 illustrates the ship location for the southern basin of Lake Michigan. These field studies were designed to give as broad a variety of conditions as possible to yield the best mean values for calculating Lake input. To understand aerosol dynamics better for the Lake environment and correlate this with PAH content, meteorological data were collected on each cruise.

## PAH CONCENTRATIONS AND FLUXES FROM THE AIR TO THE LAKE

The environmental role of PAH in air is of particular concern because they pose a potential health hazard. Many but not all PAH can be attributed to man's activities. Those that are formed by man originate almost entirely from combustion processes. Emitted high-temperature PAH vapors cool rapidly and condense on particles that are already present in the air, or form particles of nearly pure condensate. PAH are therefore subjected to the same physical processes that airborne particles are, these being dispersion by turbulence, transport by wind, and removal by sedimentation, impaction, washout, and rainout.

Values for PAH concentrations in air over Lake Michigan ranged from $0.1 \, ng/m^3$ for several of the compounds to as high as $4.2 \, ng/m^3$ for pyrene. Table 2 lists the concentrations of 12 PAH on 8 different cruises. Average annual airborne BaP concentrations from 1966 to 1970 in urban areas of the United States National Air sampling Network ranged from a low of 0.2

TABLE 1. Cruise Track Description for Lake Michigan Aerosols

| Cruise Ship | Number | Date | Cruise Track Description |
|---|---|---|---|
| Laurentian | I | 6/20-24/75 | Transects from Milwaukee to Saugatuck |
| Neeskay | II | 10/29-30/75 | Transect from Saugatuck to Milwaukee |
| — | III | 2/26-28/76 | Shoreline samples at Bailey's Harbor, Door County |
| Simons | IV | 8/3-6/76 | Transect from Milwaukee to Manitowoc, Manitowoc to Muskegon, and Muskegon to Milwaukee |
| Simons | V | 8/23-27/76 | Transect from Milwaukee to Chicago, Chicago to Benton Harbor, Benton Harbor to Kenosha, Kenosha to Holland, and Holland to Milwaukee |
| Simons | VI | 9/14-16/76 | Same as Cruise IV |
| Simons | VII | 10/14/76 | Stationary 2 miles off Milwaukee |
| Simons | VIII | 4/15-16/77 | Same as Cruise VII |
| Simons | IX | 4/18-25/77 | Same as Cruise V |
| Simons | X | 5/17-19/77 | Stationary 40 miles east of Chicago |
| Simons | XI | 6/9-10/77 | Same as Cruise X |
| Simons | XII | 6/11-17/77 | Same as Cruise X |
| Simons | XIII | 8/14-19/77 | Same as Cruise X |
| Simons | XIV | 8/20-25/77 | Same as Cruise V |
| Simons | XV | 8/26-30/77 | Same as Cruise X |

ng/m$^3$ over Hawaii to a high of 29.5 ng/m$^3$ at Altoona, Pennsylvania (16). The same study produced nonurban annual concentrations of 0.1 to 2.1 ng/m$^3$. These values fit well with overall individual PAH compounds and the values found for BaP over Lake Michigan of 0.3 to 1.8 ng/m³. It is reasonable to believe that southern Lake Michigan might lie between an urban and nonurban range, with urban-industrial aerosols being transported out over the Lake, diluted and eventually deposited into the Lake. BaP levels for cities located on Lake Michigan are listed in Table 3 (16); these levels would adequately produce those found over the Lake if these cities were assumed to be the principal sources.

Three to six-membered ring systems were detected, with the highest concentrations being seen with the three to four-ring compounds. Samples collected in the northern basin of the Lake (Cruise III) show lower levels compared with those from the southern basin. This north to south gradient may reflect the anthropogenic inputs.

Concentrations of PAH in Lake Michigan microlayer water were obtained on two cruises, the microlayer being operationally defined as the upper 300 $\mu$m thickness of water. Values for this interface ranged from 0.15 to 0.45 $\mu$g/l, and are similar to those found in other surface waters at 0.05

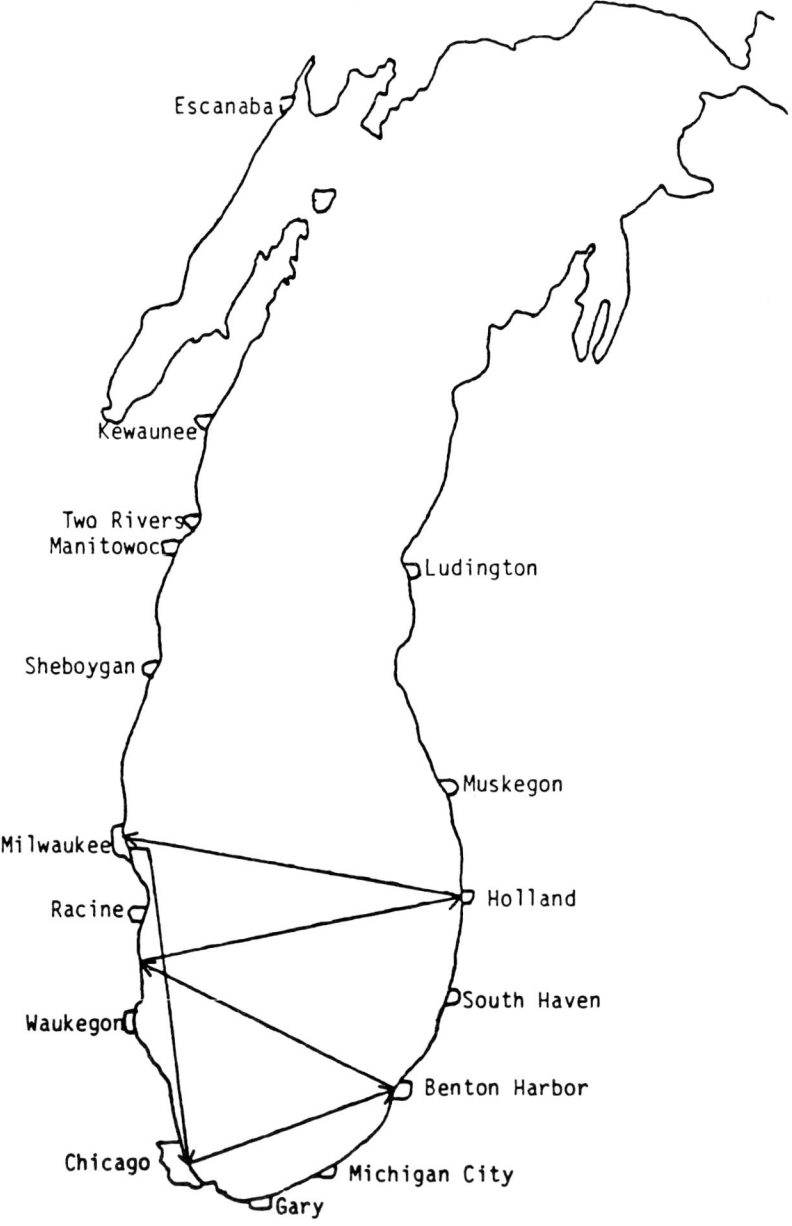

FIGURE 2. Roger Simons Lake Michigan southern cruise track, August 23-27, 1976; April 18-25, 1977; June 11-17, 1977; and August 20-25, 1977.

TABLE 2. PAH Concentrations (ng/m$^3$)$^a$ for Lake Michigan Cruises

| Com-<br>pound$^b$ | \multicolumn{11}{c}{Cruise Number} | | | | | | | | | | |
|---|---|---|---|---|---|---|---|---|---|---|---|
| | II | III | III | IV | V | V | VI | VIII | VIII | XI | XI | XII |
| Fl | 0.51 | 0.4 | 0.2 | 1.2 | — | — | 2.2 | 0.6 | 1.1 | 2.2 | — | 1.3 |
| Ph | — | 0.2 | 0.1 | 0.8 | — | 0.8 | — | 0.8 | 0.4 | 1.0 | — | 0.5 |
| An | — | 0.1 | 0.1 | 0.8 | — | 0.8 | — | 0.8 | 0.4 | 1.0 | — | 0.5 |
| Flu | 1.7 | — | 0.2 | 0.8 | 1.3 | 1.2 | — | 1.3 | 0.9 | 1.2 | 0.8 | 0.9 |
| Blf | 4.1 | 0.2 | 0.1 | — | 2.5 | 1.5 | — | 1.3 | — | 1.1 | — | 0.9 |
| Py | 4.2 | 0.1 | 0.1 | 3.4 | — | 1.8 | 0.2 | 0.4 | 0.2 | 0.5 | — | 0.4 |
| BaA | 0.8 | 0.2 | — | 0.9 | — | — | 2.5 | 0.3 | 0.2 | — | 2.2 | 0.4 |
| Per | — | 0.1 | 0.1 | 1.7 | — | — | — | — | — | — | — | — |
| Tri | — | 0.1 | 0.1 | 0.5 | — | — | — | 0.4 | 0.2 | 0.3 | — | — |
| BaP | — | — | — | — | 1.8 | — | — | — | — | — | — | 0.3 |
| O-Pp | — | — | — | — | — | 0.9 | — | — | — | — | — | — |
| BghiP | 2.0 | — | — | 1.5 | — | — | — | — | — | — | — | — |

$^a$Relative precision ±25 percent ranging, from 1.05 to 0.03 ng/m$^3$.
$^b$Fl = Fluorene         Blf = 2,3 Benzofluorene       Tri = Triphenylene
 Ph = Phenanthrene   Py = Pyrene                        BaP = Benz(a)pyrene
 An = Anthracene       BaA = Benz (a) anthracene   O-Pp = O-phenylenepyrene
 Flu = Fluoranthene   Per = Perylene                     BghiP = Benzo (ghi) perylene

TABLE 3. Annual Average Ambient BaP Concentrations (ng/m$^3$)$^a$

| Station | 1966 | 1967 | 1968 | 1969 | 1970 |
|---|---|---|---|---|---|
| Chicago | 3.3 | 3.0 | 3.1 | 3.9 | 2.0 |
| East Chicago | 6.8 | 5.7 | 1.9 | 6.8 | 5.3 |
| Hammond | 3.9 | 2.5 | 2.1 | 3.3 | 1.7 |
| Grand Rapids | N.D.$^b$ | N.D.$^b$ | 1.4 | 1.6 | 0.8 |

$^a$Reference 16.
$^b$N.D. = not determined.

to 3μg/1 (1). Fewer PAH were detected in the microlayer; however, on a comparative scale, the values represented six orders of magnitude greater than the concentrations of PAH found in the Lake air. This might suggest an atmospheric route, such as deposition to the Lake surface as a source of these high PAH levels. Direct observation of aerosols and the Lake microlayer by scanning electron microscopy provided evidence of similar particles in both compartments (4). Filters were vapor coated with gold-palladium, examined by television mode, and photographed. Cenospheres (hollow) and plerospheres (spheres packed inside larger spheres) were observed. Particles of this type are strictly anthropogenic in origin (10)

## 134  AIRBORNE PAH OVER LAKE MICHIGAN

and have been found in the fly ash of combusted coal. The presence of these particles provides a direct link between the deposition of aerosols from air to water.

The high levels of PAH in the microlayer cannot be attributed solely to atmospheric input, but may also include biological production or bubble scavenging.

If one assumes that the average particle size which PAH are associated with is 1 $\mu$m (17) and the mean wind velocity for the Lake Michigan Cruises is 4.7 m/sec, it is then possible to calculate a range of deposition velocities ($V_d$) for PAH to the water. As a lower limit of $V_d$, the wind tunnel value of $2 \times 10^{-2}$ cm/sec (13) was selected and an upper limit of $V_d$ (from field measurements) of $32 \times 10^{-2}$ cm/sec (2) was selected. These two values were used to bracket those depositions which might be seen for environmental PAH aerosols. The PAH deposition from air to the water surface can then be calculated by the following relationship:

$$D = V_d \cdot C \cdot A$$

where D = dry depositional flux; $V_d$ = deposition velocity; C = average aerosol concentration; and A = area of deposition. For dry aerosol deposition the mean concentrations of individual compounds on all cruises were selected as the best overall values, and the total surface area of Lake Michigan was taken as $5.7 \times 10^3$ m$^2$ (8). The dry flux values for individual PAH are listed in Table 4 for the northern two-thirds and the southern one-third of the Lake. The highest flux occurred for fluorene in the northern part of the Lake and the highest fluxes were seen for benzo (ghi) perylene, perylene, 2,3-benzfluorene, and fluorene in the southern sector of the Lake. High levels of benz (a) anthracene were also seen for both the northern and southern parts of the Lake and the potent carcinogen benz (a) pyrene was found at relatively high levels in the southern third of the Lake.

TABLE 4. Dry Flux ($10^5$ kg/yr) of Polycyclic Aromatic Hydrocarbons to Lake Michigan

| Compound | Northern Two-Thirds of Lake | Southern One-Third of Lake |
|---|---|---|
| Fl    | 0.72 - 11.5 | 1.6  - 25 |
| Ph    | 0.36 -  5.8 | 0.9  - 14 |
| An    | 0.24 -  3.8 | 0.9  - 14 |
| Flu   | 0.48 -  7.7 | 1.4  - 22 |
| Blf   | 0.36 -  5.8 | 2.3  - 36 |
| Py    | 0.24 -  3.8 | 1.7  - 28 |
| BaA   | 0.48 -  7.7 | 1.3  - 20 |
| Per   | 0.24 -  3.8 | 2.1  - 33 |
| Tri   | 0.24 -  3.8 | 0.43 - 6.8 |
| BaP   | —           | 1.2  - 19 |
| U-Pp  | —           | 1.1  - 17 |
| BghiP | —           | 2.8  - 44 |

If one assumes that the scavenging ratios for element mass median diameters are applicable to PAH on particles, then for a particle of a mass median diameter of 1.0 $\mu$m, the scavenging ratio is 160 (6). The concentration of PAH in rain can then be calculated using this (W), the concentration of PAH in air $\chi$, and the density of air (1200 g/m$^3$) by substituting them into the equation:

$$k = \frac{W\chi}{\rho}$$

The flux for rainout and washout from air to the water is then calculated from the relationship:

$$D' = k \cdot R \cdot A$$

where D' = wet depositional flux, k = calculated concentration of PAH in rain, R = rainfall per year to Lake Michigan, and A = area of deposition. The annual rainfall to the Lake was taken as 75.2 cm (3).

The wet flux values of individual PAH compounds to the southern and northern parts of Lake Michigan are listed in Table 5. The values are virtually the same as the upper range values for the dry deposition.

TABLE 5. Wet Flux (10$^6$ kg/yr) of Polycyclic Aromatic Hydrocarbons to Lake Michigan

| Compound | Northern Two-Thirds of Lake | Southern One-Third of Lake |
|---|---|---|
| Fl | 1.1 | 2.4 |
| Ph | 0.57 | 1.3 |
| An | 0.37 | 1.3 |
| Flu | 0.73 | 2.0 |
| Blf | 0.57 | 3.5 |
| Py | 0.37 | 2.6 |
| BaA | 0.73 | 1.8 |
| Per | 0.37 | 3.1 |
| Tri | 0.37 | 0.65 |
| BaP | — | 1.8 |
| O-Pp | — | 1.7 |
| BghiP | — | 3.3 |

## CONCLUSIONS

The conclusions drawn from this work indicated the atmosphere as a source of polycyclic aromatic hydrocarbons to Lake Michigan. Direct observations of aerosol particles provided a link between anthropogenic aerosols and deposition of these particles to the Lake surface.

Twelve PAH were identified in aerosols sampled over Lake Michigan. This constituted the first measurements of this kind on a large inland lake. Concentrations ranged from 0.1 to 4.2 ng/m$^3$; these values were found on single filters exposed for less than or equal to 24 hours. These values correlated well with those from shoreline data in the literature and added to the meager qualitative and quantitative data obtained to date on PAH in this unique environment.

Concentrations of PAHs in the Lake Michigan microlayer ranged from 0.15 to 0.45 µg/l, representing on a relative scale, $10^6$ times the concentration in air. This suggests that aerosols are a source of these compounds and that the microlayer is a repository until the PAH are removed.

The wet and dry fluxes of individual polycyclic aromatics were virtually the same, and relatively high levels of two important aromatics, benz(a)anthracene and benzo(a)pyrene, emphasize the need for data on atmospheric input to water.

It appears that PAH in aerosols originate from man-made combustion processes. Conversion of vapor phase PAH to particulates happens rapidly at ambient temperatures and pressures and by the nature of the particle sizes PAH are formed on, their residence times in the atmosphere are long and removal from air is principally by impaction. Thus fluxes of PAH from air to water are slow but significant. The atmosphere acts then as a large source of PAH and the Lake as a sink. Once the PAH reach the Lake, their buildup in the microlayer seems evident, until adsorption and sedimentation to the Lake bottom remove them from further cycling.

## ACKNOWLEDGMENTS

This work was supported by the Environmental Protection Agency Grant Number R00514201 and the Graduate School of the University of Wisconsin-Madison.

## REFERENCES

1. Adelman, J. B., and Suess, M. J. (1970): Polynuclear aromatic hydrocarbons in the water environment. *Bulletin World Health Org.* 43:479.
2. Cawse, P. A. (1974): Report AERE-R7669, Environmental and Medical Sciences Division, United Kingdom Atomic Energy Authority, Harwell, Oxfordshire.
3. Changnon, S. A., Jr. (1968): Precipitation climatology of Lake Michigan Basin, Bulletin 52, Illinois State Water Survey, Urbana.

4. Elzerman, A. W. (1976): Surface microlayer-microcontaminate interaction in fresh water lakes. p. 71. Ph.D. Thesis, Water Chemistry Program, University of Wisconsin-Madison, Madison, Wisconsin.
5. Environmental Science and Engineering, Inc., Gainesville, Florida (1976): Evolution of a method for the analysis of airborne polychlorinated biphenyls. Report prepared for the Office of Toxic Substances, Environmental Protection Agency.
6. Gatz, D. F. (1975): Pollutant aerosol deposition into southern Lake Michigan. *Water, Air, Soil Poll.* 5:239.
7. Grosjean, D. (1975): Solvent extraction and organic carbon determination in atmospheric particulate matter: the organic extraction-organic carbon analyzer (OE-OCA) technique. *Anal. Chem.* 47:797
8. Klein, D. H. (1975): Fluxes, residence times, and sources of some elements to Lake Michigan. *Water, Air, Soil Poll.* 4:3.
9. Kurz, J. L., and Olin, J. G. (1975): A new flow controller for high volume air samplers. 68th Annual meeting of the Air Pollution Control Association, Boston, Massachusetts.
10. McCrone, W. C., and Delly, J. G. (1973): *The Particle Atlas,* Ed. 2, Vol. III, Ann Arbor Science Publisher, Inc.
11. Murphy, T. J. (1976): Concentrations of phosphorus in precipitation in the Lake Michigan basin. *J. Great Lakes Res.* 2:127.
12. National Academy of Sciences (1972): *Particle Polycyclic Organic Matter,* Washington, D.C.
13. Schmel, G. A., and Sutter, S. L. (1974): Particle deposition rates on a water surface as a function of particle diameter and air velocity. *J. Res. Atm.* 3:911.
14. Strand, J. W., and Andren, A. W. (1978): Synthesis of N,N'-Bis(p-butoxy-benzylidene)-$\alpha,\alpha'$,-bi-p-toluidene and N,N'-Bis(phenylbenzylidene)-$\alpha,\alpha'$,-bi-p-toluidence and their comparison with Dexsil 300 for polycyclic aromatic hydrocarbon separations. *Anal. Chem.* 50:1508.
15. Strand, J. W., Stolzenburg, T. and Andren, A. W. (1978): A constant relative humidity—temperature chamber for the accurate weight determination of air particulate matter collected on filters. *Atm. Environ.* 12:2027.
16. U.S. Environmental Protection Agency (1975): Scientific and technical assessment report on particulate polycyclic organic matter. PB-241, p. 799.
17. Van Vaeck, L. and Van Cauwenberghe, K. (1978): Cascade impactor measurements of the size distribution of the major classes or organic pollutants in atmospheric particulate matter. *Atm. Environ.* 12:2229.
18. Winchester, J. W. and Nifong, G. D. (1971): Water pollution in Lake Michigan by trace elements from pollution aerosol fallout. *Water, Air, Soil Poll.* 1:50.

# ROOM TEMPERATURE PHOSPHORIMETRY FOR THE ANALYSIS OF ENVIRONMENTAL SYSTEMS

T. Vo-Dinh and R. B. Gammage

Health and Safety Research Division
Oak Ridge National Laboratory
Oak Ridge, Tennessee 37830

## INTRODUCTION

Room temperature phosphorimetry (RTP) is a relatively new technique gaining increasing interest among analytical spectroscopists (1, 3-5, 7, 9-11, 14-20, 24). This technique is based on the detection of the phosphorescence emitted from organic compounds adsorbed on solid substrates at room temperature.

Because of its spin-forbidden nature, phosphorescence emission exhibits typically longer decay times (milliseconds to several seconds lifetime) than the spin-allowed fluorescence process ($10^{-10}$ to $10^{-7}$ seconds lifetime). In liquid solutions at room temperature, bi- and monomolecular quenching processes usually cause nonradiative deactivation of the triplet state. The presence of oxygen, an efficient triplet quencher, is also a major contributor to the radiationless deactivation of the triplet level. The radiationless deactivation process for most molecules in the triplet state is so efficient that phosphorescence can normally be observed only when the solution is frozen into rigid matrices. Conventional methods in phosphorimetry, therefore, involve (1) careful preparation of oxygen-free solutions, (2) insertion of analyte compounds into polymer samples, or (3) use of rigid matrices of frozen organic solvent. While the first two techniques involve tedious and time-consuming preparation, the third requires experiments at low temperatures. Recently, intense phosphorescence at room temperature has been observed from various salts of polyatomic organic compounds adsorbed on solid supports, such as silica, alumina, paper, and asbestos (10). This type of phosphorescence is assumed to originate from surface-adsorbed molecules, since none could be observed from finely ground samples of free crystalline compounds. Various substances

# 140 ROOM-TEMPERATURE PHOSPHORIMETRY

of an ionic nature were found to show strong phosphorescence, especially when they were spotted onto substrate following dissolution in strongly acidic or basic solvents. As a consequence, it was believed that the ionic state of the molecules resulted in an increased molecular rigidity via adsorption to the substrate, thus reducing the effect of collisional deactivation. Hydrogen bonding was also found to be responsible for phosphorescence of adsorbed compounds at room temperature. In this laboratory, analytical applications of the RTP process have received special attention. Extensive efforts are devoted to developing the usefulness of RTP as a rapid, simple tool for monitoring trace organic pollutants. This paper presents some recent developments and illustrates some practical applications of the technique in the characterization of polynuclear aromatic hydrocarbons (PAH) in synfuel-derived samples.

## RECENT DEVELOPMENTS IN INSTRUMENTATION AND METHODOLOGY

Unlike conventional phosphorimetry, RTP does not require cryogenic equipment. Figure 1 shows the various experimental steps in an RTP assay:

1. Substrate preparation (optional pretreatment)
2. Sample delivery

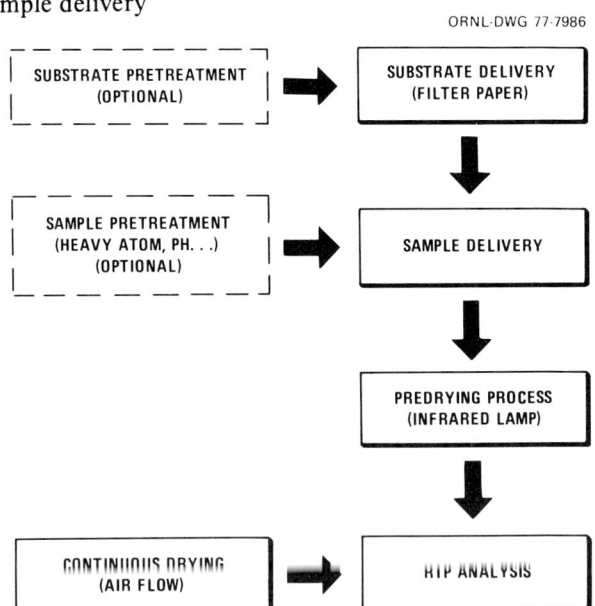

FIGURE 1. Experimental procedures for RTP.

3. Drying process
4. Spectroscopic measurement.

Three microliters of sample solution are spotted on filter paper. Predrying is achieved with infrared heating lamps. The subsequent drying process during the measurement is accomplished by blowing warm and dry air through the sample compartment. Phosphorimetric measurements are performed with, for example, a commercial Perkin-Elmer spectrofluorimeter equipped with a rotating phosphoroscope. The excitation light source is a 150-W xenon arc lamp. The detector is an R777 photomultiplier (Hamamatsu Company, Middlesex, N. J.) that has a useful photocathode spectral response from 185 to 700 nm.

The versatility of sampling procedures is one of the main advantages of the method. Various sample collection methods are possible: spotting, leaching, swipe techniques, and liquid and gas filtration. The use of filter paper or filter membranes as direct sample support greatly broadens the type of samples that can be monitored. Automation is one of the major advantages of the RTP technique. An automatic device using the RTP method for continuous analysis was previously developed for clinical applications (16). Such an automated system could be adapted to the field monitoring of large numbers of samples.

A particular process that is an invaluable aid to phosphorimetric analysis is the heavy-atom effect which enhances the triplet emission of weakly phosphorescent compounds. Since Kasha first reported on the solvent heavy-atom effect (8), it has been confirmed that the presence of heavy atoms, either as substituents (internal heavy-atom effect) or in the solvent (external heavy-atom effect), can significantly enhance the process of intersystem crossing and, therefore, the phosphorescence emission. This effect has been applied to conventional low-temperature phosphorescence (2, 6) and to RTP analyses (4, 7, 14, 15, 17) of individual compounds. The usefulness of the external heavy-atom effect has been recently extended to improve both sensitivity and selectivity of RTP analysis of complex synthetic mixtures (20) and real-life surface samples (21).

For PAH, a large variety of heavy-atom salts such as cesium iodide, sodium bromide, lead acetate, and thallium acetate have been found to be very efficient in enhancing the phosphorescence quantum yields. The presence of heavy-atom salts offers considerable advantage in quantitative analysis since the limits of detection can be decreased, in some cases by several orders of magnitude. The detection limits of the PAH investigated are in the nanogram and subnanogram range. Increasing the phosphorescence intensities of substances can be accomplished not only by adding heavy atoms to the paper but also by pretreating the substrate.

A recent study showed that some PAH compounds can be selectively affected in specific ways. Certain heavy atoms preferentially increase the RTP emission of some PAH much more than others (20). This phenomenon (selective heavy-atom perturbation) can significantly enhance the analytical figures of merit of the RTP technique, thus opening up new analytical dimensions in the assay of complex real-life mixtures.

A recent development that has also improved the selectivity of the RTP technique is the use of derivative techniques in spectral recording. This method has been shown to enhance minor spectral features of RTP spectra (22).

Another method exists for further improving the specificity of the RTP analysis. It is based on the idea of synchronous excitation (23) which has been recently applied to RTP analysis (18,19). To obtain a synchronous spectrum, both excitation and emission wavelengths are simultaneously scanned while the wavelength interval between them ($\Delta\lambda$) is kept constant. This technique can significantly enhance the selectivity of phosphorescence analyses of complex mixtures. Noteworthy is the possibility of exploiting the singlet-triplet energy splitting as an additional factor of selectivity (19). The use of a $\Delta\lambda$ value that is optimized to the singlet-triplet properties of a given compound usually renders its detection more effective than does conventional fixed excitation.

## APPLICATIONS IN ENVIRONMENTAL ANALYSIS AND POTENTIAL

The RTP technique is a sensitive method for monitoring trace organic pollutants. A wide variety of PAH compounds of biological and environmental importance is listed in Table 1. Included in this list are 7 among the 15 polyaromatic hydrocarbons that are in the Priority Pollutant list established by the Environmental Protection Agency (EPA) (12) viz, benzo(a)pyrene-BaP-, benzo(e)-pyrene-BeP-, chrysene, fluoranthene, fluorene, phenanthrene, and pyrene. The limits of optical detection (LOD) of the PNA compounds investigated are between 5 and 500 pg. Figures 2 and 3 show typical RTP spectra from two PAH compounds, phenanthrene and fluoranthene, with various heavy-atom salts. The potential of the RTP technique in multicomponent analysis is illustrated in Figure 4. This figure shows the spectra of a six-component mixture using cesium iodide as the heavy-atom perturber. The selectivity of the analysis of complex samples is further enhanced by varying the heavy-atom perturber and the excitation wavelength (20).

**TABLE 1. Limits of Optical Detection (L.O.D.) of Several PAH by RTP**

| Compound | $\lambda_{ex}$[a] (nm) | $\lambda_{em}$[b] (nm) | Heavy Atom | L.O.D. (ng) |
|---|---|---|---|---|
| Acridine | 360 | 640 | Pb(OAc)$_2$ | 0.4 |
| BaP | 395 | 698 | Pb(OAc)$_2$ | 0.5 |
| BeP | 335 | 543 | CsI | 0.01 |
| 2,3-Benzofluorene | 343 | 505 | NaI | 0.03 |
| Carbazole | 296 | 415 | CsI | 0.005 |
| Chrysene | 330 | 518 | NaI | 0.03 |
| 1,2,3,4-DBA | 295 | 567 | CsI | 0.08 |
| 1,2,5,6-DBA | 305 | 555 | NaI | 0.005 |
| Dibenzocarbazole | 295 | 475 | NaI | 0.002 |
| Fluoranthene | 365 | 545 | Pb(OAc)$_2$ | 0.05 |
| Fluorene | 270 | 428 | CsI | 0.2 |
| $\alpha$-Naphthol | 310 | 530 | NaI | 0.03 |
| Phenanthrene | 295 | 474 | NaBr | 0.007 |
| Pyrene | 343 | 595 | Pb(OAc)$_2$ | 0.1 |
| Quinoline | 305 | 505 | AgNO$_3$ | 0.04 |

Source: Reference 20.
[a] $\lambda_{ex}$ = excitation wavelength.
[b] $\lambda_{em}$ = emission wavelength.

A practical example of application of the RTP technique is the qualitative and quantitative analysis of a Synthoil sample (a coal liquid from a desulfurization process). A quick RTP screening measurement can, within a few minutes, identify and roughly quantify the most abundant constituant, pyrene (5500 ppm), in the raw Synthoil sample. Analysis of a sample that has been coarsely fractionated further identified and quantified four other PAH constituents; BaP, chrysene, fluoranthene, and phenanthrene. The identification of phenanthrene in the Synthoil sample is illustrated in Figure 5. The intensity of the main three emission peaks assigned to phenanthrene is increased in a Synthoil sample that has been enriched with that compound. The synchronous scanning technique further enhances the curvature of the two short-wavelength shoulders which belong to fluorene and its methyl-derivatives. Table 2 gives the content of PAH in the Synthoil sample analyzed by RTP.

A further demonstration of the potential of RTP is the analysis of a XAD-2 extract from a work-place air sample. Other techniques have identified classes of PAH in this sample in an experimental procedure for evaluating the analytical scheme of the Source Assessment Sampling

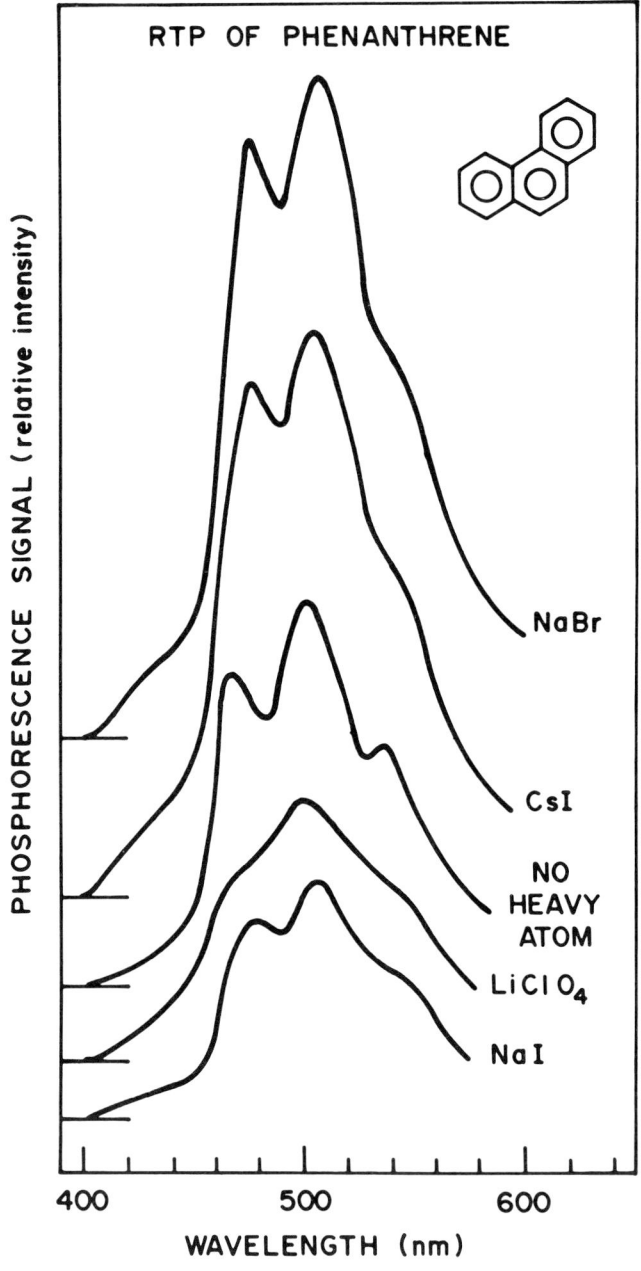

FIGURE 2. RTP spectra of phenanthrene with various heavy-atom perturbers.

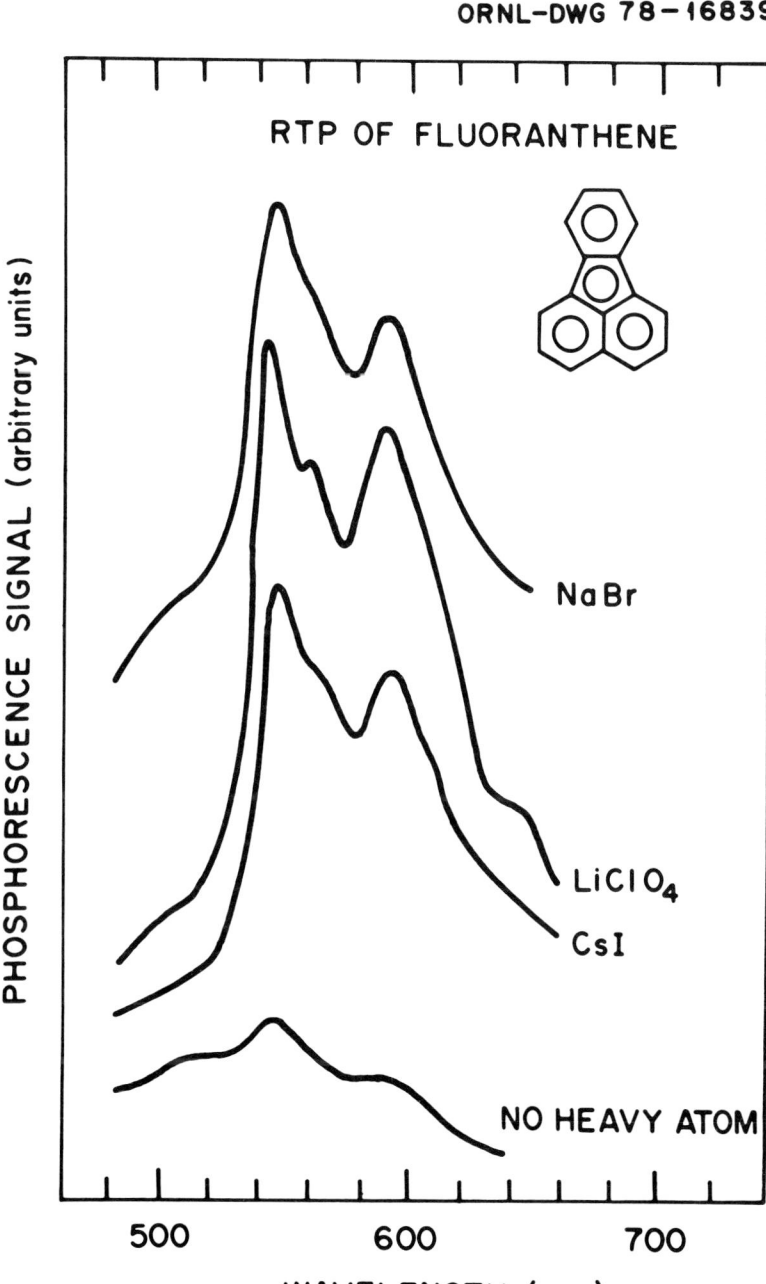

FIGURE 3. RTP spectra of fluoranthene with various heavy-atom perturbers.

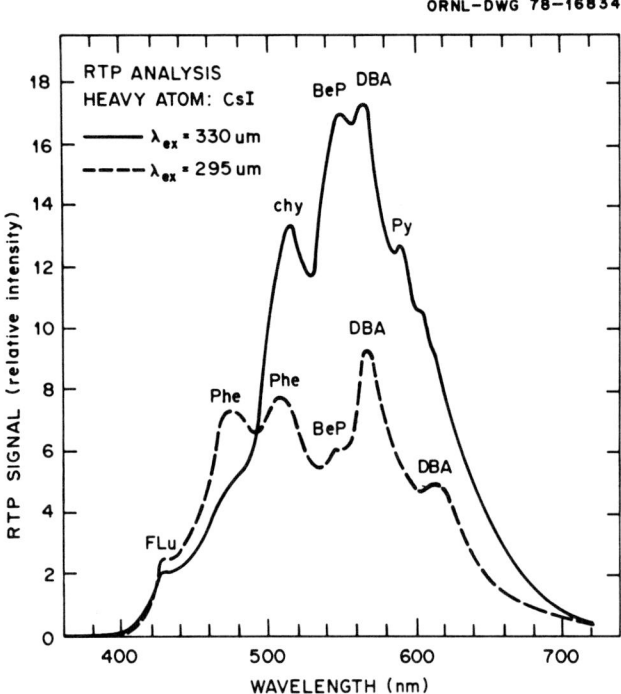

FIGURE 4. RTP spectra of a six-component mixture (Flu = fluorene; Phe = phenanthrene; Chy = chrysene; BeP = benzo(e)pyrene; DBA = 1,2,5,6-dibenzanthracene; Py = pyrene).

System (SASS) developed by the U.S. EPA (12,13). The RTP method was used to identify and quantify several major ($\sim 10^{-6}$ M) PAH compounds, such as chrysene, fluoranthene, phenanthrene, and pyrene, found at concentrations approximately tenfold higher than those of other minor constituents ($\sim 10^{-7}$ M), such as BaP, BeP, 2, 3-benzofluorene, 1, 2, 5, 6-dibenzanthracene (DBA), dibenzothiophene, and fluorene (Table 3). Two compounds, BaP and 1, 2, 5, 6-DBA, are the constituents having the most carcinogenic activity. Eight PAH compounds were on the EPA Priority Pollutants list.

Because of the extreme complexity of these fossil-fuel-derived samples, the identification of the organic components often requires a series of sophisticated, time-consuming, and relatively expensive separation and spectrochemical measurements involving techniques that are mostly laboratory oriented. Although it is often desirable to characterize as many compounds as possible, there is an equal need for simple, rapid, and reliable methods that can give within a reasonably short time period (within a few hours or one day) a qualitative or semiquantitative indication of the major components in a complex environmental sample.

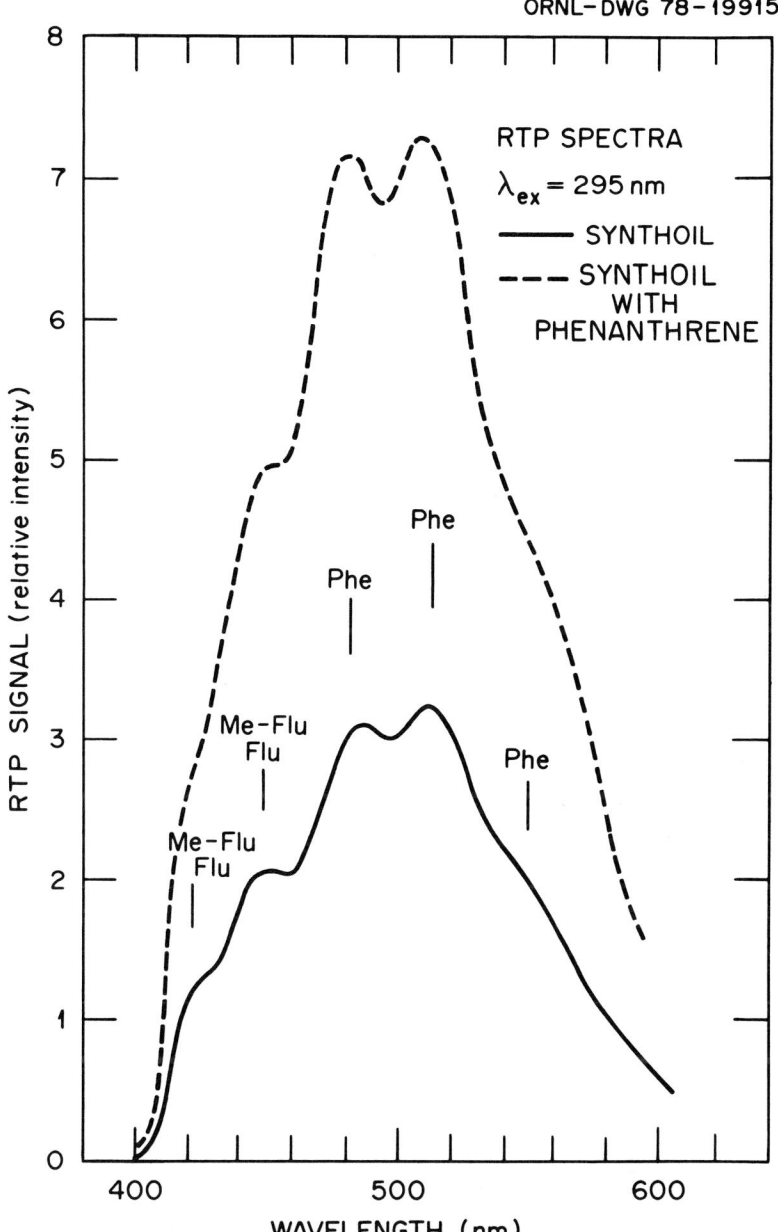

FIGURE 5. Identification of phenanthrene in a Synthoil sample (solid curve: RTP spectrum of original Synthoil sample; dashed curve: RTP spectrum of the Synthoil sample enriched with phenanthrene; Flu = fluorene; Me-Flu = methyl-derivatives of fluorene; Phe = phenanthrene).

TABLE 2. Several PAH Compounds Identified
and Quantified in a Synthoil Sample by RTP

| Compound | Concentration (ppm)[a] |
|---|---|
| Benzo(a)pyrene | 50 ± 18 |
| Chrysene | NQ[b] |
| Fluoranthene | 460 ± 160 |
| Phenanthrene | 2100 ± 700 |
| Pyrene | 5500 ± 1600 |

[a]ppm = Mg PNA/g of Synthoil.
[b]NQ: not quantified.

TABLE 3. Concentration of Some PAH Compounds
in Liquid Extract From a Work-Place
Air Sample, Analyzed by RTP

| Compound | Concentration[a] (M) |
|---|---|
| Benzo(a)pyrene | $<10^{-6}$ |
| Benzo(e)pyrene | $4 \times 10^{-7}$ |
| Benzo(b)fluorene | $3 \times 10^{-7}$ |
| Chrysene | $3.6 \times 10^{-6}$ |
| Dibenz(a,h)anthracene | $1 \times 10^{-7}$ |
| Dibenzothiophene | $3 \times 10^{-7}$ |
| Fluoranthene | $1.6 \times 10^{-6}$ |
| Fluorene | $1.5 \times 10^{-7}$ |
| Phenanthrene | $8 \times 10^{-6}$ |
| Pyrene | $2.3 \times 10^{-6}$ |

[a]Data obtained from the analysis of a diluted sample of Synthoil (1:1000 dilution in ethanol).

In the area of occupational health hazard assessment, RTP can offer several advantages as an analytical monitoring tool. The technique should provide a simple, rapid, and sensitive method for identifying major PAH components in samples from coal conversion and oil shale industries. It should find useful application in a multiphase approach to environmental analysis.

First, RTP is most suitable for "prescreening" large numbers of samples. The technique is simple and inexpensive since it requires no cryogenic equipment and uses only filter paper for the sample substrate. Unfractionated samples can be directly measured to profile one or two major components within a few minutes. Parent PAH profiling is useful for characterizing different types of fugitive emissions (6).

Second, the RTP technique also is useful if a more precise analysis is required. A longer time (2 to 4 hours) is required since the sample has to be fractionated and calibration curves for various compounds have to be measured beforehand. This type of analysis can provide a qualitative as well as a quantitative characterization of several major components (five to ten). The experimental procedures are no more complicated than those required for the EPA Level 1 Environmental Assessment (13), and the 30 percent accuracy of the RTP measurements is more than adequate for this phase of analysis.

Finally, for trace compounds (tens of ppm or less) a more precise but also more time-consuming analysis (from 1 to 2 days) can be conducted, after sample enrichment with the compound of interest. This type of measurement allows the determination of a larger number of components. At this level of sophistication, RTP can be combined with other methods to provide a complete characterization of environmental samples.

## ACKNOWLEDGMENTS

This research is sponsored by Office of Health and Environmental Research, U.S. Department of Energy, under Contract W-7405-eng-26 with Union Carbide Corporation.

## REFERENCES

1. A von Wandruska, R. M., and Hurtubise, R. J. (1976): Determination of p-aminobenzoic acid by room temperature solid surface phosphorescence. *Anal. Chem.* 48:1784–1788.
2. Boutillier, G. D., Donnell, C. M., and Rahn, R. O. (1974): Inorganic probes for the phosphorimetric determination of nucleosides. *Anal. Chem.* 46:1508–1511.
3. Bower, E.L.Y., and Winefordner, J. D. (1978): Room-temperature phosphorescence characteristics and limits of detection of several pharmaceutical compounds. *Anal. Chim. Acta.* 101:319–332.

4. Bower, E.L.Y., and Winefordner, J. D. (1979): A continuous sampling system for room temperature phosphorimetric analysis. *Appl. Spectrosc.* 33:9–12.
5. Ford, C. D., and Hurtubise, R. J. (1978): Room-temperature phosphorescence of the phthalic acid isomers, p-aminobenzoic acid, and terephthalamide adsorbed on silica gel. *Anal. Chem.* 50:610–612.
6. Gammage, R. B. (1979): Preliminary thoughts on proxy PNA compounds in the vapor and solid phase, In: *Proceedings of the Industrial Hygiene Monitoring Needs for the Coal Conversion and Oil Shale Industries,* Brookhaven National Laboratory, BNL-51002, pp. 174–188.
7. Jakovijevic, I. M. (1977): Lead or thallium salts as external heavy atoms for room-temperature quantitative phosphorescence. *Anal. Chem.* 49:2048–2050.
8. Kasha, M. (1952): Collisional perturbation of spin-orbital coupling and the mechanism of fluorescence quenching. *J. Chem. Phys.* 20:71–74.
9. Paynter, R. A., Wellons, S. L., and Winefordner, J. D. (1974): New method of analysis based on room-temperature phosphorescence. *Anal. Chem.* 46:736–738.
10. Schulman, E. M., and Walling, C. (1973): Triplet-state phosphorescence of adsorbed ionic organic molecules at room temperature. *J. Phys. Chem.* 77:902–905.
11. Seybold, P. G., and White, W. (1975): Room-temperature phosphorescence analysis: use of external heavy-atom effect. *Anal. Chem.* 47:1199–1200.
12. U.S. Environmental Protection Agency (1977): Sampling and Analysis procedures for screening of industrial effluents for priority pollutants, U.S. EPA, Cincinnati, Ohio.
13. U.S. Environmental Protection Agency (1978): *Procedures Manual: Level 1, Environmental Assessment,* 2nd ed., EPA-600/7-78-201.
14. Vo-Dinh, T., Lueyen E., and Winefordner, J. D. (1976): Heavy-atom effect on room-temperature phosphorimetry. *Anal. Chem.* 48:1186–1188.
15. Vo-Dinh, T., Lueyen, E., and Winefordner (1977): Room-temperature phosphorescence of several polyaromatic hydrocarbons. *Talanta,* 24:146–148.
16. Vo-Dinh, T., Walden, G., and Winefordner, J. D. (1977): Instrument for the facilitation of room temperature phosphorimetry with a continuous filter paper device. *Anal. Chem.* 49:1126–1130.
17. Vo-Dinh, T., and Winefordner, J. D. (1977): Room temperature phosphorimetry as a new spectrochemical method of analysis. *Appl. Spectrosc.* 13(2):261–294.

18. Vo-Dinh, T., Gammage, R. B., Hawthorne, A. R., Thorngate, J. H. (1978): Synchronous spectroscopy for analysis of polynuclear aromatic compounds. *Environ. Sci. and Technology* 12:1297–1302.
19. Vo-Dinh, T., and Gammage, R. B. (1978): Singlet-triplet energy difference as a parameter of selectivity in synchronous phosphorimetry. *Anal. Chem.* 50:2054–2058.
20. Vo-Dinh, T., and Hooyman, J. H. (1979): Selective heavy-atom perturbation for improved analyses of complex mixtures by room temperature phosphorimetry. *Anal. Chem.* 51:1915–1921.
21. Vo-Dinh, T., Gammage, R. B., and Martinez, P. R.: Identification and quantification of polynuclear aromatic compounds in Synthoil by room-temperature phosphorimetry, submitted for publication.
22. Vo-Dinh, T., and Gammage, R. B. (1979): The applicability of the second-derivative method to room-temperature phosphorescence analysis. *Anal. Chim. Acta.* 107:261–271.
23. Vo-Dinh, T. (1978): Multicomponent analysis by synchronous luminescence spectrometry. *Anal. Chem.* 50:396–401.
24. Wellons, S. L., Paynter, R. A., and Winefordner, J. D. (1974): Room-temperature phosphorimetry of biologically important compounds adsorbed on filter paper. *Spectrochim. Acta* 30A:2133–2140.

# THE INFLUENCE OF EPOXIDE HYDRATASE INHIBITION ON THE GENETIC ACTIVITY OF BENZO(a)PYRENE IN THE LIVER MICROSOME TEST USING MITOTIC GENE CONVERSION IN YEASTS AS A MUTAGENICITY TEST SYSTEM

U. Bayer*, M. Homola**, and D. Siebert**

* Institut de Recherches Scientifiques sur le Cancer
F-94800 Villejuif, France
** Institut für Landespflege Universität Freiburg,
D-7800, Freiburg, FRG

## INTRODUCTION

A great variety of different test organisms ranging from bacteria to intact mammals are used in screening tests for mutagens/carcinogens. Yeast cells take a special position within this list, in that they form a bridge in terms of biological classification that is between the procaryotic bacteria and the eucaryotic cells of higher organisms. Compared with the most frequently used bacteria (2,3,8,10,11) yeast cells offer several advantages (9,12).

In our investigations we used yeast cells for the following reasons: (1) They are eucaryotic cells, whose cell structure may better serve as a model for cells of higher organisms; (2) they can be cultured in a haploid as well as in a diploid state; (3) they are well defined genetically and offer an array of genetic endpoints, including mutations and recombinations and (4) they are single-cell microorganisms and thus combine the advantages of microorganisms with the higher relevance of eucaryotic cells (9,12,16,18,19, 20).

Mitotic gene conversion, which we have chosen as a genetic endpoint is an intragenic, nonreciprocal, recombination process in a heteroallelic locus. In the strain D4-RDII of Saccharomyces cerevisiae, mitotic gene conversion is tested in the two unlinked loci ade2 and trp5. These cells are heteroallelic in the ade 2 locus and thus require adenine for growth. After mitotic gene conversion, these cells no longer require adenine and can be selected on media lacking adenine (12,16,17). The same is true for the trp5 locus. It has been shown that for these loci mitotic gene conversion is sensitive to induction by known mutagens (12,20).

A system using liver microsome activation and yeast cells was applied in this work to study the genetic effects of benzo(a)pyrene (BaP) on the ade2 and trp5 loci. BaP has been shown to require activation by monooxygenases to genetically active epoxides which are decomposed by epoxide hydratase to inactive diols. Inhibition of the epoxide hydratase leads to an accumulation of epoxides that should be detectable by an increase in genetic effects of BaP (4,13).

Bentley et al (4), however, found both activation and inactivation of BaP after inhibition of this enzyme. These results were shown to depend on the pretreatment of the animals used as a source of microsomes and on the concentration of BaP in the system.

This study was designed to determine whether the Saccharomyces cerevisiae D4-RDII yeast strain could be applied to the specific problem of determining the genetic effects of epoxide hydratase inhibition in BaP-exposed cultures. Further, we wanted to show the utility of yeast cells as test organisms and the induction of mitotic gene conversion for the detection of genetic effects of chemical compounds in a complex metabolic situation.

## MATERIAL AND METHODS

### Chemicals

Benzo(a)pyrene was supplied by Fluka, Buchs (Switzerland) and cyclohexene-3,3-oxide was a gift from Dr. Oesch, Pharmakologisches Institut, Mainz (FRG).

### Yeast Strain

The diploid, respiration-deficient strain D4-RDII of Saccharomyces cerevisiae, heteroallelic in the two unlinked loci ade2 and trp5, was used for the induction of mitotic gene conversion (14). This strain requires adenine and tryptophane for growth. Gene conversion leads to wild-type cells no longer requiring adenine or tryptophane. Respiration deficiency increases the sensitivity of the test system against mutagens (16,18).

### Media and Culturing

Yeast cells were cultured in 5 ml of YEP medium (1 percent Difco extract yeast, 2 percent Difco bactopeptone, and 2 percent glucose) in test tubes at 25°C by shaking until they reached the late logarithmic phase, usually after 3 days. Samples (0.1 ml) from each tube were spread onto either adenine or tryptophane-deficient solid medium to eliminate cultures with a high frequency of spontaneous convertants. Media selective for convertants and for scoring survival were solid synthetic media (1.5 percent Difco bacto ager, 4 percent glucose) based on "Difco nitrogen base w/o amino acids and ammonium sulfate" (1.5 g/l) and supplemented with ammonium sulfate (1.5 g/l), amino acids, and nucleobases. All petri dishes were incubated at 25°C. Colonies were counted after 6 days.

## Treatment of Yeast Cells

Benzo(a)pyrene was dissolved in dimethylsulfoxide (DMSO). The final concentration of DMSO reached a maximum of 15 percent. Cells from the late logarithmic phase were washed twice in distilled water immediatedly before treatment. For the liver microsome test in vitro, about $5 \times 10^7$ cells were incubated with 0.5 ml of microsomes in 0.15 M phosphate buffer (pH 7.4), together with the solution of the test compound in DMSO supplemented with sodium isocitrate (8 mg), $NADP^+$ (1 mg), $MgCl_2$ (2 mg), and isocitratedehydrogenase (20 U/ml; 0.02 ml) to a final volume of 2 ml. Normally, after 8 hours the treatment was terminated by washing the cells twice in distilled water. The cells are then exposed to a "liquid holding" of 16 hours in distilled water to increase the sensitivity of the respiration-deficient strain D4-RDII (14,16,19).

## Preparation of Liver Microsomes (S-9 Fraction)

Liver microsomes were prepared from NMRI mice (a gift from Gödecke, Freiburg), which had received a single dose of 75 mg of methylcholanthrene per kg bodyweight intraperitoneally 1 day before killing. The S-9 fraction of the microsomes was frozen immediately at -80°C and thawed before the experiments.

## Statistics

An increased frequency of induced mitotic gene conversion was regarded as a positive result, when it was statistically significant ($p < 0.01$) and when it could be repeated in at least two further experiments. For statistics the tables of Kastenbaum and Bowmann (7) were used.

## RESULTS

### Activity of Dimethylsulfoxide (DMSO)

The solvent DMSO did not show any genetic activity in these and in previous experiments when used in concentrations up to 15 percent. Only a weak killing effect due to DMSO could be found (15).

### Activity of Cyclohexene-3,3-Oxide

Cyclohexene-3,3-oxide (CH33O) inhibits epoxide hydratase (4). When it was examined for genetic activity at concentrations of 2, 4, and 10 mM in the yeast test with or without liver microsomes, it did not increase the frequency of mitotic gene conversion in the strain D4-RDII of Saccharomyces cerevisiae. In no case did we find a killing effect of CH33O at the concentrations used.

### Genetic Activity of Benzo(a)Pyrene

Previous experiments have shown that BaP is not capable of inducing mitotic gene conversion in the direct test without microsomes at concentrations up to 1000 ppm (14).

The addition of liver microsomes from untreated mice led to the same results. When liver microsomes from phenobarbital-pretreated mice were used, we found a three- to fourfold increase of mitotic gene conversions in both loci. The BaP concentration was 1500 ppm. The use of methylcholanthrene-induced microsomes clearly increased the frequency of mitotic gene conversions five to eight fold over control values in both loci and at the same concentrations of BaP.

A comparison of the S-9 fraction with the 100,000 g fraction of microsomes revealed no significant differences. Therefore, we used only the S-9 fraction for the following experiments.

### Inhibition of Epoxide Hydratase

To inhibit epoxide hydratase, cyclohexene-3,3-oxide was added to the cultures at a concentration of 2 mM. When phenobarbital-induced microsomes were used, no significant effect could be found at a BaP concentration of 200 ppm. Using methylcholanthrene(MC)-induced microsomes and the same BaP concentration, we found an increased frequency of mitotic gene conversions, but the increase was not significant. Only the higher BaP concentration of 1000 ppm and the MC-induced microsomes under the same treatment conditions led to a clear and significant increase in mitotic gene conversions. This experiment was repeated four times with the same result (Table 1).

Higher concentrations of the inhibitor CH33O, 4 and 10 mM, did not lead to a further increase. A treatment time of 18 hours instead of 8 hours doubled the frequency of mitotic gene conversions but simultaneously raised the killing rate to more than 90 percent.

## Summary of the Results

Using methylcholanthrene-induced microsomes (S-9 fraction) from NMRI mice in a "yeast liver microsome" test, the inhibition of epoxide hydratase by cyclohexene-3,3-oxide led to a significant increase in induced mitotic gene conversions in the strain D4-RDII of Saccharomyces cerevisiae. This indicates that in the yeast system epoxides contribute significantly to the genetically active metabolites of BaP.

## DISCUSSION

### Results

Bentley et al (4) have shown a dual role for epoxide hydratase in BaP metabolism using the Salmonella test. When liver microsomes from

TABLE 1. Induction of Mitotic Gene Conversion in the Strain D4-RDII of Sacchromyces Cerevisiae by Benzo(a)pyrene

| Dose of BaP (ppm) | Inhibition of Epoxide Hydratase | Micro-somes | Convertants/$10^6$ Survivors | | | | Survival (%) |
|---|---|---|---|---|---|---|---|
| | | | ade2 | CFa[b] | trp5 | CFt[b] | |
| Control | - | NI[a] | 7.0 (44)[c] | - | 9.7 (61) | - | 100 |
| 200 | - | NI | 12.0 (75) | 1.7 | 17.3 (108) | 1.8 | 99.4 |
| 200 | + | NI | 13.9 (100) | 2.0 | 14.1 (102) | 1.5 | 114.9 |
| 200 | - | MC | 23.0 (81) | 3.3 | 33.2 (117) | 3.4 | 56.1 |
| 200 | + | MC | 25.4 (75) | 3.6 | 25.7 (76) | 2.7 | 47.5 |
| Control | + | MC | 10.1 (26) | - | 26.0 (67) | - | 100.0 |
| 1000 | - | MC | 35.7 (37) | 3.7 | 55.9 (58) | 2.2 | 40.3 |
| 1000 | + | MC | 78.3 (63) | 7.8 | 145.5 (114) | 5.6 | 31.3 |

[a] Treatment conditions : pH 7.4; 25°C; darkness; 8 hours; liquid holding 16 hours. Inhibition of the epoxide hydratase by cyclohexene-3,3-oxide (CH33O) 0.2 mM; liver microsomes S-9 not induced (NI) and methylcholanthrene induced (MC).
[b] CFa and CFt denote factors of relative conversion frequency over control in the respective loci.
[c] Numbers in parentheses indicate actual number of convertant colonies counted on five plates.

control mice or phenobarbital-pretreated mice were used, epoxide hydratase efficiently reduced the mutagenicity of BaP in Salmonella. Further, the inhibition of epoxide hydratase was shown to decrease the rate of mutagenicity at low concentrations of BaP but increased it at high concentrations, when the microsomes came from mice that had been pretreated with methylcholanthrene for enzyme induction. These results indicate that high BaP concentrations inhibit the oxidation of dihydrodiols to highly mutagenic dihydrodiol epoxides. The role of the epoxide hydratase is thus determined by the form of the monooxygenase in the activating system (4,6).

Our results, which show an increase in mitotic gene conversions after the inhibition of the epoxide hydratase at high concentrations of BaP, are in line with these findings. We could not, however, show the decreasing effect of the inhibition at low BaP concentrations. This may be due to the use of different inhibitiors or the difference in uptake rates of BaP for yeast as compared with Salmonella. We used cyclohexene-3,3-oxide, whereas Bentley et al (4) had used the genetically active trichloro-propeneoxide (5).

Furthermore, our results correspond with results in CHO cells, where Roszinsky-Köcher et al (13) demonstrated an increase in chromosome aberrations and sister chromatid exchanges induced by BaP after inhibition of the epoxide hydratase by cyclohexene-3,3-oxide.

In microbial or yeast tests without activation systems, BaP reveals no genetic or cytogenetic activity (10,11,13,19). The genetic activity of BaP could be detected only when the metabolic properties were known in detail and were considered in the mutagenicity test systems by the addition of microsomes. This leads to our general conclusion that there is no safe test for an unknown substance whose metabolic properties are not known and therefore are not considered in the test.

**Comparison of the Procedures**

In the Salmonella test the test substance, together with the S9 mix and the bacteria, is placed in a soft agar overlay onto the plate, which is incubated at 37°C normally for 24 hours before the colonies are counted (1).

In the yeast procedure, however, the yeast cells are put in a test tube together with the microsomes and the test substance. The treatment takes place in a definite volume of liquid at 25°C. After a definite time the treatment is terminated by washing the cells free of chemicals; the cells are then plated onto solid agar plates which are incubated for some days before the colonies are counted. The convertants are referred to $10^6$ survivors. This procedure thus allows not only the determination of genetic effects but also the determination of a toxic effect. Furthermore, the procedure allows a variation of distinct factors like pH and treatment time. As the yeast procedure also allows an exact determination of the concentration of a test compound during a given time, this test is suitable for a determination of a dose-response relationship. The yeast procedure with microsomes thus offers a better approach for further detailed studies of the relation between genetic activity and metabolism of a test substance.

## Mitotic Gene Conversion and Mutations

Recently Murthy (12) has published data on more than 200 substances that had been tested for their induction of mitotic gene conversion. In no case were there false negative or false positive results when the induction of mitotic gene conversion was compared with mutagenicity data from other test systems. This best confirms the conclusion of Zimmermann (20) that the induction of mitotic gene conversion is strongly correlated to the induction of mutations.

There are substances that induce frame-shift mutations like methylcholanthrene and those that induce base pair substitutions like methylmethane sulfonate or N-methyl-N-nitrosoguanidine (MNNG). In the Salmonella test, tester strains are used that are sensitive to only one or the other type of mutation. To avoid false negative results, therefore, a series of tester strains is necessary. In no case, however, has mitotic gene conversion revealed such a mutagen specificity.

These two points qualify mitotic gene conversion as a good genetical endpoint for mutagenicity screening.

## The Sensitivity of the System

Comparison of yeast data and Salmonella data reveal a lower sensitivity of the yeast system concerning the detection of the lowest effective dose. A comparison of the relative sensitivity, however, shows a different picture. Bentley et al (4) found a two- to fivefold increase in mutations in Salmonella after inhibition of the epoxide hydratase. Roszinky-Köcher et al (13) found an increase in chromosome aberrations and sister chromatid exchanges in Chinese hamster ovary cells (CHO cells) which was less than twofold. In our system, the increase was more than twofold in both the ade2 and the trp5 loci of the strain D4-RDII of Saccharomyces cerevisiae.

This comparison shows that, at least for the problem studied, the sensitivity of the yeast system is comparable with the relative sensitivity of the other systems.

## CONCLUSION

The data from this and other studies (12,16,18,20) present the induction of mitotic gene conversion in Saccharomyces cerevisiae as an excellent tool for mutagenicity testing. When compared with the widely used Salmonella test, the yeast system offers advantages that suggest the extended use of this test additionally or instead of the Salmonella test in the screening for mutagens/carcinogens.

## SUMMARY

Mitotic gene conversion is a mitotic, intragenic recombination process that is closely correlated to mutagenicity. It is used as a mutagenicity test system in the diploid strain D4-RDII of Saccharomyces cerevisiae to investigate the genetic properties of benzo(a)pyrene (BaP). Requiring metabolic activation, BaP induced a two- to threefold increase in mitotic gene conversion after the addition of methylcholanthrene-induced liver microsomes from mice. The inhibition of the epoxide hydratase by cyclohexenoxide led to a five- to seven-fold increase in the genetic effect, thus indicating that epoxides are the main active metabolites in this system. These results are in line with results obtained with Salmonella ("Ames test") and Chinese hamster ovary cells. From this it is concluded that all mutagenicity tests for substances requiring metabolic activation must consider specific metabolic pathways. A general test cannot be proposed.

A comparison with published data on the "Ames test" reveals advantages of the yeast system: (1) the yeast cells used are eukaryotic and diploid and theoretically more relevant for mammals; (2) mitotic gene conversion covers both mutations and recombinations in one single assay; (3) the yeast technique better allows a variation of distinct factors of the treatment and so can give more detailed information.

The yeast system therefore is proposed as an assay that should be included in screening procedures for mutagens/carcinogens.

## ACKNOWLEDGMENTS

U. Bayer is recipient of an EURATOM-grant. This work was carried out under Contract No. EC-144-77-1 ENVD of the E. C. Environmental Research Program and the Bundesministerium für Forschung und Technologie (No. MT 420).

## REFERENCES

1. Ames, B. N., Durston, W. E., Yamasaki, E., and Lee, F. D. (1973): Carcinogens are mutagens: A simple test system combining liver homogenates for activation and bacteria for detection. *Proc. Nat. Acad. Sci. USA* 70:2281-2285.
2. Ames, B. N., McCann, J., and Yamasaki, E. (1975): Methods for detecting carcinogens and mutagens with the Salmonella/mammalian-microsome mutagenicity test. *Mutation Res.* 31:347-364.
3. Ames B. N. (1979): Identifying environmental chemicals causing mutations and cancer. *Science* 204:587-593.
4. Bentley, P., Oesch, F., and Glatt, H. (1977): Dual role of epoxide hydratase in both activation and inactivation of benzo(a)pyrene. *Arch. Toxicol.* 39:65-75.
5. Callen, D. F., and Ong, T. M. (1978): Effects of the epoxide hydrase inhibitor, 1,1,1-trichloropropane-2,3-oxide on the genetic activity of aflatoxin $B_1$ metabolites in in vitro activation test systems. *Mutation Res.* 49:371-376.
6. Glatt, H. R., Vogel, K., Bentley, P., and Oesch, F. (1979): Reduction of benzo(a)pyrene mutagenicity by dihydrodiol dehydrogenase. *Nature* 277:319-320.
7. Kastenbaum, M. A., and Bowman, K. O. (1970): Tables for determining the statistical significance of mutation frequencies. *Mutation Res.* 9:527-549.
8. McMahon, R. E., Cline, J. C., and Thompson, C. Z. (1979): Assay of 855 test chemicals in ten tester strains using a new modification of the Ames test for bacterial mutagens. *Cancer Res.* 39:682-693.
9. Marquardt, H. (1968): Prüfung chemischer Verbindungen auf erbändernde Wirkung: Die Hefe als Versuchsobjekt. *Umschau Wissenschaft und Technik* 21:657-659.
10. McCann, J., Choi, E., Yamasaki, E., and Ames, B. N. (1975): Detection of carcinogens as mutagens in the Salmonella/microsome test: Assay of 300 chemicals. *Proc. Nat. Acad. Sci. USA* 72:5135-5139.
11. McCann, J., and Ames, B. N. (1976): Detection of carcinogens as mutagens in the Salmonella/microsome test: Assay of 300 chemicals: Discussion. *Proc. Nat. Acad. Sci. USA* 73:950-954.
12. Murthy, M.S.S. (1979): Induction of gene conversion in diploid yeast by chemicals: correlation with mutagenic action and its relevance in genotoxicity screening. *Mutation Res.* 64:1-17.
13. Roszinsky-Köcher, G., and Röhrborn, G. (1979): Increase in genetic activity of the polycyclic hydrocarbon benzo(a)pyrene by inhibition of the epoxide hydrase in an activation system in vitro. *Mutation Res.* 66:199-203.

14. Siebert, D. (1973): A new method for testing genetically active metabolites. Urinary assay with cyclophosphamide (Endoxan, Cytoxan) and Saccharomyces cerevisiae. *Mutation Res.* 17:307-314.
15. Siebert, D., and Kolar, G. F. (1973): Mutagenic activity of alkaryl triazenes: Induction of mitotic gene conversion by 3,3-dimethyl-1-phenyltriazene, 1-(3-hydroxyphenyl)-3,3 dimethyltriazene and by 1-(4-hydroxyphenyl)-3,3-dimethyltriazene in Saccharomyces cerevisiae. *Mutation Res.* 18:267-271.
16. Siebert, D., Bayer, U., and Marquardt, H. (1979): The application of mitotic gene conversion in Saccharomyces cerevisiae in a pattern of four assays, in vitro and in vivo, for mutagenicity testing. *Mutation Res.* 67:145-156.
17. Zimmermann, F. K., and Schwaier, R. (1967): Induction of mitotic gene conversion with nitrous acid, 1-methyl-3-nitro-1-nitrosoguanidine and other alkylating agents in Saccharomyces cerevisiae. *Molec. Gen. Genetics* 100:63-76.
18. Zimmermann, F. K. (1968): The effect of liquid holding on chemical induced lethality and mitotic gene conversion in Saccharomyces cerevisiae. *Molec. Gen. Genetics* 103:11-20.
19. Zimmermann, F. K. (1969): Genetic effects of polynuclear hydrocarbons: Induction of mitotic gene conversion. *Z. Krebsforsch* 72:65-71.
20. Zimmermann, F. K. (1971): Induction of mitotic gene conversion by mutagens. Mutation Res. 11:327-337.

# THE EFFECT OF ANTIOXIDANTS ON THE MUTAGENICITY OF BENZO(a)PYRENE AND DERIVATIVES

P. D. Sullivan, L. M. Calle, I. J. Ocasio,
J. D. Kittle, Jr., and L. E. Ellis
  Department of Chemistry
  Ohio University
  Athens, Ohio 45701

Antioxidants have been found to inhibit carcinogenesis in experimental animals (2,4,5,7,11,17-21,23,25-30,32). It has been suggested that there might exist a general mechanism(s) for this inhibition even though not every antioxidant exhibits inhibition (6). Such a mechanism(s) may involve a direct reaction of the antioxidant with a reactive metabolite of the carcinogen, or an indirect effect of the antioxidant on the in vivo activation and/or deactivation of the carcinogen (20). Our studies have been concerned with the investigation of possible direct mechanisms of antioxidant inhibition. Since many antioxidants act as free radical scavengers (24), and since free radical intermediates have been identified in the metabolism of benzo(a)pyrene (BaP) by rat liver microsomes (12,13,15), the possibility of a direct reaction between antioxidants and free radical forms of BaP was previously investigated using Electron Spin Resonance (ESR) spectroscopy (22). The results of that investigation suggested that a direct interaction between the antioxidants and BaP is not a likely mechanism for the inhibition of BaP carcinogenesis. However, our experiments did suggest that certain antioxidants may alter the in vitro metabolism of BaP.

In order to further study possible direct mechanisms, the effects of antioxidants on the mutagenicity of BaP and some of its substituted derivatives was investigated using the Ames test. Another aspect of our work has been concerned with the study of the effects of substitution on the properties and reactivity of BaP. In this regard the chemical and enzymatic oxidations of some alkyl-substituted derivatives to free radical products will also be discussed in this paper.

## Effect of Antioxidants on the Mutagenicity of Benzo(a)pyrene

As another approach towards clarifying possible direct mechanisms of inhibition by antioxidants, their effect on the mutagenicity of BaP towards *Salmonella typhimurium* Strain TA-98 was investigated using the Ames test. The antioxidants included several that have demonstrated inhibitory effects on carcinogenesis (4,11,17,21,23,25-30,32), i.e., butylated hydroxyanisole (BHA), butylated hydroxytoluene (BHT), ethoxyquin (EQ), and phenohiazine (PTH). The others which have not been tested in animals were: 4,4'-thiobis-(6-t-butyl-o-cresol) (Ethyl 736), Irganox 1010 (1010), and tetra-t-butylbiphenol (TBB), as well as commonly used antioxidants of the phenolic type such as tetramethoxythianthrene (TMTH), which was included because of its low oxidation potential; 4-hydroxy-2,2,6,6-tetramethyl piperidinooxy (NIT), included because of its known antioxidative properties (31); and phenazine methosulfate (PMS), which was used as a control since it interferes with electron transport processes and thereby inhibits the metabolic activation of BaP.

The experimental procedure for the Ames test was essentially that described by Ames (1) with some slight modifications. The hydrocarbon and antioxidant were dissolved in dimethyl sulfoxide (DMSO) and added to molten top agar at 45°C. Strain TA-98 of *Salmonella typhimurium* and an in vitro activation system containing an S-9 preparation from $\beta$-naphthoflavone-induced rat livers were then added. Mutagenicity assays were carried out using five plates per set of conditions, and the number of revertant colonies were counted after 48 hours incubation at 37°C. Spontaneous revertants were subtracted to obtain the number of revertants per plate due to added substrates.

The results from our initial experiments are shown in Table 1. Most of the antioxidants appeared to exert some inhibitory effects on the mutagenicity of BaP. An attempt was made to obtain dose response curves for the antioixdants BHA, BHT, EQ, PTH, and Vitamin C (Vit. C), with the results shown in Figure 1. These results indicate that BHT, BHA, and EQ become effective inhibitors only at relatively high concentrations and may, in fact, increase the mutagenicity of BaP at low concentrations. Vitamin C has essentially no effect, whereas PTH appears to be the most effective inhibitor of the in vitro mutagenicity of BaP, acting even at fairly low concentrations.

Studies also were carried out in which combinations of antioxidants were added to the test medium. Typical results from these studies are shown in Table 2. These results are difficult to interpret at present, particularly because of the large errors involved; however, the possibility of antagonistic effects occurring between different antioxidants is raised.

**TABLE 1.** Effect of Antioxidants on the Mutagenicity of BaP Towards *S. Typhimurium* (Strain TA-98)

| Antioxidant[a] | HIS Revertants/Plate | Percentage of BaP Control[b] |
|---|---|---|
| None | 238 ± 9 | 100 |
| PMS | 0 ± 4 | 0 ± 2 |
| TMTH | 110 ± 6 | 46 ± 5 |
| BHT | 135 ± 27 | 57 ± 13 |
| BHA | 137 ± 29 | 58 ± 14 |
| PTH | 140 ± 14 | 59 ± 8 |
| EQ | 174 ± 10 | 73 ± 7 |
| TBB | 183 ± 12 | 77 ± 8 |
| NIT | 187 ± 17 | 79 ± 10 |
| Ethyl 736 | 203 ± 30 | 85 ± 15 |
| 1010 | 226 ± 22 | 95 ± 12 |
| Vit C | 483 ± 15[c] | 89 ± 7 |

[a] The concentration of the antioxidant is 60 nmoles/plate.
[b] The concentration of BaP is 6 nmoles/plate.
[c] From a separate experiment.

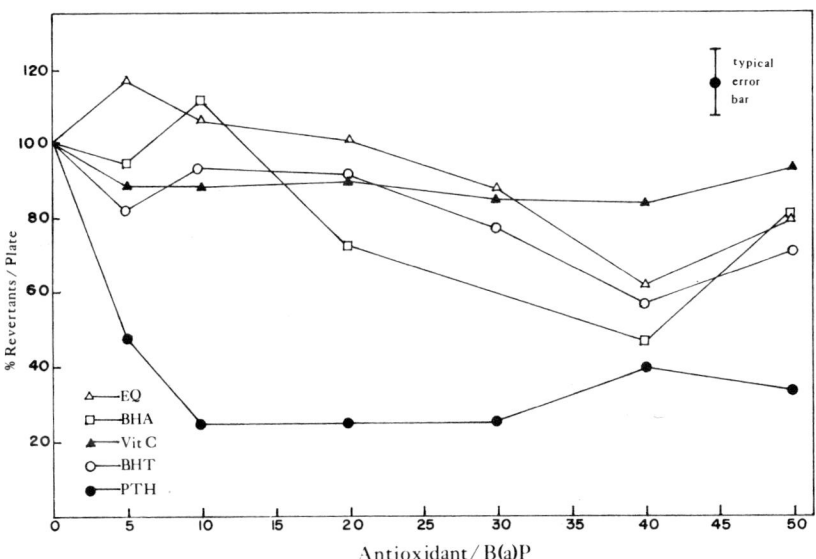

**FIGURE 1.** Dose response curves for EQ, BHA, Vit, C, BHT, and PTH.

Again, the obvious conclusion is that PTH is a more effective inhibitor than BHA, BHT, or EQ.

TABLE 2. The Effect of Combined Antioxidants[a] on the
Mutagenicity of BaP[b] Towards S. Typhimurium (Strain TA-98)

| Antioxidant | Percentage of HIS Revertants[c]/Plate | | | |
|---|---|---|---|---|
| | PTH | EQ | BHA | BHT |
| PTH | 28 ± 9 | 42 ± 18 | 60 ± 13 | 56 ± 12 |
| EQ | 42 ± 18 | 105 ± 25 | 68 ± 16 | 135 ± 17 |
| BHA | 60 ± 13 | 68 ± 16 | 73 ± 20 | 77 ± 18 |
| BHT | 56 ± 12 | 135 ± 17 | 77 ± 18 | 82 ± 17 |

[a] The concentration of the antioxidant is 60 nmoles/plate.
[b] The concentration of BaP is 6 nmoles/plate.
[c] Percentage change in the number of revertants as compared with the control value of BaP without the antioxidant.

## Effect of Substituents on the Mutagenicity of Benzo(a)Pyrene

In order to study the effects of substitution on the properties and reactivity of BaP, the mutagenicity of several substituted derivatives towards *Salmonella typhymurium* Strain TA-98 was investigated with the results shown in Table 3. Interestingly, 6-methylbenzo(a)pyrene (6-MeBaP), 8-fluorobenzo(a)pyrene (8-FBaP), 10-methylbenzo(a)pyrene (10-MeBaP), and 7-fluorobenzo(a)pyrene (7-FBaP) were found to be more mutagenic than BaP itself. These results also can be compared to the known carcinogenicity of 6-MeBaP (3), which is approximately equal to that of BaP, and 10-MeBaP (8,9,14), which is slightly less than that of BaP. No data on the carcinogenicity of 7-F and 8-F BaP are currently available.

TABLE 3. Mutagenicity of BaP and Derivatives
Towards S. Typhimurium (Strain TA-98)

| Substituted BaP Compound | Revertants/nmole |
|---|---|
| 6-Methyl | 199 ± 31 |
| 8-Fluoro | 144 ± 17 |
| 10-Methyl | 64 ± 5 |
| 7-Fluoro | 54 ± 8 |
| BaP | 41 ± 7 |
| 7-Methyl | 14 ± 2 |
| 6-Ethyl | 9 ± 1 |
| 6-Acetoxy | 8 ± 4 |
| 6-Methoxy | 6 ± 1 |
| 7,10-Dimethyl | 2 ± 1 |
| 4,5,7-Trimethyl | 1 ± 1 |

Assuming that the presence of substituents, particularly in the 7,8,9, or 10 positions and possibly also in the 6 position, will block or severely reduce the formation of a 7,8-diol-9,10-epoxide derivative, then the mutagenicity and carcinogenicity of these compounds must be attributed to other species. The next most mutagenic metabolite of BaP is the 4,5-epoxide. Hence, by analogy one might expect the mutagenicity of the substituted BaPs to have a large contribution from 4,5-epoxide derivatives. The 4,5-epoxide of BaP is not, however, carcinogenic and therefore the carcinogenicity of 6-Me and 10-MeBaP may be due to other, at present, unknown species. Further work is planned to elucidate the actual species responsible for the mutagenicity of these compounds.

## Effects of the Antioxidants BHA, BHT, EQ, and PTH on the Mutagenicity of Benzo(a)Pyrene Derivatives

The effect of the antioxidants BHA, BHT, EQ, and PTH on the mutagenicity of the 6-Me, 10-Me, 7-F, and 8-F benzo(a)pyrene derivatives was investigated. The results are shown in Table 4. The data are presented in terms of percentage revertants/plate as compared to controls without antioxidants, and the antioxidants were at concentrations 50 times that of the hydrocarbon. Again, in this case the most striking inhibition was produced by PTH, particularly against the mutagenicity of 7-F and 8-F BaP.

Among the antioxidants tested for effects on the mutagenicity of BaP and derivatives, PTH seems to be the most effective inhibitor of the mutagenicity of these compounds. PTH appears to exert some type of direct effect on the mutagenicity of BaP. Possible mechanisms for this effect might include a competitive inhibition or alteration of electron transport. Some evidence supporting these mechanisms has been obtained by means of high-performance liquid chromatography (HPLC) studies in which a qualitative decrease of BaP metabolites, produced upon incubation with rat liver microsomes in the presence of PTH, was observed. A decrease in the metabolic pathway leading to the conversion of BaP into the 6-oxy-BaP radical, via formation of 6-OH-BaP, was also observed by monitoring a decrease in the intensity of the ESR signal for the 6-oxy-BaP radical produced on incubation of BaP with rat liver microsomes and PTH (22).

Another possible inhibitory mechanism might be the result of intercalation of PTH into the DNA molecule. Such intercalation is well known for the PTH derivative, chlorpromazine (16). A number of other phenothiazines have been shown to be photomutagenic (10), indicating

TABLE 4. Effect of PTH, BHA, BHT, and EQ on the Mutagenicity of 6-MeBaP, 8-FBaP, 10-MeBaP, 7-FBaP, and BaP Towards S. Typhimurium (Strain TA-98)

| BaP Derivatives | AO/H[a] | Solvent[b] | HIS Revertants/Plate (%) | | | |
|---|---|---|---|---|---|---|
| | | | PTH | BHA | BHT | EQ |
| 6-MeBaP | 60 | 100 ± 2 | 91 ± 17 | 79 ± 8 | 64 ± 10 | 78 ± 15 |
| 8-FBaP | 50 | 100 ± 4 | 33 ± 4 | 130 ± 14 | 116 ± 14 | 130 ± 19 |
| 10-MeBaP | 50 | 100 ± 5 | 94 ± 12 | 101 ± 13 | 72 ± 11 | 107 ± 15 |
| 7-FBaP | 50 | 100 ± 4 | 20 ± 4 | 73 ± 11 | 47[c] | 85 ± 15 |
| BaP | 50 | 100 ± 6 | 69 ± 15 | 127 ± 21 | 93 ± 23 | 75 ± 21 |

[a]AO/H is the antioxidant:hydrocarbon ratio per plate.
[b]The antioxidants were added in 50 μl of DMSO.
[c]From one plate only.

that they may also intercalate with DNA. Such intercalation might block sites on the DNA, thus preventing BaP or its metabolites from binding to DNA.

## The Effect of Substituted Phenothiazines on the Mutagenicity of Benzo(A)Pyrene

Since PTH is the parent compound of a number of drugs which are widely prescribed for several medical problems including psychoses, allergies, and nausea (10), it was decided to test the inhibitory effect of several of these drugs and other substituted phenothiazines on the mutagenicity of BaP. Preliminary experiments have indicated that 2-chlorophenothiazine is more effective than PTH as an inhibitor of BaP mutagenesis and that several other derivatives have significant inhibitory effects (Table 5). These results suggest that it would be of considerable interest to carry out an epidemiological investigation to compare the cancer rates of the normal population with those of patients who have been prescribed PTH-derived drugs over a long period of time.

## Enzymatic and Chemical Oxidations of Substituted Benzo(a)Pyrene Derivatives

Another aspect of this work has involved the study of chemical and enzymatic oxidations of alkyl-substituted BaP derivatives, particularly with regard to formation of free radical products attributed to one-electron oxidation mechanisms. Results indicate that the one-electron oxidation of substituted BaPs may be considerably modified by the presence of alkyl groups. Substitution at the 6 position effectively prevents the formation of oxy radicals, except under extreme oxidizing conditions which result in the cleavage of the substituent. Steric hindrance at the 7 position makes the formation of an oxy radical somewhat more difficult, though not impossible, as shown by the ESR spectra (Figure 2) produced by oxidation with $H_2O_2$ in trifluoracetic acid. Spectrum a is from BaP, b from 10-MeBaP, c from 7-MeBaP, and d from 7,10-dimethyl BaP. The same radicals can be obtained on enzymatic oxidation of these compounds, except for 7,10-dimethyl BaP. The presence of alkyl substitutents does not affect the chemical oxidation by one-electron transfer to produce a cation radical, as shown by the ESR spectra (Figure 3) which can be produced by Thallium III oxidations of 6-ethyl, 10-methyl, and 6-methyl BaP. Certainly, this one-electron pathway may be more important for these substituted BaPs than for the parent compound.

TABLE 5. Percentage of Change in the Number of Revertants per Plate Caused by the Addition of PTH Derivatives as Compared with the Control Value for BaP

| PTH Derivatives | PTH Derivative/BaP | | | | | | | |
|---|---|---|---|---|---|---|---|---|
| | 0.25 | 0.5 | 1 | 5 | 10 | 20 | 35 | 50 |
| Phenothiazine | — | 60 ± 15 | 61 ± 10 | 48 ± 6 | 25 ± 5 | 25 ± 5 | 32 ± 5 | 34 ± 5 |
| 2-Chlorophenothiazine | 54 ± 5 | 44 ± 9 | 46 ± 9 | 25 ± 5 | 16 ± 5 | 12 ± 3 | 17 ± 3 | 20 ± 4 |
| Chlorpromazine | 49 ± 17 | 41 ± 16 | 82 ± 16 | 76 ± 22 | 81 ± 23 | a | a | a |
| Triflupromazine | 86 ± 13 | 81 ± 20 | 88 ± 24 | 71 ± 18 | 50 ± 13 | a | a | a |
| Fluphenazine | 99 ± 18 | 86 ± 15 | 90 ± 15 | 69 ± 14 | 65 ± 22 | a | a | a |
| 2-Trifluoromethyl-phenothiazine | 122 ± 22 | 113 ± 18 | 72 ± 16 | 72 ± 17 | 66 ± 16 | 68 ± 16 | 55 ± 16 | 50 ± 20 |
| Triflupromazine | 95 ± 13 | 103 ± 24 | 99 ± 16 | 85 ± 12 | 46 ± 13 | 70 ± 14 | a | a |

[a]Toxic to bacteria.

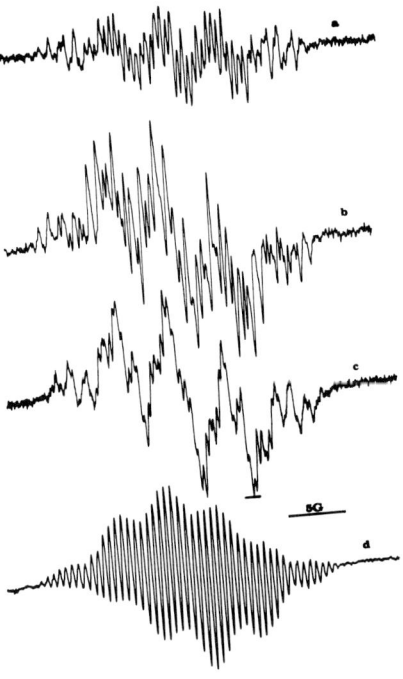

FIGURE 2. The ESR spectra obtained on oxidation with TFA in $H_2O_2$ and dichloroethane for (a) BaP, (b) 10-MeBaP at -8°C, (c) 7-MeBaP at -10°C, and (d) 7,10-DiMeBaP at -10°C.

## CONCLUDING REMARKS

Our results to date indicate that some antioxidants inhibit the mutagenicity of BaP and some of its substituted derivatives towards *Salmonella typhimurium* Strain TA-98, and that this inhibition is concentration dependent. Among the antioxidants tested, PTH is found to be the most effective inhibitor of mutagenicity. Data from the inhibition by PTH derivatives reveal that 2-chlorophenothiazine is more effective than PTH itself as an inhibitor of BaP mutagenicity. The mechanism for this inhibition is yet to be elucidated and further work is in progress to gather more information on this problem.

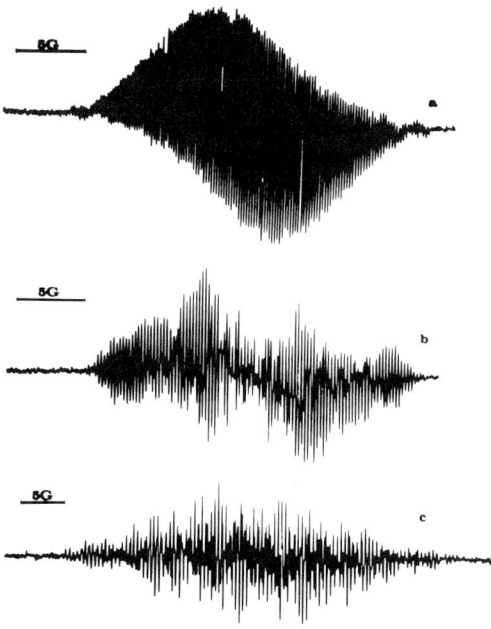

FIGURE 3. The ESR spectra obtained on oxidation with TTFA in TFA for (a) 6-EtBaP at -35°C, (b) 10-MeBaP at -16°C, and (c) 6-MeBaP at -35°C.

## ACKNOWLEDGMENTS

We would like to thank Dr. B. M. Ames for providing us with several strains of *Salmonella typhimurium*; Dr. E. Cavalieri for the 6-MeBaP, 6-MeOBaP, and 6-EtBaP samples; Dr. M. Newman for the 10-MeBaP, 7,10-dimethyl BaP, 8-FBaP, and 7-FBaP samples; and Dr. R. G. Harvey for the 7-MeBaP and 4,5,7-trimethyl BaP samples. Acknowledgment is also made to the National Institutes of Health for Grant NIH UT 2273, which has supported this research.

## REFERENCES

1. Ames, B. N., McCann, J., and Yomasaky, E. (1975): Methods for detecting carcinogens and mutagens with the Salmonella/ mammalian microsome mutagenicity test. *Mutat. Res.* 31:347-364.

2. Braun, G. G. (1975): Effect of antioxidant polyolefins on the induction of tumors by benzo(a)pyrene. *Gig. Sanit.* 6:18-22; *Chem. Abst.* 83:109590 g.
3. Cavalieri, E., Roth, R., Rogan, R., Grandjean, C., and Althoff, J. (1978): Mechanisms of tumor initiation by polycyclic aromatic hydrocarbons. In: *Carcinogenesis—A Comprehensive Survey*, Vol. 3, R. F. Freudenthal and P. W. Jones, Eds., pp. 273-284, Raven Press, New York.
4. Chan, J. T. and Black, H. S. (1978): The mitigating effect of dietary antioxidants on chemically-induced carcinogenesis. *Experimentia* 34:110-111.
5. Cumming, R. B., and Walton, M. F. (1973): Modification of the acute toxicity of mutagenic and carcinogenic chemicals in the mouse by prefeeding with antioxidants. *Food Cosmet. Toxicol.* 11:547-553.
6. Epstein, S. S., Joshi, S., Andrea, J. F., and Mantel, N. (1967): The null effect of antioxidants on the carcinogenicity of 3,4,9,10-dibenzopyrene to mice. *Life Sciences* 6:225-233.
7. Frankfort, O. S., Lipchina, L. P., and Bunto, T. V. (1967): The influence of 4-methyl-2,6-tertbutylphenol (Ionol) on the development of hepatic tumors in rats. *Bull. Exp., Biol. Med.* 8:86-88.
8. Harvey, R. G., and Dunne, B. (1978): Multiple regions of metabolic activation of carcinogenic hydrocarbons. *Nature* 273:566-568.
9. Hecht, S. S., Hirota, N., and Hoffman, D. (1978): Comparative tumor initiating activity of 10-methylbenzo(a)pyrene, 7,10-dimethylbenzo(a)pyrene and benzo(a)pyrene. *Cancer Lett.* 5:179-183.
10. Jose, J. G. (1979): Photomutagenesis by chlorinated phenothiazine tranquilizers. *Proc. Natl. Acad. Sci. USA* 76:469-472.
11. Lam, L.K.T. and Wattenberg, L. W. (1977): Effects of butylated hydroxyanisole on the metabolism of benzo(a)pyrene by mouse liver microsomes. *J. Natl. Cancer Inst.* 58:413-417.
12. Lesko, S. A., Caspary, W., Lorenzten, R., and Ts'o, P.O.P. (1975): Enzymic formation of 6-oxobenzo(a)pyrene radical in rat liver homogenates from carcinogenic benzo(a)pyrene. *Biochemistry* 14:3978-3984.
13. Lorenzten, R. J., Caspary, W. J., Lesko, S. A., and Ts'o, P.O.P. (1975): The autoxidation of 6-hydroxybenzo(a)pyrene. *Biochemistry* 14:3970-3977.
14. Miller, J. A., and Miller, E. C. Private communication.
15. Nagata, C., Inomata, M., Kodama, M., and Tagashiro, V. (1968): ESR study of the interaction between the chemical carcinogens and tissue components. III. Determination of the structure of the free radical produced either by stirring 3,4-benzopyrene with albumin or incubating it with liver homogenates. *Gann* 59:289-298.

16. Ohnishi, S., and McConnell, H. M. (1965): Interaction of the radical ion of chlorpromazine with deoxyribonucleic acid. *J. Am. Chem. Soc.* 87:2293.
17. Pamukcu, A. M., Yalciner, S., and Bryan, J. T. (1977): Inhibition of carcinogenic effects of bracken fern (*pteridium aquilinum*) by various chemicals. *Cancer* 40:2450-2454.
18. Shamberger, R. J. (1970): Relationship of selenium to cancer. I. Inhibitory effect of selenium on carcinogenesis. *J. Natl. Cancer Inst.* 44:931-936.
19. Slaga, T. J., and Bracklen, W. M. (1977): The effect of antioxidants on skin tumor initiation and aryl hydrocarbon hydroxylase. *Cancer Res.* 37:1631-1635.
20. Speier, J. L., and Wattenberg, L. W. (1975): Alterations in microsomal metabolism of benzo(a)pyrene in mice fed butylated hydroxyanisole. *J. Natl. Cancer Inst.* 55:469-472.
21. Speier, J. L., Lam. L.K.T., and Wattenberg, L. W. (1978): Effects of administration to mice of butylated hydroxyanisole by oral intubation of benzo(a)pyrene. *J. Natl. Cancer Inst.* 60:605-609.
22. Sullivan, P. D., Calle, L. M., Shafer, K., and Nettleman, M. (1978): Effect of antioxidants on benzo(a)pyrene free radicals. In: *Carcinogenesis—A Comprehensive Survey*, Vol. 3, R. F. Freudenthal and P. W. Jones, Eds.
23. Ulland, B. M., Weisburger, J. H., Yamamoto, R. S., and Weisburger, E. K. (1973): Antioxidants and carcinogenesis: BHT, but not diphenyl-p-phenylenediamine, inhibits cancer induction by N-2 fluorenylacetamide and by N-hydroxy-N-2-fluorenylacetamide in rats. *Food Cosmet. Toxicol.* 11:199-207.
24. Uri, N. (1961): Mechanisms of antioxidation. In: *Autoxidation and Antioxidants*, Vol. 1, W. O. Lundberg, Ed., Interscience, New York.
25. Wattenberg, L. W., and Leong, J. L. (1965): Effects of phenothiazines on protective systems against polycyclic hydrocarbons. *Cancer Res.* 25:365-370.
26. Wattenberg, L. W. (1972): Inhibition of carcinogenic and toxic effects of polycyclic hydrocarbons by phenolic antioxidants and ethoxyquin. *J. Natl. Cancer Inst.* 48:1425-1430.
27. Wattenberg, L. W. (1973): Inhibition of chemical carcinogen-induced pulmonary neoplasia by butylated hydroxyanisole. *J. Natl. Cancer Inst.* 50:1541-1544.
28. Wattenberg, L. W. (1974): Inhibition of carcinogenic and toxic effects of polycyclic hydrocarbons by several sulfur-containing compounds. *J. Natl. Cancer Inst.* 52:1583-1587.
29. Wattenberg. L. W. (1975): Effects of dietary constituents on the metabolism of chemical carcinogens. *Cancer Res.* 35:3326-3331.

30. Wattenberg. L. W. (1977): Inhibition of carcinogenic effects of polycyclic hydrocarbons by benzylisothiocyanate and related compounds. *J. Natl. Cancer Inst.* 58:395-398.
31. Weil, J. T., Van der Veen, J., and Olcott, H. S. (1968): Stable nitroxides as lipid antioxidants. *Nature* 219:168-169.
32. Weisburger, E. K., Evarts, R. P., and Wenk, M. L. (1977): Inhibitory effect of butylated hydroxytoluene (BHT) on intestinal carcinogenesis in rats by azoxytmethane. *Food Cosmet. Toxicol.* 15:139-141.

# MUTATION INDUCTION AT MULTIPLE GENE LOCI IN CHINESE HAMSTER OVARY CELLS: COMPARISONS OF BENZO(A)PYRENE METABOLISM BY ORGAN HOMOGENATES AND INTACT CELLS

J. H. Carver, E. P. Salazar, M. G. Knize,
D. S. Orwig, and J. S. Felton

Biomedical Sciences Division
Lawrence Livermore Laboratory
Livermore, California 94550

## INTRODUCTION*

Short-term, in vitro tests are gaining acceptance as a rapid means to assess the mutagenic and carcinogenic potential of chemicals found in the environment (3,19,21,26,27). Many of these chemicals require metabolic activation to electrophilic forms for mutagenic activity. Current mutagenesis assays rely on organ homogenates or intact cultured cells for metabolic activation. A key question is whether these activation systems duplicate, or even approximate, the response in the intact animal. The mutagenesis assay reported here seems to be a sensitive indicator of genetic damage in mammalian DNA (2). We have used four independent gene loci in our study. Because genetic loci may differ both in sensitivity and mechanism for mutational damage, the simultaneous use of several gene loci may be more effective in detecting potentially mutagenic agents or mixtures of compounds. Comparing the observed mutation induction

---

*Abbreviations in this paper include: $\alpha$-MEM, alpha minimum essential Eagle medium; AHH, aryl hydrocarbon hydroxylase; aprt, gene coding for adenine phosphoribosyltransferase, EC 2.4.2.7; ATPase, gene coding for $Na^+$ and $K^+$ activated, $Mg^{++}$-dependent ATPase, EC 3.6.1.3; 7,8 diol, 7,8 dihydroxy-7,8 dihydrobenzo(a)pyrene; 9,10 diol, 9,10-dihydroxy-9,10 dihydrobenzo(a)pyrene; 4,5 diol, 4,5 dihydroxy-4,5 dihydrobenzo(a)pyrene; hgprt, gene coding for hypoxanthine-guanine phosphoribosyltransferase, EC 2.4.2.8; HPLC, high performance liquid chromotograpy; 3-OH, 3-hydroxybenzo(a)pyrene; 9-OH, 9-hydroxybenzo(a)pyrene; tk, gene coding for thymidine kinase, EC 2.7.1.21.

with metabolite profiles produced under different conditions of in vitro metabolic activation will provide a data base leading to experiments more directly relevant to the complex in vivo bioactivation and detoxification processes.

We have compared different methods of activating benzo(a)pyrene (BaP) in an in vitro mammalian mutagenesis assay (1,2). Metabolites of BaP are known to be toxic, mutagenic, and carcinogenic (4,11,13,14,16,20,-32,33,34). The parent hydrocarbon is ubiquitous in the environment and is particularly released by processes associated with energy production, conversion, and utilization (5,9). Benzo(a)pyrene is metabolized by microsomal mixed-function oxidases, epoxide hydratase, and a variety of conjugases to form phenols, quinones, dihydrodiols, epoxides, and conjugates of the oxygenated metabolites (10,29,31,35). The 9,10-epoxides are believed to be the major mutagenic and carcinogenic metabolites of BaP (8,33).

In this paper, mutations induced by BaP are quantified at four gene loci in Chinese hamster ovary (CHO) cells: aprt, hgprt, tk, and ATPase. The mutants were selected by their resistance to 8-azaadenine (AA), 6-thioguanine (TG), 5-fluorodeoxyuridine (FUdR), or ouabain (OUA). We compare activation by liver homogenates (S9) and liver microsomes from rats induced with Aroclor® 1254, Syrian hamster embryo (SHE) cells, and kidney microsomes from male, female, and testosterone-treated female C3H/HeJ mice. Finally, a comparison is made between profiles of BaP metabolites produced by the different activation systems and the observed mammalian mutagenesis data.

## MATERIALS AND METHODS

### Preparation of Organ Homogenates and SHE Cells

For the Aroclor®-induced rat-liver material, 8-week-old outbred male rats (Simonsen albino, Sprague-Dawley derived) were injected i.p. with Aroclor® 1254 (200 mg/ml in corn oil) at 500 mg/kg body weight. Five days later, rats were sacrificed by arterial bleeding after anesthesia in a $CO_2$ atmosphere. Livers were removed under sterile conditions, washed with cold phosphate-buffered saline (PBS), and homogenized in 3 ml PBS/g wet liver weight with a Polytron homogenizer (Brinkmann Instruments) for 1 minute at setting 4. The resulting homogenate was centrifuged at 9000 x g for 15 minutes at 4° C. The supernatant (S9) was frozen and stored at -80°C. To prepare microsomes, the S9 was immediately centrifuged at 100,000 x g for 1 hour. The pellet was gently homogenized, resuspended in sterile 10 percent glycerol in PBS, and stored at -80°C. The mouse kidney microsomes were prepared from 8 to 10-week-old C3H/HeJ male and female mice. Thirty mg of testosterone pellets (kindly supplied by Dr. Richard Swank, Roswell Park Memorial Institute) were implanted subcutaneously in the backs of selected females 8 days before sacrifice. All protein concentrations were determined by the method of Lowry et al (17).

Primary Syrian hamster embryo (SHE) cells were obtained by a modification of R. J. Pienta's protocol (personal communication). Adult female hamsters were sacrificed and embryos were removed and decapitated. Embryonic tissues were dispersed into single cells by successive treatments with 0.05 percent trypsin at room temperature. Dispersed cells were incubated in flasks at 37°C for 24 hours, the debris was decanted, and the mixed cell population trypsinized and resuspended in $\alpha$-MEM culture medium containing 10 percent glycerol and 10 percent fetal calf serum. Samples were frozen in liquid nitrogen for long-term storage.

## Cell Culture and Mutagenesis Assays

The parent cells for these studies were Chinese hamster ovary (CHO) cells obtained from Dr. W. C. Dewey. A subline, CHO-AT3-2, was selected and characterized biochemically as heterozygous for both the aprt and tk loci (1), permitting single-step selection of mutants resistant to AA (aprt$^{-/-}$), TG (hgprt$^-$), FUdR (tk$^{-/-}$), or OUA (altered ATPase) (2). Approximately 18 hours before treatment with promutagen, CHO-AT3-2 cells were plated at 1 to 2 x $10^6$ cells per T25 flask. Cells were exposed for 2 hours to the promutagen in $\alpha$-MEM (serum-free) containing the activation system and the following cofactors: NADPH, 0.37 mM; NADH, 0.93 mM; NADP, 0.87 mM; glucose-6-phosphate, 6.57 mM. For activation by SHE cells, the CHO-AT3-2 cells were coincubated with lethally irradiated SHE cells and the promutagen for 48 hours at a CHO to SHE ratio of 1:1, which increased to approximately 8:1 by the end of the incubation. After either 2 hours or 48 hours of exposure, approximately 1 to 2 x $10^6$ CHO-AT3-2 cells were transferred into monolayer culture for expression of mutant (drug-resistant) phenotypes. After 3 days incubation, 3 to 5 x $10^6$ cells were placed in suspension culture for continued phenotypic expression and subsequent replating for mutant selection. Plating efficiencies ranged from 0.65 to 0.85 (data not shown). The observed mutant frequency is the ratio of mutant colonies per dish to the number of viable cells plated per dish. The induced mutant frequency is the difference between the observed mutant frequency of treated cultures and the observed spontaneous mutant frequency of untreated culture (controls).

## Assay of BaP Metabolites

For HPLC analysis, all activation systems plus cofactors were added to $\alpha$-MEM culture medium and incubated at 37°C for 15 minutes, except for kidney microsomes which were incubated for 30 minutes. To each reaction, 5.0 $\mu$Ci of random labeled [3H]BaP (diluted to a specific activity of 0.14 Ci/mmole, Amersham-Searle) was added. For analysis of SHE metabolism, the cells were incubated with BaP for 48 hours, scraped from the flask with a rubber policeman, and suspended in the culture medium for extraction. To terminate the reaction, 1 ml of cold acetone was added per ml of $\alpha$-MEM medium. The acetone-$\alpha$-MEM mixture was extracted with ethyl acetate (2 ml/ml medium), and the organic phase was removed, dried under nitrogen, and resuspended in methanol (200 $\mu$l). Twenty-five $\mu$l of this solution was used for HPLC analysis on a Hewlett Packard 1084b equipped with an RP-18 column. Metabolites were separated using a methanol-water gradient (40:60 to 90:10 in 30 minutes) at a flow rate of 4 ml/minute. Fractions were collected every 0.5 minute for 35 minutes. Peaks were identified by comparison of retention times to standards provided by IIT Research Institute, Chicago, Illinois, through the National Cancer Institute Chemical Repository. Cytochrome P-450 concentration and AHH activity was determined by the methods of Omura and Sato (25) and Nebert and Gelboin (23), respectively. The demethylase activity was determined according to Frantz and Malling (7) with measurements of formaldehyde production as described by Nash (22).

## RESULTS

### Metabolic Activation of BaP: Cell Toxicity

Figure 1 shows that the relative cell survival, or the fraction of surviving CHO-AT3-2 cells after BaP treatment, decreased with the concentration of activating protein (mg/ml). With 10 $\mu$g/ml of BaP, cell toxicity with activation by rat liver S9 and microsomes increased up to 1.50 and 0.75 mg/ml, respectively. No significant cell toxicity was observed with kidney microsomes from male mice (0.5 to 1.0 mg/ml) at BaP concentrations ranging from 20 to 150 $\mu$g/ml. Slight toxicity was seen at 100 $\mu$g/ml of BaP with kidney microsomes from untreated and testosterone-treated female mice. After 48 hours, significant toxicity ($\geq$ 20 percent relative cell survival) was observed for CHO-AT3-2 cells coincubated with SHE cells at BaP concentrations ranging from 2 to 10 $\mu$g/ml (data not shown).

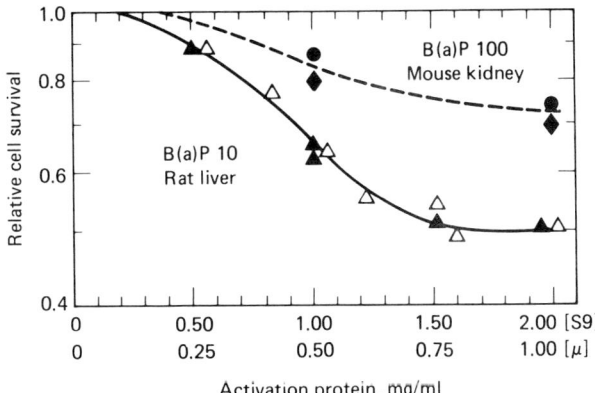

FIGURE 1. Relative cell survival of CHO-AT3-2 cells decreased with increasing concentrations of rat liver S9 (△) or microsomes (▲) or kidney microsomes from female mice untreated (●) or treated with testosterone (♦). Rat S9 and microsomes were incubated with 10 μg/ml BaP and mouse activation protein with 100 μg/ml.

## Multiple Marker Mutagenesis Assay: Drug-Resistance Markers

Figure 2 shows the increase in frequency of the induced mutations as a function of relative cell survival after treatment with BaP, activated by several metabolic activation systems. For comparison, the response to ethyl methanesulfonate (EMS) and dimethylnitrosamine (DMN) activated by rat liver S9 is given in the legend to Figure 2, expressed as a function of relative cell survival after mutagen treatment. The frequency of mutations induced after activation by rat liver microsomes was similar to that after SHE activation. Rat liver S9 produced metabolites that were apparently toxic but only slightly mutagenic to CHO-AT3-2 cells. Minimal mutagenesis was observed from BaP activated by mouse kidney microsomes. The BaP activated by kidney microsomes from male mice was neither toxic nor mutagenic; preparations from females produced some toxic responses but no significant mutation. Microsomes from females treated with testosterone slightly increased the spontaneous mutation frequencies at two out of the four loci ($OUA^R$ and $FUdR^r$). No additional toxicity or mutagenicity was observed at BaP concentrations up to 150 μg/ml.

In contrast to minimal BaP-induced mutagenesis after activation by kidney microsomes is the ability of C3H/HeJ mouse kidney microsomes to metabolize DMN to mutagenic intermediates. At ≥ 10 percent survival, preparations from male kidney yielded DMN-induced mutation frequencies approximately 3- to 22-fold higher at all four loci than those

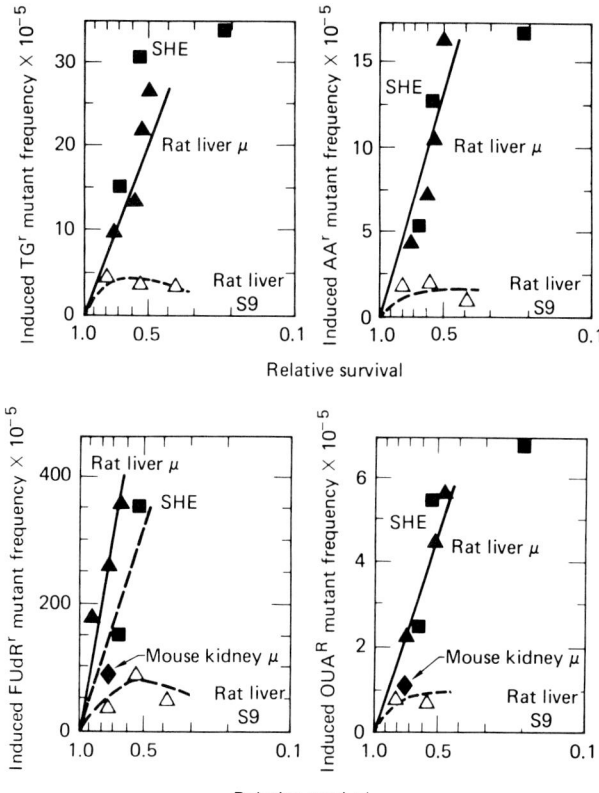

FIGURE 2. The induced frequency of AA$^r$, TG$^r$, OUA$^R$, or FUdR$^r$ mutants increased with decreasing relative cell survival (fraction of surviving CHO-AT3-2 cells) after treatment with 10 μg/ml BaP [rat liver S9 (△), microsomes (▲), and SHE (■)] and 100 μg/ml BaP [mouse kidney microsomes (◆)]. The straight-line fits of the curves for rat liver microsomes have the following slopes: AA$^r$, $4.2 \times 10^{-4}$; TG$^r$, $6.3 \times 10^{-4}$; OUA$^R$, $1.6 \times 10^{-4}$; FUdR, $2.1 \times 10^{-2}$. For comparison, the responses to EMS and DMN activated by rat liver S9 for each locus have the following slopes: AA$^r$, $3.3 \times 10^{-3}$ and $4.8 \times 10^{-4}$; TG$^r$, $4.4 \times 10^{-3}$ and $7.7 \times 10^{-4}$; OUA$^R$, $1.1 \times 10^{-3}$ and $1.6 \times 10^{-4}$; FUdR$^r$, $1.4 \times 10^{-2}$ and $1.9 \times 10^{-2}$.

of controls or with microsomes from female mice; kidney microsomes from testosterone-treated females increased DMN-induced mutations from 4- to 20-fold greater than those of controls and microsomes from untreated females (Figure 3). Differences between BaP- and DMN-induced mutation with mouse kidney microsomal activation can be seen in Table 1. Kidney microsomes were compared for a number of metabolic properties. The cytochrome P-450 content in male kidney microsomes was 8-fold higher than that in female preparations; AHH activity did not

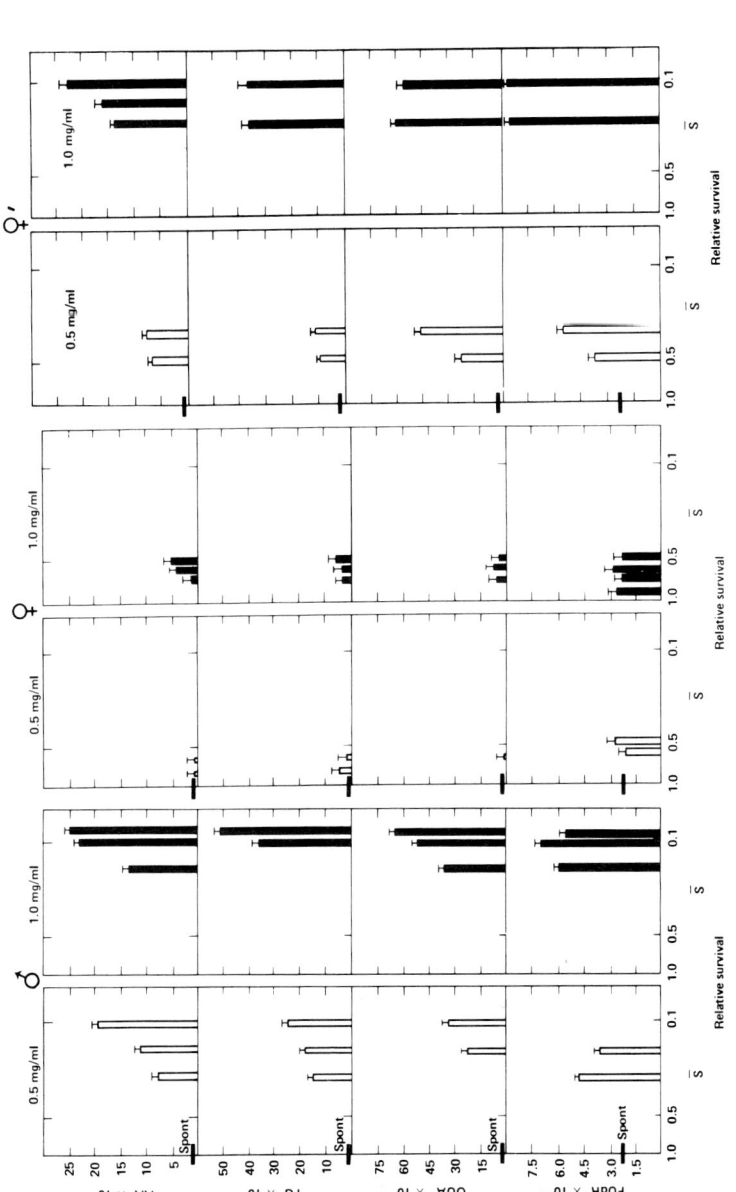

FIGURE 3. Relationship of frequencies of AA$^r$, TG$^r$, OUA$^R$, and FUdR$^r$ mutants with decreasing relative cell survival (S) after treatment with 60 to 100 mM DMN and microsomes (open bars = 0.5 mg/ml protein and closed bars = 1.0 mg/ml) from male, female, or testosterone-treated female (♀) C3H/HeJ mice. The observed, spontaneous mutation frequency at each locus is shown by the horizontal bar on the ordinate.

TABLE 1. Mouse Kidney Metabolism: Comparison of BaP and DMN

|  | Cyt. $P_{450}$[a] | DMN Demethylase[b] | AHH[c] | BaP Metabolism[d] |
|---|---|---|---|---|
| C3H ♂ | 0.39 | 2.9 | 4.0 | 11.0 |
| C3H ♀ | 0.05 | 0.4 | 4.5 | 11.4 |
| C3H ♀'[e] | 0.37 | 3.0 | 4.4 | 13.8 |

[a] Nmol P-450/mg protein.
[b] Nmol HCHO formed/mg protein/min.
[c] Pmol 3-hydroxy BaP formed/mg protein/min.
[d] Percentage of counts not found in BaP peak.
[e] Implanted with testosterone pellets.

differ in males and females. Implanting testosterone pellets in females markedly increased both cytochrome P-450 content and DMN demethylase activity but did not affect AHH activity.

## BaP Metabolite Profiles

Figure 4 shows the BaP metabolite profile obtained with 0.25 mg/ml of rat liver microsomal protein under culture conditions. Quantitation of metabolite profiles obtained with varying concentrations of rat liver S9 or microsomes or intact SHE cells is given in Table 2. Total BaP metabolism with microsomes increased with up to 0.75 mg/ml protein, then slowly decreased. Consistent levels of 3-OH, 7,8 diol, and 9,10 diol with increasing amounts of activating system indicate that steady-state levels of these compounds are formed under conditions of increasing total metabolism. In contrast, the 4,5 diol levels decrease. Tetrol or conjugate levels increase with microsomal protein concentration up to 0.75 mg/ml, implying an increase in transient levels of diol epoxides which are too reactive to be measured directly.

With rat liver S9, the total metabolism increased with increasing protein. At similar protein concentrations (0.75 mg/ml), metabolite levels are generally either similar to or higher than those measured with microsomal activation. Production of both 3-OH and 9-OH by S9 protein increases beyond the levels produced by microsomes.

In contrast to rat liver metabolism, intact SHE cells produced little or no tetrols. Amounts of water-soluble metabolites are relatively high, which may result from formation of water soluble BaP conjugates (8) during the 48-hour incubation period. The major ethyl-acetate-extractable metabolite formed is the 7,8 dihydrodiol. This intermediate is the proximal precursor to the mutagenic 7,8 dihydrodiol-9,10 epoxide.

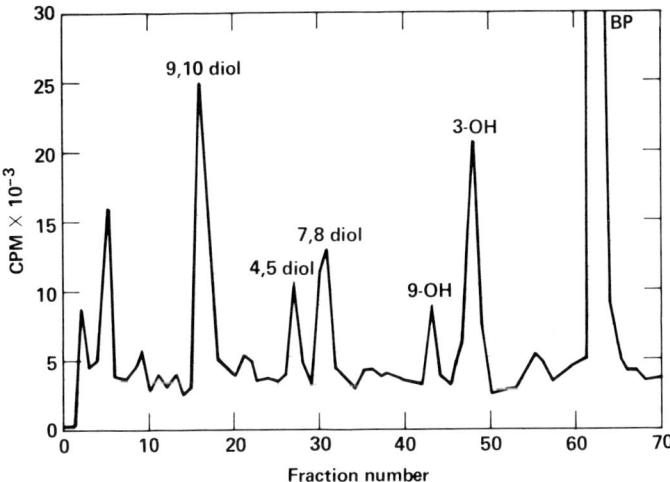

FIGURE 4. HPLC profile of Benzo(a)pyrene metabolites formed during a 15-minute incubation of [$^3$H]BaP (0.14 Ci/mmol) with 0.25 mg/ml rat liver microsomal protein. Peaks are identified by correlation of retention times with known standards.

Kidney microsomes from C3H/HeJ mice that were not induced with Aroclor were unable to generate sufficient quantities of mutagenic metabolites of BaP to yield significant mutagenesis. As seen in Table 1, total metabolite production was only 11 to 14 percent, and specific measurement of the fluorescent 3-OH product (AHH activity) was less than 5 pmol/mg protein/min, in contrast to 5,000 pmol 3-OH BaP from Aroclor®-induced rat liver microsomes (data not shown). However, cytochrome P-450 levels in male and testosterone-treated female mice are sufficient to demethylate DMN in significant quantity (Table 1). This level results in dose-dependent increases in mutation frequency at all four gene loci (Figure 3) (Felton and Carver, in preparation).

## DISCUSSION AND CONCLUSIONS

In brief, the monooxygenase system is responsible for incorporation of one atom of molecular oxygen into an aromatic substrate (in this paper BaP), resulting in an arene oxide (epoxide) intermediate; see Reference 31 for review. This reactive intermediate can, upon spontaneous rearrangement, form a phenol, be catalyzed by epoxide hydratase to a dihydrodiol or by glutathione transferase to a glutathione conjugate, or become covalently bound to nucleic acids which results in a mutagenic event. The phenols can be further conjugated to even more polar forms by UDP

## TABLE 2. Benzo(a)pyrene Metabolism in Culture

| Activation System | Protein (mg/ml) | Unident[a] | Tetrols, Conjugates[b] | Benzo(a)pyrene Metabolite Profile (cpm × 10⁻³) | | | | | | Aqueous | Unmetabolized BaP | Recovery[c] (%) |
|---|---|---|---|---|---|---|---|---|---|---|---|---|
| | | | | 9,10 diol | 4,5 diol | 7,8 diol | 9-OH | 3-OH | | | | |
| Rat Liver microsomes | 0.25 | 13.3 | 9.6 | 34.7 | 15.5 | 20.2 | 9.7 | 27.4 | | 64 | 243 | 69 |
| Rat liver microsomes | 0.50 | 12.1 | 16.2 | 35.3 | 12.1 | 22.0 | 10.8 | 30.6 | | 98 | 141 | 62 |
| Rat liver microsomes | 0.75 | 15.6 | 23.1 | 29.9 | 7.6 | 18.8 | 4.2 | 22.7 | | 122 | 86 | 56 |
| Rat liver microsomes | 1.00 | 6.9 | 10.8 | 24.7 | 7.5 | 14.5 | 17.1 | 23.0 | | 75 | 95 | 48 |
| Rat liver microsomes | 1.50 | 7.7 | 18.1 | 17.0 | 4.6 | 13.5 | 9.5 | 18.8 | | 108 | 160 | 60 |
| SHE | — | 0.01 | 0.2 | 4.5 | 7.0 | 19.5 | 0.01 | 9.0 | | 215 | 220 | 85 |
| Rat liver S-9 | 0.125 | 3.1 | 4.2 | 11.9 | 9.1 | 8.7 | 1.0 | 14.1 | | 49 | 493 | 75 |
| Rat liver S-9 | 0.250 | 8.3 | 5.8 | 18.7 | 9.6 | 17.5 | 6.5 | 23.4 | | 67 | 446 | 77 |
| Rat liver S-9 | 0.50 | 8.1 | 11.9 | 37.2 | 15.0 | 26.2 | 15.0 | 40.2 | | 95 | 255 | 65 |
| Rat liver S-9 | 0.75 | 13.8 | 13.0 | 37.6 | 14.7 | 29.5 | 18.5 | 46.9 | | 122 | 180 | 62 |

[a]Unidentified polar metabolites
[b]Presumed tetrols and sulfate conjugates
[c]Percentage of recovery equals total of organic metabolites, aqueous metabolites, and unmetabolized BP divided by radioactivity added to dish.

glucuronyl transferase or an appropriate sulfatase. The epoxide and dihydrodiol can be further metabolized by the monooxygenase system to form the very mutagenic diol epoxides. Finally, these reactive intermediates can be hydrolyzed spontaneously or be converted to tetrols by the epoxide hydratase.

Although significant differences are observed in BaP-induced mutagenesis after activation by rat liver microsomes or S9, the measured metabolite profiles are very similar. This is somewhat surprising because at any given protein concentration there is approximately 4.6 times more cytochrome P-450 in the microsomes than S9. At comparable protein concentrations, the S9 preparations produce higher levels of phenols (both 3 OH and 9-OH) than do the microsomes; these phenols may be more toxic than mutagenic to the CHO cells. This could account for the lower mutagenesis seen in Figure 2 with increasing S9 concentrations. Alternatively, recovery of metabolized BaP decreases rapidly with increasing microsomal protein concentration, suggesting more covalent binding to macromolecules (including DNA), ultimately resulting in more mutations. This loss in recovery with increasing protein concentration could also be due to covalent binding to the increased concentration of soluble proteins present in the S9. However, Krahn and Heidelberger (15) reported significant BaP-induced mutagenesis in V79 cells after activation by rat liver S9 at 2 mg/ml total protein. Similar to the Selkirk metabolite study (28), these rats were induced with 3-methylcholanthrene. The metabolite profile from liver microsomes obtained from rats induced by 3-methylcholanthrene (30) is similar to that in Figure 4, except that S9 from Aroclor®-induced rat liver homogenates shows less 3-OH in the profile.

A good comparison of covalent binding or DNA adduct formation following metabolism by microsomes and S9 may lead to a better understanding of the marked differences between the two subcellular fractions. It is apparent that steady-state levels of metabolites (such as we have measured here) may not be a good indicator of mutagenic metabolism; exact measurement of interaction with the target may be more meaningful.

The metabolite profile obtained after SHE activation in our study differs in several aspects from data previously reported (18,29). In this study, the BaP concentration and reaction time were optimized for CHO mutagenesis, i.e., 10-fold higher concentration and 48-hour incubation as opposed to 24 hour. The data in Table 2 represent a combination of metabolites in cells and in the medium; both were potential sources of metabolites for CHO cells cocultivated with SHE cells. Most significant is the high level of 7,8 diol and the low level of tetrols. In marked contrast to a previous report (18), our SHE activation yielded a larger fraction of 7,8 diol and 4,5 diol than of 9,10 diol forms and the 3-OH product exceeded the 9-OH form. The 7,8 diols, which are major metabolites, probably contributed heavily to SHE-activated BaP mutagenesis via 7,8 diol-9,10 epoxides. These diol epoxides may be more efficiently transported to the CHO target cells, with little or no accumulation of the diol epoxides and subsequent conversion to tetrols. Mutation results at the four gene loci were quantitatively similar to those obtained after BaP activation from rat liver microsomes, suggesting that both activation systems have the potential to produce significant levels of the mutagenic epoxide SHE, as evidenced by the 7,8 diol accumulation and microsomes by the presence of tetrols.

For two of the markers reported here (OUA$^R$ and TG$^r$, the BaP-induced mutation measured in CHO cells after SHE activation can be compared with results in similar assays employing a metabolizing feeder layer combined with the cells used for selection of mutant phenotypes. Huberman and coworkers (11,12) used Syrian (golden) hamster cells with V79 hamster cells. Their data for OUR$^R$ are similar to those shown in Figure 2, with results for AG$^r$ about 10-fold higher than our data for TG$^r$, consistent with the AG$^r$/TG$^r$ ratio of ten observed in our CHO system (data not shown). Newbold and colleagues (24) also used V79 cells, with BHK cells for metabolism. Their reported increase in OUA$^R$ mutants is about 20-fold higher than our results, and their data for AG$^r$ are about 25-fold above the results shown in Figure 2.

Kidney microsomes that have measurable cytochrome P-450 levels metabolize DMN sufficiently to give dose-dependent mutagenesis (Figure 3). The same microsomes metabolize very little BaP (11 to 14 percent) and produce even less potentially mutagenic products, as reflected by minimal cell toxicity and very little mutagenesis (Figure 2). Presumably the cytochrome P-450 in kidney from noninduced mice (male and testosterone-treated female) does not oxidize BaP efficiently, but does demethylate DMN efficiently.

In conclusion, we have compared profiles of BaP metabolites produced by different activation systems with the observed mutagenic activity at four gene loci in a mammalian mutagenesis system. Metabolite profiles for each activation system showed some striking similarities but also some significant differences. Although profiles of SHE-activated BaP may indeed represent metabolism that is more relevant to the intact cell than to disrupted cell homogenates (29), induced mutation levels at four separate gene loci were very similar for BaP activated by rat liver microsomes or SHE. Thus, at least for BaP, dissimilar activation systems may produce similar steady-state levels of the mutagenic epoxide intermediates. However, the frequencies of BaP mutation with S9 activation were substantially lower. Kidney microsomes from male and testosterone-treated (but not from untreated) female mice efficiently metabolized DMN, but produced insignificant amounts of BaP metabolites; mutation data paralleled these results.

The usefulness of in vitro mammalian mutagenesis tests to screen potentially mutagenic chemicals will depend in part on the applicability of the assays in detecting mutagens and promutagens from a large variety of chemical classes. For activation of BaP, rat liver microsomal preparations were as efficient as feeder layers of intact cells. Whatever the reasons for the lower efficacy of rat liver S9, it is clear from these studies that microsomal activation should be added to the usual protocol employing S9 activation of promutagens. This addition would enhance the ability of mutagenesis assays to detect mutagens that are metabolized by metabolic pathways similar to BaP.

## ACKNOWLEDGMENTS

We thank Lora Dillard and Gwen Ryan for technical assistance. The authors are especially grateful to Frederick Hatch for valuable discussions and suggestions.

Reference to a company or product name does not imply approval or recommendation of the product by the University of California or the U.S. Department of Energy to the exclusion of others that may be suitable.

This work was performed under the auspices of the U.S. Department of Energy by the Lawrence Livermore Laboratory under Contract W-7405-ENG-48 and supported by Interagency Agreements IAG-D5-E681-AN, IAG-D5-E681-AO, and IAG-D5-E681-AQ with the Environmental Protection Agency.

## REFERENCES

1. Adair, G. M., Carver, J. H., and Wandres, D. L.: Mutagenicity testing in mammalian cells, I. Derivation of a Chinese hamster ovary cell line heterozygous for the adenine phosphoribosyltransferase and thymidine kinase loci, submitted to *Mutat. Res.*
2. Carver, J. H., Adair, G. M., and Wandres, D. L.: Mutagenicity testing in mammalian cells, II. Validation of multiple drug-resistance markers having practical application for screening potential mutagens, submitted to *Mutat. Res.*
3. Clive, D. (1977): A linear relationship between tumorigenic potency in vivo and mutagenic potency at the heterozygous thymidine kinase ($TK^{+/-}$) locus of L15178Y mouse lymphoma cells coupled with

mammalian metabolism. In: *Progress in Genetic Toxicology*, D. Scott, B. A. Bridges, and F. H. Sobels, Eds., pp. 241–247, Elsevier/North Holland Biomedical Press, Amsterdam.
4. Committee on the Biological Effects of Atmospheric Pollutants (1972): *Particulate Polycyclic Organic Matter*, National Academy of Sciences, Washington, D.C.
5. Coomes, R. M. (1979): Carcinogenic testing of oil shale materials. In: *12th Oil Shale Symposium Proceedings*, pp. 100–114, Colorado School of Mines Press, Golden.
6. Felton, J. S., Olsen, L. E., Corzette, M. H., Johnson, M. E., and Carver, J. H. (1979): Strain and sex differences in a steroid inducible mouse kidney microsomal monooxygenase: Its relationship to dimethylnitrosamine metabolism and mutagenicity. *Envir. Mut.* 1:146.
7. Frantz, C. N. and Malling, H. V. (1975): Factors affecting metabolism and mutagenicity of dimethylnitrosamine and diethylnitrosamine. *Cancer Res.* 35:2307–2314.
8. Gelboin, H. V., Okuda, T., Selkirk, J. K., Nemoto, N., Yang, S. K., Rapp, H. J., and Bast, R. C. (1976): Benzo(a)pyrene metabolism: Enzymatic and liquid chromatographic analysis and application to human tissues. In: *Fundamentals in Cancer Prevention*, N. Magee, S. Takayama, T. Sugimura, Eds., pp. 167–190, University Park Press, Baltimore.
9. Guerin, M. K., (1978): Energy sources of polycyclic aromatic hydrocarbons. In: *Polycyclic Hydrocarbons and Cancer, 1, Environment, Chemistry, and Metabolism*, H. V. Gelboin and P.O.P. Ts'o, Eds., pp. 3–42, Academic Press, New York.
10. Heidelberger, C., (1975): Chemical carcinogenesis. *Annu. Rev. Biochem.* 44:79–121.
11. Huberman, E., and Sachs, L. (1976): Mutability of different genetic loci in mammalian cells by metabolically activated carcinogenic polycyclic hydrocarbons, *Proc. Nat. Acad. Sci. U.S.* 73:188–192.
12. Huberman, E., Sachs, L., Yang, S. K., and Gelboin, H. V. (1976): Identification of mutagenic metabolites of benzo(a)pyrene in mammalian cells, *Proc. Nat. Acad. Sci. U.S.*, 73:607–611.
13. Huberman, E., Sachs, L., Yang, S. K., and Gelboin, H. V. (1976): Identification of mutagenic metabolites of benzo(a)pyrene metabolism by microsomal enzymes from rhesus liver and lung. *Cancer Res.* 37:244–249.
14. Jerina, D. M. and Daly, J. W. (1974): Arene oxides: A new aspect of drug metabolism. *Science* 185:573–582.
15. Krahn, D. F. and Heidelberger, C. (1977): Liver homogenate-mediated mutagenesis in Chinese hamster V79 cells by polycylic aromatic hydrocarbons and aflatoxins. *Mutat. Res.* 46:27–44.

16. Levin, W., Wood, A. W., Yagi, H., Dansette, P. M., Jerina, D. M., and Conney, A. H. (1976): Carcinogenicity of benzo(a)pyrene 4,5,7,8- and 9,10-oxides on mouse skin. *Proc. Nat. Acad. Sci. U.S.*, 73: 243-247.
17. Lowry, O. H., Rosebrough, N. J., Farr, A. L., and Randall, R. J. (1951): Protein measurement with the folin phenol reagent. *J. Biol. Chem.* 193:265-275.
18. MacLeod, M. C., Cohen, G. M., and Selkirk, J. K. (1979): Metabolism and macromolecular binding of the carcinogen benzo(a)pyrene and its relatively inert isomer benzo(e)pyrene by hamster embryo cells. *Cancer Res.* 39:3463-3470.
19. McCann, J. and Ames, B. N. (1977): The *Salmonella* microsome mutagenicity test: Predictive value for animal carcinogenicity. In: *Origins of Human Cancer Book C Human Risk Assessment*. H. H. Hiatt, J. D. Watson, and J. A. Winsten, Eds., pp. 1431-1450, Cold Spring Harbor Laboratory, Cold Spring Harbor.
20. McCann, J., Choi, E., Yamasaki, E., and Ames, B. N. (1975): Detection of carcinogens as mutagens in the *Salmonella*/microsome test: Assay of 300 chemicals. *Proc. Nat. Acad. Sci. U.S.*, 72: 5135-5139.
21. Meselson, M., and Russell, K. (1977): Comparisons of carcinogenic and mutagenic potency. In: *Origins of Human Cancer, Book C Human Risk Assessment*, H. H. Hiatt, J. D. Watson, and J. A. Winsten, Eds., pp. 1473-1481, Cold Spring Harbor Laboratory, Cold Spring Harbor.
22. Nash, T., (1953): The colorimetric estimation of formaldehyde by means of the Hantzseh reaction. *Biochem. J.* 55:416-421.
23. Nebert, D. W. and Gelboin, H. V. (1968): Substrate-inducible microsomal aryl hydroxylase in mammalian cell culture. *J. Biol. Chem.* 243:6242-6249.
24. Newbold, R. F., Wigley, C. B., Thompson, M. H., and Brookes, P. (1977): Cell-mediated mutagenesis in cultured Chinese hamster cells by carcinogenic polycyclic hydrocarbons: Nature and extent of the associated hydrocarbon-DNA reaction. *Mutat. Res.* 43:101-116.
25. Omura, T., and Sato, R. (1964): The carbon monoxide-binding pigment of liver microsomes. *J. Biol. Chem.* 239:2370-2378.
26. Purchase, I.F.H., Longstaff, E., Ashby, J., Styles, J. A., Anderson, D., Lefevre, P. A., and Westwood, F. R. (1976): An evaluation of 6 short-term tests for detecting organic chemical carcinogens. *Br. J. Cancer* 37:873-959.
27. Rinkus, S. J., and Legator, M. S. (1979): Chemical characterization of 465 known or suspected carcinogens and their correlation with mutagenic activity in the *Salmonella typhimurium* system. *Cancer Res.* 39:3289-3318.

28. Selkirk, J. K. (1977): Divergence of metabolic activation systems for short-term mutagenesis assays. *Nature* 279:604–607.
29. Selkirk, J. K. (1977): Benzo(a)pyrene carcinogenesis: A Biochemical selection mechanism. *J. Toxicol. Environ. Health* 2:1245–1258.
30. Selkirk, J. K., Roy, R. G., Wiebel, F. J., and Gelboin, H. V., (1976): Differences in benzo(a)pyrene metabolism between rodent liver microsomes and embryonic cells. *Cancer Res.* 36:4476–4479.
31. Sims, P., and Grover, P. L. (1974): Epoxides in polycyclic aromatic hydrocarbon metabolism and carcinogenesis. *Adv. Cancer Res.* 20:165–274.
32. Wislocki, P. G., Wood, A. W., Chang, R. L., Levin, W., Yagi, H., Hernandez, O., Jerina, D. M., and Conney, A. H. (1976): High mutagenicity and toxicity of a diol epoxide derived from benzo(a)pyrene. *Biochem. Biosphys. Res. Commun.* 62(31):1006–1012.
33. Wood, A. W., Goode, R. L., Chang, R. L., Levin, W., Conney, A. H., Yagi, H., Dansette, P. M., and Jerina, D. M. (1975): Mutagenic and cytotoxic activity of benzo(a)pyrene 4,5,7,8- and 9,10-oxides and the six corresponding phenols. *Proc. Nat. Acad. Sci. U.S.* 72(8):3176–3180.
34. Yang, S. K., Gelboin, H. V., Trump, B. F., Aurrup, H., and Harris, C. C. (1977): Metabolic activation of benzo(a)pyrene and binding to DNA in cultured human bronchus. *Cancer Res.* 37:1210–1215.
35. Yang, S. K., Roller, P. P., Fu, P. P., Harvey, R. G., and Gelboin, H. V. (1977): Evidence for a 2,3-epoxide as an intermediate in the microsomal metabolism of benzo(a)pyrene to 3-hydroxybenzo(a)pyrene. *Biochem. Biophys. Res. Commun.* 77:1176–1182.

# ANALYSIS OF POLYNUCLEAR AROMATIC HYDRO-CARBONS ON THE AIRBORNE PARTICULATES OF URBAN NEW JERSEY

**A. Greenberg, R. Yokoyama, P. Giorgio, and F. Cannova**
Department of Chemical Engineering and Chemistry
New Jersey Institute of Technology
Newark, New Jersey 07102

## INTRODUCTION

The New Jersey Institute of Technology Air Pollution Laboratory is engaged in a program of monitoring selected sites in New Jersey for air pollutants once every 6 days. Substances monitored include a group of toxic vapors, metals (As, Be, Cd, Hg, Mn, Ni, and Pb), and Polynuclear Aromatic Hydrocarbons (PAH). Sites include: (1) Newark (Military Park, which is situated in a cross section of main streets in the business center); (2) Rutherford (a backyard in a residential section of a suburban town); (3) Camden (top of Rutgers-Camden library in an area having no immediate vehicular traffic, but located less than 1,000 feet from the Franklin Bridge to Philadelphia); and (4) Elizabeth (near the New Jersey Turnpike Toll Booth for Exit 13). Samples at the Newark and Camden locations are collected on glass fiber filters during a 24-hour period using high-volume samplers maintained by the New Jersey Department of Environmental Protection (N.J.D.E.P.). These filters are returned to Trenton and weights of particulate matter are determined under conditions of controlled temperature and humidity. Strips from these filters are sent to our laboratory within 2 or 3 weeks. The Rutherford and Elizabeth samples are collected on glass fiber filters using high-volume samplers maintained by the New Jersey Institute of Technology. These are usually run for periods of about 8 hours. Samples are returned promptly and kept in darkness until they are analyzed. The goal is to obtain air profiles throughout New Jersey in order to aid epidemiological studies, pinpoint sources of pollutants, and evaluate pollution from "events" such as chemical warehouse or landfill fires. Additionally, such studies will allow the profiling of ambient air prior to anticipated increases in combustion of coal, synthetic fuels, and processed solid waste. Thus, new contaminants in the air will become evident when and if they are produced as combustion by-products.

## METHODOLOGY

Glass fiber filters are normally soxhlet extracted with 150 ml of methylene chloride (all solvents employed are distilled-in-glass or HPLC-grade) for a period of 8 hours*. A 1-ml aliquot of internal standard (1-methyltriptycene in acetonitrile) is added and the solution is roto-evaporated (less than 30 C) to a volume of about 10 ml, and further concentrated to about 1 ml. This concentrate is then subjected to thin-layer chromatography on a silica gel G plate with 1:1 hexane-benzene (1). The PAH fraction is detected with UV light, scraped from the plate, and washed with 5 ml of tetrahydrofuran. This solution is subsequently concentrated to about 0.1 ml. A 10-microliter aliquot is then analyzed by reversed-phase high-performance liquid chromatography (HPLC). From the completion of soxhlet extraction, the remaining workup, including HPLC analysis, is done in one day. Sample concentration is done in a graduated amber vial.

The reversed-phase HPLC analysis relies upon a modified version of a recently published procedure employing a specific octadecylsilyl (ODS) column (2). A solvent gradient ranging linearly from 40 percent aqueous acetonitrile to 100 percent acetonitrile (15-minute gradient), and thereafter at 100 percent acetonitrile, is employed at a 0.5 ml/min flow rate throughout. Samples are monitored simultaneously by means of ultraviolet detectors (280 nm and 365 nm) and a fluorescence detector (360 nm excitation; greater than 440 nm emission). Identification of peaks is made by comparisons of retention indices with those of known mixtures, coinjection of known compounds with environmental samples, and comparison of UV absorbance ratios ($Abs_{280}/Abs_{365}$) with those of known compounds. Fluorescence peaks are used for peak identification rather than for quantitation.

## RESULTS

Figure 1 is a chromatogram obtained using a standard mixture of PAH. Benzo(a)pyrene is baseline-separated from benzo(e)pyrene, perylene, and benzo(k)fluoranthene. The latter presents a problem in separation from

---

*Some types of "binderless" glass fiber filter yield a viscous liquid (upon methylene chloride extraction and subsequent concentration) which does not permit subsequent purification. Such filters exhibit some combustion in a flame and charring in the presence of hot, concentrated sulfuric acid, and clearly have an organic binder. Soxhlet extraction with hexane seems to avoid this problem.

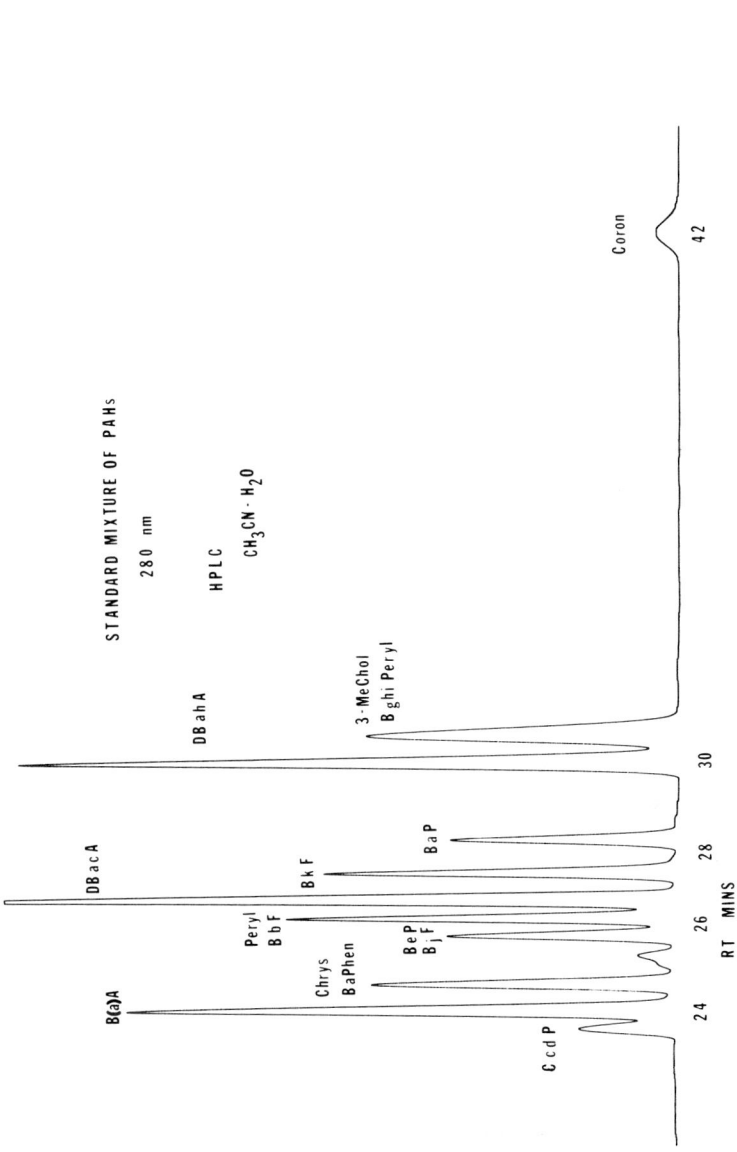

FIGURE 1. HPLC chromatogram of PAH standard; conditions are described in text [CcdP represents cyclopenteno(cd)pyrene].

BaP for many ODS columns(2). Figure 2 is a chromatogram for Rutherford, New Jersey (September 6, 1979) which is typical of the environmental samples analyzed. The benzo(a)pyrene peak appears to be homogeneous on the basis of comparison with the known (Figure 1), as well as by the near identity of the peak ratio ($Abs_{280}$ / $Abs_{365}$) with that of a pure sample.

A compound of some interest recently is cyclopenteno(cd)pyrene (CcdP) (sometimes called cyclopenta(cd)pyrene). We have obtained a sample of this compound from the Midwest Research Institute through the National Cancer Institute Chemical Repository). It is mutagenic, present in 60-fold excess relative to BaP in carbon black, and is known to be present in gasoline engine exhausts and incinerator soots (3). Our results indicate that this compound is present in concentrations less than half those of BaP. It is very photoreactive and significant decomposition may occur during collection and analysis. Table 1 lists the levels of some PAH analyzed from a sample obtained on November 17, 1979, at the Newark, New Jersey, site. The levels of PAH and their relative abundances are in reasonable agreement with other published data (4-6).

## CONCLUSIONS

The small data base presently available, involving the earlier-described analytical conditions, precludes detailed data analysis, analysis of enrichment factors and ratios, or chemical element balance (7-8). As noted by others, we find PAH levels to be lower in summer months than in winter months. This is due to a combination of factors including greater volatilization of PAH during warm weather (8-9), increased photochemical decomposition of PAH during summer months (2-3), seasonal atmospheric patterns, and increased residential heating during cold weather.

TABLE 1. PAH Levels Measured for Sample Obtained at Newark, New Jersey, Site on November 17, 1979

| Compound | Concentration ($ng/m^3$) |
| --- | --- |
| Benzo(a)pyrene | 1.5 |
| Benzo(k)fluoranthene | 1.3 |
| Beno(a)anthracene | 0.7 |
| Coronene | 1.2 |
| Benzo(ghi)perylene | 3.3 |
| Cyclopenteno(cd)pyrene | 0.5 |

# PAH IN URBAN NEW JERSEY 197

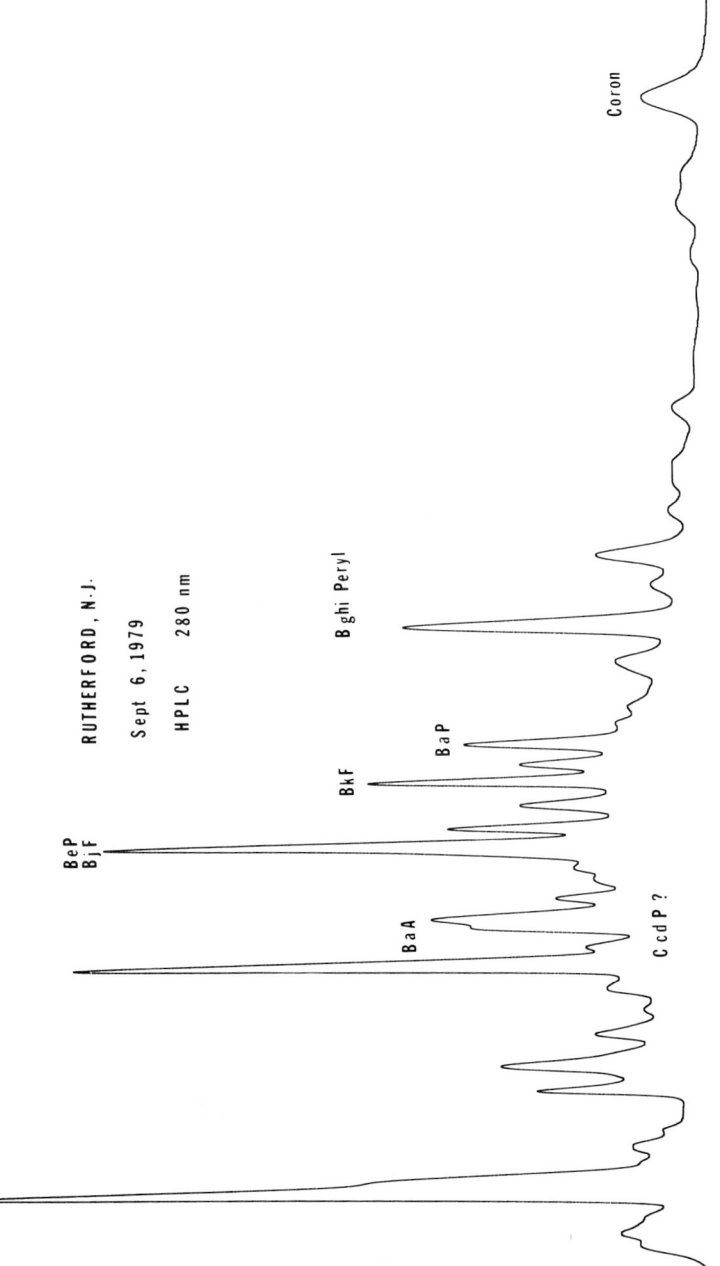

FIGURE 2. HPLC chromatogram of Rutherford, New Jersey (September 6, 1979); conditions are described in text.

## ACKNOWLEDGMENTS

The PAH analyses described here have been supported by the New Jersey Department of Environmental Protection. Analyses of toxic vapor and metals which can be compared to PAH determinations are obtained by Professors Joseph Bozzelli and Barbara Kebbekus and co-workers at the New Jersey Institute of Technology.

## REFERENCES

1. Zoccolillo, L., Liberti, A., and Brocco, D. (1972): Determination of polycyclic hydrocarbons in air by gas chromatography with high-efficiency packed columns. *Atmos. Environ* 6:715-720.
2. Ogan, K., Katz, E., and Slavin, W. (1979): Determination of polycyclic aromatic hydrocarbons in aqueous samples by reversed-phase liquid chromatography. *Anal. Chem* 51:1315-1320.
3. Wallcave, L., Nagel, D. L., Smith, J. W., and Waniska, R.D. (1975): Two pyrene derivatives of widespread environmental distribution: Cylopenta(cd)pyrene and acepyrene. *Environ. Sci. and Technol.* 9: 143-145.
4. Pierce, R. C., and Katz, M. (1975): Determination of atmospheric Isometric polycyclic arenes by thin-layer chromatography and fluorescence spectrometry. *Anal. Chem.* 47:1743-1748.
5. Fox, M. A., and Staley, S. W. (1976): Determination of polycyclic aromatic hydrocarbons in atmospheric particulate matter by high-performance liquid chromatography coupled with fluorescence techniques. *Anal. Chem.* 48:992-998.
6. Daisey, J. M., and Leyko, M. A. (1979): Thin-layer gas chromatographic method for the determination of polycyclic aromatic and aliphatic hydrocarbons in airborne particulate matter. *Anal. Chem.* 51:24-26.
7. Daisey, J. M., Leyko. M. A., and Kneip, T. J. (1979): Source identification and allocation of PAH compounds in the New York City aerosol: methods and applications. In: Polynuclear Aromatic Hydrocarbons. P. W. Jones and P. Leber, Eds., Ann Arbor Sci., Ann Arbor, Michigan.
8. Cautreels, W., and Van Cauwenberghe, K. (1978): Experiments on the distribution of organic pollutants between airborne particulate matter and the corresponding gas phase. *Atmos. Environ.* 12:1133-1141.

# REAL-TIME ANALYSIS OF GAS PHASE POLYCYCLIC AROMATIC HYDROCARBONS USING A MOBILE ATMOSPHERIC PRESSURE CHEMICAL IONIZATION MASS SPECTROMETER SYSTEM

D. A. Lane, T. Sakuma, and E.S.K. Quan
SCIEX INC.
55 Glencameron Rd. #202
Thornhill, Ontario
L3T 1P2, Canada

## INTRODUCTION

The chemistry and the pathogenic pathways of polycyclic aromatic hydrocarbons (PAH) have been the subjects of active research for more than a century.

PAH are produced primarily as a result of incomplete combustion of organic matter, and thus are believed to exist in both the vapor phase and the solid phase as an integral constituent of particulate matter. Since the quantity of such pollutants in most atmospheric samples is very low, and since they are often associated with other contaminants such as unburned fuels, phthalates, and polychlorinated biphenyls, the identification and quantitation of PAH are usually complex, time consuming, and often inaccurate because of multistep isolation and determination techniques. Most of the currently employed analytical methods involve (a) collection of particulate PAH by drawing a large volume of air through a filter, (b) extraction of the PAH collected on the filter paper with a solvent, and (c) chromatographic cleanup followed by (d) identification and quantitation using one or a combination of spectroscopic and chromatographic methods, described elsewhere (1,2).

There are a number of analytical difficulties associated with these traditional methods. The real-time analysis of PAH present in ambient air, automotive exhaust fumes, coke oven emissions, cigarette smoke, or other gaseous media cannot be achieved, mainly because of lack of selectivity, sensitivity, and mobility of the analytical instrumentation. Because of the time and physical and/or chemical manipulations employed to accomplish the analysis, the original spectrum of PAH contained in a sample could be seriously altered, and this could lead to misinterpretation of environmental problems. Lane (3,4) has shown that certain PAH deposited on the surface of particulate matter react easily with atmospheric oxidants and has described the complexity of atmospheric processes on PAH adsorbed on particulate matter. To elucidate oxidation processes, it is important that the oxidation intermediates be identified. The widely accepted high-volume sampler collects only PAH adsorbed on particulate matter and the collection process may, in fact, contribute to the decomposition or loss of the PAH. Very limited information (5) is available on vapor-phase PAH, which may be better absorbed by human tissues than particulate PAH. Thus, it is important to be able to perform real-time analyses for PAH in ambient air and emission sources without complex sample preparations.

Vo-Dinh et al (6) reported the real-time detection of naphthalene and methylnaphthalene down to 1 ppb using synchronous scanning phosphorescence techniques. Although his approach was an important step toward the realization of real-time analysis of PAH, a more sensitive instrument is required to detect higher PAH which have much lower vapor pressures.

Atmospheric-pressure chemical-ionization mass spectrometry (APCI-MS) has been shown to be a very useful tool for the ultratrace analysis of vapor-phase compounds such as ammonia, hydrogen fluoride, amines, nitrosamines, ketones, alcohols, phthalic acid esters, benzene, alkylated benzenes, phenols, explosives, and drug metabolites (7–10).

The purpose of this study was to apply this ultrasensitive APCI-MS technology to the analysis of vapor-phase PAH, and thus to eliminate the current time- and manpower-consuming techniques and to eliminate all sample manipulation procedures.

## INSTRUMENTATION

The TAGA™ (an acronym for Trace Atmospheric Gas Analyzer) instrumentation is based upon the well-established principles of chemical ionization (CI) coupled with ion detection by mass spectrometry. Chemical-ionization techniques may be employed when the ion-source pressure is maintained at or above about 1 Torr pressure. When using CI techniques, the overall sensitivity of the system increases as the ion-source pressure is increased upward from 1 Torr, and highest sensitivity is achieved when the ionization is carried out at atmospheric pressure since, at this pressure, the density of molecules and the frequency of ion-molecule collisions to form product ions are at the maximum.

To facilitate an understanding of the system, it is convenient to consider the instrument to be divided into various components as follows:

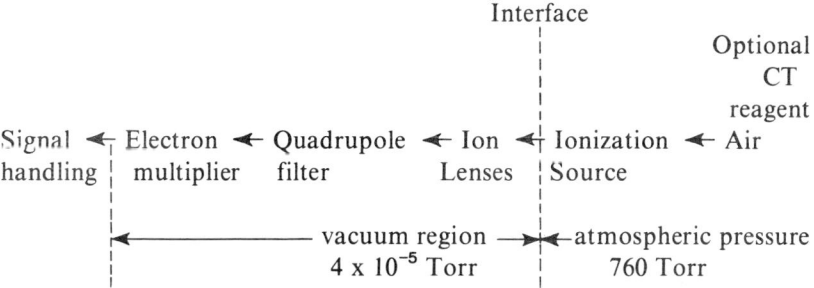

Figure 1 shows a cross-sectional schematic of the TAGA™ System. A high-capacity air pump draws a large flow of ambient air (0.1 to 5 $\ell$./sec) into a corona discharge ionization region. By choosing such flow conditions, wall-adsorption and memory-effect problems are minimized. The corona discharge is a point-to-plane source which generates large, stable currents (controllable between $10^{-7}$ and $10^{-5}$ amp) of either positive or negative reactant ions. The ions subsequently pass through a gas curtain and then through an orifice which interfaces the atmospheric-pressure

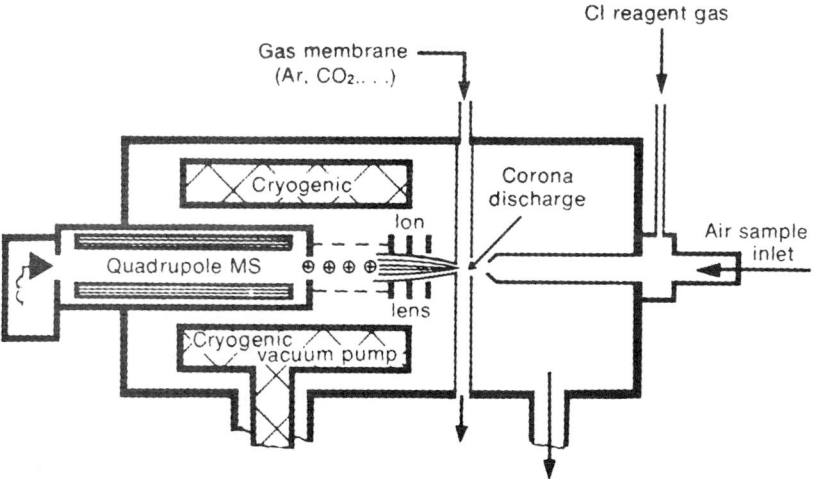

**FIGURE 1. Cross-sectional schematic of the TAGA ™ 2000 system.**

ionization region and the vacuum region which contains the mass spectrometer. The curtain gas, flowing past the orifice, serves several important functions. The gas functions primarily as a membrane through which the ions may pass while the remaining air molecules are rejected. This not only provides for efficient operation of the vacuum system but also maintains clean conditions within the mass spectrometer. An added benefit is that particulate matter in the air (which has plagued other researcher's attempts to use APCI-MS because it clogged the orifice) is directed away from the orifice by the proper selection of gas-flow conditions in this region.

The ions, after passing through the orifice, are focused by a series of electrostatic lenses into the quadrupole mass spectrometer where the ions are mass analyzed.

APCI-MS systems must handle large gas flows through the orifice interfacing the 760-Torr ion source and the $10^{-5}$ Torr pressure of the mass spectrometer. A pumping speed of 20,000 $\ell$/sec at an operating pressure of $4 \times 10^{-5}$ Torr is achieved by a small, closed-loop, two-stage helium cryo-refrigerator which is coupled to a baffled cryo-array. The TAGA™ cryo-array can operate continuously for 5 days before it is necessary to actuate an automatic overnight regeneration cycle.

## APCI-MS CHEMISTRY FOR PAH

In an ion source operated at atmospheric pressure, each ion would undergo approximately 760 times more collisions than it would if it were in a conventional 1-Torr chemical ionization source. Thus, in the TAGA™ system, there are about $10^{17}$ ion-molecule collisions per second taking place and this permits the detection of some trace compounds down to about 1 part per trillion (ppt). For ambient air analysis, APCI eliminates the sample introduction problems faced by traditional mass spectrometer systems since the air enters directly into the ion source.

In positive-mode ambient air APCI, the first reaction is the ionization of the nitrogen molecule which, in turn, transfers its charge to oxygen (since $N_2$ has a higher ionization potential than $O_2$). The $O_2^+$ ion clusters with the water vapor naturally occurring in the air and then, through a complex series of reactions, forms the hydrated proton $H_3O^+(H_2O)_n$. In the TAGA™ system, the lower proton hydrates constitute the major ions since the higher proton hydrates are declustered in the vacuum region of the system. The ability to decluster ions greatly simplifies the resulting spectra and has the added benefit of increasing the system response to the trace compounds of interest.

The trace compounds in the air then react with a proton hydrate, the net result being the transfer of a proton to the trace. The only criterion governing this reaction is that the effective proton affinity (PA) of the trace be greater than that of water. In general, this is true for any compound containing a hetero atom such as N, O, S, P, etc., and, fortunately, is also true for the PAH. In fact, it has been shown by Meot-Ner (11) that the proton affinity increases as the size of the PAH molecule increases. Since the proton affinities of the PAH (naphthalene and above) are all much greater than that of water, one would expect to see the protonated PAH ion in the spectrum.

If one were limited to the use of proton transfer reactions in ambient air APCI-MS, then the technique would be very limited indeed. This is not the case, for it is a very simple matter to alter the nature of the ionization reactions in the source, simply by adding an appropriate chemical ionization reagent gas to the carrier stream into the TAGA ™. The nature of the CI reagent added will, in turn, determine the nature of the ion-molecule chemistry which will predominate and can be selected either to highlight or to suppress a particular class of compound. If, for example, benzene is added to the inlet air stream (at a concentration of about 1 ppm), the source chemistry will be dominated by charge-transfer chemistry, and proton-transfer chemistry will be greatly suppressed simply because the added benzene ions will replace the protonated water species. Under such conditions, trace compounds which have ionization potentials (IP) lower than that of benzene will pick up the charge from benzene and will therefore be seen as the ionized parent molecular ion. Those compounds that have IPs above that of benzene will not be seen. Since all of the PAH have IPs below that of benzene, they should be seen as the parent molecular ion in the positive mode.

Thus, two methods may currently be used for the detection of PAH:

$(PAH) + H_3O^+(H_2O)_n \rightarrow (PAH)H^+ + (n+1)H_2O$   proton transfer

$(PAH) + C_6H_6^+ \rightarrow (PAH)^+ + C_6H_6$   charge transfer

## EXPERIMENTAL

All experiments were performed using a SCIEX TAGA™ 2000 APCI-MS System. The PAH were purchased from RFR Corporation and the acetone solvent and benzene CI reagent were distilled-in-glass-quality reagents. Two methods were used to add PAH vapor to the inlet air stream: one for qualitative response determination and the second for calibration and detection limit determinations.

To introduce PAH vapor for qualitative studies, a few mg of PAH was dissolved in acetone and the solution was deposited in a 19-mm-ID (22-mm-OD) glass tube. The tube was rotated to disperse the solution and the process was continued until the solvent had evaporated. Through this procedure, a 2-cm band of finely dispersed PAH was deposited on the glass tube. The glass tube was then inserted into the TAGA™ air stream line. Prepurified air from a gas bottle was passed through the tube at 0.079 ℓ/sec (10 cfh). When benzene was required as a CI reagent, ultrapure nitrogen was passed (at a flow of 1 to 10 mℓ/sec) through a Dudley bubbler containing benzene. The benzene air mixture was then bled into the air stream, upstream of the PAH tube. For each type of reaction condition (air or benzene CI), blank tests were performed before the PAH were inserted in the inlet air stream.

To obtain quantitative data and to establish detection limits in real air, a different experimental configuration was used. Outside air (indoor air contains more contaminants at higher concentrations than outdoor air!) was drawn through a glass tube 19-mm ID and 22-mm OD at a rate of 1.52 ℓ/sec. A benzene/air mixture could be added to the inlet line through a glass "Tee" in the line. Again, the benzene/air flow was adjusted to between 1 and 10 mℓ/sec.

The PAH were again dissolved in acetone and a few drops of the solution was deposited in the barrel of a 50 mℓ syringe. The syringe was rotated to coat the lower portion of the barrel with solution and taking care not to deposit solution in the needle tip. After the solvent had evaporated, the plunger was inserted into the barrel and the syringe was set aside for at least 1/2 hour to permit the vapor from the PAH to reach equilibrium with the air in the syringe. The syringe was then placed in a Sage model 341 automated syringe drive unit. The Teflon needle of the syringe (when handling such small concentrations of compounds, stainless steel needles were found to adsorb the trace and thus were not used) was passed through a second Tee piece in the inlet line and positioned so that the tip was in the center of the flow. By varying the syringe drive rate, the resulting concentration in the carrier stream could be altered. The concentration of the PAH in the line was determined from the equation:

$$C(ppb) = \frac{P_s}{P_a} \times \frac{I}{F} \times 10^9$$

where  $P_s$ is the vapor pressure of the compound in Torr at the ambient temperature
$P_a$ is the atmospheric pressure, Torr
I is the syringe injection rate (typically $<\sim 290 \times 10^{-6}$ ℓ/sec)
F is the ambient air carrier flow rate (1.52 ℓ/sec)
$10^9$ converts the concentration to parts per billion (ppb).

A plot of signal intensity (ion counts per second) vs concentration in ppb results in a linear response up to several hundred ppb before saturation begins to become apparent (a Beer's law response is observed). In the linear region, however, the slope can be used as a response factor. By dividing any observed signal for a particular PAH by the slope of its calibration curve, the concentration of that compound in the carrier may be determined. In this manner, it was possible to determine the actual concentration of PAH in the air stream when the coated glass tube "qualitative" experiments were performed.

Qualitative tests were conducted for naphthalene (Naph), fluorene (Fluor), 9-fluorenone (9-Fluor), anthracene (Anth), phenanthrene (Phen), fluoranthene (Flt) and pyrene (Pyr), whereas calibration and detection tests using the syringe technique were conducted for Naph, Fluor, Anth and Phen.

## RESULTS AND DISCUSSION

All of the PAH behaved very much as expected both under proton-transfer chemistry and under charge-transfer chemistry with benzene. Under proton-transfer chemistry, the protonated parent ion was seen together with its attendant isotope at 1 amu higher. Under benzene charge-transfer chemistry, the parent molecular ion dominated the spectrum. Because of the rather high proton affinities of the PAH, benzene charge-transfer reactions do not totally suppress the tendency of the PAH to abstract a proton.

Consequently, three peaks were occasionally seen: the $M^+$, $MH^+$ and isotope of $M^+$, and the isotope of $MH^+$. In the case of Fluor, the $(M-1)^+$ ion (the fluorenyl cation) was detected. This was the only compound for which a strong $(M-1)^+$ peak was observed. To facilitate discussion of the results each compound is considered in turn.

### Naphthalene

Because of its rather high vapor pressure ($6.69 \times 10^{-2}$ Torr at $22.5°$ C), Naph was observed through syringe injection only. In air, a rather weak $M^+$ was seen. Although the thermodynamic PA of Naph (191.3 kcal/mol) is greater than that of water, the effective PA (like that of benzene) must be less. For this reason, the Naph does not accept a proton. Figure 2(a) shows a scan from m/z = 125 to 130 and shows the weak $M^+$ peak between two impurity peaks at m/z = 127 and 129. The Naph concentration here was 25 ppb and, accepting a 3 to 1 signal-to-noise ratio as a criterion for

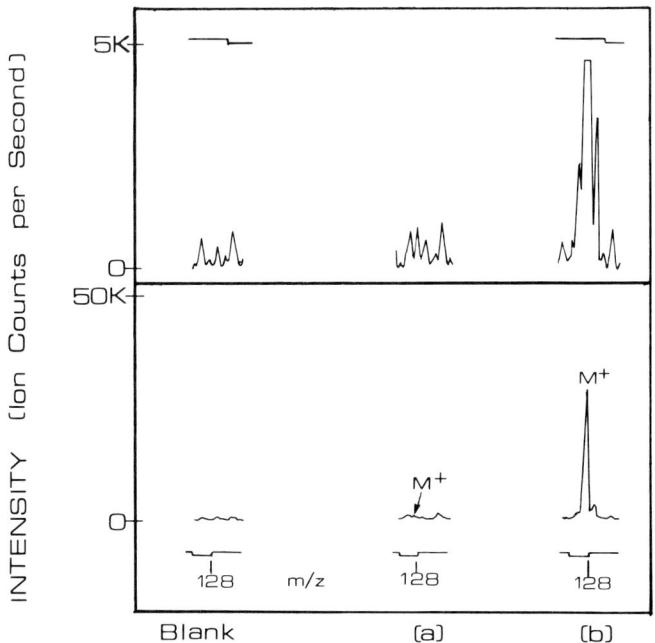

**FIGURE 2.** (a)The TAGA ™ response to 25 ppb naphthalene in air and (b) in air with benzene as the CT reagent gas. A blank scan is included in (a) showing the background signal at m/z = 128.

signal detection, the limit of detection in air is about 2 ppb with a response factor of only 26 ion counts/sec/ppb.

The scan shown in Figure 2(b) demonstrates the tremendous enhancement (about 40 times) of the parent molecular ion when using benzene charge transfer (CT) techniques. The peak at m/z = 129 is due to the $C^{13}$ isotopes of Naph, and its intensity, relative to the parent at m/z = 128, of 11.4 percent agrees well with the calculated isotope ratio of 10.96 percent. The response factor under these conditions is 970 ion counts/sec/ppb, with a detection limit of about 10 ppt or 52 ng/m$^3$.

### Fluorene

The response of the TAGA ™ to the Fluor-coated tube is shown for proton-transfer conditions in Figure 3(a) and for charge-transfer conditions in Figure 3(b). The MH$^+$ peak in Figure 3(a) is the strongest peak, as anticipated. However, there is also a relatively strong M$^+$ peak and even an (M-H)* peak of significant intensity. The M$^+$ peak is probably due to charge transfer from $O_2^+$ in the air. The (M-1)+ ion results through one or

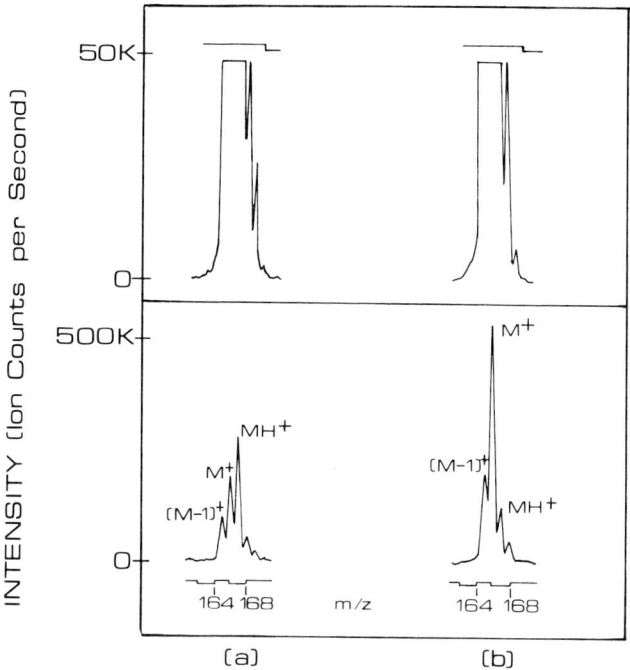

**FIGURE 3.** (a) The TAGA ™ response to 200 ppb fluorene in air and (b) the response using benzene as the CT reagent.

both of two possible mechanisms: the loss of a neutral hydrogen atom from the ionized molecule or the loss of a neutral hydrogen molecule from the protonated parent ion. The result is a very stable ion which can delocalize its charge over 7 atoms within the molecule if the standard Kekule approach is taken.

Under benzene CT conditions, the $M^+$ is enhanced over proton transfer conditions by a factor of 3, whereas the $(M-1)^+$ peak is enhanced by a factor of 2 and the $MH^+$ peak is decreased by a factor of 4. From the syringe injection of head-space vapor (vapor pressure at 22°C is 4.8 x $10^{-4}$ Torr), a detection limit of about 20 ppt was established for the charge-transfer conditions. The concentration of fluorene shown in Figure 3 was 200 ppb.

**9-Fluorenone**

The spectra obtained for 9-Fluor under the two reaction conditions are shown in Figures 4(a) and 4(b). In air, the 9-Fluor shows up very strongly as the protonated molecular ion at m/z = 181. Using benzene as a

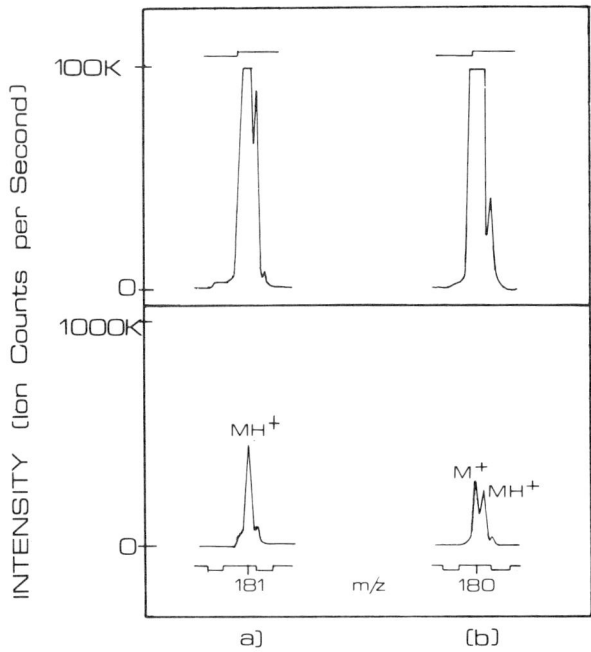

**FIGURE 4.** The trace at (a) shows the response to 9-fluorenone in air and the trace at (b) shows the response under benzene CT conditions.

charge-transfer reagent, however, the 9-Fluor results in two almost equal intensity peaks at m/z = 180 and 181, which correspond to the $M^+$ and $MH^+$, respectively. It is apparent that the proton affinity of 9-Fluor is very high (its value could not be found in the literature) since, even in the presence of the benzene charge-transfer reagent, the $MH^+$ is almost as strong as the $M^+$ ion. These two responses should provide a means for confirming the presence of 9-Fluor in the presence of ketones, aldehydes, and other compounds appearing at the same mass under proton-transfer conditions but which would not show any charge-transfer ion.

The vapor pressure at 9-Fluor could not be found in the literature and therefore it was not possible to quantitate its response. Its detection limit, however, should be less than 50 ppt.

### Phenanthrene

The air spectrum of phenanthrene, shown in Figure 5(a), demonstrates that Phen will not only accept a proton, but will also charge transfer in air. The ratio of the M$^+$ to MH$^+$ was constant for the experimental conditions and had a value of 0.69 ± .04.

When benzene was added to the system, the trace shown in Figure 5(b) was obtained. Here, the M$^+$ peak is the dominant peak, but the peak at m/z = 179 is too large to be due entirely to isotopes. For Phen, the (M+1)/M isotopic ratio was calculated to be 15.32 percent. However, in the scans, the ratio was 20.8 ± 0.6 percent, indicating that this peak was due, at least in part, to proton transfer. In view of the relatively high proton affinity of Phen (201.1 kcal/mol), this was not surprising. This ratio may be utilized to distinguish Phen from Anth as will be shown later. Benzene charge-transfer chemistry yields a signal enhancement of about a factor of 2.6.

Using the syringe injection technique, a detection limit of approximately 50 ppt (or 360 ng/m$^3$) was established for Phen which, in turn, shows that the peaks shown in Figures 5(a) and 5(b) were produced by a Phen concentration of 86 ppb. The vapor presure of Phen is 1.4 x 10$^{-4}$ Torr at 23°C.

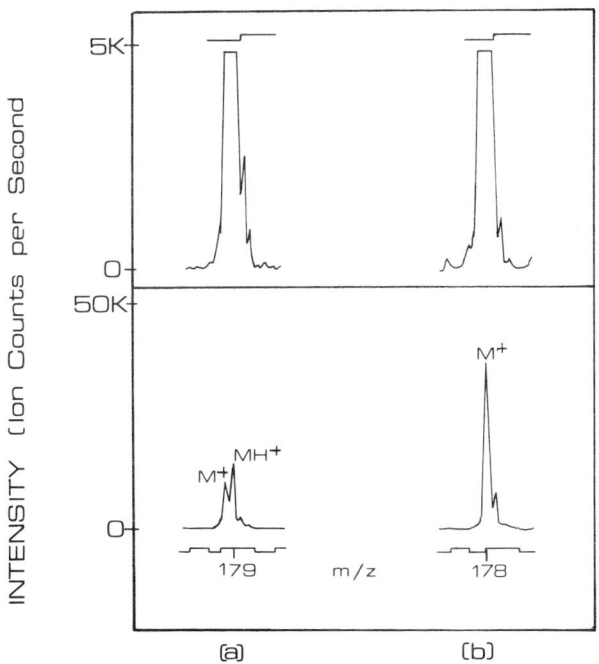

**FIGURE 5.** (a)The APCI-MS trace for 86 ppb phenanthrene in air and (b) with benzene as the charge-transfer reagent.

## Anthracene

The APCI-MS response to Anth was significantly different from that of Phen. The respective proton-transfer and charge-transfer responses are shown in Figures 6(a) and 6(b). In air, Anth shows both the M and an MH peak; the ratio of the $M^+$ to $MH^+$ is $0.67 \pm .05$ and is essentially the same as that for Phen. However, in the presence of benzene as a CT reagent, the ratio of the m/z = 179 to m/z = 178 peaks is $30 \pm 1.8$ percent (isotope effects would account for only 15.32 percent). This would suggest that benzene cannot totally suppress the proton-transfer mechanism. The fact that the Anth ratio is higher than that for Phen reflects the higher PA of Anth.

Thus, it appears that one has, in APCI-MS using benzene as a CT reagent, an empirical method for determining the PA for a particular set of PAH isomers. If the (M+1) to M ratio for a compound whose PA is not known is determined in the presence of benzene, then by comparing this ratio with the values of other isomers for which the PA is known, an approximate value of the PA may be inferred. This same sort of relation will be shown to hold for the isomers Flt and Pyr.

Quantitatively, the response of Anth (vapor pressure $2.9 \times 10^{-4}$ Torr at $22°C$) indicated that it should be detected at concentrations down to and perhaps below 50 ppt. The signals shown in Figure 5 were generated from an Anth concentration of 7 ppb.

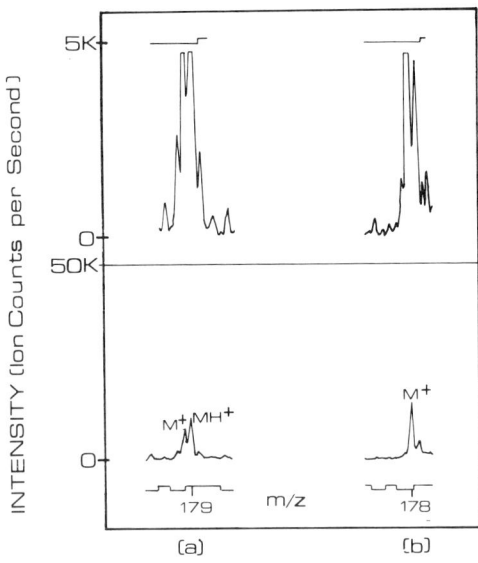

FIGURE 6. (a) The APCI-MS trace for 7 ppb anthracene in air and (b) with benzene as the CT reagent.

## Fluoranthene

Since the vapor pressure of Flt over the solid material could not be found, only a qualitative assessment of the Flt response can be made. However, since the melting point of Flt (111°C) is lower than that of Pyr (156°C), the vapor pressure of Flt is likely to be greater than that of Pyr. The response in air is shown in Figure 7(a) and shows a very weak proton-transfer peak at m/z = 203. However, the peak is very strongly enhanced by the addition of benzene vapor [Figure 7(b)] by a factor of almost 7 times.

Again, if one compares the ratio of the M+1 to M peaks in the presence of benzene, the value of 20.9 ± 0.8 percent is obtained. If only isotopes were involved, then the peak would be 17.48 percent. Thus, the proton-transfer reaction is taking place even under charge-transfer conditions.

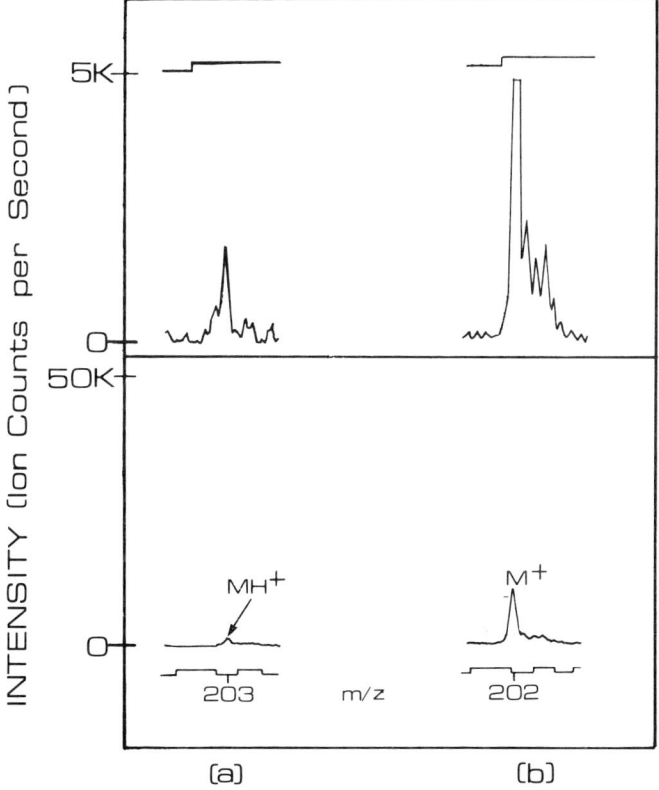

**FIGURE 7.** (a) The APCI-MS traces obtained for fluoranthene in air and (b) with benzene as a charge-transfer reagent.

## Pryene

Pyr shows up much better in air [Figure 8(a)] than does Flt. This may be attributed directly to the higher effective proton affinity of Pyr. Under the influence of benzene, the trace shown in Figure 8(b) was obtained. A measurement of the M+1 to M peak intensities under benzene CT conditions yields a value of 26.6 ± 1.6 percent. This, compared with the expected 17.48 percent due to isotope effects, would indicate that Pyr is much more successful in competing for a proton under charge transfer conditions than is Flt. Again, this indicates the higher PA of Pyr over Flt and lends more support to the concept that such an experimental test can yield approximate PA values for various PAH within a particular isomeric group.

Under the particular conditions used for the syringe injections, Pyr did not yield a response. This was not surprising since the highest concentration of Pyr possible with the syringe injection technique would be 0.92 ppt. The vapor pressure of Pyr at 20°C is $3.6 \times 10^{-6}$ Torr. Although not known precisely, it is estimated that the concentration of Pyr present in Figure 8(a) and (b) is in the vicinity of 10 to 50 ppt.

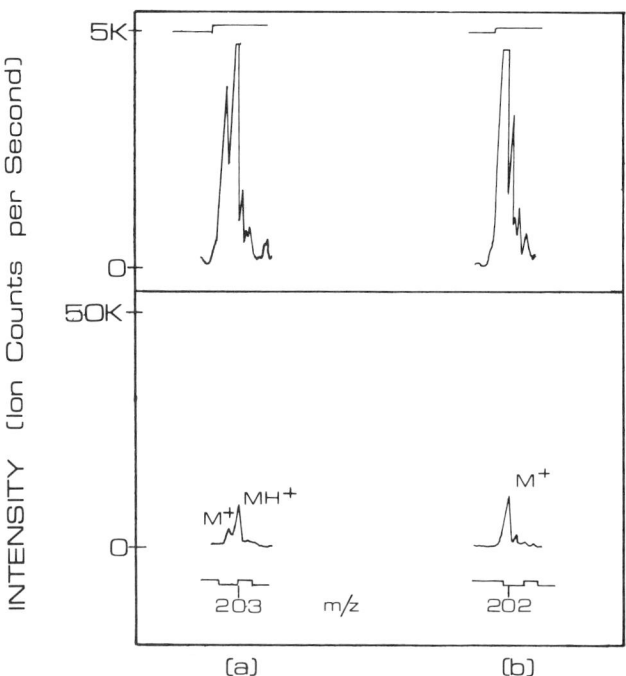

**FIGURE 8.** (a)The APCI-MS traces obtained for pyrene in air and (b) with benzene as the charge-transfer reagent.

## CONCLUSIONS

Although the information presented in this paper is very preliminary in nature, it does demonstrate the tremendous potential available in Atmospheric Pressure Chemical Ionization Mass Spectrometry for the detection of ambient, gaseous PAH in real time. Current detection limits for most of the PAH studied appear to be in the low ppt range. Work is currently under way to extend the list of compounds to include benzo(a)pyrene and its isomers as well as other PAH. Efforts are also being made to improve the methodology and to assess and, if possible, to eliminate interferences which might appear in the analysis of "real-world" samples.

The current literature contains a great paucity of fundamental information concerning the PAH commonly found associated with atmospheric particulate matter. Good vapor pressure data for PAH at normal atmospheric temperatures and pressures is almost nonexistent.

As has been shown in the case of the isomer pairs—Phen and Anth and Flt and Pyr—the ratio of the M+1 to M peaks under charge-transfer reaction conditions appears to be a useful means for empirically determining PAs of compounds within a particular isomeric series. Further work in this area is under way.

Since the PAH already have high PAs, it is likely that the oxidized species may have even higher PAs. Consequently, APCI-MS should prove itself to be an indispensable technique for tracking down the (presently) elusive PAH reaction products suspected to exist in the air.

It should be emphasized that APCI-MS is a means of performing essentially instantaneous analyses. Since there are no sample collection, manipulation, and processing procedures involved, compound losses and chemical alteration, which inherently plague traditional techniques, are virtually nonexistent.

## REFERENCES

1. National Academy of Sciences (1972): Biological effects of atmospheric pollutants: particulate polycyclic organic matter.
2. Hertz, H. S., Chesler, S. N., Eds. (1979): Trace organic analysis: A new frontier in analytical chemistry. National Bureau of Standards Special Publication 519, U. S. Government Printing Office.
3. Lane, D. A. (1975): Ph.D. Thesis, York University, Toronto, Ontario, Canada.
4. Lane, D. A., and Katz, M. (1975): The Photomodification of benzo(a)pyrene, benzo(b)fluoranthene and benzo(k)fluoranthene under

simulated atmospheric conditions. In: *Fate of Pollutants in the Air and Water Environments Part 2,* I. H. Suffet (Ed.), pp. 137–154, John Wiley and Sons, New York.
5. Boublik, T., Fried, V., and Hala, E. (1973): *The Vapor Pressures of Pure Substances.* Elsevier Scientific Publishing Co., New York.
6. Gammage, R. B., Vo-Dinh, T., Hawthorne, A. R., Thorngate, J. H., and Parkinson, W. W. (1978): A New generation of monitors for polynuclear aromatic hydrocarbons from synthetic fuel production. In: *Carcinogenesis,* Vol. 3, P. W. Jones and R. I. Freudenthal, Eds., pp. 155–174, Raven Press, New York.
7. Lovett, A. M., Reid, N. M., Buckley, J. A., French, J. B., and Cameron, D. M. (1979): Real-time analysis of breath using an atmospheric pressure ionization mass spectrometry. *Biomed. Mass Spectrom.* 6(3): 91-97.
8. Thomson, B. A., Davidson, W. R., and Lovett, A. M. (1979): Applications of a versatile technique for trace analysis—atmospheric pressure negative chemical ionization. *Environ. Health Perspectives* (accepted for publication).
9. Lane, D. A., Thomson, B. A., Lovett, A. M., and Reid, N. M. (1979): Real-time tracking of industrial emissions through populated areas using a mobile APCI mass spectrometer system. Presented at the 8th International Mass Spectrometry Conference, Olso, Norway.
10. Thomson, B. A., Sakuma, T., Fulford, J. E., Lane, D. A., Reid, N. M., and French, J. B. (1979): Fast in-situ measurement of PCB levels in ambient air to the $ng/m^3$ level using a mobile APCI mass spectrometer system. Presented at the 8th International Mass Spectrometry Conference, Oslo, Norway.
11. Meot-Ner, M., Hunter, E. P., and Field, F. H. (1979): Bonding energies and resonance stabilization in dimer ions of polycyclic aromatics. Presented at the 27th Annual Conference on Mass Spectrometry and Allied Topics, Seattle, Washington.

# RAT MAMMARY GLAND VERSUS MOUSE SKIN: DIFFERENT MECHANISMS OF ACTIVATION OF AROMATIC HYDROCARBONS

E. Cavalieri*, D. Sinha**, and E. Rogan*

*Eppley Institute for Research in Cancer
University of Nebraska Medical Center
Omaha, Nebraska 68105
**Roswell Park Memorial Institute, Buffalo, N.Y. 14263

## INTRODUCTION

It is axiomatic to assume that covalent binding of chemical carcinogens to macromolecular cellular targets is the triggering process in the series of events leading to cancer. Binding of polycyclic aromatic hydrocarbons (PAH) requires metabolic activation by monooxygenase enzymes. In recent years it has been proposed that the ultimate carcinogenic forms of PAH are vicinal diol epoxide derivatives (13,14,15,24). A considerable amount of data supporting this mechanism is available for unsubstituted, as well as alkyl-substituted, PAH. It has become increasingly evident, however, that the complexity of PAH carcinogenesis can best be understood in terms of multiple mechanisms of activation producing different ultimate metabolite(s) (3). The predominance of one mechanism over another depends upon the nature of the activating enzymes present in the target tissues of various animal species.

In addition to diol epoxides, other electrophilic intermediates, such as radical cations of benzylic carbonium ions of alkylated PAH generated via formation of reactive benzylic esters, have been postulated as important ultimate carcinogenic forms of PAH.

Hydroxylation of the benzylic carbon atom is a major pathway in the metabolism of alkyl-substituted PAH (Figure 1). These hydroxy derivatives are generally esterified with glucuronic acid, phosphate, acetate, or sulfate moieties to form stable polar compounds that can be

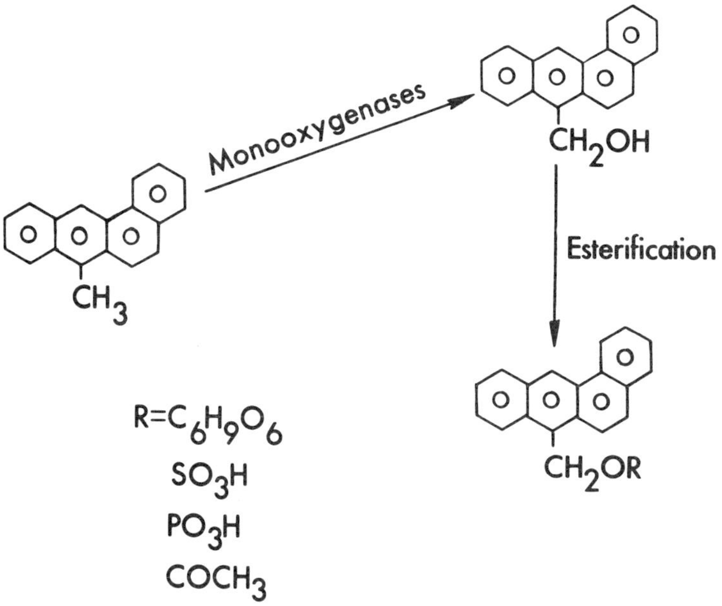

**FIGURE 1.** Hydroxylation and esterification of a benzylic carbon atom to form a possible reactive alkylating compound.

rapidly excreted. However, when a benzylic ester is formed and bears a good leaving group, it may become a potential ultimate carcinogenic metabolite by generating a reactive carbonium ion capable of binding to cellular nucleophiles (4,10,23,27). To establish the relevancy of this mechanism we have studied the carcinogenicity on mouse skin of 6-methylanthanthrene and 6,12-dimethylanthanthrene relative to the inactive parent compound anthanthrene. In this case, the metabolic activation leading to carcinogenesis for the two methyl derivatives might arise from hydroxylation of the methyl group, followed by esterification to form a reactive alkylating agent. It was previously reported that 6-methylanthanthrene and 6,12-dimethylanthanthrene were carcinogenic by subcutaneous injection in rats (16), whereas anthanthrene was inactive. We present in this paper similar results on mouse skin by initiation-promotion and repeated application.

The hypothesis that one-electron oxidation of PAH to radical cations is an important mechanism of activation leading to carcinogenesis is based on several lines of evidence (6,7,12,21,22,28). A general scheme illustrating one-electron oxidation of a hydrocarbon and subsequent binding to a nucleophile is outlined in Figure 2. Removal of one electron from the $\pi$ system generates a radical cation in which the charge distribution varies with the structure of the hydrocarbon. The position(s)

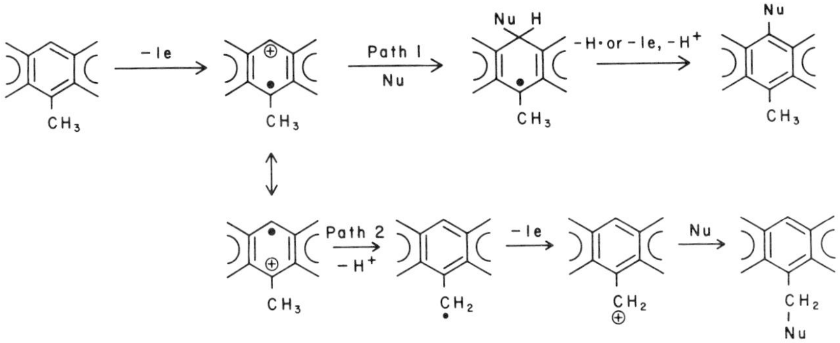

# MECHANISMS OF ACTIVATION OF AROMATIC HYDROCARBONS 217

FIGURE 2. Mechanism of trapping of a PAH radical cation by a nucleophile (Nu).

of highest charge density is the most susceptible to nucleophilic attack. In path 1, the nucleophile attacks the carbon atom on the hydrocarbon nucleus. The resulting radical may then lose a hydrogen atom or be oxidized to an arenonium ion with loss of a proton to complete the substitution reaction. When charge density at the carbon atom adjacent to the methyl group is appreciable, loss of a methyl proton (path 2) yields a benzylic radical which is rapidly oxidized to a benzylic carbonium ion with subsequent trapping by a nucleophile.

Activation of the carcinogens N-hydroxy-2-acetylaminofluorene (1,2,11) and benzo(a)pyrene (BaP) (21) via the one-electron oxidation mechanism was observed in vitro with horseradish peroxidase, as well as in vivo with mammary peroxidase (20) for the former compound. These results led us to postulate that rat mammary gland might activate PAH selectively by one-electron oxidation. With that in mind, we studied by direct application on the mammary gland the carcinogenic effect of a series of key representative PAH with different ionization potentials (IPs). The carcinogenicity of these hydrocarbons was compared with that of the same compounds in mouse skin. The results obtained are encouraging in terms of possible development of the mammary gland as a target organ in which one-electron oxidation is the selective mechanism of activation of PAH.

## RESULTS AND DISCUSSION

### Carcinogenicity of Anthanthrene, 6-Methylanthanthrene, and 6,12-Dimethylanthanthrene on Mouse Skin by Repeated Application and Initiation-Promotion

As shown in Table 1, the carcinogenic activity of the compounds was tested on CD-1 mice from the Eppley Colony. Only 6-methylanthanthrene at the high dose level elicited a carcinogenic response, while

TABLE 1. Skin Tumors in CD-1 Mice by Repeated Application of Anthanthrene Derivatives[a]

| Substance and Dose | Effective No. of Mice[b] | Tumor Incidence (%) | Survival, Mean ± S.D. (weeks) | Age at First Tumor Appearance, Mean ± S.D. (weeks) | No. of Papillomas / No. of TBA[b] | No. of Carcinomas / No. of TBA[c] |
|---|---|---|---|---|---|---|
| 0.6 μmole |  |  |  |  |  |  |
| Anthanthrene | 28 | 0 | 65 ± 12 | — | 0 | 0 |
| 6-Methyl-anthanthrene | 21 | 48 | 66 ± 7 | 46 ± 11 | 5/4 | 5/5 |
| 6,12-Dimethyl-anthanthrene | 28 | 0 | 69 ± 6 | — | 0 | 0 |
| 0.2 μmole |  |  |  |  |  |  |
| Anthanthrene | 28 | 0 | 70 ± 1 | — | 0 | 0 |
| 6-Methyl-anthanthrene | 21 | 5 | 67 ± 8 | 69 ± 0 | 1/1 | 0 |
| 6,12-Dimethyl-anthanthrene | 28 | 0 | 70 ± 0 | — | 0 | 0 |
| Dioxane-DMSO (75:25) | 30 | 0 | 70 ± 0 | — | 0 | 0 |
| Untreated Control | 30 | 0 | 70 ± 0 | — | 0 | 0 |

[a]Nine-week-old female CD-1 mice in groups of 28 were treated with the test compound at indicated doses in 33 μl dioxane-DMSO (75:25) twice weekly for 40 weeks. The test ended after 70 experimental weeks.
[b]Tumor-bearing animals.
[c]Histologically examined.

at the low dose, only one tumor was found by the end of the experiment. In an ongoing three-dose-level experiment, both 6-methylanthanthrene and 6,12-dimethylanthanthrene displayed tumor-initiating activity after 20 weeks of promotion (Table 2).

These findings are consistent with previously reported results on the carcinogenicity of subcutaneously injected 6-methyl-and 6,12-dimethyl-anthanthrene in rats (16). In this case we can state that the carcinogenicity of these compounds cannot be attributed to formation of diol epoxides, since anthanthrene does not have a bay region. We postulate that the carcinogenicity of the two methyl derivatives is elicited by possible hydroxylation of the methyl group, followed by esterification to form a reactive alkylating agent. It cannot be excluded, however, that blocking of the meso-anthracenic 6 and 12 positions as possible sites of metabolic oxidation would generate an ultimate carcinogenic epoxide in another molecular region.

TABLE 2. Tumor-Initiating Activity of Anthanthrene Derivatives on CD-1 Mouse Skin After 20 Weeks of Promotion[a]

| Initiator | Dose ($\mu$mole) | Mice with Tumors (%) | Tumors/ Mouse |
|---|---|---|---|
| Anthanthrene | 0.18 | 8 | 0.08 |
| | 0.06 | 4 | 0.04 |
| | 0.02 | 4 | 0.04 |
| 6-Methylanthan- throne | 0.18 | 42 | 0.6 |
| | 0.06 | 17 | 0.2 |
| | 0.02 | 0 | 0 |
| 6,12-Dimethyl- anthanthrene | 0.18 | 62 | 0.8 |
| | 0.06 | 8 | 0.08 |
| | 0.02 | 17 | 0.3 |
| Solvent Dioxane- DMSO (75:25) | — | 0 | 0 |

[a]Eight-week-old female CD-1 mice in groups of 24 were treated with the indicated dose given 10 times in 33 $\mu$l of dioxane-DSMO (75:25) every other day. One week after the last dose, the backs of mice were treated with 17 nmole of 12-0-tetradecanoylphorbol-13-acetate in 33 $\mu$l of acetone twice weekly for 20 weeks.

## Induction of Rat Mammary Tumors by Direct Application of PAH

A series of PAH, which had previously been tested on mouse skin (see below), was chosen for direct application in the rat mammary gland. This method has been reported and described in detail (25). As shown in Table 3, the compounds were tested at two dose levels in groups of 20 female Sprague-Dawley rats. Besides the very strong mammary carcinogen 7,12-dimethylbenz(a)anthracene (BaA-7,12-CH$_3$), which was applied at the low dose level as a positive control, the only specific mammary carcinogen used was BaP. At the high dose level it induced a 40 percent incidence of adenocarcinomas and at the low dose, 30 percent. Fibrosarcomas were also diagnosed at both dose levels. The metabolite BaP 7,8-dihydrodiol was inactive, indicating that the mechanism of carcinogenesis of BaP may be different in mammary gland from mouse skin and cannot be easily explained in terms of diol epoxide formation. Fibrosarcomas were induced by BaP-6-CH$_3$ and BaA-7-CH$_3$ at the high dose level, a finding which suggests that these compounds are weakly carcinogenic or noncarcinogenic in mammary gland. Compounds such as 5-methylchrysene (CH-5-CH$_3$) and dibenz[a,h]anthracene [DB(a,h)A], which are potent carcinogens in mouse skin (see below), were inactive in the mammary gland. BaA, which was virtually inactive on mouse skin (see below), also did not elicit mammary tumors. To our knowledge the only other PAH tested by this technique were 3-methylcholanthrene (MC), MC-1-OH, MC-1-one, and 3-methylcholanthrylene (9). Among these compounds, MC was found to be the only mammary carcinogen.

## Comparative Tumor-Initiating Activity of Benzo(a)pyrene and Benzo(a)pyrene 7,8-Dihydrodiol on Mouse Skin

The tumor-initiating activity of BaP was compared with that of the racemic BaP 7,8-dihydrodiol in two solvent systems, acetone and dioxane-DMSO (75:25) (Table 4). Both compounds were more active when applied in acetone. The tumorigenic effect of the two hydrocarbons can be considered to be about the same, although it appeared that BaP was slightly more active than its derivative when treated in acetone. These results are consistent with the previous experiments in which the two compounds were compared as tumor initiators (8,26) and as complete carcinogens (17,18). The similar carcinogenicity of BaP and BaP 7,8-dihydrodiol in mouse skin differs from the activity in the rat mammary gland (see above), in which only BaP was active.

## Carcinogenicity of PAH by Repeated Application on Mouse Skin

The carcinogenicity of BaP and BaP-6-CH$_3$ (5) was compared at two dose levels, as shown in Table 5. As in the results obtained in rat mammary gland (Table 3), the parent compound proved to be a stronger carcinogen than the methyl derivative.

In the BaA series, the activity of the various compounds as complete carcinogens was as follows (Table 6): BaA elicited a borderline effect, BaA-7-CH$_3$ was a relatively potent carcinogen, MC was very active, and BaA-7,12-CH$_3$ was most active. These results parallel those obtained in studies of direct application in mammary gland (Table 3), including MC which was previously tested by Dao et al·(9).

However, compounds such as Ch-5-CH$_3$ and DB(a,h)A, which were inactive in the rat mammary gland, were potent carcinogens on mouse skin (Table 7).

## CONCLUSIONS

In the rat mammary gland the carcinogenicity of BaP may not be understood in terms of an ultimate diol epoxide intermediate since the presumed proximate carcinogenic metabolite BaP 7,8-dihydrodiol was inactive. It may also be that the metabolic pathways for detoxification of BaP 7,8-dihydrodiol differ significantly between mouse skin and rat mammary gland.

The only PAH carcinogenic in mammary gland are those with generally low IPs (Table 8). This could mean that PAH radical cations are involved in the mechanism of tumor initiation, since the ease of formation of these intermediates is dependent on the IP of the hydrocarbon. Compounds like Ch-5-CH$_3$ and DB(a,h)A, which have relatively high IPs and thus cannot be activated by one-electron oxidation, are inactive in the mammary gland. These two hydrocarbons, however, are potent carcinogens on mouse skin.

These data tend to suggest that one-electron oxidation might be a selective mechanism of carcinogenic activation for PAH in mammary gland, while it is safe to assume that multiple mechanisms of activation are involved in mouse skin for most PAH.

## ACKNOWLEDGMENT

This work was supported by Public Health Service Contract N01CP33278 from the National Cancer Institute, NIH. We thank Dr. Gustav Granroth for evaluation of selected histopathological material and Ms. Jeanne Laubscher for valuable technical assistance.

TABLE 3. Carcinogenicity of Aromatic Hydrocarbons by Direct Application in Rat Mammary Gland[a]

| Compound | Structure | Dose[b] | No. of Rats with Tumors Total Number of Rats | Tumor Incidence (%) | No. and Type of Tumor[c] | | |
|---|---|---|---|---|---|---|---|
| | | | | | ACA(%) | FS(%) | FA(%) |
| Benzo(a)pyrene | | High | 16/20 | 80 | 8/20(40) | 10/20(50) | 2/20(10) |
| | | Low | 10/20 | 50 | ACA+FS 4/20(20) | 4/20(20) | 0 |
| | | High | 0/20 | 0 | 6/20(30) | 4/20(20) | 0 |
| Benzo(a)pyrene 7,8-dihydrodiol | | Low | 0/20 | 0 | 0 | 0 | 0 |
| | | High | 3/18 | 16.7 | 0 | 3/18(16.7) | 0 |
| 6-Methylbenzo(a)-pyrene | | Low | 0/20 | 0 | 0 | 0 | 0 |
| | | High | 0/20 | 0 | 0 | 0 | 0 |
| Benz(a)anthracene | | Low | 0/20 | 0 | 0 | 0 | 0 |

## TABLE 3. (Continued)

| Compound | Structure | Dose[b] | No. of Rats with Tumors Total Number of Rats | Tumor Incidence (%) | No. and Type of Tumor[c] | | |
|---|---|---|---|---|---|---|---|
| | | | | | ACA(%) | FS(%) | FA(%) |
| 7-Methylbenz(a)anthracene | | High | 6/20 | 30 | 0 | 6/20(30) | 0 |
| | | Low | 0/20 | 0 | 0 | 0 | 0 |
| 7,12-Dimethylbenz(a)anthracene | | Low | 20/20 | 100 | 16/20(80) | 4/20(20) | 0 |
| | | High | 0/20 | 0 | 0 | 0 | 0 |
| 5-Methylchrysene | | Low | 0/20 | 0 | 0 | 0 | 0 |
| | | High | 0/20 | 0 | 0 | 0 | 0 |
| Dibenz(a,h)anthracene | | Low | 0/20 | 0 | 0 | 0 | 0 |

[a] The compounds were "dusted" on the exposed right inguinal gland of 50-day-old female Sprague-Dawley rats. The animals were sacrificed after 20 experimental weeks.
[b] High dose: 16 μmole; low dose: 4 μmole.
[c] ACA: adenocarcinoma; FS: fibrosarcoma; FA: fibroadenoma.

TABLE 4. Tumor-initiating Activity of Benzo(a)pyrene and Benzo(a)pyrene 7,8-Dihydrodiol on CD-1 Mouse Skin[a]

| Initiator | Solvent | Mice with Tumors (%) | Tumors/Mouse |
|---|---|---|---|
| BaP | Acetone | 92 | 2.6 |
| | Dioxane-DMSO (75:25) | 60 | 1.6 |
| BaP 7,8-dihydrodiol | Acetone | 76 | 2.1 |
| | Dioxane-DMSO (75:25) | 68 | 1.8 |
| None | | 0 | 0 |

[a] Seven-week-old female CD-1 mice in groups of 25 were treated on the skin with 10 mini-doses of 0.02 μmole of initiator in 33 μl of solvent delivered every other day. One week after the last treatment, the backs of mice were painted with 17 nmole of 12-O-tetradecanoylphorbol-13-acetate in 33 μl acetone twice weekly for 30 weeks.

## MECHANISMS OF ACTIVATION OF AROMATIC HYDROCARBONS 225

TABLE 5. Skin Tumors in Swiss Mice Treated with Benzo(a)pyrene and 6-Methylbenzo(a)pyrene by Repeated Application[a]

| Compound | Dose (μmole) and Frequency of Treatment | Effective No. of Mice[b] | Tumor Incidence (%) | Survival Mean ± S.D. (weeks) | Age of First Tumor Appearance Mean ± S.D. (weeks) | No. and Type of Skin Neoplasms | | |
|---|---|---|---|---|---|---|---|---|
| | | | | | | Papillomas | Keratotic Epitheliomas | Carcinomas |
| Benzo(a)pyrene | 0.2 4x weekly for 20 weeks | 28 | 100 | 26 ± 5 | 18 ± 2 | 2 | 6 | 30 |
| | 0.1 2x weekly for 20 weeks | 28 | 79 | 42 ± 6 | 29 ± 7 | 2 | 8 | 18 |
| 6-Methylbenzo(a)-pyrene | 0.2 4x weekly for 20 weeks | 22 | 86 | 41 ± 8 | 20 ± 4 | 4 | 4 | 15 |
| | 0.1 2x weekly for 20 weeks | 23 | 30 | 50 ± 4 | 36 ± 8 | 2 | 1 | 4 |
| Control (Acetone) | — | 29 | 0 | 65 ± 11 | — | 0 | 0 | 0 |

[a] Eight-week-old female Swiss mice in groups of 30 were treated at the indicated dose dissolved in 16.7 μl acetone. All surviving mice were sacrificed after 52 experimental weeks.
[b] Histologically examined.

TABLE 6. Skin Tumors in Swiss Mice Treated with Benz(a)anthracene, 7-Methylbenz(a)anthracene, 7,12-Dimethylbenz(a)anthracene and 3-Methylcholanthrene by Repeated Application[a]

| Compound | Dose (µmole) | Effective No. of Mice[b] | Tumor Incidence (%) | Survival Mean ± S.D. (weeks) | Age of First Tumor Appearance Mean ± S.D. (weeks) | No. and Type of Skin Neoplasms | | |
|---|---|---|---|---|---|---|---|---|
| | | | | | | Papillomas | Keratotic Epitheliomas | Carcinomas |
| Benz(a)anthracene | 0.4 | 39 | 3 | 65 ± 15 | 53 ± 0 | 1 | 0 | 0 |
| 7-Methylbenz(a)-anthracene | 0.8 | 30 | 77 | 48 ± 6 | 34 ± 6 | 4 | 5 | 22 |
| | 0.4 | 30 | 67 | 48 ± 6 | 38 ± 7 | 4 | 4 | 16 |
| 7,12-Dimethyl-benz(a)-anthracene | 0.025 | 30 | 93 | 33 ± 7 | 14 ± 7 | 7 | 24 | 17 |
| | 0.0125 | 30 | 70 | 45 ± 15 | 26 ± 12 | 4 | 11 | 16 |
| | 0.0062 | 30 | 47 | 54 ± 16 | 50 ± 14 | 2 | 7 | 8 |
| 3-Methyl-cholanthrene | 0.2 | 45 | 100 | 25 ± 3 | 18 ± 3 | 3 | 8 | 55 |

[a] Seven-week-old female Swiss mice in groups of 30 (40 for benz(a)anthracene, 50 for methylcholanthrene) were treated at the indicated dose dissolved in 16.7 µl acetone. The groups were treated as follows: benz(a)anthracene, twice weekly for 30 weeks; 7-methylbenz(a)-anthracene, twice weekly for 25 weeks; 7,12-dimethylbenz(a)anthracene, twice weekly for 20 weeks. The surviving mice treated with the above compounds were sacrificed at 70, 52 and 70 experimental weeks, respectively.
[b] Histologically examined.

TABLE 7. Skin Tumors in Swiss Mice by Repeated Application of 5-Methylchrysene and Dibenz(a,h)anthracene[a]

| Compound | Dose (μmole) | Effective No. of Mice[b] | Tumor Incidence (%) | Survival Mean ± S.D. (weeks) | Age of First Tumor Appearance Mean ± S.D. (weeks) | No. and Type of Skin Neoplasms[d] | | | |
|---|---|---|---|---|---|---|---|---|---|
| | | | | | | No. of Papillomas/ No. of TBA[c] | No. of Keratotic Epitheliomas/ No. of TBA | No. of Carcinomas/ No. of TBA | No. of TBA |
| 5-Methylchrysene | 0.60 | 29 | 93 | 40 ± 9 | 27 ± 4 | 2/2 | 8/8 | 18/18 | |
| | 0.20 | 29 | 86 | 51 ± 10 | 34 ± 5 | 5/5 | 7/7 | 13/13 | |
| | 0.066 | 30 | 63 | 58 ± 4 | 40 ± 6 | 0/0 | 7/7 | 11/11 | |
| Dibenz(a,h)anthracene | 0.30 | 29 | 86 | 45 ± 9 | 31 ± 8 | 5/5 | 6/6 | 18/16 | |
| | 0.10 | 29 | 62 | 45 ± 8 | 33 ± 7 | 1/1 | 4/4 | 10/10 | |
| | 0.033 | 30 | 67 | 47 ± 7 | 34 ± 9 | 2/2 | 8/6 | 17/15 | |

[a] Eight-week-old female Swiss mice in groups of 30 were treated at the indicated dose with the test compounds in 33 μl acetone. 5-Methylchrysene was applied twice weekly for 30 weeks, and all surviving animals were killed at 52 experimental weeks. Dibenz(a,h)anthracene was given twice weekly for 20 weeks, and survivors were killed at 60 experimental weeks.
[b] Histologically examined.
[c] Tumor-bearing animals.
[d] Other skin tumors were as follows: 5-methylchrysene—0.20 μmole, 2 trichoepitheliomas, 1 fibrous histiocytoma, 1 skin granuloma; 0.066 μmole, 1 trichoepithelioma; dibenz(a,h)anthracene—0.30 μmole, 1 trichoepithelioma; 0.10 μmole, 3 basal cell carcinomas, 1 trichoepithelioma, 2 skin granulomas; 0.033 μmole, 1 basal cell carcinoma, and 2 sebaceous adenomas.

TABLE 8. Comparative Carcinogenicity of PAH in Mouse Skin and Rat Mammary Gland

| Compound | Structure | Ionization Potential[a] (eV) | Carcinogenicity in:[b] Mouse Skin | Carcinogenicity in:[b] Rat Mammary Gland |
|---|---|---|---|---|
| 5-Methylchrysene | | CA. 7.7 | +++ | — |
| Dibenz(a,h)anthracene | | 7.57[c] | +++ | — |
| Benz(a)anthracene | | 7.54 | ± | — |
| 7-Methylbenz(a)-anthracene | | 7.37 | +++ | + |
| Benzo(a)pyrene | | 7.23 | ++++ | +++ |
| Benzo(a)pyrene 7,8-dihydrodiol | | | ++++ | — |
| 7,12-Dimethylbenz-(a)anthracene | | 7.22 | +++++ | +++++ |
| 3-Methylcholanthrene | | 7.12 | ++++ | ++++ [d] |
| 6-Methylbenzo(a)pyrene | | 7.08 | +++ | + |

[a] Determined from maximum absorption of the charge-transfer complex of each compound with chloranil.
[b] Extremely active, +++++; very active, ++++; active, +++; moderately active, ++; weakly active, +; very weakly active, ±; inactive, -.
[c] Determined by polarographic oxidation (19).
[d] Reference 9.

## REFERENCES

1. Bartsch, H., and Hecker, E. (1971): On the metabolic activation of carcinogen N-hydroxy-N-2-acetylaminofluorene. III. Oxidation with horseradish peroxidase to yield 2-nitrosofluorene and N-acetoxy-N-2-acetylaminofluorene. *Biochim. Biophys. Acta* 237:567-578.
2. Bartsch, H., Miller, J. A., and Miller, E. C. (1972): N-Acetoxy-N-acetylaminofluorene and nitrosoarenes, one-electron non-enzymatic and enzymatic oxidation products of various carcinogenic aromatic acethydroxamic acids. *Biochim. Biophys. Acta* 273:40-51.
3. Cavalieri, E., Rogan, E., and Roth, R. (1980): Multiple mechanisms of activation in aromatic hydrocarbon carcinogenesis. In: *Free Radicals and Cancer*, R. A. Floyd, Ed., Marcel Dekker, Inc., New York (in press).
4. Cavalieri, E., Roth, R., and Rogan, E. (1979): Hydroxylation and conjugation at the benzylic carbon atom: A possible mechanism of carcinogenic activation for some methyl-substituted aromatic hydrocarbons. In: *Polynuclear Aromatic Hydrocarbons*, P. W. Jones and P. Lieber, Eds., pp 517-529, Ann Arbor Science Publishers, Ann Arbor, Michigan 517-529.
5. Cavalieri, E., Roth, R., Grandjean, C., Althoff, J., Patil, K., Liakus, S., and Marsh, S. (1978): Carcinogenicity and metabolic profiles of 6-substituted benzo(a)pyrene derivatives on mouse skin. *Chem.-Biol. Interact.* 22:53-67.
6. Cavalieri, E., Roth, R., and Rogan, E. G. (1976): Metabolic activation of aromatic hydrocarbons by one-electron oxidation in relation to the mechanism of tumor initiation. In: *Carcinogenesis, Vol. I. Polynuclear Aromatic Hydrocarbons: Chemistry, Metabolism and Carcinogenesis*, R. I. Freudenthal and P. W. Jones, Eds., pp. 181-190, Raven Press, New York.
7. Cavalieri, E., Roth, R., Rogan, E., Grandjean, C., and Althoff, J. (1978): Mechanisms of tumor initiation of polycyclic aromatic hydrocarbons. In: *Carcinogenesis Vol. III. Polynuclear Aromatic Hydrocarbons*. P. W. Jones and R. I. Freudenthal, Eds., pp. 273-284, Raven Press, New York.
8. Chouroulinkov, I., Gentil, A., Grover, P. L., and Sims, P. (1976): Tumor-initiating activities on mouse skin of dihydrodiols derived from benzo[a]pyrene. *Brit. J. Cancer* 34:523-532.
9. Dao, T. L., King, C., and Tominaga, T. (1971): Isolation, identification, and biological study of compounds derived from 3-methylcholanthrene by irradiation in dimethyl sulfoxide. *Cancer Res.* 31:1492-1495.

10. Flesher, J. W., and Tay, L, K. (1978): Reaction of the carcinogens 7-hydroxymethyl-12-methylbenz(a)anthracene and 7-acetoxymethyl-12-methylbenz(a)anthracene with DNA. *Res. Commun. Chem. Path. Pharmacol.* 22:345-355.
11. Floyd, R. A., and Soong, L. M. (1977): Obligatory free radical intermediate in the oxidative activation of the carcinogen N-hydroxy-2-acetylaminofluorene. *Biochim. Biophys. Acta* 498:244-249.
12. Fried, J. (1974): One-electron oxidation of polycyclic aromatics as a model for the metabolic activation of carcinogenic hydrocarbons. In: *Chemical Carcinogenesis,* Part A, P.O.P. Ts'o and J. DiPaolo, Eds., pp. 197-215, Marcel Dekker, New York.
13. Jerina, D. M., and Daly, J. W. (1976): Oxidation at carbon. In: *Drug Metabolism,* D. V. Parke and R. W. Smith, Eds., pp. 15-33, Taylor & Francis, Ltd., London.
14. Jerina, D. M., Lehr, R., Schaefer-Ridder, M., Yagi, H., Karle, J. M., Thakker, D. R., Wood, A. W., Lu, A.Y.H., Ryan, D., West, S., Levin, W., and Conney, A. H. (1977): Bay region epoxides of dihydrodiols: A concept which explains the mutagenic and carcinogenic activities of benzo(a)pyrene and benz(a)anthracene. In: *Origins of Human Cancer,* Book B, H. H. Hiatt, J. D. Watson, and J. A. Winsten, Eds., pp. 639-658, Cold Spring Harbor Laboratory, New York.
15. Jerina, D. M., Lehr, R. W., Yagi, H., Hernandez, O., Dansette, P., Wislocki, P. G., Wood, A. W., Chang, R. L., Levin, W., and Conney, A. H. (1976): Mutagenicity of benzo(a)pyrene derivatives and the description of a quantum mechanical model which predicts the ease of carbonium ion formation from diol epoxides. In: *In Vitro Metabolic Activation and Mutagenesis Testing,* F. J. DeSerres, J. R. Fouts, J. R. Bend and R. M. Philpot, Eds., pp. 159-177, North Holland Publishing Co., Amsterdam.
16. Lacassagne, A., Buu Hoi, N. P., and Zajdela, F. (1958): Relation entre structure moléculaire et activité cancérogene dans trois séries d'hydrocarbures aromatiques hexacycliques. *C. R. Acad. Sci. Paris* 246:1477-1480.
17. Levin, W., Wood, A. W., Wislocki, P. G., Kapitulnik, J., Yagi, H., Jerina, D. M., and Conney, A. H. (1977): Carcinogenicity of benzo-ring derivatives of benzo(a)pyrene on mouse skin. *Cancer Res.* 37:3356-3361.
18. Levin, W., Wood, A. W., Yagi, H., Jerina, D. M., and Conney, A. H. (1976): (±)-trans-7,8-dihydroxy-7,8-dihydrobenzo(a)pyrene: A potent skin carcinogen when applied topically to mice. *Proc. Natl. Acad. Sci. U.S.A.* 73:3867-3871.

## RESULTS AND DISCUSSION

### Evaluation of the Specificity of the Heme Effect

In previous reports (19,20) we have demonstrated the apparent ubiquity of the stimulatory response produced by hematin when tested with extrahepatic tissues. With the exception of rabbit fetal liver, we have found that additions of hematin increase the rates of monooxygenation reactions, specifically in extrahepatic tissues, and that nonezymatic pathways do not appear to be mechanistically involved in the heme effect.

Recent evaluations of a number of porphyrin derivatives and their capacities to produce effects similar to hematin are summarized in Table 1. Not only is the hematin-mediated reaction specific for extrahepatic tissues, but it is also specific for $Fe^{2+}$-protoporphyrin-IX. Neither protoporphyrin-IX nor $Co^{2+}$- or $Zn^{2+}$-protoporphyrin-IX were effective in producing similar stimulatory responses. Biliverdin, the immediate biodegradation product of heme, was also ineffective. Hemoglobin, however, was partially effective in stimulating rates of these enzyme reactions. Investigators have demonstrated that the heme moieties of the hemoglobin molecule are capable of exchanging in aqueous medium with other protein molecules (6). Such an exchange reaction may also be responsible for the limited stimulatory effects of hemoglobin in our

TABLE 1. Effect of Protoporphyrin Compounds and Derivatives on Rates of BaP-Hydroxylation with Rabbit Brain S-9 Fractions[a]

| Compound | Activity | Ratio |
|---|---|---|
| Control | 26 | 1.0 |
| Protoporphyrin-IX | 45 | 1.7 |
| $Co^{2+}$-protoporphyrin-IX | 21 | 0.8 |
| $Zn^{2+}$-protoporphyrin-IX | 17 | 0.7 |
| $Fe^{2+}$-protoporphyrin-IX-chloride (hemin) | 1606 | 61.8 |
| $Fe^{2+}$-protoporphyrin-IX-hydroxide (hematin) | 1648 | 63.4 |
| Hemoglobin | 207 | 8.0 |
| Biliverdin | 25 | 1.0 |

[a]Reaction flasks contained 1 mg/ml of rabbit brain S-9 protein and were incubated for 2 hours in the presence of cofactors and 80 $\mu$M of BaP. Each of the listed compounds was added to yield a final heme concentration of 12 $\mu$M. Protoporphyrin-IX was prepared as described (21). Activities are expressed as picomoles metabolite formed/mg protein/2 hr and values represent means of triplicate determinations.

assay system. FeCl₂, myoglobin, catalase, and horseradish peroxidase also were tested as possible substitutes for hematin, but were ineffective in eliciting any stimulatory responses.

## Time Course of the Heme Effect

The data in Table 2 illustrate the time course of the effects of hematin with rabbit brain, lung, and liver S-9 fractions employed as enzyme sources. Most notable in these experiments is the lack of effect of hematin additions on hepatic preparations when assayed at either short (15 minutes) or long (120 minutes) time intervals. Occasionally we have observed a slight (up to 20 percent) hematin-mediated stimulation of hepatic activities during short (10 to 15 minute) incubations; these results are similar to those reported by Brown and Kupfer (5). Longer incubations of hepatic preparations with hematin invariably lead to overall slight but significant decreases in enzymatic activities relative to control flasks. These phenomena are in contrast to the situation with extrahepatic tissues. In nonhepatic tissues, illustrated in Table 2 by the brain and the lung, the stimulatory effects of heme are relatively minimal at shorter incubation times but increase with time (as denoted by the ratios of activities) up to 60 minutes for lung and 120 minutes for brain.

In addition (data not shown), we have found that the initial lag period characteristic of the heme-mediated stimulation could be partially

**TABLE 2. Time Course of BaP-Hydroxylation as Assayed with S-9 Fractions of Rabbit Brain, Lung, and Liver[a]**

| Time (minutes) | Hematin (12 μM) | Rabbit Tissue | | |
|---|---|---|---|---|
| | | Brain | Lung | Liver |
| 15 | — | 6 | 64 | 560 |
| | | (1.7) | (1.3) | (1.0) |
| 15 | + | 10 | 85 | 569 |
| 30 | — | 13 | 98 | — |
| | | (3.9) | (2.0) | |
| 30 | + | 51 | 200 | — |
| 60 | — | 21 | 243 | — |
| | | (13.1) | (4.1) | |
| 60 | + | 275 | 998 | — |
| 120 | — | 46 | 681 | 1460 |
| | | (29.4) | (3.0) | (0.8) |
| 120 | + | 1350 | 2050 | 1206 |

[a]Reaction flasks contained 2 mg/ml of S-9 tissue protein and were incubated for the indicated time periods. Activities were expressed as picomoles metabolite formed/mg protein and values represent means of triplicate determinations. Values in parentheses represent ratios of activities (+heme/-heme).

circumvented either by preincubation of the complete reaction mixture (including substrate) before the addition of hematin at time zero or by the addition of 1 to 10 $\mu$M of 3-hydroxy-BaP (the major metabolite in these systems) to the complete reaction mixture together with hematin at time zero. These data, although preliminary at this time, indicate the possible requirement for metabolism of substrate as a precondition for the initiation of the hematin-dependent oxidation reactions.

## Cofactor Requirements and the Effects of Specific P-450 Modifiers

Hematin-dependent and hematin-independent reactions are both dependent on the presence of reducing equivalents in the reaction mixture, and NADPH is the preferred source of these reducing equivalents (21). Furthermore, neither the hematin-dependent nor hematin-independent oxidations occurred when heat-inactivated tissue subfractions or tissue blanks were employed, thus ruling out apparent nonenzymatic mechanisms.

Table 3 lists the relative effects of a number of known modifiers of monooxygenase systems. Both hematin-dependent and hematin-independent reactions were inhibited by classical P-450 inhibitors such as carbon monoxide, aniline, cytochrome c, $\alpha$-naphthol, $E_2$, and 7,8-benzoflavone. Metyrapone, however, produced stimulatory effects on both reaction systems, although this substance usually exhibits inhibitory effects on P-450-mediated oxidations. Liebman (15) has demonstrated a

Table 3. Effects of Potential Modifiers on BaP-Hydroxylase Activities with Rabbit Brain S-9 Fraction[a]

| Modifier | Concentration (mM) | Ratio (Experimental/Control) | |
|---|---|---|---|
| | | -Hematin | +Hematin |
| None | — | 1.00 | 1.00 |
| CO | —[b] | 0.75 | 0.02 |
| Aniline | 0.1 | 0.43 | 0.02 |
| Cytochrome c | 0.01 | 0.68 | 0.06 |
| $\alpha$-Naphthol | 0.01 | 0.77 | 0.53 |
| 17-$\beta$-Estradiol | 0.1 | 0.09 | 0.50 |
| 7,8-Benzoflavone | 0.1 | 0.29 | 0.88 |
| Metyrapone | 1.0 | 2.83 | 1.13 |

[a] All incubations were performed for 2 hours and values are expressed as ratios of means of activities (experimental/control) obtained for each experiment.
[b] CO incubations were performed in a 50% CO:50% $O_2$ atmosphere and compared with controls incubated in a 50% $N_2$:50% $O_2$ atmosphere.

similar stimulatory effect by metyrapone on rates of acetanilide hydroxylation and tricloroethylene oxidation. Thus, although this particular effect is unusual and unexplained, it is not unprecedented.

Many of the data available to date regarding the hematin effect have been derived from studies utilizing BaP as substrate. We are currently examining other substrates of P-450-mediated monooxygenation reactions and our results (not shown) indicate that the metabolism of DMBA and $E_2$ is affected in a similar stimulatory fashion, as is BaP. However, the metabolism of FAA is not enhanced by hematin additions, in rabbit brain, lung, or liver. These results, together with the apparent differential effects of some of the modifiers listed in Table 3 on hematin-dependent and hematin-independent reactions, may indicate a selective effect of hematin on specific P-450 cytochromes. This possibility is presently being examined in greater detail.

## Regulation of Cytochrome P-450 and Heme Biosynthesis

The regulation of holocytochrome P-450 pools in animal tissues appears to be subject to coordinated controls involving heme biosynthesis in the mitochondria, apocytochrome P-450 synthesis in the rough endoplasmic reticulum, and, finally, the incorporation of heme into the apoprotein, yielding a functional holoenzyme (27). However, the control mechanisms regulating this coordinated biosynthesis are not well elucidated. Reports by Correia and Meyer (9) and Correia et al (8) have demonstrated that pools of apocytochrome P-450 can be experimentally produced in the liver and that such pools can be reconstituted by exogenous heme, both in vitro (9) and in vivo (8). Other investigators have provided evidence of the existence of free pools of apocytochrome P-450 in hepatic preparations from fetal (4,26) and adult (16) rats. Negishi and Krebich (18) recently reported that, in developing rats, insertion of the heme moiety occurred in parallel with the synthesis of apoprotein such that no large quantities of apocytochrome P-450 accumulated in the livers of the prenatal animals. Such apparent discrepancies remain, at least temporarily, unresolved. Improved methodologies regarding the analytical estimation of small pools of P-450 are necessary to provide more accurate determinations of low concentrations of cytochrome P-450(s) existing in various tissues.

Other studies (23,24,28,29) have indicated that, although the oxidative capacities of fetal livers from rabbits, rats, guinea pigs, mice, and humans are low, heme biosynthetic activities occur with substantially higher rates in fetal livers than in livers from adult animals. Insofar as it has been studied in adult liver, evidence has accumulated that the availability of heme for cytochrome P-450 biosynthesis in the normal animal is not rate-limiting (3,10,22). However, only very few studies on the biosynthetic aspects of heme in extrahepatic tissues have been performed. It appears that in the adrenal gland of rats (7) and in the human fetal adrenal (23), the activities of the heme biosynthetic enzymes are comparable to the activites found in the liver. In rats, the activity of delta-aminolevulinic acid synthethase (the rate-limiting enzyme for heme biosynthesis in adults) decreases in the order: spleen, liver, adrenal, kidney, heart (24). Rabbbit brain has also been found to possess fairly high activity of this enzyme (17).

The postulation of the existence of free apocytochrome pools in tissues that possess relatively high rates of heme biosynthetic activity represents an enigma that is not readily explained in terms of presently known aspects of the properties of this hemoprotein. If such a situation does indeed exist, then additional insight will be required to understand such phenomena, especially with regard to regulatory aspects that control heme complexation with P-450 apoproteins. Interesting and possibly analogous situations have been reported for at least two other heme enzymes, tryptophan pyrrolase and heme oxygenase. Fiegelson and Greengard (11) and others (2,25) have demonstrated that tryptophan pyrrolase, a cytoplasmic enzyme catalyzing the rate-limiting step in tryptophan degradation, exists as pools of both apo- and holoenzyme. Not only are the ratios of these respective pools altered by pretreating animals with enzyme-inducing agents, but the apoproteins are subject to substrate-mediated complexation with heme to yield a functional holoenzyme. Yoshida and Kikuchi (30) have also provided evidence that heme oxygenase, an enzyme important in the catabolism of heme, has only a very low affinity for heme and that the heme moiety apparently functions simultaneously as a prosthetic group and as a substrate for the enzyme. Whether these situations are analogous to the situation regulating cytochrome P-450 activities in various tissues remains an intriguing and yet-to-be-determined possiblity.

## SUMMARY

Data have been presented indicating that the stimulatory effect of hematin on monooxygenase reactions is due to a specific action of hematin on cytochrome P-450(s). This conclusion is based partially on experiments examining the cofactor requirements, substrate specificity, and effects of specific P-450 modifiers. In addition, it is suggested that the monooxygenase systems of extrahepatic tissues may be subject to different regulatory controls that those existing in the liver and that a possible mechanism for the hematin stimulation may be the complexation of heme with pools of free apocytochrome P-450 that exist in these tissues.

## ACKNOWLEDGMENTS

These studies were supported by NICHD Grant HD-04839, National Foundation Grant CRBS-250, and NIH Grant HL-03174. C.J.O. was supported by DHEW Environmental Pathology/Pathophysiology Training Grant ES-07032.

## REFERENCES

1. Atlas, S. A., Boobis, A. R., Felton, J. S., Thorgiersson, S. S., and Nebert, D. W. (1977): Ontogenetic expression of polycyclic aromatic compound-inducible monooxygenase activities and forms of cytochrome P-450 in rabbit. *J. Biol Chem.* 242:4712-4721.
2. Badaway, A.A.B., and Evans, M. (1973): The effects of chemical porphyrogens and drugs on the activity of rat liver tryptophan pyrrolase. *Biochem. J.* 136:885-892.
3. Bhat, K. S., Sardana, M. K., and Padmanaban, G. (1977): Role of haem in synthesis and assembly of cytochrome P-450. *Biochem. J.* 164:295-303.
4. Black, O., and Bresnick, E. (1972): Ontogenetic changes of proteins of endoplasmic reticulum. *J. Cell. Biol.* 52:733-742.
5. Brown, J. E., and Kupfer, D. (1975): Interactions of heme with hepatic microsomal monooxygenase: Effect on benzpyrene hydroxylation. *Chem. Biol. Interactions* 10:57-64.
6. Bunn, H. F., and Jandl, J. H. (1968): Exchange of heme among hemoglobins and between hemoglobin and albumin. *J. Biol. Chem.* 10:465-475.
7. Condie, L. W., Tephly, T. R., and Baron, J. (1976): Studies on heme synthesis in the rat adrenal. *Ann. Clin. Res.* 9, Suppl. 17:83-88.
8. Correia, M. A., Farrell, G. C., Schmidt, R., deMontellano, P.R.O., Yost, G. S., and Mico, B. A. (1979): Incorporation of exogenous heme into hepatic cytochrome P-450 in vivo. *J. Biol. Chem.* 254:15-17.
9. Correia, M. A., and Meyer, U. A. (1975): Apocytochrome P-450: Reconstitution of functional cytochrome with hemin in vitro. *Proc. Nat. Acad. Sci. (USA)* 72:400-404.

10. Druyan, R., and Kelley, A. (1972): The effect of exogenous delta-aminolevulinate on rat liver haem and cytochromes. *Biochem J.* 129:1095-1099.
11. Fiegelson, P., and Greengard, O. (1972): Immunochemical evidence for increased titers of liver tryptophan pyrrolase during substrate and hormonal enzyme induction. *J. Biol. Chem.* 232:3714-3717.
12. Garfinkel, D. (1958): Studies on pig liver microsomes. I. Enzymatic and pigment composition of different microsomal fractions. *Arch. Biochem. Biophys.* 77:493-509.
13. Guenthner, T. M., and Nebert, D. W. (1978): Evidence in rat and mouse liver for temporal control of two forms of cytochrome P-450 inducible by 2,3,7,8-tetrachlorodibenzo-p-dioxin. *Eur. J. Biochem.* 91:449-456.
14. Klingenberg, M. (1958): Pigments of rat liver microsomes. *Arch. Biochem. Biophys.* 75:376-386.
15. Liebman, K. (1969): Effects of metyrapone on liver microsomal drug oxidations. *Molec. Pharmacol.* 5:1-9.
16. Levin, W., Alvares, A. P., and Kuntzman, R. (1970): Distribution of radioactive hemoprotein and CO-binding pigment in rough and smooth endoplasmic reticulum of rat liver. *Arch. Biochem. Biophys.* 139:230-235.
17. Muzyka, V. I. (1972): Delta-aminolevulinic acid synthetase in grey substance of brain hemispheres. *Biokhimiya* 37:1220-1223.
18. Negishi, M., and Krebich, G. (1978): Coordianted polypeptide synthesis and insertion of protoheme in cytochrome P-450 during development of endoplasmic reticulum membranes. *J. Biol. Chem.* 253:4791-4797.
19. Omiecinski, C. J., Bond, J. A., and Juchau, M. R. (1978): Stimulation by hematin of monooxygenase activity of extrahepatic tissues from rats, rabbits, and chickens. *Biochem. Biophys. Res. Commun.* 83:1004-1011.
20. Omiecinski, C. J., Chao, S. T., and Juchau, M. R. (1980): Modulation of monooxygenase activities by hematin and 7,8-benzoflavone in fetal tissues of rats, rabbits, and humans. *Devel. Pharm. Therapeut.* 1:90-100.
21. Omiecinski, C. J., Namkung, M. J., and Juchau, M. R. (1980): Mechanistic aspects of the hematin-mediated increases in brain monooxygenase activities. *Molec. Pharmacol.* (in press).
22. Rajamanickam, C., Rao, M.R.S., and Padmanaban, G. (1975): On the sequence of reactions leading to cytochrome P-450 synthesis—effect of drugs. *J. Biol. Chem.* 250:2305-2310.
23. Rifkind, A. B., Bennett, S. Forster, E. S., and New, M. I. (1975): Components of the heme biosynthetic pathway and mixed function

oxidase activity in human fetal tissues. *Biochem. Pharmacol.* 24:839-846.
24. Sardesai, V. M., Lenaghan, R., Rosenberg, J. C. (1972): Tissue delta-aminolevulinic acid synthetase activity in hemorrhagic shock. *Biochem. Med.* 6:366-371.
25. Schimke, R. T., Sweeney, F. W., and Berlin, C. M. (1965): The roles of synthesis and degradation in the control of rat liver tryptophan pyrrolase. *J. Biol. Chem.* 240:322-331.
26. Siekevitz, P. (1974): The differentiation of rat liver endoplasmic reticulum membranes: apo-cytochrome P-450 as a membrane protein. *J. Supramolec. Struct.* 1:471-489.
27. Tait, G. H. (1978): The biosynthesis and degradation of heme. In: *Heme and Hemoproteins*, F. DeMatteis and W. N. Aldridge, Eds., pp. 1-48, Springer-Verlag, New York.
28. Woods, J. S., and Dixon, R. L. (1970): Perinatal differences in delta-aminolevulinic acid synthetase activity. *Life Sci.* 9:711-719.
29. Woods, J. S., and Dixon, R. L. (1972): Studies on the perinatal differences in the activity of hepatic delta-aminolevulinic acid synthetase. *Biochem. Pharmacol.* 21:1735-1744.
30. Yoshida, T., and Kikuchi, G. (1978): Features of the reaction of heme degradation catalyzed by the reconstituted microsomal heme oxygenase system. *J. Biol. Chem.* 253:4230-4236.

# INHIBITION OF PULMONARY METABOLISM OF BENZO(A)PYRENE PRODUCED BY ACUTE TOBACCO SMOKE EXPOSURE

**W. C. Lubawy and R. S. Isaac**

Division of Pharmacodynamics and Toxicology
College of Pharmacy, University of Kentucky
Lexington, Kentucky 40506

## INTRODUCTION

Benzo(a)pyrene, BaP, is one of the 2,000 or more identified compounds in tobacco smoke and is thought to be one of the polycyclic aromatic hydrocarbons responsible for the high incidence of lung cancer in smokers (15). In order to develop maximum carcinogenic potential, BaP must be metabolized by cytochrome P-450 mediated aryl hydrocarbon hydroxylase (AHH) to reactive intermediates which then can bind covalently to nucleic acids and proteins of various tissues (7). Since metabolic activation appears to be required for BaP-induced carcinogenesis, it is important to identify factors that affect the rate at which BaP is metabolized by pulmonary tissue. This investigation concerns the immediate effect of whole smoke or the gas phase from smoke and the delayed effect of tobacco-smoke-condensate administration on the metabolism of BaP by the isolated perfused rabbit lung (IPL).

Numerous reports indicate that chronic smoke exposure increases the activity of pulmonary AHH. Welch et al (28) reported that homogenates of lung tissue obtained from rats exposed to cigarette smoke daily for 3 days possessed 12 times more BaP hydroxylase activity than lung homogenates obtained from control animals. Holt and Keast (16) and Abramson and Hutton (1) obtained similar results in mice. Akin and Benner (2) showed that in rats the largest increase in enzyme activity (enzyme induction) from smoke exposure was produced by cigarettes that delivered the highest level of total particulate matter. Similar indications of cigarette smoke producing enzyme induction have been reported with a variety of systems (6, 26). In these studies AHH activity was monitored several hours to several days following smoke exposure.

We became interested in enzyme activity changes that occur during and immediately after smoking, as shown by the rabbit isolated lung model. For several reasons we thought that the initial effect of smoke might be to decrease the rate of metabolism of BaP, with the enzyme induction commonly seen in most species being a delayed effect. A number of substances are capable of inhibiting initial drug metabolism and then subsequently inducing it, i.e., N-methyl-3-piperidyl-N'-N diphenyl carbamate (21), SKF 525A (23, 9), nikethamide, iproniazide, N-ethyl-3-piperidylbenzilate (24) and hexobarbital (18). Whole smoke could act in an analogous manner. Additionally, fresh cigarette smoke contains a wide variety of toxic gases (14) such as carbon monoxide (CO) and cyanide. Both of these gases have been shown to inhibit drug metabolism in microsomes (22, 17). Finally, we previously demonstrated that acute smoke exposure results in decreased metabolism of nicotine by the IPL (19) and therefore thought that BaP metabolism might be similarly affected.

## MATERIALS AND METHODS

### Chemicals

$^{14}$C-benzo(a)pyrene (60.7 $\mu$Ci/mmole) specifically labelled in the 7 and 10 positions was purchased from Amersham Corporation, Arlington Heights, Illinois. Commercial grade 3-methylcholanthrene (3MC) was purchased from Eastman Kodak, Rochester, New York. All chemicals for extraction were A.C.S. reagent grade. HPLC separation was performed using LiChrosolv solvents from E. Merck, Darmstadt, Germany, purchased through MCB, Cincinnati, Ohio. Tobacco-smoke condensate (TSC), supplied by Dr. John Benner of the Kentucky Tobacco and Health Research Institute, was collected by trapping whole smoke from University of Kentucky 3A1 reference cigarettes at -80° C.

### Isolated Lung

The isolated-lung procedure employed was essentially that previously described by Neimeier and Bingham (20), with minor modification. Male New Zealand rabbits weighing 2-½ to 3 kg were pretreated I.P. 24 hours earlier with 20 mg/kg 3MC, 100 mg/kg tobacco-smoke condensate (TSC), or 1 ml/kg of the corn-oil vehicle. Rabbits were anesthetized with 40 mg/kg sodium pentobarbital and given 2000 U/kg heparin I.V. As much blood as possible was drawn via cardiac tap and the cardiopulmonary unit was removed intact. The trachea, pulmonary artery, and left atrium were cannulated and the preparation was suspended in an artificial thorax. Pulmonary tissue was perfused with Krebs-Ringer bicarbonate buffer (12) fortified with 4.5 percent bovine serum albumin and 5 mM glucose. Ventilation with a warmed, humidified $CO_2$-air mixture was accomplished by an alternating negative pressure, 50 cycles per minute, applied to the thorax. The thorax and reservoir were water jacketed and kept at 37°C. Only freshly silicone-treated glassware and siliconized tubing were in contact with the perfusate, which was kept at pH 7.35 to 7.40. After a 10 to 15-minute equilibration period, 4 $\mu$Ci of 7,10-$^{14}$C-labelled BaP dissolved in 0.2 ml of acetone was slowly infused into the fluid perfusing the lung over a 1-minute period. Preliminary experiments demonstrated that the disposition of $^{14}$C-BaP was not influenced by the apparatus or the perfusate over a 2-hour period.

Experiments were terminated by draining the lung of all perfusate and flushing the lung with 15 ml of ice-cold saline. Lobes were quickly dissected free of nonpulmonary tissue, weighed, and homogenized in 20 ml of saturated NaCl at -15°C. All procedures were performed under subdued yellow lighting to minimize photooxidation.

## Assay of $^{14}$C-BaP and Metabolites

Two-ml aliquots of the perfusate-rinse mixture or the lung homogenate were extracted twice with 6 ml each of ethyl acetate:acetone (2:1). Extracts were combined, evaporated to dryness with nitrogen, and reconstituted in 0.5 ml of methanol. Aliquots of methanol were injected into a Model 6000 HPLC fitted with a $C_{18}$ $\mu$Bondapak column (Waters Associates, Milford, Massachusetts). Metabolites and unchanged $^{14}$C-BaP were eluted initially with water methanol (40:60) for 10 minutes. This was gradually changed to 50 percent each of the water methanol (40:60) and methanol:ethyl acetate (90:10) according to Program 3 of the Model 660 solvent programmer (Waters Associates). Flow rate was 0.8 ml/min and fifty 1-minute samples were collected and counted in Aquasol-2 cocktail (New England Nuclear, Boston), with corrections made using the external standard/channels ratio methods and quench curves.

Various 1-minute sample collections were pooled and numbered as follows (Figure 1). Fraction 1 was composed of samples with a retention

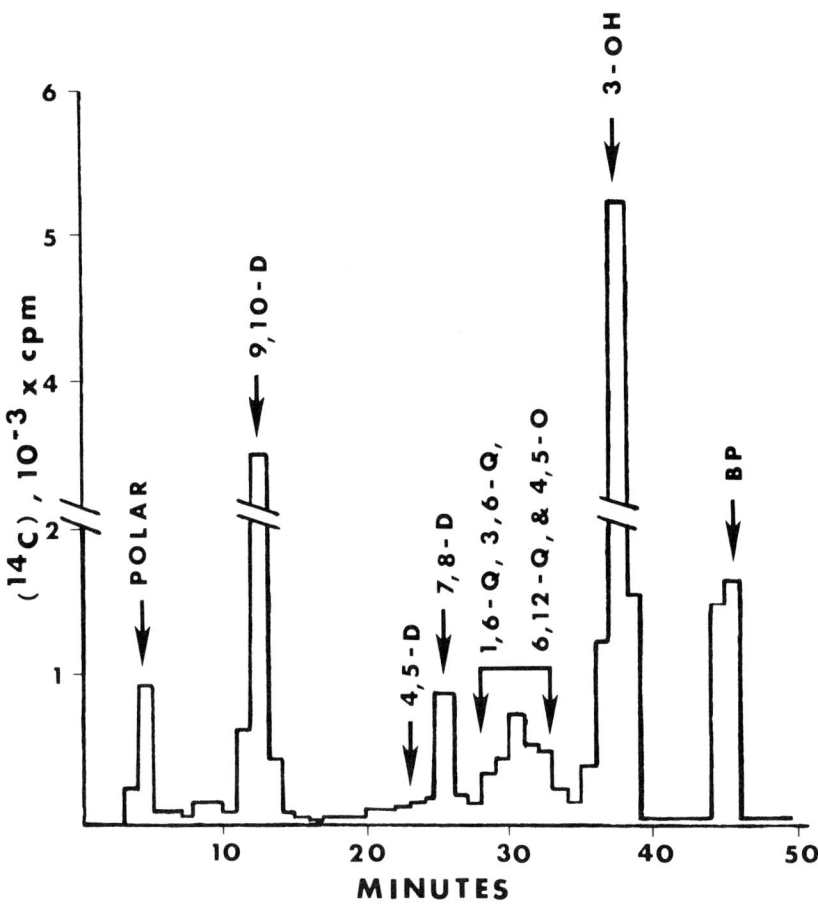

FIGURE 1. HPLC radiochromatogram of metabolites formed after addition of $^{14}$C-benzo(a)pyrene to the isolated perfused rabbit lung.

time of 3 to 11 minutes and contained the most polar compounds extracted. We believe these are tetrols and possibly extractable conjugates. Fraction 2 contained samples retained for 12 through 16 minutes. This fraction had the same retention time as a 9,10-dihydrodiol standard. Fraction 3 contained samples retained for 17 through 26 minutes, the same retention time as a 4,5-dihydrodiol standard. This fraction may contain other diols as well. Fraction 4 contained samples retained for 27 through 29 minutes, the same as a 7,8-dihydrodiol standard. Fraction 5 contained samples retained for 30 through 35 minutes. This is probably a mixture of quinones and 4,5 oxide. Fraction 6 was composed of samples retained for 36 to 43 minutes. The major peak in this group had the same retention time as a 3 phenol standard. Other phenols also may be present. Fraction 7 contained samples retained for 44 through 47 minutes which was the same retention time as $^{14}$C-BaP.

## Cigarette-Smoke and Gas-Phase Administration

In smoking experiments, 40 ml of fresh whole smoke from a nonradioactive, University of Kentucky 2R1 reference cigarette was generated into a glass reservoir by a single port-reverse smoking machine similar to that previously described (4). This puff was immediately drawn past the lung through a Y-shaped cannula, the lower arm of which was placed in the trachea. As the lung "inhaled" it would draw samples of this smoke from the reservoir through one of the upper arms of the cannula. A small vacuum applied to the other arm drew the smoke through the cannula, making it available to the lung during inhalation. The same vacuum served to clear smoke from the cannula as the lung exhaled, thereby keeping the recycling of smoke to a minimum. Smoke was in the lung for approximately 9 seconds. A 40-ml puff of smoke was generated and presented to the lung in this manner once every minute. The warmed humidified $CO_2$-air mixture was used to ventilate the lung between puffs. Respiration frequency was normally 50 cycles/min with 20 cycles/min being used during smoke exposure. About half the smoke generated was "inhaled" by the lung. Using this system, we demonstrated that the absorption of CO and nicotine into fluid perfusing the lung was linearly related to the number of puffs of smoke presented to the lung and that deposition of particulate matter was even throughout the pulmonary tissue (11).

In some experiments, only the gas phase from smoke was administered to the lung. To accomplish this a cambridge filter was placed between the smoke generating machine and the storage reservoir, thereby removing all particulate matter. The resulting gas phase was then administered to the lung in the same manner as whole smoke.

Cigarettes were conditioned at 60 percent humidity for at least 24 hours prior to smoking to ensure constant moisture content and more uniform burning characteristics. Cigarettes typically yield nine to ten 40-ml puffs when smoked completely with this apparatus.

## Statistical Analysis

Data were analyzed using the Statistical Analysis System general linear models procedure (3). The analysis was performed with and without

## 248 BaP METABOLISM SLOWED BY TOBACCO SMOKE

the arc-sine transformation for normalizing the distribution of percentage data. Results were identical in each case.

## RESULTS

### Effect of Pretreatment on $^{14}$C-BaP Metabolism by the IPL

In preliminary experiments the disappearance of $^{14}$C-BaP from the albumin buffer perfusing the IPL was determined using lungs obtained from corn-oil-, 3MC-, or TSC-pretreated rabbits (Figure 2). The disappearance was log linear for at least 15 minutes and was faster in lungs

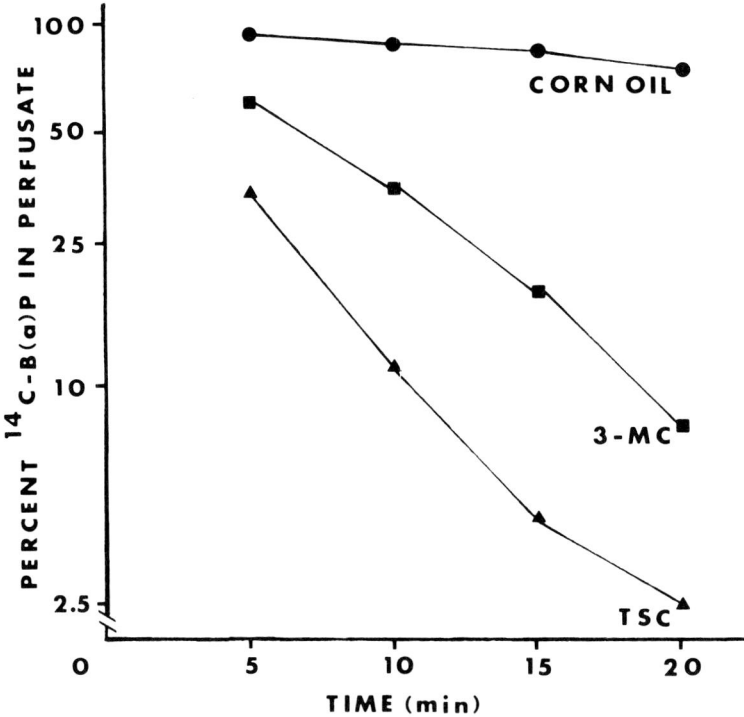

FIGURE 2. Disappearance of $^{14}$C-benzo(a)pyrene from buffer perfusing isolated rabbit lungs. Lungs were obtained from animals pretreated 24 hours earlier with corn oil, 3-methylcholanthrene (3MC), or tobacco-smoke condensate (TSC).

from 3MC- and TSC-pretreated rabbits than in lungs from rabbits pretreated with corn oil.

In a second series of experiments, the metabolic profile of radioactivity recovered from perfusate and lung after a 15-minute perfusion was determined (Table 1). The percentage of radioactivity remaining as unchanged BaP was considerably higher when lungs from corn-oil-pretreated rabbits were used, 48.3 ± 13.5 percent (mean ± S.E., N = 3), compared with that obtained with 3MC or TSC pretreatment, 10.9 ± 4.1 percent and 7.3 ± 1.7 percent, respectively.

Lungs from corn-oil-pretreated rabbits produced less metabolites than those from 3MC- or TSC-pretreated animals. For example, the nonextractable radioactivity primarily represented by water-soluble material occupied only 9.4 ± 1.2 percent of the radioactivity recovered with lungs from corn-oil-pretreated rabbits. Experiments with lungs from 3MC- and TSC-pretreated animals yielded 20.6±4.5 and 24.0±3.1 percent radioactivity as nonextractable material. This pattern continued with four of the six metabolite fractions collected from the HPLC. The two fractions in which there was no statistical difference between corn-oil- and 3MC- or TSC-pretreated animals were fraction 2 which had the same retention time as 9,10-dihydrodiol standard and fraction 6 which had the same retention time as a 3-phenol standard.

The influence of whole cigarette smoke or the gas phase from this smoke on the disposition of $^{14}$C-BaP by perfused lungs isolated from 3MC-pretreated rabbits is shown in Table 2. $^{14}$C-BaP was added to the perfusate immediately after lungs were exposed to smoke or gas phase. Experiments were terminated 20 minutes later. Whole smoke increased the percentage of unchanged $^{14}$C-BaP recovered from 7.51 ± 0.81 percent (mean ± S. E., N = 3) with no smoke to 24.64 ± 10.8, 57.05 ± 7.45, and 62.19 ± 4.61 percent with exposure to two, four, and eight puffs of smoke, respectively. When only the gas phase from eight puffs was administered, 10.36 ± 3.13 percent of radioactivity recovered was unchanged $^{14}$C-BaP. This was not significantly different from results for the control.

Lungs exposed to cigarette smoke produced less metabolites than did control lungs. This pattern was fairly consistent from nonextractable radioactivity through the six fractions obtained from the HPLC separation. Eight puffs of gas phase produced little change in the metabolite profile from that seen with no treatment.

Since some of the initial products formed from $^{14}$C-BaP are metabolized further, we investigated the effect of smoke on these processes as well. In these experiments, $^{14}$C-BaP was added to the buffer perfusing the IPL. Thirty minutes later, after most of the $^{14}$C-BaP was metabolized and levels of metabolites in perfusate reached their peak, the lung was smoked with four puffs of whole smoke and perfusion continued for another

TABLE 1. Percentage Distribution of Radioactivity from $^{14}$C-B(a)P Metabolism in Perfused Lungs Isolated from Variously Pretreated Rabbits (Mean ± S.E., N = 3)

| Treatment | Source | Nonextractable | Fraction 1 | 2 | 3 | 4 | 5 | 6 | B(a)P |
|---|---|---|---|---|---|---|---|---|---|
| Corn oil | Perfusate | 5.3±1.6 | 12.0±3.3 | 25.0±11.7 | 2.8±0.1 | 2.4±0.7 | 3.3±0.4 | 6.4±0.5 | 42.8±14.7 |
| | Lung | 10.2±1.5 | 2.6±1.1 | 10.8±3.8 | 1.4±0.3 | 2.3±0.8 | 4.8±1.2 | 18.6±6.5 | 49.3±13.2 |
| | Total | 9.4±1.2 | 4.0±0.9 | 13.6±5.6 | 1.6±0.2 | 2.3±0.8 | 4.5±1.1 | 16.3±5.0 | 48.3±13.5 |
| 3-MC | Perfusate | 12.6±4.0[a] | 19.5±4.9 | 43.4±3.2 | 4.0±0.9 | 3.8±0.1[a] | 3.3±0.7 | 6.7±0.7 | 6.8±1.7[a] |
| | Lung | 23.4±5.2[a] | 5.8±1.5[a] | 16.5±1.0 | 3.4±0.1[a] | 4.1±0.4[a] | 9.9±2.2[a] | 24.6±4.6 | 12.0±6.4[a] |
| | Total | 20.6±4.5[a] | 8.7±0.4[a] | 23.8±1.1 | 3.5±1.0[a] | 4.1±0.1[a] | 8.2±0.4 | 20.2±2.6 | 10.9±4.1[a] |
| TSC | Perfusate | 19.3±2.8[a] | 16.0±1.4 | 43.0±2.2 | 4.2±0.2 | 4.3±0.1[a] | 4.8±0.8 | 4.8±0.8 | 3.6±0.7[a] |
| | Lung | 25.6±3.4[a] | 7.6±1.0[a] | 16.1±1.5 | 3.5±0.9[a] | 4.4±0.3[a] | 9.5±0.5[a] | 24.8±1.5 | 8.6±2.1[a] |
| | Total | 24.0±3.1[a] | 9.7±0.7[a] | 22.9±1.5 | 3.7±0.7[a] | 4.4±0.2[a] | 8.3±0.6[a] | 19.8±1.4 | 7.3±1.7[a] |

[a]Significantly different from corn oil at $p < 0.05$.

TABLE 2. Effect of Cigarette Smoke on the Percentage Distribution of Radioactivity from $^{14}$C-B(a)P Metabolism in Perfused Rabbit Lungs (Mean ± S.E., N = 3)

| Number of Puffs Whole Smoke | Nonextractable | \multicolumn{7}{c}{Fraction} |
|---|---|---|---|---|---|---|---|---|
| | | 1 | 2 | 3 | 4 | 5 | 6 | B(a)P |
| 0 | 26.28±2.13 | 12.92±2.01 | 19.03±0.80 | 3.91±0.82 | 1.84±0.48 | 7.04±1.1 | 21.47±3.09 | 7.51± 0.81 |
| 2 | 9.31±1.31[a] | 2.97±0.91[a] | 26.17±5.04 | 0.89±0.62[a] | 4.98±0.32[a] | 6.8±0.90 | 24.23±2.09 | 24.64±10.8 |
| 4 | 10.00±4.68[a] | 1.55±0.41[a] | 10.87±3.73[a] | 1.38±0.46[a] | 2.05±0.48 | 3.48±0.93[a] | 13.55±3.27[a] | 57.05± 7.45[a] |
| 8 | 3.31±0.37[a] | 2.04±0.17[a] | 12.03±1.33[a] | 0.23±0.02[a] | 0.58±0.06[a] | 5.30±1.19 | 14.30±1.68[a] | 62.19± 4.61[a] |
| Gas phase only | | | | | | | | |
| 8 | 21.68±4.49 | 10.89±3.31 | 19.01±3.49 | 1.57±0.61[a] | 4.20±0.58[a] | 10.02±1.27 | 22.27±1.47 | 10.36± 3.18 |

[a]Significantly different from control (no smoke) at $p < 0.05$.

90 minutes. Results of these studies are illustrated in Figure 3. With control lungs, perfusate levels of various metabolites increased during the first 30 minutes and then slowly declined over the next 90 minutes. This was seen for fractions 2, 4, 5, and 6. Smoke exposure at 30 minutes

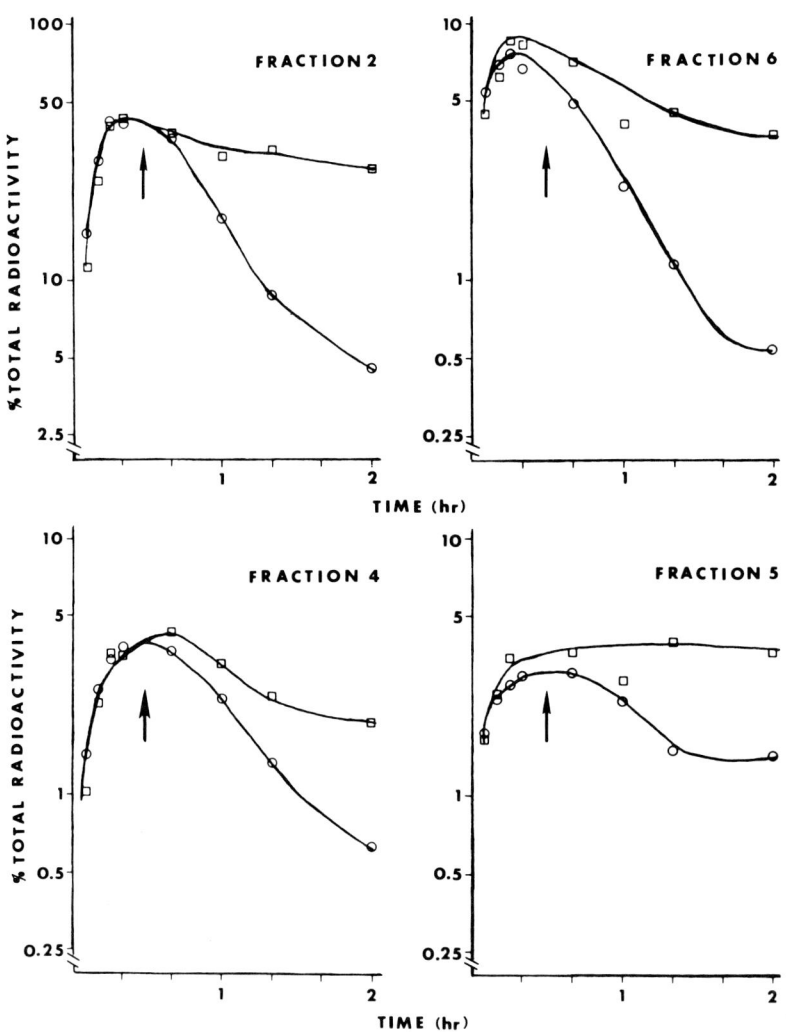

FIGURE 3. Effect of cigarette smoke on the level of $^{14}$C-benzo(a)pyrene metabolites in the buffer perfusing the isolated rabbit lung. Circle (o----o) represent control lungs. Squares (□----□) represent lungs exposed to four puffs of whole tobacco smoke (arrow) 30 minutes after addition of $^{14}$C-benzo(a)pyrene. Fraction numbers described in Methods and Materials section.

shows the disappearance of these fractions during the subsequent 90 minutes. Fractions 1 and 3 (not pictured) peaked at 30 minutes and then plateaued. Little change occurred during the subsequent 90 minutes with both control and smoke-exposed lungs.

Figure 4 shows the influence of four puffs of smoke on the appearance of nonextractable metabolites in the perfusate. As expected, smoke exposure caused the amount of these metabolites to plateau at a lower level, near 40 percent of the perfusate radioactivity, than controls which plateaued near 63 percent.

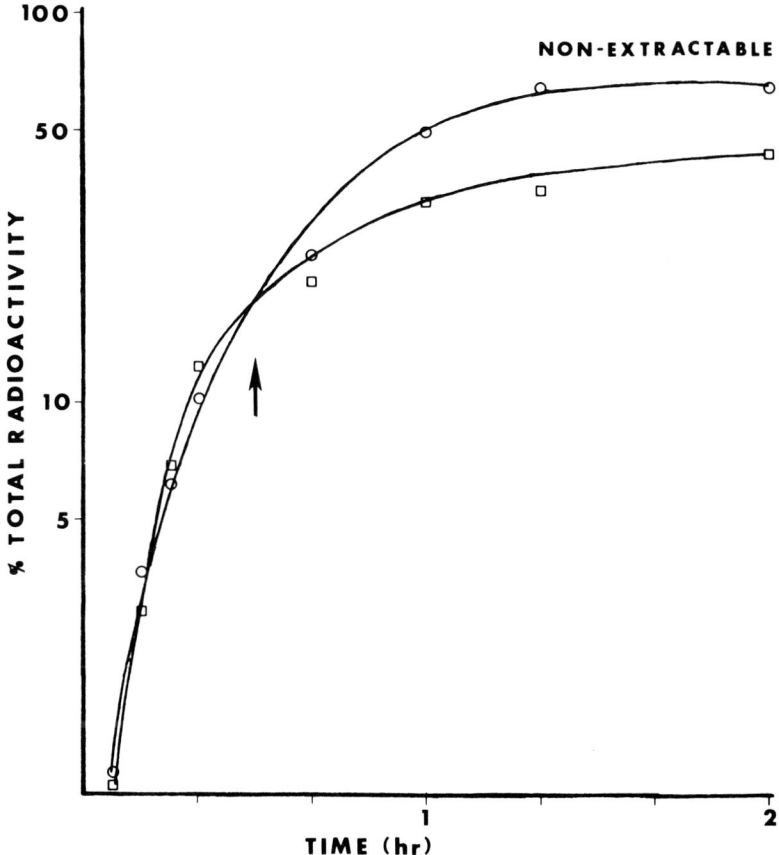

FIGURE 4. Effect of cigarette smoke on the appearance of nonextractable radioactivity in the buffer perfusing isolated rabbit lungs. Circles (o----o) represent control lungs. Squares (□----□) represent lungs exposed to four puffs of whole tobacco smoke (arrow) 30 minutes after addition of $^{14}$C-benzo(a)pyrene.

## DISCUSSION

These studies demonstrate that the immediate effect of cigarette smoke on the pulmonary metabolism of $^{14}$C-BaP is inhibition, in contrast to the delayed effect which is induction. Additionally, they demonstrate that the bioconversion of the primary metabolites of BaP is also decreased. The immediate effect of cigarette smoke reducing the metabolism of $^{14}$C-BaP is probably due to materials carried on the particulate matter of smoke, since the gas phase alone produced little effect.

The increased $^{14}$C-BaP metabolism produced by 3MC and whole-smoke-condensate administration in this study is consistent with the rabbit pulmonary enzyme induction reported by Bingham et al (5) following 3MC pretreatment and Warshawsky et al (27) followng BaP pretreatment. Smith et al (25), however, observed no increase in pulmonary BaP metabolizing ability following 3MC pretreatment. The reasons for these differences are not immediately apparent but may be related to the rabbits employed. We used male New Zealand rabbits which were the same strain used in studies where enzyme induction is seen (5, 27). Smith et al (25) used male Dutch Belt rabbits and did not observe pulmonary enzyme induction. Differences may also be due to the dosage of $^{14}$C-BaP placed in the IPL system. We added 0.07 $\mu$moles of BaP to the lung. Bingham et al (5) and Warshawsky et al (27) added 1.19 and 1.24 $\mu$moles, respectively. Smith et al (25) added 20 $\mu$moles.

The mechanism by which smoke initially inhibits $^{14}$C-BaP metabolism is unknown. It is conceivable that some of the 2,000 or more components in smoke may interfere with $^{14}$C-BaP metabolism by saturating the enzyme surface with alternative substrate. This may be analogous to hexabarbital inhibition of ethylmorphine demethylation (8) or phenothiazine inhibition of BaP hydroxylation (10).

The failure of the gas phase of smoke to dramatically alter $^{14}$C-BaP metabolism is interesting in light of the previous report of the ability of CO to bind cytochrome P-450 and inhibit xenobiotic metabolism (22). It is likely that insufficient CO is present in cigarette smoke to influence metabolism in the IPL, since very high levels are required to do this in microsomes. Additionally, the constant ventilation of the lung with air after smoke exposure would allow $O_2$ to rapidly displace CO.

The level of smoke exposure used in these experiments does not interfere with all functions of the lung. Hagedorn and Kostenbauder, using the same exposure IPL system described here, demonstrated that smoke did not influence the ability of the lung to metabolize prostaglandin F2$\alpha$ (12) or convert angiotensin I to angiotensin II (13).

These experiments indicate that the complex mixture of materials present in smoke can acutely influence the disposition of the individual components of that smoke. They also illustrate the problem involved with attempting to explain the cellular consequences of smoke exposure by examining the disposition and biologic activity of smoke components individually. However, these results must be interpreted cautiously with regard to their implication in the chronic human smoker. Because of the extreme difficulty of chronically exposing rabbits to cigarette smoke, lungs from 3MC-pretreated nonsmoked animals were used in these studies demonstrating acute smoke inhibition of BaP metabolism. Smoke exposure may result in enzyme changes that are not exactly duplicated by 3MC; therefore, lungs chronically exposed to smoke may have altered sensitivity to the inhibitory effect of acute smoke.

## ACKNOWLEDGMENTS

This investigation was supported by University of Kentucky Tobacco and Health Research Institute Project Number 098. The authors thank Dr. Robert Griffith for development of the smoke-exposure system and Ms. Ginny Maggard for invaluable assistance with surgical and assay procedures.

## REFERENCES

1. Abramson, R. K., and Hutton, J. J. (1975): Effects of cigarette smoking on aryl hydrocarbon hydroxylase activity in lungs and tissues of inbred mice. *Cancer Res.* 35:23-29.
2. Akin, F. J., and Benner, J. F. (1976): Induction of aryl hydrocarbon hydroxylase in rodent lung by cigarette smoke: A potential short-term bioassay. *Toxicol. Appl. Pharmacol.* 36:331-337.
3. Barr, A. J., Goodnight, J. H., Sall, J. P., and Helwig, J. T. (1976): A users guide to SAS 76, SAS Institute Inc., Raleigh.
4. Benner, J. F., Owens, S., Hancock, R., and Griffith, R. B. (1973): Smoking machine development and inhalation atmosphere monitoring. In: *Procedings of the University of Kentucky Tobacco and Health Workshop Conference No. 4,* pp. 494-506, Tobacco and Health Research Institute, Lexington.
5. Bingham, E., Niemeier, R., and Dalbey, W. (1976): Metabolism of environmental pollutants by the isolated perfused lung. *Fed. Proc.* 35:81-84.

6. Cohen, A. M., Uotila, P., Hartiala, J., Suolinna, E., Simberg, N., and Pelkonen, O. (1977): Metabolism and covalent binding of [$^3$H]Benzo(a)pyrene by isolated perfused lungs and short term tracheal organ culture of cigarette smoke-exposed rats. *Cancer Res.* 37:2147-2155.
7. Dipple, A. (1976): Polynuclear aromatic carcinogens. In: *Chemical Carcinogens*, C. E. Searle, Ed., pp. 245-314, American Chemical Society, Washington, D.C.
8. Feller, D. R., and Lubawy, W. C. (1973): Interactions of the hexobarbital enantiomers with rat liver microsomes. *Pharmacology* 99:129-137.
9. Fernandez, G., Villarruel, M. C., and Castro, J. A. (1978): Mechanism of the drug metabolizing enzymes' induction by 2-diethylaminoethyl-2-2-diphenylvalerate-HCl (SKF 525-A). *Toxicol. Appl. Pharmacol.* 46:315-321.
10. Goujon, F. M., and Nebert, D. W., and Gielen, J. E. (1972): Genetic expression of aryl hydrocarbon hydroxylase induction. IV. Interaction of various compounds with different forms of cytochrome P-450 and the effect of benzo(a)pyrene metabolism in vitro. *Mol. Pharmacol.* 8:667-680.
11. Griffith, R. B., Kostenbauder, H. B., and Lubawy, W. C. (1980): A tobacco smoke-perfused lung interface system. Submitted for publication.
12. Hagedorn, B., and Kostenbauder, H. B. (1977): Studies on the effect of tobacco smoke on biotransformation of vasoactive substances in the isolated perfused rabbit lung. I. Prostaglandin F$_2\alpha$. *Res. Commun. Chem. Pathol. Pharmacol.* 18:495-506.
13. Hagedorn, B., and Kostenbauder, H. B. (1978): Studies on the effect of tobacco smoke on the biotransformation of vasoactive substances in the isolated perfused rabbit lung. II. Angiotensin I. *Res. Commun. Chem. Pathol. Pharmacol.* 20:195-198.
14. Hoffman, D., Rathkamp, G., Brunnemann, K. D., and Wynder, E. L. (1973): Chemical studies on tobacco smoke. XXII. On the profile analysis of tobacco smoke. *Sci. Total Environ.* 2:157-171.
15. Hoffman, D., and Wynder, E. L. (1976): Environmental respiratory carcinogenesis. In: *Chemical Carcinogens*, edited by C. E. Searle, pp. 324-365, American Chemical Society, Washington, D.C.
16. Holt, P. G., and Keast, D. (1973): Induction of aryl hydrocarbon hydroxylase in the lungs of mice in response to cigarette smoke. *Experientia* 29:1004.
17. Kitada, M., Chiba, K., Kamataki, T., and Kitagawa, H. (1977): Inhibition by cyanide of drug oxidations in rat liver microsomes. *Japan. J. Pharmacol.* 27:601-608.

18. Lindner, E., and Beyhl, F. E. (1978): Induction of microsomal drug metabolizing enzymes caused by hexobarbital. *Experientia* 34:226.
19. McGovern, J. P., Lubawy, W. C., and Kostenbauder, H. B. (1976): Uptake and metabolism of nicotine by the isolated perfused rabbit lung. *J. Pharmacol. Exp. Ther.* 199:198-207.
20. Niemeier, R. W., and Bingham, E. (1972): An isolated perfused lung preparation for metabolic studies. *Life Sci. Part II, Biochem. Gen. Mol. Biol.* 11:807-820.
21. Noordhoek, J. (1968): Pharmacokinetics and dose-sleeping time lines of hexobarbital in mice. *European J. Pharmacol.* 3:242-250.
22. Omura, T., Sato, R., Cooper, D. Y., Rosenthal, O., and Estabrook, R. W. (1965): Function of cytochrome P-450 of microsomes. *Fed. Proc.* 24:1181-1189.
23. Serrone, D. M., and Fujimoto, J. M. (1961): The diphasic effect of N-methyl-3-piperidyl (N',N')-diphenylcarbamate HCl (MPOC) on the metabolism of hexobarbital. *J. Pharmacol. Exp. Therap.* 133:12-17.
24. Serrone, D. M., and Fujimoto, J. M. (1962): The effect of certain inhibitors in producing shortening of hexobarbital action. *Biochem. Pharamacol.* 11:609-615.
25. Smith, B. R., Philpot, R. M., and Bend, J. R. (1978): Metabolism of benzo(a)pyrene by the isolated perfused rabbit lung. *Drug Metab. Disp.* 6:425-431.
26. Vahakangas, K., Nevasaari, K., Pelkonen, O., and Karki, N. T. (1977): The metabolism of benzo(a)pyrene in isolated perfused lungs from variously treated rats. *Acta Pharmacol. et Toxicol.* 41:129-140.
27. Warshawsky, D., Niemeier, R. W., and Bingham, E. (1978): Influence of particulates on metabolism of benzo(a)pyrene in the isolated perfused lung. In: *Polynuclear Aromatic Hydrocarbons: Carcinogenesis*, Vol. 3, R. I. Freudenthal and P. W. Jones, Eds., pp. 347-360, Raven Press, New York.
28. Welch, R. M., Loh, A., and Conney, A. H. (1971): Cigarette smoke: Stimulatory effect on metabolism of 3,4-benzopyrene by enzymes in rat lung. *Life Sci.* 10:215-221.

# MANGANIC ACETATE AND HORSERADISH PEROXIDASE/HYDROGEN PEROXIDE: IN VITRO MODELS OF ACTIVATION OF AROMATIC HYDROCARBONS BY ONE-ELECTRON OXIDATION

E.G. Rogan, R. Roth*, and E. Cavalieri

Eppley Institute for Research in Cancer
University of Nebraska Medical Center
42nd and Dewey Ave.
Omaha, NE 68105

## INTRODUCTION

It has been suggested that metabolic oxidation of polycyclic aromatic hydrocarbons (PAH) to radical cations capable of binding to cellular nucleophiles might be an important in vivo pathway of carcinogen activation (7,11,17). Some of the properties of these intermediates have been investigated in the past using a system consisting of iodine as oxidant and pyridine as nucleophile and solvent (6).

We have recently completed the study of a second chemical model for one-electron oxidation, namely, Mn (III) acetate in acetic acid. This system was chosen because it was already known that PAH having ionization potentials (IPs) below about 8 eV react with Mn (III) acetate by a one-electron transfer mechanism involving intermediate PAH radical cations (1-3,12). This oxidant system has been previously studied in detail for only a few substituted benzenes and naphthalenes. At the temperatures previously employed (>90°C), secondary oxidation reactions and competing pathways involving thermal decomposition of manganic acetate to carboxymethyl radicals complicate the results. It was expected that for the PAH of interest, which have much lower IPs (<7.5 eV), clean reactions would occur at lower temperatures. Such has proven the case.

---

*Current Address: Midwest Research Institute, 425 Volker Blvd., Kansas City, MO, 64110.

This system has been quite useful in better understanding properties of PAH radical cations, as well as of considerable synthetic utility in preparing acetate derivatives of PAH.

We have also developed a new biochemical model system for studying one-electron oxidation of PAH, namely activation of these compounds by horseradish peroxidase (HRP) in the presence of $H_2O_2$ (16). This enzyme, which contains a heme group and is similar to cytochrome P-450, activates N-hydroxy-2-acetylaminofluorene (4,5,8-10) and the aromatic hydrocarbons benzo(a)pyrene (BaP) and 6-methylbenzo(a)pyrene (BaP-6-$CH_3$)(16) by one-electron oxidation. It may be that such activations in animals (14) are critical in the tumor initiation process.

These chemical and biochemical approaches are useful in the study of the chemical properties of the reactive radical cations as well as for their ability to form adducts of PAH with biological macromolecules by one-electron oxidation.

## RESULTS AND DISCUSSION

### One-Electron Oxidation of PAH by Mn (III) Acetate

Clean, dry, soldi $Mn(OAc)_3 \cdot 2H_2O$ (0.400 mmole) was added to a stirred solution or partial suspension of 0.200 mmole of hydrocarbon in 25 ml of dry acetic acid under nitrogen at 40°C. The manganic acetate was added all at once, and the progress of each reaction was followed by thin-layer chromatography (TLC). When the manganic ion had been consumed or when the reaction stopped progressing, the mixture was diluted with chloroform; washed with successive portions of water, aqueous sodium bisulfite, sodium bicarbonate, and water; and dried over $Na_2SO_4$. Crude product mixtures were initially analyzed by comparison with authentic standards on TLC and by nuclear magnetic resonance (NMR) spectroscopy. Subsequently, major products were recrystallized after isolation by column chromatography on alumina or silica gel and identified in most cases by NMR, mixed mp, and TLC comparisons with authentic compounds prepared by published procedures.

The results of these studies are given in Table 1 and are readily interpreted in terms of a mechanistic scheme analogous to that presented previously (6). Compounds such as phenanthrene and chrysene that have a relatively high IP are not oxidized by $Mn^{3+}$ at 40°C. For unsubstituted PAH with relatively low IP, aromatic acetates are derived from direct attack by acetate ion on the radical cation at positions of maximum charge density followed by $Mn^{3+}$ oxidation of the resulting radical and loss of a proton (Figure 1 and Table 1: compounds 3, 4, and 6). When positions

## TABLE 1. One-Electron Oxidation by Manganic Acetate

| Compound | Reaction Time | Product | Yield[a] (%) | Ionization Potential[b] (eV) |
|---|---|---|---|---|
| 1. Phenanthrene | 96 hr | No Rxn | — | 8.19 |
| 2. Chrysene | 96 hr | No Rxn | — | 7.8 |
| 3. Benz(a)anthracene (BaA) | 48 hr | BaA-7-OAc | 90-100 | 7.54 |
| 4. Pyrene | 96 hr | Pyrene-1-OAc | 15-20[c] | 7.50 |
| 5. BaA-7-$CH_3$ | 24 hr | BaA-7-$CH_2$OAc | 85 | 7.37 |
|  |  | BaA-7-$CH_3$-12-OAc | 10 |  |
| 6. Benzo(a)pyrene (BaP) | <10 min | BaP-6-OAc | 95 | 7.23 |
|  |  | quinones | 5 |  |
| 7. BaA-7, 12-$CH_3$ | <10 min | BaA-12-$CH_3$-7$CH_2$OAc | 50-60 | 7.22 |
|  |  | BaA-7-$CH_3$-12-$CH_2$OAc | 40-50 |  |
| 8. 3-Methylcholanthrene (MC) | <10 min | MC-1-OAc | 100 | 7.19 |
| 9. BaP-6-$CH_3$ | 10 min | BaP-6-$CH_2$OAc | 75-80 | 7.08 |
|  |  | BaP-6-$CH_3$-1-OAc[d] | 15-20 |  |
|  |  | BaP-6-$CH_3$-3-OAc[d] |  |  |
| 10. BaP-6-$CH_2$OH | 2.5 hr | BaP-6-OAc[d,e] | 85 |  |
|  |  | BaP-6-CHO[d,e] | 15 |  |
| 11. BaA-7-$CH_2$OH | 48 hr | BaA-7-OAc | 30 |  |
|  |  | BaA-7-CHO | 20 |  |
| 12. MC-1-OH | <10 min | MC-1-one | 100 |  |
| 13. MC-2-OH | <10 min | MC-1-OAc-2-OH[f] | 30 |  |
|  |  | (cis and trans) |  |  |
| 14. MC-2-one | 16 hr | MC-1-OAc-2-one[d,f] | 90 |  |
| 15. BaP-6-CHO | 48 hr | No Rxn | — |  |
| 16. BaA-7-CHO | 96 hr | No Rxn | — |  |
| 17. MC-1-one | 64 hr | No Rxn | — |  |

[a] Based on NMR analysis of the crude product prior to chromatography.
[b] Determined by charge-transfer complex with chloranil.
[c] The remaining amount was starting material.
[d] Isolated by HPLC.
[e] Separated by chromatography on MgO/celite 2:1 in benzene-chloroform.
[f] Identified by NMR and mass spectra.

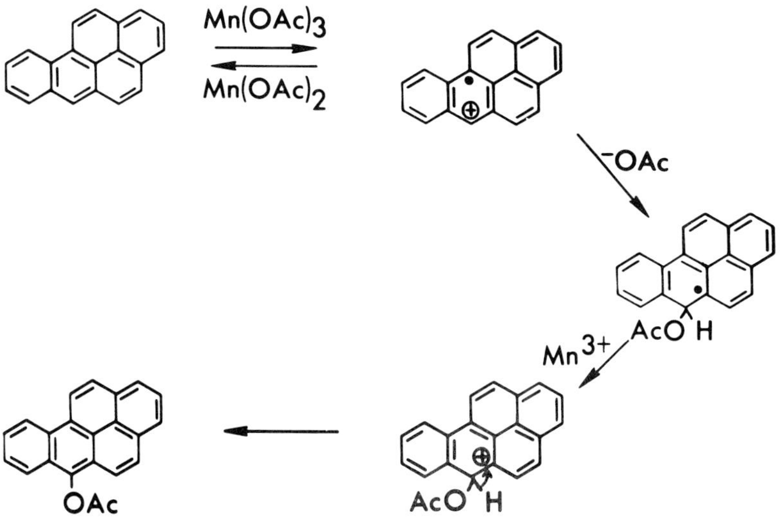

FIGURE 1. One-electron oxidation of benzo(a)pyrene by Mn(III) acetate and trapping of its radical cation by acetate ion.

corresponding to maximum charge density in the parent hydrocarbon bear a carbonyl group, a polar substituent that would drastically destabilize the adjacent positive charge, no reaction occurs (Table 1: compounds 15, 16, and 17). When, however, these sites bear a substituent having a benzylic proton, loss of that proton from the radical cation can ensue to give a radical species that is rapidly oxidized by $Mn^{3+}$. The resulting benzylic cation is trapped by acetate, giving side-chain substitution (Figure 2 and Table 1: compounds 5, 7, 8, 9, and 13). If the analogous benzylic position bears a hydroxyl substituent, loss of the hydroxyl proton from the benzylic cation may occur instead, to give the corresponding carbonyl compounds as minor products in the case of BaP and benz(a)anthracene (BaA) hydroxymethyl derivatives or the exclusive product in the case of 1-hydroxy-3-methylcholanthrene (MC-1-OH) (Table 1: compounds 10, 11, and 12). The latter compound represents a special case in which loss of the benzylic proton is favored by the coplanarity of the benzylic C-H bond with the aromatic $\pi$ system. The major products derived from the hydroxymethyl derivatives are the corresponding aromatic acetates (Table 1: compounds 10 and 11). These compounds are presumably formed initially by a pathway analogous to that of the parent hydrocarbons, i.e., direct attack by acetate on the aromatic radical cation at the position of maximum charge density followed by loss of an hydroxyl proton and formaldehyde (Figure 3). In the case of MC-2-one (Table 1: compound 14),

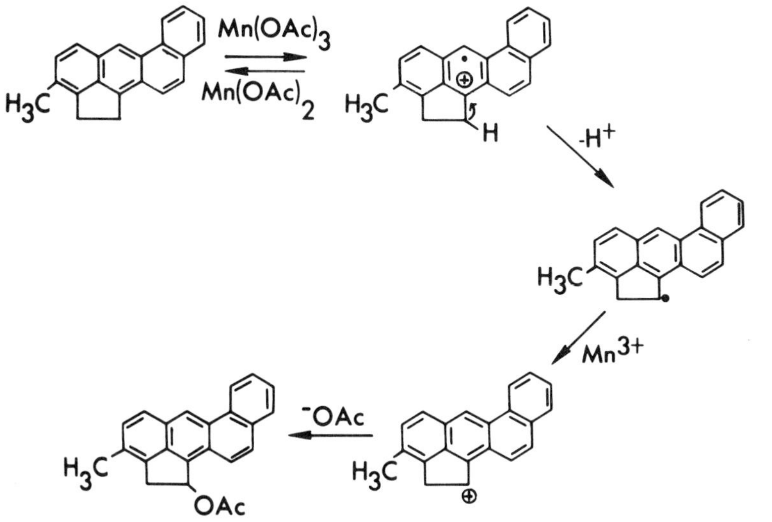

**FIGURE 2.** One-electron oxidation of 3-methylcholanthrene by Mn(III) acetate and trapping of its radical cation by acetate ion.

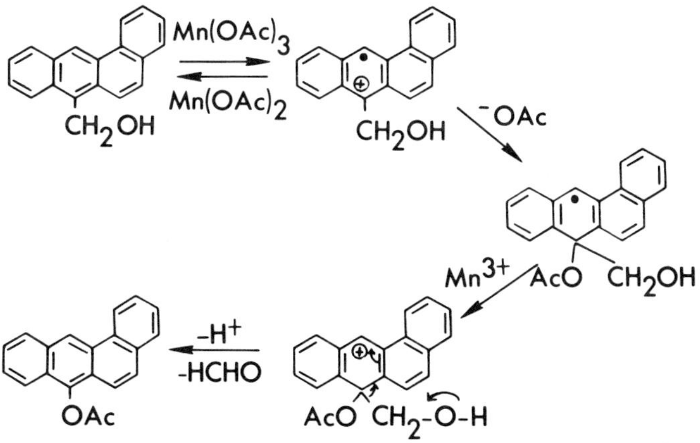

**FIGURE 3.** One-electron oxidation of 7-hydroxymethylbenz(a)anthracene by Mn(III) acetate and trapping of its radical cation acetate ion at the position of maximum charge density.

the substitution at the benzylic position occurs at a lower rate of reaction because of the presence of an adjacent carbonyl group, which destabilizes the radical cation, thus increasing the IP of the hydrocarbon.

## Binding of PAH to DNA by HRP/$H_2O_2$

HRP (EC 1.11.1.7), type II, Sigma Chemical Company, was used in studies of HRP/$H_2O_2$-catalyzed binding of PAH to DNA. The [$^{14}$C]PAH are BaA, BaP, and BaA-7, 12-$CH_3$ from Amersham (Arlington Heights, Illinois); anthracene, dibenz(a,h)anthracene, and phenanthrene from California Bionuclear (Sun Valley, California); 3-methylcholanthrene (MC) from New England Nuclear (Boston, Massachusetts); and BaA-7-$CH_3$ and BaP-6-$CH_3$, which were synthesized in our laboratory. The standard 1-ml reaction mixtures contained 0.067 M Sörenson buffer, pH 7.0; 4mM DNA; 0.2 mg HRP; 30 to 90 $\mu$M [$^{14}$C]PAH in 10 $\mu$l dimethylsulfoxide; and 5 x $10^{-4}$M $H_2O_2$. Chemicals were added in the order listed, and the reaction mixtures were incubated for 30 minutes at 37°C in the dark, unless otherwise specified. The reaction was stopped with the addition of 0.1 ml of 20 X standard saline citrate and 0.1 ml of 10 percent sodium lauryl sulfate. DNA was purified according to Rogan and Cavalieri (15) and analyzed for DNA content by absorption at 259 nm and bound [$^{14}$C]PAH by liquid scintillation counting. A minimum of duplicate experiments was performed. For each experiment, control incubations were included which did not contain $H_2O_2$. The same background levels of binding were obtained when HRP was omitted instead of $H_2O_2$.

The ability of nine [$^{14}$C]PAH to serve as substrates in HRP/$H_2O_2$-catalyzed binding to DNA has been correlated with their IPs. Table 2 shows that the levels of binding differed over about a 30-fold range. Those

Table 2. HRP/$H_2O_2$-Catalyzed Binding to DNA[a]

| Compound | [$^{14}$C]PAH ($\mu$mol/mol DNA)[b] | [$^{14}$C]PAH Bound in Controls ($\mu$mol/mol DNA) | Ionization Potential[c] (eV) |
|---|---|---|---|
| Phenanthrene | 3.8±0.8 (11) | 1.2±0.4 (7) | 8.19 |
| Dibenz(a,h)anthracene | 4.3±1.0 (10) | 1.4±0.5 (8) | 7.57[d] |
| Benz(a)anthracene | 4.0±0.5 (12) | 0.9±0.2 (8) | 7.54 |
| Anthracene | 8.8±1.6 (9) | 6.3±0.7 (4) | 7.43 |
| 7-Methylbenz(a)anthracene | 5.6±0.6 (6) | 2.6±0.2 (4) | 7.37 |
| Benzo(a)pyrene | 89.2±5.6 (8) | 1.3±0.3 (18) | 7.23 |
| 7,12-Dimethylbenz(a)-anthracene | 63.9±4.6 (12) | 1.0±0.2 (8) | 7.22 |
| 3-Methylcholanthrene | 60.6±4.1 (10) | 3.4±0.2 (4) | 7.19 |
| 6-Methylbenzo(a)pyrene | 39.8±5.3 (9) | 2.2±0.2 (7) | 7.08 |

[a]Reaction mixtures were incubated as described in text. Either $H_2O_2$ or horseradish peroxidase was omitted from control incubations.
[b]Values are corrected for controls and are given ± standard error of measurement. Number in parentheses indicates the number of determinations.
[c]Determined from maximum absorption of the charge-transfer complex of each compound with chloranil.
[d]Determined by polarographic oxidation (13).

PAH with IPs above about 7.35 eV were minimally bound, while BaP, BaP-6-CH$_3$, BaA-7,12-CH$_3$ and MC, with IPs below about 7.35 eV, were bound at significant levels.

## CONCLUSIONS

In the model system consisting of Mn (III) acetate in acetic acid as one-electron oxidant, the nucleophilic attack of the acetate ion takes place at the position(s) of maximum charge density. In the case of alkyl-substituted PAH, the attack occurs generally at the methyl group when substituted at the position of maximum charge density. These data allow us to predict the position(s) involved in covalent binding of PAH to cellular macromolecules when activation occurs via one-electron oxidation.

Study of the binding of a series of PAH to DNA catalyzed by the HRP/H$_2$O$_2$ system shows that the level of binding is influenced by the IP of the PAH. This represents a further indication (16) that HRP activates PAH by one-electron oxidation. The HRP/H$_2$O$_2$ oxidizing system can be used to form PAH adducts with biological macromolecules in vitro, which can then be compared with PAH adducts produced in vivo.

## ACKNOWLEDGMENTS

This work was supported by National Cancer Institute contract NO1 CP33278. We wish to express our thanks for the skillful technical assistance of K. Saugier and P. Katomski-Beck.

## REFERENCES

1. Andrulis, P. J. Jr., and Dewar, M.J.S. (1966): Aromatic oxidation by electron transfer. III. Oxidation of 1- and 2-methoxynaphthalene by manganic acetate. *J. Am. Chem. Soc.* 88:5483-5485.
2. Andrulis, P.J. Jr., Dewar, M.J.S., Dietz, R., and Hunt, R.L. (1966): Aromatic oxidation by electron transfer. I. Oxidations of *p*-methoxytoluene. *J. Am. Chem. Soc.* 88:5473-5478.
3. Aratani, T., and Dewar, M.J.S. (1966): Aromatic oxidation by electron transfer. II. Oxidations of aromatic ethers and amines by manganic acetate. *J. Am. Chem. Soc.* 88:5479-5482.
4. Bartsch, H., and Hecker, E. (1971): On the metabolic activation of the carcinogen *N*-hydroxy-*N*-2-acetylaminofluorene. III. Oxidation with horseradish peroxidase to yield 2-nitrosofluorene and acetoxy-*N*-2-acetylaminofluorene. *Biochem. Biophys. Acta* 237:567-578.
5. Bartsch, H., Miller, J. A., and Miller, E. C. (1972): Acetoxy-*N*-acetylaminoarenes and nitrosoarenes. One-electron non-enzymatic and enzymatic oxidation products of various carcinogenic aromatic acethydroxamic acids. *Biochem. Biophys. Acta* 273:40-51.

6. Cavalieri, E., and Roth, R. (1976): Reaction of methylbenzanthracene and pyridine by one-electron oxidation: A model for metabolic activation and binding of carcinogenic aromatic hydrocarbons. *J. Org. Chem.* 41:2679-2684.
7. Cavalieri, E., Roth, R., and Rogan, E. G. (1976): Metabolic activation of aromatic hydrocarbons by one-electron oxidation in relation to the mechanism of tumor initiation. In: *Carcinogenesis, Vol. I., Polynuclear Aromatic Hydrocarbons: Chemistry, Metabolism and Carcinogenesis,* R. I. Freudenthal and P. W. Jones, Eds., pp. 181-190, Raven Press, New York.
8. Floyd, R. A., and Soong, L. M. (1977): Obligatory free radical intermediate in the oxidative activation of the carcinogen $N$-hydroxy-2-acetylaminofluorene. *Biochem. Biophys. Acta* 498:244-249.
9. Floyd, R. A., Soong, L. M., and Culver, P. L. (1976): Horseradish peroxidase/hydrogen peroxide-catalyzed oxidation of the carcinogen $N$-hydroxy-$N$-acetyl-2-aminofluorene as affected by cyanide and ascorbate. *Cancer Res.* 36:1510-1517.
10. Floyd, R. A., Soong, L. M., Walker, R. D., and Stuart, M. (1976): Lipid hydroperoxide activation of $N$-hydroxy-$N$-acetylaminofluorene via a free radical route. *Cancer Res.* 36:2761-2767.
11. Fried, J. (1974): One-electron oxidation of polycyclic aromatics as a model for the metabolic activation of carcinogenic hydrocarbons. In: *Chemical Carcinogenesis,* Part A, P.O.P. Ts'o and J. DiPaolo, Eds., pp. 197-215, Marcel Dekker, New York.
12. Heiba, E. I., Dessau, R. M., and Koshl, W. J., Jr. (1969): Oxidation by metal salts. III. The reaction of manganic acetate with aromatic hydrocarbons and the reactivity of the carboxymethyl radical. *J. Am. Chem. Soc.* 91:138-141.
13. Pysh, E. S., and Yang, N. C. (1963): Polarographic oxidation potentials of aromatic compounds. *J. Am. Chem. Soc.* 85:2124-2130.
14. Reigh, D. L., Stuart, M., and Floyd, R. A. (1978): Activation of the carcinogen $N$-hydroxy-2-acetylaminofluorene by rat mammary peroxidase. *Experientia,* 34:107-108.
15. Rogan, E. G., and Cavalieri, E. (1974): 3-Methylcholanthrene-inducible binding of aromatic hydrocarbons to DNA in purified rat liver nuclei. *Biochem. Biophys. Res. Commun.* 58:1119-1126.
16 Rogan, E. G., Katomski, P. A., Roth, R. W., and Cavalieri, E. L. (1979): Horseradish peroxidase/hydrogen peroxide-catalyzed binding of aromatic hydrocarbons to DNA. *J. Biol. Chem.* 254:7055-7059.
17. Wilk, M., Bez, W., and Rochlitz, J. (1966): Neue Reaktionen der carcinogenen Kohlenwasserstoffe 3,4-Benzpyrene, 9,10-dimethyl-1,2-benzanthracene und 20-methylcholanthrene. *Tetrahedron* 22:2599-2608.

# COMPARATIVE METABOLISM OF DIHYDRODIOLS OF POLYCYCLIC AROMATIC HYDROCARBONS TO BAY-REGION DIOL EPOXIDES

D. R. Thakker*, W. Levin**, H. Yagi*, M. Tada*,
A. H. Conney** and D. M. Jerina*

*Section on Oxidation Mechanisms, Laboratory of Bioorganic Chemistry
National Institute of Arthritis, Metabolism, and Digestive Diseases
National Institutes of Health
Bethesda, Maryland 20205

**Department of Biochemistry and Drug Metabolism
Hoffmann-La Roche Inc.
Nutley, New Jersey 07110

## INTRODUCTION

Carcinogenicity of the Polycyclic Aromatic Hydrocarbons (PAH)[†] has been attributed to their metabolism to reactive metabolites by cytochrome P-450-dependent monooxygenases and epoxide hydrase. Rapid advances in recent years have led to the discovery that (+)-benzo(a)pyrene (BaP) 7,8-diol 9,10-epoxide-2 (Figure 1) is an ultimate carcinogenic metabolite of BaP (1). The results from our and other laboratories which led to the identification of a bay-region diol epoxide as an ultimate carcinogenic metabolite of BaP have been reviewed (2,13). In the course of this first identification of an ultimate carcinogen from any

---

[†] Abbreviations used are: BaP, benzo(a)pyrene; BA, benzo(a)anthracene; BeP, benzo(e)pyrene; DBA, dibenzo(a,h)anthracene; PAH, polycyclic aromatic hydrocarbon; (±BaP 7,8-dihydrodiol, (±)-*trans*-7,8-dihydroxy-7,8-dihydro BaP; BaP diol epoxide-1, (±)-7β, 8α-dihydroxy-9β,10β-epoxy-7,8,9,10-tetrahydro BaP; BaP diol epoxide-2, (±)7β,8α-dihydroxy-9α,10α-epoxy-7,8,9,10-tetrahydro BaP; diol epoxides of other PAH are defined in a similar manner: BA 3,4-dihydrodiol, *trans*-3,4-dihydroxy-3,4-dihydrobenzo(a)anthracene; dihydrodiols of phenanthrene, chrysene, BeP and DBA are similarly abbreviated; BA H₄-1-ol, 1-hydroxy-1,2,3,4-tetrahydro BA:HPLC, high-pressure liquid chromatography.

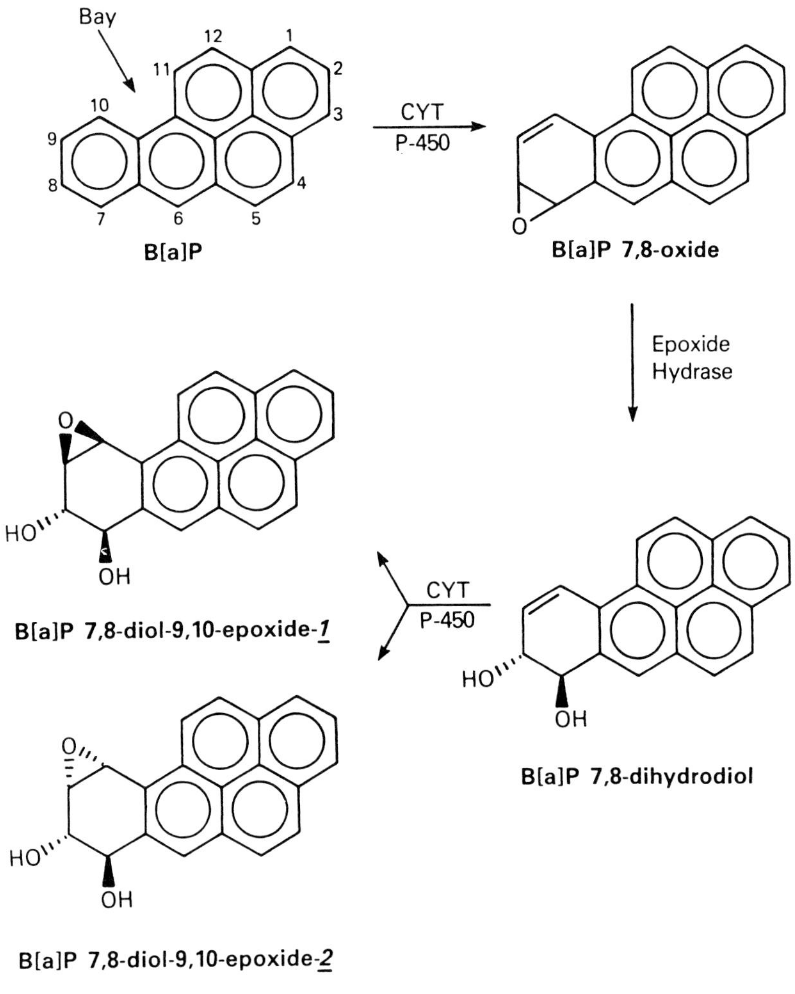

**FIGURE 1.** Metabolic activation of BaP to bay-region diol epoxides. The enantiomer of diol epoxide-2 shown is the highly tumorigenic (+)-2 enantiomer.

member of the polycyclic hydrocarbon class of carcinogens, we have developed a unified theory to explain and predict carcinogenicity of PAH. This concept, known as the bay-region theory, postulates that epoxides on saturated, angular benzo-rings, when present in the bay-region of a PAH (Figure 1), should possess high chemical reactivity and biological activity (4-6). Hence, the bay-region theory proposes that bay-region diol epoxides should be ultimate carcinogenic forms of the parent PAH. Studies on at least seven different PAH have borne out the predictions of the theory (see References 2,4-6,13). Although the bay-region theory involves no attempt to take metabolic factors into account in the prediction of relative carcinogenicity of the PAH, metabolism does play an important role as a determinant of carcinogenicity for this class of carcinogens. In the previous Battelle symposium, we had discussed comparative metabolism of a series of PAH to their corresponding dihydrodiols with bay-region double bonds in relation to their relative carcinogenicity (17). In this chapter we extend such considerations one step further and discuss the metabolism of trans dihydrodiols of several PAH to bay-region diol epoxides. The structures of the dihydrodiols and the bay-region diol epoxides under consideration are shown in Figures 2 and 3, respectively. Bay-region diol epoxides, when placed in aqueous solutions, undergo solvolysis to tetraols by trans and cis- attack of water at the benzylic position of the epoxide (Figure 4) (25). In addition, diol epoxide-1 isomers of several PAH isomerize to keto-diols at neutral or alkaline pH (21,25). Furthermore, the isomer-1 of several bay-region diol epoxides have been noted to react with methanol when present in the mobile phase used to chromatograph such diol epoxides on reverse-phase ODS columns. Generalized examples of the solvolytic and isomerization products formed from bay-region diol epoxides are shown in Figure 4. Identification of such products as metabolites may be taken as evidence for the formation of diol epoxides from a dihydrodiol.

## BaP 7,8-DIHYDRODIOL

(±)-BaP 7,8-dihydrodiol is metabolized by cytochrome P-450-dependent monooxygenases to a pair of diastereromeric bay-region diol epoxides (Figure 1) (20). With microsomes from 3-methylcholanthrene-treated rats, the dihydrodiol is metabolized almost exclusively to the 7,8-diol-9,10-epoxides (83 to 92 percent of total metabolites). The diol epoxide-2 and diol epoxide-1 (cf. Figure 1) diastereomers are formed in the ratios of 1.4-1.8 to 1. The (-)-[7R,8R]-enantiomer of BaP 7,8-dihydrodiol is metabolized by cytochrome P-448, primarily to diol epoxide-2; the ratio of diol epoxide-2 to diol epoxide-1 formed being 6 to 1 (18). In contrast, (+)-BaP-[7S,8S]-dihydrodiol is metabolized to diol epoxide-2 and diol epoxide-1 in the ratio of 1 to 22. Metabolically formed BaP 7,8-dihydrodiol consists predominantly of the (-)-[7R,8R]-enantiomer (96 percent) and hence is metabolized to (+)-diol epoxide-2 as the major product (~88 percent of total metabolites). Thus, high stereospecificity is observed in the metabolism of BaP 7,8-dihydrodiol. Interestingly, (+)-BaP 7,8-diol-9,10-epoxide-2, the major product of metabolically formed BaP 7,8-dihydrodiol, is the most tumorigenic metabolite of BaP in the newborn mouse tumor model (1).

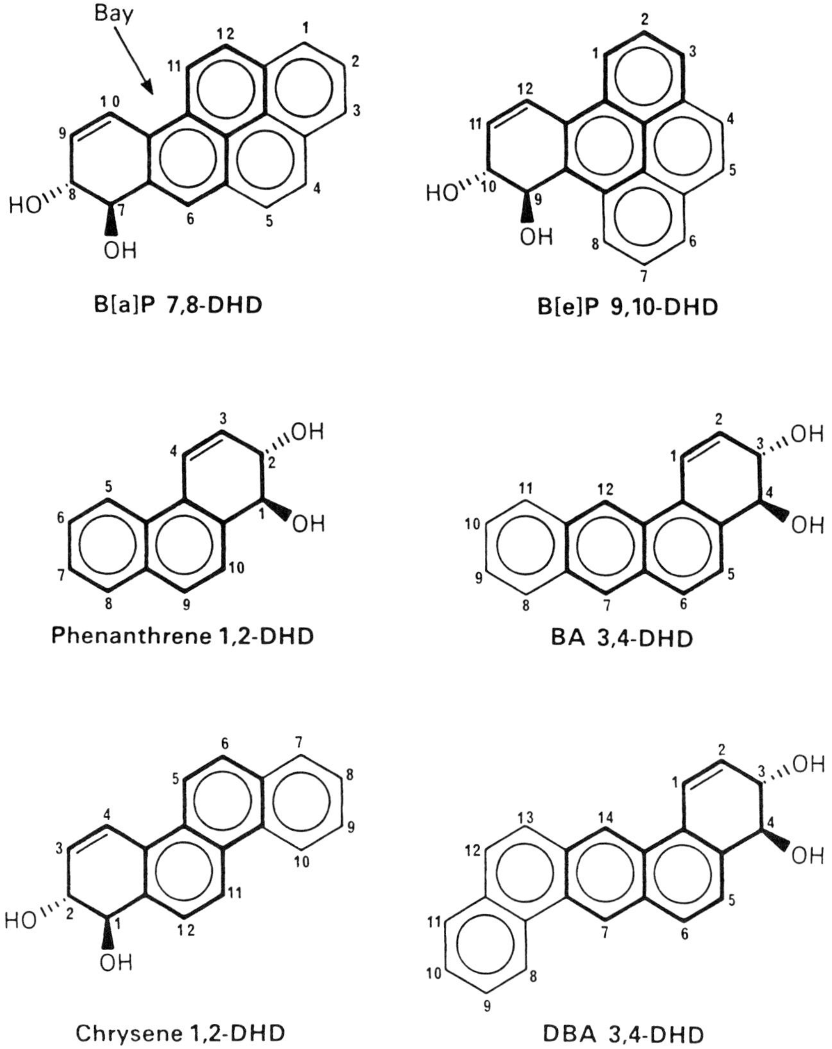

**FIGURE 2.** Dihydrodiols of several PAH with bay-region double bond. The darkened phenanthrene portion of each structure emphasizes the bay region in this figure and Figure 3.

## BENZO(a)ANTHRACENE 3,4-DIHYDRODIOL

Since unlabelled benzo(a)anthracene (BA) 3,4-dihydrodiol substrates were used, quantitation of the substrate consumed and the tetraols

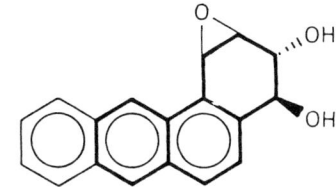
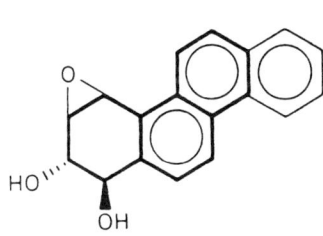
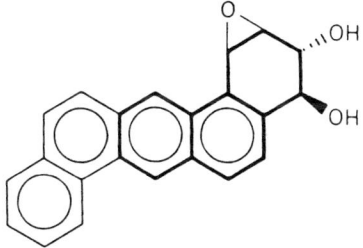

B[a]P 7,8-diol-9,10-epoxide

B[e]P 9,10-diol-11,12-epoxide

Phenanthrene 1,2-diol-3,4-epoxide

BA 3,4-diol-1,2-epoxide

Chrysene 1,2-diol-3,4-epoxide

DBA 3,4-diol-1,2-epoxide

**FIGURE 3.** Bay-region diol epoxides of several PAH. The isomers in which the oxirane oxygen is cis to the benzylic hydroxyl group are defined as diol epoxides-1, and the isomers in which these two groups are trans to each other are defined as diol epoxides-2. The structures shown do not imply either diastereomer.

formed was achieved using BA H$_4$-1-ol as an internal standard. Chromatographic mobilities of the tetraols derived from BA 3,4-diol-1,2-epoxides on solvolysis, BA 3,4-dihydrodiol, and the internal standard BA H$_4$-1-ol are shown in Figure 5a. The HPLC separation of metabolites of (±)-BA 3,4-dihydrodiol is shown in Figure 5b. Comparison of the two

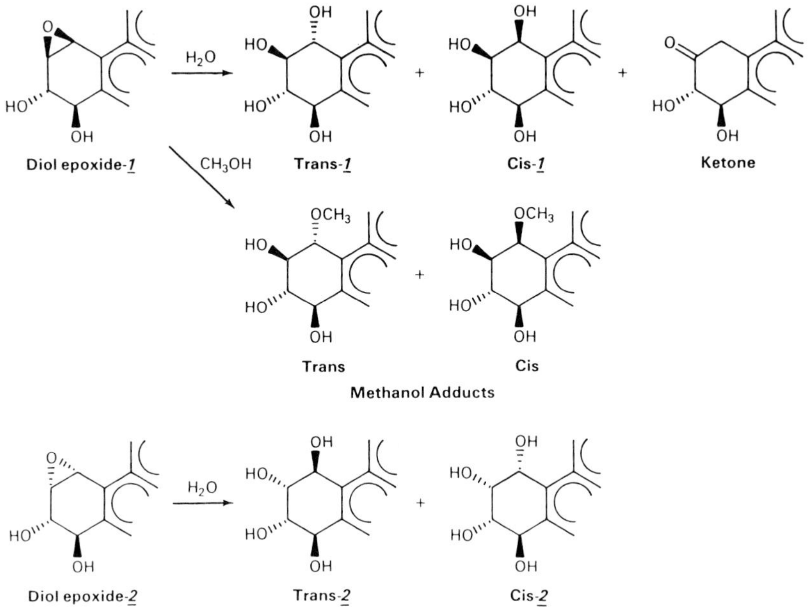

## 272 METABOLISM OF DIHYDRODIOLS OF PAH

**FIGURE 4.** Products formed upon solvolysis and isomerization of the diastereomeric bay-region diol epoxides of a PAH.

chromatograms indicates that tetraols from diol epoxide-$\underline{2}$, i.e., trans-$\underline{2}$ and cis-$\underline{2}$, are formed in much greater quantities than tetraols from diol epoxide-$\underline{1}$. In addition, two other metabolites, $M_1$ and $M_2$ are formed in significant quantities. Both these metabolites show similar UV spectra. The UV peaks for metabolites $M_1$ and $M_2$ do not necessarily reflect their true relative amounts among the metabolites of BA 3,4-dihydrodiol since their extinction coefficients, which have not yet been determined, are expected to be different from those of the tetraols of BA. When effluent was monitored at 260 nm, the peaks due to metabolites $M_1$ and $M_2$ were substantially bigger than shown in Figure 5. The HPLC profiles of metabolites formed from the (+)-and (-)-BA 3,4-dihydrodiols are also compared in Figure 5. It is apparent that both racemic dihydrodiol and the (-)-enantiomer are metabolized mainly to diol epoxide-$\underline{2}$ along with only trace amounts of diol epoxide-$\underline{1}$, as evidenced by the presence of very small amounts of trans-$\underline{1}$ and cis-$\underline{1}$ in the HPLC profile (Figure 5). The (+)-enantiomer appears to form very small amounts of tetraols (Figure 5d). The quantitation of diol epoxide metabolites and the specific activity of cytochrome P-448 toward ($\pm$)-BA 3,4-dihydrodiol as substrate are listed in Table 1. Varying amounts of the 3 peaks present between the BA 3,4-dihydrodiol substrates and BA $H_4$-1-ol are due to decomposition of BA 3,4-dihydrodiol during incubation and workup. Increased amounts of these peaks formed when the dihydrodiol was allowed to stand at room temperature in either methanol or tetrahydrofuran prior to chromatography.

## PHENANTHRENE 1,2-DIHYDRODIOL

The 1,2-dihydrodiol with the bay-region double bond, obtained from a large-scale incubation of [$^3$H]-phenanthrene with microsomes from 3-methylcholanthrene-treated rats, was found to be 93 percent enantiomerically pure (16). Absolute stereochemistry of the major enantiomer has yet to be assigned. Chromatographic mobilities of the tetraols formed from the diastereomeric phenanthrene 1,2-diol-3,4-epoxides upon hydrolysis (22) and the metabolites formed from [$^3$H]-phenanthrene 1,2-dihydrodiol are shown in Figure 6. A substantial portion of the largest peak in the metabolite profile (~19 min) was established to be phenanthrene 1,2-diol-3,4-epoxide-$\underline{2}$, based on the fact that approximately 50 percent of it was converted to trans-$\underline{2}$ tetraol on mild acid treatment of the sample. One might anticipate that more vigorous acid treatment may lead to further formation of tetraols. In addition, the three peaks which follow the largest peak also disappeared on mild acid treatment, suggesting that they also arise by chromatographic decomposition of diol epoxide. Thus, diol epoxide-$\underline{2}$ appears to be a major metabolite of the highly enantiomerically enriched phenanthrene 1,2-dihydrodiol used as a substrate in this study. The early peak of radioactivity which emerges after the solvent breakthrough is presently unidentified. Diol epoxides constitute ~63 percent of the total metabolites of phenanthrene 1,2-dihydrodiol (Table 1). The dihydrodiol was found to be a poor substrate for cytochrome P-448, since it was metabolized at approximately one-tenth of the rate of the parent hydrocarbon (Table 1).

## CHRYSENE 1,2-DIHYDRODIOL

The [$^3$H] chrysene 1,2-dihydrodiol was obtained by large-scale incubation of 6-[$^3$H]-chrysene with microsomes from 3-methylcholanthrene-pretreated rats. The dihydrodiol was found to be 86 percent enantiomerically pure (16). Absolute stereochemistry of this dihydrodiol also has yet to be assigned. Separation of the tetraols which arise on solvolysis of

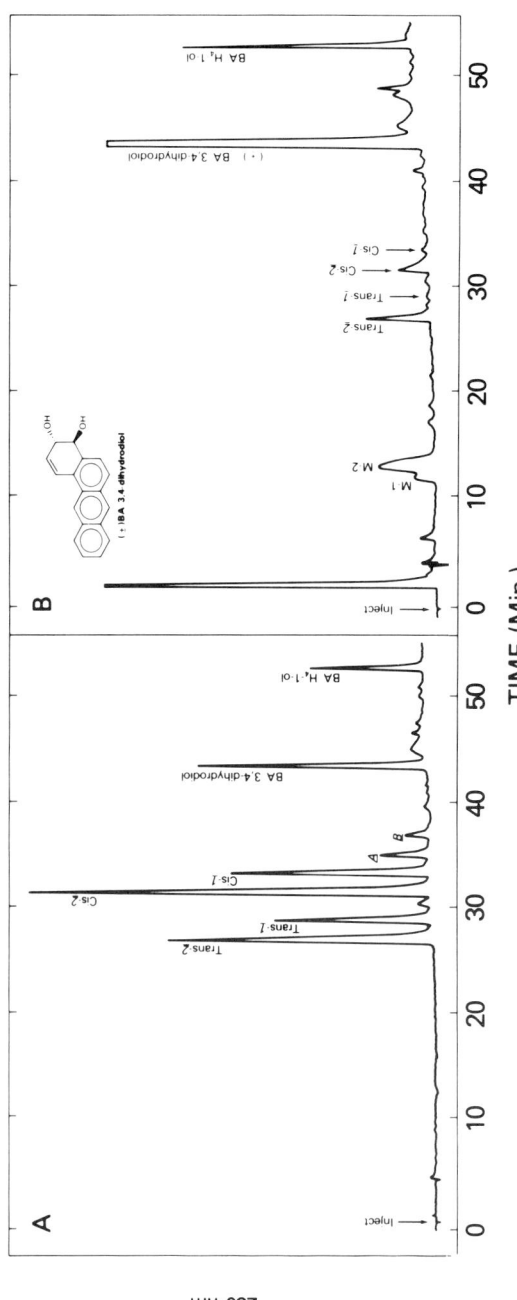

FIGURE 5. A: Chromatographic mobilities of tetraols derived from BA 3,4-diol-1,2-epoxides, BA 3,4-dihydrodiol, and an internal standard BA H₄-1-ol. Peaks A and B are presently uncharacterized solvolysis products of diol epoxide-2 and diol epoxide-1, respectively. B: Metabolites of (±)-BA 3,4-dihydrodiol formed by liver microsomes from 3-methylcholanthrene-treated, Long-Evans rats. (±)-BA 3,4-dihydrodiol (2.5 μmol) was incubated with microsomes (12.5 mg protein, 1.61 nmol cytochrome P-448/mg protein), potassium phosphate (pH 7.4, 5 mmol), MgCl₂ (0.15 mmol) and NADPH (0.05 mmol) in a total volume of 50 ml for 10 min. The products were extracted into 150 ml of ethyl acetate/acetone (2/1), the solvent was concentrated to a small volume, and the products were analyzed by HPLC on a Du Pont Zorbax ODS column (0.62 cm x 25 cm). A linear gradient of 40 percent methanol/water to 100 percent methanol at a rate of gradient change of 1 percent/min and flow rate of 1.2 ml/min was used. The effluent was monitored at 250 nm and peak areas were integrated by an Autolab System IV integrator with the BA H₄-1-ol as internal standard.

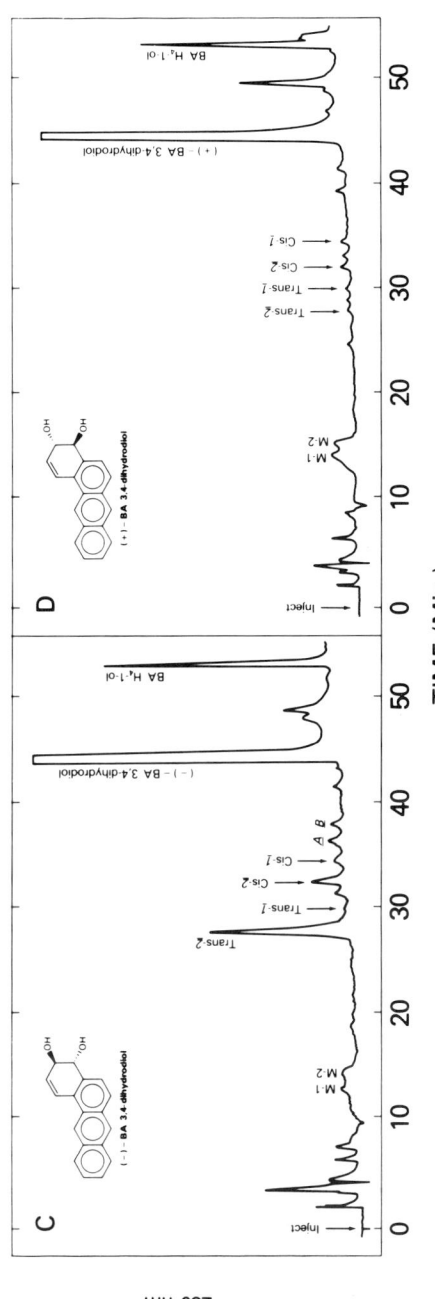

FIGURE 5 (Continued) C & D: Metabolism of (C) (−)-BA [3R,4R]- and (D) (+)-BA [3S,4S]-dihydrodiols by microsomes from 3-methylcholanthrene-treated rats. Absolute stereochemistry of the (+)- and (−)-enantiomers is as shown. The (−)- and (+)-BA 3,4-dihydrodiols (100 nmol) were incubated with microsomes from 3-methylcholanthrene-treated rats (0.5 to 1.0 mg protein) in a total volume of 2.0 ml for 10 min. Other incubation and workup conditions as well as the chromatographic conditions were similar to those described for the (±)-BA 3,4-dihydrodiol.

**TABLE 1. Metabolism of Dihydrodiols of Several PAH to Bay-Region Diol Epoxides[a]**

| Dihydrodiol | Cytochrome P-448 Activity[b] (nmol product/nmol hemoprotein/min) | Dihydrodiol[c]/PAH | Diol Epoxide-1/Diol Epoxide-2 | Total Diol Epoxides[d] (%) |
|---|---|---|---|---|
| (±)-BA 3,4-dihydrodiol | 4.4 | 0.6 | ~1:50 | <10 |
| Phenanthrene 1,2-dihydrodiol | 1.4 | 0.2 | ~1:5[e] | ~63[e] |
| Chrysene 1,2-dihydrodiol[f] | 1.4 | 0.9 | 1:1.7 | 60-66 |
| (±)-BaP 7,8-dihydrodiol | 1.4 | 0.4 | ~1:1 | 83-92 |

[a] Experimental conditions are described in appropriate figure legends. All the results were obtained with liver microsomes from 3-methylcholanthrene-treated rats.
[b] Specific activities were calculated from the total metabolites which elute before the substrate. For BA 3,4-dihydrodiol, the calculations were based on the amount of substrate consumed as determined by the peak area ratio method using Ba H$_4$-1-ol as an internal standard.
[c] Dihydrodiol/PAH denotes the ratio of specific activities of cytochrome P-448 (nmol product formed/nmol hemoprotein/min) toward the dihydrodiols and corresponding parent hydrocarbons. Specific activities of cytochrome P-448 for BA (3), phenanthracene (16), chrysene (16), BaP (3) and BaP 7,8-dihydrodiol (20) were obtained from previously published results.
[d] Total diol epoxides denotes the diol epoxides expressed as percent of total metabolites formed which elute in discrete chromatographic peaks. In the case of diol epoxides of BA the values represent the amount of diol epoxides as percent of total substrate consumed.
[e] Based on total radioactivity in tetraol peaks after acidification of the sample with mild acid. These numbers may change upon more vigorous acid treatment.
[f] The ratio of diol epoxide-1 to diol epoxide-2 and the percent of total diol epoxides in the case of metabolism of chrysene 1,2-dihydrodiol are subject to change if peak C (Figure 7) is later confirmed to be derived from diol epoxide 1 and included in the calculations. The proportion of total diol epoxides will go up to >85 percent of total metabolites if peak C is related to diol epoxide-1.

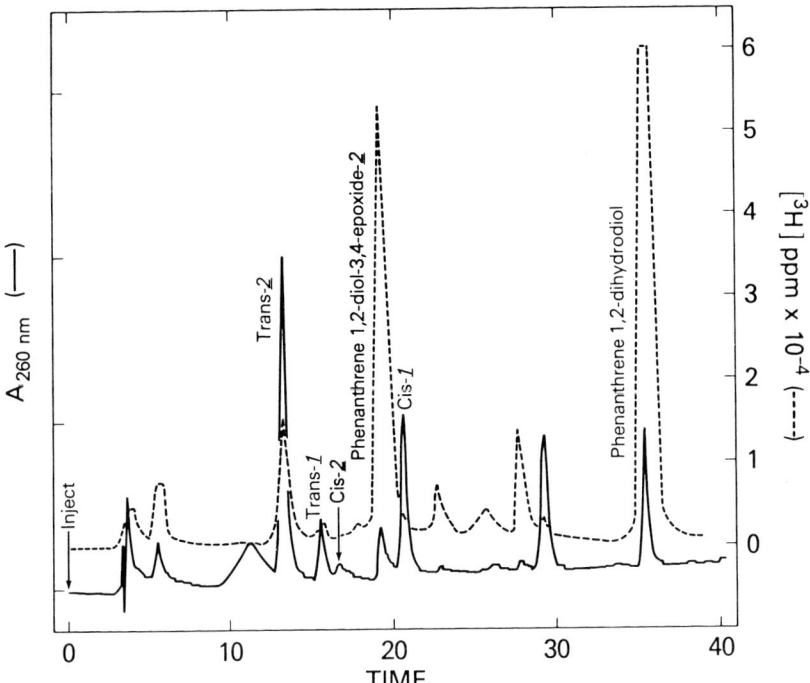

**FIGURE 6.** Chromatographic separation of phenanthrene 1,2-dihydrodiol and the tetraols which arise on solvolysis of the phenanthrene 1,2-diol-3,4-epoxides (——) along with the metabolites of [$^3$H]-phenanthrene 1,2-dihydrodiol(- - -). The peak at 29.0 min in the UV trace is possibly a keto-diol formed by isomerization of phenanthrene 1,2-diol-3,4-epoxide-1. The 1,2-dihydrodiol (100 nmol) was incubated for 10 min with microsomes from 3-methylcholanthrene-treated rats (0.25 mg protein), NADPH (2 μmol), MgCl$_2$ (6 μmol) and potassium phosphate (pH 7.4, 200 μmol) in a total volume of 2.0 ml. Products were analyzed by HPLC after extraction into 6 ml of ethyl acetate/acetone (2/1); a Du Pont Zorbax ODS column (0.62 cm x 25 cm) was used with a linear gradient of 30 percent methanol in water to 100 percent methanol at a rate of gradient change of 1 percent/min and a flow rate of 1.2 ml/min.

the chrysene 1,2-diol-3,4-epoxides (22) as well as the metabolites of [$^3$H]-chrysene 1,2-dihydrodiol are shown in Figure 7. Tetraol trans-2 is a major metabolite of the 1,2-dihydrodiol; however, small amounts of cis-2, trans-1, and cis-1 were also detected. The shoulder on cis-1 tetraols peak appears to have resulted from solvolysis of diol epoxide-1 during chromatography. The identity of three metabolite peaks (A-C) has yet to be established. Peak C, however, appears to have been derived from diol epoxide-1 (cf. Figure 3), since direct injection of chrysene 1,2-diol-3,4-epoxide-1 results in the formation of trans-1, cis-1, and two poorly separated peaks which chromatograph with peak C. Approximately 60 to

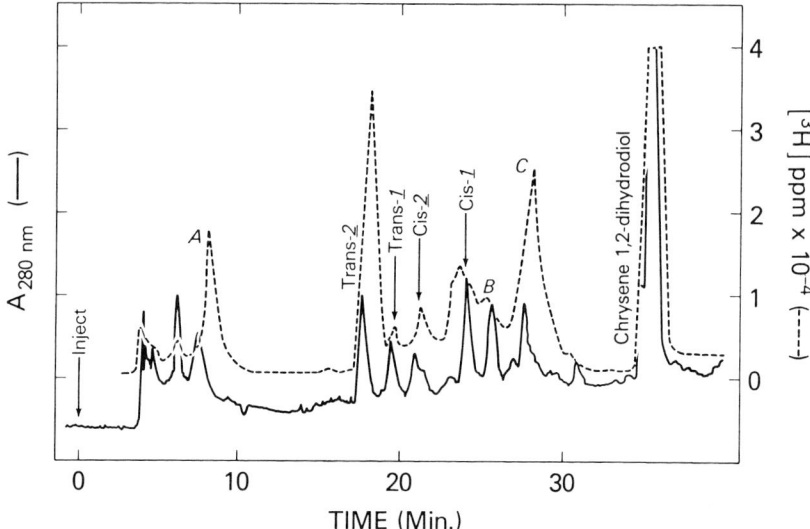

FIGURE 7. Chromatographic separation of chrysene 1,2-dihydrodiol and the tetraols which arise on solvolysis of the chrysene 1,2-diol-3,4-epoxides (———) as well as metabolites of [$^3$H]-chrysene 1,2-dihydrodiol (- - -). The 1,2-dihydrodiol (50 nmol) was incubated for 10 min with microsomes from 3-methylcholanthrene-treated rats (0.5 mg protein), NADPH (1 μmol), MgCl$_2$ (3 μmol) and potassium phosphate (pH 7.4, 100 μmol) in a total volume of 1.0 ml. The products were analyzed by HPLC after extraction into ethyl acetate/acetone (2/1). The chromatographic conditions were similar to those used for the metabolites of phenanthrene 1,2-dihydrodiol except that the gradient was started at 50 percent methanol/water.

66 percent of the total metabolism of chrysene 1,2-dihydrodiol can be accounted for by the formation of bay-region diol epoxides (Table 1). If peak C is derived from diol epoxide-1, then the diol epoxides constitute approximately 88 percent of the total metabolites of the 1,2-dihydrodiol. Although the metabolically formed chrysene 1,2-dihydrodiol of high enantiomeric purity is not as good a substrate for cytochrome P-448 as is (±)-BA 3,4-dihydrodiol, it is metabolized almost as effectively as is chrysene (Table 1).

## BENZO(e)PYRENE 9,10-DIHYDRODIOL

HPLC separation of the metabolites of (±)-BeP 9,10-dihydrodiol and the tetraols which arise on solvolysis of the BeP 9,10-diol-11,12-epoxides is shown in Figure 8. Peaks P-1 and P-2 have been identified as phenolic metabolites of the 9,10-dihydrodiol. Peak X has been identified as

4,5,9,10-tetrahydroxy-4,5,9,10-tetrahydro BeP based on its UV and mass spectra (24). Tetraols and methanol adducts, if formed at all, are only trace metabolites. In analogy to the metabolism of the bay-region dihydrodiol of BaP, i.e., BaP 9,10-dihydrodiol (19) in which the hydroxyl groups are *in* the bay-region, the 9,10-dihydrodiol of BeP is poorly metabolized at the adjacent double bond by rat liver microsomes (10,24).

## COMPARATIVE METABOLISM OF DIHYDRODIOLS OF PAH TO BAY-REGION DIOL EPOXIDES

The metabolism of dihydrodiols of several PAH to their corresponding bay-region diol epoxides is compared in Table 1. Among the several dihydrodiols studied, (±)-BA 3,4-dihydrodiol is the best substrate for cytochrome P-448. The 1,2-dihydrodiols of chrysene and phenanthrene, along with the 7,8-dihydrodiol of BaP are less active as substrate for cytochrome P-448. When compared with their respective parent hydrocarbons, chrysene 1,2-dihydrodiol is about as good a substrate as chrysene, whereas phenanthrene 1,2-dihydrodiol is metabolized at only 20 percent the rate of phenanthrene. The (±)-BaP 7,8-dihydrodiol and (±)-BA 3,4-dihydrodiol are metabolized one-third and two-thirds as efficiently as their respective parent hydrocarbons. Examination of the distribution of products between diol epoxides-1 and -2 reveals that the (-)-enantiomer of BA 3,4-dihydrodiol forms almost exclusively diol epoxide-2, which is similar to results obtained with the (-)-enantiomer of BaP 7,8-dihydrodiol (7,18). Interestingly, however, (+)-BA 3,4-dihydrodiol forms very little diol epoxides. This is in marked contrast to results obtained with (+)-BaP 7,8-dihydrodiol (18), which is metabolized predominantly to diol epoxide-1. The biosynthetic 1,2-dihydrodiol of phenanthrene (93 percent enantiomeric purity) also forms predominantly diol epoxide-2. Similarly, chrysene 1,2-dihydrodiol formed by microsomes from 3-methylcholanthrene-treated rats, which is also highly enantiomerically pure (84 percent), produces more diol epoxide-2 compared to diol epoxide-1 if peak C is excluded from the calculations. Interestingly, the (-)-[R,R]-enantiomers of BaP 7,8-dihydrodiol (9,12) and BA 3,4-dihydrodiol (11,23) are at least 5-fold more tumorigenic on mouse skin and in newborn mice than are their (+)-[S,S]-enantiomers. In both cases, the (-)-[R,R]-enantiomers are also predominantly metabolized to the more tumorigenic diol epoxide-2 isomers (1,8,11,23). The cytochrome P-448 system thus exhibits similar stereospecificity in the metabolism of the proximate carcinogens (-)-BaP 7,8-dihydrodiol and (-)-BA 3,4-dihydrodiol, as shown schematically in Figure 9.

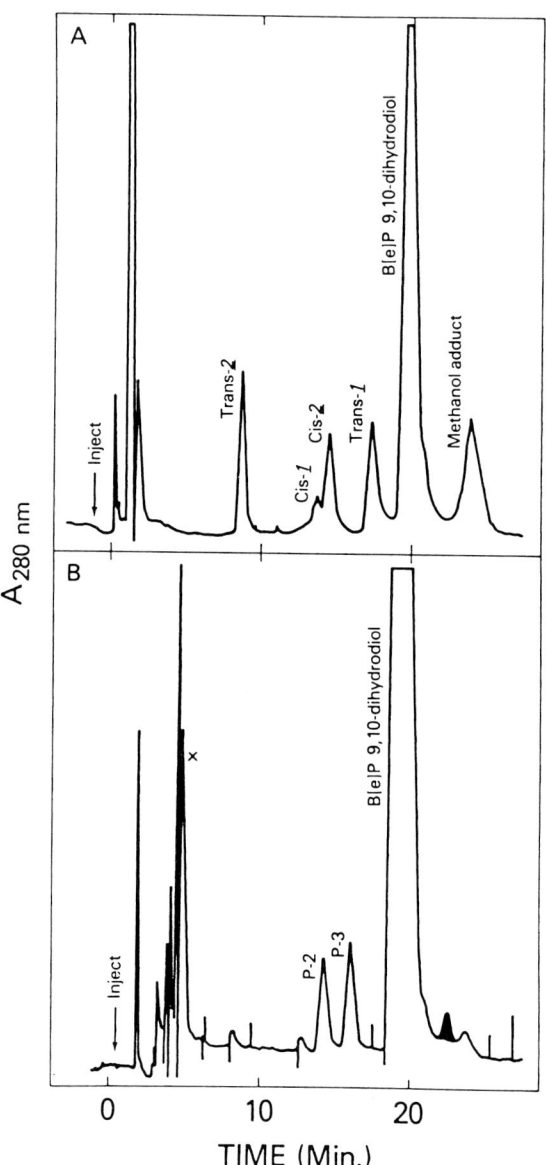

FIGURE 8. Chromatographic separation of the tetraols formed on solvolysis of BeP 9,10-diol-11,12-epoxides (A) and the metabolites of BeP 9,10-dihydrodiol (B). The (±)-BeP 9,10-dihydrodiol was metabolized by microsomes from aroclor-pretreated rats as described (10, 24). The shaded peak corresponds in retention time to a methanol adduct of the diol epoxide(s). Metabolites were chromatographed on a Du Pont Zorbax-CN column (0.62 cm x 25 cm) with 45 percent methanol/water at a flow rate of 2.0 ml/min as mobile phase.

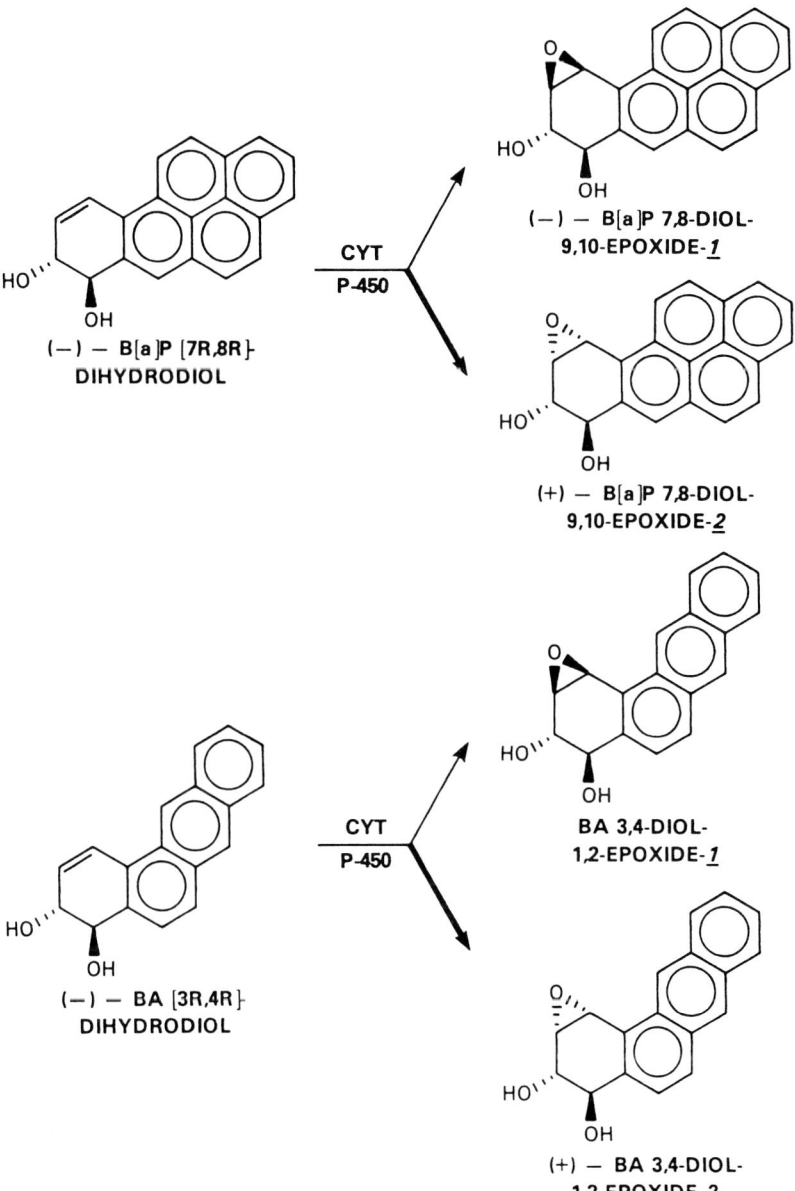

FIGURE 9. Metabolism of (-)-BaP [7R,8R]-dihydrodiol and (-)-BA [3R,4R]-dihydrodiol by rat liver microsomes. The bold arrows indicate the major metabolic pathways. Where signs of rotation are shown, absolute stereo-chemistry is indicated.

## CONCLUSIONS

The present results indicate that the dihydrodiols of various PAH are metabolized to the bay-region diol epoxides with different degrees of regiospecificity. For example, BaP 7,8-dihydrodiol as well as phenanthrene and chrysene 1,2-dihydrodiols of high enantiomeric purity are metabolized predominantly to bay-region diol epoxides (60 to 92 percent of total metabolites). In contrast, much lower amounts of bay-region diol epoxides are formed from (±)-BA 3,4-dihydrodiol (<10 percent of total metabolites). The absolute configuration of the dihydrodiol is a determining factor in the percentage of total metabolites formed as bay-region diol epoxides. The extent of metabolism of several PAH dihydrodiols to bay-region diol epoxides is summarized in Table 2. A comparison of these results with the percent conversion of the parent hydrocarbons to dihydrodiols *in the bay-region* leads to an interesting correlation. When the PAH is metabolized to a high percentage of the dihydrodiol *in the bay-region*, the dihydrodiol with a bay-region double bond is metabolized to a *high percentage* of *bay-region diol epoxides*. For example, the dihydrodiol of BaP in the bay-region, i.e., BaP 9,10-dihydrodiol, represents 20 percent of the total metabolites of BaP and the 7,8-dihydrodiol is metabolized almost exclusively at the 9,10-position by microsomes from 3-methylcholanthrene-treated rats. Similarly, chrysene is metabolized to form high amounts of the 3,4-dihydrodiol in the bay region (34 to 46 percent of total metabolites), and the 1,2-dihydrodiol of chrysene is extensively metabolized at its 3,4-position to form bay-region diol epoxides. Although the metabolism of dibenzo(a,h)anthracene (DBA)

TABLE 2. Bay-Region Metabolism of PAH and Their Dihydrodiols With Bay-Region Double Bonds[a]

| Formation of Dihydrodiol in Bay Region | Dihydrodiol in Bay Region[b] (%) | Formation of Bay Region Diol Epoxide | Bay Region[c] Diol Epoxides (%) |
|---|---|---|---|
| BaP → 9,10-DHD | ~20 | BaP 7,8-DHD → 7,8-diol-9,10-epoxide | 83-92 |
| BA → 1,2-DHD | <2 | BA 3,4-DHD → 3,4-diol-1,2-epoxide | <10 |
| Phenanthrene → 3,4-DHD | 8-10 | Phenanthrene 1,2-DHD → 1,2-diol-3,4-epoxide | ~63 |
| Chrysene → 3,4-DHD | 34-46 | Chrysene 1,2-DHD → 1,2-diol-3,4-epoxide | 60-66 |
| BeP → 9,10-DHD | trace | BeP 9,10-DHD → 9,10-diol-11,12-epoxide | trace |
| DBA → 1,2-DHD | ~13 | DBA 3,4-DHD → 3,4-diol-1,2-epoxide | high |

[a] Dihydrodiol is abbreviated as DHD. See also Figures 2 and 3.
[b] Data taken from References 3 (BaP), 15 (BA), 16 (phenanthrene and chrysene), unpublished results (BeP), and (14) (DBA). In all cases, hepatic microsomes from 3-methylcholanthrene-treated rats were the enzyme source.
[c] Data taken from References (20) (BaP 7,8-dihydrodiol) and (16) (DBA 3,4-dihydrodiol).

3,4-dihydrodiol at the bay-region 1,2-double bond has not been quantified, significant amounts of the bay-region diol epoxides are formed (14). The 1,2-dihydrodiol in the bay-region represents 13 percent of the total metabolism of DBA. In contrast, BA and its 3,4-dihydrodiol are both poorly metabolized at the bay-region (1,2-position). Phenanthrene presents an example where the extent of bay-region metabolism (3,4-position) of the hydrocarbon is intermediate (8 to 10 percent). The 1,2-dihydrodiol is metabolized to bay-region 1,2-diol 3,4-epoxides in good yield but the dihydrodiol was found to be a poor substrate for cytochrome P-448 compared with the parent hydrocarbon, having only one-fifth the activity of the parent hydrocarbon. BeP 9,10-dihydrodiol (10,24) as well as the parent hydrocarbon (unpublished results) are not metabolized at the bay-region to any significant extent. In the case of BeP 9,10-dihydrodiol, the low metabolism to bay-region diol epoxides is also believed to be due to the predominant pseudo-diaxial conformation of the bay-region hydroxyl groups (10). We have previously shown in a similar blocking effect of the diaxial bay-region hydroxyl groups BaP 9,10-dihydrodiol on metabolism of its adjacent double bond (19). The results described in this paper give a preliminary indication of a parallel relationship between the metabolism of PAH to dihydrodiols in the bay-region, and the metabolism of the dihydrodiols with bay-region double bonds to the corresponding bay-region diol epoxides. It appears that the susceptibility of the bay-region double bond to metabolic attack is not affected significantly by the presence of hydroxyl groups when they are present outside of the bay-region. Further studies with other hydrocarbons are warranted to establish whether these relationships are of a general nature. Such information can be of substantial value in light of the recent results which indicate that the bay-region diol epoxides of several PAH are ultimate carcinogenic metabolites (2,13).

## REFERENCES

1. Buening, M. K., Wislocki, P. G., Levin, W., Yagi, H., Thakker, D. R., Akagi, H., Koreeda, M., Jerina, D. M. and Conney, A. H. (1978): Tumorigenicity of optical enantiomers of the diastereomeric benzo-(a)pyrene 7,8-diol-9,10-epoxides in newborn mice: Exceptional activity of (+)-7$\beta$,8$\alpha$-dihydroxy-9$\alpha$,10$\alpha$-epoxy-7,8,9,10-tetrahydrobenzo-(a)pyrene. *Proc. Nat. Acad. Sci. USA* 75:5358-5361.
2. Conney, A. H., Levin, W., Wood, A. W., Yagi, H., Lehr, R. E. and Jerina, D. M. (1978): Biological activity of polycyclic hydrocarbon metabolites and the bay-region theory. In: *Advances in Pharmacology and Therapeutics.* Cohen, Y., Ed., pp. 43-52, Pergamon Press, Oxford.

3. Holder, G., Yagi, H., Dansette, P., Jerina, D. M., Levin, W., Lu, A.Y.H., and Conney, A. H. (1974): Effects of inducers and epoxide hydrase on the metabolism of benzo(a)pyrene by liver microsomes and a reconstituted system: Analysis by high-pressure liquid chromatography. *Proc. Nat. Acad. Sci. USA* 71:4356-4360.
4. Jerina, D. M. and Daly, J. W. (1976): Oxidation at carbon. In: *Drug Metabolism—from Microbes to Man.* D. V. Parke and R. L. Smith, Eds., pp. 13-32, Taylor and Francis, Ltd., London.
5. Jerina, D. M., Lehr, R. E., Schaefer-Ridder, M., Yagi, H., Karle, J. M., Thakker, D. R., Wood, A. W., Lu, A.Y.H., Ryan, D., West, S., Levin, W. and Conney, A. H. (1977): Bay-region epoxides of dihydrodiols: A concept which explains mutagenic and carcinogenic activity of benzo(a)pyrene and benzo(a)anthracene. In: *Origins of Human Cancer.* H. Hiatt, J. D. Watson, and I. Winsten, Eds., pp. 639-658, Cold Spring Harbor Laboratory, Cold Spring Harbor, New York.
6. Jerina, D. M., Lehr, R. E., Yagi, H., Hernandez, O., Dansette, P. M., Wislocki, P. G., Wood, A. W., Chang, R. L., Levin, W. and Conney, A. H. (1976): Mutagenicity of benzo(a)pyrene and the description of a quantum mechanical model which predicts the ease of carbonium ion formation from diol epoxides. In: *In vitro Metabolic Activation in Mutagenesis Testing.* F. J. de Serres, J. R. Fouts, J. R. Bend, and R. M. Philpot, Eds., pp. 159-177, Elsevier/North Holland Biomedical Press, Amsterdam.
7. Jerina, D. M., Yagi, H., Thakker, D. R., Karle, J. M., Mah, H. D., Boyd, D. R., Gadaginamath, G., Wood, A. W., Buening, M., Chang, R. L., Levin, W. and Conney, A. H. (1978): Stereoselective metabolic activation of polycyclic aromatic hydrocarbons. In: *Advances in Pharmacology and Therapeutics.* Vol. 9, Y. Cohen, Ed., pp. 53-62, Pergamon Press, Oxford.
8. Kapitulnik, J., Wislocki, P. G., Levin, W., Yagi, H., Jerina, D. M. and Conney, A. H. (1978): Tumorigenicity studies with diol-epoxides of benzo(a)pyrene which indicate that ($\pm$)-trans-7$\beta$,8$\alpha$-dihydroxy-9$\alpha$,10$\alpha$-epoxy-7,8,9,10-tetrahydrobenzo(a)pyrene is an ultimate carcinogen in newborn mouse. *Cancer Res.* 38:354-358.
9. Kapitulnik, J., Wislocki, P. G., Levin, W., Yagi, H., Thakker, D. R., Akagi, H., Koreeda, M., Jerina, D. M. and Conney, A. H. (1976): Marked differences in the carcinogenic activity of optically pure (+)- and (-)-7,8-dihydroxy-7,8-dihydrobenzo(a)pyrene in newborn mouse. *Cancer Res.* 38:7661-7665.
10. Lehr, R. E., Taylor, C. W., Kumar, S., Levin, W., Chang, R. L., Wood, A. W., Conney, A. H., Thakker, D. R., Yagi, H., Mah, H. D. and Jerina, D. M. (1979): Differences in metabolism provides a basis

for the low mutagenicity and carcinogenicity of benzo(e)pyrene compared to benzo(a)pyrene. In: *Polynuclear Aromatic Hydrocarbons: Third International Symposium on Chemistry and Biology—Carcinogenesis and Mutagenesis.* P. W. Jones and P. Leber, Eds., pp. 37-49, Ann Arbor Science Publishers, Inc., Ann Arbor, Michigan.

11. Levin, W., Thakker, D. R., Wood, A. W., Chang, R. L., Lehr, R. E., Jerina, D. M. and Conney, A. H. (1978): Evidence that benzo(a)anthracene 3,4-diol-1,2-epoxide is an ultimate carcinogen on mouse skin. *Cancer Res.* 38:1705-1710.

12. Levin, W., Conney, A. H. and Jerina, D. M. (1980): Evidence in D. M. and Conney, A. H. (1977): Marked differences in the tumor-initiating activity of optically pure (+)- and (-)-trans-7,8-dihydroxy-7,8-dihydrobenzo(a)pyrene on mouse skin. *Cancer Res.* 37:2721-2725.

13. Nordqvist, M., Thakker, D. R., Yagi, H., Lehr, R. E., Wood, A. W., Levin, W., Conney, A. H. and Jerina, D. M. (1980): Evidence in support of the bay-region theory as a basis for the carcinogenic activity of polycyclic aromatic hydrocarbons. In: *Molecular Basis of Environmental Toxicity.* R. S. Bhatnager, Ed., pp. 329-357, Ann Arbor Science Publishers, Ann Arbor, Michigan.

14. Nordqvist, M., Thakker, D. R., Yagi, H., Tyan, D. E., Thomas, P. E., Levin, W., Conney, A. H. and Jerina, D. M. (1979): The highly tumorigenic 3,4-dihydrodiol is a principal metabolite formed from dibenzo(a,h)anthracene by liver enzymes. *Mol. Pharmacol.* 16:643-655.

15. Thakker, D. R., Levin, W., Yagi, H., Karle, J. M., Lehr, R. E., Ryan, D., Thomas, P. E., Conney, A. H. and Jerina, D. M. (1979): Metabolism of benzo(a)anthracene to its tumorigenic 3,4-dihydrodiol, *Mol. Pharmacol.* 15:138-153.

16. Thakker, D. R., Nordqvist, M., Levin, W., Conney, A. H. and Jerina, D. M. (in preparation)

17. Thakker, D. R., Nordqvist, M., Yagi, H., Levin, W., Ryan, D., Thomas, P., Conney, A. H. and Jerina, D. M. (1979): Comparative metabolism of a series of polycyclic aromatic hydrocarbons by rat liver microsomes and purified cytochrome P-450. In: *Polynuclear Aromatic Hydrocarbons: 3rd International Symposium on Chemistry and Biology—Carcinogenesis and Mutagenesis.* P. W. Jones and P. Leber, Eds., pp. 445-472, Ann Arbor Science Publishers, Inc., Ann Arbor, Michigan.

18. Thakker, D. R., Yagi, H., Akagi, H., Koreeda, M., Lu, A.Y.H., Levin, W., Wood, A. W., Conney, A. H. and Jerina, D. M. (1977): Metabolism of benzo(a)pyrene. VI. Stereoselective metabolism of benzo(a)pryene and benzo(a)pyrene 7,8-dihydrodiol to diol epoxides. *Chem. -Biol. Interact.* 16:281-300.

19. Thakker, D. R., Yagi, H., Lehr, R. E., Levin, W., Buening, M., Lu, A.Y.H., Chang, R. L., Wood, A. W., Conney, A. H. and Jerina, D. M. (1978): Metabolism of trans-9,10-dihydroxy-9,10-dihydrobenzo(a)pyrene occurs primarily by arylhydroxylation rather than formation of a diol epoxide. *Mol. Pharmacol.* 14:502-513.
20 Thakker, D. R., Yagi, H., Lu, A.Y.H., Levin, W., Conney, A. H. and Jerina, D. M. (1976): Metabolism of benzo(a)pyrene: Conversion of (±)-trans-7,8-dihydroxy-7,8-dihydrobenzo(a)pyrene to highly mutagenic 7,8-diol-9,10-epoxides. *Proc. Nat. Acad. Sci. USA* 73:3381-3385.
21. Whalen, D. L., Montemarano, J. A., Thakker, D. R., Yagi, H. and Jerina, D. M. (1977): Changes in mechanisms and product distributions in the hydrolysis of benzo(a)pyrene-7,8-diol-9,10-epoxide metabolites induced by changes in pH. *J. Am. Chem. Soc.* 99: 5522-5524.
22. Whalen, D. L., Ross, A. M., Yagi, H., Karle, J. M. and Jerina, D. M. (1978): Stereoelectronic factors in the solvolysis of bay-region diol epoxides of polycyclic armoatic hydrocarbons. *J. Am. Chem. Soc.* 100:5218-5221.
23. Wislocki, P. G., Buening, M. K., Levin, W., Lehr, R. E., Thakker, D. R., Jerina, D. M. and Conney, A. H. (1979): Tumorigenicity of the diastereomeric benz(a)anthracene 3,4-diol-1,2-epoxides and the (+)- and (-)-enantiomers of benz(a)anthracene 3,4-dihydrodiol in newborn mice, *J. Nat. Cancer Res.* 63:201-204.
24. Wood, A. W., Levin, W., Thakker, D. R., Yagi, H., Chang, R. L., Ryan, D. E., Thomas, P. E., Dansette, P. M., Wittaker, N., Turujman, S., Lehr, R. E., Kumar, S., Jerina, D. M. and Conney, A. H. (1979): Biological activity of benzo(e)pryene: An assessment based on mutagenic activities and metabolic profiles of the polycyclic hydrocarbon and its derivatives. *J. Biol. Chem.* 254:4408-4415.
25. Yagi, H., Thakker, D. R., Hernandez, O., Koreeda, M. and Jerina, D. M. (1977): Synthesis and reactions of the highly mutagenic 7,8-diol-9,10-epoxides of the carcinogene benzo(a)pyrene. *J. Am. Chem. Soc.* 99:1609-1611.

# REPRODUCTIVE SURVIVAL AND MACROMOLECULAR SYNTHESIS IN CULTURED MAMMALIAN CELLS EXPOSED TO SINGLE METABOLITIES AND MIXTURES OF METABOLITES OF BENZO(A)PYRENE

G. D. Griffin, C. S. Dudney, P. S. Furcinitti,
T. D. Jones, and P. J. Walsh

Health and Safety Research Division
Oak Ridge National Laboratory
Oak Ridge, Tennessee 37830

## INTRODUCTION

Although it is well known that many metabolites of BaP* exhibit varying degrees of cytotoxicity when exposed to cells in vitro (3,6,13,14), the mechanisms by which these cytotoxic effects are produced are largely unknown. Many studies have shown that BaP metabolites, particularly the diol-epoxides, can covalently bind to intracellular macromolecules such as DNA, RNA, and protein (1, 7). The effect that this binding has on subsequent functioning of the macromolecules is perhaps less well characterized. The toxic effect in vitro of mixtures of BaP metabolites has not been extensively investigated, although for human in vivo exposure situations, a mixture of both single-chemical and multiple-chemical metabolites is probably the common occurrence. In order to assess human health risks from coal-combustion processes, it became important (1) to study the composite toxicity of metabolite mixtures compared with the toxicity of individual components of the mixture and (2) to investigate cellular macromolecular effects of toxic metabolites in an effort to understand molecular mechanisms involved in toxicity.

---

*The abbreviations used are: BaP, benzo(a)pyrene; BaP 4,5-oxide, benzo(a)-pyrene 4,5-oxide; 3-HOBaP, 3-hydroxybenzo(a)pyrene; 6-HOBaP, 6-hydroxybenzo(a)pyrene; BaP trans-7,8-dihydrodiol, trans-7,8-dihydroxy-7,8-dihydrobenzo(a)pyrene; BaP 1,6-quinone, benzo(a)pyrene,1,6-quinone.

## MATERIALS AND METHODS

Chinese hamster V79 cells, kindly supplied by Dr. E. Huberman, Biology Division, Oak Ridge National Laboratory, were grown and subcultured as described by Huberman et al (4), although Dulbecco's modification of Eagle's minimal essential medium was used as the culture medium. One day before an experiment was started, cells were seeded into a flask at a rather low density and allowed to grow overnight. These cells were then used for the experiment. Reproductive cell survival was assessed by ability to form colonies in 60-mm Petri dishes. The experimental procedure followed that of Huberman et al (4) except that different times of exposure to the BaP metabolites were used. The percentage of cloning efficiency for metabolite-treated cells was defined as the number of colonies in a treated dish divided by the number of colonies in an acetone control dish times 100. Metabolites of BaP (supplied by the IIT Research Institute through the National Cancer Institute Carcinogenesis Research Program, Dr. David Longfellow) were dissolved in acetone for addition to the culture.

Cell counts in treated and control cultures were determined using a hemocytometer, using Trypan Blue staining to assess viable cells. Labeling of DNA, RNA, and protein was accomplished using appropriate radioactive precursors (thymidine, uridine, and leucine respectively) obtained from radiochemical suppliers (New England Nuclear and Amersham/Searle). Usually, 5 to 10 $\mu$Ci/ml of $^3$H-labeled compound or 0.05 to 0.10 $\mu$Ci/ml of $^{14}$C-labeled chemical was added to the culture medium. The usual period of labeling was 16 to 18 hours immediately before the culture was terminated, except for the earliest 4-hour exposure time point, in which case the labeling was for 4 hours. All labeling experiments were carried out using replicate 60-mm dishes (duplicate or triplicate). Macromolecules were precipitated by cold 10 percent trichloroacetic acid, then separated and prepared for counting by standard techniques (10,11). Amounts of DNA, RNA, and protein in cell lysates were determined by colorimetric assays [diphenylamine (2), orcinol (8), and Lowry procedures (9), respectively]. Test results of the labeling studies were calculated for metabolite-treated dishes as percentage of labeling in control dishes, which were treated in the same manner except no metabolite was added.

## RESULTS AND DISCUSSION

### BaP Metabolite Mixtures

Initial experiments were conducted to study the cytotoxic effect of varying cell-exposure time to a single BaP metabolite. The results for

3-HOBaP are shown in Figure 1. There is an increasing cytotoxic effect with increasing time exposure. There is not a direct quantitative agreement between cytotoxic effect and dose if dose is defined as concentration of toxic material multiplied by exposure time, but a general correlation is evident. Other metabolites including BaP trans-7,8-dihydrodiol and BaP 4,5-oxide, showed similar effects, i.e., increasing toxicity with increased exposure times. Although a number of investigators (4,13) have chosen brief exposure times (1 to 3 hours) to measure toxicity, we selected an exposure time of 48 hours for the mixture experiments, since this may be more representative of in vivo situations. Exposure intervals of 1 to 3 hours would be short compared with a cell-cycle time of 12 hours, and significant proportions of the total cell population may not be sensitive if the lethal damage events occur in only one or two phases of the cell cycle.

The results of representative experiments in which a relatively nontoxic BaP metabolite (BaP trans-7,8-dihydrodiol) is mixed with a rather toxic metabolite (BaP 1,6-quinone) are shown in Figures 2 and 3. Figure 2 illustrates the cytotoxic effect of an equimolar mixture of the two

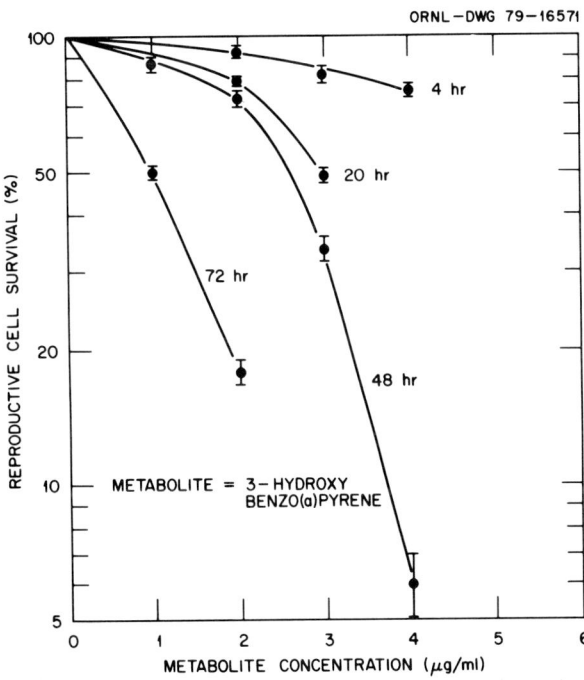

FIGURE 1. Effect of increasing exposure time on the cytotoxicity of B(a)P metabolites. Reproduction survival is defined in the Materials and Methods Section. Error bars indicate one coefficient of variation.

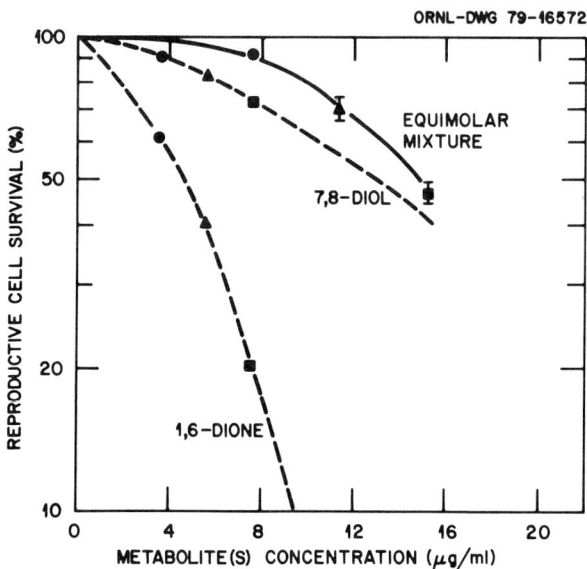

FIGURE 2. Cellular cytotoxicity caused by a mixture of B(a)P metabolites. Dotted lines show the cytotoxicity of the individual components of the mixture. Symbols (●, ▲, ■) indicate the overall concentration of the combined metabolites in the mixture and the concentration of each individual metabolite that went into making up the sum total: e.g., for the ■ symbol, the cell survival produced by 15 µg/ml of the mixture was 45 percent, while the cell survival produced by 7.5 µg/ml of each of the two components was 73 and 20 percent, respectively.

metabolites. The mixture is not as toxic as would be expected if the toxic effects of the individual components were the result of independent damage events. (See the dotted lines in Figure 2. E.g., if the toxic effects were independent, then net cell survival of mixture-exposed cells at individual component concentrations of 7.5 µg/ml should be 20 percent x 75 percent = 15 percent.) In fact, the toxic effect produced by the mixture seems to be a large reduction in cytotoxicity compared to that produced by BaP 1,6-quinone alone and a small increase in cytotoxicity compared with that produced by BaP trans-7,8-dihydrodiol alone. Figure 3 shows the cytotoxic effect of a mixture containing 2 moles of BaP trans-7,8-dihydrodiol per mole of BaP 1,6-quinone. Only in the case of the highest mixture concentration does the cytotoxicity of the mixture approach that predicted from the product of the individual cytotoxicities. At other concentrations, the cytotoxicity of the mixture is less than would be predicted.

Another mixture combination was tried, in which increasing amounts of 3-HOBaP were added to two different concentrations of BaP 4,5-oxide (Figure 4). In each case, the mixture produced the same degree of

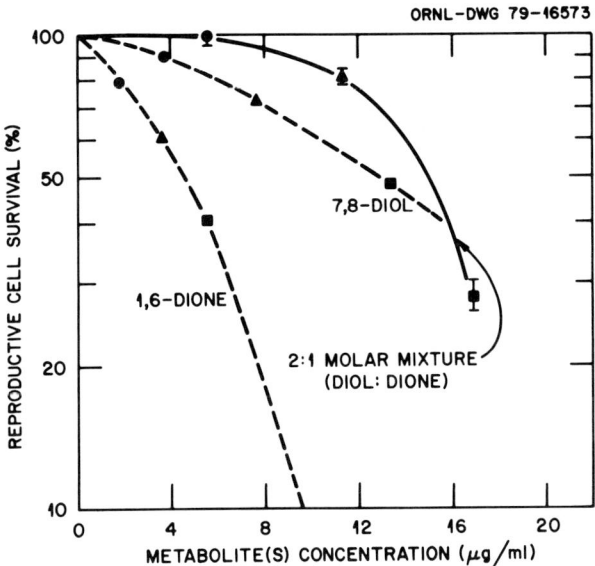

**FIGURE 3.** Cellular cytotoxicity caused by a mixture of B(a)P Metabolites. See Legend for Figure 2.

cytotoxicity as the BaP 4,5-oxide alone (either 65 percent or 45 percent cell survival, depending on the concentration of BaP 4,5-oxide) no matter what concentration of 3-HOBaP was added (up to 3 $\mu$g/ml of 3-HOBaP). These results seem to suggest that BaP 4,5-oxide is able to block effectively the toxic activity of 3-HOBaP, perhaps by acting at intracellular sites that may be sensitive to 3-HOBaP.

The results with BaP trans-7,8-dihydrodiol and BaP 1,6-quinone mixtures suggest a neutralizing effect of the less cytotoxic component ameliorating the toxicity of the more cytotoxic component. A protective effect of benzo(e)pyrene on 7,12-dimethylbenz(a)anthracene skin tumor initiation in mice has been noted in a study by Slaga et al (12) which suggests that this effect could be due to modification in the potent carcinogen metabolism by the very weak carcinogen benzo(e)pyrene. In our in vitro studies, these cells are not capable of further metabolizing the metabolites administered (5), so that the effects observed must occur at subcellular sites directly involved in cytotoxic mechanisms. The nature of these sites is not known to us.

## Macromolecular Effects of Toxic Metabolites

Two metabolites, BaP 4,5-oxide and 6-HOBaP, were used in these studies. These metabolites differ considerably in their toxicity, with BaP

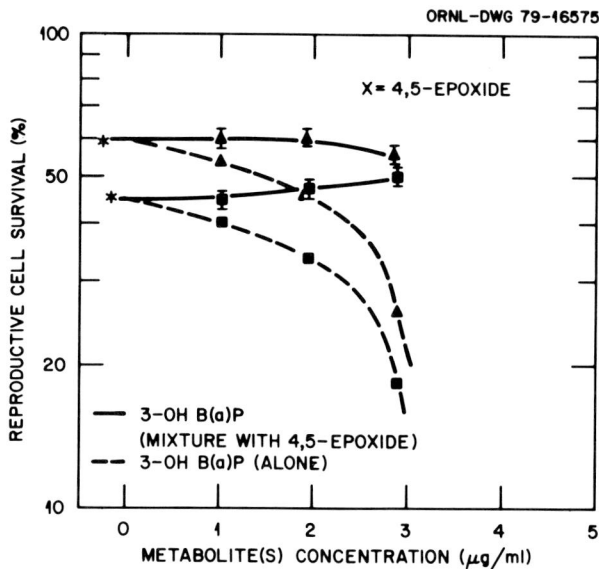

FIGURE 4. Cellular cytotoxicity caused by a mixture of B(a)P metabolites. Cells were exposed to either 0.6 or 0.8 μg/ml of B(a)P 4,5-oxide (associated cell survivals were 60 or 45 percent, respectively) and increasing concentrations (1, 2, or 3 μg/ml) of 3-HOB(a)P. Dotted lines indicate the cytotoxic effect that might be anticipated if these increasing amounts of 3-HOB(a)P were added to cells which had already been subjected to the level of cytotoxicity produced by the two concentrations of B(a)P 4,5-oxide.

4,5-oxide being much more cytotoxic and mutagenic (6). V79 cells were seeded at a rather low density ($8 \times 10^4$) in 60-mm dishes, and then exposed to the two aforementioned metabolites for varying periods of time. The cell number and uptake of radioactive precursors into certain macromolecules was determined over sequential intervals during and after the period in which the metabolite was in the culture medium.

The effect of varying times of exposure to 6-HOBaP on subsequent cell growth is shown in Figure 5. The effect of BaP 4,5-oxide is very similar to that of 6-HOBaP except for 48-hour exposure, in which the subsequent cell growth closely parallels the growth of cells exposed for 24 hours to 6-HOBaP. The fact that the growth curves for metabolite-treated cells, after an initial delay whose duration depends upon length of exposure, all become more or less parallel to the growth curve for untreated cells suggests that exponential phase growth rates for metabolite-treated cells are not grossly different from those of the control. One possible explanation for the shape of the cell-growth curves is that the BaP metabolite induces a delay in the cell cycle and, after the metabolite is removed, the cells eventually recover the normal cycle kinetics. The difference in hours between the exponential growth phases of control and treated cultures might provide an estimate of such a delay. For BaP 4,5-oxide, tentative delays of 6, 12, and 18 hours can be estimated for exposure times of 4, 12, and 48 hours, respectively. For 6-HOBaP, delays of 4, 6, 18, and 36 hours can be estimated for exposure times of 4, 12, 24, and 48 hours.

Another effect must be taken into account when interpreting these growth curves. Treatment with metabolite results in some cytotoxicity, the degree increasing with increasing exposure times. Reproductive cell-survival measurements were made as a part of this experiment and the percentage of cell survival associated with varying exposure times is given in Figure 5 (Corresponding cell-survival values for 4-, 12-, and 48-hour

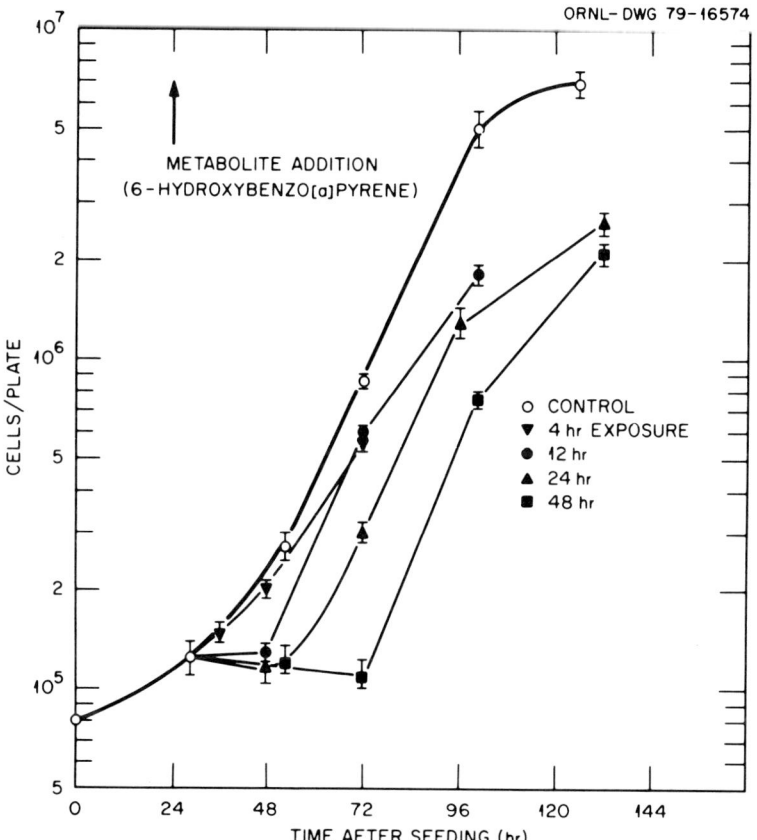

**FIGURE 5.** Cell growth following treatment with a B(a)P metabolite. Duplicate plates were exposed for the times indicated. Control plates received acetone. All cell counts indicate viable cells as counted by Trypan Blue staining. Reproductive cell survivals associated with exposure times of 4, 12, 24, and 48 hours were 87, 75, 51, and 31 percent, respectively.

exposure to BaP 4,5-oxide were 73, 58, and 28 percent respectively). If the time required for the metabolite to kill and lyse cells is short relative to the doubling time of the cells (12 hours), then the viable cells remaining might be in exponential growth a short time after the cytotoxic period. The apparent cell-growth plateau from the start of the metabolite addition to the next cell count might not be a genuine plateau but could consist of a rapid decrease in cell number, followed by exponential growth of the surviving cells. We have not determined such intermediate cell counts on the longer exposure times, and so cannot confirm or disprove this interpretation. In any case, this latter explanation cannot fully account for the results with respect to the 6-HOBaP-treated growth curves because the cell-growth plateau at the 48-hour exposure is so long that some delay in cell-cycle time must be invoked to explain the data.

To further investigate the cell-cycle kinetics of the metabolite-treated cells, and to study the effects of metabolite treatment on certain macromolecular synthesis processes, radioactive precursors for DNA, RNA, and protein were used. The results of these experiments for DNA are shown in Figure 6. The long labeling times were chosen to provide incorporation of a label proportional to the percentage of cells cycling and the cell multiplication rate, and will not provide information about

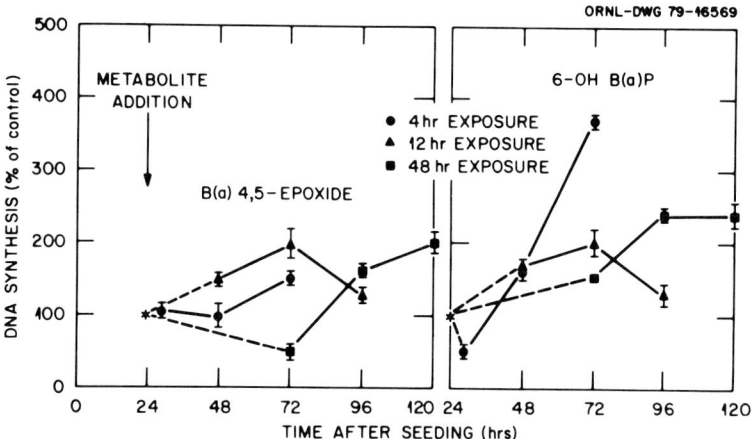

FIGURE 6. Incorporation of $^{14}$C-thymidine into DNA of cells exposed for varying times to B(a)P metabolites. The labeling time was for 16 to 18 hours except for the first time point of 4-hour exposure, in which case the labeling time was 4 hours. All data points were calculated as cpm/$\mu$g of DNA, and normalized by dividing by the corresponding incorporation in control cultures. The * indicates the point at which metabolite-treated and control cultures were separated to start the experiment.

the rate of DNA synthesis. It can be seen from Figure 6 that several effects are produced, depending on the length of exposure and the metabolite used. BaP 4,5-oxide shows no initial effect (4-hour exposure), an increased uptake compared with that of controls (12-hour exposure), and an inhibition (48-hour exposure). Results for the shorter exposure times suggest that metabolite treatment causes an increase in the percentage of cells cycling after the removal of the metabolite (see data points at 72 hours). Since cultures exposed for both 4 and 12 hours are in apparently normal exponential growth at this time, the increase in DNA labeling compared with that of controls is difficult to explain otherwise. On the other hand, 48-hour exposures result in an inhibition of DNA labeling initially (see 72-hour data point), followed by an increse to well above control levels. These data are consistent with a slowing or delay of the cell cycle during the long metabolite exposure.

The data for 6-HOBaP (Figure 6) show that this metabolite has different effects on the cell-cycle kinetics than does BaP 4,5-oxide. The most dramatic difference is in the uptake of precursor into DNA by the cultures exposed for 4 hours. For 6-HOBaP, there is an initial inhibition, followed by a greatly increased uptake. These results again suggest a delay or slowing of the cell cycle, followed by a rebound that causes an increase in the percentage of cells cycling compared with that for the controls.

Surprisingly, cultures exposed for 48 hours show increases in uptake into DNA at a time (see 72-hour data point) when the cell count seems to be static, as compared with the control cultures which are in exponential growth. Such an effect might occur if the metabolite induced a mitotic block. It may be that some of the measured uptake into DNA is occurring as a result of DNA repair, but it is difficult to distinguish this repair incorporation from incorporation into newly synthesized DNA in growing cultures.

The results of labeling RNA are shown in Figure 7. The long labeling time used will probably result in most of the label being incorporated into ribosomal RNA, rather than into the rapidly synthesized and degraded messenger RNA. Of course, transfer RNA also will be labeled, but this is only 10 to 12 percent of the total RNA. Rather dramatic differences are evident when the results from the two metabolites are compared (see the 4- and 48-hour-exposure curves particularly). The increases in incorporation into RNA, compared with control cultures, suggest increases in synthesis of ribosomal RNA. The most logical reason for this response to be elicited is in preparation for increased protein synthesis.

Indeed, measurements of protein synthesis (Table 1) indicated that it was increasing during the same time intervals as increased RNA synthesis. For both 12- and 48-hour 6-HOBaP exposures, protein synthesis was inhibited initially after metabolite removal and then gradually increased

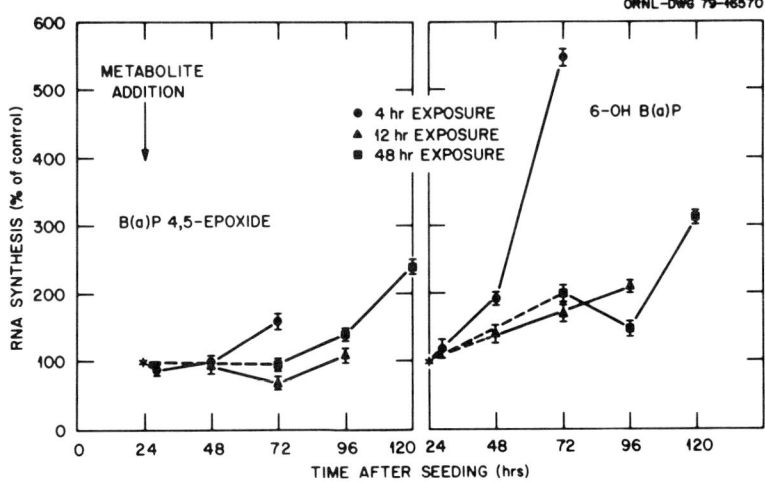

FIGURE 7. Incorporation of $^3$H-uridine into RNA of cells exposed for varying times to B(a)P metabolites. See Legend to Figure 6 for details. All data points were calculated as cpm/μg of RNA, and normalized by dividing by the corresponding incorporation in control cultures.

TABLE 1. Protein, RNA, and DNA Synthesis in V79 Cells Exposed to 6-HOB(a)P for 12 or 48 Hours

| | Precursor Incorporation Compared With Control Cultures[a] (%) | | | | | |
|---|---|---|---|---|---|---|
| | 12-Hour Exposure | | | 48-Hour Exposure | | |
| | A[b] | B | C | A[b] | B | C |
| Protein | 60 | 70 | 80 | 50 | 55 | 100 |
| RNA | 140 | 170 | 210 | 190 | 150 | 315 |
| DNA | 150 | 200 | 130 | 150 | 230 | 235 |

[a]Data derived from duplicate dishes. The coefficient of variation was 5 to 15 percent.
[b]A, B, and C refer to time intervals after removal of 6-HOB(a)P from the culture medium of 12, 36, and 60 hours and 0, 24, and 48 hours for the 12-hour exposed and 48-hour exposed cultures, respectively.

toward control levels as the cells continued to grow. The observed inhibition of protein synthesis could be due to slowed protein synthesis associated with the slowing of the cell cycle. It also may be a consequence of fewer cells in the treated cultures, although for the 12-hour exposure, the cell killing is only 25 percent, or multiple factors may be involved.

These studies have indicated that the effects of BaP metabolite addition to cells are complex, affecting many intracellular processes in different ways. Interpretation of how complex mixtures of chemicals react with cells will remain obscure until the most significant intracellular sites and processes for chemical interaction are identified.

## ACKNOWLEDGMENT

Research sponsored by the Office of Health and Environmental Research, U. S. Department of Energy under Contract W-7505-eng-26 with the Union Carbide Corporation.

## REFERENCES

1. Brookes, P., Baird, W. M., and Dipple, A. (1974): Interaction of the carcinogen 7-methylbenz(a)anthracene with DNA of mammalian cells. In: *Chemical Carcinogenesis,* P.O.P. Ts'o and H. V. Gelboin, Eds., pp. 149-157, Marcel Dekker, Inc., New York.
2. Burton, K. (1956): A study of the conditions and mechanisms of the diphenylamine reaction for the colorimetric estimation of deoxyribonucleic acid. *Biochem. J.* 62:315-323.
3. Gelboin, H. V., Huberman, E., and Sachs, L. (1969): Enzymatic hydroxylation of benzopyrene and its relationship to cytotoxicity. *Proc. Natl. Acad. Sci. U.S.A.* 64:1188-1194.
4. Huberman, E., Aspiras, L., Heidelberger, C., Grover, P. L., and Sims, P. (1971): Mutagenicity to mammalian cells of epoxides and other derivatives of polycyclic hydrocarbons. *Proc. Natl. Acad. Sci.* 68:3195-3199.
5. Huberman, E., and Sachs, L. (1974): Cell mediated mutagenesis of mammalian cells with chemical carcinogens. *Int. J. Cancer.* 13:326-333.
6. Huberman, E., Sachs, L., Yang, S. K., and Gelboin, H. V. (1976): Identification of mutagenic metabolites of benzo(a)pyrene in mammalian cells. *Proc. Natl. Acad. Sci. U.S.A.* 73:607-611.
7. Huberman, E., and Sachs, L. (1977): DNA binding and its relationship to carcinogenesis by different polycyclic hydrocarbons. *Int. J. Cancer.* 19:122-127.
8. Kerr, S. E., and Seraidarian, K. (1945): The separation of purine nucleosides from free purines and the determination of the purines and ribose in these fractions. *J. Biol. Chem.* 159:211-225.

9. Lowry, O. H., Rosebrough, N. J., Farr, A. L., and Randall, R. J. (1951): Protein measurements with the folin phenol reagents. *J. Biol. Chem.* 193:265-275.
10. Mans, R. J., and Novelli, G. D. (1961): Measurement of the incorporation of radioactive amino acids into protein by a filter-paper disk method. *Arch. Biochem. Biophys.* 94:48-53.
11. Schmidt, G., and Thannhauser, S. J. (1945): A method for the determination of deoxyribonucleic acid, ribonucleic acid, and phosphoproteins in animal tissues. *J. Biol. Chem.* 161:83-89.
12. Slaga, T. J., Jecker, L., Bracken, W. M., and Weeks, C. E. (1979): The effects of weak or non carcinogenic polycyclic hydrocarbons on 7,12-dimethylbenz(a)anthracene and benzo(a)pyrene skin tumor initiation. *Cancer Lett.* 7:51-59.
13. Wislocki, P. G., Wood, A. W., Chang, R. L., Levin, W., Yagi, H., Hernandez, O., Dansette, P. M., Jerina, D. M., and Conney, A. H. (1976): Mutagenicity and cytotoxicity of benzo(a)pyrene arene oxides, phenols, quinones, and dihydrodiols in bacterial and mammalian cells. *Cancer Res.* 36:3350-3357.
14. Wood, A. W., Wislocki, P. G., Chang, R. L., Levin, W., Lu, A.Y.H., Yagi, H., Hernandez, O., Jerina, D. M., and Conney, A. H. (1976): Mutagenicity and cytotoxicity of benzo(a)pyrene benzo-ring epoxides. *Cancer Res.* 36:3358-3366.

# BENZO(a)PYRENE ACTIVATION AND DETOXIFICATION BY HUMAN PULMONARY ALVEOLAR MACROPHAGES AND LYMPHOCYTES

M. V. Marshall*†, T. L. McLemore**,***, R. R. Martin**,
M. H. Marshall*, N. P. Wray**,***, D. L. Busbee†,
E. T. Cantrell‡, M. S. Arnott*, and A. C. Griffin*

    *The University of Texas System Cancer Center
    M. D. Anderson Hospital and Tumor Institute
    Houston, TX 77030
  **Baylor College of Medicine
    Houston, TX 77030
***The Veterans Administration Hospital
    Houston, TX 77030
  †North Texas State University
    Denton, TX 76203
  ‡Texas College of Osteopathic Medicine
    Fort Worth, TX 76107

## INTRODUCTION

Most investigations of carcinogen metabolism by human tissues have attempted to study the metabolism of benzo(a)pyrene (BaP) in mitogen-stimulated peripheral blood lymphocytes utilizing fluorometric analysis of the relatively noncarcinogenic phenolic derivatives of BaP, the major metabolites produced by the arylhydrocarbon hydroxylase (AHH) enzyme system (5,6,14,16,20-26,28). A more comprehensive approach to the study of BaP biotransformation in man would be to quantitate all metabolites produced by this enzyme system, since formation of metabolites such as the BaP 7,8-dihydrodiol-9,10-epoxide may be a determinant of cancer susceptibility (3,13,30). One of the major steps in secondary BaP metabolism is the epoxide hydrase-mediated conversion of BaP epoxides to dihydrodiols (31). Some dihydrodiols are detoxification products (4,5- and 9,10-dihydrodiol); however, the 7,8-dihydrodiol may be further metabolized to the 7,8-dihydrodiol-9,10-epoxide, the presumed ultimate carcinogen of BaP (12,13). Persons with a greater capacity for production of the 7,8-dihydrodiol might, therefore, be at greater cancer risk due to further metabolism to ultimate carcinogens (9). The concentration of toxic metabolites in the cell depends not only upon synthesis of primary metabolites (epoxides and phenols) but largely upon their further metabolism by conjugation reactions (4,8). These conjugation reactions include esterification of hydroxyl groups with sulfate and glucuronic acid or conjugation of epoxides with glutathione.

Investigation of carcinogen metabolism by cells of the human respiratory tract could be most pertinent, since these cells are directly exposed to inhaled environmental carcinogens. Initial studies of lung tissue metabolism of carcinogens were done by Cantrell and co-workers (5). These studies demonstrated that pulmonary alveolar macrophages (PAMs) obtained by bronchopulmonary saline lavage from normal cigarette smokers exhibited AHH activity (as determined fluorometrically). Later studies by McLemore et al confirmed the original observations and extended the work to include noncancer and lung cancer patients (20-24). In addition, a culture system for human PAMs was developed to study in vitro induction of AHH by benz(a)anthracene (BaA) or cigarette tars (19,23-26). It was demonstrated that in vivo induction of AHH in PAMs by cigarette smoking was positively correlated with the in vitro induction observed following incubation with various PAH inducers (23,24,26). Additional studies demonstrated that AHH activity is present in lung tissue obtained at the time of surgery from cigarette smokers (21). Recent studies by Autrup et al have demonstrated the ability of human PAMs and bronchus obtained at autopsy from lung cancer patients to metabolize free BaP (3) and BaP bound to ferric oxide particles (2). Induction of AHH activity by PAH adsorbed to asbestos has also been observed in PAMs and lymphocytes (7,19).

Free PAM metabolite profiles are similar to those of human blood lymphocytes and monocytes from healthy volunteers (10,27). However, they differ from previously reported BaP metabolite profiles in cultured human PAMs which demonstrated little or no production of BaP phenols and the production of large quantities of the 7,8-dihydrodiol following induction with BaA (3,11). These previous studies were performed on autopsy samples obtained from lung cancer patients, whereas samples used in our study were obtained from healthy volunteers.

Our research has focused on the role of human PAMs in activation and detoxification of PAH, primarily BaP. In this paper, we have compared PAMs and circulating lymphocytes from five smokers and five nonsmokers for their ability to metabolize BaP as determined by high-pressure liquid chromatography (HPLC).

## MATERIALS AND METHODS

### Study Subjects

PAMs and lymphocytes were obtained as previously described from ten normal volunteers, 21 to 38 years of age, who had not received any medication that would be expected to alter the metabolism of BaP (17,20). The subjects included five smokers and five nonsmokers. The smoking history of the five smokers is illustrated in Table 1. Procedures for preparation of PAMs, identification of metabolites, and conjugate extraction are described in reference 17.

### Lymphocytes

Procedures for preparation of lymphocytes are described in reference 20. Following 72 hours of exposure to mitogens, the cells were induced for 24 hours by the addition of 20 $\mu$M BaA. Controls were exposed to an equal volume of the solvent, i.e., acetone. After induction, the cells were centrifuged and resuspended at $1 \times 10^6$ cells/vial, based on the initial cell count. The cells were then exposed to 25 nmole $^3$H-BaP/ml (specific

**TABLE 1. Source of Cells Used for Metabolism Study**

| Subject | Sex | Race | Age | Smoking History (pack-years) |
|---------|-----|------|-----|------------------------------|
| 11S | M | Latin | 21 | 8 |
| 12S | F | White | 33 | 18 |
| 13S | M | Black | 30 | 20 |
| 14S | F | Black | 38 | 7.5 |
| 15S | F | White | 24 | 15 |
| 11NS | M | White | 23 | — |
| 12NS | M | White | 23 | — |
| 13NS | M | White | 22 | — |
| 14NS | M | White | 22 | — |
| 15NS | M | White | 23 | — |

activity 250 mCi/mmole) for 24 hours. Extraction of free metabolites and conjugates is the same as described for PAMs, except that 0.1 ml was removed for determination of cell number prior to extraction.

## Metabolite Extraction

After 24 hours of culture, 3 ml of medium was removed from culture vials and 50 nmoles $^3$H-BaP (specific activity 500 mCi/mmole) was added to the remaining 2 ml of medium, and PAMs were further incubated for 1 to 24 hours at 37° C. Metabolism was stopped by the addition of an equal volume of ethyl acetate. The ethyl acetate layers were pooled and dried over anhydrous $MgSO_4$ before evaporation under $N_2$.

## Conjugate Extraction

The aqueous layers from each subject were pooled and 1/100 volume of 1 M sodium acetate (pH 4.5) was added, followed by 1000 units/ml of crude $\beta$-glucuronidase, containing approximately 60 units/ml sulfatase activity (Sigma G0258). The samples were then incubated for 4 hours and reextracted twice with ethyl acetate as previously described (17). The extraction efficiency for BaP and several metabolites ($^3$H-4,5-dihydrodiol, $^3$H-7,8-dihydrodiol, and $^3$H-3-OH BaP) was $95.0 \pm 15.1$ percent. The total radioactivity recovered following glucuronidase treatment and the initial ethyl acetate extraction was $96.9 \pm 16.3$ percent.

## Metabolite Identification

The dried ethyl acetate extracts were dissolved in methanol prior to injection into a liquid chromatograph equipped with a 0.46 x 25-cm Zorbax $C_8$ column. Metabolites were eluted at 38° C with a gradient of 65 to 80 percent methanol/$H_2O$ containing 1 percent tetrahydrofuran at 900 psi, which is a modification of the procedure of Yang et al (31). Fluorescense and absorbance were monitored throughout the elution (excitation 320 to 385 nm, emission >408 nm) and 0.3-min samples were collected for liquid scintillation counting. Metabolites were identified by co-chromatography of authentic standards.

## Chemicals

[G-$^3$H]-benzo(a)pyrene, specific activity 20,000 to 30,000 mCi/mmole, was obtained from Amersham (Arlington Heights, Illinois). Benzo(a)pyrene was purified to >99 percent purity by HPLC prior to use.

Unlabeled BaP was obtained from Sigma (St. Louis, Missouri). Tissue culture media was obtained from Gibco (Gand Island, New York). Unlabeled benzo(a)pyrene standards were supplied by Dr. David Longfellow, National Cancer Institute, Division of Cancer Cause and Prevention. HPLC grade solvents were obtained from Burdick and Jackson or Fisher Scientific (Fairlawn, New Jersey).

## RESULTS

The ability of cultured human PAMs to metabolize BaP and to conjugate BaP metabolites was evaluated by HPLC. Here free BaP metabolites formed by PAMs during 1 hour of incubation with $^3$H-BaP are shown in Figure 1. The major metabolites observed are quinones and

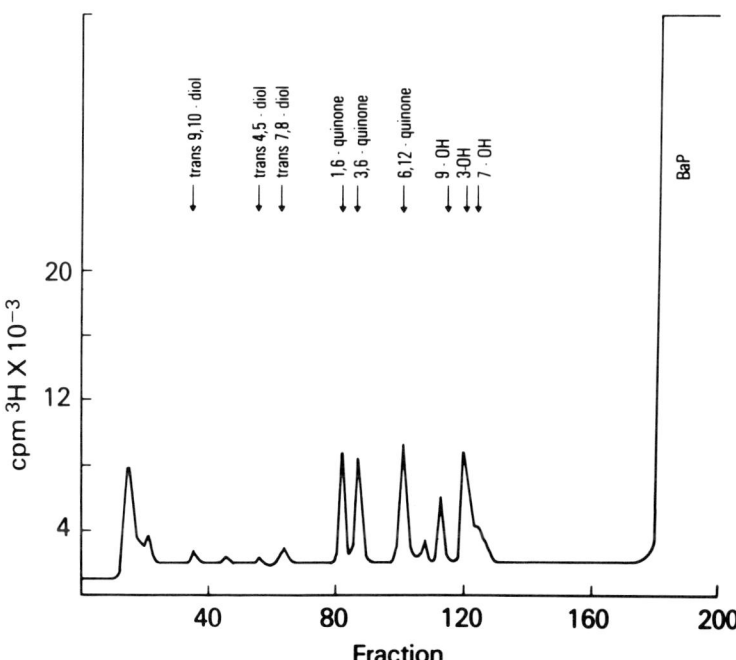

FIGURE 1. 1 hour human pulmonary alveolar macrophage metabolites. PAMs obtained from healthy volunteers were incubated for 1 hour with $^3$H-B(a)P and the metabolites were extracted as described in the Materials and Methods section. Extracts from several subjects were pooled and approximately 5 x 10$^6$ cells were analyzed by HPLC.

phenols. Noninduced BaP metabolites produced by PAMs following 24 hours of incubation with $^3$H-BaP are shown in Figure 2a. The major metabolite observed after 24 hours is the 7,8-dihydrodiol. Several unidentified minor metabolites were also observed in the tetrol-dihydrodiol region. Little or no 4,5-dihydrodiol was observed. Figure 2b shows the BaP metabolite profile from PAMs which had been induced with BaA for 24 hours prior to $^3$H-BaP addition. As in noninduced cells, the major metabolite was the 7,8-dihydrodiol. An increase was also observed in 9,10-dihydrodiol and in both 9-OH and 3-OH BaP. The peak between the 9,10- and 4,5-dihydrodiol is probably a breakdown product of BaP, whereas the peak eluting between 6,12-quinone and 9-OH BaP may be 6-OH methyl BaP (27). A greater induction was observed with lymphocytes exposed to BaA (Figure 3) compared with PAMs (Figure 2).

The sulfate and glucuronide conjugates from noninduced and BaA-induced PAMs incubated with $^3$H-BaP for 24 hours are shown in Figure 4. The major metabolites released following glucuronidase/sulfatase treatment are quinones and phenols. A new peak that eluted after 7-OH BaP was observed in both induced and noninduced metabolite profiles. An increase in the more polar conjugates was observed following BaA induction (Figure 4b).

Free BaP metabolites formed by PAMs from several individuals during 24 hours of incubation with $^3$H-BaP are shown in Figure 5. Although the number of individuals studied was small, considerable interindividual variation occurred in the extent of BaP metabolism by PAMs. Lymphocytes from these individuals were also analyzed for their ability to metabolize $^3$H-BaP (Figure 6). With the exception of one smoker (subject 11S), the relative ability to metabolize $^3$H-BaP was similar for both cell types obtained from each individual. The PAMs were able to metabolize more BaP than were lymphocytes in 24 hours, with a greater percentage of 7,8-dihydrodiol production relative to other metabolites.

Since conjugation is a principal pathway for detoxification of various BaP metabolites, the conjugation of BaP metabolites by human PAMs and lymphocytes also was evaluated (Table 2). Again, considerable interindividual variation in the conjugation of BaP metbolites was observed. The major class of known compounds released following glucuronidase/sulfatase teatment was the quinones. Conjugation was more pronounced following BaA induction of PAMs from smokers. This was not as evident in PAMs from nonsmokers, which generally conjugated less BaP metabolites than did PAMs from smokers. PAMs from several smokers were able to conjugate large quantities of 7,8-dihydrodiol. Lymphocytes from these individuals also were compared for their ability to conjugate BaP, as illustrated in Table 3. As with PAMs, the major class

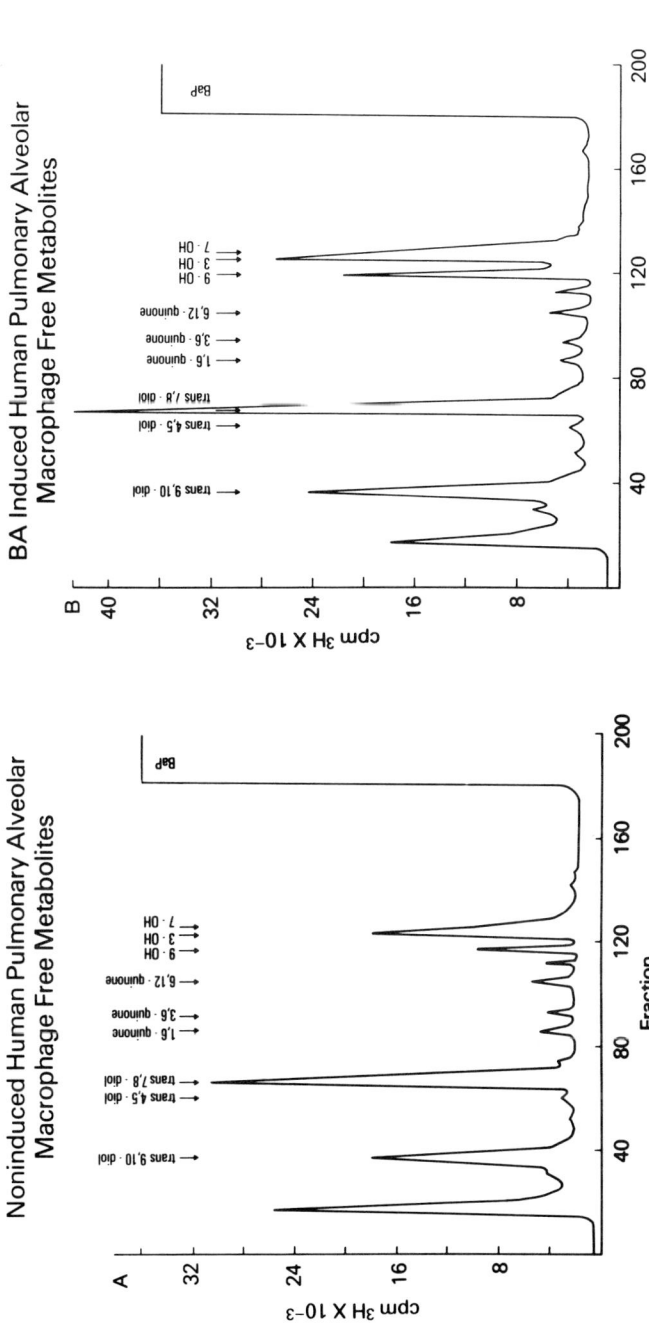

FIGURE 2. Free metabolites produced by human pulmonary alveolar macrophages. (A) PAMs obtained from healthy volunteers were cultured for 24 hours with $^3$H-BaP; and the metabolites were then extracted as described in the Materials and Methods section. Extracts from several subjects were pooled and approximately $6 \times 10^6$ cells were analyzed by HPLC. (B) PAMs were treated as above except that they were exposed to 20 µM BaA for the initial 24 hours in culture for induction of AHH activity.

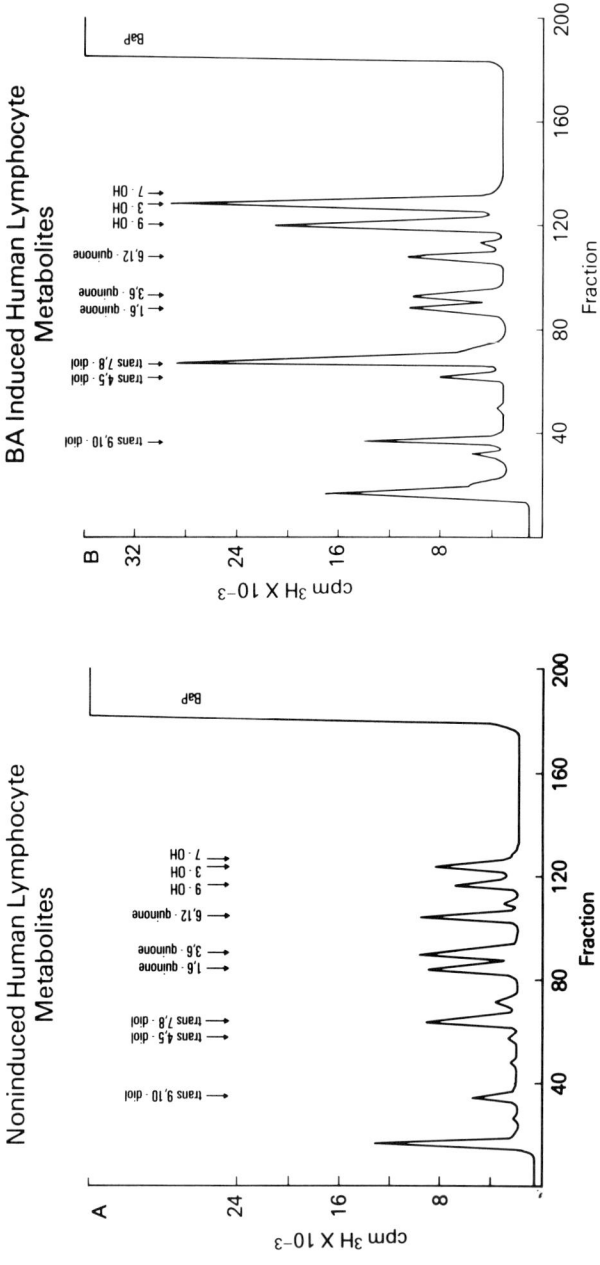

FIGURE 3. Free metabolites produced by circulating human lymphocytes. (A) Lymphocytes obtained from healthy volunteers were cultured for 3 days in the presence of mitogens, and then cultured for 24 hours following addition of 0.5 percent acetone. The lymphocytes were then exposed to $^3$H-BaP for 24 hours as described in the Materials Methods section. Extracts from several subjects were pooled and approximately $5 \times 10^6$ cells were analyzed by HPLC. (B) Lymphocytes were treated as above except that they were exposed to 20 $\mu$M BaA for 24 hours prior to $^3$H-BaP administration.

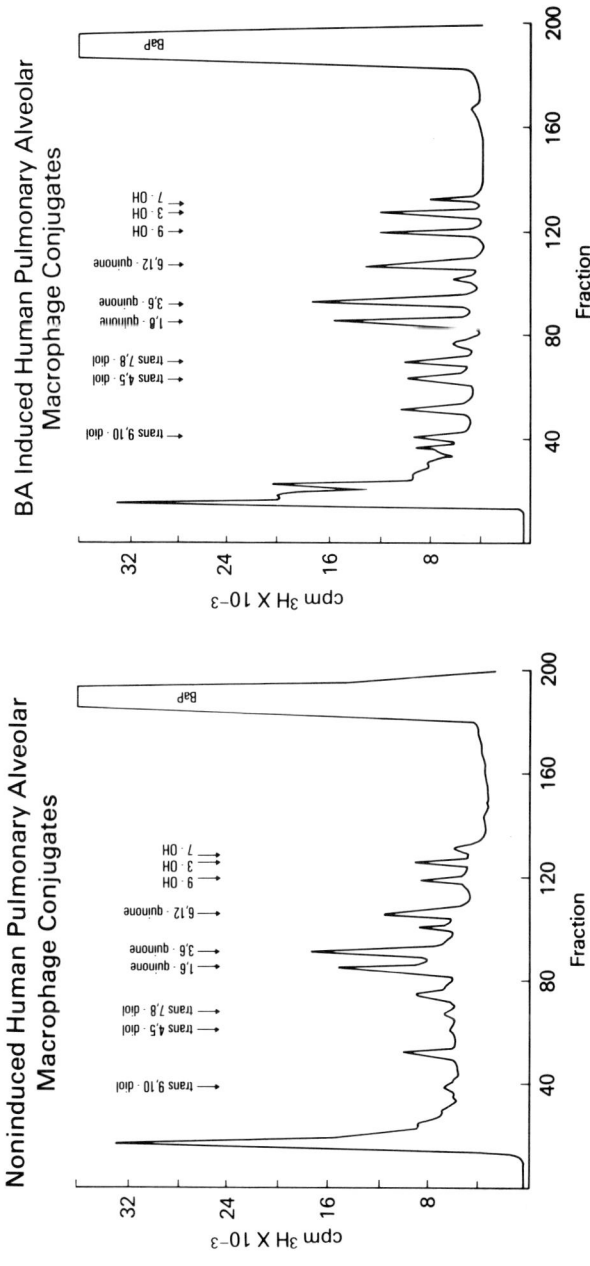

FIGURE 4. Conjugates produced by human pulmonary alveolar macrophages. (A) Metabolites produced by noninduced PAMs were hydrolyzed by a glucuronidase/sulfatase mixture and extracted as described in Materials and Methods section. Hydrolyzed conjugates from several subjects were pooled and approximately $6 \times 10^6$ cells were analyzed by HPLC. (B) Same as A except the conjugates were from BaA-induced PAMs.

TABLE 2. Conjugates of Benzo(a)pyrene Produced by Human Pulmonary Alveolar Macrophages

| Subject Number | Nonsmoker Pam Conjugates (pmoles/10⁶ cells/24 hours) | | | | | | Subject Number | Smoker Pam Conjugates (pmoles/10⁶ cells/24 hours) | | | | | |
| | Dihydrodiols | | | Quinone | Phenols | | | Dihydrodiols | | | Quinones | Phenols | |
| | 9,10 | 4,5 | 7,8 | | 9-OH | 3-OH | | 9,10 | 4,5 | 7,8 | | 9-OH | 3-OH |
|---|---|---|---|---|---|---|---|---|---|---|---|---|---|
| 11NS[a] CON[b] | 1.5 | 4.3 | 4.5 | >100. | 4.9 | <0.1 | 11S CON | <0.1 | <0.1 | 0.7 | 20.5 | <0.1 | <0.1 |
| IND[c] | 1.4 | 5.1 | 3.1 | >100. | 9.8 | 1.4 | IND | <0.1 | 0.1 | 1.3 | 21.2 | <0.1 | <0.1 |
| 12NS[a] CON | 4.4 | 8.4 | 8.3 | >100. | 8.1 | 1.9 | 12S CON | 1.3 | 0.2 | 1.3 | 4.7 | 1.0 | <0.1 |
| IND | 5.9 | 8.9 | 11.2 | >100. | 6.8 | 1.0 | IND | 1.3 | 1.0 | 2.0 | 9.3 | 2.2 | 0.4 |
| 13NS CON | <0.1 | <0.1 | 0.1 | 2.1 | <0.1 | <0.1 | 13S CON | 0.2 | 1.2 | 0.8 | <0.1 | <0.1 | <0.1 |
| IND | <0.1 | <0.1 | 0.2 | 2.9 | <0.1 | <0.1 | IND | <0.1 | 0.3 | 1.3 | 0.2 | <0.1 | <0.1 |
| 14NS CON | <0.1 | <0.1 | <0.1 | 0.6 | <0.1 | <0.1 | 14S CON | <0.1 | <0.1 | <0.1 | 14.7 | 1.6 | 2.3 |
| IND | 0.3 | <0.1 | <0.1 | 2.7 | 0.3 | 0.2 | IND | 0.1 | 0.1 | 0.9 | 19.9 | 2.0 | 2.1 |
| 15NS CON | <0.1 | <0.1 | 0.2 | <0.1 | <0.1 | <0.1 | 15S CON | <0.1 | <0.1 | 0.8 | <0.1 | <0.1 | <0.1 |
| IND | <0.1 | <0.1 | 0.2 | <0.1 | <0.1 | 0.6 | IND | 1.2 | 0.6 | 1.2 | <0.1 | 4.9 | 4.4 |

[a]Incubated 21 hours instead of 4 hours.
[b]CON—Noninduced.
[c]IND-Induced with BaA for 24 hours.

TABLE 3. Conjugates of Benzo(a)pyrene Produced by Circulating Lymphocytes

| Subject Number | Nonsmoker Lymphocyte Conjugates (pmoles/10⁶ cells/24 hours) | | | | | | | | Subject Number | Smoker Lymphocyte Conjugates (pmoles/10⁶ cells/24 hours) | | | | | | | |
|---|---|---|---|---|---|---|---|---|---|---|---|---|---|---|---|---|---|
| | Dihydrodiols | | | Quinones | Phenols | | | | | Dihydrodiols | | | Quinones | Phenols | | | |
| | 9,10 | 4,5 | 7,8 | | 9-OH | 3-OH | | | | 9,10 | 4,5 | 7,8 | | 9-OH | 3-OH | | |
| 11NS CON[a] | <0.1 | <0.1 | 0.1 | 15.9 | 2.0 | 2.3 | | | 11S CON | <0.1 | <0.1 | <0.1 | <0.1 | 1.7 | 10.7 | | |
| IND[b] | <0.1 | 0.6 | 0.9 | 26.4 | 2.3 | 1.5 | | | IND | <0.1 | <0.1 | <0.1 | <0.1 | <0.1 | <0.1 | | |
| 12NS CON | 2.3 | 3.7 | 13.0 | 24.8 | 1.8 | 2.8 | | | 12S CON | 3.5 | 1.6 | <0.1 | 17.7 | 0.9 | 1.3 | | |
| IND | 4.0 | 3.6 | 2.7 | 23.1 | 6.8 | 6.5 | | | IND | 0.5 | <0.1 | 1.7 | 10.1 | 1.1 | <0.1 | | |
| 13NS CON | <0.1 | <0.1 | <0.1 | 9.1 | 0.1 | <0.1 | | | 13S CON | 0.4 | 0.3 | 0.9 | 12.0 | 1.2 | 1.9 | | |
| IND | 0.1 | <0.1 | <0.1 | 3.6 | <0.1 | 0.1 | | | IND | 0.3 | 0.5 | 2.4 | 9.1 | 2.1 | 0.5 | | |
| 14NS CON | 0.8 | 3.5 | 5.4 | 11.2 | 3.9 | 0.9 | | | 14S CON | <0.1 | <0.1 | <0.1 | <0.1 | <0.1 | <0.1 | | |
| IND | 4.1 | 5.5 | 5.6 | 27.4 | 6.4 | 1.1 | | | IND | <0.1 | <0.1 | <0.1 | <0.1 | <0.1 | <0.1 | | |
| 15NS CON | 0.2 | <0.1 | 0.9 | 5.7 | 1.4 | 5.1 | | | 15S CON | <0.1 | <0.1 | <0.1 | <0.1 | 0.5 | 0.3 | | |
| IND | 0.4 | 2.1 | 3.3 | <0.1 | 3.4 | 3.7 | | | IND | <0.1 | 0.9 | 1.6 | 1.5 | 8.5 | 8.3 | | |

[a]Noninduced.
[b]Induced with BaA for 24 hours.

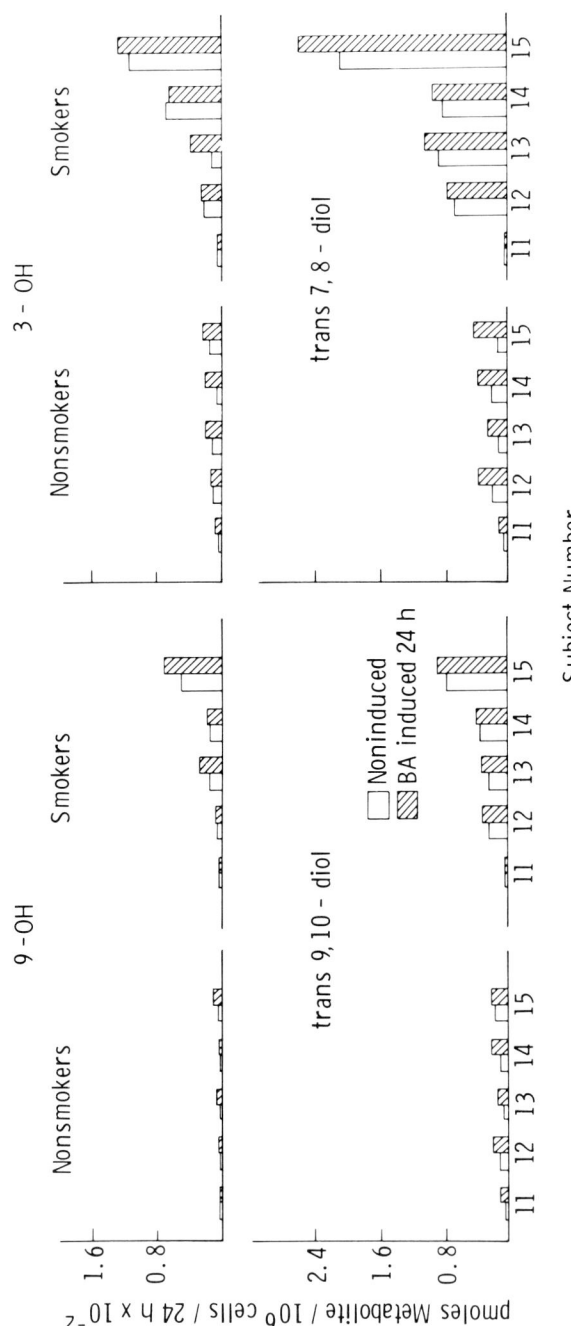

FIGURE 5. Comparison of human pulmonary alveolar macrophage metabolites from smokers and nonsmokers. Free metabolites produced by PAMs from five healthy smokers and five healthy nonsmokers were analyzed by HPLC as described in the Materials and Methods section.

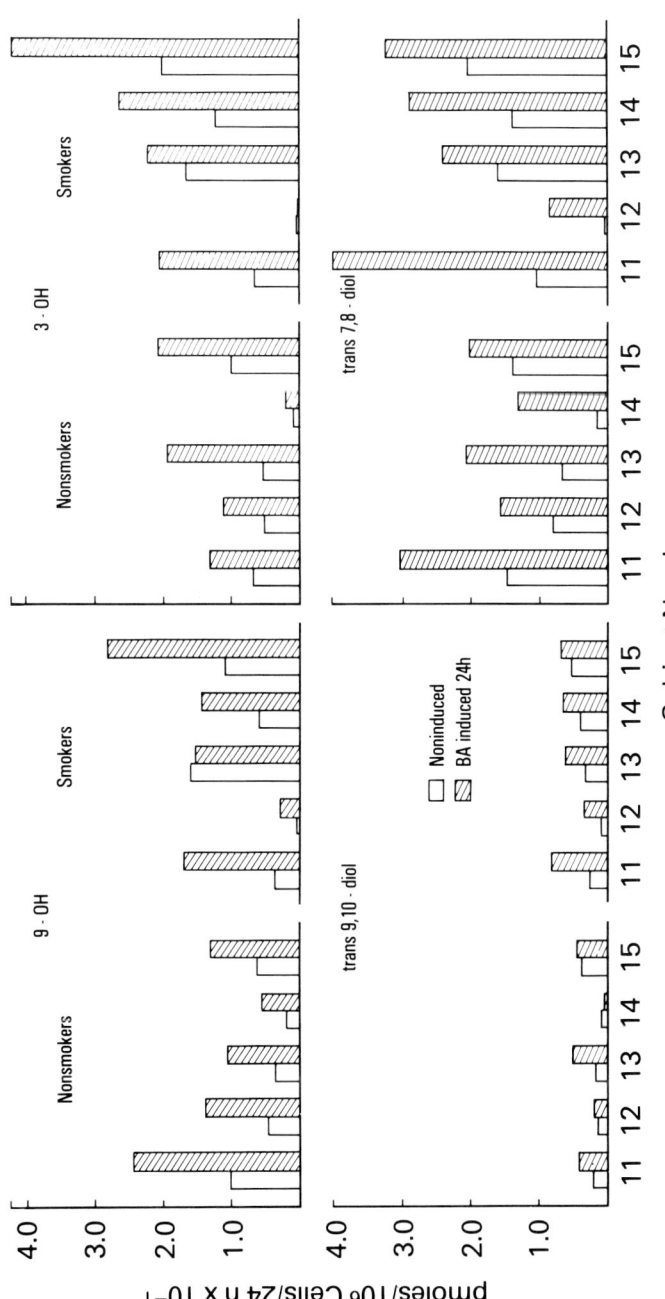

FIGURE 6. Comparison of circulating human lymphocyte metabolites from smokers and nonsmokers. Free metabolites produced by circulating lymphocytes from the same subjects described in Figure 5 were analyzed by HPLC as described in the Materials and Methods section.

of metabolites released following glucuronidase/sulfatase digestion was quinones, with several exceptions. Lymphocytes from one smoker (subject 14S) appeared to lack conjugating activity. The overall conjugating activity was higher in lymphocytes (Table 3) than in PAMs (Table 2). The lymphocytes of smokers appeared to have a lower conjugating activity than those of nonsmokers.

## DISCUSSION

Since the discovery that PAH, which have been identified in cigarette smoke, are metabolized to their active carcinogenic forms by human tissues, investigators have been intrigued by the hypothesis that carcinogen metabolism may be related to human cancer risk. This concept is supported by animal studies which demonstrate a positive relationship between the amount of AHH induction in animal tissues and tumor incidence following exposure to PAH (15). Most studies utilizing human tissues have employed peripheral blood lymphocytes to study PAH metabolizing capacities in man. It has been well documented that there is interindividual variation, with some individuals demonstrating a greater capacity for metabolism of BaP (5,21,25,26). Other studies utilizing peripheral blood lymphocytes to study cancer and noncancer patients have demonstrated distinct increases in AHH inducibility in individuals with primary lung cancer (20,22,24). Other sutdies did not confirm this increase in lung cancer patients (28). It is now apparent that the lymphocyte culture system for measurement of AHH activity has many problems which increase variability with regard to BaP metabolism (16). In addition, most studies to date have utilized the fluorometric (AHH) assay for BaP metabolism which measures primarily the phenolic metabolites of BaP. For these reasons, and lack of a prospective study, a definitive relationship between carcinogen metabolism and cancer risk in man has not been demonstrated. In the current study, we have investigated the ability of human cells to activate and detoxify BaP. Utilizing these procedures, free BaP metabolites as well as conjugates were identified by HPLC. This approach allows better quantitation of BaP metabolism in man than did previous studies.

During short term (1 hour) incubation of human PAMs with $^3$H-BaP, the major metabolites produced are phenols and quinones (17). Several unidentified metabolites were observed in 1- and 24-hour incubations. These metabolites eluted prior to the 9,10-dihydrodiol, between the 9,10-dihydrodiol and the 4,5-dihydrodiol, and also between the 6,12-quinone and 9-OH BaP. The compounds that elute prior to the 9,10-dihydrodiol consist of tetrols, triols, conjugates, and nonenzymatic product(s) (27,31).

The compound that elutes between the 9,10-dihydrodiol and the 4,5-dihydrodiol is probably a nonenzymatic degradation product of BaP, whereas the compound that elutes between the 6,12-quinone and 9-OH BaP could be 6-hydroxymethyl BaP (27).

During 24 hours of incubation, the major metabolites produced by human PAMs are the 7,8-dihydrodiol and the 9,10-dihydrodiol. A larger percentage of phenols is observed, in contrast to the observations of Autrup et al who utilized PAMs obtained at autopsy, primarily from cancer patients (3). There was a greater inducibility in lymphocytes compared with PAMs; this was also observed by Okano et al in comparing the metabolism of monocytes and lymphocytes (27). Our observation of greater BaP metabolism by PAMs compared with lymphocytes is similar to the observation by Okano et al that monocytes, the precursors of a type of macrophage, are more active than lymphocytes in metabolism of BaP (27). In short-term incubations, monocytes displayed a 2 to 5-fold greater activity than lymphocytes (27). Our data indicate that pretreatment of cells with BaA leads to a greater induction of BaP metabolism in lymphocytes compared with PAMs (Figures 5 and 6). The differences in induction and total metbolism could be related to cell size and/or an increased capacity for enzyme function. PAMs are metabolically active cells capable of ingestion and metabolism of xenobiotics, whereas lymphocytes must undergo mitogen stimulation prior to induction of the AHH enzyme system. Because of the location of PAMs in the terminal airways, they may play a crucial role in the activation or detoxification of inhaled xenobiotics, including particulates (2,3,11,17,18,21,29). PAMs metabolized more BaP than did lymphocytes and also produced more 7,8-dihydrodiol (Figures 5 and 6). Greater differences were observed in metabolites from PAMs obtained from smokers and nonsmokers than with lymphocytes. The higher activity in smoker PAMs may be due to prior induction of AHH by components of cigarette smoke. In view of the ability of PAMs to activate BaP to mutagens (11) and the greater activity compared with circulating lymphocytes, it appears that PAMs may play an important role in pulmonary carcinogenesis.

In general, BaP metabolism in cultured PAMs is reflected by BaP metabolism at 24 hours in cultured blood lymphocytes. This observation confirms previous work by McLemore et al (23-26) that there is a positive correlation between AHH activity in PAMs and lymphocytes from smokers and nonsmokers without evidence of cancer. It remains to be seen if there is a negative correlation in metabolism of BaP by PAMs and lymphocytes from lung cancer patients. Our observations confirm the report by Autrup et al (3) that the major metabolite produced by PAMs was the 7,8-dihydrodiol. Another report has demonstrated a positive correlation between the phenolic metabolites of BaP produced by various tissues from patients without evidence of cancer and a dissociation between these values in cells obtained from lung cancer patients (20,22,24). On the basis of our observations and those of Autrup et al (3), it is possible that lung cancer patients might have increased production of the 7,8-dihydrodiol, which predisposes them to greater cancer risk. Further studies will be necessary to confirm this hypothesis because many factors (smoking habits, age, general health, differences in experimental design) might also explain differences observed in the present volunteer studies and in previous studies involving lung cancer patients.

The ability to detoxify xenobiotics through conjugation might also be important in the accumulation of reactive metabolites at the target site of tumor induction (3,4,8,17). The major metabolites produced by PAMs and released following glucuronidase/sulfatase treatment are more polar than the 9,10-dihydrodiol (Figure 4). The next most plentiful metabolites are quinones which may arise spontaneously from phenols (27). From the data in Tables 2 and 3, it appears that lymphocytes conjugate more BaP metabolites than do PAMs during 24 hours of incubation.

These procedures do not allow identification of glutathione conjugates, nor do they distinguish between sulfate and glucuronide conjugates. An alternative procedure developed by Herman Autrup (1) is currently in use. This technique utilizes an alumina column to separate the different classes of BaP metabolites into unconjugated metabolites, sulfates, glucuronides, and glutathione conjugates. This technique will be utilized for all subsequent studies. In addition, $E.\ coli$ $\beta$-glucuronidase will be utilized because of optimal hydrolysis at pH 7.0 rather than pH 4.5. This should permit less nonenzymatic conversion of phenols to quinones. Additionally, the antioxidants butylated hydroxytoluene or $\alpha$-tocopherol are added to the mixture prior to extraction to minimize spontaneous formation of quinones (27).

We have analyzed the BaP metabolites and conjugates formed by PAMs and lymphocytes from several individuals. Utilizing this approach, further investigation of activation and detoxification by several human cell types could provide the basis for more precise and comprehensive studies of carcinogen and drug metabolism in the human lung, and for a better assessment of cancer risk in selected populations.

## ACKNOWLEDGMENTS

This work was supported in part by grants from The Robert A. Welch Foundation G-035, Biomedical Research Support Grant RR5511, Council for Tobacco Research U.S.A. grants 1102 and 1094, American

Cancer Society Grant PDT-149, Contract N01-CP-85671 from the National Cancer Institute, and a grant from the Veterans Administration Hospital, Houston, Texas. The assistance of Deborah Harris, Dorothy Haynes, and Debra Stanberry is greatly appreciated.

## REFERENCES

1. Autrup, H. (1979): Separation of water soluble metabolites of benzo(a)pyrene formed by cultured human colon. *Biochem. Pharmacol.* 28:1727-1730.
2. Autrup, H., Harris, C. C., Schafer, P. W., Trump, B. F., Stoner, G. D., and Hsu, T-C. (1979): Uptake of benzo(a)pyrene—ferric oxide particulates by human pulmonary macrophages and release of benzo(a)pyrene and its metabolites. *Proc. Soc. Exptl. Biol. Med.* 161:280-284.
3. Autrup, H., Harris, C. C. Stoner, G. D., Selkirk, J. K., Schafer, P. W., and Trump, B. F. (1978): Metabolism of [$^3$H]benzo(a)pyrene by cultured human bronchus and human pulmonary alveolar macrophages. *Lab. Inv.* 38:217-224.
4. Bock, K. W. (1977): Dual role of glucuronyl and sulfotransferases converting xenobiotics into reactive or biologically inactive and easily excretable compounds. *Arch. Toxicol.* 39:77-85.
5. Cantrell, E. T., Busbee, D. L., Warr, G. A., and Martin, R. R. (1973): Induction of aryl hydrocarbon hydroxylase in human lymphocytes and pulmonary alveolar macrophages, a comparison. *Life Sci.* 13:1649-1654.
6. Cantrell, E. T., Warr, G. A., Busbee, D. L., and Martin, R. R. (1973): Induction of aryl hydrocarbon hydroxylase in human pulmonary alveolar macrophages by cigarette smoking. *J. Clin. Invest.* 52:1881-1884.
7. Corson, M. A., McLemore, T. L., Mace, M. L., Marshall, M. V., Snodgrass, D. R., Martin, R. R., Wray, N. P., and Brinkley, B. R. (1979): Asbestos mediated induction of aryl hydrocarbon hydroxylase in human pulmonary alveolar macrophages. *J. Natl. Cancer Inst.* (submitted for publication).
8. Depierre, J. W., and Ernster, L. (1978): The metabolism of polycyclic hydrocarbons and its relationship to cancer. *Biochem. Biophys. Acta* 473:149-186.
9. Fishman, J., Fukushima, D. K., O'Connor, J., and Lynch, H. T. (1979): Low urinary estrogen glucuronides in women at risk for familial breast cancer. *Science* 204:1089-1091.
10. Gelboin, H. V., Selkirk, J., Okuda, T., Nemoto, N., Yang, S. K., Wiebel, F. J., Whitlock, J. P., Jr., Rapp, H. J., and Bast, R. C., Jr.

(1977): Benzo(a)pyrene metabolism: enzymatic and liquid chromatographic analysis and application to human liver, lymphocytes and monocytes. In: *Biological Reactive Intermediates.* D. J. Jallow, J. J. Kocsis, R. Snyder, and H. Vaino, Eds., pp. 98-123, Plenum Press, New York.
11. Harris, C. C., Hsu, I.-C., Stoner, G. D., Trump, B. F., and Selkirk, J. K. (1978): Human pulmonary alveolar macrophages metabolise benzo(a)pyrene to proximate and ultimate mutagens. *Nature* 272:633-634.
12. Huberman, E., and Sachs, L. (1977): DNA binding and its relationship to carcinogenesis by different polycyclic hydrocarbons. *Int. J. Cancer* 19:122-127.
13. Kapitulnik, J., Wislocki, P. G., Levin, W., Yagi, H., Thakker, D. R., Akagi, H., Koreeda, M., Jerina, D. M., and Conney, A. H. (1978): Marked differences in the carcinogenic activity of optically pure (+) -and (-) - trans, 7,8-dihydroxy-7,8-dihydrobenzo(a)pyrene in newborn mice. *Cancer Res.* 38:2661-2665.
14. Kellerman, G., Shaw, C. R., and Luyten-Kellerman, M. (1973): Aryl hydrocarbon hydroxylase inducibility and bronchogenic carcinoma *N. Engl. J. Med.* 289:934-937.
15. Kouri, R. E. (1976): Relationship between levels of aryl hydrocarbon hydroxylase activity and susceptibility to 3-methylcholanthrene and benzo(a)pyrene-induced cancers in inbred strains of mice. In: *Carcinogens, Vol. 1, Polynuclear Aromatic Hydrocarbons: Chemistry, Metabolism, and Carcinogenesis.* R. J. Freudenthal and P. W. Jones, Eds., pp. 139-151, Raven Press, New York.
16. Kouri, R. E., Imblum, R. L., Sosnowski, R. G., Slomiany, B. J., and McKinney, C. E. (1979): Parameters influencing quantitation of 3-methylcholanthrene-induced aryl hydrocarbon hydroxylase activity in cultured human lymphocytes. *J. Env. Path. and Toxicol.* 2:1079-1098.
17. Marshall, M. V., McLemore, T. L., Martin, R. R., Jenkins, W. T., Snodgrass, D. R., Corson, M. A., Arnott, M. S., Wray, N. P., and Griffin, A. C. (1979): Patterns of benzo(a)pyrene metabolism in human pulmonary alveolar macrophages. *Cancer Letters* 8:103-109.
18. McLemore, T. L., Corson, M., Mace, M., Arnott, M., Jenkins, T., Snodgrass, D., Martin, R., Wray, N., and Brinkley, B. R. (1979): Phagocytosis of asbestos fibers by human pulmonary alveolar macrophages. *Cancer Letters* 6:183-189.
19. McLemore, T. L., Jenkins, W. T., Arnott, M. S., and Wray, N. P. (1979): Aryl hydrocarbon hydroxylase induction in mitogen-stimulated lymphocytes by benzanthracene or cigarette tars adsorbed to asbestos fibers. *Cancer Letters* 7:171-177.

20. McLemore, T. L., Martin, R. R., Busbee, D. L., Richie, R. C., Springer, R. R., Topell, K. L., and Cantrell, E. T. (1977): Aryl hydrocarbon hydroxylase activity in pulmonary macrophages and lymphocytes from lung cancer and noncancer patients. *Cancer Res.* 37:1175-1181.
21. McLemore, T. L., Martin, R. R., Pickard, L. R., Springer, R. R., Wray, N. P., Toppell, K. L., Mattox, K. L., Guinn, G. A., Cantrell, E. T., and Busbee, D. L. (1978): Analysis of aryl hydrocarbon hydroxylase activity in human lung tissue, pulmonary macrophages, and blood lymphocytes. *Cancer* 41:2292-2300.
22. McLemore, T. L., Martin, R. R., Springer, R. R., Wray, N., Cantrell, E. T., and Busbee, D. L. (1979): Aryl hydrocarbon hydroxylase activity in pulmonary alveolar macrophages and lymphocytes from lung cancer and noncancer patients: a correlation with family histories of cancer. *Biochem. Genet.* 17:795-806.
23. McLemore, T. L., Martin, R. R., Toppell, K. L., Busbee, D. L., and Cantrell, E. T. (1977): Comparison of aryl hydrocarbon hydroxylase in cultured blood lymphocytes and pulmonary macrophages. *J. Clin. Inv.* 60:1017-1024.
24. McLemore, T. L., Martin, R. R., Wray, N. P., Cantrell, E. T., and Busbee, D. L. (1978): Dissociation between aryl hydrocarbon hydroxylase activity in cultured pulmonary macrophages and blood lymphocytes from cancer patients. *Cancer Res.* 38:3805-3811.
25. McLemore, T. L., Warr, G. A., and Martin, R. R. (1977): Induction of aryl hydrocarbon hydroxylase in human pulmonary alveolar macrophages and peripheral blood lymphocytes by cigarette tars. *Cancer Letters* 2:161-168.
26. McLemore, T. L., Warr, G. A., and Martin, R. R. (1977): In Vitro induction of aryl hydrocarbon hydroxylase in human pulmonary macrophages by benzanthracene. *Cancer Letters* 2:327-334.
27. Okano, P., Miller, H. N., Robinson, R. C., and Gelboin, H. V. (1979): Comparison of benzo(a)pyrene and (-)-trans-7,8-dihydro-7,8-dihydrobenzo(a)pyrene metabolism in human blood monocytes and lymphocytes. *Cancer Res.* 39:3184-3193.
28. Paigen, B., Gurtoo, H. L., Minowada, J., Houten, L., Vincent, R., Paigen, K., Parker, N. B., Ward, E., and Hayner, N. T. (1977): Questionable relation of aryl hydrocarbon hydroxylase to lung-cancer risk. *New Engl. J. Med.* 297:346-350.
29. Stoner, G. D., Harris, C. C., Autrup, H., Trump, B. F., Kingsbury, E. W., and Myers, G. A. (1978): Explant Culture of Human Peripheral Lung. I. Metabolism of benzo(a)pyrene. *Lab. Inv.* 38:685-692.

30. Wattenberg, L. W., Jerina, D. M., Lam, L. K. T., and Yagi, H. (1979): Neoplastic effects of oral administration of (+)-trans-7,8-dihydroxy-7,8-dihydrobenzo(a)pyrene and their inhibition by butylated hydroxyanisole. *J. Natl. Cancer Inst.* 62:1103-1106.
31. Yang, S. K., Roller, P. P., and Gelboin, H. V. (1977): Enzymatic mechanism of benzo(a)pyrene conversion to phenols and diols and an improved high pressure liquid chromatography separation of benzo(a)pyrene derivatives. *Biochemistry* 16:3680-3687.

# BENZO(a)PYRENE RADICALS AND OTHER DERIVATIVES FORMED IN INTERACTIONS WITH CYSTEINE/SERUM ALBUMIN

**B. N. Srinivasan and E. Fujimori**
Department of Fine Structure
Boston Biomedical Research Institute
20 Staniford Street
Boston, Massachusetts 02114

## INTRODUCTION

It is generally believed that benzo(a)pyrene (BaP), a widely occurring carcinogenic hydrocarbon, needs to be oxidized to an active product to become a mutagenic and carcinogenic chemical. BaP is oxidized by enzymatic reactions to epoxides (BaP-7,8-diol-9,10-epoxides) or radicals. It is now known that exposure of BaP to polluted air results in the formation of directly active mutagens (9). It is also possible that photooxidation of polycyclic aromatic hydrocarbons in atmospheric particulate matter produces photoactivated products (3). Besides the known radical 6-oxo-BaP (BaP-0·)(8), a BaP radical cation (BaP·$^+$) can be produced by one-electron oxidation because of the unique character at the 6-position of BaP (6,14). Recent studies also support the formation of cation radicals as the critical mechanism of BaP activation (2,11,12). The 6-position of BaP has the highest electron density (5) and hence easily undergoes one-electron oxidation, giving rise to free radicals which can bind to cellular macromolecules, protein, and DNA.

The radical cation (BaP·$^+$) is obtained by removal of one electron, whereas a neutral radical (BaP·) is formed by stripping both an electron and a hydrogen ion, from the hydrocarbon. The pK value of BaP·$^+$, whose dissociation gives rise to BaP·, has not been determined. Although no spectroscopic evidence has been obtained, both BaP·$^+$ and BaP· may have similar spectral characteristics. By comparison with the fluorescence of diphenylamine $(C_6H_5)_2NH$ and its free radical $(C_6H_5)_2N·$ (16), it is assumed that the fluorescence spectrum of BaP radicals is shifted to a shorter wavelength region. Furthermore, BaP·$^+$ is only a doublet, but BaP· can be considered as a methylene type at the 6-position with two free electrons at the same carbon atom and treated as a triplet in its ground state. Thus the two species could be differentiated by electron spin resonance (ESR) spectra. The ESR spectrum of BaP·$^+$ will give a single-line spectrum (with hyperfine structure), whereas oxidation of BaP by $I_2$ gives a poorly resolved single-line spectrum (6,14). On the other hand, BaP· will exhibit a six-line spectrum similar to the triplet ground-state molecule diphenyl-methylene (7,15). Figure 1 shows the BaP radical cation and the BaP neutral radical.

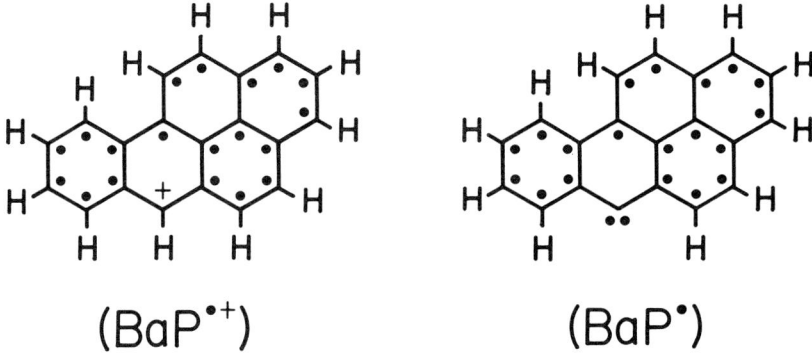

FIGURE 1. BaP radical cation (BaP·$^+$) and BaP neutral radical (BaP·).

## SERUM ALBUMIN/CYSTEINE-BaP

We have been investigating the interactions of the typical SH-protein, serum albumin, and cysteine with BaP. BaP crystals (Aldrich) were mixed with a 10 percent aqueous solution of bovine serum albumin (BSA)(Sigma) at 4°C in the dark for several days. Upon removal of BaP crystals by centrifugation, a fairly large amount of BaP was soluble in this highly concentrated protein solution. It was estimated from absorption spectrum that the concentration of total soluble (both bound and free) BaP is 118 μmole in 1.5 mmole BSA solution. This gave a value of 0.08 BaP molecule per BSA molecule. Bothorel and Desmazes (1), using up to 2 percent BSA, reported 0.063 BaP mole/mole. The solubility of BaP in 10 percent human serum albumin (HSA) was 65 μmole, or one-half of that in 10 percent BSA. Franke (4) reported 0.064 BaP mole/mole for HSA. The solubility of BaP in cysteine solution was much lower than micromolar.

In the BSA-BaP complex prepared in this study, a BaP excimer fluorescence has not been observed. This observation indicates that the BaP molecule is dispersed monomerically in the protein molecule. In contrast, by using 2-chloroethanol-water as a binary transient solvent, it was possible to trap more than 10 BaP molecules per BSA molecule without any protein denaturation (1). It was proposed that there are two binding sites: one at the protein surface with a small binding constant, and the other inside the protein and trapping about ten hydrocarbon molecules probably as molecular aggregates. BaP aggregates exhibited a new red-shifted absorption around 420 nm and a broad fluorescence (480 to 550 nm) with a peak at 505 nm, probably from excimer molecules.

## UV FLUORESCENCES, PYRENE-TYPE PRODUCTS, AND BaP NEUTRAL RADICALS

As recently reported (13), the fluorescence of the BSA-BaP and cysteine-BaP systems extends from 300 nm in the UV region to 600 nm in the visible region. These BSA/cysteine-BaP complexes exhibit two types of UV fluorescences, I and II, in the near-UV region 300 to 400 nm. The type I UV fluorescence for the BSA-BaP complex is a well-defined, structured fluorescence with maxima at 340, 357, and 378 nm and a shoulder at 330 nm. The excitation maximum is at 307 nm with an additional small band at 296 nm. For the cysteine-BaP complex, the type I UV fluorescence appears as a structureless broad spectrum in the 310 to 370 nm range with a maximum at 330 nm. The intensity of this fluorescence is directly related to the concentration of cysteine (up to 4 percent). It has also been found that the formation of the type I UV fluorescence is particularly sensitive to the presence of oxygen. As shown in Figure 2 (curve 1), this fluorescence was not clearly formed in a closed aqueous mixture (closed from atmosphere) of cysteine (5 percent) and BaP, kept at room temperature for 1 day. In an open system, however, the type I fluorescence was formed at 330 nm (Figure 2, curve 2). This result indicates the involvement of oxygen in this reaction. Figure 2 also shows the excitation spectra for the fluorescences of the type I (excitation peak at 280 nm), the type II (see below) (excitation peak at 282 nm with shoulder at 270 nm), and BaP (excitation maxima at 297, 287, and 269 nm). The latter two excitation spectra correspond to absorption bands of the second electronic-excited singlet states. The lyophilized mixture from BaP-cysteine (4 to 5 percent) in aqueous solution (kept at room temperature for 1 day), in which the UV fluorescence was observed at 330 nm, exhibited a well-separated intense six-line ESR spectrum (13). Although the fluorescence at 330 nm was not clearly observed in the cysteine concentration of 0.1 percent, the weaker six-line ESR spectrum was still observed, together with a single absorption, due to the 6-oxo-BaP radical at the position of DPPH ($\alpha,\alpha'$-diphenyl-$\beta$-picryl-hydrazyl) with g value 2.0036 (Figure 3). Both the type I UV fluorescence

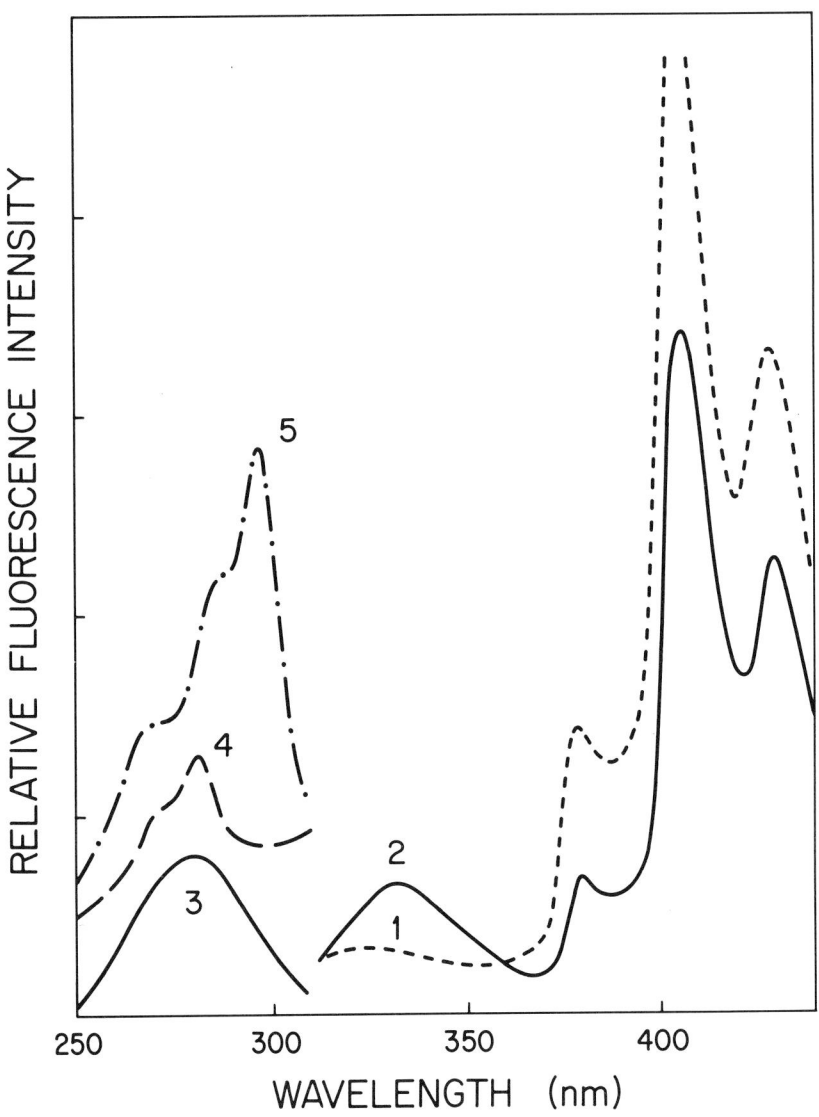

FIGURE 2. Fluorescence (1,2) and excitation (3-5) spectra of the cysteine (5%)-BaP mixture, kept for 1 day at room temperature. Curve 1, closed from atmosphere; curve 2 open to atmosphere. Curves 1 and 2 are for excitation at 280 nm. Curves 3, 4, and 5 are for the fluorescences at 330, 380, and 406 nm, respectively.

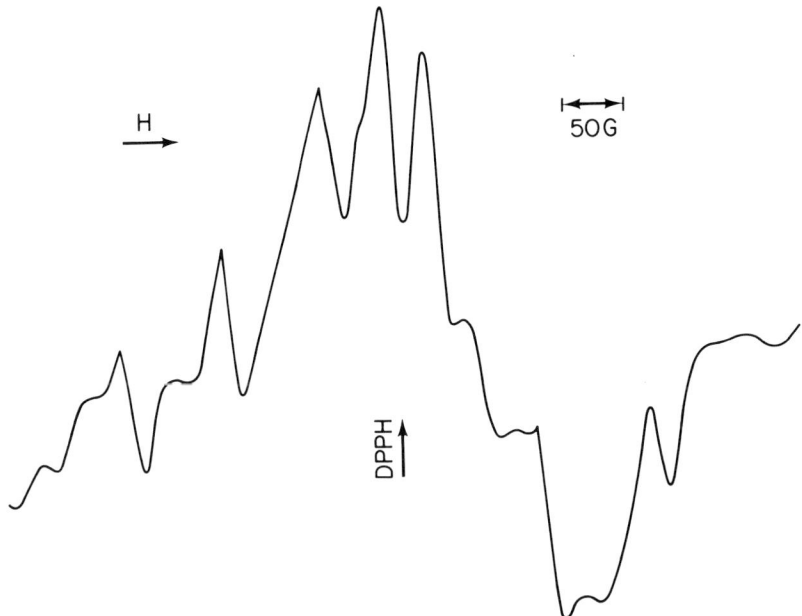

**FIGURE 3.** ESR spectrum of the cysteine (0.1%)-BaP mixture, lyophilized after standing for 1 day at room temperature.

and the six-line ESR spectrum are tentatively attributed to the BaP neutral radical (BaP·).

The other UV fluorescence type II is a sharp fluorescence at 380 nm with another expected peak at 402 nm which may be hidden under a strong visible fluorescence at 407 nm of unchanged BaP. This fluorescence has excitation maxima at about 331 and 346 nm and has been observed in both BaP complexes with BSA and cysteine. The type II UV fluorescence corresponds to the fluorescence of pyrene-like products, which is in good agreement with that of BaP-7,8 diol-9,10-epoxide (10,17). The formation of pyrene-type products was enhanced by a lower concentration of both cysteine and oxygen. Because the formation of the BaP· radical increases with the increase of cysteine and oxygen concentration, it appears that the radical formation and the pyrene-type formation are mutually exclusive and that these products are formed under different conditions. Figure 4 shows the type II UV fluorescence and excitation spectra of the aqueous mixture of cysteine and BaP (kept at room temperature for a few days), in which the free BaP was removed by benzene extraction. A species responsible for the type II fluorescence at 380 nm remained in the aqueous solution, indicating the formation of more polar-pyrene-type products than BaP.

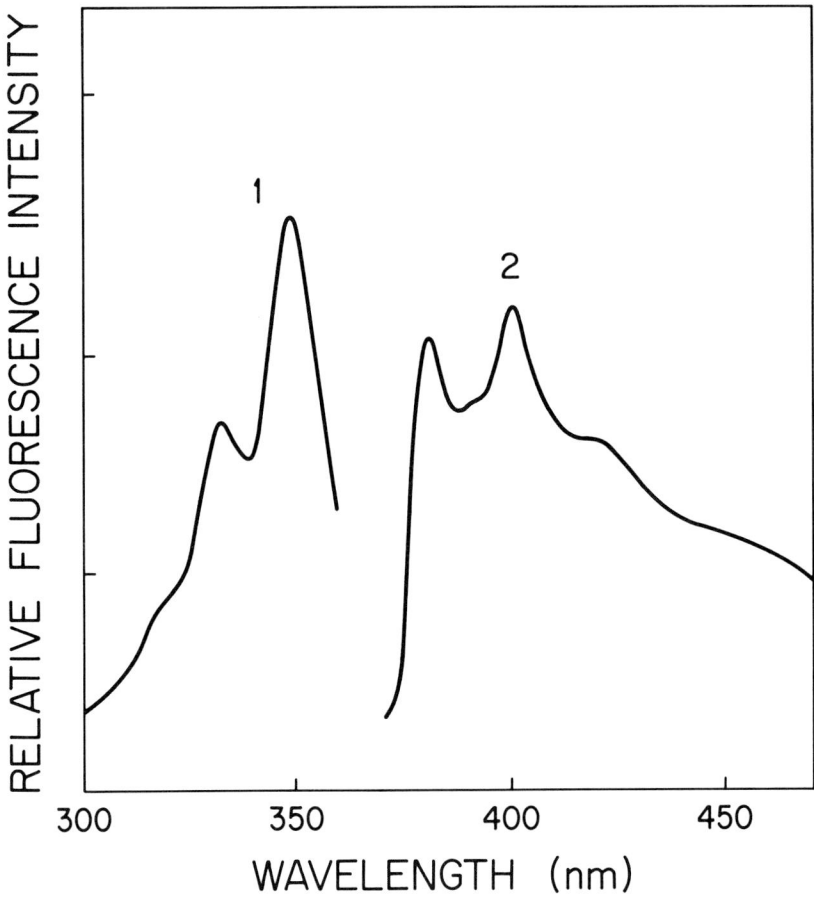

FIGURE 4. Excitation (1) and fluorescence (2) spectra of the cysteine-BaP aqueous mixture, treated with benzene after standing for a few days. Curve (1) is for the fluorescence at 400 nm and curve (2) is for excitation at 331 nm.

In contrast, a species responsible for the type I fluorescence can be extracted by benzene. This is shown in Figure 5 for the structural three-peaked (340, 357, 378 nm) type I UV fluorescence spectrum of a benzene extract from the 10 percent BSA-BaP aqueous mixture kept in the dark at 4°C. The benzene extract was washed repeatedly with water to remove suspended BSA, as tested by the absence of tryptophan fluorescence in the aqueous phase. This result indicates that a species exhibiting the type I fluorescence is as polar as BaP. The condensed benzene extract showed a broad, structural ESR signal (13). Although the concentrated aqueous solution of the BSA-BaP complex gave a six-line ESR spectrum, the

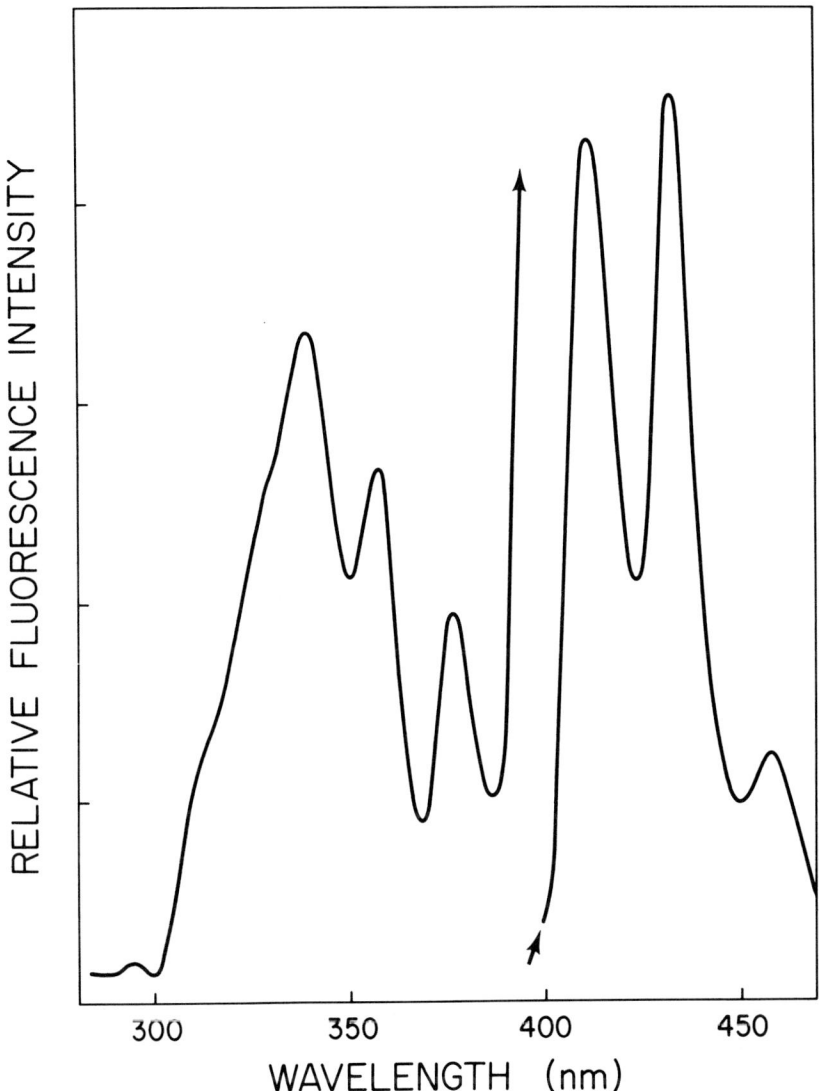

**FIGURE 5.** Fluorescence spectrum of the benzene extract from the BSA (10%)-BaP complex (excited at 270 nm).

lyophilized sample of the BSA-BaP complex showed at least two broad ESR bands on either side of the 6-oxo-BaP radical. Figure 6 shows the ESR spectra of lyophilized native and lipid-free HSA-BaP complexes. As observed with BSA, lipid-free HSA showed a higher intensity of the lower g-valued band than that in the native HSA-BaP complex. The amount of the 6-oxo-BaP radical was lower in lipid-free HSA-BaP.

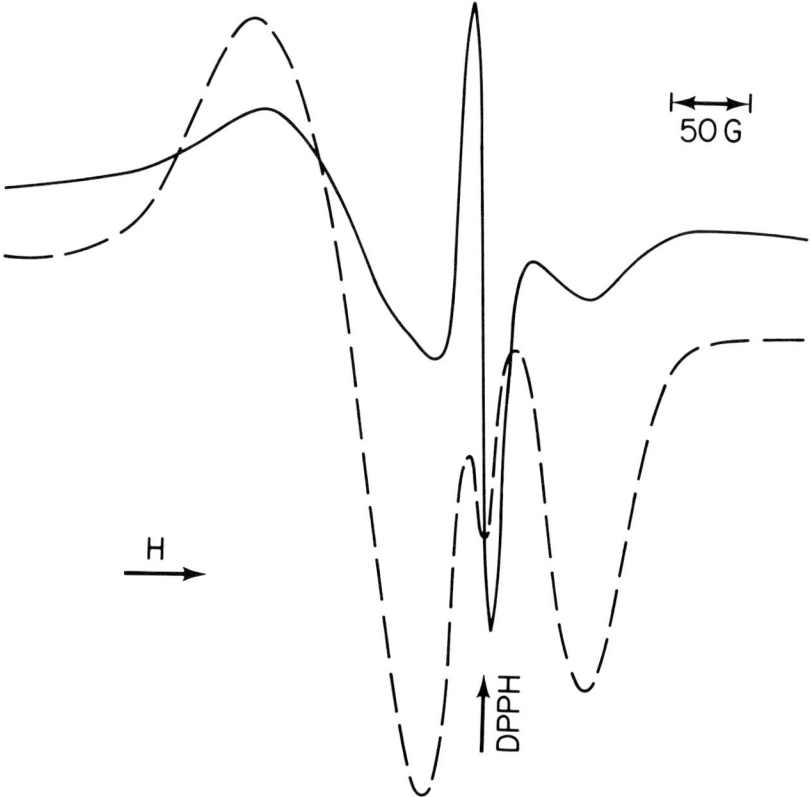

FIGURE 6. ESR spectra of the lyophilized HSA-BaP complex (—) and lipid-free HSA-BaP complex (---).

## VISIBLE FLUORESCENCES AND HYDROXY-BaP PRODUCTS

Figure 7 shows that UV(365 nm) irradiation of the BSA-BaP complex reduces the intensity of both the type I UV fluorescence (330 to 380 nm) and the visible fluorescence (400 to 460 nm). As seen in the visible fluorescence, the nonirradiated BSA-BaP complex (Figure 7, curve 1) exhibited unusually higher intensities of the second band at 432 nm and the third band at 458 nm, compared with the intensity of the first band at 407 nm. This change is indicative of the formation of hydroxy-BaP derivatives which are known to emit at 430 and 460 nm (18).

Both the nonirradiated and irradiated BSA-BaP complexes were treated with benzene. After extensive repeated benzene extraction, the nonirradiated BSA-BaP complex lost both the hydroxy-BaP derivatives

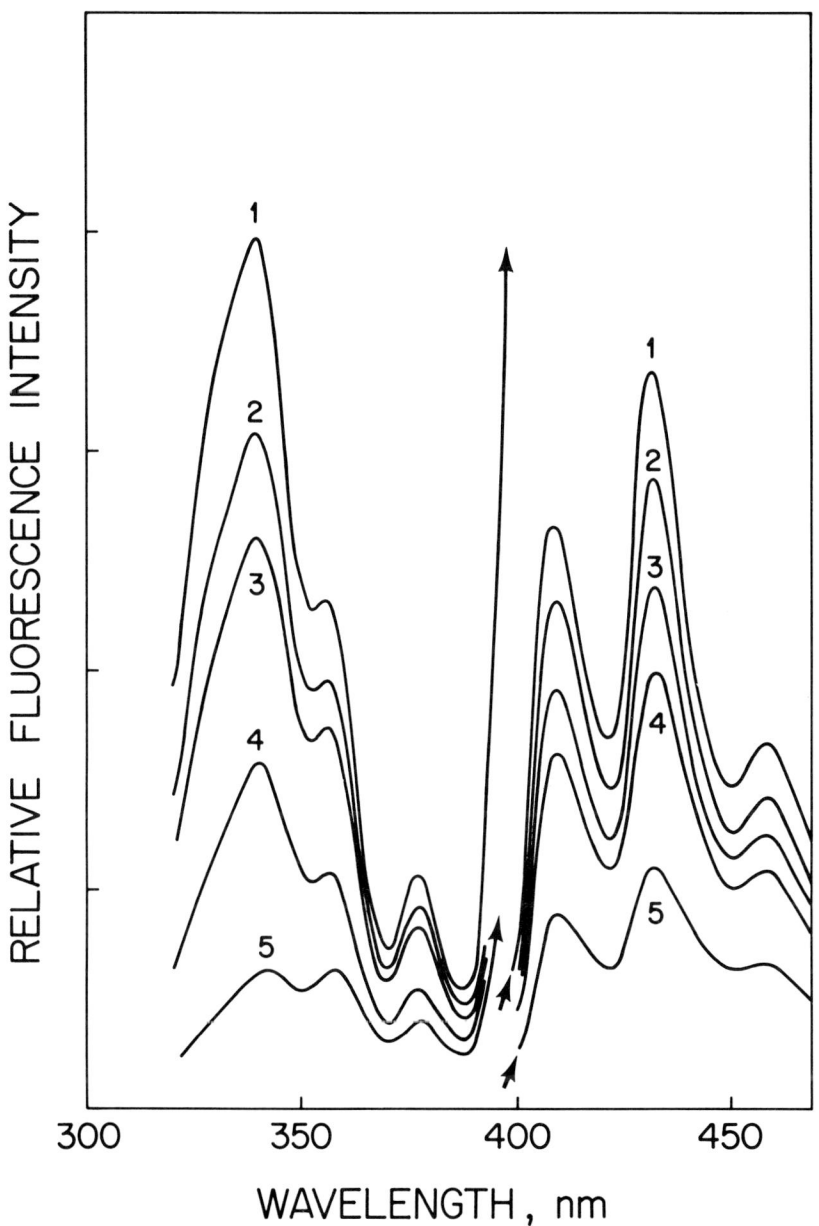

FIGURE 7. Fluorescence spectral changes in the BSA (10%)-BaP complex when irradiated with UV (365 nm) light. Curve 1 before irradiation; curves 2, 3, 4 and 5 at 2, 10, 30, and 60 minutes, respectively (excitation at 307 nm).

## 328 BaP RADICALS AND OTHER DERIVATIVES

and the type I UV fluorescent BaP product. These changes were demonstrated in Figure 8, in which a broad UV fluorescence at 345 nm appeared from the tryptophan residues when excited at 305 nm (Figure 8, curve 1). The benzene extraction could remove the type I UV fluorescence, apparently regenerating the UV fluorescence of tryptophan residues. Excitation at 389 nm gave rise to two broad fluorescence maxima at 415 and 433 nm (Figure 8, curve 2), probably because of strongly bound BaP.

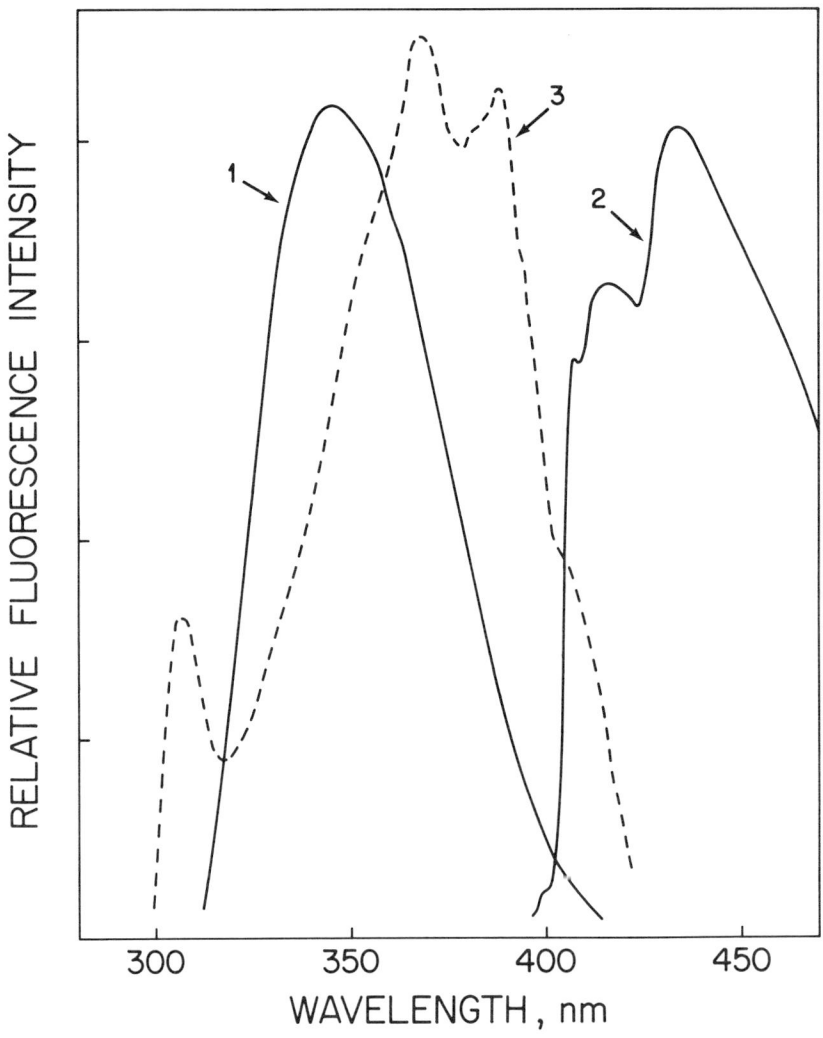

FIGURE 8. Fluorescence (1,2) and excitation (3) spectra of BSA-BaP complex after benzene extraction. Curves 1 and 2 are for excitation at 305 and 389 nm, respectively. Curve 3 is for the fluorescence at 430 nm.

The excitation maxima at 388, 369, and 350 nm for the 430 nm fluorescence also provide evidence for the binding of BaP to the protein and for the removal of hydroxy-BaP derivatives by benzene (Figure 8, curve 3). The excitation maxima for the 430 nm fluorescence of hydroxy-BaP derivatives are red-shifted to 396, 381, and 363 nm (18).

Figure 9 shows the fluorescence and excitation spectra of UV(365 nm)-irradiated BSA-BaP complex after repeated benzene extraction.

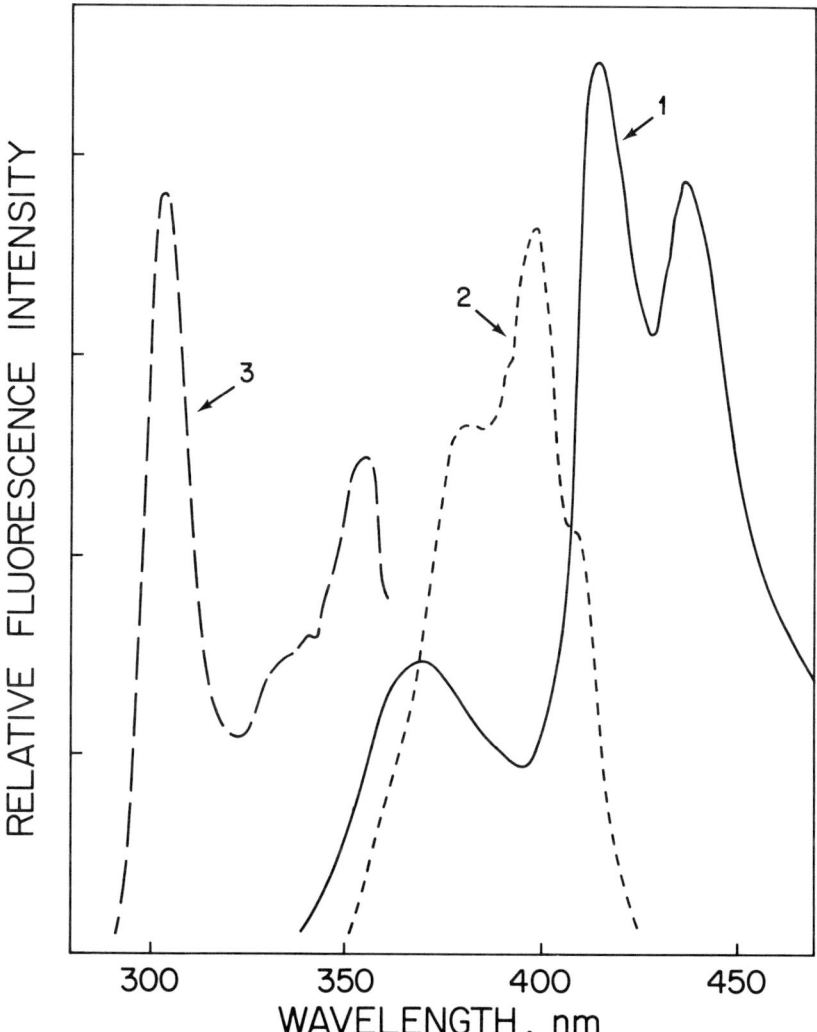

**FIGURE 9.** Fluorescence (1) and excitation (2,3) spectra of UV-irradiated BSA-BaP complex after benzene extraction. Curve 1 is for excitation at 305 nm. Curves 2 and 3 are for the fluorescences at 415 and 370 nm, respectively.

Excitation at 305 nm produced a single UV fluorescence maximum at 371 nm, probably because of photo-modified products of tryptophan residues, in addition to red-shifted intense fluorescence peaks at 415 and 437 nm of bound BaP photo-products (Figure 9, curve 1). The red-shifted excitation maxima at 398 and 382 nm for the 415 nm fluorescence correspond to those of hydroxy BaP derivatives (Figure 9, curve 2). The excitation spectrum for the 370 nm fluorescence exhibited a new peak at 355 nm as well as at 304 nm (Figure 9, curve 3).

In conclusion, the results of the UV fluorescence spectrum in the 330 to 380 nm region and the six-line ESR spectrum indicate the formation of the BaP neutral radical (BaP·) from BaP by hydrogen abstraction at the 6-position. The possible mechanism for this formation has recently been discussed (13). It is important to point out that these interactions of BaP with the SH-proteins, serum albumins, and cysteine can also produce other BaP products (pyrene-like compounds and hydroxy-BaP derivatives) as well as the radical species. Furthermore, some BaP products are firmly bound to BSA in the dark, while photooxidized BaP products are photo-bound to a greater extent. More information on BaP-protein interactions and their role in carcinogenesis awaits further investigation.

## ACKNOWLEDGMENT

This research was supported by contract (EE-77-S-02-4347) from the U.S. Department of Energy.

## REFERENCES

1. Bothorel, P., and Desmazes, J. P. (1974): Trapping of benzo(a)pyrene by bovine serum albumin, *Biochim. Biophys. Acta* 365:181-192.
2. Cavalieri, E., Roth, R., Rogan, E., Grandjean, C., and Althoff, J. (1978): Mechanisms of tumor initiation by polycyclic aromatic hydrocarbons. In: *Carcinogenisis, Vol. 3: Polynuclear Aromatic Hydrocarbons,* P. W. Jones and R. I. Freudenthal, Eds., pp. 273-284, Raven Press, New York.
3. Fox, M. A. and Olive, S. (1979): Photooxidation of anthracene on atmospheric particulate matter. *Science* 205:582-583.
4. Franke, R. (1968): Die hydrophobe Wechselwirkung von polycyclischen aromatischen Kohlenwasserstoffen mit Humanserumalbumin. *Biochim. Biophys. Acta* 160:378-395.
5. Loew, G. H., Wong, J., Phillips, J., Hjelmeland, L., and Pack, G. (1978): Quantum chemical studies on the metabolism of benzo(a)-pyrene. *Cancer Biochem. Biophys.* 2:123-130.

6. Menger, E. M., Spokane, R. B. and Sullivan, P. D. (1976): Free radicals derived from benzo(a)pyrene. *Biochem. Biophys. Res. Commun.* 71:610-616.
7. Murray, R. W., Trozzolo, A. M., Wasserman, E., and Yager, W. A. (1962): EPR of diphenylmethylene, a ground-state triplet. *J. Am. Chem. Soc.* 84:3213-3214.
8. Nagata, C., Tagashira, Y., and Kodama, M. (1974): Metabolic activation of benzo(a)pyrene: significance of the free radical. In *Chemical Carcinogenesis,* Part A, P. O. P. Ts'O and J. A. DiPaolo, Eds., pp. 87-111, Marcel Dekker, New York.
9. Pitts, J. N., Jr., Van Cauwenberghe, K. A. Grosjean, D., Schmid, J. P., Fitz, D. R., Belser, W. L., Jr., Knudson, G. B., and Hynds, P. M. (1978): Atmospheric reactions of polycyclic aromatic hydrocarbons: Facile formation of mutagenic nitro derivatives. *Science* 202:515-519.
10. Prusik, T., Geacintov, N. E., Tobiasz, C., Ivanovic, V., and Weinstein, I. B. (1979): Fluorescence study of the physico-chemical properties of a benzo(a)pyrene 7,8-dihydrodiol·9,10-oxide derivative bound covalently to DNA. *Photochem. Photobiol.* 29:223-232.
11. Rogan, E., Roth, R., and Cavalieri, E. (1978): Enzymology of polycyclic hydrocarbon binding to nucleic acids. In: *Carcinogenesis, Vol. 3: Polynuclear Aromatic Hydrocarbons,* P. W. Jones and R. I. Freudenthal, Eds., pp. 265-271, Raven Press, New York.
12. Sivarajah, K., Anderson, M. W. and Eling, T. E. (1978): Metabolism of benzo(a)pyrene to reactive intermediate(s) via prostaglandin biosynthesis. *Life Sciences* 23:2571-2578.
13. Srinivasan, B. N. and Fujimori, E. (1979): Benzo(a)pyrene-serum albumin/cysteine interactions: fluorescence and electron spin resonance studies. *Chemico.-Biol. Interact.* 28:1-15.
14. Sullivan, P. D., Calle, L. M., Shafer, K., and Nettleman, M. (1978): Effect of antioxidants on benzo(a)pyrene free radicals. In: *Carcinogenesis, Vol. 3: Polynuclear Aromatic Hydrocarbons,* P. W. Jones and R. I. Freudenthal, Eds., pp. 1-8, Raven Press, New York.
15. Trozzolo, A. M., and Gibbons, W. A. (1967): The absorption, emission and excitation spectra of diaryl-methylenes. *J. Am. Chem. Soc.,* 89:239-243.
16. Van Duuren, B. L., and Chan, T. L. (1971): Fluorescence spectrometry. In: *Spectrochemical Methods of Analysis,* J. D. Winefordner, Ed., p.423, Wiley-Interscience, New York.
17. Wade, C. G., Baker, D. E., and Bartholomew, J. C. (1978): Selective fluorescence quenching of benzo(a)pyrene and a mutagenic diol epoxide derivative in mouse cells. *Biochemistry* 17:4332-4337.
18. Yang, C. S., and Kicha, L. P. (1978): A direct fluorometric assay of benzo(a)pyrene hydroxylase. *Anal. Biochem.* 84:154-163.

# POLYNUCLEAR AROMATIC HYDROCARBONS IN NORWEGIAN DRINKING WATER RESOURCES

**B. Olufsen**

Central Institute for Industrial Research
Blindern, Oslo 3, Norway

## INTRODUCTION

"Industrial effluents and atmospheric pollutants can find their way into water supplies. It is important that attention should be given to this problem in terms of carcinogenic compounds". Ever since the World Health Organization (WHO) stated this 15 years ago (16), the occurrence and health effects of organic pollutants in drinking water have gained an increasing attention. The polynuclear aromatic hydrocarbons (PAH) play a sinister part in the group of carcinogenic environmental pollutants.

In Norway, some 95 percent of the total drinking water supply is surface water (3). This makes the sources sensitive to atmospheric fall-out of PAH originating from local activities such as electrochemical industry, from oil heaters, and from traffic or from long-range transportation (5). The overall burden of pollution in Norwegian drinking water is usually considered to be low, but "...the potential danger of continuous exposure to carcinogenic PAH through drinking water should not be overlooked, even at low concentration..." (2).

In 1971, when the WHO (17) set a standard for the amount of PAH in drinking water, recommending a maximum level of 200 ng of six specific PAH per liter of surface water, the proposed analytical method involved benzene extraction followed by thin-layer chromatography (TLC). Some authors (7) later reported analytical problems in meeting the WHO limits, and for those laboratories using gas chromatography (GC) instead of TLC, separation of some of the six compounds was not possible. High-resolution gas chromatography with glass capillary columns has somewhat changed this situation and, coupled to a mass spectrometer, it has been extremely helpful in the extended identification of compounds in the PAH fraction of environmental samples.

The work presented here is aimed at a method generally applicable to any water sample and is tailored to suit separation and identification on glass capillary columns and flame ionization detection (FID).

## DIRECT SOLVENT EXTRACTION OF WATER

Solvent extraction has been considered a general method for the isolation of PAH from water. Preliminary studies in our laboratory (12) have, however, proven this method unsatisfactory. Three successive extractions of the same river-water sample often gave the highest concentration of PAH in the third extraction, and sometimes the second extraction did. Spiking experiments also showed that the "water" was absorbing compounds and not releasing them upon extraction. We attribute this to the particulate matter present in the water. Thus liquid/liquid extraction of water containing particulates does not give quantitative or reliable data for the analysis of PAH. This is in agreement with the results obtained by Hurtubise and co-workers (10) who reported poor reproducibility in a recovery study with unfiltered retort water. Acheson et al (1) also showed reduced extraction efficiencies with increasing levels of suspended solids in the water. Upon realizing this problem, other methods including some kind of filtration of the water, had to be considered.

## ADSORPTION ON A MACRORETICULAR RESIN AND SIMPLE ELUTION WITH ORGANIC SOLVENTS

During the past few years, several papers have been published on the use of porous polymer resins as adsorbents for trace organic compounds in the analysis of water. By passing contaminated water through a column of a macroreticular resin such as Amberlite XAD-2, organic compounds are adsorbed by Van der Waals forces (8). Elution with an organic solvent releases the compounds which are further analyzed by adequate cleanup, and separated and identified by, e.g., GC. Particulate matter is retained on a plug of glass-wool and can be analyzed separately.

Several authors have reported recoveries of aromatics with the use of XAD-2 (6,11,14,15). However, with the exception of Strup et al (14), none of the authors have considered higher boiling PAH.

In an effort to include data for those compounds, recovery studies on 27 PAH were carried out with the apparatus and procedure described by Junk et al (11). Both methanol and N, N-dimethylformamide (DMF): water (9:1) were tried as solvents in the gravity elution of the column since they are soluble in water and already well established in our procedure for the analysis of PAH (4). The practical upper limit for elution was considered to be 150 ml of solvent.

DMF:water gave the best results, the values for which are given in Table 1. The data reflect absolute recoveries based on external standards. Fair recoveries are obtained for the low-molecular-weight PAH, but these decrease drastically with increasing molecular weights. Parallel to these calculations, data based on internal standards added prior to adsorption and after elution were evaluated; these indicate that no losses take place after elution. Thus, the losses that prevent good recoveries seem to occur during the adsorption/elution processes. Part of the lower boiling PAH is probably drained through the column because of their relatively higher solubility in water, while the compounds with the higher molecular weights are strongly retained by Van der Waals forces. Internal standards added prior to adsorption compensate to some extent for those losses.

TABLE 1. Absolute Recovery of PAH from Water/XAD (Including Transfer to Cyclohexane) Based on External Standards

| PAH | Percent | PAH | Percent |
| --- | --- | --- | --- |
| Naphthalene | 82.5 | Fluoranthene | 82.7 |
| 2-Methylnaphthalene | 85.3 | Pyrene | 102.4[c] |
| Biphenyl | 83.5 | Benzo(a)fluorene | 84.1 |
| Acenaphthylene | 83.1 | Benzo(b)fluorene | 78.6 |
| Acenaphthene[a] | 88.4 | 1-Methylpyrene | 80.9 |
| Dibenzofuran | 86.7 | Benz(a)anthracene | 75.7 |
| Fluorene | 80.7 | Chrysene/triphenylene | 75.3 |
| 2-Methylfluorene | 87.0 | Benzo(j)fluoranthene | 60.5 |
| Dibenzothiophene | 83.8 | Benzo(e)pyrene | 106.3[d] |
| Phenanthrene | 83.4 | Benzo(a)pyrene | 54.8 |
| Anthracene | b | Perylene | 25.7 |
| 2-Methylanthracene | 80.4 | Dibenz(a,c)anthracene | 45.6[d] |
| 1-Methylphenanthrene | 85.9 | Benzo(ghi)perylene | 20.0 |

[a]Compared with a standard containing both acenaphthene and 3-methylbiphenyl.
[b]Obscured by an unknown compound.
[c]Other compounds interfere. Suspect values.
[d]Blank value substracted. Probably still too high. FID response factor includes dibenz-(a,h)anthracene.

Strup et al (14) reported low recoveries for a spiked water sample and attribute this "...to sample losses through adsorption of the walls of the glass water reservoir...". This is a generally known problem, and an attempt was made to overcome it in this work by washing the reservoir walls with the DMF:water later used for elution. Still, low recoveries were

## 336 PAH IN WATER

obtained, suggesting that the source of the problem lies within the resin body itself. To solve this problem, three alternatives crystallized:
• To find a solvent or solvent mixture which (hot or cold) quantitatively elutes all of the interesting adsorbed PAH within reasonable volume limits
• To use several internal standards to correct for losses
• To use a continuous extraction procedure instead of the simple elution.
Further work was concentrated on the last alternative.

## ADSORPTION AND CONTINUOUS EXTRACTION OF NEVER-DRYING RESIN IN A MODIFIED SOXHLET APPARATUS

To achieve quantitative elution of adsorbed PAH from XAD-2 resins, continuous extraction was considered to be the most promising alternative. An apparatus based on a modified soxhlet extractor was constructed which allows cleaning of the polymer prior to use, adsorption both on location and in the laboratory, continuous extraction (overnight) with a low (30 ml) solvent (methanol) volume, and regeneration. The recommended (11) height-to-diameter ratio of the resin bed is maintained at 6. The apparatus is described elsewhere (13).

Two recovery tests were carried out on 30 aromatic compounds:
• Direct spiking onto the top of the resin, no water elution prior to extraction
• Direct spiking onto the top of the resin, elution with 4 liters of tap water prior to extraction.

All values were corrected with cleanup/FID response factors. A chromatogram of the selected PAH is shown in Figure 1. The chromatographic conditions used throughout this work are shown in Figure 5.

The results of the experiment using direct spiking with no water elution are shown in the uppermost part of Figure 2. With the exception of carbazole, excellent recoveries are achieved for all compounds based on internal standards. Thus, no problems arise in the absence of water.

Direct spiking followed with a water elution of 4 bed volumes/min (15) simulating adsorption of each of the PAH on a level of some 350 ppt resulted in the recoveries shown at the bottom of Figure 2. Once water is introduced in the system, recovery problems arise, now dominating the lower-boiling PAH fraction. A different aqueous solubility of each of the PAH may contribute to this picture, the choice of 3,6-dimethylphenanthrene as the internal standard in this section may be poor, or Van der Waals forces of the resin may compete with those of the water molecules, most certainly favoring one PAH over another.

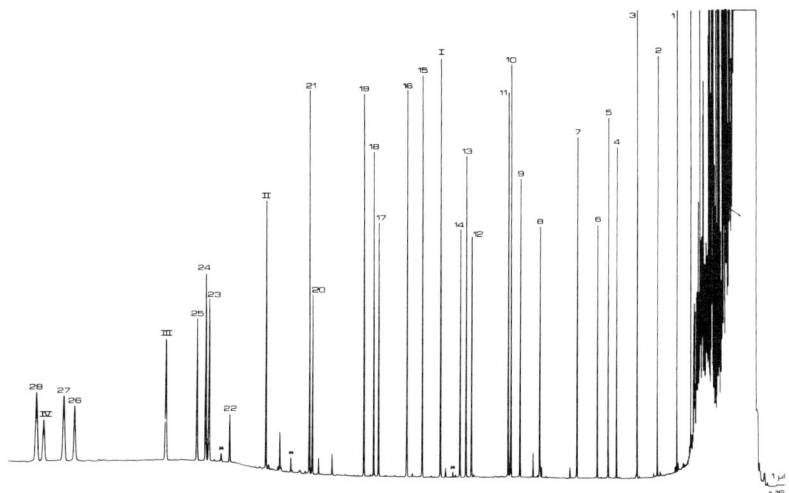

FIGURE 1. Gas chromatographic separation of PAH compounds used for cleanup/FID response factors in the recovery studies. Peak identification: (1) naphthalene, (2) 2-methylnaphthalene, (3) biphenyl, (4) acenaphthylene, (5) acenaphthene, (6) dibenzofuran, (7) fluorene, (8) 2-methylfluorene, (9) dibenzothiophene, (10) phenanthrene, (11) anthracene, (12) 2-methylanthracene, (13) 1-methylphenanthrene, (14) carbazole, (15) fluoranthene, (16) pyrene, (17) benzo-(a)fluorene, (18) benzo(b)fluorene, (19) 1-methylpyrene, (20) benz(a)anthracene, (21) chrysene/triphenylene, (22) benzo(k)fluoranthene, (23) benzo(e)pyrene, (24) benzo(a)pyrene, (25) perylene, (26) indeno(1,2,3-cd)pyrene, (27) dibenz-(a,c,/a,h)anthracenes, (28) benzo(ghi)perylene. Roman letters indicate internal standards.

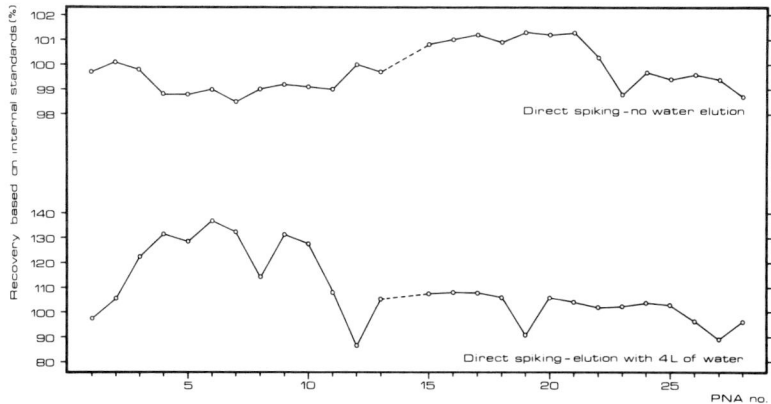

FIGURE 2. Recovery of PAH with continuous extraction of XAD-2. For peak identification, see Figure 1.

The results do reveal a method suitable for quantitative analysis of PAH in water, the only exception being carbazole. Still another internal standard should be included to compensate for losses in the lower section. Internal standards should be added *after* the completion of adsorption to prevent a continuous washing-out by water.

## THE CAPACITY OF THE RESIN

Wastewater with a high content of PAH was drained through the XAD-2 soxhlet column and the eluate was collected. The resin was analyzed for PAH as described above and the eluate was shaken twice with cyclohexane and further analyzed for PAH following the applied procedure (4). PAH was found both in the adsorbed fraction and in the eluate. However, the level of concentration was much higher in the resin (Table 2) than in the eluate, the latter needing extreme evaporation to allow

**TABLE 2.** Concentration of PAH in Wastewater from an Alumina Reduction Plant

| Peak | PAH | ppb |
|---|---|---|
| 1 | Phenanthrene | 1.8 |
| 2 | Anthracene | 0.9 |
| 3 | Methylphenanthrene/methylanthracene | 0.7 |
| 4 | 2-Methylanthracene | 0.7 |
| 5 | 4,5-Methylenephenanthrene | 0.3 |
| 6 | Methylphenanthrene/methylanthracene | 0.5 |
| 7 | 1-Methylphenanthrene | 0.4 |
| 8 | Fluoranthene | 7.5 |
| 9 | Pyrene | 6.4 |
| 10 | Ethylmethylenephenanthrene ? | 3.4 |
| 11 | Benzo(a)fluorene | 8.2 |
| 12 | Benzo(b)fluorene | 7.2 |
| 13 | 4-Methylpyrene | 0.7 |
| 14 | Methylpyrene | 4.2 |
| 15 | 1-Methylpyrene | 1.4 |
| 16 | Benzothionaphthene ? | 5.6 |
| 17 | Benzo(c)phenanthrene + ? | 2.2 |
| 18 | Benzophenanthridine ? | 1.4 |
| 19 | Benz(a)anthracene | 14.6 |
| 20 | Chrysene/Triphenylene | 27.3 |
| 21 | Benzo(b)fluoranthene | 21.2 |
| 22 | Benzo(j/k)fluoranthenes | 10.5 |
| 23 | Benzo(e)pyrene | 17.0 |
| 24 | Benzo(a)pyrene | 13.5 |
| 25 | Perylene | 3.2 |
| 26 | Indeno(1,2,3-cd)pyrene | 8.1 |
| 27 | Dibenz(a,c/a,h)anthracenes | 2.2 |
| 28 | Benzo(ghi)perylene | 8.3 |
| 29 | Anthanthrene ? | trace |

detection by GC/FID. No significant differences were observed between the chromatographic profiles of the two fractions.

Whether the PAH in the eluate is the result of aromatics extracted from particulate matter passing through the resin bed or actually excess from saturation of the polymer was not determined. In any event, the relatively low concentration of PAH in the eluate compared with the adsorbed fraction should not lead to hesitations in the general application of the method.

The chromatogram of the resin fraction is shown in Figure 3.

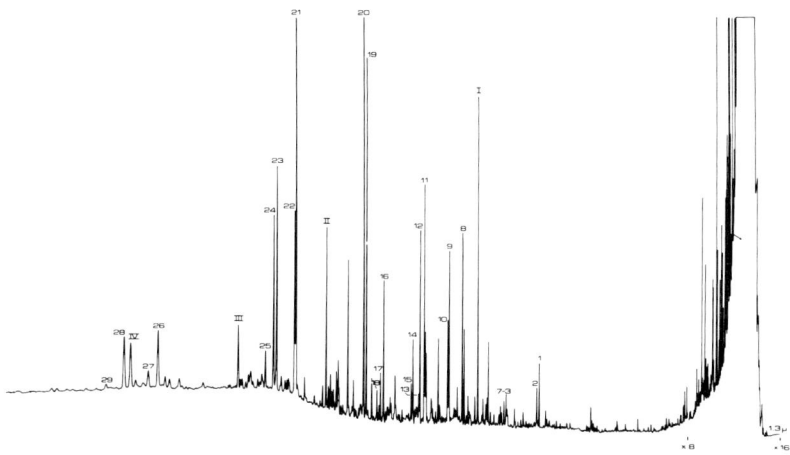

FIGURE 3. Gas chromatogram of PAH in wastewater from an alumina reduction plant. For peak identification see Table 2. Roman letters indicate internal standards.

## PAH IN DRINKING WATER RESOURCES

To examine the level of PAH in Norwegian drinking water and their raw water sources, some 20 samples have been analyzed by the continuous extraction technique. The sampling stations were close to electrochemical industry, in a typical agricultural area, or close to discharge of municipal sewage. The water samples were in most cases collected in 5-liter dark glass, stoppered bottles and adsorption carried out in the laboratory. For positive identification of PAH, several hundred liters of water was passed through the column of location.

The concentration of PAH in drinking water is shown to be extremely low. An example is shown in Table 3. The corresponding chromatogram is shown in Figure 4.

TABLE 3. Concentration of PAH in a Tap-Water Sample

| Peak | PAH | ppt |
|---|---|---|
| 1 | Naphthalene | 2.9 |
| 2 | 2-Methylnaphthalene | 1.4 |
| 3 | 1-Methylnaphthalene | 1.1 |
| 4 | Biphenyl | 0.32 |
| 5 | Acenaphthene | 0.82 |
| 6 | Dibenzofuran | 0.62 |
| 7 | Fluorene | 0.72 |
| 8 | Dibenzothiophene | 0.21 |
| 9 | Phenanthrene | 3.1 |
| 10 | Anthracene | 0.35 |
| 11 | Methylphenanthrene/methylanthracene | 0.48 |
| 12 | Methylphenanthrene/methylanthracene | 0.48 |
| 13 | 2-Methylanthracene | 0.06 |
| 14 | 4,5-Methylenephenanthrene | 0.30 |
| 15 | Methylphenanthrene/methylanthracene | 0.45 |
| 16 | 1-Methylphenanthrene | 0.37 |
| 17 | Fluoranthene | 2.6 |
| 18 | Pyrene | 1.1 |
| 19 | Ethylmethylenephenanthrene ? | 0.16 |
| 20 | Benzo(a)fluorene | <0.05 |
| 21 | Benzo(b)fluorene | <0.05 |
| 22 | 4-Methylpyrene | <0.05 |
| 23 | Methylpyrene | 0.08 |
| 24 | 1-Methylpyrene | <0.05 |
| 25 | Benz(a)anthracene | 0.49 |
| 26 | Chrysene/triphenylene | 0.79 |
| 27 | Benzo(b)fluoranthene | 0.21 |
| 28 | Benzo(j/k)fluoranthenes | 0.07 |
| 29 | Benzo(e)pyrene | 0.20 |
| 30 | Benzo(a)pyrene | <0.05 |

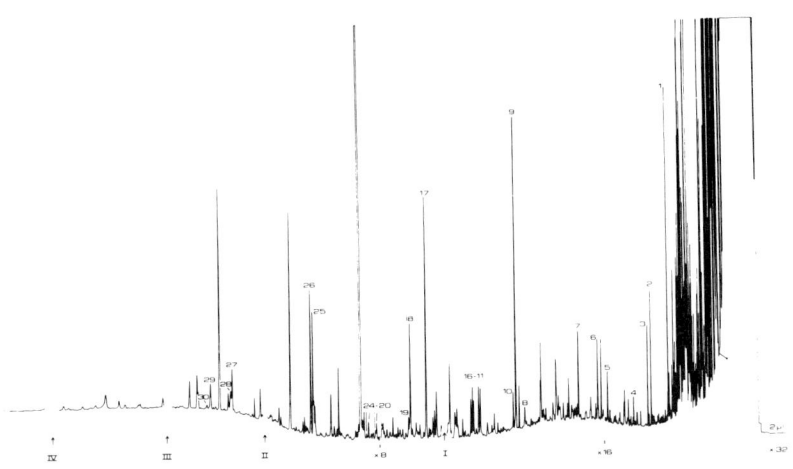

FIGURE 4. Gas chromatogram of PAH in a tap water sample, 300 liters analyzed. For peak identification see Table 3. Roman letters indicate internal standards.

For the remaining samples, only detection limits for single components are given. Based on peaks eluting in the area of fluoranthene/pyrene with a height-to-noise ratio of approximately 10 (9), the detection limits are in the range of 1 to 10 ppt. A chromatogram is shown in Figure 5.

The results reported here show that the levels of PAH in Norwegian drinking water and its sources are well below the recommended WHO standard (17). Five liters of water is also sufficient for screening tests carried out with the aim of checking for this standard.

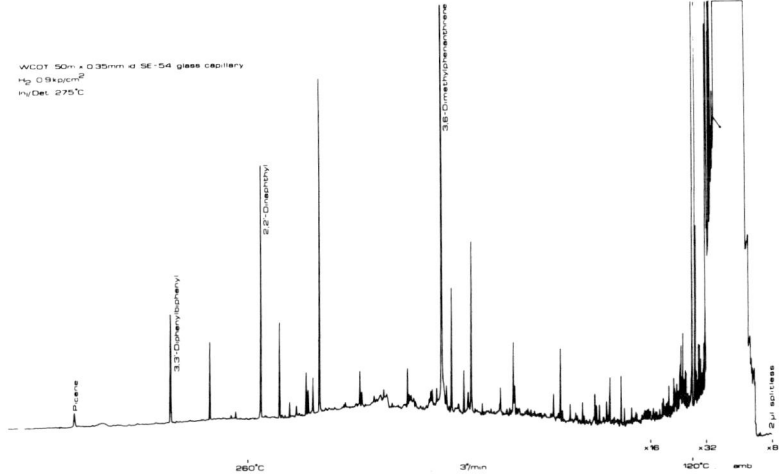

**FIGURE 5.** Gas chromatogram of a raw water sample; 5 liters analyzed. Detection limits for PAH are in the range of 1 to 10 ppt. The identified peaks are internal standards.

## CONCLUSIONS

- Direct liquid/liquid extraction of water containing particulates does not give quantitative or reliable data for the analysis of PAH.
- Simple gravity elution of PAH adsorbed on XAD-2 is not quantitative for the high-molecular-weight compounds within reasonable volume limits.
- Continuous extraction of the resin quantitatively recovers adsorbed PAH. Three internal standards are recommended.
- Concentration of single PAH in Norwegian drinking water and its raw water sources is on the ppt level.
- Five liters of water is sufficient for screening tests to check the recommended WHO standard for PAH in surface drinking water.

## ACKNOWLEDGMENT

Georg Carlberg is thanked for valuable discussions and encouragement.

## REFERENCES

1. Acheson, M. A., Harrison, R. M., Perry, R., and Wellings, R. A. (1976): Factors affecting the extraction and analysis of polynuclear aromatic hydrocarbons in water. *Water Research* 10: 207-212.
2. Basu, D. K., and Saxena, J. (1978): *Environ. Sci. Technol.* 12:795-798.
3. Berg, N., Carlberg, G. E., and Lunde, G. (1978): Analysis of water samples from the river Glomma. *Vann* 4: 1-9 (In Norwegian).
4. Bjørseth, A. (1977): Analysis of polycyclic aromatic hydrocarbons in particulate matter by glass capillary gas chromatography. *Anal. Chim. Acta* 94: 21-27.
5. Bjørseth, A., Lunde, G., and Lindskog, A. (1979): Long-range transport of polycyclic aromatic hydrocarbons. *Atmospheric Environment* 13: 45-53.
6. Chriswell, C. D., Ericson, R. L., Junk, G. A., Lee, K. W., Fritz, J. S., and Svec, H. J. (1977): Comparison of macroreticular resin and activated carbon as sorbents. *J. Amer. Water Works Asso.* 69: 699-674.
7. Crathorne, B., and Fielding, M. (1978): Analytical problems in meeting WHO limits for PAHs in water. *Proc. Analyt. Div. Chem. Soc.* 15: 155-158.
8. Dressler, M. (1979): Extraction of trace amounts of organic compounds from water with porous organic polymers. *J. Chromatogr.* 165: 167-206.
9. Grob, K., Grob, K., Jr., and Grob, G. (1975): Organic substances in potable water and in its precursor. III. The closed-loop stripping procedure compared with rapid liquid extraction. *J. Chromatogr.* 106: 299-315.
10. Hurtubise, R. J., Phillip, J. D., and Skar, G. T. (1978): *Anal. Chim. Acta* 101: 333-338.
11. Junk, G. A., Richard, J. J., Grieser, M. D., Witiak, D., Witiak, J. L., Arguello, M. D., Vick, R., Svec, H. J., Fritz, J. S., and Calder, G. V. (1974): Use of macroreticular resins in the analysis of water for trace organic contaminants. *J. Chromatogr.* 99: 745-762.
12. Olufsen, B., and Carlberg, G. E. (1979): Technical Report, Central Institute for Industrial Research, Oslo (In English).

13. Olufsen, B. (1980): Adsorption and continuous extraction of never-drying resin in a modified Soxhlet apparatus. *Anal. Chim. Acta* (113: 393-394.
14. Strup, P. E., Wilkinson, J. E., and Jones, P. W. (1978): Trace analysis of polycyclic aromatic hydrocarbons in aqueous systems using XAD-2 resin and capillary column gas chromatography-mass spectrometry analysis. In :*Carcinogenesis, Vol. 3: Polynuclear Aromatic Hydrocarbons,* P. W. Jones and R. I. Freudenthal, Eds., pp. 131-138, Raven Press, New York.
15. Webb, R. G. (1975): Isolating organic water pollutants: XAD resins, urethane foam, solvent extraction. EPA-660/4-75-003, Athens, Georgia, p. 4.
16. World Health Organization, Expert Committee on the Prevention of Cancer (1964): *Wld. Hlth. Orgn. Tech. Rep. Serv.* No. 276.
17. World Health Organization (1971): *International Standards for Drinking Water,* 3rd Edition, Geneva, Switzerland.

# STRUCTURE-CARCINOGENICITY STUDIES OF POLYCYCLIC AROMATIC HYDROCARBONS: A PATTERN RECOGNITION APPROACH

**B. Nordén, U. Edlund, and S. Wold**

Department of Organic Chemistry
Chemical Institution
University of Umeå
S-901 87 UMEÅ, Sweden

## INTRODUCTION

Structure-activity relationships (SAR) of polycyclic aromatic hydrocarbons (PAH) have been studied by numerous workers for the past 40 years. The major purposes of these studies have been to clarify the mechanism of carcinogenic action of these compounds and to predict the biological activity based on the chemical structures of PAH. These two subjects are hopefully interrelated. A theoretical prediction of the biological activity of a compound deduced from its chemical structure should constitute a piece of information valuable toward the understanding of carcinogenesis. Thus, correlations of theoretical and experimental variables with carcinogenic activities have their relevance if one is aware of the complex nature of biological systems including carcinogenic processes. When relating carcinogenic activities of PAH to their structure, a variety of different properties can be considered, for instance, hydrophobicities, conversion to appropriate metabolites, and interactions with critical receptors.

Until some years ago one thought of the hydrocarbon itself as the ultimate carcinogen. In the 50's the Pullmans (1) introduced the well-known concept of the K and L region. Some years later Mason (2) proposed a mechanism that included charge-transfer complexes as being the preferred mode of action for PAH. Birks (3) and other (4) also suggested similar types of complexes that induce cancer. These so-called "physical theories" were severely criticized in several respects by Pullman (5), who argued that their correlations were based on too few compounds. Theoretical support for the Pullman theory was given by others (6).

A few years later Dipple et al (7) proposed that a positively charged sigma complex located at the K region was the active metabolite. At present, arene epoxides are considered to be the primary metabolites of PAH (8). These, in turn, are isomerized and hydrated to other metabolites or diolepoxides, some compounds of which have their epoxide function positioned at the bay region (9). The diolepoxides and/or the carbonium ion at the bay region are believed to be the ultimate carcinogen (10). However, the crucial evidence of this theory has not yet been presented, although many indications point in this direction (11).

Correlation of single variables with the carcinogenicity of PAH is still the dominating approach (12). As pointed out earlier complex inter-relations among a number of structural properties are normally believed to determine the biological activity (13). Therefore, we believe that there will be a demand for more sophisticated techniques of statistical data analysis i.e., to base SAR on methods which can handle several variables simultaneously, such as the various methods of pattern recognition (PaRC) (14).

One advantage with PaRC methods is, if properly applied, that the probability of correlations by chance can be held to a very low level. This is not the case when correlating activity to a single variable. Here, the probability for a correlation by chance increases with the number of correlations, and when this has reached the number of PAH usually considered in SAR, one runs the risk of finding correlations by chance.

Although there are some excellent reviews on PaRC methods (15), we shall briefly discuss the general approach before looking deeper into one of the methods, the SIMCA method (16). The scope of PaRC classification studies in the study of SAR is to find mathematical descriptions of structurally similar and dissimilar substances. These descriptions can then be used to classify "unknown" substances. However, a well-formulated problem is a demand for a good result.

First the compounds are characterized by relevant variables containing structural information. These can be different quantum mechanical indices describing, for instance, the chemical reactivity. They can be other types of theoretical variables such as steric parameters. One can also include variables that are experimentally determined, log K, retention indices, spectroscopically determined variables, and so on.

The next step is a search for regularities or patterns in these variables. When dealing with structure-carcinogenicity studies, one hopefully finds different patterns for the class of carcinogens and the class of inactive compounds. The mathematical descriptions of these class patterns are determined from compound sets of known class assignment, that is, active or inactive compounds. These sets of compounds constitute the training set. Objects of unknown classification can then be put in a test set and, according to their fit (F-tests, etc.) to the earlier adapted mathematical models, one can make predictions of their class assignments, i.e., if they are active or not. Some methods also detect so-called "out-liers", i.e., compounds not described by any of the found models. It is sometimes also possible to relate external "effect" variables, such as the biological activity, to the position of a compound in the mathematical class structure. The graphical representation of the results is also a feature of PaRC methods.

In this study we have used the SIMCA method (16) of PaRC. To get a fairly large data set without too many missing observations we have concentrated on the unsubstituted PAH. We excluded compounds having three rings or less in order to avoid a trivial size factor. Five compounds having several missing variables were put in a test set. Naphtho(2,3-b)pyrene was also put in the test set because of the divergent conception about the activity of this compound. The PAH structures together with the weighted activities are shown in Table 1. Variables included in the analysis are both theoretical and experimental. A separate data analysis was made with only theoretical variables, an interesting approach for future prospective studies.

The theoretical and measured variables are listed below:

*Theoretical Variables*

- Four of the Pullman indices (1) are included: $BLE_K$, $PLE_K$ and the complex indices $(BLE + CLE_{min})_K$ and $(PLE + CLE_{min})_L$. It is not necessary to include the $CLE_{min}$ indices since they are covered by the complex indices. Sung (17) has calculated values for compounds that do not have a true K or L region.
- Resonance energies calculated by Huckel MO-method, $E_r$ (18).
- NBMO values for the sigma complexes believed to arise by $O_2$ or $O_2$-containing ions attacking the K region, $1-a_{or}$ (18).
- The energy of the highest filled (or lowest empty) molecular orbital as a measure of the electron-donor (or acceptor) properties, $|k|$ (5).
- The energy difference between the highest filled and successively unfilled energy levels, $\Delta E_{1,2}$ (19).
- The superdelocalizability indices as a measure of the reactivity of the K region, $I_K$ (6).
- The hydrophobicity estimated from hydrophobic fragmental constants, log P (20).

**TABLE 1. Compound Structures, Numbers, Activities[a], and Class Division[b]**

| Structure number | activity[a] | class |
|---|---|---|
| Training set | | |
| 1 | 1+ | |
| 2 | 1+ | 1a |
| 3 | 0 | 1b |
| 4 | 0-1+ | |
| 5 | 0 | |
| 6 | 0 | |
| 7 | 4+ | 1a |
| 8 | 2+ | 1a |
| 9 | 2+ | 1a |
| 10 | 1-2+ | |
| 11 | 1-2+ | 1a |
| 12 | 0-1+ | |
| 13 | 0 | 1b |
| 14 | 0 | 1b |
| 15 | 0 | 1b |
| 16 | 0-1+ | |
| 17 | 0 | |
| 18 | 0 | |
| 19 | 0 | 1b |
| 20 | 0 | 1b |

## TABLE 1. (Continued)

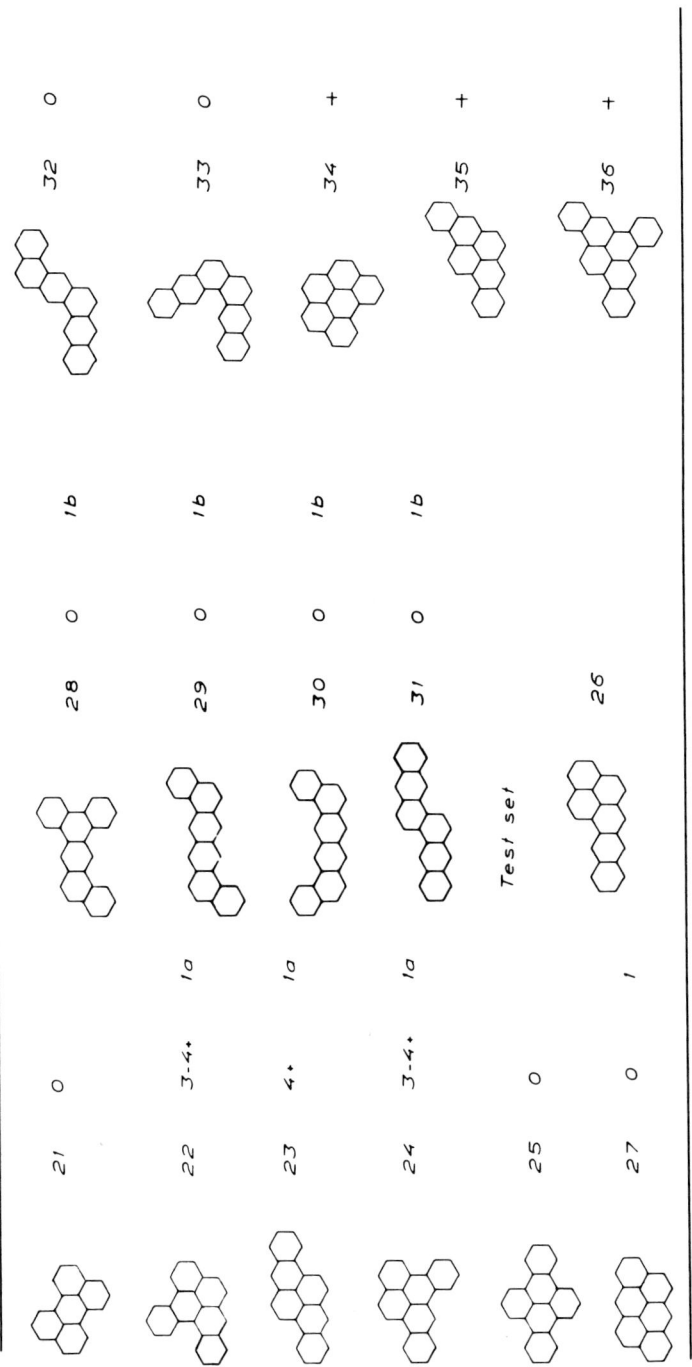

[a]See References 5, 12, 29, and 30. The activities are weighted values from these articles.
[b]Model 1: all compounds together in the first analysis. Second analysis, Models 1a and 1b. Class 1a: active compounds with moderate and high activity. Class 1b: inactive compounds predicted to have medium or high activity by Model 1.

- An asymmetric molecular structure induces an optimal distribution pattern for certain reactions, therefore we included the number of symmetry-axes, $A_s$ (21).
- The calculated charge at the most positive center of PAH radical cations, $C_{rc}$ (22).
- The calculated charge difference between the most and second most positive center of radical cations, $\Delta C_{rc}$ (22).
- A related variable with the exception that the radical cations should not be positioned at the K region, $\Delta C_{rc,non-K}$ (22).

*Measured Variables*

From the absorption spectra of Clar (23), six variables can be extracted totally, $\lambda_{max}$, log $\varepsilon$, and $\Delta$ (bandwidth) for the $\beta$- and p-bands, respectively. Some of the compounds do not have any $\alpha$-band and therefore we exclude this variable. One value for $\lambda_{max}$ is consistently taken at the longest wavelength.
- The overlapping integral determined from the quantum intensity of the fluorescence emission of tryptophan and the molar extinction coefficient of the hydrocarbon, $J_1$ (5).
- The ionization potential determined spectroscopically, $I_p$ (22).

For some compounds a few variables were not available. These values can be estimated using procedure where the mathematical model is applied, ignoring the missing values. Thereafter, the theta values can be specified by fitting the compounds to the calculated mathematical model followed by an estimation of the missing values.

Thus, we have included 15 theoretical and 8 measured variables, most likely the largest data matrix now available for the 36 PAH in the literature, without to many missing observations.

## THE SIMCA METHOD

The data of each class can be approximated by a separate principal components (PC) model (16).

$$y_{ik}^{(q)} = \alpha_i^{(q)} + \sum_{a=1}^{A} \beta_{ia}^{(q)} \cdot \theta_{ak}^{(q)} + \epsilon_{ik}^{(q)} \qquad (1)$$

The index q indicates that the data belong to class q. The data $y_{ik}$, the value of variable i measured on the object k, are described by the PC model. The parameters $\alpha_i$ and $\beta_{ia}$, which are estimated from the class training set, are specific for the variables and describe the position and direction of the class model. The parameter $\theta_{ak}$ determines the position of object k in the class. The parameters ($\alpha$, $\beta$, and $\theta$) are estimated to minimize the residuals $\varepsilon_{ik}$ in the least-square sense over all objects k and all variables i in the class. Thus, the residuals $\varepsilon_{ik}$, as describing the "noise" of the data, afford distance measures between the k:th object and the class. A refers to the number of components in the PC model of each class and thus each class can be described as a A-dimensional hyperplane in the measurement space (n-dimensional space, n is the number of variables). Figure 1a shows the situation for A+1 in three dimensions.

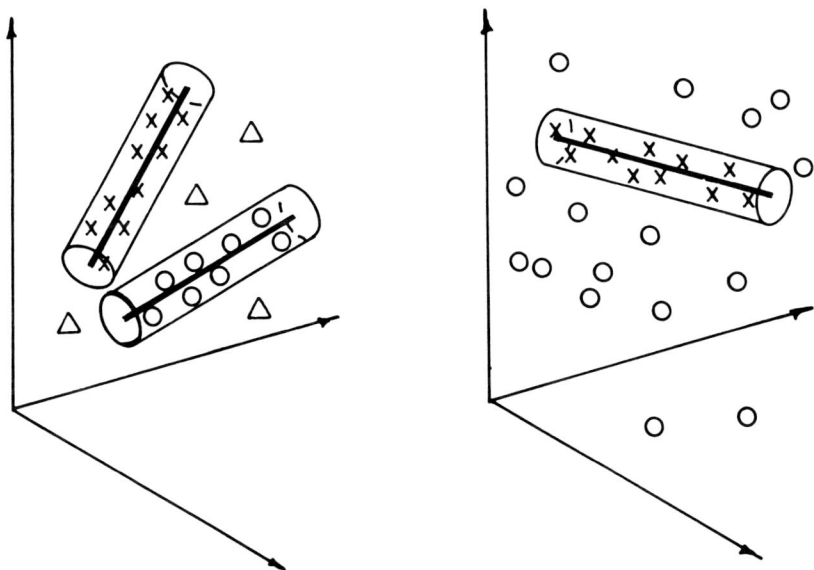

FIGURE 1(a). A three-dimensional space where each object is represented by a data point. The PC models describe the data with the dimensionality A=1 (equivalent to a line; A=2 is equivalent to a plane in space). The lines are surrounded by confidence regions that consequently enclose each class. Symbol △ represents so-called outliers not belonging to either of the two classes. (b) A three-dimensional space which shows two classes of objects representing an asymmetric data structure. Only one class can be described by a PC model (A=1, equivalent to a line). The other class is lacking systematic structure. The line is surrounded by a confidence region and the classification of new objects is made by observing whether they fall inside or outside this region. The asymmetric case of pattern recognition is a rather common situation in multivariate data analysis and causes failures because it is not recognized by most of these statistical methods and certainly not by analyzing one variable at a time.

A special case of classification is where one of two classes shows no systematic structure, while the other class is well described by the PC model. This situation indicates an asymmetric data structure and is illustrated in Figure 1b. A new object is classified according to its position inside or outside the model of the well-described class. Both Figures 1a and 1b show confidence regions around the lines (A=1).

The asymmetric case of classifications was recently found by Dunn and Wold (24). They showed that carcinogenic 4-nitroquinoline 1-oxides formed a good homogenous class, whereas the inactive molecules showed no systemic behavior in measurement space.

The principles of the method are briefly described as follows.

**Phase 1.** By using data obtained on compounds with known classifications (training set) we can calculate the *parameter* values for a separate PC model for each class. Initially, when dealing with many different variables, as in this case, it is customary to scale the data. This scaling (or regularization) procedure is done so that each variable gets the same standard deviation over the whole data set. The complexity of the PC model, i.e., the number of product terms in Equation (1) (A), is estimated using a cross-validation technique (25). We can also obtain the *modelling power* $\psi_i$, which describes how much a given variable participates in the mathematical description of the classes. The $\psi_i$ are obtained by comparing residual standard deviations with the corresponding data standard deviations.

$$\psi_i = 1 - (s_i / s_{i,y}), \tag{2}$$

where $s_i$ is the residual standard deviation of variable i over all the data in the training set, and $s_{i,y}$ is the corresponding standard deviation of the training set data y. A value of $\psi_i$ close to one indicates a high modelling power; a value of $\psi_i$ close to zero indicates a low modelling power. Variables having low modelling powers are deleted (outliers are also deleted from the training set) and the PC models of the classes are then recalculated from the reduced data set. We also get the *distances* between the classes in the measurement space, as well as the *discrimination power*; that is, how important a variable is for the separation of the classes.

**Phase 2.** Compounds with unknown class assignment and compounds with many missing observations are put in a test set. These are then classified according to the earlier adapted PC models.

**Phase 3.** Relationships between the position of an object in the class structure ($\theta$) and one or several external "effect" variables. In this case this corresponds to the prediction of the level of the carcinogenic activity.

In summary we can obtain the following information (26).
(a) The parameter values of the class PC models.
(b) A classification of compounds (both training and test sets) as active or nonactive based on their fit to the class models.
(c) The relevance of each variable, i.e., to what extent a variable participates in the description of the classes.
(d) A prediction of the level of carcinogenicity of compounds in the active class.

## RESULTS AND DISCUSSION

The data analysis was made essentially as described earlier (27). Initially, scaling of the variables of the data matrix was done. Then, in a first analysis all compounds were treated together in one single class (called Model 1). Variables with low modelling power were deleted and the Model 1 was reexamined. The deleted variables were $\log \varepsilon_p$, $\log \varepsilon_\beta$, $J_1$, $\log P$, $A_s$, $E_r$, $1-a_{or}$, $\Delta_\beta$, $C_{rc}$, $\Delta C_{rc}$ and $\Delta C_{rc, \text{non-}\kappa}$. The modelling power for the rest of the variables (twelve) are tabulated in Table 2. It is interesting to note that $\Delta E_1$, $|kl|$, $(PLE+CLE_{min})_L$, $I_K$, and $I_p$ have the highest relevance in this case. Applying one compound of equation [1] describes 68 percent of the variation (standard deviation) of the data structure having the variables mentioned above deleted. The parameters $\alpha$ and $\beta$ for the one-component PC model of class Model 1 are shown in Table 3. The $\theta$-values are shown in Table 4.

The positions of the active compounds in Model 1 were related to their carcinogenicity level, and a plot of the carcinogenic activity against $\theta_1$ is shown in Figure 2. The same result is obtained using only theoretical variables but with a lower correlation coefficient (see legend of Figure 2). If adding the inactive compounds in the same plot, some of the inactive PAH are predicted to have medium or high activity (because of their low $\theta_1$-value, i.e., ⓧ on Figure 3). One plausible interpretation might be that all these compounds penetrate some target cell, followed by a different mode of transportation for inactive and active compounds. Alternatively, all compounds are converted to specific metabolites but only the metabolites of the active compounds are appropriate for the transportation and interaction with the critical receptor.

To investigate this, we analyzed the compounds of the training set with predicted medium or high activity as two classes. This approach afforded a new training set where we have Class 1a, which contains compounds that in fact are active (compounds 2, 7-9, 11, 22-24, Table 1), and Class 1b, which contains compounds predicted to have medium or high activity but actually are inactive (compounds 3, 13-15, 19, 20, 27-31,

TABLE 2. Modelling Power for Variables of Models 1, 1a, and 1b

| Variable | Modelling Power 1 | Model 1a Only | Model 1b Only | Models 1a and 1b |
|---|---|---|---|---|
| $BLE_K$ | 0.09 | 0.21 | 0.11 | 0.16 |
| $BLE+CLE_{min})_K$ | 0.36 | 0.25 | 0.16 | 0.20 |
| $PLE_L$ | 0.45 | 0.0 | 0.71 | 0.20 |
| $(PLE+CLE_{min})_L$ | 0.56 | 0.23 | 0.59 | 0.38 |
| $\lambda_{max,p}$ | 0.48 | 0.88 | 0.45 | 0.60 |
| $\lambda_{max,}$ | 0.16 | 0.22 | 0.05 | 0.13 |
| $\Delta E_1$ | 0.67 | 0.70 | 0.54 | 0.61 |
| $\Delta E_2$ | 0.36 | 0.58 | 0.0 | 0.15 |
| $I_K$ | 0.54 | 0.25 | 0.24 | 0.24 |
| $\log P$ | | 0.15 | 0.05 | 0.10 |
| $E_r$ | | 0.13 | 0.10 | 0.11 |
| $1-a_{or}$ | | 0.28 | 0.26 | 0.27 |
| $|k|$ | 0.59 | 0.57 | 0.57 | 0.57 |
| $\Delta_\rho$ | 0.27 | 0.78 | 0.52 | 0.62 |
| $I_\rho$ | 0.54 | 0.69 | 0.49 | 0.58 |
| $\Delta C_{rc}$ | | 0.36 | 0.35 | 0.35 |
| $\Delta C_{rc,non-K}$ | | 0.15 | 0.32 | 0.23 |

Table 1). The rest of the compounds (slightly active or inactive) and the earlier test set of Model 1 constitute the new test set.

The complete data analysis was repeated, that is, to fit the PC equation separately to the two classes and to check whether some variables with low modelling power have to be deleted. Variables which did not contribute in defining the class structures were: $\log \varepsilon_p$, $\log \varepsilon_\beta$, $J_1$, $A_s$, $\Delta\beta$, and $C_{rc}$. We found two statistically well-defined classes. Measurement space consisted of 17 variables. In this 17-dimensional space a one

TABLE 3. Values for the Parameters $\alpha$ and $\beta$ of Models 1, 1b, and 1a

| Variable | $\alpha_1$ | $\beta_1$ | $\alpha_{1b}$ | $\beta_{1b}$ | $\alpha_{1a}$ | $\beta_{1a}$ |
|---|---|---|---|---|---|---|
| $BLE_K$ | 4.2E-13 | 1.8E-01 | -3.6E-01 | 1.9E-01 | -5.1E-01 | -7.4E-03 |
| $(BLE+CLE_{min})_K$ | 7.2E-13 | 2.8E-01 | -5.7E-01 | 5.1E-03 | -3.8E-01 | 2.0E-01 |
| $PLE_L$ | 3.1E-13 | 3.0E-01 | -8.5E-01 | -8.9E-02 | -2.0E-01 | 6.3E-02 |
| $(PLE+CLE_{min})_L$ | 5.3E-13 | 3.2E-01 | -8.4E-01 | -1.3E-01 | -3.0E-01 | 9.6E-02 |
| $\lambda_{max,p}$ | -7.7E-15 | -3.1E-01 | 7.3ED-01 | 4.0E-01 | -6.8E-02 | -3.2E-01 |
| $\lambda_{max,}$ | 1.8E-14 | -2.1E-02 | 6.1E-01 | -1.7E-01 | 3.7E-01 | -2.4E-01 |
| $\Delta E_1$ | 1.8E-13 | 3.3E-01 | -7.3E-01 | -3.0E-01 | -4.9E-02 | 3.6E-01 |
| $\Delta E_2$ | 4.4E-13 | 2.8E-01 | -7.3E-01 | -1.5E-01 | -2.1E-01 | 7.9E-02 |
| $I_K$ | 9.1E-13 | -3.2E-01 | 6.8E-01 | 6.5E-02 | 2.3E-01 | -3.2E-01 |
| log P | | | 5.3E-01 | -2.2E-01 | 1.3E-01 | -1.8E-01 |
| $E_r$ | | | 4.0E-01 | -2.6E-01 | 1.5E-01 | -2.4E-01 |
| $1-a_{or}$ | | | 9.8E-04 | 3.0E-01 | -4.1E-01 | -3.0E-01 |
| $|k|$ | 1.9E-13 | 3.2E-01 | -7.5E-01 | -2.4E-01 | -5.5E-02 | 3.9E-01 |
| $\Delta\rho$ | -6.9E-15 | -2.5E-01 | 6.8E-01 | 4.7E-01 | -1.9E-01 | -2.3E-01 |
| $\rho$ | 8.5E-13 | 3.2E-01 | -7.6E-01 | -2.9E-01 | 8.2E-02 | 3.4E-01 |
| $\Delta C_{rc}$ | | | -5.3E-01 | -1.0E-01 | 1.2E+00 | -1.7E-01 |
| $\Delta C_{rc,non-K}$ | | | -1.3E-01 | -2.3E-01 | 1.1E+00 | -8.7E-02 |

**TABLE 4.** $\theta$ for Model 1 and Test Set

| Training Set | $\theta$ | Test Set | $\theta$ |
|---|---|---|---|
| 3 | -3.31 | 26 | -4.37 |
| 5 | 1.37 | 32 | -1.99 |
| 6 | 7.84 | 33 | -1.84 |
| 13 | -3.41 | 34 | 0.04 |
| 14 | -5.61 | 35 | -2.40 |
| 15 | -1.35 | 36 | -0.23 |
| 17 | 2.11 | | |
| 18 | 1.96 | | |
| 19 | -0.99 | | |
| 20 | -1.03 | | |
| 21 | 0.01 | | |
| 25 | 3.76 | | |
| 27 | -3.49 | | |
| 28 | 0.08 | | |
| 29 | -2.92 | | |
| 30 | -2.20 | | |
| 31 | -2.38 | | |
| 1 | 3.76 | | |
| 2 | -0.32 | | |
| 4 | 2.62 | | |
| 7 | 1.37 | | |
| 8 | 0.21 | | |
| 9 | 2.26 | | |
| 10 | 3.57 | | |
| 11 | 0.16 | | |
| 12 | 1.85 | | |
| 16 | 1.78 | | |
| 22 | -1.57 | | |
| 23 | -3.35 | | |
| 24 | -0.04 | | |

component model (A=1) was found for class 1a, and one component model (A=1) for class 1b. Thus, we have the situation shown in Figure 1a.

The one component of class 1a describes 69 percent of the data variation (standard deviation), and the one component of class 1b describes the variation to 73 percent. The class models also afford a good classification. None of the eight compounds of class 1a is incorrectly assigned and two compounds of the inactive class are enclosed by both PC models but are closer to the active class (19 and 20). The rest of the training set (17 of 19) are correctly classified. The parameters $\alpha$ and $\beta$ are collected in Table 3, and the $\theta$-values in Table 5.

It is interesting to note that (Table 2) the variables representing the ideas of the "physical theories" also have the highest relevancy in the

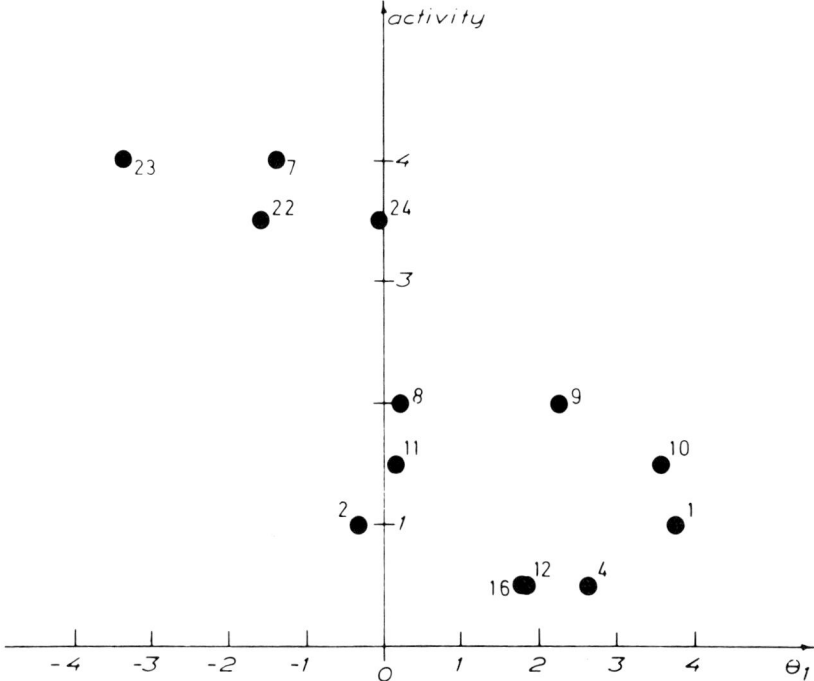

FIGURE 2. Plot of the activity versus $\theta$ showing the position of each object in the active group of compounds of Model 1. The lower the $\theta$-parameter value, the more active is the compound. Correlation coefficients: $r = 0.812$ ($p<0.001$) including both theoretical and measured variables, and $r = 0.792$ ($p<0.0001$) including only theoretical variables.

description of the classes ($\Delta_p$, $\Delta E_1$, $\lambda_{max,p}$, $I_p$). Considering only the active class, these variables still have the highest modelling power. The modelling power, when considering only class 1a or class 1b, is shown in Table 2. Further interpretation of this result is currently beyond the scope of this study.

A test of the compounds with several missing variables shows that Model 1 predicts compound 34 [benzo(ghi)perylene] to be slightly active or inactive. The other compounds (32, 33, 35, and 36) together with compound 26 are predicted to have medium or high activity. Compounds 26, 32, and 33 are enclosed by Model 1b, and compound 35 is enclosed by Model 1a, but compound 36 is detected as an outlier. Thus, only three out of the five compounds with much missing data are correctly classified.

Finally, a validation of the result had to be done to determine if the fitted PC models were consistent in their class predictions. The study was

## 358 STRUCTURE-CARCINOGENICITY STUDIES OF PAH

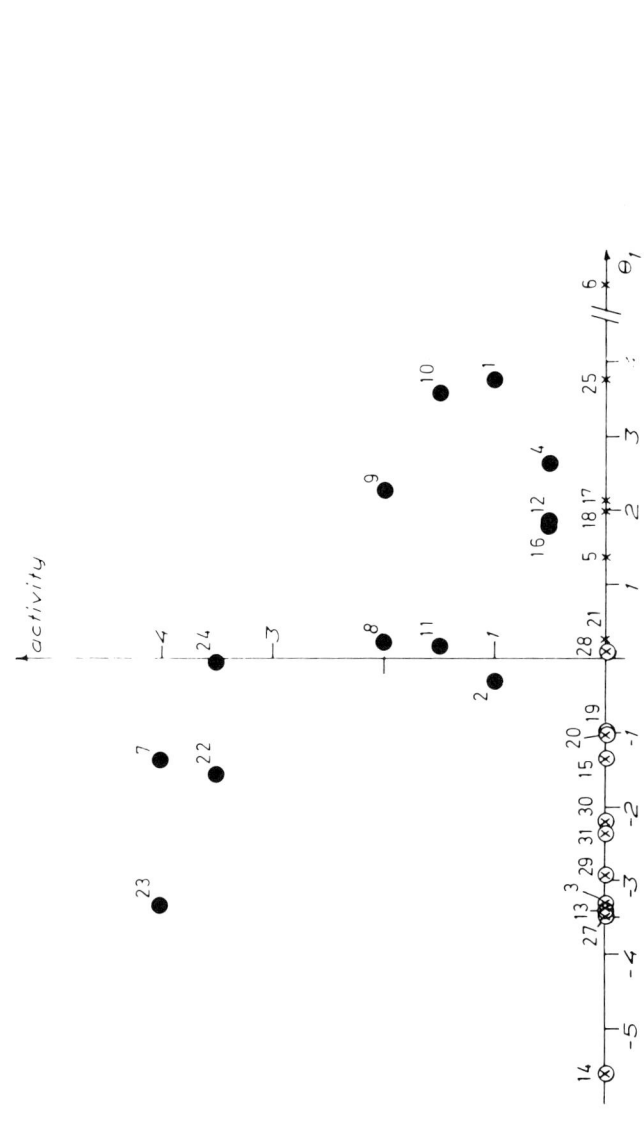

FIGURE 3. A plot of the activity versus $\theta$ (from the Model 1 analysis) showing the position of all compounds. Inactive compounds with low $\theta$-value are predicted to have medium or high activity. The second analysis is based on this first situation and hence class 1a (active compounds) will contain compounds 2, 7-9, 11, 22-24. Class 1b (inactive compounds) will contain compounds denoted by o (3,13-15, 19, 20, 27-31).

## STRUCTURE-CARCINOGENICITY STUDIES OF PAH 359

**TABLE 5. $\theta$ for Model 1a, 1b, and Test Set**

| Model 1a | $\theta$ | Model 1b | $\theta$ | Test Set | $\theta_{1a}$ | $\theta_{1b}$ |
|---|---|---|---|---|---|---|
| 2 | 1.30 | 3 | 4.33 | 1 | 4.19 | -2.93 |
| 7 | -0.66 | 13 | 0.70 | 4 | 3.21 | -1.87 |
| 8 | 1.16 | 14 | 4.18 | 5 | 2.79 | -1.13 |
| 9 | 2.87 | 15 | -2.09 | 6 | 6.89 | -3.14 |
| 11 | 0.85 | 19 | -1.38 | 10 | 2.68 | -2.89 |
| 22 | -1.41 | 20 | -1.57 | 12 | 2.83 | -2.35 |
| 23 | -3.47 | 27 | 0.60 | 16 | 1.75 | -1.75 |
| 24 | -0.65 | 28 | -3.25 | 17 | 2.19 | -2.55 |
|  |  | 29 | -.55 | 18 | 2.09 | -3.15 |
|  |  | 30 | -1.04 | 21 | 0.29 | 1.52 |
|  |  | 31 | 0.08 | 25 | 1.87 | -1.65 |
|  |  |  |  | 26 | -3.09 | 0.29 |
|  |  |  |  | 32 | -1.61 | -2.52 |
|  |  |  |  | 33 | -1.64 | -0.80 |
|  |  |  |  | 34 | 0.97 | -1.21 |
|  |  |  |  | 35 | -1.72 | -0.81 |
|  |  |  |  | 36 | -0.79 | -1.17 |

performed by leaving out a quarter of the compounds of each class and putting them into the test set. Then, the mathematical models were fitted to the remaining compounds of the class and, in the end, the compounds left out were classified. This procedure was done repeatedly so that each compound was left out once. A high prediction rate of the compounds left out indicated stable class structures.

The validation of Model 1, that is, for the prediction of the level of the active compounds, showed a correlation coefficient of 0.779 ($p < 0.01$) for the observed and predicted activity of compounds left out. The validation of Model 1a and Model 1b showed that none of the eight compounds of Model 1a was wrongly classified, and two compounds of Model 1b (compounds 19 and 20, Table 1) incorrectly classified. This classification rate (17/19) was significantly better than chance ($\chi^2 = 6.23$, $p < 0.05$).

The graphical representation of the result of Model 1a and Model 1b is shown in Figure 4. This is an eigenvector plot (28) where the 17-dimensional measurement space is reduced to two dimensions. This is done by a linear projection down on a plane. The projection is done to preserve most of the variance between the objects. This plot gives a fair idea of the original measurement space.

Another interesting approach is to let the measurement space be spanned by the objects and consider the variables as points in this space. We can do a similar eigenvector analysis as above and subsequently we can see if any groupings among the variables occur. A plot of $\beta_1$ versus $\beta_2$ from the eigenvector analysis (the slopes of the two-dimensional plane in Figure 4) is shown in Figure 5, where related variables are grouped together.

Our procedure, written as a flow diagram for the classification of PAH, is shown in Figure 6.

## SUMMARY

We have shown that it is possible to use theoretical and measured variables to (1) classify PAH as carcinogenic or nonactive compounds and (2) make predictions about the level of the activity of a given compound. The active class of PAH appears as an asymmetric case of classification, hardly detectable by considering one variable at a time.

The results indicate that two "factors" influence the carcinogenicity. The first factor influences the level of activity; the second factor in some way inhibits the carcinogenicity even where the first factor corresponds to high activity (1b).

STRUCTURE-CARCINOGENICITY STUDIES OF PAH 361

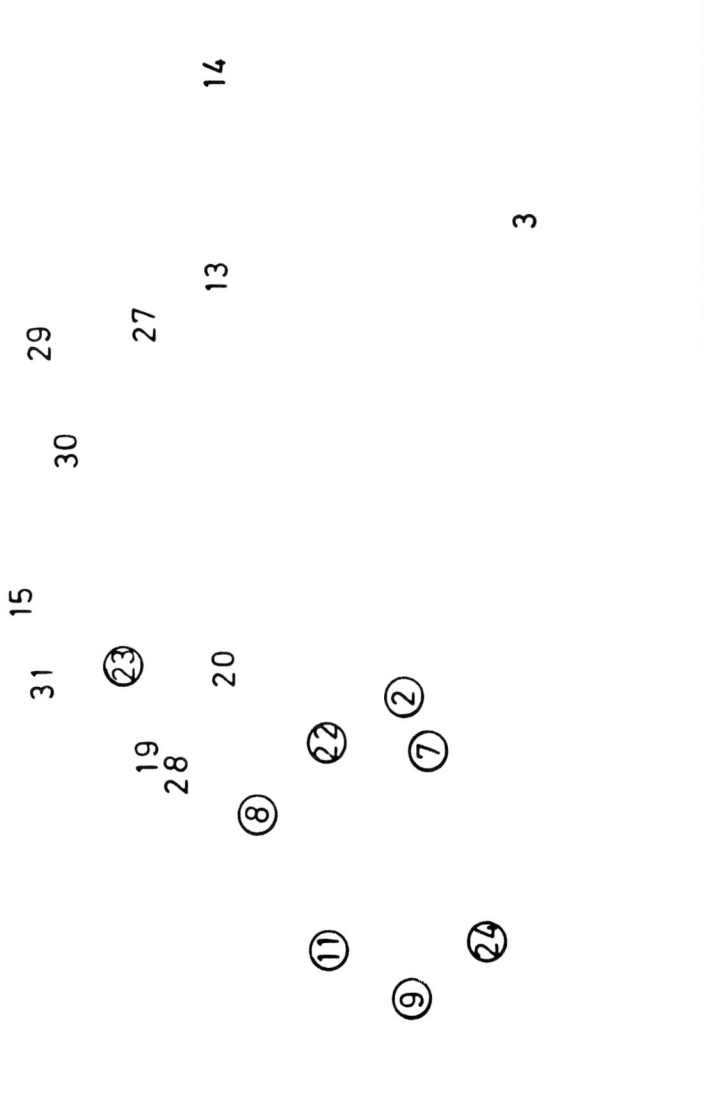

FIGURE 4. A plot of the result of an eigenvector analysis of 17 dimensions. The numbers refer to the numbering of compounds of Models 1a (numbers in circles) and 1b.

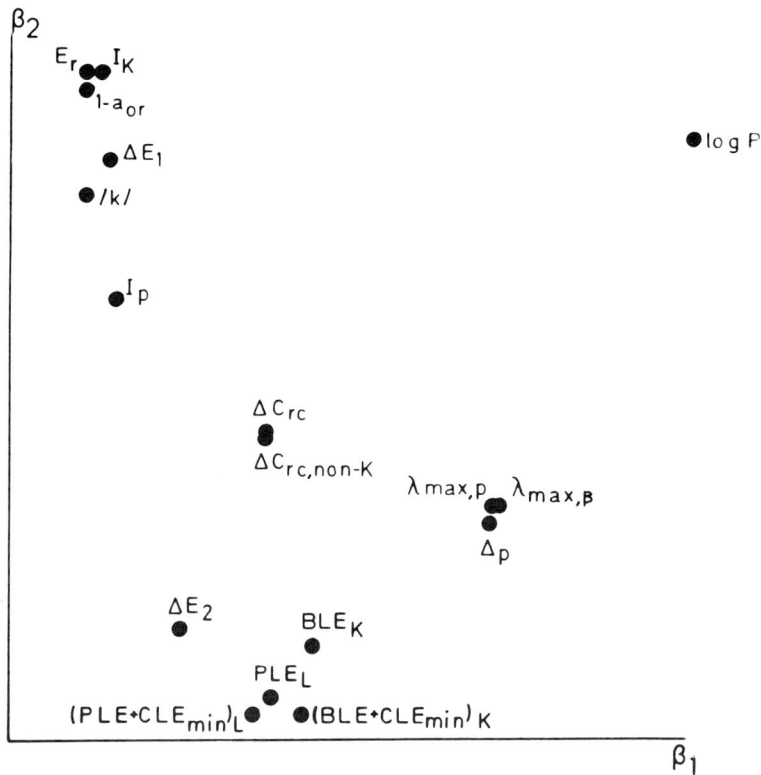

FIGURE 5. A plot of $\beta_1$ versus $\beta_2$ from the eigenvector (see text for explanation).

# STRUCTURE-CARCINOGENICITY STUDIES OF PAH 363

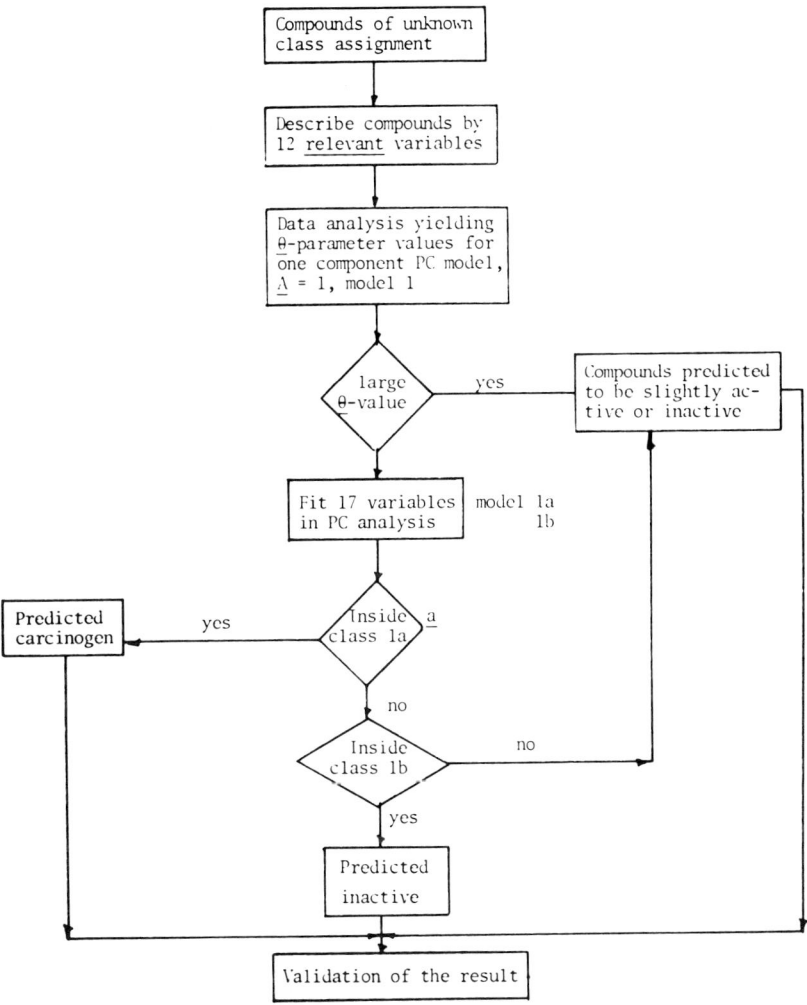

**FIGURE 6.** Our procedure for the classification of PAH. Classification rate (class 1a or 1b) in this work was 17/19 ($\chi^2$ = 6.23, p<0.05).

## REFERENCES

1. Pullman, A., and Pullman, B. (1955): Electronic structure and carcinogenic activity of aromatic molecules. *Cancer Res.* 3:117-169.
2. Mason, R. (1958): Electron mobility in biological systems and its relation to carcinogenesis. *Nature* 181:820-822.
3. Birks, J. B. (1961): A physical theory of carcinogenesis by aromatic hydrocarbons. *Nature* 190:232-235.
4. Allison, A. C., and Nash, T. (1963): Electron donation and acceptance by carcinogenic compounds. *Nature* 197:758-763.
5. Pullman, A. (1964): The theory of chemical carcinogenesis and the problem of hydrocarbon-protein interactions. *Biopolymers Symposia No. 1* pp. 47-65.
6. Mainster, M. A., and Memory, J. D. (1967): Superdelocalizability indices and the Pullman theory of carcinogenesis. *Biochem. Biophys. Acta* 148:605-608.
7. Dipple, A., Lawley, P. D., and Brooks, P. (1968): Theory of Tumor Initiation by Chemical Carcinogens: Dependence of activity on structure of ultimate carcinogen. *Eur. J. Cancer* 4:493-506.
8. Jerina, D. M., Lehr, R. E., Yagi, H., Hernandez, D., Dansette, P. M., Wislocki, P. G., Wood, A. W., Chang, R. L., Levin, W., and Conney, A. H. (1976): Mutagenicity of benzo(a)pyrene derivatives and the description of a quantum mechanical model which predicts the ease of carbonium formation from diol epoxides. In: *In Vitro Metabolic Activation in Mutagenesis Testing.* F. J. de Serres, J. R. Fonts, J. R. Bend, and R. M. Philpot, Eds., pp. 159-177, Elsevier/North Holland Biomedical Press, Amsterdam.
9. Fu, P. P., and Harvey, R. G. (1978): Molecular orbital theoretical prediction of the isomeric products formed from reactions of arene oxides and related metabolites of polycyclic aromatic hydrocarbons. *Tetrahedron* 34:857-866.
10. Wood, A. W., Levin, W., Lu, A.Y.H., Ryan, D., West, S. B., Lehr, R. E., Schaefer-Ridder, M., Jerina, D. M., and Conney, A. H. (1976): Mutagenicity of metabolically activated benzo(a)anthracene 3,4-dihydrodiol: Evidence for bay region activation of carcinogenic polycyclic hydrocarbons. *Biochem. Biophys. Res. Commun.* 72: 680-686.
11. Lehr, R. E., and Jerina, D. M. (1977): Metabolic activations of polycyclic hydrocarbons: Structure-activity relationships. *Arch. Toxicol.* 39:1-6.
12. Berger, G. D., Smith, I. A., Seybold, P. G., and Serve, M. P. (1978): Correlation of an electronic reactivity index with carcinogenicity in polycyclic hydrocarbons. *Tetrahedron Lett.* 3:231-234.

13. Berenblum, I. (1969): The problem of quantitative carcinogenesis. In: *Physicochemical Mechanisms of Carcinogenesis.* E. D. Bergmann and B. Pullman, Eds., pp. 321-324, Jerusalem.
14. Massart, D. L., Dijkstra, A., Kaufman, L. (1978): *Evaluation and Optimization of Laboratory Methods and Analytical Procedures.* Elsevier, Amsterdam.
15. Kowalski, B. R., and Wold, S. (1980): Pattern recognition in chemistry. In: *Handbook of Statistics, Vol. 2.* P. R. Krishnaiah and L. Kanal, Eds., North Holland, Amsterdam (in press).
16. Wold, S., and Sjöström, M. (1977): SIMCA: A method for analyzing chemical data in terms of similarity and analogy. In: *Chemometrics: Theory and Practice, ACS Symp. Ser. No. 52.* B. R. Kowalski, Ed., pp. 243-282.
17. Sung, S. S. (1972): Cancérologie: Essai d'application de la théorie des régions K et L à un nouveau lot d'hydrocarbures aromatiques polycyclique. Etude avec les indices complexes de réactivité. *C. R. Acad. Sci.* 274:1597-1600.
18. Scribner, J. D. (1969): Formation of a sigma complex as a hypothetical rate-determining step in the carcinogenic action of unsubstituted polycyclic aromatic hydrocarbons. *Cancer Res.* 29:2120-2126.
19. Sung, S. S. (1967): Cancérologie: Sur l'existence éventuelle d'une corrélation plus générale entre le pouvoir cancérogene d'une substance et une de ses propriétés moléculaires. *C. R. Acad. Sci.* 264:189-192.
20. Nys, G. G., and Rekker, R. F. (1974): The concept of hydrophobic fragmental constants (f-values). II. Extension of its applicability to the calculation of lipophilicities of aromatic and heteroaromatic structures. *Eur. J. Med. Chem.—Chimica Therapeutica* 9:361-375.
21. Arcos, J. C., and Argus, M. F. (1974): *Chemical Induction of Cancer, Vol. IIA.* Academic, New York.
22. Pullman, B., Pullman, A., Umans, R., and Maigret, B. (1969): A few afterthoughts. In: *Physico-chemical mechanisms of carcinogenesis.* E. D. Bergmann and B. Pullman, Eds., pp. 325-338, Jerusalem.
23. Clar, E. (1964): *Polycyclic Hydrocarbons, Vols. 1 and 2.* Academic, London.
24. Dunn, III, W. J., and Wold, S. (1978): A structure-carcinogenicity study of 4-nitroquinoline 1-oxides using the SIMCA method of pattern recognition. *J. Med. Chem.* 21:1001-1007.
25. Wold, S. (1978): Cross-validatory estimation of the number of components in factor and principal components models. *Technometrics* 20:397-405.

26. Albano, C., Dunn, III, W., Edlund, U., Johansson, E., Nordén, B., Sjöström, M., and Wold, S. (1978): Four levels of pattern recognition. *Anal. Chem. Acta. Comp. Tech. Optim.* 103:429-443.
27. Nordén, B., Edlund, U., and Wold, S. (1978): Carcinogenicity of polycyclic aromatic hydrocarbons studied by SIMCA pattern recognition. *Acta. Chem. Scand.* B 32:602-608.
28. Kowalski, B. R. (1974): Pattern recognition in chemical research. In: *Computers in Chemical and Biochemical Research, Vol. 2,* C. E. Klopfenstein and C. L. Wilkins, Eds., pp. 1-76, Academic, New York.
29. Jones, D. W., and Matthews, R. S. (1974): Carcinogenicity and structure in polycyclic hydrocarbons. *Prog. Med. Chem.* 10:159-203.
30. Dipple, A. (1976): Polynuclear aromatic carcinogens. In: *Chemical Carcinogens, ACS Monograph 173.* C. E. Searle, Ed., pp. 245-314, Washington.

# POLYCYCLIC AROMATIC HYDROCARBONS IN MARINE SEDIMENTS, BIVALVES, AND SEAWEEDS: ANALYSIS BY HIGH-PRESSURE LIQUID CHROMATOGRAPHY

**B. P. Dunn***

Environmental Carcinogenesis Unit
British Columbia Cancer Research Center
601 W. 10th Ave.
Vancouver, B.C., Canada V5Z 1L3

## INTRODUCTION

The contamination of coastal marine areas by PAH is of concern as such areas are often sources of seafoods. We have recently examined commercial shellfish from a number of countries (3), and have found that contamination of these organisms by PAH carcinogens is widespread.

The increasing utilization of coastal waters for aquaculture operations makes it imperative to develop criteria for the suitability of coastal areas for commercial seafood production, with respect to carcinogen contamination. Two major problems in determining the degree of contamination of a marine area by PAH are the questions of what type of environmental sample to measure, and which PAH compounds to quantitate. In this paper we examine the relationship between levels of a range of PAH isomers in three types of marine samples: sediments, bivalve mollusks (mussels, *Mytilus edulis*), and seaweed (*Fucus sp.*).

## METHODS AND MATERIALS

Samples of shoreline sediments, mussels, and seaweed were obtained from ten locations in the inner and outer harbour of Vancouver, British Columbia, and were stored frozen until analysis. Sample extraction and

---
*Research Scholar, Department of National Health and Welfare, Canada.

purification procedures were in general as previously described (1,2). In brief, samples were digested in alcohol and KOH, then water added and the hydrocarbons partitioned into isooctane. Interfering materials were removed by column chromatography on Florisil, followed by selective extraction of PAH from isooctane into dimethyl-sulfoxide (DMSO). Further interfering materials were removed by column chromatography on Sephadex LH-20, as described by Giger and Blumer (5), except utilizing a solvent of toluene/ethanol (1:1) instead of benzene/methanol (1:1). An internal standard of radioactively labelled benzo(a)pyrene was added to all samples before extraction to correct for losses during purification procedures.

Purified PAH extracts were concentrated into a small volume of DMSO, and aliquots of 2 to 10 microlitres were chromatographed on a Perkin Elmer HC-ODS reversed-phase liquid chromatography column. Chromatography was at 60°C and a flow rate of 0.5 ml/min. The solvent consisted of 40 percent acetonitrile, with a linear gradient of 1.4 percent acetonitrile/min beginning 6 minutes after injection. PAH were detected by means of a UV absorption monitor at a wavelength of 296 nm (absorption maximum for benzo(a)pyrene), followed by a fluorescence monitor operating with a broad excitation band (340-380 nm, selected by a Corning 7-54 filter in series with a 7-60 filter), and by measuring emission at wavelengths greater than 400 nm (selected by a Corning 3-73 UV cutoff filter, in series with a 4-76 red blocking filter to eliminate red leakage from the 7-54/7-60 filter combination). Solvents were air saturated, and a variable back-pressure restrictor (Varian Associates) was used on the output line of the second detector to maintain an internal pressure of approximately 200 psi inside both detectors to suppress bubble formation. Each detector output was fed to a separate dual pen recorder, with one pen set at 10 times the sensitivity of the other pen to increase the dynamic range of chart recorder traces which could easily be measured.

Chromatographic peaks were identified by retention time, co-chromatography of reference compounds, and by the ratio in response between UV and fluorescence detectors. Peak areas were measured by triangulation, and the amount of each component was determined by the use of response factors prepared with reference compounds. The degree of contamination of the original sample was calculated utilizing the calculated amount of a compound in an injection, the volume of an injection, the recovery of the sample per unit volume of the injected material (determined by measurement of the $^3$H-benzo(a)pyrene internal standard), and the weight of the sample.

Benz(a)anthracene and chrysene were not resolved under the chromatography conditions employed. Subsequent chromatography of representative samples under different conditions (30°C column temperature, and a linear gradient of 3 percent/min acetonitrile starting at 60 percent acetonitrile) resolved these two isomers, and indicated that they were present in approximately equal amounts. The results presented in this paper are derived from the unresolved benz(a)anthracene/chrysene peak using an averaged response factor, and are reported as the sum of the benz(a)anthracene and chrysene concentrations.

## RESULTS

Samples of sediments, mussels (*Mytilus edulis*), and seaweed (*Fucus sp.*) were obtained from seven sites, and in addition, sediments and mussels were obtained from a further three sites. The sites sampled were all in the inner and outer harbour area of Vancouver, British Columbia, and ranged from sandy recreational beaches to muddy-bottomed areas in the immediate vicinity of wharfs and docks. Analysis for benzo(a)pyrene and other PAH indicated that the degree of contamination of samples from different sites varied by as much as 100 fold for sediment samples, and 40 fold for Fucus and mussel samples.

For each type of sample (sediments, 12 samples from 10 sites; mussels, 19 samples from 10 sites; Fucus, 8 samples from 7 sites), a linear regression analysis was performed to investigate the relationship between the levels of benzo(a)pyrene and the levels of other PAH isomers in individual samples. Figure 1 shows the correlation coefficients between benzo(a)pyrene and eight other PAH, listed in order of their retention time in reversed-phase liquid chromatography. For sediments, there was a statistically significant positive correlation between the levels of benzo(a)pyrene and the levels of all other PAH measured. In mussel samples, there was a good correlation between levels of benzo(a)pyrene and levels of higher-molecular-weight PAH. In addition, there was a positive correlation between phenanthrene and benzo(a)pyrene levels in mussels, although the relationship was not quite significant at the 1 percent confidence level. For Fucus samples, benzo(a)pyrene was well correlated with higher-molecular-weight PAH isomers. However, in Fucus there was no statistically significant correlation between levels of benzo(a)pyrene and levels of phenanthrene, fluoranthene or benz(a)anthracene + chrysene.

Since levels of benzo(a)pyrene in marine samples appear to be highly correlated with levels of other high-molecular-weight PAH, it appears valid to use this compound as an indicator for the degree of PAH contamination of samples. Using benzo(a)pyrene measurements, we investigated the relationship between the degree of contamination of mussels, sediments, and seaweeds at different sites. Where duplicate or triplicate samples of one type were available from a location, the levels of carci-

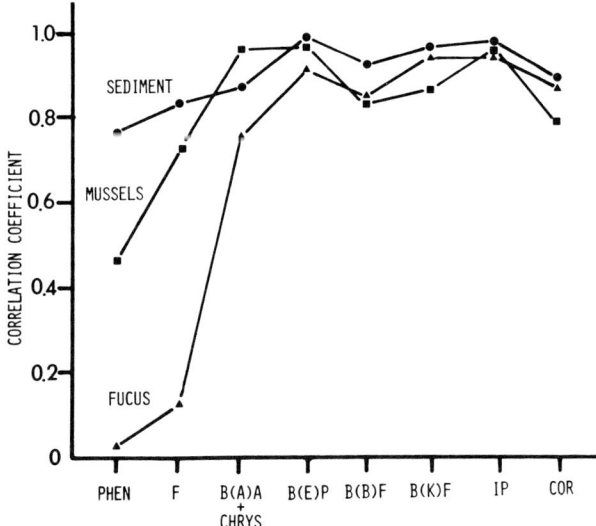

FIGURE 1. Correlation between benzo(a)pyrene and other PAH for three types of samples. PAH are listed left to right in order of increasing retention time in reversed-phase chromatography (increasing hydrophobicity). Abbreviations: PHEN, phenanthrene; F, fluoranthene; B(A)A, benz(a)anthracene; CHRYS, chrysene; B(E)P, benzo(e)pyrene; B(B)F, benzo(b)fluoranthene; B(K)F, benzo(k)fluoranthene; IP, indeno(123-cd)pyrene; COR, coronene. Correlation coefficients are statistically significant at $p = 0.01$ when greater than 0.575 (mussels, $n = 19$), 0.708 (sediments, $n = 12$), or 0.834 (Fucus, $n = 8$).

nogen were averaged. Levels in sediments were calculated on the basis of ash-free dry weight (that portion of the sample lost during ashing at 500°C) to help eliminate variability due to the different physical nature of the sediments (which ranged from coarse sand with less than 1 percent ashable organic material, to mud containing greater than 10 percent organic material).

Figure 2 shows a significant correlation between the levels of benzo(a)pyrene in sediments, and levels of benzo(a)pyrene in mussels from the same location. The one data point showing the most deviation from the relationship was at a site called Coal Harbour in Vancouver, which is in the immediate vicinity of a large pleasure boat marina and associated fueling barges. At this location (data from which are indicated with a distinctive symbol in Figures 2 to 4), levels of benzo(a)pyrene in mussels were higher than might be expected from the level of contamination of sediments. This may reflect direct contamination of the organisms by PAH arising from marina operations.

Figure 3 shows the relationship between levels of benzo(a)pyrene in sediments and levels in seaweed from the same area. In parallel with the

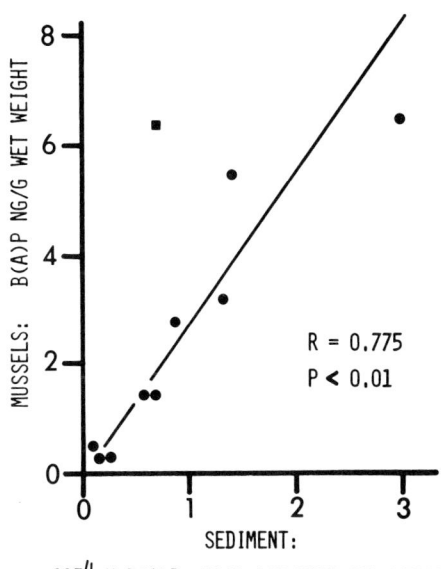

FIGURE 2. Relationship between benzo(a)pyrene levels in sediments and mussels. ■ Coal Harbour, ● Other Locations. B(A)P, benzo(a)pyrene.

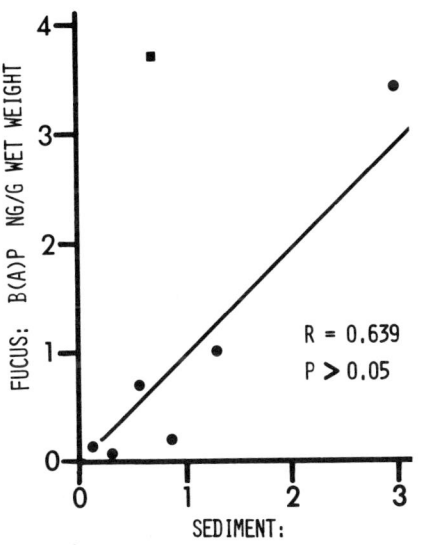

FIGURE 3. Relationship between benzo(a)pyrene levels in sediments and Fucus. ■ Coal Harbour, ● Other Locations. B(A)P, benzo(a)pyrene.

results from mussels, Fucus from the Coal Harbour site contained a disproportionately high level of carcinogen, relative to the level in sediments from that area. Including all data points, there was no statistically significant relationship between benzo(a)pyrene levels in sediments and Fucus; however, if the data from the Coal Harbour location are eliminated from the calculations as being atypical, the relationship becomes statistically significant.

Figure 4 demonstrates that there is a close relationship between the levels of benzo(a)pyrene in mussels and the levels in seaweed. On a wet weight basis, levels in Fucus are approximately half the levels in the tissues of mussels from the same location. In comparing the data from these two marine organisms, the data from the Coal Harbour location fit well with the trend established by the other locations.

A visual inspection of chromatograms of PAH isolated from sediments, mussels, and Fucus from different areas revealed no substantial differences in the types of PAH compounds encountered. This suggests that the harbour area studied either lacks point sources of PAH with distinctive fingerprints, or that if such sources exist, tidal mixing is sufficient to give an approximately uniform blend of PAH.

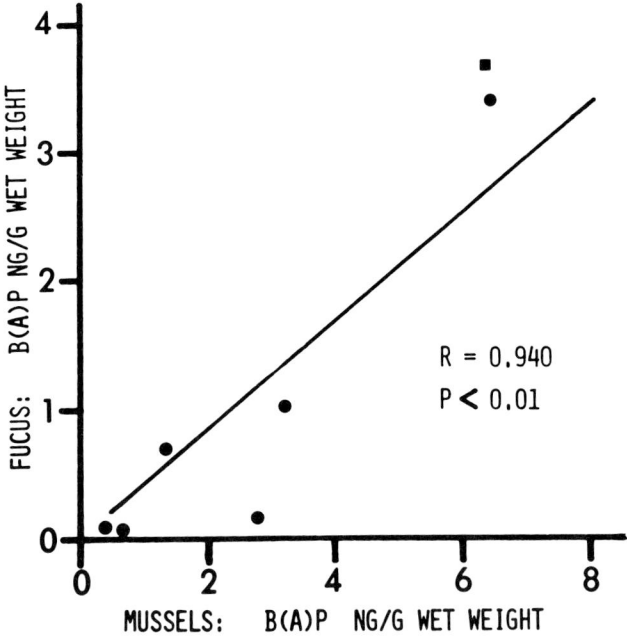

FIGURE 4. Relationships between benzo(a)pyrene levels in mussels and Fucus. ■ Coal Harbour, ● Other Locations. B(A)P, benzo(a)pyrene.

In comparing chromatograms from different sample types, there were some differences in the relative prevalence of different PAH isomers. This is demonstrated in Figures 5 and 6, which show data for seven PAH isomers ranging from fluoranthene (four fused rings) to coronene (seven fused rings). Levels of PAH in individual samples have first been normalized by dividing by the benzo(k)fluoranthene level of each sample. Where multiple samples of a particular type were available from a location, normalized data were then averaged. Finally, the normalized levels of individual PAH in mussels or Fucus were divided by the normalized level of the same isomer in sediments from the same location. Each data point represents the mussel/sediment or Fucus/sediment relative enrichment ratio of a single PAH isomer at a single location—lines on the figures join median values.

Figure 5 indicates that for mussels, with the exception of fluoranthene, as the molecular weight and hydrophobicity of PAH increase, there is a general trend towards lower levels in mussels, relative to levels in sediments. In contrast, for Fucus (Figure 6) the only general trend is over the range fluoranthene to benzo(b)fluoranthene, where there is a trend towards higher levels in organisms relative to sediment as the hydrophobicity of the PAH isomers increases. In addition to this trend, there appears also to be a selective deficit of benzo(a)pyrene in Fucus, relative to sediments.

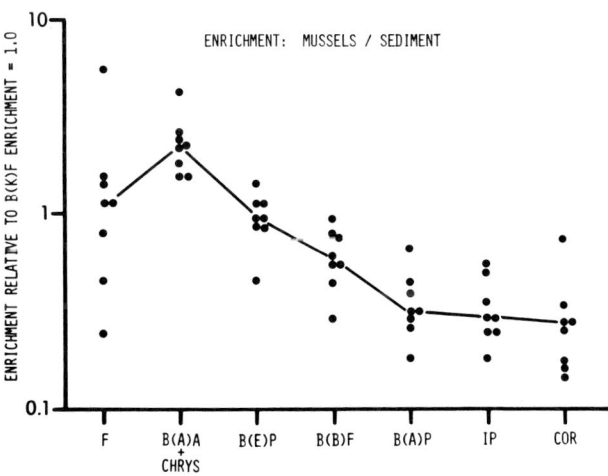

**FIGURE 5.** Relative ratios of PAH isomers in mussels, relative to sediments. For explanation of derivation of data, see text. Abbreviations: as for Figure 1 except B(A)P, benzo(a)pyrene.

## DISCUSSION

A statistically significant correlation was found for all sample types between benzo(a)pyrene and all other PAH isomers containing five or more fused rings. This correlation was extended to compounds as small as phenanthrene (three fused rings) for sediments, and fluoranthrene (four fused rings) for mussels. These data suggest that benzo(a)pyrene may be a good indicator for the presence of higher-molecular-weight, carcinogenic PAH in samples. The usefulness of benzo(a)pyrene as an indicator for lower-molecular-weight compounds appears to depend strongly on the sample type. It may be noted that although levels of phenanthrene and fluoranthrene in Fucus were not correlated with levels of benzo(a)pyrene, the levels of these two compounds were correlated with each other. This suggests that other PAH compounds may be found which are useful indicators for lower-molecular-weight isomers such as naphthalene, phenanthrene, anthracene, fluoranthrene, and pyrene.

There is evidence that the turnover of lower-molecular-weight PAH in mussels may be much more rapid than that of higher-molecular-weight compounds such as benzo(a)pyrene (4). Little is known about the half-lives of various PAH in contaminated seaweeds; however, it is unlikely that these organisms have any system for metabolizing PAH, and PAH are therefore most probably lost by partitioning into water. Lower-molecular-weight, more-water-soluble PAH would be expected to be lost much more rapidly from seaweed than higher-molecular-weight compounds. For both mussels and Fucus, a more rapid loss of lower-molecular-weight than higher-molecular-weight PAH after transient pollution incidents may explain why the levels of low-molecular-weight compounds such as phenanthrene are not well correlated with levels of benzo(a)pyrene.

In comparing different sample types from the same location, there appears to be a correlation between levels of PAH in sediments, and levels in marine organisms in the same area. Sediments are the major sink of PAH in coastal waters, and it is probably that the equilibrium of PAH in sediments with overlying water results in a "baseline" contamination of organisms at a level determined by the degree of contamination of sediments. This baseline level of PAH in organisms may be exceeded in areas where there is extensive discharge of PAH into water, such as at the Coal Harbour marina area noted in Figures 2 to 4. The data suggest that although sediments may be very useful in determining the minimum level to which organisms would be expected to be contaminated in a given area, only measurements of levels in the organisms themselves can give a true measure of how contaminated seafoods from an area would be.

Either in areas where sediments are the immediate source of PAH in organisms, or in areas where there are substantial primary souces of PAH, the pattern of PAH isomers in sediments is likely to be a relatively unbiased record of the fingerprint of PAH in an area. Accordingly, a comparison of relative levels of different PAH isomers in sediments and in marine organisms may be able to reveal characteristics of the uptake and retention of PAH by marine biota.

In comparison with sediments, both Fucus and mussels appear to be depleted in fluoranthene, relative to benz(a)anthracene/chrysene. This may reflect the fact that fluoranthene is approximately 20 times more water soluble than benz(a)anthracene, and 100 times more soluble than chrysene (6). As discussed above, lower-molecular-weight, more-water-soluble compounds may be lost more rapidly from mussels and Fucus than higher-molecular-weight isomers. Since the steady state level of a PAH isomer in an organism will reflect both its uptake rate and its half-life in the organism, rapidly discharged compounds would be expected to be depleted in an organism relative to their proportion in the PAH source which is contaminating the organism.

Although most pronounced for fluoranthene, the trend towards lower relative levels of lower-molecular-weight PAH in Fucus also may apply to benz(a)anthracene/chrysene, and benzo(e)pyrene (Figure 6). In addition,

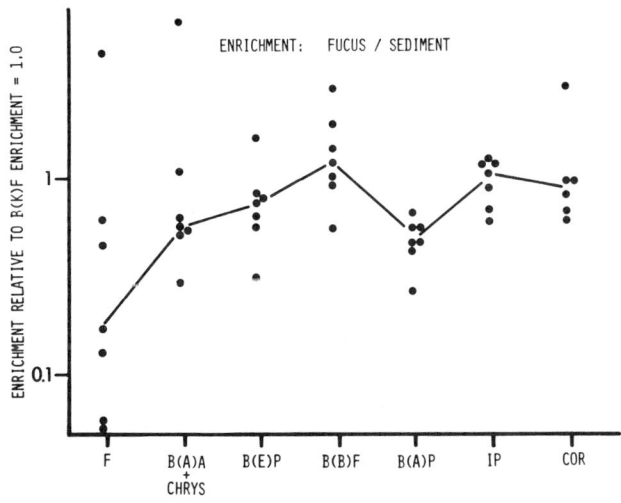

**FIGURE 6.** Relative ratios of PAH isomers in Fucus, relative to sediments. For explanation of derivation of data, see text. Abbreviations: as for Figure 1 except B(A)P, benzo(a)pyrene.

Fucus samples appear somewhat depleted in benzo(a)pyrene relative to other PAH isomers with five to seven fused rings. Preliminary and incomplete experiments in this laboratory suggest that benzo(a)pyrene may be more susceptible to photooxidation than other related PAH—the considerable exposure of seaweeds to light (necessary for photosynthesis) may result in a selective destruction of benzo(a)pyrene relative to other PAH.

The mechanisms by which seaweeds and bivalve mollusks might take up PAH from their environment are not well understood. However, two fundamentally different processes may be envisioned. In one, the PAH in sediments are attached to organic or inorganic particles, which are transferred whole to the receptor organism. In this case the organism would be expected to receive an unbiased sample of the PAH which can be measured in sediments. The other mechanism involves the partitioning of PAH from sedimentary particles into water, followed by their subsequent absorption or adsorption from water by organisms. In this latter case, the fingerprint of the PAH in the organisms would be expected to be enriched in those compounds which are most readily soluble, and transferred from particles into solution.

For seaweed samples, the pattern of enrichment of PAH relative to sediment is consistent with a nonselective uptake of PAH, with the pattern subsequently modified by the weathering away of lower-molecular-weight isomers, and possibly the selective photodestruction of benzo(a)pyrene. In view of the large surface area of seaweeds such as Fucus and the gelatinous coating of the surface, it is possible that the PAH measured in seaweeds result from sediment particles trapped on the surface, rather than PAH incorporated into the plant material itself.

For mussels, the general trend towards lower levels of higher-molecular-weight PAH, relative to levels in sediments, suggests an uptake mechanism which involves, at some point, the solution of PAH in water. Superimposed on this pattern, the depleted level of fluoranthene is probably a result of a competing phenomenon, namely the more rapid turnover and shorter half-life of lower-molecular-weight PAH in mussels.

## ACKNOWLEDGMENTS

This work was supported by National Health Research and Development Project grant number 610-1138-40, and a Research Scholar award to the author from the Department of National Health and Welfare, Canada. I thank R. Armour and J. Fee for their expert technical assistance.

## REFERENCES

1. Dunn, B. P. (1976): Techniques for determination of benzo(a)pyrene in marine organisms and sediments. *Env. Sci. Technol.* 10:1018–1021.
2. Dunn, B. P. (1980): Benzo(a)pyrene in the marine environment: analytical techniques and results. In: *Proceedings of the International Symposium on the Analysis of Hydrocarbons and Halogenated Hydrocarbons in the Aquatic Environment,* Plenum Publishing, N.Y., in press.
3. Dunn, B. P., and Fee, J. (1979): Polycyclic aromatic hydrocarbon carcinogens in commercial seafoods. *J. Fisheries Res. Board Canada.* 36:1469–1476.
4. Dunn, B. P., and Stich, H. F. (1976): Release of the carcinogen benzo(a)pyrene from environmentally contaminated mussels. *Bull. Env. Contam. Toxicol.* 15:398–401.
5. Giger, W., and Blumer, M. (1974): Polycyclic aromatic hydrocarbons in the environment: isolation and characterization by chromatography, visible, ultraviolet, and mass spectrometry. *Anal. Chem.* 46:1663–1671.
6. May, W. E., Wasik, S. P., and Freeman, D. H. (1978): Determination of the solubility behaviour of some polycyclic aromatic hydrocarbons in water. *Anal. Chem.* 50:997–1000.

# THE TISSUE HYDROCARBON BURDEN OF MUSSELS FROM VARIOUS SITES AROUND THE SCOTTISH COAST

P. R. Mackie, R. Hardy, K. J. Whittle, C. Bruce, and A. S. McGill

Ministry of Agriculture, Food and Fish
Torry Research Station
135 Abbey Road
Aberdeen, Scotland

## INTRODUCTION

The use of bivalves such as mussels (*Mytilus edulis*) to monitor various marine pollutants—heavy metals, transuranic elements, petroleum hydrocarbons and halogenated hydrocarbons—has been proposed (8) because of the alleged ability of bivalves to accumulate these and thus act as integrators of the input. An advantage of this approach, if successful, was that the repeated handling of large water samples for chemical analyses, with all its attendant problems, would become unnecessary, being replaced by a simpler tissue analysis. Analyses of animals was also considered more relevant to potential pollutant effects on marine biota.

A survey of pollutant levels around the United States coast using mussels and oysters (the so-called "mussel watch") was initiated in 1976; two previous but more limited surveys for organochlorine pesticides had already been completed (2,10).

A mussel watch has been initiated in U. K. waters and this paper reports the results of the analyses of hydrocarbon content and composition of animals collected around the coastline of Scotland.

## MATERIALS AND METHODS

### Materials

Following as far as possible the previous recomendations on mussel sampling (17), samples were collected from 27 stations around the Scottish coastline (Figure 1). At least 70 of the dominant-size individuals were selected; those that were intended for heavy metal and pesticide analyses (4) were allowed to depurate before freezing but the remainder which were to be used in this work were wrapped in clean aluminium foil and frozen immediately using solid carbon dioxide. A minimum of 12 animals per station were used in the analyses. After shucking, the tissue was divided into two fractions: (a) muscle and (b) green gland and gut; the analyses were performed on homogenous samples of the minced fractions.

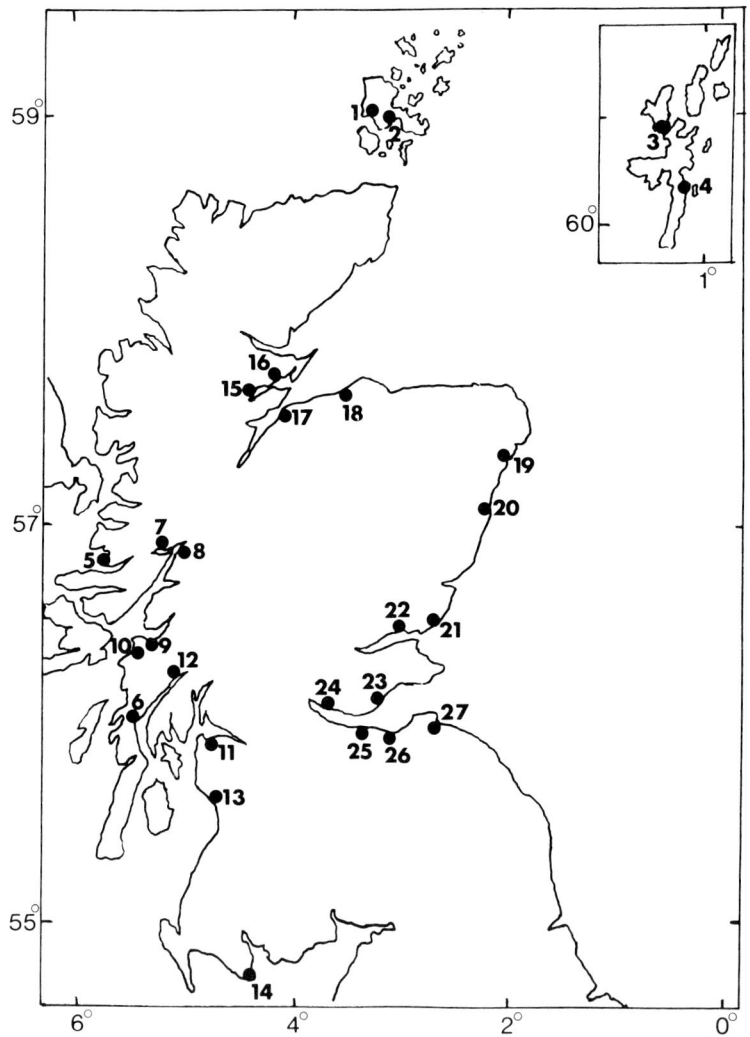

FIGURE 1. Location of sampling stations around the Scottish Coast; for station identification see Table 2.

## Methods

Precautions were taken to prevent adventitious contamination (12,13). Tissue samples (approximately 10 g) were extracted with chloroform:methanol (1) and the hydrocarbon fractions were isolated from the lipid extract by the following modification of the method of Giger and Schaffner (7). First the $n$-alkanes were isolated by silicic acid chromatography (12) and then the PAH were separated from the residual lipids by fractionation on a Sephadex LH-20 column (7). The fraction containing the PAH, i. e., the 50 to 100 ml eluant (methanol:benzene), was concentrated on a rotary film evaporator, dissolved in pentane, and applied to a second silicic acid column (7). This column was then eluted with 25 ml pentane followed by 50 ml of pentane:benzene (1:1), yielding a clean PAH fraction. This modification has two advantages over that originally specified (7) in that the alkanes are isolated intact instead of distributed over both the 0 to 50 and 50 to 100 ml Sephadex fractions and the use of pentane:benzene (1:1) produces a PAH fraction free of sterols. When methylene chloride is used as specified in the original method, sterols coelute with the PAH fraction.

Gas chromatography (GC) was performed using columns and conditions given by Mackie et al (14). PAH were analyzed quantitatively on an SE-52 glass capillary column (20 m) fitted into a Pye 104 gas chromatograph coupled to a VG Micromass 16F mass spectrometer. The source was operated in the electron impact mode and the molecular ions (or the M-1 or M-15 ions in the case of the methyl and dimethyl naphthalenes, respectively) of the compounds listed in Table 1 were determined by selective ion recording.

TABLE 1. Reproducibility of the PAH Standards in the GC/MS System

|  | Range[a] | Mean | Variance (calculated from range) |
|---|---|---|---|
| Naphthalene | 16.5-18.7 | 17.7 | 5 |
| Methyl naphthalene | 6.6-7.2 | 6.9 | 4 |
| Dimethyl naphthalene | 5.1-5.4 | 5.3 | 2 |
| Dibenzothiophene | 10.4-11.0 | 10.7 | 2 |
| Phenanthrene | 13.4-14.7 | 14.1 | 4 |
| Methyl anthracene | 17.2-18.0 | 17.6 | 2 |
| Fluoranthene | 17.8-18.7 | 18.3 | 2 |
| Pyrene | 22.1-23.1 | 22.7 | 2 |
| Benzanthracene | 4.2-4.5 | 4.3 | 3 |
| Benzo (a) pyrene | 5.8-6.1 | 5.9 | 2 |

[a] Area response per ng of compound for five injections; 20 ng injected at a sensitivity of $10^{-6}$ AFS.

## RESULTS AND DISCUSSION

Prior to introducing the results it seems appropriate to consider the accuracy of the methods employed. This has already been reported for the aliphatic hydrocarbons (12).

An assessment of the extraction and column fractionation procedures was conducted with $^{14}$C benzo(a)pyrene and it was found that 94 percent of the BaP added to tissues could be recovered. Almost all the loss of standard occurred in the final silicic acid column fractionation. Known concentrations of mixtures of PAH were then analyzed by GC/MS and the response and reproducibility determined (Table 1). The response varied from component to component, being much lower for benzanthracene and benzpyrene than for the other aromatics determined by molecular ion. As reproducibility was good, i. e., in no instance did the variance exceed 5 percent, no simple explanation can be advanced for response variability. Other experiments indicated that column absorption was not a factor here and certainly the reproducibility of the results gives some support to this. As a safeguard, after every five sample analyses a calibration run was always carried out. The response for the standards stayed remarkably constant during the course of this work.

### Paraffinic Hydrocarbons

Although the emphasis of this paper is directed towards the PAH, a brief outline of the alkane composition is presented to provide a wider understanding of hydrocarbon accumulation in the mussel. The detailed results will be discussed elsewhere.

The alkane distributions with respect to carbon number fall into four categories as shown in Figure 2. Condensed data from stations are given in Table 2.

The results from the muscle tissue are very scattered and confusing for both total alkane and phytane concentrations and also $n$-paraffin distributions. There is a tendency for high values to be associated with the more intensely populated industrialized areas, but the highest values of all were obtained from areas expected to be clean. The results for visceral tissues (green gland and gut) are much more in line with predicted input although it must be recalled that the animals were not allowed to depurate prior to freezing. Thus, any oily particles adhering to the mucus or the food in the gut would have influenced the analyses. The clean and urban sites both give mean values of 3.0 $\mu$g/g, with ranges of 0.9 to 7.1 and 0.4 to 5.7, respectively. Industrial sites have a much higher mean of 8.0 $\mu$g/g and range of 3.7 to 14.1, overlapping with the clean and urban sites. Phytane

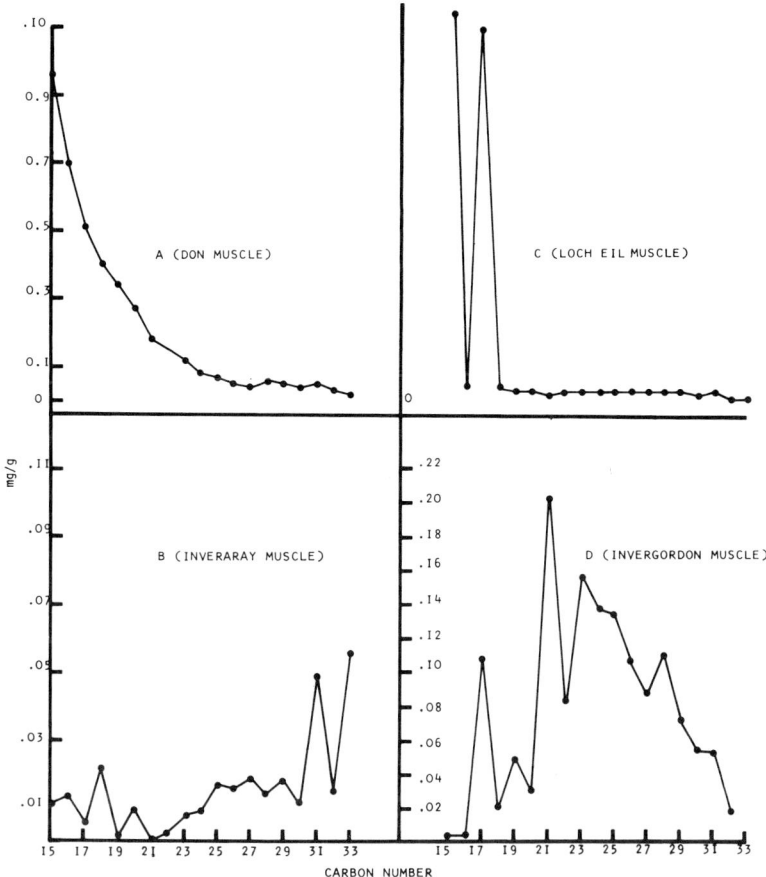

FIGURE 2. The four general types of *n*-alkane distributions found: (a) oil (Don muscle), (b) sediment (Inveraray muscle), (c) phytoplankton (Loch Eil muscle), and (d) unknown (East Invergordon muscle).

shows a similar concentration overlap: the range is 0.1 to 6.7 for industrial sites, $>.05$ to 2.4 for urban sites, 0.1 to 0.5 for clean sites, and the means are 1.7, 0.6, and 0.3, respectively.

With the exception of station 2 (Waulkmill Bay) all mussels having an "unweathered oil" type alkane distribution (Figure 2a) are in areas of industrial activity, although the reverse does not always hold true; e. g., South Queensferry (station 25) in the Firth of Forth lies in the central industrial belt of Scotland but does not possess a readily ascribable alkane profile. As indicated in the table, the alkane composition was complex in most instances, almost certainly indicating a variety of inputs, sometimes with one or other predominating.

TABLE 2. Station Identification, Description, and Condensed Alkane Data (in μg/g wet tissue)

| Station | Position | Station Type | 15 Σ 33 A[a] | 15 Σ 33 B[b] | Pristane A | Pristane B | Phytane A | Phytane B | Suspected Alkane Source/Sec (from Figure 2) |
|---|---|---|---|---|---|---|---|---|---|
| 1 | The Bush | Clean | 0.9 | 4.7 | 0.1 | 0.5 | Tr[d] | 0.2 | P[e] |
| 2 | Waulkmill Bay | " | 0.7 | 5.1 | 0.6 | 0.5 | 0.1 | 0.4 | O[f] |
| 3 | Orka Voe | " | 0.4 | 1.4 | 0.1 | 0.7 | 0.1 | 0.5 | N/K[g] |
| 4 | Brei Wick | " | 5.5[c] | — | 1.0 | — | 1.5 | — | N/K |
| 5 | Arisaig | " | 9.2 | 2.0 | 1.3 | 0.2 | 0.8 | 0.1 | O P |
| 6 | Ardrishaig | " | — | — | — | | — | | |
| 7 | Loch Eil | " | 0.7 | 2.6 | 0.1 | 0.6 | 0.1 | 0.2 | S P |
| 8 | South Fort William | Urban | 0.4 | 2.3 | Tr[f] | 0.1 | Tr | 0.1 | P S |
| 9 | Dunstaffnage | " | 0.9 | 4.5 | 0.1 | 1.0 | Tr | 0.1 | P O |
| 10 | Oban | " | 2.6 | 5.5 | 0.2 | 3.3 | 0.1 | 2.4 | P |
| 11 | Gourock | Industrial | 0.7 | 12.6 | Tr | 22.7 | Tr | 6.7 | O S |
| 12 | Inveraray | Urban | 0.3 | 1.5 | 0.1 | 0.8 | Tr | 0.1 | P S |
| 13 | Irvine | Industrial | 1.9 | 14.1 | 0.5 | 3.0 | 0.1 | 2.4 | O N/K |
| 14 | Whithorn | Clean | 1.3 | 7.1 | 0.3 | 0.2 | 0.2 | 0.2 | P N/K |
| 15 | Dingwall | Urban | 0.4 | 2.1 | Tr | 1.2 | Tr | 0.1 | S |
| 16 | East Invergordon | Industrial | 1.5 | 4.4 | 0.4 | 3.1 | 0.1 | 0.4 | P N/K |
| 17 | West Inverness | Urban | 1.5 | 5.7 | 0.2 | 0.3 | 0.1 | 0.3 | P O |
| 18 | Burghead | " | 0.5 | 2.0 | 0.1 | 0.8 | 0.1 | 0.5 | |
| 19 | Cruden Bay | " | 1.3[c] | | 0.1 | | 0.1 | | N/K |
| 20 | Don Estuary | Industrial | 4.1 | 5.4 | 0.5 | 0.9 | 0.4 | 0.8 | O |
| 21 | Carnoustie | Urban | 0.3 | 1.8 | 0.1 | 0.2 | Tr | 0.1 | S P N/K |
| 22 | Stannergate | Industrial | 0.9 | 4.3 | 0.2 | 0.8 | 0.1 | 0.6 | O P |
| 23 | Kinghorn | Urban | 0.6 | 1.0 | Tr | Tr | Tr | Tr | N/K |
| 24 | Culross | " | 1.1 | 2.7 | Tr | 4.8 | Tr | 0.5 | P N/K S |
| 25 | South Queensferry | " | 1.1 | 0.4 | Tr | 0.1 | Tr | Tr | P/S/NK |
| 26 | Joppa | Industrial | 1.5 | 3.7 | 0.1 | 0.1 | Tr | 0.1 | O/S/P |
| 27 | Tyne Mouth | Clean | — | 1.5 | — | 0.6 | — | 0.3 | N/K |

[a]Muscle.
[b]B-green gland and gut.
[c]Whole animal.
[d]Tr = < 0.05 μg/g.
[e]P phytoplankton.
[f]O - oil.
[g]N/K unknown.
[h]S - sediment.

These results give some support to the view that the mussel alkanes do provide an indication of recent or continuing pollution, although mussels are known to depurate paraffinic hydrocarbons rapidly(3).

## Polynuclear Aromatic Hydrocarbons

A typical ion chromatogram is presented in Figure 3. The mass spectrometer as used could be tuned to only four ions per run, so that

Figure 3 is a composite of two GLC runs. The aromatic distributions from the whole-mussel tissue are presented in Table 3, and their percentage distribution by ring size is shown in Figure 4. The samples were analyzed as muscle and viscera but, to present a more concise picture, these two sets of results have been combined in the appropriate ratios to give data for the whole animal. It is appreciated that Figure 4 does not give a complete picture since several compounds known to be present for example

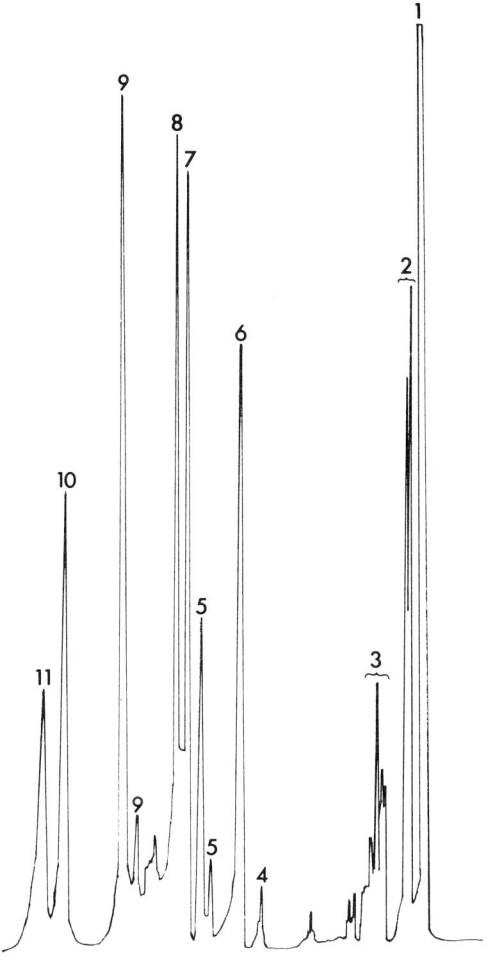

FIGURE 3. Composite ion chromatogram of PAH on SE 52 glass capillary column (20 m x 0.3 mm I. D.) temperature programmed at 5°/min from 100 to 260°C. Peaks: 1, naphthalene; 2, methyl naphthalenes; 3, dimethylnaphthalenes; 4, dibenzothiophene; 5, $^{C}1$ phenanthrenes; 6, phenanthrene; 7, fluoranthene; 8, pyrene; 9, benzanthracenes/phenanthrenes; 10, benzofluoranthenes; 11, benzo(a)pyrene.

TABLE 3. PAH Content of Whole Mussel in ng/g (Wet Tissue)[a]

| Station | Naphthalene | Methyl Naphthalenes | C2 Naphthalenes | Phenanthrene | Methyl Phenanthrenes/ Anthracenes | Pyrene | Fluoranthene | Benzanthracenes/ Phenanthrenes | Benzo(a)pyrene | Total | Dibenzothiophene[b] |
|---|---|---|---|---|---|---|---|---|---|---|---|
| 1 | 10 | 20 | 34 | 30 | 6 | 3 | 2 | 1 | 1 | 107 | 1 |
| 2 | 2 | 9 | 8 | 16 | 8 | 6 | 27 | 45 | 13 | 134 | 3 |
| 3 | TR | 2 | 7 | 9 | 49 | 8 | 13 | 25 | 6 | 119 | 3 |
| 4 | 3 | 11 | 30 | 11 | 38 | 10 | 6 | 14 | 13 | 136 | 2 |
| 5 | 3 | 16 | 27 | 7 | 4 | 2 | 5 | 3 | 1 | 68 | 4 |
| 6 | 1 | 5 | 12 | 15 | 9 | 6 | 10 | 8 | 5 | 71 | 3 |
| 7 | 1 | 2 | 5 | 18 | 12 | 161 | 201 | 514 | 255 | 1169 | 2 |
| 8 | 14 | 29 | 53 | 42 | 19 | 158 | 198 | 363 | 236 | 1112 | 3 |
| 9 | 2 | 6 | 10 | 8 | 4 | 7 | 11 | 4 | 2 | 54 | 1 |
| 10 | 4 | 7 | 47 | 32 | 31 | 9 | 12 | 31 | 22 | 195 | 3 |
| 11 | 1 | 4 | 29 | 147 | 214 | 283 | 269 | 627 | 329 | 1903 | 24 |
| 12 | TR | 1 | 9 | 12 | 6 | 3 | 6 | 14 | 13 | 64 | 3 |
| 13 | 1 | 15 | 106 | 621 | 197 | 540 | 476 | 710 | 137 | 2803 | 42 |
| 14 | 2 | 4 | 7 | 33 | 18 | 78 | 85 | 47 | 12 | 286 | 2 |
| 15 | 1 | 2 | 11 | 22 | 18 | 75 | 92 | 110 | 29 | 360 | 2 |
| 16 | 1 | 2 | 6 | 335 | 5 | 280 | 438 | 130 | 40 | 1237 | 1 |
| 17 | 196 | 44 | 50 | 8 | 63 | 5 | 7 | 5 | 22 | 400 | 28 |
| 18 | TR | 1 | 3 | 9 | 7 | 16 | 26 | 28 | 15 | 105 | 1 |
| 19 | TR | TR | 2 | 43 | 12 | 46 | 49 | 32 | 9 | 193 | 1 |
| 20 | 11 | 41 | 154 | 30 | 94 | 7 | 8 | 10 | 5 | 360 | 16 |
| 21 | 1 | 3 | 17 | 4 | 22 | 8 | 12 | 25 | 13 | 109 | 2 |
| 22 | 1 | 10 | 82 | 123 | 150 | 63 | 76 | 127 | 78 | 710 | 8 |
| 23 | TR | 1 | 3 | 15 | 7 | 29 | 35 | 30 | 9 | 129 | 1 |
| 24 | TR | 1 | 4 | 10 | 5 | 154 | 94 | 168 | 55 | 491 | 1 |
| 25 | TR | 2 | 4 | 7 | 3 | 16 | 11 | 34 | 35 | 112 | TR |
| 26 | 7 | 25 | 70 | 28 | 20 | 25 | 32 | 24 | 18 | 249 | 7 |
| 27 | 7 | 18 | 42 | 17 | 19 | 13 | 31 | 13 | 6 | 166 | 8 |

[a] For station identification see Figure 1, Table 2.
[b] The values for dibenzothiophene are not included in the total value.

dimethylphenanthrene and bnezofluoranthene were not determined. However, apart from the data for the five-ring system, which would be approximately doubled by the addition of benzofluoranthene data, the figure should give a reasonable idea of the PAH distribution with respect to ring size. None of the distribution patterns closely resemble crude oil in which the composition of the ring systems is 2>>3>>4> 5(16); however, many did possess relatively large quantities of the four-ring components, which is what one would expect if the major source is from the pyrolysis of organic compounds. At station 20 the composition had some similarities

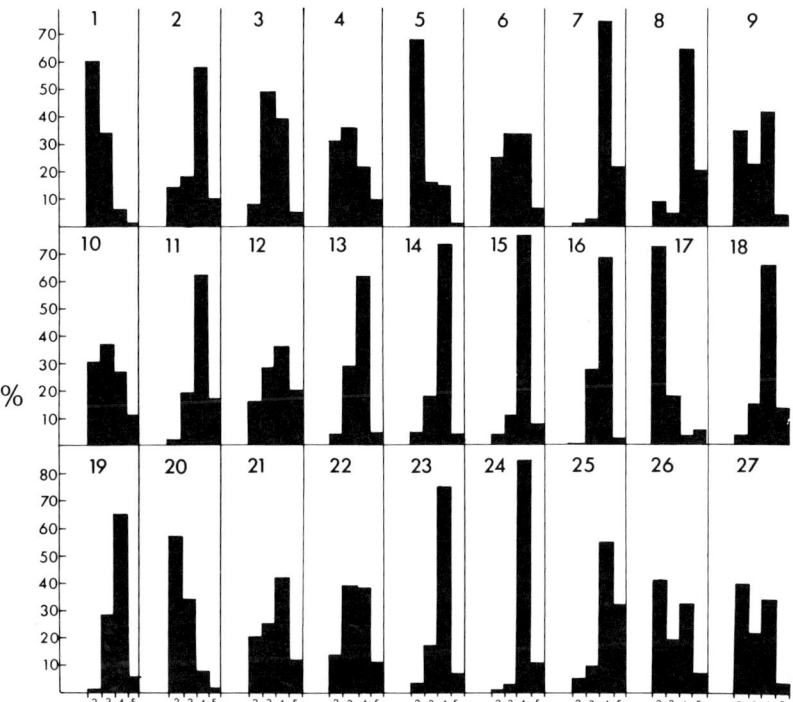

**FIGURE 4.** Percentage distribution of total PAH by ring size; for station identification see Table 2.

with crude oil and it is interesting to note that the alkane pattern of these mussels also closely resembled that of fresh oil. It was difficult to discern any similar correlations in the other samples. As with the *n*-alkanes, the best agreement between distribution and concentration of PAH with population density was observed in the results from the green gland and gut. However a similar but more variable relationship was preserved in the combined results (Table 3), the green gland and gut constituting around 20 percent of the total tissue. The PAH level in muscle relative to green gland and gut was similar to that observed for paraffinic hydrocarbons in fish muscle relative to liver (1:10) (12), the corresponding ratio for the mussel *n*-alkanes was 1:5.

The total PAH concentrations in mussels from the clean or sparsely populated sites ranged from 54 to 136, with a mean of 109 ng/g. Station 14 (Isle of Whithorn), although sparsely populated, contained three times the mean value and has been treated as aberrant. Similarly, station 7 (Loch Eil) in a remote area had a PAH concentration approximately tenfold higher than background (1169 ng/g), but it was noted during collection that the adjacent road had been asphalted some 2 weeks prior to sampling. It seemed not unreasonable to assume that the high concentration of PAH derived from runoff from the road. This possible source of PAH could also exist for the majority of the stations since most are near roadways (although not recently asphalted). An examination of fresh road asphalt of the type used in Scotland showed that the four-ring system was indeed the dominant PAH group present (Figure 5a). However, pyrene proved to be the main constituent and not benzanthracene/benzphenanthrene or fluoranthene (Figure 5b) as noted in the majority of the mussel samples.

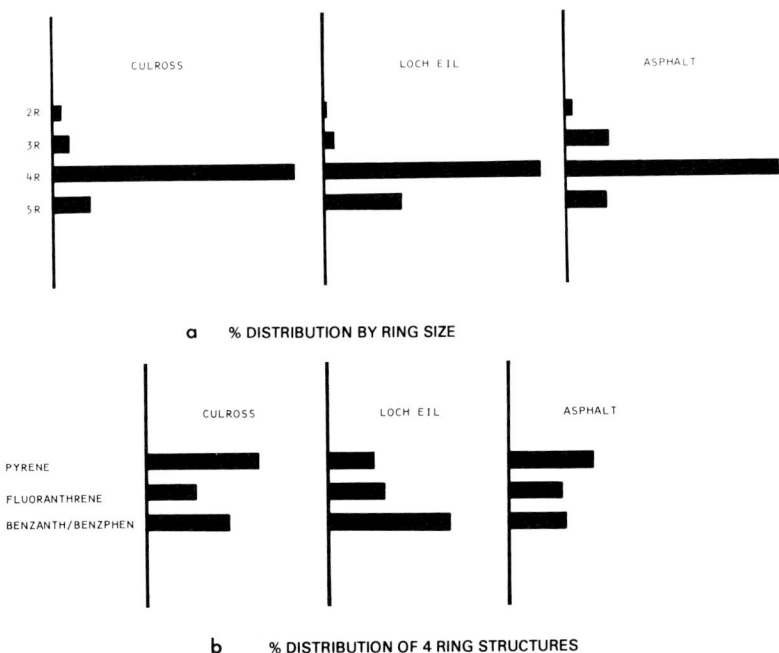

FIGURE 5. (a) Percentage distribution of PAH by ring size in mussels close to roadways compared with asphalt; (b) percentage distribution of the four-ring structures.

Differences have been noted in the uptake/depuration of benzanthracene and fluoranthene in oysters (11) and benzanthracene, benzo(a)pyrene in herring (21), and this may account for the results obtained here if indeed road water runoff is a major source of aromatic hydrocarbons. It is more likely that airborne particulate matter which has been shown to contain large amounts of chrysene/benzanthracene relative to pyrene could be the source (19). This mussel sample (Loch Eil) has not been categorized as a clean site mussel and is hence excluded from the composite data.

Samples from heavily industrialized regions such as the Forth and Clyde valleys contained 260 to 2803 (mean 1127) ng/g while smaller centers of population with no heavy industry contained 112 to 360 ng/g (mean 221 ng/g).

Comparison of the results from this survey with the original mussel watch data (8) is not possible since that study simply stated whether or not the sample was thought to contain oil, and no figures for the hydrocarbon content were given.

A survey carried out around the coast of England showed "oil" concentrations on the order of 6 to 150 $\mu$g/g (Law, unpublished data), the oil content being determined by fluorescence spectroscopy using an Ekofisk crude oil as the calibration standard. It is difficult to compare these results with those found in the present work because of the nonspecific nature of the fluorescence method; nevertheless, if it is assumed that the $n$- alkanes measured comprise ~10 percent of the crude oil, then the range of concentrations found (3 to 140 $\mu$g/g "oil") was very similar. In the English survey an agreement between hydrocarbon content and population density was noted, similar to the results presented here and those of our colleagues who determined heavy metals (4) on a survey of the Scottish coast.

It is relatively easy to make such general observations on relationships between hydrocarbon content and suspected input, but it is much more difficult to attempt to ascribe a source to the suite of PAH found. There are a number of potential sources of PAH to the marine environment. In the light of recent knowledge, marine biogenic production seems unlikely (9). The likely sources are the products of incomplete combustion and oil or oil products, and the differences between the two suites of PAH arising from these sources have been discussed in detail elsewhere, especially with regard to sedimentation (22). These hydrocarbons are available to the mussel in two forms, either dissolved or attached to particulate material present in the sea, but there is disagreement on whether or not some of these, especially naphthalenes, are available when adsorbed onto particulates (6,18). One would also expect a fractionation of the PAH in the water, with the lower members being enriched in the soluble fraction which would become more or less available to the animal, depending upon whether the soluble or particulate material represents the major source of acquisition.

Differences in the uptake and depuration of the various hydrocarbon structures by bivalves have been noted (11, 15, 20). An enrichment factor of 15:1 for methyl naphthalenes relative to the $C_{12-20}$ paraffins has been reported for oysters exposed to crude oil (20). Depuration also differs according to ring size; thus, Lee et al (11), in work on oysters, found a biological half-life of 2 days for naphthalene compared with 18 days for benz(a)pyrene. This persistence of high-molecular-weight aromatics has been noted by other workers (5). If the same factors operate in mussels, it is unlikely that the aromatic composition of the mussels would closely reflect that of the environment.

## Laboratory Experiments

In order to study uptake and depuration, mussels were exposed to silica particles (5 $\mu$) coated with a mixture of weathered Iranian and Arabian crude oils for 24 hours, followed by 24 hours depuration.

The green gland and gut showed a slight increase in total alkanes after 24 hours' exposure and a still further increase after the depuration (0.9 to 1.4 to 3.0); furthermore, the composition of the alkanes after exposure was reminiscent of the crude oil. Conversely, the muscle seemed relatively unaffected both in content (1.1 to 0.4 to 1.3) and composition of the alkanes.

The results for PAH are presented in Figure 6. An increase was noted in the concentration of the naphthalenes, especially the dimethyl naphthalenes and dibenzothiophene, both relatively major components of the oil, together with a reversal of the phenanthrene to the methyl phenanthrenes ratio. The lower concentrations observed for all compounds except the naphthalenes in the day one sample cannot be explained; the use of 12 animals per sample would seem to preclude the possibility that this was caused by biological variation. The results however are in broad agreement with those obtained by Lee et al (11) in experiments with oysters exposed to a crude oil enriched with aromatics. Analysis of a more detailed experiment is under way to clarify some of the anomalies raised by this preliminary experiment.

## CONCLUSIONS

Mussels from around the coast of Scotland have, on the whole, higher concentrations of $n$-alkanes and PAH when obtained from sites close to populous or industrial areas than those from more pristine areas. From this point of view, analysis of these animals does appear to provide a qualitative indication of input. The content and composition of hydrocarbons in the mussels almost certainly represents a balance between uptake, deposition, and depuration of individual hydrocarbon components. All work to date suggests that mussels take up oil components rapidly but have also a fairly rapid depuration, especially of the alkanes and lower aromatics, the higher PAH being considerably more persistent. Thus it is unlikely that mussels can be used as quantitative integrators of pollution by oil. The measurement of some of the components that are not readily depurated,

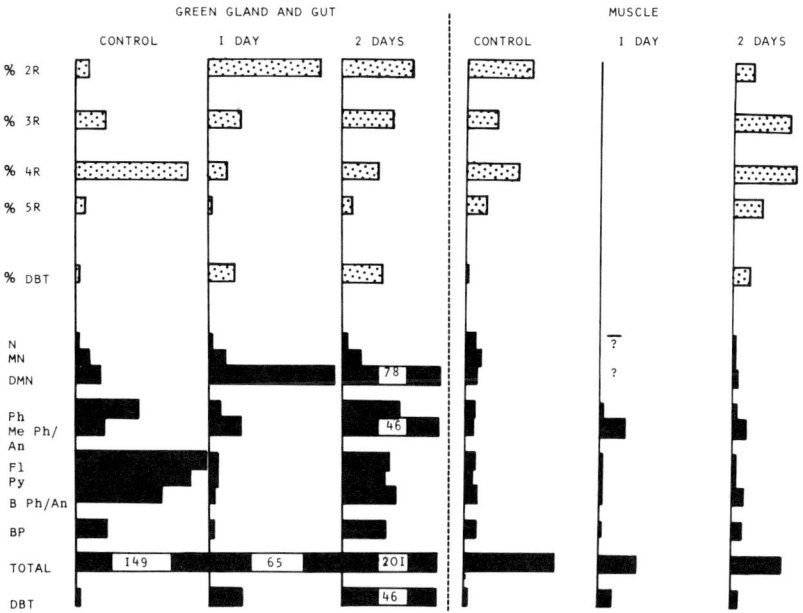

**FIGURE 6.** Laboratory exposure of mussels to oil-coated silica. Upper: percentage distribution with respect to ring size; lower: amounts of individual components in ng/g.

such as the four- and five-ring systems, may serve as indicators of past exposure, but before being sure of this, additional laboratory studies on their uptake and depuration will have to be performed. How quantitative such indications of oil will be and how representative the measurements are for other biota in that area remain unanswered. Until answers are obtained, the value of a continuing mussel watch program with regard to oil would not seem to justify its expense.

## ACKNOWLEDGMENTS

Staff of the Aberdeen Marine Laboratory of the Department of Agriculture and Fisheries for Scotland collected the mussel samples.

## REFERENCES

1. Bligh, E. G., and Dyer, W. J. (1959): A rapid method of total lipid extraction and purification. *Can. J. Biochem. Physiol.* 37: 911-917.
2. Butler, P. A. (1973): Residues in fish, wildlife and estuaries. *Pest. Monit. Bull.* 6: 238-262.

3. Clark, R. C., and Finley, J. S. (1975): Uptake and loss of petroleum hydrocarbons by the mussel. *Mytilus edulis*, in laboratory experiments. *Fishery Bull.* 73(3): 508-515.
4. Davies, I. M., and Pirie, J. M. (1978): Trace metals in mussels from the Scottish Coast. I.C.E.S. CM/1978/E33. Marine Environmental Quality Committee.
5. DiSalvo, L. H., and Guard, H. E. (1975): Hydrocarbons associated with suspended particulate matter in San Francisco Bay waters. In: *Proceedings of the 1975 Conference on Prevention and Control of Oil Pollution*, American Petroleum Institute, Washington, D. C. pp. 169-173.
6. Fossato, V. U., and Canzonier, W. J. (1976): Hydrocarbon uptake and loss by the mussel *Mytilus edulis*. *Mar. Biol.* 36: 243-250.
7. Giger, W., and Schaffner, C. (1978): Determination of polycyclic aromatic hydrocarbons in the environment by glass capillary gas chromatography. *Anal. Chem.* 50: 243-249.
8. Goldberg, E. D., Bowen, V. T., Farrington, J. W., Harvey, G., Martin, J. H., Parker, P. L., Riseborough, R. W., Robertson, W., Schneider, E., and Gamble, E. (1978): The mussel watch. *Environ. Conserv.* 5(2): 101-125.
9. Hase, A., and Hites, R. A. (1976): On the origin of polycyclic aromatic hydrocarbons in recent sediments: biosynthesis by anaerobic bacteria. *Geochim. Cosmochim. Acta* 40: 1141-1143.
10. Holden, A. V. (1973): International cooperative study of organochlorine and mercury residues in wildlife, 1969-1971. *Pest. Monit. Bull.* 7: 37-52.
11. Lee, R. F., Gardner, W. S., Anderson, J. W., Blaylock, J. W., and Barwell-Clarke, J. (1978): Fate of polycyclic aromatic hydrocarbons in controlled ecosystem enclosures. *Environ. Sci. Technol.* 12: 832-838.
12. Mackie, P. R., Whittle, K. J., and Hardy, R. (1974): Hydrocarbons in the marine environment. I. n-Alkanes in the Firth of Clyde. *Est. Coast. Mar. Sci.* 2: 359-374.
13. Mackie, P. R., Hardy, R., and Whittle, K. J. (1977): Sampling and extraction methods and their associated problems. In: *Petroleum Hydrocarbons in the Marine Environment*, A. D. McIntyre and K. J. Whittle, Eds. *Rapp. Proc.-Verb. Réun. Cons. Internat. L'Explor. Mer.* 171: 27-32.
14. Mackie, P. R., Hardy, R., and Whittle, K. J. (1978): Preliminary assessment of the presence of oil in the Ecosystem at Ekofisk after the blow-out, April 22-30, 1977. *J. Fish. Res. Bd Can.* 35: 544-551.
15. Neff, J. M., Cox, B. A., Dixit, D., and Anderson, J. W. (1976): Accumulation and release of petroleum-derived aromatic hydrocarbons by four species of marine animals. *Mar. Biol.* 38: 279-289.

16. Pancirov, R. J. (1974): Compositional data on API Reference oils used in biological studies: A No. 2 fuel oil, a Bunker C, Kuwait crude oil and South Louisiana crude oil. American Petroleum Institute Report No AID. IBA 74, Washington, D.C.
17. Phillips, D.J.H. (1976): The common mussel *Mytilus edulis* as an indicator of pollution by zinc, cadmium, lead and copper. 1. Effects of environmental variables on uptake of metals. *Mar. Biol.* 38: 59-69.
18. Roesijadi, G., Woodruff, D. L., and Anderson, J. W. (1978): Bioavailability of naphthalenes from marine sediments artifically contaminated with Prudhoe Bay crude oil. *Environ. Poll.* 15: 223-229.
19. Van Vaeck, L., and Van Cauwenberghe, K. (1978): Cascade impactor measurements of the size distribution of the major classes of organic pollutants in atmospheric particulate matter. *Atmos. Environ.* 12: 2229-2239.
20. Vaughan, B. E. (1973): Effects of oil and chemically dispersed oil on selected marine biota — a laboratory study. *Am. Pet. Inst. Publ.* 4191, Washington D. C.
21. Whittle, K. J., Murray, J., Mackie, P. R., Hardy, R., and Farmer, J. (1977): Fate of hydrocarbons in fish. *Rapp. Proc. Verb. Réun. Cons. Internat. Explor. Mer.* 171:139-142.
22. Youngblood, W. W., and Blumer, M. (1975): Polycyclic aromatic hydrocarbons in the environment: homologous series in soils and recent marine sediments. *Geochim. Cosmochim. Acta* 39: 1303-1314.

# SORPTION PROPERTIES OF POLYNUCLEAR AROMATIC HYDROCARBONS AND SEDIMENTS: HETEROCYCLIC AND SUBSTITUTED COMPOUNDS

J. C. Means*, J. J. Hassett**, S. G. Wood***, W. L. Banwart**, S. Ali***, and A. Khan**

*Chesapeake Biological Laboratory
University of Maryland
Solomons, Maryland 20688

**Department of Agronomy
University of Illinois
Urbana, Illinios 61801

***Department for Environmental Studies
University of Illinois
Urbana, Illinois 61801

## INTRODUCTION

Polynuclear aromatic hydrocarbons (PAH) have long been recognized as components of the waste streams of fossil fuel combustion and also of advanced processes such as the conversion of coal, oil shale, and tar sands into acceptable liquid and gaseous fuels (13,34,39). These pollutants enter the environment as gaseous, liquid, and solid wastes and are of concern to environmental scientists because of their toxicity and carcinogenicity (6,39).

Numerous papers have detailed the occurence of PAH and heterocyclic and substituted PAH in water and sediments (7,14,28,37). Sorption/desorption equilibria are important parameters in assessing the environmental fate and impact of organic pollutants in aquatic systems. The biological effects of organic pollutants have been shown to be greatly reduced by sorption to a sediment (31). However, few experimental data exist on the sorption properties of PAH on representative sediments (20,23,29,30,36).

The sorption of organic materials by sediments and soils is dependent on the nature of the adsorber and the adsorbate. Sorption occurs at a water-sediment interface when the forces of attraction between the sorbing compound and the substrate surface are greater than the combination of the repulsive forces between sorbate and the sorbant (41) and the forces of attraction between the sorbate and the solvent (16,22). The sorption of organic molecules by a soil or sediment has been shown to be related to a number of soil properties, including organic carbon content, clay type and content, cation exchange capacity, and pH (2,10,27,35,38,40). The chemical nature of the compound being sorbed also has an influence on the amount of adsorption (3,4,9,32).

Several investigators have reported that soil organic matter is the primary factor influencing the sorption of hydrophobic herbicides (12,15,18). Karickhoff et al (23) reported a significant positive correlation between the linear Freundlich partition coefficients (Kp) of several PAH and the organic carbon content (oc) of three local sediments:

$$Cs = KpCw^{1/n} \qquad [1]$$

where Cs equals the equilibrium concentration of the compound in the solid phase calculated on a dry weight basis; Cw equals the equilibrium solution concentration of the compound, Kp is the partition coefficient (constant), and $1/n$ is a constant related to the surface adsorption capacity of the solid phase.

They further determined that when these linear Kp values were normalized to sediment organic content, Koc = $(100)Kp/(\%$ oc), the resultant unique constant Koc exhibited a high positive correlation to the octanol-water partition coefficient (Kow) for that compound (17). The relationship was represented by the equation:

$$\log Koc = 1.00 \log Kow - 0.21 \qquad [2]$$

In earlier work, we have reported that this relationship holds for several other compounds (19,20,24,29,30,42), including some substituted and heterocyclic aromatic compounds. In this paper, we have extended our study to PAH containing heterocyclic atoms of nitrogen and sulfur and PAH substituted with oxygen- and nitrogen-containing moieties. The sorption behavior of these compounds on sediments which have been chemically and physically characterized is reported. The sorption properties of these compounds are compared with those previously reported for normal PAH.

## MATERIAL AND METHODS

### Sediment Preparation

The three soil/sediments (Table 1) were selected as a subset of 30 sediments and associated watershed soils collected to provide a range of

TABLE 1. Location and Field Notes on Selected Soil/Sediments

| Soil/Sediment | Notes |
|---|---|
| EPA-6 | Sediment sample from Antelope Creek public use area in lake behind Big Bend Dam on Missouri River, southwest of Pierre, South Dakota. Sample is a grayish clayey material with a fair silt content. |
| EPA-14 | Soil taken from a highly eroded red clay hillside southeast of the mouth of Big Sandy River and Ohio River. Point Park, Ceredo, West Virginia. |
| EPA-23 | Sediment taken from Crane Lake north of blind number 63, Sanganois Wildlife Refuge, confluence of Sangamon and Illinios Rivers. |

compositional and physical properties. All of the soils/sediments were collected from river systems and watersheds which are geographically targeted as possible sites for coal conversion facilities (26). The samples were air dried, crushed, and passed through a No. 20 sieve.

The pH, cation exchange capacity (CEC), and percentage of total nitrogen and of organic carbon as well as textural information were determined on each sample (Table 2). Soil reaction pH was determined on 1:1 and 1:2 soil-water mixtures by the Peech method (33). Cation exchange capacity was measured by the ammonium acetate method as modified by Banwart and Hassett (5). Total nitrogen was determined by the Bremner method (8), and total organic carbon was measured by the Walkley-Black method as described by Allison (1). Particle size analysis was determined by the hydrometer method of Day (11) using hydrogen peroxide to destroy the organic matter (25). Semiquantitative analysis of the clay mineral content of the less than 2-$\mu$ fraction of the clays was performed using a Phillips-Norelco X-ray diffractometer equipped with a digital computer. The montmorillonite of vermiculite (14 to 15 A), illite ($\sim$10 A), and kaolinite ($\sim$7 A) content of each soil/sediment is given in Table 2.

## Compound Purification and Radiolabelling

Dibenzothiophene (Pfaltz and Bauer), acridine (Aldrich), 2,2' biquinoline (Aldrich), and 13-H dibenzo(a,i)carbazole (Aldrich) were obtained in 99 percent pure form and tritiated using $BF_3$ - $^3H_3PO_4$ by the method of Hilton and O'Brien (21). The resulting generally labelled compounds were purified by preparative thin-layer chromatography on silica gel G plates (New England Nuclear). The final purities of all

TABLE 2. Physical-Chemical Properties of Selected Sediments and Soil

| Soil/Sediment | pH (1:1) | pH (1:2) | CEC (ml/100g) | Total N (%) | Organic C (%) | Sand (%) | Clay (%) | Silt (%) |
|---|---|---|---|---|---|---|---|---|
| EPA-6 (sed.) | 7.83 | 8.23 | 33.01 | 0.097 | 0.72 | 0.2 | 68.6 | 31.2 |
| EPA-14 (soil) | 4.54 | 4.30 | 18.86 | 0.064 | 0.48 | 2.1 | 63.6 | 34.4 |
| EPA-23 (sed.) | 6.70 | 7.10 | 31.15 | 0.195 | 2.38 | 17.3 | 69.1 | 13.6 |

Semiquantitative Determination of the Clay Minerals in the Less than 2-$\mu$ Fraction of the Sediment and Soil Samples

| Soil/Sediment | Montmorillonite and/or Vermiculite | Illite | Kaolinite |
|---|---|---|---|
| EPA-6 | 60.8 (M) | 4.6 | 3.2 |
| EPA-14 | 13.8 (V&M) | 12.8 | 37 |
| EPA-23 | 57.6 (M) | 4.2 | 7.3 |

(M) = montmorillonite.
(V) = vermiculite.

compounds studied were 99 percent or higher. Acetophenone (ICN Pharmaceuticals), $\alpha$-naphthol (ICN Pharmaceuticals), and benzidine (New England Nuclear) were obtained in $^{14}C$ labelled form. It was necessary to purify each compound by preparative thin-layer chromatography. The final purity of each compound was 99 percent or higher.

The radioactive content of samples was measured by liquid scintillation counting in a Packard 2425 Tri-Carb liquid scintillation spectrometer using Aquasol (New England Nuclear) as the counting cocktail. Soil and sediment samples processed in a Packard Model 306 sample oxidizer were analyzed using Carbo-Sorb (Packard) and Monophase 40 (Packard) as collecting media for $^{14}CO_2$ and $^3H_2O$, respectively, and Permaflour V (Packard) as the scintillation counting cocktail. All samples were counted for 10 minutes and all counts were corrected for background and counting efficiency.

## Determinations of Water Solubilities, Octanol-Water Partition Coefficients and Sorption Isotherms

Water solubilities, octanol-water partition coefficients, and sorption isothermics were determined for each compound using the procedures outlined in Means et al (29). In the case of acetophenone, $\alpha$-naphthol and benzidine, unlabelled pure compound was used to adjust the activity of the labelled compound for these experiments.

## RESULTS AND DISCUSSION

The water solubilities and octanol-water partition coefficients (Kow) at 24°C of the compounds studied are reported in Table 3. Each value is the result of at least 12 determinations. Equilibrium values were obtained for water solubilities within 48 hours and for octanol-water partition coefficients within 1 hour. The sources of variability in these measurements have been discussed previously (29). Care was taken to prevent photo-oxidation of the compound during the equilibration period. Loss of material by adsorption onto glassware was determined to be less than 2 percent. The Kow values were determined at at least two different concentrations to ensure that no concentration effects were observed.

The sorption isotherms for acetophenone, dibenzothiophene, acridine, biquinoline, and dibenzocarbazole were linear over a range of water phase concentrations on all sediments and soils tested. The linear Freundlich sorption constants (Kp) and corresponding Koc values are reported in Table 4. The isotherms for $\alpha$-naphthol and benzidine were curvilinear. The Kp values are presented in Table 4 along with their corresponding Koc values.

Acetophenone, an aromatic ketone, was sorbed weakly to two of the sediments tested. The Koc values of 25 and 29 compared well with average Koc of 35. The partition coefficients are highly correlated to the organic carbon content of the sediments (24). However, on EPA-6, a sediment with a very low carbon content (0.72 percent) but a high clay content (69 percent), an unexpectedly high Kp value was obtained. The data suggest that increased sorption can occur when the montmorillonite clay content of a sediment is high and the organic carbon content is low.

Karickoff (23) suggests that the Koc for a compound may be calculated from the Kow for that compound. The calculated Koc of 24 agrees well with the experimental Koc values (Table 5).

Dibenzothiophene was sorbed strongly to all of the sediments tested. Linear isotherms were obtained. The Koc values for the three sediments

**TABLE 3.** Water Solubilities and Octanol-Water-Partition Coefficients for Selected Aromatic Compounds

| Compound | Water Solubility | Octanol-Water Partition Coefficient (Kow) |
|---|---|---|
| Acetophenone | 5540 ppm | 38.6 |
| 4,4' Benzidine | 360 ppm (pH 5.9) | 46 (pH 5.9) |
| $\alpha$ - Naphthol | 866 ppm | 700 |
| Dibenzothiophene | 1.47 ppm | 24,000 |
| Acridine | 38.4 ppm | 4,200 |
| 2,2' -Biquinoline | 1.02 ppm | 20,179 |
| 13H-Dibenzo(a,i)carbazole | 10.4 ppb | 2,514,000 |

TABLE 4. Sorption Coefficients for Selected Aromatic Compounds

| Compound | EPA-6 | | EPA-14 | | EPA-23 | | Mean |
|---|---|---|---|---|---|---|---|
| | Kp | Koc | Kp | Koc | Kp | Koc | $Koc^a$ |
| Acetophenone | 0.682 | 94 | 0.122 | 25 | 0.679 | 29 | 35 |
| 4,4'-Benzidine[b] | 2,852 | 396,111 | 8,037 | 1,674,375 | 1,790 | 75,210 | 198,786 |
| $\alpha$-Naphthol[c] | 30 | 4,198 | 2.8 | 585 | 14 | 589 | 600 |
| Dibenzothiophene | 61 | 8,444 | 50 | 10,354 | 389 | 16,300 | 11,230 |
| Acridine | 394 | 54,798 | 335 | 69,879 | 529 | 22,243 | 48,900 |
| 2,2' Biquinoline | 281 | 38,990 | 107 | 22,270 | 430 | 18,060 | 26,500 |
| 13H-Dibenzo(a,i)-carbazole | 12,568 | 1,745,500 | 8,134 | 1,694,500 | 13,900 | 583,900 | 1,393,000 |

[a] Average of all Koc values determined for this compound on 10 to 16 sediments and soil samples.
[b] Kp (molar) and Koc values are the result of the solution of the Freundlich equation where 1/n equals the average value of 0.5.
[c] Kp (molar) and Koc values are the result of the solution of the Freundlich equation where 1/n equals 0.310, 0.593, and 0.387, respectively.

TABLE 5. Comparison of Experimentally Determined and Calculated Sorption Constants (Koc)

| Compound | Kow | Koc | Koc(calc.) |
|---|---|---|---|
| Acetophenone | 38.6 | 35 | 24 |
| Benzidine | 46.0 | N. A. | N. A. |
| Naphthol | 700 | 600 | 432 |
| Dibenzothiophene | 24,000 | 11,230 | 14,798 |
| Acridine | 4,200 | 48,900 | 2,100 |
| Biquinoline | 20,179 | 26,500 | 10,080 |
| Dibenzocarbazole | 2,514,000 | 1,393,000 | 1,257,000 |

tend to converge on an average value of 11,230 and the Kp values were highly correlated with the organic carbon content. No other soil characteristics correlated well with the Kp values. The calculated Koc (14,798) is in good agreement with the Koc determined experimentally (11,230) (Table 5). Dibenzothiophene appears to be sorbed as a neutral hydrophobic substance. The heterocyclic sulfur atom does not appear to significantly influence the sorptive behavior of the compound (20).

Three heterocyclic nitrogen-containing PAH were also studied. The isotherms obtained were all linear over a range of solution phase concentrations. The Kp values and Koc values are given in Table 4. The sorption constants indicate that these compounds are very strongly sorbed as hydrophobic substances. The amount of sorption is correlated to organic carbon content but not to other soil characteristics. However, sample EPA-23 consistently exhibits less sorption (Koc) of the nitrogen-containing compounds than the other substrates tested. Since this sediment has the highest organic carbon content of all of the substrates tested, this result is difficult to explain. No other sediment parameter would predict this decrease in sorption. Comparison of the Koc values for these three compounds calculated from the octanol-water partition coefficients shows that the actual amount of sorption is much greater than predicted (Table 5). It appears that the presence of heterocyclic nitrogen in a PAH molecule greatly enhances its ability to be sorbed to sediments. The reasons for this phenomenon are not clear and will need further study.

Two aromatic compounds substituted with polar functional groups were also studied. For $\alpha$-naphthol, the isotherms were nonlinear, as expected from the $1/n$ values of 0.310, 0.593, and 0.387. The sorption of $\alpha$-naphthol on two of the substrates (EPA-14 and EPA-23) was well correlated with organic carbon content (19), with good agreement between calculated and experimentally determined Koc values (Table 5). However, on EPA-6, the amount of $\alpha$-naphthol sorbed was much higher than would be predicted from the sediment organic carbon content. We have postulated that the organic carbon/montmorillonite ratio may be important in exploring this high degree of sorption (19,24).

Benzidine, a primary amine-substituted aromatic compound, was also studied. The isotherms were curvilinear and gave the best fit to the equation $Cs = KpCw^{0.5}$. In this case, the sorption constants were highly correlated with hydrogen ion activities, as calculated from pH measurements, with sorption of the ionized species predominating. In a more detailed study, we have determined that, when the isotherms are corrected for sorption of the neutral species, sorption of the ionized forms of benzidine is highly correlated with surface area and negatively correlated with organic carbon content. We suggest that, for compounds such as benzidine, organic carbon may be masking the active sites for adsorption of this ionizable compound (42).

## ACKNOWLEDGMENTS

This paper is Contribution No. 870 of the Chesapeake Biological Laboratory, Center for Environmental and Estuarine Studies, University of Maryland, the Institute for Environmental Studies, University of Illinois, and the Department of Agronomy, University of Illinois. This work was supported by United States Environmental Protection Agency Contract 68-03-2555.

## REFERENCES

1. Allison, L. E. (1965): Organic carbon. In: *Methods of Soil Analysis, Agronomy,* 9(2), C. A. Black, Ed., pp. 1367-1378, American Society of Agronomy, Madison, Wisconsin.
2. Bailey, G. W., and White, J. L. (1964): Review of adsorption and desorption of organic pesticides by soil colloids, with implications concerning pesticide bioactivity. *J. Agric. Food Chem.* 12: 324-332.
3. Bailey, G. W., and White, J. L. (1970): Factors influencing the adsorption, desorption, and movement of pesticides in soil. *Residue Rev.* 32: 29-92.
4. Bailey, G. W., White J. L., and Rothberg, T. (1968): Adsorption of organic herbicides by montmorillonite: role of pH and chemical character of adsorbate. *Soil Sci. Soc. Amer. Proc.* 32: 222-234.
5. Banwart, W. L., and Hassett, J. J. (1976): *Laboratory Introduction to Soil Science,* p. 118. Stipes Publishing Co., Champaign, Illinois.
6. Blumer, M. (1976): Polycyclic aromatic compounds in nature. *Sci. Amer.,* 234: 35-45.
7. Blumer, M., and Youngblood, W. W. (1976): Polycyclic aromatic hydrocarbons in the environment: Homologous series in soils and recent marine sediments. NTIS No. AD A023637. Office of Naval Research.
8. Bremmer, J. M. (1965): Total nitrogen. In: *Methods of Soil Analysis, Agronomy* 9(2), C.A. Black, Ed., pp. 1149-1237, American Society of Agronomy, Madison, Wisconsin.
9. Briggs, G. G. (1969): Molecular structure of herbicides and their sorption by soils. *Nature (London),* 223: 1288.
10. Coggins, C. W., Jr., and Crafts, A. S. (1959): Substituted urea herbicides: their electrophoretic behavior and influence of clay colloids in nutrient solution on their phytotoxicity. *Weeds* 7: 349-359.
11. Day, P. R. (1965): Particle fractionation and particle size analysis. In: *Methods of Soil Analysis, Agronomy* 9(1), C. A. Black, Ed., pp. 545-567, American Society of Agronomy, Madison, Wisconsin.
12. Doherty, P. J., and Warren, G. F. (1969): The adsorption of four herbicides by different types of organic matter and a bentonite clay. *Weed Res.* 9: 20-26.
13. Forney, A. J., Haynes, W. P., Gasior, S. J., Johnson, G. E., and Strakey, J. P., Jr. (1974): Analysis of tars, chars, gases, and water found in effluents from the synthane process. Progress Report 76. Bureau of Mines, U. S. Department of the Interior.
14. Giger, W., and Blumer, M. (1974): Polycylic aromatic hydrocarbons in the environment: isolation and characterization by chromatography, visible-ultraviolet, and mass spectrometry. *Anal. Chem.,* 46: 1663-1671.

15. Grover, R. (1971): Adsorption of picloram by soils colloids and various other adsorbents. *Weed Sci.* 19: 417-418.
16. Gustafson, R., and Paleos, J. (1971): Interactions responsible for the selective adsorption of organics on organic surfaces. In: *Organic Compounds in Aquatic Environments,* S. D. Faust and J. V. Hunter, Eds., pp. 213-237, Marcel Dekker, New York, N. Y.
17. Hamaker, J. W., and Thompson, J. M. (1972): Adsorption. In: *Organic Chemicals in the Soil Environment,* C.A.1 Goring and J. W. Hamaker, Eds., pp. 49-143, Marcel Dekker, New York, N.Y.
18. Hance, R. J. (1965): Observation on the relationship between the adsorption of diuron and the nature of adsorbent. *Weed Res.* 5: 108-114.
19. Hassett, J. J., Banwart, W. L., Means, J. C., and Wood, S. G. (1979): Sorption of $\alpha$-Naphthol: implications concerning the limits of hydrophobic sorption (1979). *Soil Sci. Soc. Amer. J.* (in press).
20. Hassett, J. J., Means, J. C., Banwart, W. L., Wood, S. G., Ali, S., and Khan, A. (1979): Sorption of Dibenzothiophene by Soils and Sediments. *J. Envir. Quality* (in press).
21. Hilton, B. D., and O'Brien, R. D. (1964): A simple techinque for tritiation of aromatic insecticides. *J. Agric. Food Chem.,* 12: 236-239.
22. Huang, P. M., Wang, T.S.C., Wang, M. K., Wu, M. H. and Shu, N. W. (1977): Retention of phenolic acids by noncrystalline hydroxy-aluminum and -iron compounds and clay minerals of soils. *Soil Sci.* 123: 213-219.
23. Karickhoff, S. W., Brown, D. S., and Scott, T. A. (1979): Sorption of hydrophobic pollutants on neutral sediments. *Water Res.* 13: 241-248.
24. Khan, A., Hassett, J. J., Banwart, W. L., Means, J. C., and Wood, S. G. (1979): Sorption of acetophenone by sediments and soils. *Soil Sci.* 128:297-302.
25. Kilmer, V. J., and Alexander, L. T. (1949): Methods of making mechanical analysis of soils. *Soil Sci.,* 68: 15-24.
26. Lindquist, A. E. (1977): Siting potential for coal gasification plants in the United States. *Bureau of Mines Circular* 8735.
27. Liu, L. C., Cibes-Viade, H., and Koo, F.K.S. (1970): Adsorption of ametryne and diuron by soils. *Weed Sci.* 18: 470-474.
28. Mackenzie, M. J., and Hunter, J. V. (1979): Sources and fates of aromatic compounds in urban stormwater runoff. *Env. Sci. Tech.* 13: 179-183.
29. Means, J. C., Hassett, J. J., Wood, S. G., and Banwart, W. L. (1979): Sorption properties of energy-related pollutants and sediments. In: *Carcinogenesis, Vol. 1, Polynuclear Aromatic Hydrocarbons,* P. W. Jones and P. Leber, Eds., pp. 327-340, Ann Arbor Science, Ann Arbor, Michigan.

30. Means, J. C., Hassett, J. J., Banwart, W. L., and Wood, S. G. Adsorption of selected polynuclear aromatic hydrocarbons by sediments and soils. *Envir. Sci. Tech.* (in review).
31. Osgerby, J. M. (1970): Sorption of un-ionized pesticides by soils. In; *Sorption and Transport Processes in Soils.* Monograph 37, Society of Chemical Industry, London.
32. Parfitt, R. L. (1969): Mechanisms of adsorption of polymers and polysaccharides by montmorillonite. Ph.D. Thesis, University of Adelaide, Adelaide, Australia.
33. Peech, M. (1965): Hydrogen ion activity. In: *Methods of Soil Analysis, Agronomy* 9(2), C. A. Black, Ed., pp. 914-926, American Society of Agronomy, Madison, WI.
34. Sharkey, A. G., Schultz, J. L., White, C., and Lett, R. (1976): Analysis of organic material in coal, coal ash, fly ash, and other fuel and emission samples. Report No. EPA-600/2-76-075. Office of Research and Development, U. S. Environmental Protection Agency, Washington, D. C.
35. Sheets, J. J., and Danielson, L. L. (1960): Herbicides in soils. In: *The Nature and Fate of Chemicals Applied to Soils, Plants, and Animals,* pp. 170-181. USDA, ARS 20-9.
36. Smith, J. H., Mabey, W. R., Bohovos, N., Holt, B. R., Lee, S. S., Chou, T. W., Bomberger, D. C., and Mill, T. (1977): Environmental pathways of selected chemicals in freshwater systems. Part II: Laboratory studies. Final Report No. EPA-600/7-78-074. Office of Research and Development, U. S. Environmental Protection Agency, Athens, GA.
37. Suess, M. J. (1976): The environmental load and cycle of polycyclic aromatic hydrocarbons. *Sci. Total Environ.,* 6: 239-250.
38. Talbert, R. E., and Fletchall, O. H. (1965): The adsorption of some s-triazines in soils. *Weeds* 13: 46-52.
39. TRW Systems and Energy, Inc. (1976): Carcinogens relating to coal conversion processes. U. S. Energy Research and Development Administration. Oak Ridge, Tennessee.
40. Upchurch, R. P., and Mason, D. D. (1962): The influence of soil organic matter on the phytotoxicity of herbicides. *Weeds* 10: 9-14.
41. Zettlemoyer, A. C., and Micale, F. J. (1971): Solution adsorption thermodynamics for organics on surfaces. In: *Organic Compounds in Aquatic Enviroments.* S. Faust and J. Hunter, Eds., pp. 165-185, Marcel Dekker, New York, N. Y.
42. Zierath, D. L., Hassett, J. J., Banwart, W. L., Means, J. C., and Wood, S. G. (1979): Sorption of benzidine by sediments and soils. *Soil Sci.* (in press).

# ANALYSIS OF PAH IN ENVIRONMENTAL SAMPLES BY HIGH TEMPERATURE STABLE GLASS CAPILLARY COLUMNS

**J. M. Meuser, F. R. Moore, P. E. Strup,**
**J. E. Wilkinson, and A. Bjorseth***

Battelle
Columbus Laboratories
Columbus, Ohio 43201

## INTRODUCTION

That PAH are widespread environmental contaminants has been firmly established (2,3,13,14). Also accepted is the fact that the major source of these PAH contaminants is incomplete combustion of organic materials (13,14). Because the process of combustion is a complex and unspecific one, the resultant mixture is also highly complex. The categorization of PAH in the complex environmental matrix, then, requires an analytical method which has a high separation efficiency. The importance of this feature is recognized when one wishes to identify the many compounds present in the sample or to distinguish between isomers, for example when one isomer is carcinogenic and the other is not.

Beyond the interest in resolving complex mixtures, however, the sensitivity of the analytical method is also of concern. Because the biological activity of PAH is frequently important and often associated with minor sample components, the analytical procedure must be well suited for the detection of low concentrations (14).

Thus far, the technique that seems to satisfy these requirements of high separation efficiency and high sensitivity is glass-capillary gas chromatography. Until recently, however, the temperature limit of these columns has been 250 to 280°C (11). Consequently, analysis time is

---

* Present address: Central Institute for Industrial Research, Blindern, Oslo 3, Norway.

prolonged and the less volatile sample components show very poor peak shape. With a few exceptions (15), compounds of very high molecular weight have remained undetected. Recently, Grob and Grob published methods and refinements for the preparation of high-quality glass capillary columns that show excellent stability to at least 350°C (10). The columns resulting from the adaptation of their techniques, coated with apolar phases (9) are inactive, show high thermostability (5), and are well suited for the analysis of PAH in environmental samples. Analysis time is reduced and even the least volatile sample components are resolved.

## EXPERIMENTAL

### Column Preparation

The method for column preparation is only slightly modified from the method published by Grob and Grob (10). After the glass column has been drawn, its preparation involves four basic steps. The inner glass surface of the column is first subjected to acid leaching. A 20 percent solution of HCl is introduced into the column, allowing 8 percent of the total length to remain empty. The ends are sealed, and the column is heated to 180°C for 16 hours. The leaching eliminates metal ions and produces a heavily hydrated surface layer, ideally 5 to 8 $\mu$m in depth (8).

Leaching is followed by dehydration. The HCl is washed out of the column with distilled water. Both ends are connected to a water aspirator and the column is heated to 280°C for 24 hours. Dehydration creates a more defined silica structure and opens the glass surface. The vacuum applied at each end is important in accomplishing a uniform dehydration throughout the column.

The defined, open surface is critical for complete silylation of the capillary column (7). This third step involves introducing a 20 percent solution of diphenyltetramethyldisilazane into the column, sealing the ends, and heating to 400°C for 8 hours. This treatment produces almost total adsorptive and catalytic inertia, along with providing the high thermostability of the column. We have found that diphenyltetramethyldisilazane produces a more inert, thermostable, and efficient column than does hexamethyldisilazane (HMDS).

After the column is washed with toluene, methanol, and ether to remove any excess silylating reagent, the glass surface is ready for the final, coating step utilizing the static coating technique (4,6,12). A solution of 0.4 percent SE-52 n-pentane is used to fill the column. One end is then sealed with a water glass plug and allowed to harden completely, 6 to 8 hours; the open end is connected to a water aspirator. The column is allowed to evacuate slowly while resting in a water bath (12).

Finally, the column is conditioned by slowly heating it overnight to 340°C and holding at the upper temperature for 2 hours. Column testing is performed using a solution described by Grob to evaluate column characteristics and a standard PAH solution to evaluate the specific separation efficiency and sensitivity of the column for PAH (5,10).

**Sample Preparation**

The samples to be analyzed are soxhlet extracted overnight with cyclohexane. Further cleanup is performed using a previous reported method of liquid-liquid extraction (1).

*Analysis*

The instrument employed for all analyses presented here is a Carlo Erba Model 2150 AC, equipped with a Grob splitless injector and a flame ionization detector. All samples were analyzed on a 15-m, SE-52 column (9), temperature programmed from 100°C to 330°C at 3 deg/min. Hydrogen is used as the carrier gas. Several advantages are obtained by doing so. First, the purity of commercial hydrogen is often much better than the purity of other carrier gases available, and column life is prolonged. Second, phase damage caused by oxygen contamination of the gas is avoided in the reducing atmosphere of hydrogen. Third, the optimum carrier gas velocity permits a shorter analysis time than when either helium or nitrogen is run at its optimum velocity. For example, the optimum velocity of hydrogen is 2-1/2 times faster than that of nitrogen (11). A portable hydrogen detector is used to ascertain that there are no leaks in the system. With this precaution we have experienced no difficulties and many advantages in using hydrogen.

## RESULTS

The chromatogram of a complex, laboratory-prepared PAH solution containing 35 PAH having a wide volatility range is shown in Figure 1. Many of the compounds frequently occur in environmental samples. Peak identification is given in Table 1. The baseline resolution of several critical isomer pairs should be noted. Phenanthrene is separated from anthracene, benz(a)anthracene and chrysene are baseline separated, and so are benzo(e)pyrene and benzo(a)pyrene. Since both benz(a)anthracene and benzo(a)pyrene have been reported to exhibit carcinogenic properties (14), it is important to separate them from their respective non-carcinogenic isomers. Additionally, one can observe two isomers of dibenzopyrene and coronene. The boiling point of coronene is reported to

### 408 THERMOSTABLE GLASS CAPILLARY COLUMNS

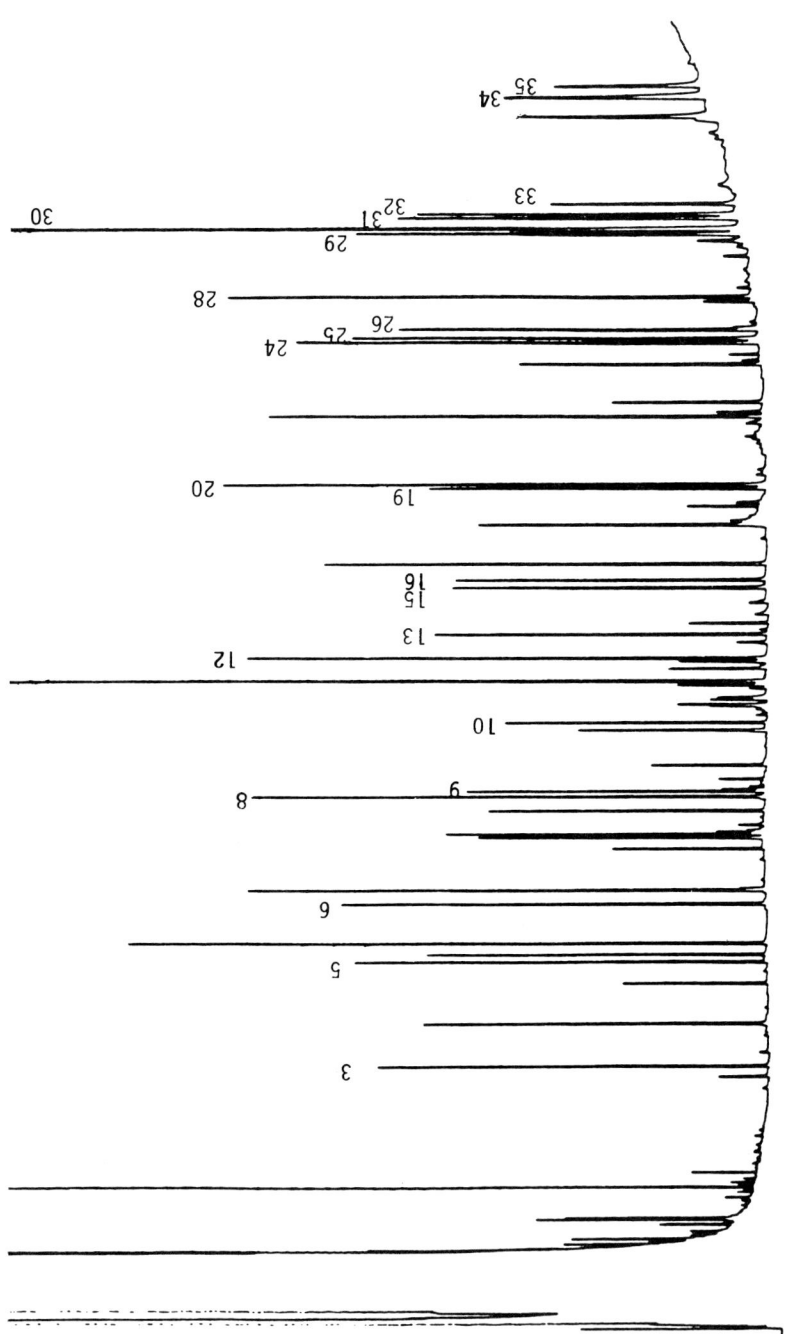

**FIGURE 1.** Glass-capillary gas chromatogram of a prepared PAH standard. Peak assignment in Table 1.

**TABLE 1. Peak Assignment**

| | |
|---|---|
| 1. Naphthalene | 19. Benz(a)anthracene |
| 2. Azulene | 20. Chrysene |
| 3. 1-Methylnaphthalene | 21. Naphthacene |
| 4. 1,5-Dimethylnapthalene | 22. Benzo(k)fluoranthene |
| 5. Acenaphthene | 23. 7,12-Dimethylbenz(a)anthracene |
| 6. Fluorene | 24. Benzo(e)pyrene |
| 7. 9,10-Dihydroanthracene | 25. Benzo(a)pyrene |
| 8. Phenanthrene | 26. Perylene |
| 9. Anthracene | 27. 9,10-Diphenylanthracene (Internal Standard) |
| 10. 2-Methylanthracene | |
| 11. 9-Methylanthracene (Internal Standard) | 28. P-quaterphenyl |
| 12. Fluoranthene | 29. Indeno(1,2,3-cd)pyrene |
| 13. Pyrene | 30. Dibenz(a,h)anthracene |
| 14. 9,10-Dimethylanthracene | 31. Picene |
| 15. Benzo(a)fluorene | 32. Benzo(g,h,i)perylene |
| 16. Benzo(b)fluorene | 33. Anthanthrene |
| 17. 1,1'-Binaphthyl | 34. Coronene |
| 18. 9-Phenylanthracene (Internal Standard) | 35. Dibenz(a,h)pyrene |

be 525°C (14). With an analysis time of 45 minutes, the excellent thermostability of the column avoids the necessity of a long isothermal period at the end of the run, which would serve to decrease the analysis time.

A chromatogram of PAH in particulates from an aluminum plant is given in Figure 2. The chromatogram reveals that the high separation efficiency and high-temperature stability is also achieved for real-world samples. Coronene and dibenzpyrene are eluted as sharp, well-chromatographed peaks.

A chromatogram of PAH in secondary alumina used to scrub the potroom exhaust from an aluminum plant is presented in Figure 3. The trace is very similar to that observed in Figure 2. Also in this case high molecular weight PAH are easily observed, even with an analysis time of less than 1 hour.

Chromatograms of PAH in samples from two different sampling points of a commercial woodburning power plant are shown in Figures 4 and 5. Again, the peak assignments are given in Table 1. However, because of the limited number of peaks and the unusual distribution, the assignment has to be regarded as tentative. The sample from the secondary collector (Figure 4) contains few compounds of higher molecular weight. In variance to this, the bottom ash (Figure 5) contains fewer low-molecular-weight and more high-molecular PAH. In both cases it is apparent that there are compounds present other than the parent PAH usually found in environmental samples (2).

# 410 THERMOSTABLE GLASS CAPILLARY COLUMNS

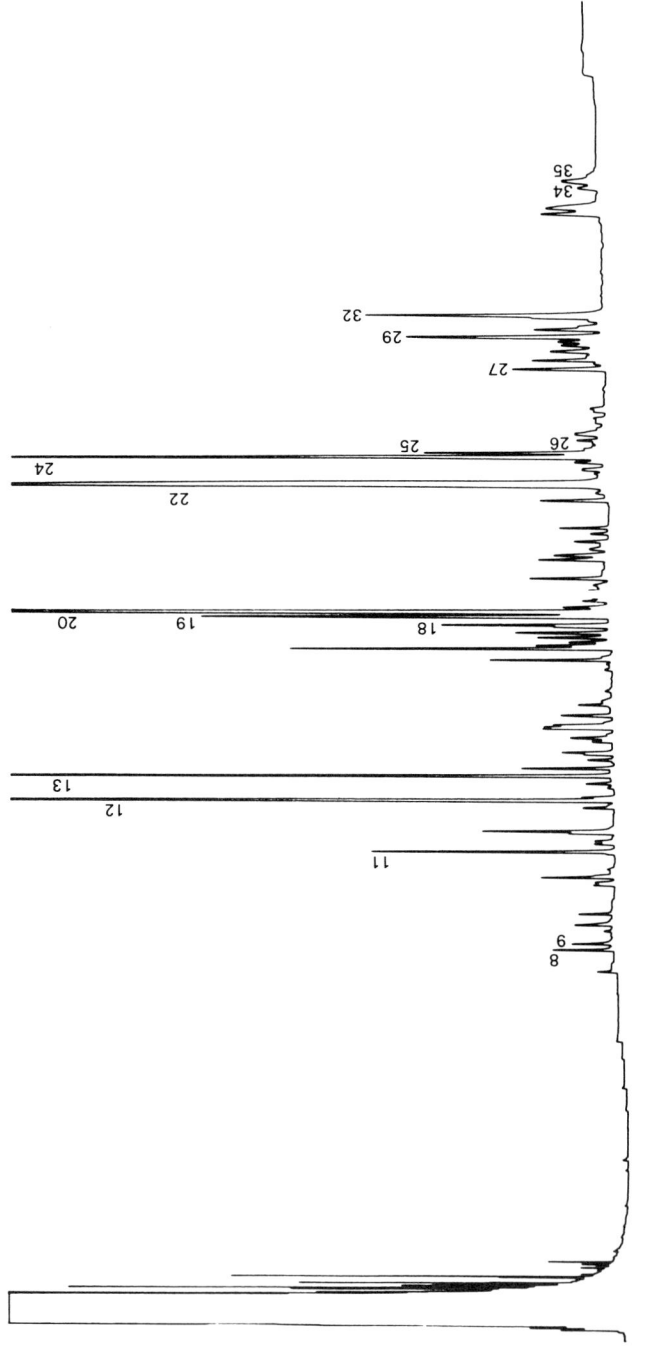

FIGURE 2. Glass-capillary gas chromatogram of PAH in particulate from an aluminum plant. Peak assignment in Table 1.

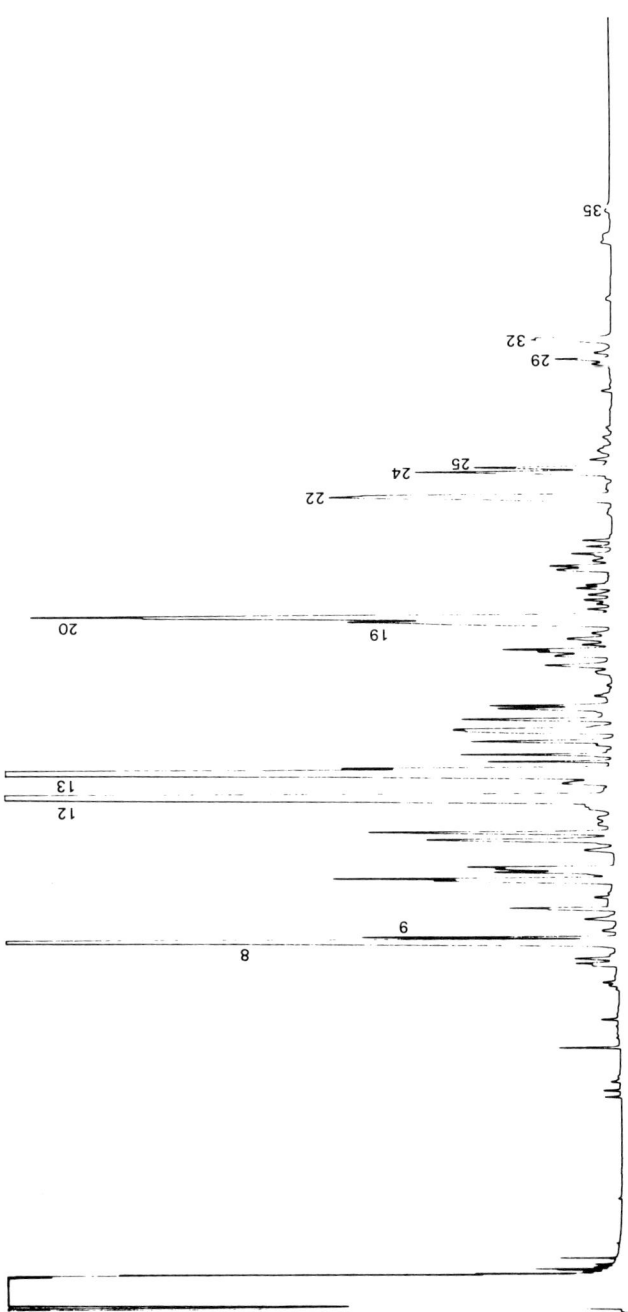

FIGURE 3. Glass-capillary gas chromatogram of PAH in secondary alumina. Peak assignment in Table 1.

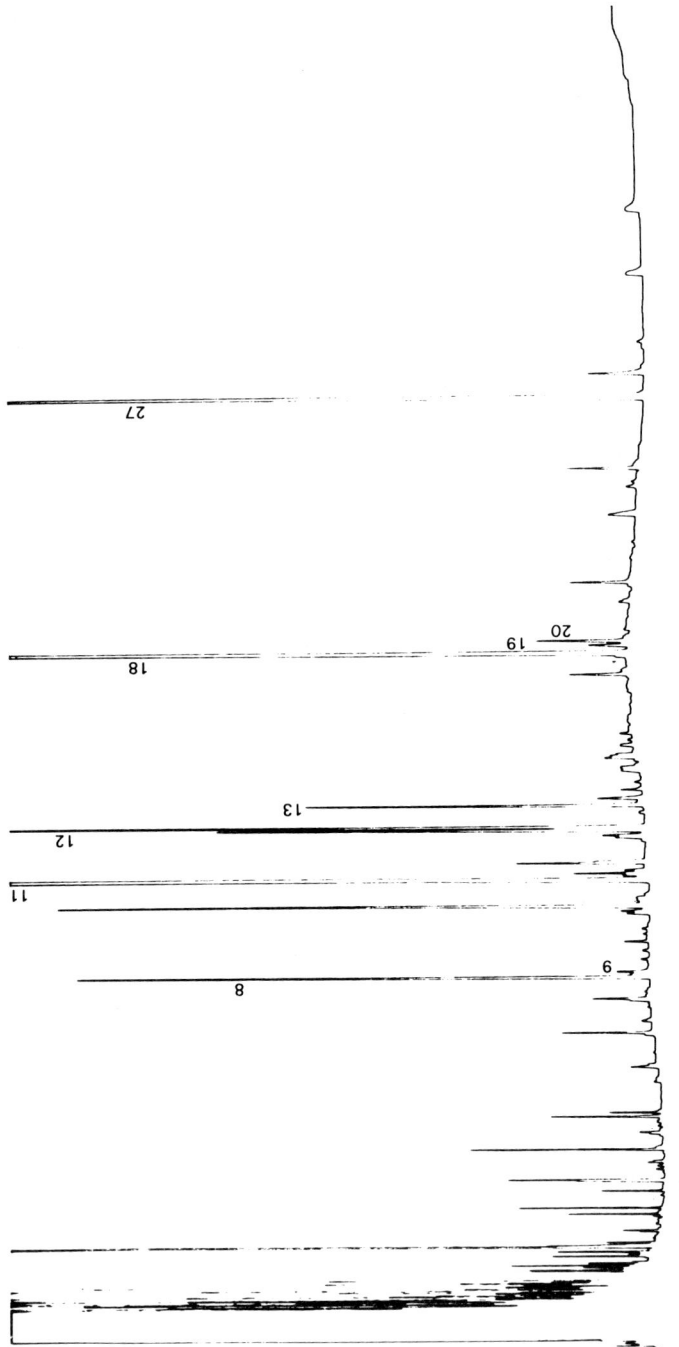

FIGURE 4. Glass-capillary gas chromatogram of PAH in fly ash from secondary collector, woodburning power plant. Peak assignment in Table 1.

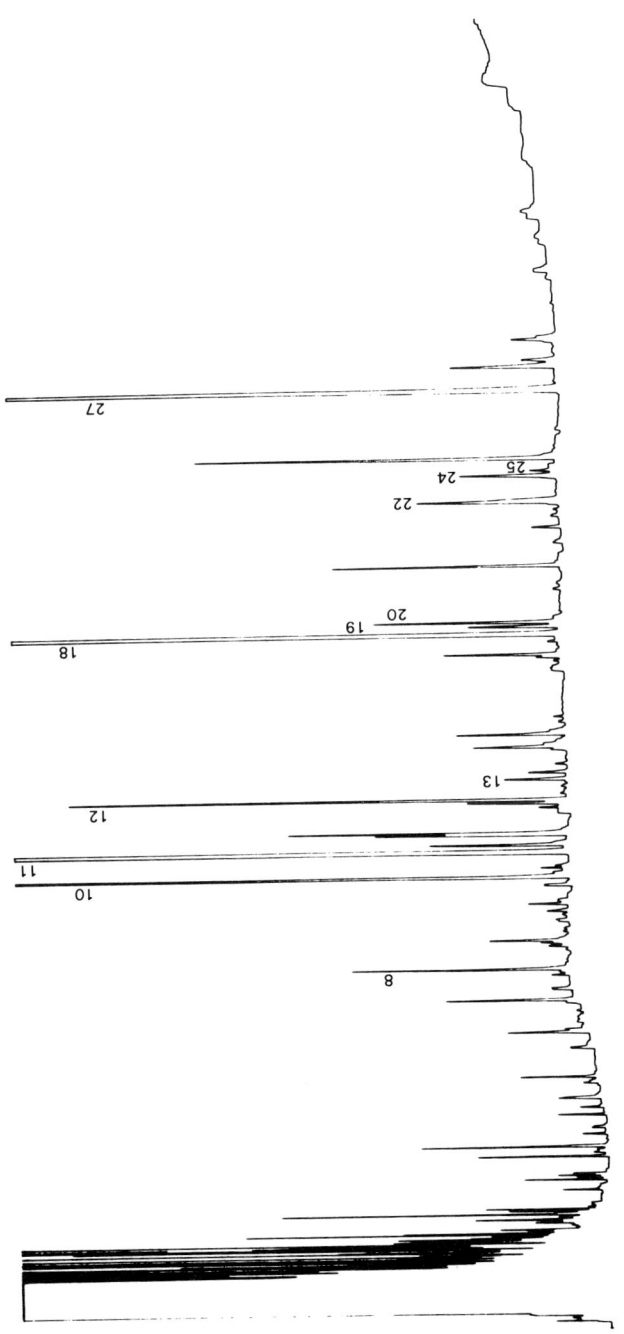

FIGURE 5. Glass-capillary gas chromatogram of PAH in bottom ash, wood-burning power plant. Peak assignment in Table 1.

## CONCLUSION

With the complexity and importance of PAH in environmental samples, the availability of a highly efficient and highly sensitive analytical technique is imperative. Glass-capillary gas chromatography, utilizing thermostable columns, appears to satisfy the requirements of such a method.

## REFERENCES

1. Bjorseth, A. (1977): Analysis of polycyclic aromatic hydrocarbons in particulate matter by glass capillary gas chromatography. *Anal. Chim. Acta* 94:21-27.
2. Bjorseth, A. (1978): Analysis of polycyclic aromatic hydrocarbons in environmental samples by glass capillary gas chromatography. In: *Carcinogenesis,* Vol. 3, P. W. Jones and R. I. Freudenthal, Eds., pp. 75-87, Raven Press, New York.
3. Bjorseth, A., Meuser, J., Moore, F., Strup, P., and Wilkinson, J. (1979): Analysis of polycyclic aromatic hydrocarbons in air and water. In: *Control of Specific (Toxic) Pollutants,* pp. 109-117, Air Pollution Control Association.
4. Giabbai, M., Shoults, M., and Bertsch, W. (1978): Static coating of glass capillary columns: Some practical observations. *Journal of HRC and CC* 1:277-278.
5. Grob, K., Jr., Grob, G., and Grob, K. (1978): Comprehensive, standardized quality test for glass capillary columns. *J. Chrom.* 156:1-20.
6. Grob, K., and Grob, K., Jr. (1977): Are we using the full range of film thickness in capillary-GLC? *Chromatographia* 10:250-255.
7. Grob, K., Grob, G., and Grob, K., Jr. (1979): Deactivation of glass capillary columns by silylation. Part I. Principles and basic technique. *Journal of HRC and CC* 2:31-35.
8. Grob, K., and Grob, G. (1979): Further evidence for the silica layer as produced by acid leaching. *Journal of HRC and CC* 2:527-528.
9. Grob, K. (1977): Gum phases for glass capillary columns; a recommendation for users, a challenge for polymer scientists. *Chromatographia* 10:625.
10. Grob, K., and Grob, G. (1976): A new, generally applicable procedure for the preparation of glass capillary columns. *J. Chrom.* 125:471-485.
11. Grob, K., and Grob, G. (1979): Practical capillary gas chromatography—a systematic approach. *Journal of HRC and CC* 2:109-117.

12. Grob, K. (1978): Static coating of glass capillary columns. Solvent selection; column filling; solvent evaporation. *Journal of HRC and CC* 1:93-94.
13. Jones, P. W., and Fruedenthal, R. I. (1978): *Carcinogenesis,* Vol. 3, Raven Press, New York.
14. National Academy of Sciences (1972): *Particulate Polycyclic Organic Matter,* Washington, D.C.
15. Stenberg, U., Alsberg, T., Blomberg, L., and Wannman, T. (1979): Gas chromatographic separation of high-molecular polynuclear aromatic hydrocarbons in samples from different sources, using temperature-stable glass capillary columns. In: *Polynuclear Aromatic Hydrocarbons,* P. W. Jones and P. Leber, Eds., pp. 313-326. Ann Arbor, Michigan.

# ON THE METABOLIC ACTIVATION OF THE BENZOFLUORANTHENES*

S. S. Hecht, E. LaVoie, S. Amin, V. Bedenko, and D. Hoffmann

Naylor Dana Institute for Disease Prevention
American Health Foundation
Valhalla, New York 10595

The benzofluoranthenes occur in the environment and are carcinogenic in experimental animals. Benzo(b)fluoranthene (BbF), benzo(j)fluoranthene (BjF), and benzo(k)fluoranthene (BkF) (see Figure 1) have been detected in gasoline engine exhaust, urban air, cigarette smoke, soil, water, and broiled and smoked foods (11). The levels of these compounds in the environment are comparable to and sometimes exceed those of benzo(a)pyrene. Some comparative data are summarized in Table 1. The benzofluoranthenes are formed in the incomplete combustion of organic matter, as are the other PAH. The temperature of combustion affects the yields of individual PAH. Higher temperatures may favor formation of certain benzofluoranthenes over benzo(a)pyrene (BaP). In the pyrolysis of n-butylbenzene, yields of benzo(b)fluoranthene and benzo(a)pyrene were comparable at 650°C while benzo(b)fluoranthene predominated at 750°C (2). The higher levels of benzo(b)fluoranthene than benzo(a)pyrene in gasoline engine exhaust may be a consequence of this temperature effect.

The carcinogenicity of the benzofluoranthenes has been assayed on mouse skin and by subcutaneous injection in mice. When applied to mouse skin (0.1 percent in acetone), benzo(b)fluoranthene and benzo(j)fluoranthene induced tumors in 85 and 100 percent of the animals, respectively, while benzo(k)fluoranthene and benzo(ghi)fluoranthene were inactive (17). Benzo(b)fluoranthene did not induce tumors at a dose of 0.01 percent, whereas benzo(a)pyrene gave 85 percent tumor-bearing

---
*Number 23 in "A Study of Chemical Carcinogenesis".

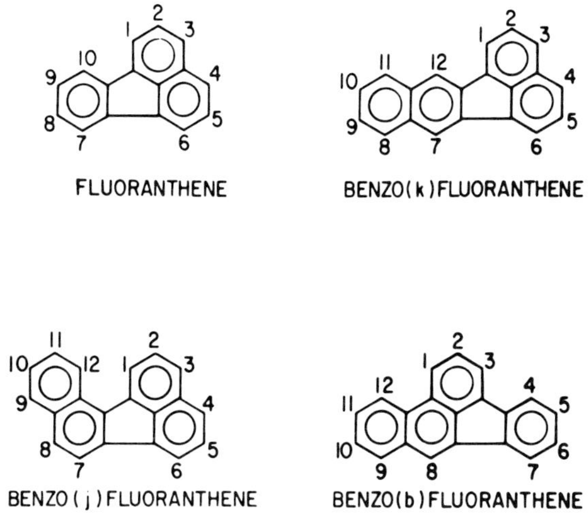

**FIGURE 1.** Structures of fluoranthene, benzo(b)fluoranthene, benzo(j)fluoranthene, and benzo(k)fluoranthene.

animals at this dose. Thus, the benzofluoranthenes are less powerful carcinogens than benzo(a)pyrene on mouse skin. Subcutaneous injection of benzo(b)fluoranthene and benzo(k)fluoranthene in mice gave sarcomas in 75 and 48 percent of the animals, respectively, compared with 84 percent with benzo(a)pyrene (13). Benzo(b)fluoranthene, benzo(j)fluoranthene, and benzo(k)fluoranthene were all mutagenic toward *S. typhimurium* TA 100 with activation (see Figure 2) (14). Despite the environmental occurrence, carcinogenicity, and mutagenicity of the benzofluoranthenes, no studies have been reported on their metabolic activation. The binding of dibenzo(a,e) fluoranthene to nucleic acids has been investigated (16).

## MATERIALS AND METHODS

### Apparatus

High-pressure liquid chromatography was performed with a Waters Associates Model ALC/GPC-204 high-speed liquid chromatograph equipped with a Model 6000A solvent delivery system, a Model 660 solvent programmer, a Model U6K septumless injector, A Model 440 UV/visible detector, and columns 1 (3.9 mm x 30 cm μBondapak/$C_{18}$, Waters, Inc., Milford, Mass.), 2(9.4 mm x 50 cm Whatman Magnum 9

TABLE 1. Some Comparative Environmental Levels of Benzofluoranthenes and Benzo(a)pyrene

| | Benzo(b)fluoranthene | Benzo(j)fluoranthene | Benzo(k)fluoranthene | Benzo(a)pyrene |
|---|---|---|---|---|
| Air in U.S. Cities ($\mu$g/1000 m$^3$)[a] | 0.1-7.4 | 0.01-4.4 | 0.03-20 | 0.2-17.0 |
| Gasoline Engine Exhaust Condensate (ppm)[b] | 64 | 17 | 54 | 31.5(72.6) |
| Cigarettes ($\mu$g/100 cig.)[c] | 0.3 | 0.6 | 0.7 | 3.9 |
| European Rivers (ng/1)[d] | 10.4-362 | 4.6-420 | 4.2-173 | 0.6-350 |

[a]Reference 10.
[b]Isolated amounts; numbers in parentheses are actual amounts determined by the isotope dilution method. Reference 9.
[c]Reference 18.
[d]References 3, 4, and 5.

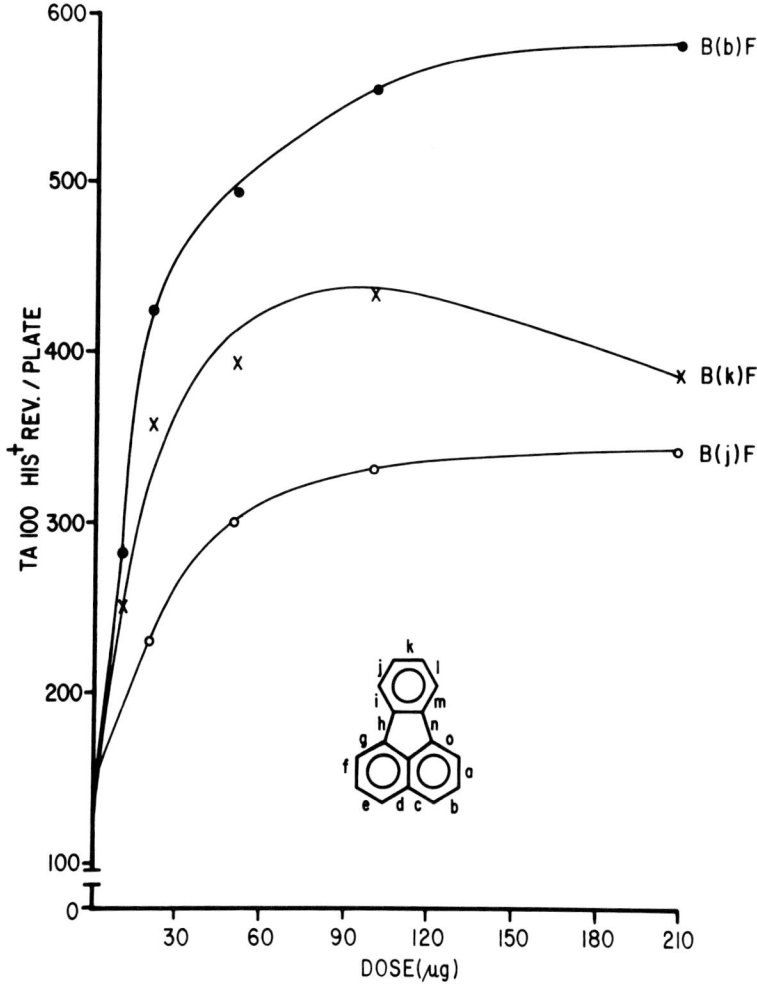

**FIGURE 2.** Mutagenicity toward *S. typhimurium* **TA 100** of benzo(b)fluoranthene, benzo(j)fluoranthene, and benzo(k)fluoranthene.

ODS, Whatman, Inc., Clifton, N.J.), and 3 (4.6 mm x 250 mm LiChrosorb RP-18, 10 μm, EM Reagents, Cincinnati, Ohio). Solvent programs were as follows: benzo(j)fluoranthene metabolites, column 1, 50 percent methanol in $H_2O$ for 20 min, then linear to 80 percent methanol in $H_2O$ in 40 min at 2 ml/min; benzo(b)fluoranthene metabolites, column 1, 50 percent methanol in $H_2O$ for 36 minutes, then linear to 80 percent methanol in 40 min at 3 ml/min; benzo(k)fluoranthene metabolites, column 3, 50 percent methanol in $H_2O$ for 15 min, then linear to 100 percent methanol in 60 min at 2 ml/min. UV spectra were run on a Cary

Model 118 spectrometer. Mass spectrometry was done with a Hewlett-Packard Model 5982A instrument.

## Chemicals

Benzo(b)fluoranthene was obtained from the Deutsche Gesellschaft für Teerverwertung. Benzo(j)fluoranthene and benzo(k)fluoranthene were synthesized (6,15), as were 9,10-dihydro-9,10-dihydroxybenzo-(j)fluoranthene, 9,10-dihydro-9,10-dihydroxybenzo(b)fluoranthene, and 8,9-dihydro-8,9-dihydroxybenzo(k)fluoranthene (1). Benzofluoranthenes and their metabolites were purified by high-pressure liquid chromatography prior to mutagenicity and carcinogenicity assays. NADP$^+$ and glucose-6-phosphate were obtained from Sigma Chemical Co., St. Louis, Missouri; 1,1,1-trichloropropene oxide (TCPO) was obtained from Aldrich Chemical Co., Milwaukee, Wisconsin; and Aroclor® 1254 was procured from Analabs, Inc., Hamden, Conn.

## Metabolism in Vitro

The in vitro metabolism studies were performed using the identical S-9 mix used in the mutagenicity assays. The S-9 mix contained, per ml, 100 μmoles of potassium phosphate buffer at pH 7.4, 8.0 μmoles of MgCl$_2$, 1.65 μmoles of KCl, 5.0 μmoles of glucose-6-phosphate, 4.0 μmoles of NADP, and 0.5 ml of S-9 fraction. Incubations were performed at 37°C for 20 minutes using a 25-ml Erlenmeyer flask to which was added 250 μg of compound in 10 μl of dimethylsulfoxide and 2 ml of S-9 mix. The effect of epoxide hydrase inhibition was examined by employing $2.0 \times 10^{-3}$ M TCPO in these incubation mixtures. The incubations were terminated by addition of 2 ml of ice-cold acetone. The mixture was then extracted (5x) with 10-ml aliquots of ethyl acetate. The ethyl acetate solution was concentrated in vacuo below 40°C prior to analysis or purification of the metabolites by reverse-phase HPLC.

For preparation of metabolite bands for mutagenicity, 2.5 mg of benzo(j)fluoranthene or benzo(b)fluoranthene was incubated with 20 ml of S-9 mix. After extraction, bands were collected from the entire sample by repetitive injections on column 1 (benzo[j]fluoranthene) or on column 2 as described below (benzo[b]fluoranthene). Each band was concentrated by evaporation under reduced pressure at 30°C and the residue was dissolved in 150 μl DMSO. For mutagenicity assays, 50 μl was added to each of 2 plates for each band.

For tumor-initiation assays, 20 mg of benzo(b)fluoranthene or benzo(j)fluoranthene was metabolized as described above. Metabolites of benzo(j)fluoranthene were separated on column 2, with a program of 50 percent methanol for 20 min, then linear to 80 percent methanol in 40 min, with a flow rate of 5 ml/min. Metabolites of benzo(b)fluoranthene were separated on column 2, with a program of 50 percent methanol for 36 min, then linear to 80 percent methanol in 40 min at a flow rate of 5 ml/min. Solvents were removed and the entire residue from each band was redissolved in 20 ml of acetone for bioassay.

### Mutagenicity Assays

Mutagenicity studies were performed as previously described (8) using *S. typhimurium* TA 100 (TA 1535/pKm 101) provided by Dr. Bruce Ames of the University of California, Berkeley. The S-9 fraction employed in both mutagenicity assays and in metabolic studies was obtained from the livers of male Fischer-344 rats weighing 300 to 500 g which had been treated 5 days prior to sacrifice with 500 mg/kg Aroclor® 1254.

### Tumor Initiation

Compounds to be tested as tumor initiators were applied to the skin of Swiss Albino female mice (Ha/ICR). Each group consisted of 20 mice. When the animals had entered the second telogen phase of hair cycle, the metabolite band to be tested was applied as a solution in acetone (100 $\mu$l) ten times on alternate days to the shaved backs of mice. Ten days after the last initiator dose, promotion was begun by application thrice weekly of 2.5 $\mu$g of tetradecanoylphorbol acetate in acetone (100 $\mu$l) for 20 weeks.

## RESULTS AND DISCUSSION

### Benzo(j)fluoranthene

A high-pressure liquid chromatogram of the metabolites of benzo(j)fluoranthene formed in vitro is shown in Figure 3. Six bands of metabolites were collected and tested for mutagenicity toward *S. typhimurium* TA 100 with activation and for tumor-initiating activity on mouse skin. The results are summarized in Table 2. Band 4 had the

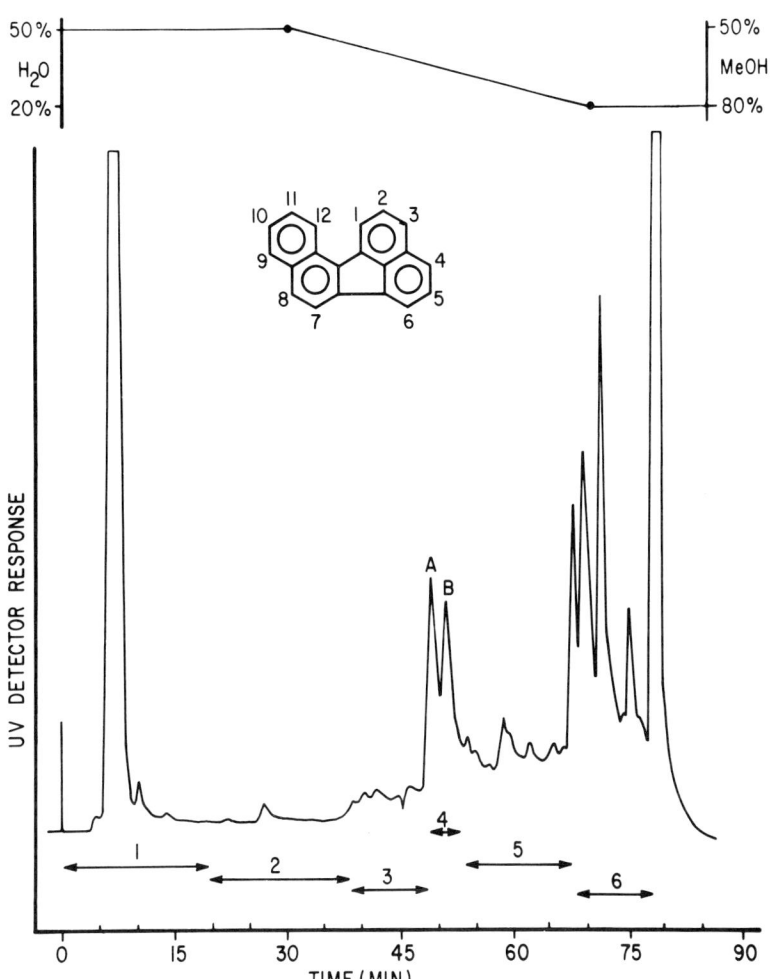

**FIGURE 3.** Metabolites formed from benzo(j)fluoranthene upon incubation with rat liver 9000 × g supernatant. Metabolite bands 1-6 are indicated at bottom. Peaks A and B are mutagenic dihydrodiols; peak B is BjF-9,10-diol (see text).

highest mutagenic and tumor-initiating activity, suggesting that this band contained proximate mutagenic or tumorigenic metabolites of benzo(j)-fluoranthene. When tested separately, each of the two peaks A and B of band 4 was mutagenic.

For structural elucidation, incubations were carried out in the presence of the epoxide hydrase inhibitor TCPO. Peaks A and B of band 4 disappeared, indicating that they were dihydrodiols. This was confirmed by mass spectrometry. There are five likely dihydrodiol metabolites of

TABLE 2. Mutagenicity and Tumor Initiating Activity of Benzo(j)fluoranthene Metabolites

| Metabolite Band[a] | S. Typhimurium TA 100[b] (His$^+$ revertants/plate) | Percentage of Tumor-Bearing[c] Animals |
|---|---|---|
| 1 | 177 ⎫ | |
| 2 | 275 ⎬ | 15 |
| 3 | 261 ⎭ | |
| 4 | 662 | 30 |
| 5 | 207 | 10 |
| 6 | 255 | 5 |

[a]See Figure 2 for metabolite bands.
[b]Metabolites formed from 2.5 mg benzo(j)fluoranthene. Each band was collected and, after concentration, redissolved in 150 μl DMSO for assay; 50 μl of each solution was applied to each of two replicate plates. DMSO control; 140 His$^+$ revertants/plate.
[c]Metabolites formed from 20 mg benzo(j)fluoranthene. Each band was collected and, after concentration, redissolved in acetone for assay. Each group consisted of 20 mice. Mice were initiated with 10 subdoses of the metabolite. Following initiation, tetradecanoylphorbol acetate (2.5 μg) was applied 3 times weekly for 20 weeks.

benzo(j)fluoranthene. The high mutagenicity of peaks A and B suggested that one of these dihydrodiols was trans-9,10-dihydro-9,10-dihydroxybenzo(j)fluoranthene (BjF-9,10-diol). To confirm this, BjF-9,10-diol was synthesized as outlined in Figure 4.

7-Methylfluoranthene was brominated with N-bromosuccinimide to give 7-bromomethylfluoranthene which was coupled with allyl magnesium bromide. Hydroboration of the resulting olefin followed by oxidation with pyridinium chlorochromate and then silver oxide gave 4-(7-fluoranthenyl)-butyric acid. Cyclization gave the desired ketone for conversion to BjF-9,10-diol by conventional methods (7). The sequence used in this synthesis for construction of the butyric acid side chain may be useful in other syntheses of PAH dihydrodiols when succinoylation reactions do not give the desired isomers or give mixtures of isomers.

The UV spectra, mass spectra, and HPLC retention volumes of the synthetic dihydrodiol and metabolite peak B of band 4 were identical, confirming the structure of this proximate mutagen as BjF-9,10-diol. The structure of the other mutagenic dihydrodiol (peak A) is not yet known. The comparative mutagenic activities toward S. typhimurium TA 100 of

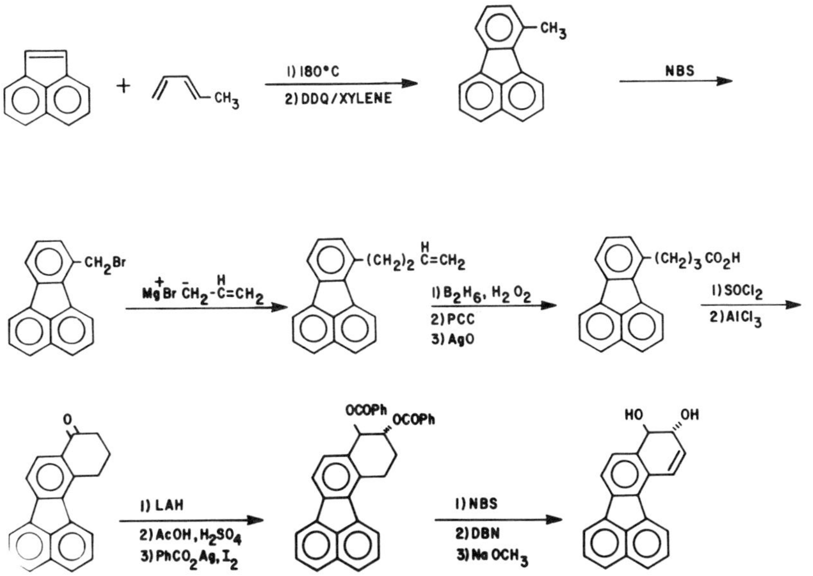

FIGURE 4. Synthesis of B(j)F-9,10-diol.

BjF-9,10-diol and benzo(j)fluoranthene are illustrated in Figure 5. The structure of BjF-9,10-diol is similar to those of dihydrodiols which are proximate forms of other carcinogenic PAH and can form bay-region dihydrodiol epoxides (12). The presumed ultimate mutagen derived from BjF-9,10-diol, 9,10-dihydro-9,10-dihydroxy-11,12-epoxybenzo(j)fluoranthene has the epoxide ring in a four-sided, "pseudo-bay region", rather than the more common three-sided bay region of such molecules as benz(a)anthracene, chrysene, and BaP.

## Benzo(b)fluoranthene

A high-pressure liquid chromatogram of the metabolites formed from benzo(b)fluoranthene by rat liver 9,000 x g supernatant is shown in Figure 6. As in the case of benzo(j)fluoranthene, bands of metabolites were collected and tested for mutagenicity toward *S. typhimurium* TA 100 with activation and for tumor-initiating activity on mouse skin. However, none of these bands showed significant mutagenic or tumorigenic activity. Only peak A, eluting at 12 min, disappeared when incubations were run in the presence of TCPO. The dihydrodiol structure of this peak was confirmed by mass spectrometry. The low mutagenicity of this dihydrodiol and its rapid elution from a reverse-phase HPLC column suggested that it was probably either the 2,3- or 4,5-dihydrodiol of benzo(b)fluoranthene.

## 426 ACTIVATION OF BENZOFLUORANTHENES

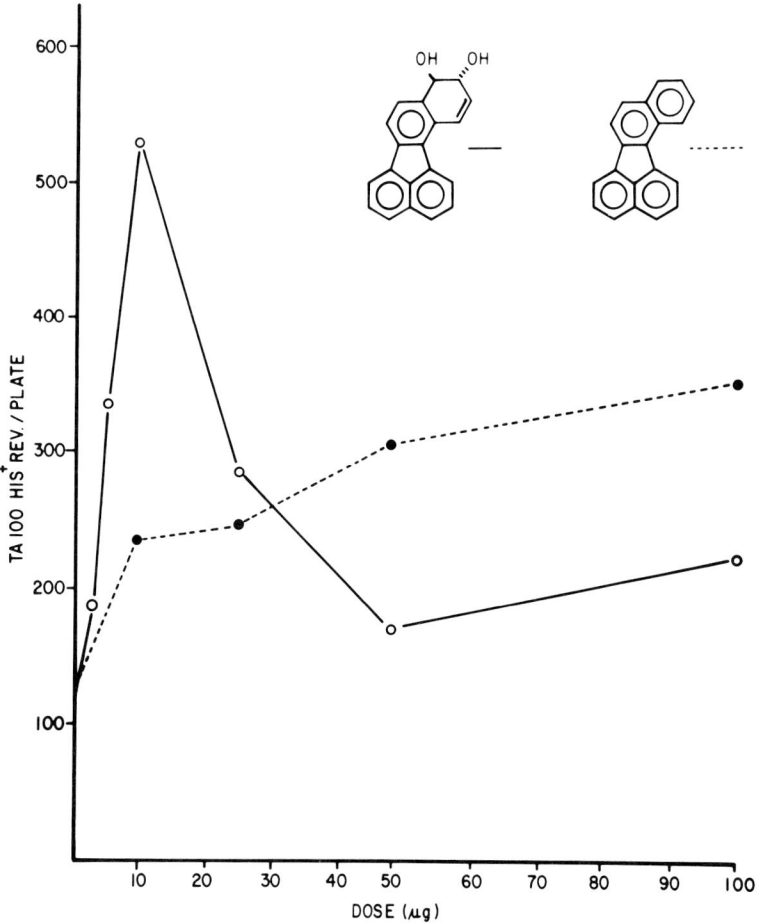

**FIGURE 5.** Mutagenicity toward *S. typhimurium* TA 100 of BjF-9,10-diol and benzo(j)fluoranthene.

We expected that trans-9,10-dihydro-9,10-dihydroxybenzo(b)fluoranthene (BbF-9,10-diol) would be the major proximate mutagen and carcinogen of benzo(b)fluoranthene since it could form a bay-region dihydrodiol epoxide (12). The experiments described above provided no evidence for metabolic formation of this compound. However, low rates of metabolism of benzo(b)fluoranthene to BbF-9,10-diol or rapid further metabolism of BbF-9,10-diol could not be excluded. Therefore, BbF-9,10-diol was synthesized as outlined in Figure 7.

Succinoylation of 6b,7,8,9,10,10a-hexahydrofluoranthene gave the desired isomer as the major product. The resulting keto-acid was reduced

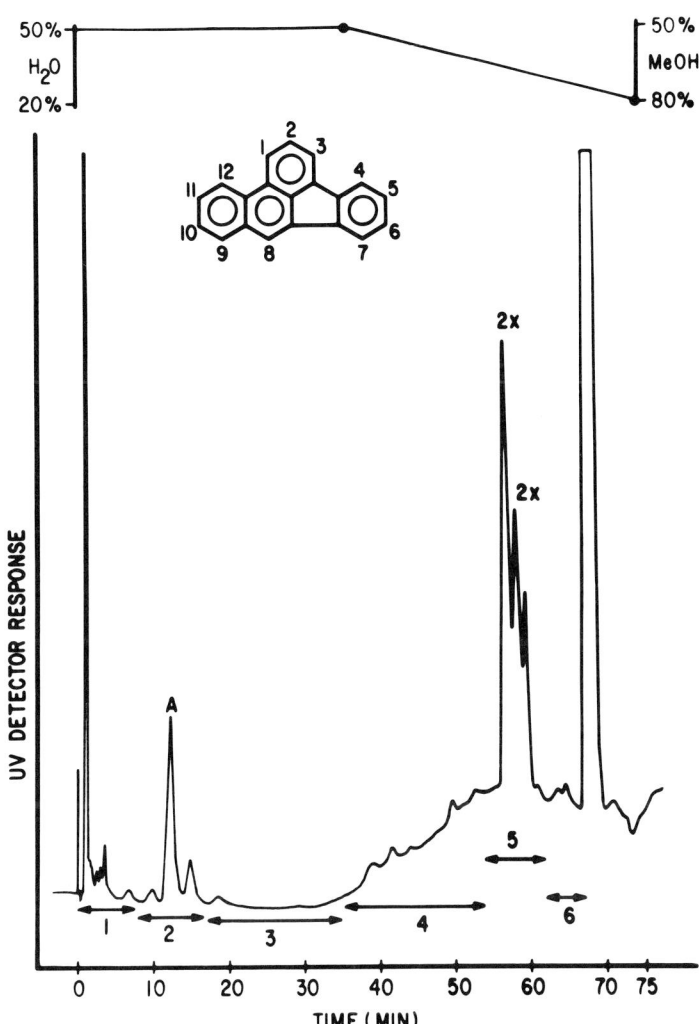

**FIGURE 6.** Metabolites formed from benzo(b)fluoranthene upon incubation with rat liver 9000 x g supernatant. Metabolite bands 1-6 are indicated at bottom. 2x = sensitivity/2. Peak A is a nonmutagenic dihydrodiol (see text).

and the corresponding ester was dehydrogenated to 4-(3-fluoranthenyl)butyric acid. Cyclization gave the desired ketone precursor. The integrity of the ring system was confirmed by conversion of an aliquot to benzo(b)fluoranthene by reduction, dehydration, and dehydrogenation. The ketone was converted to BbF-9,10-diol by the usual method.

The HPLC retention volume of BbF-9,10-diol was identical to a minor metabolite of BbF; however, their UV spectra were different. Thus,

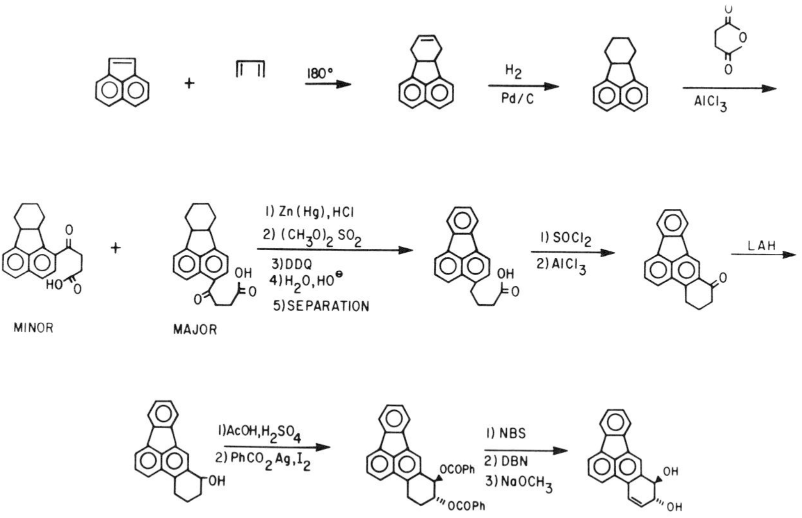

**FIGURE 7.** Synthesis of BbF-9,10-diol.

BbF-9,10-diol was not detected as a metabolite of BbF by rat liver 9,000 x g supernatant, at least under our conditions. We also failed to detect BbF-9,10-diol as a metabolite when incubations were done with 9,000 x g supernatants from Aroclor®-pretreated Sprague-Dawley rats, mice, or Syrian golden hamsters. When BbF-9,10-diol was incubated with rat liver 9,000 x g supernatant, a major polar metabolite, presumably a tetrol, was formed. No corresponding HPLC peak was observed in the metabolism of BbF, indicating that BbF-9,10-diol was not formed as a transient intermediate.

In agreement with the bay-region theory of PAH activation, BbF-9,10-diol was mutagenic toward *S. typhimurium* TA 100 with activation, as shown in Figure 8. This mutagenicity is presumably due to the formation of 9,10-dihydro-9,10-dihydroxy-11,12-epoxybenzo(b)fluoranthene as an ultimate mutagen. However, BbF-9,10-diol does not account for the mutagenicity of benzo(b)fluoranthene since it was not formed in significant quantities from benzo(b)fluoranthene using the same liver fractions employed for the mutagenicity assays. Since none of the metabolites of benzo(b)fluoranthene detected by HPLC showed significant mutagenicity, the formation of an unstable mutagen, possibly the 2,3- or 4,5-epoxide, is indicated.

### Benzo(k)fluoranthene

A high-pressure liquid chromatogram of the metabolites formed from benzo(k)fluoranthene in vitro by rat liver 9,000 x g supernatant is shown

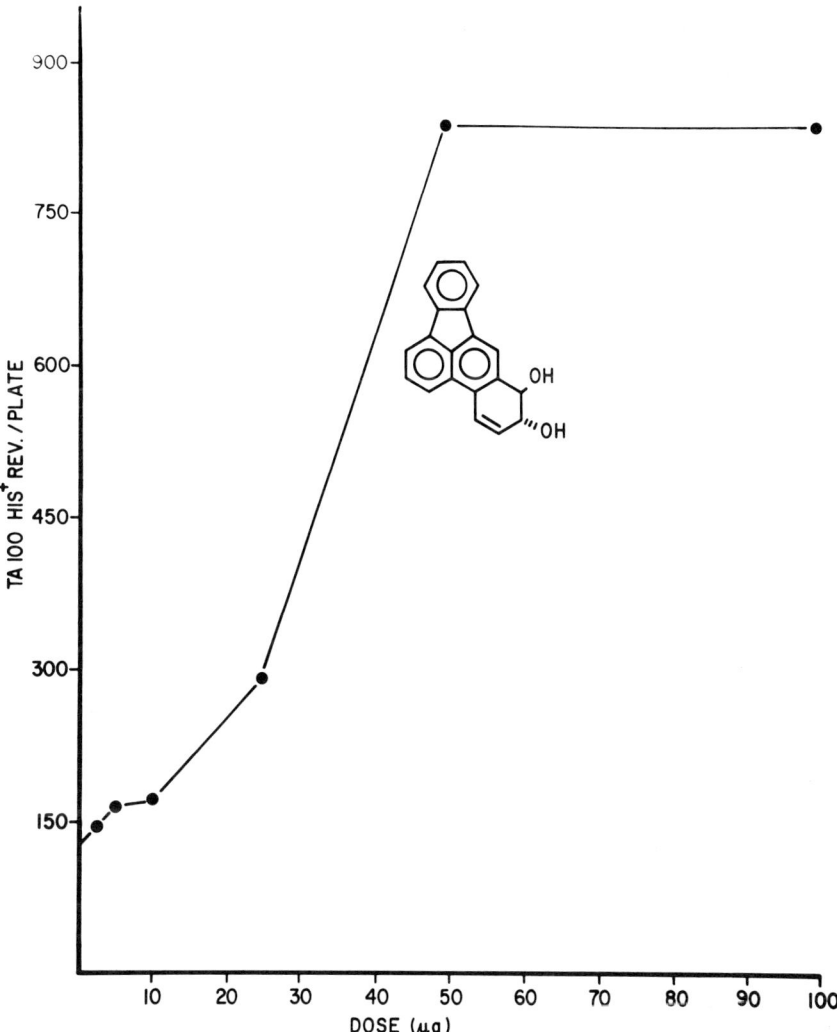

**FIGURE 8.** Mutagenicity toward *S. typhimurium* **TA 100** of B(b)F-9,10-diol.

in Figure 9. The relative simplicity of the metabolic pattern is probably a consequence of the symmetry of benzo(k)fluoranthene. The major peak, eluting at 36 min, disappeared when incubations were done in the presence of TCPO, while the minor peaks eluting at 45 to 55 min increased in intensity. Thus, the former was a dihydrodiol, while the latter were probably phenolic metabolites. The metabolic dihydrodiol was mutagenic toward *S. typhimurium* TA 100 when tested with activation. The two likely dihydrodiol metabolites of benzo(k)fluoranthene are the

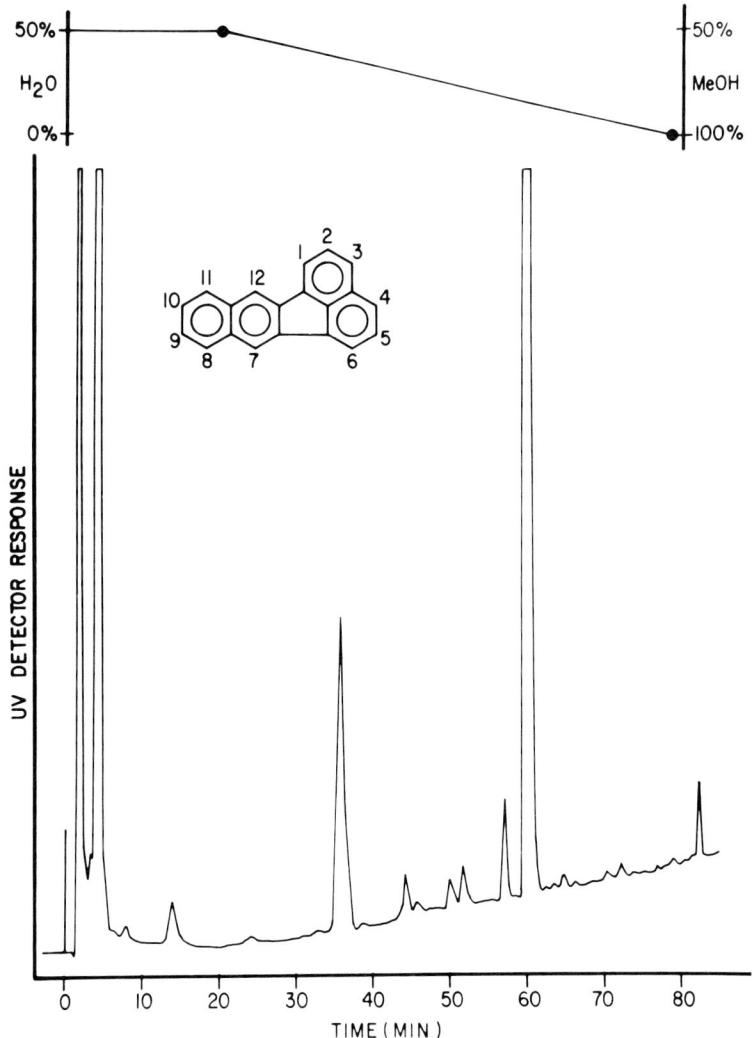

**FIGURE 9.** Metabolites formed from benzo(k)fluoranthene upon incubation with rat liver 9000 x g supernatant. The peak eluting at 36 min is B(k)F-8,9-diol (see text).

2,3- and 8,9-dihydrodiols. Only the latter can form a dihydrodiol epoxide. Therefore, we synthesized trans-8,9-dihydro-8,9-dihydroxybenzo(k)fluoranthene (BkF-8,9-diol). This synthesis was readily accomplished since the major product of succinoylation of fluoranthene was 4-(8-fluoranthenyl)-4-oxobutyric acid (6). This was cyclized to give 8-oxo-8,9,10,11-octahydrobenzo(k)fluoranthene (6), which was converted to BkF-8,9-diol by the same sequence used in the synthesis of the other two dihydrodiols.

The synthetic dihydrodiol had a UV spectrum, mass spectrum, and HPLC retention volume identical to those of the metabolic dihydrodiol. Thus, this proximate mutagen of BkF was identified as BkF-8,9-diol. The mutagenicity of synthetic BkF-8,9-diol toward *S. typhimurium* TA 100, with activation, is shown in Figure 10. These results indicate that BkF is activated, at least partially, by formation of a non-bay-region dihydrodiol epoxide as a proximate mutagen.

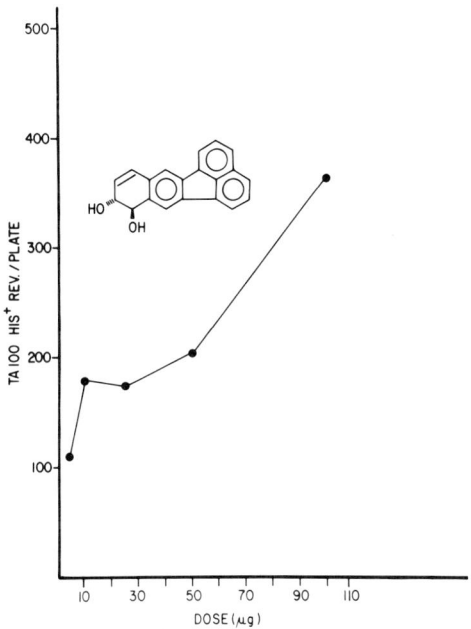

**FIGURE 10.** Mutagenicity toward *S. typhimurium* **TA 100** of **BkF-8,9-diol.**

## SUMMARY

The environmental agents benzo(j)fluoranthene, benzo(b)fluoranthene, and benzo(k)fluoranthene are mutagenic toward *S. typhimurium* TA 100 and benzo(j)fluoranthene and benzo(b)fluoranthene are carcinogenic when applied to mouse skin. Proximate mutagens of benzo(j)fluoranthene and benzo(k)fluoranthene were identified by comparison with synthetic standards as 9,10-dihydro-9,10-dihydroxybenzo(j)fluoranthene which can form a "pseudo-bay-region" dihydrodiol epoxide and 8,9-dihydro-8,9-dihydroxybenzo(k)fluoranthene which can form a non-bay-region dihydrodiol epoxide. A synthetic dihydrodiol of benzo(b)fluoranthene, 9,10-dihydro-9,10-dihydroxybenzo(b)fluoranthene, which can form a bay-region dihydrodiol epoxide, was mutagenic but was not detected as a metabolite of benzo(b)fluoranthene.

## ACKNOWLEDGMENTS

This study was supported by Grant No. ES-02030 from the National Institute of Environmental Health Sciences. Stephen S. Hecht is recipient of National Cancer Institute Research Career Development Award 5K04CA00124.

## REFERENCES

1. Amin, S., Benedenko, V., LaVoie, E., Hecht, S. S., and Hoffmann, D.: Synthesis of dihydrodiols as potential proximate carcinogens of benzofluoranthenes. *J. Org. Chem.* (submitted for publication).
2. Badger, G. M., Kimber, R.W.L., and Novotny, J. (1964): The formation of aromatic hydrocarbons at high temperatures. *Austral. J. Chem.* 17:778-786.
3. Baum, E. J. (1978): Occurrence and surveillance of polycyclic aromatic hydrocarbons. In: *Polycyclic Hydrocarbons and Cancer*, Vol. 1, H. V. Gelboin and P.O.P. T'so, Eds., pp. 45-70, Academic Press, New York.
4. Borneff, J., and Kunte, H. (1964): Carcinogenic substances in water and soil XVI. Evidence of PAH in water samples through direct extraction. *Arch. Hyg. Bakteriol.* 148: 585.
5. Borneff, J., and Kunte, H. (1965): Carcinogenic substances in water and soil XVII. About the origin and evaluation of the PAH in water. *Arch. Hyg. Bakteriol.* 149: 226.
6. Buu-Hoi, N. P., Lavit, D., and Lamy, J. (1959): A new synthesis of dibenz(a,c)anthracene, benzo(k)fluoranthene, and benzo(b)fluoranthene. *J. Chem. Soc.* 1845-1849.
7. Harvey, R. G., and Fu, P. P. (1978): Synthesis and reactions of diol epoxides and related metabolites of carcinogenic hydrocarbons. In: Polycyclic Hydrocarbons and Cancer, Vol. 1, H. V. Gelboin and P.O.P. T'so, Eds., pp. 133-165, Academic Press, New York.
8. Hecht, S. S., El Bayoumy, K., Tulley, L., and LaVoie, E. (1979): Structure-mutagenicity relationships of N-oxidized derivatives of aniline, o-toluidine, 2'-methyl-4-aminobiphenyl, and 3,2'-dimethyl-4-aminobiphenyl. *J. Med. Chem.* 22: 981-987.
9. Hoffmann, D., and Wynder, E. L. (1962): The isolation and identification of polynuclear aromatic hydrocarbons from gasoline engine exhaust condensate. *Cancer* 15: 93-102.

10. Hoffmann, D., and Wynder, E. L. (1977): Organic particulate pollutants-chemical analysis and bioassays for carcinogenicity. In: *Air Pollution*, Third Edition, Vol. II, Arthur C. Stern, Ed., pp. 361–455, Academic Press, New York.
11. International Agency for Research on Cancer (1973): *Monographs on the Evaluation of Carcinogenic Risk of Chemicals to Man*, Vol. 3, pp. 69–90, International Agency for Research on Cancer, Lyon.
12. Jerina, D. M., Yagi, H., Lehr, R. E., Thakker, D. R., Schaefer-Ridder, M., Karle, J. M., Levin, W., Wood, A. W., Chang, R. L., and Conney, A. H. (1978). In: *Polycyclic Hydrocarbons and Cancer*, Vol. 1, H. V. Gelboin and P.O.P. T'so, Eds., pp. 173–188, Academic Press, New York.
13. Lacassagne, A., Buu-Hoi, N. P., Zajdela, F., Lavit-Lamy, D., and Chalvet, O. (1963): Activité cancérogène d'hydrocarbures aromatiques polycycliques a noyau fluoranthène. *Acta. Univ. Intern. Contra Cancrum* 19:480–496.
14. LaVoie, E. J., Bedenko, V., Hirota, N., Hecht, S. S., and Hoffman, D. (1979): A comparison of the mutagenicity, tumor initiating activity and complete carcinogenicity of polynuclear aromatic hydrocarbons. In: *Polynuclear Aromatic Hydrocarbons*, P. W. Jones and P. Leber, Eds., pp. 705–721, Ann Arbor Science Publ., Ann Arbor, Michigan.
15. Nenitzescu, C. D., and Avram, M. (1950): The Polymerization of 1,2-dihydronaphtalene and dehydrogenating condensation of 1,2,3,-4-tetrahydronaphthalene. *J. Am. Chem. Soc.* 72:3486–3490.
16. Perin-Roussel, O., Ekert, B., Zajdela, F., and Jacquignon, P. (1978): Binding of dibenzo(a,e)fluoranthene, a carcinogenic, polycyclic hydrocarbon without K-region to nucleic acids in a subcellular microsomal system. *Cancer Res.* 38: 3499–3504.
17. Wynder, E. L., and Hoffmann, D. (1959): Carcinogenicity of benzofluoranthenes. *Cancer* 12: 1194–1199.
18. Wynder, E. L., and Hoffmann, D. (1967): *Tobacco and Tobacco Smoke,* Academic Press, New York.

# MULTIMEDIA HUMAN EXPOSURE AND CARCINOGENIC RISK ASSESSMENT FOR ENVIRONMENTAL PAH

J. Santodonato, D. Basu, and P. H. Howard
    Center for Chemical Hazard Assessment
    Syracuse Research Corporation
    Syracuse, New York 13210

Many regulatory approaches to reducing the exposure of humans to environmental contaminants have been media oriented. For many contaminants, this approach seems justified because the major release of the substance is by air emissions, water effluents, solid wastes, etc. However, many chemicals have multimedia environmental releases and resulting human exposures. PAH are a good example, in that humans may be exposed from air, water, food, and tobacco smoking. In the following, we review what is known about the quantitative aspects of exposure from these various sources. In addition, the data on the carcinogenic effects of PAH in experimental mammals are extrapolated to define levels of carcinogenic risk in humans. These figures are compared with estimates of the current human exposure from all media.

## HUMAN EXPOSURE FROM VARIOUS MEDIA

The quality of the monitoring data varies considerably depending upon the type of sample being considered. In general, the most extensive information is available on BaP, with significantly less information on other carcinogenic PAH and total PAH, and with scarcely any information available on nitrogen- or oxygen-containing polycyclics. To estimate human exposures to polycyclic organics, this study considers concentrations for BaP, carcinogenic PAH [BaP, benzo(j)fluoranthene, and indeno (1,2,3-cd)pyrene], and total PAH. However, it does not include the

## Ambient Air

BaP concentrations in ambient air decreased over a 10-year period from a median individual average concentration of 3.2 ng/m$^3$ in 1966 to 0.5 ng/m$^3$ in 1975 (1). In 1976, the highest value for BaP in an urban area was 2.9 ng/m$^3$ (2). These values are based upon analysis of samples taken in 34 urban sites by the EPA National Air Surveillance Network (NASN). Unfortunately, the average concentration for BaP is the only type of air monitoring data extensively available in the literature. The types and numbers of other PAH compounds monitored vary from one investigation to another. We used the 1974-75 Los Angeles monitoring data of Gordon (3) to provide estimates of the other PAH compounds (Table 1). The use of Los Angeles data is somewhat questionable because automobiles, a not-so-representative source of PAH, are major sources of air pollution in that area. However, the reported BaP concentration (3) (0.46 ng/m$^3$) is very close to the NASN median average (0.5 ng/m$^3$) for 1975 and the total PAH concentration (10.9 ng/m$^3$) is very close to the recent values (1975-1976) reported for total PAH in five Canadian cities (4) (mean concentrations, 11.1, 9.95, 18.6, 3.4, and 3.2 ng/m$^3$; the individual PAH measured were slightly different). Butler and Crossley (5) recently demonstrated that the concentrations of PAH monitored indoors and outdoors in suburban England are approximately the same; therefore, no indoor adjustment factor seems necessary. However, certain occupational or other factors may affect the atmospheric exposure for certain individuals (6).

In order to convert from the ambient concentration to inhalation intake, the amount of air breathed must be considered. The International Commission on Radiological Protection (7) indicates that 15 m$^3$ (10-year-

TABLE 1. U.S. Human Exposure to PAH Compounds Due to Inhalation of Ambient Air

|  | BaP | Carcinogenic PAH | Total PAH |
|---|---|---|---|
| Ambient Air (ng/m$^3$) | 0.5–2.9 | 2.0 | 10.9 |
| Inhalation Intake (μg/day)[a] | 0.0095–0.0435 | 0.038 | 0.207 |

[a] Assume average air breathed per day was 19 m$^3$.

old child) to 23 m$^3$ (adult man) per day are reasonable ranges; an average of 19 m$^3$ was used for deriving the values in Table 1. Using 2.5 ng/m$^3$ BaP and 15 m$^3$ of air a day, Butler and Crossley (5) concluded that a reasonable exposure to BaP from inhalation of air in England was 0.034 μg/day, which is in reasonably good agreement with the U.S. value in Table 1.

The processes that result in atmospheric contamination are clearly combustion or pyrolysis of hydrocarbon materials. Although the quantitative amounts from various sources are debatable, major sources include burning coal-refuse banks, coke production, residential fireplaces, forest fires (natural), coal-fired residential furnaces, and oil-fired commercial/institutional boilers (8,9,10).

## Water

On the basis of work by Basu and Saxena (11), the average concentrations of BaP, carcinogenic PAH, and total PAH in drinking water in 15 U.S. cities are 0.55 ng/l (range from not detected to 2.1 ng/l), 2.1 ng/l (0.2 to 11.3), and 13.5 ng/l (0.3 to 138.5), respectively. As can be seen by the range, these values vary considerably. The low concentrations somewhat reflect the extremely low water solubilities of PAH compounds. Slightly higher drinking-water values have been reported in Europe [e.g., 3 to 5 ng/l carcinogenic PAH (12,13) and 40 to 60 ng/l total PAH (13,14)], but these differences will have relatively negligible effects on the calculated daily intake values through drinking water compared with other sources. Assuming that a human consumes approximately 2 liters of water per day (7), the daily intake of PAH via drinking water would be as indicated in Table 2.

Water contamination can result in increased amounts of PAH in edible aquatic organisms [mollusks and edible algae, but probably not fish (15)] and in food stuffs irrigated by the contaminated water (16). Estimates from these sources are included in the food discussion below.

TABLE 2. U.S. Human Exposure to PAH Compounds Due to Drinking Water

|  | BaP | Carcinogenic PAH | Total PAH |
|---|---|---|---|
| Drinking Water (ng/l) | 0.55 | 2.1 | 13.5 |
| Intake (μg/day)[a] | 0.0011 | 0.0042 | 0.0270 |

[a] Assume average intake of 2 l/day.

No qualitative or quantitative data are available on sources of water contamination that result in drinking-water intake of PAH. Several reports indicate that PAH levels in water and sediments may be due to local sources such as oil spills (17), coke-oven effluents (18), and road runoff (19,20,21). However, some recent studies on the chemistry of PAH found in water sediments suggest that the main sources in the aquatic environment are anthropogenic, airborne, combustion-produced PAH (22). Thus, it is possible that the same sources that cause atmospheric contamination also contribute significantly to water contamination.

**Food**

It is difficult to evaluate the human dietary intake of PAH through foods because the amount depends not only on the food habits of the individual and the style of cooking, but also on the origin of the foods. To provide a reasonably accurate estimate of the PAH dietary intake, average concentrations of PAH in representative food items would have to be available. Although considerable information on food PAH levels has been reported, the information has been generated on grab samples in different laboratories using different analytical procedures. The available monitoring data are somewhat slanted toward samples that are likely to have higher BaP and PAH concentrations (e.g., smoked products and crops raised in "polluted environments"). In addition, very limited monitoring data are available on milk and dairy products (cheese), which make up approximately one-third of the daily diet (7) of 1600 g (milk—508 g/day; cheese—19 g/day). Table 3 summarizes the available data concerning PAH levels in food. Because of the inadequate and biased analytical data, somewhat lower ranges of values were selected for human-intake calculations (0.1 to 1.0 ppb for BaP and 1 to 10 ppb for PAH) in Table 4. Combining these ranges with average total daily food consumption of 1600 g/day (7) by man from all types of foods, the estimates of dietary PAH and BaP intake indicated in Table 4 seem reasonable.

The sources of PAH food contamination are even more complex than water-contamination sources. However, from the available data, some general preliminary conclusions can be reached. Because above-ground parts of crops usually contain ten times as much BaP as underground parts (50), atmospheric deposition may be a major source in many crops. However, root crops (carrots, potatoes, etc.) may absorb some PAH compounds from the soil (51) (levels in soil may be related to atmospheric deposition or contaminated water). Seafood, especially filter feeders (clams, oysters), is likely to have PAH concentrations that reflect the

TABLE 3. Range of PAH Levels in Food (ppb)

|  | BaP | Total PAH | References |
|---|---|---|---|
| Vegetable oils and margarine | 0.2–6.8 | 2.1–136 | 23–25 |
| Fish and other aquatic foods: |  |  |  |
| Smoked | Trace–6.6 | 5.2–162 | 26–31 |
| Non-smoked | 0 | 1.8–3.2 | 26–31 |
| Smoked meat and meat products | Trace–3.6 | 1.5–150 | 26,27,29,30, 32,33 |
| Cooked meat: |  |  |  |
| Charcoal Broiled: |  |  |  |
| Hamburger | 0–2.6[a] | 0.3–43.9[a] | 34 |
| Steak | 4.4–50.4 | 70–183.7 | 34 |
| Barbecued | 3.5–10.5 | 37.5–186.1 | 32,35,36 |
| Fruits | ND[b]–29.7[c] | — | 37,39,40 |
| Grain and cereal products | 0.1–60 | — | 38–40,42,43 |
| Sugar and adjuncts | 0.2–72.0 | — | 38,44,45 |
| Vegetables | ND–24.3[c] | — | 38,39,41 |
| Beverages | ND–21.3 | — | 24,40,46–49 |

[a] Depending upon the way the meat is cooked and its fat content.
[b] Not detected.
[c] Polluted environment.

TABLE 4. U.S. Human Exposure to PAH Compounds Due to Food Consumption

|  | BaP | Total PAH |
|---|---|---|
| Food (ppb) | 0.1–1.0 | 1.0–10.0 |
| Intake ($\mu$g/day)[a] | 0.16–1.6 | 1.6–16 |

[a] Assume average total daily food consumption of 1600 g/day.

levels in the water and this again may be due somewhat to atmospheric deposition or to local effluent sources (52–54). PAH levels in fish are likely to be low because of the considerable ability of fish to metabolize PAH, especially BaP (15). Meats and smoked products appear to have considerable PAH input from the way in which they are cooked or prepared. For example, a charcoal-broiled steak can contain as much as

50 ppb BaP (34) depending upon the way in which it is cooked (hot or cool fire). Thus, food contamination by PAH appears to result from several sources, the individual magnitude of which varies depending upon the type of food.

## TOTAL HUMAN EXPOSURE

Table 5 summarizes the calculated exposure estimates from various sources. The most significant result of these calculations is the high relative exposure likely to be provided by food. However, it should be pointed out that of the three estimates, the food calculation is the least exact because of the lack of reliable monitoring data. Nevertheless, all the estimates are probably accurate within an order of magnitude, which would still leave food as the overwhelming source of PAH exposure to humans. Even smoking one pack of cigarettes, which has been estimated to result in a BaP exposure of 0.4 $\mu$g/day (6), could result in less than the projected exposures from food. Also, food preparation can result in significant PAH exposure. For example, one charcoal-broiled steak may contain an amount of BaP equivalent to that in the smoke from 22.5 packs of cigarettes (55).

The projections of exposure in Table 5 are compatible with previous estimates. For example, Shabad and Il'nitskii (16) state that they believe the amount of carcinogenic PAH consumed by man via water is typically 0.1 percent of the amount he consumes via food. In Table 5, the calculated amount of total PAH exposure from water is 1.0 to 0.1 percent of the amount of exposure from food. Borneff (19) estimates that man's yearly PAH intake is 3 to 4 mg (8 to 11 $\mu$g/day) from fruits, vegetables, and bread; 0.1 mg (0.2 $\mu$g/day) from fats and oils of vegetable origin; and 0.05 mg (0.1 $\mu$g/day) from smoked meat or fish and drinking water. All these values are within the limits of the estimates in Table 5, but the basis for Borneff's (19) estimates is not known. The PAH intake values from

TABLE 5. Estimated Human Exposure to PAH from Various Sources ($\mu$g/Day)

| Source | BaP | Carcinogenic PAH[a] | Total PAH |
|---|---|---|---|
| Air | 0.0095–0.0435 | 0.038 | 0.207 |
| Water | 0.0011 | 0.0042 | 0.0270 |
| Food | 0.16–1.6 | b | 1.6–16 |

[a] Total of BaP, benzo(j)fluoranthene, and indeno(1,2,3-cd)pyrene.
[b] No data available.

food consumption in Table 5 are based upon the assumption that the average concentration of PAH in food will be 1 to 10 ppb. This seems to be a reasonable range based upon the available monitoring data but needs to be confirmed, and major portions of the diet (e.g., dairy products) for which no information is available need to be analyzed.

## CARCINOGENIC RISK

Epidemiological evidence concerning the health effects of PAH comes from both occupational and community air pollution studies (56). However, the effects of ingested PAH on humans have been largely ignored, despite the fact that this appears to be quantitatively the most significant route of exposure in nontobacco smokers. The absence of health-effects data concerning PAH ingestion is the most important information gap in the present epidemiology data base.

Occupational studies, especially among gas-production workers and coke-plant workers, show that long-term exposure to the products of the destructive distillation of coal is clearly associated with an elevated rate of lung cancer as well as cancer of the digestive tract and other sites (57-72). The relative risk for lung cancer is reported to range from 1.6 to as high as 33. For cancer at other sites, relative risks have been determined for kidney (7.5), stomach (1.7), bladder (1.7), blood (1.7), and skin (4.0). However, the results of most occupational studies are difficult to interpret because of the abundance of confounding variables (e.g., tobacco smoking). The greatest sources of variability probably lie in the selection of exposed and control populations, and in the procedure for estimating intensity of exposure to PAH.

Community studies, which involve much lower levels of inhalation exposure and are more problematic because of the heterogenous populations involved, attempt to detect an association between community morbidity and mortality rates and some direct or indirect index of air pollution. Several approaches have been taken in the conduct of large-scale epidemiologic studies of the association between community levels of air pollution and cancer mortality. In general, four methodological categories can be designated: (a) urban-rural comparisons (73-83); (b) migrant studies (84-87); (c) regression analyses of lung cancer death rates for residents of countries or states with respect to gross indicators of cigarette consumption, industrialization, and air pollution (88-97); and (d) retrospective and prospective analyses of data for sampled persons with individually identified exposures and background characteristics (77,78,98-101). These studies show, in certain cases, an excess of respiratory cancers and malignancies of other sites (e.g., stomach, prostrate) associated with elevated PAH levels in urban settings.

Although there is general agreement on the existence of an "urban factor" for cancer mortality, there is no consensus that the urban factor is air pollution. Alternative explanations proposed to account for the urban-rural differences in lung cancer mortality include: (a) migration to cities for medical treatment; (b) increased bacterial and viral infection in cities; (c) increased bronchitis in cities; (d) increased occupational exposure to carcinogens in cities; and (e) urban/rural variations in smoking habits.

However, studies of local variations by Winklestein and co-workers (102) in Buffalo and Hagstrom and co-workers (103) in Nashville show no consistent relationship between air pollution levels and lung cancer incidence within the urban environment. These studies should help to inhibit the tendency to equate the residual urban-rural gradient, after correcting for age and smoking, with the direct effects of air pollution. An additional finding in both studies, which could not be explained, was an association between stomach cancer and particulate air pollution. The possibility that particulates containing adsorbed PAH were cleared from the lungs by mucociliary action and subsequently swallowed seems obvious. However, the possible contribution of dietary factors was not considered in these or any other studies of PAH pollution and human cancer. In addition, fundamental differences in urban-rural dietary habits may be confounding any interpretation of cancer mortality data with regard to air pollution in general and PAH pollution in particular.

In light of our knowledge regarding the contribution of various environmental media (i.e., air, food, water) to the total PAH exposure in man, it becomes important to consider a multimedia approach to cancer-risk assessment. Furthermore, it seems that the tendency to focus on lung cancer is unwarranted if one considers both the available epidemiologic evidence and our knowledge of dietary sources of PAH.

Unfortunately, when evaluating the carcinogenic threat of PAH to man, no one has yet considered the relevance of dietary sources or their impact on gradients of cancer for sites other than the respiratory tract. This may be especially important in light of the fact that the absolute rate of digestive-tract cancer in both rural and urban populations can be considerably higher than the rate of respiratory-tract cancer. In addition, the oral administration of PAH to experimental mammals commonly produces lymphomas, leukemia, and tumors of the lung, skin, stomach, and endocrine tissues (56). Thus, the possibility is raised that the true carcinogenic impact of PAH in humans may be underestimated by the historical practice of considering inhalation as the primary route of exposure.

## DERIVATION OF AN ALLOWABLE DAILY INTAKE

In the absence of reliable dose-response information for PAH-induced cancer in man, it is possible to turn to statistical models for predicting human carcinogenic risks based on animal data. However, several problems must be dealt with in the use of such mathematical models: (a) selection of test chemicals and route of administration; (b) extrapolation from high doses in animals to low doses in man; (c) species-to-species conversion of results (i.e., rodent-to-human); and (d) partitioning of acceptable dose among various environmental media. Because man is exposed to PAH in the form of complex mixtures and by several routes of absorption, it is unlikely that an ideal animal study can be selected for application to statistical models. Nevertheless, a reasonably conservative assumption can be drawn, namely, that data from animal studies employing a single PAH of high carcinogenic activity [e.g., BaP, DB(ah)A] will provide a valid carcinogenic risk estimate for the entire class. In addition, animal studies are available which employ various routes of administration, so that it is possible to calculate allowable daily intakes (ADIs) for PAH by dietary intake, gavage, or intratracheal instillation.

The choice of an appropriate model for risk extrapolation must take into account that very little is known about the events occurring at the lower end of a carcinogenesis dose-response curve. This end of the curve, however, is of greatest interest for human-risk assessment. The choice of a linear model is often recommended because it is unlikely to underestimate the risk at low-dose levels. The linear model is considered very conservative because carcinogenic response often provides a concave dose-response curve in the low-dose range. On the other hand, because of the background incidence of cancer that exists in any large population, the linear model probably provides a very reliable estimate of risk at low doses in the environment.

For the calculation of an ADI for PAH by several routes of exposure based on stochastic effects (tumor occurrence) in animals, a modification of a model developed by the U.S. Environmental Protection Agency may be employed. The following equation is derived from a one-hit, nonthreshold-dose-response model:

$$\text{ADI} = \frac{(d \times le \div Le) \times (Le \div L)^3 \times \text{HBW} \times (\text{RL})}{\ln[(1-Pt) \div (1-Pc)] \times (\text{HBW} \div W)^{1/3}} \qquad (1)$$

where

| | | |
|---|---|---|
| ADI | = | acceptable daily intake (mg) |
| d | = | dose (mg/kg/day) |
| le | = | length of exposure (days) |
| Le | = | duration of experiment—surviving animals sacrificed at termination |
| L | = | expected lifespan for test animals |
| HBW | = | human body weight (kg)—assume 70 kg |
| RL | = | additional risk over lifetime—use 1/100,000 |
| Pt | = | proportion of test animals with tumors |
| Pc | = | proportion of control animals with tumors |
| W | = | average weight of test animals (kg). |

This model estimates the intake of a compound which would be associated with a 1/100,000 increased lifetime risk of a 70-kg human developing cancer. Most of the expressions in this equation represent more or less reasonable assumptions about the chemical basis of carcinogenesis.

The expression (d × le ÷ Le) is a conversion of the administered dose (d) to the lifespan-weighted average dose (D). Because an ADI is intended to define risk levels for lifetime exposures, such an adjustment factor is clearly desirable. This specific adjustment factor presumes that D × Le will elicit the same response as d × le. This is based on the assumption that the total cumulative dose, rather than the dose schedule, is the critical factor in generating the carcinogenic response. This assumption is currently accepted for carcinogenesis due to radiation. However, the appropriateness of this assumption for PAH, particularly because they require metabolic activation, is not known. Further, this expression does not distinguish between compounds that are rapidly eliminated from the body and those that are concentrated by and sequestered in the body for long periods of time. Nonetheless, the use of the lifespan-weighted average dose seems to be a reasonable approximation, given the current understanding of the effects of dose schedule on chemical carcinogenesis and the lack of detailed pharmacokinetic data on many of the compounds for which an ADI must be established.

The term $(Le \div L)^3$ is meant to take into account the relationship of dose (D) to the time until the appearance of neoplasms (t). Empirical observations of both chemical- and radiation-induced carcinogenesis indicate that this relationship may be expressed as:

$$t = t_o (D_o/D)^{1/3} \qquad (2)$$

Going back to Equation (1) and taking the calculated ADI as a function of $D_o$, $t_o$ equal to life expectancy (L), t equal to the experimental observation period (Le), and D equal to the experimental lifespan-weighted average dose, Equation (2) is incorporated as:

$$D_o = D \left(\frac{t}{t_o}\right)^3$$

or

$$ADI/k = (d \times le \div Le) \times (Le \div L)^3 \qquad (3)$$

where k is equal to the remaining terms in the right-hand side of Equation (1). Thus, the two expressions on the right-hand side of Equation (3) may be regarded as a lifespan-weighted average dose corrected for life expectancy.

The term $(HBW \div W)^{1/3}$ serves as an adjustment factor for applying the results from a given experimental mammal with a body weight of W to humans, where a body weight (HBW) of 70 kg is assumed. This cube root of the ratio of body weights is essentially an adjustment on a dose per unit surface area basis and may not be suitable for chemicals such as PAH which require metabolic activation.

Using the model described above, several ADIs were calculated from animal carcinogenesis data on BaP, DB(ah)A, and mixtures of PAH found in automobile exhaust. Studies involving exposure by ingestion, dermal application, and intratracheal instillation of BaP are included for the purpose of comparison, and because all routes of absorption for PAH may be relevant to environmental situations. Table 6 summarizes the data used for calculation of ADIs. The derived ADIs for BaP and other carcinogenic PAH are quite similar except where skin painting was the route of administration. This discrepancy is most likely due to the fact that mouse skin is highly sensitive to the action of carcinogenic PAH.

The most noteworthy aspect of the ADIs shown in Table 6 is that the average human diet currently results in exposure to PAH which is in excess of all calculated ADIs (see Table 7) and may in fact surpass tobacco smoking as a major contributor to human PAH body burden in certain cases (e.g., high consumption of charcoal-broiled and smoked meats). PAH levels in air and water do not in themselves appear to present an unacceptable risk based upon the ADIs calculated from animal data. Whether the PAH content of the average human diet is actually responsible for a portion of current cancer incidence in this country, however, cannot be reliably ascertained at this time. Nevertheless, these data raise the possibility that an excessive risk for PAH-induced cancer may presently exist in both urban and rural populations, and that this excess risk may be largely self-imposed.

TABLE 6. Calculation of Allowable Daily Intake (ADI) for Carcinogenic PAH

| Compound | Species | Route | Tumor Site | Dose (mg/kg/day) | Duration of Exposure (days) | Duration of Experiment (days) | Number of Animals | Number with Tumors (%) | Calculated ADI (ng/day) | References |
|---|---|---|---|---|---|---|---|---|---|---|
| BaP | CFW mouse | Dietary | Fore-stomach | 5.14 | 110 | 162 | 40 | 4(10) | 47 | 104 |
| BaP | Hamster | Intra-tracheal | Respiratory tract | 0.109 | 238 | 238 | 39 | 12(40) | 48 | 105 |
| BaP | C57BL/6J mouse | Dermal | Skin | 0.05 | 420 | 420 | 27 | 26(96) | 4.7 | 106 |
| DB(ah)A | DBA/2 mouse | Oral | Lung | 23 | 258 | 347 | 27 | 24(89) | 108 | 107 |
| PAH mixture[a] | Mouse | Dermal | Skin | 0.563 | 548 | 548 | 89 | 46(52) | 43 | 108 |
| PAH mixture[b] | Mouse | Dermal | Skin | 0.033 | 548 | 548 | 81 | 29(36) | 4.2 | 108 |

[a] Includes 0.033 mg/kg carcinogenic and 0.530 mg/kg noncarcinogenic components.
[b] Mixture of BaP, DB(ah)A, B(a)A, and B(b)F, 1:0.7:1.4:0.9.

**TABLE 7. Total Exposure to BaP in Relation to Allowable Daily Intake**

|  | Amount BaP (ng) | Relative Proportion |
| --- | --- | --- |
| Allowable Daily Intake | 48 | 1.0 |
| Water | 1.1 | 0.02 |
| Air | 9.3–43.5 | 0.20–0.91 |
| Food | 160–1600 | 3.3–33 |
| Smoking | 400 | 8.3 |

## REFERENCES

1. Faoro, R. B., and Manning, J. A. (1966-1975): Trends in benzo(a)pyrene. U.S. Environmental Protection Agency, Research Triangle Park, North Carolina.
2. Manning, J. Personal communication. U.S. Environmental Protection Agency, Research Triangle Park, North Carolina.
3. Gordon, R. J. (1976): Distribution of airborne polycyclic aromatic hydrocarbons throughout Los Angeles. *Environ. Sci. Technol.* 10:370-373.
4. Katz, M., Sakuma, T., and Ho, A. (1978): Chromatographic and spectral analysis of polynuclear aromatic hydrocarbons: Quantitative distribution in air of Ontario cities. *Environ. Sci. Technol.* 12:909-915.
5. Butler, J. D., and Crossley, P. (1979): An appraisal of relative airborne suburban concentrations of polycyclic aromatic hydrocarbons monitored indoors and outdoors. *Sci. Total Environ.* 11:53-58.
6. Bridbord, K., Finklea, J. F., Wagoner, J. K., Moran, J. B., and Caplan, P. (1976): Human exposure to polynuclear aromatic hydrocarbons. In: *Carcinogenesis, Vol. 1. Polynuclear Aromatic Hydrocarbons: Chemistry, Metabolism, and Carcinogenesis*, R. I. Freudenthal and P. W. Jones, Eds., pp. 319-324, Raven Press, New York.
7. International Commission on Radiological Protection (1974): *Report of the Task Group on Reference Man*, Pergamon Press, New York.
8. National Academy of Sciences (1972): Particulate polycyclic organic matter. Committee on Biological Effects of Atmospheric Pollutants, Washington, D.C.
9. U.S. Environmental Protection Agency (1975): Scientific and technical assessment report on particulate polycyclic organic matter (PPOM). Publication No. EPA-600/6-75-001, Washington, D.C.

10. Preliminary Assessment of the Sources, Control and Population Exposure to Airborne Polycyclic Organic Matter (POM) as Indicated by Benzo(a)pyrene B(a)P. Draft report by Energy and Environmental Analysis, Inc., VA. to U.S. Environmental Protection Agency, N. C., May 1978.
11. Basu, D. K., and Saxena, J. (1977-1978): Analysis of raw and drinking water samples for polynuclear aromatic hydrocarbons. EPA P.O. CA-7-2999-A, and CA-8-2275-B, Exposure Evaluation Branch, HERL, Cincinnati, Ohio.
12. Borneff, J. (1964): Carcinogenic substances in water and soil. Part XV: Interim Results of the Former Investigations. *Arch. Hyg. Bakt. (Berl.)* 148:1.
13. Borneff, J., and Kunte, H. (1964): Carcinogenic substances in water and soil. Part XVI: Evidence of PAH in water samples through direct extraction. *Arch. Hyg. Bakt. (Berl.)* 148:585.
14. Borneff, J., and Kunte, H. (1969): Carcinogenic substances in water and soil. Part XXVI: A routine method for the determination of PAH in water. *Arch. Hyg. Bakt. (Berl.)* 153:220.
15. Lu, P.-Y., Metcalf, R. L., Plummer, N., and Mandel, D. (1977): The environmental fate of three carcinogens: Benzo(a)pyrene, benzidine, and vinyl chloride evaluated in laboratory model ecosystems. *Arch. Environ. Contam. Toxicol.* 6(2-3):129-142.
16. Shabad, L. M., and Il'nitskii, A. P. (1970): Perspective on the problem of carcinogenic pollution in water bodies. *Gig. Sanit.* 29:19.
17. Giger, W., and Blumer, M. (1974): Polycyclic aromatic hydrocarbons in the environment: Isolation and characterization by chromatography, visible, ultraviolet, and mass spectrometry. *Anal. Chem.* 46:1663.
18. Fedorenko, Z. P. (1964): The effect of biochemical treatment of waste water of a by-product coke plant on the BP content. *Gig. Sanit.* 29:19.
19. Borneff, J. (1977): Fate of carcinogens in aquatic environments. In: *Fate of Pollutants in the Air and Water Environments, Part 2*, I. H. Suffet, Ed., pp. 393-408, J. Wiley & Sons, New York.
20. Andelman, J. B., and Suess, M. J. (1970): Polynuclear aromatic hydrocarbons in the water environment. *Bull. Wld. Health Org.* 43:479-508.
21. Suess, M. J. (1970): Presence of polynuclear aromatic hydrocarbons in coastal waters and the possible health consequences. *Rev. Int. Oceanogr. Med.* 18:181.

22. Hase, A., and Hites, R. A. (1978): On the origin of polycyclic aromatic hydrocarbons in the aqueous environment. In: *Identification and Analysis of Organic Pollutants in Water*, L. H. Keith, Ed., Ann Arbor Science, Ann Arbor, Michigan.
23. Howard, J. W., Turicchi, E. W., White, R. H., and Fazio, T. (1966): Extraction and estimation of polycyclic aromatic hydrocarbons in vegetable oils. *J. Assoc. Off. Anal. Chem.* 49:1236.
24. Swallow, W. H. (1976): Survey of polycyclic aromatic hydrocarbons in selected foods and food additives available in New Zealand. *New Zealand J. of Sci.*19:407-412.
25. Biernoth, G., and Rost, H. E. (1967): The occurrence of PAH in coconut oil and their removal. *Chem. Ind.* 45:2202-2203.
26. Malanoski, A. J., Greenfield, E. L., Barnes, C. J., Worthington, J. M., and Joe, F. L., Jr. (1968): Survey of polycyclic aromatic hydrocarbons in smoked foods. *J. Assoc. Off. Anal. Chem.* 51:114.
27. Thorsteinsson, T. (1969): Polycyclic hydrocarbons in commercially and home-smoked food in Iceland. *Cancer* 23:455.
28. Dungal, N. (1961): Can smoked food be carcinogenic? *Acta Unio Intern. Contra. Cancrum.* 17:365-366.
29. Howard, J. W., Teague, R. T., White, R. H., and Fry, B. E. (1966): Extraction and estimation of PAH in smoked foods. Part I. General method. *J. Assoc. Off. Anal. Chem.* 49:595-611.
30. Howard, J. W., White, R. H., Fry, B. E., and Turicchi, E. W. (1966): Extraction and estimation of polycyclic aromatic hydrocarbons in smoked foods. Part II. Benzo(a)pyrene. *J. Assoc. Off. Anal. Chem.* 49:611.
31. Masuda, Y., and Kuratsune, M. (1971): Polycyclic aromatic hydrocarbons in smoked fish, Katsuibushi. *Gann* 62:27.
32. Panalaks, T. (1976): Determination and identification of polycyclic aromatic hydrocarbons in smoked and charcoal-broiled food products by high pressure liquid chromatography. *J. Environ. Sci., Health* B 11:399-415.
33. Lo, M., and Sandi, E. (1978): Polycyclic aromatic hydrocarbons (polynuclears) in foods. In: *Residue Reviews*, Vol. 69, pp. 34-86, Gunther and Gunther, Eds., Springer-Verlag.
34. Lijinsky, W., and Ross, A. E. (1967): Production of carcinogenic polynuclear aromatic hydrocarbons in cooking of food. *Food and Cosmetics Toxicol.* 5:343.
35. Lijinsky, W., and Shubik, P. (1965): PH carcinogens in cooked meat and smoked fish. *Ind. Med. Surg.* 34:152-154.
36. Frethein, K. (1976): Carcinogenic polycyclic aromatic hydrocarbons in Norwegian smoked meat. *J. Agr. Food Chem.* 24:976.

37. Shiraishi, Y., Shiratori, T., and Takabatake, E. (1975): Determination of polycyclic aromatic hydrocarbons in foods. IV. 3,4-Benzopyrene in fish and shellfish. *J. Food Hyg. Soc. Japan (Shokuhin Eiseigaku Zasshi)* 16:178-181.
38. Shiraishi, Y., Shiratori, T., and Takabatake, E. (1973): Determination of polycyclic aromatic hydrocarbons in foods. II. 3,4-Benzopyrene in Japanese Foods. *J. Food Hyg. Soc. Japan (Shokuhin Eiseigaku Zasshi)* 14:173-178.
39. Kolar, L. R., Ledvina, J. T., and Hanus, F. (1975): Contamination of soil, agricultural plants, and vegetables by 3,4-benzopyrene in the Ceske Budejovice. *Cesk. Hyg.* 20:135.
40. IARC Monographs on the Evaluation of Carcinogenic Risk of the Chemical to Man, Vol. 3, Lyon, France.
41. Shiraishi, Y., Shiratori, T., and Takabatake, E. (1974): Polycyclic aromatic hydrocarbons in foods. III. 3,4-Benzopyrene in vegetables. *J. Food Hyg. Soc. Japan (Shokuhin Eiseigaku Zasshi)* 15:18-21.
42. Siddiqui, I., and Wagner, K. H. (1972): Determination of 3,4-benzpyrene and 3,4-benzofluoranthene in rain water, ground water, and wheat. *Chemosphere* 1:83-88.
43. Graf, W., and Nowak, W. (1966): Promotion of growth in lower and higher plants by carcinogenic polycyclic aromatics. *Arch. Hyg. Bakt. (Berl.)* 150:513-528. ORNL/tr.-4111.
44. Kuratsune, M. (1956): Benzo(a)pyrene content in certain pyrogenic materials. *J. Nat. Cancer Inst.* 16:1485-1496.
45. Fabian, B. (1969): Carcinogenic substances in edible fat and oil. Part VI. Further investigations on margarine and chocolate. *Arch. Hyg. Bakt. (Berl.)* 153:21-24.
46. Masuda, Y., Mori, K., Hirohata, T., and Kuratsune, M. (1966): Carcinogenesis in the esophagus. III. Polycyclic aromatic hydrocarbons and phenols in whiskey. *Gann* 57:549.
47. Vitzthum, O. G., Werkhoff, P., and Hubert, P. (1975): New volatile constituents of black tea aroma. *J. Agr. Food Chem.* 23:999-1102.
48. Kuratsune, M., and Hueper, W. C. (1958): Polycyclic aromatic hydrocarbons in coffee soots. *J. Nat. Cancer Inst.* 20:37.
49. Kuratsune, M., and Hueper, W. C. (1960): Polycyclic aromatic hydrocarbons in roasted coffee. *J. Nat. Cancer Inst.* 24:463.
50. Kolar, L. (1975): Contamination of soils and agricultural crops with the carcinogenic 3,4-benzopyrene and its causes. *Rostl. Vyroba* 21:261.
51. Shabad, L. M. (1975): Possibility of setting a permissible level of benzo(a)pyrene in the soil. *Gig. Sanit.,* 4:88.
52. Cahnmann, H. J., and Kuratsune, M. (1957): Determination of polycyclic aromatic hydrocarbons in oysters collected in polluted water. *Anal. Chem.* 29:1312.

53. Guerrero, H., Biehl, E. R., and Kenner, C. T. (1976): High-pressure liquid chromatography of benzo(a)pyrene and benzo(ghi)perylene in oil-contaminated shellfish. *J. Assoc. Off. Anal. Chem.* 59:989.
54. Dunn, B. P., and Stich, H. F. (1976): Release of the carcinogenic benzo(a)pyrene from environmentally contaminated mussels. *Bull. Environ. Contamin. Toxicol.* 15:398.
55. Lijinsky, W., and Shubik, P. (1964): Benzo(a)pyrene and other polynuclear hydrocarbons in charcoal-broiled meat. *Science* 145:153.
56. Santodonato, J., Howard, P. H., and Basu, D. K.: Multimedia health assessment document for polycyclic organic matter. *J. Environ. Pathol. Toxicol.* (in press).
57. Maxumdar, S., Redmond, C., Sollecito, W., and Sussman, N. (1975): An epidemiological study of exposure to coal tar pitch volatiles among coke oven workers. *APCA J.* 25(4):382-389.
58. Menck, H. R., and Henderson, B. E. (1976): Occupational differences in rates of lung cancer. *J. Occup. Med.* 18(12):797-801.
59. Henry, S. A., Kennaway, N. M., and Kennaway, E. L. (1931): The incidence of cancer of the bladder and prostate in certain occupations. *J. Hyg.* 31:125-137.
60. Kuroda, S. (1937): Occupational pulmonary cancer of generator gas workers. *Ind. Med. Surg.* 6:304-306.
61. Kawai, M., Amanoto, H., and Harada, K. (1967): Epidemiologic study of occupational lung cancer. *Arch. Environ. Health* 14:859-864.
62. Bruusgaard, A. (1959): The occurrence of certain forms of cancer among employees in gasworks. *Tid. for den Norske Laegeforen* 79:755-756.
63. Reid, D. D., and Buck, C. (1956): Cancer in coking plant workers. *Brit. J. Ind. Med.* 13:265-269.
64. Lloyd, J. W. (1971): Long term mortality study of steelworkers. V. Respiratory cancer in coke plant workers. *J. Occup. Med.* 13(2):53-68.
65. Doll, R. (1952): The causes of death among gas workers with special reference to cancer of the lung. *Brit. J. Ind. Med.* 9:180-187.
66. Doll, R., Fisher, R.E.W., Gammon, E. J., Gunn, W., Hughes, G. O., Tyrer, F. H., and Wilson, W. (1965): Mortality of gas workers with special reference to cancers of the lung and bladder, chronic bronchitis, and pneumoconiosis. *Brit. J. Ind. Med.* 22:1-12.
67. Doll, R., Vessey, M. P., Beasley, R.W.R., Buckley, A. R., Fear, E. C., Fisher, R.E.W., Gammon, E. J., Gunn, W., Hughes, G. O., Lee, K., and Norman-Smith, B. (1972): Mortality of gasworkers—final report of a prospective study. *Brit. J. Ind. Med.* 29:394-406.

68. Redmond, C. K., Ciocco, A., Lloyd, J. W., and Rush, H. W. (1972): Long term mortality study of steelworkers. *J.O.M.* 14(8):621-629.
69. Redmond, C. K. (1976): Epidemiological studies of cancer mortality in coke plant workers. Presented at Seventh Conference on Environmental Toxicology, Dayton, Ohio, October 13-15, 1976.
70. Redmond, C. K., Strobino, B. R., and Cypess, R. H. (1976): Cancer experience among coke by-product workers. *Ann. N.Y. Acad. Sci.* pp. 102-115.
71. Hendricks, N. V., Berry, C. M., Lione, J. G., Thorpe, J. J. (1959): Cancer of the scrotum in wax pressmen. I. Epidemiology. *A.M.A. Arch. Ind. Health* 19:524-529.
72. Hammond, E. C., Selikoff, I. J., Lawther, P. L., and Seidman, H. (1976): Inhalation of benzpyrene and cancer in man. *Ann. N.Y. Acad. Sci.* 271:116-124.
73. Mancuso, T. F., MacFarlane, E. M., and Porterfield, J. D. (1955): Distribution of cancer mortality in Ohio. *Amer. J. Public Health* 45:58-70.
74. Mancuso, T. F., and Coulter, E. J. (1958): Cancer mortality among native white, foreign-born white, and nonwhite male residents of Ohio: Cancer of the lung, larynx, bladder, and central nervous system. *J. Nat. Cancer Inst.* 20:79-105.
75. Levin, L., Morton, L., Haenszel, W., Carroll, B. E., Gerhardt, P., Handy, V. H., and Ingraham, S. C. (1960): Cancer incidence in urban and rural areas of New York State. *J. Nat. Cancer Inst.* 24(6):1243-1257.
76. Griswold, M. H. (1955): *Cancer in Connecticut, 1935-1951.* Connecticut State Department of Health, Hartford, Connecticut.
77. Haenszel, W., Loveland, D. B., and Sirken, M. G. (1962): Lung-cancer mortality as related to residence and smoking histories. I. White males. *J. Nat. Cancer Inst.* 28:947-1001.
78. Haenszel, W., and Traeuber, K. E. (1964): Lung cancer mortality as related to residence and smoking histories. II. White females. *J. Nat. Cancer Inst.* 32:803-838.
79. Hoffman, E. F., and Gilliam, A. G. (1954): Lung cancer mortality, geographic distribution in the United States for 1948-1949. *Public Health Rep.* 69:1033-1042.
80. Manos, N. E., and Fisher, G. F. (1959): An index of air pollution and its relation to health. *J. Air Pollution Control Assn.* 9(1):5-11.
81. Prindle, R. A. (1959): Some considerations in the interpretation of air pollution health effects data. *J. Air Pollution Control Assn.* 9:12-19.
82. Stocks, P., and Campbell, J. M. (1955): Lung cancer death rates among non-smokers and pipe and cigarette smokers—an evaluation

in relation to air pollution by benzpyrene and other substances. *Brit. Med. J.* 2:923-939.
83. Curwen, M. P., Kennaway, E. L., and Kennaway, N. M. (1954): The incidence of cancer of the lung and larynx in urban and rural districts. *Brit. J. Cancer* 8:181-198.
84. Eastcott, D. F. (1956): The epidemiology of lung cancer in New Zealand. *Lancet* 1:37-39.
85. Dean, G. (1959): Lung cancer among white South Africans. *Brit. Med. J.* 2:852-857.
86. Dean, G. (1964): Lung cancer in South Africans and British immigrants. *Proc. Roy. Soc. Med.* 57:984-987.
87. Menck, H. R., Casagrande, J. T., and Henderson, B. E. (1974): Industrial air pollution—possible effect on lung cancer. *Science* 183:210-212.
88. Blot, W. J., Bronton, L. A., and Fraumeni, J. F., Jr. (1977): Cancer mortality in U.S. counties with petroleum industries. *Science* 198:51-53.
89. Henderson, B. E., Gordon, R. J., Menck, H., SooHoo, J., Martin, S. P., and Pike, M. C. (1975): Lung cancer and air pollution in southcentral Los Angeles County, California, USA. *Am. J. Epidemiol.* 101:477-488.
90. Just, J., Maziarka, S., and Dlugasiewicz, M. (1969): Respiratory tract cancer death rate and the atmospheric pollution in some Polish cities. *State Communal Hyg. Inst.* 20(5):515-526.
91. Stocks, P. (1952): Epidemiology of cancer of the lung in England and Wales. *Brit. J. Cancer* 6:99-111.
92. Stocks, P. (1959): Cancer and bronchitis mortality in relation to atmospheric deposit and smoke. *Brit. Med. J.* 1:74-79.
93. Stocks, P. (1958): Air pollution and cancer mortality in Liverpool Hospital Region and North Wales. *Int. J. Air Pollut.* 1:1-13.
94. Stocks, P. (1960): The relations between atmospheric pollution in urban and rural localities and mortality from cancer, bronchitis, and pneumonia with particular reference to 3:4 benzopyrene, beryllium, molybdenum, vanadium, and arsenic. *Brit. J. Cancer* 14:398-418.
95. Stocks, P. (1966): Recent epidemiological studies of lung cancer mortality, cigarette smoking, and air pollution with discussion of a new hypothesis of causation. *Brit. J. Cancer* 20:595-623.
96. Stocks, P. (1967): Lung cancer and bronchitis in relation to cigarette smoking and fuel consumption in twenty countries. *Brit. J. Prev. Soc. Med.* 21:181-185.
97. Carnow, B. W., and Meier, P. (1973): Air pollution and pulmonary cancer. *Arch. Environ. Health* 27:207-218.

98. Buell, P., and Dunn, J. E. (1967): Relative impact of smoking and air pollution on lung cancer. *Arch. Environ. Health* 15:291-297.
99. Hammond, E. C., and Horn, D. (1958): Smoking and death rates—report on 44 months of follow-up of 187,783 men: Part II. Death rates by cause. *J. Am. Med. Assoc.* 166:1294-1308.
100. Dean, G. (1966): Lung cancer and bronchitis in Northern Ireland, 1960-1962. *Brit. Med. J.* 1:1506.
101. Hitosugi, M. (1968): Epidemiological study of lung cancer. *Inst. Public Health Bull.* 17:237-256.
102. Winkelstein, W., Jr. and Kantor, S. (1969): Stomach cancer, positive association with suspended particulate air pollution. *Arch. Environ. Health* 18:544-547.
103. Hagstrom, R. M., Sprague, H. A., and Landau, E. (1967): The Nashville air pollution study—VII: Mortality from cancer in relation to air pollution. *Arch. Environ. Health* 15:237-248.
104. Neal, J., and Rigdon, R. H. (1967): Gastric tumors in mice fed benzo(a)pyrene, a quantitative study. *Texas Rep. Biol. Med.* 25:553-557.
105. Ketkar, M., Reznik, G., Schneider, P., and Mohr, U. (1978): Investigations on the carcinogenic burden by air pollution in man. Intratracheal instillation studies with benzo(a)pyrene in bovine serum albumin in Syrian hamsters. *Cancer Letters* 4:235-259.
106. Wislocki, P. G., Change, R. L., Wood, A. W., Levin, W., Yagi, H., Hernandez, O., Mah, H. D., Dansette, P. M., Jerina, D. M., and Conney, A. H. (1977): High carcinogenicity of 2-hydroxybenzo(a)pyrene on mouse skin. *Cancer Res.* 37:2608-2611.
107. Snell, K. C., and Stewart, H. L. (1962): Induction of pulmonary adenomatosis in DBA/2 mice by the oral administration of dibenz(a,h)anthracene. *Acta Un. Int. Cancer* 19:692-694.
108. Schmähl, K., Schmidt, G., and Habs, M. (1977): Syncarcinogenic action of polycyclic hydrocarbons in automobile exhaust gas condensates. In: *Air Pollution and Cancer in Man.* V. Mohr, D. Schmähl, and L. Tomatis, Eds., pp. 53-59, Int. Agency for Research on Cancer, Scientific Publ. No. 16.

# METABOLISM AND SUBSEQUENT BINDING OF BENZO(a)PYRENE TO DNA IN PLEURONECTID AND SALMONID FISH

**U. Varanasi, D. J. Gmur and M. M. Krahn**

Environmental Conservation Division
Northwest and Alaska Fisheries Center
National Marine Fisheries Service
National Oceanic and Atmospheric Administration
2725 Montlake Boulevard East
Seattle, Washington 98112

## INTRODUCTION

A high incidence of hepatomas has been reported in some species of Pleuronectidae (20,22). Although in most cases the etiology of the pathology has not been ascertained, a close relation between pollution and occurrence of liver tumors was strongly suspected in the case of tumors occurring in English sole (*Parophrys vetulus*) from the Duwamish River (22). In laboratory studies (10,18,19,28,31), several chemicals that are known carcinogens in mammals have caused liver tumors in certain fish species, including salmonids. Polynuclear aromatic hydrocarbons (PAH) have not yet been demonstrated to induce carcinogenesis in marine fish. However, it is reported that marine fish are able to accumulate a variety of PAH from sediment (21), water (30,38), and diet (9,29,36), and that hepatic and extrahepatic tissues of these fish possess enzyme systems such as aryl hydrocarbon monooxygenases (AHM), epoxide hydrase, and glutathione-S-transferases (4,37).

Studies with freshwater trout (*Salmo trutta lacustris*) demonstrated that trout liver microsomes actively biotransformed benzo(a)pyrene (BaP) into a variety of electrophilic metabolites and catalyzed binding of activated BaP to DNA (2). Our recent study (35) with marine fish showed that liver enzymes from coho salmon (*Oncorhynchus kisutch*) and starry flounder (*Platichthys stellatus*) extensively metabolized BaP into reactive intermediates that bind to DNA. Thin-layer chromatography revealed that metabolites of BaP formed by liver extracts from these fish were characterized by a high proportion of non-K-region dihydrodiols (35).

Metabolism of BaP by liver enzymes from two species of Pleuronectidae (starry flounder and English sole) and a salmonid (coho salmon) was further investigated by high-performance liquid chromatography (HPLC). Moreover, the influence of pretreatment of these fish species with 3-methylcholanthrene (MC), BaP, and Prudhoe Bay crude oil (PBCO) on in vitro metabolic activation and covalent binding of BaP to DNA was studied. The results of these latter studies are summarized and discussed herein.

## MATERIALS AND METHODS

### Chemicals

Benzo(a)pyrene, $^3$H-BaP and $^{14}$C-BaP were purified by elution from a silica gel column as described previously (35,41), and $^3$H-BaP and nonlabeled BaP were dissolved together in absolute ethanol to yield BaP having a specific activity of 500 MCi/mmole. The standards for the oxygenated metabolites of BaP were provided by the courtesy of Dr. David G. Longfellow, NCI Carcinogenesis Research Program, Bethesda, Maryland.

### Animals

Sexually immature starry flounder, English sole, coho salmon, and male Sprague-Dawley rats (average weight $\simeq$ 125 g) were used in this study. Starry flounder were obtained from an estuary of the Columbia River, coho salmon from Manchester, Washington, and English sole from Point Pulley, Washington. Animals were injected intraperitoneally with 10 mg/kg of BaP, MC, or PBCO dissolved in corn oil and were killed 24 hours after the injection. Supernatants (10,000 x g) of liver homogenates were prepared according to previously described procedures (35).

### Covalent Binding of Activated BaP to DNA

The standard reaction mixture after optimization experiments contained 2 mg of DNA added in 2.5 ml of 0.02 M phosphate buffer (pH

7.4), 0.75 mg of NADPH added to 0.1 ml of 0.1 M EDTA (pH 7.4), and 0.2 ml of the 10,000 x g supernatant ($\simeq$ 5 mg protein). The reaction was started by adding 5 nmoles of BaP in 50 $\mu$l of ethanol. The mixture was incubated in the dark for 15 minutes at 25°C when fish liver supernatant was used and at 37°C when rat liver supernatant was used. The reaction mixture then was treated according to the procedure described previously (7,35). All optimization studies were carried out using liver supernatants isolated from the MC-pretreated fish when the temperature of the surrounding water was 8° $\pm$ 1°C. When studies on the influence of PAH pretreatment of both fish species on the in vitro binding of BaP to DNA were carried out, the water temperature was 13° $\pm$ 1°C. There was no detectable difference in the binding of metabolically activated $^3$H-BaP to DNA when liver enzymes from untreated or corn-oil-treated fish were used. Accordingly, most of the data for control fish in these studies were from untreated fish.

## Metabolism of BaP and Characterization of Ethyl Acetate-Extractable Metabolities

Metabolites were formed by incubating liver supernatants with $^3$H-BaP or $^{14}$C-BaP under the conditions described above, without the addition of DNA, and assay and extraction procedures were identical to those published previously (35).

Separations of ethyl acetate-extractable metabolites were carried out with a Hewlett-Packard 1084B high-performance liquid chromatograph using a 0.26 x 25-cm Perkin-Elmer HC-ODS 10-$\mu$m column. Both the Hewlett-Packard ultraviolet spectrometer model and a Perkin-Elmer MPF-44A fluorescence spectrometer fitted with a 20 $\mu$l, square micro flow cell were used for detection. Detector response was simultaneously recorded and integrated using two Hewlett-Packard recorder-integrators, 3385A (fluorescence) and 7985B (ultraviolet).

Acetic acid/water (0.5 percent V/V) (solvent A) and methanol (solvent B) were used for the separation of metabolites in a nonlinear gradient. Starting at 10 percent of solvent B, the gradient was changed as follows (change in %B/time): 10 minutes at 5%/min, 8 minutes at 1.25%/min, 4 minutes at 7.5%/min, to the final conditions of 100 percent solvent B, which were maintained for 2 minutes. The flow rate was 1.0 ml/min and the HPLC oven temperature was 55°C. Retention times of metabolites were matched against those of the standards detected by both UV and fluorescence spectrometry. Further characterization of metabolites present in the ethyl acetate extract was carried out by radiometry. Fractions were collected at 15-second intervals and radioactivity was determined as described previously (32). From the retention times of metabolites present in each ethyl acetate extract, most radioactive peaks were characterized and quantified.

## RESULTS

### Influence of PAH-Pretreatment on In Vitro Binding of BaP to DNA

The results in Figures 1 and 2 demonstrate that starry flounder liver supernatant (10,000 x g) was more sensitive than the liver preparations from coho salmon and English sole to variations in reaction parameters such as pH and concentrations of protein and NADPH. Reaction kinetics of binding for the English sole liver preparation were similar to those obtained with the coho salmon liver preparation. The results (Figures 1, 2) also show that when fish were held a 8°C, the binding value (pmole BaP equivalents bound/mg DNA/mg protein) obtained with the liver supernatants from MC-pretreated English sole was considerably lower than the values for both coho salmon and starry flounder. Moreover, the binding values obtained with liver supernatants from MC-pretreated coho salmon and starry flounder held at 8°C were about one-third of those obtained when the fish were held at 13°C (Table 1).

Data in Table 1 were obtained primarily with liver supernatants isolated from fish held at 13°C; therefore, valid comparisons can be made of binding values obtained after treatment of fish species with BaP, MC, or PBCO. The results in Table 1 show that the binding values obtained for the untreated English sole and starry flounder were three and eight times greater than the corresponding value for coho salmon (Table 1). Moreover, pretreatment of English sole with PBCO resulted in a 18-fold increase in binding value compared with that for untreated English sole; the increase in the binding value for PBCO-pretreated starry flounder was only fivefold.

Pretreatment of coho salmon, starry flounder, and rat with MC resulted in a substantial increase in the in vitro binding values of $^3$H-BaP to DNA compared with those obtained with liver extracts of untreated animals (Table 1). No such value was obtained for MC-pretreated English sole in this study. The binding value obtained with liver extracts of MC-pretreated starry flounder was about ten times greater than that obtained with the untreated fish; the corresponding increase in the binding value was 48- and 12-fold for coho salmon and rat (Table 1), respectively. For both coho salmon and starry flounder, the binding of BaP to DNA was slightly greater when fish were pretreated with BaP than when they were pretreated with MC.

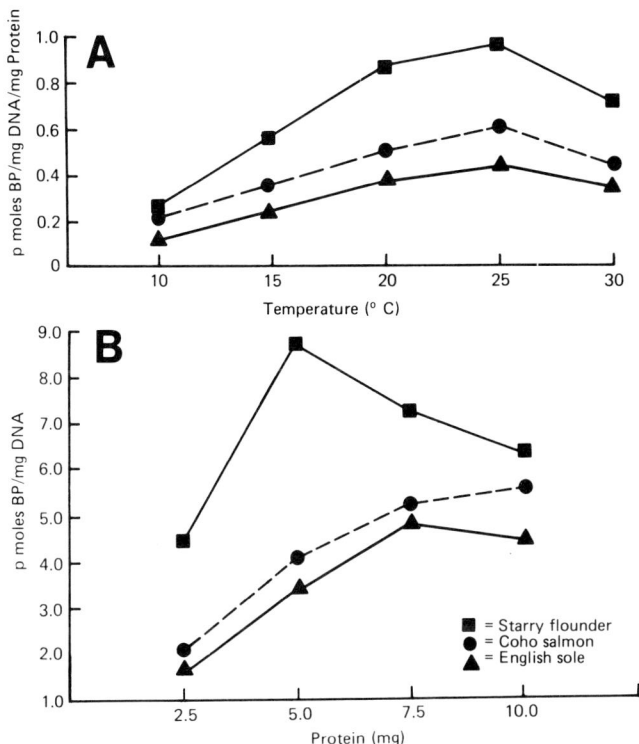

**FIGURE 1.** In vitro covalent binding of $^3$H-BaP to DNA by liver supernatants (10,000 x g) from MC-pretreated fish as a function of (a) temperature and (b) protein/reaction mixture. Data are from single experiments that gave similar results when repeated. Measurements were made in triplicate.

## Metabolism of Benzo(a)pyrene

Figure 3 depicts metabolite profiles produced by liver supernatants isolated from MC-pretreated fish species and rat. The data revealed that, for all three fish species, 9,10-dihydro-9,10-dihydroxy benzo(a)pyrene (BaP 9,10-dihydrodiol) and 7,8-dihydro-7,8-dihydroxy benzo(a)pyrene (BaP 7,8-dihydrodiol) were the major metabolites; 3-hydroxy BaP was also present in considerable amounts. The metabolite profile for the MC-treated rat revealed the presence of a high proportion of 3- and 9-hydroxy BaP and BaP 3,6-quinone, together with significant amounts of the non-K-region dihydrodiols (BaP 7,8-dihydrodiol and BaP 9,10-dihydrodiol). It should be noted that profiles of BaP metabolites obtained after incubation of liver extracts with $^{14}$C-BaP were similar to those obtained with $^3$H-BaP.

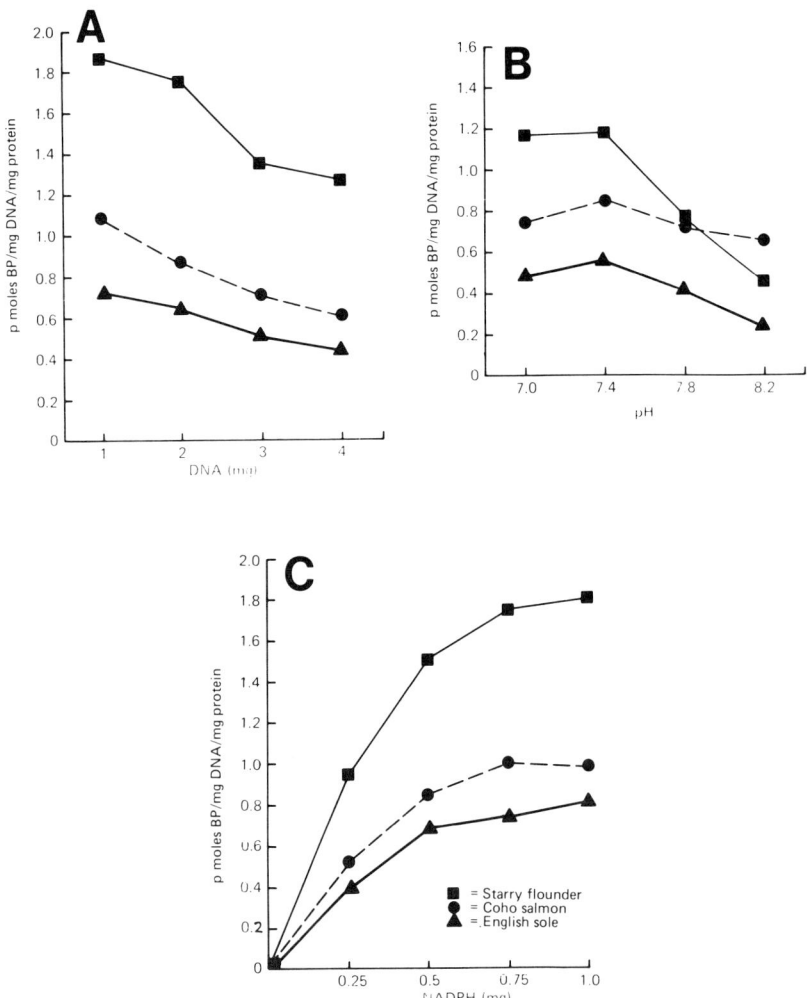

FIGURE 2. In vitro covalent binding of $^3$H-BaP to DNA by liver supernatants (10,000 x g) from MC-pretreated fish as a function of (a) DNA/reaction mixture, (b) pH, and (c) NADPH/reaction mixture. Data are from single experiments that gave similar results when repeated. Measurements were made in triplicate.

TABLE 1. In Vitro Binding of Activated BaP to DNA
Catalyzed by Liver Supernatants from Fish and Rat[a]

| Species | System[b] (Supernatant) | pmole of BaP Equivalent/mg DNA/mg Protein[c] | % of Control Value |
|---|---|---|---|
| Starry flounder | Control | 0.15 | 100 |
| | MC | 1.62 | 1,100 |
| | BaP | 1.70 | 1,100 |
| | PBCO | 0.74 | 500 |
| | MC[d] | 0.53 | —[d] |
| English sole | Control | 0.06 | 100 |
| | PBCO | 1.05 | 1,800 |
| | MC | 0.16 | —[e] |
| Coho salmon | Control | 0.02 | 100 |
| | MC | 0.97 | 4,900 |
| | BaP | 1.06 | 5,300 |
| | MC[d] | 0.30 | —[e] |
| Rat | Control | 0.06 | 100 |
| | MC | 0.69 | 1,200 |

[a]Maximum standard deviation between values from two separate experiments was 14 percent.
[b]Liver supernatants (10,000 x g) were obtained from either untreated (control) animals or those injected with 10 mg/kg of 3-methylcholanthrene (MC), benzo(a)pyrene BaP, or Prudhoe Bay crude oil (PBCO) when the water temperature for fish was 13°C.
[c]Liver supernatants ($\approx$5 mg protein) from different animals were incubated in the dark with 5 nmole of BaP, 2 mg of salmon sperm DNA, and cofactors for 15 minutes at 25°C (for fish) and 37°C (for rat). Each value is an average of two experiments and three replicate measurements using pooled liver extracts from five animals. Binding values for incubation without NADPH were less than 0.001 and are subtracted from the values reported.
[d]Liver supernatants were obtained from MC-pretreated fish when the water temperature was 8°C.
[e]No control fish were sampled at 8°C.
Source: Taken in part from Reference 35.

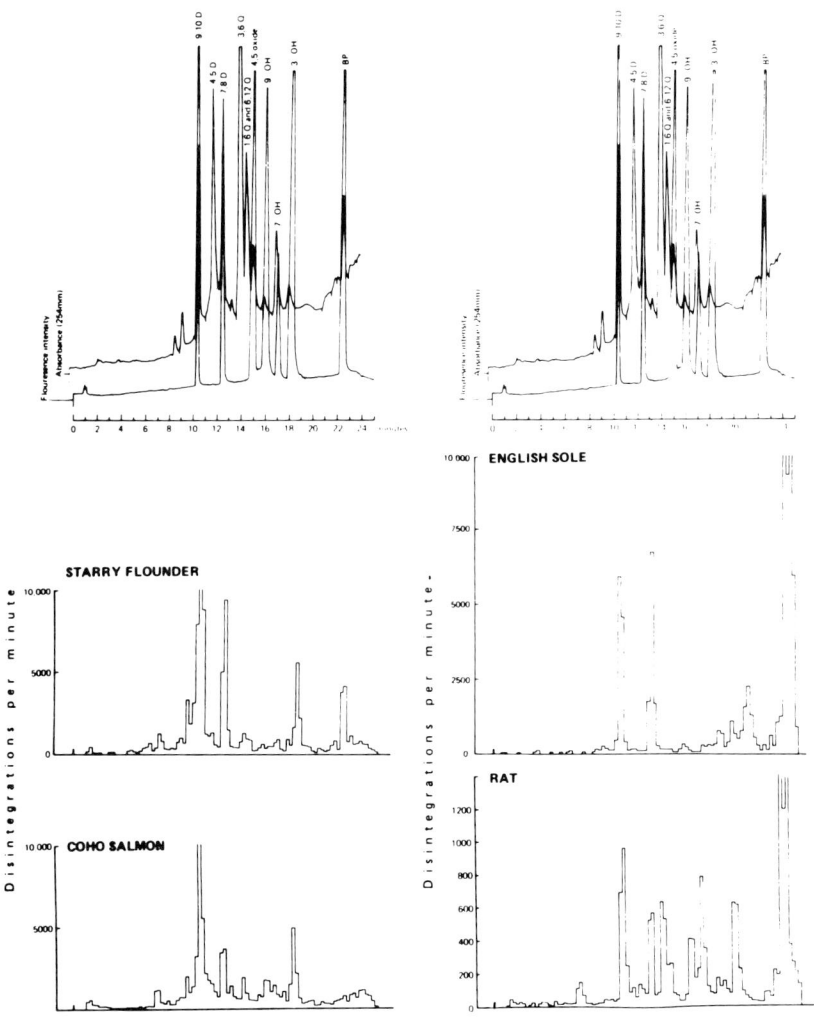

FIGURE 3. HPLC of reference BaP metabolites and ethyl acetate-extractable metabolites of $^3$H-BaP produced by liver supernatants (10,000 x g) of MC-pretreated starry flounder, coho salmon, English sole, and rat. Reference compounds and metabolites were detected by both UV and fluorescence spectrometry. Metabolites formed by incubating 5 nmole $^3$H-BaP with liver supernatants ($\simeq$5 mg protein) of fish (at 25°) and rat at (37°) for 15 minutes were separated by fractions collected at 15-second intervals. Abbreviations are 9,10-D, BaP 9,10-dihydrodiol; 4,5-D, BaP 4,5-dihydrodiol; 7,8-D, BaP 7,8-dihydrodiol; 3,6-Q, BaP 3,6-quinone; 1,6-Q, and 6,12-Q, 1,6-BaP quinone and 6,12-BaP quinone; 9,OH, 9-hydroxy BaP; 7-OH, 7-hydroxy BaP; 3-OH, 3-hydroxy BaP; BaP, benzo(a)pyrene.

Results in Table 2 show that pretreatment of starry flounder and coho salmon with MC resulted in a considerable increase in BaP metabolism catalyzed by the liver supernatants. Liver extracts from MC-pretreated starry flounder produced a larger proportion (52 percent) of ethyl acetate-extractable metabolites than did liver extracts from both coho salmon (40 percent) and rat (38 percent). Ratios of the concentrations of non-K-region dihydrodiols to monohydroxybenzo(a)pyrenes (3-hydroxy BaP and 9-hydroxy BaP) were significantly ($P<0.05$) greater in all four fish liver extracts than in the rat liver extract (Table 2). Pretreatment of both coho salmon and starry flounder with MC resulted in increases in proportions of BaP 9,10-dihydrodiol and the "prediol" components eluted before BaP 9,10-dihydrodiol, and a marked decrease in the proportion of BaP 7,8-dihydrodiol formed by the liver extracts. For example, BaP 7,8-dihydrodiol comprised 32 and 17 percent, respectively, of the ethyl acetate-extractable metabolites in incubations containing liver extracts from untreated and Mc-pretreated starry flounder; the corresponding values for BaP 9,10-dihydrodiol were 44 and 52 percent, respectively (Table 2).

**TABLE 2. Metabolism of Benzo(a)Pyrene by Liver Extracts from Fish and Rat**

| | Starry Flounder | | Coho Salmon | | Rat, |
|---|---|---|---|---|---|
| | Control | MC Pretreated | Control | MC Pretreated | MC Pretreated |
| | Percentage of Total Radioactivity[a] | | | | |
| Unreacted BaP | 40 | 13 | 59 | 5 | 35 |
| Ethyl acetate-extractable metabolites | 39 | 52 | 20 | 40 | 38 |
| Aqueous phase | 21 | 35 | 21 | 55 | 27 |
| | Ethyl Acetate-Extractable Metabolites[b] | | | | |
| | (nmole/incubation) | | | | |
| "Prediol"[c] | 0.07 | 0.34 | 0.01 | 0.30 | 0.15 |
| BP 9,10-dihydrodiol | 0.64 | 1.23 | 0.22 | 0.67 | 0.34 |
| BP 7,8-dihydrodiol | 0.46 | 0.40 | 0.28 | 0.18 | 0.19 |
| BP 3,6-quinone | tr[d] | 0.09 | 0.07 | 0.07 | 0.21 |
| 3-Hydroxy BP | 0.27 | 0.27 | 0.10 | 0.21 | 0.27 |
| 9-Hydroxy BP | tr | 0.05 | 0.04 | 0.10 | 0.24 |

[a] Data are from single experiments which gave similar results when repeated.
[b] Incubation conditions are given in the text; 5 nmole of BaP was used per each incubation. Metabolite extracts were from fish held at 13°C.
[c] Conditions for HPLC are given under Materials and Methods.
[d] Components eluted before BaP 9,10-dihydrodiol.
[e] tr, trace.

## DISCUSSION

Environmental temperature can have a significant influence on the activity of certain enzymes in poikilothermic organisms (14). For example, Stegeman (34) reported that at low temperatures ($\approx 6°C$) virtually no induction occurred in the hepatic AHM activity of *Fundulus heteroclitus* after treatment with MC, whereas the induction was significant at 16°C. These results (34) provide a possible explanation for our findings showing the low values for binding of metabolically activated BaP to DNA yielded by liver supernatants from MC-pretreated fish held at 8°C (Table 1).

An interesting finding was that the binding values for both benthic fish species—English sole and starry flounder—were greater than the binding value for coho salmon, a pelagic fish. Benthic fishes are habitually exposed to a multitude of xenobiotics [some of which are known inducers of mixed-function oxidases (MFO)] present in sediments (21); therefore, it is possible that liver preparations from these fish yield higher values for the in vitro binding of BaP to DNA than do coho salmon. However, because many environmental factors and species-specific differences influence xenobiotic metabolizing capability of aquatic organisms (4,37) the above explanation for the observed differences in binding values for pleuronectids and salmonids should be considered as tentative.

Much of the petroleum released into the marine environment probably will accumulate in bottom sediment and, as a consequence, benthic fishes will be exposed to increasing concentrations of PAH (13). The present results showing that the magnitude of increase in the in vitro binding was much greater for English sole after treatment with PBCO than for starry flounder are noteworthy. Further investigation is necessary to assess whether English sole is more sensitive to petroleum exposure than is starry flounder.

It has been proposed that the induction of hepatic AHM in fish can indicate the presence of toxic chemicals in their environment (25). Our results and those of others (2,35) show that the increase in in vitro binding of BaP to DNA on preexposure of fish or rat to PAH (MC or BaP) was greater than the increase in BaP metabolism and AHM activity. Thus, it appears that covalent binding of BaP to DNA may prove to be a more sensitive index than AHM activity to indicate pre-exposure of fish to certain toxic chemicals. Moreover, the binding value may provide a more useful index when an attempt is made to correlate metabolism with carcinogenicity of a PAH (7,26). In mammals, carcinogenic potential of a compound has been shown to correlate roughly with the extent of covalent binding to DNA (7,16,17,23,24,26,27). Accordingly, studies to assess the extent of covalent binding of a variety of pollutants to cellular DNA in various target tissues of marine organisms may give useful information as to which of these pollutants present in the marine environment may be a potential carcinogen for a particular species or tissue.

## Benzo(a)pyrene Metabolism

The preponderance of the ono-K-region dihydrodiols compared with monohydroxybenzo(a)pyrenes formed by liver enzymes of the three species of fish may be in part due to a particular species of cytochrome P-450, because non-K-region oxygenation is favored under conditions of high cytochrome $P_1$-450/P-450 ratio (26). In their studies with a freshwater salmonid, Ahokas et al (1) reported that trout liver cytochrome was closely related to cytochrome $P_1$-450 of rat liver microsomes. Metabolites of BaP formed by trout liver microsomes also were characterized by high proportions of the non-K-region dihydrodiols; however, 3-hydroxy BaP was the major metabolite (2). The differences in BaP concentrations and in subcellular fractions employed in these two studies make it difficult to compare in detail the metabolite profiles produced by liver enzymes from freshwater and saltwater salmonids.

The hepatic MFO, together with conjugating enzymes, are responsible for both activation and detoxification of PAH. Therefore, treatment of animals with chemicals that alter these enzyme systems can result in either increased or decreased toxicity of the PAH. Studies show that pretreatment of rats (2) or mice (39) with either MC or BaP resulted in a considerable increase in in vitro metabolism of BaP catalyzed by liver microsomes, together with increases in the proportions of both BaP 7,8-dihydrodiol and BaP 9,10-dihydrodiol. Although pretreatment of fish with MC in the present study resulted in a similar increase in BaP metabolism, there was a marked decrease in the proportions of BaP 7,8 dihydrodiol and a considerable increase in the proportion of "prediol" components. Moreover, preliminary data (Varanasi, Gmur, and Krahn, unpublished results) show that pretreatment of these fish with either BaP or PBCO resulted in similar alterations in the metabolite profiles. BaP 7,8-dihydrodiol is further metabolized (5,8,15) to an ethyl acetate-extractable metabolite, 7,8,9,10-tetrahydro-7,8,9,10-tetrahydroxy BaP, which is formed via BaP 7,8-dihydrodiol-9,10-epoxide, the postulated ultimate carcinogen of BaP (33). Recent studies of Cohen et al (8) demonstrated that the major ethyl acetate-extractable metabolite formed by rat trachea was BaP 9,10-dihydrodiol, whereas enzymatic hydrolysis of the water-soluble metabolites released 3-hydroxy BaP and BaP 7,8-

dihydrodiol as the major metabolites. Thus, it appears that BaP 7,8-dihydrodiol can be rapidly metabolized and conjugated, whereas BaP 9,10-dihydrodiol which is believed to be resistant to epoxidation and conjugation (5,8) may accumulate under the in vitro conditions employed in the present study. Studies with rat hepatocytes (6,15) show that BaP 9,10-dihydrodiol easily egressed in extracellular fluid, whereas BaP 7,8-dihydrodiol was equally divided between the cell and extracellular medium. Thus, extensive metabolism of BaP and the production of large proportions of BaP 9,10-dihydrodiol by liver enzymes of these fish, especially PAH-pretreated fish, may have important consequences in efficient removal of BaP from the liver and its transport to other tissues (e.g., muscle or skin) via the circulatory system.

In conclusion, the results show that liver enzymes of both salmonid and pleuronectid fish catalyzed metabolic activation and subsequent binding of BaP to DNA; BaP was largely converted to the non-K-region dihydrodiols which are precursors of biologically active diol-epoxides implicated as ultimate carcinogens. It is important therefore to investigate transport of carcinogenic PAH and their metabolites from liver to the edible tissues of fish exposed to PAH. Moreover, considering that certain fish species, such as pleuronectids, frequently develop epidermal neoplasia (41), whereas others rarely do, an evaluation of the interactions of the activated PAH with DNA in skin of fish would be of interest.

## ACKNOWLEDGMENTS

We thank Victor D. Henry for technical assistance and Gail E. Siani for helping with manuscript preparation. This work was supported in part by the Bureau of Land Management through interagency agreement with the National Oceanic and Atmospheric Administration, and which is managed by the Outer Continental Shelf Environmental Assessment Program (OCSEAP) Office.

## REFERENCES

1. Ahokas, J. T., Pelkonen, O., and Karki, N. T. (1977): Characterization of benzo(a)pyrene hydroxylase of trout liver. *Cancer Res.* 37:3737-3743.
2. Ahokas, J. T., Saarni, H., Nebert, D. W., and Pelkonen, O. (1979): The in vitro metabolism and covalent binding of benzo[a]pyrene to DNA catalyzed by trout liver microsomes. *Chem. Biol. Interact.* 25:103-111.

3. Arcos, J. C., and Argus, M. F. (1968): Molecular geometry and carcinogenic activity of aromatic compounds. New perspectives. *Adv. Cancer Res.* 11:305-471.
4. Bend, J. R., and James, M. O. (1978): Xenobiotic metabolism in marine and freshwater species. In: *Biochemical and Biophysical Perspectives in Marine Biology,* Vol. IV, D. C. Malines and J. R. Sargent, Eds., pp. 125-188, Academic Press, New York.
5. Booth, J., and Sims, P. (1976): Different pathways involved in the metabolism of the 7,8- and 9,10-dihydrodiols of benzo(a)pyrene. *Biochem Pharamacol.* 25:979-985.
6. Burke, M. D., Vadi, H., Jernstrom, B., and Orrenius, S. (1977): Metabolism of benzo(a)pyrene with isolated hepatocytes and the formation and degradation of DNA-binding derivatives. *J. Biol. Chem.* 252:6424-6431.
7. Buty, S. G., Thompson, S., and Slaga, T. J. (1976): The role of epidermal aryl hydrocarbon hydroxylase in the covalent binding of polycyclic hydrocarbon to DNA and its relationship to tumor initiation. *Biochem. Biophys. Res. Commun.* 70:1102-1108.
8. Cohen, G. M., Marchok, A. C., Nettlesheim, P., Steele, V. E., Nelson, F., Huang, S., and Selkirk, J. K. (1979): Comparative metabolism of benzo(a)pyrene in organ and cell cultures derived from rat tracheas. *Cancer Res.* 39:1980-1984.
9. Collier, T. K., Thomas, L. C., and Malins, D. C. (1978): Influence of environmental temperature on disposition of dietary naphthalene in coho salmon (*Oncorhynchus kisutch*): Isolation and identification of individual metabolites. *Comp. Biochem. Physiol.* 61C:23-28.
10. Dawe, C. J. Stanton, M. F., and Schwartz, F. J. (1964): Hepatic neoplasms in native bottom-feeding fish of Deep Creek Lake, Maryland. *Cancer Res.* 24:1194-1201.
11. Dipple, A., Lawley, P. D., and Brookes, P. (1968): Theory of tumor initiation by chemical carcinogens: Dependence of activity on structure of ultimate carcinogen. *Eur. J. Cancer* 4:493-506.
12. Holder, G. M., Yagi, H., Dansette, P. M., Jerina, D.M., Levin, W., Lu, A.Y.H., and Conney, A. H. (1974): Effects of inducers and epoxide hydrase on the metabolism of benzo(a)pyrene by liver microsomes and a reconstituted system: Analysis by high pressure liquid chromatography. *Proc. Natl. Acad. Sci. U.S.A.* 71:4356-4360.
13. Herbes, S. E., and Schwall, L. R. (1978): Microbial transformation of aromatic hydrocarbons in pristine and petroleum contaminated sediments. *Appl. Environ. Microbiol.* 35:306-316.
14. Hochachka, P. W., and Somero, G. N. (1971): Biochemical adaptation to the environment. In: *Fish Physiology,* Vol. VI, W. S. Hoar and P. J. Randall, (Eds.) pp. 99-190, Academic Press, New York.

15. Jones, C. A. Moore, B. P., Cohen, G. M., Fry, J. R. and Bridges, J. W. (1978): Studies on the metabolism and excretion of benzo(a)pyrene in isolated adult rat hepatocytes. *Biochem. Pharmacol.* 27:693-702.
16. Kinoshita, N., and Gelboin, H. V. (1972): The role of aryl hydrocarbon hydroxylase in 7,12-dimethylbenz(a)anthracene skin tumorigenesis: On the mechanism of 7,8-benzoflavone inhibition of tumorigenesis. *Cancer Res.* 32:1329-1339.
17. Kouri, R. E., Salerno, R. A., and Whitmire, C. E. (1973): Relationships between aryl hydrocarbon hydroxylase inducibility and sensitivity to chemically induced subcutaneous sarcomas in various strains of mice. *J. Natl. Cancer Inst.* 50:363-368.
18. Lee, D. J., Wales, J. H., Ayres, J. L., and Sinnhuber, R. O. (1968): Synergism between cyclopropenoid fatty acids and chemical carcinogens in rainbow trout (*Salmo gairdneri*). *Cancer Res.* 28:2312-2318.
19. Matsushima, R., and Sugimura, T. (1976): Experimental carcinogenesis in small aquarium fishes. *Prog. Exp. Tumor Res.* 20:367-379.
20. McCain, B. B., Gronlund, W. D., Myers, M. S., and Wellings, S. R. (1979): Tumors and microbial diseases of marine fishes in Alaskan waters. *J. Fish Dis.* 2:111-130.
21. McCain, B. B., Hodgins, H. O., Gronlund, W. D., Hawkes, J. W., Brown, D. W., Myers, M. S., and Vandermeulen, J. H. (1978): Bioavailability of crude oil from experimentally oiled sediments to English sole (*Parophrys vetulus*) and pathological consequences. *J. Fish. Res. Board Can.* 35:657-664.
22. McCain, B. B., Pierce, K. V., Wellings, S. R., and Miller, B. S. (1977): Hepatomas in marine fish from an urban estuary. *Bull. Environ. Contam. Toxicol.* 18:1-2.
23. Nebert, D. W., Benedict, W. F., and Gielen, J. E. (1972): Aryl hydrocarbon hydroxylase, epoxide hydrase and 7,12-dimethylbenz(a)anthracene-produced skin tumorigenesis in the mouse. *Mol. Pharmacol.* 8:374-379.
24. Nebert, D. W., Boobis, A. R.,Yagi, H. Jerina, D. M., and Kouri, R. E. (1977): Genetic differences in mouse cytochrome $P_1$-450-mediated metabolism of benzo(a)pyrene in vitro and carcinogenic index in vivo. In: *Biological Reactive Intermediates,* D. J. Jollow, J. J. Kocsis, R. Snyder, and H. Vainio, Eds. pp. 125-145, Plenum Publishing Corp., New York.
25. Payne, J. R. (1976): Field evaluation of benzopyrene hydroxylase induction as a monitor for marine petroleum pollution. *Science* 191:945-946.

26. Pelkonen, O., Boobis, A. R., and Nebert, D. W. (1978): Genetic differences in the binding of reactive carcinogenic metabolites to DNA. In: *Carcinogenesis: A Comprehensive Survey*, Vol. III, P. W. Jones and R. I. Freudenthal, Eds., pp. 383-400, Ravin Press, New York.
27. Phillips, D. H., Grover, P. L., and Sims, P. (1978): The covalent binding of polycyclic hydrocarbons to DNA in the skin of mice of different strains. *Int. J. Cancer* 22:487-494.
28. Pliss, G. B., and Khudoley, V. V. (1975): Tumor induction by carcinogenic agents in aquarium fish. *J. Natl. Cancer Inst.* 55:129-133.
29. Roubal, W. T., Collier, T. K., and Malins, D. C. (1977): Accumulation and metabolism of carbon-14 labeled benzene, naphthalene, and anthracene by young coho salmon (*Oncorhynchus kisutch*). *Arch. Environ. Contam. Toxicol.* 5:513-529.
30. Roubal, W. T., Stranahan, S. I., and Malins, D. C. (1978): The accumulation of low molecular weight aromatic hydrocarbons of crude oil by coho salmon (*Oncorhynchus kisutch*) and starry flounder (*Platichthys stellatus*). *Arch. Environ. Contam. Toxicol.* 7:237-244.
31. Sato, S., Matsushima, T., Tanaka, N., Sugimura, T., and Takashima, F. (1973): Hepatic tumors in the guppy (*Lebistes reticulatus*) induced by aflatoxin $B_1$, dimethylnitrosamine, and 2-acetylaminofluorene. *J. Natl. Cancer Inst.* 50:767-778.
32. Selkirk, J. K., Croy, R. G., Roller, P. P., and Gelboin, H. V. (1974): High-pressure liquid chromatographic analysis of benzo(a)pryene metabolism and covalent binding and the mechanism of action of 7,8-benzoflavone and 1,2-epoxy-3,3,3-trichloropropane. *Cancer Res.* 34:3474-3480.
33. Sims, P., Grover, P. L., Swaisland, A., Pal, K., and Hewer, A. (1974): Metabolic activation of benzo(a)pyrene proceeds by a diol-epoxide. *Nature* 252:326-328.
34. Stegeman, J. J. (1979): Temperature influence on basal activity and induction of mixed function oxygenase activity in *Fundulus heteroclitus*. *J. Fish. Res. Board Can.* 36:1400-1405.
35. Varanasi, U., and Gmur, D. J. (1979): Metabolic activation and covalent binding of benzo(a)pyrene to DNA catalyzed by liver enzymes of marine fish. *Biochem Pharmacol.* (in press).
36. Varanasi, U., Gmur, D. J., and Treseler, P. A. (1979): Influence of time and mode of exposure on biotransformation of naphthalene by juvenile starry flounder (*Platichthys stellatus* ) and rock sole (*Lepidopsetta bilineata)*. *Arch. Environ. Contam. Toxicol.* 8:673-692.
37. Varanasi, U., and Malins, D. C. (1977): Metabolism of petroleum hydrocarbons: Accumulation and biotransformation in marine

organisms. In: *Effects of Petroleum on Arctic and Subarctic Marine Environments and Organisms,* Vol. II, D. C. Malins, Ed., pp. 175-270, Academic Press, New York.
38. Varanasi, U., Uhler, M., and Stranahan, S. I. (1978): Uptake and release of naphthalene and its metabolites in skin and epidermal mucus of salmonids. *Toxicol. Appl. Pharmacol.* 44:277-285.
39. Wang, I. Y., Rasmussen, R. E., Petrakis, N. L., and Wang, A.-C. (1976): Enzyme induction and the difference in the metabolite patterns of benzo(a)pyrene produced by various strains of mice. In: *Carcinogenesis: A Comprehensive Survey,* Vol. I, R. I. Freudenthal and P. W. Jones, Eds., pp. 77-89, Raven Press, New York.
40. Wellings, S. R., McCain, B. B., and Miller, B. S. (1976): Epidermal papillomas in Pleuronectidae of Puget Sound, Washington. Review of the current status of the problem. *Prog. Exp. Tumor Res.* 20:55-74.
41. Yang, S. K., Gelboin, H. V., Trump, B. F., Autrup, H., and Harris, C. C. (1977): Metabolic activation of benzo(a)pyrene and binding to DNA in cultured human bronchus. *Cancer Res.* 37:1210-1215.

# BENZO(A)PYRENE-DNA ADDUCT FORMATION IN CELLS: TIME-DEPENDENT DIFFERENCES IN THE BENZO(A)PYRENE-DNA ADDUCTS PRESENT

W. M. Baird and R. U. Dumaswala

The Wistar Institute of Anatomy and Biology
36th Street at Spruce
Philadelphia, Pennsylvania 19104

## INTRODUCTION

Carcinogenic polycyclic aromatic hydrocarbons are metabolically activated and bind covalently to DNA, RNA, and protein in systems in which they induce biological effects (reviewed in 7,9,10,12,28). In both mouse skin and rodent embryo cell cultures, the carcinogenic activity of a hydrocarbon is directly correlated with the amount of the hydrocarbon bound to DNA and to a specific protein, the h-protein (1,8,11,23).

To identify the hydrocarbon metabolites responsible for this binding to cellular macromolecules, it is necessary to isolate and characterize the hydrocarbon adducts. In 1973, Baird and Brookes (3) developed a technique for isolating the hydrocarbon-DNA adducts formed in rodent embryo cell cultures. This procedure involved isolation of the DNA from [$^3$H]hydrocarbon-treated cells, enzymatic degradation of the DNA to deoxyribonucleosides, and chromatography of the adducts on columns of Sephadex LH20 eluted with methanol: water gradients. By comparing the chromatographic properties of the benzo(a)pyrene [BaP]-DNA adducts formed in cells with those of the DNA adducts formed by reaction of the 7,8-diol-9,10-epoxide of BaP [BaPDE] with DNA, Sims et al (32) demonstrated that BaPDE was the BaP metabolite responsible for the formation of the major BaP-DNA adducts in hamster embryo cell cultures. It has subsequently been found in various cultured cell lines and in mouse skin that most of the BaP-DNA adducts formed result from this diol epoxide (reviewed in 28). In cell cultures and in mouse skin the DNA adducts formed from several other hydrocarbons have been found to result from the binding of bay-region diol epoxides (18) of these hydrocarbons (reviewed in 28).

Two stereoisomers of this [BaPDE] exist (13): one syn, has the 9,10-epoxide on the same side of the ring as the 7-hydroxyl and one, anti, has the 9,10-epoxide on the opposite side of the ring from the 7-hydroxyl (6,35) (see Figure 1). Jennette et al (17) described techniques for isolating the adducts formed by both BaPDE with individual deoxyribonucleosides by high-pressure liquid chromatography (HPLC) on a reverse-phase column. In studies of the binding of the diol epoxides to DNA in aqueous solutions, adducts on the amino groups of deoxyguanosine, deoxyadenosine, and deoxycytidine are formed (17,22,24,25,26). The anti-BaPDE may also react with the N-7 of guanine under these conditions (27), although there is no evidence that this adduct is formed in cells or animal tissue (21). King et al (19) developed a method for separating the

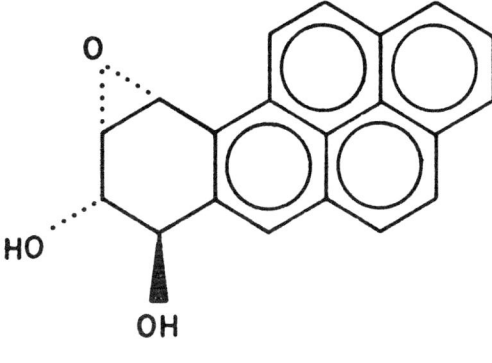

anti-BaPDE

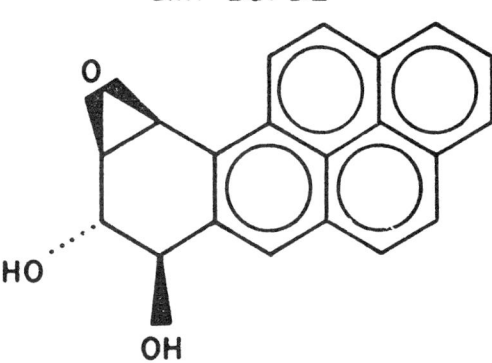

syn-BaPDE

**FIGURE 1.** Structure of the syn-and anti-forms of BaP-7,8-diol-9,10-epoxide.

BaP-DNA adducts of these isomers by eluting Sephadex LH20 columns with borate-containing methanol: water gradients, the anti-BaPDE adducts formed a complex with the borate that eluted earlier than the syn-BaPDE adducts. Using this technique, Shinohara and Cerutti (31) demonstrated that both isomers were involved in the binding of BaP to DNA in rodent cells. The major BaP-DNA adducts formed in several other cell systems were shown to result from the reaction of the anti-isomer with the amino group of guanine (2,16). Baird and Diamond (5) found that in hamster embryo cell cultures, the ratio of the syn-adducts to anti-adducts decreased the longer the cells were exposed to BaP. Ivanovic et al (14) observed that individual base adducts in the DNA of hamster embryo cells treated with BaP for 18 hours differed from those in cells treated for 21 hours and then incubated for an additional 24 hours after removal of the BaP. They proposed that these changes were probably due to selective excision of some adducts.

DNA exists in cells as chromatin, a complex of DNA and histones organized as a series of subunits called nucleosomes. A portion of the nucleosome is sensitive to digestion by the enzyme micrococcal nuclease, while the remainder, which exists as a histone octamer-DNA complex, is relatively resistant to digestion by this enzyme (33,36). Tlsty and Lieberman (34) have shown that in N-acetoxy-2-acetylaminofluorene-treated human cells, DNA repair synthesis occurs initially in the micrococcal nuclease sensitive regions. Yamasaki et al. (36) and Jahn and Litman (15) measured the release of BaPDE-DNA adducts from nuclei and chromatin by micrococcal nuclease in order to determine the proportion of the binding of BaPDE that occurs in different regions of the nucleosome. They found that 60 to 65 percent of the bound carcinogen was released.

To determine the reason for and significance of the change with time in BaP-DNA adducts in hamster embryo cells, we have used HPLC and chromatography on Sephadex LH20 columns to examine the adducts formed in these cells. We also investigated the effect of time of treatment of the culture on the proportion of the BaP-DNA adducts in the micrococcal nuclease-sensitive and-resistant regions of the nuclei.

## MATERIALS AND METHODS

### Preparation of [$^3$H]BaP-DNA

Hamster embryo cell cultures were prepared from 13-day hamster embryos (Lakeview, Newfield, New Jersey, or Leo Goodwin Institute, Fort Lauderdale, Florida) as described previously (4). Tertiary cultures were prepared in either 150 cm² (Corning Glass Works, Corning, New York) or 175 cm² (Falcon, Oxnard, California) culture flasks with 50 ml Eagle's minimum essential medium containing 10 percent fetal bovine serum (4). [G-³H]BaP (Amersham, Arlington Heights, Illinois) was diluted with unlabeled BaP to 5 Ci/mmole and cultures were exposed to 0.5 nmole [³H]BaP/ml medium and 0.1 percent DMSO for the times specified. DNA was isolated by a phenol extraction procedure and degraded to deoxyribonucleosides with the enzymes DNAse I from bovine pancreas, phosphodiesterase from *Crotalus atrox* venom and alkaline phosphatase from *E. coli* (Sigma Chemical Co., St. Louis, Missouri) as described previously (3,4).

## Preparation of BaPDE-deoxyribonucleoside Adducts

[¹⁴C]-anti-BaPDE, specific activity 29 mCi/mmole, and [³H]syn-BaPDE, specific activity 214 mCi/mmole, were provided by the Cancer Research Program of the National Cancer Institute, Division of Cancer Cause and Prevention, Bethesda, Maryland. One volume of BaPDE in tetrahydrofuran: triethylamine (19:1) was added to 7 or more volumes of water: ethanol (2:1) solutions of calf thymus DNA, type I (Sigma), polydeoxyguanylic-polydeoxycytidylic acid (Sigma), polydeoxyandenylic-thymidylic acid (Sigma), polydeoxycytidylic acid (Miles Research Products, Elkhart, Indiana), or polydeoxycytidylic-deoxyinosinic acid (Sigma) (26,29). After 18 hours, samples were extracted with ethyl acetate and nucleic acid was precipitated as described previously (5). The samples were then degraded to deoxyribonucleosides as described for BaP-DNA.

## Chromatography on Sephadex LH20 Columns

The borate elution procedure described previously (5,19,30) was modified as follows: a column (25 cm X 1.5 cm) packed with Sephadex LH20 (Pharmacia, Piscataway, New Jersey) was first equilibrated with methanol: 0.05 M sodium borate, pH 8.7 (1:1) and the enzyme-degraded DNA sample was applied. The column was then eluted, first with 60 ml of the equilibration mixture and then with a linear gradient of 100 ml methanol:0.05 M sodium borate (1:1) and 100 ml methanol:0.05 M sodium borate (8:2). One hundred and ten 2.0 ml fractions were collected and 0.5 ml samples were counted by liquid scintillation counting. Absorbance at 260 nm of deoxyribonucleosides was measured for the first

30 fractions. A marker of p-nitrobenzylpyridine was added to each column and its position detected by UV-absorption.

## High-Pressure Liquid Chromatography

Enzyme-digested DNA samples were first placed on quick-Sep LH20 columns (20) (Isolab, Akron, Ohio) eluted with 25 ml of water to remove deoxyribonucleosides, and then eluted with methanol (eight 2-ml fractions were collected) to remove BaP-deoxyribonucleoside adducts. Methanol fractions containing the BaP-DNA adducts (usually the second and third) were combined and evaporated to a low volume. Samples (50 $\mu$l) were chromatographed on an Altex Model 312 HPLC (Altex Scientific, Berkeley, California) on an Altex 25 cm X 4.6 mm Ultrasphere-octyl $C^8$ reverse-phase column at room temperature. A precolumn of CO:PELL ODS (Whatman, Clifton, New Jersey) was used for all samples, and all were passed through a fluoropore filter (Millipore, Bedford, Massachusetts) prior to chromatography. Samples were eluted for 35 minutes with methanol: water (55:45) and then for 10 minutes with a linear gradient of methanol: water (55:45 to 65:35), followed by 5 minutes at 65:35 at a flow rate of 1.0 ml/minute. One hundred and sixty 0.3 ml fractions were collected in scintillation vials and counted by liquid scintillation counting. Fractions at the end of the eluate as well as the first ten fractions contained no radioactivity; these fractions are not plotted on the figures in the Results section.

Samples of BaP-DNA adduct hydrolysis products were prepared by placing DNA samples in 0.5 N HCl at 100°C for 60 minutes (22). The acid-released material was extracted twice into equal volumes of ethyl acetate and the radioactive material in both phases was measured. The ethyl acetate-extractable material was evaporated to dryness, dissolved in methanol, and chromatographed as described above.

## Determination of Micrococcal Nuclease-Releasable Material

Nuclei were prepared from BaP-treated hamster embryo cell cultures as described by Smerdon et al. (33). Samples containing 50 $\mu$g of DNA/ml digestion buffer (33) were incubated with 5 units of micrococcal nuclease (Sigma, E.C. No. 3.1.4.7 from *Staphylococcus aureus,* Foggi strain) at 37°C. At the times specified in Figure 7, the undigested material was precipitated by the addition of an equal volume of 1 M NaCl, 1 M $HClO_4$ at 0°C for 20 min. After centrifugation, the UV-absorbance ($A_{260}$) of the acid-soluble material was determined and radioactivity was measured by liquid scintillation counting.

## RESULTS

To facilitate our studies on the effects of the length of incubation with BaP on the BaP-DNA adducts formed in hamster embryo cell cultures, a smaller LH20 column and a methanol:water gradient system similar to that described by Remsen et al (30) was used. The smaller DNA samples required in this system permitted experiments to be carried out using standard 150 or 175 cm$^2$ plastic tissue culture flasks instead of roller bottles. The LH20 chromatographic elution profiles of enzyme digests of BaP-DNA samples from cells exposed to 0.5 nmole [$^3$H]BaP/ml medium for either 5, 24, or 72 hours are shown in Figure 2. The 5-hour sample contained a marker of enzyme-digested [$^{14}$C]anti-BaPDE-DNA; the anti-adducts chromatographed with peak 1. Although not shown, a sample of [$^3$H]syn-BaPDE-DNA adducts chromatographed in the same position as peak 2 on this gradient. The 5-hour sample of BaP-DNA from hamster embryo cells contained a small amount of early eluting material, major amounts of syn-BaPDE and anti-BaPDE adducts are nearly identical at 5 material eluting later (peak 3). As can be seen in Table 1 the relative amounts of syn-BaPDE and anti-BaPDE adducts are nearly identical at 5 hours; after 24 hours of exposure, the material in peak 3 has decreased and the anti-BaPDE adduct peak is now slightly greater than the syn-BaPDE; after 72 hours of exposure, the anti-BaPDE adduct peak is considerably greater than the syn-BaPDE adduct peak.

The BaP-DNA adducts formed in cells were then examined by HPLC. The elution profiles of samples of DNA reacted with either [$^{14}$C] anti-BaPDE or [$^3$H]syn-BaPDE are shown in Figure 3. These diol epoxides were also reacted with poly dG-polydC, poly dA-dT•dA-dT, poly dC, and poly dI-dC and the elution positions of the adducts formed are given in Table 2. As reported by others (17,22,24,25,26), the major adducts with DNA are with guanine, but small amounts of cytosine and adenine adducts are also formed.

The elution profile of the 5-hour sample of [$^3$H]BaP-DNA on this gradient contained five major peaks of adducts, designated A through E in Figure 4. On the basis of the elution profiles of the BaPDE-deoxyribonucleoside adducts shown in Figure 3 and Table 2, peak C contains anti-BaPDE-dG adducts, peak D contains syn-BaPDE-dG adducts, and peak E contains dA adducts of both BaPDE's. The material in peaks A and B appear to be cytosine adducts, since peak A elutes in the same position as an anti-BaPDE-dC adduct marker and peak B elutes in the same position as a syn-BaP-dC adduct marker.

The percentage of the total amount of material that is eluted in each peak at different times is given in Table 3. As can be seen, after 24 hours of exposure, the amount of material in peaks B and E decreased, the amount in Peak C increased, and the amounts in peaks A and D were unchanged. After 72 hours of exposure, the relative amount of material decreased in peaks B and E, increased in peaks C and D, and was about the same in peak A (Table 3 and Figure 4).

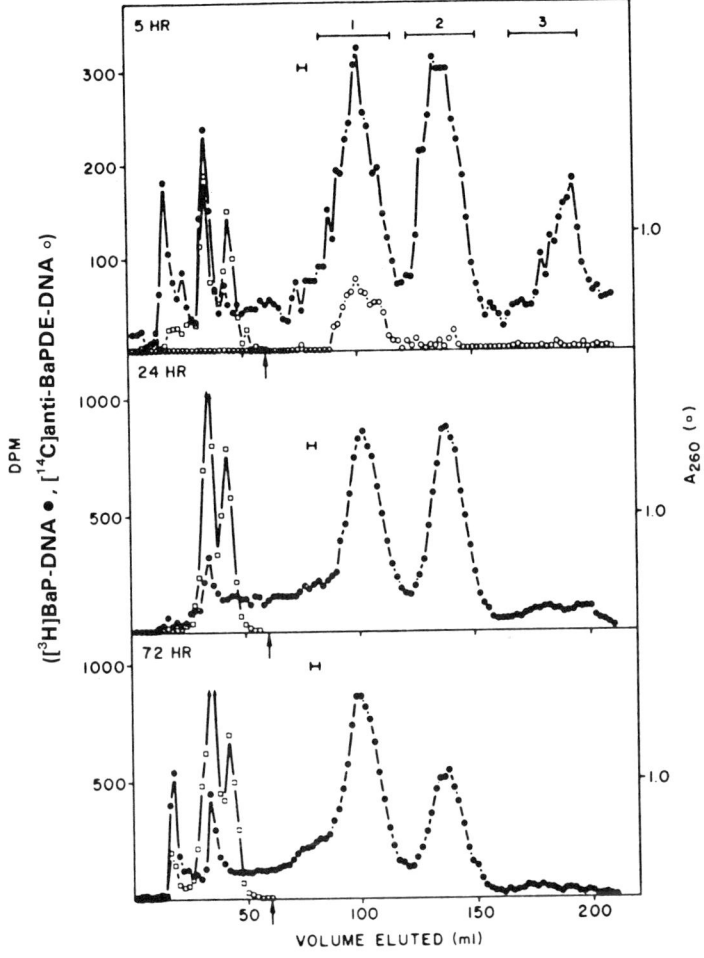

FIGURE 2. Sephadex LH20 column elution profiles of the [$^3$H]BaP-deoxyribonucleoside adducts formed in hamster embryo cell cultures exposed for various lengths of time to [$^3$H]BaP. The cells were treated with [$^3$H]BaP, and DNA samples were isolated, enzymatically degraded to deoxyribonucleosides, and chromatographed on Sephadex LH20 columns eluted with borate-containing methanol: water gradients as described in the Materials and Methods section. ● DPM/0.5 ml fraction; † start of the gradient; ⊢ elution position of p-nitrobenzylpyridine marker.

TABLE 1. Sephadex LH20 Separation of [$^3$H]BaP-DNA Adducts[a]

| [$^3$H]BaP-DNA Sample | Peaks | | | |
|---|---|---|---|---|
| | Early | 1 | 2 | 3 |
| 5 hour | 7 | 38 | 39 | 16 |
| 24 hour | 5 | 44 | 37 | 7 |
| 72 hour | 7 | 60 | 29 | 3 |

[a] Data from Figure 2 are expressed in terms of the relative amount of the radioactivity in each peak for the BaP-DNA samples shown in that figure. Elution volumes for these peaks were: early, 25 to 40 ml; peak 1, 90 to 120 ml; peak 2, 125 to 155 ml; and peak 3, 175 to 205 ml. Anti-BaPDE-DNA adducts elute in the same volume as peak 1; syn-BaPDE-DNA adducts elute in the same volume as peak 2.

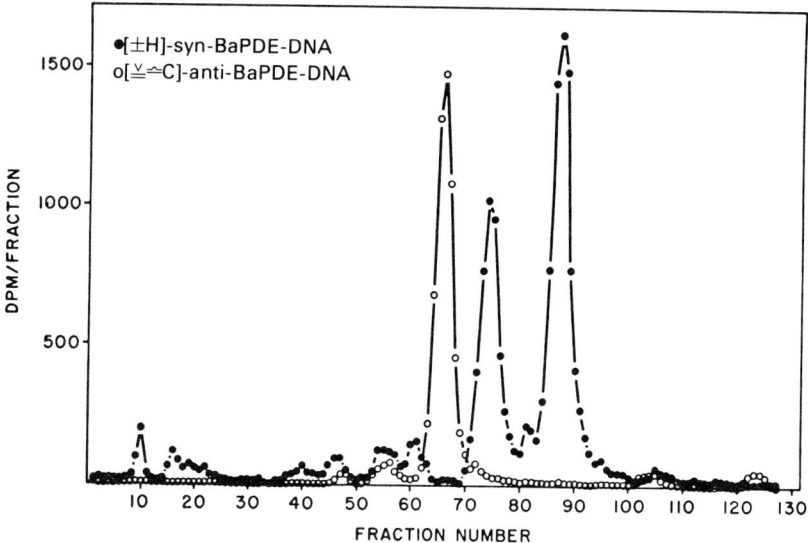

FIGURE 3. HPLC elution profiles from an Ultrasphere-octyl reverse-phase column of enzyme digests of DNA samples prepared by reaction of the DNA in aqueous; ethanol solution with [$^3$H]syn-BaPDE or [$^{14}$C]anti-BaPDE. Sample preparation and HPLC conditions are described in the Materials and Methods section.

To relate the HPLC results to the Sephadex LH20 column results, peaks 1 and 2 from an LH20 column of a 24-hour BaP-DNA sample were chromatographed on the HPLC system. To remove the large amounts of borate in these samples and reduce volume, the samples in each LH20 peak were pooled, diluted with 2 volumes of water, and passed through a $C_{18}$ Sep-pack (Waters Associates, Milford, Massachusetts) eluted with water to remove borate and then with methanol to remove the BaP-DNA

TABLE 2. HPLC Elution Positions of Anti- and Syn-BaPDE-Deoxyribonucleoside Adducts[a]

| Nucleic Acid | Anti-Adducts | | | Syn-Adducts | | | |
|---|---|---|---|---|---|---|---|
| dG-dC | 45 | <u>52</u> | 65 | 53 | <u>60</u> | 74 | 90 |
| dA-dT | 44 | <u>114</u> | <u>131</u> | 91 | 124 | <u>138</u> | |
| dC | 44 | <u>66</u> | | <u>56</u> | 77 | 90 | |
| dI-dC | <u>43</u> | <u>64</u> | | <u>60</u> | <u>108</u> | | |

[a]Adducts peaks that represented more than 25 percent of the total BaPDE adducts in the sample are underlined.

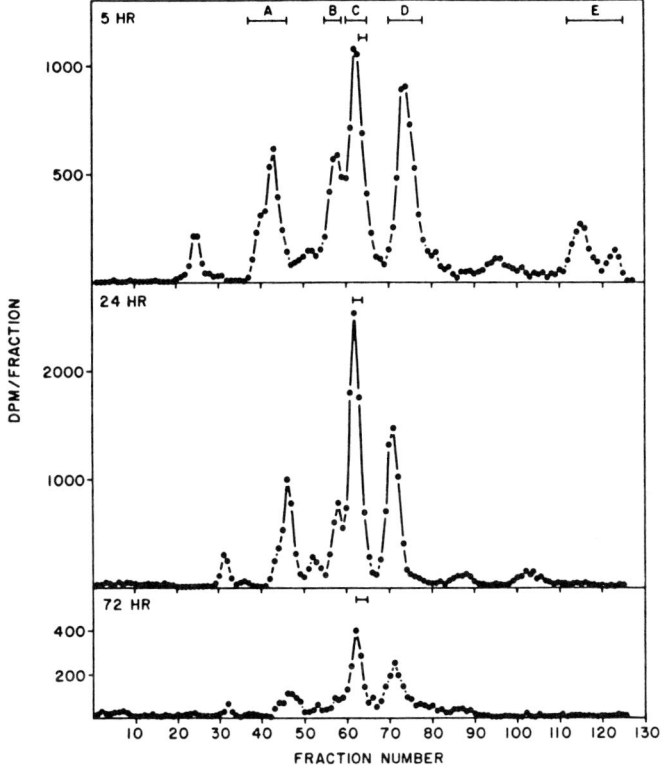

FIGURE 4. HPLC elution profiles of enzyme degraded [³H]BaP-DNA samples from hamster embryo cell cultures treated with [³H]BaP for length of time specified. Samples were prepared and chromatographed on a reverse-phase column as described in the Materials and Methods section. The relative amounts of radioactivity in the peaks designated A,B,C,D,E, are given in Table 3. ⊢⊣ indicates the elution position of the major [¹⁴C]anti-BaPDE-DNA adduct used as a marker in all columns.

TABLE 3. HPLC Separation of [$^3$H]B(a)P-DNA Adducts[a]

| [$^3$H]BaP-DNA Sample | Percentage of Radioactivity in Peak | | | | |
|---|---|---|---|---|---|
| | A | B | C | D | E |
| 5 hour | 16 | 14 | 26 | 25 | 10 |
| 24 hour | 18 | 9 | 37 | 24 | 4 |
| 72 hour | 16 | 5 | 41 | 31 | 0 |

[a] Data from Figure 4 are expressed in terms of the relative amount of the radioactivity in each peak for the (±H)BaP-DNA samples analyzed by HPLC. The peak fractions and their corresponding markers are: A, 37 to 46, anti-dC; B, 55 to 59, syn-dC; C, 60 to 65, anti-dG; D, 70 to 78, syn-dG; E, 112 to 135, syn-dA and anti-dA. The total amounts of BaP-DNA adducts in these samples were: 5 hours, 5 pmoles/mg DNA; 24 hours, 9 pmoles/mg DNA; 72 hours, 4 pmoles/mg DNA.

adducts. The methanol eluates were chromatographed by HPLC as described above; the elution profiles are shown in Figure 5. Peak 1 from the LH20 column contained [$^{14}$C]anti-BaPDE-adduct marker; this marker was added to peak 2 before HPLC. In both samples, the HPLC peaks were very broad, probably because of residual borate. However, the sample in peak 2 could be resolved into two peaks by HPLC, the major one eluting in the same position as peak D, the syn-BaPDE-dG adduct peak, and the smaller one eluting in the position of peak B. Both the [$^{14}$C]marker and the [$^3$H]BaP-DNA adducts in peak 1 eluted from the HPLC column earlier than the [$^{14}$C]marker usually does, probably because both remained complexed to the borate. The major [$^3$H]peak co-chromatographed with the [$^{14}$C]marker in this location.

The identity of the BaP diol epoxides in these DNA samples was also examined by acid hydrolysis of BaP-DNA to release tetrols, extraction of the hydrolysis products with ethyl acetate, and chromatography on HPLC. The elution positions of the hydrolysis products of [$^3$H]syn-BaPDE and [$^{14}$C]anti-BaPDE are shown at the top of Figure 6. A 5-hour sample of [$^3$H]BaP-DNA from cells and the marker of [$^{14}$C]anti-BaPDE-DNA were hydrolyzed and chromatographed; the elution profile is shown at the bottom of Figure 6. The ratios of [$^3$H]BaP tetrols eluted at the positions of the syn- and anti-tetrols are similar to those described for the 5-hour sample analyzed on borate-containing gradients on LH20 column (Figure 2, Table 1) and by HPLC (Figure 4, Table 3) based upon our identification of peaks A and C resulting from the anti-BaPDE and peaks B and D resulting from the syn-BaPDE.

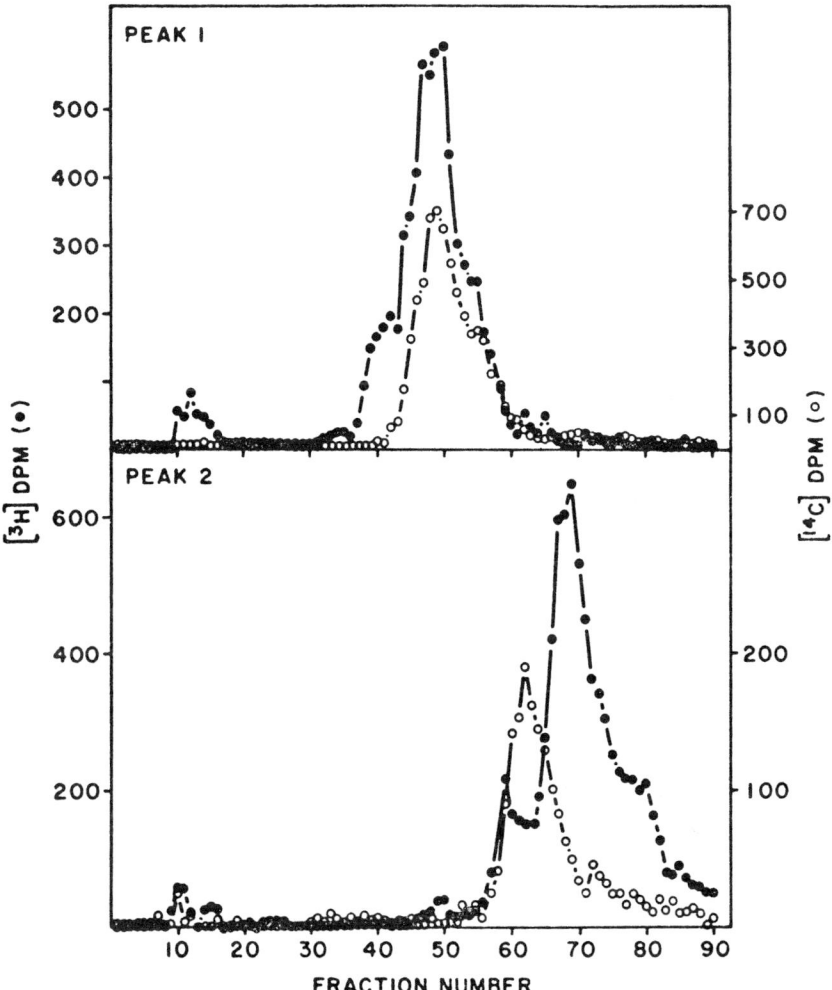

FIGURE 5. HPLC elution profiles of peaks 1 and 2 from an LH20 column of a 24-hour [$^3$H]BaP-DNA sample. The material in peaks from a column such as that shown in Figure 2 were prepared as described in the Results section and analyzed by HPLC on a reverse-phase column as described in the Materials and Methods section. ● DPM/fraction of [$^3$H]BaP-DNA adducts; o DPM/fraction of [$^{14}$C]anti-BaPDE-DNA adducts.

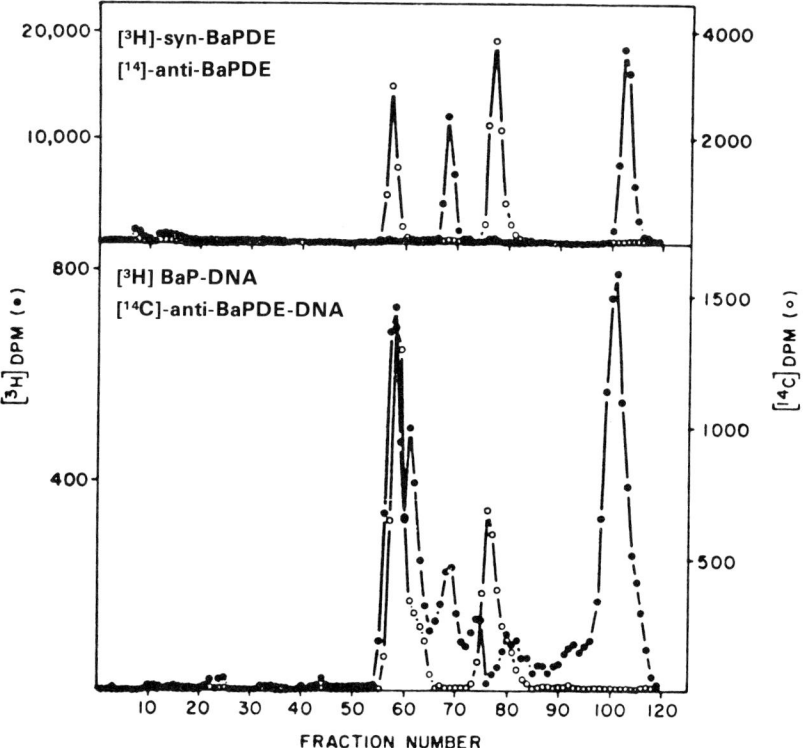

FIGURE 6. HPLC elution profiles of the hydrolysis products of syn- and anti-BaPDE, [$^{14}$C]anti-BaPDE-DNA, and [$^3$H]BaP-DNA. Samples were hydrolyzed and chromatographed on a reverse-phase column as described in the Materials and Methods section.

To establish how changes in the BaP-DNA adducts with length of exposure to BaP were related to the micrococcal nuclease sensitivity of the BaP bound in the cell nuclei, samples from [$^3$H]BaP-treated cells were digested with micrococcal nuclease (Figure 7). The time course of DNA digestion as measured by release of $A_{260}$ from these samples is shown at the top; this was identical for both the 5-hour and 24-hour samples. The lower portion of Figure 7 shows the release of tritium by micrococcal nuclease from these samples in 60 minutes; in the 5-hour sample almost 50 ± 4 percent of the radioactivity was released, whereas in the 24-hour sample, only 18 ± 3 percent was released.

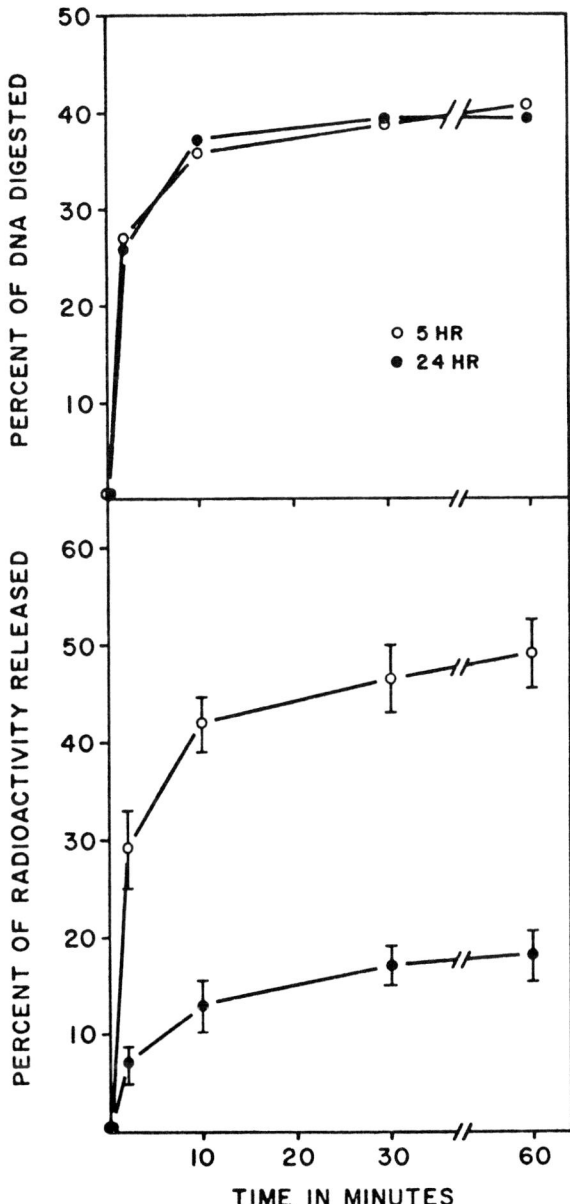

FIGURE 7. Micrococcal nuclease digestion of [$^3$H]BaP-nuclei from hamster embryo cells. Samples from cells treated for 24 hours (●) or 5 hours (o) were prepared and assayed as described in the Materials and Methods section. The top shows the percentage of the DNA digested as measured by the amount of nonprecipitable $A_{260}$ absorbing material. The bottom shows the percentage of the radioactivity that was not perchloric acid precipitable.

## DISCUSSION

The results demonstrate that the nature of the BaP-DNA adducts formed in hamster embryo cells varies with the length of time of exposure of the cells to [$^3$H]BaP. As the length of time of incubation increases, the proportion of the DNA adducts of the anti-isomer increases. This result has now been demonstrated by HPLC of the BaP-deoxyribonucleoside adducts and the acid-released BaP-tetrols as well as by chromatography on Sephadex LH20 columns eluted with borate-containing methanol:water gradients. In previous studies (5) with cultures of hamster embryo cells grown in roller bottles, after 4 hours and 6 hours of exposure to BaP most of the BaP-DNA adducts were syn-BaPDE adducts; after 24 hours or 72 hours of exposure most were anti-BaPDE adducts. In the present study with hamster embryo cells growing in stationary flasks, the BaP-DNA adducts at 5 hours resulted from equal amounts of each BaPDE isomer; after 24 hours of exposure to BaP more of the adducts resulted from the anti-isomer, but the ratio of anti to syn was less than that in roller bottles (5); by 72 hours, the anti-isomer adducts predominated in both types of culture. Thus, the culture conditions exert an important influence on the nature of the BaP-DNA adducts and on how these are altered with time of exposure.

The major changes in the individual BaP-deoxyribonucleoside adducts with time are the decrease in the adenine adducts and in the material in peak B, and the increase in the material in peak C, suggesting selective excision of some adducts as proposed by Ivanovic et al. (14). The much greater proportion of the BaP-DNA adducts in the micrococcal nuclease-sensitive regions of the chromatin at early times than at later times suggests that adducts in these regions are more rapidly excised. More rapid repair of nuclease-sensitive regions has been demonstrated for cells treated with N-acetoxy-2-acetylaminofluorene (34). If the more stable anti-BaPDE is better able to bind to nuclease-resistant regions than the syn-BaPDE, syn-BaPDE binding should occur to a greater extent in the nuclease-sensitive regions and, therefore, more would be lost as the adducts in this region are repaired. Studies are in progress to characterize the time-dependent changes that occur in formation and removal of BaP-DNA adducts in micrococcal nuclease-sensitive and nuclease-resistant regions of the chromatin.

## ACKNOWLEDGMENTS

The [$^{14}$C]anti-BaPDE and [$^3$H]syn-BaPDE were provided by the Cancer Research Program, Division of Cancer Cause and Prevention,

National Cancer Institute, Bethesda, Maryland. The hamsters from the Leo Goodwin Institute were also provided by the National Cancer Institute, Bethesda, Maryland. This work was supported by Grants CA 19948, CA 08936, and CA 21778 from the National Cancer Institute, Department of Health, Education, and Welfare.

## REFERENCES

1. Abell, C. W., and Heidelberger, C. (1962): Interaction of carcinogenic hydrocarbons with tissues. VIII. Binding of tritium-labeled hydrocarbons to the soluble proteins of mouse skin. *Cancer Res.* 22:931-946.
2. Autrup, H., Harris, C. C., Trump, B. F., and Jeffrey, A. M. (1978): Metabolism of benzo(a)pyrene and identification of the major benzo(a)pyrene-DNA adducts in cultured human colon. *Cancer Res.* 38:3689-3696.
3. Baird, W. M., and Brookes, P. (1973): Isolation of the hydrocarbon-deoxyribonucleoside products from the DNA of mouse embryo cells treated in culture with 7-methylbenz(a)anthracene-$^3$H. *Cancer Res* 33:2378-2385.
4. Baird, W. M., and Diamond, L. (1976): Effect of 7,8-benzoflavone on the formation of benzo(a)pyrene-DNA-bound products in hamster embryo cells. *Chem.-Biol. Interact.* 13:67-75.
5. Baird, W. M., and Diamond, L. (1977): The nature of benzo(a)-pyrene-DNA adducts formed in hamster embryo cells depends on the length of time of exposure to benzo(a)pyrene. *Biochem. Biophys. Res. Commun.* 77:162-167.
6. Beland, F. A., and Harvey, R. G. (1976): The isomeric 9,10-oxides of trans-7,8-dihydroxy-7,8-dihydrobenzo(a)pyrene. *J. Chem. Soc. Chem. Commun.* 84-85.
7. Brookes, P. (1977): Role of covalent binding in carcinogenicity. In: *Biological Reactive Intermediates,* D. J. Jallow, J. J. Kocsis, R. Snyder, and H. Vainio, Eds., pp. 470-480, Plenum Press, New York.
8. Brookes, P., and Lawley, P. D. (1964): Evidence for the binding of polynuclear aromatic hydrocarbons to the nucleic acids of mouse skin: Relation between carcinogenic power of hydrocarbons and their binding to deoxyribonucleic acid. *Nature* 201:781-784.
9. Diamond, L., and Baird, W. M. (1977): Chemical carcinogenesis in vitro. In: *Growth, Nutrition and Metabolism of Cells in Culture,* Vol. 3, G. H. Rothblat and V. J. Cristofalo, Eds., pp. 421-470, Academic Press, New York.

10. Dipple, A. (1976): Polynuclear aromatic carcinogens. In: *Chemical Carcinogens*, C. E. Searle, Ed., pp. 245-314, ACS Monograph Series, No. 173, American Chemical Society, Washington.
11. Duncan, M., Brookes, P., and Dipple A. (1969): Metabolism and binding to cellular macromolecules of a series of hydrocarbons by mouse embryo cells in culture. *Int. J. Cancer* 4:813-819.
12. Heidelberger, C. (1975): Chemical carcinogenesis. *Annu. Rev. Biochem.* 44:79-121.
13. Hulbert, P. B. (1975): Carbonium ion as ultimate carcinogen of polycylic aromatic hydrocarbons. *Nature* 256:146-148.
14. Ivanovic, V., Geacintov, N. E., Yamasaki, H., and Weinstein, I. B. (1978): DNA and RNA adducts formed in hamster embryo cell cultures exposed to benzo(a)pyrene. *Biochemistry* 17:1597-1603.
15. Jahn, C. L., and Litman, G. W. (1979): Accessibility of deoxyribonucleic acid in chromatin to the covalent binding of the chemical carcinogen benzo(a)pyrene. *Biochemistry* 18:1442-1449.
16. Jeffrey, A. M., Weinstein, I. B., Jennette, K. W., Grzeskowiak, K., Nakanishi, K., Harvey, R. G., Autrup, H., and Harris, C. (1977): Structures of benzo(a)pyrene-nucleic acid adducts formed in human and bovine bronchial explants. *Nature* 269:348-350.
17. Jennette, K. W., Jeffrey, A. M., Blobstein, S. H., Beland, F. A., Harvey, R. G., and Weinstein, I. B. (1977): Nucleoside adducts from the in vitro reaction of benzo(a)pyrene-7,8-dihydrodiol 9,10-oxide or benzo(a)pyrene 4,5-oxide with nucleic acids. *Biochemistry* 16:932-938.
18. Jerina, D. M., Lehr, R. E., Yagi, H., Herandez, O., Dansette, P. M., Wislocki, P. G., Wood, A. W., Chang, R. L., Levin. W., and Conney, A. H. (1976): Mutagenicity of benzo(a)pyrene derivatives and the description of a quantum mechanical model which predicts the case of carbonium ion formation from diol epoxides. In: *In Vitro Metabolic Activation in Mutagenesis Testing*, F. J. deSerres, J. R. Fouts, J. R. Bend, and R. M. Philpot, Eds., pp. 159-177, Elsevier/ North-Holland Biomedical Press, Amsterdam.
19. King, H.W.S., Osborne, M. R., Beland, F. A., Harvey, R. G., and Brookes, P. (1976): ($\pm$)-7$\alpha$,8$\beta$-dihydro-9$\beta$,10$\beta$-epoxy-7,8,9,10-tetrahydrobenzo(a)pyrene is an intermediate in the metabolism and binding to DNA of benzo(a)pyrene, *Proc. Natl. Acad. Sci. U.S.A.* 73:2679-2681.
20. King, H.W.S., Osborne, M. R., and Brookes, P. (1977): The metabolism and DNA binding of 3-methylcholanthrene. *Int. J. Cancer* 20:564-571.
21. King, H.W.S., Osborne, M. R., and Brookes, P. (1979): The in vitro and in vivo reaction at the $N^7$-position of guanine of the ultimate

carcinogen derived from benzo(a)pyrene. *Chem. Biol. Interact.* 24:345-353.
22. Koreeda, M., Moore, P. D., Wislocki, P. G., Levin, W., Conney, A. H., Yagi, H., and Jerina, D. M. (1978): Binding of benzo(a)pyrene 7,8-diol-9,10-epoxides to DNA, RNA, and protein of mouse skin occurs with high stereoselectivity. *Science* 199:778-781.
23. Kuroki, T., and Heidelberger, C. (1972): Determination of the h-protein in transformable and transformed cells in culture. *Biochemistry* 11:2116-2124.
24. Meehan, T., and Straub, K. (1979): Double-stranded DNA stereoselectively binds benzo(a)pyrene diol epoxides, *Nature* 277:410-412.
25. Meehan, T., Straub, K., and Calvin, M. (1977): Benzo(a)pyrene diol epoxide covalently binds to deoxyguanosine and deoxyadenosine in DNA. *Nature* 269:725-727.
26. Osborne, M. R., Beland, F. A., Harvey, R. G., and Brookes, P. (1976): The reaction of ($\pm$)-7$\alpha$,8$\beta$-dihydroxy-9$\beta$,10$\beta$-epoxy-7,8,9,10-tetrahydrobenzo(a)pyrene with DNA. *Int. J. Cancer* 18:362-368.
27. Osborne, M. R., Harvey, R. G., and Brookes, P. (1978): The reaction of trans-7,8-dihydroxy-anti-9,10-epoxy-7,8,9,10-tetrahydrobenzo(a)pyrene with DNA involves attack at the $N^7$-position of guanine moieties. *Chem.-Biol. Interact.* 20:123-130.
28. Phillips, D. H., and Sims, P. (1979): Polycyclic aromatic hydrocarbon metabolites: their reactions with nucleic acids. In: *Chemical Carcinogens and DNA*, Vol. 2, P. L. Grover, Ed., pp. 29-57, CRC Press, Inc., Boca Raton, Florida.
29. Pulkrabek, P., Leffler, S., Weinstein, I. B., and Grunberger, D. (1977): Conformation of DNA modified with a dihydrodiol epoxide derivative of benzo(a)pyrene. *Biochemistry* 16:3127-3132.
30. Remsen, J., Jerina, D., Yagi, H., and Cerutti, P. (1977): In vitro reaction of radioactive 7$\beta$,8$\alpha$-dihydroxy-9$\alpha$,10$\alpha$-epoxy-7,8,9,10-tetrahydrobenzo(a)pryene and 7$\beta$,8$\alpha$-dihydroxy-9$\beta$,10$\beta$-epoxy-7,8,9,10-tetrahydrobenzo(a)pyrene with DNA. *Biochem. Biophys. Res. Commun.* 74:934-940.
31. Shinohara, K., and Cerutti, P. A. (1977): Excision repair of benzo(a)pyrene-deoxyguanosine adducts in baby hamster kidney 21/C13 cells and in secondary mouse embryo fibroblasts C57BL/6J. *Proc. Natl. Acad. Sci. U.S.A.* 74:979-983.
32. Sims, P., Grover, P. L., Swaisland, A., Pal, K., and Hewer, A. (1974): Metabolic activation of benzo(a)pyrene proceeds by a diol-epoxide. *Nature* 252:326-328.
33. Smerdon, M. J., Tlsty, T. D., and Lieberman, M. W. (1978): Distribution of ultraviolet-induced DNA repair synthesis in nuclease

sensitive and resistant regions of human chromatin. *Biochemistry* 17:2377-2386.
34. Tlsty, T. D., and Lieberman, M. W. (1978): The distribution of DNA repair synthesis in chromatin and its rearrangement following damage with N-acetoxy-2-acetylaminofluorene, *Nucleic Acids Res.* 5:3261-3273.
35. Yagi, H., Thakker, D. R., Hernandez, O., Koreeda, M., and Jerina, D. M. (1977): Synthesis and reactions of the highly mutagenic 7,8-diol 9,10-epoxides of the carcinogen benzo(a)pyrene. *J. Am. Chem. Soc.* 99:1604-1611.
36. Yamasaki, H., Roush, T. W., and Weinstein, I. B. (1978): Benzo(a)pyrene 7,8-dihydrodiol-9,10-oxide modification of DNA: relation to chromatin structure and reconstitution. *Chem.-Biol. Interact.* 23:201-213.

# OXIDATIVE AND NONOXIDATIVE METABOLISM OF POLYCYCLIC AROMATIC HYDROCARBONS IN RABBIT AND CHICKEN AORTAS AND IN HUMAN FETAL SMOOTH-MUSCLE CELLS

**J. A. Bond, R. M. Kocan, E. P. Benditt, and M. R. Juchau**

Departments of Pharmacology and Pathology
School of Medicine
University of Washington
Seattle, Washington 98195

## INTRODUCTION

Atherosclerosis, a morphological form of arteriosclerosis, is a multifactorial disease process that has reached epidemic levels, particularly in Western Europe and the United States where it is the leading cause of death (2). Research in atherosclerosis, however, has been guided by general hypotheses that have tended to neglect various environmental factors that may play a critical role in the initiation and progression of atherosclerosis. Among the most notable factors are chemicals (cytotoxic and mutagenic), viruses, and radiation.

The somatic mutation-selection hypothesis for atherosclerosis (3) postulates that the requisite focal proliferation of smooth muscle cells (SMC) in the intima of the aorta may stem from a single SMC that undergoes cellular alteration, presumably a mutation, which initiates the subsequent formation of atherosclerotic plaques. This hypothesis suggests new routes for research into the nature of atherogenesis. Environmental promutagens are postulated to act as initiators and/or promoters of atherosclerotic lesions in much the same manner as they are proposed as initiators and promoters of benign and malignant neoplasms (12). Known tumor initiators such as chemical mutagens (1,22,23), radiation (10,15,16), and oncogenic viruses (9) all have caused enhanced development and increased incidences of atherosclerotic lesions in experimental animals. The hypothesized analogy between atherosclerosis and cancer has further support in light of the fact that both share common risk factors—urban living, cigarette smoke, and dietary habits (14).

Polycyclic aromatic hydrocarbons (PAH) are products derived from the combustion of such energy sources as petroleum products, wood, and coal. They are ubiquitous and abundant pollutants of air, water, and soil. One of the most widely studied methylated PAH is 7,12-dimethylbenz[a]-anthracene (DMBA), whose wide spectrum of biological activity includes production of mammary cancer (13) and adrenal necrosis (21) in rats. DMBA is considered to be the most potent mouse skin carcinogen of all tested PAH.

In most mammalian tissues, PAH are enzymatically converted to epoxide intermediates which can spontaneously rearrange to phenols, be converted enzymatically to trans-dihydrodiols via epoxide hydratase, be reduced back to the parent compound via epoxide reductase, be conjugated enzymatically or nonenzymatically with glutathione (GSH) via GSH S-epoxide transferase, or react directly with cellular macromolecules. The trans-dihydrodiols may be further oxidized to the diol-epoxides that also react with cellular macromolecules.

This paper describes the various enzyme systems in aortas of rabbits and chickens and in human fetal aortic SMC in culture which are responsible for the overall metabolism of DMBA and benzo[a]pyrene-4,5-oxide (BaP-4,5-oxide). The relative levels of these enzyme activities are important as they may be key determinants of the steady-state levels of arene oxides, diol oxides, and/or other reactive intermediates. The formation and detection of dihydrodiols and phenols [as assessed by high-pressure liquid chromatography (HPLC)] and GSH conjugates may suggest the presence in aortas of an epoxide intermediate. Epoxide and diol-epoxide intermediates are thought to mediate cellular transformation by binding to the DNA of cells.

## MATERIALS AND METHODS

### Chemicals

[$^{14}$C]-DMBA (specific activity 10 to 100 mCi/mmole) was purchased from Amersham/Searle, Arlington Heights, Illinois. [$^{14}$C]-styrene oxide was synthesized from styrene to a final specific activity of $3.25 \times 10^{-2}$ mCi/mmole. BaP-4,5-oxide was both synthesized from benzo[a]pyrene (BaP) and purified in our laboratory. DMBA and 1,2-benz[a]anthracene (BA) were supplied by Sigma Chemical Co., St. Louis, Missouri. 3-Methylcholanthrene (3-MC) was purchased from Mann Research Laboratories, New York, N.Y. [$^{14}$C]-DMBA, DMBA, BA, and 3-MC were further purified by thin-layer chromatography as previously described (5).

Hematin was obtained from Calbiochem, Los Angeles, California; 7,8-benzoflavone (7,8-BF) and 1,2-epoxy-3,3,3-trichloropropane (TCPO) from Aldrich Chemical Co., Milwaukee, Wisconsin; Aroclor 1254 from Analabs, Inc., No. Haven, Connecticut; and Aquasol and Omnifluor from New England Nuclear, Boston, Massachusetts. NADPH, NADH, glucose 6-phosphate (G6P), glucose 6-phosphate dehydrogenase (G6PD), 17-$\beta$-estradiol ($\beta$-E$_2$), GSH, and UDP-glucuronic acid (UDPGA) all were obtained from Sigma Chemical Co., St. Louis, Missouri. All other reagents and chemicals utilized were of the highest purity commercially available.

## Treatment of Animals With Inducing Agents

Animals were divided into three groups consisting of 3 to 5 controls and 3 to 5 experimental animals in each group. After a 2-day acclimatization period, animals were injected intraperitoneally with single doses of inducing agents as follows: group 1, 3-MC (40 mg/kg) in corn oil and sacrificed 48 hours later, and group 2, Aroclor 1254 (500 mg/kg) in corn oil and sacrificed 5 days later. Controls were injected with equal volumes of corn oil. At the time of sacrifice, aortas were excised and prepared as described previously (4). Proteins were determined by the method of Lowry et al (17). Values presented in the tables represent the means of 3 to 4 separate experiments.

## Assay of Metabolites of [$^{14}$C]-DMBA

Analyses of the rates of formation of the various metabolites of DMBA from aortic homogenates of rabbits and chickens were performed according to the method described by DiGiovanni et al (7) with slight modifications as described below. Aortic whole homogenates equivalent to 3 to 6 mg of protein from untreated or pretreated animals were incubated with vigorous shaking in a Dubnoff incubator at 37°C for 60 minutes in total darkness under continuous O$_2$ flow. Typical incubation flasks contained 25 nmoles of [$^{14}$C]-DMBA (1 $\mu$Ci), 120 nmoles of unlabelled DMBA, NADPH (1.2 mM, final concentration), NADH (0.71 mM, final concentration), and sufficient potassium phosphate buffer (0.1 M, pH 7.35) to bring the total volume to 2.0 ml. Reaction rates were zero order with respect to substrate and reduced nucleotide concentrations and were linear with time (up to 60 minutes) and with increasing protein concentrations in the ranges utilized. When the effects of modifiers on enzyme activities were studied, hematin, TCPO, 7,8-BF, or $\beta$-E$_2$ were added as indicated in the Results section. Sample blanks consisted of heat-inactivated (100°C; 10 minutes) tissue fractions and all samples were run in duplicate. Reactions were started by addition of substrate and terminated by additions of ethyl acetate:acetone (2:1).

## High-Pressure Liquid Chromatography

Ethyl acetate:acetone (2:1) extracts of DMBA metabolites from duplicate incubation flasks containing aortic tissue preparations were pooled. The pooled extracts were evaporated to dryness in the dark under gentle air flow and then reconstituted in 1.0 ml methanol. Eight $\mu$l of the methanolic solution were injected into a HPLC (Micromeritics Model 7000-001) for analyses as described by Bond et al (5). Retention times of the radioactive metabolites were compared with those of standard DMBA compounds (DMBA and metabolites) (8).

## Epoxide Hydratase Assay

Aortic 600 × g supernatant fractions (5 minutes) equivalent to 1.0—1.5 mg of protein were incubated for 60 minutes with vigorous shaking in a Dubnoff incubator at 37°C in the dark. Typical incubation flasks contained BaP-4,5-oxide (36 $\mu$M, final concentration) and sufficient Tris HCl buffer (0.1 M, pH 8.5) to bring the total reaction volume to 1.0 ml. Reaction rates were zero order with respect to substrate and were linear with respect to time and protein concentrations within the studied ranges. When tested as a modifier, TCPO (0-4 $\mu$M) was added as indicated in the Results section. Sample blanks consisted of heat-inactivated (100°C; 10 minutes) tissue fractions and all samples were run in triplicate.

The reaction mixture was terminated and extracted three times with dichloromethane. Dichloromethane extracts from triplicate incubation flasks containing aortic tissue preparations were pooled and evaporated to dryness in the dark under gentle air flow. The dried extract was reconstituted in 1.0 ml of methanol, and 8 $\mu$l of this methanolic solution were injected into a HPLC. The BaP-4,5-diol was separated on a 2.0 mm × 0.25-m Vydac reverse phase column operated in the isocratic mode (methanol:water 65 to 35 percent).

Samples were monitored by UV absorption and quantities of BaP-4,5-diol formed were determined by measuring the area under the curve. Final values were obtained by comparing with standard BaP-4,5-diol curves.

## Glutathione-S-Epoxide Transferase Assay

Aortic 105,000 × g (60 minutes) supernatant fractions equivalent to 1.0—1.5 mg protein were incubated for 10 minutes with vigorous shaking in a Dubnoff incubator at 37°C in the dark. Typical incubation flasks contained 1 μmole of [$^{14}$C]-styrene oxide (72,000 dpm), GSH (3.7 mM, final concentration), and sufficient Tris HCl buffer (0.1 M, pH 8.0) to bring the total rection volume to 1.0 ml. Reaction rates were zero order with respect to substrate and were linear with respect to time and protein concentrations within the ranges studied. When tested as a modifier, BaP-4,5-oxide (0-72 μM) was added as indicated in the Results section. Sample blanks consisted of heat-inactivated tissue (100°C; 10 minutes) fractions and all samples were run in triplicate.

The reaction was terminated and extracted three times with petroleum ether (b.p. 20 to 40°C). Five hundred μl of the final aqueous extract was transferred to a scintillation vial to which was added 10 ml of Aquasol. Samples were counted on a Beckman 8000 series liquid scintillation counter with an efficiency consistently greater than 85 percent.

## Cells in Culture

Human fetal abdominal SMC were obtained and cultured as previously described (5). Pretreatment of cells included additions of (1) 100 μM 7,8-BF, (2) 1 mM GSH, (3) 1 mM UDPGA, (4) 130 μM TCPO, and (5) 5 mM ATP and 2 mM Na$_2$SO$_4$. Two hours subsequent to the pretreatments, cells were incubated for 24 hours in the presence of 2 μCi [$^{14}$C]-DMBA (26 μM). Following incubations with DMBA, cell growth medium was removed from culture plates and extracted twice with ethyl acetate:acetone (2:1). An aliquot of the remaining aqueous phase was counted on a Beckman 8000 series liquid scintillation counter with an efficiency of >85 percent.

## RESULTS

### Metabolism of DMBA in Aortic Homogenates of Rabbits and Chickens

The observed effects of 3-MC and Aroclor 1254 and of modifiers on DMBA metabolism in vitro in rabbit and chicken aortic homogenates are presented in Table 1. Regardless of the inducing agent utilized, the major metabolites eluting from the HPLC cochromatographed with trans-5,6-dihydro-5,6-dihydroxy-7,12-dimethylbenz(a)anthracene (DMBA-5,6-trans-diol, trans-8,9-dihydro-8,9-dihydroxy-7,12-dimethylbenz(a)anthracene (DMBA-8,9-trans-diol), 7,12-dihydroxymethylbenz(a)anthracene (7,12-diOHMBA), 7-hydroxymethyl-12-methylbenz(a)anthacene (7-OHM-12-MBA), and 12-hydroxymethyl-7-methylbenz(a)anthracene (12-OHM-7-MBA). Total quantities of hydroxylated metabolites recovered

TABLE 1. Effects of Various Inducers in Vivo and Modifiers in Vitro on 7,12-dimethylbenz(a)anthracene Metabolism in Rabbit and Chicken Aortas[a]

### Chickens Pretreated with

| Metabolites Co-Chromatographing with | None | | Modifier | | | 3-MC | | Modifier | | | Aroclor 1254 | | Modifier | | |
|---|---|---|---|---|---|---|---|---|---|---|---|---|---|---|---|
| | None | TCPO | Hematin | 7,8-BF | $\beta$-$E_2$ | None | TCPO | Hematin | 7,8-BF | $\beta$-$E_2$ | None | TCPO | Hematin | 7,8-BF | $\beta$-$E_2$ |
| DMBA-5,6-diol | 3.9 | 1.2 | 16.8 | 0.5 | ND[b] | 16.9 | 1.4 | 16.1 | 6.1 | 4.5 | 19.2 | ND | 21.3 | 2.1 | 0.4 |
| DMBA-8,9-diol | 2.9 | 1.7 | 7.6 | ND | ND | 15.5 | 10.5 | 21.3 | 6.6 | 3.5 | 17.7 | ND | 37.6 | 6.0 | 1.0 |
| 7,12-diOH-MBA | 2.3 | 1.5 | 7.2 | 0.6 | ND | 10.4 | 10.9 | 14.6 | 2.8 | 3.2 | 15.0 | 1.9 | 30.9 | 6.0 | 2.6 |
| 7-OHM-12-MBA | 2.6 | 5.9 | 5.4 | 0.4 | ND | 12.7 | 15.7 | 17.1 | 4.7 | 7.0 | 18.9 | 9.8 | 29.9 | 9.1 | 6.3 |
| 12-OHM-7-MBA | 2.2 | 3.6 | 5.8 | ND | ND | 10.2 | 12.6 | 16.7 | 2.1 | 8.1 | 19.4 | 7.7 | 33.0 | 7.0 | 6.2 |
| Totals | 13.9 | 13.9 | 42.8 | 1.4 | ND | 65.7 | 51.1 | 85.8 | 22.3 | 26.3 | 90.2 | 19.4 | 152.7 | 30.2 | 16.5 |

### Rabbits Pretreated with

| Metabolites Co-Chromatographing with | None | | Modifier | | | 3-MC | | Modifier | | | Aroclor 1254 | | Modifier | | |
|---|---|---|---|---|---|---|---|---|---|---|---|---|---|---|---|
| | None | TCPO | Hematin | 7,8-BF | $\beta$-$E_2$ | None | TCPO | Hematin | 7,8-BF | $\beta$-$E_2$ | None | TCPO | Hematin | 7,8-BF | $\beta$-$E_2$ |
| DMBA-5,6-diol | 25.6 | 8.1 | 66.8 | 8.3 | ND | 94.1 | 6.7 | 168.7 | 30.1 | ND | 53.2 | 43.3 | 169.1 | 22.5 | 5.8 |
| DMBA-8,9-diol | 30.4 | 9.0 | 47.5 | 9.9 | ND | 52.1 | 31.9 | 108.0 | 10.8 | 2.6 | 31.1 | 17.6 | 46.1 | 11.9 | 8.3 |
| 7,12-diOH-MBA | 18.4 | 22.6 | 33.5 | 10.4 | ND | 34.6 | 28.9 | 65.5 | 34.8 | ND | 26.0 | 21.6 | 50.9 | 19.0 | 14.8 |
| 7-OHM-12-MBA | 12.7 | 6.7 | 13.5 | 5.2 | ND | 17.5 | 15.2 | 32.6 | 4.9 | 5.5 | 15.9 | 9.7 | 15.5 | 12.2 | 13.1 |
| 12-OHM-7-MBA | 6.9 | 6.4 | 14.1 | 1.6 | ND | 26.1 | 20.2 | 43.7 | 9.1 | 3.5 | 20.1 | 12.7 | 43.6 | 12.6 | 13.5 |
| Totals | 94.0 | 52.8 | 175.4 | 35.4 | ND | 222.4 | 163.2 | 418.5 | 89.7 | 11.6 | 156.3 | 104.9 | 325.2 | 78.2 | 55.5 |

[a] Incubation flasks containing aortic whole homogenates from untreated and treated rabbits and chickens were incubated for 60 minutes in the dark. Pooled organic fractions from duplicate incubation flasks were evaporated to dryness, reconstituted in methanol to 1 ml, and injected into a high-pressure liquid chromatograph for analyses. Values represent mean specific activities (pmoles/mg protein/min) (N = 3) for each of the peaks reported.

[b] ND indicates that metabolites could not be detected. Values less than 0.4 were considered not detectable.

from the HPLC column were significantly increased by pretreatment of rabbits or chickens with Aroclor 1254 or 3-MC.

The effects of hematin (9 $\mu$M), when added to incubation flasks, were quite remarkable and corroborated earlier investigations in our laboratory (19,20). Additions of hematin markedly increased (by 2- to 3-fold) the aortic metabolism of DMBA for both treated and untreated rabbits and chickens. Hematin additions appeared to mask the enhancement of monooxygenase activities produced by inducing agents. For example, in chickens the 6-fold induction observed with Aroclor 1254 pretreatment was only about 3-fold when hematin was present in reaction vessels.

Addition of 7,8-BF (130 $\mu$M) to incubation flasks containing rabbit and chicken aortic homogenates from each treatment group markedly inhibited the formation of all detectable metabolites of DMBA (50 to 80 percent inhibition). Addition of $\beta$-E$_2$ (100 $\mu$M) to incubation flasks from each treatment group decreased the formation of all detectable metabolites of DMBA (50 to 100 percent inhibition).

In rabbit aortic homogenates, additions of TCPO (13 $\mu$M), a potent epoxide hydratase inhibitor, to incubation flasks from all treatment groups resulted in decreases in the formation of DMBA-5,6-trans-diol and DMBA-8,9-trans-diol, while concomitantly decreasing or leaving unchanged the quantities of 7,12-diOHMBA, 7-OHM-12-MBA, and 12-OHM-7-MBA. However, addition of TCPO (13 $\mu$M) to incubation flasks containing aortic homogenates from untreated or pretreated chickens decreased the formation of DMBA-5,6-trans-diol and DMBA-8,9-trans-diol, while concomitantly increasing the formation of 7-OHM-12-MBA and 12-OHM-7-MBA.

## Aortic Glutathione S-Epoxide Transferase Activity in Rabbits and Chickens

The formation of glutathione conjugates as a function of time in untreated and treated rabbit and chicken aortas appeared to be linear for up to 10 minutes (Figure 1). Pretreatment of rabbits or chickens with either 3-MC (40 mg/kg) or phenobarbital (40 mg/kg) resulted in no detectable changes in the rates of formation of the conjugate.

Additions of BaP-4,5-oxide (0-72 $\mu$M) to incubation flasks containing homogenates from untreated rabbit or chicken aortas and styrene oxide as substrate resulted in a concentration-dependent inhibition of measurable conjugate formation. In both species, maximal inhibition was observed at 72 $\mu$M BaP-4,5-oxide (Table 2).

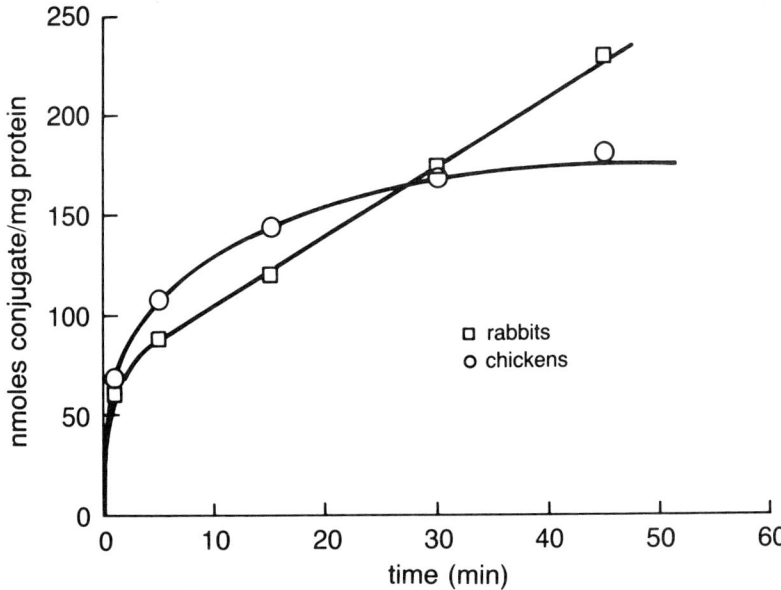

FIGURE 1. Formation of glutathione conjugates as a function of time. Rabbit or chicken aortic 105,000 × g (60 minutes) supernatant fractions were incubated for 1, 5, 15, 30 or 45 minutes. Heat-inactivated controls were run for each time period. Each point represents the mean of three separate experiments. Values presented represent final specific activities after subtraction of background and of radioactivity in the respective boiled controls.

TABLE 2. Inhibition of Glutathione S-epoxide Transferase Activity by Benzo(a) Pyrene-4,5-oxide[a,b] in Aortas of Rabbits and Chickens

| Final Concentration of BaP-4,5-oxide ($\mu$M) | Rabbits | Chickens |
|---|---|---|
| None | 100[c] | 100 |
| 4.5 | 70 | 76 |
| 9.0 | 67 | 65 |
| 18.0 | 52 | 39 |
| 36.0 | 47 | 22 |
| 72.0 | 28 | 14 |

[a]Aortic 105,000 × g (60 minutes) supernatant fractions were incubated with [$^{14}$C]-styrene oxide (1 mM), glutathione (3.7 mM), Tris buffer (pH 8.0) in the dark. Aqueous extracts were counted on a liquid scintillation counter with an efficiency of >85 percent.
[b]Substrate for reaction is [$^{14}$C]-styrene oxide (1 mM).
[c]Values expressed as percentage of control.

## Epoxide Hydratase Activity in Aortas of Rabbits and Chickens

As seen in Figure 2, the formation of detectable BaP-4,5-diol as a function of time in both rabbits and chickens was linear for 60 minutes and 90 minutes, respectively. All remaining experiments utilizing rabbit and chicken aortas were performed with a 60-minute incubation time and were linear with protein concentration.

TCPO was tested for its effects on rabbit and chicken epoxide hydratase, and as expected, small varying concentrations of TCPO (0—4 $\mu$M) produced a concentration-dependent inhibition of both rabbit and chicken epoxide hydratase. Maximal inhibition (35 percent) was observed in both species at 4 $\mu$M TCPO (Table 3).

Pretreatment of rabbits or chickens with either 3-MC (40 mg/kg) or phenobarbital (40 mg/kg) resulted in no significant alteration in enzyme activity.

## Effects of Modifiers on DMBA Metabolism in SMC in Culture

Human fetal SMC in culture were incubated with various modifiers in order to assess their effects on overall cellular metabolism of DMBA.

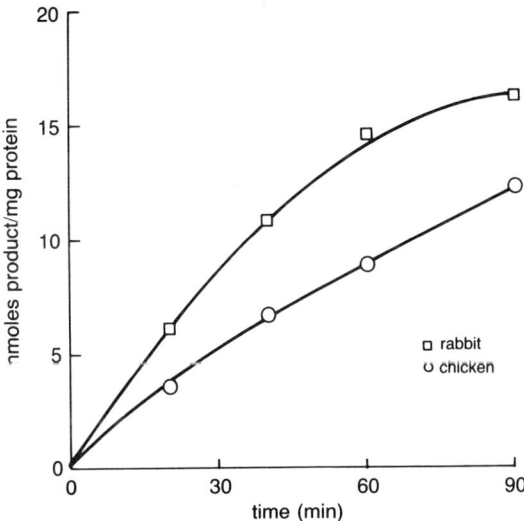

FIGURE 2. Formation of benzo(a)pyrene-4,5-diol as a function of time. Rabbit or chicken aortic 600 × g (5 minutes) supernatant fractions were incubated for 15, 45, 60, or 90 minutes. Heat-inactivated controls were run for each time period. Dichloromethane extracts were pooled, evaporated to dryness, reconstituted in 1.0 ml methanol, and injected into a HPLC for analyses. Samples were monitored by UV absorption and quantities of BaP-4,5-diol formed were determined by measuring the area under curve.

TABLE 3. Effects of 1,2-Epoxy-3,3,3-Trichloropropane (TCPO) on Epoxide Hydratase in Rabbits and Chickens[a]

| Final Concentration of TCPO ($\mu$M) | Rabbits | Chickens |
|---|---|---|
| None | 100[b] | 100 |
| 0.8 | 67.9 | 65.9 |
| 1.6 | 63.7 | 50.2 |
| 2.4 | 41.1 | 43.6 |
| 3.2 | 37.7 | 39.1 |
| 4.0 | 35.8 | 35.1 |

[a]Incubation flasks containing aortic 600 x g supernatant fractions (5 minutes) were incubated with BaP-4,5-oxide (36 $\mu$M) and Tris HCl buffer (pH 8.5) for 60 minutes in the dark. Dichloromethane extracts from triplicate incubation flasks were pooled, evaporated to dryness, reconstituted in 1.0 ml methanol, and injected into a HPLC. Samples were monitored by UV absorption and quantities of BaP-4,5-diol formed were determined by measuring the area under the curve.
[b]Values expressed as percentage of control.

Table 4 reveals the data obtained after an aliquot of the aqueous phase (after ethyl acetate:acetone extraction of growth medium) was counted from each group to which modifiers had been added. Additions of 100 $\mu$M 7,8-BF or 130 $\mu$M TCPO to cells in culture resulted in decreased quantities of dpm detected in the aqueous phase. However, additions of 1 mM GSH or 1 mM UDPGA to cells resulted in marked enhancement of recoverable dpm in the aqueous phase. Additions of ATP and $Na_2SO_4$ did not affect recoverable dpm.

## DISCUSSION

If PAH play a role in the development of atherosclerosis, it becomes of paramount importance to ascertain the capacity of aortic tissue to biotransform PAH. Previous work in our laboratory has demonstrated that enzymes present in chicken aortas catalyze the conversion of PAH to metabolites that covalently bind to DNA and produce mutations in *Salmonella typhmurium* auxotrophs (6). Similar data (unpublished) also have been obtained with rabbit aortic homogenates as the enzyme source. The work of Albert et al (1) has provided the data that demonstrate the capacity of BaP and DMBA to elicit atherosclerotic lesions in chickens. This report has provided evidence that cytochrome P-450-dependent mono-oxygenases in rabbit and chicken aortas are capable of converting an atherogenic chemical, DMBA, to potentially mutagenic intermediates,

TABLE 4. Effects of Various Modifiers Added to Human Fetal Smooth Muscle Cells in Culture on the DPM Remaining in the Aqueous Extract[a]

| Modifier | Final Concentration | Percentage of Control |
|---|---|---|
| None | -- | 100 |
| 7,8-BF | 100 $\mu$M | 76 |
| GSH | 1 mM | 129 |
| UDPGA | 1 mM | 152 |
| TCPO | 130 $\mu$M | 83 |
| ATP; $Na_2SO_4$ | 5 mM; 2 mM | 103 |

[a]Two hours subsequent to pretreatment with modifiers, human fetal smooth muscle cells were incubated with [$^{14}$C]-DMBA for 24 hours. At the end of 24 hours, cell growth medium was removed from culture plates and extracted twice with ethyl acetate:acetone (2:1). An aliquot of the remaining aqueous phase was counted on a liquid scintillation counter with an efficiency of >85 percent. Radioactivity remaining in the aqueous phase following ethyl acetate:acetone extractions from growth media to which no modifiers were added (control) was 426 ± 94 dpm, n = 4. Background radioactivity was calculated as that remaining in the aqueous phase following incubations with growth media containing no smooth muscle cells and was subtracted from control values. Other values in the table represent percentages of control values obtained.

but, also demonstrates the presence of two important nonoxidative inactivating enzymes, GSH S-epoxide transferase and epoxide hydratase.

Pretreatment of rabbits and chickens with either 3-MC or Aroclor 1254 resulted in marked increases in overall biotransformation of DMBA. The values obtained for the rates of monooxygenation were of the same order of magnitude (7) as observed for mouse skin, but about 2 to 3 orders of magnitude less than those commonly measured in hepatic preparations of chickens or rats. The inhibitory effects of 7,8-BF and $\beta$-$E_2$ on DMBA metabolism corroborated earlier investigations in our laboratory (6) and suggest a possible role of an initial activation step of PAH in the aorta to reactive intermediates. Additions of TCPO (13 $\mu$M) were effective in decreasing the formation of DMBA-dihydrodiols, while concomitantly increasing amounts of hydroxymethyl metabolites of DMBA in chickens.

Levels of epoxide hydratase in rabbit and chicken aortas were observed to be 1 to 2 orders of magnitude less than those commonly observed in rat or chicken hepatic preparations (18). Additions of TCPO yielded a predictable concentration-dependent inhibition of epoxide hydratase in both species. Earlier experiments from our laboratory (6) demonstrated the capacity of TCPO to increase the binding of DMBA reactive intermediates to DNA over that of controls.

Moderate levels of GSH S-epoxide transferase activity were detected in rabbits and chickens. Similar to epoxide hydratase, levels were 1 to 2 orders of magnitude less than those seen in hepatic preparations (11). BaP-4,5-oxide appeared to be a potent inhibitor of the conjugation reaction when styrene oxide (1 mM) was the substrate.

Because of the limited availability of human fetal SMC cells, direct measurements of epoxide hydratase and GSH S-epoxide transferase were not possible. However, we could indirectly assess the presence of these enzymes by measuring dpm in the final aqueous extract after pretreatment of cells with various modifiers. We have previously demonstrated (5) that human fetal abdominal smooth muscle cells have the capacity to biotransform BaP primarily to phenols and DMBA to dihydrodiol, hydroxymethyl, and phenolic metabolites. In this investigation, additions of GSH or UDPGA resulted in increases in recoverable dpm, while 7,8-BF and TCPO resulted in decreased recoverable dpm. ATP and $NA_2SO_4$ pretreatments did not result in any changes in recoverable dpm.

The data presented in this report revealed a potentially important aspect of the overall metabolism of PAH by aortas and cells in culture. Rabbit and chicken aortas possess the capacity to bioactivate PAH (via an inducible monooxygenase and noninducible hydrase) as well as the capacity to "inactivate" potentially reactive species of PAH via conjugation reactions. Human fetal SMC also appear to possess enzymes capable of conjugating metabolites of DMBA. The significance of aortic PAH metabolism and its relationship to atherogenesis remains to be elucidated. However, the view that the aorta is merely an inactive "tube" or "conduit" is untenable, as the aorta can now be viewed as a possible "target" organ for PAH where reactive intermediates could be generated locally. Environmental procarcinogens and promutagens could traverse the walls of the aorta, be metabolically bioactivated by aortic monooxygenases, and initiate or promote the development of lesions by means of a somatic cell mutation.

## ACKNOWLEDGMENT

This work was supported by NIH Grants NS-04939, HL-01374, and HD-04839.

## REFERENCES

1. Albert, R. E., Vanderlaan, M., Burns, F. J., and Nishizumi, M. (1977): Effects of carcinogens on chicken atherosclerosis. *Cancer Res.* 37: 2232-2235.
2. Department of Health, Education, and Welfare (1971): *Arteriosclerosis: A report by the National Heart and Lung Institute, Task*

Force on Arteriosclerosis, Vol. 2, Publication No. (NIH), pp. 72-219, U.S. Government Printing Office, Washington D.C.
3. Benditt, E. P. (1978): The monoclonal theory of atherogenesis. In: Atherosclerosis Reviews, R. Paoletti and A. M. Gotto, Jr., Eds., pp. 77-85, Raven Press, New York.
4. Bond, J. A., Omiecinski, C. J., and Juchau, M. R. (1979): Kinetics, activation, and induction of aortic monooxygenases-biotransformation of benzo(a)pyrene. Biochemical Pharmacol. 28:305-311.
5. Bond, J. A., Kocan, R. M., Benditt, E. P., and Juchau, M. R. (1979): Metabolism of benzo(a)pyrene and 7,12-dimethylbenz(a)anthracene in cultured human fetal aortic smooth muscle cells. Life Sciences 25:425-430.
6. Bond, J. A., Yang, H.-Y.L., Majesky, M. W., Benditt, E. P., and Juchau, M. R. (1979): Metabolism of benzo(a)pyrene and 7,12-dimethylbenz(a)anthracene in chicken aortas: monooxygenation, bioactivation to mutagens, and covalent binding to DNA in vitro. Toxicol. Appl. Pharmacol. (in press).
7. DiGiovanni, J., Slaga, T. J., Berry, D. L., and Juchau, M. R. (1977): Metabolism of 7,12-dimethylbenzanthracene in mouse skin homogenates analyzed with high-pressure liquid chromatography. Drug Metab. Dispos. 5:295-301.
8. DiGiovanni, J., Berry, D. L., Slaga, T. J., Jones, A. H., and Juchau, M. R. (1979): Effects of pretreatment with 2,3,7,8-tetrachlorodibenzo-p-dioxin on the capacity of hepatic and extrahepatic mouse tissues to convert procarcinogins to mutagens for Salmonella typhimurium auxotrophs. Toxicol. Appl. Pharmacol. (in press).
9. Fabricant, C. G., Fabricant, J., Litrenta, M. M., and Minick, C. R. (1978): Virus-induced atherosclerosis. J. Exp. Med. 148:335-340.
10. Gold, H. (1961): Production of arteriosclerosis in the rat. Effect of X-ray and a high-fat diet. Arch. Pathol. 71:268-273.
11. Hayakawa, T., Myokei, Y., Tagi, H., and Jerina, D. M. (1977): Purification and some properties of glutathione-S-epoxide transferase from guinea pig liver. J. Biochem. 82:407-415.
12. Heidelberger, C. (1975): Chemical carcinogenesis. Ann. Rev. Biochem. 44:79-121.
13. Huggins, C., Grand, L. C., and Brillantes, F. P. (1961): Mammary cancer induced by a single feeding of polynuclear hydrocarbons, and its suppression. Nature 189:204-207.
14. Juchau, M. R., Bond, J. A., Kocan, R. M., and Benditt, E. P. (1979): Bioactivation of polycyclic aromatic hydrocarbons in the aorta: evidence for a role in the genesis of atherosclerotic lesions. In: Polynuclear Aromatic Hydrocarbons, P. W. Jones and P. Leber, Eds., pp. 639-659, Ann Arbor Science Publishers, Ann Arbor, Michigan.

15. Lamberts, H. B., and deBoer, W.G.R.M. (1963): Contributions to the study of immediate and early X-ray reactions with regard to chemoprotection VII. X-ray induced atheromatous lesions in the arterial wall of hypercholesterolemic rabbits. *Int. J. Rad. Biol.* 6:343-350.
16. Lindsay, S., Kohn, H. I., Dakin, R. L., and Jew, J. (1962): Aortic arteriosclerosis in the dog after localized aortic X-irradiation. *Circulation Res.* 10:51-60.
17. Lowry, O. H., Rosebrough, N. J., Farr, A. L., and Randall R. J. (1951); Protein measurement with the folin phenol reagent. *J. Biol. Chem.* 193:265-275.
18. Oesch, F., Schmassmann, H., Bentley, P. (1978): Specificity of human, rat, and mouse skin epoxide hydratase toward K-region epoxides of polycyclic hydrocarbons. *Biochemical Pharmacol.* 27:17-20.
19. Omiecinski, C. J., Chao, S. T., and Juchau, M. R. (1979). Modulation of monooxygenase activities by hematin and 7,8-benzoflavone in fetal tissues of rats, rabbits, and humans. *Dev. Pharmacol. Ther. (in press).*
20. Omiecinski, C. J., Bond, J. A., Juchau, M. R. (1978): Stimulation by hematin of monooxygenase activity in extrahepatic tissues from rats, rabbits, and chickens. *Biochem. Biophys. Res. Commun.* 93:1004-1011.
21. Wheatly, D. N., Hamilton, A. G., Currie, A. R., Boyland, E., and Sims, P. (1966): Adrenal necrosis induced by 7-hydroxymethyl-12-methylbenz(a)anthracene and its prevention. *Nature* 211:1311-1213.
22. White, J., and Mider, G. B. (1941): The effect of dietary cystine on the reaction of dilute brown mice to methylcholanthrene (preliminary report). *J. Natl. Cancer Inst.* 2:95-97.
23. White, J., Mider, G. B., and Heston, W. E. (1942-3): Note on the comparison of dosage of methylcholanthrene on the production of leukemia and sclerotic lesions in strain dilute brown mice on a restricted cystine diet. *J. Natl. Cancer Inst.* 3:453-454.

# HYDROCARBON-DEOXYRIBONUCLEOSIDE ADDUCTS IN VIVO AND IN VITRO AND THEIR RELATIONSHIP TO CARCINOGENICITY

G. M. Cohen*, S. W. Ashurst*,
J. K. Selkirk**, and T. J. Slaga**

*Department of Biochemistry
University of Surrey
Guildford GU2 5XH
Surrey, England

**Biology Division
Oak Ridge, Tennessee 37830, U. S. A
Oak Ridge National Laboratory

## INTRODUCTION

Chemical carcinogenesis in many tissues, such as skin, liver, and bladder, appears to be a two-stage process (47). The first stage, tumor initiation, requires only a single application of a carcinogen and is considered to be essentially irreversible. The second-stage, tumor promotion, requires multiple applications of a second noncarcinogenic chemical and is reversible, at least in the early stages. Despite extensive investigations over many years, a detailed mechanism for tumor induction by any single chemical still eludes definition. However, certain general principles have emerged. Tumor initiation is believed to result from the interaction of the carcinogen, or a metabolically activated form, with some critical cellular macromolecule (37). Thus, delineation of the active form or forms of the carcinogen and the identification of the key cellular targets with which the carcinogen interacts to initiate tumor formation have emerged as two of the major problems in carcinogenesis.

Polycyclic aromatic hydrocarbons (PAH) are one class of carcinogenic compounds which have been extensively investigated because of their possible relationship to human lung cancer and certain occupational cancers of the scrotum, skin, and bronchus, and also their widespread occurrence in our environment [e.g. in cigarette smoke, diesel exhaust, coal tar and soot, water, and the food chain (15,25)].

PAH are metabolized to a vast array of oxidative metabolites, including oxides, phenols, dihydrodiols, and diol-epoxides (17, 28, 44, 45). Recent evidence strongly implicates one particular type of metabolite, i.e., bay-region diol-epoxides, as being the major carcinogenic and mutagenic metabolites of PAH (29, 46). These diol-epoxides are formed by a three-step activation, i.e., initial conversion of the PAH by a microsomal mixed function oxidase to an epoxide, then further metabolism to a dihydrodiol by the microsomal epoxide hydratase, followed by a second oxidation of the dihydrodiol on an adjacent olefinic double bond in the bay region to form a diol-epoxide.

Some of these oxidized metabolites may also be further metabolized to conjugated metabolites including glutathione (28, 46), glucuronic acid (12, 38) and sulfate ester conjugates (11). While the biological activity of the majority of these conjugated metabolites has not yet been determined, they are in all probability detoxification products. Thus it appears that it is the overall balance of oxidative and conjugating enzymes, both in the intact organism and in each individual tissue or cell, which determines how much of a reactive metabolite is available for interaction with critical cellular macromolecules. This is further illustrated in the simplified version of benzo(a)pyrene metabolism shown in Figure 1. The amount of the putative ultimate carcinogen $(\pm)7\alpha,8\beta$-dihydroxy-$9\beta,10\beta$-epoxy-7,8,9,10-tetrahydrobenzo(a)pyrene (BaP) 7,8-diol-9,10-oxide-interacting with critical cellular macromolecules could be determined providing all the various competing rate constants were known. Numerous factors including age, sex, nutritional factors, hormonal status, prior or concomittant administration of certain drugs or other foreign chemicals (41) may all exert a profound influence on one or several of the rate constants shown in Figure 1, thus influencing the amount of any critical carcinogen-macromolecule interaction. It should also be emphasized that some of the above factors affecting the various rate constants may be very different in different tissues.

A realistic in vivo determination of all of these various rate constants would, in the foreseeable future, appear to be an impossible task. However, it may be possible to assess the critical macromolecular binding. Although the key macromolecule(s) with which carcinogens covalently bind are not unequivocally known, a good correlation has been found to exist between the differing carcinogenicities of a series of PAH and their covalent binding DNA, but not to RNA or protein (9). Many other studies utilizing various chemical classes of carcinogens have also found a good relationship between carcinogenicity and binding to DNA

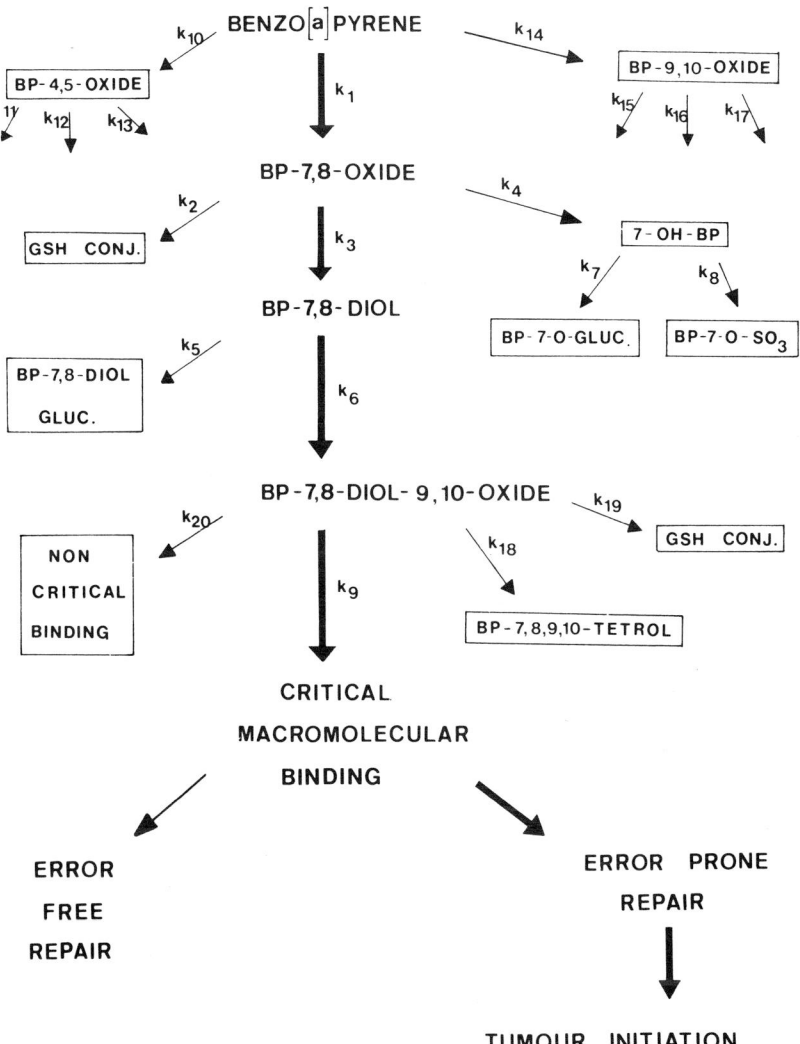

FIGURE 1. A simplified scheme of the metabolism of benzo(a)pyrene illustrating the possible effects of the multitude of different pathways on the amount of reactive metabolite(s) interacting with critical macromolecules such as DNA. If the individual rate constants $K_1 \rightarrow K_{20}$ are known, then the fraction of the dose interacting with critical macromolecules can be calculated.

(14). However, with certain classes of carcinogens, such as nitrosamines and nitrosamides, the qualitative as well as the quantitative nature of the carcinogen-DNA binding is of major importance. With these carcinogens, the quantitatively minor $O^6$-alkylation of guanine appears to be of much greater biological significance than the major adduct formed (i.e., $N^7$-alkylation of guanine) (34).

With PAH the nature of the metabolites that covalently bind to DNA appears to be markedly dependent on the nature of the metabolic activation system used. When benzo(a)pyrene is either applied to mouse skin in vivo (16, 21, 32), or incubated with hamster embryo cells (26, 40), short-term organ culture of human or bovine bronchus (21, 27), cultured human colon (4), mouse liver and lung (19), or human diploid fibroblasts (7), the major DNA or RNA adduct is derived from interaction of BaP-7, 8-diol-9,10-epoxide mainly with the C-10 of the diol-epoxide and the exocyclic amino group of guanine (27, 33, 40). In contrast to these studies, when PAH are activated by liver microsomes and the reactive metabolites trapped by reaction with exogenous DNA, the hydrocarbon-deoxyribonucleoside adducts formed are strikingly different from those obtained in vivo or using intact cells (2, 3, 30, 48); and, in the case of benzo(a)pyrene, the major adduct appears to be derived from further metabolism of 9-hydroxybenzo(a)pyrene at the K-region (e.i., 4,5-position) (31, 42).

The discrepancy between the adducts formed utilizing either microsomes or intact cells appears to be primarily due to disruption of the relative activities of the oxidative (Phase I) and conjugating enzymes (Phase II). This has been discussed by the authors in part in previous publications (2, 3) but will be further expanded in this study. We also demonstrate that it is the qualitative rather than the quantitative nature of the binding of benzo(a)pyrene to mouse skin DNA which is related to carcinogenicity.

## MATERIALS AND METHODS

Generally tritiated benzo(a)pyrene was obtained from the Radiochemical Centre, Amersham, Bucks, England. Radioactive and nonradioactive metabolites of benzo(a)pyrene were kindly provided by the National Cancer Institute, Bethesda, Maryland (U. S. A.). Glucose-6-phosphate dehydrogenase, NADP, deoxyribonuclease I, and snake venom phosphodiesterase were obtained from Boehringer Mannheim Chemical Co., Lewes, England. Glucose-6-phosphate, alkaline phosphatase, and calf thymus DNA were obtained from Sigma Chemical Company, Poole, Dorset, England. 2, 3, 7, 8-Tetrachlorodibenzo-p-dioxin (TCDD — Lot 851-144-2) was generously supplied by the Dow Chemical

Company, Midland, Michigan. 12-O-Tetradecanoylphorbol-13-acetate (TPA) was obtained from Dr. P. Borchert, University of Minnesota, Minneapolis, Minnesota.

## MICROSOME AND HEPATOCYTE MEDIATED BINDING OF [$^3$H]-BENZO(a)PYRENE TO DNA

Isolated hepatocytes from male Wistar rats were prepared as previously described (20). Microsomes from both rat liver and mouse skin ($C_{57}BL$) were prepared by differential centrifugation as described elsewhere (1). Microsomes were incubated with an NADPH-generating system and exogenous DNA essentially by the method of Grover and Sims (22) with slight modifications (1). The DNA from both hepatocytes and microsomes was isolated and purified as described elsewhere (2).

## HYDROLYSIS OF DNA AND SEPARATION OF HYDROCARBON-DEOXYRIBONUCLEOSIDE ADDUCTS

The DNA was hydrolyzed by sequential digestion with deoxyribonuclease I, snake venom phosphodiesterase, and alkaline phosphatase essentially as described by Baird and Brookes (5). The enzymic hydrolyzates were then either (a) separated on a Sephadex LH-20 column (90 x 1.5 cm) and eluted with a 30 to 100 percent methanol gradient in water as described by Baird and Brookes (32), or (b) applied to a short column of Sephadex LH-20 (0.9 x 3 cm) to remove unhydrolyzded DNA and unmodified deoxyribonucleosides; the less polar hydrocarbon-deoxyribonucleosides were then eluted with methanol and subsequently separated by high-pressure liquid chromatography (HPLC) on a Waters μBondapak octadecylsilane (ODS) reverse-phase column using an isocratic 1:1 methanol:water solvent system as previously described (1).

## TUMOR EXPERIMENTS AND IN VIVO MACROMOLECULAR BINDING

Female Sencar mice (bred at Oak Ridge National Laboratory, Oak Ridge, Tennessee) 7 to 9 weeks old were shaved 2 days prior to treatment and only those mice in the resting phase of the hair cycle were used. In the tumor experiments, groups of 30 animals received a single topical application of benzo(a)pyrene (100 nmoles), followed 1 week later by twice-weekly applications of the tumor promoter TPA (2 $\mu$g). One group of animals received a single topical application of a nontoxic dose of TCDD (1 $\mu$g in 0.2 ml acetone) 72 hours prior to receiving the carcinogen. The binding studies were carried out exactly as the tumor experiment except that $^3$H-benzo(a)pyrene was used ($\sim$ 250 $\mu$Ci/animal). After application of the carcinogen, the mice were sacrificed 3 and 24 hours later. The epidermal material was removed by the heat treatment method (36) and DNA, RNA, and protein were extracted using the method of Huberman and Sachs (24).

## RESULTS

### Hydrocarbon-Deoxyribonucleoside Adducts Following Metabolic Activation with Liver Microsomes or Intact Hepatocytes

Both isolated hepatocytes and liver microsomes from 3-methylcholanthrene pretreated rats metabolized [$^3$H]-benzo(a)pyrene to reactive metabolites which covalently bound to either endogenous or exogenous DNA, respectively. The DNA was isolated, purified, and enzymically hydrolyzed to deoxyribonucleosides, which were separated by HPLC. Striking differences were observed between the hydrocarbon-deoxyribonucleoside adducts formed following metabolic activation by either liver microsomes or isolated hepatocytes (Figure 2). The major adduct (co-chromatographing with peak B) formed following metabolic activation of benzo(a)pyrene by liver microsomes also co-chromatographed with the adduct formed following microsomal metabolic activation of 9-hydroxybenzo(a)pyrene (Figure 3a). In contrast, with isolated hepatocytes the major adduct co-chromatographing with peak C (Figure 2b) and only small amounts of adduct co-chromatographing with peak B were observed. The adduct co-chromatographing with peak (C) was similar to that obtained following reaction of either ($\pm$)BaP-7,8-diol-9,10-epoxide with DNA (Figure 2b) or microsomal activation of [$^3$H]-7, 8-dihydro-7, 8-dihydroxybenzo-(a)pyrene together with exogenous DNA (Figure 3b). Thus peak B appears to be due to a further metabolite of 9-hydroxybenzo-(a)pyrene, tentatively identified as 9-hydroxybenzo(a)pyrene 4, 5-oxide (31, 42), bound to an unknown base in DNA, while peak C is most probably derived from binding of BaP-7,8-diol-9,10-epoxide to the exocyclic amino group of guanine.

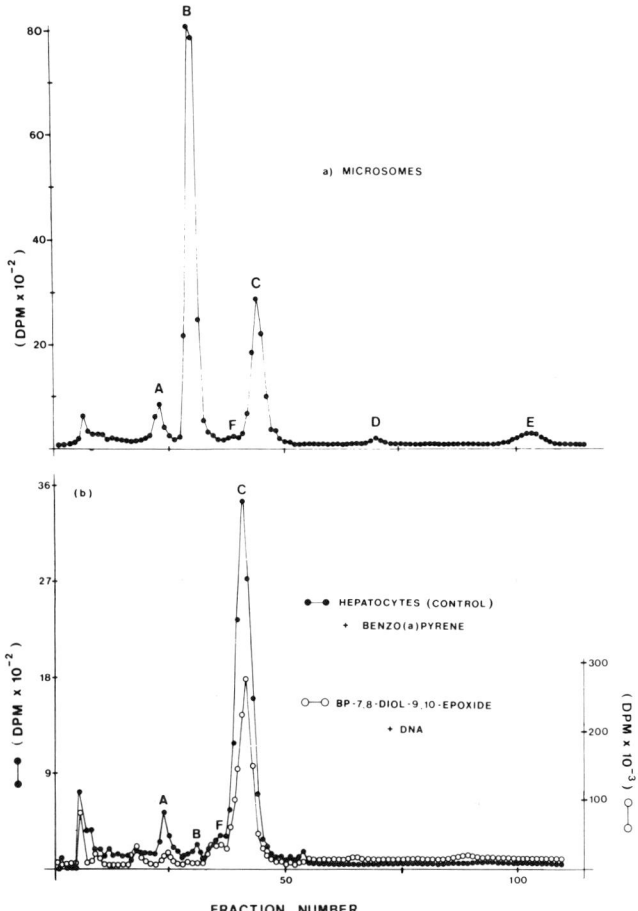

FIGURE 2. High-pressure liquid-chromatographic separation of benzo(a)pyrene-deoxyribonucleoside adducts formed following metabolic activation of [$^3$H]-benzo(a)pyrene by (a) liver microsomes or (b) hepatocytes. Liver microsomes, together with an NADPH-generating system and exogenous DNA, or isolated hepatocytes, were incubated with [$^3$H]benzo(a)pyrene. At the end of the incubation period, either the exogenous DNA or the DNA from the hepatocytes was isolated, purified, and hydrolyzed sequentially with DNase I, phosphodiesterase, and alkaline phosphatase. The hydrocarbon-deoxyribonucleoside adducts were then separated by HPLC as described in the section Materials and Methods and the amount of radioactivity in each fraction determined by liquid scintillation counting. Identification of the major DNA-bound product in hepatocytes was achieved by co-chromatography with BP-deoxyribonucleoside products formed by hydrolysis of DNA previously reacted with [$^3$H]anti-BaP-7, 8-diol-9, 10-epoxide (Figure 2b, o——o).

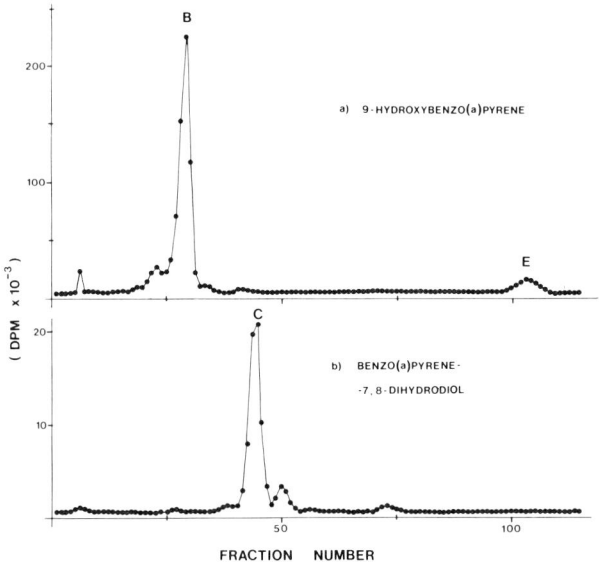

FIGURE 3. High-pressure liquid-chromatographic separation of metabolite-deoxyribonucleoside adducts formed following microsomal activation of (a) [$^3$H]9-hydroxybenzo(a)pyrene or (b) [$^3$H]7,8-dihydro-7,8-dihydroxybenzo(a)pyrene. Liver microsomes were incubated with the appropriate metabolite, an NADPH-generating system, and exogenous DNA. The DNA was isolated and purified and the adducts separated as described in the legend to Figure 2.

## Hydrocarbon-Deoxyribonucleoside Adducts Following Metabolic Activation By Mouse Skin In Vivo Or Skin Microsomes

A similar discrepancy in the nature of the adducts was also observed following examination of the adducts obtained from the DNA of mouse skin ($C_{57}$BL mice) treated in vivo with [$^3$H]-benzo(a)pyrene (100 nmoles) or after metabolic activation of the carcinogen by isolated skin microsomes (from $C_{57}$BL mice) and subsequent binding to exogenous DNA. In the in vivo study the major adduct formed co-chromatographed with peak C and was derived from BaP-7,8-diol-9,10-epoxide, in agreement with other studies (16, 21, 32) (results not shown). In contrast to this, in preliminary studies using mouse skin microsomes to activate benzo(a)pyrene, little diol-epoxide adduct was formed and other not yet fully characterized adducts were obtained (results not shown). Similar preliminary results have also been reported (8), when with mouse skin microsomes, most of the adducts were derived from benzo(a)pyrene 4, 5-oxide or from further metabolism of 9-hydroxybenzo(a)pyrene (8, 42).

## Relationship Of Hydrocarbon-Deoxyribonucleoside Adducts To In Vivo Carcinogenicity In Mouse Skin

We were interested in determining the effects of different pretreatments in vivo, which modulated the amounts of hydrocarbon-DNA binding, and their relationship to in vivo carcinogenicity of the hydrocarbon to mouse skin. As part of these studies (10) we found that topical application to mouse skin of a nontoxic dose of 2, 3, 7, 8-tetrachlorodibenzo-p-dioxin (TCDD - 1 $\mu$g) 72 hours prior to administration of an initiating dose of [$^3$H]-benzo(a)pyrene (100 nmoles) caused a marked increase in the binding of DNA, RNA, and protein (Table 1). However, despite the increased binding in DNA, a marked inhibition of carcinogenicity of benzo(a)pyrene was observed when the mice were pretreated with TCDD (Table 2). A similar inhibition of carcinogenicity was reported in CD1-mice by Di Giovanni et al (18), but in this study the binding of benzo(a)pyrene was not investigated.

In an attempt to resolve this apparent discrepancy between increase in DNA binding and inhibition of carcinogenicity, the nature of the hydrocarbon-DNA adducts was investigated further. A very striking difference was observed between the Sephadex LH 20-elution profiles of the hydrocarbon-deoxyribonucleoside adducts obtained following enzymic

**TABLE 1. In Vivo Covalent Binding of $^3$H-Benzo(a)pyrene to Mouse Epidermal DNA, RNA and Protein after Pretreatment with TCDD**

| Treatment | Time | Hydrocarbon Bound (pmoles/mg) | | |
|---|---|---|---|---|
| | (hr) | DNA | RNA | Protein |
| BP | 3 | 4.8 | 3.6 | 51 |
| BP + TCDD | 3 | 12.0 | 7.8 | 130 |
| BP | 24 | 9.0 | 8.5 | 79 |
| BP + TCDD | 24 | 25.0 | 11.3 | 205 |

Note: Sencar mice were treated with [$^3$H]-benzo(a)pyrene (BP-100 nmoles). Some mice also received 1 $\mu$g of TCDD, applied topically 3 days prior to application of $^3$H-benzo(a)pyrene.

TABLE 2. Effects of 2,3,7,8-Tetrachlorodibenzo-p-dioxin (TCDD) on the Initiation of Skin Tumors by Benzo(a)pyrene (BP)

| Pretreatment | Initiator | No. of Mice[a] | No. of Papillomas per Mouse at 15 Weeks | Percentage of Mice with Papillomas at 15 Weeks |
|---|---|---|---|---|
| Acetone | BP | 28 | 3.8 | 85 |
| TCDD | BP | 29 | 0.3 | 25 |
| Acetone | Acetone | 30 | 0.1 | 7 |

[a]Number of mice surviving after 15 weeks of promotion.
[b]The control mice received only acetone, followed by twice weekly applications of 2 $\mu$g of TPA

Note: Female Sencar mice were initiated with 100 nmoles of BP and promoted twice weekly with 2 $\mu$g of TPA for 15 weeks. Three days before initiation, some of the mice received 1 $\mu$g of TCDD.

hydrolysis of the DNA isolated from the skins of mice treated with benzo(a)pyrene with and without prior topical application of TCDD (Figure 4). When mice (Sencar) were treated with benzo(a)pyrene alone, in agreement with the author's earlier results and those of others using HPLC (21, 32), only one major radioactive hydrocarbon-deoxyribonucleoside adduct eluting between 516 and 574 ml (fractions 126 through 140, Figure 4a) was obtained. This adduct appears to be due to BaP-7, 8-diol-9,10-oxide bound to the exocyclic amino group of guanine. However, when mice were pretreated with TCDD 72 hours prior to application of benzo(a)pyrene, no detectable hydrocarbon-deoxyribonucleoside adduct corresponding to the BaP-7,8-diol-9,10-oxide adduct was observed (Figure 4b). Most of the radioactivity chromatographed with fairly polar unidentified material eluting before 84 ml (Figure 4b).

## DISCUSSION

A major difference was observed in the nature of the hydrocarbon-deoxyribonucleoside adducts formed by either skin or liver microsomes compared with those formed by mouse skin in vivo or intact hepatocytes. When liver microsomes were used to metabolically activate benzo(a)pyrene, the major hydrocarbon-deoxyribonucleoside adduct was derived from a further metabolite of 9-hydroxybenzo(a)pyrene (Figure 2a), in agreement with other results (1, 8, 30, 31, 42 and 49). However, in vivo or with intact

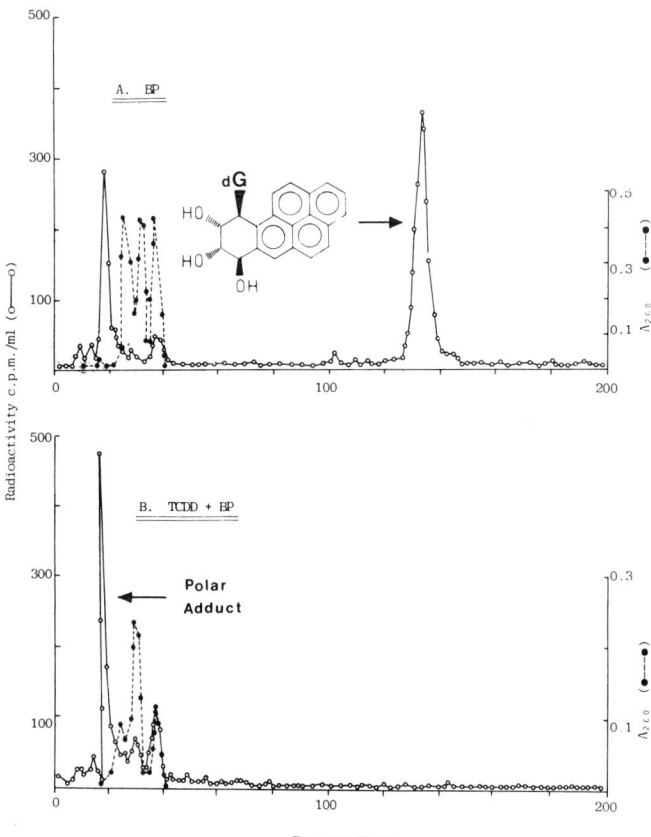

FIGURE 4. Sephadex LH-20 chromatographic elution profiles of enzyme digests from DNA isolated from mouse skin treated with (A) $^3$H-benzo(a)pyrene (100 nmoles/mouse) alone or (B) TCDD (1 μg) 72 hours before application of $^3$H-benzo(a)pyrene (100 nmoles/mouse). The DNA digests were chromatographed on an LH-20 column (90 x 1.5 cm) and eluted with a 30 to 100 percent water:methanol gradient (1,000 ml). ●----● U. V. was monitored continuously at 254 nm; o——o c.p.m./fraction for 1.0 ml sample of each fraction of 4.1 ml.

cell systems, the major adduct was BaP-7, 8-diol-9, 10-oxide bound to the exocyclic amino group of guanine (Figure 2b and 4a), in agreement with the results of many other studies utilizing intact cell systems (6, 7, 16, 19, 21, 23, 26, 27, 32, 43, 46, 48, 50, 51). The major reason for the qualitative alteration in the nature of the adducts with these different metabolic activation systems appears to be disruption on the microsomes of the normal cellular balance of oxidative and conjugating enzymes. A similar conclusion was also reached by Vähäkangas et al (50) who observed different adducts when either rat lung microsomes or perfused lung was used to metabolically activate benzo(a)pyrene. The absence in isolated microsomes of both essential cofactors necessary for glutathione and glucuronic acid conjugation and also some of the enzymes responsible for sulfate conjugation causes some chemicals to be metabolized in a different way to their in vivo metabolic fate. Thus, in the microsomal activation of benzo(a)pyrene, a second oxidation of 9-hydroxygenzo(a)pyrene is favored resulting in the formation of 9-hydroxybenzo(a)pyrene 4,5-oxide and subsequent binding to DNA, whereas in intact cells the phenol is converted primarily to its glucuronic acid or sulfate ester conjugate (13, 38). These results illustrate certain important limitations of various short-term test systems, utilizing microsomes or 9,000 g supernatant fractions as metabolic activation systems, and the subsequent extrapolation of the results to the in vivo situation.

It is interesting to note that in this study binding of the putative ultimate carcinogen derived from benzo(a)pyrene, i.e., BaP-7,8-diol-9,10-oxide to the exocyclic amino group of guanine in tissues susceptible (mouse skin - Figure 4a) and relatively resistant (isolated hepatocytes -Figure 2b) to benzo(a)pyrene-induced carcinogenesis (25), was observed. To interpret these observations, the covalent binding found in these studies is compared with other data available in the literature, together, if possible, with the known biological consequences for the tissue or cell (Table 3). Although it may be a little early to generalize, it would appear that the binding of BaP-7,8-diol-9,10-oxide to the exocyclic amino group of guanine may be an essential feature of biological systems in which benzo(a)pyrene exerts its carcinogenic, mutagenic or cell-transformation effects. Some support for this hypothesis was obtained in the inhibition of benzo(a)pyrene tumorigenesis by TCDD (Table 2) and the absence of significant amounts of BaP-7,8-diol-9,10-oxide bound to the exocyclic amino group of guanine (Figure 4b). More data are obviously required to substantiate this hypothesis and it is possible that the biological effects are due to binding of the diol-epoxide or other metabolites to other cellular macromolecules or to other bases in the DNA. However, in a recent study, Phillips et al (43) have shown that following topical application of either benzo(a)pyrene, 7,12-dimethylbenz-(a)anthracene, or 3-methylcholanthrene to skins of mice either susceptible, intermediate or relatively resistant to hydrocarbon-induced skin carcinogenesis, the amounts and disappearance of the different diol-epoxide deoxyribonucleoside adducts were similar. Thus it is possible that the presence of the BaP-7,8-diol-9,10-oxide bound to the exocyclic amino group of guanine or the processing of this damage by the cell is an essential, but by itself an insufficient, feature to result in tumor formation. Other, as yet less well-characterized processes such as fidelity of DNA repair, rates of cell

**TABLE 3. Biological Systems Where Adducts Derived from Benzo(a)pyrene-7,8-diol-9,10-epoxide Have Been Detected**

| Biological System | Concentration or Dose of Benzo(a)pyrene | Biological Observed Effect | Pmoles Bound/mg DNA | Reference |
|---|---|---|---|---|
| Mouse skin | 0.1 - 1 µ moles | Low to high susceptibility to skin carcinogenesis | 1.4 → 34.2 | 10, 43 |
| Mouse salivary gland epithelial cells | 4 µM | Foci of hyperplastic epithelium | 32.6 | 51 |
| Rat tracheal epithelial cells | 8 µM | Transformation | 1.9 | 48 |
| Chinese hamster V.79 cells | 4 µM | Mutagenesis | 13.1 | 39 |
| Hamster embryo cells | 0.5 µM | Transformation | 11.9 | 6 |
| Cultured human bronchus | 1.5 µM | — | 2 → 151 | 23 |
| Cultured human colon | 1.5 µM | — | 0.08 → 1.6 | 4 |
| Mouse lung (A/J) | 50 nmoles | Pulmonary adenomas | 0.3 | 19 |
| Isolated perfused rat lung | 2.5 µM | — | 0.11 | 50 |
| Mouse liver (A/J) | 50 nmoles | Relatively resistant to carcinogenesis | 1.1 | 19 |
| Isolated rat hepatocytes | 55 µM | Relatively resistant to carcinogenesis | 5.2 | 2 |
| Rat liver | 2 mg/kg | Relatively resistant to carcinogenesis | 0.5 | 35 |

Note: Data compiled from the literature of systems of widely differing susceptibilities to benzo(a)pyrene. The evidence that it is the BP-7,8-diol-9,10-oxide bound to DNA in most of these systems is fairly good. Caution must be exercised in the extrapolation of these data because of many factors such as differing doses. Results not corrected for how much of the binding is due to the specific adduct or differences in repair or removal of the adducts.

division, immune competence, and tumor promotion are also of critical importance in determining whether specific carcinogen-DNA interactions ultimately manifest themselves as tumors. However, the results in this study clearly illustrate that it is the qualitative rather than the quantitative nature of the binding of benzo(a)pyrene to DNA (Figure 4, Table 1) which bears a much closer relationship to its tumorigenicity.

## SUMMARY

The qualitative rather than the quantitative nature of hydrocarbon-DNA interactions appears to be of major importance in carcinogenesis. Pretreatment of mouse skin with 2,3,7,8-tetrachlorodibenzo-p-dioxin causes both a marked inhibition of the tumorigenicity of benzo(a)pyrene accompanied by an increase in the gross covalent binding of benzo(a)pyrene to epidermal DNA. However, this inhibition of tumor formation was associated with a specific decrease in the formation of a $(\pm)7\alpha, 8\beta$-dihydroxy-$9\beta,10\beta$-epoxy-7,8,9,10-tetrahydrobenzo(a)pyrene (BaP-7, 8-diol-9,10-oxide) adduct bound to the exocyclic amino group of deoxyguanosine. It is proposed that the formation or the cellular processing of this adduct may be an essential but by itself inadequate event in the induction of carcinogenesis or mutagenesis by benzo(a)pyrene. The qualitative nature of hydrocarbon-deoxyribonucleoside adducts is markedly dependent on cellular integrity. When microsomes from rat liver are used to metabolically activate benzo(a)pyrene, the quantitatively major DNA-binding metabolite is derived from a further metabolite of 9-hydroxybenzo(a)pyrene. However, in vivo or in intact cells, where both conjugating enzymes, together with their necessary cofactors, and oxidative enzymes are present, the major hydrocarbon-deoxyribonucleoside adduct is derived from BaP-7,8-diol-9,10-oxide. It is hypothesized that a major reason for these observed differences in the nature of the hydrocarbon-deoxyribonucleoside adducts is the loss of conjugating enzymes or their necessary cofactors in microsomal preparations, which in turn results in certain metabolites being converted to reactive metabolites not normally seen in the intact organism.

## ACKNOWLEDGMENTS

This work was supported in part by grants from the Medical Research Council and the Cancer Research Campaign of Great Britain. S. W. Ashurst was in receipt of a Science Research Council studentship.

Research sponsored also by NIH Grant CA-20076 and the Office of Health and Environmental Health, U. S. Department of Energy, under contract W-7405-eng-26 with Union Carbide Corporation.

## REFERENCES

1. Ashurst, S.W., and Cohen, G.M. (1979): Magnesium ions affect the quantitative but not the qualitative microsome mediated binding of benzo(a)pryene to DNA. *Chem. Biol. Interactions* 28: 279-289.
2. Ashurst, S. W., and Cohen, G. M. (1980): *Chem. Biol. Interactions* 29: 117-127.
3. Ashurst, S. W., Mehta, R., and Cohen, G. M. (1979): Importance of conjugation reactions in determining the qualitative nature of polycyclic aromatic hydrocarbon-DNA interactions. *Medical Biology* 57: 313-320.
4. Autrup, H., Harris, C. C., Trump, B. F., and Jeffrey, A. M. (1978): Metabolism of benzo(a)pyrene and identification of the major benzo(a)pyrene-DNA adducts in cultured human colon. *Cancer Res.* 38: 3689-3696.
5. Baird, W. M., and Brookes, P. (1973): Isolation of the hydrocarbon-deoxyribonucleoside products from the DNA of mouse embryo cells treated in culture with 7-methylbenz(a)anthracene-$^3$H. *Cancer Res.* 33: 2378-2385.
6. Baird, W. M., and Diamond, L. (1976): Effect of 7, 8-benzoflavone on the formation of benzo(a)pyrene-DNA-bound products in hamster embryo cells. *Chem. Biol. Interactions* 13: 67-75.
7. Baird, W. M., and Diamond, L. (1978): Metabolism and DNA binding of polycyclic aromatic hydrocarbons by human diploid fibroblasts. *Int. J. Cancer* 22: 189-195.
8. Boobis, A., and Nebert, D. W. (1977): Genetic differences in the metabolism of carcinogens and in the binding of benzo(a)pyrene metabolites to DNA. In: *Advances in Enzyme Regulation* 15: 339-362. G. Weber, Ed. Pergamon Press, Oxford and New York 1977.
9. Brookes, P., and Lawley, P. D. (1964): Evidence of the binding of polynuclear aromatic hydrocarbons to the nucleic acids of mouse skins: relation between carcinogenic power of hydrocarbons and their binding to DNA. *Nature* 202: 781-784.
10. Cohen, G. M., Bracken, W. M., Iyer, R. P., Berry, D. L., Selkirk, J. K., and Slaga, T. J. (1979): Anticarcinogenic effects of 2,3,7,8-tetrachlorodibenzo-p-dioxin on benzo(a)pyrene and 7,12-dimethyl-benz(a)anthracene-tumor initiation and its relationship to DNA binding. *Cancer Res.* 39: 4027-4033.

11. Cohen, G. M., Haws, S. M., Moore, B. P., and Bridges, J. W. (1976): Benzo(a)pyren-3-yl hydrogen sulfate, a major ethyl acetate-extractable metabolite of benzo(a)pyrene in human, hamster and rat lung cultures. *Biochem. Pharmacol.* 25: 2561-2570.
12. Cohen, G. M., and Moore, B. P. (1977): The metabolism of benzo(a)pyrene, 7,8-dihydro-7,8-dihydroxybenzo(a)pyrene and 9,10-dihydro-9,10-dihydroxybenzo(a)pyrene by short-term organ cultures of hamster lung. *Biochem. Pharmacol.* 26: 1481-1487.
13. Cohen, G. M., Moore, B. P., and Bridges, J. W. (1977): Organic solvent soluble sulphate ester conjugates of monohydroxybenzo(a)pyrenes. *Biochem. Pharmacol.* 26: 551-553.
14. Colburn, N. H., and Boutwell, R. K. (1968): The binding of β-propiolactone and some related alkylating agents to DNA, RNA, and protein of mouse skin. *Cancer Res.* 28: 653-660.
15. Committee on Biological Effects of Atmospheric Pollutants (1973), National Academy of Sciences, Washington, D. C.
16. Daudel, P., Duquesne, M., Vigny, P., Grover, P. L., and Sims, P. (1975): Fluorescence spectral evidence that benzo(a)pyrene-DNA products in mouse skin arise from diol-epoxides. *FEBS Lett.* 57: 250-253.
17. DePierre, J. W., and Ernster, L. (1978): The metabolism of polycyclic hydrocarbons and its relationship to cancer. *Biochem. Biophys. Acta.* 473: 149-186.
18. Di Giovanni, J., Juchau, M. R., Berry, D. L., and Slaga, T. J. (1979): 2,3,7,8-tetrachlorodibenzo-p-dioxin: potent anticarcinogenic activity in CD-1 mice. *Biochem. Biophys. Res. Commun.* 86: 577-584.
19. Eastman, A., Sweetenham, J., and Bresnick, E. (1978): Comparison of in vivo and in vitro binding of polycyclic hydrocarbons to DNA. *Chem. Biol. Interactions* 23: 345-353.
20. Fry, J. R., Jones, C. A., Wiebkin, P., Belleman, P., and Bridges, J. W. (1976): The enzymic isolation of adult rat hepatocytes in a functional and viable state. *Anal. Biochem.* 71: 341-350.
21. Grover, P. L., Hewer, A., Pal, K., and Sims, P. (1976): The involvement of a diol-epoxide in the metabolic activation of benzo(a)pyrene in human bronchial mucosa and in mouse skin. *Int. J. Cancer* 18: 1-6.
22. Grover, P. L., and Sims, P. (1968): Enzyme catalysed reactions of polycyclic hydrocarbons with DNA and protein in vitro. *Biochem. J.* 110: 159-160.
23. Harris, C. C., Autrup, H., Connor, R., Barrett, L. A., McDowell, E. M., and Trump, B. F. (1976): Interindividual variation in binding

of benzo(a)pyrene to DNA in cultured human bronchi. *Science* 194: 1067-1069.
24. Huberman, E., and Sachs, L. (1977): DNA binding and its relationship to carcinogenesis by different polycyclic hydrocarbons. *Int. J. Cancer* 19: 122-127.
25. International Agency for Research on Cancer (1973): *IARC Monographs on the Evaluation of Carcinogenic Risk of the Chemical to Man.* Vol. 3, Lyon.
26. Ivanovic, V., Geacintov, N. E., Yamasaki, H., and Weinstein, I. B. (1978): DNA and RNA adducts formed in hamster embryo cell cultures exposed to benzo(a)pyrene. *Biochemistry* 17: 1597-1603.
27. Jeffrey, A. M., Weinstein, I. B., Jennette, K. W., Grzeskowiak, K., Nakanishi, K., Harvey, R. G., Autrup, H., and Harris, C. (1977): Structures of benzo(a)pyrene-nucleic acid adducts formed in human and bovine explants. *Nature* 269: 348-350.
28. Jerina, D. M., and Daly, J. W. (1974): Arene oxides: a new aspect of drug metabolism. *Science* 185: 573-582.
29. Jerina, D. M., Lehr, R. E., Yagi, H., Hernandez, O., Dansette, P. M., Wislocki, P. G., Wood, A. W., Chang, R. L., Levin, W., and Conney, A. H. (1976): Mutagenicity of benzo(a)pyrene derivatives and the description of a quantum mechanical model which predicts the ease of carbonium ion formation from diol epoxides. In: *In Vitro Metabolic Activation in Mutagenesis Testing,* F. J. de Serres, J. R. Bend, and R. M. Philpot, pp. 159-177, Eds. Elsevier, Amsterdam.
30. King, H.W.S., Thompson, M. H., and Brookes, P. (1975): The benzo(a)pyrene deoxyribonucleoside products isolated from DNA after metabolism of benzo(a)pyrene by rat liver microsomes in the presence of DNA. *Cancer Res.* 34: 1263-1269.
31. King, H.W.S., Thompson, M. H., and Brookes, P. (1976): The role of 9-hydroxybenzo(a)pyrene in the microsome mediated binding of benzo(a)pyrene to DNA. *Int. J. Cancer* 18: 339-344.
32. Koreeda, M., Moore, P. D., Wislocki, P. G., Levin, W., Conney, A. H., Yagi, H., and Jerina, D. M. (1978): Binding of benzo(a)pyrene 7,8-diol-9,10-epoxides to DNA, RNA and protein of mouse skin occurs with high stereoselectivity. *Science* 199: 778-781.
33. Koreeda, M., Moore, P. D., Yagi, H., Yeh, H.J.C., and Jerina, D. M. (1976): Alkylation of polyguanylic acid at the 2-amino group and phosphate by the potent mutagen $(\pm)$-$7\beta,8\alpha$-dihydroxy-$9\beta,10\beta$-epoxy-7,8,9,10-tetrahydrobenzo(a)pyrene.*J. Am. Chem. Soc.* 98: 6720-6722.
34. Loveless, A. (1969): Possible relevance of $O^6$-alkylation of deoxyguanosine to mutagenicity of nitrosamines and nitrosamides. *Nature* 223: 206-208.

35. Lutz, W. K., Viviani, A., and Schlatter, C. (1978): Nonlinear dose-response relationship for the binding of the carcinogen benzo(a)pyrene to rat liver DNA in vivo. *Cancer Res.* 38: 575-578.
36. Marrs, J. M., and Voorhees, J. J. (1971): A method of bioassay of an epidermal chalone-like inhibitor. *J. Invest. Dermatol.* 56: 174-181.
37. Miller, J. A. (1970): Carcinogenesis by chemicals: an overview—G. H. A. Clowes Memorial Lecture. *Cancer Res.* 30: 559-576.
38. Nemoto, N., and Gelboin, H. V. (1976): Enzymatic conjugation of benzo(a)pyrene oxides, phenols and dihydrodiols and UDP-glucuronic acid. *Biochem. Pharmacol.* 25: 1221-1226.
39. Newbold, R. F., Wigley, C. B., Thompson, M. H., and Brookes, P. (1977): Cell-mediated mutagenesis in cultured Chinese hamster cells by carcinogenic polycyclic hydrocarbons: nature and extent of the associated hydrocarbon-DNA reaction. *Mutation Res.* 43: 101-116.
40. Osborne, M. R., Beland, F. A., Harvey, R. G., and Brookes, P. (1976): The reaction of ($\pm$)-7$\alpha$, 8$\beta$-dihydroxy-9$\beta$, 10$\beta$-epoxy-7, 8, 9, 10-tetrahydrobenzo(a)pyrene with DNA. *Int. J. Cancer* 18: 362-368.
41. Parke, D. V., and Smith, R. L. (1977): *Drug Metabolism—From Microbe to Man.* Taylor and Francis Limited, London.
42. Pelkonen, O., Boobis, A. R., Yagi, H., Jerina, D. M., and Nebert, D. W. (1978): Tentative identification of benzo(a)pyrene metabolite-nucleoside complexes produced in vitro by mouse liver microsomes. *Mol. Pharmacol.* 14: 306-322.
43. Philips, D. H., Grover, P. L., and Sims, P. (1978): The covalent binding of polycyclic hydrocarbons to DNA in the skin of mice of different strains. *Int. J. Cancer* 22: 487-494.
44. Selkirk, J. K. (1977): Benzo(a)pyrene carcinogenesis: a biochemical selection mechanism. *J. Tox. Environ. Health* 2: 1245-1258.
45. Sims, P., and Grover, P. L. (1974): Epoxides in polycyclic aromatic hydrocarbon metabolism and carcinogenesis. *Advan. Cancer Res.* 20: 165-274.
46. Sims, P., Grover, P. L., Swaisland, A., Pal, K., and Hewer, A. (1974): Metabolic activation of benzo(a)pyrene proceeds by a diol-epoxide. *Nature* 252: 326-328.
47. Slaga, T. J., Sivak, A., and Boutwell, R. K. Eds. (1978): *Mechanisms of Tumor Promotion and Cocarcinogenesis,* Raven Press, New York.
48. Steele, V., Marchok, A. C., Ashurst, S. W., and Cohen, G. M. In preparation.
49. Thompson, M. H., King, H.W.S., Osborne, M. R., and Brookes, P. (1976): Rat liver microsome-mediated binding of benzo(a)pyrene metabolites to DNA. *Int. J. Cancer* 17: 270-274.
50. Vähäkangas, K., Nebert, D. W., and Pelkonen, O. (1979): The DNA binding of benzo(a)pyrene metabolites catalysed by rat lung

microsomes in vitro and in isolated perfused rat lung. *Chem. Biol. Interactions* 24: 167-176.
51. Wigley, C. B., Thompson, M. H., and Brookes, P. (1976): The nature of benzo(a)pyrene binding to DNA in an epithelial cell culture system. *Europ. J. Cancer* 12: 743-745.

# RECOGNITION OF DNA DAMAGE IN VITRO AND IN VIVO

S. M. D'Ambrosio[*,**], F. B. Daniel[*,†], D. Brash[**],
R. E. Gibson[**], N. J. Joyce[**], W. Kim[‡],
D. Mhaskar[**], J. Ortman[**], K. Schmidt[**],
A. Wani[**], D. Witiak[‡], R. W. Hart[*,**,‡]

>  [*]Department of Pharmacology
>  [**]Department of Radiology
>  School of Medicine
>  The Ohio State University
>  Columbus, Ohio 43210
>  [†]U.S. Environmental Research Center
>  Cincinnati, Ohio 45268
>  [‡]Department of Medicinal Chemistry
>  The Ohio State University
>  Columbus, Ohio 43210

## INTRODUCTION

Many prescreen assays used to detect carcinogenic agents depend upon the ability of the agent to induce DNA damage. Such damage is usually determined indirectly by measuring DNA repair, misrepair, or mutation (13). However, not all DNA damage induced by chemical and physical carcinogens is repaired. For example, $N$-acetoxy-acetylaminofluorene (AAAF) induces two types of base modifications in cellular DNA with the resultant C8 adduct to guanine being repairable, and the N7 adduct not being repairable (1). Methylating agents such as methylmethane sulfonate (MMS) induce a multitude of DNA base modifications including the poorly repaired $O^6$-methylguanosine (29). Polynuclear aromatic hydrocarbons (PAH) such as 7,12-dimethlybenz(a)anthracene (DMBA) also induce a variety of different types of base modifications, some of which are readily repaired and others of which are not repaired

(19). Unscheduled DNA synthesis (UDS), an indirect and frequently used measure of DNA repair, depends upon: (a) the ability of a cell to repair DNA damage; (b) an agent to induce repair; (c) cellular nucleotide pools; and (d) patch sizes. Therefore, the main limitation of the UDS assay is that it will not detect unrepaired DNA damage nor be a reliable measure of DNA repair in vivo due to the afore-cited factors. Since unrepaired DNA damage may be responsible for the biological effects of many chemical and physical carcinogenic agents, the indirect measure of such damage is inappropriate in vivo.

The only direct measures of DNA damage are those that utilize damage-specific enzymes or antibodies which detect damage, or methods measuring the binding of radiolabeled carcinogen to cellular DNA (2). However, the use of radiolabeled carcinogen is often too expensive or the carcinogen too difficult to radiolabel with the specific activity required for detection. Endonucleases, on the other hand, have been used successfully to detect and quantitate, with a high degree of sensitivity, damage induced in cellular DNA by ultraviolet (UV) radiation (4), alkylators (24,25), and PAH (15,35). The most commonly used endonucleases have been those obtained from *Aspergillus orzae* (32) and *Micrococcus luteus* (4). A single-strand endonuclease obtained from *A. orzae* has been reported to recognize distortions in DNA resulting from treatment of cells and purified DNA with various chemical (11,17,34,35) and physical (15,33) agents. Such DNA damage resulting from UV irradiation, and chemical binding to DNA is thought to cause distortions in DNA resulting in single-strand regions recognizable by the S-1 endonuclease. The crude preparation of UV-endonuclease obtained from *M. luteus* exhibits specific activity toward DNA damaged by UV-irradiation (4) (recognizing over 95 percent of the thymine dimers formed). Other endonuclease and $N$-glycosidase activities have been reported in *M. luteus* that act upon DNA damaged by alkylating agents (24), and through depurinations (18).

Since endonucleases have been used successfully in directly measuring the extent and repair of DNA damage induced by physical and chemical carcinogens in vitro, it was our objective to utilize endonucleases to detect such damage induced in vivo. S-1 and *M. luteus* endonucleases were used to detect damage induced by DMBA in vitro in Syrian hamster embryo (SHE) cells, and in vivo in rat mammary gland cells.

## MATERIALS AND METHOD

### Cell Culture

SHE cultures were obtained by the disassociation of 12-day old embryos. Cultures were maintained in modified minimal essential medium with Earles salts (MEM-E) (GIBCO), supplemented with 10 percent fetal calf serum (FCS) (FLOW), 100 units/ml penicillin (Sigma), 100 µg/ml streptomycin (Sigma), and 2.5 units/ml Fungazone (GIBCO). Cultures were maintained at 37°C in a humid 5 percent $CO_2$ atmosphere and frozen at Passage 2 in liquid nitrogen. All cells were grown to confluency, at which time they were trypsinized and seeded to a density of 300,000 in 60-cm-diameter Corning plastic tissue culture dishes. Approximately 6 hours after seeding, cells were incubated in the modified MEM-E containing 3 percent calf serum (Flow). Eighteen hours later, DNA was labeled by incubating "strased" cells in modified MEM-E containing 10 percent FCS and either 10 µCi/ml of [$^{14}$C]thymidine (New England Nuclear, specific activity 250 Ci/mol) or 20 µCi/ml of [$^{3}$H]thymidine (Amersham Searle Corp., specific activity 6.5 Ci/mM) (6,7). Additional label was added, as above, 24 hours later and treated or not treated with DMBA at the doses and for the periods of time indicated in the legends of the figures and charts.

Normal human skin fibroblasts (NHSF) RG1530SU were obtained from the foreskin of an apparently normal human male. Cells in Passage 18 were grown in MEM-E and labeled with [$^{3}$H]thymidine or [$^{14}$C]thymidine as described above.

## Cell Treatment

SHE cells prelabeled with [$^{14}$C]thymidine or [$^{3}$H]thymidine were treated in modified MEM-E containing 10 percent FCS with thin-layer, chromatographically purified DMBA (Eastman). Purity was checked by high-pressure liquid chromatography (HPLC) and found to be greater than 99 percent DMBA. DMBA was dissolved in dimethylsulfoxide (DMSO) and added to the culture media, so that the final concentration of DMSO was less than 0.1 percent. DMSO was added to untreated cultures as a control.

NHSF were washed twice with phosphate buffered saline (PBS). The PBS was poured off the plate and cells were irradiated with 100 $J/m^2$ UV radiation (254-nm) from a GE germicidal lamp at a fluence of 10 $J/m^2$/sec. $J/m^2$/sec.

**Extraction of DNA.** Following incubation with DMBA or irradiation with UV, cells were washed twice with a NaCl-ethylenediaminotetraacetic (salt-EDTA) solution (26) and scraped into 1 ml of this solution on ice. After collecting the cells by centrifugation, cells were resuspended into TEN buffer (40 mM Tris, 10 mM EDTA, 100 mM NaCl, pH 8.0). Proteinase K (100 µg per ml) and Sarkosyl (0.2 percent) were added and the samples were incubated at 45 C for 10 minutes and 37 C for 60 minutes. Phenol saturated with NET buffer (40 mM NaCl, 2 mM EDTA,

20 mM Tris, pH 8.0) was added and samples were rolled gently for 1 hour. After removal of the aqueous layer, samples were again rolled with phenol. The aqueous layer was washed twice with ether, and placed into a dialysis bag, and dialyzed against two changes of NET buffer.

**Endonuclease Reaction.** *M. luteus* crude lysate was prepared according to the procedure of Carrier and Setlow (4). S-1 endonuclease was prepared from δ-amylase powder (Sigma Chem. Co.) according to the procedure of Vogt (32).

Five microliters of $^3$H-labeled DNA was incubated with 62 µl of 5 mM KPO$_4$ buffer (pH 7.4) and 8 µl of reaction buffer (TEM) (0.5 M Tris, 0.1 M EDTA, 0.1 M mercaptoethanol), and the *M. luteus* enzyme was incubated at 37° C for 30 minutes according to the procedure of Shackleton and Roberts (27). Thirty microliters of $^{14}$C-labeled DNA was incubated with 32 µl of the phosphate and 8 µl of reaction buffers and enzyme in 30 µl of phosphate buffer. Samples were incubated at 37° C for 30 minutes and layered on top of a 5 to 20 percent alkaline sucrose gradient.

In order to determine the extent of S-1 endonuclease recognition of DMBA-damaged DNA, 5 µl $^3$H- or 30 µl $^{14}$C-labeled DNA was incubated with 5 or 30 µl of S-1 Buffer A (1 mM ZnCl$_2$, 80 mM NaAcetate, 100 mM NaCl, pH 4.5). S-1 endonuclease in S-1 Buffer B (1 mM ZnCl$_2$, 40 mM NaAcetate, 100 mM NaCl, pH 4.5) was added to a final volume of 100 µl. Samples were incubated at 37° C for 30 minutes and then layered on top of a 5 to 20 percent alkaline sucrose gradient.

The reactions were terminated by layering the sample on top of a 0.2-ml solution containing 0.5 M NaOH and 0.01 M EDTA on top of a 5 to 20 percent alkaline sucrose gradient (0.5 M NaCl, 0.1 M EDTA and 0.2 M NaOH). After centrifugation for 100 minutes at 45,000 rpm and 20° C in the SW 56 rotor of a Beckman L5-50 ultracentrifuge, the weight-average molecular weight (Mw) was determined as previously described (23,16).

*Treatment of Animals*

Fifty-day-old Sprague Dawley rats were injected with 5.12 mg pure DMBA (as determined by HPLC) in 0.5 ml, 95 percent DMSO:5 percent PBS (v/v). Twenty-four hours later the animals were injected with 0.5 ml of a mixture of ketamine, HCl, and promazine as an anesthetic. An incision was made near the nipple of each mammary gland, the skin pulled back and the underlying tissue in the area surrounding the nipple was carefully dissected. This tissue was placed on a sterile glass cutting block and minced by cross cutting with scalpels. The minced tissue was then placed into a 1-ml homogenizer (Wheaton Dounce) and given five to six strokes with the loose piston. After checking for intact nuclei under a phase contrast microscope, nuclei were collected by centrifuging at 1500 rpm for 5 minutes, resuspended into 1 ml of TEN buffer, and DNA was extracted as described above. After dialysis with two changes of NET buffer, 150 $\mu$l was placed into sterile test tubes. Ten milliliters of S-1 endonuclease buffer C (5 mM $ZnSO_4$, 40 mM NaAcetate, 100 mM NaCl, pH 3.95) and 40 $\mu$l of S-1 endonuclease (50 units) in S-1 endonuclease buffer B were added and samples were incubated at 37°C for 30 minutes. The reaction was terminated and the DNA was centrifuged as described above.

**Detection of Nonradiolabeled DNA.** Recently, a highly sensitive and quantitative method has been developed to assay DNA sedimented through alkaline sucrose gradients. This method, which has been described elsewhere (3), involves: (a) neutralization of the gradient fractions with HCl; (b) precipitation of the DNA after the addition of lysozyme as a carrier with trichloroacetic acid (TCA); (c) removal of the TCA, lipid, sucrose, and salts by washing the precipitated DNA with redistilled ethanol; (d) after drying, the reaction of DNA with diaminobenzoic acid at 60 C° for 30 minutes; and (e) the addition of 1 N HCl, and the determination of fluorescence at 405 nm (excitation) and 500 nm (emission). The relative fluorescence of each gradient fraction is used to determine the weight-average and number-average molecular weights of the DNA sedimented.

**Isolation of Radiolabeled DNA.** *SHE DNA.* The monolayers of 10 to 20 150-mm-diameter Corning plastic tissue culture plates were treated with 50 $\mu$l of an acetone solution containing [$^3$H]DMBA (Amersham Searle) having a specific activity of 100 to 120 $\mu$Ci/mole. Twenty-four hours following addition of the hydrocarbon, the medium was changed and cells were incubated for the times indicated in the charts and figures. DNA was harvested from the cells after rinsing the plates twice with PBS, by adding 4 ml of lysing solution (8 M urea, 0.035 M sodium dodecyl sulfate (SDS), 0.01 M EDTA, 0.24 M sodium phosphate, pH 6.8). The viscous layer was removed and 2 ml of isoamyl alcohol per 50 ml lysate was added, homogenized for 30 seconds in a Waring blender, and loaded onto a 6-g hydroxyapatite (Bio-Rad HTP-DNA-grade) column. The DNA was eluted from the column as previously described (10) and after dialysis the radioactivity and the quantity of DNA were determined. The DNA was hydrolysed with DNase, snake-venom phophodiesterase, and alkaline phosphotase as previously described(10).

*Rat Mammary DNA.* Fifty-day-old Sprague-Dawley rats were injected as described above with 5.12 mg of [$^3$H]DMBA (Amersham Searle) having a specific activity of 16.2 Ci/mM and a chemical and radiochemical purity greater than 98 percent as determined by HPLC. Rats were killed 24 hours after treatment and the mammary glands were removed and quickly frozen in liquid nitrogen and stored at $-70°$ C until the DNA was to be isolated. Tissue samples were weighed and kept on ice before being cut into smaller pieces. The tissue fragments were placed into a 50-ml sterile polpypropylene tube, and fat free mammary gland cells were prepared using a modification of a previously described procedure (21). Collagenase (1.8 mg/ml) in PBS was added at 5 ml per gram of tissue, and agitated in a 37° C shaking water bath for 60 minutes. An additional 10 ml of PBS was added to the sample, which was then centrifuged for 15 minutes at 200 x g. The supernatant containing the fat cells was carefully aspirated off and counted for radioactivity. The epidermal cell pellet was washed with 20 ml of PBS. Approximately two volumes of 1.5 percent SDS was added to each sample, followed by autodigested pronase (final concentration of 100 $\mu$g/ml). Samples were incubated at 37° C with gentle shaking for 60 minutes, at which time they were made 6 percent in para-amino salicylic acid (PAS) and 1 percent NaCl by the addition of solid PAS and NaCl. One volume of Kirby's phenol was added to each sample, followed by shaking for 60 minutes at room temperature. After centrifugation (60 minutes at 450 x g), the aqueous layers were removed and 1.0 ml of Tris-EDTA-PAS-NaCl was added to the phenol layer which was reextracted. Following recentrifugation, the aqueous layers were combined. These aqueous DNA preparations were mixed by vortexing, and a small aliquot was removed from each for determination of radioactivity. The nucleic acids were precipitated by adding 1.5 volumes of 2-ethoxyethanol and were stored at 0° C for several hours. The precipitated DNA was rinsed two times with 5.0 ml of absolute ethanol, and two times with 5.0 ml of anhydrous ether. After removal of the excess ether under a stream of dry nitrogen, the DNA was dissolved at 25° C in a minimal volume of 0.1 M Tris, 0.01 M EDTA, and a pH of 7.3. RNase A (Sigma, heat-treated) was added to a final volume of 50 $\mu$g per ml and incubated in a 37° C water bath for 30 minutes with shaking. Pronase was added to a final volume of 50 $\mu$g per ml and samples were shaken at 37° C for 30 minutes. The aqueous fraction was extracted with one volume chloroform:isoamyl alcohol 24:1, (v/v) twice. After making the solution 1 percent NaCl, the DNA was precipitated with 1.5 volumes of 2-ethoxyethanol. The precipitated DNA was washed two times with absolute ethanol, two times with ether, and then dried under nitrogen. After weighing the DNA, it was dissolved in a minimal volume of 0.3 M Tris and 0.05 mM $MgCl_2$ (pH 7.4) and shaken at room temperature overnight to completely dissolve the DNA. DNase I (50,000 units) was added to each sample and incubated for 60 minutes at 37°C with shaking. Aliquots were removed for liquid scintillation counting and the amount of DNA was determined using diaminobenzoic acid (9,10).

## Chemical Synthesis

DMBA-5,6-oxide was prepared as previously described (8) and was purified by crystallization from ether-n-hexane under reduced light and $N_2$ atmosphere. The DMBA-5,6-oxide exhibited an NMR spectrum (acetone-$d_6$; 90 MHz) $\delta$ 2.92 (s, 3H, CH$_3$), 4.58 (d, 1H, CH-O-, J = 4.13 Hz), 5.03 (d, 1H, CHO-, J = 4.13 Hz), 7.4 to 8.3 (m, 8H, aromatic). UV (methanol) max 274.5 nm. The oxide was recrystallized 2 to 3 times, dried under reduced pressure, and stored at 0°C over CaCl$_2$ in an amber bottle prior to use. The compound was found to be greater than 95 percent pure as determined by NMR.

## RESULTS

### Conditions for Endonuclease Recognition

In order to determine the optimal conditions for recognition of damaged DNA by S-1 and *M. luteus* endonucleases, the DNA from UV-irradiated, NHSF- or DMBA-treated SHE cells was incubated with various concentrations of S-1 and *M. luteus* endonucleases. We used UV radiation to induce DNA damage since much is already known about the types (thymidine dimers) and extent of damage induced. The S-1 endonuclease in concentrations greater than 20 units/$\mu$g of DNA resulted in large numbers of nonspecific single-strand breaks in unirradiated DNA. Although no such breaks were induced when 2.5 units/$\mu$g of DNA was used, little difference in the sedimentation of control and UV-irradiated DNA was observed (Table 1). Differences in the sedimentation profiles between UV-irradiated and control DNA were apparent at concentrations between 2.5 and 20 units/$\mu$g of DNA (Table 1). Thus the optimal concentration of S-1 endonuclease for recognition of damage induced by UV radiation was observed to be 7.5 units/$\mu$g of DNA. Since 100 J/m$^2$ of UV radiation produces approximately 0.3 percent dimers (5) and 25 UV-endonuclease-sensitive sites per $10^8$ daltons DNA, it appears that S-1 endonuclease recognizes 16 percent of the total number of dimers induced by UV radiation.

**TABLE 1. Effect of S-1 Endonuclease Concentration on UV and Control DNA[a]**

| S-1 Endonuclease (units/μg DNA) | UV (J/m$^2$) | Mw (10$^8$ Daltons) | 1/Mw (10$^8$ Daltons) | Breaks (10$^8$ Daltons)[b] |
|---|---|---|---|---|
| 2.5 | 0 | 0.25 | 3.90 | -0.5 |
| 2.5 | 100 | 0.29 | 3.40 | |
| 5.0 | 0 | 0.18 | 5.35 | 1.77 |
| 5.0 | 100 | 0.14 | 7.12 | |
| 10.0 | 0 | 0.18 | 5.49 | 2.87 |
| 10.0 | 100 | 0.11 | 8.26 | |
| 20.0 | 0 | 0.17 | 5.85 | 3.77 |
| 20.0 | 100 | 0.10 | 9.62 | |

[a] Normal human skin fibroblasts were irradiated or not irradiated with 100 J/m$^2$ UV. Cells were collected and the DNA extracted and incubated with S-1 endonuclease as described in "Methods".
[b] Δ/Mw = [1/Mw(UV) - 1/Mw(control)].

Using the above data we proceeded to determine the extent of recognition by S-1 endonuclease of damage induced by DMBA. We used SHE cells for this study because: (a) they readily activate PAH into metabolites that bind to cellular DNA (9) and (b) much is known about the binding of DMBA to SHE DNA (9). Since the concentration of S-1 endonuclease appears to be somewhat critical for recognition of DNA damage, we determined the effects of endonuclease concentration upon DNA obtained from SHE cells treated with 5 μg/ml DMBA (Table 2). We observed that the optimal concentration of enzyme needed for maximal recognition of DMBA-induced DNA damage in vitro appears to be between 2.5 and 5.0 units/μg of DNA. The number of endonuclease-sensitive sites induced also appears to decrease as the number of units of S-1 endonuclease increases, probably caused by nonspecific degradation of control, as well as DMBA-damaged DNA. The maximal number of endonuclease-sensitive sites induced by 5.0 units/μg of DNA was approximately 9 per 10$^9$ Daltons DNA.

To determine the extent of recognition by S-1 endonuclease of damage induced by DMBA in vitro, we compared the number of endonuclease-sensitive sites to the amount of radiolabeled DMBA bound to SHE DNA (Table 3). As shown in Table 3, 1.0 μM of DMBA induced approximately 9.0 S-1 endonuclease-sensitive sites and the binding of 65.6 radiolabeled DMBA molecules per 10$^9$ Daltons DNA. This indicates that

**TABLE 2. Effect of S-1 Endonuclease Concentration on the Recognition of DMBA-Damaged DNA**[a]

| Concentration ($\mu$l) | Breaks per $10^9$ Daltons[b] |
|---|---|
| 2.5 | 9.0 |
| 5.0 | 9.0 |
| 7.5 | 6.0 |
| 10.0 | 2.0 |

[a]NHSF or SHE cells were treated in culture with 15 $\mu$M 5,6-epoxide DMBA for 15 minutes or with 5 $\mu$g/1 DMBA for 24 hours, respectively. DNA was extracted with phenol as described in "Methods".

[b]The number of breaks per $10^9$ daltons ($\Delta 1/M_w$) was determined following sedimentation of DNA through alkaline sucrose gradients. $\Delta 1/M_w = [1/M_w \text{ enzyme} - 1/M_w \text{ no enzyme}]$ treated$-[1/M_w \text{ enzyme}]$ control.

**TABLE 3. Comparison of Endosite Assay with DNA Binding Assay DMBA-Treated SHE Cells**[a]

| Time[b] (hours) | Endosites[c] per $10^9$ Daltons | Binding[d] Molecules per $10^9$ Daltons |
|---|---|---|
| 0 | 8.3±.8 | 73.8±14.6 |
| 3 | 9.3±.9 | 65.4±13.0 |
| 24 | 9.1±.9 | 57.6±11.4 |

[a]Cells were incubated with 1 $\mu$M DMBA for 24 hours.
[b]Following treatment, cells were washed with MEM-E and collected for assay at the times indicated.
[c]Endonuclease prepared from *M. luteus* was used.
[d]Cells were treated with 1 $\mu$M [$^3$H]DMBA.

the S-1 endonuclease recognized approximately 14 percent of the total amount of damage induced by DMBA.

Endonuclease obtained from *M. luteus* has been shown to detect damage induced by various alkylating agents (24), acetoxy-acetyaminofluorene (25), X-radiation (24), depurinations/depyrimidations (18) and UV radiation (4). We determined whether endonuclease obtained from *M. luteus* would recognize damage induced by DMBA in SHE cells.

A typical alkaline sucrose gradient profile of DNA from SHE cells treated with DMBA and incubated with and without *M. luteus* endonuclease is shown in Figure 1. DNA from cells treated with DMBA and incubated with endonuclease sediments slower than does DNA not incubated with endonuclease. DNA from control cells incubated with and without endonuclease co-sedimented, indicating that the *M. luteus* endonuclease did not introduce a large number of nonspecific breaks. Table 4 shows the effect of increasing concentrations of *M. luteus* endonuclease in inducing breaks (i.e., recognition of damage) in DNA from cells treated with 5,6-epoxide-DMBA or DMBA.

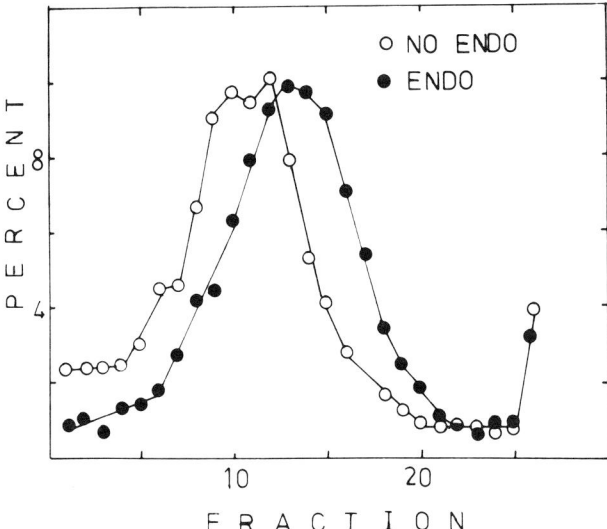

**FIGURE 1.** Recognition of DMBA-damaged SHE DNA by *M. luteus* endonuclease. SHE cells were incubated with 5 µg/ml DMBA for 24 hours. The cells were collected and the DNA isolated as described in "Methods". $^3$H-labeled DNA was incubated with 30 µl of the *M. luteus* endonuclease (●), while $^{14}$C-labeled DNA was not (o). After 30 minutes incubation at 37°C, DNAs were layered on top of a 5 to 20 percent alkaline sucrose gradient as described under "Methods".

Figure 2 compares the extent of recognition by S-1 and *M. luteus* endonuclease as a function of DMBA concentration. As shown in this figure, there appears to be a linear relationship between the number of endonculease-sensitive sites and the concentration of DMBA used. A similar relationship has been observed with the binding of radiolabeled DMBA to SHE DNA (9).

The efficacy of using endonucleases to determine the initial levels of damage induced, and to monitor removal was determined by comparing some of the more standard DNA repair assays (UDS, BubR photolysis)

**TABLE 4. Effect of** *M. Luteus* **Endonuclease Concentration Upon the Recognition of 5,6-Epoxide-DMBA and DMBA-Damaged DNA**[a]

| Concentration (µl) | 5,6-Epoxide-DMBA (breaks per $10^9$ Daltons)[b] | DMBA (breaks per $10^9$ Daltons) |
|---|---|---|
| 0.0 | 0.0 | 0.00 |
| 2.5 | 8.0 | NT |
| 5.0 | 13.0 | 1.3 |
| 7.5 | 21.0 | NT |
| 10.0 | 31.0 | 1.1 |
| 15.0 | NT[c] | 3.7 |
| 20.0 | 29.0 | NT |
| 30.0 | NT | 3.9 |

[a]SHE cells were incubated with 5 µg/ml DMBA for 24 hours.
[b]See Table 2.
[c]NT not tested.

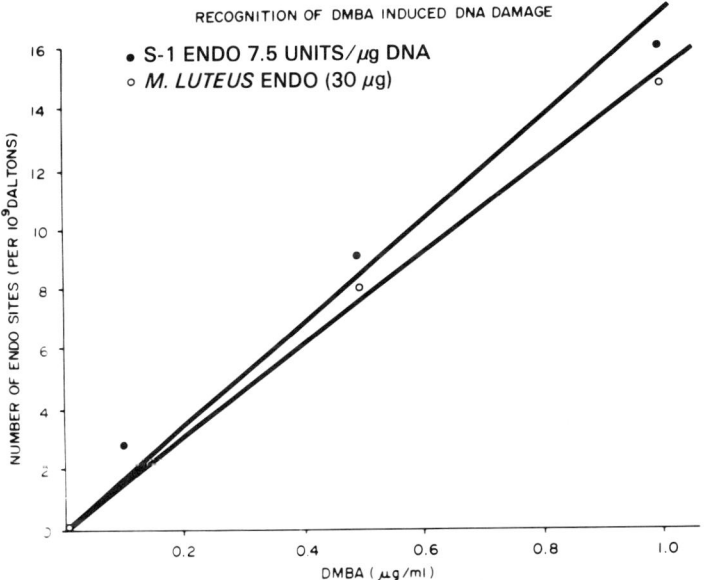

**FIGURE 2.** Relationship between the number of S-1 and *M. luteus* endonuclease-sensitive sites induced and the concentration of DMBA. SHE cells were incubated with various concentrations of DMBA for 24 hours. The DNA was extracted as described in "Methods" and either incubated with 30 µl *M. luteus* or with 7.5 units/µg DNA S-1 endonuclease. Reaction samples were then layered on top of alkaline sucrose gradients and the number of endonuclease-sensitive sites was determined as described under "Methods".

to the loss of endonuclease-sensitive sites and radiolabeled DMBA bound to DNA (Table 3). As shown in Table 3, few, if any, endonuclease-sensitive sites and radiolabeled DMBA bound to DNA are lost over a 24-hour period. Over the same period of time, little UDS is induced and there appear to be very few BudR photolabile sites induced (6,7). These data indicate, as shown by these four methods, that SHE cells repair very little of the DNA damage induced by DMBA. Thus, the endonuclease-sensitive site assay is at least equivalent to other, more expensive (using radiolabeled DMBA) ones, and as predictive as other (UDS and BudR photolysis) assays for detecting the initial and persistent DNA damage induced by chemical carcinogens.

## Recognition of DNA Damage in Vivo

Having determined the conditions for optimal endonuclease recognition of DMBA-induced DNA damage in vitro, we utilized the S-1 endonuclease to determine the extent of DNA damage induced in rat mammary DNA in vivo. In these experiments, 50-day-old Sprague-Dawley rats, shown by previous studies (28) to be most susceptible to the induction of mammary tumors, were injected with 5.12 mg (20 mM) of DMBA. The mammary glands were dissected, the DNA isolated and: (a) incubated with S-1 endonuclease to determine the number of endonuclease-sensitive sites; and (b) the level of radiolabeled DMBA bound to DNA was determined. Using a recently developed fluorescent assay to detect nanogram quantities of DNA after sedimentation through alkaline sucrose gradients (3), we compared the sedimentation profiles of mammary DNA obtained from control and DMBA-treated rats incubated with 20 units/$\mu$g DNA S-1 endonuclease. As shown in the top panel of Figure 3, the DNA from rats treated and not treated with DMBA co-sedimented as expected since DMBA does not introduce alkaline labile breaks in DNA. When this same DNA is incubated with S-1 endonuclease, we observe that control DNA sediments further down the gradient than does DNA obtained from the treated animal. From these data we calculated that approximately 4.5 endonuclease-sensitive sites per $10^9$ Daltons of DNA are induced into the rat mammary DNA. In a parallel experiment, animals were injected with 5.12 mg of radiolabeled DMBA and the amount bound within a 24-hour period was determined. From this experiment, 4.0 mM of radiolabeled DMBA was observed to be bound per mole of DNA phosphate, or 12.0 DMBA molecules were bound per $10^9$ Daltons of DNA. These results (Table 5), although preliminary, would indicate that S-1 endonuclease recognizes approximately 37 percent of the damage induced in vivo by DMBA. Other

# DNA DAMAGE 535

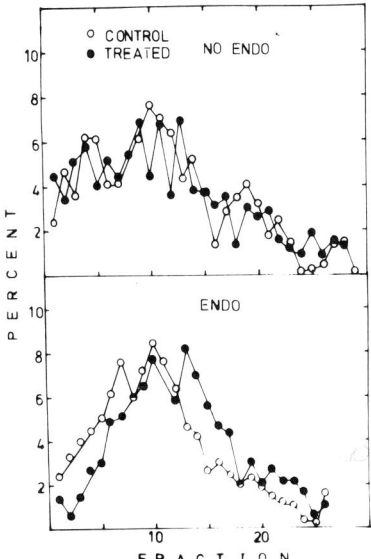

FIGURE 3. Recognition of DMBA-damaged DNA in rat mammary cells. Fifty-day-old Sprague-Dawley rats were injected with 5.12 mg of DMBA as described in "Methods". The individual glands were dissected and the DNA extracted and incubated with endonuclease as described under "Methods". Controls (o) were injected with a 0.5 ml solution of 95 percent DMSO:5 percent PBS while treated animals (•) were injected with 5.12 mg DMBA in the DMSO:PBS solution. Top panel shows the sedimentation patterns of DNA not incubated with endonuclease while the bottom panel shows the sedimentation patterns of DNA incubated with 20 units/µg DNA of S-1 endonuclease.

TABLE 5. Comparison of Endosites to the Amount of DMBA Bound to DNA[a] in Vivo

| Endosites[b] per $10^9$ | DMBA Bound[c] | |
|---|---|---|
| | mM/mole DNA $PO_4$ | Per $10^9$ Daltons |
| 4.5 | 4.0 | 12.0 |

[a] Animals were treated with 5.12 mg DMBA or [$^3$H]DMBA for 24 hours.
[b] DNA was extracted and incubated with 20 S-1 units/µg DNA. Endonuclease reaction in 200 µl as described in "Methods". The number of endosites (per $10^9$ Daltons) + $(1/M_{w(treated)} - 1/M_{w(control)})$ endo - $1/M_{w(treated)} + M_{w(control)})$ no endo.
[c] The amount of $^3$H-labeled DMBA bound was determined by pooling the DNA from all mammary glands. The data are the average for two animals.

experiments are in progress to substantiate these observations of recognition of damage in vivo.

## DISCUSSION

It is well established that the ultimate character of many of the known chemical carcinogens is that of an electrophile and that these electrophiles react readily with cellular DNA. Such damage to DNA resulting in base modification, base mispairs, and pertubations in DNA replication, have been strongly implicated in the initiation of the carcinogenic event (12). That damage to cellular DNA (at least for UV-induced cancer) is responsible for cellular transformation and has been uniquely demonstrated by Hart et al (14).
environmental carcinogens. All of these methods, although used in making regulatory policy, have certain serious limitations in being able to predict the absolute carcinogenic potential to man. Some bacterial (20) and mammalian (31) cell culture assays are based upon the ability of a carcinogen to induce DNA damage expressed as a mutation. Such assays do not always predict, with a great deal of certainty, which compounds are carcinogenic in man. Epidemiological determinations require large, exposed populations and/or exposure to the potent carcinogen over long periods of time and over an appropriate control population in order to determine risk. Long-term animal studies usually are very costly and, due to the use mainly of selectively inbred rodent strains, DNA repair capacity may not accurately reflect the effect of exposure to man. Other assays utilizing DNA repair are based upon the idea that such agents either induce UDS (30) or inhibit DNA replication (22). These assays may also not be an accurate determination of the harmful effects of carcinogenic agents to man since: (a) not all carcinogenic agents induce DNA repair in all cell systems, (b) the repair that is induced may not be an accurate monitor of the removal of "carcinogenic" lesions, and (c) such assays may reflect other parameters not related to DNA damage and/or repair. Thus it is important to use a method that is able to detect carcinogenic damage directly in vivo. Such an assay must be: (a) sensitive in being able to detect small amounts of damage in small biological samples, (b) quantitative in being able to determine the extent of damage induced, and (c) able to distinguish between carcinogens and noncarcinogens.

The studies reported in this paper describe the development of a system to detect DNA damage in specific organs in vivo. This assay utilizes: (a) enzymes isolated from *M. luteus and A. orzae* to detect damage in cellular DNA induced by DMBA or other carcinogens (not reported here), (b) alkaline sucrose sedimentation of nonradiolabeled DNA for quantitation of the breaks induced by enzymes recognizing such damage, and (c) detection of nonradiolabeled DNA following sedimentation. DNA damage induced by DMBA causes base modifications which result in base mispairings and distortions in the double helix. Supposedly, these distortions result in localized denatured regions which are digested by single-strand specific endonucleases like the S-1 enzyme obtained from *A. orzae*. The base modifications resulting from DMBA binding to cellular DNA are probably recognized by either damage-specific endonucleases and/or N-glycosidases present in the *M. luteus* extract.

Our data demonstrates the usefulness of using endonucleases, alkaline sucrose gradient sedimentation, and fluorescent detection of nonradiolabeled DNA to detect initial levels of DNA damage and and as a tool for monitoring their removal. Although the S-1 endonuclease recognizes approximately 16 percent of the total number of dimers induced by UV radiation, and approximately 16 percent of the total amount of damage induced by DMBA in SHE cells, such data indicate the usefulness of using endonuclease as a direct measurement of DNA damage. Other studies in progress in our laboratories are demonstrating the sensitivity of alkaline sucrose gradient sedimentation coupled to the DNA-fluorescent assay in quantitating alkaline labile breaks induced by ethylnitrosourea (in preparation). In comparison to other assays (i.e., radiolabeled carcinogen) for direct measurement of DNA damage, the endonuclease-sensitive site assay is quicker, equally as quantitative, and less expensive. Further, the endonuclease-sensitive assay is more sensitive and quicker than the more standard in vitro assays utilizing UDS or pertubations of DNA replication as an indirect measure of damage.

The usefulness of the endonuclease-sensitive site assay for monitoring DNA damage in vivo was demonstrated by the detection of DMBA-induced DNA damage in rat mammary gland cells. These preliminary studies indicate that the S-1 endonuclease is able, with a high degree of sensitivity, to detect DNA damage that would otherwise be impossible when utilizing UDS and DNA replication assays, and quicker and less expensive than long-term animal studies or studies with radiolabeled compounds. Other studies are under way to correlate the extent of recognition in vivo with: (a) the amount of radiolabeled carcinogenic and noncarcinogenic analogs bound to DNA, and (b) the induction of mammary tumors.

## ACKNOWLEDGMENTS

We wish to thank L. Carter, M. A. Drum, C. C. Kung, and R. Crane for their technical assistance, K. Lamb for his computer assistance, and

Drs. Stoica and Gould for advice and assistance with the animal experiments.

This research was supported by Grant No. 505337 from the U.S. EPA, and Contract No. CA21371 from NCI.

## REFERENCES

1. Amacher, D. E., and Lieberman, M. W. (1977): Removal of acetoxyaminofluorene from the DNA of control and repair deficient human fibroblasts. *Biochem. Biophys Res. Commun.* 74: 285–290.
2. Brash, D. E., and Hart, R. W. (1978): DNA damage and repair in vivo. *J. Environ. Path. Tox.* 2: 79–114.
3. Brash, D. E., and Hart, R. W. (1979): A sensitive method for analyzing nonradiolabeled DNA from alkaline sucrose gradients (in preparation).
4. Carrier, W. L., and Setlow, R. B. (1970): Endonuclease from *Micrococcus luteus* which has activity toward ultraviolet-irradiated deoxyribonucleic acid: purification and properties. *J. Bacterial.* 102: 178–186.
5. Carrier, W. L., and Setlow, R. B. (197): The excision of pyrimidine dimers (the detection of dimers in small amounts). *Methods Enz.* 21: 230–237.
6. D'Ambrosio, S. M., Daniel, F. B., and Hart, R. W. (1979): Cellular repair of DNA damage induced by 7,12-dimethylbenz(a)anthracene and its fluoro analogs in vitro. In: *Polynuclear Aromatic Hydrocarbons*, P. W. Jones and P. Leber, Eds., pp. 793–803, Ann Arbor Sci. Pub., Ann Arbor, Michigan.
7. D'Ambrosio, S. M., Daniel, F. B., Hart, R. W., Gazer, F. D., and Witiak, D. T. (1979): DNA repair in Syrian hamster embryo cells treated with 7,12-dimethylbenzy(a)anthracene and its weakly carcinogenic-5-fluoro analog. *Cancer Lett.* 6: 255–261.
8. Dansette, P., and Jerina, D. M. (1974): A facile synthesis of arene oxides at the K region of polycyclic hydrocarbons. *J. Am. Chem. Soc.* 96: 1224–1227.
9. Daniel, F. B., Cazer, F. D., D'Ambrosio, S. M., Hart, R. W., Kim, W. H., and Witiak, D. T. (1979): Comparative metabolism and DNA binding of 7,12-dimethylbenz(a)anthracene and its weakly carcinogenic 5-fluoro analog. *Cancer Lett.* 6: 263–272.
10. Daniel, F. B., Wong, L. K., Orvaec, C. T., Cazer, F. D., Wang, C.L.A., D'Ambrosio, S. M., Hart, R. W., and Witiak, D. T. (1979): Biochemical studies on the metabolism and DNA-binding of DMBA and some of its monofluoro derivatives of varying carcinogenicity.

In: *Polynuclear Aromatic Hydrocarbons*, P. W. Jones and P. Leber, Eds., pp. 855–884, Ann Arbor Sci. Pub., Ann Arbor, Michigan.
11. Heflich, R. H., Domez, D. J., Maher, V. M. and McCormick, J. J. (1977): Reactive derivatives of benzo(a)pyrene and 7, 12-dimethylbenz(a)anthracene cause S-1 nuclease-sensitive sites in DNA and "UV-like" repair, *Biochem. Biophys. Res. Commun.* 77: 634–641.
12. Hart, R. W., Daniel, F. B., Davis, M. T., and Lewis, N. J. (1976): A rational evaluation of the use of DNA damage as an environmental carcinogenesis/mutagenesis prescreen. In: *A Rational Evaluation of Pesticidal vs Mutagenic/Carcinogenic Action*, R. W. Hart, H. E. Kraybill and F. J. DeSerres, Eds., pp. 99–118, DHEW publication no. (NIH) 78-1306.
13. Hart, R. W., Hall, K., and Daniel, F. B. (1979): DNA repair and mutagenesis in mammalian cells. *Photochem. Photobiol.* 79: 131–154.
14. Hart, R. W., Setlow, R. B., and Woodhead, A. D. (1977): Evidence that pyrimidine dimers in DNA can give rise to tumors. *Proc. Nat. Acad. Sci. USA*, 74: 5574–5578.
15. Heflich, R. H., Mahoney-Leo, E., Maher, V. M., and McCormick, J. J. (1979): Removal of thymine-containing pyrmidine dimer from UV-light-irradiated DNA by S-1 endonuclease. *Photochem. Photobiol.* 30: 247–250.
16. Howland, G. P., Hart, R. W., and Yette, M. L. (1975): Repair of DNA strand breaks after gamma-irradiation of protoplasts isolated from cultured wild carrot cells. *Mutat. Res.* 27: 81–87.
17. Karran, P., Higgins, N. P., and Strauss, B. (1977): Intermediates in excision repair by human cells: Use of S-1 nuclease and benzylated napthoylated cellulose to reveal single-strand breaks. *Biochem.* 16: 4483–4490.
18. Luval, J. (1977): Two enzymes are required for strand incision in repair of alkylated DNA. *Nature* 269: 829–832.
19. Maher, V. M., McCormick, J. J., Grover, P. L., and Sim, P. (1977): Effect of DNA repair on the cytotoxicity and mutagenicity of polycyclic hydrocarbon derivatives in normal and xeroderma pigmentosum human fibroblasts. *Mutat. Res.* 43: 117–138.
20. McCann, J., Choi, E., Yamasaki, E., and Ames, B. N. (1975): Detection of carcinogens as mutagens in the *Salmonella*/microsome test: Assay of 300 chemicals. *Proc. Nat. Acad. Sci. U.S.A.* 72: 5135–5139.
21. Moon, R. C., Jass, D. H., and Young, S. (1969): Preparation of fat cell "free" rat mammary gland. *J. Histochem. Cytochem.* 17: 182–186.
22. Painter, R. B. (1977): Rapid test to detect agents that damage human DNA, *Nature* 265: 650–651.

23. Regan, J. D., and Setlow, R. B. (1974): Two forms of repair in the DNA of human cells damaged by chemical carcinogens and mutagens. *Cancer Res.* 34: 3318–3325.
24. Riazuddin, S., and Grossman, L. (1977): *Micrococcus luteus* correndonuclease: 1. Resolution and purification of two endonucleases specific for DNA containing pyrimidine dimers. *J. Biol. Chem.* 252: 6280–6286.
25. Riazuddin, S., Grossman, L., and Mahler, I. (1977): *Micrococcus luteus* correndonuclease III. Evidence for involment in repair in vivo of two endonucleases specific for DNA containing pyrimidine dimers. *J. Biol. Chem.* 252: 6294–6298.
26. Setlow, R. B., Regan, J. D., German, J., and Carrier, W. L. (1969): Evidence that xeroderma pigmentosum cells do not perform the first step in the repair of ultraviolet damage to their DNA. *Proc. Nat. Acad. Sci. U.S.A.* 64: 1035–1041.
27. Shackleton, J., and Roberts, J. J. (1978): Repair of alkylated DNA in chinese hamster cells measured by loss of enzyme-sensitive sites in isolated DNA. *Biochem. Biophy. Res. Comm.* 81: 1168–1174.
28. Simpson-Herren, L., and Griswald, P. (1970): Studies of the kinetics of growth and regression of 7,12-dimethylbenz(a)anthracene-induced mammary adenocarcinoma in Sprague-Dawley rats. *Cancer Res.* 30: 813–818.
29. Singer, B. (1979): N-Nitroso alkylating agents: Formation and persistance of alkyl derivatives in mammalian nuclic acids as contributing factors in carcinogenesis. *J. Natl. Cancer Inst.* 62: 1329–1339.
30. Stan, R.H.C. and Stich, H. F. (1975): DNA repair synthesis of cultured human cells as a rapid bioassay for chemical carcinogens. *J. Can.* 16: 284–291.
31. Trosko, J. E., and Hart, R. W. (1976): DNA mutation frequencies in mammals. *Interdiscipl. Topic Geront.* 9: 168–197.
32. Vogt, V. M. (1973): Purification and further properties of single-strand specific nuclease from *Aspergillus orzae. Eur. J. Biochem.* 33: 192–200.
33. Wiegard, R. D., Godson, G. H., and Rudding, C. M. (1975): Specificity of S-1 nuclease from *Aspergillus orzae. J. Biol. Chem.* 250: 8848–8855.
34. Yamaski, H., Pulkrrabek, P., Grunberger, D., and Weinstein, I. B. (1977): Differential excision from DNA of C-8 and $N^2$ guanosine adducts of N-acetyl-2-aminofluorene by single-strand-specific endonuclease. *Cancer Res.* 37: 3756–3760.

35. Yamaski, H., Roush, T. W., and Weinstein, B. I. (1978): Benzo(a)pyrene-7,8-dihydrodiol-9-10-oxide modification of DNA: Relation to chromatin structure and reconstitution. *Chem. Biol. Interact.* 23: 201–213.

# THE PROBLEM OF PAH DEGRADATION DURING FILTER COLLECTION OF AIRBORNE PARTICULATES— AN EVALUATION OF SEVERAL COMMONLY USED FILTER MEDIA

F. S.-C. Lee, W. R. Pierson, and J. Ezike*

Engineering and Research Staff
Research
Ford Motor Company
Dearborn, Michigan 48121

## INTRODUCTION

The atmospheric fate and chemical pathways of polynuclear aromatic hydrocarbons (PAH) have drawn much attention recently. Results of several studies have shown that some PAH compounds can react readily with nitrogen oxides (6,15), sulfur oxides (7), sulfuric acid aerosols (10), or ozone (9,15) or other oxidant gases even in the absence of light. Many of the PAH oxidation products, e.g., nitro-PAH and certain hydroxy-PAH species, are active in the Ames test (15). Thus the possibility exists that these products are responsible for the bioassay responses reported for the polar non-PAH fractions of the organic extracts from particulate matter in ambient air or source plumes (3,5,15).

Given these circumstances, it seems natural to ask whether the collected PAH in contact with the filter might oxidize during or after sampling. The extent of the oxidation and the nature of the product species then ought to depend on the chemical and physical properties of the filter medium. If so, then chemical measurements or bioassays of samples collected on certain filter types might be seriously affected.

The foregoing considerations led to the present investigation. Our experiments show that substantial degradation of BaP and of BaA occurs on certain filters. The extent of degradation depends critically on filter type, polytetrafluoroethylene (PTFE Teflon) membrane filters consistently yielding the best PAH recoveries of the five commonly used filter types tested. Similar filter dependences appear to exist also for PAH

---

*Present address: Department of Chemistry, Atlanta University, Atlanta, Georgia 30314.

compounds collected in real-world samples (that is, atmospheric aerosols, vehicle exhaust, air in a vehicle tunnel).

Experiments with BaP revealed similarly extensive degradation on filter surfaces—again heavily dependent on filter type—even during storage in the dark with no air stream passing through the filter. Again, the PTFE membrane filter type consistently minimized the degradation problem. Even so, some loss was always encountered and thus it is not clear that spurious effects in filter sampling can ever be truly eliminated.

The picture that emerges from this study is that extensive dark oxidation of filter-collected PAH does occur on most filter types; that the filter medium itself assumes a leading role in this degradation; that the atmospheric reactions hitherto advanced (e.g., atmospheric formation of nitro derivatives, Ref. 15) need not be invoked to explain the presence of direct-acting mutagens (15) in filter-collected atmospheric aerosol samples; and that lightly loaded filters are especially susceptible to the problems described, owing to their proportionately greater contact with the filter. It follows, of course, that misconceptions regarding the chemical properties and biological effects of airborne particulate organic material may currently exist. Gas-phase PAH determinations involving adsorption on polymeric materials may be subject to similar problems, though the surface reactivities of these materials remain to be explored.

## EXPERIMENTAL

The filter-exposure experiments employed radiotracer techniques in combination with high-performance liquid chromatography (HPLC). This approach was chosen because of its specificity and sensitivity in product separation and quantification. It has been widely used in biomedical research, though its potential for environmental studies has not yet been fully evaluated. Experience in the present study indicates that the method is well suited to the investigation of the chemical fate of trace organic pollutants under environmental conditions.

### Material and Apparatus

All of the reagent and PAH compounds were obtained commercially. [7,10-$^{14}$C]BaP, [12-$^{14}$C]BaA, and [G-$^3$H]BaP were purchased from Amersham with specific activities of 60.7 mCi/mmole, 54.5 mCi/mmole, and 15 Ci/mmole, respectively. Original solutions were each diluted to about 1$\mu$Ci/ml and stored in either benzene or cyclohexane solutions. They were periodically purified by a silica-gel column to ensure radio-chemical purity.

**TABLE 1. Filter Media Tested in This Study**

| Filter Identification | Description of Material and Structure | Manufacturer |
|---|---|---|
| Gelman A | Glass fiber Type A | Gelman Instrument Co. |
| 2500 QAO | Silica (quartz) fibers | Pallflex |
| T60A20 | Micro glass fibers with Teflon binder on fibers | Pallflex |
| Fluoropore (0.2 or 0.5 $\mu$m) | PTFE membrane bonded to polyethylene net | Millipore |
| Zefluor (0.2 or 1 $\mu$m) | PTFE membrane supported by PTFE fibers | Ghia Corp. |

The types of filters tested in this study are listed in Table 1, along with their material and structure descriptions and manufacturers. For convenience, the commercial trade names of these filters will be used throughout the text. Unless otherwise indicated, all of the filters were used directly without further treatment. Nominal pore size of all membrane filters was 0.2 $\mu$m unless otherwise indicated.

HPLC analyses were performed on a Waters HPLC system consisting of two Model 6000 pumps, a Model 660 gradient programmer, a Model 710 WISP sample processor, and a Model 440 dual micro UV detector. A Schoeffel Model FS-970 LC spectrofluorometer was used for HPLC/fluorescence measurements. All solvents used were UV-grade distilled-in-glass quality obtained from Burdick and Jackson.

Radioactivity assays were performed on a Beckman LS-7000 Liquid Scintillation Counter (LSC). The scintillation cocktails used were Insta-Gel (for aqueous samples) and Insta-Fluor (for organic samples), both obtained from Packard.

## Sampling and PAH Exposure

Standard high-volume samplers were used for air samplings. They were equipped with covered shelters which also shielded the filter samples from sunlight. Filters of either 8 by 10-inch or 142-mm disc sizes were used. Urban air samples were collected on the roof of our research building in Dearborn, Michigan, in a suburban area. Diesel particulates were collected from an exhaust dilution tube described elsewhere (11).

Since only relative values obtained from parallel runs are compared in this study, sampling details other than particulate mass, sampling time, and air volume were not recorded. Highway and rural air samples were collected in the Allegheny Tunnel of the Pennsylvania Turnpike and its vicinity in southwestern Pennsylvania. Details of sampling conditions will be published separately and general conditions are described in a previous paper (13).

For BaP and BaA sampling exposure studies, the blank or particulate preloaded filters to be tested in each set of parallel runs were each spiked with an equal amount of $^{14}$C-BaP or $^{14}$C-BaA. The spiking was performed by placing solution drops of 1 to 3-$\mu$l sizes across the filter with a microsyringe until the total volume was doped on the filter surface. The spiking process was controlled so that about the same surfce area of each filter was covered by $^{14}$C-BaP or $^{14}$C-BaA solution droplets (about 10 $\mu$g/cm$^2$; the spiked area for BaP could be checked with a black light). Unless otherwise indicated, the amount spiked was ~0.12 $\mu$Ci (0.5 $\mu$g) of $^{14}$C-BaP or ~0.1 $\mu$Ci (0.4 $\mu$g) of $^{14}$C-BaA per filter. After spiking, the filters were dried in air and subsequently either used for sampling exposure or stored. In the latter case, the filters were each wrapped with aluminum foil and stored at room temperature.

Some sampling was carried out in which the spiked filter was placed downstream from another filter in the sampler. The upstream filter served to remove the particulates, thus allowing only gases to reach the downstream filter. This tandem filter set for each of the parallel runs was arranged so that approximately the flow rate and face velocity were the same for each filter set. Thus, for instance, filters with low pressure drops (T60A20, Gelman A and Tissuquartz) were covered by filters with high pressure drops (Zefluor and Fluoropore) or vice versa.

## HPLC and Radio-Liquid-Chromatographic (RLC) Analyses

Filter samples after sampling or storage were soxhlet extracted for at least 12 hours with dichloromethane or other solvents as indicated. The extract was then concentrated by rotary evaporation and nitrogen gas blowdown before HPLC and RLC analysis.

Several different columns were used for reverse-phase HPLC analysis, as illustrated in the respective chromatograms presented in later sections. All of the normal-phase HPLC analyses were performed under the same chromatographic conditions on a Waters Radial Compression Column (Radial-PAK B). The sample was injected while the column was eluted with n-heptane. During the subsequent 10 min, the eluent was changed gradually from heptane to chloroform using a solvent gradient preset in the programmer (Curve 9, a nonlinear gradient with ~5 percent chloroform up to ~5 min and then nearly exponential progression to 100 percent chloroform in the second half of the gradient period).

A typical RLC assay involved the following steps. First, an aliquot of the extract was counted in the LSC to obtain the total extractable $^{14}$C activity for the filter. This allowed calculation of $^{14}$C recoveries after sample treatment. The remainder of the extract was then concentrated to a volume of 500 $\mu$l or less. An aliquot of the concentrate was withdrawn into a 100- or 200-$\mu$l syringe. A known volume of the concentrate in the syringe was placed in the LSC vial and counted directly, while the remaining portion was injected into the HPLC. The eluent from the column along with the separated peaks was collected at 0.5 to 2-min intervals and each fraction was counted in the LSC. After quenching corrections (Beckman H number external quenching correction method), the activities associated with BaP and other peaks were used for yield and recovery calculations.

$^3$H-BaP was used along with $^{14}$C-BaP in some of the runs. The double-labelling runs were carried out in order to assess the possible losses or conversions of BaP during analysis as distinguished from the sampling/exposure step. In these runs, the $^3$H-BaP was added to the filter after sampling. The spiked filter was then extracted immediately. The $^{14}$C and $^3$H activities associated with the products or BaP were then counted separately using a channel-ratio method.

All the RLC chromatograms were obtained by plotting the observed radioactivities associated with each collected fraction of the HPLC eluent against their respective retention times. Since they are shown for qualitative comparisons, detailed quantitative yield calculations for the $^{14}$C products are not reported here.

## RESULTS AND DISCUSSION

### BaP and BaA Recoveries

Recoveries of $^{14}$C-BaP spiked on different types of filter media after sampling varying amounts of air are quantitatively compared in Table 2. These experiments were carried out with the tandem filter arrangement described earlier. Because of the limited number of Hi-Vol samplers available during the experiment, only two or three parallel runs could be performed each time. The results listed in a given column in the table represent runs made simultaneously under identical conditions. Thus, for runs within each set, errors associated with pollutant concentration

TABLE 2. Percentage of Recovery of $^{14}$C-BaP Spiked on Blank Filters After Sampling Pre-Filtered Urban Air—Dependence of $^{14}$C-BaP Recovery on Filter Types and Sampling Conditions

| | Sampling Conditions | | | | | |
|---|---|---|---|---|---|---|
| Set No. | 1 | 2 | 3 | 4 | 5 | 6 |
| Time (hr) | 8 | 17 | 18 | ~50 | 65 | 30 |
| Air Vol (m$^3$) | 35 | 80 | 140 | 250 | 1,200 | 145 |
| μg BaP Spiked[a] | 0.5 | 0.5 | 0.5 | 0.5 | 0.5 | 5 |
| Filter Type[b] | % BaP Recovery (% of Total Extractable $^{14}$C Radioactivities)[c] | | | | | |
| Glass Fiber | 35 | - | - | 11 | 2.6 | 75 |
| Tissuquartz | - | - | 1.3 | - | - | 62 |
| T60 A20 (Pallflex) | - | 50 | - | 40 | - | 86 |
| Fluoropore | 60 | 55 | 65 | 71 | - | 90 |
| Zefluor | - | - | - | - | 45 | - |

[a]The amount of BaP spiked on each filter was ~10 ng/(cm$^2$ surface area) for the first five sets of runs and was ~100 ng/cm$^2$ for Set 6.
[b]All the filters used were 142 mm diam (~100 cm$^2$ sample area) except for the 1,200 m$^3$ air exposure runs for which 8 = 10-inch filters were used (406 cm$^2$ sample area) at flow rates of 3 to 5 cfm.
[c]CH$_2$Cl$_2$ extracts.

changes or meteorological factors were essentially eliminated. Even between sets, the meteorological effects are probably small for sets 1 through 4 which were sampled under similar conditions within the 3 summer months of 1978. Sets 5 and 6 were run in summer 1979 and winter 1978, respectively. Therefore, they are discussed separately.

Throughout the paper, $^{14}$C-BaP or BaA recovery data will be expressed as the percentage of recovery of $^{14}$C-BaP or BaA activity relative to the total extracted $^{14}$C activity. This was chosen instead of absolute recovery (percentage of recovery from original $^{14}$C-BaP or BaA spiked) in order to offset errors caused by losses during analysis. In a typical $^{14}$C-BaP run, the total extractable $^{14}$C activities accounted for about 80 percent of the $^{14}$C-BaP activity originally spiked. Of the missing 20 percent, about 5 to 10 percent was found to remain in the filter because of incomplete solvent extraction. The other 10 to 15 percent was probably lost either during analysis or through sublimation during sampling. In contrast, the total recoverable $^{14}$C activity in a $^{14}$C-BaA run generally accounted for only 30 to 40 percent of the activity originally spiked. The much higher loss here as compared with the BaP runs can be reasonably attributed to BaA sublimation, consistent with the expectation from vapor-pressure considerations. A similar observation has also been reported elsewhere (1).

Typical RLC chromatograms run on a normal-phase HPLC column

are shown in Figure 1 (set 5, Table 2). In addition to $^{14}C$ activities, RLC chromatograms showing $^3H$ activities and fluorescence responses are also included in the figure. The latter two traces will be discussed later. Most of the runs listed in Table 2 were run on reverse-phase columns. Typical reverse-phase RLC runs comparing recoveries of Fluoropore or Gelman A filter series are shown in Figures 2 and 3, respectively.

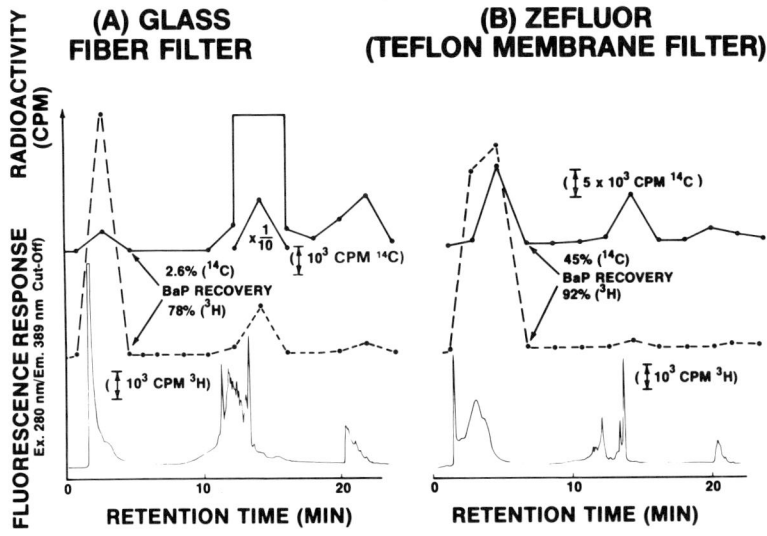

FIGURE 1. Radio-$^{14}C$, radio-$^3H$, and fluorescence-liquid chromatograms comparing the degradations of $^{14}C$-BaP on a blank glass fiber and a blank Teflon membrane filter after being exposed to urban air. Samples were extracted with $CH_2Cl_2$. HPLC runs were performed on a Waters normal-phase radial compression column with 2 ml/min flow rate. See the experimental section in the text for other chromatographic details. Retention-time calibrations: BaP = 3.3 min; $N_2$-BaP = 12 min; OH-BaP = 14 to 15 min; BaP-diones and BaP-dihydrodiols = 20 to 25 min. These are the two parallel runs listed as set 5 in Table 2.

A comparison of $^{14}C$ BaP recoveries from runs in sets 1 through 4 shows that in every case the Gelman A and Tissuquartz filters gave substantially lower recoveries than did T50A2 and Fluoropore filters. Of the latter two types, Fluoropore exhibited the higher recovery. Similar results are seen in set 5, where another Teflon membrane filter (Zefluor) gives much higher recovery than the Gelman A filter. As indicated in Table 1, T60A20 is a glass-fiber filter impregnated with PTFE Teflon. Although the detailed structure of the filter is not described by the manufacturer, it is believed that most but not all of the filter glass surface is covered with Teflon. The order of the BaP recoveries, i.e., Teflon membrane > T60A20 > glass fiber (Gelman A) or silica fiber (Tissu-

## 550 PAH DEGRADATION DURING SAMPLING

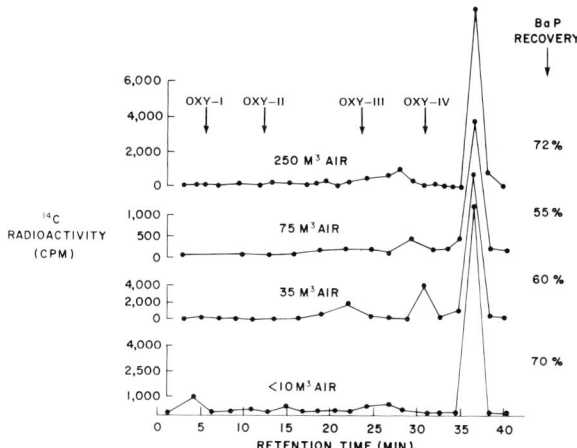

FIGURE 2. Radio-liquid chromatogram (reverse-phase) showing the recoveries of $^{14}$C-BaP on Fluoropore filters after being exposed to varying amounts of prefiltered urban air. Extracted with $CH_2Cl_2$. Analyzed on a 5 $\mu$m HI-EFF Micropart $C_{18}$ column obtained from Applied Science Laboratories, Inc.; linear gradient of 40 percent $CH_3CN$ in water to 100 percent $CH_3CN$ in 50 minutes. See Figure 6 for detailed peak designations.

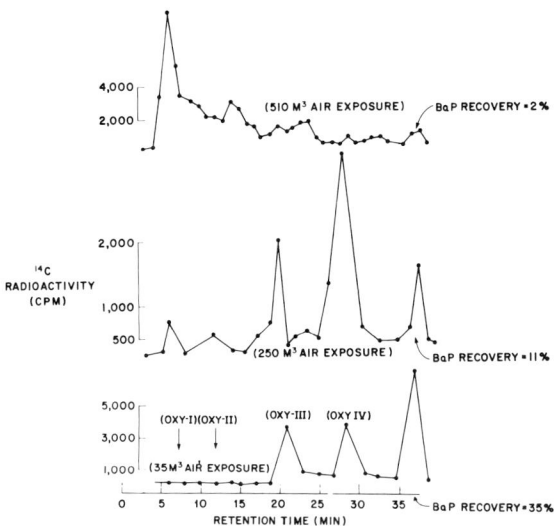

FIGURE 3. Radio-liquid chromatogram (reverse-phase) showing the recoveries of $^{14}$C-BaP on glass-fiber filters after being exposed to varying amounts of prefiltered urban air. $CH_2Cl_2$ extracts. Chromatographic conditions same as in Figure 2. Peak designations same as in Figure 6.

## PAH DEGRADATION DURING SAMPLING 551

quartz), thus suggests that the degradation of BaP was enhanced by the glass or silica surfaces of the filter. In addition to the filters listed here, we have tested two research-stage filters, viz., a Teflon-impregnated glass fiber filter (TX60A40, Pallflex) and a high-purity processed glass-fiber filter (Pallflex). The respective recoveries obtained for these two filters were 52 and 6 percent versus 65 percent for Fluoropore, in accord with the general observation recounted above. Experiments were also carried out to pretreat the Gelman A filter surface before air exposure. As shown in Figure 4, little difference in BaP recovery was noticed when untreated filters were compared with heated and acid-washed filters or filters preexposed to a particulate-free air stream.

Similar results were observed for $^{14}C$-BaA; as seen in Figure 5, BaA recoveries were much higher with the Zefluor filter than with Gelman A filters.

The foregoing results strongly suggest that the surface properties of the filter play a major role in determining the extent of PAH degradation during air exposure. These surface properties may include several complex and interrelated factors such as catalysis, specific surface area, affinity for water and other vapors, surface pH, or surface polar functional groups. Although the mechanisms are not understood, it is interesting to note that the observed order of increasing BaP recovery is in

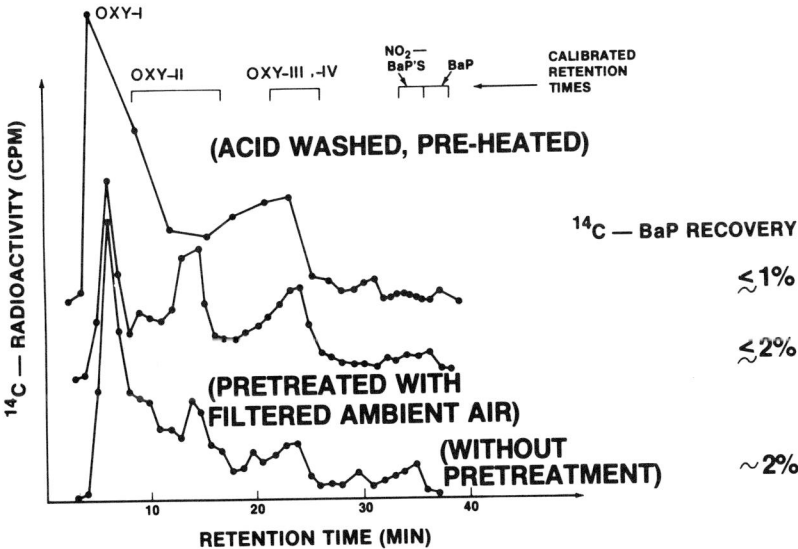

**FIGURE 4.** Radio-liquid chromatograms (reverse-phase) comparing the recovery of $^{14}C$-BaP after air sampling using treated and untreated glass-fiber filters. Samples were extracted with $CH_2Cl_2$. HPLC runs were performed on a Waters $\mu$ Bondapak $C_{18}$ column; linear gradient of 40 percent $CH_3CN$ in water to 100 percent $CH_3CN$ in 30 minutes; solvent flow rate = 1 ml/min. See Figure 6 for peak designations.

## 552 PAH DEGRADATION DURING SAMPLING

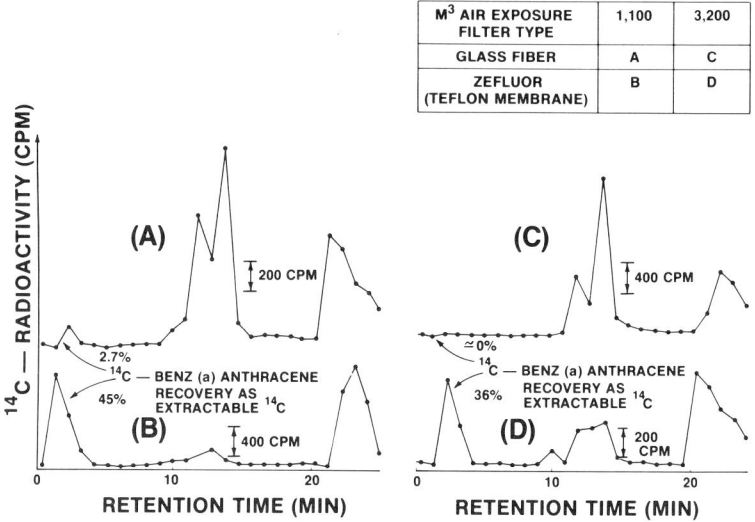

FIGURE 5. Radio-liquid chromatogram showing the degradations of $^{14}$C-BaA on different types of filters after being exposed to urban air. Extracted with $CH_2Cl_2$. Chromatographic conditions same as in Figure 1.

accordance with the order of decreasing specific surface areas of the filter, i.e., Tissuquartz > glass fiber > T60A20 > Fluoropore. The large surface areas associated with silica or glass-fiber filters not only enhance their catalytic activities, but also enhance the uptake of water and other vapors (2). These factors may all be responsible for the large PAH degradations observed on these filters.

It has been reported that gaseous oxidants such as sulfur oxides and nitrogen oxides may be converted to sulfate and nitrate on the glass-fiber-filter surface (14). It is possible that these converted oxides in turn cause the oxidation of PAH. A much higher vapor affinity of glass-fiber filter as compared with Teflon membrane filter could be the reason why much more non-$^{14}$C-associated fluorescing material, particularly in the chloroform-eluted region (~11 to 13 min in the fluorescence traces in Figure 1), existed in the glass-fiber filter than in the Zefluor filter after air exposure. The $^3$H/$^{14}$C double-labeling technique was employed in this pair of runs, with $^3$H-BaP spiked onto the filter before extraction but after sampling to assess losses caused by oxidation after air exposure. The observation of more $^3$H-BaP conversion in the glass-fiber run (78 percent recovery versus 92 percent recovery for Zefluor, Figure 1) implies that the glass-fiber filter or its adsorbed material causes additional BaP degradation during extraction.

In all of the runs discussed above, an equal amount of $^{14}$C-BaP (0.5

$\mu$g) was spiked on each filter surface before sampling. In Set 6 runs, however, a tenfold higher amount of $^{14}$C-BaP, i.e., 100 $\mu$g/cm$^2$, was spiked on each filter. Recoveries were higher at the higher dose for every filter, although the order of their recovery remained the same (see Table 2). Again, the results suggest that the surface area or some commensurate surface property such as number of reaction sites (hence, e.g., the $^{14}$C-BaP/filter surface area ratio), rather than the volume of air exposure, governs the extent of BaP degradation.

Several experiments were also carried out in which $^{14}$C-BaP spiked on particulate-laden filters rather than blank Tissuquartz filters was exposed to an air stream. Results from these experiments are listed in Table 3 along with the results obtained from blank-filter runs made under similar conditions. Higher but still poor $^{14}$C-BaP recoveries were obtained in the particulate-laden-filter runs. Comparing runs 4 and 5 with runs 1, 2, and 3, one finds that the recovery increases regularly as the particulate loading on the filter increases. In the extreme case, the recovery in run 1 is 16 times higher than that in runs 4 and 5, even though run 1 had actually entailed exposure to four times more air.

Qualitatively, these results show that $^{14}$C-BaP adsorbed on the surface of air particulate is more oxidation-resistant than $^{14}$C-BaP adsorbed directly on the surface of Tissuquartz filters. Quantitative statements are difficult to make because of the limited number of runs and because the $^{14}$C-BaP spiked filters in runs 1 through 3 presumably was at least partly adsorbed on the filter surface rather than completely on the particulates, precluding quantitative assessment of BaP recovery from particulates alone. Nevertheless, the data do demonstrate clearly that the physical and chemical nature of the surface onto which the PAH are adsorbed can result in large variations in the reactivities of PAH toward oxidation.

**TABLE 3. Percentage of Recovery of $^{14}$C-BaP Spiked on Tissuquartz Filters Preloaded with Air Particulates Under Different Sampling Conditions**

| Run No.[a] | 1 | 2 | 3 | 4 | 5 |
|---|---|---|---|---|---|
| m$^3$ of Air Passing Through the Filter Before Spiking | 1,800 | 300 | 300 | 0 | 0 |
| Particulates Preloaded on the Filter Before Spiking (mg) | 300 | 50 | 50 | 0 | 0 |
| Air Passing Through the Spiked Filter (Exposure) | 5,600 | 1,200 | 1,200 | 1,500 | 1,500 |
| BaP Recovery (%) | 26 | 6.1 | 4.5 | 1.6 | 1.8 |

[a]Runs 2 and 3 and runs 4 and 5 are two sets of parallel runs each sampled under identical sampling conditions.

## Product Identification and $^{14}C$ Radioactivity Balance

A detailed product identification is beyond the scope of this work and was not performed. Some of the expected BaP oxidation products were identified and several others were calibrated in our HPLC procedure. A run made for product identification is shown in Figure 6. About 10 to 15 $\mu$g of BaP was spiked on a Gelman A filter which was subsequently exposed to 500 m$^3$ of air using the tandem filter already described.

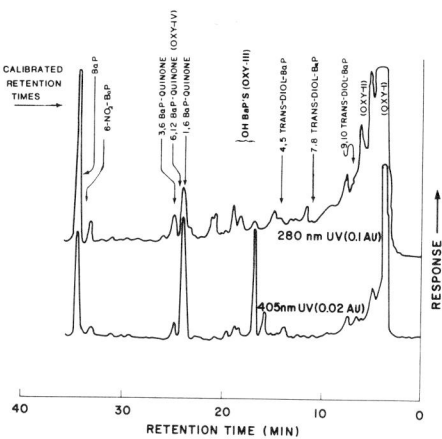

FIGURE 6. HPLC calibrations of possible oxy-BaPs produced by air oxidation of BaP spiked on a Gelman A filter. Sample was extracted in CH$_2$Cl$_2$. HPLC was performed on a column system consisting of two Whatman Partisil 5 $\mu$ODS columns in series; linear gradient of 40 percent CH$_3$CN in water to 100 percent CH$_3$CN in 30 minutes; solvent flow 0.8 ml/min.

As shown in the figure, both 3,6- and 1,6-BaP-quinones (Oxy-IV) were tentatively identified in this run on the basis of their respective retention times and UV absorption ratios at 280 and 405 nm. Oxy-III most likely consists of several isomeric hydroxy-BaPs, again judging from the retention times and UV absorption data. Hydroxy-BaP isomers calibrated in this run include 1-, 3-, 4-, and 7-OH-BaPs. The last two isomers had retention times identical to those of Oxy-III. The first two isomers, however, were found to degrade readily into isomeric BaP-quinones, with the extent of degradation depending on the cleanliness of the column during analysis. Their retention times were both very close to that of Oxy-III but varied slightly from run to run, probably because of the in-column conversion problems described above. The identification of hydroxy-BaPs as major products is also supported by the retention-time calibrations given in normal-phase columns, as shown in Figure 1.

The retention times for three of the BaP-dihydrodiols were also calibrated in this run. Their identifications were complicated by the interfering peaks and by their small yields. As seen in Figures 2 and 3, they were at most, minor products since no $^{14}C$ activity peaks were observed in the region. Similarly, the absence of noticeable $^{14}C$ activity peaks at the calibrated retention time for 6-nitro-BaP excludes it as a significant product. The other two major products remaining unidentified are Oxy-I and Oxy-II. The other minor peaks in Figure 6 were due to vapor pickup and their identities were not pursued further.

It is seen from Figure 3 that in the Gelman A filter runs the recovery of $^{14}C$-BaP decreases and the yields of $^{14}C$-oxidation products increase regularly as the volume of air exposure increases from 35 through 250 to 510 $m^3$. The two major products, $^{14}C$-BaP-quinones (Oxy-IV) and $^{14}C$-hydroxy-BaP (Oxy-III), were both further oxidized along with parent $^{14}C$-BaP to the highly polar products, Oxy-I, as more and more air flowed through the filter. This is consistent with the mechanism of BaP degradations caused by gaseous oxidations. In contrast, such a trend of air-volume dependence is not observed in the Fluoropore runs (Figure 2); there, within the data scatter, the recovery is in fact constant ($\pm$ 15 percent) in going from 10 to 250 $m^3$ of air. Furthermore, the total yield of the oxidation products observed in the Fluoropore runs is, on the average, less than 10 percent and does not seem to vary systematically with air volume. Because of the low flow rates associated with Fluoropore filters and thus the long sampling times, we did not carry out the experiments to air exposures beyond 250 $m^3$. However, a run was made under similar conditions using a 1-$\mu$m pore size Zefluor filter (Set 5, Table 2), and 45 percent of the $^{14}C$-BaP was still recovered even after the sampling of 1200 $m^3$ of air, as opposed to the almost complete conversion in the Gelman A run made simultaneously.

In runs shown in Figures 2 and 3, the total chromatographically accounted-for $^{14}C$ activity is, on the average, about 75 to 85 percent of what was extracted. The fate of the missing 15 to 25 percent of the extractable $^{14}C$ is not clear at present. It is possible that some of the $^{14}C$ was not eluted from the column. The column was routinely cleaned with a flow of tetrahydrofuran and chloroform after each radioactive run, and some $^{14}C$ activity was indeed observed in the solvent eluents. More $^{14}C$ total activity and $^{14}C$-hydroxy-BaPs were observed in normal-phase column runs than in reverse-phase column runs—for instance, the total $^{14}C$ activity recovery in runs shown in Figure 1 is 85 to 90 percent, versus 75 to 85 percent for runs shown in Figures 2 and 3—suggesting column losses of certain $^{14}C$ species in reverse-phase runs.

The near-constant recovery of $^{14}C$-BaP in the series of runs shown in Figure 2 is a strong indication that oxidation by gases is greatly reduced when a filter with inert surfaces such as Teflon is used. This is further discussed later.

## Conversion of $^{14}$C-BaP During Analysis and Storage

This set of experiments was carried out to assess: (1) the degradation of BaP during analysis, (2) the extent of BaP degradation in contact with the filter during storage, and (3) the stability of BaP adsorbed on air particulates during storage. Results from these experiments are summarized in Table 4.

TABLE 4. $^{14}$C-BaP Recovery from Filters after Storage

| Storage Time (days)[a] | <1 | 3-4 | 8-14 | 20 | 50 |
|---|---|---|---|---|---|
| Filter Type | % $^{14}$C-BaP Recovery (% of Total Extractable $^{14}$C Radioactivities) | | | | |
| Glass Fiber | 85 | - | 45 | 30 | - |
| Tissuquartz | 92 | - | 40 | 20 | 0 |
| T60A20 (Pallflex) | 85 | 70<br>73 | 70 | - | - |
| Fluoropore | 90 | 60 | 94 | - | - |
| Zefluor | - | 78 | 83 | - | - |
| Ambient Air Particulates Collected on Tissuquartz Filters[b] | 100[b]<br>85 | 100 | 90<br>76 | 76 | 52 |

[a] Samples were stored at room temperature in the dark (wrapped in aluminum foil).
[b] Percent recovery for each run was normalized to the recovery of one of the fresh sample runs (>1 day storage) as 100 percent. See text for details.
[c] $^{14}$C-BaP spike ≃ 0.4 μg. Extracted in $CH_2Cl_2$.

The first column in Table 4 lists runs analyzed immediately after spiking. The 8 to 15 percent $^{14}$C-BaP deficits in all the blank filter runs in the first column occurred, it should be noted, even without any significant air exposure or storage. Such deficits were most likely caused by reactions of $^{14}$C-BaP with filter materials, oxygen, or other trace impurities in the system and were to be expected in view of the trace amounts of $^{14}$C-BaP used in our study.

When $^{14}$C-BaP recoveries after storage are compared for different types of filters, it is seen that the data follow the same trends as those observed in the sampling exposure runs reported earlier. Thus, while the recoveries in Gelman A and Tissuquartz runs decrease regularly with increasing storage time, such dependence is not observed in the Teflon membrane filter runs; in fact, the recoveries in the latter runs seem to remain constant (79 ± 14 percent) within a 2-week period. Thus, in contrast to the glass or silica fiber media, the surfce activities of the two Teflon membrane filters seem to be limited, and could be essentially "consumed" by a small fraction of the spiked $^{14}$C-BaP in the first few days, after which further oxidation became insignificant. Here again, the results support the hypothesis that air oxidation of PAH requires an active surface, and thus can be minimized when an inert surface such as Teflon is used for collection.

In the sampling exposure runs discussed in the earlier sections, the roximately typical experimental duration was app3 to 4 days, including sampling, extraction, and other analytical steps. From the discussions above, it is seen that at least part of the $^{14}$C-BaP losses observed there can be attributed to reactions of BaP with the trace "inherent" filter activities rather than to actual air exposure. For Fluoropore filter runs, the highest possible recovery for the sampling runs was probably 75 to 80 percent, i.e., recoveries observed after 3 to 4 days of storage (or $^{14}$C-BaP/filter surface contact) as listed in Table 4. Consequently, for the Fluoropore runs shown in Figure 2, *the existence of an air stream going through the filter had no measurable effect (less than 10 percent) on BaP degradation.* Again, then, PAH degradation can be minimized with an inert filter.

The stability of BaP adsorbed on air particulates during prolonged storage was also studied. In this experiment, an 8 by 10-in. Tissuquartz filter was used to first collect about 300 mg of urban air particulates. The loaded filter was then marked off into eight sections (without physical separations) and each was then spiked with about 0.4 $\mu$g $^{14}$C-BaP ($\sim$10 ng/cm$^2$). The whole filter was then exposed to approximately 5600 m$^3$ of air using the procedure described earlier. After exposure, the filter was divided into eight sections. One section was analyzed immediately, while all the others were put into storage. The air-exposure step was carried out to destroy all the possible $^{14}$C-BaP adsorbed directly on the silica fiber of the filter and thus allow the study of only particulate-adsorbed BaP.

Results from the study are listed in the last row in Table 4. All the recovery data were normalized to the recovery of one of the fresh sample runs (immediately after 5600 m$^3$ air exposure) as 100 percent. A comparison of this and the blank Tissuquartz runs clearly shows that the rate of $^{14}$C-BaP degradation is much slower in the former case. Within experimental error, it is seen that the recoveries here are nearly constant for 2 weeks; and even after 52 days of storage, more than half of the $^{14}$C-BaP was still recovered, as opposed to the complete conversion in the blank filter case. Thus, at least under "static" conditions such as these, the particulate-adsorbed PAH seem to be quite stable toward oxidation in the absence of light. The last conclusion cannot be applied directly to the stability of BaP during air sampling because of the presence of oxidant gases and the "dynamic" situations involved. We are currently addressing this problem by the collection and analysis of PAH in real-world samples, as discussed in the next section.

## Comparisons of Airborne Particulates Collected on Different Filter Media

Results from the model experiments described above show that the extent of PAH degradation during sampling depends strongly on the type of filter used. The use of Teflon membrane filters in place of the commonly used glass-fiber filters was found to suppress greatly such degradation problems. This result should be directly applicable to the sampling of volatile PAH which exist partly or wholly in the vapor phase in the sample source but subsequently condense, during sampling, on the collection medium, i.e., a filter or an absorbent trap. The result is also pertinent for the sampling of particulate-associated PAH for two reasons. First, surface-initiated reactions may still occur because of the direct contact between the filter and the particulates. Second, PAH may migrate between particulate and filter surface because of sublimation/readsorption during sampling. Significant sublimation losses of PAH during high-volume sampling have been observed for nonvolatile PAH such as BaA, BaP, and benzo(g,h,i)perylene and are found to vary mostly with sampling conditions but not so much with equilibrium vapor pressure (1). In our experiments, BaP and BaA losses ascribable to sublimation ranged from 10 to 15 percent (BaP) to 60 to 70 percent (BaA), as stated earlier. Thus, even for particulate-associated PAH with extremely low vapor pressures, the above-cited filter effects may still be significant.

An investigation is under way to compare various types of airborne particulates collected on different filter types. Some preliminary results are reported below. The data to be compared here are: (1) the percent by weight of organics extracted by dichloromethane, (2) the percent by weight of polar materials extracted by acetonitrile, and (3) the estimated relative concentration of gross PAH.

The filters chosen for this comparative study were Zefluor, Gelman A, and T60A20. Zefluor was selected as the representative Teflon membrane filter because of its low background impurities and low affinity for organic vapors as compared with Fluoropore. Background contaminations for these filters are compared in Figure 7. Among the four filters tested, Zefluor gives the lowest impurity content in its blank filter extract. Fluoropore is relatively clean in the polar fractions but its nonpolar impurities are, in fact, the highest, presumably originating from its polymeric backing material. These contaminants may potentially interfere with the PAH fluorescence measurements, and substantial amounts of flourescing material from extracts of blank Fluoropore filters have indeed been reported (16).

Airborne particulates from the four types of sources listed in Table 5 have been compared. The mass emission rates mg/m$^3$ air) measured from

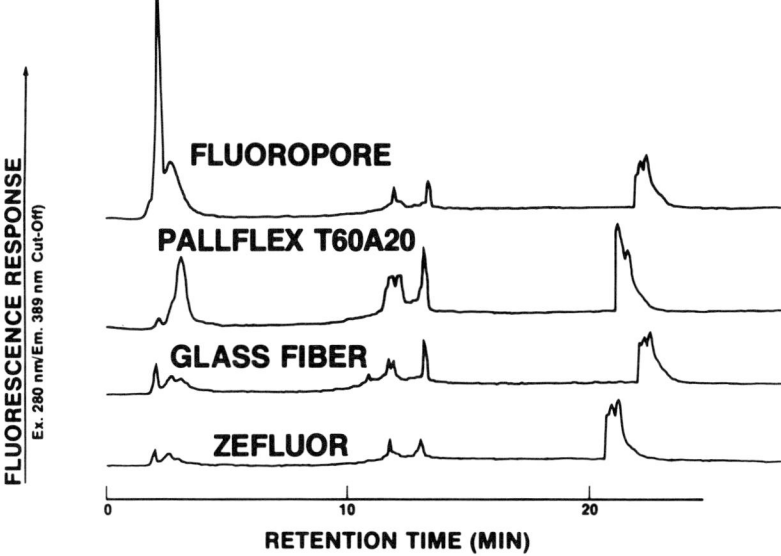

FIGURE 7. HPLC assay of fluorescence impurities in the blank $CH_2Cl_2$ extracts of several commonly used filter media. Chromatographic conditions same as in Figure 1. Extracts were obtained by soxhlet extraction of a 142-mm Fluoropore disc and an 8 x 10-inch size of each of the other three filters. The Fluoropore fluorescence response observed should be multiplied by a factor of 3.26 when quantitative comparisons of these filters are made on a per-unit area basis.

different filter runs agree in general. This was expected for Zefluor and Gelman A filters since both are reported to have > 99.9 percent filtration efficiencies (2,8). Our results also indicate that the loss of particulates due to filter penetration is not significant for T60A20 under our sampling conditions.

The normal-phase liquid chromatograms for extracts of these particulates are illustrated and compared in Figure 8. The first group of peaks eluted by heptane represents nonpolar fluorescent material including mostly multiring aromatics and PAH. The two major peaks comprising most of the PAH are labelled as PAH-1 and PAH-2. The area under PAH (1 + 2) should be proportional to the amount of gross PAH in the injected sample. The second group of peaks represents chloroform-elutable material. The two major peaks are labelled $X_1$ and $X_2$. Interestingly, $X_1$ was observed only in vehicle-exhaust-related samples and potentially can be used as a source marker for exhaust emissions. Peak Y represents acetonitrile-elutable materials which were mostly highly polar oxygenated compounds. For our present purpose, the latter two groups of peaks need not be discussed further.

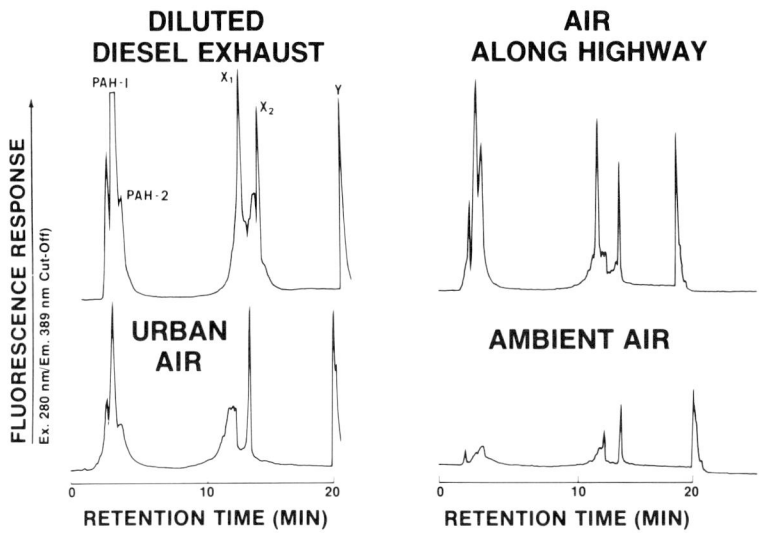

FIGURE 8. HPLC/fluorescence profiles comparing $CH_2Cl_2$ extracts of airborne particulates collected from different sources. Chromatographic conditions same as in Figure 1.

Results of these experiments are listed in Table 5. Two sets of runs (sets A and B) were conducted for each particulate type. Each set consists of two parallel runs sampled on different filter types. Each particulate sample was soxhlet extracted first with dichloromethane and then with acetonitrile. The extracts were each analyzed gravimetrically in order to evaluate the solvent extractable mass as a percentage of total particulate mass. Finally, each extract was analyzed by HPLC and the aggregate amount of PAH and other nonpolar fluorescent compounds was estimated from the integrated fluorescence response in the PAH region of the chromatogram.

It is apparent in Table 5 that, within a given set, there are large differences in percentage of extractables among the various filter types. For the dichloromethane extracts, there does not seem to be a persistent pattern to these differences. But acetonitrile extracted less material from the Zefluor filters than from the T60A20 or the glass-fiber filters. This observation implies that the chemical nature of the organic extractable material is susceptible to the influence of the filter medium itself.

The simplest inference is that chemical change is promoted somehow (perhaps catalytically) by the T60A20 and Gelman A filter surfaces. Such a filter artifact problem would be more serious for lightly loaded filters than for heavily loaded ones since the percentage of material undergoing reactions would be larger in the former case. Another possibility is

**TABLE 5.** Percentage of Extractable Organics Associated with Various Airborne Particulates Collected with Different Filter Media

| Source of Particulates[a] | Code of Parallel Runs | Filter Type | Extractable Mass (% of particulate mass) | | PAH and Other Non-polar Fluorescent Materials (normalized % $CH_2Cl_2$-extractable)[c,d] |
|---|---|---|---|---|---|
| | | | $CH_2Cl_2$ | $CH_3CN$ (after $CH_2Cl_2$)[b] | |
| Urban air (Dearborn, Michigan) | A | Zefluor | 9.5 | 5 | 0.154 |
| | | Glass fiber | 9.1 | 8 | 0.105 |
| | B | Zefluor | 10 | 5 | 0.33 |
| | | Glass fiber | 12 | 12 | 0.25 |
| Highway air (collected at Allegheny Tunnel, Pennsylvania) | A | Zefluor | 35 | 2 | 0.48 |
| | | T60A20 | 28 | 17 | 0.33 |
| | B | Zefluor | 26 | 0 | 0.28 |
| | | T60A20 | 26 | 9 | 0.14 |
| Rural ambient air (near Allegheny Tunnel, Pennsylvania) | A | Zefluor | 29±11 | 2±2 | 0.06 |
| | | T60A20 | 17±12 | 7±1 | 0.03 |
| | B | Zefluor | 35±2 | 9±5 | 0.10 |
| | | T60A20 | 5±1 | 10±3 | 0.06 |
| Diesel exhaust (collected in dilution tube) | A | Zefluor | 60 | 1.5 | 0.79 |
| | | T60A20 | 61 | 1.8 | 0.73 |
| | B | Zefluor | 66 | 2.7 | 0.62 |
| | | T60A20 | 60 | 3.0 | 0.69 |

[a]The particulate loadings on filter samples were ~ 30 to 100 mg for urban air; 20 to 100 mg for highway air; 3 to 20 mg for ambient air; and 100 to 200 mg for exhaust samples.

[b]The sample was soxhlet extracted with dichloromethane, followed by acetonitrile

[c]Estimated from relative fluorescence responses of sample and mixed PAH standard. Insofar as the PAH abundance distribution differs between sample and standard or between sample pairs, these numbers are only approximate.

pickup, by adsorption or reaction, of heavy vapor molecules by the T60A20 or Gelman A filter surfaces.

The first hypothesis is supported by the PAH/nonpolar fluorescence yields listed in the last column of Table 5. With the exception of diesel samples collected in the dilution tube, the yield of PAH/nonpolar fluorescence materials is always higher for the Zefluor filter than for the other filters of the same set, consistent with the notion of filter-catalyzed degradation. A vapor pickup mechanism, by contrast, would not explain the data since the problem to be explained is that the amount of PAH/nonpolar fluorescence material is higher on the Zefluor filter, the one that is supposed to have the poorest ability to pick up vapors, than on the T60A20 and Gelman A.

It is interesting to note that the diesel samples collected in the dilution tube do not seem to follow the trends displayed by the other three particulate types tested. The nonpolar fluorescence yields, as well as the dichloromethane and acetonitrile extractable figures, are almost equal between Zefluor and T60A20 runs. A plausible explanation for this seemingly discordant result is that the greater dilution-tube sample masses and the much shorter sampling exposure ($\leq 1$ hour versus $\geq 1/2$ day for the other particulate samples) would tend to mask the artifact formation process; in addition, the dilution-tube gas composition differs very much from the atmosphere. This makes the point that filter-type differences may or may not show up, depending on the regime in which they are evaluated; hence, any evaluation, or any extrapolation from one regime to another (e.g., dilution tube to atmosphere), must be conducted with this in mind.

## ACKNOWLEDGMENTS

The authors wish to thank D. Schuetzle for his valuable discussions and suggestions. Thanks are also due to F. C. Ferris, T. J. Prater, M. C. Paputa, and other colleagues in our laboratory who have participated in the collaborative studies.

## REFERENCES

1. Barton, S. C., Johnson, N. D., and Das, B. S. (1979): A feasibility study of alternatives to high-volume sampling for labile constituents of atmospheric particulates. Report for Air Resources Branch, Ministry of Environment, Canada, by Ontario Research Foundation, Ontario, Canada.

2. Benson, A. L., Coletta, G. C., and Levins, P. L. (1977): Evaluation of filter media for quantitative collection of particulate matter from engine exhaust. Report to U.S. E.P.A. by Arthur D. Little, Inc., Report No. ADL 75994.
3. Daisey, J. M. (1979): Organic compounds in urban aerosols. Paper presented at the Conference on Aerosols: Anthropogenic and Natural - Sources and Transport, January.
4. Fox, M. A., and Olive, S. (1979): *Science* 205:582-583.
5. Huisingh. J., Bradow, R., et al (1978): Application of bioassay to the characterization of diesel particle emissions: Part II. Application of a mutagenicity bioassay to monitoring light duty diesel particulate emissions. Paper presented at Symposium on Application of Short-Term Bioassays in the Fractionation and Analysis of Complex Environmental Mixtures, EPA, Williamsburg, Virginia.
6. Jäger, J. (1978): *J. Chromatography* 152:575-578.
7. Jäger, J., and Rakovic, M. (1974): *J. Hyg. Epidemiol. Microbiol. Immunol.* 18:137.
8. John, W., and Reischl, G. (1978): *Atmospheric Environment* 12: 2015-2019.
9. Lane, D. A., and Katz, M. (1977): The photomodification of BaP, BbF and BkF under simulated atmospheric conditions. In: *Fate of Pollutants in Air and Water Environments, Part 2, Chemical and Biological Fate of Pollutants in the Environment,* Suffet, I. H., Eds., John Wiley and Sons, New York.
10. Lee, F. S.-C., Prater, T. J., and Ferris, F. (1979): PAH emissions from a stratified-charge vehicle with and without oxidation catalyst: sampling and analysis evaluation. In: *Polynuclear Aromatic Hydrocarbons,* Vol. 3, Jones, P. W. and Leber, P., Eds., pp. 83-110, Ann Arbor Science Publishers, Inc., Ann Arbor, Michigan.
11. Mckee, D. E., Ferris, F. C., and Goeboro, R. E. (1978): Unregulated emissions from a PROCO engine powered vehicle. Paper No. 780592, Society of Automotive Engineers Meeting, Detroit.
12. National Academic Science Report (1972): Biological effects of atmospheric pollutants: Particulate polycyclic organic matter. U. S. Government Printing Office, Washington, D. C.
13. Pierson, W. R., et al (1978): *J. Air Poll. Control Assoc.* 28:123-132.
14. Pierson, W. R., Hammerle, R. H., and Brachaczek, W. W. (1976): *Anal. Chem.* 48:1808-1811.
15. Pitts, J. N., Jr., Van Cauwenberghe, K. A., Grosjean, D., Schmid, J. P., Fitz, D. R., Belser, W. L., Jr., Knudson, G. B., and Hynds, P. M. (1978): *Science* 202:515-519.
16. Robertson, D. J., Groth, R. H., Garder, D. G., and Glastris, E. G. (1979): *J. Air Poll. Control Assoc.* 29:143-146.
17. Tebbens, B. D., Mukai, M., and Thomas, J. F. (1971): *J. Am. Indus. Hyg. Assoc.* 365-372 June.

# PROXY METHODS AND COMPOUNDS FOR WORKPLACE MONITORING OF POLYNUCLEAR AROMATIC HYDROCARBONS

R. B. Gammage* and A. Bjørseth**,
 *Health and Safety Research Division
 Oak Ridge National Laboratory
 Oak Ridge, Tennessee 37830
 **Battelle's Columbus Laboratories
 Central Institute for
 Industrial Research
 Blindern, Oslo 3 NORWAY

What is a proxy compound or substance when one is referring to polynuclear aromatic hydrocarbon (PAH) compounds? It is a measurable parameter or compound that is in some manner representative of the very large and complex group of PAH compounds that compose nearly all real-life samples. The representation may be either direct or indirect, as will be shown later. Alternative words for proxy that often appear in the literature include indicator, surrogate, or signature.

Since one is often interested in biological impact, the proxy should ideally be the limiting toxicant in terms of dose-effect. In workplace monitoring this is usually not too practical, even if one were fortunate enough to know the identity of the limiting toxicant. The choice of proxy is usually governed by factors of expediency, or habit in the case of benzo(a)pyrene (BaP). Also the choice is driven in large part by economic pressures to find intermediate, appropriate, and low-cost technological solutions to the determination of PAH exposures.

With health risk in mind, there is justifiable reason for the widespread use of BaP as a direct proxy. It is the classically studied PAH and its metabolism to ultimate carcinogens, such as 7,8-dihydrodiol and 7,8-diol-9,10-epoxides, is comparatively well understood (1). The occurrence of BaP is ubiquitously linked to complex PAH emissions. Some individuals would go so far as to say that "it is frequently necessary to know only the BaP content in environmental monitoring. This is because BaP is a representative carcinogen in the environment, and BaP content is highly correlated with the levels of other PAH compounds (2)".

One of the purposes of this paper is to show that BaP is not a universally acceptable proxy for PAH compounds. What might be adequate for representing PAH compounds in the outdoors air can be quite inadequate for monitoring the quality of workplace environments. The BaP content might be too variable, or its concentration too low, compared with that of other PAH compounds, in order to warrant its selection as an indicator. One also needs to recognize that BaP analyses are time consuming as well as expensive. The logic is developed that as complete as possible an unbiased analysis of the working environment is necessary prior to the selection of monitoring techniques and choice of the proxy compounds or indicator substances.

If the industrial hygienist needs rapid warning of dangerous levels of PAH-containing fumes of PAH-contaminated surfaces, then more indirect and quickly measurable proxies have to be sought. These might include a coexistent and easy to measure product such as carbon monoxide, the total concentration of airborne particulates, or filtered gross fluorescence as a measure of surface contamination. Again, if one were concerned with vapors of PAH compounds at ambient temperatures, the extremely low equilibrium vapor concentration of BaP, $10^{-7} g/m^3$ (3), makes it unsuitable as an indicator. The most appropriate proxy would be a volatile, low-boiling compounds having a high equilibrium vapor equilibrium vapor concentration.

These problems and decisions are very important because of the wide and growing variety of working environments that require occupational health control of PAH compounds. Workers in synfuel technologies (gasification and liquefaction involving coal, shale oil or tar sands), roofers, roadworkers, workers in the reduction of aluminum and ferroalloy industries, foundry workers, and coke-oven workers are all at risk from PAH exposure. To minimize hazards to employees in these old and newly emerging technologies, better approaches toward characterization and monitoring of PAH are needed. As for proxy PAH compounds, preliminary thoughts were expressed at a recent symposium (4).

## INDIRECT, NEAR-REAL-TIME PROXY METHODS FOR PAH

The industrial hygienist sometimes needs to evaluate a hazardous situation quickly in terms of the potential human exposure to PAH. PAH measurement needs to be made in real time. Since no established techniques exist for monitoring selected PAH compounds in real time, one is forced to look for indirect indicators that can be measured very quickly.

## Airborne Particulates

Carbon monoxide is proposed as the leading candidate for a proxy compound in coal gasification plants (5). It would be used as an indicator gas for monitoring in designated confined areas, so that it could be used as a basis for an alarm mechanism in selected areas, and as a potential index of workers' exposure to an array of toxic chemicals. Unfortunately, knowledge based on data of gas stream composition (6) would seem to preclude its use as an indicator of PAH-bearing heavy oils. Background levels of CO from nonplant sources, and possible fluctuations in these levels, are high enough to preclude indicator monitoring of heavy oils at their permissable exposure limit [0.2 mg/$m^3$ threshold limit value—time weighted average concentration (7)].

A gas, such as CO, is intrinsically unsuitable as a proxy for particulate matter. The gas, for example, disperses more readily than the particles of oil or tar mists can be produced in the absence of CO. A more valid but still indirect proxy method is to measure the total concentration of airborne particulate matter and attempt to correlate this parameter with one that relates to the PAH content, such as the benzene-soluble fraction. With this in mind, NIOSH conducted a study, with the American Iron and Steel Institute, on the respirable particulate fractions of coke-oven emissions (8).

Three hundred samples of aerosols were collected. Each was analyzed for particulate loading, benzene solubles, and BaP. Correlations between each possible pairing were analyzed statistically. The highest correlation coefficient of the logarithms was 0.85 between particulate and benzene-soluble concentrations. No allowance was made for differences among the five plants visited in the study, the different sampling locations within a plant, or the day to day variations in the emissions; each factor was believed to have a significant effect on the data. Even better correlation could be anticipated if the data are standardized for particular locations within a given plant. Several types of rapid reading and portable aerosol sensors are available commercially. They are designed to give mass concentration loadings and use the principle of either beta attentuation, light scattering, or the piezoelectric balance. Rapid measurement of particulate loading might give a rapid indication of the PAH-bearing and OSHA-controlled benzene solubles. This potentially promising approach to indicator monitoring is as yet untested, but it is worth factoring into health and environmental testing and evaluation programs at synfuel facilities.

## Skin Contaminants

Skin contamination in situ cannot be measured and quantified directly in terms of the concentration of carcinogenic constituents. One has to resort to a rather indirect method of measurement, i.e., total fluorescence induced by near-ultraviolet light. The gross fluorescence becomes the indicator medium and it is observed either with the naked eye or with a photomultiplier tube. The former technique is qualitative and tells one only whether or not fluorescent contamination is present on the skin. The information is restricted to a yes or no answer. Detection of the filtered fluorescence with a photomultiplier tube improves the sensitivity and adds the potential element of quantification. It remains to be seen, however, whether this crude fluroescent indicator is relatable to biologically significant exposures.

Two types of survey instrument are in the developmental stage at Oak Ridge National Laboratory. Each is intended for the improved detection and quantification of fluorescing surface contaminants. A spill spotter (9) irradiates an object with modulated near-ultraviolet light transmitted through a telephoto lens. It is designed for examining all types of surfaces, independent of the lighting conditions. A luminoscope, employing fiber optics lightguides, and a stethescopic cap pressed against the skin is more specifically designed to monitor small areas of skin contaminated by oils and tars (10). Both instruments will measure the intensity of filtered fluorescent light and use the measured intensity as an indicator of the level of surface contamination. Preliminary results for a prototype luminoscope indicate that raw-coal distillates or recycle solvents can be measured in amounts as small as 1 nl/cm$^2$ (10).

## Rapid Analytical Screening

The industrial hygienist collects many specimens of tar or oil in the form of collected particulate matter or wipe samples. These must be evaluated expeditiously for the presence of PAH. Remedial action, such as decontamination after accidental exposure, may be required based on where, and in what amounts, PAH are found. Quick evaluation also aids in decisions for further, more specific analyses. Once again, there is a general need to resort to the proxy concept.

A naphthalene-sensitized, gross fluorescence spot test is available (11) for estimating general levels of PAH in organic solvents. Results can be obtained quickly enough (within a few minutes) for the analysis to be placed in the category of near-real-time. With the fluorescence being viewed by the naked eye alone, only order-of-magnitude estimates of PAH content are possible. If quenching effects are absent, the lower limit of detection of a 1 $\mu$l spot on filter paper is in the range of 1 to 10 pg PAH. No doubt, improvements with regard to reproducibility and sensitivity could be obtained if a photomultiplier tube were incorporated. Because no more specific information is forthcoming, other than the overall PAH content, this is essentially an indirect proxy technique.

A variant of this approach uses the addition of chromogenic or fluorogenic reagents to induce fluorescent color changes in aromatic amines that are cancer-suspect agents (12). The chemical spot tests were developed as a "swipe" technique for detecting a variety of primary aromatic amines and related compounds on painted, metal, and concrete surfaces. Detection limits were usually less than 200 ng/cm$^2$ of the surface being tested. This test attracts additional attention because of the recent finding that the primary aminoarenes can carry a large fraction of the overall mutagenicity of the total PAH in synfuel products (13).

## DIRECT PROXY METHODS

### Real-Time Techniques

Volatile PAH are of concern because there is the possibility of both chronic and acute exposures to their vapors. The vapor composition in headspace above an oil spill, for example, should be dominated by the most volatile and abundant constituents. A proxy compound for the spectrum of volatile PAH should, therefore, be such a constituent. Naphthalene, or one of its methyl derivatives, is an attractive candidate for a proxy. These compounds are quite volatile and they are ubiquitous and abundant in synthetically derived oil and tar. Naphthalene can comprise as much as 10 weight-percent of coal tar products (14).

A second derivative, wavelength-modulated, ultraviolet-absorption spectrometer in portable form (with the added power of a microcomputer to correct for interfering compounds) has been developed at Oak Ridge National Laboratory (15). Its purpose is demonstrated in Figure 1; the presence of naphthalene and 2-methylnaphthalene is shown in headspace above a solvent-refined coal (SRC) light oil at ambient temperature. The detection sensitivity is high (Figure 2), with the lower limit for naphthalene being in the few ppb range (16). A prototype instrument will be evaluated for use as an area monitor and leak detector at the low-Btu coal gasifier located at the University of Minnesota at Duluth (UMD) during the coming year.

For rapid screening of liquid or solid samples, a room-temperature phosphorescence (RTP) technique (17) shows promise as a chemical spot test, giving compound-specific information. A liquid sample, such as a swipe sample dissolved in an organic solvent, is spotted on filter paper. Phosphorescence is selectively induced by treatment with a heavy atom perturber such as lead acetate. It is possible to identify and roughly quantify a major PAH compound without any fractionation. A 15 µg, raw

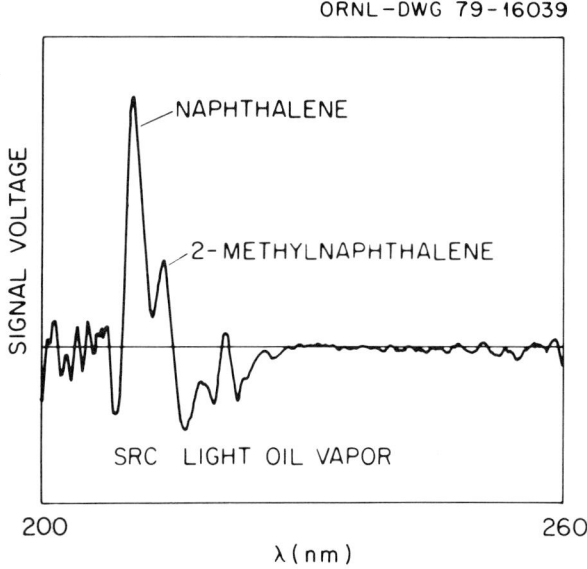

**FIGURE 1.** Second-derivative UV-absorption spectrum of aromatic head-space vapors above a solvent refined coal light oil.

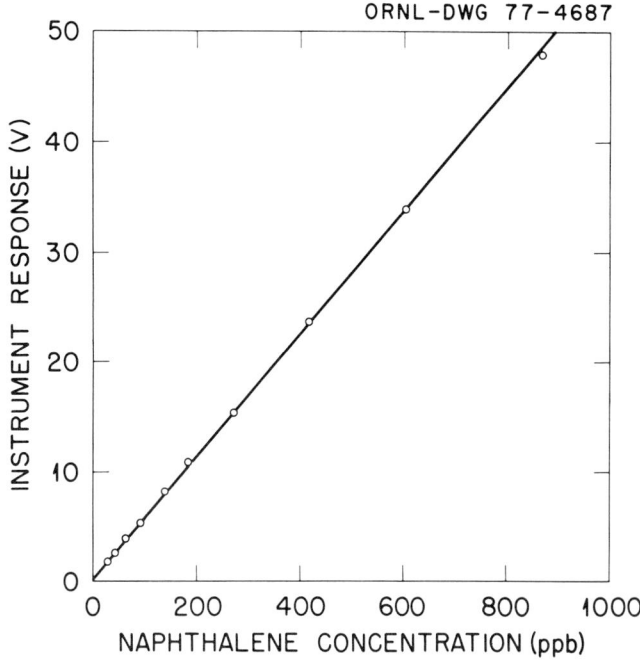

**FIGURE 2.** Analytical curve for naphthalene vapor.

## PROXY METHODS FOR WORKPLACE MONITORING 571

Synthoil coal liquefaction tar, for example, was dissolved in methylene chloride before spotting on filter paper. The RTP spectrum shown in Figure 3 is specific to pyrene in the amount of 5500 ± 1600 ppm. Independent chromatographic analysis of the Synthoil gave a pyrene concentration of 4,300 ppm. This ability to measure (within a few minutes) a major component in an unfractionated real-life sample opens the door to simple and rapid PAH indicator monitoring.

**Slower But More Precise Techniques**

More sophisticated analytical techniques are mentioned briefly in this paper. The reason is that several techniques are gaining widespread use for rapidly fractionating and analyzing either single or small numbers of PAH compounds that can indicate carcinogenic potential. Within 1 to 4 hours, complex mixtures of PAH can be resolved into individual components and quantitatively determined with high sensitivity.

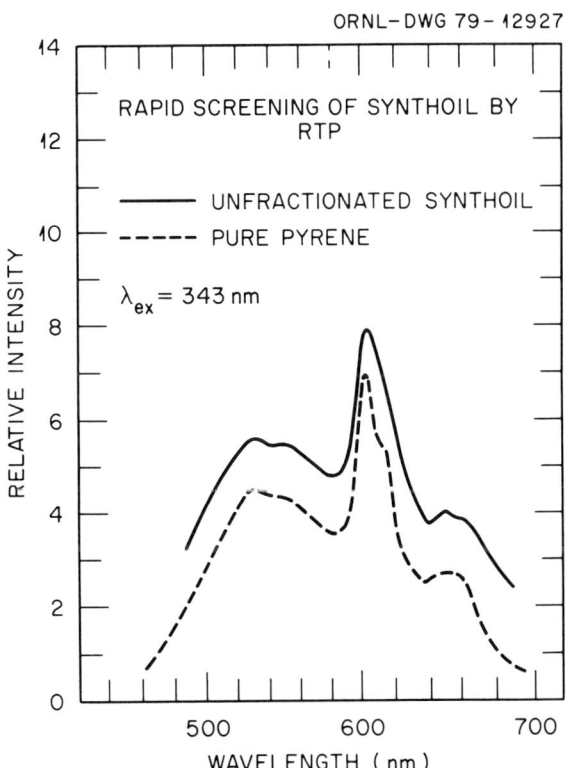

**FIGURE 3.** Identification of pyrene in unfractionation Synthoil by selective room-temperature phosphorescence.

The highest separation efficiency is offered by a glass capillary gas chromatograph ($GC^2$) with flame ionization and/or electron capture detectors (18,19). When interfaced with a mass spectrometer (MS), very high sensitivity and selectivity are obtained. The instrument can separate pairs of isomers, such as phenanthrene/anthracene, benz(a)anthracene/chrysene, and benzo(a)pyrene/benzo(e)pyrene. Usually, a computerized $GC^2$-MS allows identification of different PAH compounds at a level below 1 ng. $GC^2$-MS has recently been applied to characterize PAH in work environments (20).

The sensitivity, rapidity, and accuracy of high-performance liquid chromatography (HPLC) has been demonstrated for analysis of BaP and other PAH in workplace air (20,21). The final step of detection is measurement of the ultraviolet absorbance ratio at two different wavelengths (21). First, of course, an analysis has to be made of PAH standards to determine the characteristic elution volume and absorbance ratio for each PAH peak. Aliquots of benzene-soluble, air-filter extract have been directly analyzed without any additional separation. Concentrations of PAH, such as fluoranthene, pyrene, and BaP, can be calculated by reference to simple calibration curves. The chromatographic procedure is complete within 30 minutes (21).

In the category of major PAH component analysis, a one-dimensional dual-band, thin-layer chromatography (TLC) technique is valuable (2). Analysis on the order of ten PAH compounds in real-life samples is possible at concentrations of each PAH of 1 ng/ml. The same technique is simpler yet for the analysis of BaP alone. Speed is such that 20 to 30 samples can be analyzed per day.

In the same vein, a rapid TLC separation and in situ fluorescent analyzer is being integrated with an automatic isolation procedure (23).

## PROXY PAH COMPOUNDS

There are several specific PAH compounds that can act as proxies for a spectrum of PAH compounds. However, distinction needs to be made between PAH found in occupational environments and those occurring in the air of urban and rural environments.

In a given urban, out-of-doors environment, mixing of PAH from various local sources apparently takes place to a thorough enough extent that there is a high degree of correlation among the PAH species. Any one compound is then a fairly good index of the concentrations of the others (24). BaP is usually chosen to be the representative carcinogen. Other PAH compounds that are fairly stable against photochemical decomposition, such as phenanthrene, chrysene, or triphenylene, but not more reactive linear compounds such as benz(b)anthracene and dibenz(a,h) anthracene, would also be adequate proxy compounds. These major, stable PAH compounds in aerosols from industrial sources are also known to survive transport over long journeys with, at times, surprisingly little dilution (25).

In the workplace, a greater care should be exercised in the selection of proxy PAH. The thorough mixing of PAH from various emission sources is not as likely in the workplace as it is in urban or rural atmospheres. Certainly, it becomes more questionable to concentrate on a single proxy without a thorough knowledge of the PAH profiles and their variation at different plant locations. In, for example, a coal gasification or liquefaction plant, products varying considerably in PAH concentration and PAH profile are encountered. Fugitive emissions of PAH may likewise vary considerably at different points within a plant, and also depend upon the process conditions and coal feedstock.

These types of difficulties manifested themselves in the NIOSH-American Iron and Steel Institute coke-oven study (8) mentioned earlier. The correlation coefficients of the logarithms of the BaP concentration versus the particulate concentration and the benzene solubles concentration were 0.7 and 0.6, respectively. This poor degree of correlation strongly suggests that the compositions of the benzene solubles and particulates in gross samples may vary considerably. There remains a need for collection and characterization of fugitive emissions that takes into account plant location, process conditions, and feedstock.

The concept proposed by Bjørseth (26) is recommended as a first step for resolving these problems; parent PAH profiles (PPP) of emissions are determined. From the nature and constancy of these profiles, a proxy compound or compounds can be chosen which will be indicative of the concentration of the carcinogenic constituents.

The common practice for measuring BaP as a single proxy without foreknowledge of the PPP should be discouraged. The BaP may be present only as a trace or minor constituent and, therefore, be difficult to analyze. The PPP shown in Figure 4 serve to illustrate this point. In the aluminum plant emissions, benzopyrenes are abundant, while in the Søderberg paste plant they are nearly 2 orders of magnitude lower in concentration compared to other major PAH compounds. Only in the former case is BaP a suitable indicator compound.

Other types of samples containing BaP at concentrations 2 orders of magnitude less than the most abundant PAH are creosote (27) and Synthoil (28).

It is reasonable to select a proxy compound having carcinogenic activity, but it need not exhibit the carcinogenic activity directly; cocarcinogenic compounds such as pyrene and fluoranthene are active in enhancing the tumorogenic effects of BaP several fold (29). Since these compounds are almost universally abundant in real-life PAH mixtures,

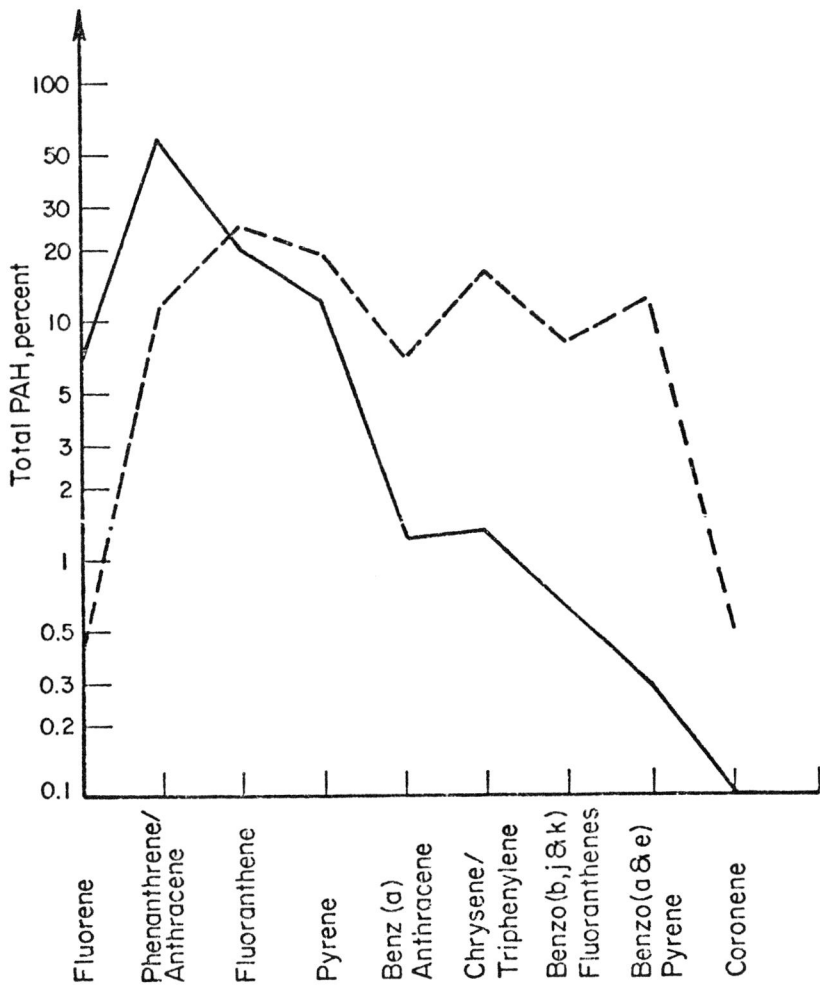

FIGURE 4. Parent PAH profile of PAH compounds in particulate matter from an aluminum plant (---) and a Soderberg paste plant(—).

they should be viewed as promising PAH proxy compounds in situations where the selection of BaP would be unwise.

Detailed studies on PAH in fugitive workplace emissions are needed to establish baseline data. With this knowledge, proxy compound selection can be made more objectively than at present. Monitoring by proxy should then give results more reflective of the carcinogenic potential of the exposure.

## REFERENCES

1. Yang, S. K., Roller, P. P., and Gelboin, H. V. (1978): Benzo(a)pyrene metabolism: Mechanism in the formation of epoxides, phenols, dihydrodiols, and the 7,8-diol-9,10-epoxides. In: *Carcinogenesis, Vol. 3: Polynuclear Aromatic Hydrocarbons*, P. W. Jones and R. I. Freudenthal, Eds., pp. 285-301, Raven Press, New York.
2. Matsushita, H. (1978): Analytical methods for monitoring polycyclic aromatic hydrocarbons in the environment. In: *Polycyclic Hydrocarbons and Cancer*, H. V. Gelboin and P. O. Ts'o, Eds., Volume 1, pp. 71-81, Academic Press, New York.
3. Pupp, C., Lao, R. C., Murray, J. J., and Pottie, R. F. (1974): Equilibrium vapor concentrations of some polycyclic aromatic hydrocarbons, $As_4O_6$ and $SeO_2$ and the collection efficiencies of these air pollutants. *Atmos. Environ.* 8:915-925
4. Gammage, R. B. (1979): Preliminary thoughts on proxy PNA compounds in the vapor and solid phase. In: *Assessing the Industrial Hygiene Monitoring Needs for the Coal Conversion and Oil Shale Industries*, O. White, Jr., Ed., pp. 173-188, Report No. BNL 51002.
5. Criteria for a recommended standard: Occupational exposures in coal gasification plants (1978). DHEW (NIOSH) publication No. 78-191.
6. Attari, A. (1973): Fate of trace constituents of coal during gasification. Publication NTIS PB-223-001, Institute of Gas Technology, Chicago, Illinois.
7. Threshold limit values for chemical substances and physical agents in the workroom environment with intended changes for 1978: American conference of governmental industrial hygienists.
8. Burg, J. R., and Teass, A. W. (1975): A statistical analysis of 300 personal samples of respirable particulate coke-oven emissions. Memorandum to the Deputy Director, NIOSH, dated December 18, 1975.
9. Schuresko, D. D., and Jones, G., Jr. (1979): A portable fluorometric monitor to detect PNA contamination of work area surfaces. In: *Assessing the Industrial Hygiene Monitoring Needs for the Coal Conversion and Oil Shale Industries*, O. White, Jr., Ed., pp. 214-222, Report No. BNL 51002.
10. Vo-Dinh, T., and Gammage, R. B. (1980), to be presented at the American Industrial Hygiene Conference, Houston, TX, May 18-23, 1980.
11. Smith, E. M., and Levins, P. L. (1978): Sensitized fluorescence for the detection of polycyclic aromatic hydrocarbons. Publication No. EPA-600/7-78-182.

12. Weeks, R. W., Jr., Dean, B. J., and Yasuda, S. K. (1976): Detection limits of chemical spot tests toward certain carcinogens on metals painted, and concrete surfaces. *Anal. Chem.* 48:2227-2233.
13. Ho, C.-h., Clark, B. R., Guerin, M. R., Ma, C. Y., Rao, T. K. (1979): Aromatic nitrogen compounds in fossil fuels - A potential hazard. In: *Preprints of papers, Division of Fuel Chemistry, American Chemical Society,* Vol. 24, No. 1, pp. 281-291.
14. Major components of German high-temperature conversion process coal tar (1977). In : *Environmental Health and Control Aspects of Coal Conversion: An Information Overview,* Volume 1, pp. 4-102, Publication No. ORNL/EIS-94.
15. Hawthorne, A. R. (1979): DUVAS: A field portable second-derivative UV-absorption spectrometer for monitoring PNA vapors. In:*Assessing the Industrial Hygiene Monitoring Needs in the Coal Conversion and Oil Shale Industries,* O. White, Jr., Ed., pp. 191-202, Report No. BNL 51002.
16. Hawthorne, A. R., and Thorngate, J. H. (1979): Application of second-derivative UV-absorption spectroscopy to polynuclear aromatic compound analysis. *Appl. Spectrosc.* 33:301-305.
17. Vo-Dinh, T., Gammage, R. B., and Martinez, P. R. (1980): Identification and quantification of polynuclear aromatic compounds in Synthoil by room-temperature phosphorimetry. *Anal. Chim. Acta* (in press).
18. Bjørseth, A. (1978): Analysis of polycyclic aromatic hydrocarbons in environmental samples by glass capillary gas chromatography. In: *Carcinogenesis, Vol. 3: Polynuclear Aromatic Hydrocarbons,* P. W. Jones and R. I. Freudenthal, Eds., pp. 75-83, Raven Press, New York.
19. Bjørseth, A., and Eklund, G. (1979): Analysis of polynuclear aromatic hydrocarbons by glass capillary gas chromatography using simultaneous flame ionization and electron capture detection. *J. High Resol. Chrom. & CC.* 2:22-26.
20. Bjørseth, A., and Eklund, G. (1979): Analysis of polynuclear aromatic hydrocarbons in working atmospheres by computerized gas chromatography-mass spectrometry. *Anal. Chim. Acta* 105:119-128.
21. Lynch, M., and Weiner, E. (1979): HPLC: high-performance liquid chromatography. *Env. Sci. & Techn.* 13:666-671.
22. Boden, H. (1976): The Determination of benzo(a)pyrene in coal tar pitch volatiles using HPLC with selective UV detection. *J. Chrom. Sci.* 14:391-395.
23. Swanson, D., Morris, C., Hedgecoke, R., Jungers, R., Thompson, R., and Bumgarner, J. E. (1978): A rapid analytical procedure for the analysis of benzo(a)pyrene in environmental samples. *Trends in Fluorescence* 1:22-27.

24. Baum, E. J. (1978): Occurrence and surveillance of polycyclic aromatic hydrocarbons. In: *Polycyclic Hydrocarbons and Cancer,* Volume 1, H. V. Gelboin, and P.O.P. Ts'o, Eds., Academic Press, New York.
25. Lunde, G., and Bjørseth, A. (1977): Analysis of polycyclic aromatic hydrocarbons in long-range transported aerosols. *Nature* 268:518-519.
26. Bjørseth, A. (1979): Determination of polynuclear aromatic hydrocarbons in the working environment. In: *Polynuclear Aromatic Hydrocarbons,* P. W. Jones and P. Leber, Eds., pp. 371-381, Ann Arbor Science Publishers, Inc., Ann Arbor, Michigan.
27. Lijinsky, W., Domsky. I., Mason, G., Ramahi, H. Y., and Safari, T. (1963): The chromatographic determination of trace amounts of polynuclear hydrocarbons in petrolatum, mineral oil, and coal tar. *Anal. Chem.* 35:952-956.
28. Analytical chemistry division annual progress report (1977), p. 56, ORNL-5360; and Coal technology annual interim report for fiscal year ending June 20 (1976), p. 96, ORNL-5208.
29. Hoffmann, D., Schmeltz, I., Hecht, S. S., and Wynder, E. L. (1978): Tobacco carcinogenesis. In: *Polycyclic Hydrocarbons and Cancer,* H. V. Gelboin, and P.O.P. Ts'o, Eds., Volume I, pp. 85-117, Academic Press, New York.

# MUTAGENIC MATERIAL IN AIR PARTICLES IN A STEEL FOUNDRY

C. Kaiser\*, A. Kerr\*\*, D. R. McCalla\*, J. N. Lockington\*\*, and E. S. Gibson\*\*.

\*Department of Biochemistry
McMaster University
Hamilton, Ontario, Canada

\*\*DOFASCO Ltd.
Hamilton, Ontario, Canada

## INTRODUCTION

A recent study published by Gibson et al (4) indicates that lung cancer mortality amongst foundry workers at DOFASCO, Ltd. was significantly higher than that found elsewhere in the plant or in a control population. The risk increased with length of employment and appeared to be concentrated in certain occupational groups. Possible sources of carcinogens include metal fumes, pyrolysis products formed from the organic materials used as mould and core binders, dusts, or some combination of these. Exposures to all airborne materials were well below the ACGIH threshold limit values (1978). Preliminary data indicate that there was no direct correlation between the apparent lung cancer risk in various parts of the foundry and the levels of total benzene-soluble material, "marker" polycyclic aromatic hydrocarbons, total or respirable particulate matter, various metal fumes or silica.

Possible explanations for the excess lung cancer in the foundry are (a) the presence of unidentified carcinogens in the air in high risk areas (b) synergism between various contaminants, or (c) different exposures in the 1940's when employees in the study started work. The objective of this study is to examine the possibility that unidentified carcinogens are present by using as a primary tool the bacterial mutagenicity assay developed by B. N. Ames and his colleagues (1). This assay measures the ability of the test material to induce mutations in specially devised strains of *Salmonella typhimurium*. Since many mutagens and carcinogens require metabolic "activation" by mammalian enzymes, rat liver extract, referred to as the 9000 x g supernatant fraction (S-9), is added to the test plates. The Salmonella/S-9 assay has been subjected to the most thorough validation studies of any of the short-term assays for potential carcinogens, and has proved to have good predictive value for many classes of organic compounds (5, 6, 10) including aromatic hydrocarbons and amines, but is less effective with chlorinated hydrocarbons and some simple nitrosamines (12). The principal advantages of microbial assays are their speed and simplicity which permit the processing of the large number of samples required for detection and characterization of potentially carcinogenic materials in the environment. One must, however, be aware of the possibility of sampling artifacts (9) and of possible interactions between individual components of complex mixtures (7, 11, 13).

There are many data on the levels and identities of polycyclic aromatic hydrocarbons in various environments; however, use of mutagenicity assays permits examination of organic mixtures for potentially hazardous material without the necessity of making assumptions about the chemical nature of the compounds involved. Such assays have given positive results with particulate material collected from urban air (e.g., 2, 3, 8, 14, 15, 16) and exhaust from internal combustion engines (17). Work done to date has been concentrated on the particulate fraction of foundry air which is known to include condensed polycyclic aromatic hydrocarbons. At this stage, results of preliminary work are presented, describing some studies on the use of various sampling devices and on optimization of the assays, followed by data which show that a diverse range of mutagenic compounds is present in foundry air particulates.

## MATERIALS AND METHODS

Tritiated benzo(a)pyrene ($^3$H-BaP) was obtained from New England Nuclear. NADP and glucose-6-phosphate were purchased from Sigma Chemical Company. Other chemicals and solvents used were of reagent grade.

### Sampling

Three types of sampling devices were used. In all of these, particulate material was collected on tared type A/E glass fiber filters (Gelman) which are 99.9 percent efficient using the 0.3 μm DOP test. High-volume sampling (1.5 m$^3$/min) was carried out using 8 x 10-inch filters in a conventional apparatus (Soiltest Canada Ltd. or General Steelwares). The face velocity was about 30 m/min. Medium-volume samples (28 1/min or 0.028 m$^3$/min) were collected on a 115-mm-diameter filter in a glass cassette at a face velocity of 3.0 m/min, using a carbon-vane vacuum pump. Low-volume samples (2 1/m) were collected on 37-mm-diameter filters in polystyrene cassettes, using Mine Safety Appliance personal sampling pumps. Again, the face velocity was about 3.0 m/min.

## Extraction

Filters were weighed and placed in cellulose thimbles in a glass soxhlet apparatus (protected from light) and extracted with reagent-grade methanol for 16 hours. The methanol was carefully removed in a rotary evaporator at 30 C and the residue dissolved in 0.5 or 1 ml dimethysulfoxide (DMSO) for mutagen assay, or in cyclohexane for liquid:liquid extraction. In the sections that follow, some of the results are presented in the form of the number or revertants induced by the extract from a given weight of particulate material.

## Mutagenicity Assays

The assays were carried out as described by Ames et al (1). The 2-nitronaphthalene (Aldrich Chemical Co.) and 2-acetylaminofluorene, kindly supplied by Dr. J. A. Miller, University of Wisconsin, were included as positive controls in every set of assays as direct and indirect acting mutagens, respectively, and benzo(a)pyrene (Sigma) was included periodically. The bacteria used were tested regularly for resistance to ampicillin and sensitivity to crystal violet to make sure that they retained the pKM101 plasmid and the deep rough mutation (1). Solvent blanks and extracts of blank filters were run periodically and were consistently found to be nonmutagenic. The S-9 preparations used were made, as described by Ames et al (1), from male rats which had been injected 5 days previously with 500 mg/kg of Aroclor 1254 (Monsanto) in corn oil.

Bacterial colonies were routinely counted using a New Brunswick Scientific "Biotran II" Automated Colony Counter, the performance of which was checked regularly with manual counts.

## Liquid:Liquid Extraction

To the flask containing the residue from the methanol extract, 100 ml cyclohexane was added which was then transferred to a glass separatory funnel. The flask was rinsed thoroughly with 100 ml water which was also transferred to the separatory funnel and used to extract the "aqueous fraction", which was then taken to dryness in a rotary evaporator at 50 C. The acidic and basic compounds were then sequentially extracted from the cyclohexane solution (designated the "original" cyclohexane solution) as follows. First, the solution was extracted with 100 ml of 5 percent aqueous sodium carbonate and then washed with 25 ml of water which was added to the sodium carbonate solution. The latter was neutralized and adjusted to pH 1 with dilute HCl and back-extracted with 100 ml cyclohexane, which was then evaporated to give the "acidic fraction". Second, the "original" cyclohexane solution was extracted with 5 percent aqueous acetic acid to remove basic material, after which it was washed with 25 ml water which as added to the dilute acetic acid layer. The pH of the latter was adjusted to pH 11 with 20 percent NaOH and back-extracted with cyclohexane, which was then evaporated to give the "basic fraction". Finally, the cyclohexane layer remaining from the last aqueous wash was evaporated to give the "neutral fraction". When the performance of this extraction procedure was monitored using radioactive BaP, over 90 percent of the radioactivity was recovered in the neutral fraction, as expected.

## RESULTS AND DISCUSSION

### Sampling

Although most of the data on the mutagenicity of air particulates currently available have been obtained using samples collected with high-volume samplers, the work of White (18) indicated that these devices, which operate at face velocities of about 30 m/min, may give low recoveries of important polycyclic hydrocarbons. The reality of potential losses is illustrated by an experiment we performed in which 100 $\mu$g of $^3$H-BaP was applied as a solution in ethanol to a 25-cm$^2$ area of an 8 x 10-inch Type A/E glass fiber filter. After the ethanol had evaporated completely, the filter was placed in a special holder downstream, separated from another filter, and subjected to an air flow of 30 m/min for 24 hours in a high-volume sampler located in a nonindustrial area. After this time, only 54 percent of the original radioactive material was recovered from the filter. All the applied radioactivity was recovered from control filters which were not exposed to the air flow. However, when the time during which 100 µg of $^3$H-BaP was exposed to the air stream was decreased to 3 hours, the loss of radioactivity was too small to be detected.

Chromatography of the material recovered after 24 hours on 20 percent acetylated cellulose thin-layer plates showed that most of the radioactivity was found in B(a)P, with a small amount of material having a lower Rf value. There was no evidence for the presence of 6-nitrobenzo(a)pyrene (9). The ambient temperatures during these experiments were from about 20 to 27 C. Thus, while BaP placed directly on the filter may behave differently from BaP in particulate matter, these experiments suggest that long sampling times may lead to poor recovery of BaP, but short sampling times may not. Fortunately short sampling times suffice to collect enough foundry particulates for mutagenicity assays. To test directly the adequacy of high-velocity samplers operated for short periods for this study, we compared the recovery to total particulate material and mutagenic activity using high volume collectors with the results obtained simultaneously using two other devices which operate at the lower velocity of about 3.0 m/min (see Materials and Methods section). The amounts of foundry particulate material per m$^3$ of air collected in the three devices were similar (2 to 6 mg/m$^3$), with a tendency to higher yields with the high-velocity samplers. Total mutagenic activity recovered per mg particulate or per m$^3$ of air was also similar using the three devices. This suggests that there are not excessive losses of total mutagenic material with the high volume high velocity samplers and is also consistent with what would be expected for BaP on the basis of the experiments with $^3$H-BaP described above. Obviously in all of these experiments we are looking at materials having relatively low vapor pressures since the evaporation step involved later in the procedure will result in the loss of any highly volatile materials.

Silver-backed 37-mm filters proved to be unsuitable for collection of material for mutagenicity tests since the residue from methanol extracts was very toxic to the test bacteria. Extracts of blank filters were not toxic so it is possible that the silver-backing may have reacted with some component present in the atmosphere to form a toxic material. This problem was not encountered when unbacked glass-fiber filters were used.

## Extraction

Preliminary experiments with filters on which particulate material had been collected showed that a 16 hour extraction with methanol in a soxhlet extractor resulted in efficient recovery of mutagenic material. No further mutagenic material was recovered by elution with either toluene:hexane:isopropanol (70:10:20; v/v) or with water. Methanol has been used to extract mutagenic material from air particulates by Dehen et al (3) and Tokiwa et al (16).

**Optimization of the Mutagenic Assays**

Considerable literature now exists concerning the need for careful optimization of the amount of S-9 preparation used for each type of sample being tested. Figure 1 shows the results of a series of experiments in which the amounts of extract from air particulates and the S-9 were systematically varied. Clearly, larger amounts of S-9 preparation are required for maximum activity when the amount of extract is increased. However, when the amount of extract applied to the plate was greater than that obtained from 500 $\mu$g of particles, the response was nonlinear so that it is important to keep doses low enough to be in the linear range.

In work to date we have employed strains TA98 and TA100, which between them detect a wide variety of mutagens. Initial experiments showed that strain TA98 gave the stronger positive response and this strain was used in the experiments reported here.

**Levels and Nature of the Mutagenic Activity**

The levels of mutagenic activity found in samples obtained from the same location near the pouring floor where molten steel is cast into preformed molds vary over at least a threefold range and are much higher than those of urban Hamilton air samples. In addition, much of the mutagenic material in the foundry samples required metabolic activation, whereas the mutagens from urban air samples were largely "direct acting" (i.e., were mutagenic even when no S-9 preparation was added).

Considerable qualitative variation was apparent when extracts from different samples of particulates were individually subjected to liquid-liquid fractionation to give "aqueous", "acidic", "basic", and "neutral" fractions which were then assayed for mutagenic activity. Figure 2 shows results from samples taken on two successive days in the summer of 1979. On July 4 the level of total mutagens was relatively high and active material was present in all fractions. Indeed, the basic and aqueous fractions had more activity than the neutral fraction. These results indicate that mutagens of several diverse chemical classes were present. Also, about 20 percent of the total mutagenic activity was "direct acting".

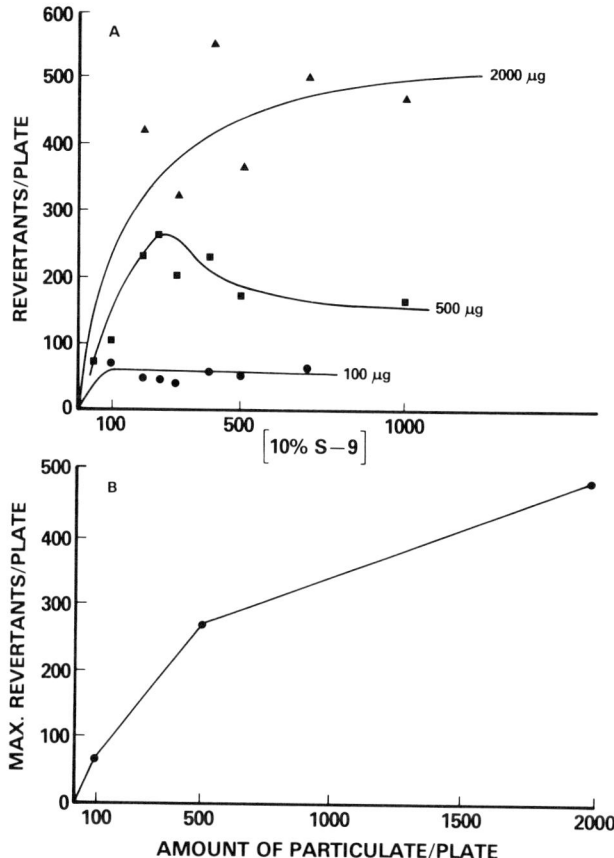

FIGURE 1. (A) The effect of various amounts of S-9 preparation on the response of *Salmonella typhimurium* TA98 to material extracted from foundry particulates. The horizontal axis gives the volume of S-9 mix [10 percent of which was S-9 preparation (1)] that was added to each plate. The numbers on the curves show the weight of particulate material equivalent to the extract applied to each plate. (B) Yield of revertants with various amounts of extract measured using the optimal amount of S-9 mix for each concentration. Data from Figure 1A.

Since PAH would appear in the neutral fraction and require metabolic activation by the S-9 preparation, it appears that these compounds can account for only a small fraction (maximum about 8 percent) of the total mutagenic activity obtained on July 4. In contrast, on July 5, the level of mutagenic material was lower and a larger proportion of the activity was found in the neutral fraction. However, 47 percent of the activity was direct acting.

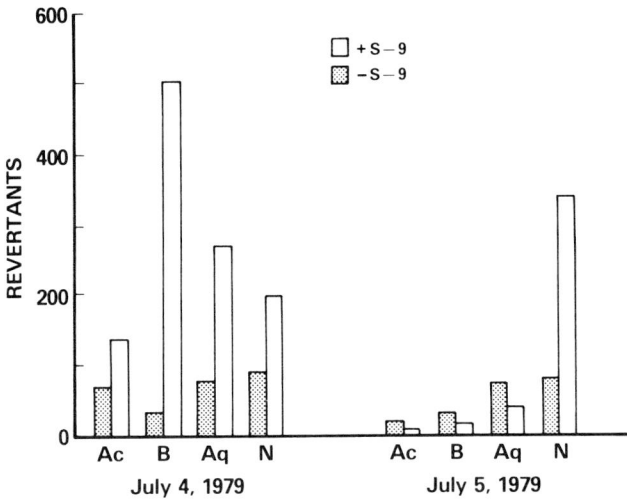

FIGURE 2. Mutagenic activity in various fractions of extract from foundry air particulates detected with *Salmonella typhimurium* TA98. The solid bars represent direct acting material (i.e., material which induced mutations in the absence of added S-9 mix), while open bars represent activity measured in the presence of 500 µl per plate of S-9 mix containing 10 percent S-9 prepartion. Ac = acidic fraction (see Materials and Methods section); B = basic fraction; Aq = aqueous fractions, and N = neutral fraction.

This type of day-to-day variation has been observed on other occasions and is probably related to the activities in the foundry on the particular days involved. On July 4, the pouring floors were working at capacity and there was extensive shake-out of sand from castings, whereas on July 5 there was less pouring and only a limited amount of shake-out. Other workers have demonstrated the presence of mutagens other than PAH in particulate material from urban air particulates (9, 15).

Our current efforts are directed toward assessing the amount and nature of particulate mutagenic material in various parts of the foundry having various levels of lung cancer risk, trying to pinpoint the source of the various classes of compounds present, and characterizing the non-PAH mutagens.

## SUMMARY

Methanol extracts of particulates obtained from steel foundry air were mutagenic in the Salmonella/S-9 test (Ames assay). Samples collected on type A/E glass fiber filters in devices operating at face velocities of 30 and 3 m/min yielded samples having similar levels of mutagenicity per $m^3$ of air. The levels of mutagenic activity found in extracts of foundry particulates showed considerable day-to-day variation and were many times higher than those found in particulates from outside urban air. Liquid:liquid extraction was used to separate the total extract into several fractions. Considerable amounts of activity were found in the aqueous, basic, and acidic fractions as well as in the neutral fraction. Also, up to 47 percent of the total activity measured was mutagenic without metabolic activation by the S-9 fraction of rat liver. It is concluded that foundry air particulates contain mutagens belonging to several chemical classes and that at least under some conditions PAH represent a relatively small proportion of the total mutagenic material present.

## ACKNOWLEDGMENT

This study is supported by a grant from the Ministry of Labour, Province of Ontario, Canada. We thank Dr. B. N. Ames for supplying the *Salmonella* strains used.

## REFERENCES

1. Ames, B. N., McCann, J., and Yamasaki E. (1975): Methods for detecting carcinogens and mutagens with the *Salmonella*/ mammalian microsome mutagenicity test. *Mutatian Res.* 31: 374-390.
2. Commoner, B., Madyastha, R., Bronsdon, R. B., and Vithayathil, A.J. (1978): Environmental mutagens in urban air particulates. *J. Toxicol. & Environ. Health* 4: 59-77.
3. Dehnen, W., Pitz, N., and Romingas, R. (1977): The mutagenicity of airborne particulate pollutants. *Cancer Lett.* 4: 5-12.
4. Gibson, E. S., Martin, R. H., and Lockington, J. N. (1977): Lung cancer mortality in a steel foundry, *J. Occup. Med.* 19: 807-12.
5. McCann, J., Choi, E., Yamasaki, E., and Ames, B. N. (1975): Detection of carcinogens as mutagens in the *Salmonella*/ microsome test: assay of 300 chemicals. *Proc. Nat. Acad. Sci. USA* 72: 5135-5139.
6. McCann, J., and Ames, B. N. (1976): Detection of carcinogens as mutagens in the *Salmonella*/microsome test: assay of 300 chemicals: Discussion. *Proc. Nat. Acad. Sci. USA* 73: 950-954.

7. Nagao, M., Yahagi, T., Honda, M., Seino, Y., Matsushima, T., and Sigimura, T. (1977): Demonstration of mutagenicity of aniline and o-toluidine by Norharman. *Proc. Japan Academy* 53: Ser. B, 34-37.
8. Pitts, Jr., J. N., Grosjean, D., Mischke, T. M., Simmon, V. F. and Poole D. (1977): Mutagenic activity of airborne particulate organic pollutants. *Toxicol Lett.* 1: 65-70.
9. Pitts Jr., J. N., Van Cauwenberghe, K. A., Grosjean, D., Schmid, J. P., Fitz, D. R., Belser Jr., W. L., Knudson, G. B., and Hynds, P. M. (1978): Atmospheric reactions of polycyclic aromatic hydrocarbons: facile formation of mutagenic nitro derivatives. *Science* 202: 515-518.
10. Purchase, I. F. H., Longstaff, E., Ashby, J., Styles, J. A., Anderson, D., Lefevre, P. A., and Westwood, F. R. (1976): Evaluation of six short-term tests for detecting organic chemical carcinogens and recommendations for their use. *Nature* 264: 624-627.
11. Salamone, M. F., Heddle, J. A., and Katz, M. (1979): The use of the *Salmonella*/microsomal assay to determine mutagenicity in paired chemical mixtures. *Can. J. Genet. Cytol.* 21: 101-1207
12. Simon, V. F., (1979): In vivo mutagenicity assays of chemical carcinogens and related compounds with *Salmonella typhimurium*. *J. Natl. Cancer Inst.* 62: 893-9
13. Stoltz, D. R., Stavric, B., Iverson, F., Bendall, R., and Klassen, R. (1979): Suppression of naphthylamine mutagenicity by amaranth. *Mutation Res.* 60: 391-393.
14. Talcott, R., and Wei, E. (1977): Airborne mutagens bioassayed in *Salmonella typhimurium*. *J. Natl. Cancer Inst.* 58: 449-451.
15. Teranishi, K., Hamada, K., and Watanabe, H. (1978): Mutagenicity in *Salmonella typhimurium* mutants of the benzene-soluble organic matter derived from airborne particulate matter and its five fractions. *Mutation Res.* 56: 273-280.
16. Tokiwa, H., Morita, K., Takeyoshi, H., Takahashi, K., and Ohnishi, Y. (1977): Detection of mutagenic activity in particulate air pollutants. *Mutation Res.* 48: 237-248.17.
17. Wang, Yi Y., Rappaport, S. M., Sawyer, R. F., Talcott, R. E., and Wei, E. T. (1978): Direct-acting mutagens in automobile exhaust, *Cancer Lett.* 5: 39-47
18. White, L. D. (1975): The collection, separation, identification and quantitation of selected polynuclear aromatic hydrocarbons and metals in coal tar and coke oven emissions. *Ph.D. Thesis, University of Cincinnati*.

# COMPARISON OF SIM GC/MS AND HPLC FOR THE DETECTION OF POLYNUCLEAR AROMATIC HYDROCARBONS IN FLY ASH COLLECTED FROM STATIONARY COMBUSTION SOURCES

**S. G. Zelenski, G. T. Hunt, and N. Pangaro**
GCA/Technology Division
Burlington Road
Bedford, Massachusetts 01730

## INTRODUCTION

Polynuclear aromatic hydrocarbons (PAH) have long been recognized as hazardous compounds. Recent availability of more specific, sensitive detectors and high-resolution separation techniques are providing much-needed information on the distribution of low levels of these compounds in the environment. The two most common analytical methods currently employed are selected ion mass spectrometry (SIMS), coupled with capillary gas chromatography (CGC) (1-3) and high performance liquid chromatography (HPLC) coupled with fluorescence and/or ultraviolet detection (4,5).

The complexity of the PAH isolation scheme depends upon the source of the sample. Samples with potentially high levels of interfering compounds such as anaerobic sediments are usually subjected to more extensive and time-consuming cleanup than relatively "clean" samples such as fly ash from high-temperature combustion. However, common to all procedures is an extraction step followed by some separation of the PAH mixture into subgroups or individual constituents.

While the advantages of both CGC-SIMS and HPLC have been touted, our preliminary experiences indicated that without extensive cleanup either method alone suffered from the presence of interfering compounds. Further concern over the reliability of the methods was generated after a review of literature revealed the extreme variability of reported PAH data for similar combustion sources. CGC-SIMS was at the greatest disadvantage because more unquestioned reliance is usually placed on the "magical black box" of mass spectrometry.

As both methods were readily available in our laboratory and we were about to embark on an extensive investigation of PAH in a variety of multimedia environmental samples, we felt that a comparison of the two methods for identification and quantitation of PAH in typical samples should be initiated. We were aware of other investigations comparing GC-MS and HPLC (6), but many advances in computer data processing as well as separation technology had occurred since those comparisons were made. In addition, many reports had appeared since 1978 relying solely on one or the other method for identification and quantitation of PAH in environmental samples.

We chose four different combustion fly ash samples for analysis. The samples were extracted and prepared and identical aliquots were distributed for CGC-SIMS and for HPLC fluorescence analysis. Spiked samples were also prepared using a precleaned representative fly ash sample to develop information on the recovery of a range of PAH from a combustion source particulate. Results were not compared until the analyses were complete to prevent any analyst bias when subjective judgments were required.

## EXPERIMENTAL PARAMETERS

### Reagents

PAH standards were obtained from the following sources:
Aldrich Chemical, Metuchen, New Jersey
Chem Services, Westchester, Pennsylvania
RFR Corporation, Hope, Rhode Island
Supelco, Inc., Bellefonte, Pennsylvania
Analabs, North Haven, Connecticut.

Toluene and methanol used in sample extractions were Burdick and Jackson distilled-in-glass. These were checked for potential contamination prior to use. Methylene chloride and pentane used in the chromatography fractionation were also distilled-in-glass from the same supplier. Methanol used in HPLC analysis was prefiltered using a 0.45-$\mu$ Millipore filtering apparatus. Water utilized in HPLC analysis was distilled/deionized and filtered in a similar manner. Silica Gel G (Davison Grade 950, 60 to 200 mesh) used in column chromatography was precleaned and activated at 100°C prior to use.

## Instrument Conditions

### HPLC

All HPLC separations were performed using a Du Pont 850 liquid chromatographic system. A four-step gradient sequence of methanol/water was used to optimize isomeric PAH separations. Operating parameters are listed in Table 1. All quantitative measurements were made utilizing a Perkin-Elmer 650-10S fluorescence spectrophotometer. Detector outputs were interfaced to a Spectra-Physics Minigrator/Recorder to permit measurements on the basis of peak areas. Operating parameters are listed in Table 1.

**TABLE 1. Summary of Instrumentation Operating Conditions**

| HPLC | | Fluorescence | GC/MS |
|---|---|---|---|
| Du Pont 850 HPLC with gradient elution | | Perkin-Elmer 650-10S detector | Hewlett-Packard 5985 GC/MS/DS |
| | | | SE-54 capillary column (30 m) |
| Column, Zorbax ODS $C_{18}$, 25.0 cm x 4.7 mm | | Wavelength (nm)<br>Excitation 300<br>Emission 400 | Flow Rate, 2.5 ml/min<br>UHP helium<br>Column Temperature, 100°C held |
| Flow Rate, 1.5 ml/min | | Slit Width, 15 nm | for 2 min, 3°/min to 265°C and hold |
| Solvent, methanol/water | | Output, 1 volt to Spectra Physics mini-grator | Injection, splitless for 30 sec |
| Step Gradient | | | Volume, 1.0 $\mu$l |
| Methanol (%) | Time (min) | Chart Speed, 40 cm/hr | Temp, 275 ° |
| 1. 80-85 | 20 | | MS Conditions |
| 2. 85-90 | 15 | | |
| 3. 90-hold | 10 | | Electron energy, 70 V |
| 4. 90-95 | 10 | | Source temp, 200° |
| Oven Temperature, 40°C | | | |

### GC/MS

A Hewlett-Packard 5985, GC/MS quadrupole mass spectromter was utilized for all quantitative measurements. A 30-meter SE-54 capillary column was used for separation of PAH mixtures. For chromatography and instrument operating conditions, see Table 1.

## PROCEDURES

Four separate samples were obtained for analyses. It is worthy to note that each of them was a bulk particulate collected from various types of stationary combustion sources. A listing of pertinent information relating to sample origin, particulate sizing, and combustion type is given in Table 2. Since these samples were available in bulk quantities, subsequent analyses and quantitative measurements were facilitated. Each of these samples was analyzed in duplicate.

As a further check on the procedure, two spiked fly ashes were also analyzed. The multiple-step analytical procedure is outlined in Figure 1. At all points in the procedure, precautions were taken to shield all samples and standards from exposure to interior lighting.

Prior to the column chromatography of the sample extracts, the efficiency of the fractionation procedure was evaluated using a standard PAH mixture. Column recovery data are given in the experimental results.

## RESULTS AND DISCUSSION

To check the extraction and chromatography steps, a spiked fly ash sample was included in the analyses for PAH. While 98 percent of the spiked d-8 naphthalene was recovered, varying amounts of other PAH were found after extraction from the fly ash. Recoveries range from 2 percent for 9,10-dimethylanthracene to 82 percent for fluoranthene. While these results may help to explain the variability of reported PAH results as noted earlier, they would not affect this experiment as aliquots of the same solution were sent for HPLC and GC-MS analysis.

Figure 2 shows the HPLC-fluorescence chromatogram of the extract of the particulates from the hopper of a baghouse on a bituminous coal-fired combustion source. Separations are achieved on a $C_{18}$ reverse-phase column and detection is by fluorescence. Good resolution of all components is demonstrated except between the pairs of chrysene/benz(a)-anthracene and benzo(a)pyrene/benzo(e)pyrene. The concentrations range from 50 ng/g particulate to 6200 ng/g particulate. It appears that HPLC coupled with fluorescence detection is capable of good resolution and high sensitivity when analyzing a fly ash sample for PAH.

However, an examination of Table 3 reveals a different conclusion. Selected ion scanning GC-MS results are now shown alongside HPLC results. Many of the HPLC identifications are not confirmed by GC-MS, even though the HPLC quantitation is usually within the detection limit for GC-MS. This suggests the fallibility of the HPLC method in assigning

TABLE 2. Particulate Data Summary Table

| Particulate Collection Device | Fuel Type | Combustion Type | Control Device Temp (°C) | Particulate Sizing | Sampling Method |
|---|---|---|---|---|---|
| Electrostatic precipitator | Bituminous coal | Pilot Plant | 177-204 | 18% ≤ 3μ | SASS train cyclones |
| Baghouse fabric filter | Bituminous coal | Utility | 1243 | Median diameter 2.9 μ | Anderson impactor |
| Electrostatic precipitator | Oil | Utility | 343° inlet, 163-171 outlet | — | — |
| Baghouse fabric filter | Bituminous coal | Pilot Plant | 177-182 | 86.1 ≤ 3μ | SASS train cyclones |

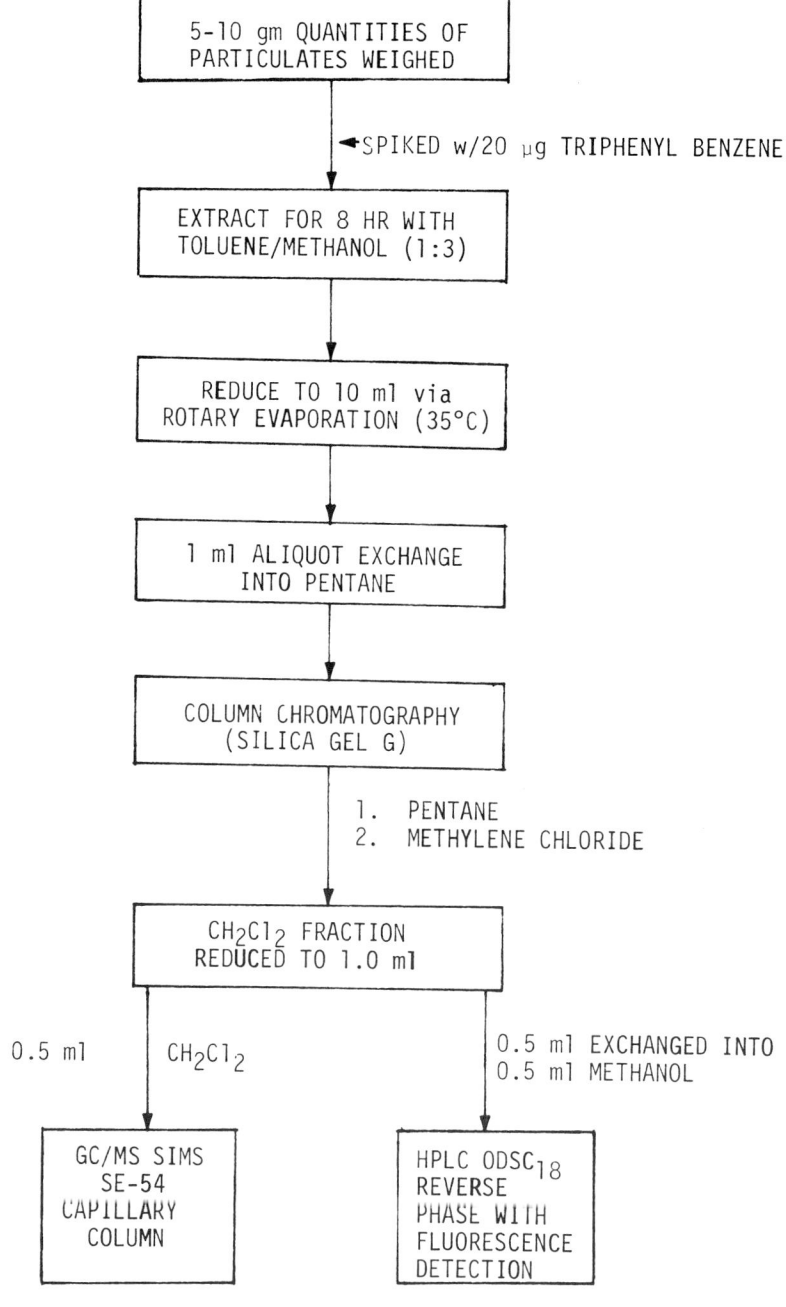

FIGURE 1. Diagram of the method used for PAH analysis.

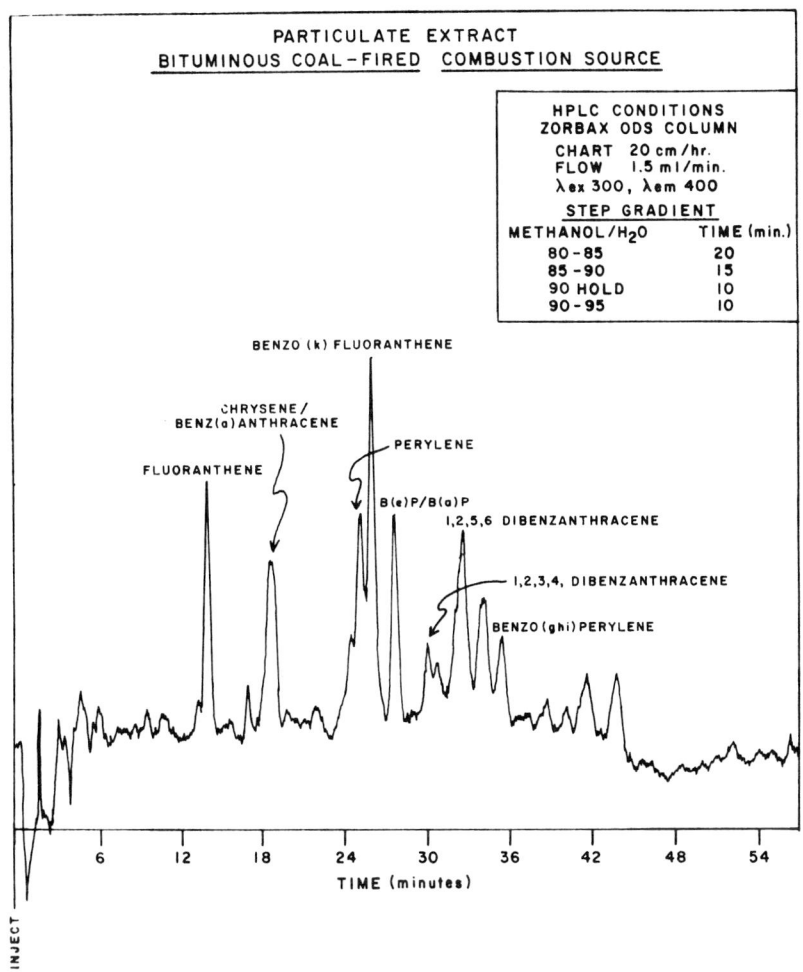

**FIGURE 2.** HPLC chromatograph of PAH in particulates from a bituminous coal-fired combustion source.

identifications based on the specific fluorescence detector and retention times. It is also interesting that when GC-MS confirms an HPLC identification, the quantitation provided by GC-MS is usually considerably higher.

A possible explanation may be the nonlinearity of one or both of the calibration curves used for quantitation. A significant error may be introduced by utilization of the linear regression analysis to smooth the calibration curve. We are currently investigating the significance of this possibility.

TABLE 3. Comparison of Results From GC-MS Analysis and From HPLC Analysis of Different Flyash Samples

| | Particulate (ng/g) | | | | | | | |
|---|---|---|---|---|---|---|---|---|
| | A (Baghouse) | | B (ESP) | | C (Baghouse) | | D (ESP) | |
| Component | HPLC | GC-MS | HPLC | GC-MS | HPLC | GC-MS | HPLC | GC-MS |
| Anthracene | D[a] | 10 | ND[b] | 1830 | ND[a] | 26 | ND[b] | 10 |
| Fluoranthene | 6185 | 427 | 193 | 460 | 61 | 10 | 44 | 10 |
| Pyrene | D[a] | 318 | D[a] | 500 | ND[b] | 10 | D[a] | 14 |
| | | <10 | | <10 | | <10 | | <10 |
| Benz(a)anthracene | 616 | | 31 | | 5.2 | | ND[b] | |
| Chrysene | | 3400 | | 10 | | <10 | | <10 |
| Benzo(a)pyrene | | <10 | | <10 | | <10 | | <10 |
| | 144 | | 6.6 | | 0.1 | | 3.3 | |
| Benzo(e)pyrene | | <10 | | 7.4 | | <10 | | <10 |
| Benzo(k)fluoranthene | 96 | <10 | 6.4 | <10 | ND[b] | <10 | ND[b] | <10 |
| Perylene | D[a] | <20 | D[a] | <20 | ND[b] | <20 | ND[b] | <20 |
| 1,2,3,4-Dibenzoanthracene | 50 | <20 | 3.1 | <20 | ND[b] | <20 | ND[b] | <20 |
| 1,2,5,6-Dibenzoanthracene | 510 | <20 | 42 | <20 | ND[b] | <20 | ND[b] | <20 |
| Benzo(g,h,i)perylene | 280 | <20 | 98 | <20 | 8.2 | <20 | 65 | <20 |
| Indeno(1,2,3-c,d)pyrene | ND[b] | <20 | D[a] | <20 | D[a] | <20 | ND[b] | <20 |
| Anthanthrene | D[a] | <50 | ND[b] | <50 | ND[b] | <50 | ND[b] | <50 |
| Coronene | ND[b] | <100 | D[a] | <100 | ND[b] | <100 | ND[b] | <100 |
| 2-Methyl anthracene | ND[b] | 281 | D[a] | 759 | D[a] | 10 | ND[b] | <10 |
| 9-Methyl anthracene | ND[b] | <10 | ND[b] | 1000 | ND[b] | <10 | ND[b] | <10 |

[a] D detected, but not quantitated.
[b] ND not detected, <1 ng/g particulate.

## CONCLUSIONS

HPLC with fluorescence detection provides comparable resolution and increased sensitivity over GC-MS in the analysis of PAH. However, in complex environmental samples, HPLC may provide misleading information on the presence of PAH in the sample. Conversely, while selected ion scanning GC-MS theoretically provides a more selective detector, it lacks the sensitivity of HPLC-fluorescence analyses. It appears that the ideal analytical scheme would include a prescreening with HPLC-fluorescence to establish the concentration ranges and types of PAH present, while GC-MS would provide confirmation of identification and quantitation in its usable range.

## REFERENCES

1. Giger, W., and Schaffner, C. (1978): Determination of polycyclic aromatic hydrocarbons in the environment by glass capillary gas chromatography. *Anal. Chem.* 50(2):243–249.
2. Bjørseth, A. (1978): Analysis of polycyclic aromatic hydrocarbons in environmental samples by glass capillary gas chromatography. In: *Carcinogenesis*, Vol. 3, P. W. Jones and R. I. Freudenthal, Eds., pp. 75–83, Raven Press, New York.
3. Lao, R. C., Thomas, R. S., and Monkman, J. L. (1975): Computerized gas-chromatographic-mass spectrometric analysis of polycyclic aromatic hydrocarbons in environmental samples. *J. Chromat.* 112:681-700.
4. Ogan, K., Katz, E., and W. Slavin (1979): Determination of polycyclic aromatic hydrocarbons in aqueous samples by reversed-phase liquid chromatography. *Anal. Chem.* 51(8):1315–1320.
5. Smillie, R. D., Wang, D. T., and Meresz, O. (1978): The use of a combination of ultraviolet and fluorescence detectors for the selective detection and quantitation of polynuclear aromatic hydrocarbons by high pressure liquid chromatography. *J. Environ. Sci.Health* A13: 47–59.
6. Thomas, R. S., Lao, R. C., Wang, D. T., Robinson, D., and Sakuma, T. (1978): Determination of polycyclic aromatic hydrocarbons in atmospheric particulate matter by gas chromatograph-mass spectrometry and high pressure liquid chromatography. In: *Carcinogenesis*, Vol. 3, P. W. Jones and R. I. Freudenthal, Eds., pp. 9–19, Raven Press, New York.

# CHEMICAL CHARACTERIZATION OF POLYNUCLEAR AROMATIC HYDROCARBONS IN AIRBORNE EFFLUENTS FROM AN EXPERIMENTAL FLUIDIZED BED COMBUSTOR

R. L. Hanson, R. L. Carpenter, and G. J. Newton

Inhalation Toxicology Research Institute
Lovelace Biomedical and Environmental Research Institute
P. O. Box 5890
Albuquerque, New Mexico 87115

## INTRODUCTION

Fluidized-bed coal combustors are under development to provide a more efficient and environmentally acceptable method for electricity generation, using coal as the fuel. These combustors (FBC) operate at lower combustion temperatures than pulverized-coal combustors, resulting in lower emissions of nitrogen oxides. Limestone or dolomite can be used in the bed to reduce sulfur oxides in the stack gas.

The lower combustion temperature for FBC may result in emission of organic compounds as vapors and/or associated with particles. Several polynuclear aromatic hydrocarbons (PAH) have been found in fly ash from pulverized-coal combustion (15) and organic extracts of the fly ash from pulverized coal combustion have been found to be mutagenic (5). Therefore, there has been interest in the organic emissions from FBC.

Organic analysis of the effluents from FBC have been reported by the Inhalation Toxicology Research Institute (ITRI) (3,7), Battelle's Columbus Laboratories (BCL) (2,13,14), and Argonne National Laboratory (ANL) (8). BCL has sampled effluents from its atmospheric pressure (AFBC) 6-inch FBC and effluents from the Exxon Miniplant pressurized FBC (PFBC). Argonne National Laboratory operates both an AFBC and a PFBC.

BCL has been conducting environmental assessments of various effluents and found organic compounds in the flue gas from the PFBC (the Exxon Miniplant) (2,14). The majority of organic compounds found (1740 $\mu g/m^3$) was low-molecular-weight $C_6$-$C_{12}$ hydrocarbons, while only 58

$\mu g/m^3$ was hydrocarbons greater than $C_{12}$. The presence of oxygen, nitrogen, or sulfur heterocycles was not reported.

Vapor-phase hydrocarbons concentrations from the 6-inch AFBC at BCL were reported as 85 and 360 ppm(C) (13). Polycylic organic matter (POM) associated with particles amounted to 72 and 5 $\mu g/m^3$. Organic class analyses indicated the presence of aliphatic esters, phthalate esters, nitriles, hydroxyls, and complex carbonyls. A Tenax®-GC adsorbent sample of the flue gas contained reduced sulfur compounds with gas chromatographic retention times similar to those of benzothiophene and dibenzothiophene.

Research at ANL has dealt with characterization of effluents from their PFBC (8). Particulate samples were collected on a woven-wire-mesh filter at a temperature of 150 to 170°C, and exhaust gases were collected on a glass-fiber filter after cooling to 40 to 50°C. The particulate samples from the woven-wire-mesh filter were found to contain from 0.8 to 20 mg of extractable organic material per gram of fly ash. An organic carbon determination of two samples gave 1.0 and 2.7 percent organic carbon, compared with 0.15 percent extractable organics. The cooled effluent collected on the glass fiber filter produced a yellow-orange oily material. The condensible organic vapors contained about 25 times greater total organics than the extractable organics obtained from particulate samples. Recent studies at ANL indicate that most of the condensible materials are produced during the PFBC start-up. Direct acting mutagens were found in the extracts of particulate samples (10). Both direct and indirect mutagens were found in the condensible samples collected on glass fiber filters. Samples extracted by dimethyl sulfoxide (DMSO) gave about twice the mutagenic activity when compared to extracts obtained from a combination of several organic solvents.

The Inhalation Toxicology Research Institute (ITRI) has been involved in a collaborative research program with the Morgantown Energy Technology Center (METC) to characterize airborne effluents from their 18-inch diameter experimental AFBC. Chemical characterization of airborne effluents from the AFBC at METC has been reported for combustion of Montana Rosebud subbituminous coal and Texas lignite (3,4,7,11,12). For both fuels the vapor phase organic concentration was found to be greater than the particle-associated organics. Differences in organic composition of airborne effluents were found with different fuels burned in this FBC (7). Vapor-phase and particle-associated PAH from the combustion of three coals and Paraho oil shale in the AFBC at METC are compared. Emphasis at ITRI has been on the quantitation of total organics and the identification of PAH rather than on broad chemical-class characterization of the organics in the airborne effluents from the METC-AFBC.

## MATERIALS AND EXPERIMENTAL METHODS

Effluent samples were obtained from an experimental AFBC operated by METC. The AFBC was sampled on seven separate occasions (Table 1) during combustion studies burning Montana Rosebud subbituminous coal, Texas lignite, Western Kentucky bituminous coal, and Paraho oil shale. This unit, shown schematically in Figure 1, is an 18-inch ID AFBC. Samples were collected before and after each of three effluent cleanup devices (cyclone 1, cyclone 2, and bag filter). The complete sampling system is described in detail elsewhere (3). A small portion of effluent sample (15 to 25 l/min) was withdrawn (duct to sample velocity ratios of ~1.2:1) and diluted with 100 l/min of dry cool air. The diluted effluent was conveyed to a chamber from which aerosol samples were obtained. Vapor-phase samples for organic characterization were taken immediately downstream from the diluter by means of an adsorbent trap preceded by a silver membrane filter (Selas Flotronics Membranes).

The adsorbent trap contained Tenax®-GC (2,6-diphenyl-p-phenylene oxide polymer, 60/80 mesh) which was chosen as an organic adsorbent because of its thermal stability and demonstrated ability to collect PAH from combustion effluents (1,9). For the first four sampling periods (Table 1) the Tenax® traps held approximately 1 g of Tenax® and a sample flow rate of about 2 l/min was used. For the last three sampling periods a larger trap containing about 5 g of Tenax® was used at a flow rate of about 10 l/min.

A concentric electrostatic precipitator (CESP) was used to capture particles passing through the sampling system. The material collected by the CESP was also analyzed for organics. For the last three sampling periods (Table 1), a high-volume sampling system was used to collect larger

**TABLE 1.** Summary of Coal Types and Operating Condition of FBC Sampling

| Coal Type | Bed Material | Bed Temperature (°C) | Sampling Period |
|---|---|---|---|
| Montana Rosebud[a] | Limestone | 782-849 | 1 |
| Montana Rosebud[a] | Limestone | 813-841 | 2 |
| Texas Lignite | $SiO_2$ | 771-943 | 3 |
| Texas Lignite | $SiO_2$ and limestone | 768-843 | 4 |
| Western Kentucky[b] | Limestone | 827-860 | 5 |
| Montana Rosebud[a] | Limestone | 771-907 | 6 |
| Paraho Oil Shale | None | 706-836 | 7 |

[a]Subbituminous coal.
[b]Bituminous coal.

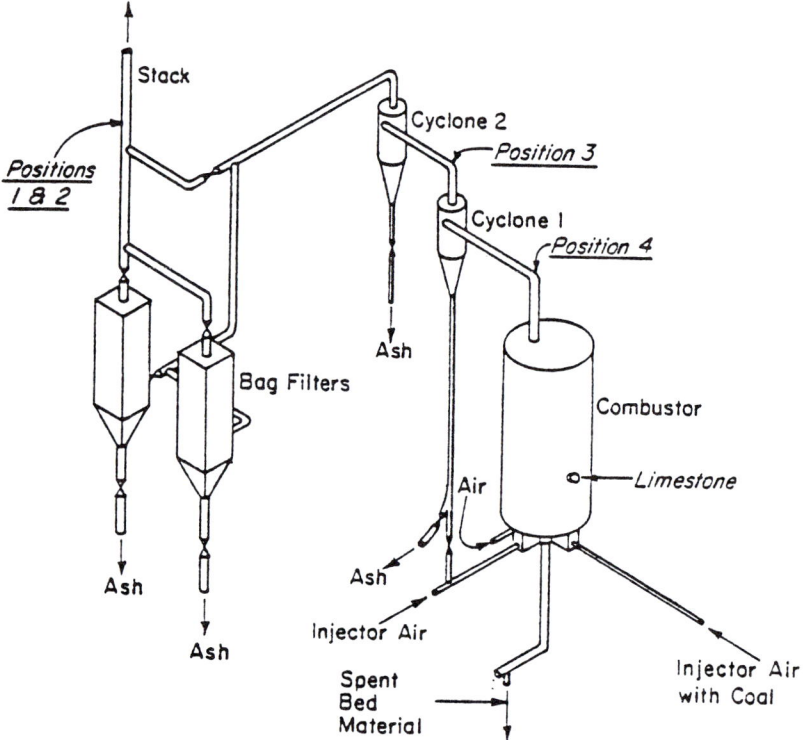

**FIGURE 1.** Schematic of the experimental fluidized-bed combustor.

quantities of effluent particles. The high-volume sampling probe consisting of an 8 × 10-inch-high volume filter (2830 l/min) was placed in the flue opposite the smaller probe. In addition, samples of cyclone 1 ash, cyclone 2 ash, and bag filter ash were obtained and analyzed.

Tenax was preextracted with n-pentane and dried prior to use. Silver-membrane filters and glass-fiber filters were used as supplied. Extraction of blank filters resulted in no detectable peaks in the gas chromatograms. All solvents were glass distilled.

Tenax®samples were soxhlet extracted for 6 hours using n-pentane as the solvent. Particulate samples were extracted for 1 hour in sealed amber bottles in an ultrasonic bath (Bransonic B-52, 55 kHz, 200 watts) using dichloromethane. Solvents with low boiling points were chosen to minimize thermal conversion reactions during extraction. The extracts were concentrated by rotary evaporation to about 1 ml. When gravimetric determinations of extractable material were desired, the remaining solvent was removed by a stream of dry nitrogen to near dryness and samples were placed in covered beakers overnight while the remaining solvent evaporated.

Extracts were analyzed by gas chromatography (GC) and gas chromatography-mass spectrometry (GC-MS). These techniques provided quantitation of total gas-chromatographable hydrocarbons, GC profile analysis, and tentative identification of many components.

For quantitation and profile analysis, a hydrogen flame ionization detector was used on a Varian 3700 gas chromatograph. Three glass capillary columns were used for these analyses: 30 m WCOT Dexsil 300, 25 m WCOT OV-101, and 30 m WCOT SP-2100. Temperature programs were: 60°C for 5 min, 5°C/min to 270°C, hold at 270°C for 20 to 30 min and 60°C for 5 min, 4°C/min to 270°C, hold at 270°C for 20 to 30 min. Splitless injection was used with sample volumes of 1 to 3 $\mu$l. To calibrate the gas-chromatographic system, a known PAH mixture was used as a standard. The average response factor was calculated by dividing the peak areas obtained (Varian CDS-111) by the total weight of the standards injected. Extracts were quantitated by dividing total peak areas by the average response factor.

A Finnigan Model 4000 GC-MS and INCOS data system supplied with the NBS mass spectral library was used for gas chromatography-mass spectrometric identifications. Either a 30 m WCOT Dexsil 300 or a 30 m WCOT SP-2100 glass capillary column was used with splitless injection of 1 $\mu$l of sample. The temperature program was: 60°C for 5 min, 6°C/min to 270°C, hold at 270°C for 20 min. Tentative identifications were based upon retention characteristics compared with those of standards and the fit to the spectra in the NBS mass spectral library. Absolute identification of some isomeric compounds was not possible because of the unavailability of standards. Mass spectral fits of greater than 800 (1000 is a perfect fit) were achieved for most of the compounds.

## RESULTS

### Vapor-Phase Organic Concentrations of FBC Effluents

Concentrations of vapor-phase organics collected on Tenax® are listed in Table 2 for different fuels burned in the FBC. During the sampling the fuels were burned under differing conditions, and this is reflected in the range of values.

Table 3 lists compounds tentatively identified in Tenax® vapor-phase samples from Montana Rosebud subbituminous coal effluents and Paraho oil shale effluents taken after the bag filter. More oxygenated compounds were found in the vapor phase of oil shale airborne effluents than in the vapor phase of the subbituminous coal effluents. This may have been a result of the greater oxygen content of the kerogen in oil shale.

TABLE 2. Range of Vapor-Phase Organic Concentrations (mg/m$^3$) for FBC Effluents from Various Fuels

| Sampling Location | Fuels | | | |
|---|---|---|---|---|
| | Montana Rose Bud Subbituminous Coal | Texas Lignite | Western Kentucky Bituminous Coal | Paraho Oil Shale |
| After bag filter | 0.084-20 | 0.4-5.1 | 0.2-1.3 | 0.42-4.0 |
| Before bag filter | 0.007-21 | 1.0-12 | 0.2-7.6 | 0.5 -1.8 |
| Between cyclones | 1.1-1.3 | 1.1-4.9 | 0.4-4.5 | N.D.[a] |
| Above bed | 0.28-1.4 | 1.6-33 | 1.61 | N.D. |

[a]N.D. = not determined.

TABLE 3. Tentative Identification of Vapor Phase Organic Compounds Sampled After the Bag Filter

| Montana Rosebud Subbituminous Coal | | Paraho Oil Shale | |
|---|---|---|---|
| Molecular Weight | Tentative Identification | Molecular Weight | Tentative Identification |
| 118 | Benzofuran | 106 | Benzaldehyde |
| 128 | Naphthalene | 128 | Naththalene |
| 134 | Benzothiophene | 142 | 1-Methylnaphthalene |
| 142 | 1-Methylnaphthalene | 142 | 2-Methylnaphthalene |
| 142 | 2-Methylnaphthalene | 154 | Biphenyl |
| 152 | Acenaphthylene or biphenylene | 156 (4 Peaks) | Dimethylnaphthalene isomers |
| 154 | Acenaphthene or biphenyl | 168 | Dibenzofuran |
| | | 174 | Hydroxynaphthalenedione |
| 168 | Dibenzofuran | 178 | Phenanthrene |
| 178 | Phenanthrene | 180 | 9-Fluorenone |
| 184 | Dibenzothiophene | 182 | Methyldibenzofuran |
| | | 184 | Dibenzothiophene |
| | | 190 | Methylphenylhexanone |
| | | 202 | Fluoranthene |
| | | 202 | Pyrene |
| | | 220 | 2,6-Bis(1,1-Dimethylethyl)-4-methylphenol |
| | | 236 | Trimethylphenyldihydroindene |

## Extractable Organic Concentrations of FBC Particles

Ash particles were collected and analyzed from the bag filter, primary cyclone, secondary cyclone, above the bed, between the cyclones, before the bag filter, and in the stack after the bag filter. Comparisons of average dichloromethane extractable organics determined gravimetrically are shown in Figure 2. Results indicated an increase of at least a factor of four in extractable organics as the exhaust stream cools from cyclone 1 to the bag filter.

## Comparison of Vapor-Phase and Particle Extractable Concentrations in FBC Airborne Effluents

For the FBC studies, particles in the exhaust stream were collected on silver-membrane filters in front of Tenax® and also by the CESP on the exhaust line of the aerosol sampling chamber. Comparative results for CESP and Tenax® samples for Texas lignite combustion are given in Figure 3 and those for Tenax® prefilters and Tenax® for Paraho oil shale combustion are given in Figure 4. For both fuels the vapor-phase organic concentration is higher (97.3 percent for Texas lignite and 89 percent for Paraho oil shale after the bag filter) than the extractable-particle-associated organic concentration in the FBC exhaust.

## Qualitative Identification of PAH in Extracts of Particle and Ash

Several airborne effluent samples from each fuel type have been analyzed by GC-MS. For purposes of illustration, tables containing all compounds tentatively identified in a certain type of sample for a given fuel have been prepared. This approach illustrates the diversity and similarity of the samples. Tables 4 through 7 list the tentative identifications of components in dichloromethane extracts of bag filter ash from the four fuel types burned in the FBC. Table 8 lists the compounds tentatively identified from a CESP sample taken above the bed while burning Texas lignite.

## DISCUSSION

### Effect of Fuel Type on Vapor Phase and Particle Associated Extractable Organic Compounds

The range of vapor-phase organic concentrations for the four fuel types overlapped. The highest concentrations and greatest variety of vapor-phase organics were found when Montana Rosebud subbituminous coal and Texas lignite were burned. Oxygen and sulfur PAH were found in the extracts of FBC airborne effluent samples from all four fuels. Nitrogen containing compounds were found in extracts of effluent samples from

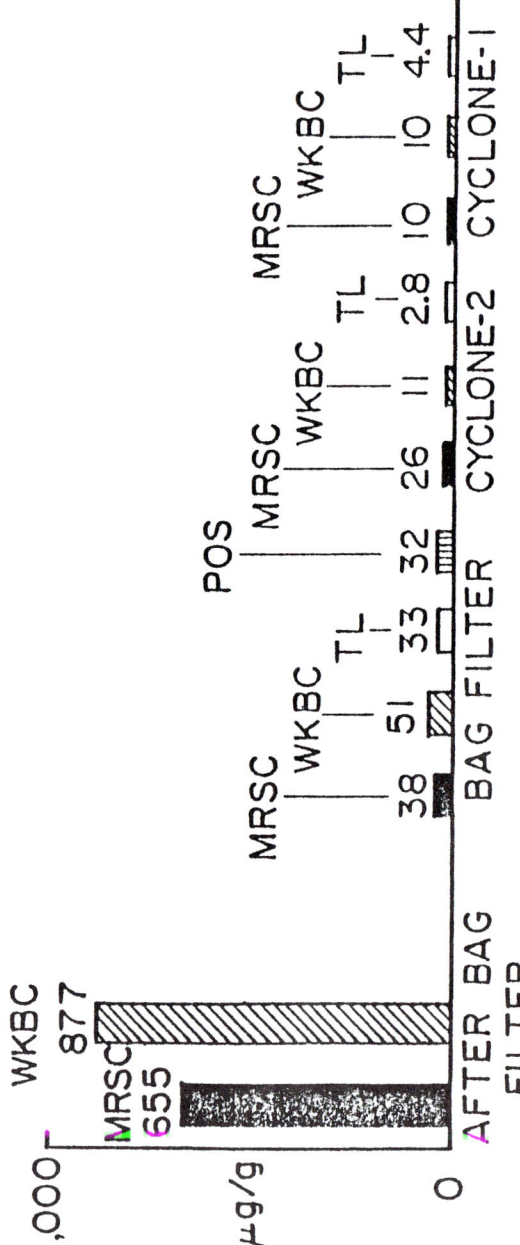

FIGURE 2. Gravimetric dichloromethane extractables from FBC particles. Combustion of Montana Rosebud Subbituminous (MRSC), Western Kentucky Bituminous Coal (WKBC), Texas Lignite (TL), and Paraho Oil Shale (POS).

PAH IN EXHAUST OF FBC 607

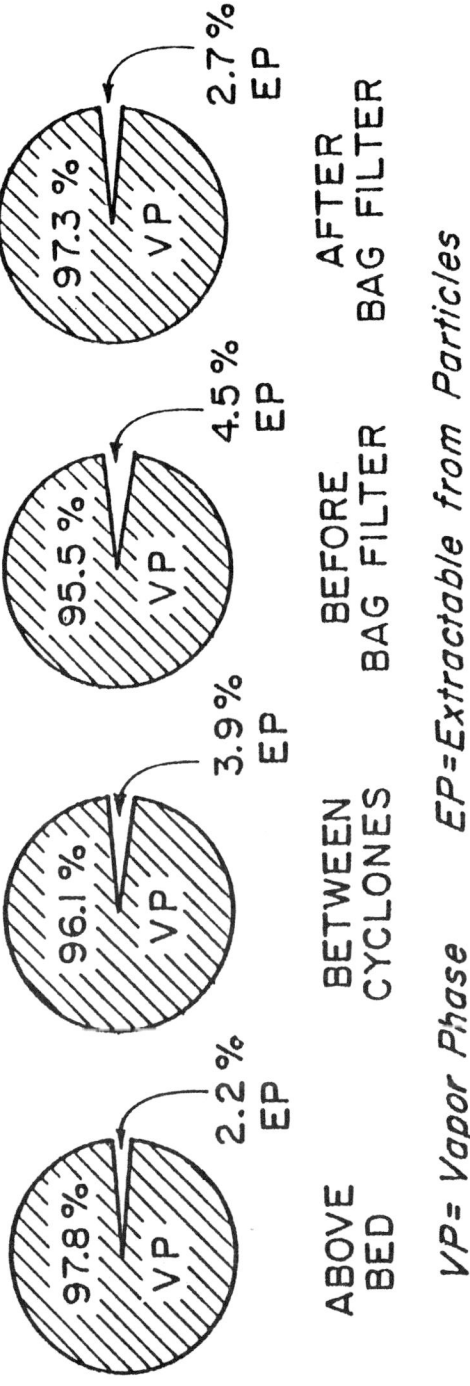

FIGURE 3. Comparison of PAH effluent concentrations for FBC of Texas lignite.

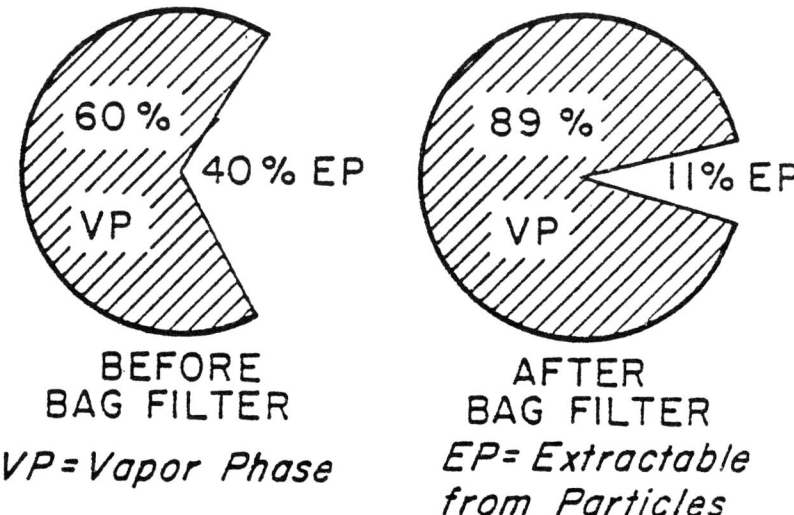

FIGURE 4. Comparison of PAH effluent concentrations for FBC of Paraho oil shale.

Texas lignite and Montana Rosebud subbituminous coal combustion, but not in those from FBC combustion of Western Kentucky bituminous coal and Paraho oil shale.

## Effect of Sampling Position on the Adsorption of Vapor-Phase Organics by Particles

The vapor-phase organic concentration was at least 20 times particle-associated (dichloromethane extractable) organic concentrations at all four sampling locations during combustion of Texas lignite (Figure 3). For Paraho oil shale, concentrations were almost equal before the bag filter but about eight times greater for the vapor phase after the bag filter (Figure 4). This indicates that the exhaust cleanup system is more efficient in removing organic matter associated with particles than vapor phase organics.

Adsorption of vapor-phase organics by particles appeared to occur as the exhaust cooled. The bag filter ashes had higher concentrations of extractable organics per gram than the cyclone ashes (Figure 2). High-volume-filter samples taken after the bag filter has about 17 times greater extractable organics than the bag filter ash, suggesting that the particles that passed through the bag filter had significantly higher concentrations of extractable organics per unit mass than the particles collected by the effluent cleanup devices. By sampling before the bag filter, an indication of

TABLE 4. Tentative Identifications of Compounds for Dichloromethane Extracts of Bag Filter Ashes While the FBC Was Burning Montana Rosebud Subbituminous Coal

| Molecular Weight | Tentative Identification |
|---|---|
| 128 | Naphthalene |
| 131 | Methylindolizine isomer |
| 142 (2 peaks) | Methylnaphthalene isomers |
| 152 | Acenaphthylene or biphenylene |
| 154 | Phenylpyridine |
| 156 | Dimethylnaphthalene isomer |
| 157 | Ethylquinoline isomer |
| 168 | Dibenzofuran |
| 178 | Phenanthrene |
| 178 | Anthracene |
| 178 (2 peaks) | Ethynediyl bis-benzene isomers |
| 179 (3 peaks) | Fluorene-9-imine, benzo(h)quinoline, phenanthridine, acridine or biphenylcarbonitrile |
| 180 (4 peaks) | Methylfluorene isomers, fluorenone isomers, or benzo(c)cinnoline |
| 181 | 5,6-Dihydrophenanthridine |
| 182 | Dimethylbiphenyl or dimethylacenaphthene |
| 182 | Xanthene and methyldibenzofuran isomers |
| 184 | Dibenzothiophene |
| 192 (4 peaks) | Methylphenanthrene isomers |
| 194 | Anthracenone or phenanthrenone |
| 196 (2 peaks) | Xanthrenone or phenazinol |
| 202 | Fluoranthene |
| 202 | Pyrene |
| 204 (3 peaks) | Phenylnaphthalene isomers, dihydrofluoranthene isomers, or dihydropyrene isomers |
| 208 (2 peaks) | 9,10-Anthraquinone and 9,10-phenanthroquinone |
| 216 (4 peaks) | Benzofluorene, methylpyrene isomers, or methylfluoranthene isomers |
| 218 (3 peaks) | Phenanthroimidazole isomers or anthraimidazole isomers |
| 220 (2 peaks) | Methylphenylcinnoline isomers or methylphenylquinoxaline isomers |
| 228 (3 peaks) | Benzo(ghi)fluoranthene, benzo(c)phenanthrene, benz(a)anthracene, chrysene, or triphenylene |
| 230 | Terphenyl isomer |
| 230 (3 peaks) | Benz(de)anthracene-7-one or dibenzo(ch)-naphthyridine isomers |
| 234 (2 peaks) | Benzonaphthothiophene isomers |
| 240 | Methylbenzo(ghi)fluoranthene isomer |
| 240 | Benzothienobenzothiophene |

## TABLE 4 (Continued)

| Molecular Weight | Tentative Identification |
|---|---|
| 252 | Benzofluoranthene isomer |
| 252 | Benzo(e)pyrene, benzo(a)pyrene, or perylene |
| 254 (2 peaks) | Binaphthalene isomers |
| 256 | $S_8$ Molecular sulfur |
| 258 | Benz(a)anthracene-7,12-dione or benzo(c)phenanthrenedione |

TABLE 5. Tentative Identifications of Compounds for Dichloromethane Extracts of Bag Filter Ashes While the FBC Was Burning Texas Lignite

| Molecular Weight | Tentative Identification |
|---|---|
| 118 | Benzofuran |
| 128 | Naphthalene |
| 132 | Ethenylbenzaldehyde |
| 134 | Benzothiophene |
| 135 | Benzisothiazole |
| 136 | Dihydrobenzothiophene |
| 142 (2 peaks) | Methylnaphthalene isomers |
| 143 | Methylquinoline isomer |
| 148 | Methylbenzo(b)thiophene isomers |
| 152 | Acenaphthylene or biphenylene |
| 154 | Acenaphthene or biphenyl |
| 160 | Phenylthiophene isomer |
| 168 | Dibenzofuran |
| 178 (2 peaks) | Phenanthrene and anthracene |
| 184 | Dibenzothiophene |
| 192 (3 peaks) | Methylphenanthrene isomers |
| 202 (2 peaks) | Fluoranthene and pyrene |
| 204 | Phenylnaphthalene isomer |
| 218 | Phenanthroimidazole |

the compounds adsorbed from the vapor-phase organics by the bag filter ash could be obtained by comparing with extracts from bag filter ash collected at the same time. The extract of bag filter ash contained compounds with molecular weights of 152, 154, and 168. Extracts of the particles collected before the bag filter had phenanthrene and anthracene (molecular weight 178) as the lowest molecular weight products. Extracts of Tenax®samples taken after the bag filter contained compounds of molecular weight 118 to 184 including 152, 154, and 168 (Table 3). Thus, the bag filter ash appeared to adsorb vapor-phase organics that had lower molecular weights than the organics found on the particles collected before

TABLE 6. Tentative Identification of Compounds for Dichloromethane Extracts of Bag Filter Ashes While the FBC Was Burning Western Kentucky Bituminous Coal

| Molecular Weight | Tentative Identification |
| --- | --- |
| 154 | Biphenyl or acenaphthene |
| 168 | Dibenzofuran |
| 168 | Methylbiphenyl or methylacenaphthene-isomer |
| 178 | Phenanthrene |
| 180 | 9-Fluorenone or benzo(c)cinnoline |
| 182 | Diphenylmethanone |
| 182 (2 peaks) | Methyldibenzofuran isomers |
| 184 | Dibenzothiophene |
| 202 | Fluoranthene or pyrene |
| 204 (2 peaks) | Phenylnaphthalene isomers |

TABLE 7. Tentative Identification of Compounds for Dichloromethane Extracts of Bag Filter Ashes While the FBC Was Burning Paraho Oil Shale

| Molecular Weight | Tentative Identification |
| --- | --- |
| 178 | Phenanthrene or anthracene |
| 202 | Fluoranthene |
| 202 | Pyrene |
| 208 | 9,10-Phenanthroquinone, or 9,10-anthraquinone |
| 228 | Triphenylene, chrysene, benz(a)anthracene, or benzo(c)phenanthrene |
| 230 | Terphenyl isomer |
| 230 (2 peaks) | Benz(de)anthracene-7-one or dibenzo(ch)-naphthyridine isomers |
| 248 | 2,5-Bis(1,1-dimethylpropyl)-2,5-cyclohexadiene-1,4-dione |
| 252 | Benzofluoranthene isomer |
| 252 | Benzo(e)pyrene, benzo(a)pyrene, or perylene |
| 258 | Benz(a)anthracene-7,12-dione or benzo(c)phenanthrenedione |

the bag filter, but these lower molecular weight compounds were not completely removed from the vapor phase by the bag filter.

After exhaust leaves the stack of the FBC it will be cooled and diluted by ambient air. Samples of exhaust particles collected by the CESP were diluted and cooled to near ambient temperature. These particles have been found to have a much higher extractable organic content than particles collected by the exhaust cleanup devices (7). Adsorption of organics does occur, but the quantity adsorbed has been calculated to be less than a monolayer on the surface of particles (7).

**TABLE 8. Tentative Identification of Compounds for Dichloromethane Extract of Concentric Electrostatic Precipitator Sample Taken Above the Bed While Texas Lignite Was Burning in the FBC**

| Molecular Weight | Tentative Identification |
|---|---|
| 128 | Naphthalene |
| 142 (2 peaks) | Methylnaphthalene isomers |
| 154 | Acenaphthene or biphenyl |
| 156 (2 peaks) | Dimethylnaphthalene isomers |
| 168 | Dibenzofuran |
| 178 | Ethynediyl bis-benzene or phenanthrene |
| 180 | 1,1-Dimethylethyl-4-methoxyphenol |
| 220 | 2,6-Bis(1,1-dimethylethyl)-2,5-cyclohexadiene-1,4-dione |
| 220 | 2,6-Bis(1,1-dimethylethyl)-4-methylphenol |
| 234 | 2,6-Bis(1,1-dimethylethyl)-4-ethylphenol |
| 248 | 2,5-Bis(1,1-dimethylpropyl)-1,5-cyclohexadiene-1,4-dione |

## Comparison of Hydrocarbons in FBC Effluents

The data on hydrocarbon concentrations from the various experimental FBC can be summarized with respect to vapor phase and particle associated hydrocarbon concentrations. Vapor phase hydrocarbons for four experimental FBC are compared in Figure 5. The PFBC at ANL has the highest reported vapor-phase concentration at 30 mg/m$^3$. Only single values have been reported by Battelle's Columbus Laboratories (BCL) for its AFBC and the Exxon PFBC. The wide range of vapor-phase hydrocarbon concentrations from various fuels at the METC-AFBC are illustrated by the values from Table 2 for sampling after the bag filter. The particle-associated hydrocarbon concentrations are compared in Figure 6. The concentrations are reported per gram of particles collected, since the different particulate cleanup devices employed can greatly affect the particulate mass loading and these devices have not been optimized in these experimental FBC. Again, the ANL-PFBC has been reported to have the highest concentration at 5 mg/g. The AFBC at BCL and METC have very similar particle-associated concentrations of organics. The values used from the METC-AFBC were from high-volume-filter samples given in Figure 2, collected after the bag filter.

Many of the same PAH have been reported in the flue gas from the Exxon-PFBC and BCL-AFBC as have been found in samples from the METC-AFBC. Some mutagenicity has been reported for suspended particles from the Exxon-PFBC, the ANL-PFBC, and the METC-AFBC (6,10,14). Some of the PAH identified in FBC samples are known mutagenic compounds.

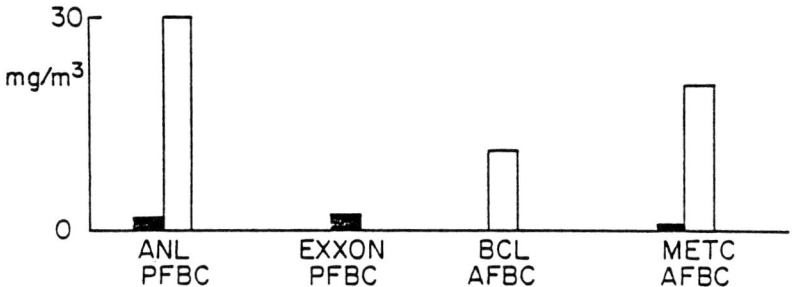

**FIGURE 5.** Vapor-phase hydrocarbons in FBC effluents. Dark bars represent the lowest value reported; clear bars represent the highest value reported.

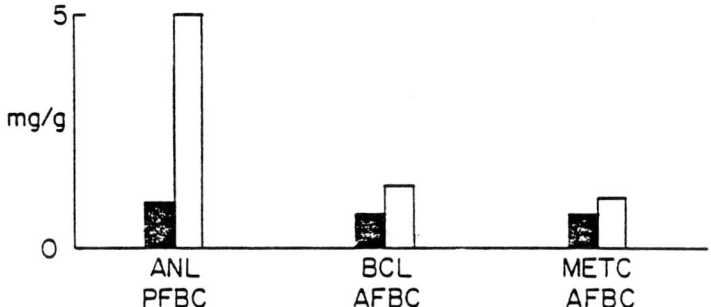

**FIGURE 6.** Range of hydrocarbon concentrations for FBC particles. Dark bars represent the lowest value reported; clear bars represent the highest value reported.

## CONCLUSIONS

A great deal has been learned from the sampling and characterization efforts. The reactive nature of exhaust gases and high temperatures resulted in a sampling strategy involving aerosol extraction and dilution to cool the exhaust. The ITRI program has included the use of biological testing to evaluate the toxicity and mutagenicity of FBC exhaust samples. Chemical characterization efforts have been focused by results from biological testing. Because of sample size limitations, fractionation by column chromatography or HPLC have not been used in support of chemical characterization and biological testing of FBC exhaust samples.

The PAH associated with particulate emissions have been found to change with fuel type burned in the FBC. The concentrations of extractable organics have been found to depend on sampling location in the FBC exhaust stream. As the effluent stream cooled, the concentration of extractable organics increased. Higher extractable organic concentrations have been found in the particulate samples collected with the sampling system than were found in the samples collected by the effluent cleanup devices, presumably as a result of the cooling and dilution used in the sampling system.

PAH have been quantitated and identified in particulate and vapor phase samples from an experimental FBC. Similar chemical analysis have been applied to fly ash collected by the effluent cleanup devices. We found the following from these studies: (1) In general, vapor phase PAH concentrations exceed particle-borne concentrations by a factor of 20. (2) Ash collected by the cleanup devices contains lower extractable PAH concentrations than particles collected by the sampling system. (3) The PAH concentration for particles increased as they approached the stack.

The PAH associated with the particles collected after the bag filter and the ash collected by the bag filter contained higher molecular weight compounds (five rings) than the vapor-phase samples collected after the bag filter. Oxygen and sulfur PAH were found in effluent samples from all four fuels burned in this FBC. Nitrogen PAH were found in effluent samples from combustion of Montana Rosebud subbituminous coal and Texas lignite.

The study of the transfer of vapor-phase PAH to particles at the bag filter during combustion of Montana Rosebud subbituminous coal indicated that compounds with molecular weights of 152, 154, and 168 were present on the bag filter ash and the vapor-phase samples after the bag filter but not in particulate samples collected before the bag filter.

Measurements of mutagenic activity made at ITRI showed that some FBC effluent particles are mutagenic (6). Should reduction in the quantity of mutagens emitted by FBC be needed, our data suggest that vapor-phase cleanup of the effluents will be important. The finding of mutagenic activity in extracts of particulate samples from the FBC should serve as an indication of potential health risk. However, this must be related to other sources and a possible relationship between mutagenic activity and human exposure must be established before any evaluation of risk is possible.

## ACKNOWLEDGMENTS

This Research was performed under U.S. Department of Energy Contract EY-76-C-04-1013.

The authors acknowledge the excellent cooperation and assistance received from the staff of the Morgantown Energy Technology Center, especially that from W. Wallace, J. Kovach, D. Green, U. Grimm and M. Mazza. The assistance of C. H. Hobbs, E. B. Barr, S. Weissman, R. Peele, R. Tamura, D. Horinek, S. Rothenberg, C. R. Clark, and A. Sanchez of the staff of the Inhalation Toxicology Research Institute is also acknowledged.

## REFERENCES

1. Adams, J., Menzies, K. and Levins, P. (1977): Selection and Evaluation of Sorbent Resins for the Collection of Organic Compounds, U.S. Environmental Protection Agency Report, EPA-600/7-77-044.
2. Bliss, C., Ed. (1977): *Proceedings of the Fifth International Conference on Fluidized Bed Combustion*, December 12-14, 1977, Washington, D.C.
3. Carpenter, R. L., Weissman, S. H., Newton, G. J., Hanson, R. L., Peele, E. R., Maaza, M. H., Kovach, J. J., Green, D. A., Grimm, U. (1978). *Characterization of aerosols produced by an experimental fluidized bed coal combustor operated with sub-bituminous coal*, Lovelace Biomedical and Environmental Research Institute Unclassified Report, LF-57.
4. Carpenter, R. L., Newton, G. J., Rothenberg, S. J., and Denee, P. B. (1980): Respirable aerosols from fluidized bed coal combustion. II. Physical characterization of fly ash. *Environ. Sci. & Technol.* (in press).
5. Chrisp, C. E., Fisher, G. L. and Lammert, J. E. (1978): Mutagencity of filtrates from respirable coal fly ash, *Science* 199:73.
6. Clark, C. R. and Hobbs, C. H.: Mutagenicity of effluents from an experimental fluidized bed coal combustor (submitted to *Environmental Mutagenesis*).
7. Hanson, R. L., Carpenter, R. L., Newton, G. J., and Rothenberg, S. J. (1979): Studies of organic material present in the exhaust stream of an experimental fluidized bed coal combustor, *Environmental Science and Health* A14(4):223.
8. Isaacson, H. R., and Haugen, D. A. (1978): Toxicology of samples of FBC process and effluent streams, p. 39, ANL-CENI-7306.
9. Jones, P. W., Giammar, R. D., Strup, P. E., and Stanford, T. B. (1976): Efficient collection of polycyclic organic compounds from combustion effluents, *Environ. Sci. Technol* 10:806.
10. Kubitschek, H. E. and Haugen, D. A. (1979): Biological activity of effluents from fluidized bed combustion of high-sulfur coal. In: *Proceedings of the Park City Environmental Health Conference*, held April 4-7, 1979, at Park City, Utah (in press).
11. Mazza, M. H., Green, D. A., Paris, M. W., and Newton, G. J. (1978): Mineral characterization of fluidized-bed combustion aerosol ash--Montana Rosebud sub-bituminous coal. MERC/TPR-78/1.
12. Mei, J. S., Grimm, U., and Halow, J. S. (1978): Fluidized-bed combustion test of low-quality fuels--Texas lignite and lignite refuse, MERC/RI-78/3.

13. Merryman, E. L., Levy, A., Felton, G. W., Liu, K. T., Allen, J. M., and Nack, H. (1977): Method for analyzing emissions from atmospheric fluidized-bed combustor. U.S. Environmental Protection Agency Report EPA-600/7-77-034.
14. Murthy, K. S., Howes, J. E., Nack, H., and Hoke, R. C. (1979): Emissions from pressurized fluidized-bed combustion processes, *Environ. Sci. Technol.* 13:197.
15. Sucre, L., Jennings, W., Fisher, G. L., Raabe, O. G., and Olechno, J. (1979): Polynuclear aromatic hydrocarbons associated with coal combustion. In: *Trace Organic Analysis: A New Frontier in Analytical Chemistry*, H. S. Hertz and S. N. Chesler, Eds., *NBS Special Publication* 519, p. 109.

# USE OF PAH TRACERS DURING SAMPLING OF COAL FIRED BOILERS

T. W. Sonnichsen*, M. W. McElroy**, A. Bjørseth***

*KVB, Inc.
A Research-Cottrell Company
17332 Irvine Boulevard
Tustin, California 92680
**Electric Power Research Institute
3412 Hillview Avenue
Palo Alto, California 94303
***Central Institute for
Industrial Research
Blindern, Oslo 3 NORWAY

## INTRODUCTION

Polynuclear Aromatic Hydrocarbons (PAH) are emitted in trace quantities from the combustion of pulverized coal. KVB has completed a program under sponsorship of the Electric Power Research Institute (EPRI) to quantify PAH emissions from coal-fired utility boilers and to determine the influence of several operational factors on these emissions, including fuel and combustion conditions, particulate emission characteristics and control devices.

As part of this program, PAH tracers were injected into the sampling train concurrent with the collection of "real" samples. It was recognized that the behavior of the tracer PAH compounds would not necessarily simulate that of the real PAH, which had been subjected to particulate-matter and boiler-surface contact prior to entry into the sampling system. Tracers included deuterated forms of PAH compounds, characteristic of coal combustion emissions, and other forms not found in stack gases. The original intent of the tracer injections had been to verify the accuracy and precision of the sampling and analytical procedures. It could also provide a possible means (if necessary) to correct losses due to systematic errors. Prior to this effort, no known comprehensive evaluation of tracer recoveries from sampling systems in the field had been made.

### 618 USE OF PAH TRACERS DURING SAMPLING

The PAH sampling program was conducted in three phases. The first was a laboratory evaluation of the sample train and analytical methodology. The second and third consisted of field measurements of PAH compounds from two coal-fired utility boilers. This paper will discuss only the results of the tracer evaluations and the potential reasons for the relatively poor recovery levels observed. The PAH emission measurements from the coal combustion will be documented in an EPRI report (1) to be issued in early 1980.

## SAMPLING AND ANALYTICAL PROCEDURES

### PAH Sampling System

The sampling train used for PAH collection is shown schematically in Figure 1. The system consisted of an Aerotherm high-volume stack sampler (HVSS) modified to include an organic adsorbent module and gas temperature conditioner both developed by Battelle (2), and a PAH tracer injection system. A dual parallel filter system was also added to allow increased sampling times under heavy particulate loadings. The modified sampling train allowed sample flow rates as high as 3 to 4 scfm and total sample volumes approaching 1000 scf. This system is similar to the EPA

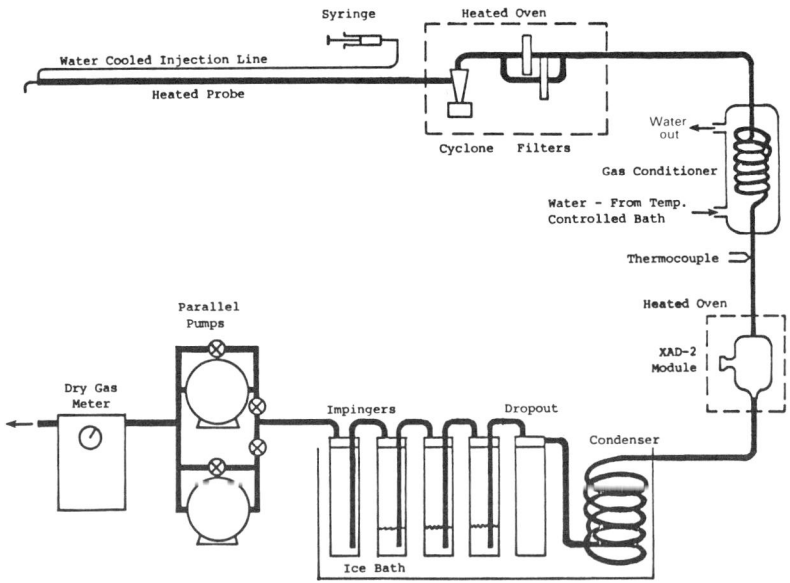

**FIGURE 1. PAH sampling system.**

SASS train widely used for emission testing and PAH measurement from combustion systems.

The sampling system consisted of a heated probe, cyclone, and filter maintained at stack gas temperatures, followed by a Battelle gas conditioner and organic module. The gas conditioner consisted of a glass coil enclosed in circulating water. During sampling, the flow rate of the circulating water was controlled to maintain exit gas temperatures to the organic module at 130 to 140°F. The organic module contained a bed of granular XAD-2 adsorbent positioned vertically to eliminate the possibility of gas channeling and bypassing the XAD-2 granules. The module was housed in a heated oven maintained at 135°F. The balance of the system included the impingers for water dropout, pumps, flow meters, and controls characteristic of the HVSS system.

A water-cooled injection line was added to allow probe tip injection of the PAH tracers. Two alternate injection points located upstream of the cyclone and the gas conditioner were used also during the last phase of the test program. The rate of injection was controlled by a syringe pump. Tracers generally were injected through the sample run.

Following sampling, the sampling train was thoroughly cleaned. Dry particulate matter from the cyclone and filters was sealed in Teflon bottles. Repeated washings of the probe, cyclone, filter housing and gas conditioner were made using methylene chloride, acetone, and water. The organic modules were sealed. The bulk of the samples were transported by KVB directly to the analytical laboratory for analysis to minimize potential shipping losses.

**Analytical Method**

The analyses reported in this paper were conducted by Battelle-Columbus Laboratories (BCL). Limited analyses by other laboratories were included as a cross-check of the BCL results. Included in this section is a discussion of the BCL procedures used.

The PAH analytical procedures consisted of extraction and concentration, liquid chromatographic cleanup and analysis and quantification by gas chromatography/mass spectrometry (GC/MS). Limited comparisons were also made using glass capillary gas chromatography with flame ionization detection.

Extractions were carried out using only distilled-in-glass solvents to avoid contamination. XAD-2 samples were soxhlet extracted with methylene chloride for 24 hours. Filters were extracted with methylene chloride by ultrasonic agitation for 1 hour. Water samples were extracted with methylene chloride, separated and the methylene chloride dried over magnesium sulfate. Extractions from various segments of the sampling train were combined as directed by KVB.

The extracts were spiked with a 5 $\mu$g standard solution of 9-methylanthracene, 9-phenylanthracene, and 9, 10-diphenylanthracene. Samples were reduced in volume to 10 ml by use of a rotating evaporator. Fifteen ml cyclohexane was added to allow a solvent exchange for proper liquid chromatography (LC) separation and the solution was reduced to 0.5 ml by Kuderna-Danish evaporation.

The samples were then liquid chromatographed on silica gel to remove interfering species. The PAH were eluted with 75 ml of 20 percent methylene chloride in petroleum ether followed by 30 ml of 50 percent methylene chloride in petroleum ether. Kuderna-Danish techniques were used again to reduce the fraction to approximately 1 ml.

The PAH compounds were then quantitatively analyzed by chemical ionization (CI) GC/MS using a 6-foot by 1/4-inch glass column filled with 3 percent OV-3 on 100/120 mesh Supelcoport. The column was temperature-programmed from 160 to 320°C at 4°C/min, holding the upper temperature until completion of the analysis. The GC/MS system consisted of a Finnigan Model 9610 gas chromatograph coupled with a Finnigan Model 4000 mass spectrometer.

The sensitivity of the system was improved by using single ion monitoring. The molecular ions of up to eight compounds were monitored simultaneously. Quantification was achieved by rationing the ion current of the particular molecular ion to that of an appropriate internal standard and then applying a calibration factor to minimize the differences in ionization efficiencies of the PAH compound and the internal standard. The use of single ion monitoring also minimized interferences from extraneous compounds not removed during the LC cleanup.

## PAH Tracers

Tracer solutions were prepared by KVB using material supplied from Merck and Co., Inc. The PAH material was weighted and dissolved in methylene chloride. Tracer solutions were made with concentrations that would provide the desired quantity of PAH tracer with a total injection of approximately 20 ml of methylene chloride.

Listed in Table 1 are the PAH tracers used in the program together with the PAH compounds identified by Battelle, the laboratory internal standards, and the symbols used in this report to identify the tracer compounds. The PAH materials have been divided into three groups covering a wide range of molecular weights spanning the normal spectrum of PAH compounds.

## TABLE 1. PAH Analysis

| PAH Compound | BCL Internal Standard | KVB Tracer Compound | Symbol |
|---|---|---|---|
| Anthracene/phenanthrene<br>Methyl anthracene<br>Fluoranthene<br>Pyrene<br>Methyl pyrene/fluoranthene | 9-Methylanthracene | $D_{12}$ anthracene<br>$D_{10}$ pyrene | A<br>X |
| Benzo(c)phenanthrene<br>Chrysene/benzo(a)anthracene<br>Methyl chrysenes<br>7,12-Dimethyl benz(a)anthracene | 9-Phenylanthracene | $D_{12}$ chrysene<br>$\beta,\beta'$-binaphthyl | C<br>B |
| Benzo fluoranthenes<br>Benz(a)pyrene<br>Benz(e)pyrene<br>Perylene<br>Methyl benzo pyrenes<br>3-Methyl cholanthrene<br>Indeno (1,2,3-cd) pyrene<br>Benzo(g,h,i)perylene<br>Dibenzo(a,h)anthracene<br>Dibenzo(c,g)carbazole<br>Dibenz(a1 and ah)pyrenes<br>Coronene | 9-10-Diphenylanthracene | $D_{12}$ Perylene<br>P-quaterphenyl | P<br>Q |

A summary of the tracers used, the quantities injected (Inj), quantities analyzed (Anl), and the total recovery factor (F) is presented in Table 2. Initially, only one PAH tracer, p-quaterphenyl, was used. As the uncertainty of the recovery of the tracer became apparent, additional tracers were added to a total of six compounds. Note that the quantity of tracers injected increased from nominally 1 $\mu$g to greater than 10 $\mu$g. Tracers were increased and quantities were changed in an effort to improve the relatively poor recovery of the tracers. A laboratory verification program of Battelle's analytical techniques, conducted as part of this study in response to the low levels of tracer recovered, indicated that these quantities were well in excess of the 100 ng detection limit for the total sample. As given in Table 2, the tracer recovery was very widespread with recovery factors ranging from 0.03 to 14.20. (Note that a recovery factor of 1.00 corresponds to 100 percent recovery of the tracer.) The wide variability of these recoveries indicated severe problems, including the possible "disappearance" of PAH tracers by some unknown factor and also problems with the sampling and analytical methodology.

## RESULTS

### Laboratory Evaluation

The first phase of the program consisted of a laboratory evaluation of the tracer preparation, sampling train operation, and the analytical

# 622  USE OF PAH TRACERS DURING SAMPLING

### TABLE 2. Summary of PAH Tracer($\mu$) Injections and Recoveries

| Test No. | $D_{12}$ anthracene ||| $D_{12}$ chrysene ||| $D_{12}$ perylene ||| $D_{10}$ pyrene ||| $\beta,\beta'$-Binaphthyl ||| p-quarterphenyl |||
|---|---|---|---|---|---|---|---|---|---|---|---|---|---|---|---|---|---|---|
| | Inj. | Anl. | f | Inj. | Anl. | f | Inj. | Anl. | f | Inj. | Anl. | f | Inj. | Anl. | f | Inj. | Anl. | f |
| Laboratory Combustor |||||||||||||||||||
| 1  | — | — | — | — | — | — | — | — | — | — | — | — | — | — | — | 9.8 | 8.9 | 0.91 |
| 2  | — | — | — | — | — | — | — | — | — | — | — | — | — | — | — | 3.3 | 1.5 | 0.45 |
| 3  | — | — | — | — | — | — | — | — | — | — | — | — | — | — | — | 0.6 | 0.5 | 0.83 |
| 10 | — | — | — | — | — | — | — | — | — | — | — | — | — | — | — | 0.4 | 0.2 | 0.50 |
| 11 | — | — | — | 1.5 | 1.8 | 1.20 | — | — | — | — | — | — | — | — | — | 0.3 | 0.3 | 1.00 |
| 12 | — | — | — | 1.5 | 1.9 | 1.27 | — | — | — | — | — | — | — | — | — | 1.1 | 0.3 | 0.27 |
| 13 | — | — | — | 1.5 | 1.2 | 1.00 | — | — | — | — | — | — | — | — | — | 1.1 | 1.0 | 0.91 |
| 14 | — | — | — | 1.2 | 1.2 | 1.00 | 1.2 | 0.6 | 0.50 | — | — | — | — | — | — | 1.3 | 0.8 | 0.62 |
| 15 | — | — | — | 1.3 | 0.3 | 0.23 | 1.3 | ND | 0 | — | — | — | — | — | — | 1.4 | 0.2 | 0.14 |
| 16 | — | — | — | 1.5 | 0.3 | 0.20 | 1.5 | ND | 0 | — | — | — | — | — | — | 1.6 | 0.1 | 0.06 |
| 17 | — | — | — | 1.2 | ND | — | 1.3 | ND | 0 | — | — | — | — | — | — | 1.3 | 0.1 | 0.08 |
| 18 | — | — | — | 0.7 | ND | — | 0.7 | ND | 0 | — | — | — | — | — | — | 0.7 | ND | 0 |
| 19 | — | — | — | 2.1 | 0.8 | 0.38 | 2.1 | 0.2 | 0.10 | — | — | — | — | — | — | 2.2 | ND | 0 |
| 20 | — | — | — | 1.8 | 0.1 | 0.06 | 1.8 | ND | 0 | — | — | — | — | — | — | 1.9 | 0.2 | 0.11 |
| 21 | — | — | — | 1.8 | 0.4 | 0.22 | 1.8 | ND | 0 | — | — | — | — | — | — | 1.9 | 0.3 | 0.16 |
| 22 | 1.5 | 4.5 | 3.0 | 1.5 | 8.7 | 5.80 | 1.5 | 2.1 | 1.40 | — | — | — | — | — | — | 1.4 | 0.4 | 0.29 |
| 23 | 1.9 | 2.6 | 1.37 | 1.9 | 6.9 | 3.63 | 1.9 | 0.4 | 0.21 | — | — | — | — | — | — | 1.8 | 1.1 | 0.61 |
| 200,000-Lb-Steam/Hr B&W Boiler |||||||||||||||||||
| 24 | 2.3 | 0.2 | 0.09 | 2.3 | 0.2 | 0.09 | 2.3 | 0.1 | 0.04 | — | — | — | — | — | — | 2.2 | ND | 0 |
| 25 | 2.3 | 0.9 | 0.39 | 2.3 | 2.0 | 0.87 | 2.4 | 0.1 | 0.04 | — | — | — | — | — | — | 2.2 | 0.5 | 0.23 |
| 26 | 4.4 | 0.2 | 0.05 | 4.4 | 0.6 | 0.14 | 4.4 | 0.2 | 0.05 | — | — | — | — | — | — | 4.3 | 0.2 | 0.05 |
| 27 | 2.3 | 2.4 | 1.04 | 2.3 | 5.3 | 2.30 | 2.3 | 0.2 | 0.08 | — | — | — | — | — | — | 2.2 | 0.6 | 0.27 |
| 28 | 2.2 | 0.2 | 0.09 | 2.2 | 0.2 | 0.09 | 2.2 | 0.1 | 0.05 | — | — | — | — | — | — | 2.1 | ND | 0 |
| 29 | 2.3 | 1.0 | 0.43 | 2.2 | 3.1 | 1.41 | 2.3 | 0.1 | 0.04 | — | — | — | — | — | — | 2.2 | 0.5 | 0.23 |
| 30 | 1.6 | 0.1 | 0.06 | 1.5 | 0.1 | 0.07 | 1.6 | 0.1 | 0.06 | — | — | — | — | — | — | 1.5 | 0.1 | 0.07 |
| 31 | 1.6 | 0.4 | 0.25 | 1.6 | 0.7 | 0.44 | 1.6 | 0.1 | 0.06 | — | — | — | — | — | — | 1.5 | 0.4 | 0.27 |
| 32 | — | — | — | — | — | — | — | — | — | — | — | — | — | — | — | — | — | — |
| 33 | 1.1 | 0.2 | 0.18 | 1.0 | 0.9 | 0.90 | 1.7 | ND | 0 | — | — | — | — | — | — | 1.0 | 1.5 | 1.50 |
| 34 | 0.8 | 8.9 | 11.13 | 0.8 | 3.4 | 4.25 | 0.8 | 1.0 | 1.25 | — | — | — | — | — | — | 0.8 | 2.8 | 3.50 |
| 35 | 1.2 | 0.2 | 0.17 | 1.1 | 0.9 | 0.82 | 1.2 | ND | 0 | — | — | — | — | — | — | 1.1 | 0.4 | 0.36 |
| 36 | 2.9 | 3.2 | 1.10 | 2.9 | 0.5 | 0.17 | 2.9 | ND | 0 | — | — | — | — | — | — | 2.8 | 0.1 | 0.04 |
| 37 | 1.8 | 1.9 | 1.06 | 1.8 | 1.6 | 0.89 | 1.8 | ND | 0 | — | — | — | — | — | — | 1.8 | 0.3 | 0.17 |
| 575-MM CE Boiler |||||||||||||||||||
| 39 | 10.9 | 1.9 | 0.17 | 5.9 | 0.3 | 0.05 | 5.9 | 0.2 | 0.03 | 10.3 | 2.1 | 0.20 | 6.4 | 0.03 | 0.05 | 6.4 | 2.1 | 0.33 |
| 40 | 6.7 | 1.0 | 0.15 | 3.6 | 1.3 | 0.26 | 3.6 | 0.3 | 0.08 | 6.3 | 1.2 | 0.21 | 3.9 | 1.0 | 0.26 | 3.9 | 3.2 | 0.82 |
| 41 | 8.1 | 2.2 | 0.27 | 4.4 | 1.2 | 0.27 | 4.4 | 0.6 | 0.14 | 7.6 | 2.1 | 0.28 | 4.7 | 1.3 | 0.28 | 4.7 | 3.8 | 0.81 |
| 42 | 13.0 | 1.7 | 0.13 | 7.1 | 1.9 | 0.27 | 7.0 | 0.6 | 0.06 | 12.3 | 2.4 | 0.20 | 7.6 | 1.6 | 0.21 | 7.6 | 3.6 | 0.47 |
| 43 | 12.8 | 0.7 | 0.06 | 6.9 | 2.6 | 0.38 | 6.9 | 0.5 | 0.07 | 12.0 | 0.8 | 0.07 | 7.4 | 0.4 | 0.05 | 7.5 | 7.9 | 1.05 |
| 44 | 12.3 | 16.2 | 1.32 | 6.7 | 8.7 | 1.30 | 6.7 | ND | 0 | 11.6 | 16.5 | 1.42 | 7.2 | 11.3 | 1.57 | 7.2 | 20.5 | 2.85 |
| 45 | 11.2 | 1.6 | 0.14 | 6.4 | 29.0 | 4.45 | 6.4 | 3.2 | 0.50 | 11.1 | 1.9 | 0.17 | 6.9 | 26.0 | 3.96 | 6.9 | 98.1 | 14.20 |
| 46 | 12.7 | 0.9 | 0.07 | 6.9 | 1.9 | 0.28 | 6.9 | 0.7 | 0.10 | 35.8 | 3.7 | 0.10 | 24.6 | 5.4 | 0.22 | 24.7 | 11.3 | 0.46 |
| 47 | 12.8 | 0.6 | 0.05 | 6.9 | 2.0 | 0.29 | 6.9 | 0.6 | 0.09 | 36.2 | 1.6 | 0.04 | 24.8 | 0.7 | 0.03 | 24.9 | 12.4 | 0.50 |
| 48 | 8.4 | 0.8 | 0.10 | 4.5 | 1.7 | 0.38 | 4.5 | 0.8 | 0.18 | 8.0 | 5.3 | 0.66 | 4.9 | 5.2 | 1.06 | 5.0 | 13.4 | 2.68 |
| 49 | 11.8 | 2.2 | 0.19 | 6.4 | 1.3 | 0.20 | 6.4 | 0.3 | 0.05 | 12.1 | 11.8 | 0.98 | 7.5 | 4.5 | 0.60 | 7.5 | 11.0 | 1.47 |
| 50 | 12.4 | 0.5 | 0.04 | 6.7 | 0.7 | 0.10 | 6.7 | 0.3 | 0.05 | 11.6 | 0.7 | 0.06 | 7.2 | 1.0 | 0.14 | 7.2 | 4.0 | 0.56 |
| 51 | 12.0 | 0.2 | 0.02 | 6.5 | 0.5 | 0.08 | 6.5 | ND | — | 11.0 | 6.7 | 0.61 | 6.8 | 2.8 | 0.41 | 6.8 | 3.1 | 0.46 |
| 52 | 12.0 | 0.2 | 0.02 | 6.5 | 0.9 | 0.14 | 6.5 | 0.1 | 0.02 | 11.2 | 0.2 | 0.02 | 6.8 | ND | 0 | 6.8 | 0.2 | 0.03 |
| 53 | 11.5 | 0.5 | 0.04 | 6.2 | 0.4 | 0.07 | 6.2 | 0.8 | 0.13 | 10.8 | 2.0 | 0.19 | 6.7 | 0.6 | 0.09 | 6.7 | 0.6 | 0.09 |
| 54 | 12.9 | 1.6 | 0.12 | 6.9 | 6.9 | 1.00 | 6.9 | 3.2 | 0.46 | 12.1 | 3.7 | 0.31 | 7.4 | 6.1 | 0.82 | 7.5 | 11.2 | 1.49 |
| 56 | 11.7 | 2.0 | 0.17 | 6.4 | 1.5 | 0.23 | 6.3 | 0.5 | 0.08 | 11.0 | 1.1 | 0.10 | 6.8 | 0.8 | 0.12 | 6.8 | 1.5 | 0.22 |

procedures. Standard tracer solutions were prepared and injected while drawing (a) hot air and (b) flue gas from a laboratory combustor.

Several aliquots of tracer solutions with one, two and three tracer compounds were sent to BCL for analysis. These samples were subjected to the same workup and analytical procedures as described previously for field samples. As a cross-check, a limited number of samples were analyzed by Denver Research Institute (DRI) using no workup and chromatographic quantification of only the specific PAH compounds. A summary of the relative recoveries from these analyses is presented in Table 3. Although there is significant scatter in the data, the results indicate reasonably close inter-laboratory agreement and recoveries near 100 percent.

Two sets of check-out tests were made while drawing hot air at 350°F. The first set used a single tracer (p-quaterphenyl), the second a double tracer (p-quaterphenyl and $D_{12}$ chrysene). For all tests, the PAH tracers were injected into the sampling trains through the probe tip set up (illustrated in Figure 1) at concentrations and rates comparable to the later field tests. The recovery factors are shown in Figure 2 in bar graph form. The initial set of test results showed fair (0.50) to excellent (0.80) recovery.

TABLE 3. Relative Recoveries of Tracer Injection Solution (Laboratory Tests)

|  | Sample 58 | | Sample 68 | | Sample 106 | |
| --- | --- | --- | --- | --- | --- | --- |
|  | BCL | DRI | BCL | DRI | BCL | DRI |
| $D_{12}$ chrysene | — | — | 69<br>87<br>90<br>160<br>79 | 102 | 247 | — |
| $D_{12}$ perylene | — | — | — | — | 115 | — |
| p-quaterphenyl | 88<br>75<br>104 | 102 | 60<br>72<br>88<br>105 | 102 | 144 | — |

The second set of tests showed excellent recovery for the $D_{12}$ chrysene, while the p-quaterphenyl recoveries were mixed. The bar graph in Figure 2 has been separated into the recovery found in the heated portion of the train (probe, cyclone and filter) and the cold portion (gas conditioner and adsorber). Note that the fraction depositing in the heated portion varied among tests, representing 5 to 80 percent of the total injection. The results of these tests under ideal sampling conditions were encouraging and

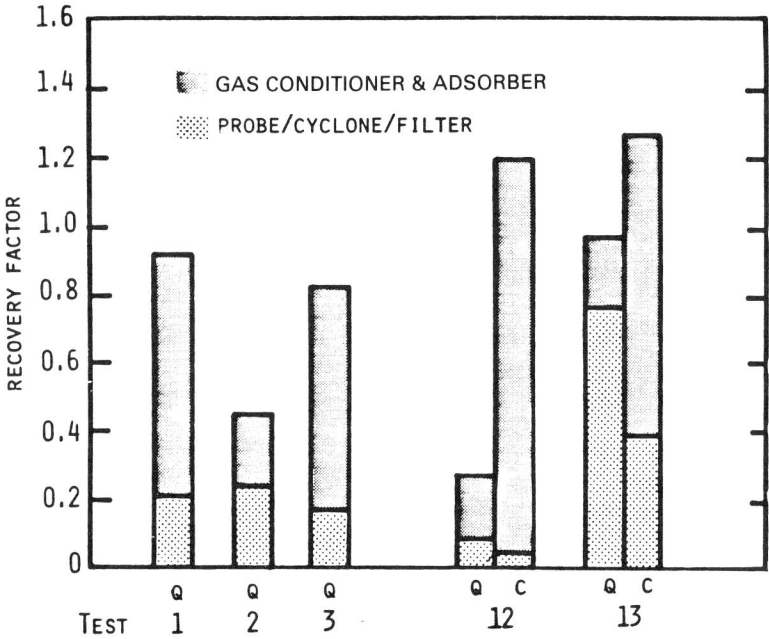

FIGURE 2. Tracer recovery for hot-air tests.

suggested that the system would be workable for the more complex conditions with particulate-laden stack gases.

The balance of the laboratory tests involved injecting PAH tracers while sampling flue gas from the KVB $4 \times 10^6$ Btu/hr pulverized-coal boiler upstream and downstream from a baghouse (BH). Single (p-quaterphenyl), triple (plus $D_{12}$ chrysene and $D_{12}$ perylene) and quadruple (plus $D_{12}$ anthracene) were used during these tests. The objective was to verify the use of the sampling and analytical techniques under conditions similar to those expected during the follow-on field programs. The results of these tests, again in bar graph form, are presented in Figures 3 and 4.

Several observations can be made from these results:
- Recoveries ranged from near zero to greater than five. (Note that the high recoveries were limited to the quadruple tracer, which may indicate an error in its preparation.)
- Significant species-to-species variations are evident for nearly all tests. In general, $D_{12}$ chrysene showed the best recovery whereas $D_{12}$ perylene had the poorest.
- No effect of particulate loading was apparent for the tests conducted upstream and downstream from the laboratory baghouse.

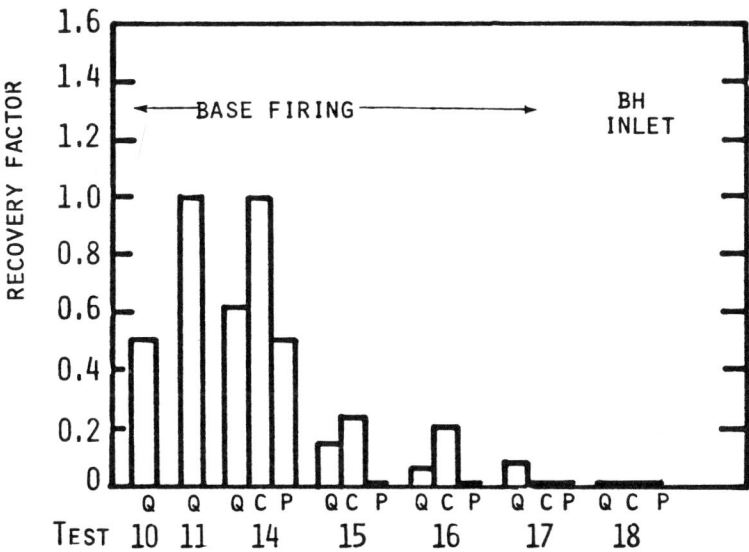

FIGURE 3. Tracer recoveries for laboratory combustor tests.

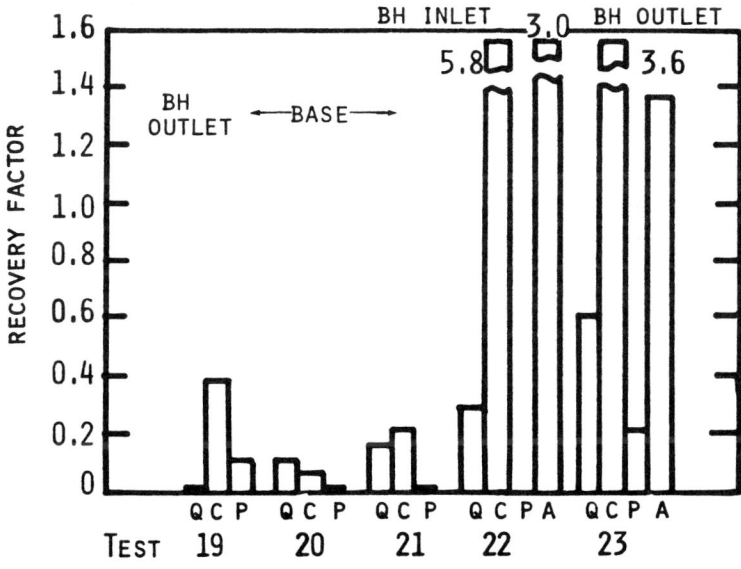

FIGURE 4. Tracer recoveries for laboratory combustor tests.

Unfortunately, breakdown of these data into recovery from the various portions of the train was not possible, so only the total recoveries are presented.

## Field Test at a Baghouse

The second phase of the program involved PAH sampling from a small utility boiler (210,000 lb steam/hr) equipped with a high-efficiency (greater than 99.9 percent) fabric filter baghouse for particulate control. The boiler fired low-sulfur (0.6 percent) western coal. The objective of these tests was to characterize PAH emissions at the inlet and outlet of the baghouse.

The quadruple tracer solution was injected simultaneously with the sampling. The recovery factors for these tests are presented in Figures 5 and 6 for the baghouse inlet and outlet, respectively. As before, the unshaded bars represent total recovery and the shaded areas represent collection from the hot section of the train (associated with the particulate catch) and the cold section (associated with the gaseous collection). With the exception of Tests 34 and 35, these tests were conducted with stack gas particulate and gaseous consitituents characteristic of full-load operation and at temperatures near 300° F. The other tests were made at a reduced load with comparable stack gas temperatures.
Examination of the recovery factors shows several distinct trends.

- The recovery factors of the tracers within a single test are not constant, despite injection as a single solution. In general, the recovery factors

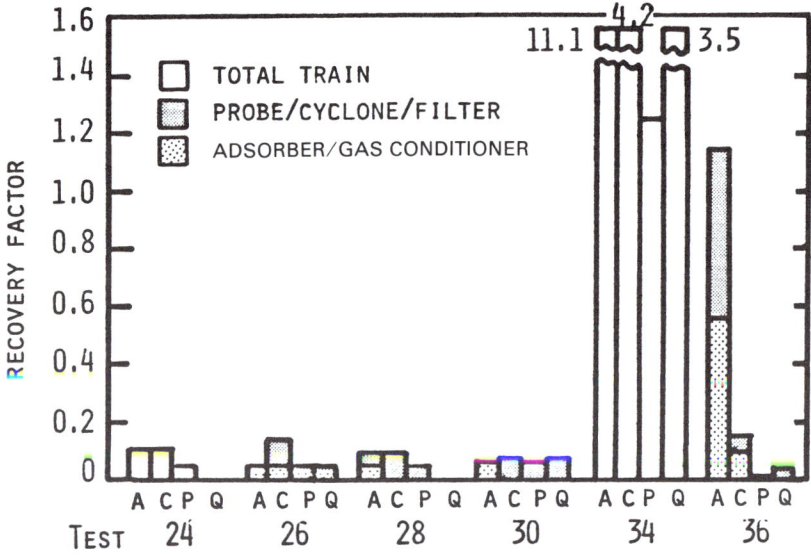

FIGURE 5. Tracer recoveries at field baghouse.

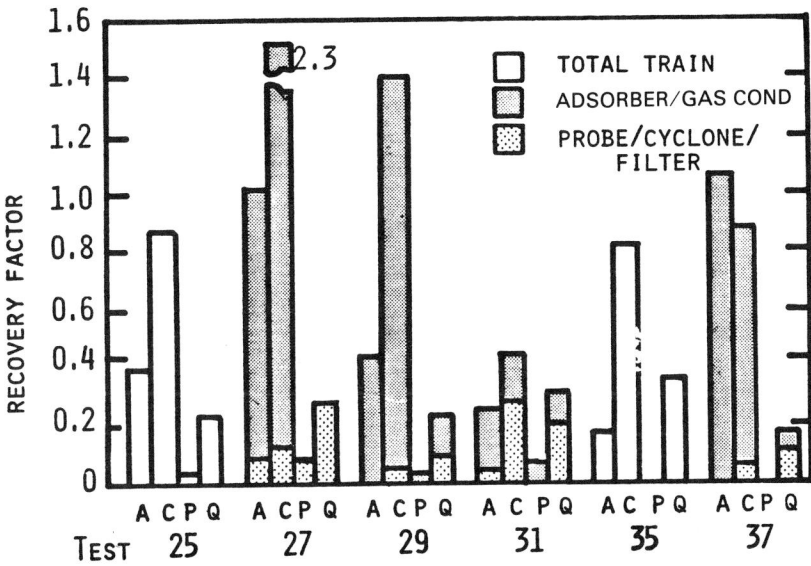

FIGURE 6. Tracer recoveries at field baghouse.

tend to be reduced at the heavier POM species ($D_{12}$ perylene and p-quaterphenyl) compared to the two lighter tracers. The tracer showing the worst recovery is the $D_{12}$ perylene.
- Greater recovery of tracers for the tests conducted at the baghouse outlet is evident. With the exception of a single set of tests, each outlet test showed a higher recovery than in the corresponding inlet test.
- Significant variations in recovery factors are present for similar test condition and sample locations.
- Greater recovery is evident in the adsorber section of the sampling train.

The second trend is especially significant in that it suggests higher recovery at conditions of reduced particulate concentration.

## Field Test Program on an Electrostatic Precipitator

The third and final phase of the program was the characterization of PAH emissions from a large coal-fired utility boiler (575 MW) equipped with a high-efficiency electrostatic precipitator (ESP). The unit fired high-sulfur (2.3 percent) eastern coal with temperatures at the ESP inlet and outlet near 240°F.

The objective of the program was to characterize the PAH emissions at the ESP inlet and outlet. Again, tracers were injected during sampling. Based on the results of the baghouse tests, a series of diagnostic tests of the sampling system were also conducted. These tests included (a) increased concentrations of tracer compounds, (b) injection of the tracers during various levels of particulate build-up on the filters, and (c) injection in the sampling system downstream from the hot section, that is, past the area of particulate collection. To facilitate these tests, two additional PAH compounds were added to the set of tracers previously used. These six PAH compounds contained two compounds in each of the three groups given in Table 1. These tracers were injected together in one solution or split into two solutions of three tracers to conduct separate diagnostic tests during the same sampling run.

Additional changes from the baghouse program included (a) the injection location and (b) laboratory cross checks. Due to the absence of the cooling water for the probe tip injection system, the point of the injection was moved just upstream of the cyclone, thereby removing the injection line from contact with the hot flue gases. This change should not have affected tracer recovery appreciably. In addition, the analytical program was expanded to include a limited number of analyses by A. D. Little (ADL) as a cross-check of the analytical methodology.

The relative recoveries of the four tracer solutions used during these tests, expressed in percent, are given in Table 4. Solution A, the base solution containing all six compounds and used for the bulk of the tests, was analyzed by both labs. Solution B contained a higher concentration of all six compounds while C and D were the two triple tracer solutions used in the diagnostic tests. The BCL analyses of $D_{12}$ chrysene, $D_{12}$ perylene, $D_{10}$ pyrene and $\beta,\beta'$-binaphthyl showed excellent recoveries. The $D_{10}$ anthracene recoveries were consistently low whereas the p-quaterphenyl recoveries were high. Results of the ADL analyses were comparable for all compounds except $D_{12}$ chrysene and $D_{10}$ pyrene, which had lower recoveries than in the BCL analyses.

**TABLE 4.** Relative Recoveries of Tracer Injection Solutions (ESP Tests)

|  | ADL[a] | BCL[b] | | | |
|---|---|---|---|---|---|
|  | A | A | B | C | D |
| $D_{10}$ anthracene | 57.5 | 39.8 | 85.1 | 40.5 | — |
| $D_{12}$ chrysene | 12.6 | 118 | 93.6 | 119 | — |
| $D_{12}$ perylene | 83.0 | 84.4 | 80.9 | 107 | — |
| $D_{10}$ pyrene | 47.6 | 97.4 | 93.5 | — | 93.8 |
| $\beta,\beta'$-binaphthyl | 120 | 89.6 | 84.6 | — | 76.0 |
| p-quaterphenyl | 171 | 196 | 142 | — | 178 |

[a]ADL—A. D. Little.
[b]BCL—Battelle Columbus Laboratories.

The recovery factors for the tests conducted at the inlet and outlet of the ESP are presented in Figures 7 and 8, respectively. With the exception of Tests 44 and 45, all analyses were made by BCL. Several trends are apparent from these data:

- With the exception of the ADL analysis, the tracer recoveries are generally poor — about 20 percent. The difference in the ADL data has not been explained.
- Significant differences in the recoveries of the PAH compounds in the same solution are evident. The recoveries of $D_{12}$ chrysene, $\beta,\beta'$-binaphthyl, and p-quaterphenyl were consistently higher than those of the remaining compounds.
- Comparable recoveries are evident for the ESP inlet and outlet samples. This is contrary to the results of the field baghouse tests.
- Use of the higher concentrations of tracers in Tests 46 and 47 did not improve recoveries.
- The bulk of the recoveries was from the particulate catches, again unlike the baghouse results.

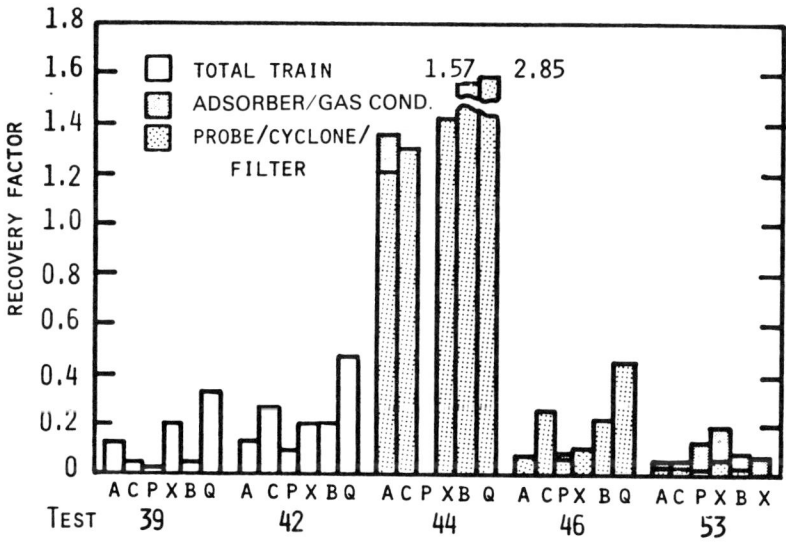

FIGURE 7. Tracer recoveries at ESP inlet.

The results of the diagnostic tests are presented in Figure 9. Spike tests were conducted during which Solution C was injected during the initial

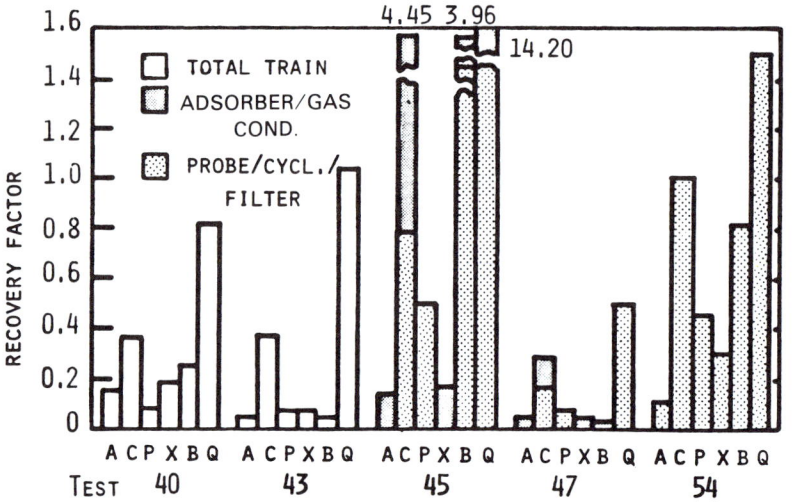

FIGURE 8. Tracer recoveries at ESP outlet.

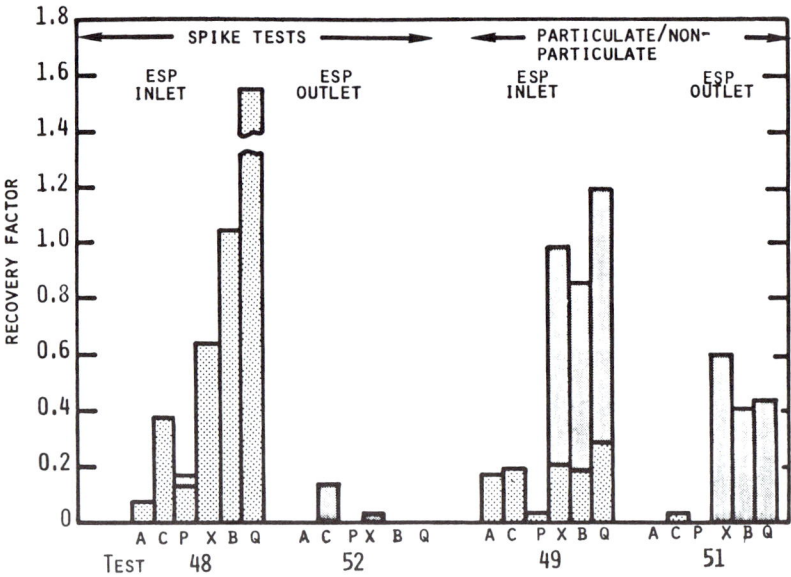

FIGURE 9. Tracer recoveries for diagnostic tests.

part of the sampling with presumably low particulate levels and Solution D was injected near the end of the test. As shown, the spike tests conducted at the ESP inlet showed enhanced recovery for the second PAH set with nearly all PAH tracers recovered in the particulate catch. The single test at the ESP outlet showed poor recovery for all tracers.

The balance of the diagnostic tests involved a split in the injection location. While Solution C was injected upstream of the cyclone, as before, Solution D was simultaneously injected just upstream of the gas conditioner. This test was designed to separate injection under particulate and nonparticulate environments. As shown in Figure 9, the test conducted at the ESP inlet showed excellent recovery for the downstream injection with poor recovery for upstream injection. (The occurrence of Solution D tracer in the probe/filter analysis was probably due to an error during cleanup.) The results at the ESP outlet showed a similar trend of enhanced recovery for PAH tracers not subjected to the particulate-laden filters. These data support the previous results of the baghouse program as to the potential interference of particulate matter.

## DISCUSSION

The results of the tracer injection tests generally showed poor recovery for the tracer compounds. While the recoveries for PAH tracer material from standard solutions and hot air tests approached 100 percent, the recovery for tracers injected during sampling of flue gas streams, with few exceptions, was below 30 percent. Significant variations in tracer recoveries were apparent both between PAH compounds in the same injection solution and repeated tests conducted at as nearly identical test conditions as possible. These results indicate significant problems with the accuracy and precision of current sampling methodology used to quantify PAH emissions. The factors contributing to these problems can only be speculated on.

Of primary concern as causes of the poor recoveries were such controllable factors as (a) the procedures used for tracer solution preparation and subsequent injection, and (b) the analytical preparation and methodology. The results of the tracer solution analyses and hot air tests would appear to demonstrate that the preparation and injection procedures of the tracer solutions were made satisfactorily. As for the second concern, the poor recoveries prompted an extensive evaluation of the accuracy and precision of the BCL methodology. The results of these tests generally showed the CI GC/MS procedures to provide accurate, repeatable analyses of laboratory-prepared PAH samples. Not evaluated, however, was the efficiency of the extraction procedures when particulate matter was present. There is growing evidence that conventional extraction of PAH from fly ash is not completely efficient (3), and this remains a potential source of the poor recovery levels.

Another factor that could have contributed to the variable recoveries observed is the influence of uncontrollable items associated with particulate and stack gas conditions. These items could affect both tracer and real PAH compounds in several ways including (a) chemical alteration by surface reactions and/or (b) irreversible adsorption on fly ash surfaces. Considering the complex composition and morphology of the particulate matter, any number of chemical phenomena are possible. These unknown factors pose a significant sampling and analytical complication to PAH measurement.

Operation and make-up of the sample train could similarly affect recoveries. Filter and adsorber surfaces could act as a catalyst surface altering the PAH structure. Operation of the train at reduced temperatures (below the water dew point) could result in condensation of sulfuric acid and subsequent degradation of the PAH compounds.

In light of these and other potential interfering factors, a significant degree of uncertainty exists in PAH measurement procedures. Much additional work under controlled laboratory sampling conditions using realistic field conditions (e.g., particulate composition, stack temperature, etc.) is needed. It is strongly recommended that repeated measurements be made during these evaluations due to the high variability observed in these results.

This high degree of variability between the recoveries of the tracer compounds under identical test conditions raises questions as to the similarity of the behavior of PAH compounds in general. If the tracer results reflect the behavior of PAH compounds, then highly irregular recoveries of real PAH compounds could be expected. The use of tracers as a means of correcting sampling losses is therefore not recommended until these phenomena are better understood.

Earlier investigations of PAH sampling train performance using clean, laboratory-generated flue gas (without particulate matter) and limited field measurements using conventional procedures should be reexamined. Even the results of this study should be viewed with caution due to the uncertainties present.

## REFERENCES

1. KVB, Inc. (1980): Report to Electric Power Research Institute, Characterization of polycyclic organic matter in power plant effluents (to be issued).
2. Jones, P. W., et al (1977): Measurement of polycylic organic materials and other hazardous organic compounds in stack gases - state of the art. NTIS PB 274 013.
3. Griest, W. H., et al (1979): Identification and quantification of polynuclear organic matter (POM) on particulates from a coal-fired power plant, EPRI Report EA-1092.

# BINDING OF BaP DIOL EPOXIDE TO CHROMOSOMAL HISTONE PROTEINS

### A. Kootstra*, M. C. Cronen**, and T. J. Slaga*

*University of Tennessee—Oak Ridge
Graduate School of Biomedical Sciences,
and Biology Division
Oak Ridge National Laboratory
Oak Ridge, Tennessee 37830

**Southern Colleges University Union Student
Centre College
Danville, Kentucky. Winter 1979.

### INTRODUCTION

In order to understand how a chemical such as benzo(a)pyrene [BaP] exerts its carcinogenic effect, a fundamental knowledge is required concerning the interaction of metabolites of BaP with the macromolecules of the eukaryotic cell and the consequences of such interaction at the molecular level. Since evidence suggests that DNA is an important target of the ultimate carcinogenic form of BaP, the BaP diol-epoxide (6, 14, 25-27, 34), we have studied the binding of BaP diol epoxide (anti-isomer) to chromatin in vitro, in order to obtain some information about the interaction of this carcinogen with the various components of chromatin (16, 17).

Chromatin at the primary level of organization, consists of a linear array of nucleosomal core particles which are linked to one another by a variable DNA spacer or linker region, and this structure, which has the appearance of beads on a string, is produced through the interaction of histone proteins with eukaryotic DNA (10, 18, 24, 37).

The histone proteins, which are a class of basic proteins, can be divided into two groups: the very lysine-rich histones H1 (and H5 in avian erythrocytes) and the inner histones H2A, H2B, H3, and H4. The very lysine-rich histones are associated with the linker DNA connecting adjacent nucleosomes and also seem to interact with the nucleosomal core particle (23, 32). Current evidence suggests that these very lysine-rich histones are important molecules in the generation of higher order packaging of the nucleosomal core particles into the thicker interphase chromatin fibers (22, 29). The inner histones, the second group, are responsible for the structural arrangement of the nucleosomal core particle DNA. Biophysical and biochemical evidence has shown that the nucleosomal core particles has the dimensions of a flat disk about 110 A in diameter and 57 A in height (10, 11, 30). The nucleosomal core particle contains two of each of the four inner histones, an octamer of H2A, H2B, H3, and H4, around which 140 (146 from more recent analysis) base pairs of DNA double helix are coiled.

Comparative primary sequence analysis of the inner histones has shown that these proteins have been highly conserved during the evolutionary process, the arginine-rich histones, H3 and H4, more so than the moderately lysine-rich histones H2A and H2B (7, 9, 15, 33). These results, together with the observation from reconstitution experiments which suggested that the arginine-rich histones H3 and H4 alone are capable of generating the nucleosomal structure, imply that the histones H3 and H4 are the essential structural components of the nucleosomal core particles, while histones H2A and H2B may play a somewhat different role (5, 28).

Although the histone proteins are highly conserved, they are also subject to extensive postsynthetic side-chain modifications such as phosphorylation, acetylation, methylation, and ADP-ribosylation (1, 8, 12, 31). The exact biological role of these specific enzymatic side-chain modifications is unknown, but it has been suggested that the phosphorylation and dephosphorylation of histone H1 has been linked with chromosomal condensation during cell division (3). The analysis of the positions of acetylation has shown that a relatively small number of lysine side-chain modifications occurred and these are clustered in the N-terminal portions of the inner-histone molecules. Since the N-terminal regions of these histones are thought to interact with the DNA phosphate groups, enzymatic modification of these N-terminal regions may change the histone-DNA interaction and subsequently influence the structure and function of chromatin (1, 31). In view of the importance of histone modifications, the modification of histone proteins, in situ, by potential carcinogens (4, 13, 17, 21, 35) may well have important consequences in the initial molecular events that lead to malignant transformation of the eukaryotic cell. Such histone modifications by carcinogens could be viewed as epigenetic events. The data presented in this paper suggest that the epsilon amino acid side-chains of lysine residues may well be important targets for the BaP diol epoxide molecules, and that there appears to be some sterospecificity in the binding of the carcinogen to chromatin histone protein in vitro.

## MATERIALS AND METHODS

### Isolation of Soluble Chromatin

Chicken erythrocyte nuclei were isolated as described previously (17). The nuclear suspension was diluted with isolation buffer [10 mM tris (pH 7.0), 10 mM NaCl, 3 mM $MgCl_2$, 0.5 percent Nonidet P40, 0.1 mM PMSF] to a concentration of $1.4 \times 10^9$ nuclei/ml. The suspension was made up to 1.0 mM $CaCl_2$, and the digestion by micrococcal nuclease (60 units/ml) at 37°C was stopped after 10 minutes by addition of ethylenediaminetetraacetic acid (EDTA), 10 mM final concentration. The digested nuclear suspension was dialyzed (Spectrapor) for 24 hours at 4°C, with three changes of 0.2 mM EDTA (pH 7.0), 0.1 mM PMSF. The lysed nuclear suspension was centrifuged at 10,000 x g for 10 minutes at 4°C, and the supernatant solution containing solubilized chromatin was concentrated with an Amicon PM10 membrane. Removal of subnucleosomal material, mononucleosomes, and up to pentanucleosomes was achieved by centrifugation of the solubilized chromatin through 5 to 25 percent sucrose gradients containing 0.2 mM EDTA (pH 7.0), 0.1 mM PMSF, and 20 mM NaCl (16, 17). Oligonucleosomal fractions (for convenience these oligonucleosomes are referred to as chromatin) were pooled and dialyzed against 10 mM tris (pH 7.0), 0.1 mM PMSF at 4°C for 12 hours.

### Dissociation and Reaction with Carcinogen

The dialyzed oligonucleosomes were concentrated to 7.3 $A_{260}$ units/ml, and 5-ml aliquots were made up to either 0.65 or 2.0 M in NaCl concentration by slow addition of solid NaCl with gentle stirring. The solutions were kept at 4°C overnight and equilibrated at 37°C for 10 minutes. The carcinogen, $^{14}C$-labeled BaP diol-epoxide anti-isomer (specific activity, 29.4 mCi/mmole), which was dissolved in tetrahydrofuran:triethylamine (19:1 v/v), was added to the chromatin solutions at a DNA:carcinogen ratio of 100:1 (w/w) while the organic solvent concentration was <1.0 percent. The reaction was carried out at 37°C for 30 minutes under yellow light, after which the chromatin solutions were dialyzed against 0.2 mM EDTA (pH 7.0) until the radioactive background was constant. The dialyzed soluble carcinogen-labeled chromatin solutions were isolated and lyophilized. The lyophilized chromatin was solubilized in sodium dodecyl sulfate (SDS) sample application buffer (19) to a concentration of 1.0 mg/ml with respect to the histone proteins.

## Trypsin Digestion and Analysis by SDS Polyacrylamide Gel Electrophoresis

Chromatin dissolved in 5 mM tris/HCl pH 8.0 (10 $A_{260}$ units/ml) was digested with trypsin-treated L-1-tosylamido-2-phenylethyl chloromethyl ketone (Worthington Biochemical Corp.), as described by Whitlock and Simpson (36). The digestion was terminated by the addition of SDS (0.1 percent final concentration) and the samples were immediately lyophilized. The extent of tryptic digestion was routinely monitored by 20 percent SDS polyacrylamide slab gel electrophoresis.

## SDS-polyacrylamide Electrophoresis

Fifteen or 20 percent SDS-slab gel electrophoresis was performed essentially as described by Laemmli (19), and 8 $\mu$l of the protein sample (8 $\mu$g/slot) was loaded. Electrophoresis was performed at 100 V/gel. The gels were stained for 20 minutes in 20 percent acetic acid, 30 percent isopropanol, and 0.01 percent Coomassie blue (8); rinsed in water; destained in 10 percent acetic acid; and then photographed. Densitometer scans of the stained protein bands were obtained with a Zeineh soft laser scanning densitometer.

## Fluorography of Stained Slab Gels

For fluorography, the slab gels were treated with dimethyl sulfoxide and 2,5-diphenyloxazole (scintillation grade) as described by Laskey and Mills (20). The films (Kodak X-ray film XRP-5) were exposed at -70°C. The developed films were photographed and scanned as described above.

## RESULTS

### Labeling of Histone Proteins as a Function of NaCl Concentration

Figure 1 shows the densitometer tracings of stained histone proteins obtained from chromatin reacted with BaP diol-epoxide (anti-isomer) in

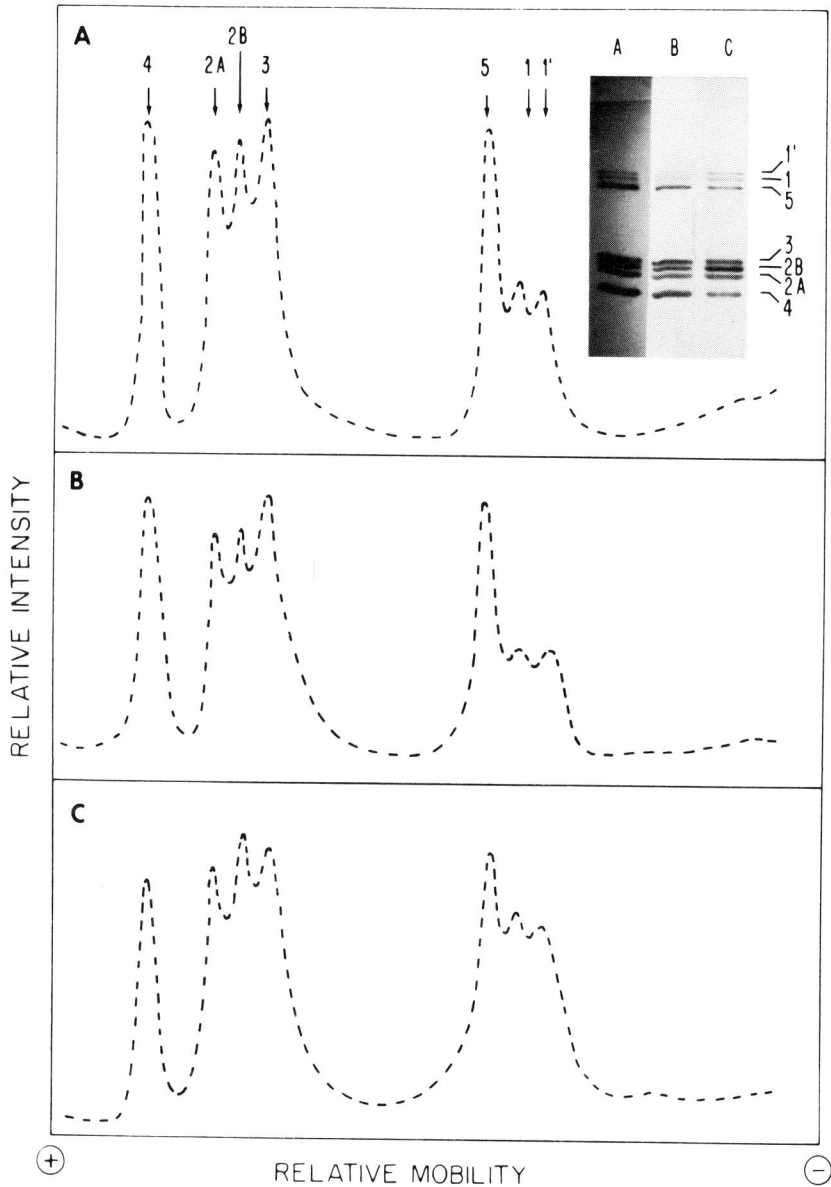

FIGURE 1. Densitometer tracing of carcinogen-labeled nucleosomal histone proteins after the gels were stained. Histones from oligonucleosomes were labeled (A) in the absence of NaCl, (B) in the presence of 0.65M NaCl, or (C) in the presence of 2.0M NaCl. Inset, stained gel.

the absence of NaCl (A) and in the presence of 0.65M NaCl (B) and 2.0M NaCl (C). Although all the histones were recovered after dialysis, the corresponding fluorogram (Figure 2), which indicates the relative amount of radioactive carcinogen bound to the histone proteins, revealed that in the control chromatin, after the reaction with the carcinogen, histones H3 and H2B bound most of the carcinogen, while histone H4 was associated with a relatively small amount of the carcinogen. The fluorogram also showed the presence of an unidentified labeled molecule which migrated just above histone H4.

When chromatin was labeled under 0.65M NaCl dissociating conditions, histone H5, but neither of the H1 molecules, was extensively modified by the carcinogen. This was in contrast to the labeling pattern under nondissociating conditions. Futhermore, under these conditions of labeling (0.65M NaCl), histone H2A was also modified by the carcinogen, although not to the same extent as histones H3 and H2B (Figure 2B).

When chromatin was dissociated in the presence of 2.0M NaCl, BaP diol-epoxide modified all the histones (data not shown). The soluble chromatin fraction after dialysis showed that all the very lysine-rich histones were modified but only histones H3 and H2A were associated with the carcinogen (Figure 2C). Analysis of the histone proteins in the precipitated fraction revealed that the very lysine-rich histones, H1', H1, and H5, also were labeled but that the distribution of the label in the inner histones was different. Here the histones H2B and H4 were found to contain the carcinogen. The nature of this differential precipitation of chromatin after dialysis is currently under investigation.

## Tryptic Digestion of Carcinogen-Bound Chromatin

The control chromatin labeled with $^{14}$C-BaP diol-epoxide as shown in Figure 1A, in which histones H3 and H2B contained most of the bound carcinogen (Figure 2A), was digested by trypsin which removes the N-terminal regions of the inner histones. The kinetics of digestion monitored by SDS poly-acrylamide gel electrophoresis (Figure 3) and the stained gels were treated for fluorography (Figure 3, at the bottom). The analysis of the fluorogram shows that the relative intensity, which is related to the amount of labeled carcinogen present, was not significantly changed during the tryptic digestion. Furthermore, the position of the modified proteins remained unaltered during the tryptic digestion and remains associated with the positions in which the intact H3 and H2B molecules migrate. These results therefore imply that the modified molecules are resistant to tryptic digestion.

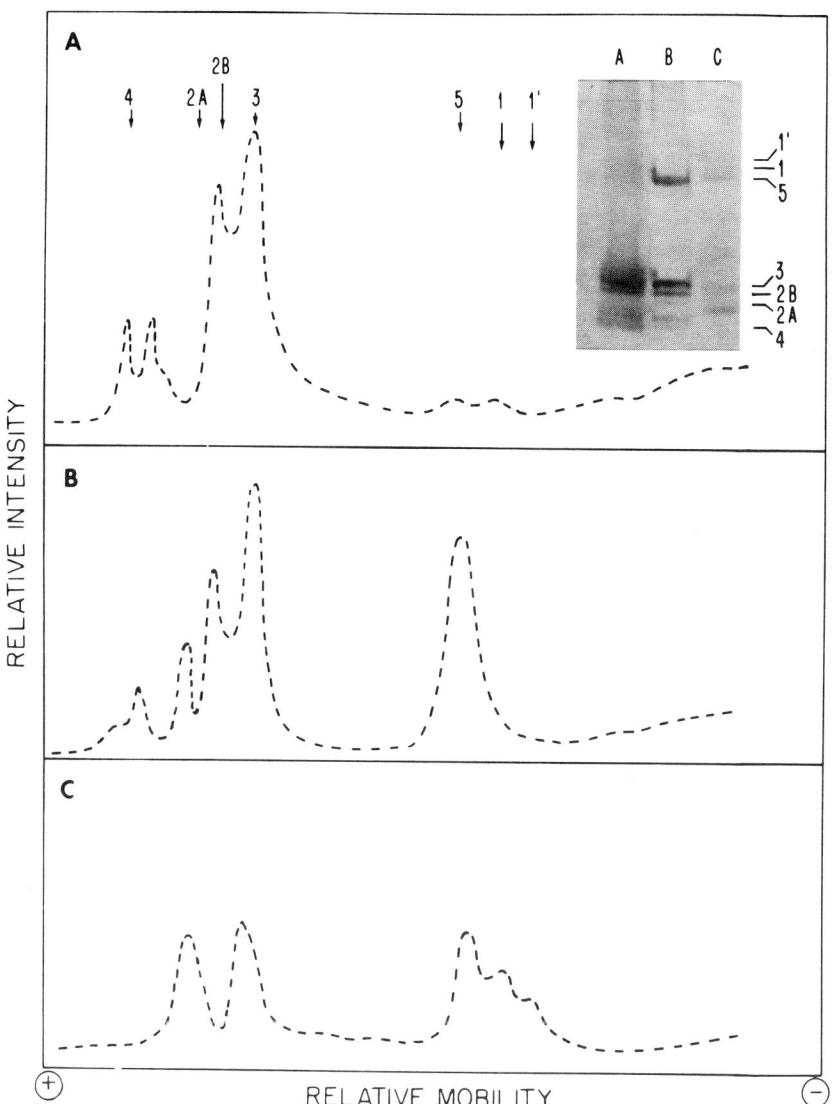

FIGURE 2. Densitometer tracing of the developed X-ray film of the stained gel shown in Figure 1 after treatment for fluorography. A, B, and C refer to the conditions of labeling as described in Figure 1. Inset, the developed fluorogram; the apparent increase in mobility of histones H3 and H2A (inset right-hand side) was caused by a slight deformation of the gel as a consequence of the treatment for fluorography. The developed X-ray film was matched with the stained protein bands in the gel to ensure proper identification.

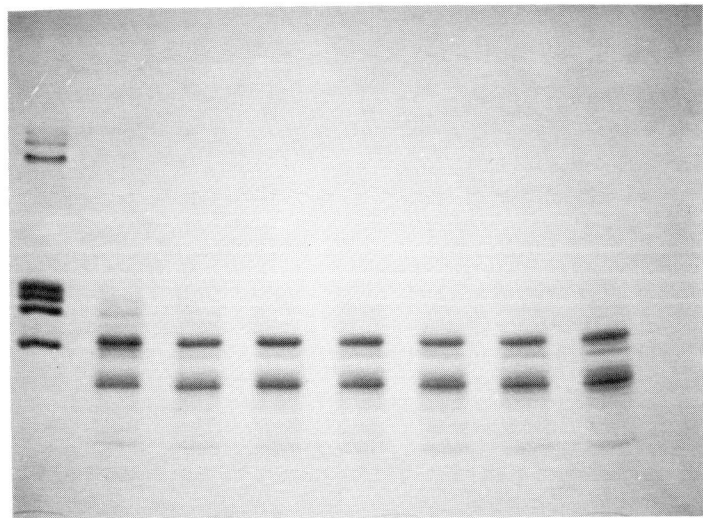

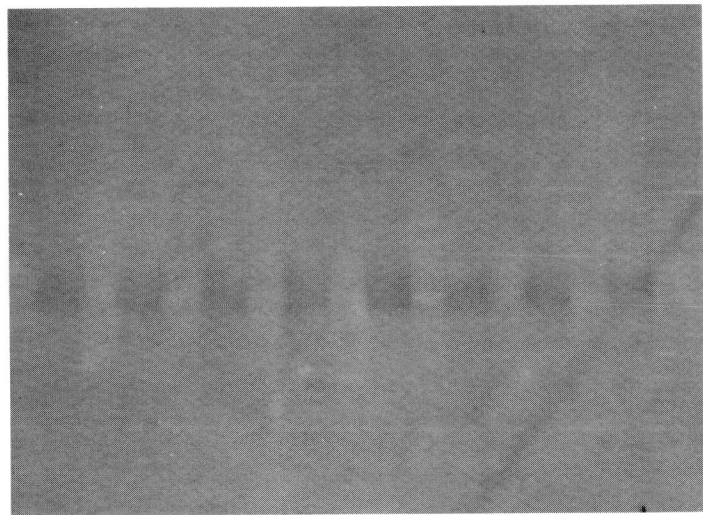

FIGURE 3. Oligonucleosomes, labeled with BaP diol-epoxide (anti-isomer) as described in Figure 1 (A) were digested by trypsin. Aliquots(2.0 $A_{260}$ units) were removed at 0, 20, 40, 60, 80, 100, 120, and 140 minutes and the proteins were subjected to electrophoresis on 20 percent SDS polyacrylamide gels. The stained gel is shown in the top panel. The gel was then treated for fluorography and the bottom panel shows the developed X-ray film. The position of the carcinogen-modified histones H3 and H2B was not affected by the tryptic digestion.

## DISCUSSION

The experimental evidence presented here shows that BaP diol-epoxide (anti-isomer), when reacted with chicken erythrocyte chromatin in vitro, has a differential affinity for the histone proteins. Under nondissociating conditions and at neutral pH histones, H3 and H2B were predominantly labeled, whereas H4 bound the carcinogen to a lesser extent. The unidentified carcinogen-labeled molecule that migrated just above H4 could be modified H4 or a low-level nonhistone protein; however, from current experiments it appears that this may be a breakdown product of H3. The relatively low level of binding to H5 and H1 in native chromatin was somewhat surprising, since these histones are thought to be more exposed in the chromatin structure as judged from their rapid disappearance during tryptic digestion. However, since they do become accessible to the carcinogen when chromatin is dissociated in 2.0M NaCl, the potentially reactive groups of the H5 and both H1 molecules appear to be masked by their interaction with DNA. Dissociation of chromatin in 0.65M NaCl now makes the H5 a target molecule, but neither of the H1 molecules were labeled under these conditions. This suggests that both histone H1 molecules still are complexed with the DNA in a manner that prevents the carcinogen from reacting with the H1 histones, while histone H5 appears to be dissociated from the complex. At the same time, histone H2A has become accessible to the carcinogen, under 0.65M NaCl dissociating conditions. This indicates that histone H5 when bound to chromatin may be responsible for protecting histone H2A from the attack by the carcinogen. This observation would be in conformation with current evidence suggesting that the very lysine-rich histones are in proximity to histone H2A (2).

Furthermore, tryptic digestion of chromatin in which histone H3 and H2B were mainly modified showed that the carcinogen-modified proteins were resistant to tryptic digestion. This suggests that the modified amino acid side chains involved in the binding of the carcinogen are the epsilon amino groups of lysine residues and/or the side chains of arginine residues. Therefore, such modifications of the side chains of lysine residue may produce subtle changes in the structure of chromatin, in a way similar to that postulated for the enzymatic side-chain modifications in vivo, and as a consequence render regions of chromatin available for transcription which were not previously transcribed.

Thus, it seems that the differential affinity of BaP diol-epoxide (anti-isomer) for chicken erythrocyte chromosomal histone proteins in vitro is a function of the three-dimensional arrangement of the nucleosomal subunit structure and that the presence of the very lysine-rich histones prevents the carcinogen from interacting with histone H2A. The influence of other chromosomal proteins on the binding of BaP diol-epoxide to nucleosomal histone proteins (35) would confirm the suggestions that the specificity of binding of a carcinogen to chromatin histone proteins can be altered by the presence of other proteins and hence the three-dimensional structure of chromatin may well confer specific constraints with respect to the binding of potential carcinogens.

## ACKNOWLEDGMENT

This research is supported by NIH Grant CA 20076 and the Office of Health and Environmental Research, U. S. Department of Energy, under Contract W-7405-eng-26 with the Union Carbide Corporation.

## REFERENCES

1. Allfrey, V. G. (1975): Post-synthetic modifications of histone structure. In: *Chromatin and Chromatin Structure*, H. J. Li and R. Eckhardt, Eds., pp. 167-191, Academic Press.
2. Bonner, W. M., and Stedman, J. D. (1979): Histone H1 is proximal to histone 2A and to A24. *Proc. Natl. Acad. Sci. USA* 76:2190.
3. Bradbury, E. M., Inglis, R. J., and Matthews, H. R. (1974): Control of cell division by very lysine-rich histones phosphorylation. *Nature* 247:257.
4. Bresnick, E., Vaught, J. B., Chaung, A.H.L., Stoming, T. A., Bockman, D., and Mukhtar, H. (1977): Nuclear aryl hydroxylase and interaction of polycyclic hydrocarbons with nuclear components. *Arch. Biochem. Biophys.* 181:257.
5. Camerini-Otero, R. D., Sollner-Webb, B., and Felsenfeld, G. (1977): Super coiling energy and nucleosome formation: The role of the arginine-rich histone kernel. *Nucl. Acids Res.* 4:1159.
6. Daudel, P., Duquesne, M., Vigny, P., Grover, P. L., and Sims, P. (1975): Fluorescence spectral evidence that benzo(a)pyrene-DNA products in mouse skin arise from diol-epoxides. *FEBS Lett.* 57:250.
7. DeLange, R. J., and Smith, E. L. (1975): Histone function and evolution as viewed by sequence studies. In: *The Structure and Function of Chromatin*, Ciba Foundation Symposium 28, F. W. Fitzsimons and G.E.W. Wolstenholme, Eds., pp. 59-76, Elsevier, Amsterdam.
8. Dixon, G. H., Candido, E.P.M., Honda, B. M., Louie, A. J., MacLeod, A. R., and Sung, M. T. (1975): The biological role of post-synthetic modifications of basic nuclear proteins. In: *The*

*Structure and Function of Chromatin*, Ciba Foundation Symposium 28, F. W. Fitzsimons and G.E.W. Wolstenholme, Eds., pp. 59-76, Elsevier, Amsterdam.
9. Elgin, S.C.R., and Weintraub, H. (1975): Chromosomal proteins and chromatin structure. *Annu. Rev. Biochem.* 44:725.
10. Felsenfeld, G. (1978): Chromatin. *Nature* 271:115.
11. Finch, J. T., Lutter, L. C., Rhodes, D., Brown, R. S., Rushton, B., Levitt, M., and Klug, A. (1977): Structure of nucleosome core particles of chromatin. *Nature* 269:29.
12. Giri, C. P., West, M. H., and Smulson, M. (1978): Nuclear protein modification and chromatin subunit structure. I. Differential poly-(adenosine diphosphate)ribosylation of chromosomal proteins in nuclei versus isolated nucleosomes. *Biochem.* 17:3495.
13. Jungman, R. A., and Schweppe, J. S. (1972): Binding of chemical carcinogens to nuclear proteins of rat liver. *Cancer Res.* 32:952.
14. King, H.W.S., Thompson, M. H., Tarmy, E. M., Brooks, P., and Harvey, R. G. (1976): On the nature of the product which results when 7,8-dihydro-7,8-dihydroxybenzo(a)pyrene is metabolized *in vivo* and bound to DNA. *Chem. Biol. Interact.* 13:349.
15. Kootstra, A., and Bailey, G. S. (1978): Primary structure of histone H2B from trout (Salmo trutta) testes. *Biochem.* 171:2504.
16. Kootstra, A., Slaga, T. J., and Olins, D. E. (1979): Interaction of benzo(a)pyrene diol-epoxide with nuclei and isolated chromatin. *Chem. Biol. Interact.* (in press).
17. Kootstra, A., Slaga, T. J., and Olins, D. E. (1979): Studies on the binding of BaP diol-epoxide to DNA and chromatin. In: *Polynuclear Aromatic Hydrocarbons*, P. W. Jones and P. Leber, Eds., pp. 819-834, Ann Arbor Science Publishers, Inc.
18. Kornberg, R. D. (1977): Structure of chromatin. *Annu. Rev. Biochem.* 46:931.
19. Laemmli, U. M., (1970): Cleavage of structural proteins during the assembly of the head of bacteriophage T4. *Nature* 227:680.
20. Laskey, R. A., and Mills, A. D. (1975):Quantitative film detection of $^3$H and $^{14}$C in polyacrylamide gels by fluorography. *Eur. J. Biochem.* 56:335.
21. Metzger, G., and Werbin, H. (1979): Evidence of N-acetoxy-N2-acetyl-amino fluorene induced covalent-linked binding of some nonhistone proteins to DNA in chromatin. *Biochem.* 18:655.
22. Müller, U., Zentgraf, H., Eicken, I., and Keller, W. (1978): Higher order structure of siminian virus 40 chromatin. *Science* 201:406.
23. Noll, M., and Kornberg, R. D. (1977): Action of micrococcal nuclease on chromatin and the location of histone H1. *J. Mol. Biol.* 109:383.

24. Olins, A. L., and Olins, D. E. (1974): Spheroid chromatin subunits (ν-bodies). *Science* 183:330.
25. Osborne, M. R., Thompson, M. H., Tarmy, E. M., Beland, F. E., Harvey, R. G., and Brookes, P. (1976): The reaction of 7,8-dihydro-7,8-dihydroxybenzo(a)pyrene-9,10-oxide with DNA in relation to the benzo(a)pyrene DNA products isolated from cells. *Chem. Biol. Interact.* 13:343.
26. Sims, P., Groer, P. L., Swaisland, A., Pal, K., and Hewer, A. (1974): Metabolic activation of benzo(a)pyrene proceeds by a diol-epoxide. *Nature (London)* 252:326.
27. Slaga, T. J., Bracken, W. M., Viaje, A., Levin, W., Yagi, H., Jerina, D. M., and Conney, A. H. (1977): Comparison of the tumor-initiating activities of benzo(a)pyrene arene oxides and diol-epoxides. *Cancer Res.* 37:4130.
28. Sollner-Webb, B., Camerini-Otero, R. D., and Felsenfeld, G. (1976): Chromatin structure as probed by nucleases and proteases: evidence for the central role of histones H3 and H4. *Cell* 9:179.
29. Shätling, W. H. (1979): Role of histone H1 in the conformation of oligonucleosomes as a function of ionic strength. *Biochem.* 18:596.
30. Suau, P., Kneale, G. G., Braddock, G. W., Bladwin, J. P., and Bradbury, E. M. (1977): A low resolution model for the chromatin core particle by neutron scattering. *Nucl. Acid Res.* 4:3769.
31. Sung, M. T., Harford, J., Bundman, M., and Vidalakas, G. (1977): Metabolism of histones in avian erythroid cells. *Biochem.* 16:279.
32. Varshavsky, A. J., Bakayev, V. V., and Georgiev, G. P. (1976): Heterogeneity of chromatin subunits in vitro and location of histone H1. *Nucleic Acids Res.* 3:477.
33. Von Holt, C., Strickenland, W. N., Brandt, W. F., and Strickenland, M. S. (1979): More histone structures. *FEBS Lett.* 100:201.
34. Weinstein, I. B., Jeffrey, A. M., Jennette, K. W., Blobstein, S. H., Harvey, R. G., Harris, C., Autrup, H., Kasai, H., and Nakanishi, K. (1976): Benzo(a)pyrene diol-epoxides as intermediates in nucleic acid binding in vitro and in vivo. *Science* 193:592.
35. Whitlock, J. P. (1979): The conformation of the chromatin core particle is ionic strength-dependent. *J. Biol. Chem.* 254:5684.
36. Whitlock, J. P., and Simpson, R. T. (1977): Localization of the sites along nucleosomal DNA which interact with the $NH_2$-terminal histone regions. *J. Biol. Chem.* 252:6516.
37. Woodcock, C.L.F. (1973): Ultrastructure of inactive chromatin. *J. Cell Biol.* 59:368a.

# METABOLISM OF 6-, 7-, 8-, AND 12-METHYLBENZ(a)—ANTHRACENES AND HYDROXYMETHYLBENZ(a)ANTHRACENES

S. K. Yang*, M. W. Chou*, and P. P. Fu**

*Department of Pharmacology, School of Medicine, Uniformed Services
University of the Health Sciences
Bethesda, Maryland 20014
**National Center for Toxicological Research
Jefferson, Arkansas 72079

Methylbenz(a)anthracenes (MBAs) have been found in cigarette smoke condensate (11), stack gas, and roofing tar extracts (27). Among the twelve monomethylbenz(a)anthracenes (see Figure 1), 6-, 7-, 8-, and 12-methylbenz(a)anthracene (6-, 7-, 8-, and 12-MBA) were found to be carcinogenic when injected subcutaneously or painted on the skin of rats and mice [see Arcos and Argus (1) and Newman (22) and references therein]. It was found that 7-MBA was the most active, 6-, 8-, and 12-MBA were less active, and the other MBAs were either inactive or very slightly active (1, 22, 26). The molecular basis for the differences in the carcinogenicity of the twelve MBAs is not known.

Recent studies on 7-MBA and 7,12-dimethylbenz(a)anthracene (7,12-DMBA) indicated that the 3,4-dihydrodiol metabolites (3, 10, 28, 29, 33) upon further metabolism possess higher mutagenic activity (12, 17-20), tumor-initiating activity (6, 25), in vitro DNA binding activity (4), and malignant cellular transformation activity (19, 20) than the corresponding parent hydrocarbons and their dihydrodiol derivatives. The 3,4-dihydrodiol-1,2-epoxide has been implicated in binding to DNA of mammalian cells in culture (2, 8, 13, 21, 31, 32), to DNA of mouse skin pretreated with the hydrocarbons (2, 30), and to DNA in a rat liver microsome-mediated in vitro system (13). These results and the results obtained from studies of unsubstituted polycyclic aromatic hydrocarbons (PAH) (15, 16, 36) are generally in support of the bay-region theory (15, 16).

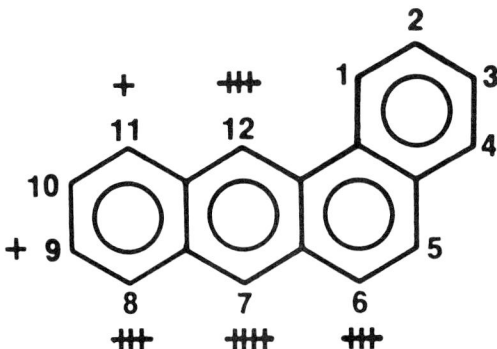

FIGURE 1. Relative carcinogenic activity of monomethylbenz[a]anthracenes tested in mouse skin on a scale of zero to ++++(26).

Jerina and Daly (14) have proposed that a methyl or alkyl substituent can substantially block ring metabolism of aromatic hydrocarbons to an arene oxide at the methyl-substituted double bond, thereby directing more of the total metabolism toward other ring positions. Jerina and Daly (14) have interpreted the animal testing data (26) for the enhanced carcinogenicity of the 6-, 7-, 8-, 9-, and 12-MBAs and the lack of enhanced carcinogenicity for the 1-, 2-, 3-, and 4-MBAs relative to that of the unsubstituted benz(a)anthracene (BA) based on the basis of this hypothesis. The lack of enhanced carcinogenicity of 5-MBA was interpreted as representing inhibition of formation of the key 3,4-dihydrodiol metabolite (14).

We have undertaken the metabolism studies of the twelve individual MBAs. Structural identification of the MBA metabolites is fundamental to the detailed understanding of activation and detoxification pathways. The in vitro metabolism studies and the subsequent mutagenicity and carcinogenicity testing of the MBA metabolites will provide insight in establishing the molecular basis for the differences in carcinogenicity among the twelve MBAs. This report describes the in vitro rat liver microsomal metabolism studies of 6-, 7-, 8-, and 12-MBAs and their hydroxymethylbenz[a]anthracene (OHMBA) derivatives. The potential proximate carcinogenic 3,4-dihydrodiol metabolites have been found to be formed from each of the hydrocarbons. Metabolites are isolated and purified by reversed-phase and normal-phase high-performance liquid chromatography (HPLC). Most of the metabolites have been characterized by their chromatographic properties in reversed-phase and normal-phase HPLC, by ultraviolet and fluorescence spectrophotometry, by mass spectrometry, and by their inability to form cis acetonides. A trans-8,9-dihydrodiol is found to be one of the major metabolites of 8-MBA. The formation of this trans-8,9-dihydrodiol indicates that the presence of a methyl group in a PAH does not necessarily block the enzymatic oxygenation at the methyl-substituted formal double bond. This finding is at variance with the hypothesis proposed previously (7,14).

## MATERIALS AND METHODS

### Materials

Reference standards, BA phenols, and dihydrodiols were obtained from the Chemical Repository of the National Cancer Institute. Dr. Melvin S. Newman has generously provided us with some 6-, 7-, and 12-MBAs which enabled us to initiate the studies described herein. 8-OHMBA was synthesized from 8-bromomethyl-BA by a procedure similar to that of Flesher et al (9). The 6-, 7-, and 12-OHMBAs were similarly prepared from the corresponding bromomethyl derivatives.

### In Vitro Incubation of the Hydrocarbons with Rat Liver Microsomes

Liver microsomes from 3-methylcholanthrene-pretreated (MC-microsomes) and phenobarbital-pretreated (PB-microsomes) male Sprague-Dawley rats were prepared as described (5). Metabolites were obtained by in vitro incubation of the hydrocarbon (8 $\mu$mol added in 4 ml methanol) in the dark at 37°C for 60 min in a 100-ml reaction mixture (pH 7.5) containing 5 mmol of tris-HCl, 0.3 mmol of magnesium chloride, 10 units of glucose-6-phosphate dehydrogenase (type XII, Sigma), 10 mg NADP$^+$, 65 mg of glucose-6-phosphate, and 100 mg protein equivalent of liver microsomes from 3-methylcholanthrene or phenobarbital pretreated rats. Substrate and its metabolites were extracted with acetone (100 ml) and ethyl acetate (200 ml), and the organic phase was dehydrated with anhydrous magnesium sulfate and evaporated to dryness under reduced pressure. The residue was redissolved in 1 ml of tetrahydrofuran (THF)-methanol (1:1) for the reversed-phase HPLC separation of metabolites. Large-scale incubations were also carried out in order to isolate a sufficient amount of metabolites for further characterization.

### High-Performance Liquid Chromatography

A Spectra-Physics model 3500B liquid chromatograph was fitted with a Du Pont 4.6 mm x 25 cm Zorbax ODS column. The sample injected onto the ODS column was eluted at ambient temperature with a 40-min linear gradient of methanol-water (1:1) to methanol at a solvent flow rate of 0.8 ml/min (reverse-phase HPLC). Each of the chromatographic peaks combined from repeated reversed-phase HPLC runs was further purified by an isocratic system (normal-phase HPLC) on a Du Pont 6.2 mm x 25 cm Zorbax SIL column with a different ratio of THF-hexane-methanol as the eluting solvent. This latter normal-phase HPLC system was used to confirm the purity of the metabolites and was also used to separate overlapping metabolites that were inseparable in the reversed-phase system (5). Each of the metabolites described in this report was eluted as a single chromatographic peak on both HPLC systems. Metabolites remaining in aqueous phase of the extraction contained less than 2 percent of the total metabolites and were not analyzed. Only unconjugated metabolites were separated and identified in this study.

## Physicochemical Properties of the Metabolites

Ultraviolet absorption spectra were measured in methanol on a Cary 118C spectrophotometer. Uncorrected fluorescence spectra were measured in methanol and in 0.1 N NaOH on a Perkin-Elmer model 44A spectrofluorometer. Mass spectral analysis was performed on a Finnigan 4000 gas chromatograph-mass spectrometer-data system by electron impact with a solid probe at 70 eV and 250°C ionizer temperature. Tests for the cis/trans isomer of the dihydrodiol metabolites were carried out by dissolving each of the dihydrodiol metabolites in anhydrous acetone in the presence of anhydrous copper sulfate, and the reaction product was isolated by reversed-phase HPLC and analyzed by mass spectrometry (35). All dihydrodiol metabolites were found to be trans isomers by their inability to form vicinal cis acetonide (35).

## RESULTS AND DISCUSSION

The metabolism studies described in this report were carried out under conditions in which a large percentage of the hydrocarbon was metabolized. Owing to recycling of the primary metabolites for further metabolic oxygenations and the relatively higher microsomal enzyme activity of the liver microsomes from 3-methylcholanthrene-pretreated rats, a greater variety and a relatively larger amount of metabolites were produced. The amount of metabolites that can be isolated and purified by the reversed-phase and normal-phase HPLC has greatly facilitated the task of structural identification.

The ultraviolet absorption and fluorescence spectra of BA phenols and dihydrodiols and many known metabolites of 7,12-dimethylbenz[a]anthracene (7, 12-DMBA) (5) have greatly aided the structural identification of the MBA metabolites. The identification of OHMBA metabolites not only provides an understanding of the metabolic pathways of the OHMBA itself but also facilitates the identification of products formed from the extensive metabolism of each MBA. Quantification of metabolites formed from each MBA cannot be made at this time because the molar extinction coefficients of the majority of the MBA metabolites are not known. Synthesis of radio-labeled MBAs is currently under way so that we will be able to compare the quantitative formation of various metabolites. Studies of 7,12-DMBA metabolism (5) indicated that the ultraviolet (at 254 nm) absorption profile by HPLC qualitatively parallels the amount of metabolites formed. Thus the relative amounts of metabolites formed can be approximately assessed within each chromatogram.

It should be pointed out, and this is apparent from the results shown in Figures 2 through 4 and 6, that in addition to 4-phenols other phenolic metabolites are also formed and they are eluted closely with 4-phenols on the reversed-phase HPLC. The identities of these phenolic metabolites are currently under investigation.

## Metabolism of 6-Methylbenz[a]anthracene

The reversed-phase HPLC separation of 6-MBA is shown in Figure 2 (solid curve). The identities of the metabolites contained in each chromatographic peak are also indicated in Figure 2. The identified metabolites formed from the microsomal metabolism of 6-MBA with MC-microsomes are (in increasing retention time): 6-OHMBA-10,11-diol, 6-MBA-3,4-diol, 6-MBA-8,9-diol, 6-MBA-10,11-diol, 6-OHMBA, and 4-OH-6-MBA. A major chromatographic peak which is eluted at about 12.5 min (Figure 2) has not been identified. The potentially key carcinogenic metabolite, 6-MBA-3,4-diol, is formed in a relatively minor amount. However, when 6-MBA was incubated with PB-microsomes, there was substantially decreased formation of 4-OH-6-MBA with concomitant increased formation of 6-MBA-3,4-diol. This was probably due to a greater extent of enzymatic hydration reaction of the 6-MBA-3,4-epoxide intermediate to form the 6-MBA-3,4-diol in PB-microsomes. As a consequence of 6-OHMBA being a very minor metabolite, very little or none of the secondary metabolites were formed (Figure 2). When PB-microsomes were used in the in vitro incubation, substantially increased formation of 6-OHMBA with concomitant increased formation of secondary metabolites was observed. In general, the percentage of MBA metabolized is considerably higher in MC-microsomes than that in PB-microsomes when the same amount of liver microsomal protein is used.

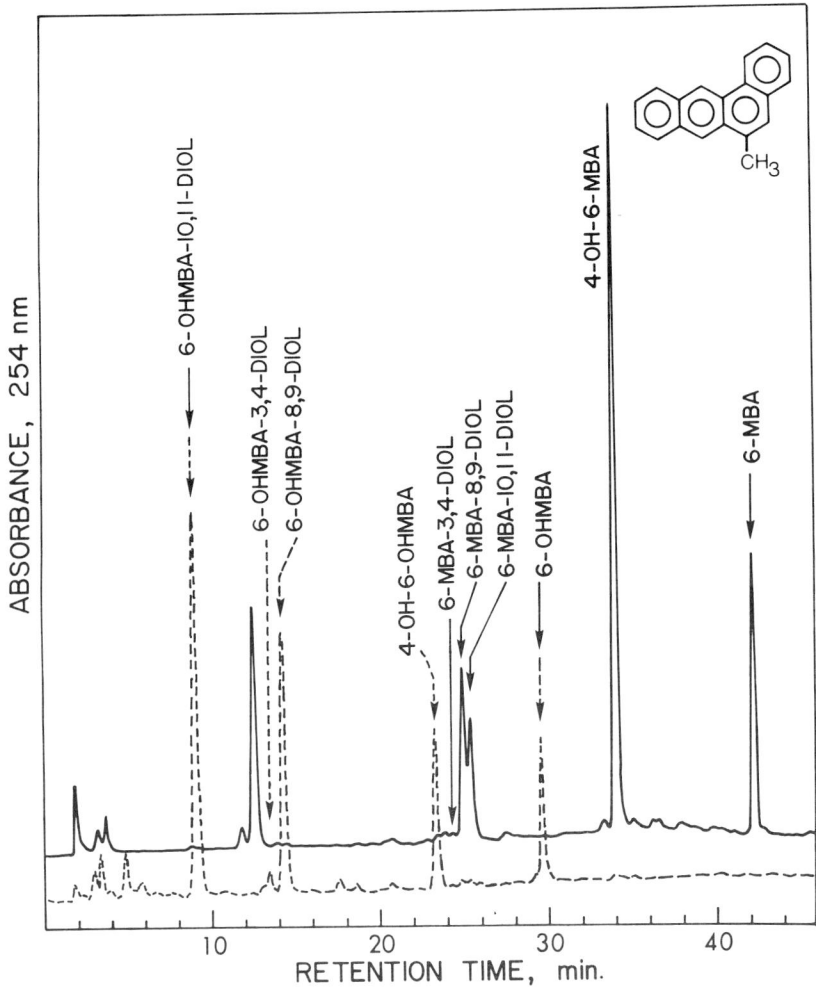

FIGURE 2. Reversed-phase HPLC separation of the metabolites of 6-methylbenz[a]anthracene (solid curve) and 6-hydroxymethylbenz[a]anthracene (broken curve). The identities of the materials contained in the unmarked chromatographic peaks are yet unknown.

A small amount of 6-OHMBA-10,11-diol was detected. This metabolite of 6-MBA may be formed by further hydroxylation at the methyl group of 6-MBA-10,11-diol which is one of the major metabolites (Figure 2). This is a possible pathway since we have found that a 7,12-DMBA-3,4-diol can be further metabolized at the methyl groups to form three other 3,4-diols (5).

There was no evidence that a 5,6-diol is formed from 6-MBA. Presumably, the 5,6-double bond of 6-MBA is inaccessible to the microsomal enzymes because of steric hindrance from the 6-methyl group. Although a 1,2-diol was found as a metabolite from the microsomal metabolism of 7-MBA (19,30) and 8-MBA (see below), respectively, 6-MBA-1,2-diol has not been found as a metabolite of 6-MBA.

## Metabolism of 6-Hydroxymethylbenz[a]anthracene

The major type of metabolites formed from the microsomal metabolism of 6-OHMBA in MC-microsomes (Figure 2, broken curve) are similar to those formed from 6-MBA. Namely, they are 6-OHMBA-10,11-diol, 6-OHMBA-3,4-diol, 6-OHMBA-8,9-diol, and 4-OH-6-OHMBA.

## Metabolism of 7-Methylbenz[a]anthracene

The identified metabolites formed from the microsomal metabolism of 7-MBA in MC-microsomes are (in elution order, Figure 3, solid curve); 7-OHMBA-5,6-diol, 7-OHMBA-8,9-diol, 7-OHMBA-10,11-diol, 7-OHMBA-3,4-diol, 7-MBA-5,6-diol, 7-MBA-8,9-diol, 4-OH-7-OHMBA, 7-MBA-3,4-diol, 7-MBA-10,11-diol, 7-OHMBA, and 4-OH-7-MBA. Although an intensive search was made, the 7-MBA-1,2-diol reported (30) as a minor metabolite has not been found in this study. Some of the metabolites formed from the microsomal metabolism of 7-MBA have been reported by other investigators (10,23,24,28,30). The dihydrodiol and phenolic metabolites of 7-OHMBA may be formed either by further metabolism at the ring positions of 7-OHMBA or by hydroxylation of the free methyl group of the corresponding 7-MBA metabolites.

Unlike the 5,6- and 8,9-diols of other MBAs (Figures 2, 4, and 6), 7-MBA-5,6-diol and 7-MBA-8,9-diol are eluted at much shorter retention times on the reversed-phase HPLC (Figure 3). The comparatively shorter retention times suggest that the hydroxyl groups of these two diols are in quasi-diaxial conformation due to steric interactions with the 7-methyl group. The hydroxyl groups of a bay-region BA-1,2-diol have been

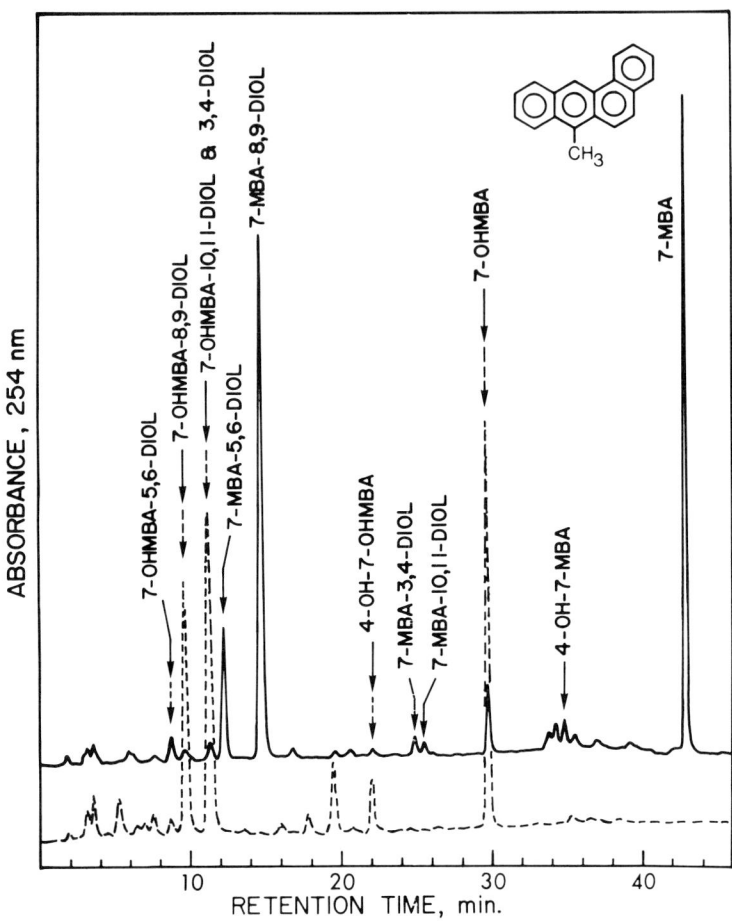

**FIGURE 3.** Reversed-phase HPLC separation of the metabolites of 7-methylbenz[a]anthra©ene (solid curve) and 7-hydroxymethylbenz[a]anthracene (broken curve). The identities of materials contained in the unmarked chromatographic peaks are yet unknown.

proven to be in quasi-diaxial conformation by X-ray crystallographic and NMR analysis (37). This phenomenon was also indicated by the relatively longer retention times of the 7-MBA 8,9-, 1,2-, and 5,6-diols on a normal-phase HPLC (18,30). As in the case of 7,12-DMBA (5,29), the formation of 5,6-diol and 8,9-diol as major metabolites from 7-MBA indicated that metabolic oxygenations at positions peri to the 7-methyl group are not inhibited.

Although 7-MBA-3,4-diol is one of the relatively minor metabolites, it has been found to be the most mutagenic in the mutagenicity test among all 7-MBA diols (30). It has high tumor-initiating activity in mouse skin (6) and causes malignant cellular transformation (19). The 7-MBA-3,4-diol may be the precursor of a reactive metabolite that binds to cellular DNA (30). The reports by Sims and co-workers (6,18,19,30,31) indicate that 7-MBA-3,4-diol is the key metabolite on the metabolic pathways in forming the most reactive metabolite of 7-MBA.

## Metabolism of 7-Hydroxymethylbenz[a]anthracene

The metabolites formed from the microsomal metabolism of 7-OHMBA in MC-microsomes are also formed from 7-MBA. The identified metabolites (in elution order, Figure 3 broken curve) are: 7-OHMBA-5,6-diol, 7-OHMBA-8,9-diol, 7-OHMBA-10,11-diol, 7-OHMBA-3,4-diol, and 4-OH-7-OHMBA. Many of these metabolites have not been reported previously (24). It is of interest to note that the 5,6-diol and 8,9-diol, both with a hydroxyl group peri to the 7-hydroxymethyl group, also have shorter retention times than the 3,4- and 10,11-diols. Thus the hydroxyl groups in 7-OHMBA-5,6-diol and 7-OHMBA-8,9-diol probably also adopt quasi-diaxial conformations.

## Metabolism of 8-Methylbenz[a]anthracene

The separation of 8-MBA metabolites by a reversed-phase HPLC is shown in Figure 4 (solid curve). The identified metabolites formed from the microsomal metabolism of 8-MBA in MC-microsomes are (in increasing retention time): 8-OHMBA-3,4-diol, 8-OHMBA-10,11-diol, 8-OHMBA-5,6-diol, 8-MBA-1,2-diol, 4-OH-8-OHMBA, 8-MBA-8,9-diol, 8-MBA-10,11-diol, 8-MBA-5,6-diol, 8-MBA-3,4-diol, 8-OHMBA, 3-OH-8-MBA, 2-OH-8-MBA, 4-OH-8-MBA (major component) and 1-OH-8-MBA (minor component), and 8-CHO-BA (8-formyl-BA). A metabolite peak with retention time of 19.2 min (Figure 4) has a UV absorption spectrum characteristic of that of BA-5,6-epoxide but its identity has not been confirmed by mass spectral analysis. The 8-MBA-1,2-diol is eluted with a retention time much shorter than that of the other 8-MBA diols. The hydroxyl groups of this bay-region diol most likely adopt quasi-diaxial conformation similar to the hydroxyl groups in the bay region BA-1,2-diol (37) and 7-MBA-1,2-diol (28).

Two of the potential proximate carcinogens, 8-MBA-3,4-diol and 8-OHMBA-3,4-diol, constitute about one-fourth of the total metabolites (Figure 4, solid curve). These two 3,4-diols may be the key metabolites responsible for the carcinogenicity of 8-MBA. However, the amount of

3,4-diol formations may not be the determining factor for the carcinogenicity of 8-MBA. The carcinogenic activity of 8-MBA may depend on how efficiently these 3,4-diols are further metabolized by the cells in vivo to the potentially reactive 3,4-diol-1,2-epoxides. It can be expected that the 8-MBA-3,4-diol, because of its higher lipid solubility, can penetrate across cell membranes more readily than 8-OHMBA-3,4-diol.

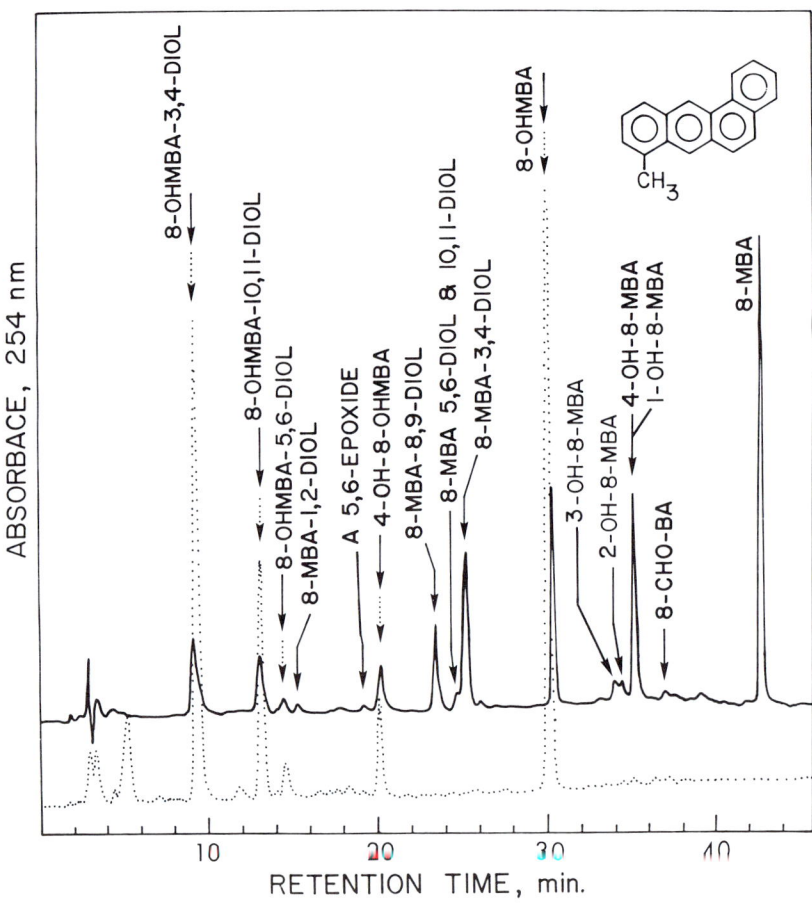

FIGURE 4. Reversed-phase HPLC separation of the metabolites of 8-methylbenz-[a]anthracene (solid curve) and 8-hydroxymethylbenz(a)anthracene (dotted curve). There are minor amounts of other unidentified phenolic metabolites which co-chromatographed with the 8-MBA phenols. The identities of materials contained in the unmarked chromatographic peaks are yet unknown.

The most unusual finding in the metabolism studies of 8-MBA is the identification of 8-MBA-8,9-diol. The 8,9-diol is one of the major metabolites and it is optically active (34). The proposed metabolic pathways of 8-MBA for the formation of the 8,9-diol are shown in Figure 5. This is the first example which indicates that the presence of a methyl

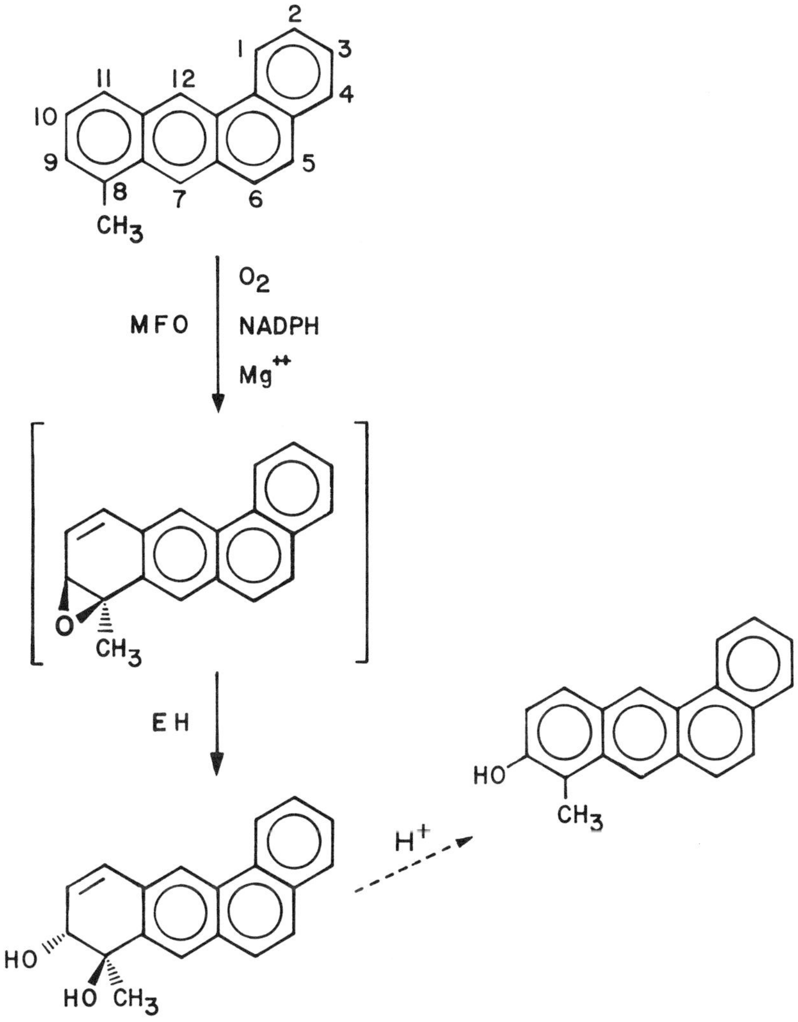

FIGURE 5. The proposed metabolic pathways of 8-MBA at the 8 and 9 positions. The acid-catalyzed dehydration of 8-MBA-8,9-diol produced 9-OH-8-MBA (broken arrow). The abbreviations are: MFO, mixed-function oxidases; EH, epoxide hydratase; NR, nonenzymatic rearrangement. Structures shown do not imply absolute stereochemistry.

group in a PAH does not sterically block the enzymatic formation of a diol at the methyl-substituted aromatic double bond. Thus the previous conclusion that the microsomal system does not expoxidize aromatic double bonds substituted by one or two methyl groups (7) apparently cannot be generalized to include all methyl-substituted PAHs. In contrast to the earlier suggestion (14), it is clear from our results (Figure 4) that the presence of a methyl group does not necessarily direct more of the total metabolism toward other ring positions.

## Metabolism of 8-Hydroxymethylbenz[a]anthracene

All the identified metabolites of 8-OHMBA formed in MC-microsomes (Figure 4, dotted curve) have been found to be formed from the microsomal metabolism of 8-MBA. Although an 8,9-diol was formed from 8-MBA, no evidence has been found for the formation of 8-OHMBA-8,9-diol. Thus, the hydroxymethyl group of a hydroxymethyl-PAH may block the oxidative metabolism at the hydroxymethyl-substituted formal double bond. It is interesting to note that the 8-OHMBA-3,4-diol is one of the most abundant metabolites of 8-OHMBA and it is also one of the major metabolites of 8-MBA (Figure 4). Thus both 8-MBA-3,4-diol and 8-OHMBA-3,4-diol may be further metabolized to form the potentially reactive 3,4-diol-1,2-epoxides and therefore may contribute to the overall carcinogenicity observed for 8-MBA.

## Metabolism of 12-Methylbenz[a]anthracene

The only study on the metabolism of 12-MBA was reported by Sims (23), in which rat liver homogenates were used as the enzyme source. Sims (23) has tentatively identified some 12-MBA metabolites including 12-MBA-5,6-diol, 12-MBA-8,9-diol, and several other phenolic derivatives. The efficient separation of metabolites by the reversed-phase HPLC (Figure 6, solid curve) has allowed us to identify many more metabolites that were formed by the incubation of 12-MBA with rat liver microsomes. The identified metabolites formed from the microsomal metabolism of 12-MBA in MC-microsomes are (in elution order, Figure 6): 12-OHMBA-5,6-diol, 12-OHMBA-10,11-diol, 12-OHMBA-3,4-diol, 12-MBA-10,11-diol, 4-OH-12-OHMBA, 12-MBA-8,9-diol, 12-MBA-3,4-diol, 12-MBA-5,6-diol, 12-OHMBA, and 4-OH-12-MBA. The possible bay-region 12-MBA-1,2-diol metabolite has not been detected. The bay-region 12-MBA-1,2-epoxide intermediate may be formed but may be too unstable to be hydrated enzymatically to 12-MBA-1,2-diol (3).

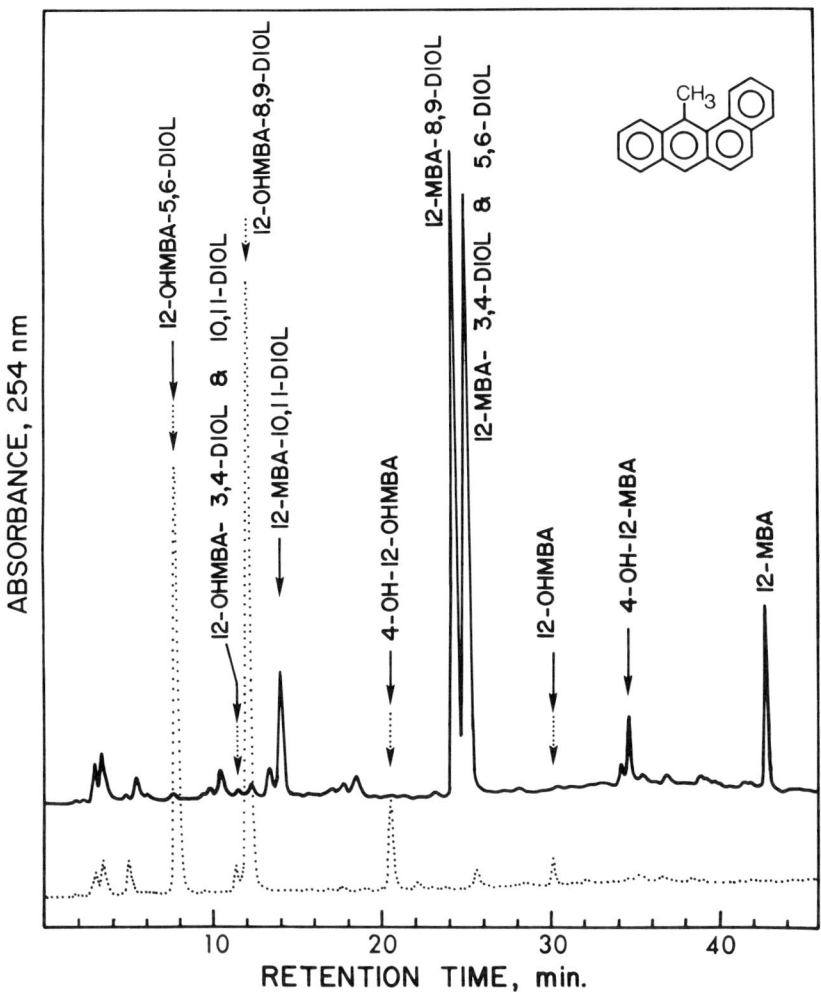

FIGURE 6. Reversed-phase HPLC separation of the metabolites of 12-methylbenz[a]anthracene (solid curve) and 12-hydroxymethylbenz[a]anthracene (dotted curve). The identities of materials contained in the unmarked chromatographic peaks are yet unknown.

Similar to the metabolism of 6-MBA (Figure 2), the hydroxymethyl derivative 12-OHMBA is also a very minor metabolite of 12-MBA (Figure 6, solid curve). Consequently, the secondary metabolites derived from further metabolism of 12-OHMBA are all minor products. These minor 12-OHMBA metabolites may also be derived from further metabolism of the 12-MBA metabolites bearing a free methyl group.

However, when liver microsomes from PB-pretreated rats were used in the in vitro incubation of 12-MBA, a substantial amount of 12-OHMBA was formed. Consequently, the 12-OHMBA diols and phenols were formed at a level relatively higher than those formed in liver microsomes from 3-MC-pretreated rats.

12-MBA-3,4-diol and 12-MBA-5,6-diol were co-chromatographed on the reversed-phase HPLC (Figure 6). These two metabolites were resolved (retention time 4.9 and 7.8 min, respectively) on the normal-phase HPLC with THF-hexane (1:1) as the elution solvent at a flow rate of 2 ml/min. The co-chromatographed 12-OHMBA-10,11-diol and 12-OHMBA-3,4-diol (Figure 6) were resolved on the reversed-phase HPLC by using water-methanol (1:1) as the elution solvent in which the 10,11-diol was eluted slightly earlier.

The 12-MBA-10,11-diol is eluted with a much shorter retention time than those of the other 12-MBA diols (Figure 6). This phenomenon is probably due to the steric interaction from the 12-methyl group which rendered the hydroxyl groups of 12-MBA-10,11-diol to adopt quasi-diaxial conformation (as noted in previous sections). The enzymatic formations of 10,11-diol from 12-MBA and 7,12-DMBA (5,29) thus indicate that metabolic oxygenation of 10,11-double bonds are not inhibited by the peri 12-methyl group.

## Metabolism of 12-Hydroxymethylbenz[a]anthracene

Sims reported (23) the tentative identification of an 8,9-diol and 3- and 4-phenols as metabolites of 12-OHMBA. Figure 6 (dotted curve) indicates that 12-OHMBA-5,6-diol, 12-OHMBA-3,4-diol, and 12-OHMBA-10,11-diol are also formed from the microsomal incubation of 12-OHMBA in MC-microsomes. These secondary metabolites have all been found from the metabolism of 12-MBA (Figure 6, solid curve). Unlike the 12-MBA-10,11-diol which is eluted with a much shorter retention time than the other 12-MBA diols, the 12-OHMBA-10,11-diol does not exhibit similar chromatographic property.

## CONCLUDING REMARKS

A majority of metabolites described above have not been reported previously. The identification of metabolites formed from the metabolism of parent hydrocarbons is the first step toward the understanding of carcinogenesis induced by chemicals that require metabolic activation. The carcinogenic activities of the twelve MBAs have been assessed mainly by the testing of tumor formation on the skin of mice and rats (1,22,26).

Thus further studies on the various metabolites formed from skin microsomes or skin homogenates of mice and rats are fundamental to the understanding of the factors that underlie the differences in carcinogenicity of the twelve MBAs tested in the skin of mice and rats. Initiation of malignant tumor formations at the target site(s) by the MBAs may be due to the formation of some key metabolites (e.g., 3,4-dihydrodiols) which may be further metabolized to the potentially reactive and electrophilic metabolites (e.g., 3,4-dihydrodiol-1,2-epoxides). Thus the in vitro and in vivo testing of the various metabolites by a variety of methods will provide further information on the activation pathways. We are currently preparing sufficient quantities of the metabolites for biological testing of mutagenicity, carcinogenicity, in vitro DNA binding activity, and hydrocarbon-DNA binding adduct formation. These studies should provide further understanding of the MBA-induced carcinogenesis.

## REFERENCES

1. Arcos, J. C., and Argus, M. F. (1974): *Chemical Induction of Cancer,* Vol. IIA, pp. 30–33, Academic Press, New York.
2. Bigger, C.A.H., Tomaszewski, J. E., and Dipple, A. (1978): Differences between products of bindings of 7,12-dimethylbenz[a]anthracene to DNA in mouse skin and in rat liver microsomal system. *Biochem. Biophys. Res. Commun.* 80: 229-235.
3. Chou, M. W., Easton, G. D., and Yang, S. K. (1979): Metabolism of 7,12-dimethylbenz[a]anthracene and its methyl-hydroxylated metabolites: Formation of phenolic metabolites at the 2-positions. *Biochem. Biophys. Res. Commun.* 88: 1085-1091.
4. Chou, M. W., and Yang, S. K. (1978): Identification of four trans-3,4-dihydrodiol metabolites of 7,12-dimethylbenz[a]anthracene and their in vitro DNA binding activities upon further metabolism. *Proc. Natl. Acad. Sci., USA* 75: 5466-5470.
5. Chou, M. W., and Yang, S. K. (1979): Combined reversed-phase and normal-phase high-performance liquid chromatography in the purification and identification of 7,12-dimethylbenz[a]anthracene metabolites. *J. Chromatography* 185: 635–654.
6. Chouroulinkov, I., Gentil, A., Tierney, B., Grover, P. L., and Sims, P. (1977): The metabolic activation of 7-methylbenz[a]anthracene in mouse skin: high tumour-initiating activity of the 3,4-dihydrodiol. *Cancer Lett.* 3: 247-253.

7. Daly, J. W., Jerina, D. M., and Witkop, B. (1972): Arene oxides and the NIH shift: the metabolism, toxicity and carcinogenicity of aromatic compounds. *Experientia* 28: 1129-1149.
8. Dipple, A., Tomaszewski, J. E., Moschel, R. C., Bigger, C.A.H., Nebzydoski, J. A., and Egan, M. (1979): Comparison of metabolism-mediated binding to DNA of 7-hydroxymethyl-12-methylbenz[a]-anthracene and 7,12-dimethylbenz[a]anthracene. *Cancer Res.* 1154-1158.
9. Flesher, J. W., Soedigdo, A., and Kelley, D. R. (1967): Syntheses of metabolites of 7,12-dimethylbenz[a]anthracene. 4-hydroxy-7,12-dimethylbenz[a]anthracene, 7-hydroxymethyl-12-methylbenz[a]anthracene, their methyl ethers, and acetoxy derivatives, *J. Med. Chem.* 10: 932-936.
10. Grover, P. L., Hewer, A., and Sims, P. (1974): Metabolism of polycyclic hydrocarbons by rat-lung preparations, *Biochem. Pharmacol.* 23: 323-332.
11. Hoffman, D., Schmeltz, I., Hecht, S. S., and Wynder, E. L. (1978): Tobacco carcinogenesis. In: *Polycyclic Hydrocarbons and Cancer*, Vol. 1, H. V. Gelboin and P.O.P. Ts'o, Eds., pp. 85-117, Academic Press, New York.
12. Huberman, E., Chou, M. W., and Yang, S. K. (1979): Identification of 7,12-dimethylbenz[a]anthracene metabolites that lead to mutagenesis in mammalian cells. *Proc. Natl. Acad. Sci.* USA 76: 862–866.
13. Ivanovic, V., Geacintov, N. E., Jeffrey, A. M., Fu, P. P., Harvey, R. G., and Weinstein, I. B. (1978): Cell and microsomes mediated binding of 7,12-dimethylbenz[a]anthracene to DNA studied by fluorescence spectroscopy. *Cancer Lett.* 4: 131-140.
14. Jerina, D. M., and Daly, J. W. (1977): Oxidation at carbon. In: *Drug Metabolism—From Microbe to Man*, D. V. Parke and R. L. Smith, Eds., pp. 13-32, Taylor and Francis, Ltd, London, England.
15. Jerina, D. M., Lehr, R., Schaefer-Ridder, M., Yagi, H., Karle, J. M., Thakker, D. R., Wood, A. W., Lu, A.Y.H., Ryan, D., West, S., Levin, W., and Conney, A. H. (1977): Bay-region epoxides of dihydrodiols: a concept explaining the mutagenic and carcinogenic activity of benzo[a]pyrene and benzo[a]anthracene. In: *Origins of Human Cancer*, H. H. Hiatt, J. D. Watson, and J. A. Winsten, Eds., pp. 639-658, Cold Spring Harbor Laboratory.
16. Lehr, R. E., Yagi, H., Thakker, D. R., Levin, W., Wood, A. W., Conney, A. H., and Jerina, D. M. (1978): The bay-region theory of polycyclic aromatic hydrocarbon-induced carcinogenicity. In: *Carcinogenesis, Vol. 3: Polynuclear Aromatic Hydrocarbons*, P. W. Jones and R. I. Freudenthal, Eds., pp. 231–241, Raven Press, New York.

17. Malaveille, C., Bartsch, H., Tierney, B., Grover, P. L., and Sims, P. (1978): Microsome-mediated mutagenicities of the dihydrodiols of 7,12-dimethylbenz[a]anthracene: high mutagenic activity of the 3,4-dihydrodiol. *Biochem. Biophys. Res. Commun.* 83: 1468–1473.
18. Malaveille, C., Tierney, B., Grover, P. L., Sims, P., and Bartsch, H. (1977): High microsome-mediated mutagenicity of the 3,4-dihydrodiol of 7-methylbenz[a]anthracene in *S. Typhimurium* TA 98. *Biochem. Biophys. Res. Commun.* 75: 427–433.
19. Marquardt, H., Baker, S., Tierney, B., Grover, P. L., and Sims, P. (1977): The metabolic activation of 7-methylbenz[a]anthracene: the induction of malignant transformation and mutation in mammalian cells by non-K-region dihydrodiols. *Int. J. Cancer* 19: 828–833.
20. Marquardt, H., Baker, S., Tierney, B., Grover, P. L., and Sims, P. (1978): Induction of malignant transformation and mutagenesis by dihydrodiols derived from 7,12-dimethylbenz[a]anthracene. *Biochem. Biophys. Res. Commun.* 85: 357–362.
21. Moschel, R. C., Baird, W. M., and Dipple, A. (1977): Metabolic activation of the carcinogen 7,12-dimethylbenz[a]anthracene for DNA binding. *Biochem. Biophys. Res. Commun.* 76: 1092–1098.
22. Newman, M. S. (1976): Carcinogenic activity of benz[a]anthracenes. In: *Carcinogenesis, Vol. 1, Polynuclear Aromatic Hydrocarbons: Chemistry, Metabolism, and Carcinogenesis*, R. I. Freudenthal and P. W. Jones, Eds., pp. 203–207, Raven Press, New York.
23. Sims, P. (1967): The metabolism of 7- and 12-methylbenz[a]anthracene and their derivatives. *Biochem. J.* 105: 591–598.
24. Sims, P. (1970): Studies on the metabolism of 7-methylbenz[a]anthracene and 7,12-dimethylbenz[a]anthracene and its hydroxymethyl derivatives in rat liver and adrenal homogenates. *Biochem. Pharmacol.* 19: 2261–2275.
25. Slaga, T. J., Gleason, G. L., DiGiovanni, J., Sukumaran, K. B., and Harvey, R. G. (1976): Potent tumor-initiating activity of the 3,4-dihydrodiol of 7,12-dimethylbenz[a]anthracene in mouse skin. *Cancer Res.* 39: 1934–1936.
26. Stevenson, J. L., and Von Haam, E. (1965): Carcinogenicity of benz[a]anthracene and benzo[c]phenanthrene derivatives. *Amer. Ind. Hyg. Assoc. J.* 26: 475–478.
27. Thomas, R. S., Lao, R. C., Wang, D. T., Robinson, D., and Sakuma, T. (1978): Determination of polycyclic aromatic hydrocarbons in atmospheric particulate matter by gas chromatography-mass spectrometry and high-pressure liquid chromatography. In: *Carcinogenesis, Vol. 3: Polynuclear Aromatic Hydrocarbons*, P. W. Jones and R. I. Freudenthal, Eds., pp. 9–19, Raven Press, New York.

28. Tierney, B., Abercrombie, B., Walsh, C., Hewer, A., Grover, P. L., and Sims, P. (1978): The preparation of dihydrodiols from 7-methylbenz[a]anthracene. *Chem. Biol. Interactions* 21: 289–298.
29. Tierney, B., Hewer, A., MacNicoll, A. D., Gervasi, P. G., Rattle, H., Walsh, C., Grover, P. L., and Sims, P. (1978): The formation of dihydrodiols by the chemical or enzymic oxidation of benz[a]anthracene and 7,12-dimethylbenz[a]anthracene. *Chem.-Biol. Interactions* 23: 243–257.
30. Tierney, B., Hewer, A., Walsh, C., Grover, P. L., and Sims, P. (1977): The metabolic activation of 7-methylbenz[a]anthracene in mouse skin. *Chem.-Biol. Interactions* 18: 179–193.
31. Vigny, P., Duquesne, M., Coulomb, H., Lacombe, C., Tierney, B., Grover, P. L., and Sims P. (1977): Metabolic activation of polycyclic hydrocarbons: Fluorescence spectral evidence is consistent with metabolism at the 1,2- and 3,4-double bonds of 7-methylbenz[a]anthracene. *FEBS Lett.* 75: 9–12.
32. Vigny, P., Duquesne, M., Coulomb, H., Tierney, B., Grover, P. L., and Sims, P. (1977): Fluorescence spectral studies on the metabolic activation of 3-methylcholanthrene and 7,12-dimethylbenz[a]anthracene in mouse skin. *FEBS Lett.* 82: 278–282.
33. Yang, S. K., Chou, M. W., and Roller, P. P. (1979): Potential proximate carcinogens of 7,12-dimethylbenz[a]anthracene: characterization of two metabolically formed trans-3,4-dihydrodiols. *J. Am. Chem. Soc.* 101: 237–239.
34. Yang, S. K., Chou, M. W., Weems, H. B., and Fu, P. P. (1979): Enzymatic formation of an 8,9-diol from 8-methylbenz[a]anthracene. *Biochem. Biophys. Res. Commun.* 90: 1136–1141.
35. Yang, S. K., McCourt, D. W., Gelboin, H. V., Miller, J. R., and Roller, P. P. (1977): Stereochemistry of the hydrolysis products and their acetonides of two stereoisomeric benzo[a]pyrene 7,8-diol 9,10-epoxides. *J. Am. Chem. Soc.* 99: 5124–5130.
36. Yang, S. K., Roller, P. P., and Gelboin, H. V. (1978): Benzo[a]pyrene metabolism: mechanism in the formation of epoxides, phenols, dihydrodiols, and the 7,8-diol-9,10 epoxides. In: *Carcinogenesis, Vol. 3: Polynuclear Aromatic Hydrocarbons*, P. W. Jones and R. I. Freudenthal, Eds., pp. 285–301, Raven Press, New York.
37. Zacharias, D. E., Glusker, J. P., Fu, P. P., and Harvey, R. G. (1979): Molecular structures of the dihydrodiols and diol epoxides of carcinogenic polycyclic aromatic hydrocarbons. X-ray crystallographic and NMR analysis. *J. Am. Chem. Soc.* 101: 4043–4051.

# DNA BINDING OF 7,12-DIMETHYLBENZ(a)ANTHRACENE (DMBA) AND RELATED COMPOUNDS

R. C. Moschel, C.A.H. Bigger, W. R. Hudgins, and A. Dipple

Chemical Carcinogenesis Program
NCI Frederick Cancer Research Center
Frederick, Maryland 21701

## INTRODUCTION

Although initial reports of the diol-epoxide mechanism for activation of benz(a)anthracene suggested that the key metabolite was an 8,9-diol 10,11-oxide (5,36), subsequent investigation of the benz(a)anthracene series showed that the important DNA binding and biologically active metabolites were diol epoxides formed in the angular 1,2,3,4-ring (29,39,41), i.e., "bay region" diol-epoxides (18,19,22). In the case of the most potent carcinogen in this series, 7,12-dimethylbenz(a)anthracene (DMBA), the first indication of the route of metabolic activation involved came from Baird and Dipple (2). They showed that while hamster embryo cell DMBA-DNA adducts were highly photosensitive, as were the parent hydrocarbon (DMBA) and 9,10-dimethylanthracene (a model for the hydrocarbon residue which would result if binding of DMBA to DNA had occurred through a saturated 1,2,3,4-ring), models for binding of DMBA through a saturated 8,9,10,11-ring (i.e., 8,9,10,11-tetrahydro-7,12-dimethylbenz(a)anthracene) or through a saturated 5,6-double bond (e.g., 5,6-dihydro-7,12-dimethylbenz(a)anthracene) were not similarly photosensitive. They concluded that, in the hamster cell mediated binding of DMBA to DNA, metabolic activation of DMBA occurred either in the 1,2,3,4-ring, resulting in retention of the 9,10-dimethylanthracene chromophore, or through a route which would leave the tetracyclic aromatic structure of the parent hydrocarbon intact.

In order to distinguish between these two possibilities, a comparison of the fluorescence properties of isolated DMBA-deoxyribonucleoside adducts and the same model compounds was undertaken (29). Mouse embryo cell DMBA-DNA adducts were shown to exhibit the same chromatographic properties and photosensitivity as the hamster cell products, and their fluorescence spectra were similar to, but not superimposable on, the spectra for 9,10-dimethylanthracene. After acid treatment, however, both the excitation and emission spectra for the adducts underwent a significant change and the resulting spectra were almost indistinguishable from those of 9,10-dimethylanthracene. These findings on the photosensitivity and fluorescence properties of the DMBA-DNA adducts derived from rodent cell cultures indicated that the bound hydrocarbon was fully saturated in the 1,2,3,4-ring, and suggested that metabolic activation of DMBA for DNA binding occurred through a diol epoxide in the 1,2,3,4-ring (29). Subsequent reports of similar studies from other laboratories were consistent with these conclusions (16,40). Bigger et al (4) have demonstrated that DMBA binds to the DNA of mouse skin (a target tissue for DMBA carcinogenesis) by the same mechanism, although with liver homogenates it does not.

The first direct evidence for DMBA binding through a diol-epoxide was provided by Nebzydoski and Dipple (11), who examined the effect of 1,1,1-trichloropropene-2,3-oxide (TCPO) on the binding of DMBA to DNA in mouse embryo cell cultures. TCPO, a known inhibitor of the enzyme epoxide hydrase (31) and, therefore, of diol formation, effectively inhibited binding of DMBA to DNA under conditions where the overall metabolism of the carcinogen was not inhibited. Also, the major DMBA-DNA product isolated by chromatography on Sephadex LH 20 eluted with a methanol-water gradient was resolved into two separate components in a methanol-sodium borate solution gradient, suggesting that, as is known for benzo(a)pyrene binding (1,20), two stereoisomeric diol-epoxides may be involved in the binding of DMBA to DNA.

The possibility that oxidation of one of the methyl groups of DMBA might precede formation of a diol-epoxide in the angular ring (16) has also been examined by comparing the cell-mediated binding of 7-hydroxymethyl-12-methylbenz(a)anthracene and of DMBA to DNA (12). The DNA binding of the hydroxymethyl-compound was less efficient than that of DMBA, and the resulting nucleoside adducts were separable from those derived from DMBA by Sephadex LH 20 column chromatography. Thus, it was concluded that 7-hydroxymethyl-12-methylbenz(a)anthracene is not an intermediate in the binding of DMBA to DNA in the mouse embryo cell system. In addition, examination of the analyses of DMBA binding to DNA in mouse skin (4) or in hamster embryo cells (2) shows that binding of DMBA to DNA through a 7-hydroxymethyl-metabolite could not account for a significant amount of hydrocarbon-nucleoside adduct formation in either of these systems. Similar conclusions have been reached by MacNicoll et al (23).

Since the 7-hydroxymethyl-compound is a carcinogen (13), experiments to probe the mechanism through which this compound binds to

DNA were carried out (12). As was observed with DMBA, the DNA binding of 7-hydroxymethyl-12-methylbenz(a)anthracene in mouse embryo cell cultures was inhibited by TCPO but the presence of TCPO had little effect on the total metabolism of the hydrocarbon. Furthermore, chromatography of the hydroxymethyl-compound nucleoside adducts on Sephadex LH 20 columns using a sodium borate solution-methanol gradient caused the adducts to elute earlier in the gradient in the presence of borate than in its absence. These observations indicate that 7-hydroxymethyl-12-methylbenz(a)anthracene, like DMBA, is activated for DNA binding through a diol-epoxide mechanism in mouse embryo cell cultures, and an examination of the fluorescence spectra of the hydroxymethyl-compound nucleoside adducts (12) confirmed that the diol epoxide of this carcinogen is formed in the 1,2,3,4-ring. The recent isolation of the trans-3,4-dihydrodiol of DMBA and of 7-hydroxymethyl-12-methylbenz(a)anthracene in several laboratories (6,7,23,35,37,42), and subsequent studies of the biological activity of these diols (3,15,23,24,-27,28,32-34) suggest that the biological activity of DMBA arises through the same route of activation as does its DNA binding.

Although the fluorescence spectra of the nucleoside adducts derived from DMBA and 7-hydroxymethyl-12-methylbenz(a)anthracene were distinguishable (12), the spectra exhibited some very interesting similarities. In particular, their respective emission spectra lacked resolution and appeared at much longer wavelengths than anticipated. Since both types of adducts were derived from hydrocarbons which have a methyl substituent in the bay region (i.e., a methyl group attached to carbon-12) it seemed that these fluorescence spectral characteristics might be general for nucleoside adducts wherein a methyl substituent is present in the bay region. In order to examine this possibility, our previous studies of the fluorescence of DMBA- and 7-hydroxymethyl-12-methylbenz(a)anthracene-nucleoside adducts have been elaborated and extended to include an examination of the fluorescence properties of nucleoside adducts derived from 7-methylbenz(a)anthracene (30). This hydrocarbon has no substituent in the bay region but its DNA binding and biologically active metabolite is a bay region diol-epoxide (i.e., a 3,4-diol-1,2-epoxide) (8,25,26,38,39).

## RESULTS AND DISCUSSION

Complete excitation and emission spectra in methanol solution at 25°C for DMBA-, 7-hydroxymethyl-12-methylbenz(a)anthracene- and 7-methylbenz(a)anthracene-deoxyribonucleoside adducts, and for the corresponding fluorescent models for these PAH where the 1,2,3,4-ring is saturated (i.e., 9,10-dimethylanthracene, 9-hydroxymethyl-10-methylanthracene, and 9-methylanthracene) are presented in Figure 1.

The spectra for all the model compounds (Figure 1B) are very similar to one another, with the maxima for 9,10-dimethylanthracene appearing at slightly longer wavelengths than for 9-hydroxymethyl-10-methylanthracene, and the maxima for the latter compound being at slightly longer wavelengths than for 9-methylanthracene. The spectra for the 7-methylbenz(a)anthracene adducts (Figure 1A) are very similar to those of 9-methylanthracene. However, while the spectra of the nucleoside adducts from the two 7,12-disubstituted benz(a)anthracenes are very similar to one another, the spectra for these types of adducts differ significantly from those of the 7-methylbenz(a)anthracene adducts. Not only are the spectra of the 7,12-disubstituted adducts shifted to the red with respect to the model compounds, their emission spectra lack resolution and appear as broad diffuse peaks.

Figure 2 shows that resolution in the emission spectra for these adducts is much improved when fluorescence measurements are made at 77° K in ethylene glycol/water solutions (17), although their fluorescence still occurs at a longer wavelength than that of the substituted anthracenes (Figure 1). The fluorescence of the 7-methylbenz(a)anthracene adducts at 77° K is not significantly different from that at room temperature. Figure 2 also shows that the adducts derived from the two 7,12-disubstituted hydrocarbons behave similarly in response to acid treatment which, by analogy with the known reactions of benzo(a)pyrene nucleoside adducts (21), leads to cleavage of the hydrocarbon-nucleoside linkage. After treatment with 0.6 N HCl at 60°C for 10 minutes, their emission spectra (Figure 2) show good resolution even at room temperature, and the resulting spectra are almost indistinguishable from those of the similarly substituted anthracene model compounds (Figure 1). In contrast, identical acid treatment results in only slight changes in the fluorescence of the 7-methylbenz(a)anthracene adducts.

These data indicate that the unusual fluorescence properties of the DMBA- and 7-hydroxymethyl-12-methylbenz(a)anthracene-nucleoside adducts are probably a consequence of the 12-methyl group in the bay region. The fluorescence properties of the 7-methylbenz(a)anthracene adducts are very similar to those of 9-methylanthracene, which serves as a model for the fluorophore which would result if the interaction with DNA occurred through a bay region diol-epoxide. In contrast, the adducts derived from the two hydrocarbons with a methyl substituent in the bay region exhibit fluorescence properties which are significantly different from those of the correspondingly substituted anthracenes. The adducts' emission lacks resolution and occurs at a much longer wavelength than the emission of the model compounds. In both cases, resolution is

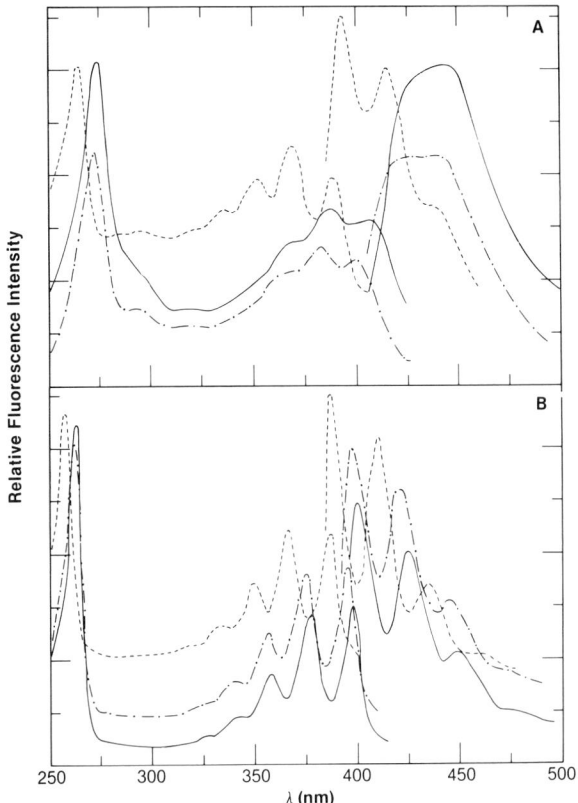

FIGURE 1. Fluorescence excitation and emission spectra of hydrocarbon-deoxyribonucleoside adducts and substituted anthracene derivatives in methanol solution. (A) 7-methylbenz(a)anthracene-deoxyribonucleoside adducts (———) (excitation spectrum determined with $\lambda_{em}$ at 415 nm; emission spectrum determined with $\lambda_{ex}$ at 264 nm), DMBA-deoxyribonucleoside adducts (———) (excitation spectrum determined with $\lambda_{em}$ at 440 nm; emission spectrum determined with $\lambda_{ex}$ at 274 nm), and 7-hydroxymethyl-12-methylbenz(a)anthracene-deoxyribonucleoside adducts (—·—) (excitation spectrum determined with $\lambda_{em}$ at 430 nm; emission spectrum determined with $\lambda_{ex}$ at 273 nm). (B) 9-methylanthracene (———) (excitation spectrum determined with $\lambda_{em}$ at 410 nm; emission spectrum determined with $\lambda_{ex}$ at 258 nm), 9-hydroxymethyl-10-methylanthracene (—·—) (excitation spectrum determined with $\lambda_{em}$ at 421 nm; emission spectrum determined with $\lambda_{ex}$ at 262 nm), and 9,10-dimethylanthracene (———) (excitation spectrum determined with $\lambda_{em}$ at 425 nm; emission spectrum determined with $\lambda_{ex}$ at 263 nm). Spectra were measured in quartz tubes using a Perkin Elmer MPF 3 fluorescence spectrophotometer and phosphorescence accessory and are uncorrected for lamp response and photomultiplier sensitivity. Reprinted from Reference 30 with permission of Elsevier/North Holland Scientific Publishers Ltd.

## Relative Fluorescence Intensity

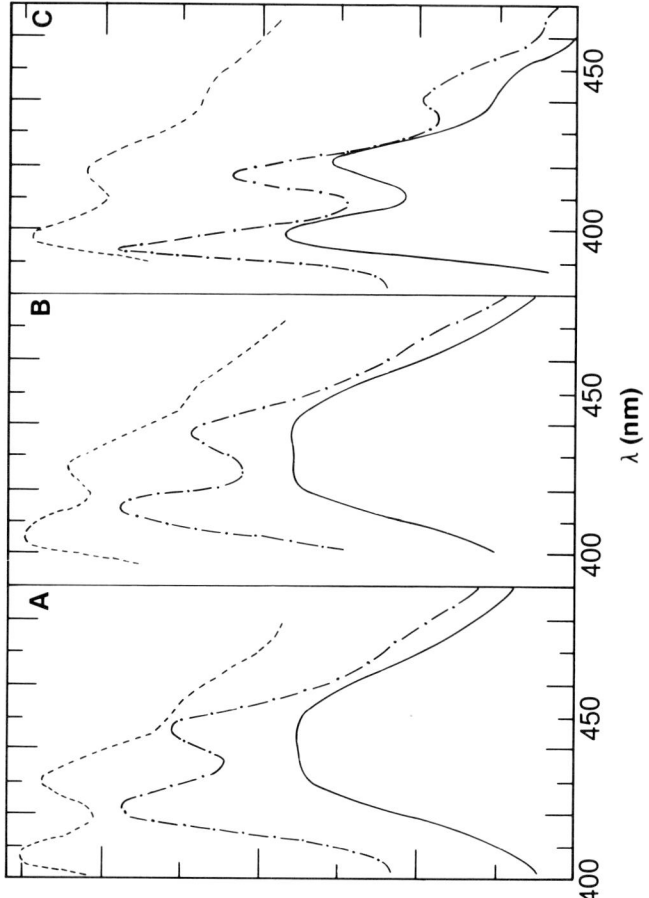

FIGURE 2. Fluorescence emission spectra for hydrocarbon-deoxyribonucleoside adducts in ethylene glycol/water (1:1) solution. (A) DMBA-deoxyribonucleoside adducts at 298 K (———) ($\lambda_{ex}$ 275 nm), at 77 K (—·—) ($\lambda_{ex}$ 275 nm), and at 298 K following treatment with 0.6 N HCl at 60°C for 10 minutes (— — —) ($\lambda_{ex}$ 270 nm). (B) 7-hydroxymethyl-12-methylbenz(a)anthracene-deoxyribonucleoside adducts at 298 K (———) ($\lambda_{ex}$ 273 nm), at 77 K (—·—) ($\lambda_{ex}$ 274 nm), and at 298 K following treatment with 0.6 N HCl at 60°C for 10 minutes (— — —) ($\lambda_{ex}$ 269 nm). (C) 7-methylbenz(a)anthracene-deoxyribonucleoside adducts at 298 K (———) ($\lambda_{ex}$ 267 nm), at 77 K (—·—) ($\lambda_{ex}$ 267 nm), and at 298 K following treatment with 0.6 N HCl at 60°C for 10 minutes (— — —) ($\lambda_{ex}$ 265 nm). Spectra were measured in quartz tubes using a Perkin Elmer MPF 3 fluorescence spectrophotometer and phosphorescence accessory and are uncorrected for lamp response and photomultiplier sensitivity. Reprinted from Reference 30 with permission of Elsevier/North Holland Scientific Publishers Ltd.

improved by lowering the temperature of the solutions to 77° K, and spectra which are very similar to those of the model compounds can be obtained by acid hydrolysis of the adducts.

These spectral properties, which are peculiar to the adducts formed from the bay region-substituted hydrocarbons, indicate some interaction between the 12-methyl group and the attached nucleoside, and it is quite possible that this substituent might cause the conformation of nucleoside and hydrocarbon in these adducts to differ from that of adducts where the hydrocarbon moiety lacks a methyl group in the bay region.

This observation is of interest because DMBA, which is substituted in the bay region, is the most potent carcinogen among the benz(a)anthracenes. 5-Methylchrysene, which is substituted in one of its bay regions, is the most potent of the monomethylchrysenes (14), and the more active cyclopenta(a)phenanthrenes are also substituted in the bay region at the 11-position (9). In addition, recent studies by Dipple and Hayes (10) indicate that DMBA-DNA adducts are poorly excised in mouse embryo cell cultures and this too might result from the conformation imposed on these adducts by the bay region substituent.

## ACKNOWLEDGMENT

These studies were supported by Contract No. NO1-CO-75380 with the National Cancer Institute, NIH, Bethesda, Maryland, 20014.

## REFERENCES

1. Baird, W. M., and Diamond, L. (1977): The nature of benzo(a)pyrene-DNA adducts formed in hamster embryo cells depends on the length of time of exposure to benzo(a)pyrene. *Biochem. Biophys. Res. Commun.* 77:162–167.
2. Baird, W. M., and Dipple, A. (1977): Photosensitivity of DNA-bound 7,12-dimethylbenz(a)anthracene. *Int. J. Cancer* 20: 427–431.
3. Bartsch, H., Malaveille, C., Tierney, B., Grover, P. L., and Sims, P. (1979): The association of bacterial mutagenicity of hydrocarbon-derived "bay-region" dihydrodiols with the Iball indices for carcinogenicity and with the extents of DNA-binding on mouse skin of the parent hydrocarbons. *Chem.-Biol. Interactions* 26: 185–196.
4. Bigger, C.A.H., Tomaszewski, J. E., and Dipple, A. (1978): Differences between products of binding of 7,12-dimethylbenz(a)anthracene to DNA in mouse skin and in a rat liver microsomal system. *Biochem. Biophys. Res. Commun.* 80: 229–235.

5. Booth, J., and Sims, P. (1974): 8,9-Dihydro-8,9-dihydroxybenz(a)anthracene 10,11-oxide: A new type of polycyclic aromatic hydrocarbon metabolite. *FEBS Lett.* 47: 30–33.
6. Chou, M. W., Easton, G. D., and Yang, S. K. (1979): Metabolism of 7,12-dimethylbenz(a)anthracene and its methyl-hydroxylated metabolites: Formation of phenolic metabolites at the 2-positions. *Biochem. Biophys. Res. Commun.* 88: 1085–1091.
7. Chou, M. W., and Yang, S. K. (1978): Identification of four *trans*-3,4-dihydrodiol metabolites of 7,12-dimethylbenz(a)anthracene and their *in vitro* DNA-binding activities upon further metabolism. *Proc. Natl. Acad. Sci. U.S.A.* 75: 5466–5470.
8. Chouroulinkov, I., Gentil, A., Tierney, B., Grover, P. L., and Sims, P. (1977): The metabolic activation of 7-methylbenz(a)anthracene in mouse skin: High tumor-initiating activity of the 3,4-dihydrodiol. *Cancer Lett.* 3: 247–253.
9. Coombs, M. M., Blatt, T. S., and Croft, C. J. (1973): Correlation between carcinogenicity and chemical structure in cyclopenta(a)phenanthrenes. *Cancer Res.* 33: 832–837.
10. Dipple, A., and Hayes, M. (1979): Differential excision of carcinogenic hydrocarbon-DNA adducts in mouse embryo cell cultures. *Biochem. Biophys. Res. Commun.* (in press).
11. Dipple, A., and Nebzydoski, J. A. (1978): Evidence for the involvement of a diol-epoxide in the binding of 7,12-dimethylbenz(a)anthracene to DNA in cells in culture. *Chem.-Biol. Interactions.* 20: 17–26.
12. Dipple, A., Tomaszewski, J. E., Moschel, R. C., Bigger, C.A.H., Nebzydoski, J. A., and Egan, M. (1979): Comparison of metabolism-mediated binding to DNA of 7-hydroxymethyl-12-methylbenz(a)anthracene and 7,12-dimethylbenz(a)anthracene. *Cancer Res.* 39: 1154–1158.
13. Flesher, J. W., and Sydnor, K. L. (1971): Carcinogenicity of derivatives of 7,12-dimethylbenz(a)anthracene. *Cancer Res.* 31: 1951–1954.
14. Hecht, S. S., Bondinell, W. E., and Hoffman, D. (1974): Chrysene and methylchrysenes: Presence in tobacco smoke and carcinogenicity. *J. Natl. Cancer Inst.* 53: 1121–1133.
15. Huberman, E., Chou, M. W., and Yang, S. K. (1979): Identification of 7,12-dimethylbenz(a)anthracene metabolites that lead to mutagenesis in mammalian cells. *Proc. Natl. Acad. Sci. U.S.A.* 76: 862–866.
16. Ivanovic, V., Geacintov, N. E., Jeffrey, A. M., Fu, P. P., Harvey, R. G., and Weinstein, I. B. (1978): Cell and microsome-mediated binding of 7,12-dimethylbenz(a)anthracene to DNA studied by fluorescence spectroscopy. *Cancer Lett.* 4: 131–140.
17. Ivanovic, V., Geacintov, N. E., and Weinstein, I. B. (1976): Cellular binding of benzo(a)pyrene to DNA characterized by low temperature fluorescence. *Biochem. Biophys., Res. Commun.* 70: 1172–1179.

18. Jerina, D. M., and Daly, J. W. (1977): Oxidation at carbon. In: *Drug Metabolism*, D. V. Parke and R. L. Smith, Eds., pp. 13–32, Taylor and Francis Ltd., London.
19. Jerina, D. M., Yagi, H., Lehr, R. E., Thakker, D. R., Schaefer-Ridder, M., Karle, J. M., Levin, W., Wood, A. W., Chang, R. L., and Conney, A. H. (1978): The bay-region theory of carcinogenesis by polycyclic aromatic hydrocarbons. In: *Polycyclic Hydrocarbons and Cancer*, Vol. 1, H. V. Gelboin and P. O. P. Ts'o, Eds., pp. 173–188, Academic Press, New York.
20. King. H.W.S., Osborne, M. R., Beland, F. A., Harvey, R. G., and Brookes, P. (1976): (±) $7\alpha$, $8\beta$-dihydroxy-$9\beta$, $10\beta$-epoxy-7,8,9,10-tetrahydrobenz(a)pyrene is an intermediate in the metabolism and binding to DNA of benzo(a)pyrene. *Proc. Natl. Acad. Sci. U.S.A.* 73: 2679–2681.
21. Koreeda, M., Moore, P. D., Yagi, H., Yeh, H.J.C., and Jerina, D. M. (1976): Alkylation of polyguanylic acid at the 2-amino group and phosphate by the potent mutagen (±)-$7\beta$-$8\alpha$-dihydroxy-$9\beta$-$10\beta$-epoxy-7,8,9,10-tetrahydrobenzo(a)pyrene. *J. Am. Chem. Soc.* 98: 6720–6722.
22. Lehr, R. E., Yagi, H., Thakker, D. R., Levin, W., Wood, A. W., Conney, A. H., and Jerina, D. M. (1978): The bay region theory of polycyclic aromatic hydrocarbon-induced carcinogenicity. In: *Carcinogenesis, Vol. 3: Polynuclear Aromatic Hydrocarbons*, P. W. Jones and R. I. Freudenthal, Eds., pp. 231–241, Raven Press, New York.
23. MacNicoll, A. D., Burden, P. M., Ribeiro, O., Hewer, A., Grover, P. L., and Sims, P. (1979): The formation of dihydrodiols by the chemical or enzymic oxidation of 7-hydroxymethyl-12-methylbenz(a)anthracene and the possible role of hydroxymethyl dihydrodiols in the metabolic activation of 7,12-dimethylbenz(a)anthracene. *Chem.-Biol. Interactions* 26: 121–132.
24. Malaveille, C., Bartsch, H., Tierney, B., Grover, P. L., and Sims, P. (1978): Microsome-mediated mutagenicities of the dihydrodiols of 7,12-dimethylbenz(a)anthracene: High mutagenic activity of the 3,4-dihydrodiol. *Biochem. Biophys. Res. Commun.* 83: 1468–1473.
25. Malaveille, C. Tierney, B., Grover, P. L., Sims, P., and Bartsch, H. (1977): High microsome-mediated mutagenicity of the 3,4-dihydrodiol of 7-methylbenz(a)anthracene in *S. typhimurium* TA 98. *Biochem. Biophys. Res. Commun.* 75: 427–433.
26. Marquardt, H., Baker, S., Tierney, B., Grover, P. L., and Sims, P. (1977): The metabolic activation of 7-methylbenz(a)anthracene: The induction of malignant transformation and mutation in mammalian cells by non-K-region dihydrodiols. *Int. J. Cancer.* 19: 828–833.

27. Marquardt, H., Baker, S., Tierney, B., Grover, P. L., and Sims, P. (1978): Induction of malignant transformation and mutagenesis by dihydrodiols derived from 7,12-dimethylbenz(a)anthracene. *Biochem. Biophys. Res. Commun.* 85: 357–362.
28. Marquardt, H., Baker, S., Tierney, B., Grover, P. L., and Sims, P. (1979): Comparison of mutagenesis and malignant transformation by dihydrodiols from benz(a)anthracene and 7,12-dimethylbenz(a)anthracene. *Br. J. Cancer* 39: 540–547.
29. Moschel, R. C., Baird, W. M., and Dipple, A. (1977): Metabolic activation of the carcinogen 7,12-dimethylbenz(a)anthracene for DNA binding. *Biochem. Biophys. Res. Commun.* 76: 1092–1098.
30. Moschel, R. C., Hudgins, W. R., and Dipple, A. (1979): Fluorescence of hydrocarbon-deoxyribonucleoside adducts. *Chem.-Biol. Interactions* 27: 69–79.
31. Oesch, F., Kaubisch, N., Jerina, D. M., and Daly, J. W. (1971): Hepatic epoxide hydrase. Structure-activity relationships for substrates and inhibitors. *Biochemistry* 10: 4858–4866.
32. Pal, K., Grover, P. L., and Sims, P. (1979): The induction of sister chromatid exchanges by dihydrodiols derived from 7,12-dimethylbenz(a)anthracene and 3-methylcholanthrene. *Cancer Lett.* 7: 45–49.
33. Slaga, T. J., Gleason, G. L., DiGiovanni, J., Sukumaran, K. B., and Harvey, R. G. (1979): Potent tumor-initiating activity of the 3,4-dihydrodiol of 7,12-dimethylbenz(a)anthracene in mouse skin. *Cancer Res.* 39:1934–1936.
34. Slaga, T. J., Huberman, E., DiGiovanni, J., and Gleason, G. (1979): The importance of the bay region diol-epoxide in 7,12-dimethylbenz(a)anthracene skin tumor initiation and mutagenesis. *Cancer Lett.* 6: 213–220.
35. Sukumaran, K. B., and Harvey, R. G., (1979): Synthesis of trans-3,4-dihydroxy-3,4-dihydro-7,12-dimethylbenz(a)anthracene, a highly carcinogenic metabolite of 7,12-dimethylbenz(a)anthracene. *J. Am. Chem. Soc.* 101: 1353–1354.
36. Swaisland, A. J., Hewer, A., Pal, K., Keysell, G. R., Booth, J., Grover, P. L., and Sims, P. (1974): Polycyclic hydrocarbon epoxides: The involvement of 8,9-dihydro-8,9-dihydroxybenz(a)anthracene 10,11-oxide in reactions with the DNA of benz(a)anthracene-treated hamster embryo cells. *FEBS Lett.* 47: 34–37.
37. Tierney, B., Hewer, A., MacNicoll, A. D., Gervasi, P. G., Rattle, H., Walsh, C., Grover, P. L., and Sims, P. (1978): The formation of dihydrodiols by the chemical or enzymic oxidation of benz(a)anthracene and 7,12-dimethylbenz(a)anthracene. *Chem.-Biol. Interactions* 23: 243–257.

38. Tierney, B., Hewer, A., Walsh, C., Grover, P. L., and Sims, P. (1977): The metabolic activation of 7-methylbenz(a)anthracene in mouse skin. *Chem.-Biol. Interactions.* 18: 179–193.
39. Vigny, P., Duquesne, M., Coulomb, H., Lacombe, C., Tierney, B., Grover, P. L., and Sims, P. (1977): Metabolic activation of polycyclic hydrocarbons. *FEBS Lett.* 75: 9–12.
40. Vigny, P., Duquesne, M., Coulomb, H., Tierney, B., Grover, P. L., and Sims, P. (1977): Fluorescence spectral studies on the metabolic activation of 3-methylcholanthrene and 7,12-dimethylbenz(a)anthracene. *FEBS Lett.* 82: 278–282.
41. Wood, A. W., Levin, W., Lu, A. Y. H., Ryan, D., West, S. B., Lehr, R. E., Schaefer-Ridder, M., Jerina, D. M., and Conney, A. H. (1976): Mutagenicity of metabolically activated benzo(a)anthracene 3,4-dihydrodiol: Evidence for bay region activation of carcinogenic polycyclic hydrocarbons. *Biochem. Biophys. Res. Commun.* 72: 680–686.
42. Yang, S. K., Chou, M. W., and Roller, P. (1979): Potential proximate carcinogens of 7,12-dimethylbenz(a)anthracene: Characterization of two metabolically formed *trans*-3,4-dihydrodiols. *J. Am. Chem. Soc.* 101: 237–239.

# BENZO(e)PYRENE DIHYDRODIOLS AND DIOL EPOXIDES: CHEMISTRY, MUTAGENICITY AND TUMORIGENICITY

R. E. Lehr*, S. Kumar*, W. Levin**, A. W. Wood**, R. L. Chang**, M. K. Buening**, A. H. Conney**, D. L. Whalen†, D. R. Thakker‡, H. Yagi‡, and D. M. Jerina‡

*Department of Chemistry
 University of Oklahoma
 Norman, Oklahoma 73019
**Department of Biochemistry and Drug Metabolism
 Hoffman-LaRoche, Inc.
 Nutley, New Jersey 07110
†Laboratory of Chemical Dynamics
 University of Maryland-Baltimore County
 Baltimore, Maryland 21228
‡Laboratory of Bioorganic Chemistry
 National Institute of Arthritis, Metabolism
  and Digestive Diseases
 National Institutes of Health
 Bethesda, Maryland 20014

## INTRODUCTION

The very weak carcinogenicity of benzo(e)pyrene (BeP) contrasts markedly with the high carcinogenicity of its much studied isomer, benzo(a)pyrene (BaP). Studies (8,11,13,16) of the mutagenicity and metabolism of BeP and several of its derivatives suggest that a possible basis for its low biological activity might reside in the failure of BeP to be metabolized to the bay-region diol epoxides. Pertinent findings included (Figure 1): (i) at best trace (<0.5 to 1.0 percent) metabolism by rat liver microsomes of BeP to the non-K-region dihydrodiol (BeP 9,10-dihydrodiol) (11): (ii) metabolism of BeP 9,10-dihydrodiol by rat liver microsomes at the aromatic nucleus, with little if any formation of BeP 9,10-diol-11,12-epoxides, so that bay-region diol opoxides would not have

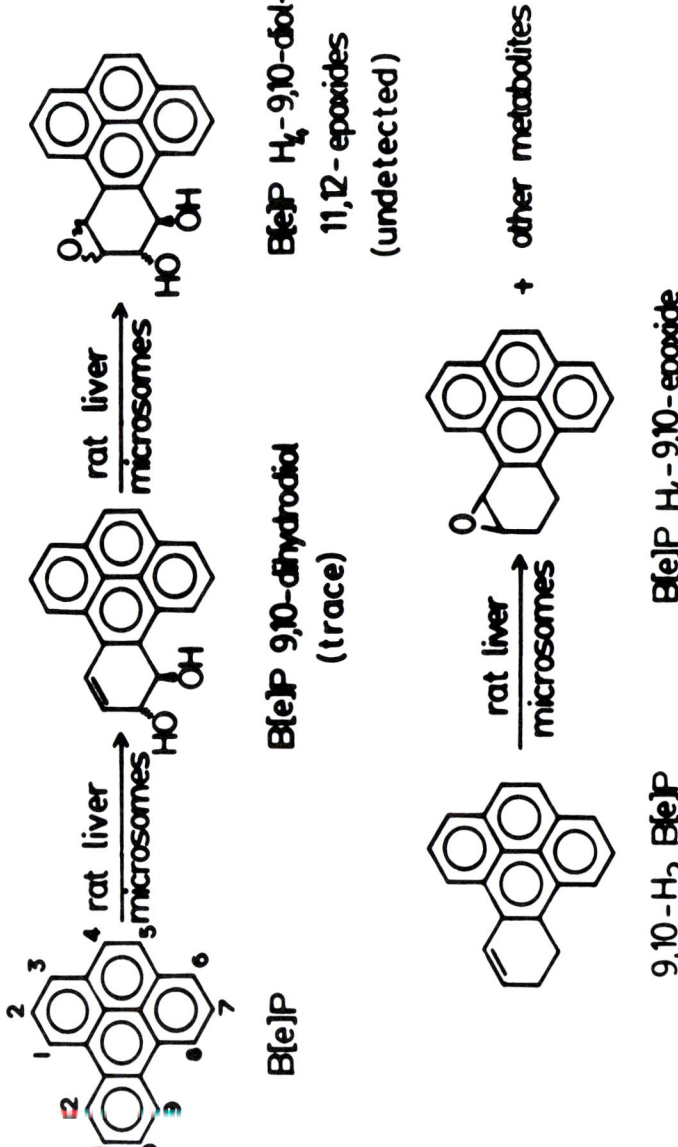

FIGURE 1. Aspects of the metabolism of BeP and derivatives of BeP.

resulted even if BeP 9,10-dihydrodiol had been produced in a significant amount (8,13,16); (iii) metabolic activation of BeP 9,10-dihydrodiol, producing species of negligible mutagenicity, as opposed to metabolic activitation of 9,10-$H_2$BeP, resulting in highly mutagenic species (8,16); (iv) BeP $H_4$-9,10-epoxide that was inherently highly mutagenic—this bay-region epoxide was shown to be a metabolite of 9,10-$H_2$BeP (8,16).

Tumorigenicity studies of BeP, its dihydrodiols, and other derivatives, which were incomplete at the time of the last Symposium (8), have now been completed (1) and are reported in the following section. The diastereomeric BeP 9,10-diol-11,12-epoxides have been prepared and their synthesis, mutagenic properties, and hydrolysis rates are also discussed.

## TUMOR INITIATION BY BeP AND ITS DERIVATIVES

The tumorigenicities of BeP, BeP 4,5-dihydrodiol (the K-region diol), BeP 9,10-dihydrodiol, BeP $H_4$-9,10-diol, and 9,10-$H_2$BeP have been determined in an initiation-promotion experiment on mouse skin, and the tumorigenicities of BeP, BeP 4,5-dihydrodiol, and BeP 9,10-dihydrodiol have been determined in newborn mice (1). The results of these exposures on mouse skin at an initiating dose of 2.5 micromoles are illustrated in Figure 2 and are representative of results at lower and higher doses than those shown. Of the compounds tested, only 9,10-$H_2$BeP was significantly tumorigenic, inducing tumors in 67 percent of the mice, with an average of 1.43 tumors per mouse. BeP, its K-region diol, and its non-K-region diol were all essentially inactive, inducing tumors in no more than 15 percent of the mice at any of the doses (1 to 6 micromoles) tested, with a maximum of 0.21 papillomas per mouse.

In newborn mice, neither BeP, BeP 4,5-dihydrodiol or BeP 9,10-dihydrodiol induced a significant number of pulmonary tumors (data not presented). BeP and BeP 4,5-dihydrodiol also failed to induce a significant number of hepatic tumors (data not presented). However, BeP 9,10-dihydrodiol, though weakly tumorigenic, induced a significantly higher number of hepatic tumors than were obtained in control animals. At a dose of 2.8 micromoles, BeP 9,10-dihydrodiol induced hepatic tumors in 61 percent of the male mice versus an incidence of 11 percent in control animals. The average number of hepatic tumors in male mice treated with BeP 9,10-dihydrodiol was 0.82 per mouse versus 0.11 per mouse in the control group (data not presented).

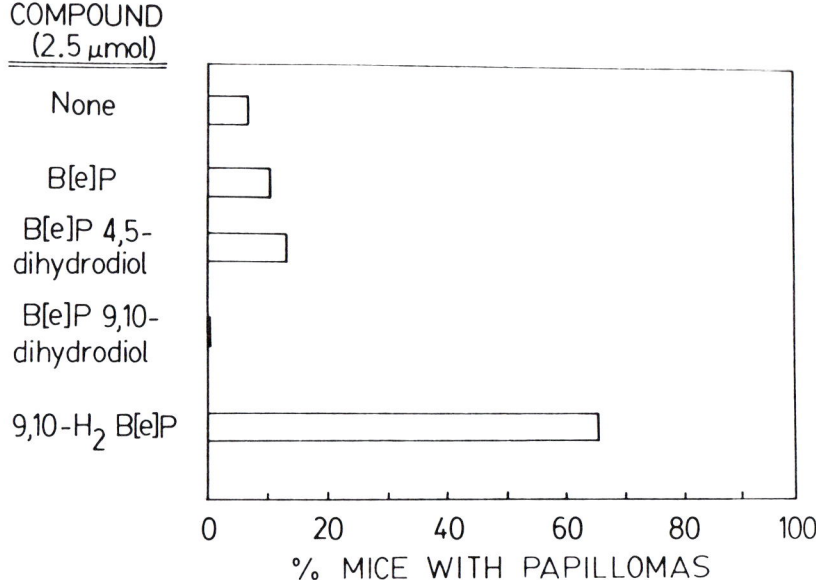

**FIGURE 2.** Tumorigenicity of BeP and derivatives of BeP on mouse skin. Female CD-1 mice (7 to 8 weeks old) were treated topically with the indicated compounds in 200 µl of acetone containing 10 percent DMSO. Seven days after application of each compound the animals were treated with TPA (16 nmoles/200 µl acetone) twice weekly for 35 weeks.

## PREPARATION OF THE DIASTEREOMERIC BeP DIOL EPOXIDES

Typically, diol epoxides are prepared from the corresponding dihydrodiol either directly with m-chloroperoxybenzoic acid, or in a two-step sequence consisting of initial bromohydrin formation with N-bromoacetamide (NBA), followed by cyclization of the bromohydrin with base. Generally, these reactions are stereospecific, with the peracid procedure affording the pure trans-diol epoxide (diol epoxide-2 series, see Figure 3) and the bromohydrin route affording the pure cis-diol epoxide (diol epoxide-1 series) (5,9,14,17,18). This has been explained on the basis of the directing influence of the quasiequatorial allylic hydroxyl group of the dihydrodiol (17,18). When this approach was applied to BaP 9,10-dihydrodiol, in which the hydroxyl groups are forced into a predominantly quasidiaxial conformation in order to relieve the nonbonded bay-region interaction of the C-10 hydroxyl group with the hydrogen atom at C-11, stereoselectivity dropped markedly, with both faces of the olefinic double bond being attacked by the electrophilic reagent in both of the

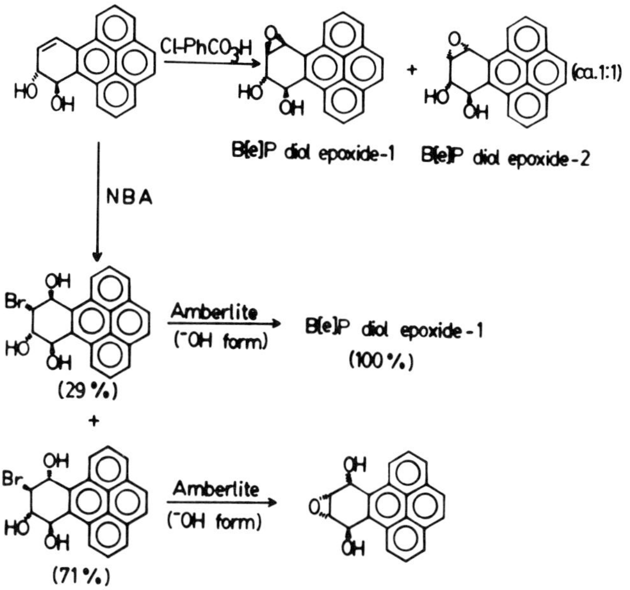

FIGURE 3. Synthetic routes to BeP diol epoxides.

above procedures (12). Purification of the diol epoxides was difficult but feasible, in part because of the relative stability of the diol epoxides in this case. Therefore, we were not surprised when an attempted epoxidation of BeP 9,10-dihydrodiol, in which the hydroxyl groups are also quasidiaxial, yielded an apparent mixture of the desired diol epoxides (6,8), which resisted our initial attempts of separation. We were surprised, however, by a subsequent report by Harvey et al (2), which claimed a 79 percent yield of pure BeP diol epoxide-2 upon treatment of BeP 9,10-dihydrodiol with m-chloroperoxybenzoic acid under conditions comparable to ours. Recently, we found that the BaP diol epoxides could be separated and isolated by high-pressure liquid chromatography (HPLC) on a Du Pont Zorbax SIL column eluted with 40 percent tetrahydrofuran in hexane, and a detailed examination of the diol epoxides formed from BeP 9,10-dihydrodiol was made which confirmed our initial observations. Specifically, treatment of BeP 9,10-dihydrodiol with m-chloroperoxybenzoic acid produces almost equal amounts of BeP diol epoxides-1 and -2 (Figure 3). Structural assignment was based upon analysis of the NMR spectra of the pure diastereoisomers and of the tetraacetates of their hydrolysis products (19). These results were further confirmed through treatment of BeP 9,10-dihydrodiol with NBA to form a pair of bromotriols which were individually cyclized to diol epoxides with base.

Cyclization to form the oxirane requires backside displacement of bromide ion by alkoxide ion. Thus, the finding that the major bromotriol isomer afforded only the BeP 9,12-diol-10,11-epoxide shown (Figure 3) established the relative stereochemistry of both bromotriol precursors, as indicated. Treatment of the minor bromotriol with base gave (exclusively) the expected diol epoxide isomer-1. Similar results were obtained during preparation of the diol epoxides of triphenylene, which is structurally similar to BeP in that the benzo-ring has two bay regions. Again, this result is in contrast to the report by Harvey et al (2).

## MUTAGENICITY OF BeP DIOL EPOXIDES

The inherent mutagenicities of the BeP diol epoxide diastereomers and the tetrahydroepoxide toward *S. typhimurium* strain TA 98 and Chinese hamster V79 cells were determined. These are compared with the values obtained for the corresponding BaP bay-region diol epoxide diastereomers and tetrahydroepoxide (Table 1). Noteworthy is the very

**TABLE 1.** Mutagenicity of BaP and BeP Diol Epoxides and Tetrahydro-Epoxide in Bacterial and Mammalian Cells[a]

| | Mutation Frequency | |
|---|---|---|
| Compound[b] | S. Typhimurium TA 98[c] | Chinese Hamster V 79 Cells[d] |
| **BeP** | | |
| 9,10-diol-11,12-epoxide | | |
|    isomer-1 | 174 | 0.4 |
|    isomer-2 | 215 | 1.2 |
| $H_4$-9,10-epoxide | 6250 | 22 |
| **BaP** | | |
| 7,8-diol-9,10-epoxide | | |
|    isomer-1 | 5300 | 18 |
|    isomer-2 | 2800 | 100 |
| $H_4$-9,10-epoxide | 3750 | 22 |

[a] Data taken from Wood et al., manuscript in preparation.
[b] Isomer-1 and isomer-2 refer to diol epoxides in which the oxirane oxygen is cis and trans to the benzylic hydroxyl group, respectively (see Figure 3).
[c] Histidine revertants/nmole/plate.
[d] 6-Azaguanine-resistant colonies/nmole/$10^5$ surviving cells.

low mutagenicity, in both systems, of the BeP diol epoxides relative to the BeP tetrahydroepoxide, and the BaP diol epoxides and tetrahydroepoxide. Thus, in Strain TA 98, BeP $H_4$-9,10-epoxide is 29 to 36 times as mutagenic as the BeP diol epoxides. With the Chinese hamster V79 cells, BeP $H_4$-9,10 epoxide is 18 to 55 times as mutagenic as the BeP diol epoxides. These results contrast with those for the BaP benzo-ring epoxides, wherein the mutagenicities of the BaP 7,8-diol-9,10-epoxides are very high and comparable to that of BaP $H_4$-9,10-epoxide. Additionally, BeP diol epoxide-2 is more mutagenic than BeP diol epoxide-1 in both bacterial and mammalian cells, whereas there is a reversal in relative mutagenicities of the BaP diol epoxides in the two systems.

## RATES OF HYDROLYSIS OF BeP DIOL EPOXIDES

The rates of hydrolysis of the BeP diol epoxides are recorded in Table 2, where they are compared with the corresponding BaP and triphenylene

**TABLE 2. Rates of Hydrolysis of BeP Diol Epoxides and Tetrahydroepoxides**

| Compound | $k_H$ (sec$^{-1}$ mol$^{-1}$) | $k_o$ (sec$^{-1}$)$^a$ × 10$^5$ |
|---|---|---|
| BeP[b] | | |
|    9,10-diol-11,12-epoxide | | |
|       isomer-1 | 9.7 | ~1.3 |
|       isomer-2 | 14.0 | 2.0 |
|    $H_4$-9,10-epoxide | 2,800 | ~9.8 |
| BaP[b] | | |
|    7,8-diol-9,10-epoxide | | |
|       isomer-1 | 510 | 420 |
|       isomer-2 | 1,400 | 13 |
| Triphenylene[b] | | |
|    1,2-diol-3,4-epoxide | | |
|       isomer-1 | 13.6 | 0.5 |
|       isomer-2 | 20.0 | 1.9 |
|    $H_4$-1,2-epoxide | 3,200 | 13 |
| BeP $H_4$-9,10-epoxide[c] | 1,700 | |
| BaP $H_4$-9,10-epoxide[c] | 12,000 | |

$^a$Small amounts of buffer were added in high-pH runs to hold pH constant.
$^b$In 10 percent dioxane/water, 0.1 M NaClO$_4$, at 25°.
$^c$In 25 percent dioxane/water, 0.1 M NaClO$_4$, at 25°

diol epoxides, and with the bay-region tetrahydroepoxide of each molecule. Both in the acid-catalyzed ($k_H$) and spontaneous ($k_o$) regions, the BeP and triphenylene diol epoxides are significantly less reactive than their corresponding tetrahydroepoxides. Tetrahydroepoxides typically solvolyze faster than the corresponding diol epoxides. For example, $k_H$ for BaP $H_4$-9,10-epoxide is 15 to 29 times that of the BaP 7,8-diol-9,10-epoxides in 25 percent dioxane/water (14). While strict comparisons are not possible because of the slightly different solvent system used, it appears that the difference in reactivity between the tetrahydroepoxide and diol epoxides in the BeP system is much larger, since in 10 percent dioxane/water the value of $k_H$ for BeP $H_4$-9,10-epoxide is 200 to 289 times the value of $k_H$ for the BeP diol epoxides. Triphenylene diol epoxides are similarly retarded in rate relative to the triphenylene tetrahydroepoxide ($k_H$ ratios of 160 and 235 for the tetrahydroepoxide relative to the diol epoxide-1 and -2 isomers, respectively).

Also of note, is that $k_H$ and $k_o$ values for isomer-2 exceed those for isomer-1, both for BeP and for triphenylene. This contrasts with the results for BaP, wherein $k_H$ for isomer-2 exceeds that of isomer-1, but $k_o$ for isomer-2 is much smaller than $k_o$ for isomer-1. A comparison of $k_H$ values for BaP and BeP $H_4$-9,10-epoxide hydrolysis in 25 percent dioxane/water reveals that the BaP isomer is seven times as reactive as the BeP isomer.

## DISCUSSION

Previous efforts to explain the relative solvolysis rates of diol epoxides and tetrahydroepoxides have explored the possible effects of conformation of the oxirane ring on reactivity (14). Thus, for BaP, molecular models indicate that two conformations for BaP $H_4$-9,10-epoxide are possible (Figure 4), and that the conformation in which the benzylic oxirane hydrogen atom is tilted away from interaction with the bay-region hydrogen atom at C-11 (the lower conformation) should be favored. In the latter conformation, the benzylic C-O bond is better positioned to overlap with the pi-system of the aromatic rings, and ring opening is expected to be faster from that conformation. This may help explain why BaP $H_4$-9,10-epoxide hydrolyzes faster (15 to 29 times) than the BaP 7,8-diol-9,10-epoxides, since the predominant conformation of each of the latter molecules (Figure 3), as indicated by Nuclear Magnetic Resonance (NMR) spectroscopy, is that in which the benzylic oxirane C-O bond is less favorably disposed for ring opening. Steric and polar substituent effects of the hydroxyl groups in the solvolysis of diol epoxides are not yet sufficiently understood and also may contribute to

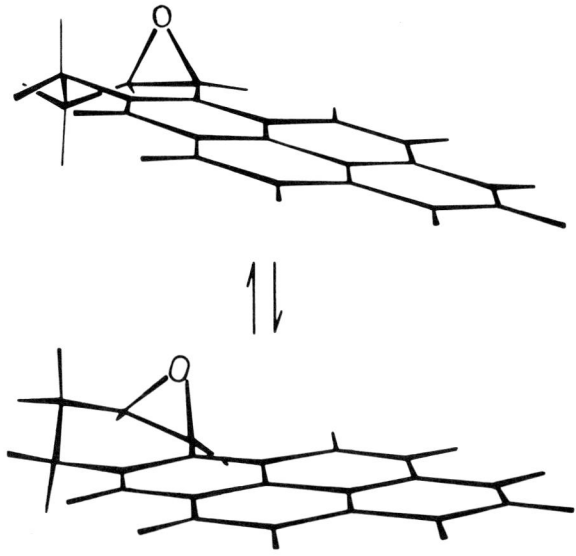

B[a]P H$_4$ 9,10-epoxide conformers

**FIGURE 4.** BaP H$_4$-9,10-epoxide conformational isomers. Steric interaction between the benzylic epoxide hydrogen and the bay-region aromatic hydrogen is possible only in the top structure.

the attenuation of their reduced solvolytic reactivities.

Conformational effects may also help explain the relative reactivity of the BeP 9,10-diol-11,12-epoxides. Thus, since both the benzylic hydroxyl group and the oxirane ring form part of bay regions, the BeP diol epoxides are locked into conformations (Figure 6) in which the hydroxyl groups are quasidiaxial, and only the BeP diol epoxide-2 has the oxirane ring optimally arranged for ring opening. The values of $k_H$ and $k_o$ are larger for BeP diol epoxide-2 relative to BeP diol epoxide-1, although the effect is not large.

More difficult to rationalize is the considerable attenuating effect of the hydroxyl groups of the BeP 9,10-diol-11,12-epoxides on reaction rate, as reflected in their $k_H$ hydrolysis rates relative to BeP H$_4$-9,10-epoxide. This effect is also probably linked to the quasidiaxial conformation of the hydroxyl groups in the BeP diol epoxides, and may reflect hindrance by the axial hydroxyl groups to solvation of the incipient carbonium ion that forms as rupture of the benzylic oxirane C-O bond occurs.

The very weak mutagenicity of the BeP 9,10-diol-11,12-epoxides relative to BeP H$_4$-9,10-epoxide is also intriguing. This may be related to

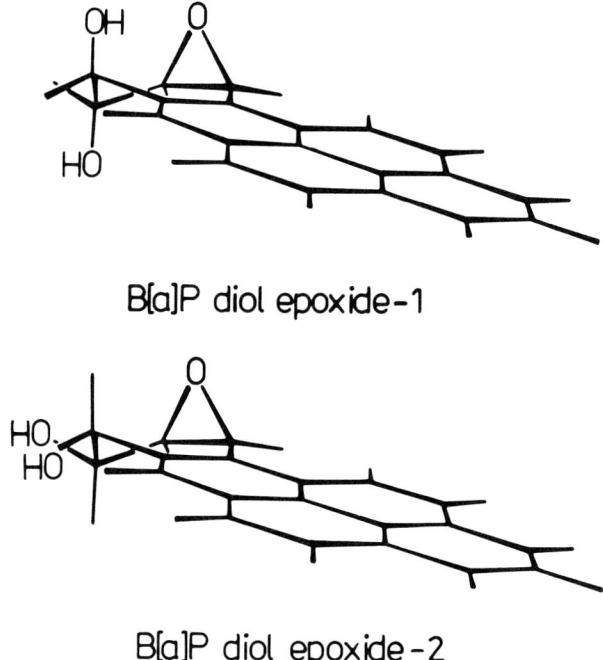

**FIGURE 5.** The major contributing conformational isomers for BaP 7,8-diol-9,10-epoxides-1 and -2.

the quasidiaxial conformation of the hydroxyl groups, since the relative difference in mutagenic activities of the diol epoxide and tetrahydroepoxides is larger than has been observed with analogous PAH derivatives wherein the hydroxyl groups are not locked into a quasidiaxial conformation. The apparent relationship between hydrolysis rate and mutation frequency for the BeP derivatives is striking. For the BeP derivatives studied, the more chemically reactive ($k_H$ and $k_o$) epoxide is also the more mutagenic, in mammalian as well as in bacterial cells. Large differences in $k_H$ values between molecules are paralled by corresponding large differences in mutation frequency. The use of hydrolysis rates as a guide to relative mutagenicity of analogous derivatives from different PAH is, however, more problematic. Thus, BaP $H_4$-9,10-epoxide solvolyzes seven times faster than BeP $H_4$-9,10-epoxide in 25 percent dioxane/water, but in *S. typhimurium* TA 98, BaP $H_4$-9,10-epoxide is only 60 percent as mutagenic as BeP $H_4$-9,10-epoxide. Quantitative comparisons of the mutagenicity of derivatives of different PAH are complicated by the differential response of different mutant strains of bacteria to different PAH. For example, in *S. typhimurium* TA

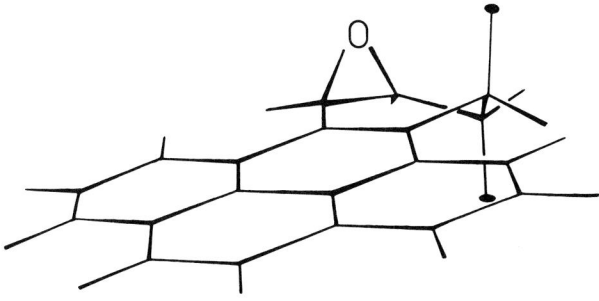

B[e]P diol epoxide - 1

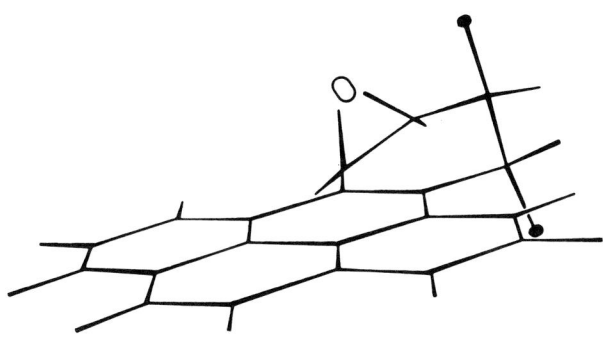

B[e]P diol epoxide - 2

**FIGURE 6.** Conformations of the BeP 9,10-diol-11,12-epoxides-1 and -2.

100, BaP $H_4$-9,10-epoxide is more mutagenic than BeP $H_4$-9,10-epoxide (8,16).

The much greater tumorigenicity, on mouse skin, of 9,10-$H_2$BeP relative to BeP, BeP 9,10-dihydrodiol, BeP 4,5-dihydrodiol, and BeP $H_4$-9,10-diol is not unexpected, and is probably mediated by metabolism to the highly reactive and mutagenic BeP $H_4$-9,10-epoxide (16). The significant, but weak, activity of BeP 9,10-dihydrodiol in inducing hepatic tumors in newborn mice could possibly reflect a small amount of metabolism to the BeP 9,10-diol-11,12-epoxides, but a determination of the tumorigenicity of the BeP diol epoxides awaits testing of the pure compounds.

## CONCLUSIONS

The bay-region theory (3,4,7) has been proven highly effective for the eight PAH which have been adequately studied (10,13,15). It predicted that, of the various possible benzo-ring tetrahydroepoxides for a given PAH, those in which the oxirane ring formed part of a bay region should be the most chemically reactive and biologically active. For BeP, initial studies suggested that insufficient metabolism to the key diol epoxides was a likely basis for the minimal biological activity of BeP. Since BeP $H_4$-9,10-epoxide was highly mutagenic, as had been anticipated by quantum chemical calculations (3,7), and the epoxide ring in the BeP 9,10-diol-11,12-epoxides occupied a bay-region position electronically analogous to that in BeP $H_4$-9,10-epoxide, high reactivity and biological activity of the BeP diol epoxides appeared likely. The present results document the first case in which the hydroxyl groups of a diol epoxide have a highly significant effect both on chemical reactivity and on inherent mutagenicity. The strongly attenuating effect of the hydroxyl groups may be a consequence of the novel quasidiaxial conformation into which the hydroxyl groups of the BeP diol epoxides are locked. Thus, the conformation of the hydroxyl groups appears to be important not only in metabolism of dihydrodiols, as shown by the virtual lack of conversion of BeP 9,10-dihydrodiol to BeP 9,10-diol-11,12-epoxides by rat liver microsomes, but also in affecting the chemical reactivity and biological activity of the diol epoxides themselves; this should be recognized as a significant factor when attempting to predict properties of diol epoxides substituted on analogous benzo-rings of other PAH.

## ACKNOWLEDGMENT

This investigation was supported in part by Grant No. 1 R01-CA-22985-02, awarded to R. E. Lehr by the National Cancer Institute, DHEW.

## REFERENCES

1. Buening, M. K., Levin, W., Wood, A. W., Chang, R. L., Lehr, R. E., Taylor, C. W., Yagi, H., Jerina, D. M., and Conney, A. H. (1980): Tumorigenic activity of benzo(e)pyrene derivatives on mouse skin and in newborn mice. *Cancer Research* (in press).
2. Harvey, R. G., Lee, H. M., and Shyamasundar, N. (1979): Synthesis of the dihydrodiols and diol epoxides of benzo(e)pyrene and triphenylene. *J. Org. Chem.* 44:78.
3. Jerina, D. M., Lehr, R. E., Yagi, H., Hernandez, O., Dansette, P. M., Wislocki, P. G., Wood, A. W., Chang, R. L., Levin, W., and Conney, A. H. (1976): Mutagenicity of benzo(a)pyrene derivatives and the description of a quantum mechanical model which predicts the ease of carbonium ion formation from diol epoxides. In: *In Vitro*

*Metabolic Activation in Mutagenesis Testing*, F. J. de Serres, J. R. Fouts, J. R. Bend and R. M. Philpot, Eds., pp. 159-177, Elsevier, Amsterdam.
4. Jerina, D. M., and Daly, J. W. (1977): Oxidation at carbon. In: *Drug Metabolism*, D. V. Parke and R. L. Smith, Eds., pp. 13-32, Taylor and Francis, Ltd., London.
5. Lehr, R. E., Schaefer-Ridder, M., and Jerina, D. M. (1977): Synthesis and reactivity of diol epoxides derived from non-K-region trans-dihydrodiols of benzo(a)anthracene. *Tetrahedron Lett.* 539.
6. Lehr, R. E., Taylor, C. W., Kumar, S., Mah, H. D., and Jerina, D. M. (1978): Synthesis of the non-K-region and K-region trans-dihydrodiols of benzo(e)pyrene. *J. Org. Chem.* 43:3462.
7. Lehr, R. E., Levin, W., Wood, A. W., Conney, A. H., Yagi, H., and Jerina, D. M. (1978): Quantum chemically based predictions of polycyclic aromatic hydrocarbon carcinogenicity: the bay region theory. In: *Structural Correlates of Carcinogenesis and Mutagenesis. A Guide to Testing Priorities?*, I. M. Asher and C. Zervos, Eds., pp. 129-135, HEW Publ. No. 78-1046.
8. Lehr, R. E., Taylor, C. W., Kumar, S., Levin, W., Chang, R., Wood, A. W., Conney, A. H., Thakker, D. R., Yagi, H., Mah, H. D., and Jerina, D. M. (1979): Differences in metabolism provide a basis for the low mutagenicity and carcinogenicity of benzo(e)pyrene compared to benzo(a)pyrene. In: *Polynuclear Aromatic Hydrocarbons: Third International Symposium on Chemistry and Biology-Carcinogenesis and Mutagenesis*, P. W. Jones and P. Leber, Eds., pp. 37-49, Ann Arbor Science Publishers, Inc., Ann Arbor, Michigan.
9. McCaustland, D. J., Fischer, D. L., Duncan, W. P., Ogilvie, E. J., and Engel, J. F. (1976): Labeled metabolites of polycyclic aromatic hydrocarbons. VI. Trans-7,8-dihydrobenzo(a)pyrene-7,8-diol-G-$^3$H and ($\pm$)-7$\alpha$,8$\beta$-dihydroxy-9$\beta$,10$\beta$-epoxy-7,8,9,10-tetrahydrobenzo(a)pyrene-G-$^3$H. *J. Labeled Comp. and Radiopharm.* 12:583.
10. Nordqvist, M., Thakker, D. R., Yagi, H., Lehr, R. E., Wood, A. W., Levin, W., Conney, A. H., and Jerina, D. M. (1979): Evidence in support of the bay-region theory as a basis for the carcinogenic activity of polycyclic aromatic hydrocarbons. In: *Molecular Basis of Environmental Toxicity*, R. S. Bhatnagar, Ed., p. 329. Ann Arbor Science Publishers, Inc., Ann Arbor, Michigan.
11. Selkirk, J. K., and MacLeod, M. C. (1979): Metabolism and macromolecular binding of benzo(a)pyrene and its noncarcinogenic isomer benzo(e)pyrene in cell cultures. In: *Polynuclear Aromatic Hydrocarbons: Third International Symposium on Chemistry and Biology-Carcinogenesis and Mutagenesis*, P. W. Jones and P. Leber, Eds., pp. 21-36, Ann Arbor Science Publishers, Inc., Ann Arbor, Michigan.

12. Thakker, D. R., Yagi, H., Lehr, R. E., Levin, W., Buening, M., Lu, A.Y.H., Chang, R. L., Wood, A. W., Conney, A. H., and Jerina, D. M. (1978): Metabolism of trans-9,10-dihydroxy-9,10-dihydrobenzo(a)pyrene occurs primarily by aryl-hydroxylation rather than formation of a diol epoxide. *Mol. Pharm.* 14:502.
13. Thakker, D. R., Nordqvist, M., Yagi, H., Levin, W., Ryan, D., Thomas, P., Conney, A. H., and Jerina, D. M. (1979): Comparative metabolism of a series of polycyclic aromatic hydrocarbons by rat liver microsomes and purified cytochrome P-450. In: *Polynuclear Aromatic Hydrocarbons: Third International Symposium on Chemistry and Biology-Carcinogenesis and Mutagenesis*, P. W. Jones and P. Leber, Eds. p. 455, Ann Arbor Science Publishers, Inc., Ann Arbor, Michigan.
14. Whalen, D. L., Ross, A. M., Yagi, H., Karle, J. M., and Jerina, D. M. (1978): Stereoelectronic factors in the solvolysis of bay-region diol epoxides of polycyclic aromatic hydrocarbons. *J. Am. Chem. Soc.* 100:5218.
15. Wood, A. W., Levin, W., Chang, R. L., Yagi, H., Thakker, D. R., Lehr, R. E., Jerina, D. M., and Conney, A. H. (1979): Bay-region activation of carcinogenic polycyclic hydrocarbons. In: *Polynuclear Aromatic Hydrocarbons: Third International Symposium on Chemistry and Biology-Carcinogenesis and Mutagenesis*, P. W. Jones and P. Leber, Eds., p. 531, Ann Arbor Science Publishers, Inc., Ann Arbor, Michigan.
16. Wood, A. W., Levin, W., Thakker, D. R., Yagi, H., Chang, R. L., Ryan, D. E., Thomas, P. E., Dansette, P. M., Whittaker, N., Turujman, S., Lehr, R. E., Kumar, S., Jerina, D. M., and Conney, A. H. (1979): Biological activity of benzo(e)pyrene, an assessment based on mutagenic activities and metabolic profiles of the polycyclic hydrocarbon and its derivatives. *J. Biol. Chem.* 254:4408.
17. Yagi, H., Hernandez, O., and Jerina, D. M. (1975): Synthesis of ($\pm$)-7$\beta$,8$\alpha$-dihydroxy-9$\beta$,10$\beta$-epoxy-7,8,9,10-tetrahydrobenzo(a)pyrene, a potential metabolite of the carcinogen benzo(a)pyrene with stereochemistry related to the antileukemic triptolides. *J. Am. Chem. Soc.* 97:6881.
18. Yagi, H., Thakker, D. R., Hernandez, O., Koreeda, M. and Jerina, D. M. (1977): Synthesis and reactions of the highly mutagenic 7,8-diol-9,10-epoxides of the carcinogen benzo(a)pyrene. *J. Am. Chem. Soc.* 99:1604.
19. Yagi, H., Thakker, D. R., Lehr, R. E., and Jerina, D. M. (1979): Benzo-ring diol epoxides of benzo(e)pyrene and triphenylene. *J. Org. Chem.* 44:3439.

# A STUDY OF THE 7,12-DIMETHYLBENZ(a)-ANTHRACENE (DMBA) BAY REGION INVOLVEMENT IN THE PRODUCTION OF CARCINOGEN AND MUTAGEN METABOLITES

Y. M. Sheikh*, M. N. Inbasekaran*, F. B. Daniel*'**, F. D. Cazer*†,
R. W. Hart‡, and D. T. Witiak*

*Division of Medicinal Chemistry
College of Pharmacy
The Ohio State University
Columbus, Ohio 43210
**Environmental Protection Agency
Cincinnati, Ohio 45268
†Radiochemistry Laboratory
Comprehensive Cancer Center
The Ohio State University
Columbus, Ohio 43210
‡Department of Radiology
College of Medicine
The Ohio State University
Columbus, Ohio 43210

7,12-Dimethylbenz(a)anthrancene (DMBA) is a potent carcinogen (2,4,28,38) on mouse skin. A single intravenous or intragastric dose of DMBA induces mammary tumors (21,40), lung tumors (47,48), ovarian tumors (27,45), and tumors of Zymbal's gland (18). Consequently, considerable effort has been directed towards the understanding of the bioactivation of DMBA (5,11,15,20,22,23,29-31,44,46,52,53) and the subsequent covalent binding of the active metabolites to cellular macromolecules (6). Recently, it was shown that incubation of DMBA in vitro with phenobarbital-induced rat liver microsomes results in ring-A activation to furnish DMBA-trans-3,4-dihydrodiol, 7OHM,12MBA-trans-3,4-dihydrodiol, 12OHM,7MBA-trans-3,4-dihydrodiol, and 7OHM,12O-HMBA-trans-3,4-dihydrodiol (Figure 1). These 3,4-dihydrodiols have higher mutagenic and DNA binding activities (5,53)

# 690 BAY-REGION METABOLITES OF DMBA

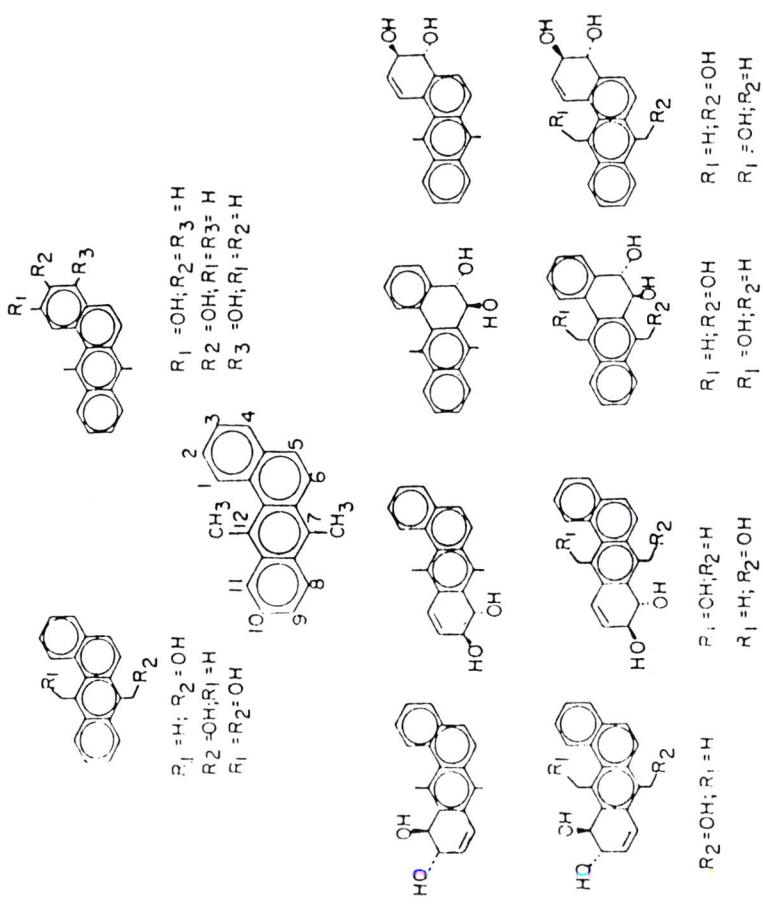

FIGURE 1. Structures of known DMBA metabolites.

than DMBA. Similarly, DMBA-trans-3,4-dihydrodiol and DMBA-trans-5,6-dihydrodiols were as active as DMBA in a tumor frequency assay, whereas the related 8,9- and 10,11-dihydrodiols were virtually inactive (15). 2-, 3-, and 4-Hydroxy-DMBA and their 7- and 12-hydroxymethyl derivatives have also been reported as metabolites (6). In connection with our work on bioactivation of DMBA in rat liver microsomes [3-methylcholanthrene (3MC)-induced] and Syrian hamster embryo cells (9) we failed to detect DMBA-trans-3,4-dihydrodiols. To explore further DMBA, fluoro-, and reduced-analogue metabolism, we employed Sprague Dawley rats, using various inducers and inhibitors of the hepatic mixed-function oxygenase system. The significance of 3,4-dihydrodiol metabolite formation, structural constraints, and enzyme-induction requirements will be discussed.

Cytochromes P-450, components of the mixed-function oxidase system, are responsible for the metabolism of numerous foreign compounds. Separation and characterization of highly purified forms of cytochromes P-450 from phenobarbital-, 3MC-, and polychlorinated biphenyl-induced rat liver microsomes have been reported (36). Different amino acid compositions and partially characterized amino acid sequences, together with immunological studies of these hemoproteins, demonstrated that cytochromes P-$450_a$, P-$450_b$, and P-$450_c$ have different primary and tertiary structures. Thus, these enzymes arise from separate genes and not by posttranslational modification of one primary gene product (3). The major phenobarbital-inducible cytochrome P-$450_b$ hydroxylates testosterone at the 16 position and effectively demethylates benzphetamine. Alternatively, the less inducible cytochrome P-$450_a$ hydroxylates testosterone at the $7\alpha$-position and has low activity for the metabolism of benzphetamine and benzo(a)pyrene. Conversely, the major 3MC-inducible cytochrome P-$450_c$ has very high catalytic activity for the metabolism of benzo(a)pyrene (BaP).

The topology of epoxide hydrase and BaP monooxygenase in the endoplasmic reticulum of rat liver has recently been investigated (37). It has been shown that epoxide hydratase may be buried deeply in the hydrophobic phase of the membrane of the hepatic endoplasmic reticulum. The specific activities of epoxide hydratase and BaP monooxygenase in rough and smooth microsomes from control and phenobarbital-treated animals are approximately the same, whereas after 3MC-treatment BaP monooxygenase is four times higher and epoxide hydratase is twice as high in the rough vessicles. Distribution studies reveal that the BaP monooxygenase complex and epoxide hydratase may form a complex of unique stoichiometry in the membrane of microsomes from control rats. Apparently, such a complex does not form following 3MC-induction (37).

In 3MC-treated $B_6$ mice the induced aryl hydrocarbon hydroxylase (AHH) activity in vitro is more sensitive to inhibition by $\alpha$-napthoflavone and more resistant to inhibition by metyrapone (1,35). Furthermore, species differences in substrate specificity of hepatic cytochrome P-448 from polycylic hydrocarbon-treated animals is well recognized (43).

We have employed inhibitors (1,14,19) of cytochrome P-450 ($CS_2$, $\alpha$-napthoflavone metyrapone, and SKF-525A) to determine whether administration of phenobarbital to male Sprague Dawley rats induces DMBA-12-methyl-hydroxyl- and 3,4-dihydrodiol-forming activities that are distinctly different from those responsible for the formation of DMBA-trans-5,6-dihydrodiol, DMBA-trans-8,9-dihydrodiol, and DMBA-trans-10,11-dihydrodiol. We also have employed various fluoro-analogues in order to explore: (1) their comparative metabolic profiles (9) using different enzyme inducers; (2) relationships of formation of various dihydrodiols with carcinogenic (41) or mutagenic activity (16); and (3) the effects of various inhibitors of the AHH system in order to characterize similarities or dissimilarities in DMBA and analogue metabolism. Additionally, we have synthesized and studied the mutagenic activity of various bay-region reduced analogues (23) in the Ames assay with and without metabolic activation.

## MATERIALS AND METHODS

### Hydrocarbons

DMBA was purchased from Eastman Chemical Company, checked for purity by high pressure liquid chromatography (HPLC) prior to use, and, when necessary, purified by column chromatography over silica gel (Silica Gel-60; E. Merck) using hexane-benzene (1:1) as eluent. The 2OH- and 4OH-DMBA derivatives were gifts of Professor Melvin S. Newman, Department of Chemistry, The Ohio State University. 3OH-DMBA was prepared according to known procedures (34) and independently converted to DMBA-trans-3,4-dihydrodiol by a method similar to the one published by Sukumaran and Harvey (42). Crystalline DMBA-trans-5,6-dihydrodiol was prepared according to published methods (17). The 1,2,3,4-tetrahydro-trans-3,4-diol analogue (50) was prepared from 1,2-dihydro-DMBA (23). The completely saturated bay-region analogue of DMBA was prepared as previously reported from these laboratories (23,50). The 2F and 3F analogues were prepared as reported (39), 4F-DMBA was a gift of Professor Newman, 5F-DMBA was prepared by a modification of the reported synthesis (33), and 6F-DMBA was a gift of Professor Wolfgang Girke, Institute für Organische Chemie der Universität Frankfurt M. (13). Authentic samples of 7OHM,12M,3OHBA, 12OHM,7M,3OHBA and 7-OHM,12MBA-trans-5,6-dihydrodiol and 12OHM,7MBA-trans-5,6-dihydrodiol were prepared by incubation of

3OH-DMBA and DMBA-trans-5,6-dihydrodiol with phenobarbital-induced rat liver S-10 fraction.

## Generation of Metabolites

Liver S-10 fractions were obtained from pretreated male Sprague Dawley rats weighing 150 to 200 g (intraperitoneal dose of phenobarbital was 100 mg/kg/day for 4 days; that of 3MC was 20 mg/kg/day for 2 days; and that of $\beta$-naphthoflavone was 75 mg/kg/day for 5 days).

Metabolites of DMBA were obtained by incubation in vitro of the substrate with liver S-10 fraction prepared by homogenizing rat liver with an equal weight of 0.1 M phosphate buffer, pH 7.3, and centrifuging the homogenate at 9,000 and 10,000 g, respectively, for a period of 1 hour each. The S-10 fractions were frozen and stored at -70°C. All incubations were carried out at 35 to 37°C in 6 ml of potassium phosphate [0.1 M Tris + 5.0 mM $MgCl_2$ + 5.0 $\mu$M $MnCl_2$ + 0.012 M (DL)-isocitrate, 4.4 units isocitrate dehydrogenase, 5.0 mM NADP, 1.0 ml S-10 fraction, and 1 mg DMBA or F-analogue substrate in 0.3 ml acetone] for a period of 2.5 hours with constant shaking. Inhibitors were preincubated for a period of 10 minutes in incubation buffer containing S-10 fraction. After addition of DMBA, incubation was continued for a period of 2.5 hours. At the end of the incubation periods, acetone (10 ml) and ethyl acetate (20 ml) were added. The mixture was vortexed and centrifuged, and the supernatant was dried over sodium sulfate (15 g) and evaporated under reduced pressure. The residue was dissolved in a mixture of acetone: ethyl acetate (1:2; 0.5 ml) with warming and analyzed by HPLC.

The stability of these S-10 fractions was examined at various times. We have found that phenobarbital- and $\beta$-naphthoflavone-induced and uninduced S-10 fractions are stable at -70°C for over 3 months. However, storage of phenobarbital-induced S-10 fractions for 1 week at 0°C resulted in loss of 3,4-dihydrodiol and stimulation of 7-methyl hydroxyl forming activity. For all inhibitor studies, two controls were utilized: one which did not contain inhibitor and one which did not contain substrate but had a maximum inhibitor concentration present. We did not find substantial variation in metabolic profiles for substrate concentrations of 0.7 to 1.3 mg per incubation.

## Isolation of Metabolites

A Laboratory Data Control (LDC) chromatography accessory module containing LDC gradient master, constrametric pumps, and spectormonitor III and equipped with Whatman Partisil PXS 10/25 ODS column (length = 25 cm; diam = 4.6 mm packed with Spherisorb ODS 5 $\mu$) and a Whatman precolumn (length = 7 cm, diam = 4.6 mm packed with CoPell$^{Tm}$ ODS 10 $\mu$) was employed for HPLC purposes. The column was eluted at ambient temperature with methanol in water (25 percent v/v) to 100 percent methanol at a solvent flow of 1.2 ml/min. Chromatographic peaks were collected and combined from 5 to 6 runs. The solvent was removed by evaporation with nitrogen and the residue examined spectrophotometrically.

Metabolites were quantified by area measurements of HPLC peaks using a UV detector at 271 nm and are the average of two to three incubations. Reliability of the quantification was checked by incubation of generally tritiated DMBA with phenobarbital and $\beta$-napthoflavone S-10 fractions. Peaks were collected by HPLC and counted and found to be within ±6 percent of the area measurements.

## Characterization of Metabolites

Ultraviolet and visible absorption spectra of the metabolites were measured in methanol on a Beckman UV 526 spectrophotometer. Fluorescence spectra of the metabolites were measured in methanol using an Aminco SPF-500 Spectrofluorometer. Data obtained were compared with reported (5,6,9,24,25,41,42,44,45,53) values for physical constant for DMBA and BA, for DMBA and BA metabolites, and for related fluoro analogues and their respective metabolites. The curves in Figure 6 are tentative.

## RESULTS AND DISCUSSION

### DMBA Metabolism

The comparative HPLC profiles of DMBA metabolites derived from phenobarbital-, $\beta$-napthoflavone, 3MC-induced, and uninduced S-10 fractions are presented in Figure 2. Retention times using reverse-phase HPLC of identified metabolites are listed in Table 1. The order of total diol production after 2.5 hours of incubation with the various S-10 fractions was 3MC-induced>$\beta$-naphthoflavone-induced>phenobarbital-induced> uninduced (Table 2). Only in the case of $\beta$-naphthoflavone-induced S-10 rat liver fraction was a substantial quantity (~10 percent) of 7-OHM,12MBA-trans-10-11-dihydrodiol detected. Alternatively, DMBA-trans-5,6-dihydrodiol production was more predominant in phenobarbital-induced (~4 percent) than in $\beta$-naphthoflavone-induced (~2.5 percent) S-10 fractions. $\beta$-Naphthoflavone- and 3MC-induced S-10 fractions most closely resembled each other relative to DMBA metabolism. These data

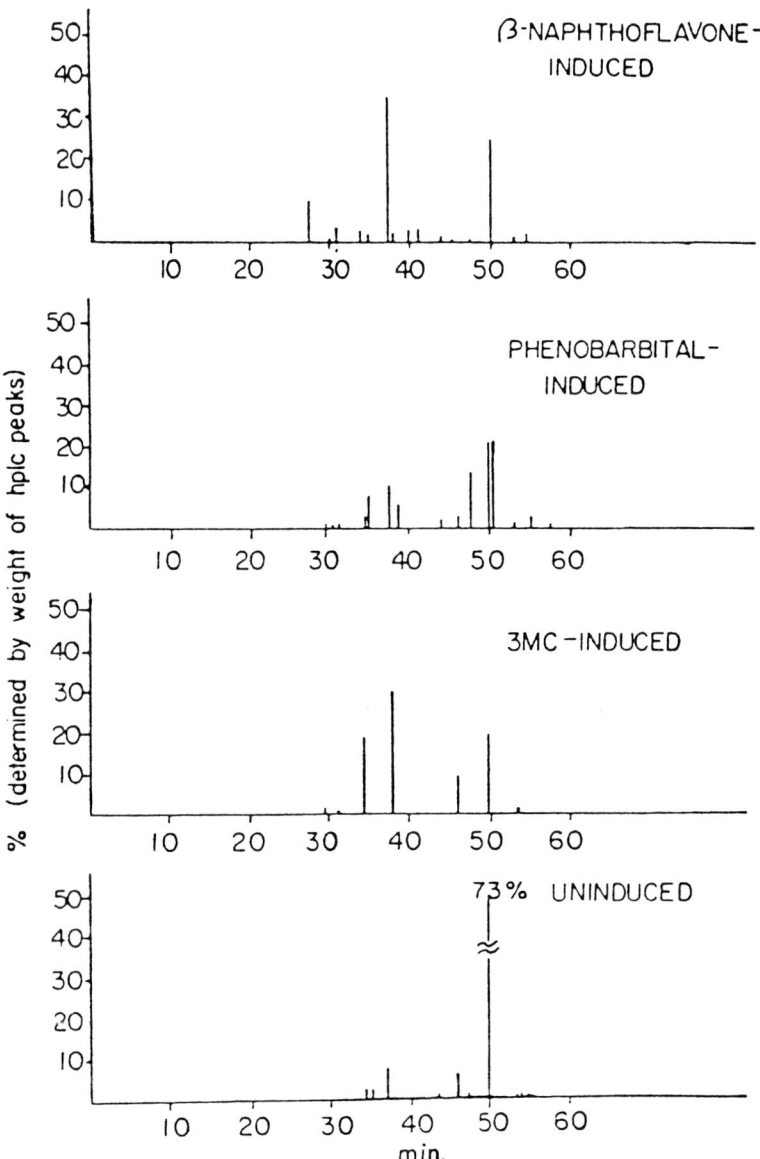

FIGURE 2. Metabolism of DMBA using induced and uninduced Sprague Dawley rat liver S-10 fractions.

TABLE 1. Retention Times of DMBA Metabolites on Partisil ODS Column; Methanol-Water Gradient (25 to 100 percent); Linear; 60 Minutes

| Metabolites | Retention Time (min) |
|---|---|
| 4OH-DMBA | 55.25 |
| 3OH-DMBA | 54.10 |
| 2OH-DMBA | 53.10 |
| 9 (or 10)OH-DMBA | 53.0 |
| 7OHM,12MBA | 50.6 |
| 12OHM,7MBA | 51.55 |
| DMBA-trans-3,4-dihydrodiol | 47.4 |
| DMBA-trans-5,6-dihydrodiol | 34.75 |
| DMBA-trans-8,9-dihydrodiol | 37.75 |
| DMBA-trans-10,11-dihydrodiol | 38.75 |
| 7OHM,12OHMBA | 38.0 |
| 7OHM,12MBA-trans-3,4-dihydrodiol | 35.30 |
| 12OHM,7MBA-trans-3,4-dihydrodiol | 35.30 |
| 12OHM,7MBA-trans-8,9-dihydrodiol | 31.20 |
| 7OHM,12MBA-trans-8,9-dihydrodiol | 30.05 |
| 7OHM-12MBA-trans-10,11-dihydrodiol | 28.10 |
| 12OHM,7MBA-trans-10,11-dihydrodiol | 28.15 |
| 7OHM,12MBA-trans-5,6-dihydrodiol | 29.05 |
| 12OHM,7OHM-BA-trans-5,6-dihydrodiol | 25.0 |

TABLE 2. Generation of Various Dihydrodiol Metabolites on Incubation of DMBA with Rat Liver Induced and Uninduced S-10 Fractions

| Metabolites | Percentage[a] of Total Metabolites | | | |
|---|---|---|---|---|
| | Uninduced | 3MC-Induced | Phenobarbital-Induced | $\beta$-Naphthoflavone Induced |
| DMBA-trans-3,4-dihydrodiol | 2 | ND[b] | 15.0 | ND |
| DMBA-trans-5,6-dihydrodiol | ND | 20 | 4.0 | 2.5 |
| DMBA-trans-8,9-dihydrodiol | 13.5 | 30 | 10.1 | 35 |
| DMBA-trans-10,11-dihydrodiol | ND | ND | ND | ND |
| 7OHM,12MBA-trans-3,4-dihydrodiol and/or 12-OHM isomer | 3 | ND | 9.1 | ND |
| 7OHM,12MBA-trans-5,6-dihydrodiol | ND | ND | ND | ND |
| 12OHM isomer | ND | ND | ND | ND |
| 7OHM,12MBA-trans-8,9-dihydrodiol | ND | 2 | ND | ND |
| 12OHM isomer | ND | <1 | <1 | <1 |
| 7OHM,12MBA-trans-10,11-dihydrodiol | ND | <1 | ND | 10 |

[a]Error of determination ~±6 percent of number shown.
[b]ND = not detected.

are consistent with the observation that $\beta$-naphthoflavone resembles 3MC induction of cytochromes P-450 and epoxide hydratase (12,14,19,49). Of considerable importance is the observation that phenobarbital-induced S-10 fraction afforded relatively large amounts of DMBA-trans-3,4-dihydrodiol (~15 percent) and either a mixture (~9 percent) or one of the other of 7-OHM,12-MBA-trans-3,4-dihydrodiol and 12-OHM,7-MBA-trans-3,4-dihydrodiol, whereas $\beta$-naphthoflavone-induced S-10 fraction produced no detectable quantities of these metabolites. Equally significant is the observation that uninduced S-10 fraction afforded small amounts of these presumed precarcinogenic 3,4-dihydrodiol metabolites. DMBA-trans-8,9-dihydrodiol was consistently detected in relatively large amounts (10 to 35 percent) with all S-10 preparations.

## Effect of Inhibitors on DMBA-Dihydrodiol Formation

As expected, all AHH inhibitors ($CS_2$, $\beta$-naphthoflavone, metyrapone, and SKF-525A) blocked formation of total DMBA metabolites in all S-10 fractions studied. However, these inhibitors exhibited markedly different effects on the relative percentage of production of the total of the various metabolites. Metyrapone (0 to 0.71 $\mu M/ml$), $CS_2$ (0 to 1.54 $\mu M/ml$), and SKF-525A (0 to 0.28 $\mu M/ml$) inhibit total dihydrodiol [sum of 3,4- 5,6-, 8,9-, and 10,11-dihydrodiols plus 7,12-bis-(hydroxymethyl)-BA] formation in phenobarbital S-10 fraction in a dose-related fashion (Figure 3). On the other hand, $\alpha$-naphthoflavone (0 to 0.29 $\mu M/ml$), an inhibitor of cytochrome P-450 (1,12), stimulated total diol formation in $\beta$-naphthoflavone-induced S-10 fraction, whereas metyrapone and SKF-525A had no effect (Figure 4). Furthermore, in the case of uninduced S-10 fraction, $\alpha$-naphthoflavone (0 to 0.57 $\mu M/ml$) and SKF-525A (0 to 0.57 $\mu M/ml$) produced a marked stimulation of total diol formation in a dose-related manner (Figure 4).

All inhibitors [$CS_2$ (0 to 2.14 $\mu M/ml$); metyrapone (0 to 0.71 $\mu M/ml$); SKF-525A (0 to 0.57 $\mu M/ml$); $\alpha$-naphthoflavone (0 to 0.57 $\mu M/ml$)] provided a marked decrease (65 to 100 percent) in the production of DMBA-trans-3,4-dihydrodiol and 7- or 12-methyl hydroxylated DMBA-trans-3,4-dihydrodiols in phenobarbital-induced S-10 fraction (Figure 5). DMBA-trans-8,9-dihydrodiol formation in phenobarbital-induced S-10 fraction also was stimulated by metyrapone (0 to 0.71 $\mu M/ml$) but was not affected by $CS_2$ (0 to 1.54 $\mu M/ml$) (Figure 6).

These, as well as several other differential effects of inhibitors on DMBA metabolism in phenobarbital-induced S-10 fraction, serve to emphasize the multiple processes involved in formation of DMBA metabolites. Whereas the combined percentage of 7- and 12-methyl hydroxylated metabolites is stimulated by $CS_2$ (0 to 1.54 $\mu M/ml$), metyrapone (0 to 0.71 $\mu M/ml$), and SKF-525A (0 to 0.57 $\mu M/ml$), the relative amount of each of these two compounds is affected dissimilarly by $CS_2$ when compared with the latter two agents (Figures 7, 8, and 9). $CS_2$ stimulated production of the 12-hydroxylated analogue relative to the 7-hydroxylated material. The converse was observed for both metyrapone

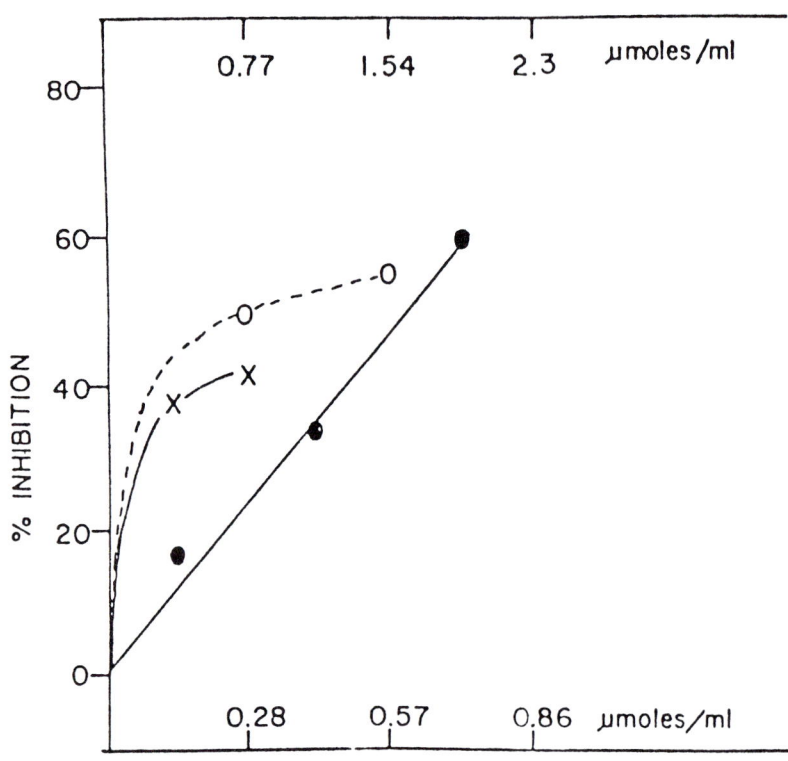

FIGURE 3. Effect of various inhibitors [O = $CS_2$ (top scale); X = SKF - 525A; ● = metyrapone] on the production of total DMBA dihydrodiols (3,4-; 5,6-; 8,9-; 10,11-) and 7,12-bis-hydroxy methylbenz(a)anthracene.

and SKF-525A. Thus, the enzyme complexes responsible for the formation of 3,4-dihydrodiols and hydroxylation of the 7- and 12-methyl groups of DMBA are likely different. The observation that DMBA-trans-8,9-dihydrodiol is a major metabolite, whereas DMBA-trans-3,4-dihydrodiol is not detected from DMBA in β-naphthoflavone- and 3MC-induced S-10 fractions suggests involvement of different enzyme systems in the production of these dihydrodiols (Figure 2, Table 2). Both metyrapone (0 to 0.57 μM/ml) and SKF-525A (0 to 0.57 μM/ml) stimulated the formation of 7-methyl-hydroxylated DMBA and inhibited 7-OHM, 12MBA-trans-10,11-dihydrodiol formation in β-naphthoflavone-induced S-10 fraction. SKF-525A also inhibited formation of 7-OHM,12MBA-trans-8,9-dihydrodiol in this fraction (Figures 10 and 11). Whereas SKF-525A does not inhibit, and metyrapone may cause minor stimulation of formation of DMBA-trans-8,9-dihydrodiol in β-naphthoflavone-induced S-10 fraction, in uninduced S-10 fractions these two metabolic inhibitors

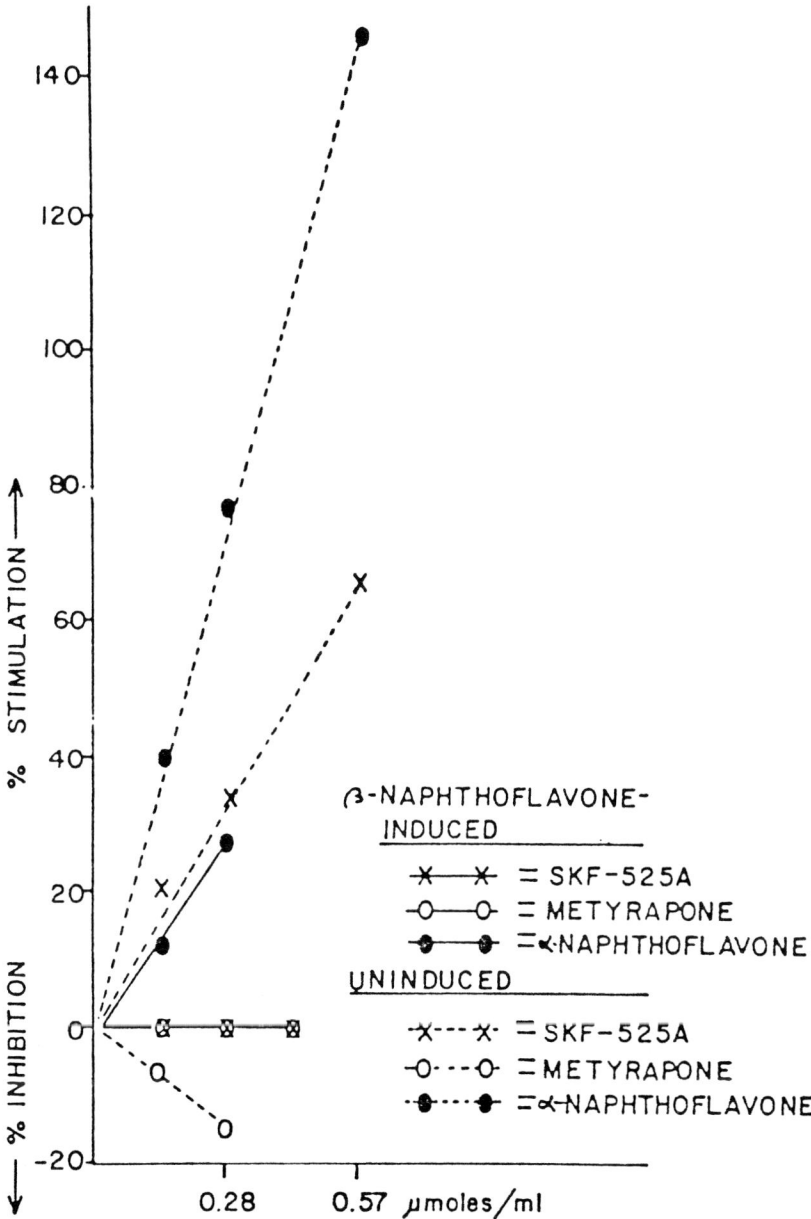

FIGURE 4. Effect of various inhibitors on the production of total DMBA diols in β-naphthoflavone-induced and uninduced rat liver S-10 fraction. Stimulation due to α-napthoflavone is relative to 7OHM, 12MBA.

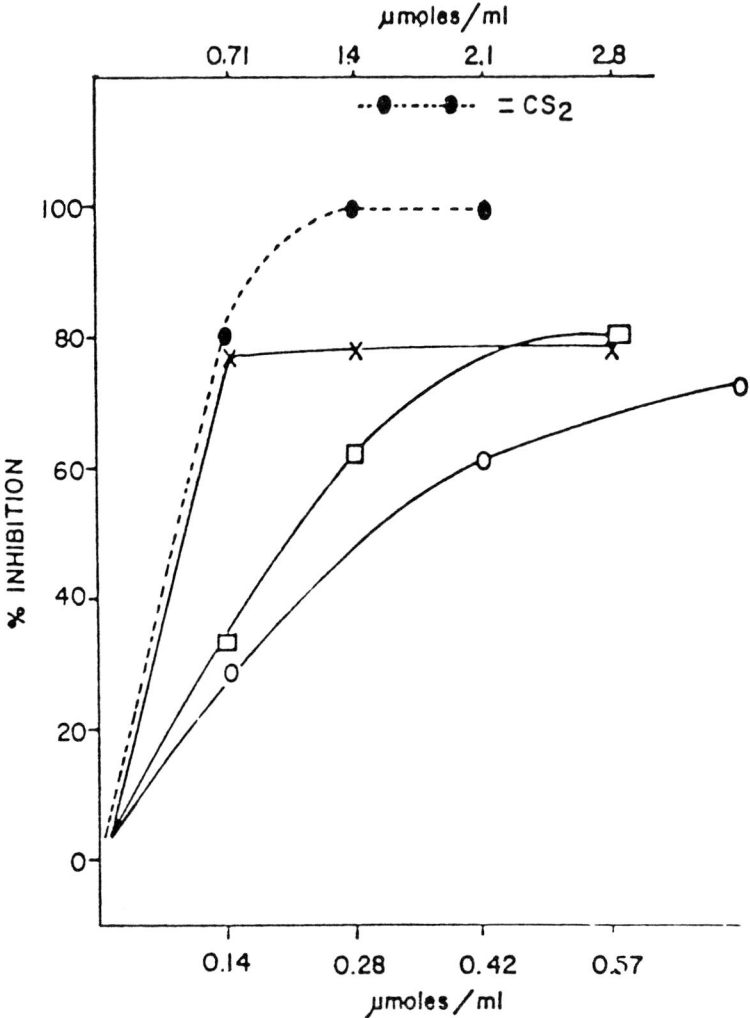

FIGURE 5. Effect of inhibitors [● = $CS_2$ (top scale); O = metyrapone; X = SKF - 525A; □ = α-naphthoflavone] on the formation of DMBA trans-3,4-dihydrodiol or the total of 7OHM, 12MBA- and 12OHM, 7MBA-trans-3,4-dihydrodiols. (Respective curves are virtually superimposable. Only the curve for DMBA-trans-3,4-dihydrodiol is shown.)

stimulated formation of this metabolite (Figure 12). It thus appears that DMBA-trans-8,9-dihydrodiol, in part, originates differently in phenobarbital- and β-naphthoflavone-induced and uninduced S-10 fractions.

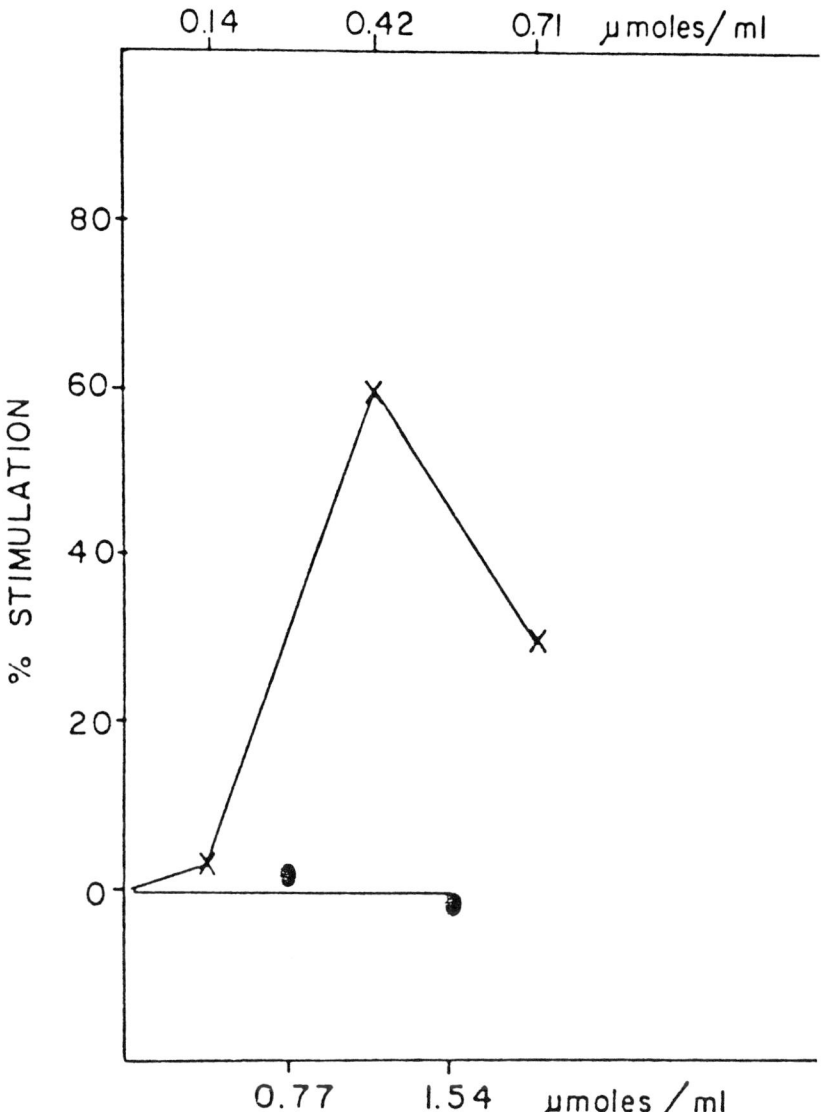

**FIGURE 6.** Effect of $CS_2$ (●, bottom scale) and metyrapone (X) on the production of DMBA-trans-8,9-dihydrodiol in phenobarbital-induced S-10 fractions.

## Metabolism of 2-, 3-, 4-, 5-, and 6-Fluoro-DMBA Analogues

Aryl fluorine bonds are of greater energy than aryl hydrogen bonds (eg, 128 kcal/mole in fluorobenzene and 112 kcal/mole in benzene) (8).

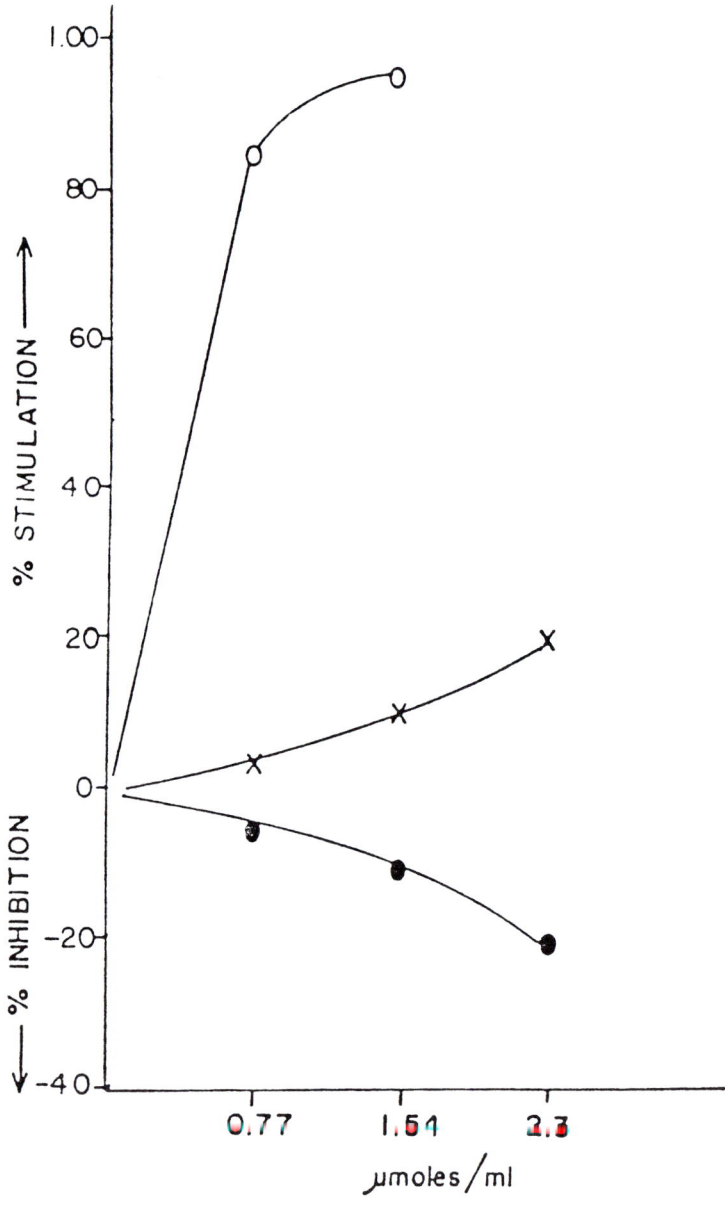

FIGURE 7. Effect of CS$_2$ on the formation of the sum of 7OHM, 12MBA and 12OHM, 7MBA (O) in relation to total metabolite production (tentative). Effect of CS$_2$ on the relative production of 12OHM, 7MBA(X) in relation to 7OHM, 12MBA (●). All data obtained employing phenobarbital-induced S-10 fractions.

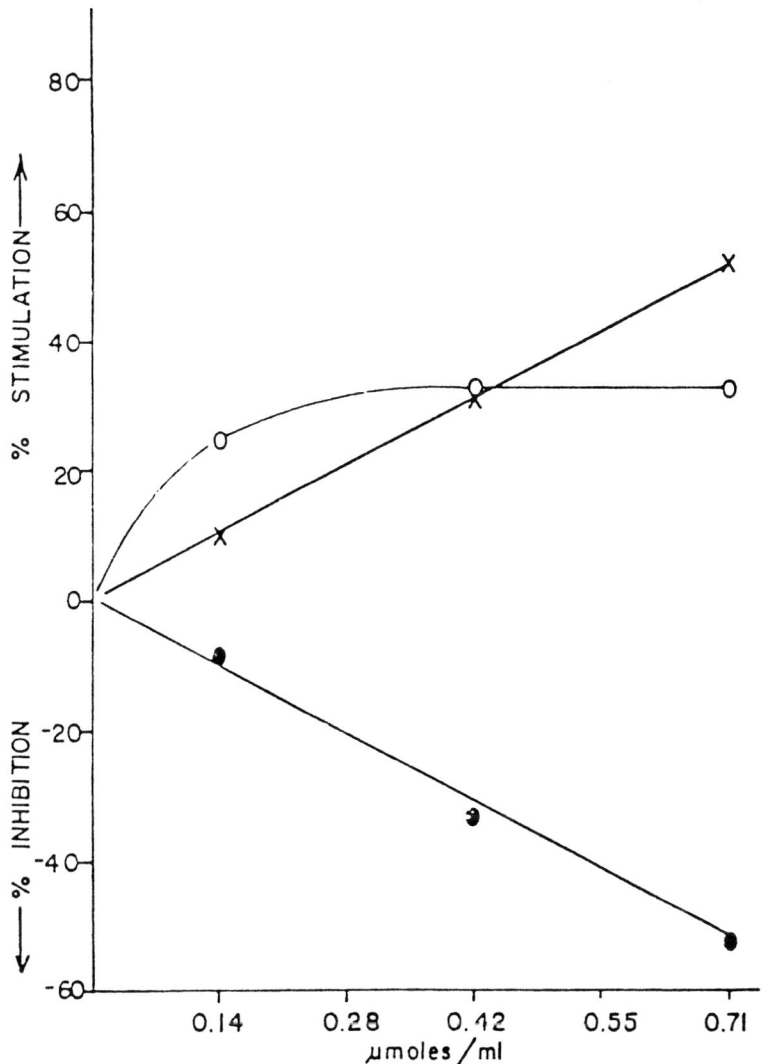

FIGURE 8. Effect of metyrapone on the formation of the sum of 7OHM, 12MBA and 12OHM, 7MBA (O) in relation to total metabolite production. Effect of metyrapone on the relative production of 12OHM, 7MBA (X) in relation to 7OHM 12MBA (●). All data obtained employing phenobarbital-induced S-10 fractions.

Additionally, insertion of F into aromatic molecules blocks biofunctionalization of bonds carrying this halogen (32). Earlier studies revealed that F-substitution for hydrogen in DMBA blocks or substantially reduces metabolism at that position (9). When inserted at position 2 or 5, DNA

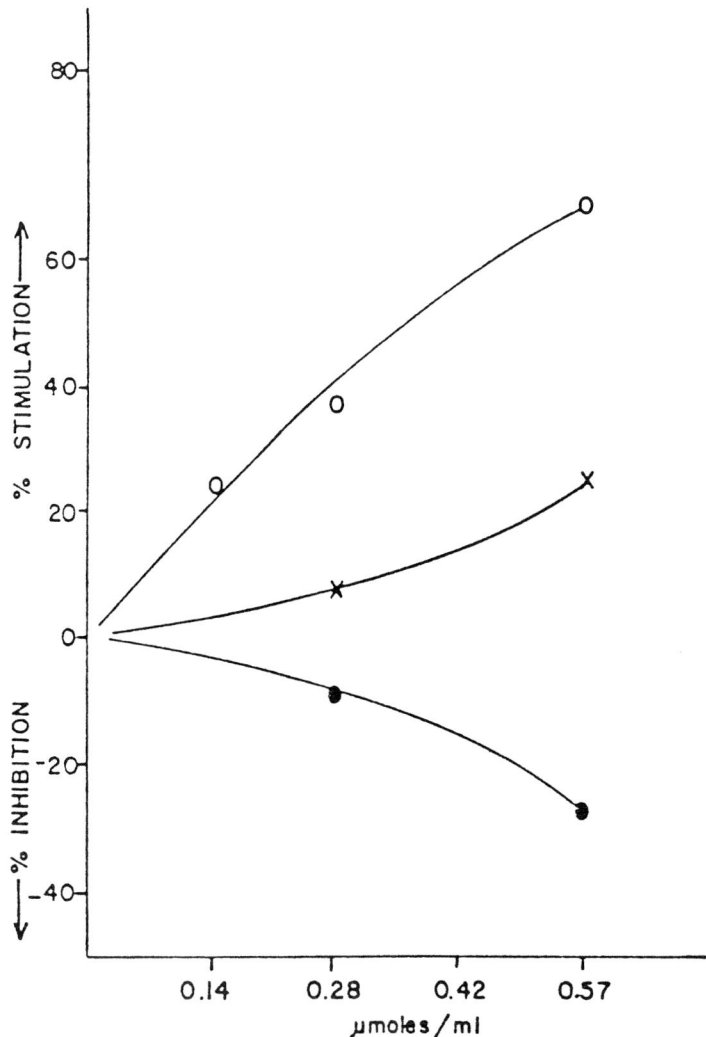

FIGURE 9. Effect of SKF-252A on the formation of the sum of 7OHM, 12MBA and 12OHM, 7MBA (O) in relation to total metabolite production. Effect of SKF-525A on the relative production of 12OHM, 7MBA (X) in relation to 7OHM, 12MBA(●). All data were obtained while employing phenobarbital-induced S-10 fractions.

covalent bonding to the hydrocarbon in mammalian-cultured cells was substantially reduced. However, at the presumably less critical site (11) F-substitution did not reduce covalent bonding to this macromolecule (9). These results correlated with the carcinogenic and mutagenic potentials of these compounds as determined by three independent groups of inves-

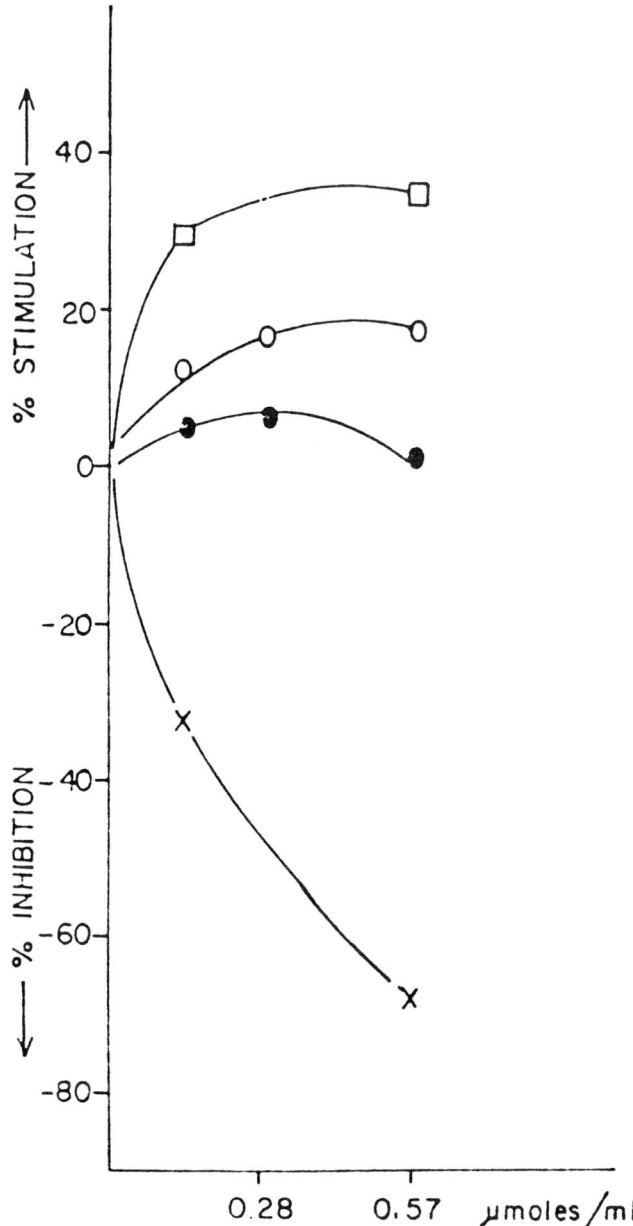

FIGURE 10. Effect of metyrapone on the production of 7OHM, 12MBA (■), DMBA-trans-8,9-dihydrodiol (O) (tentative), 7OHM, 12MBA-trans-10,11-dihydrodiol (●), and 7OHM, 12MBA-trans-8,9-dihydrodiol (X) in β-naphthoflavone-induced S-10 fractions.

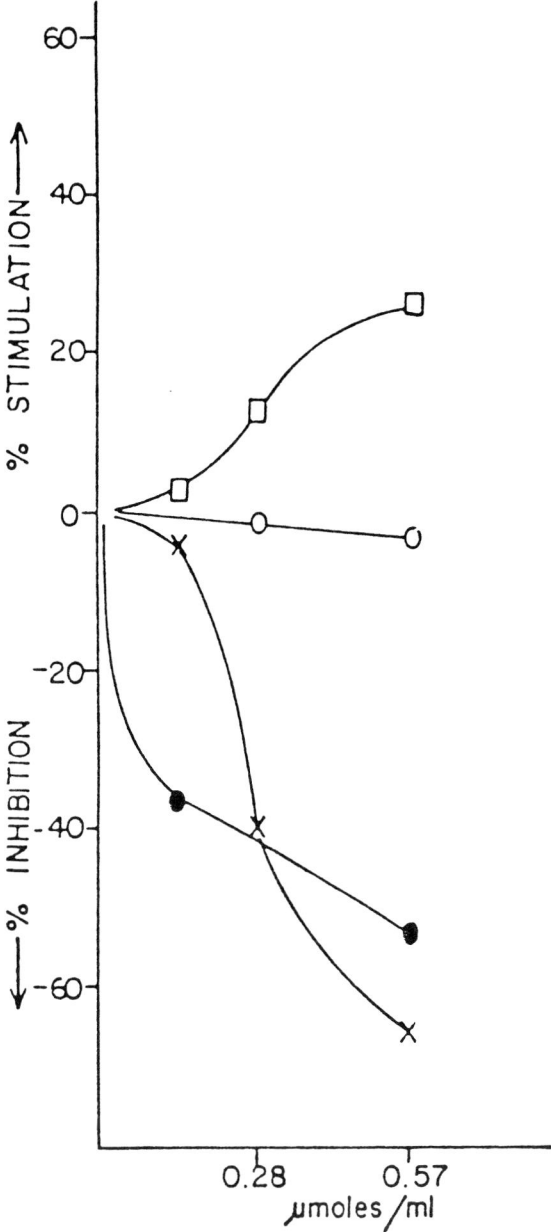

FIGURE 11. Effect of SKF-525A on the production of 7OHM, 12MBA (□), DMBA-trans-8,9-dihydrodiol (O), 7OHM, 12MBA-trans-10,11-dihydrodiol (●) and 7OHM, 12MBA-trans-8,9-dihydrodiol (X) in $\beta$-naphthoflavone-induced S-10 fractions.

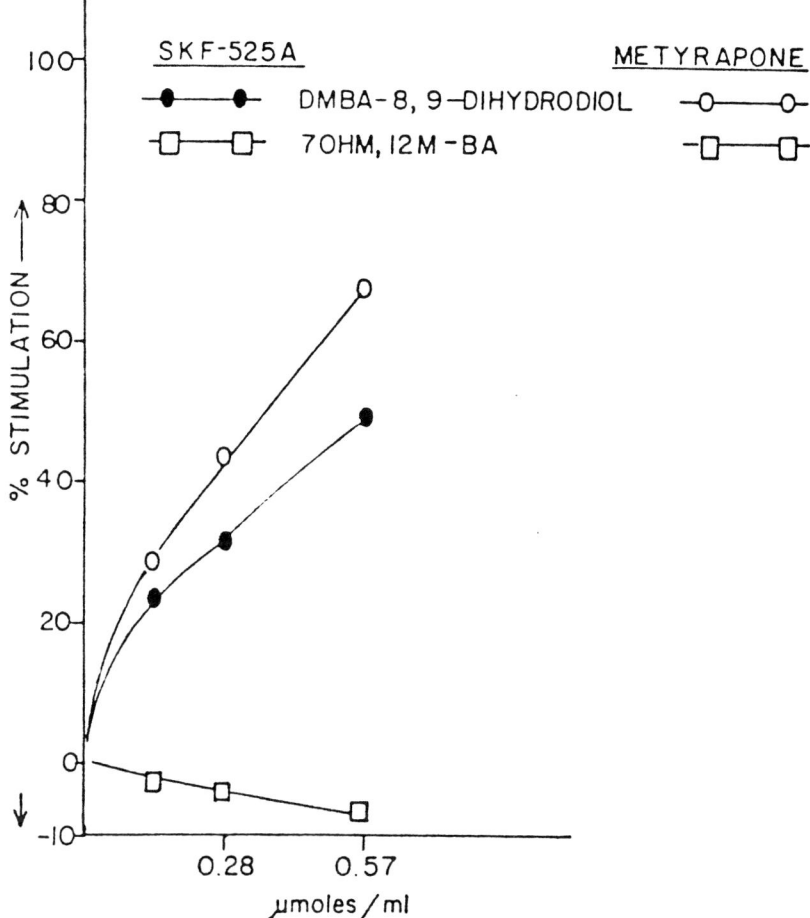

FIGURE 12. Effect of SKF-525A and metyrapone on DMBA-trans-8,9-dihydrodiol and 7OHM, 2MBA formation in uninduced S-10 fraction.

tigators (9,10,32,41). Lack of carcinogenic and mutagenic activity of the 1- and 2-F-DMBA analogues is attributable to lack of A-ring bioactivation and is consistent with the Jerina-Daly bay region theory (24-26). However, the reported carcinogenic and mutagenic activity of 4-F-DMBA (16,32,41) and the weak carcinogenic and mutagenic activity of 5-F-DMBA (16,32,41) suggests alternative metabolic routes and subsequent macromolecular bonding.

To examine further the effects of F on DMBA metabolism, 2-, 3-, 4-, 5-, and 6-F-DMBA analogues were investigated for metabolism in phenobarbital-, 3MC-, and β-napthoflavone-induced as well as in uninduced Sprague Dawley rat liver S-10 fractions. For purposes of characterizing similarities and dissimilarities in DMBA and F-analogue metabolism, various AHH inhibitors were employed.

## Characterization of Fluoro-Analogues Metabolites

As was the case with DMBA, each of the F-analogues (2-, 3-, 4-, 5-, and 6-F) yielded different HPLC metabolic profiles, depending upon the nature of the inducer employed for the preparation of the S-10 fraction (Table 3). Striking qualitative and quantitative dissimilarities in metabolic profiles were observed (Figures 13 through 17). However, irrespective of the S-10 fraction employed, trans-8,9-dihydrodiol metabolites were the major compounds formed in all preparations (Table 4). For 2-, 4-, 5- and 6-F analogues, the relative amount of trans-8,9-dihydrodiol formed from the various preparations was $\beta$-naphthoflavone-induced $>$ 3MC-induced $>$ phenobarbital-induced $>$ uninduced. However, for 3-F-DMBA the order was 3MC-induced $>$ $\beta$-naphthoflavone-induced $>$ phenobarbital-induced$>$uninduced (Table 4). Interestingly, only in the case of 6-F-DMBA was a substantial amount of the corresponding trans-10,11-dihydrodiol [3MC-induced (5.3 percent); $\beta$-naphthoflavone-induced (15 percent)] produced (Table 4). When the total of the corresponding trans-10,11-dihydrodiol and 7OHM,12M-trans-10,11-dihydrodiol (Tables 5 and 6) was measured, more was produced from 3- and 6-F-DMBA analogues in both 3MC- and $\beta$-naphthoflavone-induced S-10 fractions. Apparently, F-substitution at positions 3 and 6 in DMBA renders the 10,11 double bond more susceptible to attack by the cytochrome P-450 enzyme-system-generated oxenoid species.

The largest amounts (14 and 8 percent, respectively) of trans-5,6-dihydrodiols were formed from 3F- and 4F-DMBA analogues in 3MC-induced S-10 fractions. Only for phenobarbital- and $\beta$-naphthoflavone-induced S-10 fractions were detectable quantities of trans-5,6-dihydrodiol metabolites observed (Table 4). Additionally, metabolism of 4F-DMBA in 3MC-induced S-10 fraction furnished detectable amounts of 7OHM,12M-trans-5,6-dihydrodiol (Table 5). The carcinogenicity of 4F-DMBA and the lack of this activity for 2F- and 3F-DMBA may in part emanate from the altered kinetics of the reactions of the corresponding trans-5,6-epoxides with cellular macromolecules.

Metabolism of all F analogues in phenobarbital-induced S-10 fractions invariably furnished monohydroxy derivatives less polar than their corresponding 7-methyl hydroxylated analogues. The UV and fluorescence spectra for these monohydroxy compounds were virtually identical to 12OHM,7M-BA (Table 7), and thus they are most likely the 12OHM,7M-BA-fluoro analogues.

TABLE 3. Retention Times (min) of Characterized Fluoro-DMBA Metabolites on Partisil ODS Column; Methanol-Water Gradient (25 to 100 percent); 60 Min; Linear

| Metabolites | 2F | 3F | 4F | 5F | 6F |
|---|---|---|---|---|---|
| 7OHM,12MBA | 50.3 | 50.6 | 50.9 | 51.5 | 50.9 |
| 12OHM,7MBA | 50.9 | 51.5 | 53.75 | 53.75 | 53.75 |
| DMBA-trans-3,4-dihydrodiol | ND[a] | ND | ND | ND | 47.75 |
| DMBA-trans-5,6-dihydrodiol | 36.25 | 34.35 | 33.1 | ND | ND |
| DMBA-trans-8,9-dihydrodiol | 38.25 | 38.35 | 40.0 | 40.0 | 40.0 |
| DMBA-trans-10,11-dihydrodiol | ND | ND | ND | 40.60 | 40.25 |
| 7OHM,12OHMBA | ND | 40.6 | 42.0 | 41.85 | 43.75 |
| 7OHM,12MBA-trans-10,11 dihydrodiol | 29.35 | 29.70 | 30.85 | 31.25 | 29.0 |
| 7OHM,12-M-BA-trans-3,4-dihydrodiol | ND | ND | ND | ND | 37.05 |
| 12OHM,7MBA-trans-3,4-dihydrodiol | ND | ND | ND | ND | 37.05 |
| 7OHM,12MBA-trans-8,9-dihydrodiol | 31.25 | 33.10 | 33.75 | 33.75 | 32.20 |
| 12OHM,7MBA-trans-8,9-dihydrodiol | 32.5 | 35.60 | ND | ND | ND |
| 7OHM,12MBA-trans-5,6-dihydrodiol | ND | ND | 28.10 | ND | ND |

[a]ND = Not detected.

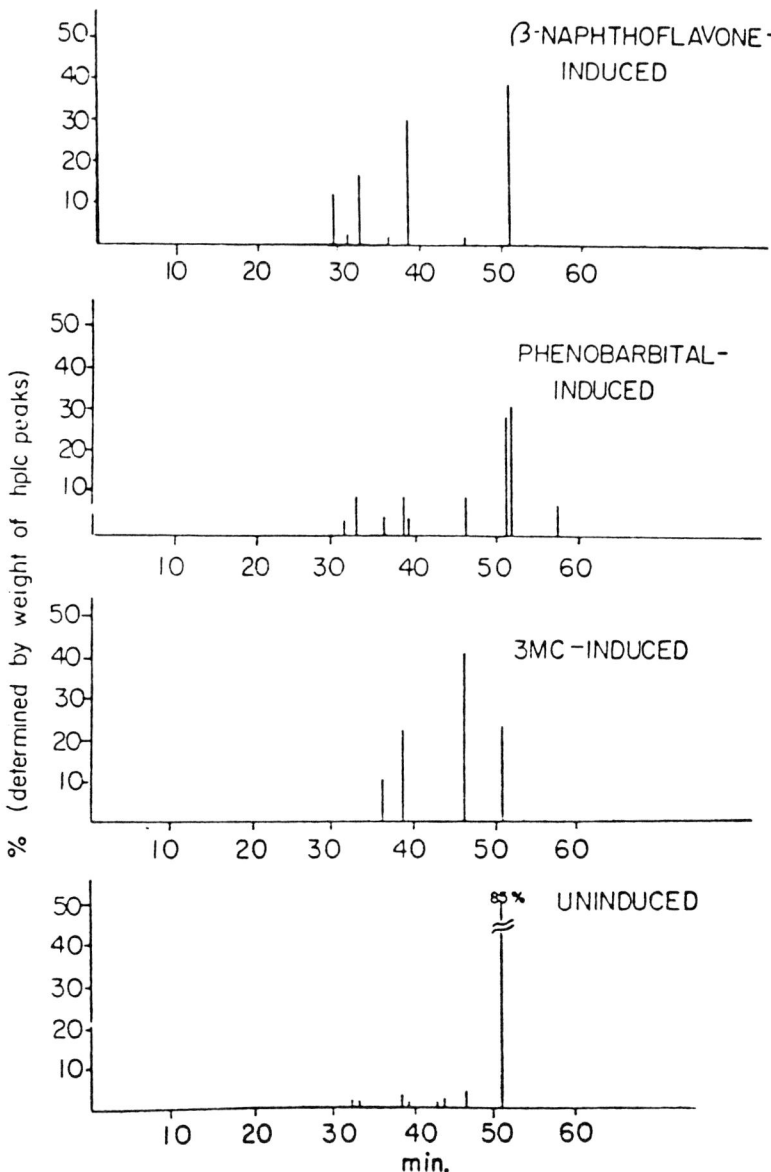

FIGURE 13. Metabolite profiles of 2F-DMBA generated in induced and uninduced male Sprague Dawley rat liver S-10 fraction.

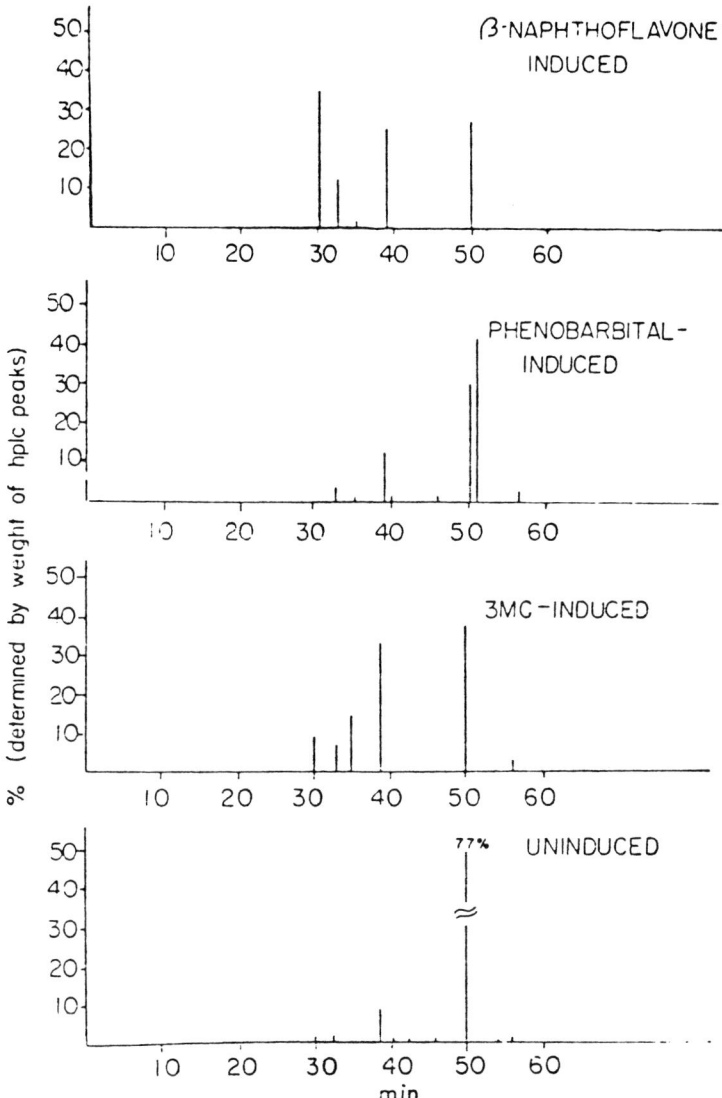

**FIGURE 14.** Metabolite profiles of 3F-DMBA generated in induced and uninduced male Sprague Dawley rat liver S-10 fraction.

Additionally, the corresponding 7,12-bishydroxymethyl benzanthracenes were produced in phenobarbital-induced S-10 fraction (Table 7). Retention times for all fluoro metabolites are presented in Table 3. Again, these data are consistent with the assigned structures.

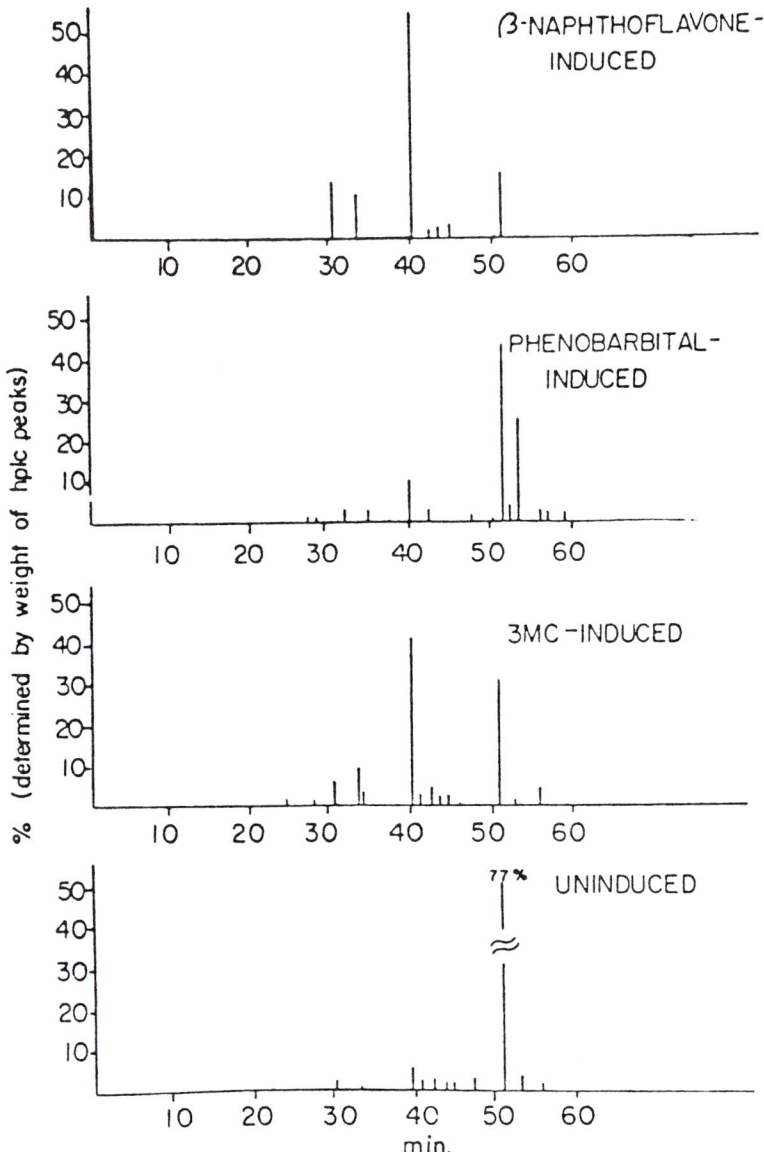

FIGURE 15. Metabolite profiles of 4F-DMBA in generated induced and uninduced male Sprague Dawley rat liver S-10 fraction.

An HPLC profile of 6F-DMBA metabolite generated in phenobarbital-induced S-10 fraction is shown in Figure 18. The metabolite appearing at a retention time of 37.25 minutes exhibited a UV spectrum (shown in Figure 19) having fine structure between 452 and 350 nm with λ maximum at 270

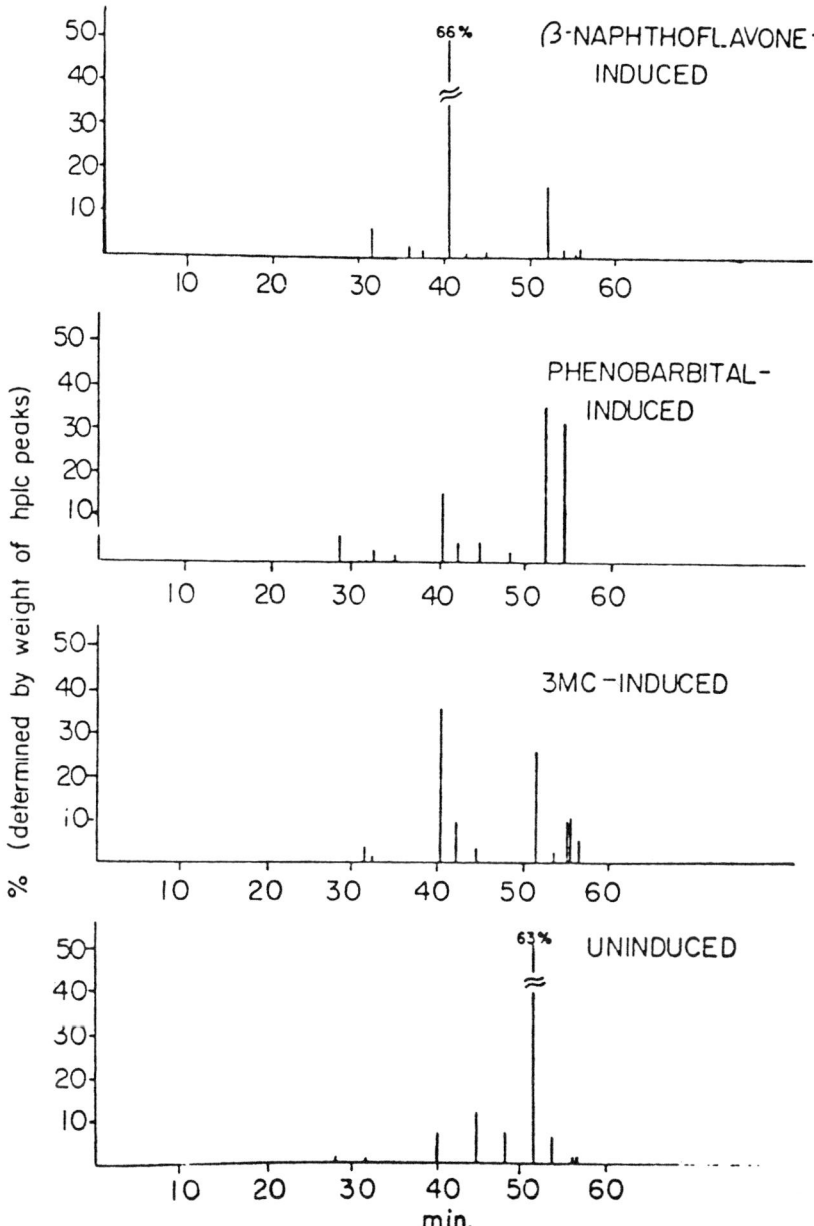

FIGURE 16. Metabolite profiles of 5F-DMBA generated in induced and uninduced male Sprague Dawley rat liver S-10 fraction.

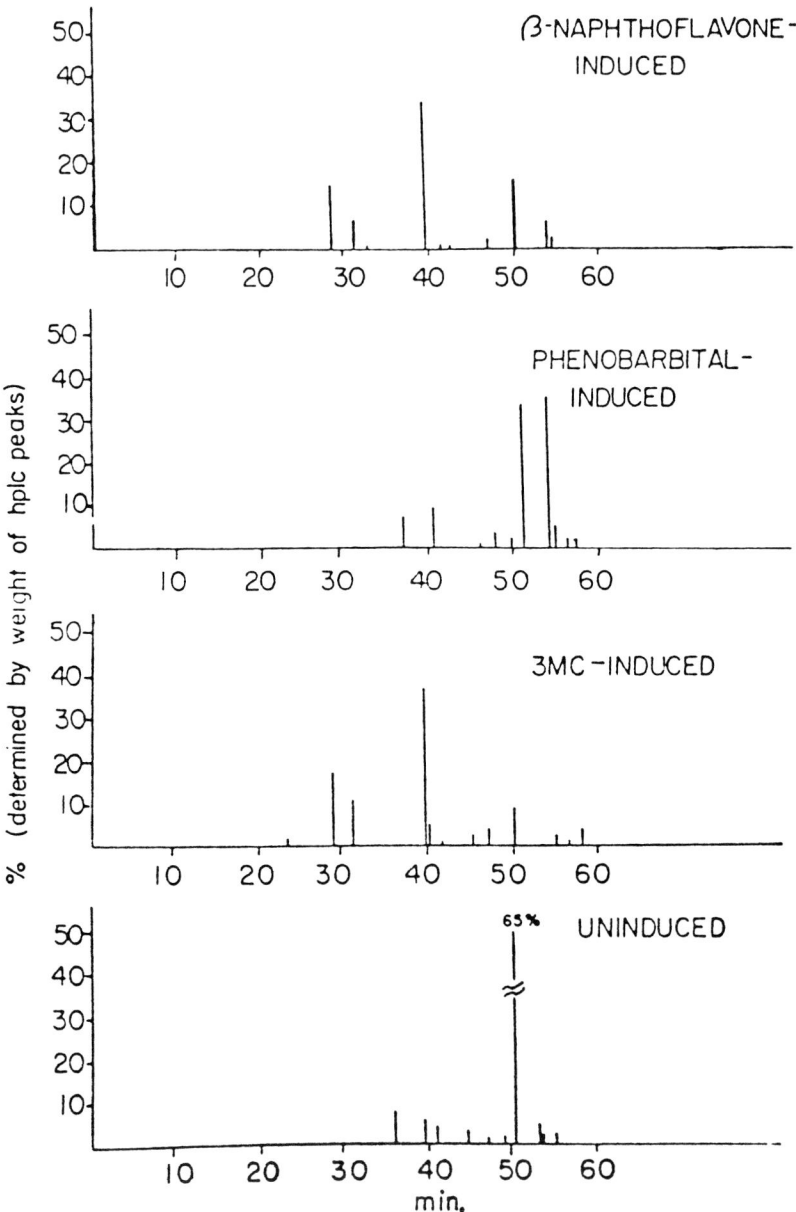

FIGURE 17. Metabolite profiles of 6F-DMBA generated in induced and uninduced male Sprague Dawley rat liver S-10 fraction.

TABLE 4. Dihydrodiol Metabolites Obtained from Various Fluoro-DMBA Analogues in Induced and Uninduced Rat Liver S-10 Fraction

| Metabolites, Respective Trans-Dihydrodiols | Phenobarbital Induced | | | | | 3MC Induced | | | | | β-Naphthoflavone Induced | | | | | Uninduced | | | | |
|---|---|---|---|---|---|---|---|---|---|---|---|---|---|---|---|---|---|---|---|---|
| | 2F | 3F | 4F | 5F | 6F | 2F | 3F | 4F | 5F | 6F | 2F | 3F | 4F | 5F | 6F | 2F | 3F | 4F | 5F | 6F |
| 3,4 | -[a] | - | - | ND[b] | 0.8*[c] | - | - | - | ND | ?[d] | - | - | - | - | ND | - | -[d] | -[d] | ND | 1.8*[c] |
| 5,6 | 3.7 | <1 | ND | - | - | ND | 14 | 8 | - | - | 1.7 | ND | ND | ND | - | ND | ?[d] | ?[d] | - | - |
| 8,9 | 8.4 | 12.6 | 10 | 10 | 9*[c] | 21 | 33 | 40 | 35 | 34 | 29.5 | 25 | 54 | 66 | 37 | 7 | 9.6 | 0.5 | 6.5 | 6.1 |
| 10,11 | ND | ND | ND | ND | ND | ND | ND | ND | ND | 5.3 | ND | ND | ND | ND | 15 | ND | ND | ND | ND | ND |

[a] - = Not detected or expected.
[b] ND = Not detected.
[c] * = Tentative identification.
[d] ? = Uncertain whether detected.

TABLE 5. Percent Methyl Hydroxylated Dihydrodiols Produced from Fluoro-DMBA Analogues in Induced and Uninduced S-10 Fractions

| Analogue, F-DMBA | 7OHM,12MBA-3,4 Dihydrodiol | | | | 7OHM,12MBA-5,6 Dihydrodiol | | | | 7OHM,12MBA-8,9 Dihydrodiol | | | | 7OHM,12MBA-10,11 Dihydrodiol | | | |
|---|---|---|---|---|---|---|---|---|---|---|---|---|---|---|---|---|
| | P[a] | 3MC[b] | NAP[c] | UN[d] | P | 3MC | NAP | UN | P | 3MC | NAP | UN | P | 3MC | NAP | UN |
| 2F | ND[e] | ND | ND | ND | ND | ND | ND | ND | 8.44 | ND | 16.3 | 2.7 | ND | ND | 12.6 | ND |
| 3F | ND | ND | ND | ND | ND | ND | ND | ND | 3.62 | 6.8 | 9.0 | 2.4 | ND | 8 | 27 | 2.4 |
| 4F | ND | ND | ND | ND | ND | 0.4 | ND | ND | 2.7 | 2.3 | 10 | 1.0 | ND | ND | 13 | 2.3 |
| 5F | ND | ND | ND | ND | ND | ?[f] | ND | ND | ND | ?[f] | 2.4 | ND | ND | ?[f] | 6 | 1.0 |
| 6F | 7 | ND | 3*[g] | 7.2*[g] | ND | ?[f] | ND | ?[f] | ND | 11 | 7.3 | ND | ND | 18 | 12.0 | 1.8 |

[a] P = Phenobarbital-induced
[b] 3MC = 3-Methylcholanthrene-induced
[c] NAP = β-Naphthoflavone-induced
[d] UN = Uninduced
[e] ND = Not detected
[f] ? = Detection uncertain
[g] * = Characterization questionable.

TABLE 6. Total of Trans-10,11-Dihydrodiols and 7 Methyl Hydroxylated Trans-10,11-Dihydrodiols using Various F-DMBA Analogues in 3MC- and β-Naphthoflavone-induced S-10 Fractions.

| 3MC Induced | | Percentage of Total Metabolities | | | | |
|---|---|---|---|---|---|---|
| | | 2F | 3F | 4F | 5F | 6F |
| DMBA-trans-10,11-dihydrodiol | | - | - | - | - | 5.3 |
| 7OHM,12MBA-trans-10,11-dihydrodiol | | - | 8 | - | - | 18 |
| | Total | | 8 | | | 23.3 |
| **β-Naphthoflavone Induced** | | | | | | |
| DMBA-trans-10,11-dihydrodiol | | - | - | - | - | 8 |
| 7OHM,12MBA-trans-10,11-dihydrodiol | | 12.6 | 27 | 13 | 6 | 12 |
| | Total | 12.6 | 27 | 13 | 6 | 20 |

TABLE 7. UV Maxima of 12-Methyl Hydroxylated Metabolites Derived from Various F-DMBA Analogues

| Compound | UV Maxima |
|---|---|
| 2F-12OHM,7MBA | 294, 283, 272, 263 |
| 3F-12OHM,7MBA | 304(sh), 294.5, 284, 273, 263, 238 |
| 4F-12OHM,7MBA | 305(sh), 296, 285.5(sh), 276, 263 |
| 5F-12OHM,7MBA | 295.4, 284.7, 275(sh), 263 |
| 6F-12OHM,7MBA | 305(sh), 293, 284.5, 275, 263 |
| 3F-7,12-Bis-OHM-BA | 302(sh), 292.5, 282, 271, 261 |
| 5F-7,12-Bis-OHM-BA | 295.5, 284, 273, 264 |

nm characteristic of DMBA-trans-3,4-dihydrodiol. These data, together with the fluoroscence spectrum (Figure 20) and comparative HPLC retention times (29.35 minutes for 7OHM,12M-BA-trans-3,4-dihydrodiol) strongly, suggest this metabolite to be 6F-7OHM,12M-BA-trans-3,4-dihydrodiol or its 12OHM,7M isomer. Thus, these data establish the second example of bay-region trans-3,4-dihydrodiol formation in a hydrocarbon having the DMBA nucleus. No corresponding trans-3,4-dihydrodiol could be detected from 5-F-DMBA in any of our rat liver S-10 preparations.

## Effect of Inhibitors on Fluoro-Analogue Metabolism

$CS_2$ (0 to 2.1 μM/ml) treatment of phenobarbital-induced S-10 fraction stimulated the combined production of the respective 7OHM,12M- BA and

# 718 BAY-REGION METABOLITES OF DMBA

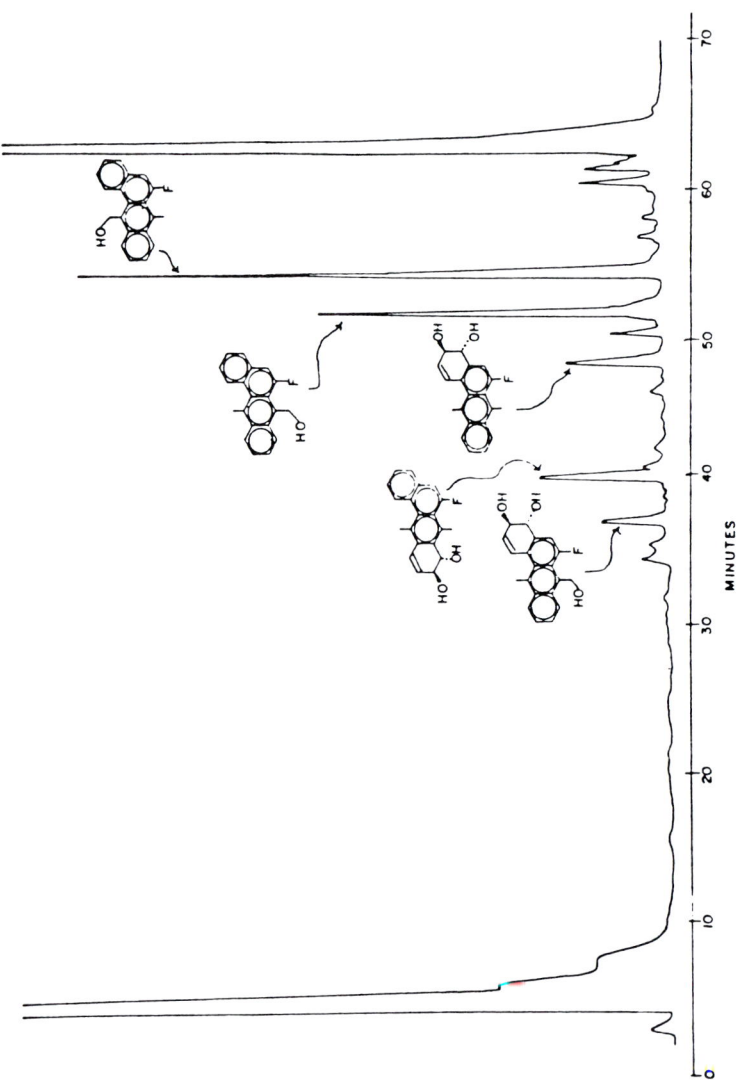

FIGURE 18. HPLC profile of 6F-DMBA metabolites generated in phenobarbital-induced S-10 fraction.

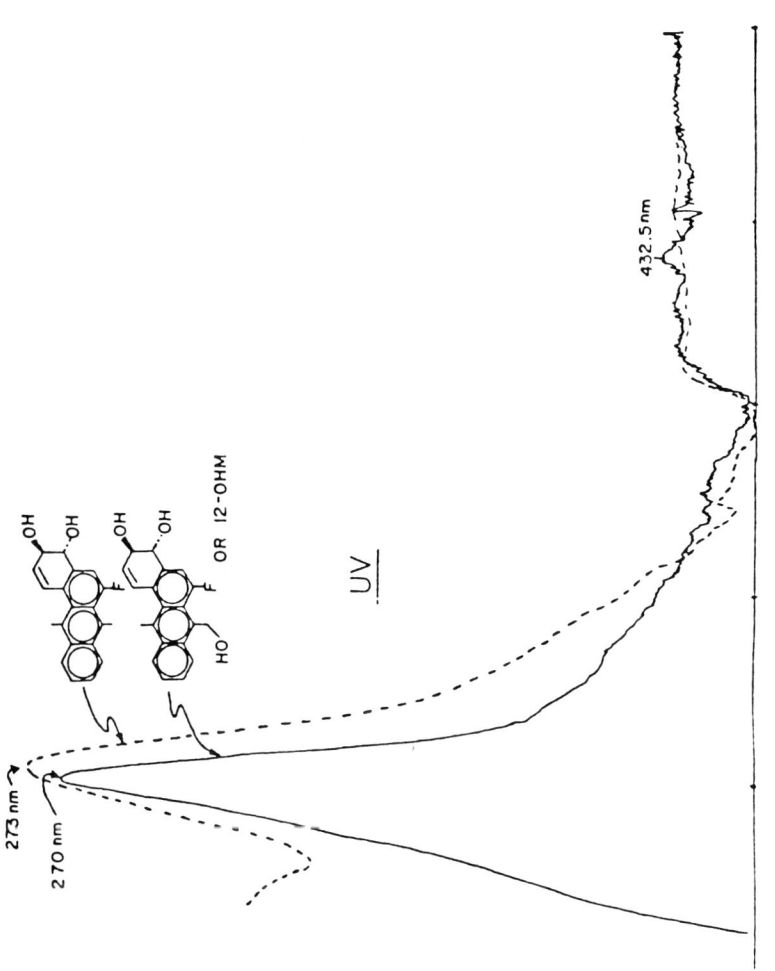

FIGURE 19. Ultraviolet spectra of presumed 6F-DMBA-trans-3,4-dihydrodiols.

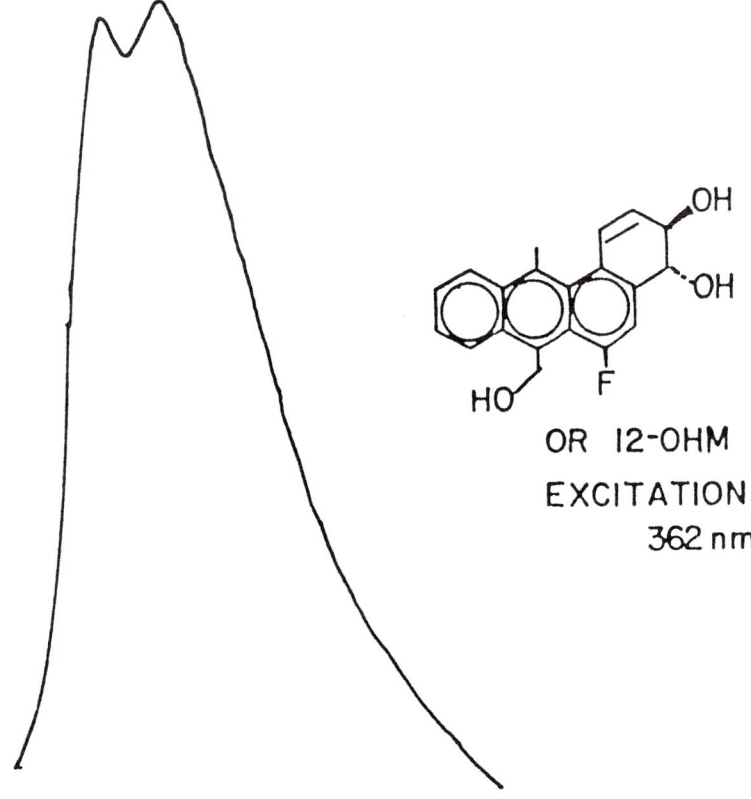

FIGURE 20. Florescense spectrum of presumed 6F-DMBA-trans-3,4-dihydrodiol.

12OHM, 7M-BA F-containing metabolites. However, formation of 7OHM,12M-BA was consistently inhibited in relation to the corresponding 12OHM,7M-BA derivatives in all the F-DMBA analogues studied (Figures 21, 22, and 23). Conversely, treatment of phenobarbital-induced S-10 fraction with metyrapone (0 to 0.71 $\mu$M/ml) stimulated the formation of 5F-7OHM,12M-BA and inhibited the production of 5F-12OHM,7M-BA (Figure 24). These results are consistent with the observations described for DMBA and reaffirm that 7- and 12-methyl hydroxylations are independent processes. $CS_2$ (0 to 2.1 $\mu$M/ml) treatment of phenobarbital-induced S-10 fraction also inhibited the production of 6F-DMBA-trans-3,4-dihydrodiol (tentatively identified) and the corresponding 7- and 12-methyl hydroxylated derivatives (Figure 25). Preincubation of $\beta$-napthoflavone-induced S-10 fraction with $CS_2$ (0 to 2.1 $\mu$moles/ml) quantitatively inhibited the formation of 3F-7OHM,12MBA-trans-10,11-dihydrodiol and 3F-7OHM,12MBA-trans-8,9-dihydrodiol.

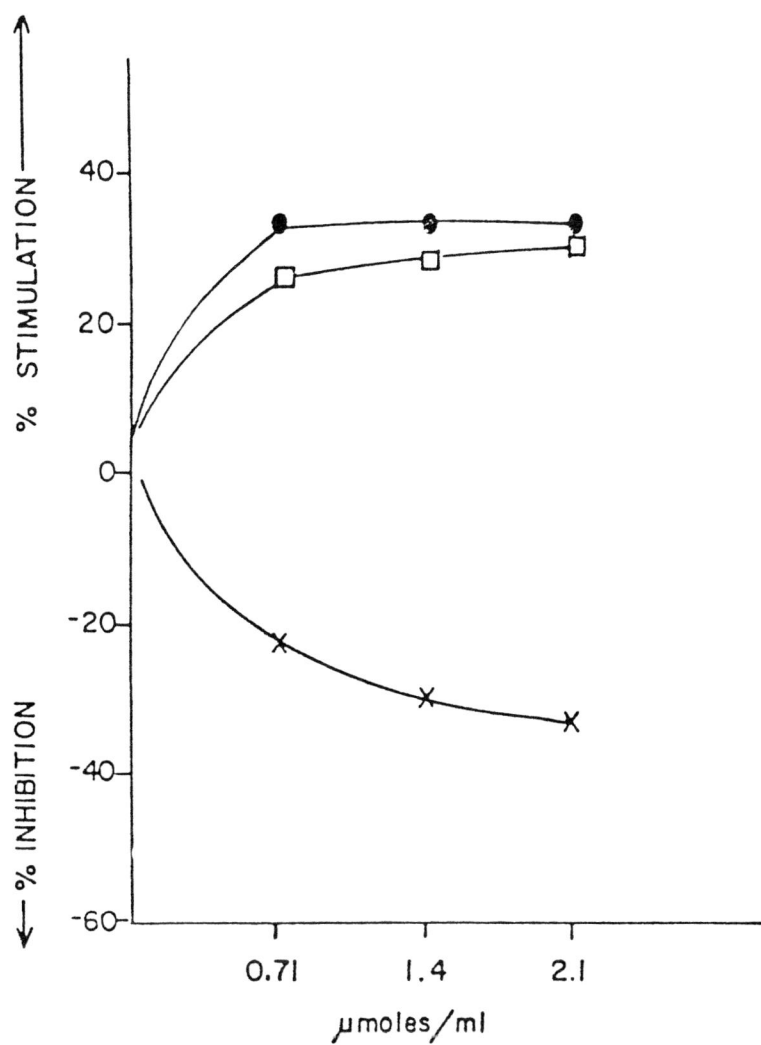

FIGURE 21. Effect of $CS_2$ on the production of 3F-DMBA metabolites [sum of 7OHM, 12M- and 12OHM,7MBA relative to total metabolites = ●12OHM, 7MBA (□) production relative to 7OHM,12MBA (X) production] in phenobarbital-induced S-10 fraction.

However, formation of 3F-DMBA-trans-8,9-dihydrodiol was inhibited to a relatively small extent, and the production of 3F-7OHM,12MBA was not affected (Figure 26).

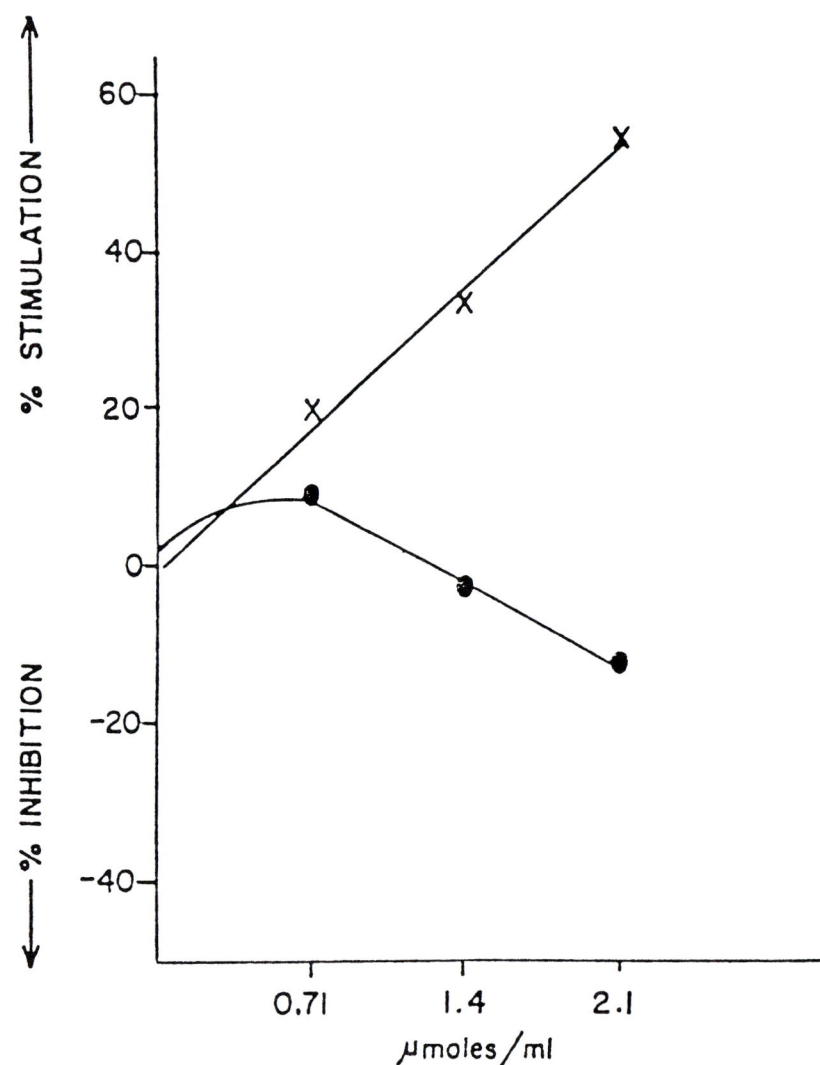

FIGURE 22. Effect of CS$_2$ on the production of 5F-DMBA metabolites [12OHM, 7MBA (X) relative to 7OHM, 12MBA(•)] in phenobarbital-induced S-10 fractions.

## CONCLUSIONS

Our data indicate that:
1. AHH in induced and uninduced S-10 fractions is inhibited differently by CS$_2$, α-napthoflavone, metyrapone, and SKF-525A.

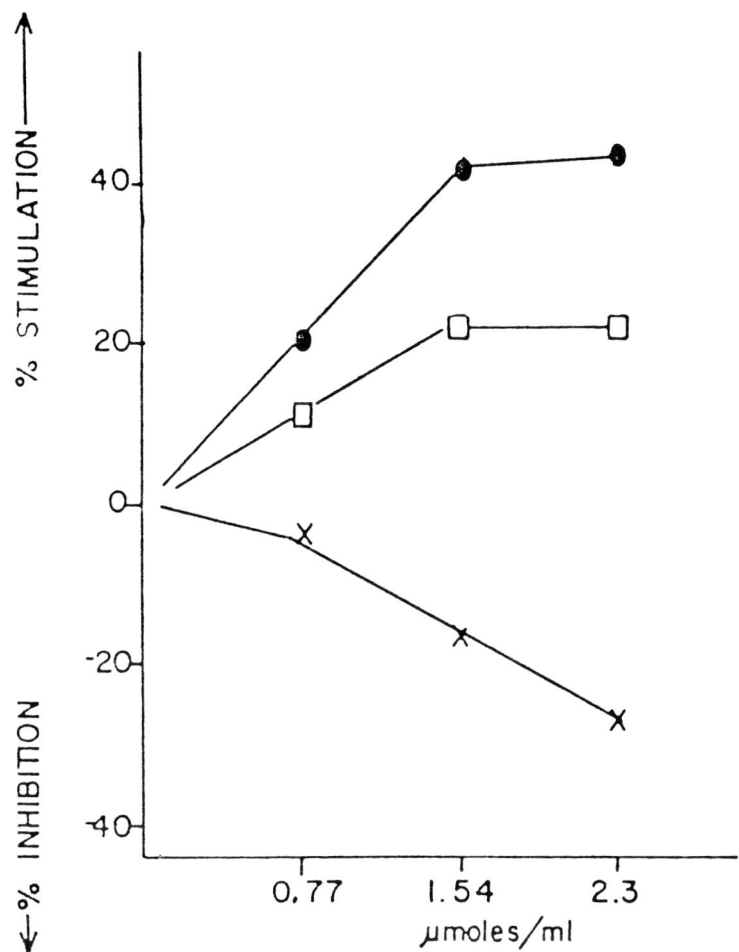

FIGURE 23. Effect of $CS_2$ on the production of 6F-DMBA metabolite [sum of 7OHM, 12M- and 12OHM,7MBA relative to total metabolite = ●; 12OHM, 7MBA (□) relative to 7OHM, 12MBA (X)].

2. DMBA-trans-3,4-dihydrodiol formation and DMBA-7- and 12-methyl hydroxylations are unrelated and independent processes.
3. DMBA-trans-8,9-dihydrodiol formation in part arises differently from phenobarbital-, 3MC-, and β-napthoflavone-induced and uninduced S-10 fractions from Sprague Dawley rat liver.
4. DMBA-trans-5,6-dihydrodiol formation does not appear to be related to trans-3,4-or trans-8,9-dihydrodiol formation.
5. These studies confirm our earlier conclusion that substitution of F for hydrogen on DMBA blocks metabolism at the site of substitution (9).

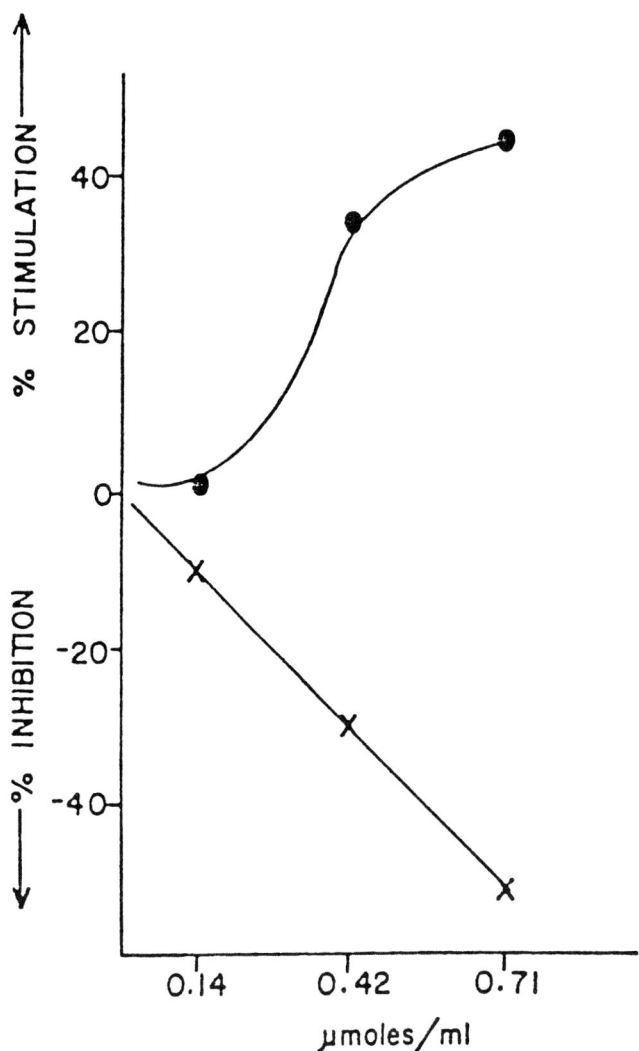

FIGURE 24. Effect of metyrapone on the production of 5F-DMBA metabolites [7OHM, 12MBA (●) relative to 12OHM,7MBA (X)] in phenobarbital-induced S-10 fractions.

6. Substitution of hydrogen at the 3 or 6 positions of DMBA by F enhances formation of the corresponding trans-10,11-dihydrodiol and 7-methyl hydroxylated derivative when compared with DMBA, perhaps because of increased nucleophilicity of the 10,11-position in 3MC- and $\beta$-naphthoflavone-induced S-10 fraction.

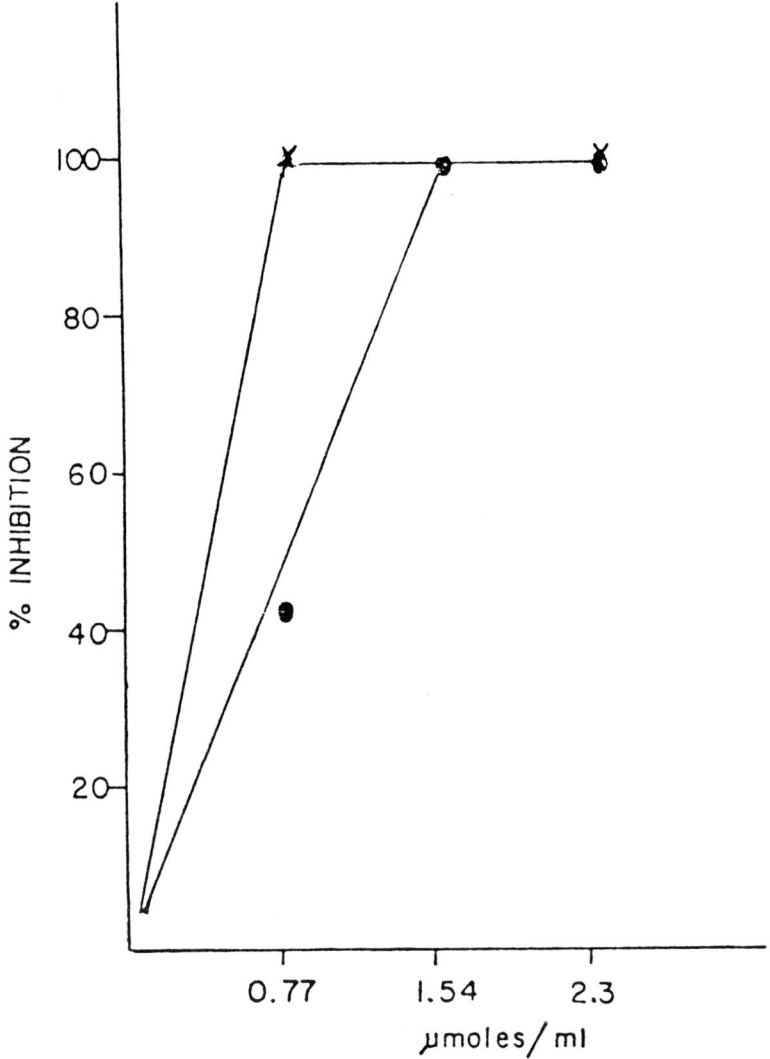

FIGURE 25. Effect of $CS_2$ on the production of 6F-DMBA metabolites (sum of 7OHM, 12M- and 12OHM, 7MBA relative to total metabolite = X; 6F-DMBA-trans-3,4-dihydrodiol = ●) in phenobarbital-induced S-10 fractions.

7. Carcinogenicity and mutagenicity of 4-F-DMBA may, in part, emanate from the reaction of the non-bay-region metabolite 4F-5,6-oxide with intracellular macromolecules (e.g., K-region covalent bonding to DNA). Thus, more than one metabolic step may be required for the binding of DMBA to DNA [W. M. Baird and L. Diamond (1979): *Chem. Biol. Interactions* 20:181].

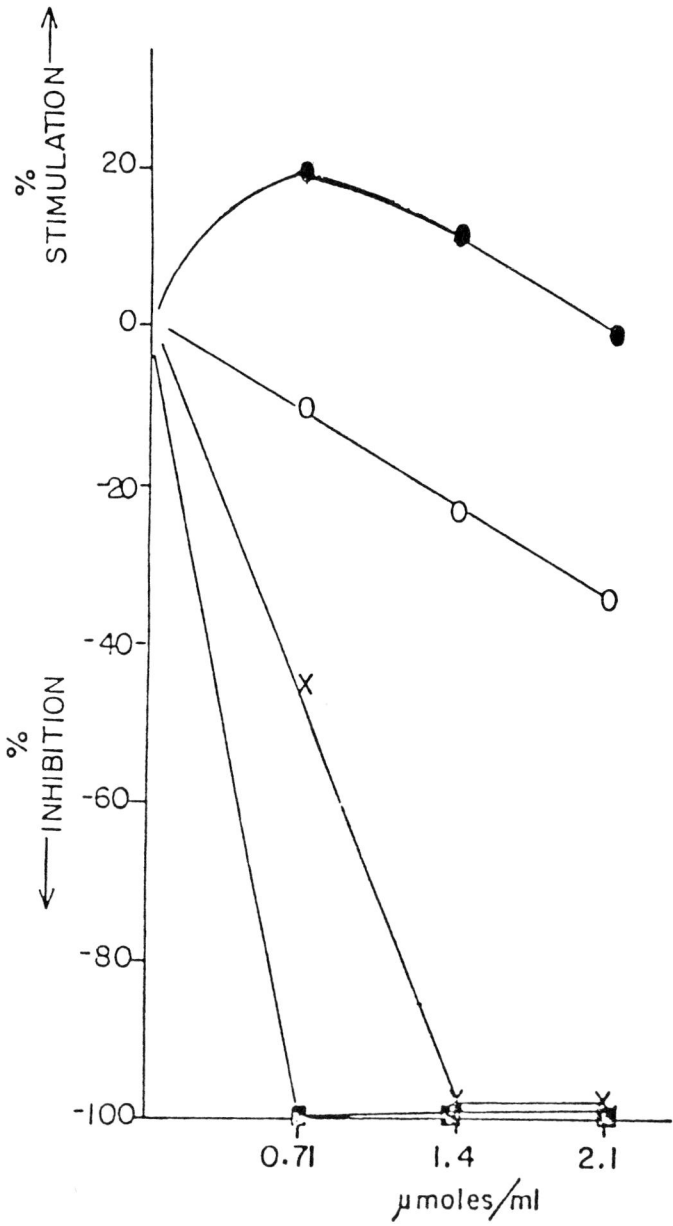

FIGURE 26. Effect of $CS_2$ on the production of 3F-DMBA metabolites (3F,7OHM, 12MBA = ●; 3F-DMBA-trans-8,9-dihydrodiol = O; 3F, 7OHM, 12MBA-trans 10,11-dihydrodiol = X; 3F, 7OHM,12MBA-trans-8,9-dihydrodiol = ■) in $\beta$-naphthoflavone-induced S-10 fractions.

8. Since 6F- and not 5F-DMBA yielded trans 3,4-dihydrodiols, this may account for the relatively higher carcinogenicity of 6F-DMBA.
9. Owing to considerable variation in metabolism for DMBA and its F analogues (both in nature and quantity of metabolites formed) in phenobarbital-, 3MC-, and β-naphthoflavone-induced and uninduced S-10 fractions and the marked difference in effects of the various metabolic inhibitors on metabolite production in these systems, caution should be exercised when labeling one or more metabolites or bioactivation pathways responsible for carcinogenesis or mutagenesis. In fact, none of the induction systems may represent potentiation of the native condition, but rather a more complex phenomenon wherein new enzyme systems are produced.

## ACKNOWLEDGMENTS

We gratefully acknowledge support of this work from the National Cancer Institute, Public Health Service Grant No. CA-21371 and U.S. Environmental Protection Agency Grant R-804201.

## REFERENCES

1. Atlas, S. A., and Nebert, D. W. (1976): Genetic association of increases in napthalene acetanilide and biphenyl, hydroxylations with inducible aryl hydrocarbon hydroxylase in mice. *Arch. Biochem. Biophys.* 175: 495-506.
2. Bock, F. G. (1964): Early effects of hydrocarbons on mammalian skin. *Prog. Exp. Tumor Res.* 4:126-168.
3. Botelho, L. H., Ryan, D. E., and Levin, W. (1979): Amino acid compositions and partial amino acid sequences of three highly purified forms of liver microsomal cytochrome P-450 from rats treated with polychlorinated biphenyls, phenobarbital or 3-methylcholanthrene. *J. Biol. Chem.* 254(13), 5635-40.
4. Boutwell, R. G. (1967): Some biological aspects of skin carcinogenesis. *Prog. Exp. Tumor Res.* 4:207-250.
5. Chou, M. W., and Yang S. K. (1978): Identification of four trans-3,4-dihydrodiol metabolites of 7,12-dimethylbenz(a)anthracene and their in vitro DNA-binding activities upon further metabolism. *Proc. Natl. Acad. Sci., U. S. A.* 75(11):5466-5470.
6. Chou. M. W., Easton, G. D., and Yang, S. K. (1979): Metabolism of 7,12-dimethylbenz(a)anthracene and its methyl hydroxylated metabolites: Formation of phenolic metabolites at the 2-position. *Biochem. Biophys. Res. Commun.* 88(3):1085-91.

7. Coon, M. J., and Vatsis, K. P. (1978): Biochemical studies on chemical carcinogenesis. Role of multiple forms of liver microsomal cytochrome P-450 in the metabolism of benzo(a)pyrene and other foreign compound In *Polycyclic Hydrocarbons and Cancer,* Vol. 1, pp. 335-360, Academic Press, Inc., New York.
8. Cox, J. O., and Pilcher, G. (1970): In *Thermochemistry of organic and organometallic compounds,* Academic Press, Inc., New York.
9. Daniel F. B., Wong, L. K., Oravec, C. T., Cazer, F. D., Wang, C.L.A., D'Ambrosio, S. M., Hart, R. W., and Witiak, D. T. (1979): In *Polynuclear* Aromatic Hydrocarbons, P. W. Jones and P. Leber, Eds., Ann Arbor Science Publishers, pp. 855-883.
10. Daniel, F. B., Cazer, F. D., D'Ambrosio, S. M., Hart, R. W., Kim, W. H., and Witiak D. T. (1979): Comparative metabolism of 7,12-dimethylbenz(a)anthracene and its weakly carcinogenic 5-fluoro analog. *Cancer Lett.* 6:263-72.
11. Dipple, A., Tomaszewski, Moschel, R. C., Bigger, C.A.H., Nebzydoski, J. A., and Egan, M. (1979): Comparison of metabolism mediated binding to DNA of 7-hydroxymethyl-12-methylbenz(a)-anthracene. *Cancer Res.* 39:1154-58.
12. Gelboin, H. V., Wieble, F. J., and Kinoshita, N. (1972): Microsomal aryl hydrocarbon hydroxylases: On their role in polycyclic hydrocarbon carcinogenesis and toxicity and the mechanism of enzyme induction. In *Biological Hydroxylation Mechanism.*, G. S. Boyd and R.M.S. Smellie, Eds., Academic Press, Inc., New York, pp. 103-133.
13. Girke, W., and Bergmann, E. D. (1976): On systhesis of 6-fluorobenz(a)anthracene derivative. *Chem. Ber.* 109:1038-1045.
14. Greiner, J. W., Kramer, R. E., Colby, H. D. (1978): Interaction of metyrapone with adrenal microsomal cytochrome P-450 in the guinea pig. *Biochem. Pharmacology* 27:2147-51.
15. Grover P. L., and Chouroulinkov, I. (1979): Tumor-initiating activities on mouse skin of diol metabolites of 7,12-dimethylbenz(a)-anthracene (DMBA) and 3-methylcholanthrene (3MC). *Proc. Am. Assoc. Cancer Res.* 20:66.
16. Harvey, R. B., and Dunne, F. B. (1978): Multiple region of metabolic activation of carcinogenic hydrocarbons. *Nature* 273:566-568.
17. Harvey, R. G., Goh, S. H., and Cortez, C. (1975): K-region oxides and related oxidized metabolites of carcinogenic aromatic hydrocarbons. *J. Am. Chem. Soc.* 97:3468.
18. Heimann, R., Huson, J. O., and Counc, A. (1968). Tumors developing in oophrectomized Sprague Dawley rats after a single gastric instillation of 7,12-dimethylbenz(a)anthracene. *Cancer Res.* 28:309-313.

19. Hildebrandt, A. G., (1972): The binding of metyrapone to cytochrome P-450 and its inhibitory action on microsomal hepatic mixed function oxidation reaction, in *Bioligical Hydroxylation Mechanism*, G. S. Boyd and R.M.S. Smellie, Eds., Academic Press, Inc., New York, pp. 79-102.
20. Huberman, E., Chou, M. W., and Yang, S. K. (1979): Identification of 7,12-dimethylbenz(a)anthracene metabolites that lead to mutagenesis in mammalian cells. *Proc. Natl. Acad. Sci. U.S.A.* 76:862-66.
21. Huggins, C., Grand, L. C., and Brillantes, F. B. (1961): Mammary cancer induced by a single feeding of polynuclear hydrocarbons and its suppression. *Nature* 189:204-207.
22. Ivanovic, V., Geacintov, N. E., Jeffrey, A. M., Fu, P. P., Harvey, R. G., and Weinstein, I. B. (1978): Cell and microsomes mediated binding of 7,12-dimethylbenz(a)anthracene to DNA studied by fluorescence spectoscopy. *Cancer Lett.* 4:131-40.
23. Inbasekaran, M. N., Witiak, D. T., Barone, K., and Loper, J. C. (1979): On synthesis and mutagenicity of bay region reduced analogues of 7,12-dimethylbenz(a)anthracene. *J. Med. Chem.* (in print).
24. Jerina, D. M., and Daly, J. W., (1976): Oxidation at carbon. In *Drug Metabolism*, D. V. Parke and R. L. Smith, Eds., pp. 15-33, Taylor and Francis, London.
25. Jerina, D. M., Lehr, R., Schaefer-Ridder, M., Yagi, H., Karle, J. M., and Thakker, D. R., Wood, A. W., Lu, A.Y.H., Ryan, D., West, S., Levin, W., and Conney, A. H. (1977): On bay region epoxides of dihydrodiols: A concept explaining the mutagenic and carcinogenic activity of benz(a)pyrene and benzo(a)anthracene. In *Origin of Human Cancer*, Cold Spring Harbor Cancer Book B.
26. Jerina, D. M., Lehr, R. E., Yagi, H., Hernandez, O., Dansette, P., Weslocki, P. G., Wood, A. W., Chang, R. L., Levin, W., and Conney, A. H. (1976): Mutagenicity of benz(a)pyrene derivative and the description of a quantum mechanical model which predicts the ease of carbonium ion formation of diol epoxides. *In Vitro Metabolic Activation in Mutagenesis Testing*, p. 159, Elsevier, Amsterdam.
27. Jull, J. W., Hawryluk, A., and Russell, A. (1968): Mechanism of induction of ovarian tumors in the mouse by 7,12-dimethylbenz(a)anthracene. Tumor induction in organ culture. *J. Nat. Cancer Inst.* 40:687-706.
28. Kennaway, E. L. (1930): Further experiments on cancer-producing substances. *Biochem. J.* 24:497-504.
29. Malaveille, C., Bartsch, H., Tierney, B., Grover, P. L., and Sims. P. (1978): Microsome-mediated mutagenesis of the dihydrodiols of 7,12-dimethylbenz(a)anthracene: high mutagenic activity of the 3,4-dihydrodiol. *Biochem. Biophys. Res. Commun.* 83, 1468-73.

30. Marquardt. H., Baker, S., Tierney, B., Grover, P. L., and Sims, P. (1978): Induction of malignant transformation and mutagenesis by dihydrodiols derived from 7,12-dimethylbenz(a)anthracene. *Biochem. Biophys. Res. Commun.* 85:357-62.
31. Moschel, R. C., Barid, W. M., and Dipple, A. (1977): Metabolic activiation of the carcinogen 7,12-dimethylbenz(a)anthracene for DNA binding. *Biochem. Biophys. Res. Comm.* 76:1092-98.
32. Newman, M. S. (1976): Carcinogenicity activity of benz(a)anthracenes in: *Carcinogenesis,* Vol. 1, R.I. Freudenthal and P.W. Jones, Eds., pp. 203-207, Raven Press, New York.
33. Newman, M. S., Fikes, L. E., Hashem, M. M., Kammar, R., Sankaran, V. (1978): On synthesis and carcinogenic activity of 5-fluoro-7(oxygenated methyl)-12--methylbenz(a)anthracene. *J. Med. Chem.* 21(10):1076-1078.
34. Newman, M. S., Khanna, M. S., Kanakarajan, K., and Kumar, S. (1978): Synthesis of 1-,2-,3-,4-,6-,9- and 10-hydroxy-7,12-dimethylbenz(a)anthracenes. *J. Org. Chem.* 43(13): 2553.
35. Oesch, F. (1972): Mammalian epoxide hydrase-inducible enzymes catalysing the inactivation of carcinogenic and cytotoxic metabolites derived from aromatic and olefinic compounds. *Xenobiotica* 3:305-40.
36. Ryan, D. E., Thomas, P. E., Korzeniowski, D., and Levin, W. (1979): Separation and characterization of highly purified forms of liver microsomal cytochrome P-450 from rats treated with polychlorinated biphenyls, phenobarbital, and 3-methyl chalanthrene. *J. Biol. Chem.* 254(4):1365-1374.
37. Seidegard, J., Moron, M. S., Erickson, L. C., DePierre, J. W. (1978): The topology of epoxide hydratase and benzpyrene monoxygenase in the endoplasmic reticulum of rat liver. *Biochem. Biophys. Acta* 543:29-40.
38. Shear, M. J. (1938): Studies in carcinogenesis methyl derivatives of 1:2-benzanthracene. *Am. J. Cancer* 33:499-537.
39. Sheikh, M. Y., Cazer, F. D., Hart, R. W., and Witiak, D. T. (1979) A facile preparation of 2- and 3-fluoro-7,12-dimethylbenz(a)anthracene. *J. Org.*
40. Shellabarger, C. J. (1976): Modifying factors in rat mammary gland carcinogenesis. In *Biology of Radiation Carcinogenesis,* J. M. Yuhas, R. W. Tennant, and J. D. Regan, Eds., pp. 31-42, Raven Press, New York.
41. Slaga, T. J., Bracken, M., Gleason, G. L., DiGiovanni, J., Berry, D. L., Tuckau, M. R., and Harvey, R. G. (1978): Tumor initiating activities of various derivatives of benz(a)anthracene and 7,12-dimethylbenz(a)-anthracene in mouse skin. In Polynuclear Aromatic Hydrocarbon, P. W. Jones and P. Leber., Eds., Chap. 49, Ann Arbor Science Publishers, Inc., Ann Arbor, Michigan, 1979.

42. Sukumaran, K. B., Harvey R. G. (1979): Synthesis of trans-3,4-dihydroxy-3,4-dihydro-7,12-dimethylbenz(a)anthracene. A highly carcinogenic metabolite of 7,12-dimethylbenz(a)anthracene. *J. Am. Chem. Soc.* 101:1353.
43. Thorgeisson, S. S., Atlas, S. A., Boobis, A. R., and Felton, J. S. (1979) Species differences in the substrate specificity of hepatic cytochrome P-488 from polycyclic hydrocarbon treated animals. Biochem. *Pharmacol.* 8:217-26.
44. Tierney, B., Hewer A., MacNicoll, A. D., Gervasi, P. G., Rattle, H., Walsh, C., Grover, P. L., and Sims, P. (1978): The formation of dihydrodiols by the chemical or enzymic oxidation of benz(a)-anthracene. *Chem. Biol. Interact.* 23:243-257.
45. Uematsu, K., and Huggins, C. (1968): Induction of leukemia and ovarian tumors in mice by pulse-dose of polycyclic aromatic hydrocarbons. *Mol. Pharmacol.* 4:411-426.
46. Vigny P., Duquesne, M., and Coulomb, H. (1977), Fluorescence spectral studies on the metabolic activation of 3-methylcholanthrene and 7,12-dimethylbenz(a)anthracene in mouse skin. *FEBS Lett.* 82(2): 278-282.
47. Walters, M. A. (1966): The induction of lung tumors by the injection of 9,10-dimethyl-1,2-benzanthracene (DMBA) into newborn suckling and young adult mice. A dose response study. *Brit. J. Cancer* 20:148-60.
48. Walters, M. A., and Roe, F.J.C. (1966): The time appearance of lung tumors in mice injected when newly born with 9,10-dimethyl-1,2-benzanthracene (DMBA). *Brit. J. Cancer* 20:161-7.
49. Whitlock, J. P. and Gelboin, H. V. (1979): Aryl hydrocarbon (benzo(a)pyrene) hydroxylase induction in cells in culture. *Pharmac. Ther.* 4:587-599.
50. Witiak, D. T., Inbasekaran, Cazer, F. D., Daniel, F. D., and Hart, R. W. Presented to the Division of Medicinal Chemistry, 176th National ACS Meeting in Honolulu, Hawaii, April 2-6, 1979 MEDI-72.
51. Wolf, C. R., Smith, B. R., Ball, L. M., Serabjit-Singh, C., Bend, J. R., and Philpot, R. M. (1979): The rabbit pulmonary monoxygenase system: Catalytic differences between two purified forms of cytochrome P-450 in the metabolism of benz(a)pyrene. *J. Biol. Chem.* 254(9):3658-3663.
52. Yang, S. K., Chou, M. W., and Roller, P. P. (1979): Potential proximate carcinogens of 7,12-dimethylbenz(a)anthracene: Characterization of two metabolically formed trans-3,4-dihydrodiol. *J. Amer. Chem. Soc* 101:237-39.
53. Yang, S. K., and Dower, W. V. (1975): Metabolic pathways of 7,12-dimethylbenz(a)anthracene in hepatic microsomes. *Proc. Natl. Acad. Sci. U.S.A* 72:2601-2605.

# METABOLISM OF 7,12-DIMETHYLBENZ[A]-ANTHRACENE: QUANTITATION OF METABOLITE FORMATIONS IN RAT LIVER MICROSOMES AND A RECONSTITUTED ENZYME SYSTEM CONTAINING HIGHLY PURIFIED CYTOCHROME P-450 OR P-448

S. K. Yang*, M. W. Chou*, P. G. Wislocki**, and A.Y.H. Lu**

*Department of Pharmacology, School of Medicine, Uniformed Services University of the Health Sciences, Bethesda, Maryland 20014 and
**Department of Animal Drug Metabolism, Merck Sharp & Dome Research Laboratories, Rahway, New Jersey 07065

7,12-Dimethylbenz(a)anthracene (DMBA) is one of the most potent carcinogenic polycyclic aromatic hydrocarbons (PAH). DMBA, Like other PAH, is biologically inactive and its biological properties such as mutagenicity and carcinogenicity require metabolic activation by the drug-metabolizing enzyme systems. The metabolic pathways and products of DMBA have been studied intensively by many investigators in the past 15 years (1,3-9,14,16,18,26,29-32,37,40,41). This report summarizes our recent findings on the identification of 37 compounds as DMBA metabolites and the quantitative aspects of in vitro DMBA metabolism in rat liver microsomes and in a reconstituted enzyme system containing highly purified cytochrome P-450 or P-448. The results revealed that all methyl groups and unsubstituted ring carbons of DMBA and its hydroxymethy derivatives 7-methyl-12-hydroxymethylbenz(a)anthracene (7-M-12-OHMBA), 7-hydroxymethyl-12-methylbenz(a)anthracene (7-OHM-12-MBA), and 7,12-dihydroxymethylbenz(a)anthracene (7,12-diOHMBA) are involved in metabolic oxygenations. Prior treatment of rats with chemicals such as phenobarbital and 3-methylcholanthrene markedly altered the selectivity of the region that was oxygenated by the liver microsomal enzymes. This region selectivity also was observed using purified cytochrome P-450 or P-448 in the reconstituted enzyme system.

## HPLC SEPARATION AND THE IDENTIFICATION OF DMBA METABOLITES

Incubations of large amounts of DMBA, 7-M-12-OHMBA, 7-OHM-12-MBA, and 7,12-diOHMBA in vitro with liver microsomes and subsequent separation by reversed-phase and normal-phase HPLC systems have allowed us to identify 37 separate DMBA metabolites. Metabolites were characterized by physicochemical methods such as NMR spectrometry, ultraviolet and fluorescence spectrophotometry, mass spectrometry, and spectropolarimetry (6-8,40,41).

### Reversed-Phase HPLC

The separation of metabolites by reversed-phase HPLC on an analytical ODS column is shown in Figure 1. The UV absorption profiles are products of DMBA (Figure 1A), 7-M-12-OHMBA (Figure 1B), 7-OHM-12-MBA (Figure 1C), and 7,12-diOHMBA (Figure 1D), obtained from in vitro incubation (60 minutes at 37°C) with liver microsomes (1 mg protein/ml incubation mixture) from phenobarbital-pretreated rats. Chromatographic peaks are numbered in Figure 1 and their identities are indicated in Table 1. Chromatographic peaks that were not numbered were observed in control microsomes containing no NADPH. Many minor peaks in Figure 1 were found to be due to very small amounts of metabolites. The identities of these minor metabolites were determined from pooled fractions that were collected from repeated chromatographic runs.

It can be seen from Figure 1A that a large number of metabolites were produced when DMBA was incubated with rat liver microsomes. The complexity is due to the occurrence of further metabolism of the primary and secondary metabolites such as 7-M-12-OHMBA, 7-OHM-12-MBA, and 7,12-diOHMBA. The hydroxymethyl derivatives can be enzymatically oxygenated at the ring positions to form dihydrodiol and phenolic products that are structurally similar to those derived from DMBA. For example, peaks A3, A10, A11, and A20 (Figure 1A and Table 1) each contain a trans 3,4 diol. These trans-3,4-diols differ with regard to the degree of hydroxylation at the 7- and 12-methyl groups.

HPLC analysis of metabolites formed from in vitro incubations of 7-M-12-OHMBA (Figure 1B), 7-OHM-12-MBA (Figure 1C), and 7,12-diOHMBA (Figure 1D) has greatly simplified our task of identifying the metabolites formed from DMBA (Figure 1A). Because all three hydroxymethyl derivatives (peaks A13, A22, and A23 in Figure 1A) retain the ring structure of DMBA, further metabolism of these three hydroxymethyl

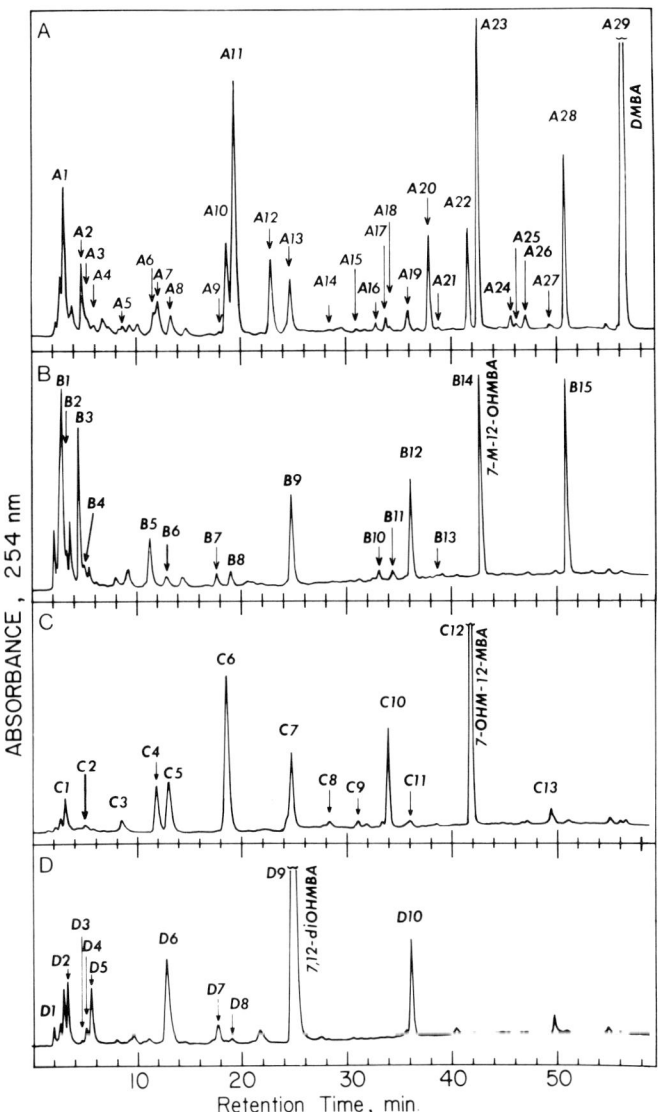

**FIGURE 1.** Reversed-phase HPLC separation of metabolites of DMBA (A), 7-M-12-OHMBA (B), 7-OHM-12-MBA (C), and 7,12-diOHMBA (D) formed by prolonged in vitro incubation with liver microsomes from PB-treated rats. Chromatographic peaks are numbered and their identities are indicated in Table 1. HPLC was carried out on a DuPont Zorbax ODS column (4.6 mm x 25 cm). The column was eluted with water-methanol (1:1) for 10 minutes and followed by a 40-minute linear gradient of water-methanol (1:1) to methanol at a solvent flow-rate of 0.8 ml/min.

**TABLE 1.** Identification of Metabolites Contained in the Chromatographic Peaks of Figure 1

| Peak | Identity[a] | Peak | Identity[a] |
|---|---|---|---|
|  | 7-M-12-COOH-BA[b] | A13,B9, C7,D9 | 7,12-diOHMBA |
|  | 7-COOH-12-MBA[b] | A14,C8 | 2-OH-7-OHM-12-MBA |
| D2 | 7,12-diOHMBA-5,6-diol | A15,C9 | 3-OH-7-OHM-12-MBA |
| A2,B3 | 7-M-12-OHMBA-5,6-diol | A16,B10 | 4-OH-7-M-12-OHMBA |
| D3 | 7,12-diOHMBA-10,11-diol | A17,C10 | 4-OH-7-OHM-12-MBA |
| A3,B4, C2,D4 | 7,12-diOHMBA-3,4-diol | A18,B11 | 3-OH-7-M-12-OHMBA |
| A4,D5 | 7,12-diOHMBA-8,9-diol | A19,B12, C10,D11 | 7-OHM-12-CHO-BA |
| A5,C3 | 7-OHM-12-MBA-10,11-diol | A19,C11 | 7-CHO-12-OHMBA |
| A6,B5 | 7-M-12-OHMBA-8,9-diol | A20 | DMBA-3,4-diol |
| A7,C4 | 7-OHM-12-MBA-5,6-diol | A21,B13 | 2-OH-7-M-12-OHMBA |
| A8,B6, C5,D6 | 4-OH-7,12-diOHMBA | A22,C12 | 7-OHM-12-MBA |
| A8,C5 | 7-OHM-12-MBA-8,9-diol | A23,B14 | 7-M-12-OHMBA |
| A9,B7 | 7-M-12-OHMBA-10,11-diol | A24 | 2-OH-DMBA |
| D7 | 2-OH-7,12-diOHMBA | A25 | 3-OH-DMBA |
| A10,C6 | 7-OHM-12-MBA-3,4-diol | A26 | 4-OH-DMBA |
| B8,D8 | 3-OH-7,12-diOHMBA | A27,C13 | 7-CHO-12-MBA |
| A11 | DMBA-5,6-diol | A28,B15 | 7-M-12-CHO-BA |
| A11,B8 | 7-M-12-OHMBA-3,4-diol | A29 | DMBA |
| A12 | DMBA—8,9-diol |  |  |
| A12 | DMBA-10,11-diol |  |  |

[a]All dihydrodiols are trans isomers (8).
[b]The carboxylic acid derivatives were isolated by a gravity-flow silica gel column (8).

metabolites to more polar products is thus expected. Theoretically, metabolites found in Figures 1B, 1C, and 1D should all be present in Figure 1A. This was found to be the case for most metabolites (Table 1), with only peaks D2, D3, D7, and D8 in Figure 1D not found directly as metabolites of DMBA. This was not surprising in view of the observation (Figure 1D) that metabolites D2, D3, D7, and D8 were all minor products of 7,12-diOHMBA. Thus these minor metabolites were probably formed from 7,12-DMBA during the metabolism of DMBA, but the amounts were too small to be detected in Figure 1A.

Carboxylic acid derivatives of DMBA (Table 1) were not isolated directly by reversed-phase HPLC shown in Figure 1. They were isolated by a procedure (8) that specifically isolates the carboxylic acid derivatives from a complex mixture of metabolites.

## Normal-Phase HPLC

Many metabolites were either partially overlapped or completely overlapped on the reversed-phase HPLC (Figure 1 and Table 1). The unresolved metabolites were completely separated from one another on an isocratic normal-phase HPLC system using a DuPont Zorbax SIL (silica gel) column. Mixtures of varying ratio of THF-hexane-methanol were used as the elution solvents (8). Examples of the separation on normal-phase HPLC are shown in Figure 2. Peaks A11, A12, and A19 in Figure 1A, each containing two metabolites, were completely resolved (Figures 2B and 2C). Furthermore, some closely eluted metabolites in Figure 1 also were separated completely by the normal-phase HPLC (Figure 2A).

The elution order of the metabolites on the reversed-phase HPLC was generally reversed on the normal-phase HPLC when the elution solvents in normal-phase HPLC did not contain methanol (Figures 1 and 2). A general rule was not apparent for the elution orders of metabolites in the normal-phase HPLC when methanol was added to the elution solvents (Figures 1 and 2 and Table 1).

## Optical Activity of Dihydrodiol Metabolites

Formation of optically active dihydrodiol metabolites indicates the stereospecific or stereoselective properties of the rat liver microsomal mixed-function oxidases and epoxide hydra(ta)se (34,35,42,43). We obtained several highly purified dihydrodiols in sufficient quantity for optical rotation measurement (Table 2). In contrast to the predominantly (-) dihydrodiol enantiomers formed from benzo(a)pyrene (34,35,42,43), the dihydrodiols formed from DMBA have both (+) and (-) optical rotations. It is also interesting to note that the specific rotation of DMBA-trans-5,6-diol is +257 degrees, whereas the 7-OHM-12-MBA-trans-5,6-diol has a specific rotation of -29 degrees. The results in Table 2 indicate that the dihydrodiols are formed stereospecifically or stereoselectively from DMBA by rat liver microsomal mixed-function oxidases and epoxide hydra(ta)se.

## METABOLIC PATHWAYS

The identified metabolites listed in Table 1 clearly indicate that all methyl groups and ring carbons numbering 1 to 6 and 8 to 11 of DMBA, 7-M-12-OHMBA, 7-OHM-12-MBA, and 7,12-diOHMBA are involved in metabolic oxygenations. The involvement of $C_7$ carbons in metabolism

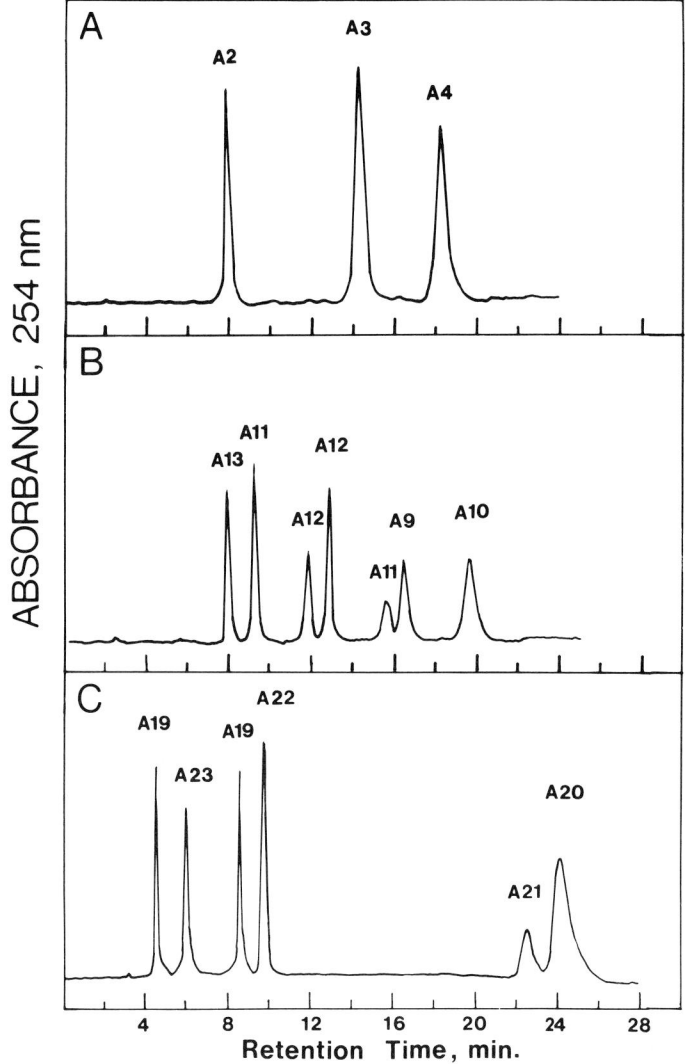

FIGURE 2. Normal-phase HPLC separation of some DMBA Metabolites. The chromatographic peaks are numbered as in Figure 1A and their identities are indicated in Table 1. HPLC was performed on a DuPont Zorbax SIL column (6.2 mm x 25 cm). The ratios of THF-hexane-methanol for the elution solvent are: A, 40:55:5; B, 30:67.5:2.5; and C, 25:75:0. The solvent flow rate was 2 ml/min. In B: the earlier eluted peak A11 is DMBA-trans-5,6-diol and the later eluted peak A11 is 7-M-12-OHMBA-trans-3,4-diol; the earlier eluted peak A12 is DMBA-trans-10,11-diol and the later eluted peak A12 is DMBA-trans-8,9-diol. In C: the earlier eluted peak A19 is 7-OHM-12-CHO-BA and the later eluted peak A19 is 7-CHO-12-OHMBA.

TABLE 2. Optical Activity of Some DMBA Metabolites

| Metabolite[a] | $[\alpha]_D^{25}$, deg |
|---|---|
| DNBA-trans-3,4-diol | −175 (0.8 mg/ml, methanol)[b] |
| DMBA-trans-8,9-diol | + 25 (0.99 mg/ml, methanol) |
| DMBA-trans-5,6-diol | +257 (1.52 mg/ml, methanol) |
| 7-OHM-12-MBA-trans-3,4-diol | −105 (2.0 mg/ml, methanol) |
| 7-OHM-12-MBA-trans-8,9-diol | + 21 (2.11 mg/ml, methanol) |
| 7-OHM-12-MBA-trans-5,6-diol | − 29 (0.68 mg/ml, methanol) |
| 7-OHM-12-MBA-trans-10,11-diol | − 97 (1.55 mg/ml, methanol) |

[a]The dihydrodiols of DMBA were obtained from the in vitro incubation of DMBA, whereas the dihydrodiols of 7-OHM-12-MBA were obtained from in vitro incubation of 7-OHM-12-MBA. The enzyme source in each case was liver microsomes from phenobarbital-pretreated male Sprague-Dawley rats.
[b]A specific rotation of -138 degrees was reported previously (40) from a sample obtained with a different microsomal preparation.

may be inferred by the detection of 2-phenols (6) or may be demonstrated by the metabolism of the trans-3,4-diols to 3,4-diol-1,2-epoxides, although the latter has not been reported.

Nine metabolites that were derived from the oxidation of the methyl groups of DMBA were identified. These are the three hydroxymethyl derivatives, two methyl aldehydes, two hydroxymethyl aldehydes, and two methyl carboxylic acids (Table 1). Both 7-CHO-12-MBA and 7-M-12-CHO-BA were found to possess moderate to weak skin-tumor-initiating activity in mice (12,15). Thus oxidation of DMBA to methyl aldehydes may contribute to the overall carcinogenicity of DMBA. The biological effects of the further oxidations of the methyl aldehydes to hydroxymethyl aldehydes and carboxylic acid derivatives are not known.

The metabolic pathways for the formation of dihydrodiols and phenolic metabolites at the ring positions are proposed in Figure 3. None of the epoxide intermediates have been detected directly. However, it has been well established that epoxides (arene oxides) are metabolites of PAH. In addition to the hydration reaction catalyzed by the microsomal epoxide hydra(ta)se, epoxide intermediates can also rearrange nonenzymatically to phenolic products. Many possible phenolic metabolites indicated in Figure 3 have not been found. These phenols may have been formed in extremely small amounts and thus were not detectable. Although four 2-phenols were found (6), none of the four possible trans-1,2-diol metabolites were detected. The absence of trans-1,2-diols may

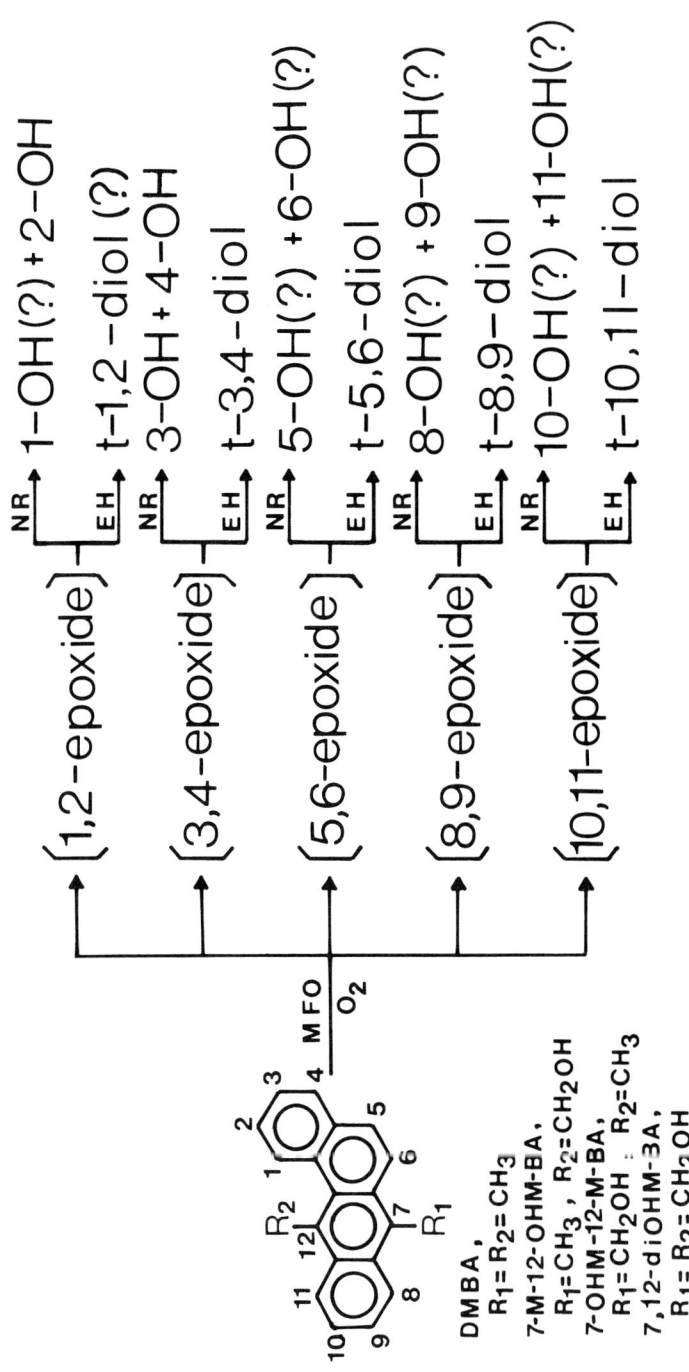

FIGURE 3. Proposed pathways of metabolic oxidations of DMBA and its hydroxymethyl derivatives at the ring positions. Epoxide intermediates (in brackets) have not been directly detected. Question marks in parenthesis indicate possible formation of that compound but it was not found in this investigation. MFO = mixed-function oxidases, EH = epoxide hydra(ta)se, NR = nonenzymatic rearrangement, t = trans. Further metabolism of the phenolic and dihydrodiol metabolites at other ring positions is not indicated.

have been due to the instability of the 1,2-epoxide intermediates. Thus 1,2-epoxide intermediates may isomerize spontaneously to 1- and 2-phenols (10) before they could be hydrated by epoxide hydra(ta)se.

Dihydrodiols bearing a free methyl group can be further hydroxylated at the methyl group to form monohydroxymethyl and dihydroxymethyl dihydrodiols. For example, we have found that DMBA-trans-3,4-diol can be metabolized by rat liver microsomes to form 7-OHM-12-MBA-trans-3,4-diol, 7-M-12-OHMBA-trans-3,4-diol, and 7,12-diOHMBA-trans-3,4-diol. In addition, both 7-M-12-OHMBA-trans-3,4-diol and 7-OHM-12-MBA-trans-3,4-diol can be further metabolized to form 7,12-diOHMBA-trans-3,4-diol (8).

Among these many and diverse metabolites, the trans-3,4-diols are considered most important since they are believed to be on the metabolic pathways by which the most reactive, mutagenic, and carcinogenic metabolites are formed. According to the "bay region" theory (21,22), formations of 1,2-epoxides from the trans-3,4-diols can lead to the formation of $C_1$ carbonium ion intermediates which are thought to be the ultimate carcinogenic metabolites of DMBA. Although direct evidence on the metabolism of trans-3,4-diols to the 3,4-diol-1,2-epoxides has not been available, indirect evidence has been reported suggesting that 3,4-diol-1,2-epoxide(s) is involved in binding to DNA of mammalian cells in culture (13,19,27), to DNA of mouse skin that has been treated with DMBA (2,38), and to DNA in a rat liver microsome-mediated in vitro system (2,7,19). DMBA-trans-3,4-diol was found to be highly mutagenic to *Salmonella typhimurium* tester strain TA 100 (24) and highly active in inducing malignant cell transformation (25). The mutagenicity of three of the four trans-3,4-diols has been tested in a cell-mediated mutagenicity assay with Chinese hamster V79 cells, and only DMBA-trans-3,4-diol was found to be more mutagenic than DMBA itself (18). A racemic DMBA trans-3,4-diol has been found to be a potent tumor initiating agent (33). The mutagenicity of 7-OHM-12-MBA-trans-3,4-diol and 7-M-12-OHMBA-trans-3,4-diol was found to be less than that of DMBA but was more mutagenic than those of 7-OHM-12-MBA and 7-M-12-OHMBA respectively (18). In an in vitro system, all three of the above-mentioned trans-3,4-diols produced products that bound to DNA more extensively than DMBA upon further metabolism with rat liver microsomes (7). These results thus indicate that there may be multiple activation pathways in the formation of ultimate carcinogenic metabolites of DMBA.

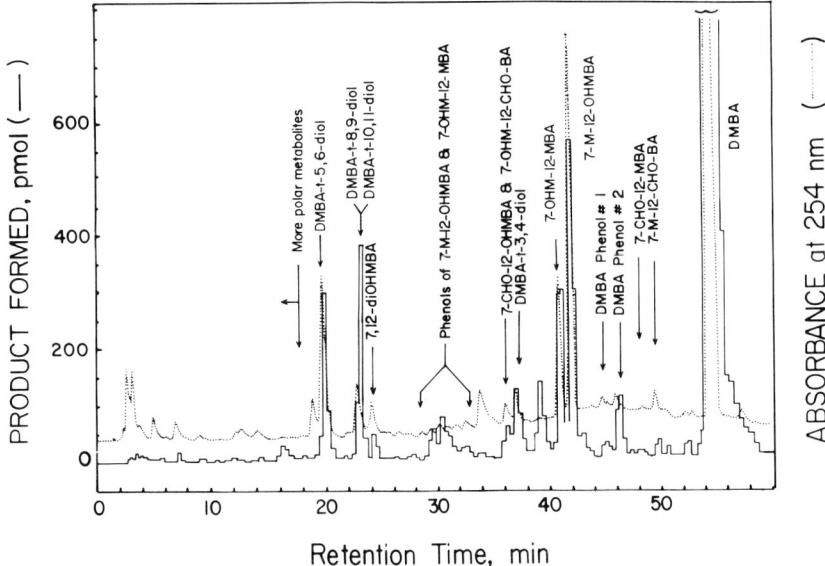

**FIGURE 4.** DMBA metabolites profile on a reversed-phase HPLC. The bar curve is the radioactivity profile of a sample obtained from the in vitro incubation of [$^{14}$C]DMBA with low concentration (0.05 mg protein/ml reaction mixture) of liver microsomes from PB-pretreated rats. The dotted curve is from a sample obtained from the microsomal incubation of DMBA with PB-microsomes (1 mg protein/ml reaction mixture) which was used as UV markers. The identities of metabolites contained in the chromatographic peaks of the dotted curve are mostly known (see Table 1 and Figure 1). The HPLC conditions were identical to that indicated in the legends of Figure 1.

## Effect of Enzyme Induction on DMBA Metabolism in Rat Liver Microsomes

The large number of metabolites formed from prolonged incubation with high concentrations of liver microsomes (Figure 1) was primarily due to extensive further metabolism of the primary metabolites of DMBA such as hydroxymethyl derivatives and the dihydrodiols and phenols of DMBA. When DMBA was incubated under initial rate conditions (i.e., shorter incubation time and lower enzyme concentrations), many of these secondary metabolites such as diols of 7-OHM-12-MBA, 7-M-12-OHMBA, and 7,12-diOHMBA were either not formed or were formed at very low levels (Figure 4). Incubation of DMBA with liver microsomes from untreated or treated rats under initial rate conditions in which DMBA was present at the saturation level led to the formation of mainly primary metabolites (Table 3). The formation of most of the metabolites

**TABLE 3.** Quantitative Determination of DMBA Metabolism in Liver Microsomes from Untreated (Control), PB- and MC-Pretreated Male, Immature Long-Evans Rats [a]

| Metabolite | Specific Activity (% of total)[b] | | |
|---|---|---|---|
| | Control | PB | MC |
| DMBA-3,4-diol | 0.096 (8.9) | 0.146 (9.2) | 0.084 (2.2) |
| DMBA-5,6-diol | 0.072 (6.7) | 0.127 (8.0) | 0.140 (3.6) |
| DMBA-8,9-diol and DMBA-10,11-diol[c] | 0.196 (18.1) | 0.133 (8.3) | 1.720 (44.4) |
| DMBA phenol #1[d] | 0.018 (1.7) | 0.016 (1.0) | 0.215 (5.5) |
| DMBA phenol #2[e] | 0.105 (9.7) | 0.089 (5.6) | 0.131 (3.4) |
| 7-OHM-12-MBA | 0.239 (22.1) | 0.249 (15.7) | 0.898 (23.2) |
| 7-M-12-OHMBA | 0.061 (5.6) | 0.543 (34.1) | 0.040 (1.0) |
| 7,12-diOHMBA | 0.015 (1.4) | 0.038 (2.4) | 0.062 (1.6) |
| Phenols of 7-M-12-OHMBA and 7-OHM-12-MBA[f] | 0.069 (6.4) | 0.088 (5.5) | 0.294 (7.6) |
| 7-CHO-12-MBA | 0.005 (0.5) | 0.030 (1.9) | N.D.[h] |
| 7-M-12-CHO-BA | 0.009 (0.8) | 0.046 (2.9) | N.D. |
| 7-OHM-12-CHO-BA and 7-CHO-12-OHMBA | 0.109 (10.1) | 0.010 (0.6) | 0.142 (3.7) |
| More polar metabolites[g] | 0.071 (6.6) | 0.045 (2.8) | 0.119 (3.1) |
| Unidentified | 0.015 (1.4) | 0.032 (2.0) | 0.027 (0.7) |
| Total | 1.08 (100) | 1.59 (100) | 3.87 (100) |
| % DMBA Metabolized | 0.4 | 1.14 | 2.46 |

[a] Seventy nmoles of [dimethyl-$^{14}$C]DMBA (in 0.04 ml methanol; specific activity 17.5 mCi/mmole) was incubated for 5 minutes at 37°C in a 1-ml reaction mixture containing 0.05 mg of liver microsomal protein from untreated, PB- and MC-pretreated rats, 0.1 nmole of potassium phosphate buffer (pH 7.0), 3 μmoles of MgCl$_2$, and 0.5 μmole of NADPH. DMBA and its metabolites were extracted with acetone (1 ml) and ethyl acetate (2 ml) and the organic phase was dehydrated with anhydrous MgSO$_4$, evaporated to dryness under a stream of nitrogen, and redissolved in 0.05 ml of tetrahydrofuran-methanol (1:1) for reversed-phase HPLC analysis (Figure 4).
[b] The specific activity is expressed as nmole products formed/min/nmole of P-450 or P-448. The % of total in the parentheses indicates the fraction of total metabolites formed. The P-450 or P-448 contents of the microsomes are: control microsomes, 0.88 nmole P-450/mg protein; PB-microsomes, 2.0 nmoles P-450/mg protein; MC-microsomes, 1.8 nmoles P-448/mg protein. The data are averaged values of duplicate samples which agree within 10 percent of the values shown.
[c] The chromatographic peaks containing DMBA-8,9-diol and DMBA-10,11-diol were analyzed by normal-phase HPLC (Figure 2) and the ratio was found to be 7:3. The ratio of these two metabolites varies depending upon the conditions of the incubation such as the concentrations of the substrate and microsomes and the length of incubation. DMBA-8,9-diol is, however, the major metabolite of the two under all conditions tested.
[d] 2-OH-DMBA and 3-OH-DMBA are the predominant metabolites.
[e] 4-OH-DMBA is the predominant metabolite.
[f] All phenolic metabolites derived from further metabolism of 7-OHM-12-MBA and 7-M-12-OHMBA (see Table 1 and Figure 2).
[g] This includes all the metabolites that are eluted earlier than DMBA-5,6-diol (see Table 1).
[h] Not detectable.

under our experimental conditions was proportional to incubation time.

Pretreatment of rats with phenobarbital (PB) resulted in an increase in total metabolism of DMBA in liver microsomes over that observed in untreated rats by 1.47-fold when activity is expressed as nmole/min/nmole P-450 and 3.34-fold when activity is expressed as nmole/min/mg protein. Like the metabolism of benzo(a)pyrene and other PAH, the total metabolism of DMBA in rat liver microsomes was markedly enhanced by methylcholanthrene (MC)-pretreatment, resulting in an increase of 3.58-fold when activity is expressed as nmole/min/nmole P-450 and 7.33-fold when activity is expressed as nmole/min/mg protein.

In control microsomes, metabolites formed from hydroxylation and further oxidation of the methyl groups of DMBA constitute more than one-third of the total metabolites (Table 3). Dihydrodiol formation of the 3,4-, 5,6-, 8,9-, and 10,11-double bonds accounts for 28 percent of the total metabolites. Hydroxylation at the 7-methyl group of DMBA is about 4-fold higher than that at the 12-methyl group. As shown in Tables 3 and 4, pretreatment of rats with either PB or MC not only increases the rate of total metabolism of DMBA but also alters the metabolite patterns. When the formation of specific metabolites in control microsomes and treated microsomes is compared (Table 4), it is clear that prior treatment of rats with PB causes increased conversion of DMBA to DMBA-3,4-diol, DMBA-5,6-diol, and 7-M-12-OHMBA but decreased conversion of DMBA to 8,9- and 10,11-diols. In contrast, MC pretreatment leads to

TABLE 4. Effect of Enzyme Induction on the Liver Microsomal Metabolism of 7,12-Dimethylbenz(a)anthracene

| Metabolite | Ratio of Specific Activities[a] | |
|---|---|---|
| | PB/C | MC/C |
| DMBA-3,4-diol | 1.52 | 0.88 |
| DMBA-3,4-diol and 3-OH-DMBA and 4-OH-DMBA[b] | 1.17 | 1.53 |
| DMBA-5,6-diol | 1.76 | 1.94 |
| DMBA 8,9-diol and DMBA-10,11-diol | 0.68 | 8.78 |
| 7-OHM-12-MBA | 1.04 | 3.75 |
| 7-M-12-OHMBA | 8.9 | 0.66 |
| All Metabolites | 1.3 | 3.1 |

[a]Data from Table 3.
[b]The amount of 3-OH-DMBA is an estimated value from DMBA phenol #1 of Table 3.

decreased formation of DMBA-3,4-diol and 7-M-12-OHMBA but increased formation of DMBA-5,6-diol, DMBA-8,9-diol, DMBA-10,11-diol, and 7-OHM-12MBA (see Table 3).

It is of interest to note that the formation of one of the potentially important metabolites, DMBA-3,4-diol, is stimulated by PB pretreatment but decreased by MC pretreatment (Table 4). However, the decreased formation of DMBA-3,4-diol in MC microsomes is accompanied by a substantial increase in the formation of DMBA 3- and 4-phenols. Thus, the metabolism of DMBA at the 3,4-double bond is still greater in MC microsomes than that in control microsomes. The decreased formation of DMBA-3,4-diol in MC microsomes may be due to the lower level of expoxide hydrase as compared with that in control microsomes.

## DMBA Metabolism in a Reconstituted System Containing Purified Cytochrome P-450 or P-448

Since enzyme induction markedly alters the metabolic pattern of DMBA metabolism, a reconstituted system containing purified rat liver cytochrome P-450 or P-448 was used to further examine the regioselectivity of the two different forms of cytochrome. Cytochrome P-450 (fraction C) is PB inducible and is the predominant form in PB microsomes (36,39), whereas cytochrome P-448 is the predominant form in MC microsomes (36). As would be expected from the results obtained in microsomes, cytochrome P-448 was much more active than cytochrome P-450 in catalyzing the metabolism of DMBA (Table 5). In the absence of epoxide hydrase, phenols were the major products in DMBA metabolism and accounted for 30 percent of the total metabolism in the cytochrome P-450 system and 46 percent in the cytochrome P-448 system. Cytochrome P-450 catalyzed the oxygenation of the methyl group both at the 7- and 12-positions, but cytochrome P-448 preferentially oxidized the 7-methyl group. In the presence of purified epoxide hydrase, more diols were formed at the expense of phenols and accounted for 35 to 40 percent of the total metabolism. However, the extent of increase in diol formation in the presence of epoxide hydrase appears to depend on the particular metabolic pathway and the form of cytochrome P-450 used. In the cytochrome P-450 system, the formation of DMBA-5,6-diol, DMBA-8,9- and 10,11-diols, and DMBA-3,4-diol was increased 6.6-, 2.5-, and 3.5-fold, respectively. In the cytochrome P-448 system, the formation of DMBA-5,6-diol and DMBA-3,4-diol was increased 3-fold, but the production of DMBA-8,9- and 10,11-diols was increased by a remarkable 36-fold. The reason for the large increase in DMBA-8,9- and 10,11-diol formation in the system containing both cytochrome P-448 and epoxide hydrase is unknown, but it is consistent with the high level of DMBA-8,9- and 10,11-diol production in MC microsomes (Tables 3 and 4).

## TABLE 5. Quantitative Determination of DMBA Metabolism in a Reconstituted Enzyme System Containing Purified Cytochrome P-450 or P-448[a]

| | Specific Activity (% of total)[b] | | | |
|---|---|---|---|---|
| | nmole/min/nmole P-450 | | nmole/min/nmole P-448 | |
| Metabolite | -EH | +EH | -EH | +EH |
| DMBA-3,4-diol | 0.02 (0.5) | 0.07 (1.5) | 0.39 (1.0) | 1.01 (1.8) |
| DMBA-5,6-diol | 0.20 (5.4) | 1.33 (29.5) | 0.91 (2.2) | 2.67 (4.8) |
| DMBA-8,9-diol and DMBA-10,11-diol[c] | 0.07 (1.9) | 0.18 (4.0) | 0.50 (1.2) | 17.9 (32.4) |
| DMBA phenol #1[d] | 0.23 (6.1) | 0.03 (0.7) | 5.4 (13.2) | 4.21 (7.6) |
| DMBA phenol #2[e] | 0.16 (4.3) | 0.10 (2.2) | 5.9 (14.6) | 2.70 (4.9) |
| 7-OHM-12-MBA | 0.98 (26.3) | 1.10 (24.4) | 14.6 (36.0) | 12.6 (22.8) |
| 7-M-12-OHMBA | 0.62 (16.7) | 0.58 (12.9) | 1.3 (3.2) | 0.65 (1.2) |
| 7,12-diOHMBA | 0.04 (1.1) | 0.05 (1.1) | 0.25 (1.2) | 0.65 (1.2) |
| Phenols of 7-M-12-OHMBA and 7-OHM-12-MBA[f] | 0.76 (20.4) | 0.24 (5.3) | 7.60 (18.8) | 5.35 (9.7) |
| 7-CHO-12-MBA | N.D.[h] | 0.12 (2.7) | 0.52 (1.3) | 0.34 (0.6) |
| 7-M-12-CHO-BA | 0.02 (0.5) | 0.04 (0.9) | 0.29 (0.7) | 0.11 (0.2) |
| 7-OHM-12-CHO-BA and 7-CHO-12-OHMBA | 0.29 (7.8) | 0.30 (6.7) | 1.52 (3.8) | 2.95 (5.4) |
| More polar metabolite[g] | 0.13 (3.5) | 0.17 (3.8) | 0.67 (1.7) | 3.0 (5.5) |
| Unidentified | 0.21 (5.5) | 0.20 (4.4) | 0.64 (1.6) | 1.03 (1.9) |
| Total | 3.72 | 4.51 | 40.5 | 55.2 |
| % DMBA Metabolized | 1.6 | 2.0 | 3.6 | 5.0 |

[a]Seventy nmoles of [$^{14}$C]DMBA (in 0.04 ml methanol; specific activity 17.5 mCi/nmole) was incubated in a 1-ml reaction mixture containing 0.1 mmole potassium phosphate buffer (pH 7.0), 1170 units of NADPH-cytochrome c reductase, 3 μmoles of MgCl$_2$, 0.02 mg of dilauroylphosphatidylcholine, 0.5 μmole of NADPH, 0.05 nmole of cytochrome P-450 (fraction C) or 0.02 nmole cytochrome P-448, and epoxide hydra(ta)se (0 and 6 units). NADPH-cytochrome c reductase was purified from PB-treated rats by the method of Yasukochi and Masters (44). One unit of reductase is defined as the amount of enzyme catalyzing the reduction of 1 nmole of cytochrome c per minute at 22°C. Cytochrome P-450 from PB-treated rats and cytochrome P-448 from MC-treated rats were purified by the method described by West et al (39). Epoxide hydra(ta)se was purified from PB-treated rats as previously described by Lu et al (23). One unit of hydra(ta)se is defined as the amount of enzyme catalyzing the hydration of 1 nmole of styrene oxide per minute at 37°C. The reaction mixtures containing cytochrome P-450 (fraction C) or P-448 were incubated for 6 and 3 minutes, respectively. The reactions were stopped by vortexing for 30 seconds after addition of 1 ml of acetone and 2 ml of ethyl acetate. More than 99.5 percent of the radioactivity was found in the organic phase.
[b]The specific activity is expressed as nmole products formed/min/nmole of P-450 or P-448. The % of total in the parentheses indicates the fraction of total metabolites formed. The data are averaged values of duplicate samples which agree within 10 percent of the values shown.
[c]DMBA-8,9-diol is the major metabolite.
[d]2-OH-DMBA and 3-OH-DMBA are the predominant metabolites.
[e]4-OH-DMBA is the predominant metabolite.
[f]All phenolic metabolites derived from further metabolism of 7-OHM-12-MBA and 7-M-12-OHMBA (see Table 1 and Figure 2).
[g]This includes all the metabolites that are eluted earlier than DMBA-5,6-diol (see Table 1).
[h]Not detectable.

In addition to cytochrome P-450 (fraction C) and P-448, three other cytochrome P-450 fractions (A, B, and D) isolated from PB-treated rats (39) were also used to evaluate the positional specificity of various forms of cytochrome P-450. Although fractions A, B, and D were not purified, it is clear from Table 6 that different forms of cytochrome P-450 show considerable regioselectivity in DMBA metabolism as judged by the oxidation of the methyl group at the 7- and 12-positions. It should be noted that the preferential attack at the 7-methyl group by purified cytochrome P-448 is consistent with the results obtained with MC microsomes. In general, the regioselectivity of microsomes and the reconstituted system is very similar, as has also been demonstrated in the study of benzo(a)pyrene metabolism (11,17).

TABLE 6. Specificity of Hydroxylation at the Methyl Groups of 7,12-Dimethylbenz(a)anthracene in Rat Liver Microsomes and in a Reconstituted Enzyme System Containing Purified Cytochrome P-450 or P-448

| Enzyme System | Specificity, nmole/min/nmole P-450 or P-448 | | Ratio |
|---|---|---|---|
| | 7-OHM-12-MBA | 7-M-12-OHMBA | |
| Liver Microsomes[a] | | | |
| Control | 0.24 | 0.06 | 3.92 |
| PB | 0.25 | 0.54 | 0.46 |
| MC | 0.90 | 0.04 | 22.4 |
| Reconstituted System[b] | | | |
| From PB-treated rats: | | | |
| P-450 (A) | 0.43 | 0.32 | 1.33 |
| P-450 (B) | 0.36 | 0.51 | 0.71 |
| P-450 (C) | 1.08 | 0.44 | 2.43 |
| P-450 (D) | 1.75 | 1.18 | 1.49 |
| From MC-treated: | | | |
| P-448 | 10.0 | 0.53 | 18.9 |

[a] Data from Table 3.
[b] The experimental conditions were the same as indicated in the legends of Table 5 except that 0.1 nmole of cytochrome P-450 or P-448 was used and the incubation time was 5 minutes (for all incubation mixtures containing cytochrome P-450) and 3 minutes (for the incubation mixture containing cytochrome P-448), respectively. Cytochrome P-450 fractions A, B, C, D were isolated from PB-treated rats by the method of West et al (39). Fraction C is PB-inducible and is the major cytochrome P-450 species in PB-treated rats. Unless otherwise stated, purified cytochrome P-450 from PB-treated rats refers to fraction C. Cytochrome P-448 is the major cytochrome species in MC-treated rats.

## REFERENCES

1. Baird, W. M., Chemerys, R., Chern, C. J., and Diamond, L. (1978): Formation of glucuronic acid conjugates of 7,12-dimethylbenz(a)anthracene phenols in 7,12-dimethylbenz(a)anthracene-treated hamster embryo cell cultures. *Cancer Res.* 38:3432-3437.
2. Bigger, C.A.H., Tomaszewski, J. E., and Dipple, A. (1978): Differences between products of bindings of 7,12-dimethylbenz(a)anthracene to DNA in mouse skin and in a rat liver microsomal system. *Biochem. Biophys. Res. Commun.* 80:229-235.
3. Booth, J., Keysell, G. R., and Sim, P. (1973): Formation of glutathione conjugates as metabolites of 7,12-dimethylbenz(a)anthracene by rat-liver homogenates. *Biochem. Pharmacol.* 22:1781-1791.
4. Boyland, E., and Sim, P. (1965): Metabolism of polycyclic compounds. The metabolism of 7,12-dimethylbenz(a)anthracene by rat-liver homogenates. *Biochem. J.* 95:780-787.
5. Boyland, E., Sims, P., and Huggins, C. (1965): Induction of adrenal damage and cancer with metabolites of 7,12-dimethylbenz(a)anthracene. *Nature* 207:816-817.
6. Chou, M. W., Easton, G. D., and Yang, S. K. (1979): Metabolism of 7,12-dimethylbenz(a)anthracene and its methyl-hydroxylated metabolites: formation of phenolic metabolites at the 2-positions. *Biochem. Biophys. Res. Commun.* 88:1085-1091.
7. Chou, M. W., and Yang, S. K. (1978): Identification of four trans-3,4-dihydrodiol metabolites of 7,12-dimethylbenz(a)anthracene and their in vitro DNA binding activities upon further metabolism. *Proc. Natl. Acad. Sci. U.S.A.* 75:5466-5470.
8. Chou, M. W., and Yang, S. K. (1979): Combined reversed-phase and normal-phase high-performance liquid chromatography in the purification and identification of 7,12-dimethylbenz(a)anthracene metabolites. *J. Chromatography* 185:635-654.
9. Chouroulinkov, I., Gentil, A., and Sims, P. (1973): Some biological effects of 7-hydroxymethyl-12-methylbenz(a)anthracene in mice. *Biomedicine* 19:438-441.
10. Daly, J. W., Jerina, D. M., and Witkop, B. (1972): Arene oxides and the NIH shift: the metabolism, toxicity and carcinogenesis of aromatic compounds. *Experientia* 28:1129-1149.
11. Deutsch, J., Leutz, J. C., Yang, S. K., Gelboin, H. V., Chiang, Y. L., Vatsis, K. P., and Coon, M. J. (1978): Regio- and stereoselectivity of various forms of purified cytochrome P-450 in the metabolism of benzo(a)pyrene and (-)trans-7,8-dihydroxy-7,8-dihydrobenzo(a)pyrene as shown by product formation and binding to DNA. *Proc. Natl. Acad. Sci., U.S.A.* 75:3123-3127.

12. DiGiovanni, J., Slaga, T. J., Berry, D. L., Harvey, R. G., and Juchau, M. R. (1978): Effects of 7,8-benzoflavone on skin tumor-initiating activities of various 7- and 12-substituted derivatives of 7,12-dimethylbenz(a)anthracene in mice. *J. Natl. Cancer Inst.* 61:135-140.
13. Dipple, A., Tomaszewski, J. E., Moschel, R. C., Bigger, C.A.H., Nebzydoski, J. A., and Egan, M. (1979): Comparison of metabolism-mediated binding to DNA of 7-hydroxymethyl-12-methylbenz(a)-anthracene and 7,12-dimethylbenz(a)anthracene. *Cancer Res.* 39:1154-1158.
14. Flesher, J. W., Soedigdo, S., and Kelley, D. R. (1967): Syntheses of metabolites of 7,12-dimethylbenz(a)anthracene, 4-hydroxy-7,12-dimethylbenz(a)anthracene, 7-hydroxymethyl-12-methylbenz(a)anthracene, their methyl ethers, and acetoxy derivatives. *J. Med. Chem.* 10:932-936.
15. Flesher, J. W., and Sydnor, K. L. (1971): Carcinogenicity of derivatives of 7,12-dimethylbenz(a)anthracene. *Cancer Res.* 31:1951-1954.
16. Gentil, A., Lasne, C., and Chouroulinkov, I. (1974): Metabolism of 7,12-dimethylbenz(a)anthracene by hamster liver homogenates. *Xenobiotica* 4:537-548.
17. Holder, J., Yagi, H., Dansette, P. M., Jerina, D. M., Levin, W., Lu, A.Y.H., and Conney, A. H. (1974): Effects of inducers and epoxide hydrase on the metabolism of benzo(a)pyrene by liver microsomes and a reconstituted system: analysis by high pressure liquid chromatography. *Proc. Natl. Acad. Sci., U.S.A.* 71:4356-4360.
18. Huberman, E., Chou, M. W., and Yang, S. K. (1979): Identification of 7,12-dimethylbenz(a)anthracene metabolites that lead to mutagenesis in mammalian cells. *Proc. Natl. Acad. Sci., U.S.A.* 76:862-866.
19. Ivanovic, V., Geacintov, N. E., Jeffrey, A. M., Fu, P. P., Harvey, R. G., and Weinstein, I. B. (1978): Cell and microsome mediated binding of 7,12-dimethylbenz(a)anthracene to DNA studied y fluorescence spectroscopy. *Cancer Lett.* 4:131-140.
20. Jerina, D. M., Dansette, P. M., Lu, A.Y.H., and Levin, W. (1977): Hepatic microsomal epoxide hydrase: A sensitive radiometric assay for hydration of arene oxides of carcinogenic aromatic hydrocarbons. *Mol. Pharmacol.* 13:342-351.
21. Jerina, D. M., Lehr, R., Schaefer-Ridder, M., Yagi H., Karle, J. M., Thakker, D. R., Wood, A. W., Lu, A.Y.H., Ryan, D., West, S., Levin, W., and Conney, A. H. (1977): Bay-region epoxides of dihydrodiols: a concept explaining the mutagenic and carcinogenic activity of benzo(a)pyrene and benzo(a)anthracene. In: *Origins of*

*Human Cancer,* H. H. Hiatt, J. D. Watson, and J. A. Winsten, Eds., pp. 639-658, Cold Spring Harbor Laboratory.
22. Lehr, R. E., Yagi, H., Thakker, D. R., Levin, W., Wood, A. W., Conney, A. H., and Jerina, D. M. (1978): The bay region theory of polycyclic aromatic hydrocarbon-induced carcinogenicity. In: *Carcinogenesis, Vol. 3: Polynuclear Aromatic Hydrocarbons,* P. W. Jones and R. I. Freudenthal, Eds., pp. 231-241, Raven Press, New York.
23. Lu, A.Y.H., Ryan, D., Jerina, D. M., Daly, J. W., and Levin, W. (1975): Liver microsomal epoxide hydrase: Solubilization, purification and characterization. *J. Biol. Chem.* 250:8283-8288.
24. Malaveille, C., Bartsch, H., Tierney, B., Grover, P. L., and Sims, P. (1978): Microsome-mediated mutagenicities of the dihydrodiols of 7,12-dimethylbenz(a)anthracene: high mutagenic activity of the 3,4-dihydrodiol. *Biochem. Biophys. Res. Commun.* 83:1468-1473.
25. Marquardt, H., Baker, S., Tierney, B., Grover, P. L., and Sims, P. (1978): Induction of malignant transformation and mutagenesis by dihydrodiols derived from 7,12-dimethylbenz(a)anthracene. *Biochem. Biophys. Res. Commun.* 85:357-362.
26. Morreal, C. E., Alks, V., and Spiess, A. J. (1976): Identification of 8,9-dihydro-8,9-dihydroxy-7,12-dimethylbenz(a)anthracene as a rat liver metabolite of 7,12-dimethylbenz(a)anthracene. *Biochem. Pharmacol.* 25:1927-1930.
27. Moschel, R. C., Baird, W. M., and Dipple, A. (1977): Metabolic activation of the carcinogen 7,12-dimethylbenz(a)anthracene for DNA binding. *Biochem. Biophys. Res. Commun.* 76:1092-1098.
28. Oesch, F., Jerina, D. M., and Daly, J. (1970): A radiometric assay for hepatic epoxide hydrase activity with 7-$^3$H-styrene oxide. *Biochem. Biophys. Acta.* 227:685-691.
29. Sims, P. (1970): Studies on the metabolism of 7-methylbenz(a)anthracene and 7,12-dimethylbenz(a)anthracene and its hydroxymethyl derivatives in rat liver and adrenal homogenates. *Biochem. Pharmacol.* 19:2261-2275.
30. Sims, P. (1973): Epoxy derivatives of aromatic polycyclic hydrocarbons. The preparation and metabolism of epoxides related to 7,12-dimethylbenz(a)anthracene. *Biochem. J.* 131:405-413.
31. Sims, P. (1970): Qualitative and quantitative studies on the metabolism of a series of aromatic hydrocarbons by rat-liver preparations. *Biochem. Pharmacol.* 19:795-818.
32. Sims, P., and Grover, P. L. (1974): Epoxides in polycyclic aromatic hydrocarbon metabolism and carcinogenesis. *Advan. Cancer Res.* 20:165-274.

33. Slaga, T. J., Gleason, G. L., DiGiovanni, J., Sukumaran, K. B., and Harvey, R. G. (1979): Potent tumor-initiating activity of the 3,4-dihydrodiol of 7,12-dimethylbenz(a)anthracene in mouse skin. *Cancer Res.* 39:1934-1936.
34. Thakker, D. R., Yagi, H., Akagi, H., Koreeda, M., Lu, A.Y.H., Levin, W., Wood, A. W., Conney, A. H., and Jerina, D. M. (1977): Metabolism of benzo(a)pyrene VI. Stereoselective metabolism of benzo(a)pyrene and benzo(a)pyrene 7,8-dihydrodiol to diol epoxides. *Chem. Biol. Interaction* 16:281-300.
35. Thakker, D. R., Yagi, H., Levin, W., Lu, A.Y.H., Conney, A. H., and Jerina, D. M. (1977): Stereospecificity of microsomal and purified epoxide hydrase from rat liver. *J. Biol. Chem.* 252:6328-6334.
36. Thomas, P. E., Korzeniowski, D., Ryan, D., and Levin, W. (1979): Preparation of monospecific antibodies against two forms of rat liver cytochrome P-450 and quantitation of these antigens in microsomes. *Arch. Biochem. Biophys.* 192:524-532.
37. Tierney, B., Hewer, A., MacNicoll, A. D., Gervasi, P. G., Rattle, H., Walsh, C., Grover, P. L., and Sims, P. (1978): The formation of dihydrodiols by the chemical or enzymic oxidation of benz(a)anthracene and 7,12-dimethylbenz(a)anthracene. *Chem.-Biol. Interactions* 23:243-257.
38. Vigny, P., Duquesne, M., Coulomb, H., Tierney, B., Grover, P. L., and Sims, P. (1977): Fluorescence spectral studies on the metabolic activation of 3-methylcholanthrene and 7,12-dimethylbenz(a)anthracene in mouse skin. *FEBS Lett.* 82:278-282.
39. West, S. B., Huang, M. T., Miwa, G. T., and Lu, A.Y.H. (1979): A simple and rapid procedure for the purification of phenobarbital-inducible cytochrome P-450 from rat liver microsomes. *Arch. Biochem. Biophys.* 193:42-50.
40. Yang, S. K., Chou, M. W., and Roller, P. P. (1979): Potential proximate carcinogens of 7,12-dimethylbenz(a)anthracene: characterization of two metabolically formed trans-3,4-dihydrodiols. *J. Am. Chem. Soc.* 101:237-239.
41. Yang, S. K., and Dower, W. V. (1975): Metabolic pathways of 7,12-dimethylbenz(a)anthracene in hepatic microsomes. *Proc. Natl. Acad. Sci., U.S.A.* 72:2601-2605.
42. Yang, S. K., McCourt, D. W., Leutz, J. C., and Gelboin, H. V. (1977): Benzo(a)pyrene diol epoxides: Mechanism of enzymatic formation and optically active intermediates. *Science* 196:1199-1201.
43. Yang, S. K., Roller, P. P., and Gelboin, H. V. (1977): Enzymatic mechanism of benzo(a)pyrene conversion to phenols and diols and

improved high-pressure liquid chromatographic separation of benzo-(a)pyrene derivatives. *Biochemistry* 16:3680-3687.
44. Yasukochi, Y., and Masters, B.S.S. (1976): Some properties of a detergent-solubilized NADPH-cytochrome *c* reductase purified by biospecific affinity chromatography. *J. Biol. Chem* 251:5337-5344.

# COMPARISON OF THE SKIN TUMOR-INITIATING ACTIVITIES OF DIHYDRODIOLS, DIOL-EPOXIDES, AND METHYLATED DERIVATIVES OF VARIOUS POLYCYCLIC AROMATIC HYDROCARBONS

T. J. Slaga*, R. P. Iyer*, W. Lyga**, A. Secrist III**,
G. H. Daub†, and R. G. Harvey‡

* Biology Division, Oak Ridge National Laboratory, Oak Ridge, Tennessee 37830;
** Department of Chemistry, The Ohio State University, Columbus, Ohio 43210;
†Department of Chemistry, University of New Mexico, Albuquerque, New Mexico 87131 and
‡Ben May Laboratory for Cancer Research, University of Chicago, Chicago, Illinois 60637

## INTRODUCTION*

---

* The abbreviations used are: PAH, polycyclic aromatic hydrocarbons; BaP, benzo(a)pyrene; 1-12MBaP, 1 through 12 methylbenzo(a)pyrene; BaA, benz(a)anthracene; 7-MBaA, 7-methylbenz(a)anthracene; 12-MB(a)A, 12-methylbenz(a)-anthracene; DMBaA, 7,12-dimethylbenz(a)anthracene; 5,7,12-TMBaA, 5,7,12-trimethylbenz(a)anthracene; DB(a,h)A, dibenz(a,h)anthracene; BeP, benzo(e)pyrene; 3-MC, 3-methylcholanthrene; 3,11-DMC, 3,11-dimethylcholanthrene; 5-F,12-MBaA, 5-fluoro-12-methylbenz(a)anthracene, 5-F, 7-MBaA, 5-fluoro-7methylbenz(a)anthracene; 5,7-DMBaA, 5,7-dimethylbenz-(a)anthracene; 7-MBaA 3,4-dihydrodiol, (±)trans-3,4-dihydroxy 3,4 dihydro 7-MBaA; DB(a,h)A 3,4-dihydrodiol, (±)trans-3,4-dihydroxy-3,4-dihydro-DB(a,h)A; chrysene 1,2-dihydrodiol, (±)trans-3,4-dihydroxy-3,4-dihydrochrysene; BeP 9,10-dihydrodiol, (±)-trans-9,10-dihydroxy 9,10-dihydrobenzo(e)pyrene; DB(a,c)A 10,11-dihydrodiol, (±trans-10,11-dihydroxy-10,11-dihydro-DB(a,c)A; 7-MBaA 3,4-diol-1,2epoxide, trans-3,4-dihydroxy-anti-1,2-epoxy-1,2,3,4-tetrahydro-7-MBaA; DB(a,h)A 3,4-diol-1,2-epoxide, trans-3,4-dihydroxy-anti-1,2-epoxy-1,2,3,4-tetrahydro-DB(a,h)A; chrysene 1,2-diol-3,4-epoxide, trans-1,2-dihydroxy-anti-3,4-epoxy-1,2,3,4-tetrahydrochrysene; chrysene 3,4-diol-1,2-epoxide, trans-3,4-dihydroxy-anti-1,2-epoxy-1,2,3,4-tetrahydro-chrysene; BeP 9,10-diol-11,12-epoxide, trans-9,10-dihydroxy-anti-11,12-epoxy-9,10-11,12-tetrahydro-BeP; DB(a,c)A 10,11-diol-12,13-epoxide, trans-10,11-dihydroxy-anti-12,13-epoxy-10,11, 12,13-tetrahydro-DB(a,c)A; TPA, 12-0-tetra-decanoylphorbol-13-acetate.

Current information indicates that at least some of the PAH are metabolized by microsomal monooxygenase systems to reactive arene oxides (17, 20, 33). Jerina and coworkers (21, 22) have proposed a theory which predicts that "bay region" diol-epoxides of PAH are important in their carcinogenic activity. A bay region occurs in a PAH when an angularly fused benzo ring is present (Figure 1). There is now direct evidence from tumorigenicity studies that bay region diol-epoxides of BaP (23,24,34,35) and BaA (29,39,40) are ultimate carcinogenic metabolites. In addition, recent studies have shown that certain benzo ring dihydrodiols (immediate precursors of bay region diol-epoxides) of the carcinogens BaP, BaA, 7-MBaA, DMBaA, DB(a, h)A, chrysene, and 3-MC are tumorigenic in mice (5-7, 23, 24, 28-31,34-36,39,40).

Recent studies have revealed that a substitution of a methyl or fluoro group not only in the 1 and 2 positions of DMBaA, which are in the bay region benzo ring, but also in the 5 position almost completely blocked the skin-tumor initiating and V79 mutagenic activites of DMBaA (18,38). The 5 position is part of the K-region and is also in a peri relationship to the adjacent angular benzo ring (Figure 1). Hecht and co-workers (15,16) reported that a methyl or fluoro substitution in the 12 position of 5-methylchrysene (which is peri to the adjacent benzo ring) was also less active as a skin-tumor initiator and a mutagen than 5-methylchrysene. In addition, they reported that fluoro substitution in the bay-region benzo ring of 5-methylchrysene in the 1 and 3 positions also diminished activity as a tumor initiator and a mutagen (15,16).

In this paper we compare the skin-tumor-initiating activities in mice of dihydrodiols and diol-epoxides of 7-MBaA, DB(a,h)A, DB(a,c)A, chrysene, BeP, and methylated and fluoro derivatives of various PAH to better understand the role of the bay region and the adjacent peri position in PAH carcinogenesis. Our results show that the 3,4-dihydrodiol of 7-MBaA, the 1,2-dihydrodiol of chrysene, and the 9,10-dihydrodiol of BeP are more potent than the corresponding parent hydrocarbons, whereas the 3,4-dihydrodiol of DB(a,h)A is less active. In general, the bay-region diol-epoxides of the above hydrocarbons were very weak skin-tumor initiators. Substitution of a methyl or fluoro group in the bay-region benzo ring as well as in the adjacent peri position is very effective in blocking the skin-tumor-initiating activities of BaP, DMBaA, 7-MBaA, and 12-MBaA.

## MATERIALS AND METHODS

### Chemicals

TPA was obtained from Dr. P. Borchert, University of Minnesota, Minneapolis, Minnesota. DMBaA and 3-MC were purchased from

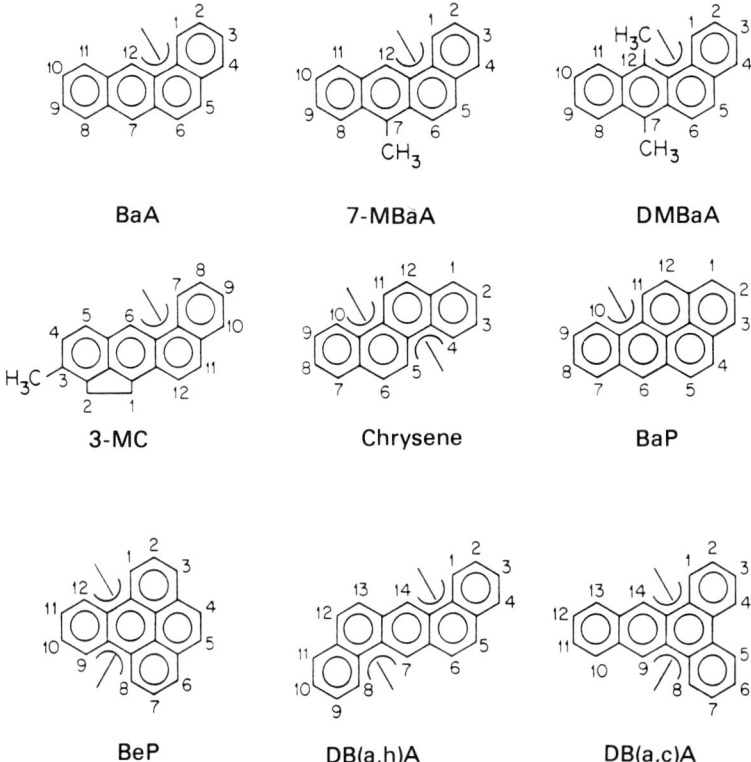

**FIGURE 1.** Structures of BaA, 7-MBaA, DMBaA, 3-MC, chrysene, BaP, BeP, DB(a,h)A, and DB(a,c)A. The arrows gives the location of the bay regions of the various PAH.

Sigma Chemical Co., St. Louis, Missouri; DB(a,h)A and DB(a,c)A from Eastman Kodak, Rochester, New York, and chrysene, BeP, and BaP from Aldrich Chemical Co., Milwaukee, Wisconsin. 3,11-DMC, 5-F,7-MBaA, 5-F,12-MBaA, 5-F-DMBaA, 5,7-DMBaA, and 5,7,12-TMBaA were a generous gift from Dr. M. Newman, The Ohio State University, Columbus, Ohio.

The various methyl derivatives of BaP were synthesized and purified as described in the literature (1-4,8,9,32). The 7,8-diol-9-methyl BaP was synthesized starting from 9 methyl-7-keto-7,8,9,10-tetrahydrobenzo(a)-pyrene as recently described (19). 7-methylbenz(a)anthracene was synthesized from 7-bromo-BaA through reaction with phenyllithium to afford 7-lithio-BaA followed by methylation with methyl iodide (26). The dihydrodiols and diol epoxides of chrysene (10,11), BeP (14), DB(a,h)A (27), DB(a,c)A (13), and 7-MBaA (26) were synthesized by the methods described. The diol-epoxides were the racemic anti-isomers (i.e., the epoxide oxygen and the benzylic hydroxyl group are located on opposite faces of the ring system) in all cases. As pointed out by Yagi et al (42), the synthetic BeP diol-epoxide is obtained as a mixture of the anti and syn stereoisomers. The pure anti BeP 9,10-diol-11,12-epoxide employed in these studies was freed of the syn-isomer by a method to be reported separately. All of the PAH were consistently prepared under yellow light immediately before use.

## Tumor Experiments

Female Sencar mice were originally obtained from Dr. R. K. Boutwell, Madison, Wisconsin, and are presently raised in Oak Ridge, Tennessee. Mice 7 to 9 weeks old were shaved with surgical clippers 2 days before treatment and only those in the resting phase of the hair cycle were used. In the tumor experiments, groups of 30 animals received a single topical application of the test compound, followed 1 week later by twice weekly applications of TPA. The incidence of papillomas was recorded weekly and they were removed at random for histological verification. DMBA, 12-MBaA, 3-MC, 7-MBaA, DB(a,h)A, DB(a,c)A, chrysene, and BeP and their dihydrodiols and methyl derivatives were applied topically in acetone. The diol-epoxides of the above PAH were applied topically in tetrahydrofuran (distilled over $LiAlH_4$ and stored over sodium wire). The dose levels of the various parent hydrocarbons used in this study were based on previously determined dose-response studies (manuscript in preparation).

## RESULTS AND DISCUSSION

### Carcinogenicity of PAH Dihydrodiols and Diol-Epoxides

The skin-tumor-initiating activities of the 3,4-dihydrodiol of 7-MBaA and DB(a,h)A, their diol-epoxides, and the corresponding parent hydrocarbons are compared in Table 1. After 15 weeks of TPA promotion, 7-MBaA-3,4-dihydrodiol at a dose of 400 nmoles had induced tumors in all the mice with an average of 6.8 papillomas per mouse, whereas 400 nmoles of 7-MBaA induced tumors in only 83 percent of the mice with an average of 3.8 papillomas per mouse. While the 3,4-dihydrodiol of 7-MBaA was approximately twice as potent as 7-MBaA, the 3,4-dihydrodiol of DB(a,h)A was found to be about half as potent as DB(a,h)A (Table 1). The bay-region diol-epoxides of 7-MBaA and DB(a,h)A were

**TABLE 1.** Comparison of the Skin-Tumor-Initiating Activities of Some Dihydrodiols and Diol-Epoxides of 7-MBaA and DB(a,h)A[a]

| Initiator | No. of Mice at 15 Weeks | Papillomas per Mouse at 15 Weeks | Mice With Tumor (%) at 15 Weeks |
|---|---|---|---|
| 7-MBaA | 29 | 3.8 | 83 |
| 7-MBaA 3,4-dihydrodiol | 28 | 6.8 | 100 |
| 7-MBaA 3,4-diol-1,2-epoxide | 29 | 0.3 | 22 |
| DB(a,h)A | 29 | 1.4 | 50 |
| DB(a,h)A 3,4-dihydrodiol | 29 | 0.7 | 37 |
| DB(a,h)A 3,4-diol-1,2-epoxide | 29 | 0.1 | 7 |
| Control (only TPA promotion) | 30 | 0.1 | 6 |

[a]7-MBaA and DB(a,h)A and their derivatives were applied at a dose of 400 nmoles and 100 nmoles, respectively, and followed 1 week later by twice weekly applications of 2 $\mu$g of TPA.

found to be very weak when tested as skin-tumor initiators. Chouroulinkov et al (7) recently reported that the 3,4-dihydrodiol of 7-MBaA was more active as a skin-tumor initiator than 7-MBaA but was less active than reported in the present study. The 3,4-dihydrodiol of DB(a,h)A was approximately half as active as a skin-tumor initiator as DB(a,h)A. Buening et al (5) also reported that the 3,4-dihydrodiol of DB(a,h)A was less active as a skin-tumor initiator than DB(a,h)A.

Table 2 shows the skin-tumor-initiating activities of the weak carcinogenic PAH cyrysene and some of its dihydrodiols and diol-epoxides. At a dose of 2 $\mu$moles, chrysene-1,2-dihydrodiol induced papillomas in 88 percent of the mice with an average of 2.8 papillomas per mouse, whereas the parent hydrocarbon initiated tumors in 73 percent with an average of 1.6 papillomas per mouse. The 3,4-dihydrodiol of chrysene was essentially inactive as a skin-tumor initiator since the TPA-only controls induced the same level of tumor response (0.1 papilloma per mouse). Similar results to ours were also reported by Levin et al (28) for the tumorigenicity of chrysene-1,2- and 3,4-dihydrodiol. The bay-region diol-epoxide (chrysene-1,2-diol-3,4-epoxide) induced tumors in 38 percent of the mice with an average of 0.5 papilloma per mouse, whereas chrysene 3,4-diol-1,2-epoxide initiated only 0.2 papilloma per mouse.

The skin-tumor-initiating activities of two other weak carcinogenic PAH and some of their dihydrodiols and diol-epoxides are shown in Table 3. The BeP-9,10-dihydrodiol, which is the immediate precursor to

**TABLE 2. Skin-Tumor-Initiating Activities of Several Dihydrodiols and Diol-Epoxides of the Weak Carcinogen Chrysene[a]**

| Initiator | No. of Mice at 15 Weeks | Papillomas per Mouse at 15 Weeks | Mice With Tumors (%) at 15 Weeks |
|---|---|---|---|
| Chrysene | 29 | 1.6 | 73 |
| Chrysene 1,2-dihydrodiol | 28 | 2.8 | 83 |
| Chrysene-1,2-diol-3,4-epoxide | 29 | 0.6 | 38 |
| Chrysene 3,4-dihydrodiol | 30 | 0.1 | 11 |
| Chrysene 3,4-diol-1,2-epoxide | 29 | 0.2 | 7 |
| Control (only TPA promotion) | 30 | 0.2 | 10 |

[a]Chrysene and the various dihydrodiols and diol-epoxides were applied at a dose of 2 μmoles and followed 1 week later by twice weekly applications of 2 μg of TPA.

the bay-region diol-epoxide induced a significant number of tumors (0.5 papilloma per mouse at a dose of 2 μmoles), whereas the parent hydrocarbon did not (Table 3). It is of interest to point out that the bay-region diol-epoxide of BeP was also ineffective as a tumor initiator. The 10,11-dihydrodiol of DB(a,c)A and the corresponding diol-epoxide

**TABLE 3. Skin Tumor-Initiating Activities of Some Dihydrodiols and Diol-Epoxides of the Weak Carcinogens BeP and DB(a,c)A[a]**

| Initiator | No. of Mice at 15 Weeks | Papillomas per Mouse at 15 Weeks | Mice With Tumors (%) at 15 Weeks |
|---|---|---|---|
| BeP | 29 | 0.2 | 17 |
| BeP 9,10-dihydrodiol | 28 | 0.6 | 32 |
| BeP 9,10-diol-11,12-epoxide | 29 | 0.1 | 6 |
| DB(a,c)A | 28 | 0.5 | 27 |
| DB(a,c)A 10,11-dihydro-diol | 29 | 0.1 | 10 |
| DB(a,c)A 10,11-diol-12,13-epoxide | 29 | 0.1 | 10 |
| Control (only TPA promotion) | 30 | 0.1 | 10 |

[a]BeP and DB(a,c)A and various dihydrodiols and diol-epoxides were applied at a dose of 2 μmoles and followed 1 week later by twice weekly applications of 2 μg of TPA.

(non-bay-region) were found to be essentially inactive as skin-tumor initiators.

It is of interest to compare the experimentally determined activities of the various dihydrodiols in Tables 1 through 3 with the relative activities MO theoretically predicted of the corresponding diol-epoxide metabolites. According to the bay-region theory (21,22), the bay-region diol-epoxides of carcinogenic PAH are characterized by exceptional reactivity predictable by $\Delta E_{deloc}$, the calculated difference in $\beta$-delocalization energies between the diol-epoxide structure and the related ring-open benzylic carbonation structure. Listed in Table 4 are the values of $\Delta E_{deloc}$ of the diol-epoxides, considered to be the metabolically activated forms of the dihydrodiols in this table. To compare the carcinogenic activities of the dihydrodiols which were assayed at different dosages, the data are expressed in terms of an arbitrary "tumorigenic index" = (papillomas per mouse) x percentage of mice with tumors)/dose ($\mu$moles). Correlation between the tumorigenic index and $\Delta E_{deloc}$ is generally good, the sole exception being the lower activity of BeP-9,10-dihydrodiol in relation to chrysene-3,4-dihydrodiol. The weak activity of the former may be, in part, a consequence of its existence in the diaxial conformation (14,41), whereas all the remaining less sterically restricted dihydrodiols in Table 4 are shown by NMR evidence (10,11,13,26,27) to exist predominantly in the diequatorial conformation. Although the biological consequences of this conformational difference are unknown, it is conceivable that the diaxial hydroxy groups of BeP-9,10-dihydrodiol sterically interfere with its enzmatic epoxidation and subsequent reaction of the resulting diol-epoxide with DNA. This suggests that the steric as well as the

TABLE 4. Comparison Between the Observed Tumorigenic Activities of PAH Dihydrodiols and the MO Theoretically Predicted Reactivities of the Corresponding Diol-Epoxide Metabolites

| Dihydrodiol | Tumorigenic Index[a] | $\Delta E_{deloc}/\beta$ |
|---|---|---|
| Chrysene-3,4 | 0.5 | 0.526 |
| DB(a,c)A-10,11 | 0.5 | 0.544 |
| BeP-9,10 | 10 | 0.714 |
| Chrysene-1,2 | 116 | 0.640 |
| DB(a,h)A-3,4 | 259 | 0.738 |
| 7-MBaA-3,4 | 1700 | 0.766[b] |

[a]Tumorigenic index = (papillomas per mouse) × (percentage of mice with tumors)/dose ($\mu$moles) under the experimental conditions described.
[b]This value of $\Delta E_{deloc}$ is calculated for BaA 3,4-dihydrodiol, since the effect of methyl substitution cannot be accurately estimated by simple perturbation theory.

electronic properties of the active dihydrodiol and diol-epoxide metabolites should be taken into account in any theory of carcinogenesis.

Although the dihydrodiols of the various PAH used in this study which are the immediate metabolic precursors of the bay-region diol-epoxides were active as tumor initiators, the various bay-region anti-diol-epoxides tested were extremely weak as tumor initiators except for the bay region anti-diol epoxide of chrysene, which had approximately one-third the activity of the parent hydrocarbon. It is of interest to point out that the only bay-region diol-epoxide tested so far that had more activity than the parent hydrocarbon is related to the weak carcinogen BaA. The bay-region ($\pm$)-anti-diol-epoxide of BaA was found to have approximately six times the activity of the parent hydrocarbon (29,39). The bay-region ($\pm$)-anti-diol-epoxide of BaP was found to have about one-third the activity of BaP (35). When the (+) and (-) enantiomeric forms of the anti-diol-epoxide of BaP were tested as skin-tumor initiators, the (+)-anti-diol-epoxide of BaP was approximately as active as BaP, whereas the (-)-anti-diol-epoxide was essentially inactive (34). The reasons for the very weak activities of the various diol-epoxides tested in the study in relation to the corresponding dihydrodiols are presently unknown. However, it is very likely that it is primarily a consequence of the indiscriminant interaction of these relatively reactive molecules with water, proteins, and other nucleophiles before they can reach the critical cellular target (presumably DNA).

## Carcinogenicity of Monomethylated Derivatives of BaP

The skin-tumor-initiating activities of various monomethylated derivatives of BaP are compared in Table 5. It is quite evident that methyl substitution at the 7,8,9 and 10 positions of BaP eliminates tumor-initiating activity. Likewise, 9-methyl BaP-7,8-dihydrodiol was found to be inactive. This is consistent with the concept of bay-region activation of BaP at these positions and in contrast to the observation by Harvey et al, who found 7-MBaP to have moderate tumorigenic activity when injected subcutaneously into rats (12). Methyl or fluoro substitution in the bay-region benzo ring of DMBaA also effectively eliminates its carcinogenic activity (18,38). 2-MBaP has weak tumor-initiating activity under our test conditions in contrast to the high sarcomogenic activity of this derivative observed by Lacassagne et al (25). The tumor-initiating activities of 3-M BaP and 4-M BaP derivatives were found to be similar to those of BaP, whereas those of 5-M BaP and 6-M BaP were much less.

However, the most striking observation of all is the fact that 11-MBaP has about three times the tumor-initiating activity of BaP. It should be

**TABLE 5. Skin Tumor-Initiating Activities of Various Monomethylated Derivatives of BaP After TPA Promotion**[a]

| Initiator | No. of Mice at 15 Weeks | Papillomas per Mouse at 15 Weeks | Mice With Tumors (%) at 15 Weeks |
|---|---|---|---|
| Control (only TPA promotion) | 29 | 0.1 | 6 |
| BaP | 30 | 2.2 | 67 |
| 1-MBaP | 29 | 3.62 | 80 |
| 2-MBaP | 29 | 0.55 | 38 |
| 3-MBaP | 29 | 2.62 | 76 |
| 4-MBaP | 30 | 2.03 | 67 |
| 5-MBaP | 30 | 0.37 | 27 |
| 6-MBaP | 29 | 0.35 | 24 |
| 7-MBaP | 27 | 0 | 0 |
| 8-MBaP | 28 | 0.03 | 3 |
| 9-MBaP | 29 | 0 | 0 |
| 10-MBaP | 29 | 0 | 0 |
| 11-MBaP | 30 | 5.6 | 90 |
| 12-MBaP | 29 | 1.7 | 69 |
| 9-MBaP 7,8-dihydrodiol | 28 | 0.07 | 7 |
| BaP 7,8-dihydrodiol | 30 | 2.53 | 77 |

[a]BaP and its monomethylated derivatives were applied topically at a dose of 200 nmoles and were followed 1 week later by twice weekly applications of 2 $\mu$g of TPA.

noted that, in the case of the 11-methyl derivative, the methyl group is attached to the carbon which forms part of a bay region but is occupying a position distal to the site of activation. A similar situation arises in the case of BaA, where substitution of the methyl group at the distal carbon of the bay region makes the compound highly tumorigenic (e.g., 12-methyl BaA or 7,12-dimethyl BaA. Likewise, 5-methylchrysene is more tumorigenic than chrysene (15). Although they were not tested, one would expect 3,6-methylcholanthrene to be more tumorigenic than 3-MC and 7- or 14-methyl DB(a,h)A to be more tumorigenic than DB(a,h)A. One explanation that might seem tangible, at least in the BaP series, is that the substitution at the 11 position enhances the stereoselectivity of enzymatic epoxidation in favor of the highly tumorigenic anti-diol-epoxide over the less carcinogenic syn-isomer. It has been shown, for example, that enzymatic epoxidation of the 9,10 double bond in BaP is not entirely stereospecific but rather stereoselective (21). The noncarcinogenicity, in mouse skin, of the syn-isomer has been related to its high reactivity and consequent inability to reach a target site.

An alternative explanation is that the steric interference to planarity of the fused ring system introduced by the 11-methyl group destabilizes aromatic conjugation, thereby conferring more olefinic character on the 7,8- and 9,10-bonds of the benzo ring, and enhancing their susceptibility to enzymatic oxidation relative to other molecular regions. Metabolism studies will be required to determine which, if either, of these interpretations is valid.

## Importance of an Unsubstituted Peri Position in PAH Carcinogenicity

A comparison of the skin-tumor-initiating activities of 3-MC, DMBaA, 7-MBaA, and 12-MBaA with those of analogous PAH bearing a methyl or fluoro group in the peri positions adjacent to the bay region benzo ring is shown in Table 6. It is quite evident that a substituted peri position in the above PAH (positions 11,5,5, and 5, respectively, of 3-MC, DMBaA, 7-MBaA, and 12-MBaA) drastically decreases their skin-tumor-initiating activities. Likewise, Table 5 shows that 6-M BaP (peri

TABLE 6. Effects of Substitution in the Peri Position on the Skin-Tumor-Initiating Activities of 3-MC, DMBaA, 7-MBaA, and 12-MBaA[a]

| Initiator | Dose (nmoles) | No. of Mice at 16 Weeks | Papillomas per Mouse at 16 Weeks | Mice With Papillomas (%) at 16 Weeks |
|---|---|---|---|---|
| 3-MC | 100 | 29 | 4.5 | 92 |
| 3,11-DMC | 100 | 30 | 0.2 | 10 |
| DMB(a)A | 10 | 28 | 8.6 | 100 |
| 5-F DMB(a)A | 10 | 29 | 0.2 | 15 |
| 5,7,12-TMB(a)A | 10 | 28 | 0.4 | 34 |
| 7-MB(a)A | 400 | 30 | 3.1 | 86 |
| 5-F,7-MB(a)A | 400 | 29 | 0.4 | 33 |
| 5,7-DMB(a)A | 400 | 29 | 0.6 | 41 |
| 12-MB(a)A | 400 | 29 | 1.6 | 73 |
| 5-F,12-MB(a)A | 400 | 29 | 0.2 | 10 |
| Control (TPA only) | | 29 | 0.1 | 6 |
| Control (3-MC only) | | 30 | 0 | 0 |

[a]The PAH were applied at the doses shown and followed 1 week later by twice weekly applications of 2 $\mu$g of TPA.

position) is much less active as a skin-tumor initiator than BaP. From these results, as well as published results, we suggest that an unsubstituted peri position adjacent to an angular benzo ring is important for carcinogenic activities (15,18,37,38). We previously reported that substitution of a methyl or fluoro group in the bay region as well as in the peri position of DMBaA almost completely counteracted the skin-tumor-initiating activities of DMBaA, whereas substitution of a fluoro in the 9 and 11 position had no effect on the activity of DMBaA (38). Hecht and co-workers (15) also found that a methyl or fluoro substitution in the 12 position of 5-methylchrysene (also a peri position) drastically decreased its skin-tumor-initiating activity.

It is quite clear that an unsubstituted peri position adjacent to an angular benzo ring, while not an absolute prerequisite for carcinogenic activity, contributes importantly to such activity. While substitution in this molecular region may potentially affect any stage in the process of metabolic activation and covalent bonding of an activated metabolite to nucleic acids, the most obvious predictable effect is on the preferred conformation of the bay-region dihydrodiol and diol-epoxide metabolites. The latter have been demonstrated by NMR and X-ray crystallographic evidence (41) to exist predominantly in a conformation in which both hydroxyl groups are oriented equatorially in the absence of methyl or other peri substituents (or fused aromatic rings in this region). Such groups force the hydroxyl functions through steric interaction to adopt the alternative diaxial conformation. This effect is illustrated for DMBaA and 5-M-DMBaA in Figure 2 for the DMBaA, 5-methylchrysene, and BaP and their methyl-substituted derivatives in Figure 3. It

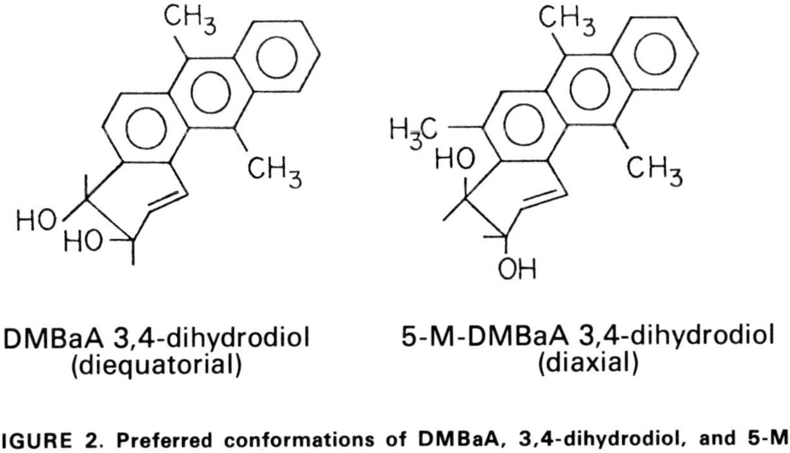

DMBaA 3,4-dihydrodiol
(diequatorial)

5-M-DMBaA 3,4-dihydrodiol
(diaxial)

FIGURE 2. Preferred conformations of DMBaA, 3,4-dihydrodiol, and 5-M-MBaA, 3,4-dihydrodiol, based on Zacharias et al (40).

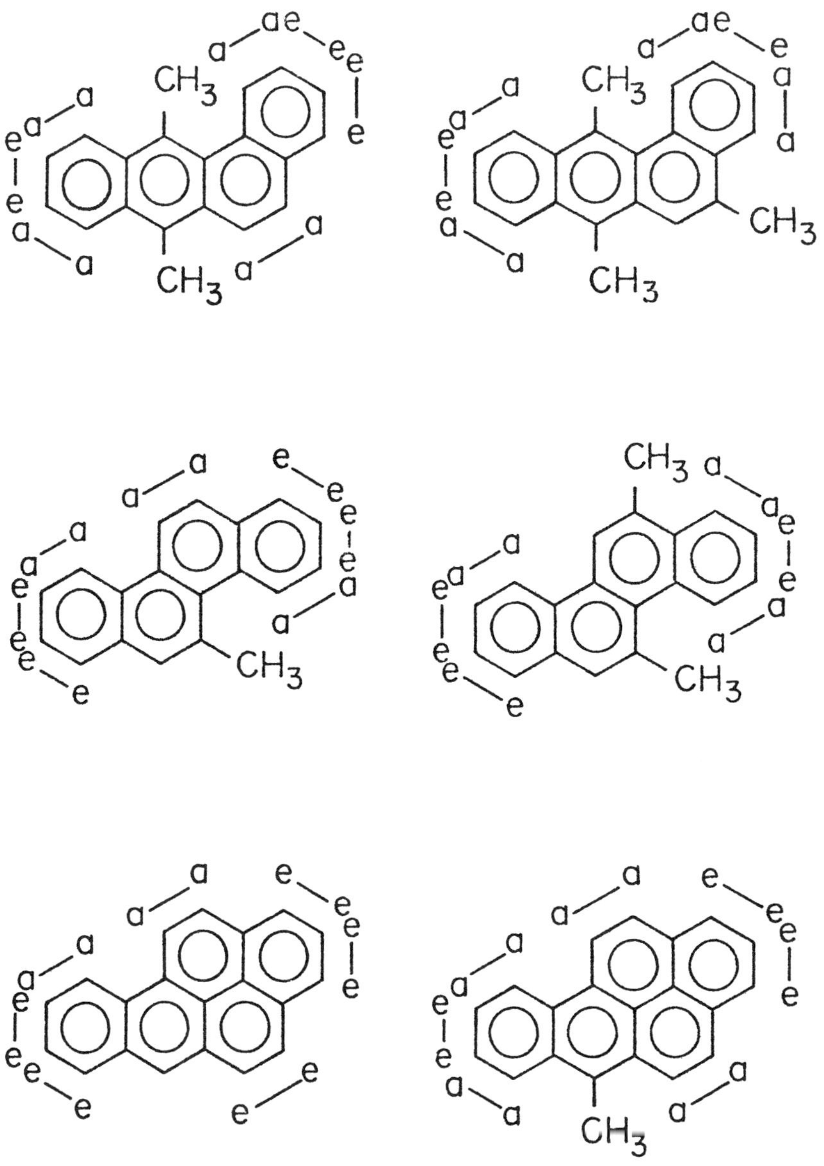

FIGURE 3. Conformations expected for the various dihydrodiols of DMBaA, 5-methylchrysene, benzo(a)pyrene, and their peri-substituted methyl analogues. The letter "a" indicates that at that site a trans-dihydrodiol would be diaxial, while "e" designates a preferred diequatorial conformer.

is probable that the diaxial hydroxyl groups sterically interfere with enzymatic epoxidation of these hindered dihydrodiols and the subsequent reaction of the resulting diol-epoxide metabolites with DNA. This is consistent with a similar effect suggested earlier in this paper to explain the unexpectedly low tumorigenic activity of the BeP-9,10-dihydrodiol.

## CONCLUSION

The data presented and discussed in this report allow us to suggest three important generalizations about PAH carcinogenicity in mouse skin. First, bay-region dihydrodiols which are the immediate precursors of bay-region diol-epoxides are the principal determinants of PAH carcinogenicity as proposed by Jerina and co-workers (21,22). This is supported by the finding that substitution of a methyl or fluoro group into the positions involved in a bay-region diol-epoxide drastically decreases its carcinogenicity (Figure 1). Second, an unhindered peri position adjacent to an angular benzene ring is necessary for carcinogenic activity. Third, a methyl group attached at the distal carbon of the bay region makes the compound more tumorigenic than the parent PAH.

## ACKNOWLEDGMENTS

This research was sponsored by NIH Grants CA20076 and CA-11968, American Cancer Society Grant BC-132D; and the Office of Health and Environmental Research, U.S. Department of Energy, under contract W-7405-eng-26 with the Union Carbide Corporation.

We thank Greta Gleason, Gerald Mills, Leroy Hardin, and Linda Ewald for their help in the tumorigenesis studies.

## REFERENCES

1. Adelfang, J. L., and Daub, G. H. (1955): The synthesis of 10-methyl-3,4-benzopyrene and 8,10-dimethyl-3,4-benzopyrene. *J. Am. Chem. Soc.* 77:3297-3300.
2. Adelfang, J. L., and Daub, G. H. (1957): The synthesis of 2-methyl-3,4-benzopyrene, and 8,9-dimethyl-3,4-benzopyrene. *J. Am. Chem. Soc.* 79:1751-1754.
3. Adelfang, J. L., and Daub, G. H. (1958): The synthesis of 5,8-dimethyl-3,4-benzopyrene, 5,10-dimethyl-3,4-benzopyrene and 5,8,10-methyl-3,4-benzopyrene. *J. Am. Chem. Soc.* 80:1405-1409.
4. Adelfang, J. L., and Daub, G. H. (1958): 9,10-dimethyl-3,4-benzopyrene. *J. Org. Chem.* 23:749.

5. Buening, M. K., Levin. W., Wood, A. W., Chang, R. L., Yagi, H., Karle, J. M., Jerina, D. M., and Conney, A. H. (1979): Tumorigenicity of the dihydrodiol of dibenzo(a,h)anthracene on mouse skin and in newborn mice. *Cancer Res.* 39:1310-1314.
6. Chouroulinkov, I., Gentil, A., Grover, P. L., and Sims, P. (1976): Tumor-initiating activities on mouse skin on dihydrodiols derived from benzo(a)pyrene. *Br. J. Cancer* 34:523-532.
7. Chouroulinkov, I., Gentil, A., Tierney, B., Grover, P., and Sims, P. (1977): The metabolic activation of 7-methylbenz(a)anthracene in mouse skin: High tumor-initiating activity of the 3,4-dihydrodiol. *Cancer Letters* 3:247-253.
8. Comp, J. L., and Daub, G. H. (1958): The synthesis of 7-methyl-, 10-methyl-, 6,7-dimethyl- and 7,10-dimethyl-3,4-benzopyrene. *J. Am. Chem. Soc.* 80:6049-6052.
9. Doyle, W. C., and Daub, G. H. (1955): The synthesis of 1-methyl-3,4-benzopyrene and 1,8-dimethyl-3,4-benzopyrene. *J. Am. Chem. Soc.* 80:5252-5255.
10. Fu, P. P., and Harvey, R. G. (1978): Synthesis of the chrysene bay-region anti-diol-epoxide from chrysene. *J.C.S. Chem. Commun.* 585-586.
11. Fu, P. P., and Harvey, R. G.: Synthesis of the dihydrodiols and diol-epoxides of chrysene from chrysene. *J. Org. Chem.* (in press).
12. Harvey, R. G. and Dunne, F. G. (1978): Multiple regions of metabolic activation of carcinogenic hydrocarbons. *Nature (London)* 273:566-568.
13. Harvey, R. G., and Fu, P. P.: Synthesis of oxidized metabolites of dibenz(a,c)-anthracene. *J. Org. Chem.* (in press).
14. Harvey, R. G., Lee, H. M., and Shyamasundar, N. (1979): Synthesis of the dihydrodiols and diol-epoxides of benzo(e)pyrene and triphenylene. *J. Org. Chem.* 44:78-83.
15. Hecht, S. S., Hirota, N., Loy, M., and Hoffmann, D. (1978): Tumor-initiating activity of fluorinated 5-methylchrysenes. *Cancer Res.* 38:1694-1698.
16. Hecht, S. S., Loy, M., Mazzarese, R., and Hoffmann, D. (1978): Synthesis and mutagenicity of modified chrysenes related to the carcinogen, 5-methyl-chrysene. *J. Med. Chem.* 21:38-44.
17. Heidelberger, C. (1975): Chemical carcinogenesis. *Ann. Rev. Biochem.* 44:79-121.
18. Huberman, E., and Slaga, T. J. (1979): Mutagenicity and tumor-initiating activity of fluorinated 7,12-dimethylbenz(a)anthracene. *Cancer Res.* 39:411-414.
19. Iyer, R. P., Lyga, W., Secrist III, A., Daub, G. H., and Slaga, T. J.:

Comparative tumor initiating activity of methylated benzo(a)pyrene derivatives in mouse skin (submitted to *Cancer Research*).
20. Jerina, D. M., and Daly, J. W. (1974): Arene oxides: A new aspect of drug metabolism. *Science* 185:573-582.
21. Jerina, D. M., and Daly, J. W. (1976): Oxidation at carbon. In: *Drug Metabolism*, D. V. Parke and R. L. Smith, Eds., pp. 15-33, Taylor and Frances Ltd., London.
22. Jerina, D. M., Lehr, R. E., Yagi, H., Hernandez, O., Dansette, P. M., Wislocki, P. G., Wood, A. W., Chang, R. L., Levin, W., and Conney, A. H. (1976): Mutagenicity of benzo(a)pyrene derivatives and the description of a quantum mechanical model which predicts the ease of carbonium ion formation from diol epoxides. In: *In Vitro Metabolic Activation and Mutagenesis Testing*, F. J. deSerres, J. R. Fouts, J. R. Bend, and R. M. Philpot, Eds., pp. 159-177, Elsevier/ North Holland Biomedical Press, Amsterdam.
23. Kapitulnik, J., Levin, W., Conney, A. H., Yagi, H., and Jerina, D. M. (1977): Benzo(a)pyrene 7,8-dihydrodiol is more carcinogenic than benzo(a)pyrene in newborn mice. *Nature* 266:378-380.
24. Kapitulnik, J., Wislocki, P. G., Levin, W., Yagi, H., Jerina, D. M., and Conney, A. H. (1978): Tumorigenicity studies with diol-epoxides of benzo(a)pyrene which indicate that $(\pm)$-trans-7$\beta$, 8$\alpha$-dihydroxy-9$\alpha$, 10$\alpha$-epoxy-7,8,9,10-tetrahydrobenzo(a)pyrene is an ultimate carcinogen in newborn mice. *Cancer Res.* 38:354-358.
25. Lacassagne, A., Zajdela, F., Buu-Hoi, N. P., Chalvet, O., et Daub, G. H. (1968): Activité cancérogène élevée des mono-, di-, et triméthyl-benzo(a)pyrenes. *Int. J. Cancer* 3:238-248.
26. Lee, H. M., and Harvey, R. G.: Synthesis of biologically active metabolites of 7-methylbenz(a)anthracene. *J. Org. Chem.* (in press).
27. Lee, H. M., and Harvey, R. G.: Synthesis of biologically active metabolites of dibenz(a)anthracene. *J. Org. Chem.* (submitted for publication).
28. Levin, W., Buening, M. K., Wood, A. W., Chang, R. L., Thakker, D. R., Jerina, D. M., and Conney, A. H. (1979): Tumorigenic activity of 3-methyl-cholanthrene metabolites on mouse skin and in newborn mice. *Cancer Res.* 39:3549-3553.
29. Levin, W., Thakker, D. R., Wood, A. W., Chang, R. L., Lehr, R. E., Jerina, D. M., and Conney, A. H. (1978): Evidence that benzo(a)-anthracene 3,4-diol-1,2-epoxide is an ultimate carcinogen on mouse skin. *Cancer Res.* 38:1705-1710.
30. Levin, W., Wood, A. W., Chang, R. L., Slaga, T. J., Yagi, H., Jerina, D. M., And Conney, A. H. (1977): Marked differences in the tumor-initiating activity of optically pure (+)- and (-)-7,8-dihydroxy-7,8-dihydro-benzo(a)pyrene in mouse skin. *Cancer Res.* 37:2721-2725.

31. Levin, W., Wood, A. W., Chang, R. L., Yagi, H., Mah, H. D., Jerina, D. M., and Conney, A. H. (1978): Evidence for bay region activation of chrysene 1,2-dihydrodiol to an ultimate carcinogen. *Cancer Res.* 38:1831-1834.
32. Patton, J. W., and Daub, G. H. (1957): The synthesis of 2-methyl-3,4-benzo-pyrene, and 2,8-dimethyl-3,4-benzopyrene. *J. Am. Chem. Soc.* 79:709-711.
33. Sims, P., and Grover, P. L. (1974): Epoxides in polycyclic aromatic hydrocarbon metabolism and carcinogenesis. *Advan. Cancer Res.* 20:165-274.
34. Slaga, T. J., Bracken, W. B., Gleason, G., Levin, W., Yagi, H., Jerina, D. M., and Conney, A. H. (1979): Marked differences in the skin tumor initiating activities of the optical enantiomers of the diastereomeric benzo(a)pyrene 7,8-diol-9,10-epoxides. *Cancer Res.* 39:67-71.
35. Slaga, T. J., Bracken, W. M., Viaje, A., Levin. W., Yagi, H., Jerina, D. M., and Conney, A. H. (1977): Comparison of the tumor-initiating activities of benzo(a)pyrene arene oxides and diol-epoxides. *Cancer Res.* 37:4130-4133.
36. Slaga, T. J., Gleason, G. L., DiGiovanni, J., Sukumaran, K. B., and Harvey, R. G. (1979): Potent tumor-initiating activity of the 3,4-dihydrodiol of 7,12-diemethyl-benz(a)anthracene in mouse skin. *Cancer Res.* 39:1934-1936.
37. Slaga, T. J., Gleason, G. L., and Hardin. L. (1979): Comparison of the skin tumor initiating activity of 3-methylcholanthrene and 3,11-dimethylcholanthrene in mice, *Cancer Letters* 7:97-102.
38. Slaga, T. J., Huberman, E., DiGiovanni, J., Gleason, G., and Harvey, R. G. (1979): The importance of the "bay region" diol-epoxide in 7,12-dimethylbenz(a)anthracene skin tumor initiation and mutagenesis. *Cancer Letters* 6:213-220.
39. Slaga, T. J., Huberman, E., Selkirk, J. K., Harvey, R. G., and Bracken, W. M. (1978): Carcinogenicity and mutagenicity of benz-(a)anthracene diols and diol-epoxides. *Cancer Res.* 38:1699-1704.
40. Wood, A. W., Levin, W., Chang, R. L., Lehr, R. E., Schaefer-Redder, M., Karle, J. M., Jerina, D. M., and Conney, A. H. (1977): Tumorigenicity of five dihydrodiols of benz(a)anthracene on mouse skin: exceptional activity of benz(a)anthracene-3,4-diol. *Proc. Natl. Acad. Sci. U.S.A.* 74:3176-3179.
41. Zacharias, D. E., Glusker, J. P., Fu, P. P., and Harvey, R. G. (1979): Molecular structures of the dihydrodiols and diol epoxides of carcinogenic polycyclic aromatic hydrocarbons. X-ray crystallographic and NMR analysis. *J. Am. Chem. Soc.* 101:4043-4051.

42. Yagi, H., Thakker, O. R., Lehr, R. E., Jerina, D. M. (1979): Benzo-ring diol epoxides of benzo(e)pyrene and triphenylene. *J. Org. Chem.* 44:3439-3442.

# LIQUID CHROMATOGRAPHIC DETERMINATION OF BENZO[a]PYRENE IN DIESEL EXHAUST PARTICULATE: VERIFICATION OF THE COLLECTION AND ANALYTICAL METHODS

S. J. Swarin* and R. L. Williams**

*Analytical Chemistry Department
**Environmental Science Department
  General Motors Research Laboratories
  Warren, Michigan 48090

## INTRODUCTION

Polynuclear aromatic hydrocarbons (PAH) are widespread contaminants of the environment. One particular PAH, benzo(a)pyrene (BaP), has been identified as a potential carcinogen and as a general indicator of combustion-related air pollution (1). Consequently, many analytical techniques have been described in the literature (2-9) for the characterization of complex mixtures of PAH and BaP, particularly in environmental samples. However, none of these techniques has been developed into a reliable, sensitive, and routine procedure for the quantitative determination of BaP in automobile exhaust particulate.

As part of our effort to obtain a data base for comparing the BaP emission rates of diesel automobiles with those of catalyst and noncatalyst gasoline automobiles (10), we developed such a method. It is based on high-performance liquid chromatography (LC) to separate BaP from other exhaust particulate components and on fluorescence detection to measure BaP with specificity and sensitivity. In this paper the method is described in some detail, including the experiments used to verify the collection and analytical methods. These experiments dealt with sampling-rate variations, sampling-time variations, filter-storage effects, effects of additional exposure to diesel exhaust gases, comparison of various extraction solvents on spiked-filter samples, and verification of the LC-fluorescence analysis by fluorescence scanning and by analyzing spiked samples.

## EXPERIMENTAL

### Particulate Emissions Measurement Apparatus

Exhaust particulate emissions were collected while a production car equipped with a 5.7-L diesel engine was driven on an electric chassis dynamometer. The entire exhaust was transferred via a 1.5-m-long stainless steel tailpipe extension to a 0.3-m-diameter dilution tunnel, where it was mixed with prefiltered air (Figure 1). The tunnel is assembled in sections joined with O-ring seals and is fabricated of clear polyacrylate, except for the short entrance and transition sections. An annular ring in the entrance section of the tunnel, coincident with the discharge end of the tailpipe extension, creates additional turbulence that promotes mixing of the exhaust gas and dilution air. The total flow rate in the tunnel is controllable over the range 19 to 34 $m^3$/min by means of a variable-sheave pulley on the fan drive. In the experiments described here, the flow rate was 19 $m^3$/min. Figure 2 is a photograph of the apparatus.

During a test, two filter samples were collected simultaneously at the sampling section. The probes for the filter assemblies were 3.0-cm-I.D. stainless steel, sized for isokinetic sampling at a 0.170 $m^3$/min sampling rate and 19 $m^3$/min flow in the dilution tunnel. Because the sampling rate was varied in some experiments, the sampling was not always isokinetic, but departures from isokinetic flow are assumed to be inconsequential in

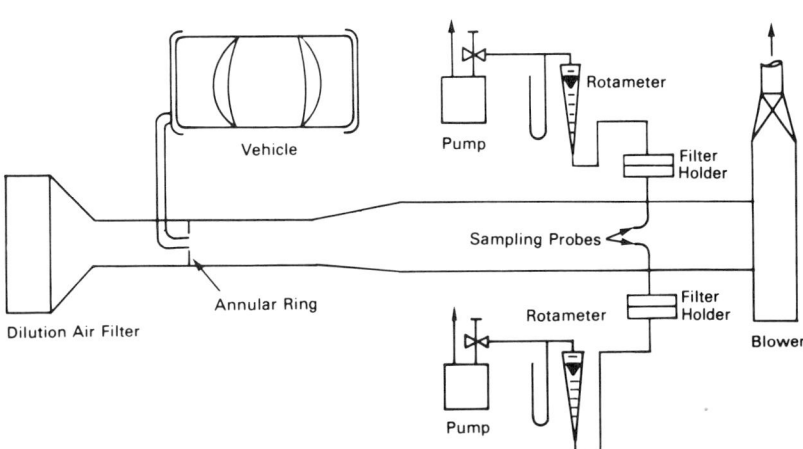

**FIGURE 1.** Schematic of the particulate sampling system.

**FIGURE 2.** Dilution tunnel for measuring particulate exhaust emissions.

view of the submicrometer size of the diesel particulates (11). A photograph of one of the filter assemblies is presented in Figure 3. The filters were 102-mm-diameter (effective diameter 94 mm) Dexiglas glass-fiber

## 774 BaP IN DIESEL EXHAUST

filters, generally pretreated by heating at 500°C for 2 hours. These filters are efficient for collecting particles of all sizes (efficiency $\geq$ 99.97 percent at 0.3 µm), free from extractable material and stable for thermal analysis, but they must be stored at constant temperature and humidity to ensure constant weight.

**FIGURE 3. Probe and filter holder for particulate sampling.**

## Liquid Chromatography Apparatus

A DuPont Model 830 liquid chromatograph (DuPont Instruments, Wilmington, Delaware) equipped with a DuPont Model 833 flow controller and fixed-wavelength (254 nm) ultraviolet detector was used. A pneumatically actuated Valco valve (Houston, Texas) with a 27-$\mu$l loop was used for sample injection. The chromatographic effluent passed through an Aminco-Keirs spectrophosphorimeter (American Instrument Co., Silver Spring, Maryland) which was used for fluorescence detection. This detector was equipped with an 18-$\mu$l flow cell supplied by the manufacturer (Part No. J4-7476). A Perkin Elmer Model 2-1 liquid chromatograph (Perkin Elmer, Norwalk, Connecticut) equipped with a Rheodyne Model 7105 syringe-loading sample injection valve (Berkeley, California) was also used with the fluorescence detector.

## Reagents

"Distilled-in-glass"-grade acetonitrile, hexane, methylene chlorine, methanol, and benzene were obtained from Burdick and Jackson (Muskegon, Michigan). Deionized water was also used. Absolute ethanol was obtained from U.S. Industrial Chemicals (Louisville, Kentucky).

PAH standards were obtained from various sources and were used without purification except for BaP, which was obtained from Eastman Kodak (Rochester, New York) and recrystallized twice from benzene-methanol before use. Stock solutions of PAH standards were prepared at $\mu$g/ml concentrations in methanol and stored in the dark. Appropriate dilutions were prepared to obtain concentrations of pg/$\mu$l for LC standards.

The chromatographic column was 25 cm x 4.6 mm Zorbax ODS (DuPont Instruments) preceded by a 4 cm x 3.2 mm RP-18 microparticulate precolumn (Altex Scientific, Berkeley, California).

## Procedure

The filter samples were extracted (soxhlet) for 3 hours (60 cycles) with a solution of benzene and ethanol (80:20). Solvent was removed from the extracted material at room temperature with a stream of nitrogen. The extracted material was brought to constant weight (24°C and 45 percent relative humidity) and weighed. The extracts were then re-extracted with two 4-ml portions of a solution of hexane and methylene chloride (65:35) using ultrasonic stirring at room temperature. This step removed the highly polar organics and inorganics which are soluble in benzene-ethanol solution. The combined hexane-methylene chloride extracts were then evaporated to dryness at 40°C in the dark under a stream of nitrogen using a Kontes evaporative concentrator (Vineland, New Jersey). The nitrogen purge that is part of the evaporative concentrator was not used; instead nitrogen was directed to flow over the surface of the solutions. The extracted material was redissolved in a measured volume (nominally 1.0 ml) of acetonitrile for final analysis.

A 27-$\mu$l aliquot of this solution was injected into the LC instrument using the Zorbax ODS column with a solution of acetonitrile and deionized water (90:10) as the mobile phase at a 1.0 ml/min flow rate. The mobile phase reservoir was continuously purged with argon to eliminate oxygen. The fluorescence detector was operated with excitation and emission wavelengths of 383 nm and 430 nm, respectively. Quantitation was made by comparing the peak height of the BaP fluorescence signal to that of external standards of known concentration. Best results were obtained when the sample was "bracketed" by injecting, in duplicate, BaP standards with concentrations just below and above that in the sample. These standards were injected immediately after the injection of the sample (i.e., 2.0, 2.8, 3.6, and 4.4 minutes after injection of the sample) so that their elution immediately followed that of the sample on the same LC pump stroke.

## RESULTS AND DISCUSSION

The method we developed for the determination of BaP in diesel exhaust particulate follows the scheme: sample collection on filters, extraction of the filters to remove the solubles (including BaP), and then LC separation and fluorescence detection. However, the development and verification of the method was performed in the reverse order of this sequence because testing of extraction efficiency depends on confirmed LC separation and fluorescence detection, and testing of sample collection depends on confirmed extraction efficiency. The Results and Discussion of this report considers the analytical method in this reversed order.

### Verification of the LC Separation and Detection Method

Although LC with fluorescence detection has frequently been cited as a suitable analytical method for the determination of PAH in complex environmental samples, very few data have been reported on the suitability of this technique for the determination of BaP in automobile exhaust samples. We performed several experiments designed to confirm that BaP was truly the species separated and detected by the LC procedure. These experiments cover the details of (1) chromatographic resolution, (2) fluorescence scanning, (3) analysis of spiked samples, and (4) storage of sample solutions.

*Chromatographic Resolution*

In this experiment, known PAH compounds were injected into the LC under the conditions given in the Procedure section, and their retention times and fluorescence intensities were recorded. Table 1 shows the results of this study, normalized to the same quantity of BaP.

Of the 19 compounds tested, none interfered with BaP chromatographically. The results for the compounds which eluted closest to BaP, dibenz(a,c)anthracene and dibenz(a,h)anthracene, are shown in Figure 4, along with those for benzo(k) fluoranthene and 9,10- dimethylbenz(a) anthracene, which have some fluorescence at the analytical wavelengths. As shown in Figure 4, all of these potential interferences are well resolved from BaP chromatographically. The relative retention times reported in Table 1 are in close agreement with those reported by Das and Thomas (2).

TABLE 1. Relative Retention Times and Fluorescence Intensities of Known PAH versus BaP

| Compound | Relative Retention | Relative Fluorescence Intensity |
|---|---|---|
| Phenanthrene | 0.377 | 0 |
| Anthracene | 0.402 | 20 |
| Fluoranthene | 0.461 | 2 |
| Pyrene | 0.529 | 0 |
| Triphenylene | 0.564 | 0 |
| Benz(a)anthracene | 0.611 | 0 |
| Chrysene | 0.622 | 0 |
| Benz(b)anthracene | 0.755 | 0 |
| Benzo(b)fluoranthene | 0.830 | 5 |
| Benzo(k)fluoranthene | 0.867 | 120 |
| Benzo(e)pyrene | 0.871 | 0 |
| Perylene | 0.882 | 50 |
| 9,10-Dimethylbenz(a)anthracene | 0.921 | 20 |
| Dibenz(a,c)anthracene | 0.963 | 0 |
| Benzo(a)pyrene | 1.000 | 100 |
| Dibenz(a,h)anthracene | 1.068 | 0 |
| Benzo(ghi)perylene | 1.457 | 15 |
| Dibenzo(a,l)pyrene | 1.614 | 0 |
| Anthanthrene | 1.751 | 60 |
| Dibenzo(a,h)pyrene | 2.425 | 2 |

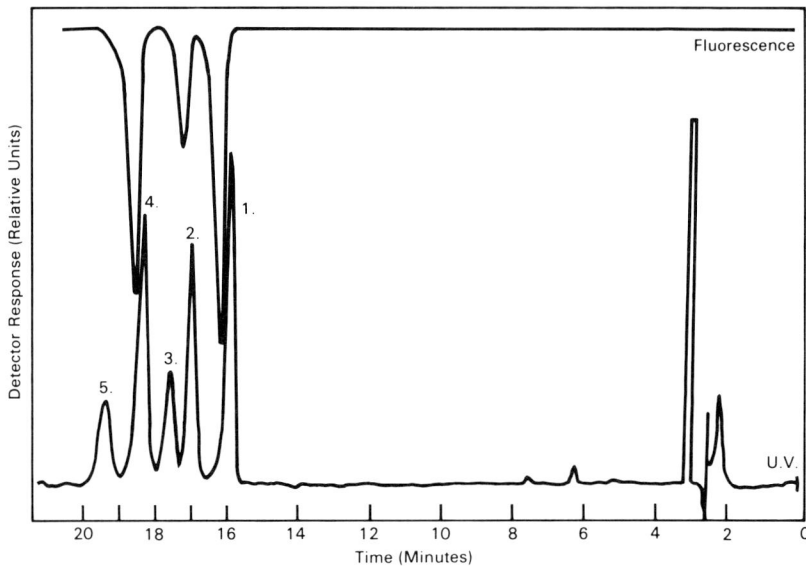

**FIGURE 4.** HPLC separation of known PAH with UV and fluorescence detection. Chromatographic conditions: 4.6-mm I.D. x 25-cm Zorbax ODS column, 90:10 (v/v) $CH_3CN:H_2O$ flow rate 1.0 ml/min. PAH separated: (1) benzo(k)fluoranthene, (2) 9,10 dimethylbenz(a) anthracene, (3) dibenz(a,c)anthracene, (4) benzo(a)pyrene, and (5) dibenz(a,h) anthracene. Approximately 25 ng of each component, except 9,10 dimethylbenz(a) anthracene, 50 ng. U.V. detector (at 254 nm) attenuation x 1. Fluorescence detector (383 nm excitation, 430 nm emission) attenuation x 30.

We also confirmed their observation that Zorbax ODS is unique in its separating ability for PAH. Attempts at complete resolution of the PAH in the BaP region using other $C^{18}$ reverse phase columns (eg $\mu$Bondapak-$C^{18}$, Partisil ODS-2, Lichrosorb-RP18, and Chromegabond-$C^{18}$) and various mobile phases were not successful.

*Fluorescence Scanning*

In this experiment designed to confirm the identity of the BaP peak in particulate samples, the emission and excitation scanning capabilities of the spectrofluorimeter were used. A sample of diesel exhaust particulate extract, prepared as described in the Procedure section, produced the liquid chromatogram shown in Figure 5A. A second aliquot of the same sample was then injected, and the flow was stopped at points A, B, C, and D so that excitation and emission spectra could be obtained. These points were established by precisely measuring the volume of eluted mobile phase i.e., 17.8, 18.1, 18.4, and 18.9 ml eluted. The emission spectra obtained at these points with an excitation wavelength of 383 nm are shown in Figure 5B. These spectra exactly match those obtained when a

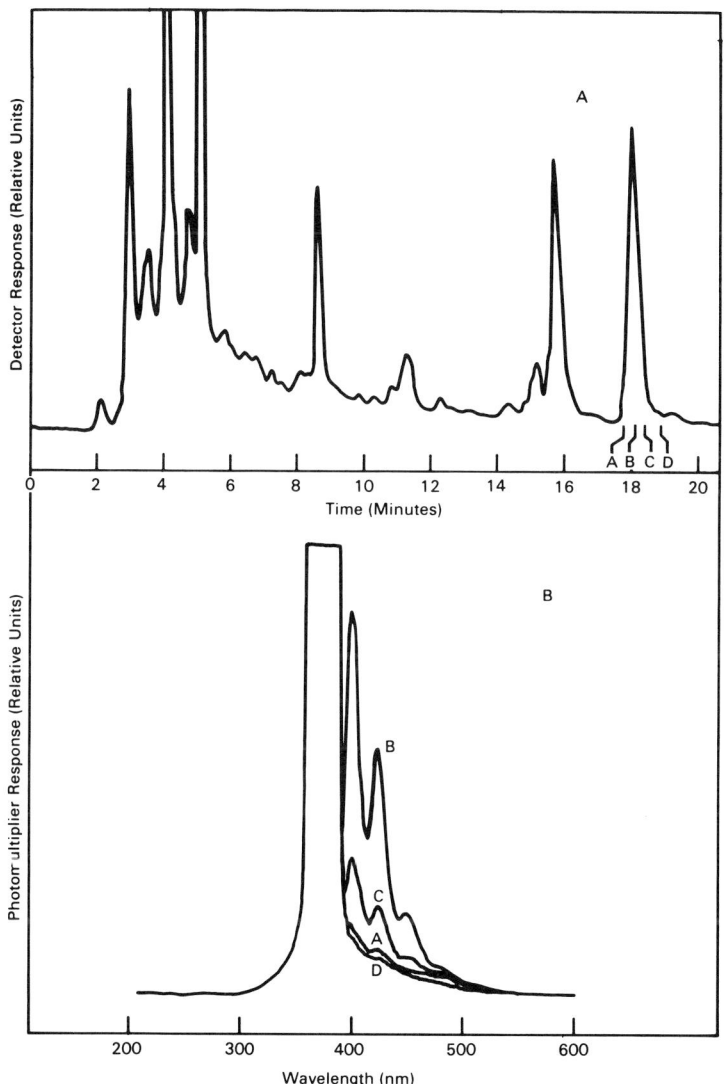

FIGURE 5. (A) HPLC separation of diesel exhaust particulate extract showing points where flow was stopped for fluorescence scanning. (Chromatographic conditions same as Figure 4.). (B) Fluorescence emission spectra (at 383 nm excitation) taken at points A, B, C, and D in part (A).

BaP standard is run under the same conditions. Although fluorescence emission spectra are not unique (in the sense that fingerprints are unique), the exact match of spectra is strong evidence that the BaP peak is free of contamination.

*Analysis of Spiked Samples*

This experiment was designed to confirm the identy of the BaP peak and utilized a series of samples of diesel exhaust particulate extracts which were analyzed and then "spiked" with BaP standard and reanalyzed. It should be emphasized that in these experiments the final acetonitrile solutions of diesel exhaust particulate extracts were spiked, so that these experiments relate only to the final chromatographic analysis and not to the recovery of BaP from the original filter sample. Table 2 shows the results of these spiking experiments.

The satisfactory recovery of the "spikes" shown in Table 2 is another indication that the chromatographic separation and fluorescence detection is selective for BaP. When the results of all of these experiments are considered together, the final analytical measurement is evidently selective for BaP.

TABLE 2. Determination of BaP in Spiked Samples

| Sample | BaP (pg/$\mu$l) | | | BaP Determined (pg/$\mu$l) |
|---|---|---|---|---|
| | Original | Added | Total | |
| A | 24.0 | 14.0 | 38.0 | 38.0 |
| B | 3.7 | 2.7 | 6.4 | 6.1 |
| C | 0.54 | 0.93 | 1.47 | 1.50 |

*Storage of Sample Solutions*

One of the objectives of this work was to develop a method for the determination of BaP that was rapid and required as little sample pretreatment as possible. However, this method yields a final acetonitrile solution of the sample which is quite complex (see Figure 5A), and the possibility of BaP decomposition or reaction in this solution needs to be considered. In an experiment designed to study the potential loss of BaP, acetonitrile solutions from several diesel exhaust particulate collections and extractions were divided into two aliquots and stored, one under normal laboratory conditions and the other protected from light by wrapping with duct tape. As shown in Figure 6 there was indeed considerable reaction or decomposition of BaP in this complex acetonitrile solution when it was exposed to light. This decomposition is quite irreproducible and exhibits unusual concentration dependence, as shown in Figure 6. However, when the samples were stored in the dark, or when the samples were analyzed immediately after preparation, good BaP results were obtained.

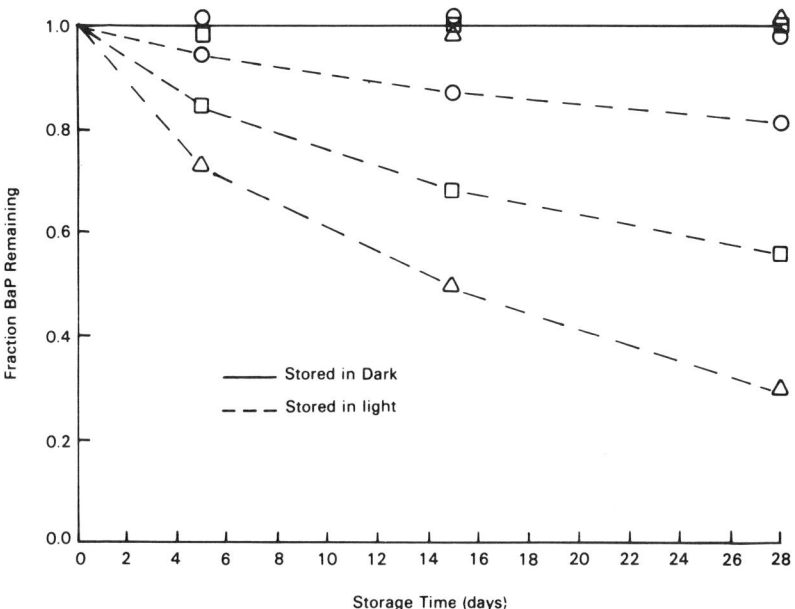

FIGURE 6. Effects of storage on acetonitrile solutions of diesel exhaust particulate extracts —— stored in dark, --- stored in light. Initial solution concentration: O 50 pg/$\mu$l, □ 500 pg/$\mu$l, △ 200 pg/$\mu$l.

## Verification of the Extraction of BaP From Filters

Another important step in the analysis was the recovery of BaP from the original filter sample. We chose to use a 3-hour soxhlet extraction with a solution of benzene and ethanol (80:20) in an attempt to give the optimum amount of organic material in a reasonable time. It was felt that the mixed solvent would extract the nonpolar hydrocarbons and PAH as well as acids, bases, esters, aldehydes, and ketones. During the course of this work, the U.S. Environmental Protection Agency recommended that methylene chloride be used as the extracting solvent (12). To test the efficiencies of these two solvent systems for the extraction of BaP, a series of extractions was performed using 20-mg portions of diesel exhaust particulate which had been collected in replicate tests using membrane filters. The bulk particulate was wrapped in a fiber-glass filter paper and extracted for 3 hours (60 cycles) using either benzene-ethanol (80:20) or methylene chloride. The sample work-up after extraction was that given in the Procedure section. Table 3 shows the results.

**TABLE 3. Comparison of Quantities of BaP Extracted with Benzene-Ethanol and Methylene Chloride**

| | | ng BaP/mg Particulate | |
|---|---|---|---|
| Trial | Description | Benzene-Ethanol | Methylene Chloride |
| 1 | 3-hr soxhlet | 8.93 | 6.28 |
| 2 | 3-hr soxhlet | 9.41 | 5.50 |
| 2A | Second 3-hr soxhlet with fresh solvent | 0.75 | 0.38 |
| 3 | 3-hr soxhlet | 7.98 | 4.94 |
| 3A | Second 3-hr soxhlet with the other solvent | 0.04 | 3.63 |

Trials 1, 2, and 3 in Table 3 show that methylene chloride extracted only 60 to 70 percent as much of the BaP from diesel exhaust particulate as benzene-ethanol. (Similar experiments using cyclohexane and methanol showed that these solvents extracted only about 40 percent as much of the BaP as benzene-ethanol.) Trial 2A, Table 3, shows that one 3-hour extraction was sufficent to remove $\geq$ 90 percent of the BaP extractable by a given solvent. Finally, Trial 3A (Table 3) confirms that, after one 3-hour soxhlet extraction with methylene chloride, considerable benzene-ethanol-extractable BaP was left on the diesel exhaust particulate. This also indicates that the low results for methylene chloride are not due to decomposition of the BaP in methylene chloride. All of these results confirm the higher efficiency of benzene-ethanol as the extraction solvent.

To further verify the extraction of BaP from filters, several experiments were performed in which the diesel exhaust particulate was "spiked" with BaP before extraction. These experiments were performed in parallel with those shown in Table 3, so that a "total expected" BaP value could be calculated from the benzene-ethanol results in Table 3 plus the spike. Table 4 shows these results.

**TABLE 4. Recovery of BaP from Spiked Samples of Diesel Exhaust Particulate Extracted with Benzene-Ethanol and Methylene Chloride**

| Trial | Description | ng BaP/mg Particulate | | | |
|---|---|---|---|---|---|
| | | Benzene-Ethanol | | Methylene Chloride | |
| | | Total Expected | Found | Total Expected | Found |
| 1 | 3-hr soxhlet | 13.2 | 13.9 | 12.5 | 8.16 |
| 2 | 3-hr soxhlet | 15.1 | 14.8 | 16.6 | 14.3 |
| 3 | 3-hr soxhlet | 14.8 | 14.2 | 14.8 | 12.1 |
| 3A | Second 3-hr soxhlet with the other solvent | — | 0.08 | — | 3.05 |

The benzene-ethanol results in Table 4 show the excellent recovery of the spikes obtained with this solvent. They also show the excellent recovery obtained for the entire analytical procedure. The inconsistency in the methylene chloride results in Table 4 show that this solvent is unsuitable for this analysis because of the inconsistency of the results; the BaP found varies from 65 to 86 percent of the total additional BaP after a methylene chloride extraction. Finally, Trial 3A again shows that benzene-ethanol effectively extracts additional BaP after a methylene chloride extraction.

## Verification of the Filter Sampling Technique

Although filter sampling is commonly used to collect samples of particulate emissions from a variety of sources (13-17), very little is known about changes in the particles which might occur during the collection process, particularly when the source is a diesel-powered automobile. This part of our report describes a series of experiments designed to examine some of the factors that might cause changes in diesel particulate during the collection and analysis of the filter samples. These experiments include (1) sampling-rate variations, (2) sampling-time variations, (3) additional exposure to diesel exhaust gases, and (4) stability during filter-sample storage.

### Sampling-Rate Variations

Sampling rate is one factor that could change the quantity and composition of diesel particulate collected on a filter. A set of nine filters was collected from the dilution tunnel at sampling rates ranging from 0.062 to

0.276 m³/min, each with a sampling period of 10.0 minutes. The concentration of particulate in the sample stream was calculated from the mass on the filter and the volume of diluted exhaust was sampled. The extractable percentage (benzene-ethanol solvent) and the BaP concentration in the particulate were also determined. Table 5 shows the results of the sampling rate experiment.

TABLE 5. Effect of Sampling Rate on Diesel Exhaust Particulate

| Sample | Sampling Rate[a] (m³/min) | Particulate Mass Concentration (mg/m³) | Extractable Percentage | BaP Concentration (ng/mg) |
|---|---|---|---|---|
| 1 | 0.062 | 9.91 | 32.8 | 8.0 |
| 2 | 0.092 | 9.70 | 38.9 | 9.4 |
| 3 | 0.120 | 9.81 | 37.9 | 11.3 |
| 4 | 0.149 | 9.76 | 35.6 | 9.8 |
| 5 | 0.175 | 9.81 | 36.6 | 9.9 |
| 6 | 0.200 | 10.09 | 36.7 | 9.8 |
| 7 | 0.229 | 10.17 | 36.9 | 11.4 |
| 8 | 0.252 | 9.93 | 34.4 | 10.8 |
| 9 | 0.276 | 9.79 | 36.2 | 11.2 |
| | | 9.89 ± 0.16[b] | 36.2 ± 1.8 | 10.2 ± 1.1 |

[a]Sampling period 10.0 minutes.
[b]One standard deviation.

If filter efficiency is low, we would expect the efficiency to increase with loading of particles on the filter. However, for a filter with high efficiency, the loading rate on the filter will be proportional to the sampling rate. The mass concentration of particulate in the stream sampled was calculated (see Table 5) for each filter used in this study. The mass concentration was 9.89 ± 0.16 mg/m³, and there was no discernible trend with sampling rate, which confirms the high efficiency of the Dexiglas filters.

For the range of sampling rates studied, the extractable percentage was 36.2 ± 1.8, and the BaP content was 10.2 ± 1.1 ng/mg. Neither the extractable percentage, which would reflect adsorption, nor the BaP concentration, which would reflect chemical reactions, showed any dependence on sampling rate.

*Sampling-Time Variations*

A set of eight filter samples was collected from the dilution tunnel at a constant sampling rate of 0.20 m$^3$/min, but the sampling period was varied from 1 to 25 minutes. Table 6 shows the concentration of particulate in the sample stream, the extractable percentage, and the BaP concentration. The range of sampling times covers the practical range commonly used in vehicle testing.

**TABLE 6. Effect of Sampling Period on Diesel Exhaust Particulate**

| Sample | Sampling Period[a] (min) | Particulate Mass Concentration (mg/m$^3$) | Extractable Percentage | BaP Concentration (ng/mg) |
|---|---|---|---|---|
| 1 | 1 | 10.03 | 43.0 | 7.5 |
| 2 | 3 | 9.77 | 43.4 | 9.9 |
| 3 | 5 | 9.76 | 38.5 | 10.3 |
| 4 | 7 | 9.23 | 37.7 | 10.6 |
| 5 | 10 | 9.14 | 37.5 | 9.3 |
| 6 | 15 | 9.07 | 40.1 | 10.4 |
| 7 | 20 | 9.43 | 40.7 | 9.1 |
| 8 | 25 | 9.27 | 38.5 | 9.2 |
|   |   | 9.46 ± 0.35[b] | 39.9 ± 2.3 | 9.5 ± 1.0 |

[a]Sampling rate, 0.20 m$^3$/minute.
[b]One standard deviation.

Any change in collection efficiency that might occur during extended sampling periods would lead to changes in the apparent mass concentration in the diluted exhaust. The mass concentration was 9.46 ± 0.35 mg/m$^3$, with no dependence on the duration of the sampling period. Processes such as adsorption and chemical reaction could affect the composition of the particulate on the filter, because particles collected in the initial layer of particulate are exposed to gaseous emission components transmitted by particle layers collected later. As can be seen in Table 6, the extractable percentage and the BaP concentration were again constant with no dependence on the duration of the sampling period.

Therefore, variations in flow rate and in the duration of the sampling period have no effect on the particle mass concentration, the extractable percentage, or the BaP concentration. Over the practical range of these two variables we found no evidence for filter efficiency changes, absorption, or chemical reaction.

*Additional Exposure to Diesel Exhaust Gases*

A set of four identical filter samples was collected from the dilution tunnel to study the effect of additional exposure of particulate to diesel exhaust gases. The sampling period for these filters was 10 minutes during which about 20 mg of particulate was collected. To complete the experiment, two filter holders were connected in series for additional sampling. A clean filter was placed in the first holder to remove the particulate. Two of the previously loaded filters were placed one at a time in the second holder for exposure to diluted exhaust gas. One was exposed for 1 minute, and a second filter was exposed for 10 minutes. The extractable percentage and the BaP concentration in the particulate were determined on the four filters (Table 7, Set 1). This experiment was repeated using lighter loaded filters, collected under different driving conditions (Table 7, Set 2).

In both sets the additional exposure to diesel exhaust gases caused no significant change in mass on the filter, in the extractable percentage or in the BaP concentration. Again, it appears that interactions between the diesel exhaust gases and particles in this sampling system have reached a steady state before the particles are collected on the Dexiglas filter.

TABLE 7. Effect of Additional Exposure to Diesel Exhaust Gases

| Sample | | Total Particulate (mg) | Additional Exposure (min) | Extractable Percentage | BaP Concentration (ng/mg) |
|---|---|---|---|---|---|
| Set 1 | 1 | 19.88 | 0 | 33.3 | 8.9 |
| | 2 | 19.83 | 0 | 33.0 | 8.2 |
| | 3 | 20.01 | 1 | 32.7 | 8.6 |
| | 4 | 20.39 | 10 | 30.6 | 8.8 |
| Set 2 | 5 | 7.53 | 0 | 30.2 | 4.9 |
| | 6 | 7.43 | 1 | 31.9 | 5.2 |
| | 7 | 7.52 | 10 | 31.3 | 5.4 |

## Stability During Filter-Sample Storage

Eleven additional replicate filter samples were collected from the dilution tunnel for a filter storage study. The amount of particulate collected on each filter was 7.5 ± 0.2 mg. These filter samples were stored in a room at 24 ± 0.5°C and 45 ± 1 percent relative humidity and with normal fluorescent light. Periodically, up to 150 days after collection, filters were randomly selected for analysis. Table 8 lists the extractable percentage and the BaP concentration.

Although the extractable percentage had a standard deviation of 4.0 percent, it showed no systematic change with storage time. The total BaP loss in 150 days was 67 percent and the BaP concentration decreased linearly with time at a rate of 0.046 ng/mg/day, as shown in Figure 7. Although long-term storage did cause a decrease in the BaP concentration, the loss in 20 days of storage was less than one standard deviation normally found for BaP in replicate samples analyzed on the same day. To avoid excessive loss of BaP by unidentified processes, the storage conditions should be carefully controlled, and good practice would dictate minimizing storage time.

**TABLE 8. Effect of Filter-Sample Storage of Diesel Exhaust Particulate**

| Sample | Storage Period (days) | Extractable Percentage | BaP Concentration (ng/mg) |
|---|---|---|---|
| 1 | 9 | 34.1 | 12.1 |
| 2 | 9 | 32.9 | 11.1 |
| 3 | 9 | 32.2 | 11.1 |
| 4 | 9 | 32.0 | 10.9 |
| 5 | 28 | 38.8 | 9.8 |
| 6 | 28 | 39.2 | 9.0 |
| 7 | 28 | 40.0 | 10.1 |
| 8 | 91 | 44.6 | 7.7 |
| 9 | 91 | 40.1 | 6.3 |
| 10 | 147 | 36.5 | 5.5 |
| 11 | 147 | 37.8 | 4.3 |
|  |  | 37.1 ± 4.0[a] |  |

[a]One standard deviation.

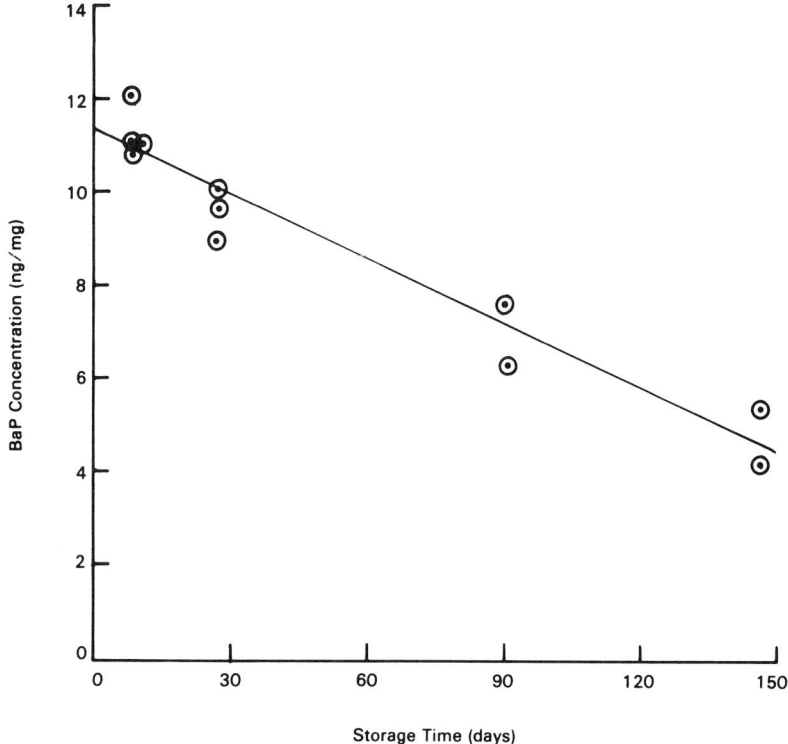

**FIGURE 7.** Effect of filter sample storage on BaP concentration.

## CONCLUSIONS

A rapid, reliable, and sensitive method for the determination of BaP in diesel exhaust particulates has been developed. The detection limit of the method is 5 pg/injection or 200 pg/filter, and the range is to 3000 ng/filter. The precision for the total analysis (from sample collection, through extraction, chromatography, and detection) is ± 10 percent (one standard deviation) at the 10 ng/mg level. The method has been verified by fluorescence scanning and analysis of spiked samples.

A solution of benzene and ethanol (80:20) has been shown to extract more BaP from the diesel exhaust particulate matrix than methylene chloride, cyclohexane, or methanol.

These results suggest that diesel exhaust particles can be efficiently sampled from a dilution tunnel onto Dexiglas filters without changing their extractable percentage or their BaP content. Variations in sampling rate and sampling time have shown no evidence for changes in the filtering efficiency with the thickness of the particulate layer and no evidence of adsorption-desorption during the filtering process, which suggests that the particles and gases had equilibrated before the mixture reached the filter. Furthermore, reexposing the filtered particulate to diesel exhaust gases produced no evidence of chemical reactions that might cause loss of BaP associated with the particles.

Filter storage of diesel particulate in laboratory light does produce a loss of BaP as does storage of diesel exhaust particulate solutions. However, good practice would dictate minimizing storage time between collection and analysis and protecting exhaust particulate solutions from light.

## ACKNOWLEDGMENT

Guidance and advice from the late C. R. Begeman was appreciated. The authors also appreciate the assistance of K. G. Kennedy, G. M. Simon, and especially A. I. Ricci who performed many of the experiments discussed here. The authors also appreciate the assistance of S. G. Anderson and J. R. Collins who operated the emissions facility.

## REFERENCES

1. National Academy of Sciences (1972): Particulate polycyclic organic matter. Committee on Biologic Effects of Atmospheric Pollutants, Division of Medical Sciences, National Research Council, NAS, Washington, D.C.
2. Das, B. S., and Thomas, G. H. (1978): Fluorescence detection in high performance liquid chromatographic determination of polycyclic aromatic hydrocarbons. *Anal. Chem.* 50:967.
3. Doran, T., and McTaggart, N. G. The combined uses of high efficiency liquid chromatography and capillary gas chromatography for the determination of polycyclic aromatic hydrocarbons in automotive exhaust condensates and other hydrocarbon mixtures. *J. Chromatogr. Sci.* 12:715.
4. Seymour, M. J., Cooper, C. V., and Teass, A. W. (1976): HPLC-fluorometric determination of benzo(a)pyrene in airborne particulate coke-oven emissions. Paper presented at American Chemical Society Meeting, San Francisco, California.
5. Jentoft, R. E., and Gouw, T. H. Analysis of polynuclear aromatic hydrocarbons in automobile exhaust by supercritical fluid chromatography. *Anal. Chem.* 48:2195.

6. Bjorseth, A. (1977): Analysis of polycyclic aromatic hydrocarbons in particulate matter by glass capillary gas chromatography. *Anal. Chem. Acta* 94:21.
7. Grimmer, G. (1977): *Air Pollution and Cancer in Man,* IARC No. 16, Lyon, pp. 29.
8. Golden, C., and Sawicki, E. (1978): Determination of benzo(a)pyrene and other polynuclear aromatic hydrocarbons in airborne particulate material by ultrasonic extraction and reverse phase HPLC with fluorescence techniques. Paper presented at Pittsburgh Conference, Cleveland, Ohio; submitted to *Analytical Letters.*
9. Seizinger, D. E. (1978): Automated fluorescence measurement of BaP levels in diesel exhaust. Paper presented at Symposium on Polynuclear Aromatic Hydrocarbons, Battelle, Columbus, Ohio.
10. Williams, R. L., and Swarin, J. (1979): Benzo(a)pyrene emissions from gasoline and diesel automobiles. Paper No. 790419, March 1, 1979, Society of Automotive Engineers, Inc., Warrendale, Pennsylvania 15096.
11. Groblicki, P., and Begeman, C. R. (1979): Particle size variation in diesel car exhaust. Paper No. 790421, March 1, 1979, Society of Automotive Engineers, Inc., Warrendale, Pennsylvania 15096.
12. Zweidinger, R. B., Tejada, S. B., Dropkins, D., Huisingh, J., and Claxton, L. (1978): Characterization of extractable organics in diesel exhaust particulate. Paper presented at Symposium on Diesel Particulate Emissions Measurement Characterization, Ann Arbor, Michigan, May, 1978.
13. Gelman, C., and Meltzer, T. (1979): Membrane filters in air analysis. *Anal. Chem.* 51:22A.
14. Witz, S., and MacPhee, R. (1977): Effect of different types of glass filters on total suspended particulates and their chemical composition. *J. Air Pollut. Constr. Assn.* 27:239.
15. Liu, B., and Lee, K., (1976): Efficiency of membrane and nuclepore filters for submicrometer aerosols. *Environ. Sci. and Technol.* 10:345.
16. Magee, W., Jonas, L., and Anderson, W. (1973): Aerosol filtration by fibrous filter mats: velocity-dependent relationships. *Environ. Sci. and Techno.* 7:1131.
17. Pich, J. (1966): Theory of aerosol filtration. In *Aerosol Science,* C. N. Davies, Ed., Academic Press, New York.

# NORMAL- AND REVERSE-PHASE LIQUID CHROMATOGRAPHIC SEPARATIONS OF POLYCYCLIC AROMATIC HYDROCARBONS

S. A. Wise, W. J. Bonnett, and W. E. May

Organic Analytical Research Division
Center for Analytical Chemistry
National Bureau of Standards
Washington, DC 20234

## INTRODUCTION

High performance liquid chromatography (HPLC) is presently an extremely useful technique for the analysis of complex mixtures of polycyclic aromatic hydrocarbons (PAH). Several HPLC methods for the determination of PAH in air particulates (1,2,11,12), automobile exhausts (7), marine sediments (16) and biota (3), and aqueous samples (9) have been reported.

At present, HPLC does not provide the high separation efficiency of capillary gas chromatography (GC); however, HPLC does offer several advantages over GC for the determination of PAH. The use of UV absorption and fluorescence detection in HPLC offers high sensitivity and specificity. In addition, various column packing materials are available that provide special selectivity for PAH isomers that are difficult to separate by GC. Finally, HPLC provides an excellent prefractionation technique for the isolation of specific groups of PAH prior to GC analysis.

Both normal-phase and reverse-phase HPLC can be utilized for the separation of PAH. The majority of HPLC analyses of PAH are performed using octadecylsilane ($C_{18}$) reverse-phase packing materials. Numerous $C_{18}$ materials are commercially available; however, these materials differ significantly in their selectivity characteristics for PAH. The retention and selectivity characteristics for PAH on several commercial $C_{18}$ columns are reported in the paper.

The retention data for over 80 PAH on two $C_{18}$ columns with different selectivity and on an amino silane ($NH_2$) column are reported. These retention characteristics were utilized to design the sequential analysis of a complex mixture of PAH extracted from urban air particulate matter. A separation of the PAH based on the number of condensed aromatic rings was first accomplished on the $NH_2$ column in the normal phase mode. Each fraction obtained from this separation was then analyzed by GC-MS and/or reverse-phase HPLC on a $C_{18}$ column with fluorescence detection for individual compound identification.

## EXPERIMENTAL

### Calculation of Retention Indices and Selectivity Ratios

The elution volumes of the solutes were measured simultaneously with elution volumes of standards [benzene, naphthalene, phenanthrene, benz(a)anthracene, and benzo(b)chrysene], and the retention index, I, was calculated as described previously by Popl et al (10) using the following equation:

$$\log I_x = \log I_n + \frac{\log R_x - \log R_n}{\log R_{n+1} - \log R_n} \quad (1)$$

where x represents the solute, n and n+1 represent the lower and higher standards, and the R values are the corresponding corrected retention volumes. The standards were assigned the following values (log I): benzene (1), naphthalene (2), phenanthrene (3), benz(a)anthracene (4), and benzo(b)chrysene (5). The retention indices for PAH with retention greater than benzo(b)chrysene were determined from extrapolation of a plot of log R vs log I for the standards. The retention indices are summrized as log I values.

The selectivity ratio, $\alpha$, was calculated as follows:

$$\alpha = \frac{V_{R2} - V_o}{V_{R1} - V_o} \quad (2)$$

where $V_{R1}$ and $V_{R2}$ are the elution volumes of the two PAH and $V_o$ is the unretained volume of the column.

## Chromatographic Columns and Conditions

Retention indices for 17 PAH were determined on eight different $C_{18}$ columns. This study did not attempt to evaluate the selectivity of all commercially available $C_{18}$ columns; the following columns were used: LiChrosorb RP-18 (BrownLee, Rheodyne, Inc., Berkeley, CA), MicroPak CH-10 and MicroPak MCH-10 (Varian Associates, Walnut Creek, CA), Nucleosil 10 $C_{18}$ (Macherey-Nagel, Duren, West Germany), Partisil 5 ODS (Whatman, Clifton, NJ), Radial Pak A (Waters Associates, Milford, MA), Vydac 201TP reverse phase (The Separations Group, Hesperia, CA), and Zorbax ODS (Du Pont Co., Wilmington, DE).

After the preliminary evaluation of the above columns, retention indices were determined for 80 compounds on two $C_{18}$ columns, which represent two different selectivities, using isocratic mobile phases of 70 to 85 percent acetonitrile in water. The retention indices for the same PAH were determined in the normal phase mode on a $\mu$Bondapak $NH_2$ column (Waters Associates) with hexane as the mobile phase.

## RESULTS AND DISCUSSION

The majority of the HPLC separations of PAH reported in the literature have used $C_{18}$ reverse-phase columns. Several of these papers (3,7,16) have also included retention data for a number of PAH on specific $C_{18}$ columns. A comparison of these data indicated that the various $C_{18}$ columns exhibit different selectivities (not just efficiencies) in the separation of PAH. Nice and O'Hare (8) have reported differences in the selectivity of several $C_{18}$ columns for steroid separations. To evaluate the differences in selectivity of $C_{18}$ columns for PAH, the retention indices of 17 PAH (14 of which are on the U.S. Environmental Protection Agency Priority Pollutant list) were determined on eight different columns from seven commercial sources. The results of this study are shown in Table 1. Using the retention data from the experiments summarized in Table 1, the selectivity factors, $\alpha$, for several PAH pairs were calculated. These results are summarized in Table 2.

For nearly complete resolution of two components on these $C_{18}$ columns ($R_s$ = 1.0), the log I values must differ by approximately 0.06 unit, depending on the efficiency of the particular column. The data in Table 1 indicate that, of the columns tested, column G was successful in partially resolving all but two of the 17 PAH [benzo(e)pyrene and benzo(j)fluoranthene]. Ogan et al (9) recently reviewed the literature regarding the separations of three groups of PAH isomers that have traditionally been difficult to separate, i.e., (1) benz(a)anthracene and

TABLE 1. Column Selectivities for Reverse-Phase Separations of PAH (Log I)

| Compound | A[b] | B[b] | C[c] | D[c] | E[b] | F[b] | G[a] | H[b] |
|---|---|---|---|---|---|---|---|---|
| Naphthalene | 2.00 | 2.00 | 2.00 | 2.00 | 2.00 | 2.00 | 2.00 | 2.00 |
| Fluorene | 2.73 | 2.70 | 2.70 | 2.71 | 2.76 | 2.76 | 2.73 | 2.74 |
| Phenanthrene | 3.00 | 3.00 | 3.00 | 3.00 | 3.00 | 3.00 | 3.00 | 3.00 |
| Anthracene | 3.14 | 3.19 | 3.12 | 3.11 | 3.11 | 3.11 | 3.24 | 3.14 |
| Fluoranthene | 3.42 | 3.44 | 3.43 | 3.42 | 3.45 | 3.43 | 3.38 | 3.44 |
| Pyrene | 3.62 | 3.69 | 3.60 | 3.59 | 3.61 | 3.63 | 3.56 | 3.66 |
| Chrysene | 4.00 | 4.04 | 3.98 | 3.98 | 3.96 | 3.97 | 4.10 | 3.99 |
| Benz[a]anthracene | 4.00 | 4.00 | 4.00 | 4.00 | 4.00 | 4.00 | 4.00 | 4.00 |
| Benzo[j]fluoranthene | 4.31 | 4.33 | 4.40 | 4.40 | 4.41 | 4.39 | 4.24 | 4.37 |
| Benzo[b]fluoranthene | 4.40 | 4.41 | 4.47 | 4.47 | 4.47 | 4.46 | 4.30 | 4.45 |
| Benzo[k]fluoranthene | 4.48 | 4.49 | 4.48 | 4.52 | 4.53 | 4.47 | 4.45 | 4.52 |
| Benzo[e]pyrene | 4.40 | 4.43 | 4.50 | 4.50 | 4.48 | 4.48 | 4.25 | 4.48 |
| Benzo[a]pyrene | 4.63 | 4.66 | 4.65 | 4.64 | 4.64 | 4.66 | 4.52 | 4.68 |
| Dibenz[a,h]anthracene | 4.78 | 4.74 | 4.89 | 4.90 | 4.92 | 4.88 | 4.69 | 4.85 |
| Benzo[b]chrysene | 5.00 | 5.00 | 5.00 | 5.00 | 5.00 | 5.00 | 5.00 | 5.00 |
| Benzo[ghi]perylene | 5.05 | 5.05 | 5.18 | 5.17 | 5.14 | 5.10 | 4.71 | 5.16 |
| Indeno[1,2,3-cd]pyrene | 5.03 | 5.04 | 5.14 | 5.09 | 5.13 | 5.10 | 4.83 | 5.13 |

[a]90 percent acetonitrile/10 percent water.
[b]80 percent acetonitrile/20 percent water.
[c]70 percent acetonitrile/30 percent water.

TABLE 2. Selectivity Factors (α) for PAH on Different C₁₈ Columns

| PAHs | A | B | C | D | E | F | G | H |
|---|---|---|---|---|---|---|---|---|
| Anthracene/phenanthrene | 1.12 | 1.22 | 1.10 | 1.08 | 1.07 | 1.08 | 1.42 | 1.11 |
| Fluoranthene/anthracene | 1.27 | 1.24 | 1.29 | 1.24 | 1.23 | 1.28 | 1.20 | 1.27 |
| Pyrene/fluoranthene | 1.19 | 1.27 | 1.14 | 1.12 | 1.11 | 1.16 | 1.29 | 1.18 |
| Benz[a]anthracene/pyrene | 1.38 | 1.32 | 1.37 | 1.33 | 1.38 | 1.32 | 1.81 | 1.31 |
| Chrysene/benz[a]anthracene | 1.00 | 1.06 | 0.99 | 0.98 | 0.96 | 0.97 | 1.27 | 0.99 |
| Benzo[e]pyrene/benz[a]anthracene | 1.55 | 1.72 | 1.53 | 1.49 | 1.35 | 1.48 | 1.77 | 1.53 |
| Benzo[a]pyrene/benzo[e]pyrene | 1.29 | 1.34 | 1.14 | 1.12 | 1.11 | 1.15 | 1.89 | 1.18 |
| Benzo[b]fluoranthene/benzo[j]fluoranthene | 1.10 | 1.11 | 1.07 | 1.06 | 1.04 | 1.06 | 1.12 | 1.07 |
| Benzo[k]fluoranthene/benzo[b]fluoranthene | 1.10 | 1.12 | 1.01 | 1.04 | 1.17 | 1.29 | 1.05[a] | 1.32 |
| Benzo[ghi]perylene/dibenz[a,h]anthracene | 1.32 | 1.40 | 1.27 | 1.24 | 1.17 | 1.29 | 1.40 | 1.32 |
| Indeno[1,2,3-cd]pyrene/benzo[b]chrysene | 1.02 | 1.04 | 1.10 | 1.11 | 1.10 | 1.11 | 0.66 | 1.10 |

[a] A different G column had a value of 0.86; see text for discussion of this separation.

chrysene; (2) benzo(k)fluoranthene, benzo(b)fluoranthene, benzo(a)pyrene, and benzo(e)pyrene; and (3) benzo(ghi)perylene and indeno(1,2,3-cd)pyrene. They described a method to separate all of these isomers using a column that contained material similar to that in column G.

The column G packing possesses a unique selectivity for PAH separations. A chromatogram illustrating the separation of 18 PAH (including several of the more difficult to separate isomers mentioned above) is shown in Figure 1. The retention of benzo(b)chrysene on column G was significantly greater than that on all other columns tested.

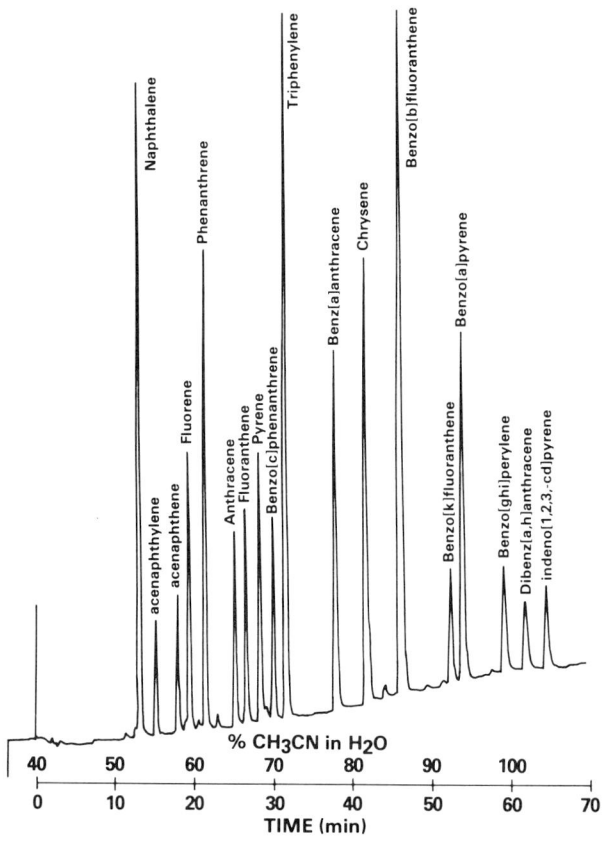

FIGURE 1. Reverse-phase liquid chromatographic separation of polycyclic aromatic hydrocarbons. Column: Vydac 201TP Reverse Phase. Detection: UV absorbance at 254 nm. Conditions: linear gradient from 40 to 100 percent acetonitrile in water at 1%/min at 1 ml/min.

As a result, the benzo(ghi)perylene and the indeno(1,2,3-cd)pyrene elute prior to the benzo(b)chrysene only on material G. In addition, the greatest resolution of benz(a)anthracene and chrysene was obtained on this column. The other columns compared in Table 1 all exhibit similar selectivities, with only slight variations in the elution order of benz(a)anthracene and chrysene. Column G exhibited the greatest resolution (based on $\alpha$ values) for all the PAH pairs listed in Table 2 except fluoranthene/anthracene and benzo(ghi)perylene/dibenz(a,h)anthracene. The benzo(a)pyrene/benzo(e)pyrene selectivity ratio for column G was significantly greater than that for any of the other columns evaluated in this study.

The separation of benzo(ghi)perylene and dibenz(a,h)anthracene on columns containing packing material G exhibits some unique properties. Ogan and co-workers (9) reported that the selectivity for these two compounds varied with the percentage of acetonitrile in the mobile phase. They reported no resolution for these two PAH at 70 percent acetonitrile and separation at higher percentages of acetonitrile with dibenz-(a,h)anthracene eluting first. We have found differences in the elution order of these compounds under identical conditions on columns from different lots of packing material G.

Chromatographic retention data, particularly on two columns of different polarity, are often useful for identification of unknown compounds. In GC the relative retention and the Kovats retention index (5) have been employed to tabulate and present retention data for intra- and interlaboratory use. Lee et al (6) recently described a new GC retention index system using PAH as standards rather than the n-alkanes used in the Kovats System. In LC, Popl and co-workers (10) reported retention indices of PAH on silica using PAH as standards.

Retention data are shown in Table 3 for over 80 PAH on two $C_{18}$ reverse-phase columns and a normal phase $NH_2$ column. However, as indicated in Table 1, all $C_{18}$ packing materials do not necessarily provide the same retention index for a given PAH. As indicated previously, material G has a unique selectivity for PAH. Thus, the retention indices are reported for column G and for column H, which was representative of the selectivity of the other $C_{18}$ columns tested.

We have found the retention data for PAH on $NH_2$ and $C_{18}$ columns to be useful for the analysis of complex mixtures of PAH as described previously (16). In the normal phase, the $NH_2$ column separates PAH based on the number of condensed aromatic rings and on some steric considerations. The addition of a methyl group to the PAH results in only a small change in the retention on the $NH_2$ column. On reverse-phase $C_{18}$ columns, separations are based largely on the solubility of the PAH in the mobile phase; thus, the addition and position of methyl groups on a PAH

TABLE 3. Logarithm of the Retention Index for Polycyclic Aromatic Hydrocarbons

| Compound | Normal Phase, NH₂ Column[a] | Reverse Phase | |
|---|---|---|---|
| | | Column G[b] | Column H[c] |
| Naphthalene | 2.00 | 2.00 | 2.00 |
| Acenaphthene | 2.10 | 2.64 | 2.51 |
| Acenaphthylene | 2.59 | 2.26 | — |
| Fluorene | 2.55 | 2.70 | 2.75 |
| 1-Methylfluorene | 2.64 | 3.19 | 3.26 |
| 2-Methylfluorene | 2.59 | 3.28 | 3.34 |
| Anthracene | 2.94 | 3.20 | 3.14 |
| 2-Methylanthracene | 3.01 | 3.71 | 3.69 |
| 9-Methylanthracene | 3.02 | 3.39 | 3.53 |
| 9,10-Dimethylanthracene | 3.08 | 3.63 | 3.90 |
| Phenanthrene | 3.00 | 3.00 | 3.00 |
| 1-Methylphenanthrene | 3.02 | 3.38 | 3.51 |
| 2-Methylphenanthrene | 3.00 | 3.72 | 3.71 |
| 3-Methylphenanthrene | 3.12 | 3.32 | 3.46 |
| 9-Methylphenanthrene | 3.02 | 3.31 | 3.51 |
| 9-Ethylphenanthrene | 2.97 | 3.53 | 3.87 |
| 9-n-Propylphenanthrene | 2.94 | 3.80 | 4.26 |
| 9-iso-Propylphenanthrene | 2.91 | 3.63 | 4.09 |
| 3,6-Dimethylphenanthrene | 3.11 | 3.57 | 3.91 |
| 1,8-Dimethylphenanthrene | 3.11 | 3.79 | 3.99 |
| 9-Methyl-10-ethylphenanthrene | 3.11 | 3.71 | 4.15 |
| 9,10-Dimethyl-3-ethylphenanthrene | 3.14 | 4.02[d] | 4.64 |
| Benzo[a]fluorene | 3.51 | 3.72 | 3.76 |

TABLE 3. (Continued)

| Compound | Normal Phase, NH$_2$ Column[a] | Reverse Phase | |
|---|---|---|---|
| | | Column G[b] | Column H[c] |
| 9-Methylbenzo[a]fluorene | 3.38 | 3.84 | 3.87 |
| Benzo[b]fluorene | 3.54 | 3.84 | 3.77 |
| 4,5-Methylenephenanthrene | 3.10 | 3.16 | 3.35 |
| Pyrene | 3.37 | 3.48 | 3.65 |
| 1-Methylpyrene | 3.46 | 3.90 | 4.13 |
| 2,7-Dimethylpyrene | 3.47 | 4.40 | 4.77 |
| 1-Ethylpyrene | 3.41 | 4.07[d] | 4.45 |
| Fluoranthene | 3.51 | 3.37 | 3.43 |
| Benz[a]anthracene | 4.00 | 4.00 | 4.00 |
| 1-Methylbenz[a]anthracene | 3.90 | 4.14 | 4.38 |
| 2-Methylbenz[a]anthracene | — | 4.09 | 4.40 |
| 3-Methylbenz[a]anthracene | — | 4.39 | 4.53 |
| 5-Methylbenz[a]anthracene | 4.04 | 4.28 | 4.48 |
| 6-Methylbenz[a]anthracene | 4.03 | 4.10 | 4.39 |
| 7-Methylbenz[a]anthracene | — | 4.14 | 4.35 |
| 8-Methylbenz[a]anthracene | 4.03 | 4.19 | 4.39 |
| 9-Methylbenz[a]anthracene | 4.08 | 4.39 | 4.53 |
| 11-Methylbenz[a]anthracene | 3.91 | 4.13 | 4.41 |
| 12-Methylbenz[a]anthracene | — | 4.10 | 4.35 |
| 7,12-Dimethylbenz[a]anthracene | 3.90 | 4.19[d] | 4.83 |
| Chrysene | 4.01 | 4.10 | 3.99 |
| 1-Methylchrysene | 4.07 | 4.43 | 4.46 |
| 2-Methylchrysene | 4.08 | 4.52 | 4.52 |
| 3-Methylchrysene | 4.12 | 4.29 | 4.42 |

## TABLE 3. (Continued)

| Compound | Normal Phase, NH$_2$ Column[a] | Reverse Phase | |
|---|---|---|---|
| | | Column G[b] | Column H[c] |
| 4-Methylchrysene | 3.95 | 4.18 | 4.35 |
| 5-Methylchrysene | 3.94 | 4.14 | 4.35 |
| 6-Methylchrysene | 4.10 | 4.14 | 4.36 |
| Triphenylene | 4.07 | 3.70 | 3.83 |
| Naphthacene | 3.95 | 4.51[d] | — |
| Benzo[c]phenanthrene | 3.64 | 3.64 | 3.91 |
| Benzo[ghi]fluoranthene | 3.84 | 3.95[d] | 4.07 |
| 20-Methylcholanthrene | 4.31 | 4.78 | 5.23 |
| Benzo[b]fluoranthene | 4.48 | 4.29 | 4.46 |
| Benzo[j]fluoranthene | 4.56 | 4.24 | 4.37 |
| Benzo[k]fluoranthene | 4.45 | 4.42 | 4.52 |
| Benzo[a]pyrene | 4.38 | 4.53 | 4.68 |
| 6-Methylbenzo[a]pyrene | 4.55 | 4.67 | 5.03 |
| Benzo[e]pyrene | 4.46 | 4.28 | 4.48 |
| Perylene | 4.61 | 4.33 | 4.50 |
| Benzo[ghi]perylene | 4.83 | 4.73 | 5.16 |
| Anthanthrene | 4.80 | 4.93 | 5.38 |
| Indeno[1,2,3-cd]pyrene | 4.90 | 4.83 | 5.13 |
| Dibenz[a,c]anthracene | 4.93 | 4.40[d] | 4.73 |
| Dibenz[a,h]anthracene | 4.94 | 4.72[d] | 4.85 |
| Benzo[b]chrysene | 5.00 | 5.00 | 5.00 |
| Picene | 5.03 | 5.10 | 5.31 |
| Biphenyl | 2.16 | 2.37 | 2.38 |
| o-Terphenyl | 2.50 | 3.20 | 3.31 |

TABLE 3. (Continued)

| Compound | Normal Phase, NH$_2$ Column[a] | Reverse Phase | |
|---|---|---|---|
| | | Column G[b] | Column H[c] |
| m-Terphenyl | 3.12 | 3.37 | 3.56 |
| p-Terphenyl | 3.28 | 3.98 | 3.75 |
| m-Quaterphenyl | 4.06 | 3.85 | 4.49 |
| p-Quaterphenyl | 4.50 | 5.04[d] | — |
| m-Quinquephenyl | 5.00 | 4.16 | 5.29 |
| 1,1'-Binaphthyl | 2.99 | 3.45 | 3.92 |
| 1,2'-Binaphthyl | 3.29 | 3.55[d] | 4.13 |
| 2,2'-Binaphthyl | 4.01 | 3.82 | 4.23 |
| 9,9'-Biphenanthryl | 4.64 | 4.18 | 5.30 |
| 9,9'-Bianthryl | 4.30 | 3.91[d] | 4.44 |

[a] n-hexane as mobile phase.
[b] Mixtures of 70 to 85 percent acetonitrile in water as mobile phase.
[c] 80 percent acetonitrile in water as mobile phase.
[d] Log I determined on a different column G.

have a significant effect on the retention. As an example of the differences in normal- and reverse-phase retention, triphenylene, benz(a)anthracene, chrysene, and naphthacene (all four condensed aromatic rings) have retention indices between 3.95 and 4.07 on the $NH_2$ column and they would elute as one broad peak. (Approximately 0.10 to 0.12 unit differences in log I values are required for complete separation on the $NH_2$ column). On $C_{18}$ column G, these same compounds are easily separated with indices of 3.70, 4.00, 4.10, and 4.55. In GC, triphenylene and chrysene are difficult to separate (6), except using liquid crystal stationary phases (4). A detailed discussion of HPLC retention as related to PAH structure will be reported elsewhere (14).

Normal- and reverse-phase HPLC were used to analyze a complex mixture of PAH extracted from urban air particulates. The PAH were first separated on the $NH_2$ column in the normal-phase mode. The chromatogram obtained for this sample is shown in Figure 2 with the

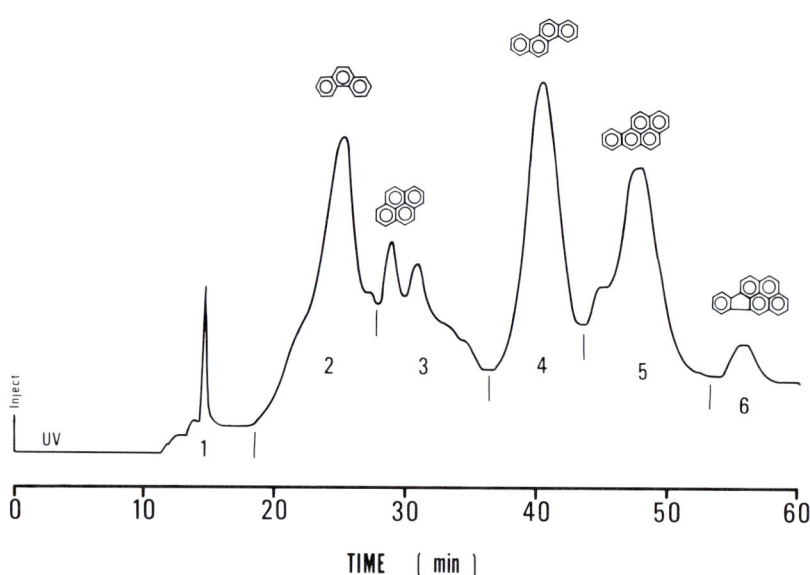

FIGURE 2. Normal-phase liquid chromatogram of PAH extracted from urban air particulates. Column: semi-preparative μBondapak $NH_2$. Detection: UV absorbance at 254 nm. Conditions: n-hexane at 1 ml/min. Numbers indicate fractions collected for subsequent analysis and the PAH structures indicate elution times for these representative compounds.

elution times of several representative PAH indicated. The PAH fractions labeled in Figure 2 were collected, concentrated, and subsequently analyzed by reverse-phase HPLC and GC-MS (15). A reverse-phase chromatogram of fraction 4 is shown in Figure 3A. This fraction contains the four condensed ring PAH benz(a)anthracene, chrysene, triphenylene, and their alkyl derivatives. The labeled compounds were identified on the basis of their fluorescence emission spectra and their chromatographic retention data. The separation of five of the six possible methylchrysene isomers is shown in Figure 3B.

Fraction 5 contains the five-ring isomeric PAH, i.e., benzo(e)pyrene, benzo(a)pyrene, perylene, benzo(j)fluoranthene, benzo(b)fluoranthene, and benzo(k)fluoranthene. As indicated by the data in Table 3, these six compounds are not completely resolved on either of the two $C_{18}$ columns.

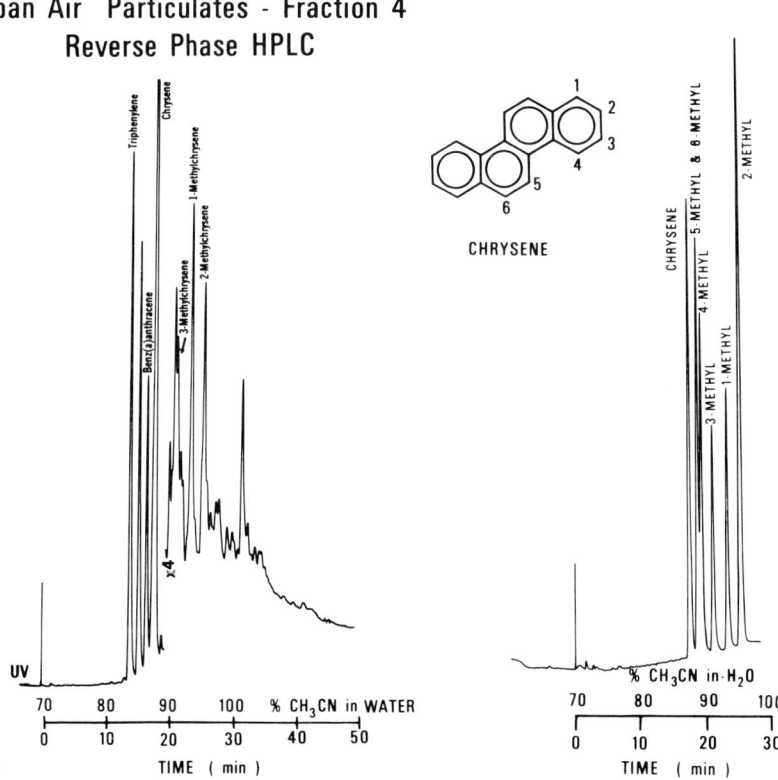

FIGURE 3. (A) Reverse-phase liquid chromatographic analysis of Fraction 4 from urban air particulates (see Figure 2). (B) Reverse-phase liquid chromatographic separation of standards of chrysene and methylchrysene isomers. Column: Vydac 201TP Reverse Phase. Detection: UV absorbance at 254 nm. Conditions: linear gradient from 80 to 100 percent acetonitrile in water at 1%/min at 1 ml/min.

However, utilizing selective fluorescence detection, all of these isomers can be measured. Figure 4 is a reverse phase chromatogram of fraction 5. These compounds were also identified using fluorescence spectroscopy and retention data.

In summary, $C_{18}$ materials from different sources were found to provide different selectivities for PAH separations. Retention indices for 80 PAH are reported on two different $C_{18}$ columns and an $NH_2$ column. These retention indices, although not unique for all $C_{18}$ or $NH_2$ columns, are useful as an aid for the separation and identification of individual PAH in complex mixtures.

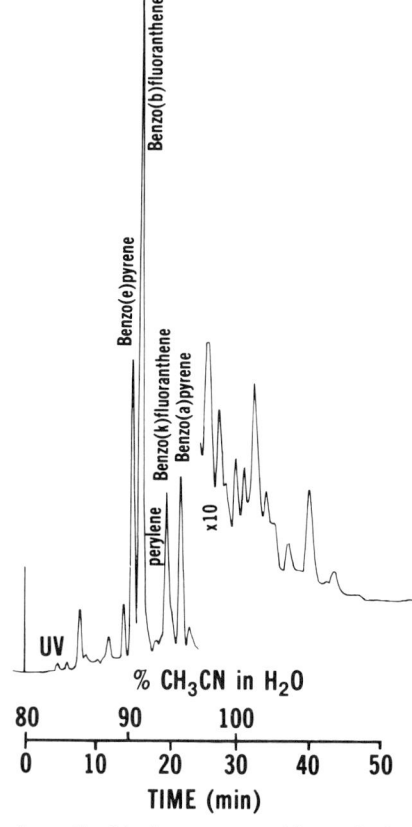

FIGURE 4. Reverse phase liquid chromatographic analysis of Fraction 5 from urban air particulates (see Figure 2). Column: Vydac 201TP Reverse Phase. Detection: UV absorbance at 254 nm. Conditions: linear gradient from 80 to 100 percent acetonitrile in water in 30 min at 1 ml/min.

## ACKNOWLEDGMENTS

The authors acknowledge partial financial support from the Office of Energy, Minerals, and Industry and the Office of Air, Land, and Water Use within the Office of Research and Development of the U. S. Environmental Protection Agency. Identification of any commercial product does not imply recommendation or endorsement by the National Bureau of Standards, nor does it imply that the material or equipment identified is necessarily the best available for the purpose.

## REFERENCES

1. Dong, M. Locke, D. C., and Ferrand, E. (1976): High-pressure liquid chromatographic method for routine analysis of major parent polycyclic aromatic hydrocarbons in suspended particulate matter, *Anal. Chem.* 48: 368–372.
2. Fechner, D. and Seifert, B. (1978): Determination of polycyclic aromatic hydrocarbons in dust deposits by high-performance liquid chromatography using multiwavelength detection, *Z. Anal. Chem.* 292(3): 199–202.
3. Hanus, J. P., Guerrero, H., Biehl, E. R., and Kenner, C. T. (1979): High-pressure liquid chromatographic determination of polynuclear aromatic hydrocarbons in oysters. *J. Assoc. Off. Anal. Chem.* 62: 29–35.
4. Janini, G. M., Muschik, G. M. Schroer, J. A., and Zielinski, W. L., Jr. (1976): Gas-liquid chromatographic evaluation and gas chromatography/mass spectrometric application of new high-temperature liquid crystal stationary phases for polycyclic aromatic hydrocarbon separations. *Anal. Chem.* 48: 1879–1883.
5. Kovats, E. (1958): Gas chromatographische Charakterisierung organischer Verbindungen, *Helv. Chim. Acta* 41:1915–1932.
6. Lee, M. L., Vassilaros, D. L., White, C. M., and Novotny, M. (1979): Retention indices for programmed temperature capillary-column gas chromatography of polycyclic aromatic hydrocarbons. *Anal. Chem.* 51: 768–774.
7. Nielsen, T. (1979): Determination of polycyclic aromatic hydrocarbons in automobile exhaust by means of high-performance liquid chromatography with fluorescence detection. *J. Chromatogr.* 170: 147–156.
8. Nice, E. C., and O'Hara, M. J. (1978): Selective effects of reversed-phase column packings in high-performance liquid chromatography of steroids. *J. Chromatogr.* 166: 263–267.

9. Ogan, K., Katz, E., and Slavin, W. (1979): Determination of polycyclic aromatic hydrocarbons in aqueous samples by reversed-phase liquid chromatography. *Anal. Chem.* 51: 1315–1320.
10. Popl, M., Dolansky, V., and Mostecky, J. (1976): Influence of the molecular structure of aromatic hydrocarbons on their adsorptivity on silica gel. *J. Chromatogr.* 117: 117–127.
11. Smillie, R. D., Wang, D. T., and Meresz, O. (1978): The use of a combination of ultraviolet and fluorescence detectors for the selective detection and quantitation of polynuclear aromatic hydrocarbons. *J. Environ. Sci. Health* A13: 47–58.
12. Thomas, R. S., Lao, R. C., Wang, D. T., Robinson, D., and Sakuma, T. (1978): Determination of polycyclic aromatic hydrocarbons in atmospheric particulate matter by gas-chromatography—mass-spectrometry and high-pressure liquid chromatography. In: *Carcinogenesis, Vol. 3: Polynuclear Aromatic Hydrocarbons,* P. W. Jones and R. I. Freudenthal, Eds., pp. 9–19, Raven Press, New York.
13. Wheals, B. B., Vaughan, C. G., and Whitehouse, M. J. (1975): Use of chemically modified microparticulate silica and selective fluorimetric detection for the analysis of polynuclear hydrocarbons by high-pressure liquid chromatography, *J. Chromatogr.* 106: 109–118.
14. Wise, S. A., Bonnett, W. J., and May, W. E. (1979): The retention characteristics of polycyclic aromatic hydrocarbons in normal- and reverse-phase liquid chromatography, manuscript in preparation.
15. Wise, S. A., Chesler, S. N., Guenther, F. R., Hilpert, L. R., and Parris, R. M. (1979): The determination of polycyclic aromatic hydrocarbons in urban air particulates by liquid chromatography and gas-chromatography/mass spectrometry, manuscript in preparation.
16. Wise, S. A., Chesler, S. N., Hertz, H. S., Hilpert, L. R., and May, W. E. (1977): Chemically bonded amino silane stationary phase for the high-performance liquid chromatographic separation of polynuclear aromatic compounds, *Anal. Chem.* 49: 2306–2310.

# GAS-CHROMATOGRAPHIC PROFILE-ANALYSIS OF PAH METABOLITES FROM RAT LIVER MICROSOMES AND CELLS IN CULTURE‡

**J. Jacob\*, G. Grimmer\*, and A. Schmoldt\*\***

*Biochemisches Institut für Umweltcarcinogene
D-2070 Ahrensburg/Holst.
Sieker Landstraβe 19
West Germany

\*\*Pharmakologisches Institut d. Universitat Hamburg
D-2000 Hamburg 20,
Martinistraβe 52
West Germany

For some polycyclic aromatic hydrocarbons (PAH) it is now well established that only (or predominantly) the bay-region epoxides or the vicinal trans-dihydrodiol epoxides are involved in malignant cell transformation. This has been confirmed for benzo(a)pyrene (BaP), some benz(a)anthracenes, 5-methylchrysene, and 3-methylcholanthrene (2, 7, 12, 13, 14). All these PAH are oxidized in other molecular regions also. It seems that several isoenzymes participate in this oxidation, two groups of which were until recently distinguished commonly as cytochrom P450 and cytochrom P448. It is, however, an open question whether there are different hydratase isoenzymes which convert the epoxides to transdihydrodiols, and if so, how many of them exist. The great interest in the metabolism of PAH in mammals which arises from this complex situation requires an effective and very sensitive method for the detection of metabolite profiles; and it is the high-pressure liquid chromatography (HPLC) technique which generally is used for this purpose (10). We decided on a gas-liquid chromatography (GLC) method, mainly for two reasons:

---

‡Fourth Communication: Inventory and Biological Impact of Polycyclic Carcinogens in the Environment.

(1) The resolution is generally higher in GLC, yielding about 2000 to 2500 theoretical plates per meter; 100,000 theoretical plates can be reached with capillary columns 50 meters in length.

(2) The method can be combined with mass spectrometry, by which well-known metabolites can be identified readily and unknown ones can be characterized.

## MATERIALS AND METHODS

### PAH

PAH used for incubation or induction were purchased from the Community Bureau of Reference, BCR, Directorate General XII, Commission of the European Communities, Brussels. The purity of all PAH was better than 99.0 percent.

### Rats

Wistar rats, male, 200±15 g; diet—Altromin.

### Solvents

All solvents were re-distilled; some of them were refined by conc. sulfuric acid. Ethylacetate was distilled over $P_2O_5$ after treatment with charcoal.

### Induction

Male Wistar rats were pretreated with compounds to be tested by intraperitoneal injection as follows: PAH or 5,6-benzoflavone, 40 mg/kg x day (3 times); phenobarbital, 80 mg/kg x day (3 times); PCB or 3,3',4,4'-tetrachlorobiphenyl, 200 mg/kg unique doses 3 days before decapitation.

### Microsomes

The preparation of microsomes was performed according to the method of Kutt and Fouts (6). The determination of microsomal protein was performed using the Lowry-method, with bovine serum albumine as a standard. Cytochrome P-450 contents were determind according to Omura and Sato (8) by registering the spectrum by means of the Aminco-DW-2-photometer.

*Incubation*

Incubations for the oxidative metabolism of PAH in vitro contained in a total volume of 2 ml were done with: 50 mM tris/HCL (pH 7.4); 0.15 mM KCl; 5 mM $MgCl_2$; 0.2 mM NADH; 1 mg microsomal protein; 20 µg PAH (dissolved in acetone); 0.5 mM NADPH; 8 mM isocitrate, and 50 µg isocitrat-dehydrogenase. Incubations were performed at 37 C under shaking; after 30 minutes they were stopped, 10 ml of acetone was added and they were stored at -20°C until investigation. Controls were incubated without NADPH. To obtain detectable amounts of metabolites, even in cases where PAH are slowly oxidized, relatively high protein concentrations and longer incubation times than usually used for obtaining theoretical maximum turnover rates were chosen. For comparison, however, all PAH were incubated under the same conditions.

*Preparation for GLC*

After evaporation of acetone in a rotary evaporator at 35°C, 10 ml of water was added and the sample was brought to pH 3 with acetic acid. The sample was extracted twice, each time with 20 ml of ethylacetate. After evaporation of the solvent the residue was dissolved in 1 ml of isopropanol and fractionated on a 10-g Sephadex LH 20 column with isopropanol as an elutant. In the case of benz(a)anthracene the first 57 ml were rejected and the fraction 57 to 160 ml was collected. Dihydrodiols required about 15 percent larger elution volumes, and phenols required about 30 percent larger elution volumes than the base PAH. To elute higher oxidized metabolites quantitatively, 3 bed-volumes are recommended. The elution profile of various PAH have been reported elsewhere (3). After careful evaporation of the solvent the residue was dissolved in 3 µl of dimethylformamide and 5 µl of Trisil (Pierce) and kept for 60 minutes at 60° C. After adding another 5 µl of N,O-bis (trimethylsilyl) trifluoroacetamide (Pierce), the mixture was allowed to stand for 30 minutes and 1 to 2 µl of the reaction mixture was used in the GLC.

## GLC

Perkin-Elmer F 20 FE Instrument. Packed 10-m glass columns were used (2 mm inner diameter, packed with Supelcoport coated with OV 101; $N_2$ as carrier gas, 20 ml $N_2$/min). Column, injection, and detection temperatures were 250°C. Injection and connection between the column and detection were all-glass. Alternatively, 25-m glass capillaries coated with SE 30 were used (carrier gas: Helium). To reduce the burden of the FID by silylation reagent, the FID-electrode was removed for 2 minutes after injection. Within this time no metabolites appeared in the fractogram.

## Mass Spectrometry

The above-mentioned glass capillary column coated with SE 30 was used for gas-liquid chromatography/mass spectrometry (GLC/MS) operating with splitless injection. The column was heated from 100°C injection temperature to 200°C rapidly at 40°C/min., and afterwards at 1°C/min. from 200 to 250°C. A Varian-MAT 112 S instrument with solvent separation was used. Ten minutes after sample injection the separator was opened (ratio 1:35). The ion source temperature was 250°C, 70 ev.

## RESULTS

The metabolite profile of BaA after incubation with rat liver microsomes of nonpretreated animals was readily established (Figure 1); only two dihydrodiols were detected.

In contrast, the profile changed dramatically after induction with 3,3' 4,4'-tetrachlorobiphenyl, which induced the cytochrom P-448 system (Figure 2).

Mass spectra of all peaks in Figure 2 were recorded and phenols, dihydrodiols, triols, and tetrahydrotetrols were found. All OTMS-ethers of hydroxyl groups containing metabolites gave intense parent peaks by which they were readily characterized. Metabolites which possessed more than one OTMS-group after silylation were readily recognized by the fragment m/e 147. Although this method has been used before by other authors (1, 11), no evidence for the presence of triols and tetrols has been given previously (Figures 3 through 6).

We applied this method to two problems: (a) the induction of monooxygenases in the rat liver by various substances, predominantly various PAH (4); and (b) the metabolism of PAH in fetal hamster lung cells in culture using BaA as substrate (5).

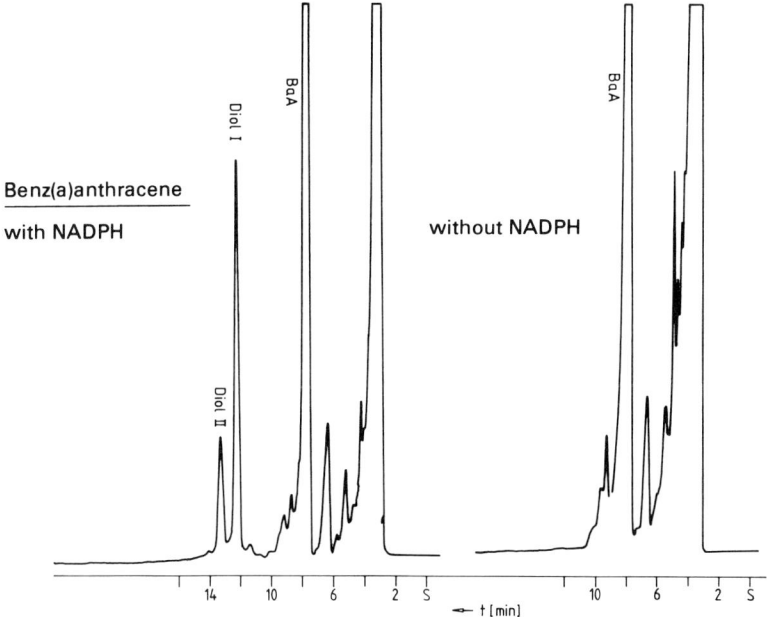

FIGURE 1. Metabolic profile of benz(a)anthracene after incubation with rat liver microsomes; no induction.

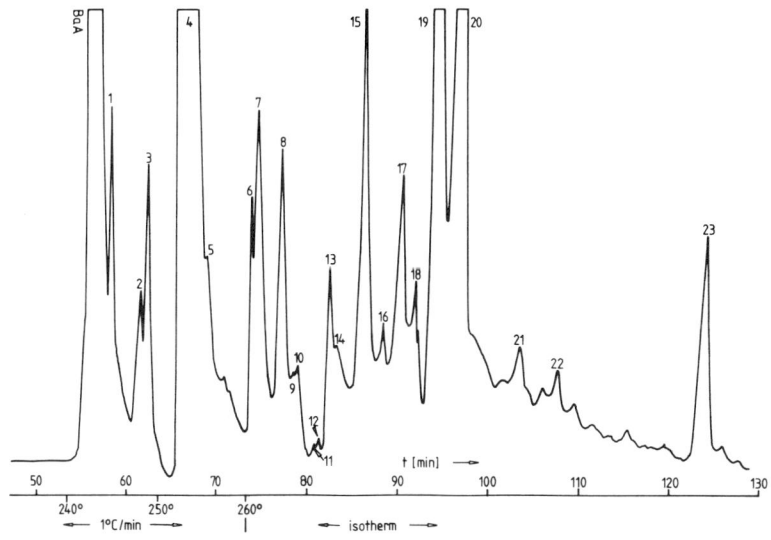

FIGURE 2. Metabolic profile of benz(a)anthracene after incubation with induced rat liver microsomes.

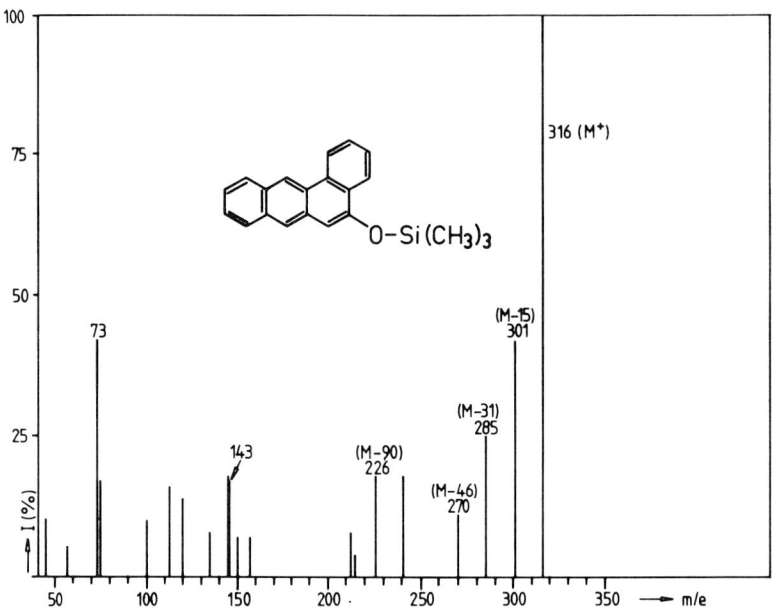

FIGURE 3. Mass spectrum of silylated benz(a)anthracene—phenol.

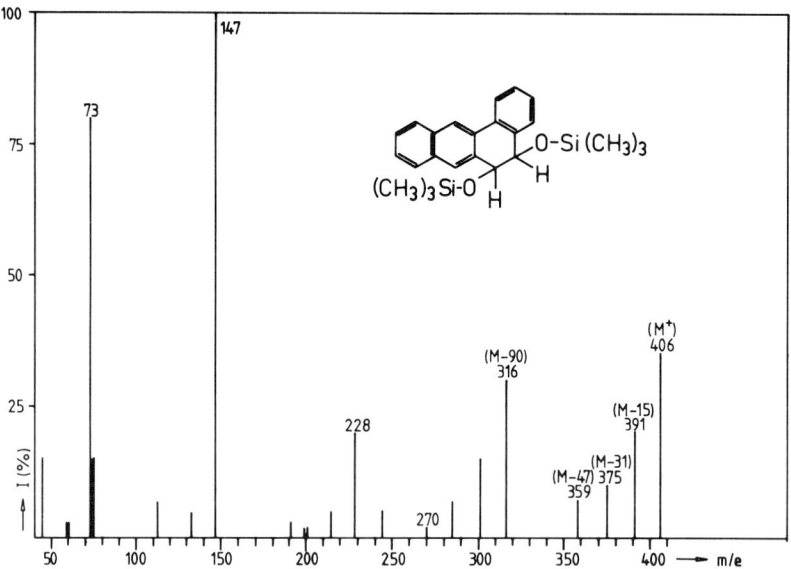

FIGURE 4. Mass spectrum of silylated benz(a)anthracene—dihydrodiol.

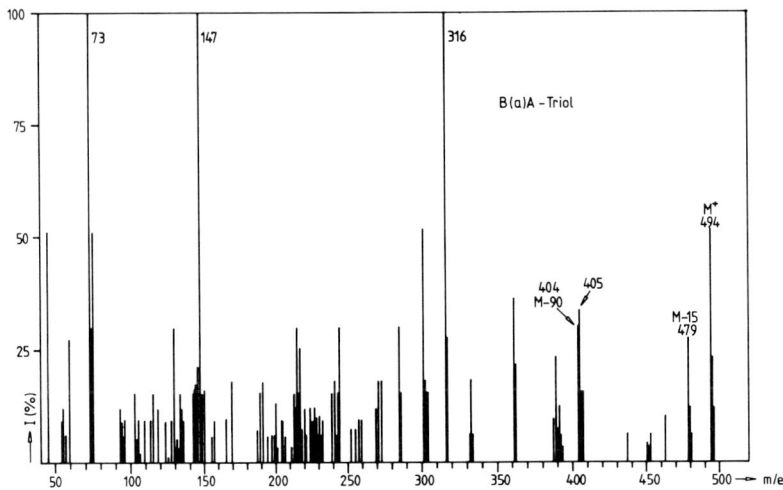

**FIGURE 5.** Mass spectrum of silylated benz(a)anthracene—triol.

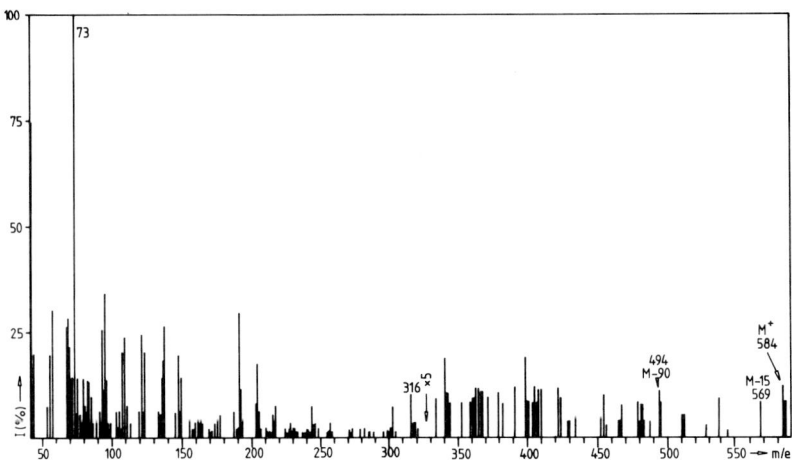

**FIGURE 6.** Mass spectrum of silylated benz(a)anthracene—tetrol.

Table 1 shows the metabolic rates of BaA as substrate in rat liver microsomes without and with pretreatment using various inducers. The induction rates differed significantly. Pyrene seemed to inhibit the oxidation slightly; benzo(e)pyrene was not an inducer; benzo(k)fluoranthene was a medium inducer; and BaP, benzo(b)- and (j)fluoranthene, BaA and, (more than all other PAH investigated) chrysene were potent inducers. This means that the inducibility of PAH does not correlate with

**TABLE 1. Metabolic Rates of BaA as Substrate in Rat Liver Microsomes without and with Induction**[a]

| Inducer | Metabolized BaA (nmol/30 minutes) | Metabolized BaA (induced/not induced) |
|---|---|---|
| None | 6.1 | 1.0 |
| 5,6-Benzoflavone | 27.0 | 4.4 |
| PCB | 38.1 | 6.2 |
| 3,3',4,4'-Tetrachlorobiphenyl | 35.6 | 5.8 |
| Phenobarbital | 44.2 | 7.2 |
| Benz(a)anthracene | 39.9 | 6.5 |
| Pyrene | 2.4 | 0.4 |
| Chrysene | 55.9 | 9.1 |
| Benzo(a)pyrene | 30.7 | 5.0 |
| Benzo(e)pyrene | 9.8 | 1.6 |
| Benzo(b)fluoranthene | 25.2 | 4.1 |
| Benzo(j)fluoranthene | 39.9 | 6.5 |
| Benzo(k)fluoranthene | 20.9 | 3.4 |

[a]40 mg PAH/kg rat, 3 times; standard conditions, 20 $\mu$g PAH, 30 minutes incubation; Wistar-rat.

their carcinogenic potency in the epicutaneous dropping test. To investigate if a distinct inducer influences the oxidation of various PAH substrates to the same or to a different extent, the metabolic rates of some PAH in rat liver microsomes of animals which were pretreated with benzo(k)fluoranthene were studied. The results are presented in Table 2, showing great differences in regard to the induction factors. Fluoranthene and pyrene, which were well metabolized already in noninduced animals, were not considerably better metabolized after benzo(k)fluoranthene pretreatment (f = 2.1). With the exception of benzo(ghi)perylene, in which no metabolites could be detected, all PAH which were metabolized to 2 percent and less showed drastic increases in metabolic rates after induction with benzo(k)fluoranthene. The results may be summarized as follows. (1) Different PAH induce the metabolism of a certain PAH (e.g., benz(a)anthracene) to a different extent. Some PAH are good inducers, others are not. It can not be excluded that some PAH inhibit the oxidation. (2) Some weak carcinogenic and even noncarcinogenic PAH are potent inducers. (3) A certain inducer stimulates the metabolism of various PAH to a different extent.

In addition to rat liver microsomal oxidation, we applied the GLC metabolite-profile analysis to an investigation of BaA metabolism in fetal hamster lung cells in culture. Cells in culture were treated with BaA for 6 and 22 hours, respectively, and the medium and cells were analyzed separately. The results are given in Table 3.

TABLE 2. Metabolic Rates of Various PAH without Inducer and after Pretreatment with Benzo(k)fluoranthene[a]

| Substrate | Without Inducer (nMol metabolites/ 30 minutes) | Induction with Benzo(k)fluor- anthene (nMol metabolites/ 30 minutes) | Metabolic Rate (induced/not induced) |
|---|---|---|---|
| Fluoranthene | 20.8 | 43.7 | 2.1 |
| Pyrene | 22.3 | 46.8 | 2.1 |
| Chrysene | 0.7 | 21.8 | 31.1 |
| Benz(a)anthracene | 6.1 | 20.7 | 3.4 |
| Benzo(b)fluoranthene | 5.2 | 49.4 | 9.5 |
| Benzo(j)fluoranthene | 2.2 | 16.5 | 7.5 |
| Benzo(a)pyrene | 1.6 | 37.0 | 23.1 |
| Benzo(e)pyrene | 0.8 | 45.8 | 57.3 |
| Indeno(1,2,3-cd)pyrene | 0.7 | 38.5 | 55.0 |
| Benzo(ghi)perylene | 0 | 0 | — |

[a] 40 mg PAH/kg rat, 3 times; standard conditons; 20 $\mu$g PAH, 30 minutes incubation Wistar-rat.

TABLE 3. Concentration of Benz(a)anthracene and Metabolites in Cells and Medium after 6 hours and 22 hours

| Incubation | | | Detected Without Arylsulfatase Treatment ($\mu$g) | | Detected After Arylsulfatase Treatment 10 minutes ($\mu$g) | | Total Metabolic Rate (%) |
|---|---|---|---|---|---|---|---|
| 6 hr | Cells | BaA | 0.5 | 15.5 | Not measured | | 22.5 |
|  | Medium | BaA | 15.0 |  |  |  |  |
| 22 hr | Cells | BaA | 0.8 | 17.1 | 0.9 | 19.5 | 56.0 |
|  |  | Metabolites | 2.5 |  | 2.6 |  |  |
|  | Medium | BaA | 8.0 |  | 9.1 |  |  |
|  |  | Metabolites | 5.8 |  | 6.9 |  |  |

[a] Incubation (20 $\mu$g substrate, 2.5 x 10⁶ cells).

More than 50 percent of the BaA was metabolized with 22 hours. Only about 10 percent of the unconverted BaA was found inside the cells, but the content of metabolites there was remarkably high. Treatment with arylsulfatase increased only the metabolites in the medium, i.e., sulfates were obviously transferred rapidly from the cells into the external medium. We do not presently have an explanation for the slight increase of BaA in the medium after sulfatase treatment. From a total of 20 $\mu$g of

BaA incubated, 19.5 μg was recovered (97.5 percent); since about $2.5 \times 10^6$ cells were exposed which converted 10 μg B(a)A, it can be calculated that 4 picograms of BaA per cell was oxidized within a period of 22 hours. This agrees with the results of Pelkonen et al (9).

## CONCLUSIONS

(1) PAH metabolite profiles can be recorded readily by means of GLC. Mass spectrometry allows for characterization and identification.

(2) Different PAH induce the monoxygenase-catalyzed oxidation of a certain PAH substrate to different extents.

(3) PAH which are inactive in the epicutaneous dropping test can be good inducers in the monoxygenase system.

(4) A certain inducer stimulates the metabolism of various PAH to different extents.

(5) Fetal hamster lung cells in tissue metabolize BaA; rates of 4 picograms of BaA per cell within 22 hours have been observed.

## ACKNOWLEDGMENTS

This research was sponsored by the Umweltbundesamt Berlin, by order of the Ministry of Internal Affairs, who is gratefully acknowledged by the authors. We also thank Mr. K. Naujack for technical assistance and the preparation of the drawings.

## REFERENCES

1. Borgen, A., Darvey, H., Castagnoli, N., Crocker, T.T., Rasmussen, R.E., and Wang, I.Y. (1973): Metabolic conversion of benzo(a)pyrene by Syrian hamster liver microsomes and binding of metabolites to desoxyribonucleic acid. *J.Med.Chem.* 16:502-506.
2. Bettencourt, A., Lhoest, G., Roberfroid, M., and Mercier, M. (1977): Gaschromatographic and mass fragmentographic assays of carcinogenic polycyclic hydrocarbon epoxide hydrase activity. *J. Chromatogr.* 134:323-330.
3. Grimmer, G. (1979): Nachweismethoden für PAH. In: *Luftqualitätskriterien für ausgewählte polyzyklische Kohlenwasserstoffe.* Umweltbundesamt Berichte 1/79, p. 40. Erich Schmidt Verlag, Berlin.
4. Jacob, J., Grimmer, G., and Schmoldt, A. (1979): Metabolitenprofile von polyzyclischen aromatischen Kohlenwasserstoffen nach Vorbehandlung mit verschiedenen Induktoren mikrosomaler Monoxy-

genasen der Rattenleber. *Hoppe-Seyler's Z. Physiol. Chem.* 360: 525-534.
5. Jacob, J., Grimmer, G., Richter-Reichhelm, H.-B., and Emura, M. (1979): Gaschromatographische Profilanalyse von PAH-Metaboliten aus Mikrosomen-Präparationen und Gewebekulturen. VDI-Kommission Reinhaltung der Luft. Hannover-Meeting.
6. Kutt, H. and Fouts, J.R. (1971): Diphenylhydantoin metabolism by rat liver microsomes and some of the effects of drug or chemical pretreatment on the diphenylhydantoin metabolism by rat liver microsomal preparations. *J. Pharmacol. Exp. Ther.* 176:11-26.
7. Lehr, R. E., Yagi, H., Thakker, D.R., Levin, W., Wood, A.W., Conney, A.H., and Jerina, D.M. (1978): The bay-region theory of polycyclic aromatic hydrocarbon-induced carcinogenicity. In: *Carcinogenesis. Vol. 3: Polynuclear Aromatic Hydrocarbons,* P.W. Jones and R.I. Freudenthal, Eds., pp. 231-241, Raven Press, New York.
8. Omura, T. and Sato, R. (1964): The carbon monoxide binding pigment of liver microsomes. *J. Biol. Chem.* 239:2370-2378.
9. Pelkonen, O., Korhonen, P., Jouppila, P. and Karki, N. (1976): Induction of aryl hydrocarbon hydroxylase in human fetal liver cell and fibroblast cultures by polycyclic hydrocarbons. *Life Sci.* 16:1403-1410.
10. Selkirk, J.K. (1978): Analysis of benzo(a)pyrene metabolism by high-pressure liquid chromatography. In: *Adv. in Chromatography* 16. J. C. Giddings, E. Groshka and P. R. Brown, Eds.
11. Stoming, T.A. and Bresnick, E. (1973): Gas chromatographic assay of epoxide hydrase activity with 3-methylcholanthrene-11,12-oxide. *Science* 181:951-952.
12. Thakker, D.R., Levin, W., Stoming, T.A., Conney, A.H., and Jerina, D.M. (1978): Metabolism of 3-methylcholanthrene by rat liver microsomes and a highly purified monoxygenase system with and without epoxide hydrase. In: *Carcinogenesis Vol. 3: Polynuclear Aromatic Hydrocarbons,* P.W. Jones and R.I. Freudenthal, Eds., Raven Press, New York.
13. Wood, A.W., Levin, W., Chang, R.L., Lehr, R.E., Schaefer-Ridder, M., Karle, J.M., Jerina, D.M., and Conney, A.H. (1977): Tumorigenicity of five dihydrodiols of benz(a)anthracene on mouse skin: Exceptional activity of benz(a)-anthracene-3,4-dihydrodiol. In: *Proc. Natl. Acad. Sci.* U.S.A. 74:3176-3179.
14. Yang, S.K., Roller, P.P., and Gelboin, H.V. (1978): Benzo(a)pyrene metabolism: Mechanism in the formation of epoxides, phenols, dihydrodiols, and the 7,8-diol-9,10-epoxides. In: *Carcinogenesis Vol. 3: Polynuclear Aromatic Hydrocarbons,* P.W. Jones and R.I. Freudenthal, Eds., Raven Press, New York.

# EXTRACTION AND RECOVERY OF POLYCYCLIC AROMATIC HYDROCARBONS FROM HIGHLY SORPTIVE MATRICES SUCH AS FLY ASH*

**W. H. Griest, J. E. Caton, M. R. Guerin, L. B. Yeatts, Jr. and C. E. Higgins**

Analytical Chemistry Division
Oak Ridge National Laboratory
Oak Ridge, Tennessee 37830

Extraction and handling recoveries are an important but sometimes overlooked part in the quantitation of polycyclic aromatic hydrocarbons (PAH). Assumption of the reproducibility of the PAH recoveries in a given extraction and isolation procedure can lead to erroneous analytical results when actual recoveries deviate because of varying sample matrix characteristics or inadvertant errors in procedural execution. Highly sorptive sample matrices may present unexpectedly difficult extraction problems.

One particularly adsorptive sample matrix is fly ash from coal-fired electric power generating plants. However, the few published reports (4,5,7,11,12,15) of fly ash analysis for PAH employed Soxhlet solvent extraction without any apparent regard for extraction and handling recoveries. Further, we are aware of only one worker (14) who has considered Soxhlet extraction recoveries of PAH from fly ash. This paper presents the results from our initial studies of the extraction and recovery behavior of PAH, using fly ash as an exemplary adsorptive sample matrix.

## DETERMINATION OF THE EXTRACTION RECOVERIES OF PAH

Radiolabeled PAH tracers, such as Carbon-14-labeled benzo(a)-pyrene ($^{14}$C-BaP), are a conveniently measured indicator of PAH extraction recoveries. The tracer is applied to the sample prior to

---

*Research sponsored by the Electric Power Research Institute and the U.S. Environmental Protection Agency Under Union Carbide Corporation contract W-7405-eng-26 with the U.S. Department of Energy.

extraction, and a small portion of the extract or isolate obtained from treatment of the sample matrix is counted by liquid scintillation spectrometry to determine the recovery of the tracer. The assumption is made that the applied tracer models the behavior of the native PAH in the sample matrix.

Employing ultrasonic extractions of fly ash with benzene we have found (9,10) $^{14}$C-BaP extraction recoveries from most fly ashes to be quite low. These recoveries are generally less than 30 percent although much higher recoveries have been obtained occasionally with apparently less sorptive fly ash samples. In contrast, $^{14}$C-BaP recoveries from the extraction of air particulate pads, water, or freshwater stream sediment typically range from 80 to 100 percent (8). The implication of these findings is that uncorrected measurements of BaP in fly ash may be low by a factor of three, from incomplete extraction alone, and possibly even lower still from uncorrected losses in isolation and handling.

The low PAH extraction recoveries from fly ash do not appear to be a simple case of incomplete application of extraction methodology or the failure of any one particular extraction method. Soxhlet extraction, solvent refluxing, and other traditional extraction methods are even less effective than ultrasonic solvent agitation and they produce lower $^{14}$C-BaP extraction recoveries. The limitation seems to lie, rather, in a fundamental physical/chemical interaction between the PAH and the fly ash matrix. We have found that the individual $^{14}$C-BaP recoveries from sequential extractions of fly ash decrease rapidly with each succeeding extraction step (Table 1) and that by the third ultrasonic solvent extraction step, the incremental increases in recovery are so low that further extraction would not be expected to substantially improve the total recovery. The trend suggests that a limiting extraction recovery factor is being approached.

TABLE 1. Extraction Recoveries of $^{14}$C-BaP and $^{3}$H-BaP from Fly Ash[a]

| Extraction No. | Recovery (%) Avg. ± Std. Dev. | | |
|---|---|---|---|
| | $^{14}$C-BaP | $^{3}$H-BaP | $^{3}$H-BaP / $^{14}$C-BaP |
| 1 | 10.5 ± 0.917 | 7.08 ± 2.18 | 0.674 |
| 2 | 4.45 ± 0.384 | 3.32 ± 0.860 | 0.746 |
| 3 | 2.24 ± 0.093 | 1.67 ± 0.603 | 0.746 |
| TOTAL | 17.2 | 12.1 | 0.72[b] |

[a]Three g of fly ash B spiked with $4 \times 10^4$ dpm of each tracer and ultrasonically extracted in 10 ml of benzene for 30 sec at room temperature with a Branson model 75 sonifier. Three extractions performed on four replicates.
[b]Average ratio.

Comparison of the results for $^{14}$C-BaP and $^3$H-BaP in Table 1 also illustrates another factor influencing recovery determination: the radiolabel of the tracer. For PAH, we are limited to isotopes of C and H, but $^{14}$C- and $^3$H-labeled tracers are readily available from commercial sources. The data in Table 1 show the ultrasonic solvent extraction recoveries of $^{14}$C-BaP and $^3$H-BaP tracers applied to the same fly ash sample. The results suggest that the use of a $^3$H-labeled PAH tracer may cause a low bias in the recovery determination relative to that defined by the $^{14}$C-labeled PAH tracer. This effect might be attributed to $^3$H-$^1$H exchange between the $^3$H-BaP tracer and the fly ash surface, and may be catalyzed by the tight binding of the tracers to the fly ash. Presumably, a similar exchange would be observed for $^2$H-labeled tracers. We do not expect an exchange of $^{14}$C - $^{12}$C to occur with the fly ash because of the attendant disruption of the PAH aromatic ring system and the possibly limited availability of exchangeable free C from fly ash.

An accounting of the unextracted tracer is important to recovery studies to determine whether the unextracted tracer remains with the spiked matrix or is lost to the container walls and other surfaces. An activity balance for the tracer can be constructed by determining the tracer activity left on these solid surfaces. Using a host matrix oxidation technique (2,10), oxidizable carbonaceous material in a solid sample is converted to $CO_2$, which is trapped in a solution for $^{14}$C-activity measurement by conventional liquid scintillation spectroscopy methods. Application of this approach to ultrasonic solvent-extracted fly ash (Set 1, Table 2) indicates that the unextracted $^{14}$C-BaP tracer is indeed left with the fly ash. However, no information as to the chemical speciation of extracted and unextracted tracer is obtained with this method.

Interestingly, examination of a spiked, but nonextracted fly ash control sample (Set 2, Table 2) yielded an unexpectedly low recovery which may

**TABLE 2. Activity Balance Accounting of $^{14}$C-BaP Tracer Applied to Fly Ash[a]**

| Set | Sample | $^{14}$C-BaP Recovery (%) Average ± Std. Dev. |
|---|---|---|
| 1 | Benzene Extract of Spiked Ash | 25.2 ± 0.03 |
|   | Spiked and Extracted Ash | 73.6 ± 3.6 |
|   | Total Accounted | 98.8 |
| 2 | Spiked Ash Control | 86.5 ± 2.8 |
| 3 | Spiked and Sonicated Ash Control | 98.3 ± 3.6 |
|   | Crushed Glass Container | 1.3 ± 0.2 |
|   | Total Accounted | 99.6 |

[a]Three g of ash B spiked with 4.1 × 10$^4$ dpm of $^{14}$C-BaP and triplicate samples (1) extracted two times for 30 sec at room temperature with 10 ml of benzene, using a Branson model 350 sonifier, (2) left unextracted, and (3) extracted once [as in (1)] but benzene evaporated onto ash. See Reference (8) for other details.

underline the difficulty in uniformly spiking a highly sorptive matrix. The low recovery was not a result of adsorption of tracer by the container walls (Set 3, Table 2). However, ultrasonication of the control fly ash in benzene (as in the extraction procedure for Set 1) and evaporation of the benzene extract back onto the fly ash allowed a quantitative accounting of the $^{14}$C-BaP applied to the fly ash. We believe that this result indicates that tracer application to the fly ash by pipetting a tracer solution to the ash does not result in a completely homogeneous distribution of tracer throughout the fly ash. The upper layers of ash may sorb the tracer as it is pipetted onto the sample and this highly labeled ash may not be satisfactorily dispersed by simple stirring. However, the process of ultrasonicating a fly ash slurry does seem to evenly distribute this labeled fly ash throughout the bulk sample.

## CHARACTERIZATION OF PAH SORPTION

The radio-labeled tracer technique also is useful for conducting studies of the nature of interactions influencing the extraction behavior of PAH from a host matrix. We have conducted an examination of the extraction behavior of other PAH from fly ash by spiking separate aliquots of a fly ash with different $^{14}$C-labeled PAH tracers and by performing separate ultrasonic extractions in benzene and recovery determinations by liquid scintillation spectroscopy. Plots of the cumulative extraction recoveries of four PAH and one paraffin from a fly ash are shown in Figure 1. The most striking result is the apparent inverse dependence of both the rate of extraction and total recovery upon the PAH ring system size. We have observed this same trend using several other fly ash samples and extraction methods. The observation that an n-paraffin is extracted more rapidly and completely than a large PAH ($C_{32}$ for the paraffin versus $C_{20}$ for BaP) suggests a sorption mechanism dependent upon aromatic properties of the sorbate.

The causes of this sorption effect are not clear, however, micron- and submicron-size fly ash paraticles are enriched with potentially complexing elements such as antimony, tin, or aluminum (3,6), and we speculate that a strong $\pi$-complex or formal bonding of the larger PAH may occur with the fly ash. The only other accessible study (14) of PAH sorption on ash suggested that the carbon content of the fly ash was the decisive factor in PAH extraction. The carbon content of their ashes was considerably higher than ours (up to 25.6 percent versus <1 percent for our samples) and may have been the controlling factor in their studies. However, the higher levels of PAH spikes applied to their ashes (133 $\mu$g/g versus 10 to 100 ng/g in our work calculated from the tracers specific activity) and the

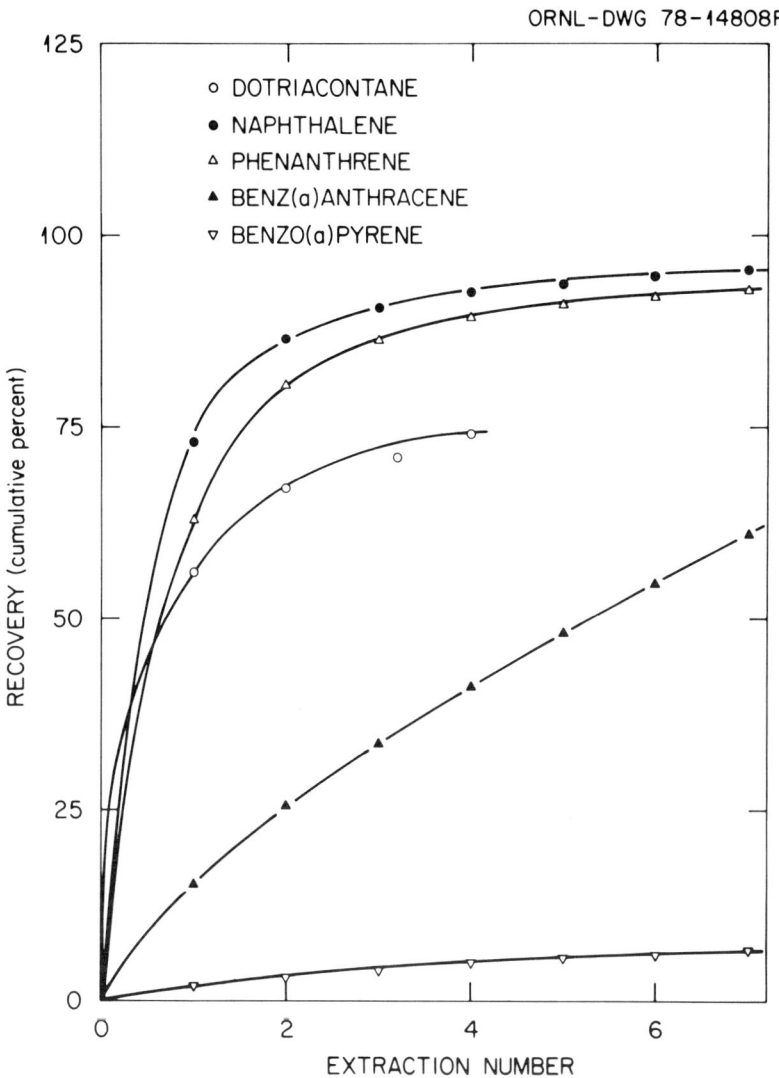

FIGURE 1. Ultrasonic Solvent Extraction Recoveries of PAH and a paraffin from Fly Ash. Three g of fly ash A was spiked with $4.5 \times 10^4$ dpm of tracer and extracted 4 to 7 times with 10 ml of benzene for 30 min at room temperature with a Sondgen Automatic Cleaner (8).

unknown matrix effects of the 400 to 500°C annealing of their ashes to vary the carbon content complicate comparison of their results with ours.

## 824 EXTRACTION AND RECOVERY OF PAH

The nature of the sorption of PAH by a highly sorptive matrix such as fly ash can be probed by a frontal elution chromatographic method. For this approach fly ash is packed into a stainless steel column (6.4 mm OD, 4.6 mm ID, and 25 cm long) in much the same way that one would dry-pack a column for liquid chromatography. After equilibration with a given solvent, a solution containing ng/ml quantities of $^{14}$C-labeled PAH in the equilibrating solvent is pumped across the column, and the solution eluted from the column is monitored for $^{14}$C-activity. Results of such studies are illustrated in Figure 2 where the affinity of fly ash for BaP is contrasted

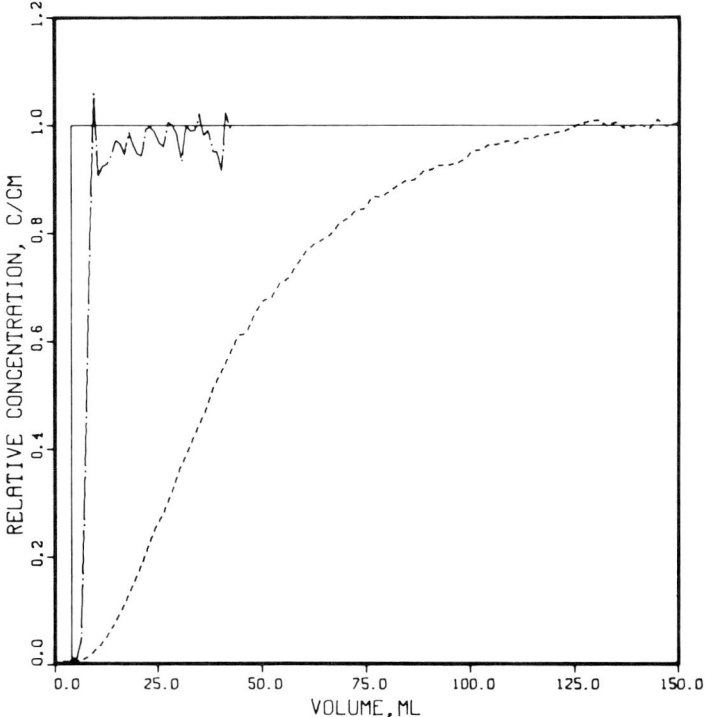

FIGURE 2. Comparison of the elution of dotriacontane (DTC) and benzo(a)pyrene from a column packed with fly ash. A 25 cm × 4.6 mm ID column was packed with 3 g of fly ash B. Tracers in benzene or hexane were pumped through the column at 1.2 ml/min and collected in 2 ml fractions for analysis by liquid scintillation spectrometry. Relative concentration is expressed as the ratio of $^{14}$C-activity per unit volume being eluted from the column to $^{14}$C-activity per unit volume being pumped onto the column. ——— Ideal elution with no sorption, —·—DTC in hexane; ---- BaP in benzene.

with affinity of fly ash for dotriacontane (DTC). The profile of $^{14}$C-activity eluted from the fly ash column when $^{14}$C-labeled DTC in hexane (24 ng/ml) is pumped onto the column (dashed line) approximates the profile one would expect to see if the fly ash showed no affinity for the DTC (solid line). In contrast, when $^{14}$C-labeled BaP in benzene (3.3 ng/ml) is pumped across a column packed with fly ash, the profile of $^{14}$C-activity eluted from the column (dotted line) indicates that there is a significant interaction between the BaP and the fly ash. Preliminary analysis of this profile (after Reference 1) indicates that at equilibrium approximately 25 ng of BaP are associated with each gram of fly ash packed in the column. If the concentration of BaP is increased the volume of solution required to saturate the fly ash (approximately 125 ml in Figure 2) is proportionally decreased. Thus for BaP concentrations on the order of 1 $\mu$g/ml, the profile of eluted $^{14}$C-activity approaches that shown by DTC.

The equilibrium sorption coefficient of approximately 25 ng of BaP per g of fly ash corresponds to a surface coverage of approximately $4.3 \times 10^{-5}$ m$^2$/g, assuming to a first approximation a rectangular surface area occupied by BaP and the bond lengths and bond angles used in other studies (17), and a flat orientation with monolayer coverage. In comparison, the surface area of similarly collected fly ash ranges from 0.3 m$^2$/g (16) to 1.4 m$^2$/g (estimated from data in Reference 9). Even when allowing an order of magnitude of uncertainty in the surface area estimation, it is evident that the BaP sorption is taking place at highly localized "hot spots" on the ash, rather than occurring in uniform layers over the fly ash surface. The sites responsible for this sorption, the nature of the interaction, and the speciation of the sorbed BaP are being investigated in our current studies.

## RECOVERY OF EXTRACTED PAHs

Once PAH have been extracted from the host matrix, they must be purified and concentrated prior to analysis. Losses of PAH can occur from handling manipulations, incomplete liquid-liquid partitioning, irreversible adsorption on column chromatography packings, and during solvent concentration steps. Although polyaromatics usually are recovered in reasonable yields (70 to 100 percent), extensive losses of diaromatics can occur. These losses are traceable to evaporative concentration of solvents, particularly during the final stages of concentration. We have found (8) that as much as 80 percent of the diaromatics can be lost, most likely from codistillation with the solvent. Use of radiolabeled tracers is not totally satisfactory in correcting for losses because the individual recoveries of the different diaromatics can vary considerably. For example, the recovery of acenaphthene may be twice that of naphthalene (8). Dual tracer approaches, using a $^{14}$C-labeled tracer and a $^{3}$H-labeled tracer, which together span the range of volatility of the diaromatics, also can be unsuccessful because of the $^{3}$H – $^{1}$H exchanging activity of adsorbents commonly used in PAH isolation, such as alumina and silica gel (8).

We have found (13) that the diaromatic recovery problem apparently can be solved by placing a small (0.5-ml) adsorbent resin cartridge downstream of the concentration flask in a flowing-nitrogen evaporative concentration apparatus. The diaromatics are lost from the solvent concentration flask during solvent evaporation but, as shown in Table 3, they are effectively trapped in the sorbent. The choice of solvent, sorbent, mass of sorbate, or nitrogen flush time within the defined limits appear to have minimal effect upon the trapping of the diaromatics.

The trapped diaromatics are recovered by flushing the adsorbent cartridge with a small volume (0.3 to 0.5 ml) of solvent or by thermal desorption onto a cryogenically cooled gas chromatographic column. Although the diaromatics are readily flushed from the polymeric resins with organic solvents, they are strongly adsorbed on the carbonaceous Ambersorb and are poorly recovered. Thermal desorption is best carried out with the Tenax resin because of its superior thermal stability over that of the other resins. Studies suggest that the method of "downstream trapping" during evaporative concentration may also offer a simple means of separating and recovering the diaromatic fraction from the polyaromatic fraction of an aromatic isolate.

TABLE 3. Trapping of Naphthalene During Evaporative Concentration in Various Solvents[a]

| Solvent | Recovery by Sorbent[b] (%) | | | |
|---|---|---|---|---|
| | Tenax | XAD-2 | Porapak Q | Ambersorb XE-340 |
| Ether | 92 | 92 | 94 | 96 |
| Methylene chloride | 96 | 94 | 94 | 96 |
| Methanol | 95 | 98 | 97 | 94 |
| Hexane | 92 | 90 | 93 | 94 |
| | 93[c] | 95[d] | | |
| | 94[e] | | | |
| Benzene | 91 | 88 | 93 | 91 |
| Cyclohexane | 92 | 98 | 95 | 92 |

[a]One microgram of naphthalene (spiked with $^{14}$C-naphthalene) in 50 ml of solvent. Five minutes extra vacuum and nitrogen treatment after solvent was evaporated.
[b]One-half ml sorbent in Pasteur pipet downstream from concentration flask.
[c]One hour additional vacuum and nitrogen flush after evaporation of solvent.
[d]One milligram naphthalene.
[e]Sixteen hours additional vacuum and nitrogen treatment after evaporation of solvent.

## CONCLUSIONS

The highly sorptive nature of some potentially environmentally significant materials such as fly ash may seriously hinder quantitative extraction of their sorbed organic content. Radiolabeled tracers offer a convenient means of probing the sorptive nature of such matrices and of obtaining the corrections for extraction and handling recoveries which are necessary for quantitative analysis.

## ACKNOWLEDGMENTS

This research was sponsored jointly by the Electric Power Research Institute and the U.S. Environmental Protection Agency under Union Carbide Corporation contract W-7405-eng-26 with the U.S. Department of Energy.

The able assistance of G. M. Henderson in conducting these experiments is gratefully acknowledged.

## REFERENCES

1. Adams, J., Menzies, K. and Levins, P. (1977): Selection and evaluation of sorbent resins for the collection of organic compounds. EPA-600/7-77-044, U. S. Environmental Protection Agency, Office of Research and Development, Washington, DC.
2. Caton, J. E., Maskarinec, M. P., Henderson, G. M., Harvey, R. W. and Guerin, M. R. (1979): The determination of tracer compounds by liquid scintillation counting after preparation of the samples by oxidation of the host matrix. Paper No. 474 presented at *The 1979 Pittsburgh Conference on Analytical Chemistry and Applied Spectroscopy*, Cleveland, Ohio, March 5-9, 1979.
3. Campbell, J. A., Laul, J. C., Nielson, K. K. and Smith, R. D. (1978): Separation and chemical characterization of finely-sized fly ash particles. *Anal. Chem.* 50:1032-1040.
4. Cuffe, S. T. and Gerstle, R. W. (1967): Emissions from coal-fired power plants: comprehensive summary. U. S. Department of Health, Education, and Welfare, Durham, NC.
5. Cuffe, S. T., Gerstle, R. W., Ornig, A. A., and Schwarts, C. H. (1964): Air pollutant emissions from coal-fired power plants; Report No. 1. *J. Air Poll. Control Assoc.* 14:353-362.
6. Davidson, R. L., Natusch, D.F.S., Wallace, J. R. and Evans, C. A. (1974): Trace elements in fly ash. Dependence of concentration on particle size. *Environ. Sci. & Tech.* 8:1107-1113.

7. Diehl, E. K., du Braeuil, F., and Glenn, R. A. (1967): Polynuclear hydrocarbon emission from coal-fired installations. *J. Eng. Power* 89(A):276-282.
8. Griest, W. H. (1978): Multicomponent polycyclic aromatic hydrocarbon analysis of inland water and sediment. *Proceedings of the International Symposium on Analysis of Hydrocarbons and Halogenated Hydrocarbons in the Aqueous Environment,* McMaster University, Hamilton, Ontario, Canada, May 25-27, in press.
9. Griest, W. H. and Guerin, M. R. (1979): *Identification and quantification of polynuclear organic matter (POM) on particulates from a coal-fired power plant.* EPRI EA-1092 Project 1057-1, Electric Power Research Institute, Palo Alto, CA.
10. Griest, W. H., Yeatts, L. B. and Caton, J. E. (1979): Recovery of polycyclic aromatic hydrocarbons sorbed on fly ash for quantitative analysis. *Anal. Chem.*, in press.
11. Guerrini, R. and Pennacchi, A. (1975): Evaluation of polynuclear hydrocarbons in stack gases from combustion of fuel oil and coal. *Rivista dei Combusti bili* (Milan) 29:349-357.
12. Hangebrauch, R. P., von Lehmden, D. J. and Meeker, J. E. (1967): Sources of polynuclear hydrocarbons in the atmosphere. U. S. Department of Health, Education, and Welfare, Cincinnati, OH.
13. Higgins, C. E., and Guerin, M. R. (1979): Recovery of diaromatics during evaporative concentration. Presented at the *Southeastern Regional Meeting of the American Chemical Society,* Roanoke, Virginia, October 24-26.
14. Jager, J. (1969): Behavior of polycyclic aromatic hydrocarbons adsorbed on solid carriers. Part I. Extraction of polycyclic aromatic hydrocarbons from solid carriers. *Ceskoslov. Hyg.* 14:135-142.
15. Kolar, L (1969): Carcinogenic 3,4-benzopyrene in the fly ash of lignite. *Rostl. Vyroba* 15:1103-1110.
16. Mather, B. (1961): Nature and distribution of particles of various sizes in fly ash. Technical Report 6-583, Corps of Engineers, Vicksburg, MISS. Quoted in Ray, S. S. and Parker, F. G. (1977): Characterization of ash from coal-fired power plants. EPA-600/7-77-010, U. S. Environmental Protection Agency, Office of Energy, Minerals Industry, Research Triangle Park, NC, p. 38.
17. Shipman, L. L. (1978): Ab inito quantum mechanical characterization of the ground electronic state of benzo (a) pyrene. Implications for the mechanism of polynuclear aromatic hydrocarbon oxidation to epoxides by cytochrome P-450. In: *Carcinogenesis, Vol. 3: Polynuclear Aromatic Hydrocarbons,* P. W. Jones and R. I Freudenthal, eds. pp. 139-44, Raven Press, NY.

# THE VOLATILITY OF PAH AND POSSIBLE LOSSES IN AMBIENT SAMPLING

**R. C. Lao and R. S. Thomas**

Chemistry Division
Air Pollution Control Directorate
Environment Canada
Ottawa, Canada, K1A 1C8

## INTRODUCTION

Air sampling, surveys, and data based on classical air filtration procedures have neglected consideration of possible losses due to volatility of the compounds sampled. A number of hazardous air pollutants, among them the polycyclic aromatic hydrocarbons (PAH), possess vapor pressures which raise serious doubts about the validity of published ambient air quality data for these compounds. Saturated or equilibrium vapor concentrations (EVCs) for some of the PAH have been determined by one of the authors (12), who found values equal to or greater than the concentrations commonly found in ambient air. Even when the EVCs are apparently reduced by effects of adsorption on airborne particulate matter, the existence and relative importance of the EVC of environmental PAH permits a qualitative estimation of the volatility and corresponding inefficiency of collection.

A number of methods have been developed for sampling PAH from the ambient atmosphere. Since its original application in 1960 (14), collection by high-volume sampling with glass-fiber filters remained the most widely used technique. Although the high-volume sampling method is very efficient in collecting airborne particulates, studies have clearly demonstrated that considerable evaporation losses of PAH may occur in the presence of air flows (12) and that PAH present in either very fine particulate form or in the vapor phase will evade collection on the filtration medium. Recently, the use of backup adsorbents or a tandem arrangement of silver membrane and glass-fiber filter has been evaluated.

The results show a marked improvement in PAH collection efficiency (3,8).

Over the past two decades, possible losses and/or disappearance of benzo(a)pyrene (BaP) and other PAH during high-volume sampling have been investigated. Unfortunately, most of the data are unsystematic and generally the results have been misinterpreted (12). Commins and Lawther (1) first drew attention to the implicit assumption of air sampling that PAH are present in air as particulates and are not volatile at normal sampling temperatures. These authors found that at 100°C a relatively large percentage of BaP disappears from a glass disk and concluded that this loss was attributable to volatilization. Since they could not detect any appreciable volatilization at room temperature, they speculated that disappearance must be due to oxidation or molecular rearrangement processes.

In a later study, Commins (2) observed that the disappearance of low-molecular weight PAH from impregnated glass-fiber filters occurs when air is passed through the filter at room temperature, but again no losses were measurable from spiked air samples. On the basis of these data, Commins concluded, therefore, that PAH must be stabilized by adsorption effects of the particulate matter.

Rondia (13) observed the volatility of BaP at 140°C but failed to detect the loss at 100°C. He also observed the disappearance of some PAH from impregnated glass-fiber filter in a closed container. Rondia thus dismissed all previous measurements of phenanthrene or PAH with less than three rings as invalid, and questioned the validity of pyrene and fluoranthene determination.

From his own experimental results, Rondia concluded that the disappearance of PAH could be due to volatilization but could not be attributable to photooxidation as had been postulated. Further, he asserted that the boiling- and melting-points temperatures are the primordial factor in determining the rate of disappearance, not the physical state of the PAH dispersed over glass fibers (thin film or crystalline).

The results of Rondia and Commins cannot be objectively compared because of the disparity of the amounts of PAH used in their experiments. Both workers observed disappearances of PAH from the impregnated glass-fiber filters but could not agree on the circumstances that caused the phenomenon. The details of measured loss rate for some PAH compared to their EVC have been discussed in a previous paper (12).

A common source of possible systematic error introduced in the collection of PAH from the ambient air is the interaction of these compounds with airborne oxidants and other reactive substances (6,9,11,15). Experiments which hope to demonstrate the loss of PAH associated with a particulate sample by merely passing air through the filter medium and measuring the remaining substances cannot effectively evaluate the vapor pressure of PAH adsorbed on particulate matter (12). Since the number and structure of potential products that might be formed from PAH are not generally known, the alternative of a qualitative screening for compounds such as quinones, expoxides, or acids is not a satisfactory approach at the concentration levels of ambient PAH samples. One method for carrying out this type of investigation would be to isotopically tag the initial PAH load. This study has just begun and further work is urgently needed.

In addition to the sampling processes, particular attention must be given to the avoidance of PAH losses at each stage in the sample preparation and analysis procedures. Solvent extraction is the most common technique for recovering the PAH from the sampling matrix. This step is followed by clean-up and concentration operations prior to any instrumental analysis. The usual solvent extraction and clean-up techniques require a long residence time and some decomposition, photooxidation, or artifact formation may occur. Serious manipulative losses of PAH due to evaporation can happen during the concentration process, particularly in situations where extracts are allowed to evaporate to dryness.

This preliminary study is designed to assess the possible loss of PAH by vaporization and highlight some basic problems in the application of the high-volume sampling techniques for PAH. Although this method may have shortcomings, because of the necessity to use the existing equipment for sampling large volumes of air on a routine basis for purposes of evaluating the impact of long-range transport of air pollutants, it is imperative that the limitations and applications of this technique be defined.

## EXPERIMENTAL

Aliquots of cyclohexane (2 ml) solutions of PAH at various concentrations were placed in a 50-ml pyrex beaker 40-mm high and 40-mm ID. The beakers were cleaned with chromic acid to smooth the surface of the glass and remove potential adsorbents. The breaker was kept in the fume hood at room temperature (20°C), with indirect fluorescent lighting during the daytime hours. Starting times were recorded when the 2 ml of PAH solution was poured into the beaker. After a set interval of time, cyclohexane was added to the beaker, the residue was redissolved, and the concentration of remaining PAH was determined by gas chromatography. Each determination was performed in duplicate.

Aliquots of PAH (1 ml) solution were added dropwise to a 47-mm Gelman-A glass-fiber disk placed over the mouth of a pyrex beaker and were retained in a fume hood. After a set interval of time the filter was placed in a 150-ml beaker, extracted by agitation with cyclohexane, and made up to volume; the concentration of PAH was then determined by GC and compared with that of original prepared solution.

A Perkin-Elmer Model 900 GC equipped with FID and Dexsil 300 column was used for the measurement of PAH. Details of PAH standards, instrumental calibrations, and analytical procedures have been documented (7).

All filters and apparatus were extracted with pure cyclohexane and the extracts were analyzed by GC-FID to ensure freedom from PAH contamination. Syringes (10 $\mu$l) manufactured by Hamilton Company were used for injection. Sample volume injected varied from 0.2 to 5.0 $\mu$l depending on sample concentrations. Results were obtained from an average of three to five injections within 5 percent deviation. Disappearance-rate curves were drawn on the basis of best fit derived from the least-square methods.

## RESULTS AND DISCUSSION

For the first set of experiments, 0.1 $\mu$g BaP was dissolved in 2 ml of cyclohexane. Results are given in Table 1 and plotted in Figure 1.

TABLE 1. Disappearance of 0.1 $\mu$g BaP from Pyrex Beakers

| Run | Time (hr) | Missing BaP ($\mu$g) | Recovered BaP ($\mu$g) | BaP Recovered (%) |
|---|---|---|---|---|
| 1 | 5  | 0.019 | 0.081 | 81 |
|   | 7  | 0.034 | 0.066 | 66 |
|   | 8  | 0.046 | 0.054 | 54 |
|   | 9  | 0.043 | 0.057 | 57 |
|   | 24 | 0.056 | 0.044 | 44 |
|   | 30 | 0.058 | 0.042 | 42 |
|   | 48 | 0.063 | 0.037 | 37 |
| 2 | 5  | 0.017 | 0.083 | 83 |
|   | 7  | 0.027 | 0.073 | 73 |
|   | 8  | 0.038 | 0.062 | 62 |
|   | 9  | 0.041 | 0.059 | 59 |
|   | 30 | 0.044 | 0.056 | 56 |
|   | 48 | 0.050 | 0.050 | 50 |

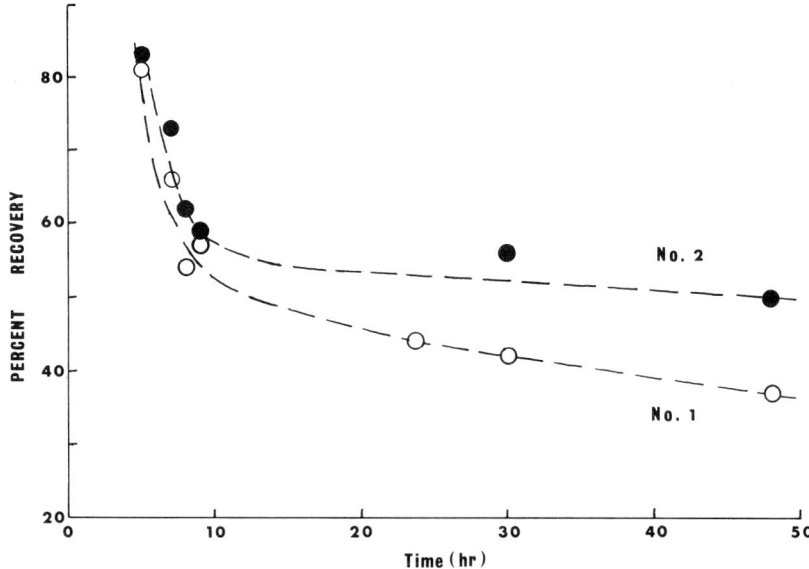

FIGURE 1. Disappearance of 0.1 μg BaP from Pyrex beakers.

Although the amount of BaP recovered from the parallel runs is not always directly related to time, there is a definite trend which indicates that the rate of disappearance is fast for the first 10 hours and levels off thereafter. The disparity in the results is partly explained by experimental errors due to the limitations of injection volume and the minute initial amounts of BaP used. In addition, variations between the surfaces of each beaker will also contribute to the difference. No definite reaction rate for disappearance can be predicted, as could be expected if the chemical kinetics were established in a controlled atmosphere. However, for the purpose of exposing the parameters of current sample treatment techniques, the plots provide an indication of expected losses.

The rates of disappearance of PAH with increased initial concentrations were then determined and the results are listed in Table 2 and illustrated in Figures 2, 3, and 4. For pressures in excess of $10^{-4}$ atm, the rate of diffusion of the vapor in the air controls the rate of sublimation and it is independent of the initial concentration (5). The evaporation process must be a zero-order reaction, even though, in Figure 2, a first-order reaction would seem to be a closer fit. The situation is obvious if log percentage of PAH recovery is plotted against time, as in Figures 3 and 4. In these two examples, a straight line can be drawn and the reaction rate is a simple function of the concentration. The disappearance phenomenon then is related to more than a simple evaporation effect. Specifying the

## TABLE 2. Disappearance of PAH from Pyrex Beakers

| Run | PAH | Quantity ($\mu$g) | Time (hr) | Missing PAH ($\mu$g) | Recovered PAH ($\mu$g) | PAH Recovered (%) |
|---|---|---|---|---|---|---|
| 1. | BaP | 10 | 18 | 0.4 | 9.6 | 96 |
| | | | 25 | 0.6 | 9.4 | 94 |
| | | | 41 | 1.0 | 9.0 | 90 |
| | | | 65 | 1.5 | 8.5 | 85 |
| | | | 91 | 2.6 | 7.4 | 74 |
| | | | 117 | 2.9 | 7.1 | 71 |
| | | | 141 | 2.6 | 7.4 | 74 |
| | | | 165 | 3.3 | 6.7 | 67 |
| | | | 189 | 3.0 | 7.0 | 70 |
| | | | 285 | 3.8 | 6.2 | 62 |
| | | | 309 | 4.5 | 5.5 | 55 |
| | | | 323 | 4.2 | 5.8 | 58 |
| | | | 357 | 4.9 | 5.1 | 51 |
| | | | 419 | 5.3 | 4.7 | 47 |
| 2. | BaP | 8 | 89 | 2.5 | 7.5 | 75 |
| | | | 137 | 3.0 | 7.0 | 70 |
| | | | 185 | 3.8 | 6.2 | 62 |
| | | | 257 | 4.3 | 5.7 | 57 |
| | | | 305 | 4.9 | 5.1 | 51 |
| | | | 473 | 6.6 | 3.4 | 34 |
| 3. | Fluoranthene | 10 | 9 | 1.0 | 9.0 | 90 |
| | | | 24 | 2.5 | 7.5 | 75 |
| | | | 48 | 4.9 | 5.1 | 51 |
| | | | 72 | 6.5 | 3.5 | 35 |
| | | | 105 | 7.7 | 2.3 | 23 |
| | | | 150 | 8.7 | 1.3 | 13 |
| 4. | Pyrene | 10 | 9 | 1.6 | 8.4 | 84 |
| | | | 24 | 3.0 | 7.0 | 70 |
| | | | 48 | 5.2 | 4.8 | 48 |
| | | | 72 | 6.4 | 3.6 | 36 |
| | | | 105 | 7.7 | 2.3 | 23 |
| | | | 146 | 8.7 | 1.3 | 13 |
| | | | 157 | 8.9 | 1.9 | 11 |
| | | 50 | 6 | 1.0 | 9.0 | 90 |
| | | | 24 | 2.1 | 7.9 | 79 |
| | | | 31 | 3.5 | 6.5 | 65 |
| | | | 79 | 4.8 | 5.2 | 52 |
| | | | 103 | 5.5 | 4.5 | 45 |
| | | | 168 | 7.5 | 2.5 | 25 |
| 5. | 3,4 Benz-acridine | 10 | 0 | 0.0 | 10 | 100 |
| | | | 25 | 5.2 | 4.8 | 48 |
| | | | 92 | 7.2 | 2.8 | 28 |
| | | | 120 | 7.5 | 2.5 | 25 |
| | | | 163 | 8.6 | 1.4 | 14 |
| | | | 195 | 9.1 | 0.9 | 9 |
| | | | 240 | 9.3 | 0.7 | 7 |
| | | | 280 | 9.6 | 0.4 | 4 |

percentage loss rather than the loss per unit throughout conditions is an unfortunate error of most previous collection efficiency studies (12). It is also surprising to find out in Table 2 that the absolute amount of pyrene in micrograms lost over a period of time T is of the same order whether the starting amount is 10 or 50 μg. Since oxidation or molecular rearrangements both are first-order reactions, the similarity of losses

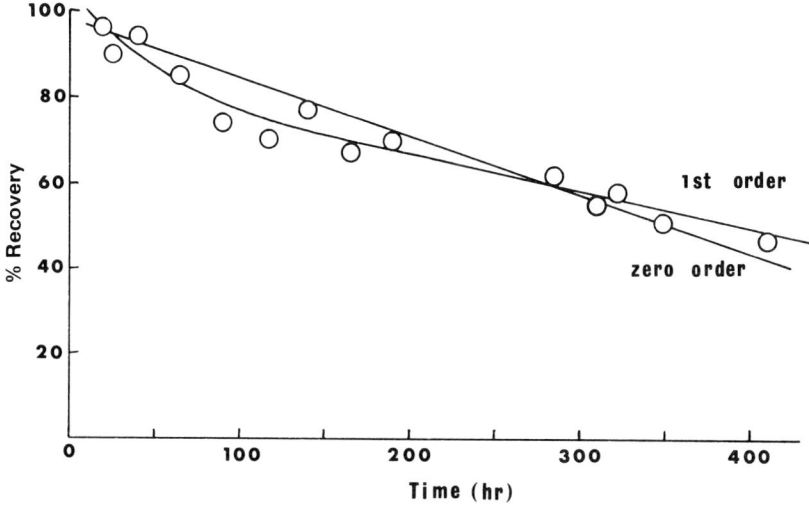

FIGURE 2. Disappearance of 10 μg BaP from Pyrex beaker.

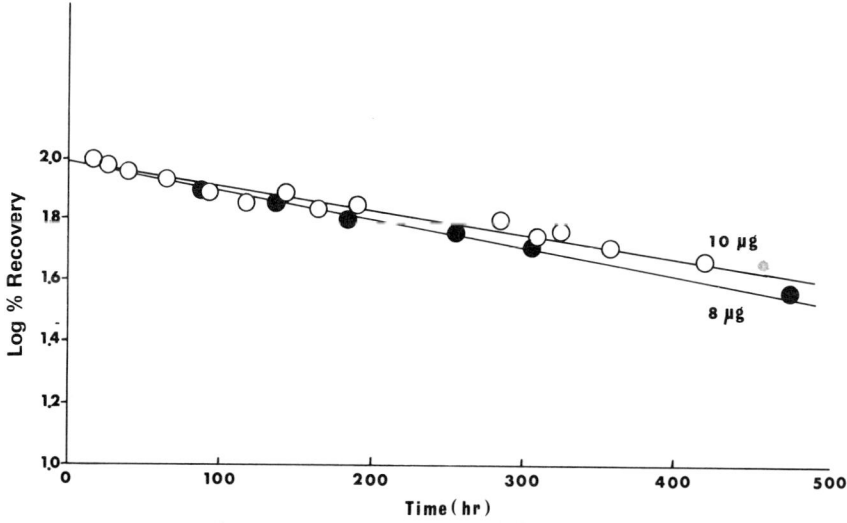

FIGURE 3. Disappearance of 10 and 8 μg BaP from Pyrex beakers.

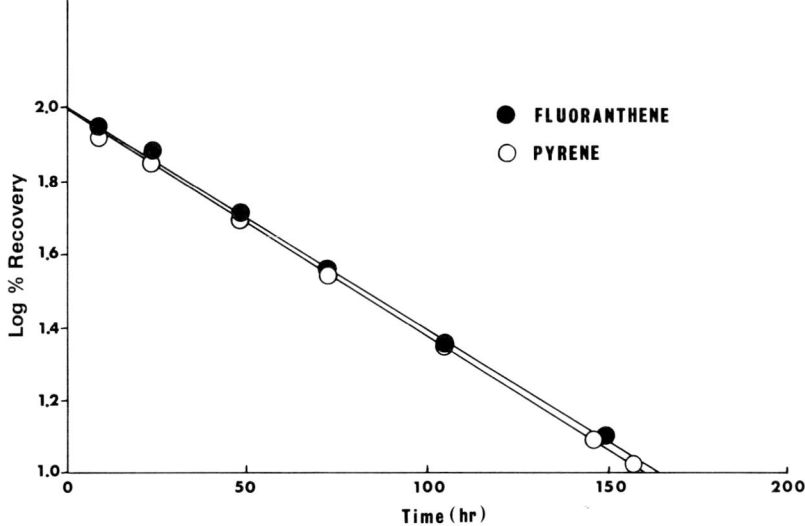

FIGURE 4. Disappearance of fluoranthene and pyrene from pyrex beakers.

indicates that the PAH disappearance mechanism can be limited to either evaporation or desorption. The conclusion is supported by the direct evidence that there are no oxidation and rearrangement products in GC runs for the exposed PAH investigated in this work, unless all the by-products have retention times identical to those of the parent compounds, which is impossible using the Dexsil 300 column.

Another aspect of the volatility problem follows from the work of Rondia (13), who observed losses at room temperature of PAH of low molecular weight, such as fluoranthene, phenanthrene, and pyrene, which had been adsorbed on airborne particulates. Only at elevated temperatures did he observe a significant decrease in quantity of BaP and BeP recovered.

Ten 49-mm diameter disks were cut from the exposed surface of a glass-fiber sheet used to sample ambient air. All the disks were spiked with a measured concentration of a BaP and benzo(k) fluoranthene (BkF) blend. Four pairs of disks were fitted in standard holders and had a measured flow of air passed through them over several intervals of time. The remaining pair of disks served as a control. Table 3 gives the results of BaP and BkF determinations by GC. No losses were observed for either PAH, and this confirms the previous conclusions (12) that particulates are able to stabilize adsorbed PAH and that high-volume sampling is almost completely efficient at normal ambient sampling temperatures for the collection of BaP, the most notorious of the PAH.

**TABLE 3. Disappearance of BaP and BkF from Used Glass-Fiber Filters**

| Run | Time (hr) | Total Air flow (l/cm$^2$) | BkF ($\mu$g) | BaP ($\mu$g) |
|-----|-----------|---------------------------|--------------|--------------|
| 1   | —         | 0                         | 1.15         | 2.05         |
| 2   | 24        | 1341.4                    | 1.18         | 1.72         |
| 3   | 49        | 2738.6                    | 1.12         | 2.38         |
| 4   | 72        | 4024.0                    | 1.17         | 2.23         |
| 5   | 120       | 6706.8                    | 1.13         | 1.82         |

## CONCLUSIONS

From these results, several conclusions can be drawn to establish some of the characteristics of the volatility of PAH and their potential losses during ambient sampling.

1. PAH that are collected on glass-fiber filters having sufficient particulate loading are adsorbed on the particulates and stabilized. The flow of air through the filter should not affect the composition and amount of all PAH but only the most volatile compounds.
2. From the measurements of vapor pressure (12) and lattice energy (26 to 29 kcal/mol for $C_{20}H_{12}$ isomers), the equilibrium between gaseous and solid BaP requires that EVC of BaP be reached before any solid appears. For BaP at 25°C the EVC value is about 136 ng/m$^3$. However, the EVC must be regarded as the upper limit since PAH are either adsorbed on particulates or intercolated in the carbon structure of the particle. What is important is that at higher sampling temperatures, PAH have a tendency to desorb from the particulates or dissolve in any entrained oily matrix and, therefore, sampling at 100°C appears theoretically less efficient than that at 25°C by approximately a factor of 8000. This factor could partially explain the seasonal difference in the ambient PAH levels in winter and summer.
3. The experiments to determine disappearance of PAH from glass-fiber filters show that the total amount of PAH which disappeared seemed to be constant per unit time. This implies that the process is not simply a function of evaporation which would produce constant

rate of disappearance but is a function of surface phenomenon. If PAH, when adsorbed on airborne particulate, have a similar heat of adsorption to that of simple aromatic hydrocarbons on activated charcoal, then, for low coverage on the adsorbing surface, $\Delta H_{ads}$ will be on the order of 12 to 15 kcal/mol. Such PAH would have negligible vapor pressure at the ambient temperature. Since $\Delta H_{ads}$ is an integrated value at a high coverage factor, when most of the surface sites are covered, the process of evaporation begins and some loosely bound PAH are also released.

Because a similar phenomenon was observed during the Pyrex beaker experiments, it can be concluded that contradictory published results may be attributed to the difference in the total initial amount of PAH used.

4. Under the conditions of ambient temperature and pressure, the bulk of PAH exist in the solid-state form and are adsorbed on the surface of particulate matter, preferentially on smaller particles (10). The high-volume sampling technique can be used without losses of BaP isomers or PAH with higher molecular weight.

5. Care should be taken in using the various chromatographic techniques for calibration or quantitative determinations, since PAH may be adsorbed on the active sites of a column-support material. This may account for the observed losses and for the different responses from compound to compound.

## REFERENCES

1. Commins, B. T., and Lawther, P. J. (1958): Volatility of 3,4-Benzoyrene in relation to the collection of smoke samples. *Brit. J. Cancer* 12: 351–354.
2. Commins, B. T. (1962), *Natl. Cancer Inst. Monograph* No. 9: 225–233.
3. Jackson, J. O., and Cupps, J. A. (1978): Field evaluation and comparison of sampling matrices for PAH in occupational atmospheres. In: *Carcinogenesis,* Vol. 3, P. W. Jones and R. I. Freudenthal, Eds., pp. 183–191, Raven Press, New York.
4. Jones, P. W., Strup, P. E., and Giammar, R. D. (1976): Improved measurement techniques for PAH in combustion effluents. In: *Carcinogenesis,* Vol. 1, P. W. Jones and R. I. Freudenthal, Eds., pp. 241–251, Raven Press, New York.
5. Jost, W. (1952): *Diffusion in Solids, Liquids and Gases,* Academic Press, New York.

6. Katz, M., and Lane, D. A. (1977): The photomodification of BaP, BbF and BkF under simulated atmospheric conditions. In: *Advances in Environmental Science and Technology, Fate of Pollutants in the Air and Water Environments,* Vol. 9, I. H. Suffet, Eds., pp. 137-54, Wiley Interscience, New York.
7. Lao, R. C., Thomas, R. S., and Monkman, J. L. (1977): Application of GC-MS to the analysis of PAH in environmental samples. In: *Carcinogenesis* Vol. 1, P. W. Jones and R. I. Freudenthal, Eds., pp. 271-281, Raven Press, New York.
8. Lingren, J. L., Krauss, H. J., and Fox, M. A. (1978): A comparison of two techniques for the collection and analysis of PAH in ambient air. In: *Proceedings, 29th Pittsburgh Conference,* No. 446.
9. McGinnes, P. R., and Smoeyink, V. L. (1974): Determination of the fate of PAH in natural water systems. *University of Illinois, Water Research Center,* Report No. 80, NTIS PB 232-168.
10. Natusch, D.F.S., and Tomkins, B. A. (1978): Theoretical considera-tion of the adsorption of PAH vapor onto fly ash in a coal-fired power plant. In: *Carcinogenesis,* Vol. 3, P. W. Jones and R. I. Freudenthal, Eds., pp. 145-154, Raven Press, New York.
11. Pitts, J. N. (1979): Photochemical and biological implications of the atmospheric reactions of amines and BaP. *Phil. Trans. R. Soc. London,* A290: 551-576.
12. Pupp, C., Lao, R. C. Murray, J. J., and Pottie, R. F. (1974): Equilibrium vapor concentrations of some PAH and the collection efficiencies of these in air pollutants. *Atmos. Envir.* 8: 915-925.
13. Rondia, D. (1965): Sur la volatilite des PAH. *Intl. J. Air and Water Poll.* 9: 113-121.
14. Sawicki, E.W.C., Elbert, T. W., and Fox, F. T. (1960): BaP content of the air of American communities, *Am. Ind. Hyg. Assoc. J.* 21: 443-451.
15. Thomas, J. F., Mukai, M., and Tebbens, B. D. (1968): Fate of airborne BaP. *Environ. Sci. and Tech.* 2: 33-39.

# SOLVENT EXTRACTION OF POLYNUCLEAR AROMATIC HYDROCARBONS

**W. K. Robbins**
Exxon Research and Engineering Company
Florham Park, New Jersey 07932

Since polynuclear aromatic hydrocarbons (PAH) generally represent a small fraction of a sample, the PAH must be concentrated and isolated from interfering organics prior to their measurement. This paper compares several extraction systems and focuses on the effects of extraction conditions as determined by the measurement of $K_D$ (distribution coefficients) for a number of representative compounds.

Systematic relationships have been shown to exist between $K_D$ and a number of extraction parameters such as PAH molecular weight, alkyl substitution of PAH, solvent strength, temperature, and extractant dilution. These relationships are consistent with a simple thermodynamic model of two-phase equilibria. Furthermore, they simplify the design of extraction procedures, since the measurement of one or two $K_D$ values may be sufficient to define the extraction characteristics for a given system.

The objective of a PAH extraction is to optimize the recovery of PAH while minimizing the coextraction of interfering materials. The recovery, or extraction efficiency (%E) for each compound is determined by four factors related by the equation

$$\%E = 100 \times \left[1 - \left(\frac{1}{1 + K_D\left(\frac{V_E}{V_S}\right)}\right)^n\right] \qquad (1)$$

where $V_E$ and $V_S$ are the volumes of extractant and solvent respectively, n is the number of times the solvent phase is extracted with fresh extractant, and $K_D$ the distribution coefficient, is a characteristic of a compound which can be determined experimentally.

A graphical representation of this relationship (Figure 1), combined with measured $K_D$ values for PAH and interfering materials, makes it possible to compare both the selectivity and the efficiency for a number of potential procedures (1). The optimum is achieved when $K_D$ for the PAH is maximized relative to that for the interfering material. As can be seen, n (which is the slope in these plots) has an exponential effect on %E. Consequently, for multiple extractions the optimum conditions occur where $K_D (V_E/V_S)$ for the PAH is $> 1$ while that for the interference is $< 1$. Once the $K_D$'s for model PAH have been determined and the effect of extraction conditions established, one can design systems for the extraction of the PAH as a class from a variety of complex matrices.

## MEASUREMENTS OF $K_D$'s

For dilute well-behaved systems of two defined, immiscible phases, the partition coefficient ($K_P$) equals the distribution coefficient ($K_D$) which is defined as the ratio of the concentration of PAH in each phase at equilibrium:

$$\ln K_P = \ln K_D = \ln \frac{(C_{PAH})_E}{(C_{PAH})_S} \qquad (2)$$

The $K_D$ data for model PAH are determined by measuring the concentration of PAH in an aliquot of the solvent phase before and after equilibration with the extractant. Prior to the measurements, the two phases are preequilibrated and separated. A known amount of PAH is added to a small portion of the preequilibrated solvent phase to give a solution of approximately 1 $\mu g/ml$. An aliquot of this solution is then equilibrated against an equal volume of extractant. The concentration of the PAH in the solvent phase before and after equilibration is used to calculate $K_D$ as

$$K_D = \frac{A_i - A_F}{A_F} \qquad (3)$$

where $A_i$ and $A_F$ are the initial and final measurements respectively.

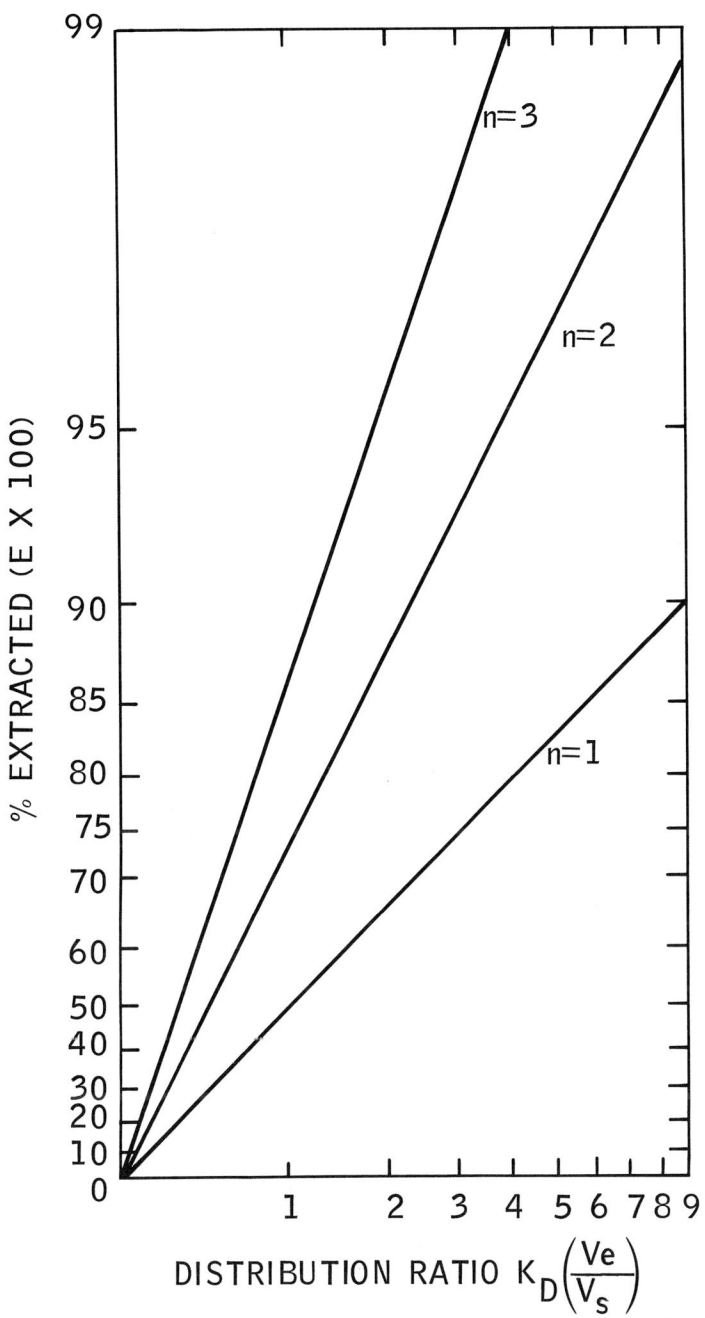

FIGURE 1. Plot of percentage extracted versus distribution ratio.

In general, this approach may be used with any measurement technique that has adequate selectivity and sensitivity. In the present study, either UV absorbance or radioassay techniques were applied to the calculation of the $K_D$ values for a number of PAH (Table 1). As expected, nearly identical results were obtained with UV and radioassay measurements of $K_D$ for benzo(a)pyrene (BaP). Although UV measurements are possible for clean model systems, the radioassay measurements are preferred because they can be applied to complex or colored matrices.

## $K_D$'s FOR PAH IN EXISTING SYSTEMS

Some of the characteristics of PAH extractions are illustrated by their $K_D$'s in four existing analytical procedures (Table 2). The U. S. Food &

**TABLE 1. PAH Compounds Studied**

| $^{14}$C-PAH (Radioassay) | Unlabelled PAH (UV Absorption) |
|---|---|
| Toluene | Pyrene |
| Anthracene | 1-Methylpyrene |
| Phenanthrene | 7 Methylbenz(a)anthracene |
| Benz(a)anthracene | Benzo(a)pyrene |
| 7,12 Dimethylbenz(a)anthracene | Benzo(g,h,i)perylene |
| Benzo(a)pyrene | Coronene |
| Dibenz(a,h)anthracene | |

**TABLE 2. Distribution Coefficients ($K_D$) for PAH Extraction Systems**

| Compound | Iso-Octane DMSO | Iso-Octane DMSO/$H_3PO_4$ (4:1) | Cyclohexane[a] DMF/$H_2O$ (9:1) | Cyclohexane NMP/ACID (4:1) |
|---|---|---|---|---|
| $^{14}$C Toluene | -- | 0.17 | 0.43 | 0.28 |
| $^{14}$C Anthracene | 6.2 | 0.83 | 2.2 (2.4) | 2.0 |
| $^{14}$C Phenanthrene | 4.3 | 0.77 | 2.5 (2.7) | 2.1 |
| Pyrene | -- | 1.0 | 3.4 (3.3) | 3.0 |
| -1 Methyl | -- | 0.33 | 2.2 | 2.2 |
| $^{14}$C Benz(a)anthracene | 20 | 1.9 | 5.0 (4.3) | 5.6 |
| -7 Methyl | | 0.50 | 2.6 | 2.8 |
| -7,12 Dimethyl | 6.0 | 0.33 | 1.7 | 1.5 |
| $^{14}$C Benzo(a)pyrene | 22 | 1.8 | 5.5 (6.9) | 6.1 |
| $^{14}$C Dibenz(a,h)anthracene | 55 | 3.3 | 12. (8.7) | 14.4 |
| Benzo(g,h,i)perylene | -- | 2.9 | 8.4 (7.4) | 10.2 |
| Coronene | -- | 2.6 | 10. (9.3) | 20 |

[a] Values in parenthesis are from reference 7.

Drug Administration (FDA) developed an iso-octane/dimethylsulfoxide (DMSO) extraction procedure for isolation of PAH from a number of matrices in the early 1960's (2,3,4). Later FDA procedures used iso-octane with $DMSO/H_3PO_4$ (4:1) (5,6). Grimmer and co-workers developed a similar selective extraction procedure based on cyclohexane and DMF-$H_2O$ (9:1) which is widely used in Europe (7,8). Earlier this year, we reported the use of N-methyl pyrrolidone (NMP) in the preparation of wastewater extracts (9). The system consisted of cyclohexane and NMP diluted with 5 percent phosphoric acid in water. The ratio of NMP to dilute $H_3PO_4$ (which will be simply called "acid" hereafter) was 4:1.

The results obtained are in good agreement with the limited literature values. Taken collectively, the results suggest parallel behavior between these extraction pairs. Upon closer study, a number of factors can be observed to have systematic effects on the $K_D$ for PAH.

For these defined systems, the $K_D$'s are best discussed in terms of the Nernst distribution law:

$$\ln K_D = \frac{\mu_S - \mu_E}{RT} \tag{4}$$

where $\mu_S$ and $\mu_E$ are the standard state chemical potentials in the solvent and extractant phases respectively. If the standard chemical potentials ($\mu°$'s) in Equation (4) are taken to be saturated solutions, it may be shown that the distribution coefficient is a measure of the relative solubility of the PAH in each phase. With this in mind, the data in Table 2 were scrutinized.

## MOLECULAR STRUCTURE EFFECTS

For each system, the distribution coefficients increase with the molecular weight of the PAHs. A plot of $\ln K_D$ against molecular weight of the unsubstituted PAH is nearly linear from toluene (one ring) through coronene (six rings). A better fit of the data is obtained if the PAH are grouped as linear plus cata-condensed or bent plus peri-condensed as illustrated in Figure 2. When the substituted PAH are compared with their parent analogs, a decrease in $K_D$ is observed (Figure 3).

The log-linear relationship between $K_D$ and molecular weight or alkyl substitution is similar to that found for series of homologous compounds in liquid partition chromatography (10), in partition-gas chromatography (11), and, to a certain extent, in adsorption chromatography (11), and, to a certain extent, in adsorption chromatography (12). Increasing the

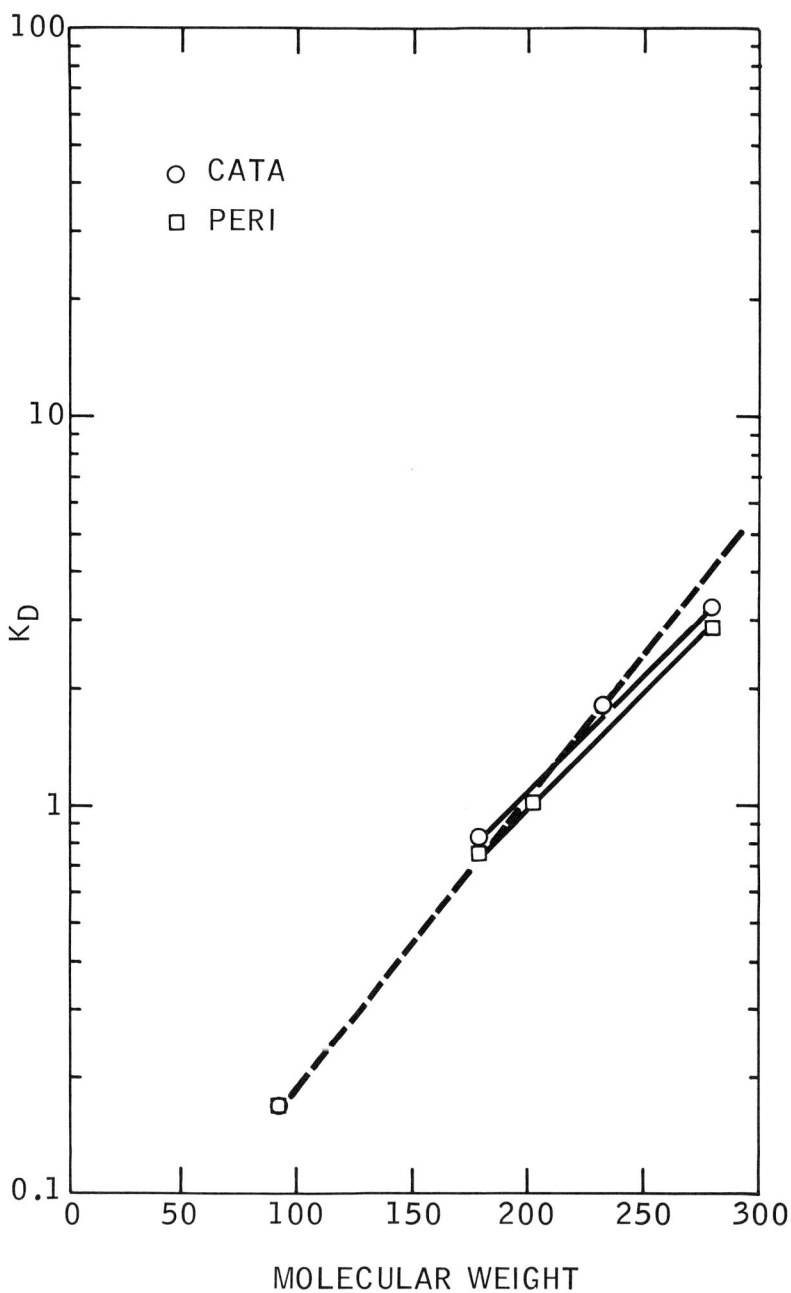

FIGURE 2. $K_D$'s for PAH — $DMSO/H_3PO_4$ (4:1) versus iso-octane.

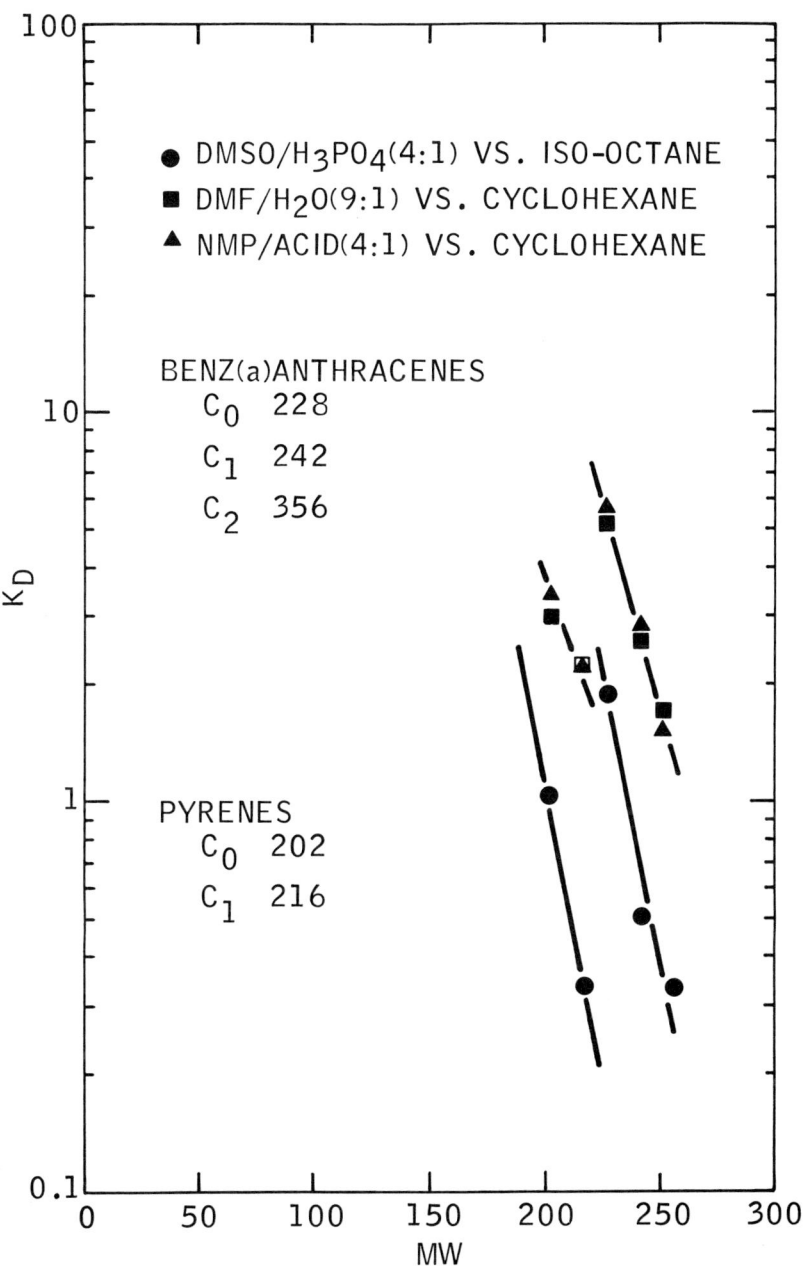

FIGURE 3. Effect of alkyl substitution on $K_D$'s.

polarizability of the molecular $\pi$-cloud increases the solubility of the PAH in the extractant phase, while increasing the alkyl substitution increases the solubility of the PAH in the solvent phase. The regular response of $K_D$ to these molecular parameters allows separations to be designed for specific compounds even when those compounds themselves are not available.

## EFFECTS OF SOLVENT-PHASE COMPOSITION

When the solvent phase is changed, $K_D$ changes as suggested by Equation (4) (Table 3). This reflects the differences in PAH solubility in the solvents. Size, shape, and aromaticity, which all contribute to solvency, are incorporated in the Hildebrandt treatment of regular solutions (13). With the exception of heptane, the pattern of implied BaP solubility is consistent with the Hildebrandt solubility parameters. A similar solvent dependency appears in the data reported for pyrene in DMSO systems (14).

## EFFECT OF TEMPERATURE ON THE DISTRIBUTION

The effect of temperature on the PAH distributions was studied by determining the $K_D$ for BaP at various temperatures. The concentration of BaP was measured in the solvent phase which had been preequilibrated with the extractant at the selected test temperature. The distribution was then accomplished at that temperature and the concentration determined again. Results were obtained at 20°C (room temperature) and 75°C (steam bath) for the DMSO/iso-octane pair. A data point at 0°C (ice bath) was obtained when a mixed aromatic solvent was equilibrated against NMP-acid. When the data obtained are plotted as ln $K_D$ against $1/T$ in °K, the linear relation predicted by Equation (1) is obtained (Figure 4).

**TABLE 3. Comparison of Aliphatic Solvent Effects When Extracted with Model Systems**

|  | $(K_D)$B(a)P NMP-Acid | $(K_D)$B(a)P DMSO | $(K_D)$ Pyrene DMSO[a] |
|---|---|---|---|
| Pentane | 8.0 | - | 9.0 |
| Hexane | 9.8 | - | 11 |
| Heptane | 8.9 | 9.3[b] | 5.1 |
| Iso-Octane | 18 | 22 | 8.7 |
| Cyclohexane | 6.1 | 11 | 4.6 |

[a] From Reference 14.
[b] From Reference 2.

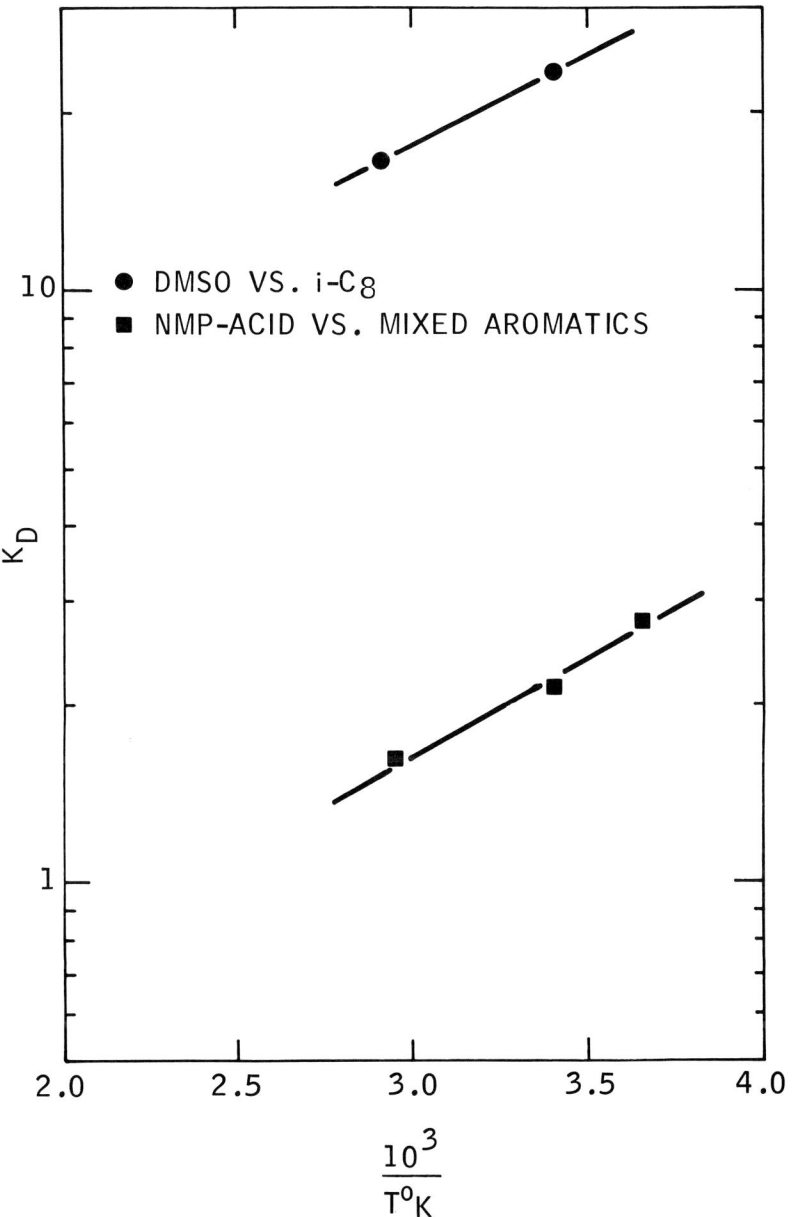

**FIGURE 4.** Effect of equilibration temperature on $K_D$ for BaP.

Although the temperature sensitivity of the distribution is small, the data illustrate the fact that the distribution is affected by temperature in a consistent manner. In many analyses, multiple extractions are performed where the $K_D$'s are close to 1. In such cases, the temperature effect is raised exponentially to become an important factor in determining the percentage extracted. For those extractions performed at elevated temperatures, the percentage of recovery may be significantly lower than predicted from room temperature $K_D$'s.

## EFFECTS OF DILUENTS ON $K_D$

Solvent strength was shown earlier to affect $K_D$ for the PAH. It is well established that solvents may be mixed to attain a desired strength (13). Likewise, a diluent added to the extractant phase will affect its strength. For example, the addition of $H_3PO_4$ to the DMSO systems substantially reduces $K_D$ as shown in Table 2. The addition of a diluent, therefore, provides a means of controlling the $K_D$'s for the PAH.

For most PAH extraction pairs, the effect of a diluent in either phase on $K_D$ is systematic, although not predictable in all respects. The origin of this effect can best be discussed in terms of activity coefficients.

The relation between the partition and distribution coefficients, when generalized, is written:

$$\ln K_P = \ln \left[ \frac{(C_{PAH})_E \gamma_E}{(C_{PAH})_S \gamma_S} \right] = \ln K_D + \ln \left( \frac{\gamma_E}{\gamma_S} \right) \quad (5)$$

where $\gamma_E$ and $\gamma_S$ are the PAH activity coefficients in the extractant and solvent respectively. This may be rearranged to

$$\ln K_D = \ln K_P - \ln (\gamma_{PAH})_E + (\gamma_{PAH})_S \quad (6)$$

For the extractions discussed to this point, the latter two terms are negligible. The presence of a diluent in a phase changes both the activity and the concentration of the PAH.

In general, the activity coefficient is a function of all the solute species which, in the present case, are the analyte A and a diluent M. For moderately dilute solutions in which there are no chemical interactions between the solute species and the concentration of the analyte is very low, the activity coefficient may be approximated by

$$\ln \gamma_{PAH} = K_i\, C_M \tag{7}$$

where $K_i$ is an interaction parameter and $C_M$ is the concentration of the diluent (15).

The effect of the diluent on $\ln K_D$ can then be expressed as the function

$$\ln K_D = \ln K_p - (K_i)_E\, (C_M)_E + (K_i)_S\, (C_M)_S \tag{8}$$

Thus, if the diluent affects only one phase, a linear plot should exist between $\ln K_D$ and the diluent concentration.

## EFFECTS OF DILUENT ADDED TO THE EXTRACTANT

In general, the linear relationship between $\ln K_D$ of the PAH and the concentration of diluent have been observed. The addition of 1 to 50 percent diluent to DMF, DMSO, or NMP was studied in detail for BaP (Figure 5). Above 10 percent dilution, all three systems exhibited the expected linearity. Over the same range, almost parallel responses were observed for $\ln K_D$ of toluene, phenanthrene, benzo(a)pyrene, and coronene when acid was added to the NMP/iso-octane pair (Figure 6). The deviations below 10 percent acid are discussed separately (*vide infra*).

As implied, the addition of diluents causes the $\ln K_D$ for the entire class of PAH to shift; for example, the peri-condensed PAH exhibit that shift in the parallel line observed for three levels of $H_3PO_4$ in DMSO (Figure 7).

A comparison of $H_2O$, acid, and $H_3PO_4$ as diluents illustrates the fact that $K_i$, the interaction parameter in Equation (8), is unique for each diluent (Table 4). In all cases, an equal volume percent of $H_3PO_4$ has a lesser effect than $H_2O$ (for example, Figure 8). The choice of diluent is generally made on the basis of factors other than the shift in $\ln K_D$. For example, DMF and NMP will hydrolyze in the presence of strong acid. Although DMF-$H_2O$ systems have been used successfully on a number of matrices (15), NMP-$H_2O$ systems occasionally resulted in emulsions; however, good phase separation is obtained when the NMP is diluted with acid. The dilute acid results in a minimal amount of NMP hydrolysis. In

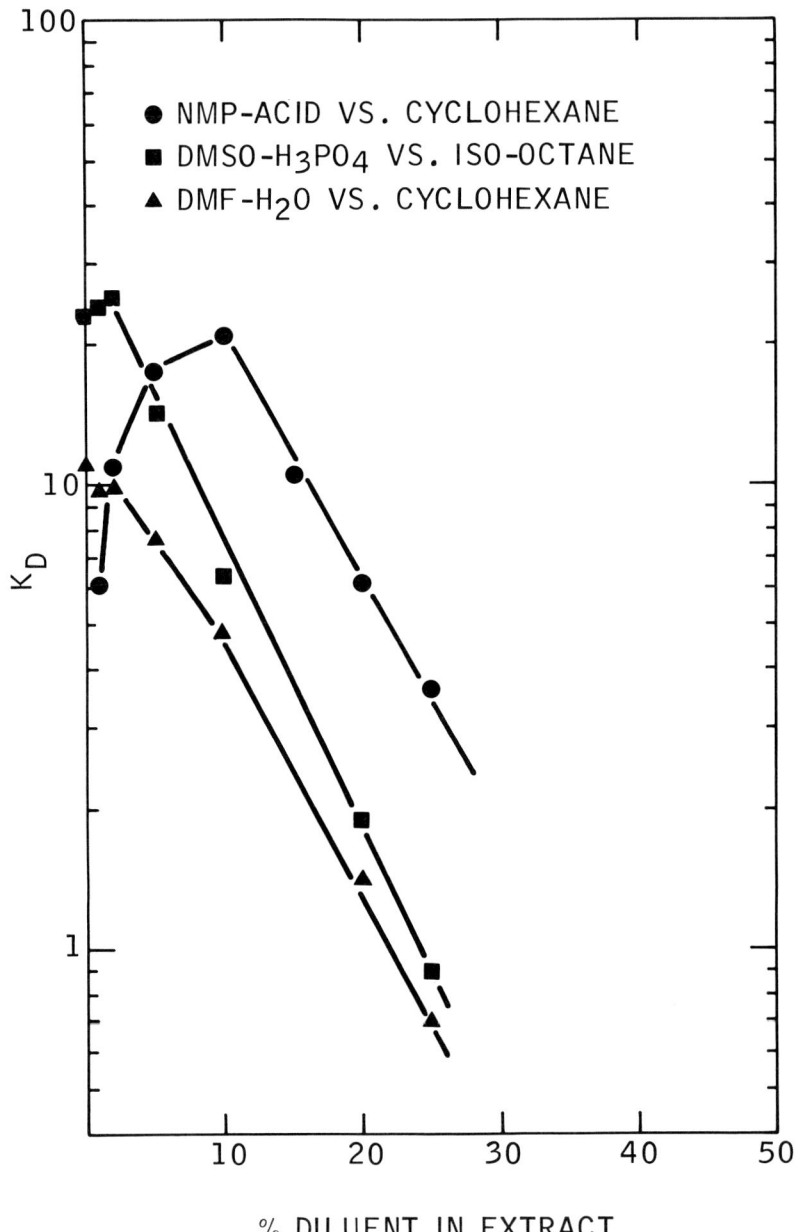

FIGURE 5. Effect of diluents of $(K_D)B(a)P$ in three systems.

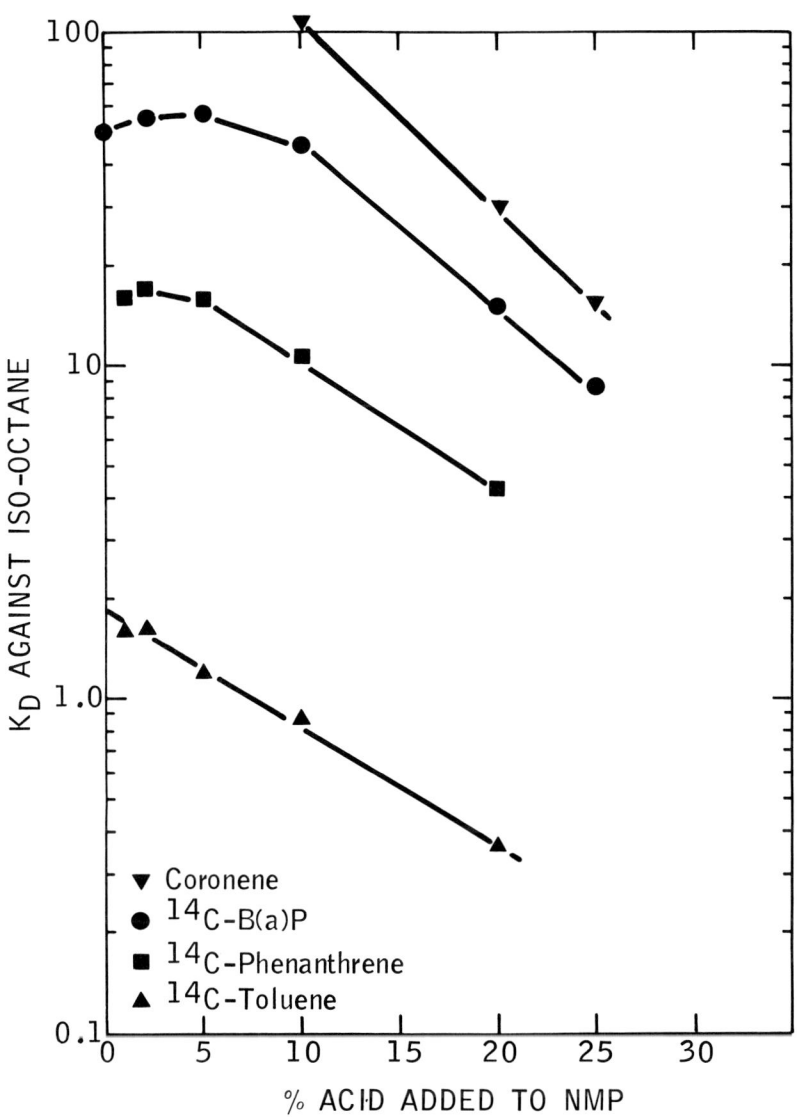

**FIGURE 6.** Effect of acid dilution on the NMP/Iso-Octane pair for 1 to 7 ring compounds.

principle, other diluents could be used in the same manner to control the $K_D$'s for the PAH as a class, but the three described here appear to give adequate flexibility to the extraction systems.

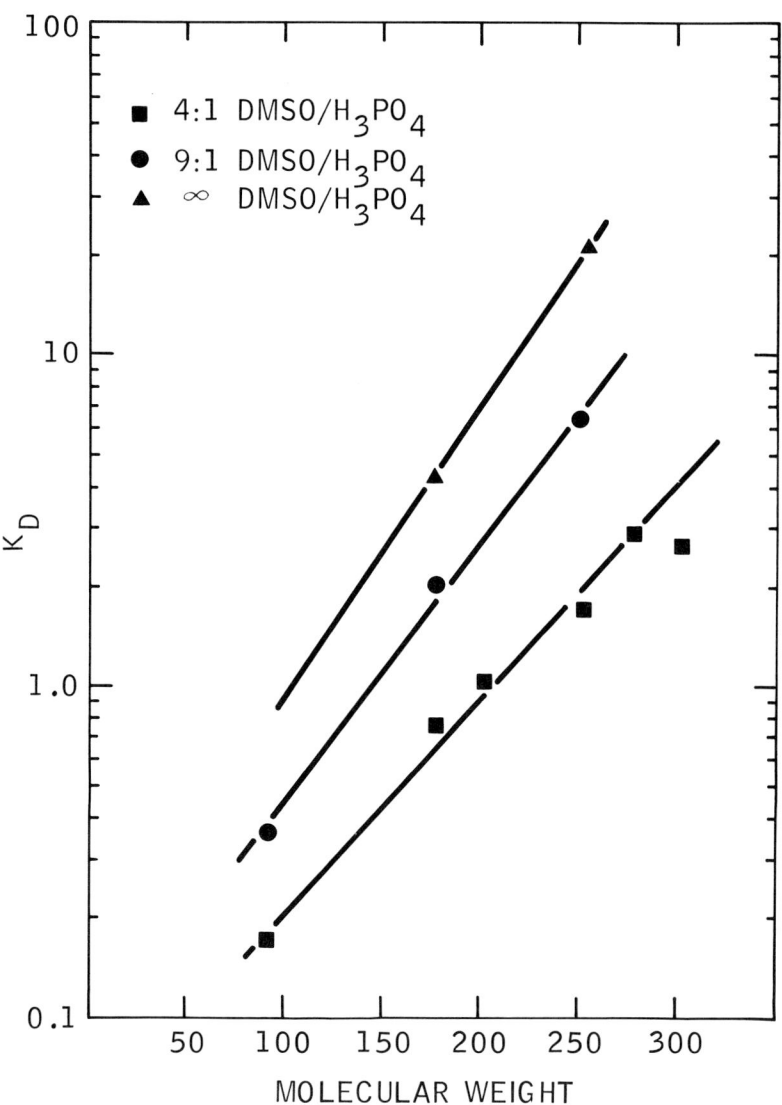

FIGURE 7. $K_D$'s versus MW -- Peri-Condensed PAH in $DMSO/H_3PO_4$ systems.

## EFFECTS OF DILUENT ADDED TO THE SOLVENT PHASE

The addition of a diluent to the solvent phase gives the predicted linear plot of ln $K_D$ versus percentage of dilution (Figure 9). Since benzene is a better solvent for PAH than iso-octane, the $K_D$ decreases with dilution

TABLE 4. Comparison of Diluents on $(K_D)$BaP

| Diluent (%) | DMSO/Iso-Octane | | DMF/Cyclohexane | | NMP/Cyclohexane | |
|---|---|---|---|---|---|---|
| | $K_D$ ($H_2O$) | $K_D$ ($H_3PO_4$) | $K_D$ ($H_2O$) | $K_D$ ($H_3PO_4$) | $K_D$ ($H_2O$) | $K_D$ Acid |
| 0  | 22   | 22   | 11   | 11  | Miscible | Miscible |
| 5  | --   | --   | --   | --  | 14       | 18       |
| 10 | 5.6  | 6.4  | 5.5  | 6.8 | 17       | 21       |
| 20 | 0.83 | 1.8  | 1.4  | --  | 5.8      | 6.1      |
| 25 | 0.56 | 0.77 | 0.70 | --  | --       | --       |

(negative slope). If iso-octane were added to a cyclohexane-NMP pair, a positive slope would be expected.

## EFFECTS OF DILUENT ON BOTH PHASES

The low values of $K_D$ for the cyclohexane-NMP pair at low levels of acid (Figure 5) are attributed to the miscibility of NMP and cyclohexane. NMP and cyclohexane are miscible at room temperature; at low concentrations of acid (~1 percent) two phases appear, but there is substantial comingling of the two reagents in each phase. In terms of Equation (8), the addition of diluent in this case is simultaneously affecting both phases. Some fraction of the diluent added to the extractant phase is being distributed into the solvent phase. At some finite concentration, the solvent phase is essentially saturated and further additions affect only the extractant phase.

## UNIQUE PROPERTIES OF NMP AS AN EXTRACTANT

The data presented in the preceding sections indicate that NMP has the greatest affinity for PAH of the reagents tested. From the physical properties of these polar aprotic solvents, one might expect each to participate in a similar polarization interaction with the PAH. As discussed, each shows a linear decrease in ln $K_D$ with diluent addition above 10 percent (Figure 5). At lower concentrations, however, NMP exhibits a deviation for BaP. When studied in detail, the effect of acid added to the NMP-isooctane pair has been found to depend on molecular weight (Figure 6). The ln $K_D$ for toluene drops off as expected, but the ln $K_D$'s for the PAH do not. As the number of rings increases, more diluent must be added before ln $K_D$ is significantly affected. These data suggest that the NMP is forming a complex with the PAH.

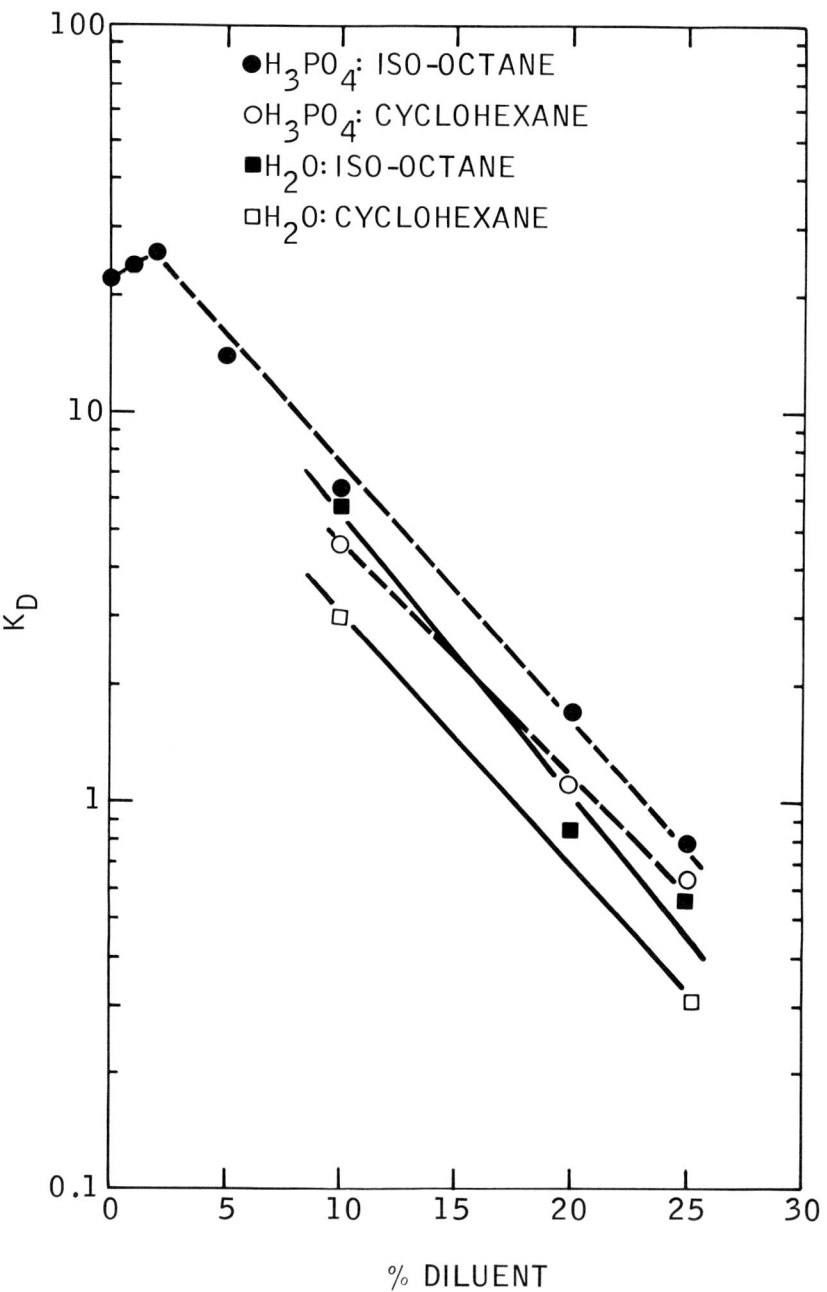

FIGURE 8. Response of $(K_D)B(a)P$ for DMSO to percentage of $H_2O$ or $H_3PO_4$.

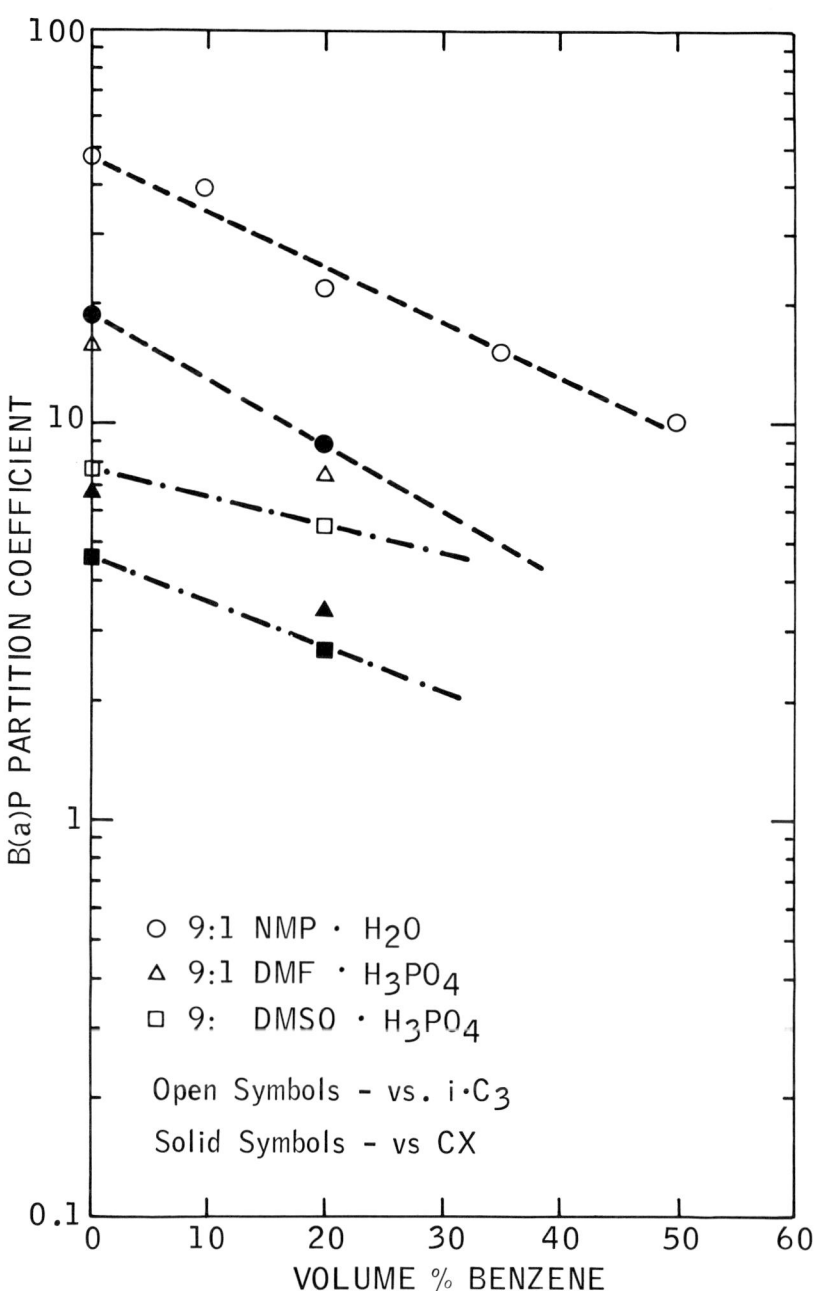

FIGURE 9. Effect of benzene on $(K_D)B(a)P$.

**858 SOLVENT EXTRACTION OF PAH**

This apparent complex may be similar to those observed between PAH and purines such as caffeine or tetramethyl uric acid. Since NMP is a tertiary lactam with a structure similar to that of the purines, its apparent complex formation with PAH may be due to the unique electron distribution of that structure.

**Caffeine**  **NMP**  **Tetramethyl Uric Acid**

Purine complexation has been used for the isolation of PAH in countercurrent extractions (16). For the larger PAH, stable 1:1 or 2:1 complexes have been isolated as solids (17). Detailed studies have indicated that the purine complexes were due to a polarization interaction between the PAH and the purines (18,19). A similar phenomenon could explain the unique characteristics of NMP for PAH extractions. Because of its unique ability to form transient polarization complexes, it has a wider range of flexibility than other extractants.

## COMPARISON WITH OTHER EXTRACTION SYSTEMS

A literature review indicates that over 30 reagents have been used for PAH extraction. Since PAH are often associated with petroleum products, characteristic solvency and selectivity of a large number of extractants have been examined for process characteristics (20).

For analytical purposes, the list could be reduced to nine reagents by rejecting those reagents which were reactive, toxic, not readily available, or known to be nonselective (Table 5). In general, the remaining reagents have high dielectric constants (>32). With the exceptions of methanol and ethylene glycol, they may also be classified as polar aprotic solvents. These nine practical reagents are classified in the table on the basis of the measurement of the distribution coefficients for BaP. The results show that DMF, DMSO, and NMP have $K_D$'s substantially larger than those of the other reagents.

TABLE 5. Practical Extraction Reagents for PAH Analyses

| | High $K_D$ | | | Low $K_D$ | |
|---|---|---|---|---|---|
| | ($K_D$ - BaP) | | | ($K_D$ - BaP) | |
| | Cyclohexane | Iso-Octane | | Cyclohexane | Iso-Octane |
| DMSO | 11 | 22 | Acetonitrile | 1.3 | -- |
| DMF | 9.4 | 24 | Nitromethane | -- | 1.6 |
| NMP | Miscible | 50 | | | |
| Sulfolane | -- | 19 | Ethylene Glycol | 0.07 | -- |
| Propylene Carbonate | 8.2 | -- | Methanol | 0.82 | -- |

Since distribution coefficients are equilibrium constants, reagents with similar physical properties might be expected to behave in the same way. For example, DMF and DMSO exhibit parallel responses to changes in extraction conditions. In the absence of H-bonding or complex formation, a general relationship exists between extraction systems:

$$\ln K_D = a \ln K'_D + b \qquad (9)$$

where a and b are constants. This relationship has been studied in detail by Hansch and co-workers (21). For the PAH extractions, this relationship means that the systematic relationships established for DMF, DMSO, and NMP can be extrapolated to the other polar aprotic solvents. For example, sulfolane or propylene carbonate would be expected to exhibit behavior parallel to the three reagents studied.

The polar aprotic reagents with lower $K_D$'s (acetonitrile, nitromethane) would also be expected to exhibit parallel behavior. The low $K_D$ values of these reagents limit the number of parameters that can be varied for use with complex samples. Thus, although they may exhibit adequate selectivity, they may be limited to relatively simple matrices.

The aprotic solvents (methanol and ethylene glycol) not only have low $K_D$ but are not selective. Consequently, they offer no advantage for PAH extractions.

## SUMMARY

The general extraction characteristics of PAH have been established by the determination of the distribution coefficients in a number of extraction systems. The results of parameter studies demonstrate that the PAH as a group respond systematically to changes in extraction conditions. Consistent with thermodynamic theory, near linear relationships between ln $K_D$ and a number of parameters are observed. Thus, ln $K_D$

    (1) decreases with volume percent diluent added to an extractant
    (2) decreases with volume percent aromatic added to the solvent phase
    (3) decreases with $1/T°K$
    (4) increases with molecular weight
    (5) decreases with alkyl substitution.

The systematic relationship between PAH and the parameters simplifies design of extraction procedures, since the measurement of just one or two $K_D$ values defines the extraction characteristics of a matrix. Furthermore, the selectivity and efficiency may be adjusted over a wide range by the addition of aromatics to the solvent phase or by adjusting the concentration of diluent in the extractant phase. This flexibility allows the extraction conditions to be selected for many specific matrix problems.

## REFERENCES

1. Robbins, W. K. (1979): *Anal. Chem.*, 51:1860-1861.
2. Haenni, E. O., Howard, J. W., and Joe, F. L. (1962): J.A.O.A.C. 45:67.
3. Haenni, E. O., Joe, F. L., Howard, J. W., and Leibel, R. L. (1962): J.A.O.A.C. 45:59.
4. Howard, J. W., and Haenni, E. O., (1963): J.A.O.A.C. 46:933.
5. Howard, J. W., Haenni, E. O., and Joe, F. L. (1965). J.A.O.A.C. 48:304.
6. Code of Federal Regulations, 1977, U.S. Food and Drug Administration, Section 172,886, Petroleum Wax, Title 21, Chapter 1, pl. 414.
7. Grimmer, G., and Böhnke, H. (1976): Chromatographia 9:30.
8. Grimmer, G., and Hildebrandt, A. (1972): J.A.O.A.C. 55:631.
9. Robbins, W. K., and Searl, T. D. (1979): The determination of polynuclear aromatic hydrocarbons in wastewater from coal liquefaction processes by the GC/UV technique. ASTM D-19 Symposium on Analytical Methods Related to Waters Associated with Production of Fuels from Alternate Sources, Pittsburgh, Pennsylvania, June 4.
10. Colin, H., Ward, N., and Guichon, G. (1978): J. Chrom. 149:169.
11. Walraven, J. H., Laddon, A. W., and Kaslmans, (1968): *Chromatographia,* Vol. 1, 195.
12. Snyder, L. R., (1968): *Principles of Adsorption Chromatography,* Dekker, New York.

13. Freiser, H. (1973): Solvent extraction. In: *An Introduction to Separation Science,* B. L. Karger, L. R. Snyder, and C. D. Horvath Eds.
14. Natusch, D.F.S., and Tomkins, B. A., (1978): *Anal. Chem.* 50:1429.
15. Long, F. A., and McDevit, W. F. (1952): *Chem. Rev* 51:119.
16. Mold, J. D., Walker, T. B., and Veasly, L. G. (1963): *Anal. Chem.* 35:2071.
17. Weil-Mutherbe, H. (1946): Biochem. J. 40:351.
18. VanDuuren, B. L. (1964): *J. Phys. Chem.* 68:2544.
19. Boyland, E., and Green B. (1962): *British J. Cancer* 16:347.
20. Deal, C. H., and Derr, E. L. (1964): *I & EC Process Design & Development* 3:394.
21. Leo, A., Hansch, C., and Elkins, D. (1971): *Chem. Rev.* 71:525.

# A COMPARISON OF EXTRACTION TECHNIQUES FOR POLYNUCLEAR AROMATIC HYDROCARBON ANALYSIS OF INDUSTRIAL EFFLUENTS AND NATURAL WATERS

**R. D. Smillie and D. T. Wang**

Ontario Ministry of the Environment
Laboratory Services Branch
P.O. Box 213
Rexdale, Ontario Canada M9W 5L1

## INTRODUCTION

The analysis for polynuclear aromatic hydrocarbons (PAH) is an important activity of environmental laboratories. In analyzing a wide array of environmental samples, the need for a systematic study and comparison of extraction techniques and procedures was strongly indicated. This paper gives an account of such investigations, using aqueous samples.

Four extraction techniques were evaluated for the analysis of PAH from industrial process streams and effluents, and from natural waters. Manual liquid-liquid (shake-out) extraction and continuous liquid-liquid extraction were investigated. Also examined were the applications of a modified Likens-Nickerson apparatus which operates on the principle of continuous solvent extraction of the steam distillate, and a microadsorption column packed with 10 $\mu$m $C_{18}$ reverse-phase material.

Organic-free water spiked with PAH, industrial effluent samples, and a natural-water sample were analyzed for PAH following extraction using these various methods. The extraction procedures and subsequent clean-ups were geared towards high-performance liquid-chromatographic (HPLC) fluorescence detection.

## EXPERIMENTAL

All solvents used were of high purity and the extraction and concentration equipment was thoroughly cleaned prior to use. Samples were analyzed in parallel with blanks. In general, organic-free water and solvents were free from PAH, but occasionally the extraction and concentration apparatus became contaminated. When contamination was detected, the problem was rectified by recleaning the glassware.

## High-Performance Liquid Chromatography

The analytical method for PAH determination utilized HPLC with a $C_{18}$ reverse-phase column. The eluent was a mixture of acetonitrile (75 percent) and water (25 percent v/v) applied at a pressure of 105 kg/cm$^2$ and a flow rate of 1.0 ml/min. Variable wavelength excitation and emission fluorescence detection were used, as described elsewhere (4) in detail.

## Samples

A sufficient quantity of aqueous sample was obtained to ensure that each of the four extraction procedures could be completely evaluated. Generally, a 32-liter sample comprising eight 4-liter bottles was required.

### *Filtered Samples*

After the sample was filtered, the sample bottle was rinsed with organic-free water, and this aqueous washing was filtered. This process was repeated twice. The glass sample bottle was then rinsed with the appropriate organic solvent and the solvent washings were pooled. The aqueous filtrate was also pooled to ensure homogeneity of the dissolved PAH fraction. The dissolved PAH fraction thus consisted of the filtrate extract and the rinsed bottle washings.

### *Nonfiltered Samples*

The complete sample from the collection bottle was added to the extraction apparatus, the sample bottle was rinsed twice with organic-free water, and the aqueous washings were added to the sample. The sample bottle was rinsed twice with the appropriate solvent and the washings were added to the rest of the solvent used for the particular extraction.

## Filtration

Half of the sample was filtered through preextracted 0.45 $\mu$m membrane filters. When the particulate-matter loading was high, preextracted "Celite" was used to aid filtration. Spent filters were soxhlet extracted for 24 hours with benzene. The extracts were concentrated by rotary evaporation to near dryness and made up with acetonitrile-water for HPLC analysis.

## Clean-up

When a clean-up procedure was warranted, the extract was chromatographed on preextracted, deactivated "Florisil" with cyclohexane, then concentrated and prepared for HPLC analysis.

## Extraction Procedures

### Manual Liquid-Liquid Extraction

Twenty-five ml of methylene chloride was added to a 1-liter sample in a 2-liter separatory funnel to saturate the solution. The aqueous sample was then extracted once with 100 ml and twice with 50 ml of methylene chloride. The combined extracts were concentrated by rotary evaporation to near dryness and made up with acetonitrile-water for HPLC analysis. When extracting a 4-liter sample, two 2-liter separatory funnels were used in parallel, and the extraction procedure remained the same as with a 1-liter sample.

### Continuous Liquid-Liquid Extraction

The continuous liquid-liquid extractor was designed specifically for solvents lighter than water. The extractor had a volume of 5 liters and the diffuser tube utilized a coarse glass frit. Generally, 4-liter samples were taken, so the remaining volume was made up with organic-free water. Approximately 250 ml of benzene was used, at a distillation rate of 5 ml/min in the 24-hour extraction procedure. The extract was concentrated to near dryness and made up with acetonitrile-water for HPLC analysis.

## Steam Distillation Extraction

Using a modified Likens-Nickerson apparatus (3), 4 liters of sample was steam distilled on a condenser at a rate of 6 ml/min, and the condensate was continuously extracted with 250 ml methylene chloride at an approximate distillation rate of 3 ml/min. After 24 hours, the solvent was concentrated by rotary evaporation to near dryness, and the extract was then made up with acetonitrile-water for HPLC analysis. Other types of continuous steam distillation/extraction equipment have been described in the literature (2, 5) and are commercially available.

## Microadsorption Column

Aqueous filtered samples were passed at 4 to 6 ml/min through a cartridge microadsorption column (4.6 mm x 10 cm) packed with 10 $\mu$m $C_{18}$ reverse-phase material. The pressures required to achieve the flow rate were 20 to 35 kg/cm$^2$. At pressures greater than 35 kg/cm$^2$ (500 psi), the microadsorption column developed leaks at the cartridge inlet line. This column was eluted with 10 ml of acetonitrile, which was concentrated to 1.0 ml for HPLC analysis.

# PAH RECOVERY STUDIES

### Organic-Free Water Spiked With PAH

Organic-free water was spiked at concentrations of 2, 4, 6 and 8 ng/l of individual PAH in acetonitrile. Recoveries for benzo(k)fluoranthene (BkF), benzo(a)pyrene (BaP), dibenz(a,h)anthracene (DBA), benzo(ghi) perylene (BP), and o-phenylenepyrene (OPP) were excellent, with the exception of the microadsorption column, as can be seen in Table 1. The percent recoveries are the grand average of the four spike concentrations.

The glass vessels that were used in these spiking experiments were washed with the extracting solvent as part of the normal extraction procedure. An exception to this procedure was the microcolumn extraction. The original intent was to use the spent microcolumn in place of the analytical column and elute directly to the fluorescent detector. However, the resolution of the PAH was poor, so this approach was abandoned.

**TABLE 1. Percentage of Recovery of PAH from Organic-Free Water**

| | Extraction Technique | | | |
|---|---|---|---|---|
| Compound | Liquid-Liquid | Continuous Liquid-Liquid | Steam Distillation | Adsorption Column |
| BkF | 96.0 | 93.4 | 96.1 | 55.7 |
| BaP | 75.7 | 84.5 | 94.0 | 43.1 |
| DBA | 94.1 | 92.9 | 88.2 | 42.2 |
| BP | 99.7 | 88.9 | 92.6 | 39.8 |
| OPP | 96.7 | 93.3 | 89.8 | 37.1 |
| Avg. | 92.4 | 90.6 | 92.1 | 43.6 |

The organics were eluted from the microcolumn with a small amount of acetonitrile, concentrated, and analyzed by HPLC. After all the water had been pumped through the microcolumn, the glass vessel was extracted with methylene chloride. An average recovery of 39.0 percent was obtained for the 5 PAH in this study, bringing the total average recovery to 82.6 percent.

There were serious disadvantages to using the microadsorption column. The flow rate inevitably slowed down with time, possibly because the column became plugged or the packing material swelled. This extraction technique was abandoned at this stage because the time required to pump 1 to 4 liters through the adsorption column was excessive, the sample container had to be extracted separately, and the overall recovery of the PAH was the poorest of the 4 methods evaluated.

## Steel Mill Process Stream

The steel mill process stream used in this study was a "flushing liquor" which consisted of an aqueous solution of the gases emitted from a coking oven. The suspended-solids level in the sample was 50 mg/l and consisted primarily of flakes of carbon. When the sample was filtered through a 0.45 $\mu$m membrane filter, oily deposits formed on the filtration equipment and oily material passed through the filter. This may have resulted in the samples being somewhat nonhomogeneous.

As the filtered sample was quite concentrated, a 10-$\mu$l direct aqueous injection was possible. The chromatogram is illustrated in Figure 1. The concentrations of the dissolved PAH obtained by the various extraction techniques are listed in Table 2. The results from each of the methods evaluated were quite compatible, with manual liquid-liquid extraction

## 868 PAH EXTRACTION METHOD COMPARISON

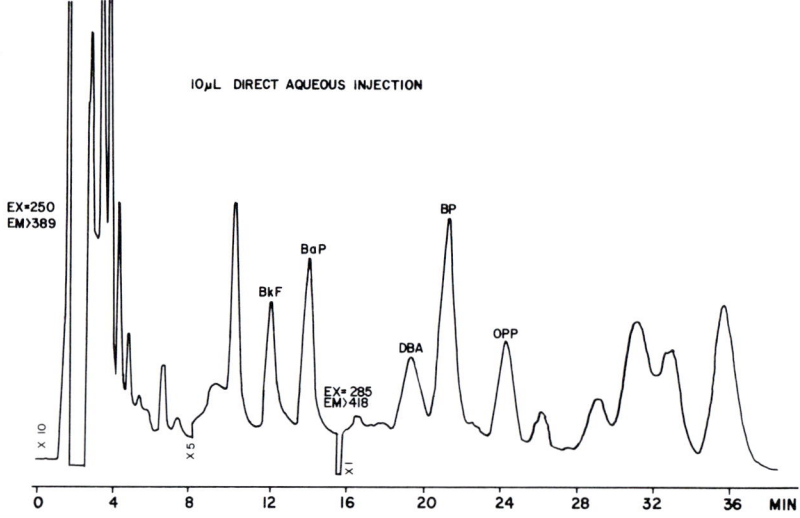

**FIGURE 1.** Fluorescence high-pressure liquid chromatogram of a 10 μl direct aqueous injection of a filtered steel mill process stream sample.

**TABLE 2.** Steel Mill Process Stream Filtrate PAH Concentration in μg/l

| Compound | Extraction Technique | | | |
| --- | --- | --- | --- | --- |
| | Direct Aqueous Injection | Liquid-Liquid | Continuous Liquid-Liquid | Steam Distillation |
| BkF | 0.08 | 0.10 | 0.11 | 0.18 |
| BaP | 0.23 | 0.40 | 0.29 | 0.35 |
| DBA | 0.03 | 0.04 | 0.03 | 0.03 |
| BP | 0.09 | 0.12 | 0.12 | 0.03 |
| OPP | 0.09 | 0.13 | 0.12 | 0.05 |
| 5 PAH Total | 0.52 | 0.79 | 0.67 | 0.64 |

Detection limit = .05 ng/l.

producing the highest PAH concentrations. It is worth noting that in the steam distillation extract, the "heavier" PAH, BP, and OPP were not recovered as efficiently, in comparison with the other methods. Anthanthrene and coronene (molecular weights of 276 and 300, respectively) were both recovered by the steam distillation technique, but not nearly as efficiently as the "lighter" PAH.

A comparison of the extraction techniques was made between the filtered and nonfiltered samples. The filtered samples comprised PAH associated with the particulate material, bottle and filter equipment washings, and the filtrate (dissolved PAH). These results are presented in Table 3. The variations in the filtered extraction procedures were small, in that the PAH concentrations from the particulate material and solvent washings were constant. The variation was due to the dissolved PAH extraction technique differences. There were substantial differences among the extraction methods for the nonfiltered sample. Continuous liquid-liquid extraction gave the highest PAH concentration, while the liquid-liquid shake-out gave the lowest. There may have been slight variability in the nonfiltered sample, because it was not possible to pool the entire sample as it was with the filtered sample. Concentrations of up to 470 $\mu$g/l of BaP had been detected in previous nonfiltered samples of this process stream. Table 4 shows how the PAH are distributed in the sample.

**Steel Mill Effluent**

The steel mill process stream (flushing liquor) underwent treatment in the plant, including a tar decanter and a two-stage ammonia still; free and fixed ammonia were removed; and the effluent was subjected to biotreatment. The effluent, at this stage, contained 600 to 1300 mg/l of suspended solids. A sample was collected at this point in the process and had a suspended solids value of 1140 mg/l. Filtration of the sample was extremely slow. Microscopic examination of the particulate material revealed fine particles ranging between 1 and 3 $\mu$m in size. A filter aid (Celite) was used to assist filtration.

The concentration of the PAH in the filtrate was low. Extraction-method comparisons of the dissolved PAH fraction are given in Table 5. The continuous liquid-liquid and steam distillation procedures recovered more PAH; however, there were no significant differences in efficiency among the methods. Of the two procedures mentioned above, the steam distillation technique provided a "cleaner" extract, as shown in Figure 2. In general, when working in the parts per trillion concentration range, the concentrated extracts were subjected to a column chromatography cleanup prior to analysis.

Comparing the PAH concentration from the filtered sample (particulate matter and filtrate) with PAH in the nonfiltered sample, all extraction procedures yielded higher concentrations for the nonfiltered sample. These results are presented in Table 6. Again, the variation in the PAH concentration of the filtered sample resulting from the extraction techniques was small. The rationale is the same as stated previously.

TABLE 3. Comparison of Total Filtered and Nonfiltered POAH Concentrations from a Steel Mill Process Stream in μg/l

| Compound | Direct Aqueous Injection, Filtered | Liquid-Liquid | | Continuous Liquid-Liquid | | Steam Distillation | |
|---|---|---|---|---|---|---|---|
| | | Filtered | Nonfiltered | Filtered | Nonfiltered | Filtered | Nonfiltered |
| BkF | .52 | .54 | .23 | .55 | .70 | .62 | .50 |
| BaP | 1.56 | 1.73 | .72 | 1.62 | 2.30 | 1.68 | 1.10 |
| DBA | .18 | .19 | .10 | .18 | .38 | .18 | .08 |
| BP | .56 | .59 | .33 | .59 | 1.08 | .50 | .13 |
| OPP | .63 | .67 | .35 | .66 | 1.05 | .59 | .26 |
| 5 PAH Total | 3.45 | 3.72 | 1.73 | 3.60 | 5.51 | 3.57 | 2.07 |

Detection limit = .05 ng/l.

**TABLE 4. Distribution of PAH in the Steel Mill Process Stream**

| PAH Associated with | | |
|---|---|---|
| Particulate (>0.45 μ) | Washings[a] | Filtrate[b] |
| 63.1 percent | 17.8 percent | 19.1 percent |

a Bottle and filtration equipment.
b Dissolved PAH is an average value for the three extraction methods evaluated.

**TABLE 5. Treated Steel Mill Effluent Filtrate PAH Concentration in ng/l**

| | Extraction Technique | | |
|---|---|---|---|
| Compound | Liquid-Liquid | Continuous Liquid-Liquid | Steam Distillation |
| BkF | 0.3 | 0.5 | 0.5 |
| BaP | 0.4 | 0.8 | 0.9 |
| DBA | nd | 0.1 | 0.1 |
| BP | 0.1 | 0.3 | 0.2 |
| OPP | 0.1 | 0.4 | 0.3 |
| 5 PAH Total | 0.9 | 2.1 | 2.0 |

Detection limit = 0.05 ng/l.

The filtered sample PAH concentrations were lower, possibly because extraction of the PAH from the particulate matter on the filters and filter aid may not have been complete. This was not surprising considering the total surface area of particulate matter on which the PAH could be adsorbed. The PAH associated with the particulate fraction comprised 98.2 percent of the total PAH concentration. Acheson et al (1) have discussed the effect of suspended solids upon PAH extraction efficiency.

The continuous liquid-liquid extraction of the nonfiltered sample produced the highest PAH value, while the liquid-liquid shake-out produced the lowest. This was most likely because there was much longer solvent contact with the former extraction procedure than with the latter.

The "cleanest" nonfiltered extract was obtained by the steam distillation method. This is illustrated in Figure 3, which compares the continuous liquid-liquid and steam distillation extract chromatograms. The manual liquid-liquid extract chromatogram was similar to that of the continuous liquid-liquid extract chromatogram.

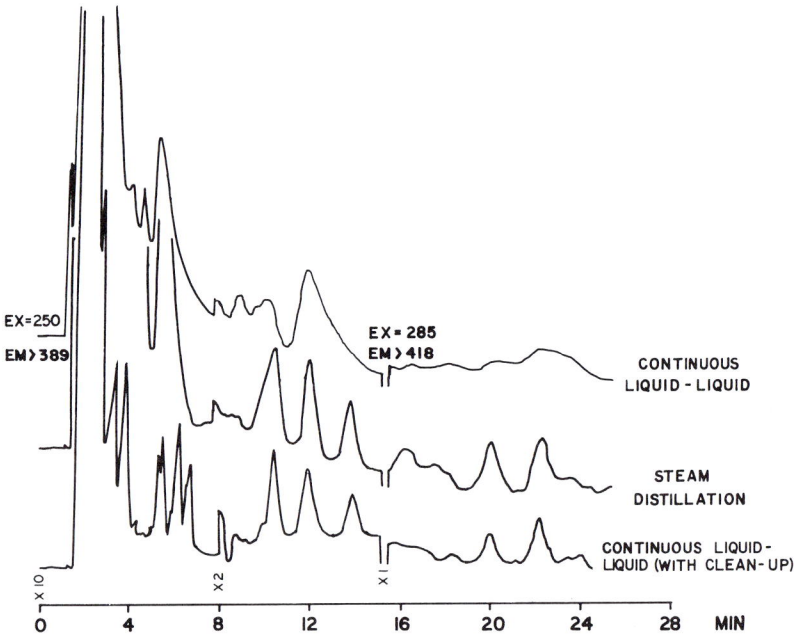

**FIGURE 2.** Comparison of chromatograms of steel mill effluent filtrate extracts from continuous liquid-liquid extraction, with and without clean-up, and steam distillation extraction.

## Receiving Waters

The particular body of water sampled received discharges from two steel mills, a sewage treatment plant and a number of primary and secondary industries. Heavy marine traffic, plus atmospheric deposition also contributed to the PAH concentration in this water body. The sample was collected approximately 0.5 meter below the water surface. Sample filtration was relatively easy in that the suspended solids level was 10 mg/l. Microscopic examination of this material showed particles 5 $\mu$m and larger. Numerous diatoms were observed.

The recoveries of the dissolved PAH in the filtrate by the three extraction procedures are given in Table 7. There was little difference in the recoveries among the three methods. Also, there was little difference between the PAH concentrations in the filtered and nonfiltered sample (Table 8). This was probably due in part to the relatively low suspended-solids value. As with the nonfiltered effluent sample, the continuous liquid-liquid technique provided the highest PAH recovery. In the receiving waters, the PAH associated with the particulate matter was 85.6 percent of the total PAH concentration.

TABLE 6. Comparison of Total Filtered and Nonfiltered PAH Concentrations from a Steel Mill Effluent in ng/l

Extraction Technique

| Compound | Liquid-Liquid | | Continuous Liquid-Liquid | | Steam Distillation | |
|---|---|---|---|---|---|---|
| | Filtered | Nonfiltered | Filtered | Nonfiltered | Filtered | Nonfiltered |
| BkF | 15.3 | 20 | 15.5 | 25 | 15.5 | 30 |
| BaP | 42.4 | 62 | 42.8 | 75 | 42.9 | 78 |
| DBA | 4.7 | 6.5 | 4.8 | 10 | 4.8 | 10 |
| BP | 17.1 | 20 | 17.3 | 34 | 17.2 | 19 |
| OPP | 13.1 | 17 | 13.4 | 33 | 13.3 | 23 |
| 5 PAH Total | 92.6 | 127.5 | 93.8 | 177 | 93.7 | 160 |

Detection limit = .05 ng/l.

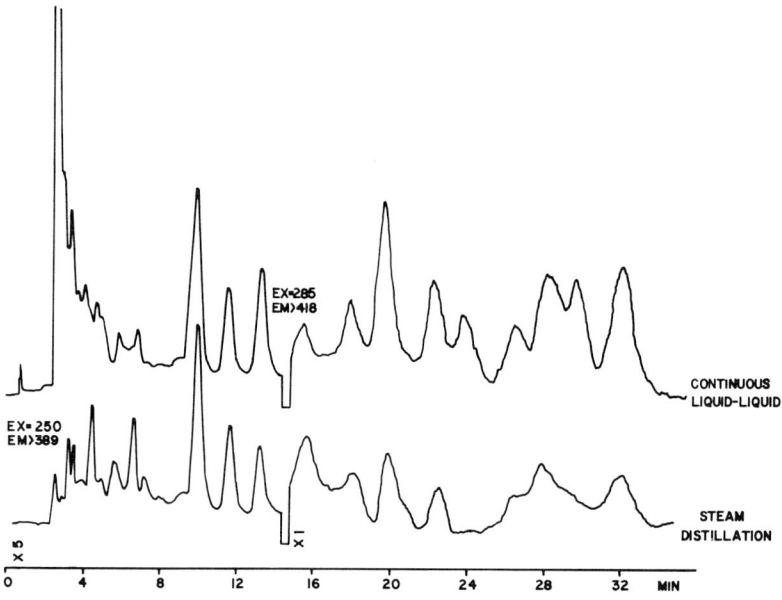

**FIGURE 3.** Comparison of chromatograms of nonfiltered steel mill effluent extractions obtained by continuous liquid-liquid and steam distillation extraction techniques.

**TABLE 7.** Receiving-Water Filtrate PAH Concentration in ng/l

| | Extraction Technique | | |
|---|---|---|---|
| Compound | Liquid-Liquid | Continuous Liquid-Liquid | Steam Distillation |
| BkF | 0.6 | 0.5 | 0.5 |
| BaP | 1.3 | 1.2 | 1.0 |
| DBA | 0.4 | 0.5 | 0.4 |
| BP | 0.8 | 0.6 | 0.5 |
| OPP | 0.5 | 0.1 | 0.5 |
| 5 PAH Total | 3.6 | 2.9 | 2.9 |

Detection limit = 0.05 ng/l.

TABLE 8. Comparison of Total Filtered and Nonfiltered PAH Concentrations from the Receiving Waters in ng/l

| Compound | Extraction Method | | | | | |
|---|---|---|---|---|---|---|
| | Liquid-Liquid | | Continuous Liquid-Liquid | | Steam Distillation | |
| | Filtered | Nonfiltered | Filtered | Nonfiltered | Filtered | Nonfiltered |
| BkF | 3.6 | 3.8 | 3.5 | 6.3 | 3.5 | 4.0 |
| BaP | 7.3 | 7.3 | 7.2 | 12.0 | 7.0 | 9.0 |
| DBA | 1.0 | 1.0 | 1.1 | 1.2 | 10.0 | 1.1 |
| BP | 4.8 | 5.8 | 4.7 | 7.3 | 4.5 | 7.2 |
| OPP | 5.3 | 6.0 | 4.8 | 8.5 | 5.3 | 6.0 |
| 5 PAH Total | 22.0 | 23.9 | 21.3 | 35.3 | 21.3 | 27.3 |

Detection limit = .05 ng/l.

## CONCLUSIONS

The microadsorption column had serious drawbacks in that samples had to be filtered prior to concentration of the filtrate. A relatively expensive pumping system was needed, as pressures of greater than 35 kg/cm$^2$ (500 psi) were required at times. This method was the most time consuming and the microadsorption columns were expensive.

The steam distillation extraction technique provided the cleanest extract as determined by HPLC. As expected, the higher molecular weight PAH, because of their relative lack of volatility, were not recovered as well as the lower molecular weight PAH. Recoveries of more water soluble heterocyclic PAH would probably be less efficient applying this method. Heat-labile compounds would also be affected by this method.

The liquid-liquid shake-out and the continuous liquid-liquid extraction are both EPA-approved methods for the extraction of base-neutral priority pollutants. In the three sets of nonfiltered samples, continuous liquid-liquid extraction recovered the most PAH, followed by the steam distillation, with the manual liquid-liquid extraction recovering the least PAH, by comparison. There were no significant differences among the methods for the three sets of filtered samples, primarily because the PAH associated with particles composed the bulk of the materials and was common to all extraction procedures.

In summary, the continuous liquid-liquid extraction technique is preferred over the liquid-liquid shake-out. There is some degree of subjectivity in the shake-out, specifically the effort put into the shaking, whereas this variable does not exist with the other method. Moreover, this extraction procedure yields higher PAH values for the same sample.

If filtration of a sample is required, there is no basic difference in the dissolved PAH recoveries by the three methods. If filtration is not required, there are significant differences. The findings in this study indicate continuous liquid-liquid extraction to be the method of choice for PAH extraction of nonfiltered samples.

## REFERENCES

1. Acheson, M. A., Harrison, R. M., Perry, R., and Wellings, R. A. (1975): Factors affecting the extraction and analysis of polynuclear aromatic hydrocarbons in water. *Water Research* 10: 207-212.
2. Lewars, E. G. (1979): *Aldrichimica Acta* 12: 22.
3. Likens, S. T., and Nickerson, G. B. (1964): *Proc. Am. Soc. Brew. Chem.* 5.

4. Smillie, R. D., Wang, D. T., and Meresz, O. (1978): The use of a combination of ultraviolet and fluorescence detectors for the selective detection and quantitation of polynuclear aromatic hydrocarbons by high-pressure liquid chromatography. *J. Environ. Sci. Health* A13: 47–59.
5. Veith, G. D., and Kimus, L. M. (1977): An exhaustive steam-distillation and solvent-extraction unit for pesticides and industrial chemicals. *Bull. Environ. Contamin. and Tox.* 17: 631–636.

# MICROSOMAL UPTAKE OF BENZO(A)PYRENE: EFFECT OF ADSORPTION TO ASBESTOS, HEMATITE, SILICA, AND CARBON BLACK

**J. R. Lakowicz and D. R. Bevan**

Gray Freshwater Biological Institute and Department of Biochemistry
University of Minnesota
P.O. Box 100
Navarre, Minnesota 55392

## SUMMARY

Particulate matter is known to increase the carcinogenic potency of polynuclear aromatic hydrocarbons. We used fluorescence spectroscopy to quantify the rates of microcrystalline and particle-adsorbed benzo(a)pyrene (BaP) uptake by rat liver microsomes. Under our experimental conditions, the times for 50 percent transfer of BaP from the particle-adsorbed states were 1.5, 4, 13, and 48 min for anthophyllite, chrysotile, hematite, and silica, respectively, and 132 minutes for BaP microcrystals. BaP was not released into microsomes from carbon black. The tenfold faster transport of BaP into microsomes from the asbestos particles when compared with the nonfibrous mineral particulates may partially explain the enhanced carcinogenic potency of particle-adsorbed PAH and the co-carcinogenic effect of cigarette smoking and asbestos inhalation.

The rate-limiting step for BaP transport into microsomes appears to be the rate of desorption from the surface of the particles. Our evidence for this mechanism is the independence of the BaP transport rates from (1) the concentrations of both particles and microsomes, (2) particle-microsome binding, and (3) the effects of the particles on the integrity of the microsomes. The rates of BaP desorption appear to be correlated with the capacity of the particles to adsorb BaP in the monomeric state. The fluorescence spectral data for anthophyllite and silica-bound BaP shows the former to have a superior ability to adsorb BaP in spite of its 35-fold smaller surface area. However, we caution that the spectra of particle-adsorbed BaP need not correlate with the BaP adsorptive capacity because the emission of adsorbed BaP molecules may be quenched. The spectral methods reported in this paper should facilitate the investigation of the particle-adsorbed state and the effects of PAH-surface interactions on the cellular availability of polynuclear aromatic hydrocarbons.

## INTRODUCTION

Carcinogenesis by polynuclear aromatic hydrocarbons (PAH) requires metabolic activation (13,36). The required aryl hydrocarbon hydroxylase activity occurs in the microsomal fraction of cells (26,37,47). Factors which alter the availability of the PAH for metabolic activation can potentially increase or decrease the incidence of cancer, possibly as a result of altering the amounts of the more carcinogenic metabolites which are formed (13).

Particulate matter is known to increase the carcinogenic potency of PAH. For example, intratracheal instillation of BaP results in only a low incidence of lung cancer unless hematite is also instilled (32,33). Other co-carcinogenic particulates include asbestos (28), aluminum and titanium oxide (40), and India ink (27). In humans, cigarette smoking and asbestos inhalation are known to be highly co-carcinogenic (34).

The mechanisms of particle-PAH co-carcinogenesis are not understood. The major route of PAH entry into the lungs is via inhaled particulates which contain adsorbed PAH. At present, the site of PAH elution is not known, but regardless of the site of elution, particles which rapidly release adsorbed PAH could increase the effective dose of carcinogens in the lungs by elution of these compounds prior to clearance of the particles from the lungs. In earlier studies (19,21) we demonstrated that adsorption of BaP to fibrous and nonfibrous mineral particulates greatly enhances its rate of uptake into vesicles of dipalmitoyl-L-$\alpha$-phosphatidycholine, which may be regarded as a model for lung surfactant. In this paper and in a companion study (18) we described the effects of particulates on the rates of BaP transport by rat liver microsomes.

## MATERIALS AND METHODS

### Source and Physical Properties of Particulates

Amorphous silica was obtained from Analabs. By nitrogen adsorption, its surface area was 381 m$^2$/g and the average particle size was 2 $\mu$m

(20). Hematite was obtained from Ventron Corporation (8.0 m$^2$/g) and carbon black from Fisher (31.1 m$^2$/g). Microscopic examination indicated the hematite particles to be of varying size, with most being less than 2 μm, and the carbon particles to be about 25 μm in diameter. Anthophyllite and Canadian chrysotile were standard samples supplied by the International Union Against Cancer (Johannesburg). The reported surface areas are 11.8 and 26.8 m$^2$/g, respectively. The particle size distributions are heterogeneous, with the average size being about 2 μm for both anthophyllite and chrysotile. For more detailed information see Timbrell (41).

## Preparations of Particulates with Adsorbed BaP

Particulates containing adsorbed BaP were prepared by mixing the particulate with a benzene solution of BaP, followed by evaporation of the benzene under reduced pressure. Unless otherwise indicated, 0.3 mg of BaP was added for each gram of particulate (57,500 M$^{-1}$ cm$^{-1}$ at 299.4 nm). The volume of benzene added was 20 ml/g of particulate. After the sample appeared dry it was kept under vacuum for 30 to 45 min at 85°C. Material which clung to the walls of the container was discarded. These samples were stored in the dark under an argon atmosphere and were used within 1 week of their preparation. The fluorescence emission spectra of BaP extracted from the particles using benzene was identical to that of the starting material.

Aqueous dispersions of BaP crystals were prepared by evaporation of a benzene solution of BaP to dryness, addition of buffer, and sonication for 30 minutes at 40 watts using a Cole-Parmer Model 8845-2 bath-type sonicator. Microscopic examination of these preparations revealed a heterogeneous size distribution of the crystals with 90 percent being less than 15 μm and 50 percent less than 5 μm. We refer to this preparation as being microcrystalline.

## Preparation of Rat Liver Microsomes

Microsomes were prepared according to Ames et al (1). Five days after inducing rats by peritoneal injection of Aroclor 1254, they were sacrificed and the livers removed. The livers were homogenized in 0.15 M KCl, and the homogenate was centrifuged at 9000 x g for 10 minutes. The supernatant was centrifuged at 105,000 x g to pellet the microsomes, and the microsomal pellet was resuspended in buffer at a concentration of 0.6 mg protein/ml.

## Fluorescence Spectral Data

Fluorescence spectral data were obtained using a computerized, photon-counting spectrofluorometer (SLM Instruments, Inc., 1101 East Huey Road, Urbana, Illinois). The following instrumental conditions were used: an excitation wavelength of 296 nm; excitation and emission bandpasses of 8 and 4 nm, respectively; a Corning 7-54 excitation filter; emission filters 0-52 and 2 mm of 1 M $NaNO_2$. The emission filters did not transmit light below 395 nm. Without these filters a moderate background signal was observed at wavelengths below 395 nm. Some background signal was also seen above 395 nm, typically less than 10 percent of the total intensity. Because a similar intensity and spectral distribution for this background was observed for all particulates, we conclude that it results from stray light scattered off the turbid suspension of particulates. These backgrounds were quantified using particles without BaP and subtracted from the spectra exhibited. The fluorescence of the microsomes was minor in relation to that resulting from the BaP and did not interfere with our observations.

Spectra for the aqueous suspension of particles with adsorbed BaP were obtained using a 2 x 2-cm cuvette, 4 cm high, which was thermostated and positioned on a magnetic stirrer so that the stirring bar remained below the light path. Continuous stirring kept the particles in suspension, and the fluorescence intensity of the adsorbed BaP did not fluctuate or lose intensity over the course of the spectral scans. The exciting beam impinged upon the sample at an angle of 20 degrees, and the emission was observed from this illuminated surface at 90 degrees to the exciting light. In spite of the turbidity of the suspensions of particles, the scattered light did not appear to interfere with our measurements. In addition, BaP emission from microsome-bound BaP could be observed even in opaque suspensions of hematite and carbon black.

## Measurement of Microsomal Uptake of BaP

BaP uptake into microsomes was quantified by the increase in fluorescence intensity at 405 nm, which occurred upon transfer of BaP from the surface of the particle into the microsomes. In particular, we assumed

$$\frac{\% \text{ Benzo(a)pyrene transferred}}{\text{to microsomes}} = \frac{I(t) - I_o}{I_\infty - I_o} \times 100$$

where $I_o$, $I(t)$, and $I_\infty$ are the fluorescence intensities before addition of microsomes, at time t after addition, and after complete transfer of BaP to the microsomes, respectively. For all the BaP uptake kinetics reported here we used 5 μg of BaP and 1 ml of rat liver microsomes, which was equivalent to 0.6 mg of microsomal protein. The buffer used was 0.1 M potassium phosphate, pH = 7.7, containing 3 mM $MgCl_2$ and 0.1 mM EDTA. An amount of particulate (16.7 mg) containing 5 μg of adsorbed BaP was suspended in 10 ml of buffer and dispersed by sonication in a bath-type sonicator for 30 minutes at room temperature. In addition to dispersing the particles, this procedure facilitated equilibration of BaP with the aqueous phase. After measurement of the initial fluorescence spectrum and intensity ($I_o$), microsomes were added in 1 ml of buffer to initiate the reaction. Complete BaP transfer to the bilayers was obtained by heating the sample to 50°C for 60 minutes. The final fluorescence intensity ($I_\infty$) was measured after reequilibration at the experimental temperature of 25°C. BaP uptake rates from the microcrystalline state, and from BaP microcrystals in the presence of particulates, were obtained in a similar manner, except that the 5 μg of BaP microcrystals was suspended in 10 ml of aqueous buffer. BaP uptake into vesicles of dipalmitoyl-L-α-phosphatidylcholine (DPPC) was measured in an identical fashion, except that 10 mg of DPPC vesicles was substituted for the microsomes (21).

## Effect of Particle-Microsome Binding on the Rates of BaP Uptake

Two procedures were adopted for studying the effect of binding of the microsomes to the particles on the kinetics of BaP uptake. In one procedure, 16.7 mg of particles with adsorbed BaP (5 μg) and 16.7 mg of particulates without BaP were suspended in 10 ml of buffer. BaP uptake was initiated by addition of 1 ml of microsomes. We assumed that microsomes would bind randomly to both the labeled and the unlabeled particles. Hence, if binding were important for BaP uptake, a decreased uptake rate would be observed in the presence of excess unlabeled particles.

In the second procedure, we investigated the effect of binding by preincubating the microsomes with particles which did not contain BaP. In particular, we incubated microsomes with the unlabeled particles (16,7 mg) for 30 minutes at 25°C. These microsomes were subsequently added to the 10 ml suspension of particles which contained adsorbed BaP. If binding were important for BaP uptake we would expect a decreased uptake rate under these conditions in which the microsomes were bound to particles not containing BaP. However, both procedures yielded equivalent BaP uptake rates.

## Measurement of Microsomal Integrity in the Presence of Particulates

Microsomal integrity was assayed by lipid peroxidation activity (8,46). NADPH and oxygen are consumed in lipid peroxidation, and we quantified this activity by NADPH oxidation and subsequent loss of fluorescence. The fluorescence of NADPH was convenient because we found it was possible to quantify consumption of NADPH even in the optically dense particulate suspension used in the BaP uptake measurements.

The reported measurements were made in 0.1 M potassium phosphate, pH = 7.7, 25° C, but equivalent activities were obtained in this same buffer when it also contained 3 mM $MgCl_2$ and 0.1 mM EDTA. The same front-fact illumination was used as for the BaP uptake measurements. The instrumental conditions were: excitation wavelength and filter, 340 nm and Corning 7-54 respectively; emission filters, Corning 3-144 and 2 mm of 1 M $NaNO_2$; emission wavelength, 464 nm. The 10 ml of assay mixture contained 16.7 mg of particles and 0.6 mg of microsomal protein. To simulate the conditions used in the BaP uptake studies, this mixture was stirred for 30 minutes at 25° C prior to initiation of lipid peroxidation by addition of 0.5 ml of $5 \times 10^{-4}$ M NADPH. The activity was obtained from the loss of NADPH fluorescence which occurred during the first 2 minutes. At this time, 50 $\mu$l of 0.2 M ADP in 5 mM $FeCl_3$ was added, and the fluorescence intensity was monitored for an additional 2 minutes. Essentially identical activities were obtained during both incubation periods, probably as a result of our use of phosphate buffer (46).

In control experiments we showed that deoxycholate and p-chloromercuribenzoate inhibited the lipid peroxidase activity, as did boiling of the microsomes. In addition, NADH did not support the lipid peroxidation activity to the same extent as did NADPH. These experiments demonstrated that the activity we measured had the properties of the NADPH-dependent lipid peroxidation described by Wills (46) and was not a nonspecific oxidation of NADPH. Upon incubation of particles and NADPH in the absence of microsomes, the fluorescence intensity of NADPH was constant. This control indicates that the particles themselves did not adsorb the NADPH, catalyze reduction of the NADPH, or cleave the phosphodiester bond. Cleavage of this bond results in an approximate fourfold increase in the fluorescence yield of NADPH.

## RESULTS

### Microsomal Uptake of Benzo(a)pyrene

The effects of particles on the microsomal uptake rates of BaP are shown in Figure 1. Several points are worthy of mention.

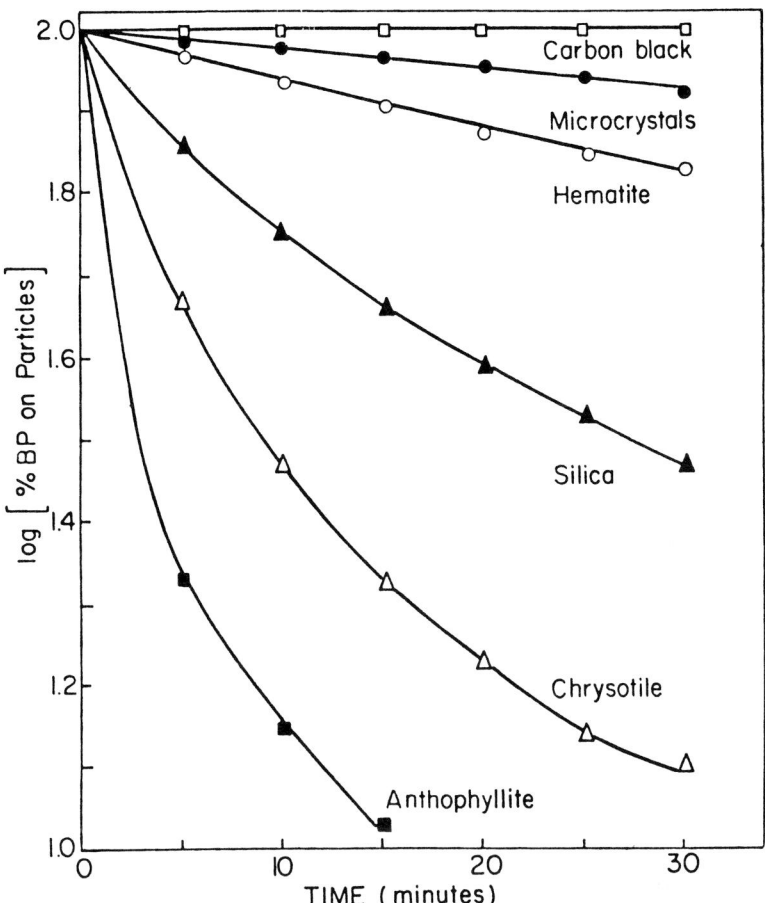

**FIGURE 1. Microsomal uptake of benzo(a)pyrene.**

1. Adsorption of BaP to all particles, except carbon black, resulted in enhanced microsomal uptake when compared with BaP microcrystals.
2. BaP is not released into microsomes from carbon black.
3. Adsorption of BaP to the particles is necessary for enhanced transport. Although not shown, simple mixtures of BaP microcrystals and particles displayed uptake rates identical to those of microcrystals alone.
4. Finally, the asbestos particles, anthophyllite and chrysotile, are about tenfold more effective than the nonfibrous particles, silica and hematite, in transporting BaP into microsomes. The apparent first-order rate constants for BaP transfer are 0.60, 0.18, 0.071,

0.014, and 0.006 min$^{-1}$ for anthophyllite, chrysotile, silica, hematite, and microcrystals, respectively.

We judged the initial transfer of BaP to follow first-order kinetics. However, a small percentage of the total BaP, perhaps 10 percent, may undergo slower transfer into microsomes. More detailed investigations of the transport kinetics would be required to quantify these different rate constants. We suspect that particle-adsorbed BaP may exist as several populations with different desorption rates from the surfaces. In addition, the solvent evaporation procedure we used to "adsorb" BaP to the surface of the particles is likely to result in some crystalline aggregates of BaP. However, further analysis is unlikely to alter our basic observation, which is the rapid transport of BaP from the surface of fibrous mineral particulates.

## Mechanism of Particle-Enhanced Microsomal Uptake of Benzo(a)pyrene

We investigated the role of particle-microsome binding on the microsomal uptake rates of BaP. We reasoned that if binding were important for BaP transport then the presence of particles in the reaction mixture which do not contain BaP should decrease its microsomal uptake rate. A decreased rate would be observable under conditions where the amount of unlabeled particulates is adequate to bind a significant fraction of the microsomes. To increase further the potential sensitivity of this experiment, we preincubated the microsomes with unlabeled particles prior to their addition to the BaP-particulate suspensions.

We determined that 16.7 mg of anthophyllite, chrysotile, or hematite is adequate to bind essentially all the microsomes under our experimental conditions, and 16.7 mg of silica binds approximately one-half of the microsomes. We preincubated microsomes with each of these particulates and added these complexes to particles containing adsorbed BaP. The preincubation with unlabeled particles had no significant effect on the BaP uptake rates. We conclude that microsome-particle binding is not correlated with the microsomal uptake rates of BaP. We note that a rapid reequilibration of the microsomes between the particles with and without adsorbed BaP would invalidate our conclusion, but such redistribution must occur in less than 30 seconds. Otherwise, the initial uptake rates we observed would have been different.

We investigated the possibility that disruption of the microsomes was responsible for particle-enhanced uptake of BaP. Microsomal integrity was assayed by lipid peroxidation activity. This activity is lost upon addition of deoxycholate, and therefore provides a measure of the integrity of the microsomal membranes. This activity was quantified fluorometrically by the oxidation of NADPH to NADP$^+$, a measurement we could make in the presence of particulates. Anthophyllite, silica, and hematite has no significant effect on the rate of lipid peroxidation (Table 1), but chrysotile was found to have a significant inhibitory effect. However, no correlation was found between the rates of BaP uptake and disruption of the microsomes. That is, anthophyllite and chrysotile show nearly identical enhancements of uptake, but anthophyllite does not disrupt the microsomes and chrysotile does. Likewise, the enhancements of anthophyllite, hematite, and silica all differ, yet none of these particles significantly affect the integrity of the microsomes. The disruptive effects of chrysotile are easily understood in terms of its unique surface chemistry (38). We also quantified the rates of BaP uptake under conditions where both the particle and microsome concentrations were increased twofold. By increasing both concentrations we were able to keep the particle-to-microsome and BaP-to-microsome ratios constant. If collisional encounters between particles and microsomes were responsible for transfer of BaP, then the BaP uptake rate should increase fourfold under these conditions. The observed rates for all particles were identical to those shown in Figure 1. These observations indicate that collisional encounters

**TABLE 1. Microsomal Lipid Peroxidase Activity in the Presence of Particulates**

| | Specific Activity (nmoles NADPH oxidized/min/ mg protein) | Activity with No Particles (%) |
|---|---|---|
| Particulates[a]: | | |
| None | 25 | (100) |
| Anthophyllite | 21 | 84 |
| Chrysotile | 9 | 36 |
| Silica | 23 | 92 |
| Hematite | 34 | 136 |
| Controls: | | |
| Deoxycholate (0.3%) | 5 | 20 |
| 100°C, 5 min | 2 | 8 |
| p-chloromercuribenzoate (1 mM) | 8 | 32 |
| NADH (no NADPH) | 9 | 36 |

[a]16.7 mg of each particulate was used.

between particles and microsomes are not of mechanistic significance for the transfer of BaP from particles to microsomes.

Because of the low water solubility of BaP in aqueous solutions (3.2 $\mu$g/l), only a small fraction of the total BaP could be dissolved in the aqueous phase of our 10-ml samples (6,19). Thus it appears unlikely that the rate-limiting step for BaP uptake into microsomes would be its rate of entry into the microsomal membranes. The zero-order dependence of the BaP uptake rate on both particle and microsome concentration probably indicates that the desorption of BaP from the surface of the particles is the rate-limiting step for BaP uptake into microsomes. Moreover, the rates of BaP uptake into DPPC vesicles were identical to those observed for microsomes, and these BaP uptake rates were relatively independent of the BaP-to-particle weight ratios (Figure 2). Thus it appears that uptake is limited by desorption and that the rates of BaP desorption were characteristic of the particle and not of the membrane into which the BaP was being transported.

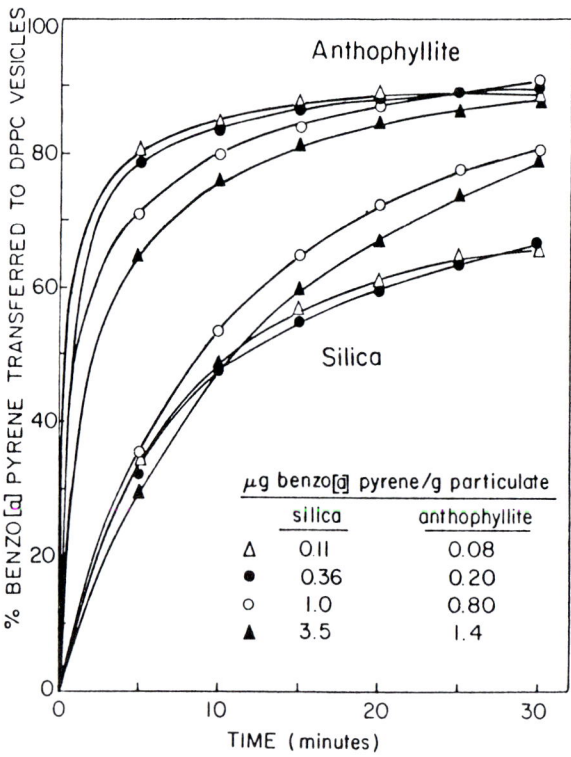

FIGURE 2. Effects of the benzo(a)pyrene-to-particle weight ratio on its uptake rate into vesicles of dipalmitoyl-L-$\alpha$-phosphatidylcholine.

## Fluorescence Spectra of Benzo(a)pyrene; A Means to Investigate the Adsorbed State

As shown in Figure 1, the uptake of BaP from microcrystalline dispersions is slow. Hence, if the binding capacity of the particles for BaP is exceeded during the solvent evaporation process, we expected these preparations to contain BaP microcrystals and thus show slow rates of BaP transport. Thus we expected particles with a high capacity to adsorb BaP, but with a low affinity for BaP (unlike carbon black), to be most effective in transporting BaP into cells. Fluorescence spectroscopy was useful in investigating the nature of the adsorbed state. This is because adsorbed and microcrystalline BaP have markedly different spectral characteristics.

When dissolved in solvents the emission spectra of BaP are highly structured. When in the crystalline state, the emission occurs at longer wavelengths and is without structure (17,21). The long wavelength emission is a result of an excited state charge transfer complex, called an excimer, between adjacent BaP molecules. As a result of the sensitivity of BaP fluorescence to the presence of neighboring BaP molecules, these spectra can potentially reveal the surface distribution of BaP on particles.

Both the structured and unstructured components of BaP fluorescence are present in aqueous suspensions of BaP microcrystals. These components are illustrated most clearly in the normalized spectra (Figure 3B). Following a tenfold dilution of this aqueous dispersion, one observes an increase in the intensity of structured emission (Figure 3A). We attribute this increase to the additional BaP which is in the nonmicrocrystalline state.

We note that the intensity of the excimer emission, which we attribute to the BaP microcrystals, does not vary significantly following dilution of the BaP suspension. This invariance is a result of the large percentage of the total BaP which remains in the microcrystalline state in buffer at equilibrium. Using the solubility of BaP in buffer of 3.2 $\mu$g BaP/l (19), we calculate that only 0.6 percent of the 5 $\mu$g of BaP dissolves in the 10 ml of buffer. This small amount of BaP does not significantly alter the amount of BaP in suspension or the intensity of the excimer emission.

We note that the 0.6 percent of solubilized BaP accounts for approximately 50 percent of the total BaP emission. This excessive contribution of the dissolved component is a result of the low relative quantum yield of BaP microcrystals. This observation has important implications for any attempt to interpret the emission spectra of aqueous suspension of BaP, in the presence or absence of particles. These spectra can be completely dominated by the water-solubilized component, and this component can vary depending upon whether adequate time was allowed for the BaP to dissolve in the aqueous phase. In addition, variation of the concentration of such suspensions results in changes in the overall appearance of the emission spectra as a result of the percentage of BaP which is solubilized.

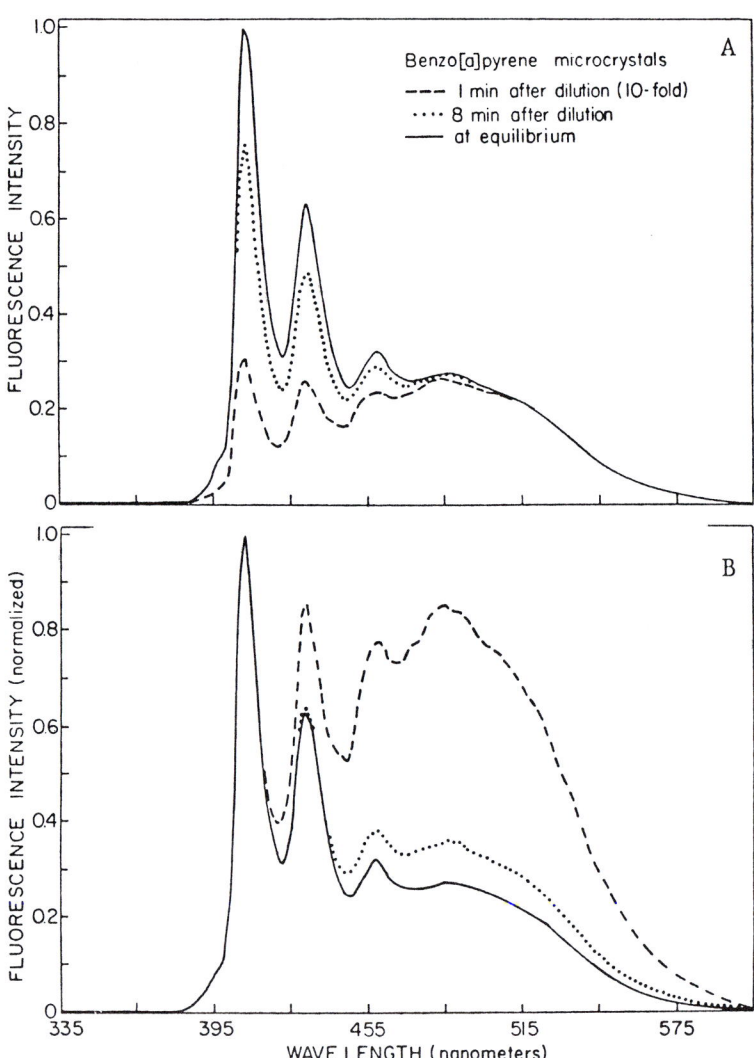

FIGURE 3. Fluorescence emission spectra of aqueous suspensions of benzo(a)-pyrene microcrystals. BaPmicrocrystals were suspended in buffer (0.01 M tris, 0.05 M KCl, pH = 7.5) at a concentration of 5µg/ml. Emission spectra were recorded at the times indicated following tenfold dilution into additional buffer.

## Fluorescence Spectra of Benzo(a)pyrene Adsorbed to Silica and Anthophyllite

We examined the fluorescence spectra of BaP adsorbed to anthophyllite and silica. We chose anthophyllite because of its rapid rate of BaP desorption, and we chose silica because of its much slower BaP desorption rate. No spectra information could be obtained for suspensions of BaP adsorbed to hematite. In this instance the adsorbed BaP was nonfluorescent, and the emission spectra of these suspensions resulted only from the BaP which dissolved in the aqueous phase.

For anthophyllite and silica the BaP emission spectra were examined over a range of BaP-to-particle weight ratios; the spectra of these suspensions were resolved into the solubilized and particle-adsorbed components (Figure 4). We noted the following:

1. In general, increasing BaP-to-particle weight ratios yielded more intense excimer emissions, probably as a result of the formation of BaP aggregates at the higher weight ratios.
2. From the total emission spectra of these suspensions we observed that the solubilized BaP was the major source of the structured emission for the silica suspensions, whereas the solubilized BaP accounts for only a small fraction of the structured emission in the anthophyllite suspensions.
3. Subtraction of the solubilized BaP component yielded the emission spectra of the adsorbed BaP (the filtering effects of the particles were accounted for in this subtraction procedure—unpublished data). These adsorbed spectra clearly indicated more monomer fluorescence for BaP adsorbed to anthophyllite than for BaP adsorbed to silica. These spectral observations may indicate that anthophyllite is superior to silica in its ability to adsorb BaP in the monomeric state, and we suspect that it is this adsorbed state which is rapidly desorbed and thus available for rapid transport into membranes.

Caution must be exercised in the interpretation of the fluorescence spectra of particle-adsorbed BaP. The emission of adsorbed BaP may be totally quenched on some particles, as we have observed for hematite (19). These invisible BaP molecules may be rapidly desorbed, but the emission spectra of the particle-BaP suspensions may be dominated by a small amount of BaP microcrystals, and thus show a high percentage of excimer emission. Thus, the fluorescence emission spectra of these nonideal suspensions may not be representative of the total BaP population, and one need not find a correlation between the intensity of monomer emission and the rates of BaP transport into membranes.

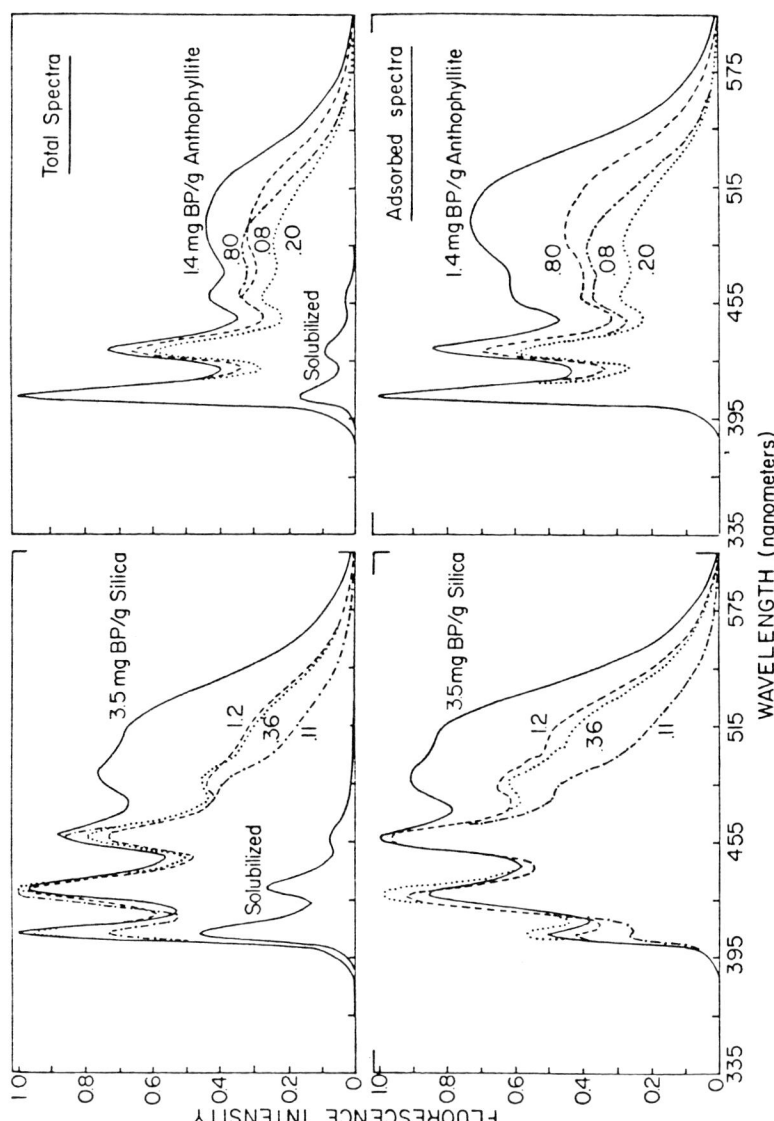

FIGURE 4. Emission spectra of benzo(a)pyrene bound to silica and anthophyllite.

## DISCUSSION

We have described a fluorescence spectroscopic method to quantify the rate of BaP uptake into microsomes. This technique allows continuous observation of the uptake within seconds after mixing the microsomes with the BaP. Direct chemical methods of measuring BaP uptake, which would require separation of the particles and microsomes followed by analysis for BaP, are complicated by the length of time required for separation, by binding between the microsomes and the particles, and by losses of the hydrophobic PAH to the walls of the containers. It should be noted that our technique measures the rate of uptake even in the opaque suspensions of particulates. Clearly, our understanding of the metabolism of PAH by the microsomal fraction of cells will be aided by a convenient means to quantify the rate of PAH transport into this cellular fraction. Furthermore, this method could be useful in bacterial mutagenesis tests and in the characterization of particle-PAH preparations which are to be used for carcinogenesis testing in animals.

### Microsomal Availability of Particle-Adsorbed BaP and Carcinogenic Potency

The known co-carcinogenic effects of particles and PAH which are observed in experimental animals correlate well with the microsomal availabilities of BaP described in this paper. For example, hematite with adsorbed BaP is more carcinogenic than simple mixtures of BaP and hematite (14). Similar results were found with asbestos-adsorbed BaP and asbestos-BaP mixtures (28). Our results indicate that BaP must be adsorbed to the particle for enhanced transport to occur. Carbon black with adsorbed BaP has been found to be less carcinogenic than BaP alone (5,39), although, depending on particle size, co-carcinogenic effects with other PAH have been observed (27). The former results agree with our own, in that carbon black did not release BaP into microsomes. We suspect that the observed variability in the co-carcinogenic effects of BaP and carbon particles is a result of the variable surface polarity possible with this particulate (24) and the effects of particle size on the relative rates of lung clearance and BaP elution. In addition, we demonstrated that fibrous minerals are superior to nonfibrous minerals in their ability to transport carcinogens into microsomes. We suggest that this may provide a partial explanation for the co-carcinogenic effects of asbestos inhalation and cigarette smoking (34).

## Other Mechanisms of Particle-PAH Co-Carcinogenesis

As indicated above, the microsomal availabilities of our BaP particle preparations correlate with the carcinogenic potencies of similar preparations which were tested in experimental animals. However, we point out that mechanisms other than particle-enhanced transport may be involved in particle-PAH co-carcinogenesis.

1. The co-carcinogenic effects of smoking and asbestos inhalation may result from the tissue damage caused by the inhaled particles. For example, carcinogen testing in animals has shown that cell trauma, such as burns, freezing with ice, surgical incisions, and scarification with sandpaper all increase the sensitivity of tissues to chemical carcinogens (3,7). In humans, lung tumors frequently originate in scar tissue (10,29). These lung scars can result from healed infarcts (11), tuberculosis (31), and particle-induced lung damage (12,30,43). Thus, asbestos-induced lung damage may increase the sensitivity of lung tissues to the chemical carcinogens present in cigarette smoke.
2. Inhaled particulates can inhibit the clearance of smoke particles from the lungs, and therby increase the total exposure to carcinogens by increasing the time available for elution of carcinogens from the smoke particles. As indicated above, inhaled particulates are known to increase the amount of scar tissue in the lungs. These scars interfere with the upward ciliary clearance of particles, and in humans these sites are frequently the origin of tumors (15). Additionally, particles are kown to be concentrated in these damaged tissues (4). In animals, asbestos inhalation inhibits the lung clearance of titanium oxide particles (9). Thus asbestos inhalation by humans could increase the residence times of smoke particles in the lungs and thereby increase the total exposure level of lungs to the carcinogens present in cigarette smoke.
3. Alternatively, smoking-induced damage of lung tissue may increase the susceptibility of lung tissue to the inherent carcinogenicity of asbestos. Smoking is known to damage lung tissue (2) and such damage results in the formation of scar tissue and inhibition of the clearance rates of the inhaled particles (42). In addition, smoking increases the fibrinogenic activity of the blood (22,25) and the occurrence of microemboli in the bloodstream. These emboli lodge in the capillaries of the lungs, again resulting in scarification (44). Perhaps the scar tissue which results from cigarette smoking is highly susceptible to the inherent carcinogenic effects of asbestos.

4. And finally, both smoking and asbestos inhalation could alter the metabolic profiles of the PAH and thereby increase the amounts of carcinogenic metabolites which are formed. Smoking is known to increase the aryl hydrocarbon hydroxylase levels in trachea and lungs (2,23,35). Since only a few PAH metabolites seem to be the ultimate carcinogens (16,45), the extent of carcinogenesis is likely to be critically dependent upon the PAH metabolic profiles. These profiles could also be altered by the cytotoxic and irritant effects of asbestos and by the enhanced microsomal availability of asbestos-adsorbed PAH. Perhaps the smoking-induced increase in AHH activity and the asbestos-enhanced availability of the PAH may act in a cooperative manner in inducing cell transformation.

In summary, our results seem to be relevant to carcinogen testing in animals, but alternative mechanisms of co-carcinogenesis may also be important in humans. We hope our methods, which allow the PAH delivery rates to be quantified, will facilitate the design of experiments which further elucidate the mechanism of particle-PAH co-carcinogenesis, and thereby increase our understanding of the multiple etiology of human cancer.

## ACKNOWLEDGMENTS

We thank the Freshwater Foundation, and especially its founder Mr. Richard Gray, Sr., without whose assistance this work would not have been possible. In addition we thank Mr. Steven Riemer for his excellent technical assistance. These studies were supported by Grant BC-261 from the American Cancer Society. D. R. Bevan was supported in part by a Postdoctoral Fellowship (CA-6405) from the National Cancer Institute. This work was done during the tenure of an Established Investigatorship (to J. R. Lakowicz) of the American Heart Association.

## REFERENCES

1. Ames, B. N., McCann, J., and Yamasaki, E. (1975): Method for detecting carcinogens and mutagens with the Salmonella/mammalian-microsome mutagenicity test. *Mutat. Res.* 31:347-364.
2. Auerbach, O., Stout, A. P., Hammond, E. C., and Garfinkel, L. (1961): Changes in bronchial epithelium in relation to cigarette smoking and in relation to lung cancer. *New Engl. J. Med.* 265:253-267.

3. Berenblum, I. (1944): Irritation and carcinogenesis. *Arch. Pathol.* 38:233-244.
4. Blenkinsopp, W. K. (1968): Particle accumulation in the lung as a possible factor in the aetiology of lung cancer. *J. Path. Bacteriol.* 96:297-304.
5. Davis, B. R., Whitehead, J. K., Gill, M. E., Lee, P. N., Butterworth, A. D., and Roe, F.J.R. (1975): Response of rat lung to 3,4-benzopyrene administered by intratracheal instillations with infusine with or without carbon black. *Br. J. Cancer* 31:443-452.
6. Davis, W. W., Krahl, M. E., and Clowes, G.H.A. (1942): A nephelometric method for determination of solubilities of extremely low order. *J. Am. Chem. Soc.* 64:101-107.
7. Deelman, H. T. (1927): The part played by injury and repair in the development of cancer, with some remarks on the growth of experimental cancers. *Proc. R. Soc. Med.* 20:19-20.
8. Ernster, L. and Nordenbrand, K. (1967): Microsomal lipid peroxidation. In: *Meth. Enzymol.*, Vol. 10, R. W. Estabrook and M. E. Pullman, Eds., pp. 574-580, Academic Press, New York.
9. Ferin, J. and Leach, L. J. (1976): The effect of amosite and chrysotile asbestos on the clearance of $TiO_2$ particles from the lung. *Environ. Res.* 12:250-254.
10. Finke, W. (1956): Chronic pulmonary disease as a possible etiologic factor in lung cancer. *Int. Rec. Med.* 169:61-72.
11. Hanbury, W. J., Cureton, R.J.R., and Simon, G. (1954): Pulmonary infarcts associated with bronchogenic carcinoma. *Thorax* 9:304-312.
12. Harington, J. S. (1974): Fibrogenesis. *Environ. Health Perspect.* 9:271-279.
13. Heidelberger, C. (1975): Chemical carcinogenesis. In: *Annu. Rev. Biochem.*, Vol. 44, E. E. Snell, P. D. Boyer, A. Meister, and C. C. Richardson, Eds., pp. 79-121, Annual Reviews, Inc., Palo Alto.
14. Henry, M. C., Port, C. D., and Kaufman, D. G. (1975): Importance of physical properties of benzo(a)pyrene-ferric oxide mixtures in lung tumor induction. *Cancer Res.* 35:207-217.
15. Hilding, A. C. (1957): Ciliary streaming in the bronchial tree and the time element in carcinogenesis. *New Engl. J. Med.* 256:634-640.
16. Kapitulnik, J. Wislocki, P. G., Levin, W., Yagi, H., Thakker, D. R., Akagi, H., Koreeda, M., Jerina, D. M., and Conney, A. H. (1978): Marked differences in the carcinogenic activity of optically pure (+)- and (-)-trans-7,8-dihydroxy-7,8-dihydrobenzo(a)pyrene in newborn mice. *Cancer Res.* 38:2661-2665.
17. Lakowicz, J. R., and Bevan, D. R. (1980): Effects of asbestos, iron oxide, silica, and carbon black on the microsomal availability of benzo(a)pyrene. *Biochemistry* 18:5170-5176.

18. Lakowicz, J. R., and Bevan, D. R. (1980): Effects of adsorption of benzo(a)pyrene to asbestos and non-fibrous mineral particulates upon its rate of uptake into phospholipid vesicles and rat liver microsomes. In: *Proceedings of the International Workshop on In Vitro Effects of Mineral Dusts* (in press).
19. Lakowicz, J. R., and Bevan, D. R. (1980): unpublished observations.
20. Lakowicz, J. R., Englund, F., and Hidmark, A. (1978): Particulate enhanced membrane uptake of 1,2-benzanthracene observed by fluorescence spectroscopy. *Biochim. Biophys. Acta.* 543:202-216.
21. Lakowicz, J. R., and Hylden, J. L. (1978): Asbestos-mediated uptake of benzo(a)pyrene observed by fluorescence spectroscopy. *Nature* 275:446-448.
22. Levine, P. H. (1973): An acute effect of cigarette smoking on platelet function. *Circulation* 48:619-623.
23. McLemore, T., Martin, R. R., Toppel, K. L., Busbee, D. L., and Cantrell, E. T. (1977): Comparison of aryl hydrocarbon hydroxylase induction in cultured blood lymphocytes and pulmonary macrophages. *J. Clin. Invest.* 60:1017-1024.
24. Medalia, A. I., and Rivin, D. (1976): Carbon blacks. In: *Characterization of Powder Surfaces*, G. D. Parfitt and K.S.W. Sing, Eds., pp. 279-351, Academic Press, New York.
25. Mustard, J. F., and Murphy, E. A. (1963): Effect of smoking on blood coagulation and platelet survival in man. *Br. Med. J.* 1:846-849.
26. Nebert, D. W., and Gelboin, H. V. (1968): Substrate-inducible microsomal aryl hydroxylase in mammalian cell culture. *J. Biol. Chem.* 243:6242-6249.
27. Pylev, L. N. (1961): Experimental induction of lung cancer in rats by intratracheal administration of 9,10-dimethyl-1,2-benzanthracene. *Bull. Exp. Biol. Med. (USSR)* 52:1316-1319.
28. Pylev, L. N., and Shabad, K. M. (1973): Some results of experimental studies in asbestos carcinogenesis. In: *Biological Effects of Asbestos*, P. Bogovski, J. C. Gilson, V. Timbrell, and J. C. Wagner, Eds., pp. 99-106, International Agency for Research on Cancer, Lyon.
29. Raeburn, C., and Spencer, H. (1957): Lung scar cancers. *Br. J. Tub. Dis. Chest* 51:237-245.
30. Reeves, A. L., Puro, H. E., and Smith, R. G. (1974): Inhalation carcinogenesis from various forms of asbestos. *Environ. Res.* 8:178-202.
31. Ripstein, C. B., Spain, D. M., and Bluth, I. (1968): Scar cancer of the lung. *J. Thorac. Cardiovasc. Surg.* 56:362-370.
32. Saffiotti, U., Cefis, F., and Kolb, L. H. (1968): A method for the experimental induction of bronchogenic carcinoma. *Cancer Res.* 28:104-124.

33. Saffiotti, U., Cefis, F., Kolb, L. H., and Shubik, P. (1965): Experimental studies of the conditions of exposure to carcinogens for lung cancer induction. *J. Air Pollut. Control Assoc.* 15:23-25.
34. Selikoff, I. J., Hammond, E. C., and Churg, J. (1968): Asbestos exposure, smoking and neoplasia. *J. Amer. Med. Assoc.* 188:22-26.
35. Simberg, N., and Uotila, P. (1978): Stimulatory effect of cigarette smoking on the metabolism and covalent binding of benzo(a)pyrene in the trachea of the rat. *Int. J. Cancer* 22:28-31.
36. Sims, P., and Grover, P. L. (1974): Epoxides of PAH metabolism and carcinogenesis. *Adv. Cancer Res.* 20:166-274.
37. Sims, P., Grover, P. L, Swaisland, A., Pal, K., and Hewer, A. (1974): Metabolic activation of benzo(a)pyrene proceeds by a diol epoxide. *Nature* 252:326-328.
38. Speil, S., and Leineweber, J. P. (1969): Asbestos minerals in modern technology. *Environ. Res.* 2:166-208.
39. Steiner, P. E. (1956): The conditional biological activity of the carcinogens in carbon blacks, and its elimination. *Cancer Res.* 14:103-110.
40. Stenback, F., Rowland, J., and Sellakumar, A. (1976): Carcinogenicity of benzo(a)pyrene and dusts in hamster lung instilled intratracheally with titanium oxide, carbon, and ferric oxide. *Oncology* 33:29-34.
41. Timbrell, V. (1970): Characteristics of the International Union Against Cancer standard reference samples of asbestos. In: *Pneumoconiosis, Proc. Int. Conf.*, H. A. Shapiro, Ed., pp. 28-36, Oxford University Press, Cape Town.
42. Tremer, H. M., Falk, H. L., and Kotin, P. (1959): Effect of air pollutants on ciliated mucus-secreting epithelium. *J. Natl. Cancer Inst.* 23:979-997.
43. Viswanathan, P. N., Dogra, R.K.S., Shanker, R., and Zaidi, S. H. (1973): Pulmonary fibrogenic response of guinea pigs to amosite dust. *Int. Arch. Arbeitsmed.* 3l:51-59.
44. Wartman, W. B., Jennings, R. B., and Hudson, B. (1951): Experimental arterial disease, I. The reaction of the pulmonary artery to minute emboli of blood clots. *Circulation* 4:747-755.
45. Weinstein, I. B., Jeffrey, A. M., Jannette, K. W., Blobstein, S. H., Harvey, R. C., Harris, C., Antrup, H., Kasai, H., and Nakanishi, K. (1976): Benzo(a)pyrene diol epoxides as intermediates in nucleic acid binding in vitro and in vivo. *Science* 193:592-595.
46. Wills, E. D. (1969): Lipid peroxide formation in microsomes. *Biochem. J.* 113:315-324.
47. Yang, S. K., McCourt, D. W., Leutz, J. C., and Gilboin, H. V. (1977): Benzo(a)pyrene diol epoxides: Mechanism of enzymatic formation and optically active intermediates, *Science* 196:1199-1201.

# CORRELATIONS OF MUTAGENIC ACTIVITY WITH POLYNUCLEAR AROMATIC HYDROCARBON CONTENT OF VARIOUS MINERAL OILS

M. Hermann*, J. P. Durand*, J. M. Charpentier*, O. Chaudé*, M. Hofnung**, N. Pétroff*, J-P.. Vandecasteele*, and N. Weill**
  *Institut Français Du Pétrole
  1 et 4 avenue de Bois Préau, 92506 Rueil Malmaison, France
  **Institut Pasteur
  28, rue du Dr. Roux, 75724 Paris Cedex 15, France

We have previously described techniques permitting the use of the Ames test (1) for the determination of the mutagenicity of mineral oils (2). We applied them to the study of the mutagenicity of oils of various origins. In the present work, we used high-pressure liquid chromatography (HPLC) for separation and analysis of the polynuclear aromatic hydrocarbons (PAH) of some of these oils. At the same time, mutagenicity determinations have been performed on pure PAH compounds identified in these oils, as well as on fractions resulting from a partial separation of the PAH portion into classes of related compounds. This approach was used in an attempt to identify the most important compounds or groups of compounds responsible for the mutagenic activity of these oils. The results for three representative products obtained from petroleum, namely a used solvent-refined crankcase oil, an hydrotreated oil, and a petroleum distillate are given. In addition, a general assessment of the mutagenicity of the various types of oils from the petroleum industry and from other origins has been performed in relation with their total PAH content.

## MATERIALS AND METHODS

### Oil Samples and Chemicals

Oil No. 0 was a white oil of medicinal quality; oils Nos. 1 and 2 were, respectively, a special highly refined steel-hardening oil and the corresponding oil used under inert atmosphere. They were kindly supplied by Dr. Limasset, INRS Nancy. Oils Nos. 3 and 4 were, respectively, a highly refined crankcase oil and the corresponding used oil obtained in a special wearing test (Petter engine). Petroleum distillates Nos. 5 and 7 were petroleum industry samples obtained by vacuum distillation (distillation ranges 380 to 500°C and 300 to 430°C, respectively). Oils Nos. 6 and 8 were solvent-refined samples obtained respectively from petroleum distillates Nos. 5 and 7. Oil No. 9 was an experimental hydrotreated petroleum residue (distillation range 400 to 550°C). Oil No. 10 was a crude oil obtained by shale pyrolysis under nitrogen atmosphere. Oil No. 11 was an anthracenic oil from coal tar distillation, kindly supplied by Dr. Ferrand, CERCHAR, Creil, France. Oil No. 12 was a steam-cracking residue distillating above 260°C. Sulfur contents of these oils were very low for oils No. 0 to No. 4 and for oil No. 9. It was equal to 2.5, 1.4, 1.6, 0.7, and 2.3 percent for oils Nos. 5, 6, 7, 8 and 12, respectively. It was not determined but should be relatively high in oils Nos. 10 and 11.

PAH were from commercial sources, mainly Aldrich and K&K except for benzo(b)fluoranthene which was a gift from CERCHAR, Creil, France.

**Analytical Methods**

*Determination of Total PAH Content*

Total PAH content of oils was determined by a previously described method (2) of extraction of the PAH with dimethyl sulfoxide (DMSO) from cyclohexane solutions of oils. Reextraction in cyclohexane was then performed by salting out of the PAH from the DMSO solution. PAH content of oil was obtained from the weight of the residue after cyclohexane distillation.

*Determination of Individual PAH by HPLC*

The method, which is described in detail elsewhere by J. P. Durand and N. Petroff (9,10), is summarized in Table 1. It allows a rapid and direct determination of the main PAH of an oil sample. On the microsilica particles modified by DMSO adsorption, PAH retention increases with the number of rings and decreases with the degree of alkylation or hydrogenation. This allows a good separation at the level of the more condensed PAH. Sulfur-containing PAH have a behavior similar to that of corresponding non-sulfur-containing PAH, making detailed identification more difficult in the case of oils with a high sulfur content.

**TABLE 1.** High Pressure Liquid Chromatography Data[a]

| | |
|---|---|
| Instrument | Varian Model 8500 with Valco Model 7000 sample injector valve |
| Column | Microsilica particles (Lichrosorb SI60 5 $\mu$m Merck). Modified by DMSO adsorption on a 4.2-mm (ID) × 30-cm column |
| Solvent | Heptane or isooctane + 0.3% DMSO - flow rate: 1 ml × min$^{-1}$ |
| Detectors | (1) UV detector LDC spectromonitor I $\lambda$ # 290 nm<br>(2) Fluorescence detector Dupont Model 836 excitation filter: 320 to 385 nm; emission filter > 408 nm |
| Injection | Oil/cyclohexane (1/1): 10 to 30 $\mu$l or DMSO extracts (10 $\mu$l). |
| Time | 20 to 60 min |

[a] The conditions indicated are those used for analytical determinations.

Identification and quantitative determination of individual PAH have been done by reference to authentic compounds when available. In other cases, identifications are proposed from mass spectrometric and UV adsorption spectrophotometric determinations on fractions obtained from semipreparative HPLC separations described below.

*PAH Fractionation by HPLC*

Two types of fractionation have been conducted on the same type of microsilica columns modified by DMSO adsorption using the solvents listed in Table 1.

- Preparative fractionations were performed on large columns (35-mm ID x 50 cm) containing 150 g of microsilica (Lichroprep SI60, 15 to 25 $\mu$m from Merck), allowing the introduction of 100 to 200 mg of PAH extracts obtained by the DMSO extraction of oils described above, and dissolved in 30 ml of cyclohexane. Weighing of the fractions obtained was possible in this case. In the instrument used (Chromatospac Prep 100 from Jobin et Yvon), detection was done by refractometry.
- Semipreparative fractionations were performed on 9-mm ID x 30-cm columns containing 9 g of microsilica (Spherosil XOA 600, 5 to 10 $\mu$m from Prolabo France), allowing the introduction of about 5 mg of PAH extracts obtained as above and diluted in 500 $\mu$l of cyclohexane.

The instrument used was the same as for analytical HPLC (Table 1), the ultraviolet detector ($\lambda$ = 290 nm) allowing an adequate monitoring of the fractionation. However, weighing of the fractions obtained was not possible.

**Mutagenicity Determinations**

Mutagenicity determinations have been performed using the two previously described procedures allowing the utilization of the

*Salmonella*/mamalian microsome test devised by Ames (1) for petroleum products (2). The first procedure, used for individual PAH or PAH extracted by DMSO as described above, is based on the utilization of DMSO solutions of the compounds to be tested. When applied to mineral oils, this procedure, which required prior extraction of PAH, offered the advantage of allowing at the same time the determination of the PAH content. The second procedure, being carried out directly on oil emulsified with a detergent was, as a consequence, much faster and also more reliable with some types of oils and was usually preferred. Mutagenicity determinations could then be carried out on individual PAH, DMSO extracts of oils, fractions from preparative and semipreparative HPLC separations, and whole oils.

## RESULTS

### Mutagenicity Determinations of Individual PAH

The mutagenicity of benzo(a)pyrene (BaP) dissolved in DMSO and in medicinal oil with and without emulsification is shown in Figure 1. The

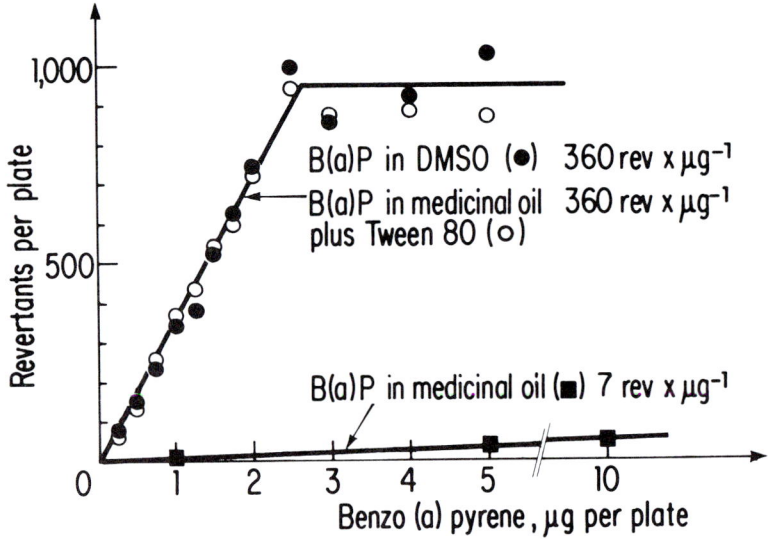

FIGURE 1. Influence of the mode of addition of BaP on the dose-response curves in the Ames test. The strain used was TA 98 and the procedure was carried out as previously described (2) in the presence of liver enzymes. BaP was added as a solution in DMSO or medicinal oil and the total volume of solvent (DMSO or oil) was adjusted to 100 μl. Spontaneous revertants have been substracted (usually 40 per plate).

results illustrate the necessity of an adequate mode of addition of BaP to obtain expression of its mutagenicity. Dissolution in DMSO and emulsification with detergents gave equally satisfactory results. The dose-response curve of BaP exhibited a large straight portion which allowed a good determination of mutagenic activity. Similar results were obtained with other pure PAH.

## Mutagenicity of Used and Hydrotreated Oils

*Mutagenicity Determinations*

Dose-response curves exhibited by used oil No. 4, derived from a nonmutagenic, highly refined oil (oil No. 3) and by an hydrotreated oil (oil No. 9) are presented in Figure 2. For oil No. 4, the curve was very similar to

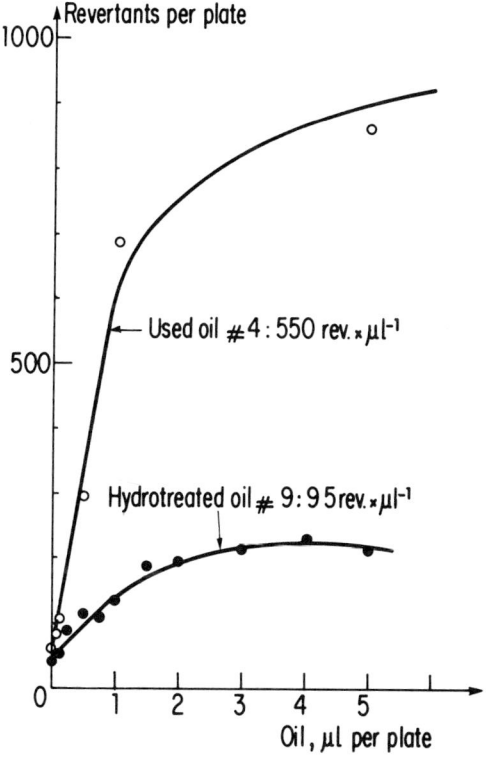

FIGURE 2. Dose-response curves in the Ames test of used oil and hydrotreated oil. The general procedure was that previously described, utilizing strain TA 98 in the presence of liver enzymes (2). Oils were added using the emulsification method with Tween 80. Spontaneous revertants have not been substracted.

that of BaP (Figure 1), but for oil No. 9, it levelled off more rapidly and the linear part used for the calculation of mutagenicity was shorter. The mutagenicity of hydrotreated oil No. 9 (95 rev. x $\mu l^{-1}$) was also lower than that of used oil No. 4 (550 rev. x $\mu l^{-1}$). Both curves were obtained using the emulsification method. The method involving extraction with DMSO (DMSO extraction method) gave similar but often lower results especially for used oils (200 rev. x $\mu l^{-1}$ for oil No. 4), probably because of incomplete extraction by DMSO.

## Detailed Study of Used Oils

The analytical profile of used oil No. 4, shown in Figure 3, in comparison with that of the corresponding unused oil (oil No. 3), indicates the formation, during the wear of oil, of a great number of PAH which have been identified when possible and subjected as pure compounds to mutagenicity determinations. At the same time, semipreparative fractionations were conducted on DMSO extracts of oil No. 4 and mutagenicity determinations were carried out on the fractions obtained. The main chemical structures observed in this oil were pyrene, fluoranthene, and higher PAH under condensed forms and methyl

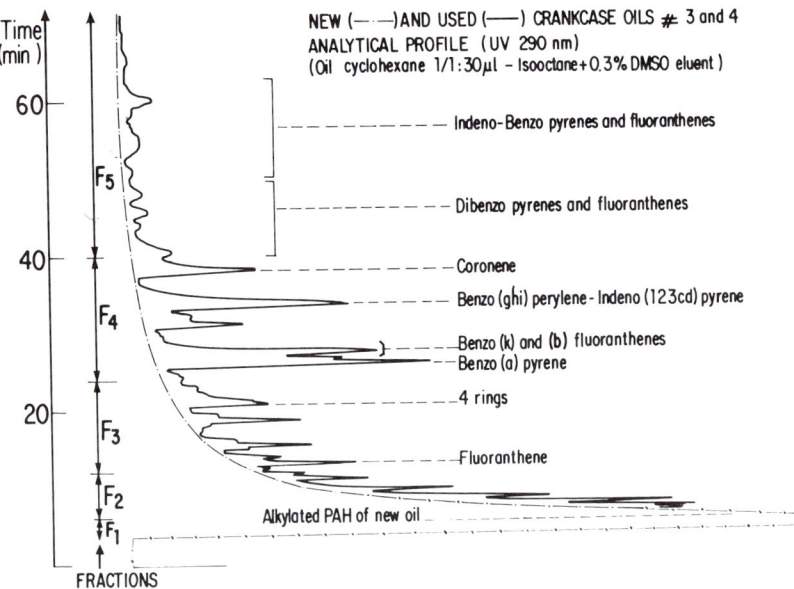

FIGURE 3. Analytical profile of used crankcase oil. The fractionation performed in the corresponding semipreparative procedure is shown (fractions $F_1$ to $F_5$).

derivatives. From the results (summarized in Table 2), it can be concluded that the components responsible for the mutagenic activity are 4-, 5-, and 6-ring PAH, most of which have been identified. BaP is, in this case, the most important identified individual contributor to the mutagenic activity of the oil, but its contribution is only 5.6 percent of the total. Another point which deserves comment is that the sum of the mutagenicities of the fractions amounts to only about 50 percent of that of the initial DMSO extract. However, the sum of the contributions of individual PAH present in fraction $F_4$, most of which have been identified, seems large enough to account for the mutagenicity of this fraction.

*Detailed Study of Hydrotreated Oils*

A detailed study, similar to that performed with used oil, has been conducted with hydrotreated oil No. 9. The results (see Figure 4 and Table 3) first indicated that the characteristic PAH structures of this oil were pyrene and higher PAH in their most condensed forms and their alkyl and hydrogenated derivatives. No BaP was present in this oil. It appeared that the main contributors to the mutagenicity of oil No. 9 were partially hydrogenated structures which were present in large quantities (52 percent by weight of the DMSO extract) and localized in fraction $F_2$. A smaller contribution of the PAH of the less abundant fraction $F_4$ (15 percent by weight of the DMSO extract) was also observed. For this oil, the sum of the mutagenicity of the individual fractions was close to that of the DMSO extract. However, the sum of the mutagenicities of the identified PAH of fraction $F_4$ appeared too high in comparison with the mutagenicity of the whole fraction, suggesting that, in this case, the mutagenicities of individual PAH were partially inhibited when measured as a mixture in the whole fraction.

**Mutagenicity of Petroleum Distillates**

*Mutagenicity Studies*

A typical dose-response curve obtained with petroleum distillate No. 5 is presented in Figure 5. The main characteristic of this curve, which differed markedly from those obtained for other types of oils such as used oil No. 2 and for pure PAH, was that the maximum increase in the number of revertants did not exceed twice the number of spontaneous revertants, thus making these oils appear only slightly mutagenic. However, the initial part of the curve indicated a fairly high mutagenic

TABLE 2. Fractionation and Mutagenicity of Used Oil No. 4

| | Fractionation of DMSO Extracts of Oil[a] | | Contribution of Individual PAH to Oil Mutagenicity | | |
|---|---|---|---|---|---|
| Fractions | Mutagenicity[b] (%) | Identified or Proposed[c] PAH Structures | Specific Mutagenicity of Pure Compound (rev × μg$^{-1}$) | Concentration in Oil[d] (μg × l$^{-1}$) | Contribution to Oil Mutagenicity (%) |
| F$_1$ | 0.13 (0.05) | Alkylated PAH of unused oil (3) | | | |
| F$_2$ | 0.27 (0.1) | id. | | | |
| F$_3$ | 16.1 (5.9) | Fluoranthene (1) | 20 | 68,000 | <0.25 |
| | | Benzo(mno)fluoranthene (2) | | | |
| | | Cyclopentapyrene (2) | | | |
| | | Pyrene (1) | <0.02 | 1,000 | ~0 |
| | | Other 4 rings PAH and methyl derivatives (3) | | | |
| F$_4$ | 12.8 (4.7) | Perylene (1) | 130 | 47,000 | 1.1 |
| | | Benzo(a)pyrene (1) | 500 | 62,000 | 5.6 |
| | | Benzo(k)fluoranthene (1) | 20 | 1,700 | <0.01 |
| | | Benzo(b)fluoranthene (1) | 40 | 57,000 | 0.45 |
| | | Indenofluoranthene (2) | | | |
| | | Indeno(1,2,3,c,d)pyrene (1) | 90 | 20,000 | 0.3 |
| | | Benzo(ghi)perylene (1) | 35 | 33,000 | 0.2 |
| | | Coronene (1) | 200 | 33,000 | 1.2 |
| | | Methyl derivatives of above PAH (3) | | | |
| F$_5$ | 15.7 (5.7) | Dibenzopyrenes (3) | | | |
| | | Dibenzofluoranthenes (3) | | | |
| | | Indenobenzo(a)pyrene (2) | | | |
| | | Indenobenzofluoranthene (2) | | | |
| ΣF$_1$–F$_5$ | 45 | | | | |

[a] The DMSO extract of oil No. 4 represented about 15 g/l of oil. Its mutagenicity was about 30 rev × μg$^{-1}$ of extract (or 200 rev × μl$^{-1}$ of oil versus 550 rev × μl$^{-1}$ for the whole oil). Fractionation of oil was done on this extract by the semi-preparative procedure.
[b] Mutagenicity is expressed as a percentage of that of DMSO extract and (in parentheses) of that of oil.
[c] Complete identification was performed when authentic compounds were available (compounds noted 1). For other individual compounds, proposed structures were derived from mass and UV adsorption spectra and chromatographic behavior (2). For groups of compounds, they were based on chromatographic behavior and molecular weights from mass spectra (3).
[d] PAH determination was done by the analytical procedure on whole oil with calibration with the pure compounds.

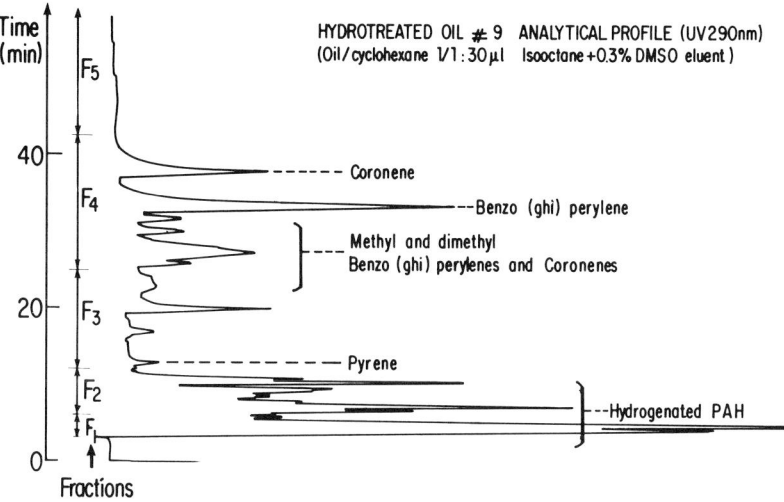

FIGURE 4. Analytical profile of hydrotreated oil. The fractionation performed in the corresponding preparative procedure is shown (fractions $F_1$ to $F_5$).

activity for these oils. The results thus suggested that petroleum distillates contained, besides mutagenic PAH, inhibitory compounds which rapidly prevented the expression of mutagenic activity when higher amounts of oil were used. Figure 6 confirms this hypothesis, showing that in a mixture of oil No. 5 and BaP, the dose-response curve of BaP is severely damped in comparison with that of BaP dissolved in medicinal oil. The initial slope of the curve, however, allows the calculation of a mutagenic activity for the mixture of oil No. 5 and BaP which is approximately equal to the sum of that of the two components. These results and the observations reported above prompted a detailed analysis of the mutagenic properties of PAH in a mixture compared with those of the individual compounds. In our first experiments, we recorded the number of revertants obtained with a constant amount of BaP in the presence of increasing amounts of petroleum distillate No. 5. The inhibition of BaP mutagenicity was observed as expected (see Figure 7.) However, these experiments also revealed that lower amounts of petroleum distillate enhanced instead of inhibiting BaP mutagenic activity. These phenomena of enhancement and subsequent inhibition of BaP mutagenicity have been observed with other petroleum distillates, but only at a lesser extent with hydro-treated oils such as No. 9 or used oils such as No. 4. Other individual PAH such as anthanthrene or 1,2 benzanthracene behaved like BaP in this type of experiment. A systematic study of the enhancement and inhibition of BaP mutagenicity by individual PAH, including sulfur-containing PAH, has been carried out in order to elucidate the nature of the compounds present

TABLE 3. Fractionation and Mutagenicity of Hydrotreated Oil No. 9

| | Fractionation of DMSO Extracts of Oil[a] | | | Contribution of Individual PAH to Oil Mutagenicity | | |
|---|---|---|---|---|---|---|
| Fractions | Weight Balance (%) | Mutagenicity[b] (%) | Identified or Proposed[c] PAH Structures | Specific Mutagenicity of Pure Compound (rev × $\mu g^{-1}$) | Concentration in Oil[d] (g × $1^{-1}$) | Contribution to Oil Mutagenicity (%) |
| $F_1$ | 27.5 | 0.04 | Highly hydrogenated PAH (3) | | | |
| $F_2$ | 52 | 87.4 | Trimethylene pyrene (2) Methyl and dimethyl Derivatives of pyrene (3) Other hydrogenated PAH (3) | | | |
| $F_3$ | 3.5 | 6.3 | Pyrene (1) Slightly Hydrogenated, 6 and 7 rings PAH (3) | <0.02 | 20,000 | ~0 |
| $F_4$ | 15 | 10.7 | Benzo (ghi) perylene (1) Coronene (1) Methyl and dimethyl derivatives of above PAH (3) | 35 200 | 162,000 54,000 | 5.9 11.3 |
| $F_5$ | 2 | 0.95 | More condensed PAH? | | | |
| $F_1$-$F_5$ | | 105.4 | | | | |

[a] The DMSO extract of oil No. 9 represented 2.5 g/l of oil Its mutagenicity was about 100 rev × $\mu l^{-1}$ of oil (versus 95 rev × $\mu l^{-1}$ for the whole oil).
[b] Fractionation of was done on the DMSO extract by the preparative procedure allowing a weight balance of the fractions. Expressed as a percentage of the mutagenicity of the DMSO extract which is, in this case, equivalent to that of oil.
[c] Complete identification was performed when authentic compounds were available (compounds noted 1). For other individual compounds, proposed structures were derived from mass and UV adsorption spectra and chromatographic behavior (2). For groups of compounds, they were based on chromoatographic behavior and molecular weights from mass spectra (3).
[d] PAH determination was done by the analytical procedure on whole oil with calibration with the pure compounds.

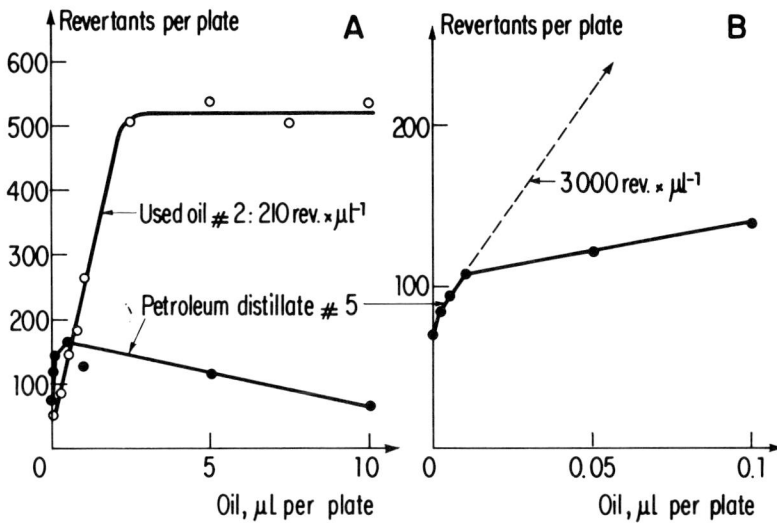

**FIGURE 5.** Dose-response curves in the Ames test for used oil No. 2 and petroleum distillate No. 5. A: curves obtained in the range 0 to 10 μl of oil; B: curve obtained for petroleum distillate No. 5 in the range 0 to 0.1 μl of oil. Other conditions are as for Figure 2.

in petroleum distillate No. 5 which were responsible for the phenomena observed. Forty-four PAH containing from one to seven rings, with or without methyl side chains, were tested. An enhancement effect was observed with naphtalene which was larger with anthracene. When higher PAH were used, an inhibition effect, not seen with anthracene, was observed following the initial enhancement effect. With benzo(e)pyrene for example, the enhancement effect was observed at a lower ocncentration than for anthracene but was partly masked because of the strong subsequent inhibition effect. Probably for this reason, with still higher PAH, only the inhibition effect could be seen. Thus both effects of enhancement and inhibition increased with the size of the PAH. The intensity of these effects varied between PAH of a given number of rings. Sulfur-containing PAH also exhibited these properties.

*Analytical Profile and Fractionation of Petroleum Distillates*

Sulfur-containing PAH were major constituents of the PAH fraction of petroleum distillate No. 5. The complexity of this fraction, illustrated in the analytical profile of Figure 8 entails a lesser predominance of individual compounds, and thus, fewer individual components could be identified. The major constituents of petroleum distillate No. 5 were alkyl

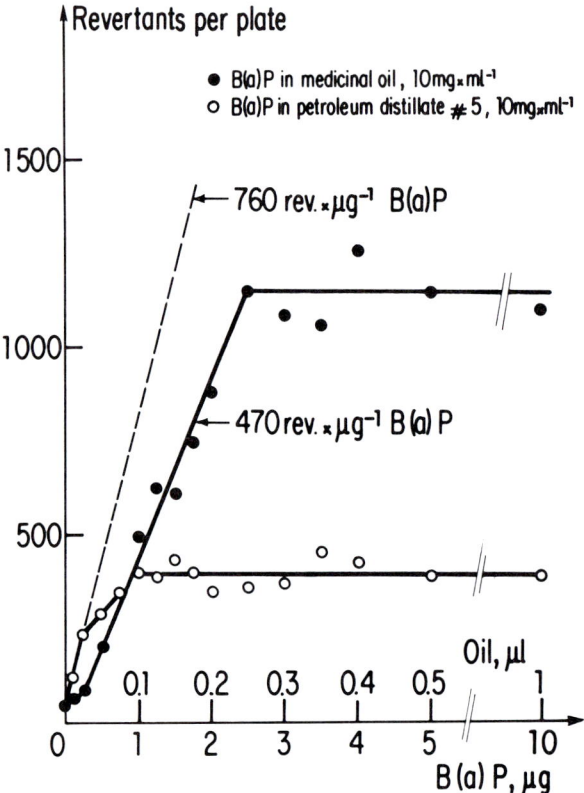

**FIGURE 6.** Dose-response curves in the Ames test of BaP dissolved in medicinal oil or in petroleum distillate No. 5. BaP was dissolved either in medicinal oil or in petroleum distillate No. 5 at a final concentration of 10 mg × ml$^{-1}$. Other conditions are as for Figure 2.

dibenzothiophene derivatives. Smaller amounts of non-sulfur-containing PAH were also present, including a low quantity (1 mg × l$^{-1}$ of oil) of BaP. The results of the fractionation experiment (shown in Table 4) indicate that fraction $F_2$ contained most of the mutagenic activity. The major constituents of this fraction, which represented 70 percent by weight of the DMSO extract, were moderately alkylated dibenzothiophene derivatives. It is worth noting that fraction $F_1$ which contained more highly alkylated derivatives of the same sulfur PAH was not mutagenic. The same fractionation experiment shows that the sum of the mutagenicities of the fractions is here much lower than that of the DMSO extract. The mutagenicity enhancement effect described above appears to be a very likely explanation of this result.

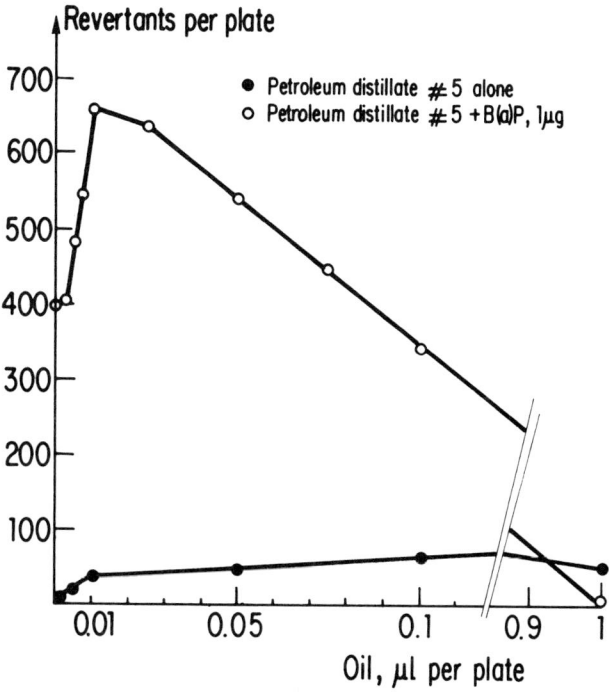

FIGURE 7. Effect of addition of petroleum distillate No. 5 on the number of revertants obtained in the presence of 1 µg of BaP. A constant amount (1 µg) of BaP as a DMSO solution and increasing amounts of petroleum distillate No. 5 were added to the Petri plates and the corresponding numbers of revertants recorded, after substraction of the numbers of revertants obtained in the presence of petroleum distillate No. 5 alone, which are also shown for comparison. Other conditions are as for Figure 2, except that spontaneous revertants have been substracted.

## General Survey of the Mutagenicities of Oils of Various Origins

Besides the detailed study of three representative types of products obtained from petroleum, a general survey of some important characteristics of mineral oils of various origins has been performed. Oils from nonpetroleum sources (anthracenic and shale oils) have been included for comparison. The results, presented in Table 5, illustrate the good correlation observed between mutagenic activity and PAH content (in all cases 20 to 40 revertants per µg of PAH extract were obtained). On the contrary, there was no correlation between mutagenic activity and BaP content. This last point confirms and extends earlier observations (3,4). The highest BaP contents have been found in used oils, which reflects the origin of the PAH of these oils. It should be pointed out, when considering

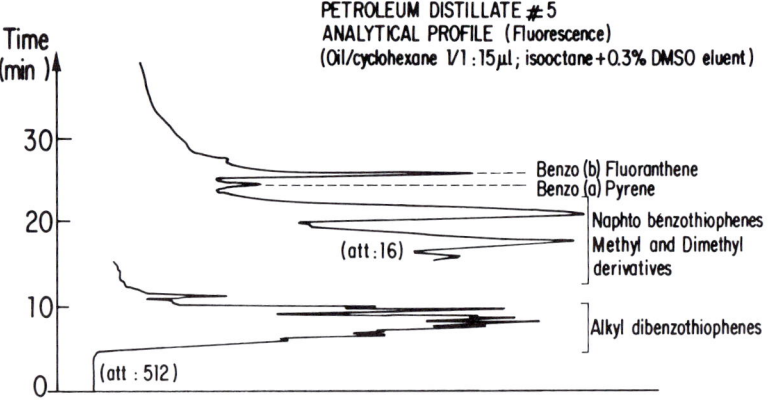

FIGURE 8. Analytical profile of petroleum distillate No.5.

all the results presented here concerning used oils, that our samples (Nos. 2 and 4) were prepared by unusually severe wearing treatment.

## DISCUSSION

The first conclusion to be drawn from this study is that the mutagenic activity of mineral oils can usually be ascribed to a complex mixture of compounds with an average mutagenic activity, rather than to any single compound. The chemical structures of these mutagenic compounds have been found to vary with the origin of oil; they were moderately alkylated PAH in the case of petroleum distillates; partially hydrogenated PAH in the case of hydrotreated oils; and 4-, 5-, and 6-ring PAH in used oils. The diversified nature of the compounds responsible for the mutagenicity of oils explains the good correlation observed between mutagenicity and PAH content. The correlation between mutagenicity results presented above and animal carcinogenicity of mineral oils as estimated from literature data when available (4,5,6) is also good. Unfortunately, these data are not available for used oils for which relatively moderate but definite mutagenic activities have been found.

The good correlation observed between PAH content and mutagenic activity suggests that the determination of the latter is satisfactory, even in the more difficult case of petroleum distillates where the accuracy of the determinations is lower. The effects of enhancement and inhibition of mutagenicity demonstrated here may increase or decrease the mutagenic activities measured in oils with respect to those of individual PAH, or may also, in other cases, approximately cancel each other according to the nature of the compounds involved. These effects explain to a large extent, the difficulties met in trying to balance the mutagenicities of fractions or

TABLE 4. Fractionation and Mutagenicity of Petroleum Distillate No. 5

| Fractions[a] | Weight Balance (%) | Mutagenicity[b] (%) | Proposed PAH Structures[c] |
|---|---|---|---|
| $F_1$ | 8.2 | <0.2 | Alkyl dibenzothrophene derivatives (5 or more $CH_2$ or $CH_3$ groups) Alkyl fluorene derivatives (7 or more $CH_2$ or $CH_3$ groups) (3) |
| $F_2$ | 73.3 | 15 | Alkyl dibenzothrophene derivatives 2,3 or 4-$CH_2$ or $CH_3$ groups) Alkyl fluorene derivatives (4, 5 or 6 $CH_2$ or $CH_3$ groups) (3) Alkyl naphtobenzothrophene derivatives 1, 2, or 3 $CH_2$ or $CH_3$ groups (3) |
| $F_3$ | 10.1 | 4.8 | 4-ring PAH (chrysene, benzanthracene, etc.) |
| $F_4$ | 4.7 | 0.3 | Naphtobenzothrophenes (3) 5-ring PAH (benzo(a)pyrene, benzofluoranthenes) (1) |
| $F_5$ | 4.7 | 0.25 | 6 or more rings PAH |
| $\Sigma F_1 - F_5$ | | ~21 | |

[a] The DMSO extract of petroleum distillate No. 5 represented 78 g/l of oil. Its mutagenicity was about 2800 rev×$\mu l^{-1}$ for oil (versus 3400 rev× $\mu l^{-1}$ for the whole oil). Fractionation was done by the preparative procedure.
[b] Expressed as a percentage of the mutagenicity of the DMSO extract which is, in this case, equivalent to that of oil.
[c] Complete identification was performed when authentic compounds were available (compounds noted 1). For other individual compounds, proposed structures were derived from mass and UV adsorption spectra and chromatographic behavior (2). For groups of compounds, they were based on chromatographic behavior and molecular weights from mass spectra (3).

TABLE 5. Mutagenicity and PAH Content of Various Oils

| Reference | Origin | Mutagenicity[a] (rev. × μl$^{-1}$) | BaP Content[a] (μg × l$^{-1}$) | PAH Content[a] (g × l$^{-1}$) | Rev. × μg$^{-1}$ PAH |
|---|---|---|---|---|---|
| 0 | White (medicinal quality) | <1 | <5 | 0.64 | |
| 1 | Hardening, new | <0.5 | <5 | 2.6 | |
| 2 | Hardening, used | 250 | 25,000 | 8.3 | 30 |
| 3 | Crankcase, new | <30 | <5 | nd(2)[b] | |
| 4 | Crankcase, used | 550 | 62,000 | 15(2)[b] | |
| 5 | Petroleum distillate | 2,800 | 840 | 78 | 36 |
| 6 | Solvent-refined (derived from No. 5) | 340 | <10 | 20 | 17 |
| 7 | Petroleum distillate | 1,200 | <10 | 54 | 22 |
| 8 | Solvent-refined (derived from No. 7) | 230 | <10 | 9.7 | 24 |
| 9 | Petroleum distillate hydrotreated | 95 | <5 | 2.5[b] | 38 |
| 10 | Shale | 6,000 | nd(1)[b] | | |
| 11 | Anthracenic | 15,500 | 200,000 | ~700 | 22 |
| 12 | Steam cracking residue | 19,500 | 330,000 | ~600 | 33 |

[a] All values are expressed per unit volume of oil. Corresponding weight values may be calculated using a mean density value of 0.85 (oil No. 0 to oil No. 9) and 1 (oil No. 10 to No. 12).
[b] nd - no determination (1) poor separation of B(a)P; (2) presence of additives alters PAH determination; (3) DMSO extraction impossible.

individual compounds with those of their mixtures. Similar effects have been reported in the case of harman and norharman (7).

The most interesting point regarding these effects, observed in a test of mutagenicity, is that they may also be relevant to the carcinogenic properties of the compounds involved. The enhancement effect, in fact, is reminiscent of the co-carcinogenic activity of naphtalene and alkyl naphthalenes in tobacco carcinogenis (8). The question actually raised here is that of the origin of these effects. Our working hypothesis is that they occur at the level of the enzymatic system of activation of the PAH.

## REFERENCES

1. Ames, B. N., McCann, J., and Yamasaki, E. (1975): Methods for detecting carcinogens and mutagens with the *Salmonella*/mammalian microsome mutagenicity test. *Mutation Research* 17: 347-364.
2. Hermann, M., Chaude, O., Weill, N., Bedouelle, H., and Hofnung, M. (1980): Adaptation of the *Salmonella*/mammalian microsome test to the determination of the mutagenic properties of mineral oils, *Mutation Research* (in press).
3. Payne, J. F., and Martins, I. (1978): Crankcase oils: are they a major mutagenic burden in the aquatic environment. *Science* 200: 329-330.
4. Medical Research Council, London (1968): The carcinogenic action of mineral oils: a chemical and biological study.
5. Catchpole, W. M., MacMillan, E., and Powell, H. (1971): Specifications for cutting oils with special reference to carcinogenicity. *Journal of the Institute of Petroleum* 57: 247-260.
6. International Agency for Research on Cancer (1973): *Monographs on the evaluation of carcinogenic risk of the chemicals to man. Certain polycyclic aromatic hydrocarbons and heterocyclic compounds.* Vol. 3, World Health Organization, I.A.R.C.
7. Nagao, M., Yahagi, T., and Sugimura, T. (1978): Differences in effects of norharman with various classes of chemical mutagens and amounts of S-g. *Biochemical and Biophysical Research Communications* 83: 373-377.
8. Schmeltz, I., Tosk, J., Hilfrich, J., Hirota, N., Hoffmann, D., and Wynder, E. L. (1978): Bioassays of naphtalene and alkyl naphtalenes for co-carcinogenic activity. Relation to tobacco carcinogenesis. In: *Carcinogenesis. Vol. 3: Polynuclear Aromatic Hydrocarbons.* P. W. Jones and R. I. Freudenthal, Eds., pp. 47-60, Raven Press, New York.

9. Durand, J. P. (1978): Séparation des hydrocarbures aromatiques polycycliques condensés dans les huiles minérales d'origine pétrolière par chromatographie en phase liquide. Dosage du benzo(a)pyrene. CNAM Thesis, Paris (available from Institut Francais du Pétrole).
10. Durand, J. P. and Petroff, N. (1980): Determination of benzo(a)pyrene and other PAH in petroleum oils by direct liquid chromatography. *Journal of Chromatography* (accepted for publication).

# HIGH ARYL HYDROCARBON HYDROXYLASE INDUCIBILITY IS POSITIVELY CORRELATED WITH OCCURRENCE OF LUNG CANCER

D. Busbee[*,†], T. McLemore[**,†], R. R. Martin[**], N. Wray[**,†], M. Marshall[*,‡], and E. Cantrell[#]

> [*]Department of Biology/Genetics Center
> North Texas State University
> Denton, Texas 76203
> [**]Department of Medicine
> Baylor College of Medicine
> Houston, Texas 77030
> [†]Veterans Administration Hospital
> Houston, Texas 77030
> [‡]Biochemistry Department
> M. D. Anderson Hospital
> Houston, Texas 77030
> [#]Pharmacology Department
> Texas College of Osteopathic Medicine
> Fort Worth, Texas 76107

Cancer occurrence is well correlated to persons chronically exposed to polynuclear aromatic hydrocarbons (1), to persons who habitually ingest smoked foods (2), and to persons with a history of cigarette smoking (3-5). Susceptibility of humans and experimental animals to carcinogenesis by polynuclear aromatic hydrocarbons (PAH) is generally considered to be a function of enzymatic activation of the PAH during their metabolism by the microsomal monooxygenase enzyme, aryl hydrocarbon hydroxylase (AHH). A major, unstable, carcinogenic intermediate formed during metabolism of the PAH, benzo(a)pyrene (BaP), by AHH has been reported to be $7\beta,8\alpha$-dihydrodiol-$9\alpha,10\alpha$-epoxy-7,8,9,10-tetrahydrobenzo(a)pyrene (6-11). This compound is mutagenic, apparently as a function of its capacity to bind cellular macromolecules, including DNA, RNA, and proteins (7, 10-12). Although a recent report by Poland and

Glover (13) suggests it to be unlikely that 2,3,7,8-tetrachlorodibenzo-p-dioxin (TCDD) is oncogenic as a function of its covalent binding to DNA, the binding of PAH carcinogens to cellular macromolecules appears requisite to their carcinogenic action and is considered to be a function of PAH activation by AHH.

Genetic control of AHH inducibility has been reported for mice (14) and for man (15,16). Shortly after initial papers indicated AHH to be inducible in the human cultured lymphocyte system (17,18), Kellermann et al (19) reported that a population of lung cancer patients fell into the upper level of AHH inducibility for a healthy control population. Kellermann inferred that an assessment of lung cancer risk might be possible, using individual levels of AHH inducibility as the criterion for risk determination. Subsequent publications addressing the relationship between AHH inducibility and chemically related cancer have varied widely, with some laboratories showing a weak correlation between the increased incidence of high AHH inducibility and lung cancer occurrence (20-22) and some groups reporting the absence of such a correlation (23,24). The variability of AHH data generated using the cultured lymphocyte system has produced general pessimism on the usefulness of AHH determinations in man and has significantly deterred consideration of this system as a potential screening tool to assess lung cancer risk.

In this paper we present data showing a significant correlation between increased levels of high AHH inducibility and lung cancer occurrence. We further propose an explanation for the discordant data obtained by investigators in this area of research.

## INDUCTION OF AHH IN PULMONARY ALVEOLAR MACROPHAGES

Cantrell et al (25) demonstrated that pulmonary alveolar macrophages (PAMs) freshly lavaged from the lungs of healthy, volunteer cigarette smokers had significantly higher levels of AHH than did cells obtained from a similar group of nonsmokers. The increased AHH levels were presumed to be due to induction of the PAM enzyme system in situ by cigarette-smoke tars. This was further documented by studies demonstrating that AHH levels in PAMs from a nonsmoker increased significantly over a 30-day period following initiation of cigarette smoking. Within 3 months after cessation of smoking, AHH activity of PAMs decreased to the presmoking levels (29).

After the development of techniques for culturing human PAMs, AHH was demonstrated to be induced in vitro in these cells by PAH, such as benzanthracene (BaA) (28), or by crude cigarette tar extracts (26,27). For a given individual, the levels of PAM AHH induced in culture by either BaA or cigarette tar extracts were equivalent. In situ and in vitro induction of AHH in PAMs was studied in groups of noncancer and primary lung cancer patients, all of whom were cigarette smokers. Comparisons of fresh PAM AHH levels and BaA-induced PAM AHH levels showed correlations for both lung cancer and noncancer patients (lung cancer R=0.971, P<0.001; noncancer R=0.685, P<0.005; Figure 1). A similar comparison of fresh PAM AHH levels and BaA-induced fold induction of AHH levels in PAMs demonstrated a positive correlation for both lung cancer (R=0.908, P<0.001) and noncancer (R=0.876, P<0.001) patients (Figure 2). These data indicate that, for PAMs, it is reasonable to express BA-induced AHH levels either as absolute values or as fold induction over normal controls, and also suggest that BA-induced AHH levels in cultured PAMs accurately reflect what the fresh PAM AHH levels would be, induced in situ in cigarette smokers, whether they have cancer or not.

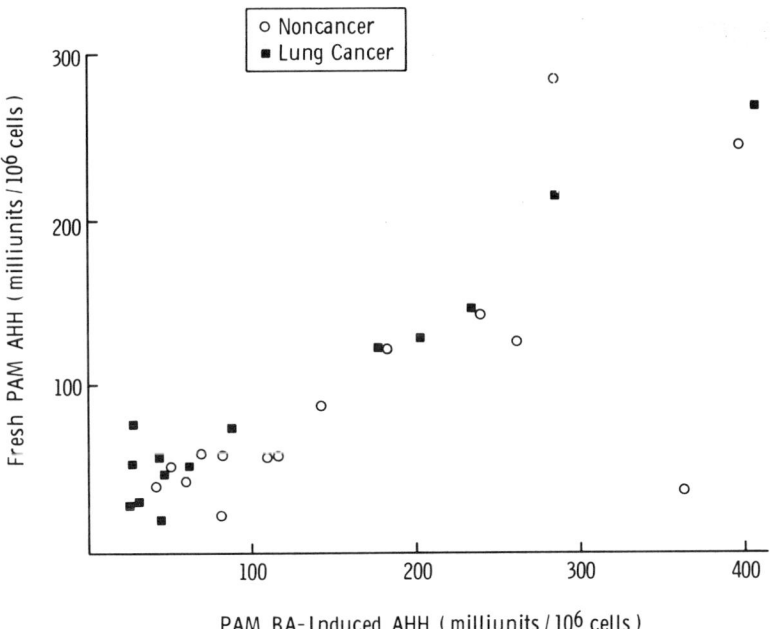

**FIGURE 1.** Comparison of AHH in fresh and cultured BA-induced PAM from individual noncancer or lung cancer patients (R=0.685) and P<0.005 for noncancer patients and R=0.971 and P<0.001 for lung cancer patients). Data from McLemore et at (33).

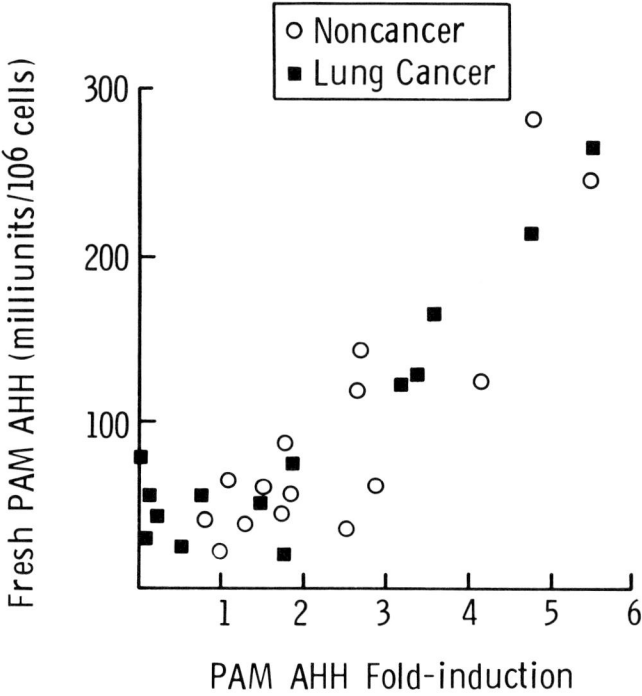

FIGURE 2. Comparison of AHH activity in fresh PAM and AHH fold induction in cultured PAM from individual patients (R=0.876 and P<0.001 for noncancer patients and R=0.908 and P<0.001 for lung cancer patients). Data from McLemore et al (33).

## COMPARISON OF AHH IN CULTURED LYMPHOCYTES, IN PULMONARY ALVEOLAR MACROPHAGES, AND IN LUNG TISSUE

To determine whether lymphocyte AHH levels are representative of enzyme levels in PAMs induced in situ, PAM and lymphocyte AHH activities were compared in individuals. Initial studies were performed in healthy volunteers and demonstrated that, for smokers and nonsmokers, a positive correlation was observed when AHH levels in PAMs freshly lavaged from the lung and in cultured peripheral lymphocytes were simultaneously compared for a given individual (29). These studies were followed by comparisons of fresh PAM AHH levels and cultured lymphocyte AHH induction in lung cancer and noncancer patients (30). The data showed a positive correlation for PAM and lymphocyte AHH levels in noncancer patients, but a lack of correlation (dissociation) of these values in patients with primary lung cancer.

Analyses of AHH levels in both healthy and patient populations were followed by studies comparing the in vitro induction of AHH in cultured PAMs and cultured lymphocytes from healthy volunteers (26). A positive correlation was demonstrated between BaA-induced AHH activity in cultured PAMs and cultured lymphocytes from both smokers and nonsmokers. This relationship was seen when enzyme levels were expressed as either absolute AHH activity or as fold induction. Subsequent studies compared the in vitro induction of AHH in PAMs and lymphocytes from lung cancer and noncancer patients. When AHH levels were compared in cultured PAMs and lymphocytes from noncancer patients, a positive correlation was observed regardless of whether the data were expressed as absolute BaA-induced levels or as fold induction (Figures 3,4, and 5). A lack of correlation was observed for AHH levels in these tissues from lung cancer patients (Figures 3,4 and 5).

Cultured lymphocyte AHH induction, fresh PAM AHH activity, and AHH levels in fresh, excised lung tissue were compared for individual noncancer patients, all of whom were cigarette smokers (31). AHH values for these tissues demonstrated a positive correlation ($R=0.987$, $P<0.001$; Figure 6). When AHH values for these tissues were compared in an age-matched group of lung cancer patients, all of whom were cigarette smokers, multiple linear regression analysis of enzyme levels showed an absence of correlation ($R=0.701$, $P>0.25$). These data suggest that the capacity of tissues to be induced for AHH may be compromised in lung cancer patients. The data also demonstrate that, although one or more of the tissues may not respond to AHH-inducing agents in lung cancer patients, another of the tissues may still be useful for determining the genetically specified capacity of the person to be induced for AHH. The data suggest that the lymphocyte is typically the tissue affected by the presence of lung cancer and that lung tissue and PAMs appear to retain a better correlation between AHH levels than do PAMs and lymphocytes or lung tissue and lymphocytes. Figure 6 also shows that consistently low AHH levels of all three tissues are more typical of noncancer patients than of cancer patients. Four out of seven noncancer patients were in the low-AHH category for PAMs, lymphocytes, and lung tissue. Only one out of seven of the lung cancer patients fell into the low AHH category for all three tissues. None of the lung cancer patients fell into the high category for all three tissues. Because of the low number of patients for which three tissues were available, further investigation of these observations is required.

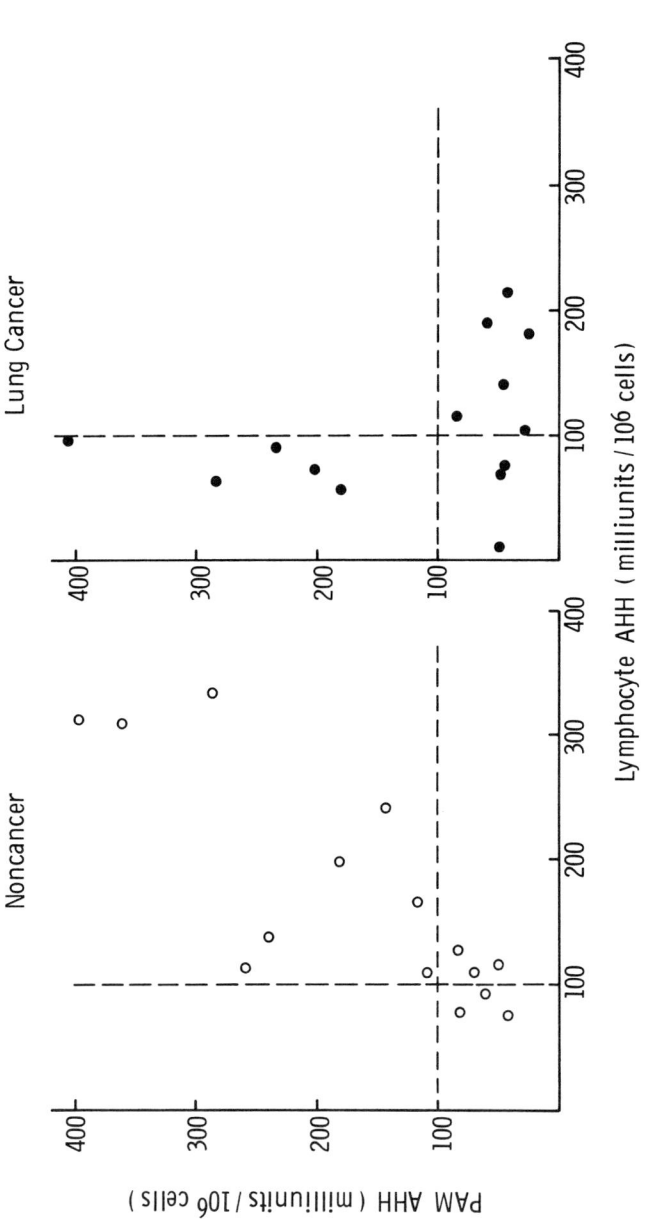

FIGURE 3. Comparison of AHH levels in BA-induced PAM and lymphocytes from individual noncancer or lung cancer patients ($R=0.801$ and $P<0.001$ for noncancer patients and $R=0.306$ and $P>0.3$ for lung cancer patients). Data from McLemore et al (33).

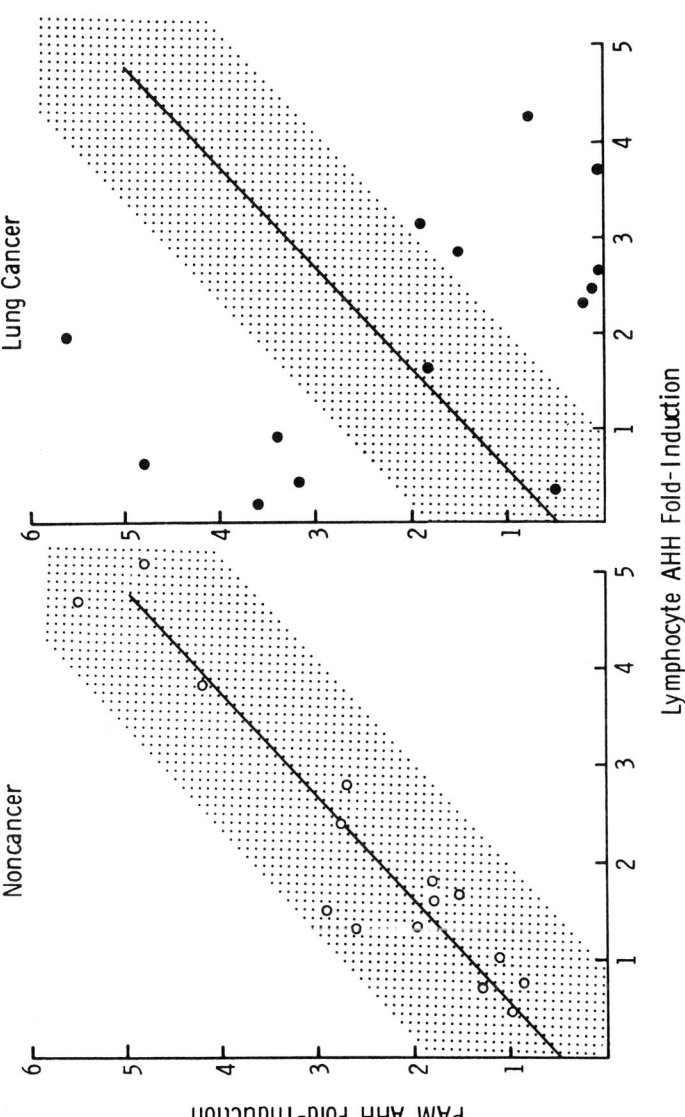

FIGURE 4. Comparison of AHH fold-induction values in PAM and lymphocytes from noncancer or lung cancer patients (R=0.942 and P<0.001 for noncancer patients and R=0.625 and P<0.02 for lung cancer patients). The regression line and the mean ±2 S.E. of estimate lines for the noncancer patient values appear on both graphs. Data from McLemore et al (33).

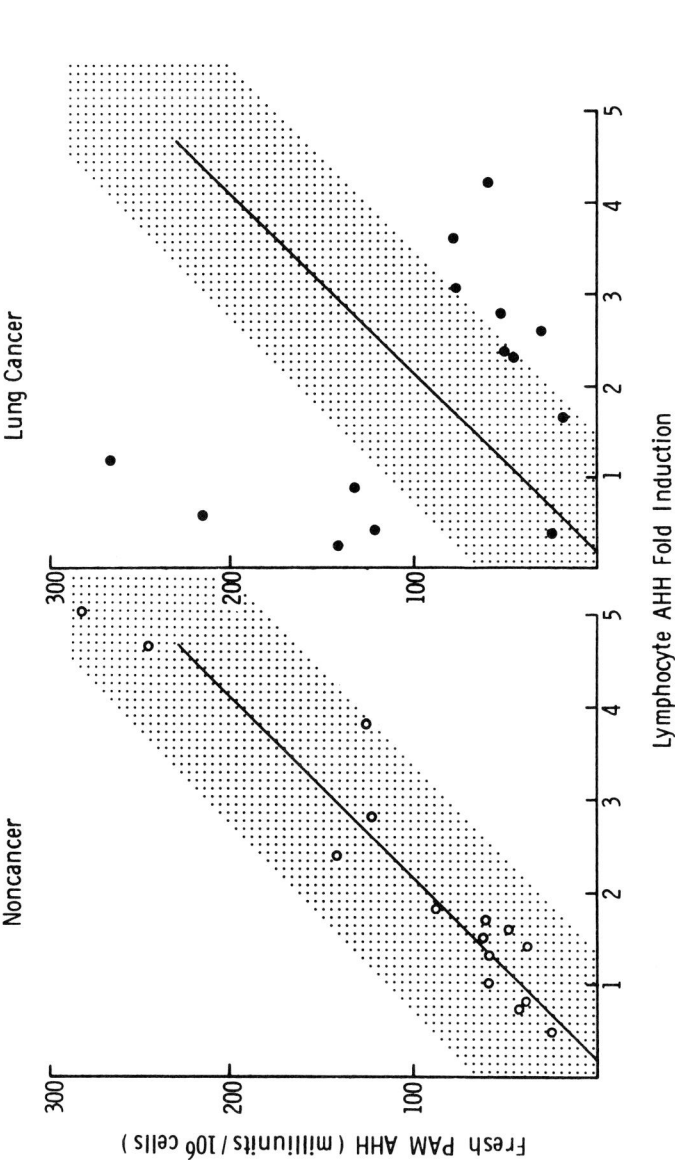

FIGURE 5. Comparison of AHH in PAM freshly lavaged from the lung and AHH fold induction in cultured lymphocytes from individual noncancer or lung cancer patients (R=0.948 and P<0.001, for noncancer patients and R=−0.466 and P>0.05 for lung cancer patients). Fold induction, (BA-induced AHH activity/noninduced AHH activity) -1. The regression line and a shaded area representing the mean ±2 S. E. of estimate lines for noncancer patients are superimposed on both graphs. Data from McLemore et al (33).

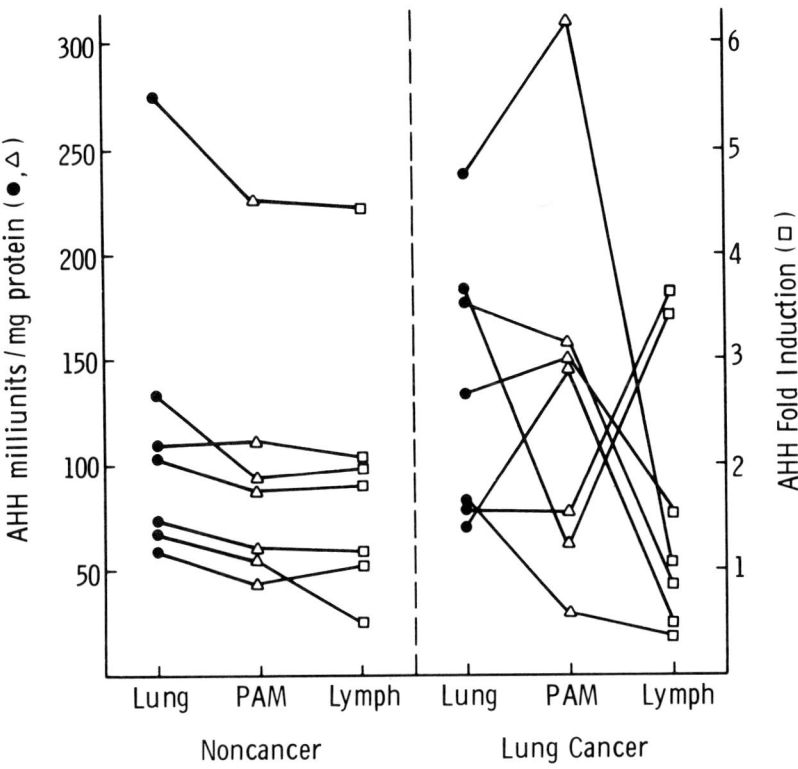

FIGURE 6. Simultaneous comparison of AHH in fresh PAMs, fresh lung tissue, and cultured lymphocytes from individual patients with and without lung cancer. ● = AHH activity in fresh lung tissue; ■ = AHH activity in fresh PAMs; △ = cultured lymphocyte AHH fold induction. Connecting lines represent enzyme values for individual patients. R=0.987, P<.001 for noncancer and R=0.701, P>.25 for lung cancer patients by multiple regression analysis. Data from McLemore et al (31).

## FURTHER ANALYSIS OF RELATIONSHIPS BETWEEN AHH LEVELS IN PAMS AND LYMPHOCYTES FROM LUNG CANCER PATIENTS

Extensive family histories were compiled, and work and health records were obtained, on a patient group examined in a double blind study. This population included 64 lung cancer and 51 noncancer patients. An initial report by McLemore et al (32) showed a positive correlation between PAM and lymphocyte AHH values in noncancer patients (R=0.915, P<0.001), while a dissociation was observed for these enzyme values in individuals from a lung cancer population (linear regression not appropriate for this set of values) (Figures 7, and 8). In this group, 69 percent of noncancer patients showed low AHH levels for both lymphocytes and PAMs, while 19 percent of the lung cancer patients were in the low-AHH category for both tissue types. Conversely, 31 percent of the noncancer patients showed high for PAM and lymphocyte AHH values. No lung cancer patients showed high AHH in both tissues, but 81 percent showed high PAM AHH or lymphocyte AHH levels. Since either of the two tissues should reflect the genetic capacity of the individual to be induced for AHH, comparison of AHH levels between lung cancer and noncancer patients was expressed in the form of combined data in Figure 9. When the highest AHH value for PAM or for lymphocytes in noncancer individuals was compared with the highest AHH value for PAM or for lymphocytes in lung cancer patients, the lung cancer patient population is

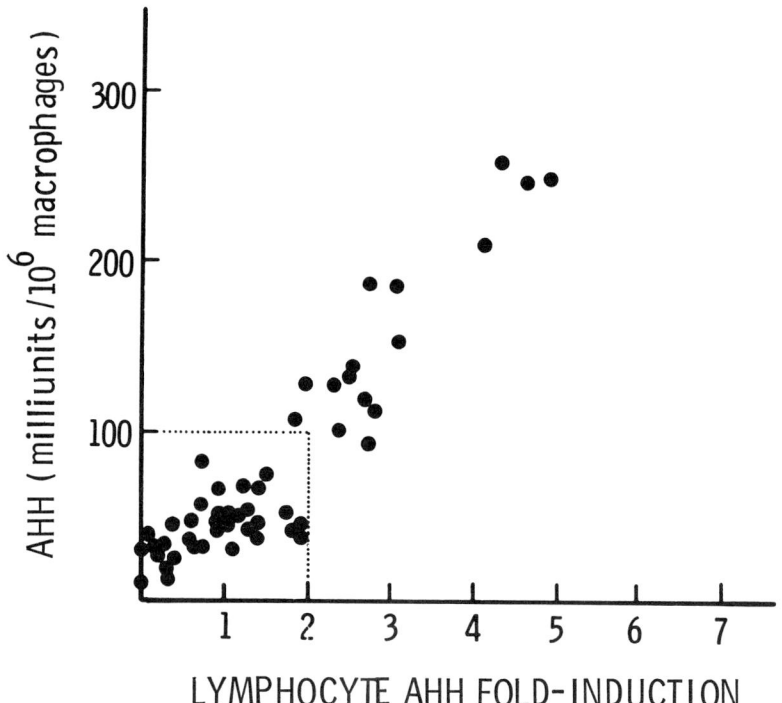

FIGURE 7. Comparison of AHH levels in lymphocytes grown in short-term tissue culture in medium containing 10 μmoles of benzanthracene, with AHH levels induced in situ in pulmonary alveolar macrophages by exposure to cigarette smoke. Tissues were obtained on the same day from noncancer patients who were cigarette smokers. AHH values in lymphocytes are expressed as fold induction rather than as absolute values. AHH levels in PAMs are expressed as milliunits/$10^6$ cells, where 1 unit is the amount of fluorescence equivalent to 1 pmole of 3-hydroxybenzo(a)pyrene produced per minute of incubation in the assay procedure. Data from McLemore et al (32).

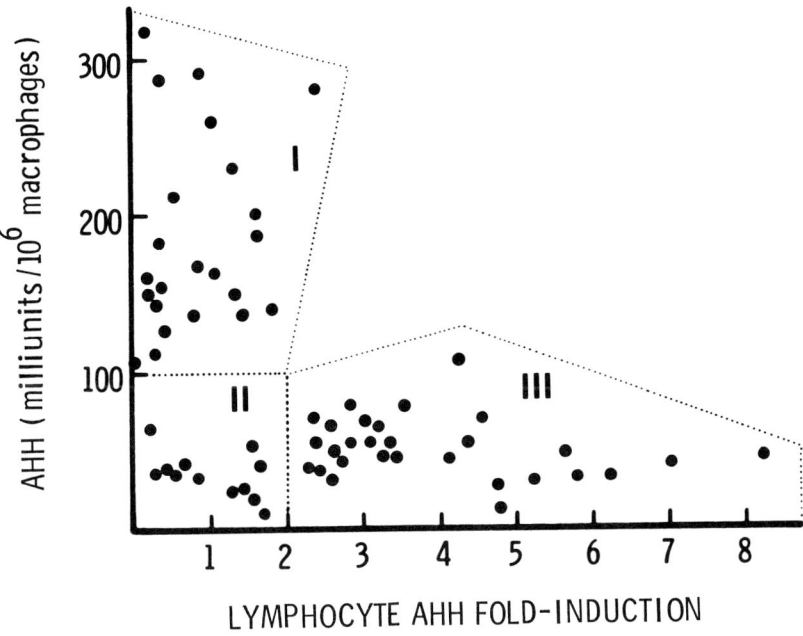

FIGURE 8. Comparison of AHH levels in lymphocytes grown in short-term tissue culture in medium containing 10 μmoles of benzanthracene, with AHH levels induced in situ in pulmonary alveolar macrophages by exposure to cigarette smoke. Tissues were obtained on the same day from lung cancer patients who were cigarette smokers. AHH values in lymphocytes are expressed as fold induction rather than absolute values. AHH levels in PAMs are expressed as milliunits/$10^6$ cells, where 1 unit is the amount of fluorescence equivalent to 1 pmole of 3-hydroxybenzo(a)pyrene produced per minute of incubation in the assay procedure. Data from McLemore et al (32).

distinctly skewed to the right. This suggests that a significantly larger percentage of persons who are highly inducible for AHH are found in the lung cancer patient population ($P<0.001$).

The lung cancer patient group shown in Figure 8 can be divided into three subpopulations: (I) persons with high PAM and low lymphocyte AHH levels, (II) persons with low AHH levels in both tissues, and (III) persons with high lymphocyte and low PAM AHH levels. No patient in group II had any reported history of cancer of any type among close family members, i.e., mother, father, siblings, grandparents, and siblings of mother and of father. Family histories of cancer were found in 9.5 percent of persons in group I and in 39.3 percent of persons in group III. Noncancer patients showed a 20.3 percent family history of cancer. Cancer types included those listed by Lynch et al (35) for the SBLA cancer family

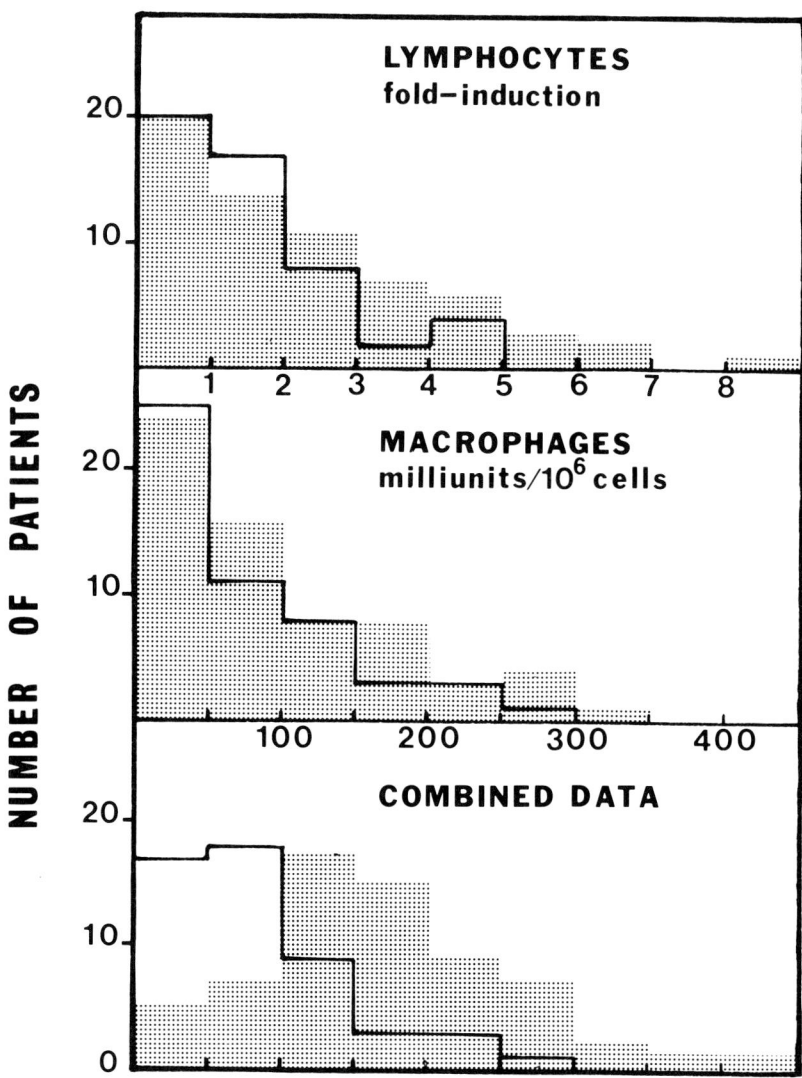

FIGURE 9. Comparisons of cultured lymphocyte AHH levels (top) from lung cancer (stippled) and noncancer (open) patients, pulmonary alveolar macrophage AHH levels (center) from lung cancer (stippled) and noncancer (open) patients, the highest AHH levels (bottom) in either cultured lymphocytes or pulmonary alveolar macrophages of lung cancer patients (stippled) and noncancer patients (open). Data from McLemore et al (34).

syndrome. The range of AHH activity for noncancer patients with and without family histories of cancer was not measureably different.

An additional difference was seen in the age of cancer patients. Persons in group I (9.5 percent family histories of cancer) averaged 5 to 6 years older at the time of initial diagnosis than persons in group II or III. The age comparisons (age at initial diagnosis) combine with the family history of cancer to suggest that a familial component exists among the lung cancer population in this study. Further investigation will be required to delineate the exact relationships between family histories of cancer, lung cancer incidence, age at initial diagnosis, and carcinogen-metabolizing capability of these individuals.

## AHH AND ANTIPYRINE ELIMINATION

The oxidation and elimination of numerous categories of drugs has been documented to be a function of AHH (36). Antipyrine is a good example; clearance of this drug from the plasma may be positively correlated with AHH inducibility in cultured lymphocytes obtained from healthy normal subjects (R=0.947, P<0.001; 36). Studies of antipyrine clearance in lung cancer patients and in a healthy control group demonstrated a decreased antipyrine half-life in lung cancer patients (Figure 10). The comparison of both antipyrine half-life and the metabolic clearance rate in healthy control subjects and in bronchogenic carcinoma patients with histologically different cancer types was reported by Kellermann et al (37). The metabolic clearance rate was higher for any of the lung cancer types than for the control group. The half-life of antipyrine in plasma was correspondingly lower for the lung cancer patient group than for the healthy control group. These data show that increased rates of plasma antipyrine clearance may be positively correlated with both high AHH inducibility and with the occurrence of bronchogenic carcinoma. The data suggest that high AHH inducibility may thus be correlated with the occurrence of bronchogenic carcinoma. These data provide a theoretical link between lung cancer occurrence and high AHH inducibility that is not directly dependent on use of the cultured lymphocyte system to measure AHH levels in cancer patients.

## CONCLUDING REMARKS

Data presented here provide evidence that two separate populations of lung cancer patients (examined by two different laboratory goups using different research plans) contain a greater number of persons who are highly inducible for AHH than do their respective noncancer control groups. Attempts to determine a correlation between lung cancer occurrence and increased AHH levels have been made by several

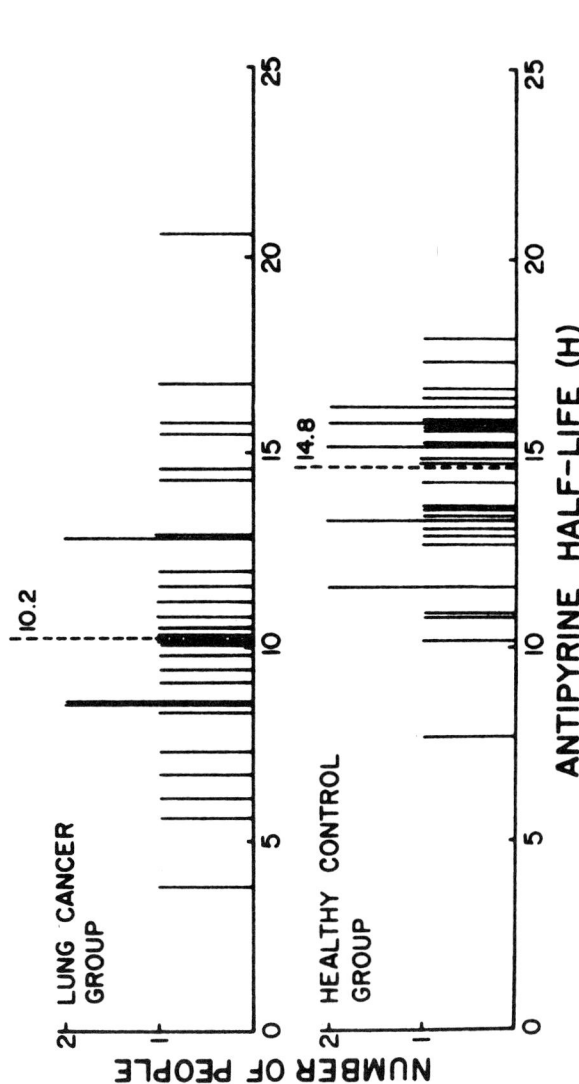

FIGURE 10. Distribution of antipyrine saliva half-lives in 31 male patients with bronchogenic carcinoma and in 31 healthy matched control subjects. Data from Kellermann et al (37).

laboratory groups. Some have reported a weak correlation between lung cancer concurrence and the increased incidence of high AHH inducibility in lung cancer populations (19,22). Other groups report no correlation between AHH and lung cancer occurrence (23,24,38). These studies have in common the use of AHH analysis in a single tissue type, the cultured lymphocyte. A high coefficient of variability is reported by all laboratories doing lymphocyte AHH studies. In addition to the difficulties encountered in assaying for AHH in cultured lymphocytes from healthy controls, our data show that lymphocytes are not inducible in 23 out of 64 (35.9 percent) lung cancer patients known to be inducible for AHH in another tissue. Analysis of AHH levels in those persons automatically skews data from the lung cancer population to the left, so that the lung cancer population AHH values are not distinguishable from those of a control group. This provides a fairly simple explanation for the difficulty encountered in correlating AHH inducibility with bronchogenic carcinoma occurrence. An examination of AHH inducibility in two or more tissues indicates that more people who are highly inducible for AHH may be found in a lung cancer population than is typical for a control group.

## REFERENCES

1. Hammond, E. C., Selikoff, I. J., Lawther, P. L., and Seidman, H. (1976): Inhalation of benzopyrene and cancer in man. *Annals N. Y. Acad. Sci.* 271:116-124.
2. Falk, H. L. and Kotin, P. (1965): Chemistry, host entry, and metabolic fate of carcinogens. *Clin. Pharm. Ther.* 4:88-96.
3. Hammond, E. C., Selikoff, I. J., and Seidman, H. (1975): Multiple interactions of cigarette smoking. Extrapulmonary cancer. *In:* P. Bucalossi, U. Vernonesi, and N. Cascinelli, Eds., *Cancer Epidemiology, Environmental Factors,* Vol. 3, pp. 147-150, Proceedings of the XI International Cancer Congress, Florence, October 20-26, 1974. Excerpta Medica, Amsterdam.
4. Steinfeld, J. L. (1971): *The health consequences of smoking: A report of the Surgeon-General:* Department of Health, Education and Welfare Publication No. (HSM) 71-7513, pp. 239-244.
5. Sterling, T. D. (1975): A critical reassessment of the evidence bearing on smoking as the cause of lung cancer. *Amer. J. Publ. Hlth.* 65:939-953.
6. Kapitulnik, J., Levin, W., Conney, A. H., Yagi, H., and Jerina, D. M. (1977): Benzo(a)pyrene 7,8-dihydrodiol is more carcinogenic than benzo(a)pyrene in newborn mice. *Nature* 266:378-380.

7. Nakanishi, K., Kasi, H., Cho, H., Harvey, R. G., Jeffrey, A. M., Jennette, K. W., and Weinstein, I. B. (1977): Absolute configuration of a ribonucleic acid adduct formed in vivo by metabolism of benzo(a)pyrene. *J. Am. Chem. Soc.* 99:258-260.
8. Thakker, D. R., Yagi, H., Lu, A. Y. H., Levin, W., Conney, A. H., and Jerina, D. M. (1976): Metabolism of benzo(a)pyrene: conversion of ($\pm$)-trans-7,8-dihydroxy-7,8-dihydro-benzo(a)pyrene to highly mutagenic 7,8-diol-9,19-epoxides. *Proc. Natl. Acad. Sci.* 73:3381-3385.
9. Wood, A. W., Levin, W., Lu, A. Y. H., Yagi, H., Hernandez, O., Jerina, D. M., and Conney, A. H. (1976): Metabolism of benzo(a)pyrene and benzo(a)pyrene derivatives to mutagenic products by highly purified hepatic microsomal enzymes. *J. Biol. Chem.* 251:4882-4890.
10. Yang, S. K., Gelboin, H. V., Trump, B. F., Autrup, H., and Harris, C. C. (1977): Metabolism activation of benzo(a)pyrene and binding to DNA in cultured human bronchus. *Cancer Res.* 37:1210-1215.
11. Yang, S. K., McCourt, D. W., Leutz, J. C., and Gelboin, H. V. (1977): Benzo(a)pyrene diol epoxides: Mechanisms of enzymatic formation and optically active intermediates. *Science* 196:1199-1200.
12. Baird, W. M., Harvey, R. G., and Brookes, P. (1975): Comparison of the cellular DNA-bound products of benzo(a)pyrene with the products formed by the reaction of benzo(a)pyrene-4,5-oxide with DNA. *Cancer Res.* 35:54-57.
13. Poland, A. and Glover, E. (1979): An estimate of the maximum in vivo covalent binding of 2,3,7,8-tetrachlorodibenzo-p-dioxin to rat liver protein, ribosomal RNA, and DNA. *Cancer Res.* 39:3341-3344.
14. Nebert, D. W., Goujon, F. M., and Gielen, J. E. (1972): Aryl hydrocarbon hydroxylase induction by polycyclic hydrocarbons: Simple autosomal trait in the mouse. *Nature (New Biol.)* 236:107,110.
15. Kellermann, G., Luyten-Kellermann, M., and Shaw, C. R. (1973): Genetic variation of aryl hydrocarbon hydroxylase in human lymphocytes. *Amer. J. Hum. Genet.* 25:347-331.
16. Atlas, S. A., Vessell, E. X., and Nebert, D. W. (1976): Genetic control of interindividual variations in the inducibility of aryl hydrocarbon hydroxylase in cultured human lymphocytes. *Cancer Res.* 36:4619-4630.
17. Whitlock, J. P., Cooper, H. L., and Gelboin, H. V. (1972): Aryl hydrocarbon (benzopyrene) hydroxylase is stimulated in human lymphocytes by mitogens and benz(a)anthracene. *Science* 177:618-619.
18. Busbee, D. L., Shaw, C. R., and Cantrell, E. T. (1972): Aryl hydrocarbon hydroxylase induction in human leukocytes. *Science* 178:315.

19. Kellermann, G., Shaw, C. R., and Luyten-Kellermann, M. (1973): Aryl hydrocarbon hydroxylase inducibility and bronchogenic carcinoma. *New Eng. J. Med.* 289:934-937.
20. Coomes, M., Muijsson, I., Mason, W., Cantrell, E., Anderson, D., and Busbee, D. (1976): Aryl hydrocarbon hydroxylase and 16$\alpha$-hydroxylase in the cultured human lymphocyte system. *Biochem. Genet.* 14:671.
21. Guirgis, H. A., Lynch, H. T., Mate, T., Harris, R. E., Wells, I., and Caha, L. (1976): Aryl hydrocarbon hydroxylase activity in lymphocytes from lung cancer and normal control. *Oncology* 33:105-109.
22. Gahmberg, C. G., Sekki, A., Kosunen, T. U., Holsti, L. R., and Makela, O. (1979); Induction of aryl hydrocarbon hydroxylase activity and pulmonary carcinoma. *Int. J. Cancer* 23:302-305.
23. Paigen, B., Gurtoo, H. L., Minowada, J., Houten, L., Vincent, R., Paigen, K., Parker, N. B., Ward, E., and Hayner, N. T. (1977): Questionable relationship of aryl hydrocarbon hydroxylase to lung cancer risk. *New Eng. J. Med.* 297:346-350.
24. Ward, E., Paigen, B., Steenland, K., Vincent., R., Minowada, J., Gurtoo, H., Sartori, P., and Havens, M. B. (1978): Aryl hydrocarbon hydroxylase in persons with lung or laryngeal Cancer. *Int. J. Cancer* 22:284-389.
25. Cantrell, E. T., Warr, G. A., Busbee, D. L., and Martin, R. R. (1973): Induction of aryl hydrocarbon hydroxylase in human pulmonary alveolar macrophages by cigarette smoking. *J. Clin. Invest.* 52:1881-1884.
26. McLemore, T. L., Martin, R. R., Toppell, K. L., Busbee, D. L., and Cantrell, E. T. (1977): Comparison of aryl hydrocarbon hydroxylase induction in cultured blood lymphocytes and plumonary macrophages. *J. Clin. Invest.* 60:1017-1024.
27. McLemore, T. L., Warr, G. A., and Martin, R. R. (1977): Induction of aryl hydrocarbon hydroxylase in human pulmonary alveolar macrophages and peripheral lymphocytes by cigarette tars. *Cancer Lett.* 2:161-168.
28. McLemore, T. L. and Martin, R. R. (1977) In Vitro induction of aryl hydrocarbon hydroxylase in human pulmonary alveolar macrophages by benzanthracene. *Cancer Lett,* 2:327-334.
29. Cantrell, E., Busbee, D., Warr, G., and Martin, R. (1973): Induction of aryl hydrocarbon hydroxylase in human lymphocytes and plumonary alveolar macrophages—a comparison. *Life Sci.* 13:1649-1654.
30. McLemore, T. L., Martin, R. R., Busbee, D. L., Richie, R. C., Springer, R. R., Toppell, K. L., and Cantrell, E. T. (1977): Aryl hydrocarbon hydroxylase activity in pulmonary macrophages and lymphocytes from lung cancer and noncancer patients. *Cancer Res.* 37:1175-1181.

31. McLemore, T. L., Martin, R. R., Pickard. L. R., Springer, R. R., Wray, N. P., Toppell, K. L., Mattox, K. L., Guinn, G. A., Cantrell, E. T., and Busbee, D. L. (1978): Analysis of aryl hydrocarbon hydroxylase activity in human lung tissue, pulmonary macrophages, and blood lymphocytes. *Cancer* 41:2200-2292.
32. McLemore, T. L., Martin, R. R., Springer, R. R., Wray, N. P., Cantrell, E. T., and Busbee, D. L. (1979): Aryl hydrocarbon hydroxylase activity in pulmonary alveolar macrophages and lymphocytes from lung cancer and noncancer patients: A correlation with family histories of cancer. *Bichem. Genet.* 17(9):795-806.
33. McLemore, T., Martin, R., Wray, N., Cantrell, E., and Busbee, D. (1978): Dissociation between aryl hydrocarbon hydroxylase activity in cultured pulmonary macrophages and blood lymphocytes from lung cancer patients. *Cancer Res.* 38:3805-3811.
34. McLemore, T. L., Martin, R. R., Wray, N. P., Cantrell, E. T., and Busbee, D. L.: Reassessment of the relationship between aryl hydrocarbon hydroxylase and lung cancer. *Cancer* (in press).
35. Lynch, H. Y., Mulcahy, G. M., Harris, R. E., Guirgis, H. A., and Lynch, J. F. (1978): Genetic and pathologic findings in a kindred with hereditary sarcoma, breast cancer, brain tumors, leukemia, lung, laryngeal, and adrenal cortical carcinoma. *Cancer* 41:2055-2064.
36. Kellermann, G. and Luyten-Kellermann, M. (1978): Benzo(a)pyrene metabolism and plasma elimination rates of phenacetin, acetanilide, and theophylline in man. *Pharmacology* 17:191-200.
37. Kellermann, G., Luyten-Kellerman, M., Jett, J., Moses, H., and Fontana, R. (1978): Aryl hydrocarbon hydroxylase in man and lung cancer. Human genetic variation in response to medical and environmental agents: Pharmacogenetics and energetics. *Human Genet.* (Suppl. 1):161-168.
38. Jett, J., Moses, H., Branum, E., Taylor, W., and Fontana, R. (1978): Benzo(a)pyrene metabolism and blast transformation in peripheral blood mononuclear cells from smoking and nonsmoking populations and lung cancer patients. *Cancer* 41:192-200.

# 2,3,7,8-TETRACHLORODIBENZO-P-DIOXIN (TCDD)-INDUCED ALTERATIONS IN OXIDATIVE AND NONOXIDATIVE BIOTRANSFORMATION OF PAH IN MOUSE SKIN: ROLE IN ANTICARCINOGENESIS BY TCDD

J. DiGiovanni,* G. S. Kishore,* T. J. Slaga**, and R. K. Boutwell*

*McArdle Laboratory for Cancer Research
University of Wisconsin
Madison, WI 53706
**Biology Division
Oak Ridge National Laboratory
Oak Ridge, TN 37830

## INTRODUCTION

Carcinogenic polycyclic aromatic hydrocarbons (PAH)* require metabolic activation to highly reactive intermediates that are capable of interacting covalently with cellular macromolecules (13). The interaction with DNA, a target for activated PAH intermediates, has received considerable attention in recent years and is presently considered a critical event in the initiation of chemical carcinogenesis. Current evidence suggests that "bay-region" diol-epoxides may be ultimate carcinogenic forms for some PAH. This is supported by recent data concerning benzo(a)pyrene (BaP)- and 7-12-dimethylbenz(a)anthracene (DMBA)-epoxides that produce a DNA-binding product similar to that isolated in vivo after treatment with the parent hydrocarbon (2,8,11,15,25,31,35,36). In mouse skin, the 7,8-dihydrodiol of BaP is the only metabolite found to be equally potent or more potent as a tumor initiator than BaP itself

---

*Abbreviations used are: PAH, polycyclic aromatic hydrocarbons; TCDD, 2,3,7,8-tetrachlorodibenzo-p-dioxin; DNA, deoxyribonucleic acid; RNA, ribonucleic acid; BaP, benzo(a)pyrene; DMBA, 7,12-dimethylbenz(a)anthracene; MCA, 3-methylcholanthrene; AHH, aryl hydrocarbon hydroxylase; p-NP, p-nitrophenol; TPA, 12-0-tetradecanoylphorbol-13-acetate.

(6,21,34). The potent carcinogenic activity of (±)-trans-7β, 8α-dihydroxy-9α, 10α-epoxy-7,8,9,10-tetrahydrobenzo(a)pyrene (BaP-diol-epoxide) in newborn mice and mouse skin lends further support to this concept (18,32). Additional evidence has accumulated in support of bay-region diol-epoxides as potential ultimate carcinogenic forms for benz(a)anthracene, 7-methylbenz(a)anthracene, chrysene, and dibenz(a,h)anthracene (7,24,33,38-42).

Many factors are known to affect the metabolism of PAH to electrophilic intermediates such as diol-epoxides, as well as detoxification products, and consequently affect carcinogenic or mutagenic activity. These factors include route of administration, species, sex, age, strain, diet, temperature, time of day, season, and the previous or concurrent administration of other drugs or environmental chemicals. 2,3,7,8-Tetrachlorodibenzo-p-dioxin (TCDD) is a highly toxic and widely studied chemical found as an environmental contaminant (30). This compound is known to markedly influence enzyme pathways responsible for both the activation and inactivation of PAH. TCDD is an extremely potent inducer of microsomal monooxygenase activity with properties similar to 3-methylcholanthrene (MCA) (28). In addition, TCDD is capable of inducing aryl hydrocarbon hydroxylase (AHH, E.C. 1.14.14.2) of hepatic and extrahepatic tissues transplacentally (3,4) in genetically "nonresponsive" mice (29) and in mouse skin (9,27,29). UDP-Glucuronyltransferase (E.C. 2.4.1.17) activities were also increased in hepatic (23,26) and renal (12) microsomal preparations, as well as glutathione-S-transferase (E.C. 2.5.1.18) activities in hepatic cytosolic preparations (1) from rats pretreated with TCDD.

The present investigation was designed to extend our earlier work on the inhibitory effect of TCDD on tumor initiation by various PAH (9). We have analyzed the effect of treating female mice with TCDD at various times before and after application of three different PAH tumor initiators. In addition, experiments were performed to investigate the effects of TCDD on the oxidative and nonoxidative biotransformation pathways for PAH in mouse epidermis in relation to the anticarcinogenic effect. The two-stage system of mouse-skin tumorigenesis provides a particular advantage for studying this phenomenon since biochemical changes can be directly correlated with changes in the tumor response. In addition, systemic metabolism of the PAH is not likely to affect the overall response since only very small doses (e.g., 2.56 μg) of the carcinogen are applied locally.

## MATERIALS AND METHODS

### Chemicals

DMBA, BaP, MCA, p-nitrophenol (p-NP) glucose-6-phosphate, glucose-6-phosphate dehydrogenase, uridine-5′-diphospho-glucuronic acid, and NADPH were purchased from the Sigma Chemical Company, St. Louis, Missouri. [$^3$H]DMBA (78 Ci/mmole) was obtained from Amersham/Searle, Arlington Heights, Illinois, and purified (>99 percent) by thin-layer chromatography with benzene/ethanol (9:1) prior to use. 7-[$^{14}$C]Styrene oxide (9.2 mCi/mmole, 98 percent pure by gas-liquid chromatography) was supplied by California Bionuclear Corporation, Sun Valley, California. 3-Hydroxybenzo(a)pyrene was a generous gift from the Carcinogenesis Program of the National Cancer Institute. TCDD was a gift from the Dow Chemical Company, Midland, Michigan (98.6 percent pure by gas-liquid chromatography, Lot No. 851-144-2). 12-0-Tetradecanoylphorbol-13-acetate (TPA) was purchased from Dr. Peter Borchert, University of Minnesota, Minneapolis, Minnesota. Styrene oxide was obtained from the Aldrich Chemical Company, Milwaukee, Wisconsin.

### Animals

Female CD-1 mice, purchased from Charles River Mouse Farms, North Wilmington, Massachusetts, were utilized for all biochemical experiments and for tumor experiments as noted. In addition, female Sencar mice (originally obtained from Dr. R.K. Boutwell, McArdle Laboratory for Cancer Research, Madison, Wisconsin, and currently raised at the Oak Ridge National Laboratory, Oak Ridge, Tennessee) also were utilized for tumor experiments where indicated. At 7 to 9 weeks of age, the backs of the mice were carefully shaved with surgical clippers. Mice were allowed to rest for 2 days and only those in the resting phase of the hair cycle were used. For topical applications, chemicals were applied to the shaved area in 0.2 ml of acetone and control animals were treated with an equal volume of acetone. For some experiments, TCDD was administered as a single intraperitoneal injection (0.1 or 1 $\mu$g per mouse). Special precautions were observed for the handling of TCDD and TCDD-treated animals. The solutions of TCDD were prepared by introducing sufficient acetone in TCDD-containing sealed glass ampules to make saturated solutions (0.09 mg TCDD/ml acetone). The solutions were mixed with corn oil (for intraperitoneal injections) or acetone (for topical application) in appropriate concentrations such that less than 0.5 ml or exactly 0.2 ml, respectively, was administered to each mouse. Animals treated with TCDD were kept in rooms separate from the control animals to prevent possible contamination of the controls.

## Tumor Experiments

Each experimental group contained 30 preshaved mice. Mice were initiated with either DMBA (10 nmoles), BaP (100 nmoles) or MCA (100 nmoles). One week after initiation, mice received applications of either 3.4 nmoles (Sencar) or 17 nmoles (CD-1) of TPA twice weekly. Promotion was continued for 18 to 20 weeks and the incidences of papillomas and carcinomas were observed and recorded weekly. Papillomas and carcinomas were removed at random for histologic verification. The tumor response is presented as the average number of papillomas per mouse.

## Enzyme Assays

Mice were killed by cervical dislocation and Nudit Cream (supplied by Helena Rubenstein, Inc.) was applied to the shaved area of the back. After 5 minutes, the Nudit Cream was thoroughly washed off under cold, running water and the skins were removed and placed on ice. The whole skins were placed dermis side down on a cold glass plate and the epidermis was scraped off with 15 strokes of a razor blade. The epidermal material from four skins was suspended in 1 ml of 0.05 M Tris—0.25 M sucrose (pH 7.5) and homogenized with a polytron PT10 homogenizer for 45 seconds (at setting 6). Microsomal fractions were obtained by centrifuging the whole homogenate at 9000 × g for 20 minutes and then centrifuging the supernatant at 105,000 × g for 1 hour. Microsmal pellets were suspended once in Tris-sucrose buffer, resedimented, and then resuspended in the same buffer.

The AHH assay was a modification of that described by Juchau et al (17). Assays were performed in semidarkness and contained, in a total volume of 1 ml: 0.4 to 1.0 mg microsomal protein, 3 $\mu$moles magnesium chloride, 7.4 $\mu$moles glucose-6-phosphate, 2 units glucose-6-phosphate dehydrogenase, 2.4 $\mu$moles NADPH, 100 nmoles BaP, and 30 $\mu$moles potassium phosphate (pH 7.4). Incubations were carried out for 30 mimutes at $37°C$ in an atmosphere of 100 percent $O_2$. Specific activity is expressed as pmoles of 3-hydroxybenzo(a)pyrene formed per milligram of microsomal protein per minute of incubation.

UDP-Glucuronyltransferase activities were measured using a modification of previous methods (14,37) with p-NP as substrate. Typical incubation mixtures contained, in a final volume of 0.5 ml: 250 nmoles p-NP, 1 $\mu$mole uridine-5'-diphospho-glucuronic acid, 37.5 $\mu$moles Tris HCl (pH 7.5), and 0.2 to 0.8 mg microsomal protein. Incubations were carried for 2 hours at 37°C and reactions were terminated by adding 2.5 ml of 0.5 mM TCA. The disappearance of color was monitored at 400 nm as described (14). Specific activity is expressed as nmoles of p-nitrophenylglucuronide formed per milligram of microsomal protein per minute of incubation.

Microsomal epoxide hydrase and soluble glutathione-S-transferase activities were measured using a slight modification of the procedures described by James et al (16) with 7-[$^{14}$C]styrene oxide as substrate. Specific activity for epoxide hydrase is expressed as nmoles of 7-[$^{14}$C]styrene glycol formed per milligram of microsomal protein per minute of incubation and glutathione-S-transferase activities are expressed as nmoles of S-(2-hydroxy-1-phenylethyl) glutathione formed per milligram of cytosol protein per minute of incubation.

Protein was determined by the Lowry method (22) using bovine serum albumin as standard. Linearity with respect to protein concentration and time and saturating concentrations of substrate were established using epidermal microsomal or cytosolic fractions for the assays as described.

## Effect of TCDD on [$^3$H]DMBA Disappearance from Mouse Epidermis

TCDD was applied topically to the shaved backs of CD-1 mice (1 μg/mouse) 3 days prior to application of 2.56 μg of [$^3$H]DMBA (10 μCi). Control animals were pretreated with acetone before application of [$^3$H]DMBA. After topical application of [$^3$H]DMBA, five mice each from control and TCDD-pretreated groups were sacrificed at the following times: 5 minutes, 30 minutes, 1 hour, and 2, 4, 12, 15 and 24 hours. Epidermal homogenates were prepared as described above and centrifuged at 800 × g for 15 minutes. The sediment was suspended in Tris-sucrose buffer and resedimented. The original supernatant was combined with the wash and centrifuged at 10,000 × g for 20 minutes to obtain combined lysosomal and mitochondrial fraction (L + M). The L + M sediment was suspended with buffer and resedimented. The supernatant fraction was combined with the wash and centrifuged at 105,000 × g for 1 hour to give microsomes. The microsomal pellet was suspended and resedimented and the wash combined with the supernatant fraction. The 800 × g, L + M, and microsomal pellets were resuspended in appropriate volumes of buffer, and aliquots of these as well as the 105,000 × g supernatant (cytosol fraction) were solubilized with 1 ml of Soluene tissue solubilizer (Packard Instrument Co.) at 40°C for 1 hour. Samples were counted with the addition of toluene + PPO (Research Products International, Elk Grove, Illinois) in a Beckman 7000 scintillation counter.

## RESULTS AND DISCUSSION

### Effects of Pretreatment with TCDD on Tumor Initiation with DMBA, BaP, and MCA

The effects of single topical applications of TCDD at various times relative to initiation with DMBA and BaP in female mice are summarized in Figure 1. TCDD was applied 10, 5, 3, or 1 day(s) or 5 minutes before

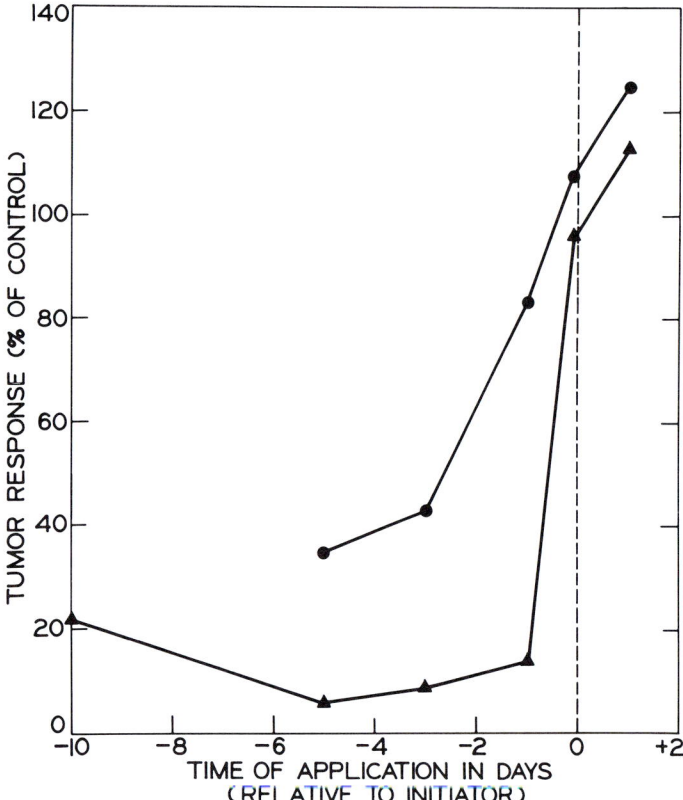

FIGURE 1. Summary of the time-dependent inhibitory effects of TCDD on tumor initiation by DMBA (▲) and BaP (●). Animals were initiated with 10 nmoles of DMBA or 100 nmoles of BaP and promoted 1 week later with twice weekly applications of TPA for 20 weeks. TCDD (1 µg per mouse, topically) was applied at various times relative to the initiator (abscissa). The tumor response (ordinate) represents the average number of papillomas per mouse (at 20 weeks) expressed as a percentage of the acetone-treated control group for each initiator. Thirty female CD-1 mice were used for each experimental group except for experiments when TCDD was given 1 hour after DMBA or BaP where female Sencar mice were employed.

initiation with DMBA and BaP (except at 10 days) and 1 day after initiation with these PAH. The dose of TCDD utilized for all tumor experiments, 1 µg/mouse, produced no visible toxic manifestations. All animals appeared normal in terms of weight gain and morphologic and histologic characteristics of the skin after 20 weeks of promotion with TPA. The application of TCDD 5 minutes before or 1 day after initiation with 10 nmoles of DMBA produced little or no effect on the average number of papillomas per mouse at the 20th week of promotion. In contrast, when TCDD was applied 10, 5, 3, and 1 day(s) before DMBA, a marked reduction in papilloma formation was observed. The results when BaP (100 nmoles) was used as the initiator were nearly identical except that the magnitude of inhibition was less than that observed with DMBA as the initiator. The inhibitory effect of TCDD was therefore highly dependent on the time of pretreatment.

In light of recent work on the effects of TCDD on MCA-induced subcutaneous sarcomas (19,20) suggesting a potential enhancing effect in certain mouse strains, we determined the effect of topically administered TCDD on MCA-initiated skin papillomas in female Sencar mice. Female Sencar mice were originally developed using the selective breeding methods described by Boutwell (5). Sencar mice are approximately 20 to 80 times more sensitive than CD-1 mice to skin tumorigenesis by DMBA (R. K. Boutwell and T. J. Slaga, unpublished observations). The results with MCA are given in Table 1. For these experiments, mice were initiated with 100 nmoles MCA and promoted twice weekly with 3.4 nmoles of TPA. This resulted in an average of 4.2 papillomas/mouse at the end of an 18-week promotion period. When TCDD was applied topically (1 µg/mouse) 3 days prior to initiation with MCA, a 57 percent reduction in papilloma formation was observed. In accord with the tumor experiments

**TABLE 1.** Effect of Single Topical Doses of TCDD on Tumor-Initiation by MCA

| Treatment (µg/mouse) | Treatment Time[a] | No. of Surviving Mice at 18 Weeks[b] | Papillomas per Mouse | Percentage of Control[c] |
|---|---|---|---|---|
| Acetone | -5 min | 30 | 4.2 | 100 |
| TCDD (1 µg) | -3 days | 29 | 1.8 | 43 |
| TCDD (1 µg) | -5 min | 30 | 4.1 | 98 |
| TCDD (1 µg) | +1 day | 30 | 4.5 | 107 |

[a]A minus sign indicates time of application prior to initiation with MCA whereas a plus sign indicates time of application after initiation.
[b]Thirty females Sencar mice were used per experimental group. Mice were initiated with 100 nmoles of MCA. One week after initiation, mice received 3.4 nmoles of TPA twice weekly for 18 weeks.
[c]The average number of papillomas per mouse expressed as a percentage of the acetone-pretreated control group.

using DMBA and BaP as initiators, topical applications of TCDD 5 minutes before or 1 day after initiation with MCA had no effect on the final number of papillomas per mouse. Similar to the effect of TCDD on tumor initiation with BaP, the magnitude of inhibition with MCA as the initiator was less than that observed with DMBA as the initiator. These results as well as data to be published elsewhere* indicate that female Sencar mice respond in a manner identical to that of female CD-1 mice under the influence of TCDD.

Table 2 illustrates the effects of single intraperitoneal doses of TCDD on tumor initiation by MCA and DMBA. In experiment 1 (Table 2), animals were initiated with 100 nmoles of MCA. TCDD ($1\mu g$/mouse) was injected intraperitoneally either 3 days or 5 minutes before initiation. Control animals received an equal volume of the corn oil vehicle. When TCDD was injected 3 days before MCA application, an approximately 59 percent inhibition in papilloma formation was observed. Injection of TCDD 5 minutes before MCA had no effect on the tumor response. In the

TABLE 2. Effect of Single Intraperitoneal Doses of TCDD on Tumor-Initiation by MCA and DMBA

| Treatment ($\mu g$/mouse) | Treatment Time[a] | No. of Surviving Mice at 18 Weeks[b] | Papillomas per Mouse | Percentage of Control[c] |
|---|---|---|---|---|
| | | Experiment 1 | | |
| Acetone (topical) | -3 days | 30 | 3.4 | 100 |
| Corn oil | -3 days | 29 | 3.9 | 100 |
| TCDD (1 $\mu g$) | -3 days | 29 | 1.6 | 41 |
| TCDD (1 $\mu g$) | -5 min | 28 | 3.6 | 92 |
| | | Experiment 2 | | |
| Acetone (topical) | -3 days | 29 | 6.1 | 100 |
| Corn oil | -3 days | 29 | 6.3 | 100 |
| TCDD (1 $\mu g$) | -3 days | 29 | 0.5 | 8 |
| TCDD (0.1 $\mu g$) | -3 days | 28 | 0.8 | 13 |

[a] A minus sign indicates time of application of TCDD prior to initiation with the hydrocarbon. Mice in experiment 1 were initiated with 100 nmoles of MCA and mice in experiment 2 were initiated with 10 nmoles of DMBA

[b] Thirty female Sencar mice were utilized per experimental group. One week after initiation, mice received 3.4 nmoles of TPA twice weekly for 18 weeks.

[c] The average number of papillomas per mouse expressed as a percentage of the intraperitoneal corn-oil-treated control groups.

---

*DiGiovanni, J., Berry, D. L., Gleason, G. L., and Slaga, T. J. Time-dependent inhibition by 2,3,7,8-tetrachlorodibenzo-p-dioxin of skin tumorigenesis with polycyclic hydrocarbons (submitted for publication).

second set of experiments, TCDD was injected (at two different dose levels) intraperitoneally 3 days prior to initiation with DMBA. Intraperitoneal doses of 1 $\mu$g and 0.1 $\mu$g TCDD produced 92 and 87 percent inhibition, respectively, in DMBA-initiated papilloma formation. We previously demonstrated a dose-response relationship for the inhibitory effect of topically applied TCDD (9). The effect of the two intraperitoneal dose levels is in good agreement with results of these earlier experiments. Furthermore, intraperitoneal injection of TCDD (3 days before initiation with either DMBA or MCA) was effective at inhibiting papilloma formation compared with topical application.

## Effects of TCDD on Epidermal AHH, EPoxide Hydrase, UDP-Glucuronyltransferase, and Glutathione-S-transferase

Induction of microsomal enzyme systems in several tissues has been postulated as a mechanism for the anticarcinogenic effects of a wide variety of compounds, including chlorinated hydrocarbons (10 and references therein). However, these studies have not demonstrated conclusively that the inhibitory actions are a result of induction within the target tissue. During the course of our experiments with TCDD, it was observed that the time course of the inhibitory effect on tumor initiation correlated with the magnitude as well as the time course for induction of monooxygenase activity in the skin (9). Application of TCDD to the skins of genetically inducible as well as "noninducible" mouse strains results in marked increases in AHH activity in this tissue (29). A single topical application of TCDD (0.3 $\mu$g) to Swiss Webster CD-1 mice produced an approximately 30-fold increase in AHH activity of the skin 3 days later (27). Figure 2A illustrates the effect of a 1 $\mu$g topical dose of TCDD on epidermal AHH activity from CD-1 mice. AHH activity was increased 12-fold 24 hours after TCDD application, and after 72 hours a 21-fold stimulation was observed. AHH activity also was elevated at 5 and 10 days after TCDD. Figures 2B and 2C depict the effects of TCDD on epidermal epoxide hydrase and glutathione-S-transferase activities, respectively, using 7-[$^{14}$C]styrene oxide as substrate. Topical pretreatment with TCDD (1$\mu$g/mouse) did not appear to significantly alter epidermal epoxide hydrase or glutathione-S-transferase activities at any of the time points observed. Figure 2D illustrates the effect of topical pretreatment with TCDD (1$\mu$g/mouse) on epidermal UDP-glucuronyltransferase activity. Twenty-four hours after treatment with TCDD, there was an approximately 1.4-fold stimulation of UDP-glucuronyltransferase activity, and by 72 hours, an approximately 2.3-fold stimulation was observed and maintained at 5 days and 10 days after TCDD. The time course for the

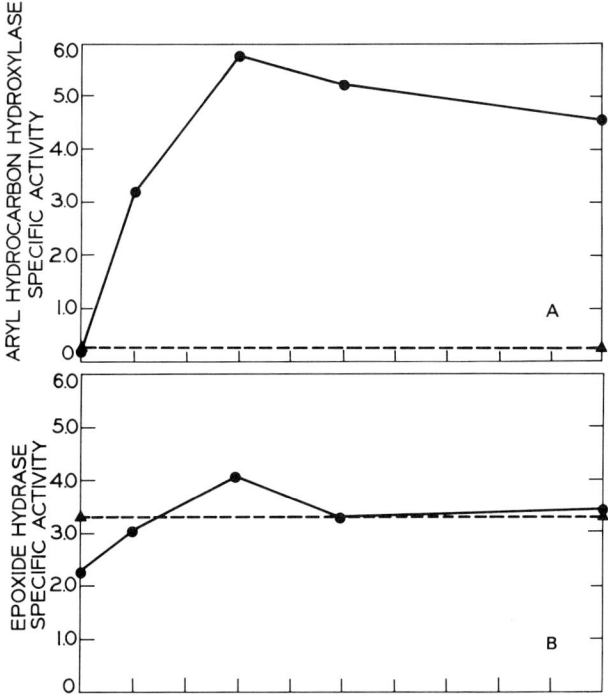

FIGURE 2. Effect of TCDD on mouse epidermal AHH (panel A), epoxide hydrase (panel B), glutathione-S-transferase (panel C) and UDP-glucuronyltransferase activity (panel D). Twenty-five female CD-1 mice were utilized for each time point, and each curve represents an average of two separate experiments run in triplicate. Animals were treated topically with 1 µg TCDD and sacrificed at 5 minutes, 1, 3, 5, and 10 days. Microsomal and cytosolic fractions were prepared as described under Materials and Methods. Specific activities are expressed as follows: AHH, pmoles of 3-hydroxybenzo(a)pyrene per milligram of protein per minute of incubation × $10^{-2}$, epoxide hydrase, nmoles of 7-[$^{14}$C]styrene glycol per milligram of protein per minute of incubation; glutathione-S-transferase, nmoles of S-(2-hydroxy-1-phenylethyl)glutathione per milligram of protein per minute of incubation; UDP-glucuronyltransferase, nmoles of p-nitrophenyl-glucuronide per milligram of protein per minute of incubation. The average control value for the 10-day experiment is represented by a dashed line.

inhibitory effects of TCDD on DMBA, BaP (Figure 1), and MCA (Table 1) tumor initiation correlated with the time course for the induction of epidermal AHH and UDP-glucuronyltransferase.

Subcutaneous injection of TCDD was shown to stimulate mouse-skin AHH activity in both C57BL/6 and DBA/2 mice (20). Table 3 demonstrates that intraperitoneal injections of either 1 or 0.1 µg of TCDD per mouse produced a 17- or 14-fold stimulation, respectively, of epidermal

AHH activity 3 days after injection. UDP-Glucuronyltransferase activities also were stimulated approximately equally (1.5-fold) at either dose of TCDD.

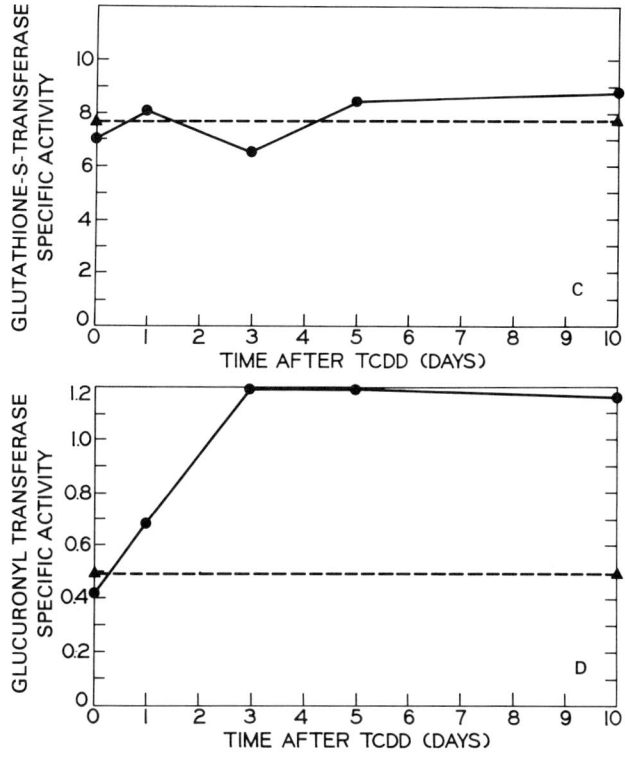

FIGURE 2. (Continued)

TABLE 3. Effect of Single Intraperitoneal Doses of TCCD on Epidermal AHH and UDP-Glucuronyltransferase[a]

| Pretreatment | AHH[b] | UDP-Glucuronyltransferase[c] |
| --- | --- | --- |
| Corn oil | 45 | 0.41 |
| TCDD (1 µg) | 755 | 0.65 |
| TCDD (0.1 µg) | 613 | 0.62 |

[a]Each value represents an average of two separate experiments run in triplicate. Fifteen mice were used per treatment group.
[b]Specific activities for AHH are expressed as pmoles 3-hydroxybenzo(a)pyrene formed per milligram of microsomal protein per minute of incubation. Animals were pretreated 72 hours prior to sacrifice.
[c]Specific activities for UDP-glucuronyltransferase are expressed as nmoles p-nitrophenyl-glucuronide formed per milligram of microsomal protein per minute of incubation.

## Effect of TCDD on [$^3$H]DMBA Disappearance from Mouse Epidermis

The correlation between the time course for induction of epidermal AHH and UDP-glucuronyltransferase by TCDD and the time course for the inhibitory effect on tumor initiation with various PAH suggested that pretreatment with the dioxin may give rise to increased rates of inactivation of PAH in mouse epidermis. In support of this concept was the finding that the quantity of total [$^3$H]DMBA bound to DNA and RNA of the epidermis in the presence and absence of TCDD-pretreatment correlated well with the tumor response under similar conditions (9). Furthermore, a recent study by Cohen et al* indicated that although the total binding of BaP to epidermal DNA was increased in mice pretreated with TCDD, the major hydrocarbon-deoxyribonucleoside adduct present in the DNA of control mice was absent (i.e., BaP-diol-epoxide bound to the exocyclic amino group of guanine).

Figure 3 illustrates the effects of TCDD-pretreatment (1 $\mu$g/mouse, topically) on the disappearance of [$^3$H]DMBA from various mouse epidermal subcellular fractions. Epidermal homogenates were fractionated into four components as follows: 800 × g sediment, L+M sediment, microsomes, and cytosol. Radioactivity associated with each fraction was determined as described under Materials and Methods. The data in Figure 3 represent an average of two separate experiments. Consistently, little difference could be observed in the disappearance of radioactivity between control and TCDD-pretreated animals for the 800 × g sediment (Figure 3a), L + M (Figure 3b), and the microsomal (Figure 3d) fraction. However, in both experiments, there was an approximately 2-fold difference (TCDD-pretreated greater than control) in radioactivity associated with the 105,000 × g supernatant (cytosolic fraction, Figure 3c) at all time points.

To further analyze the difference in radioactivity associated with the cytosolic fraction between TCDD-pretreated and control animals, the distinction was made between acid-precipitable (5 percent TCA) and acid-soluble radioactivity. These results are depicted in Figure 4 and are plotted as the ratio of free (acid-soluble DPM) to total DPM (i.e., acid-soluble + precipitable) in the top panel and as the ratio of bound DPM (acid precipitable) to total DPM in the bottom panel. Both of these values are plotted as a function of time after application of [$^3$H]DMBA. Two hours after application of [$^3$H]DMBA the ratio of free radioactivity to the total cytosolic radioactivity was much greater in the TCDD-pretreated animals

---

*Cohen, G. M., Bracken, W. M., Iyer, P. R., Berry, D. L., Selkirk, J. K., and Slaga, T. J. Anticarcinogenic effects of 2,3,7,8-tetrachlorodibenzo-p-dioxin on benzo(a)pyrene and 7,12-dimethylbenz(a)anthracene tumor initiation and its relationship to DNA binding (submitted for publication).

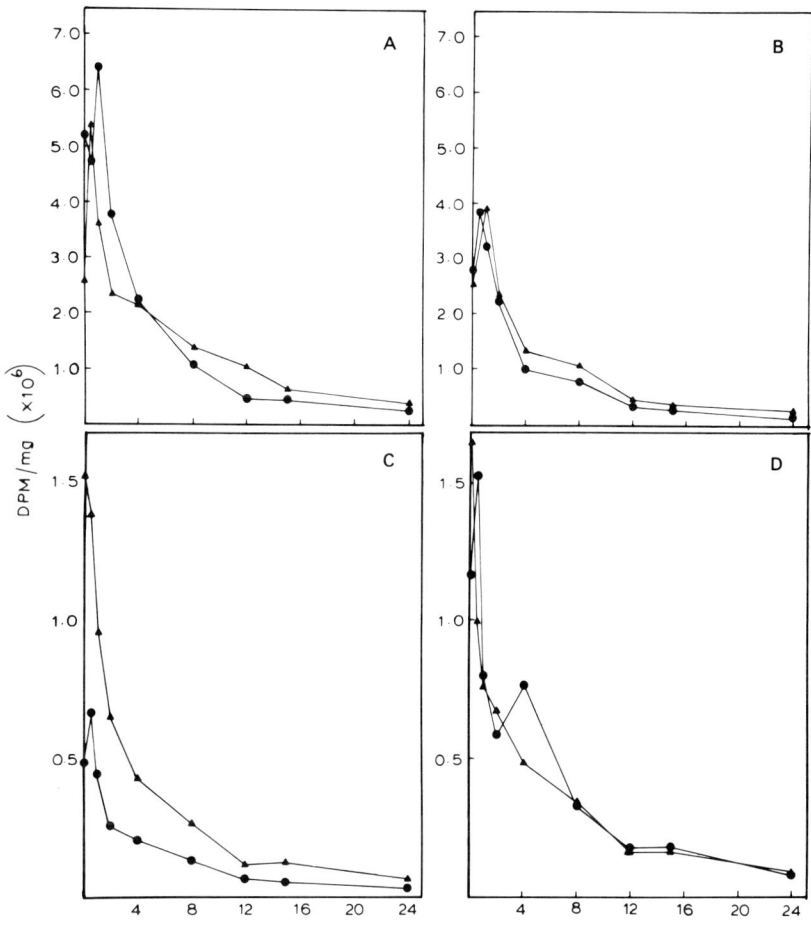

FIGURE 3. Effect of TCDD on the disappearance of [$^3$H]DMBA from mouse epidermis. Five female CD-1 mice were utilized for each time point and each curve represents an average of two separate experiments. Control (●) mice were treated with 0.2 ml of acetone and a second group were treated with 1 μg of TCDD (▲) in 0.2 ml of acetone. Three days later, 10 nmoles of DMBA (10 μCi) was applied to animals of both groups and mice in each group were sacrificed at 5 minutes, 30 minutes, 1 hour, and 2, 4, 8, 12, 15, and 24 hours. Subcellular fractions were prepared as described under Materials and Methods and the radioactivity associated with each fraction was determined. Values in the graph are expressed as DPM per milligram of protein ($\times 10^6$). Profile A, 800 × g sediment; profile B, L+M fraction; profile C, cytosol fraction; and profile D, microsomal fraction.

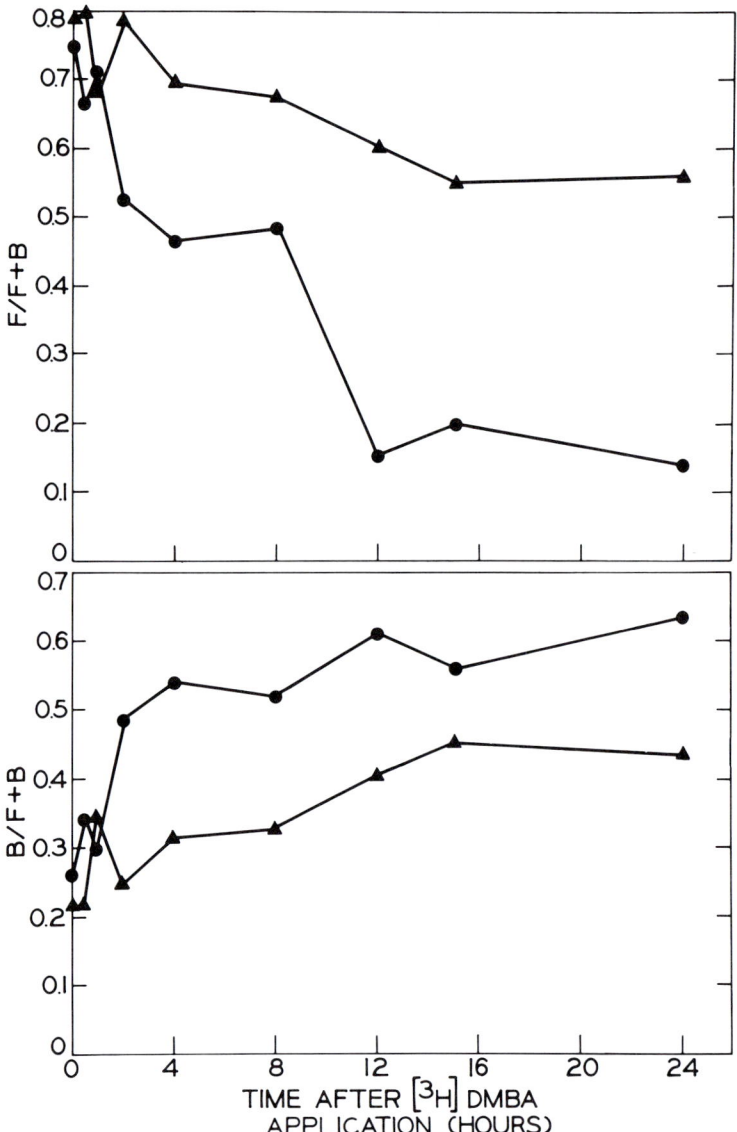

FIGURE 4. Effect of TCDD on disappearance of [$^3$H]DMBA from mouse epidermal cytosol. Aliquots of epidermal cytosol fractions (105,000 × g supernatant) from each time point in Figure 3 were treated with 5 percent TCA. Acid-soluble as well as acid-precipitable radioactivity was determined. The results are plotted as a ratio of acid-soluble (F) to acid-soluble + acid-precipitable (i.e., total DPM, F + B) radioactivity in profile A and as a ratio of acid-precipitable (B) to the total cytosol counts (F + B) in profile B. Control (●); TCDD pretreated (▲).

when compared with the control animals. The difference was approximately 4-fold by 24 hours. In contrast, the quantity of acid-precipitable radioactivity remains greater in control than TCDD-pretreated animals at time points of 2 hours or greater. These findings suggested that TCDD pretreatment led to an altered pattern of [$^3$H]DMBA metabolism in epidermal cells.

The results presented above, coupled with the observed increases in epidermal AHH and UDP-glucuronyltransferase, suggested a possible mechanism for the ability of TCDD to inhibit tumor initiation by PAH. Under the influence of TCDD, epidermal cells may be programmed to inactivate PAH carcinogens more efficiently and therefore eliminate them more readily. Further analysis of the profile of water-soluble metabolites as well as rates of clearance and excretion from epidermal cells may shed further light on the inhibitory effect of TCDD.

## SUMMARY

The findings of this work are summarized below:

1. Topical doses of TCDD, producing no visible toxic manifestations, effectively inhibited tumor initiation by DMBA, BaP, and MCA. The inhibitory effect was highly dependent on the time of treatment relative to the initiator. Maximal inhibition was observed when TCDD was given 3 to 5 days before initiation. Little or no effect was observed when TCDD was given 5 minutes before or 1 day after initiation.
2. Intraperitoneal injections of TCDD (0.1 or 1 $\mu$g/mouse) were as effective as topical applications for inhibiting DMBA and MCA tumor initiation. The inhibitory effects of intraperitoneally injected TCDD also were dependent on the time of treatment relative to the initiator.
3. Topical application of TCDD, at the doses utilized in tumor experiments, markedly induced epidermal AHH activity to maximal levels (21-fold over controls) 3 days after treatment. In addition, epidermal UDP-glucuronyltransferase activities also were maximally stimulated (2 to 3-fold over control) by 3 days following TCDD treatment. TCDD treatment had little or no effect on epidermal epoxide hydrase or glutathione-S-transferase activities.
4. The disappearance from mouse epidermis of [$^3$H]DMBA (applied 3 days after topical application of TCDD) was similar to that observed in the epidermis of control animals, with the exception of radioactivity associated with the cytosol fraction. Two hours after application of the

hydrocarbon, approximately 2-fold more radioactivity was associated with the cytosol from TCDD-pretreated animals than from control animals. Analysis of acid-soluble and acid-precipitable radioactivity in the cytosol indicated an approximately 4-fold difference in the amount of acid-soluble counts (TCDD-pretreated greater than control). Conversely, the quantity of acid-precipitable counts was greater in control (approximately 2-fold) cytosol.
5. The time course fo the inhibitory effects of TCDD on tumor initiation by DMBA, BaP, and MCA correlated with the time course for induction of epidermal AHH and UDP-glucuronyltransferase enzymes. The results presented in this investigation suggested that under the influence of TCDD, epidermal cells may be programmed to inactivate PAH carcinogens more efficiently.

## ACKNOWLEDGMENTS

Research was supported by NIH grants CA-22484, CA-07175, CA-09020, CA-20076, and by the Department of Energy under Contract W-7405-eng-26 with the Union Carbide Corporation.

## REFERENCES

1. Baars, A. J., Jansen, M., and Breimer, D. D. (1978): The influence of phenobarbital, 3-methylcholanthrene and 2,3,7,8-tetrachlorodibenzo-p-dioxin on glutathione-S-transferase activity in rat liver cytosol. *Biochem. Pharmacol.* 27:2487-2494.
2. Baird, W. M., and Dipple, A. (1977): Photosensitivity of DNA-bound 7,12-dimethylbenz(a)anthracene. *Int. J. Cancer* 20:427-431.
3. Berry, D. L., Slaga, T. J., Wilson, N. M., Zachariah, P. K., Namkung, M. J., Bracken, W. M., and Juchau, M. R. (1977): Transplacental induction of mixed function oxygenases in extra-hepatic tissues by 2,3,7,8-tetrachlorodibenzo-p-dioxin. *Biochem. Pharmacol.* 26:1383-1388.
4. Berry, D. L., Zachariah, P. K., Namkung, M. J., and Juchau, M. R. (1976): Transplacental induction of carcinogen-hydroxylating systems with 2,3,7,8-tetrachlorodibenzo-p-dioxin. *Toxicol. Appl. Pharmacol.* 36:569-584.
5. Boutwell, R. K. (1964): Some biological aspects of skin carcinogenesis *Prog. Exptl. Tumor Res.,* 4:207-250.
6. Chouroulinkov, I., Gentil, A., Grover, P. L., and Sims, P. (1976): Tumour-initiating activities on mouse skin of dihydrodiols derived from benzo(a)pyrene. *Br. J. Cancer* 34:523-532.

7. Chouroulinkov, I., Gentil, A., Tierney, B., Grover, P., and Sims, P. (1977): The metabolic activation of 7-methylbenz(a)anthracene in mouse skin: High tumor-initiating acitivity of the 3,4-dihydrodiol. *Cancer Lett.* 3:247-253.
8. Daudel, P., Duquesne, M., Vigny, P., Grover, P. L., and Sims, P. (1975): Fluorescence spectral evidence that benzo(a)pyrene-DNA products in mouse skin arise from diol epoxides. *FEBS Lett.* 57:250-253.
9. DiGiovanni, J., Berry, D. L., Juchau, M. R., and Slaga, T. J. (1978): 2,3,7,8-Tetrachlorodibenzo-p-dioxin: Potent anticarcinogenic activity in CD-1 mice. *Biochem. Biophys. Res. Commun.* 86:577-584.
10. DiGiovanni, J., Slaga, T. J., Berry, D. L., and Juchau, M. R. (1980): Inhibitory effects of environmental chemicals on polycyclic aromatic hydrocarbon carcinogenesis. In: *Modifiers of Chemical Carcinogenesis,* T. J. Slaga, Ed., pp. 145-168, Raven Press, New York.
11. Dipple, A., and Nebzydoski, J. A. (1978): Evidence for the involvement of a diol-epoxide in the binding of 7,12-dimethylbenz(a)-anthracene to DNA in cells in culture. *Chem.-Biol. Interactions* 20:17-26.
12. Fowler, B. A., Hook, G.E.R., and Lucier, G. W. (1977): Tetrachlorodibenzo-p-dioxin induction of renal microsomal enzyme systems: Ultrastructural effects on pars recta ($S_3$) proximal tubule cells of the rat kidney. *J. Pharmacol. Exptl. Therapeut.* 203:712-721.
13. Heidelberger, C. (1975): Chemical carcinogenesis. *Ann. Rev. Biochem.* 44:79-121.
14. Hollmann, S., and Touster, O. (1962): Alterations in tissue levels of uridine diphosphate glucose dehydrogenase, uridine diphosphate glucuronic acid pyrophosphatase and glucuronyltransferase induced by substances influencing the production of ascorbic acid. *Biochem. Biophys. Acta* 338-352.
15. Ivanovic, V., Geacintov, N. E., Jeffrey, A. M., Fu, P. P., Harvey, R. G., and Weinstein, I. B. (1978): Cell and microsome mediated binding of 7,12-dimethylbenz(a)anthracene to DNA studied by fluorescence spectroscopy. *Cancer Lett.* 4:131-140.
16. James, M. O., Fouts, J. R., and Bend, J. R. (1976): Hepatic and extrahepatic metabolism, in vitro, of an epoxide (8-$^{14}$C-styrene oxide) in the rabbit. *Biochem. Pharmacol.* 25:187-190.
17. Juchau, M. R., Pedersen, M. G., and Symms, K. G. (1972): Hydroxylation of 3,4-benzpyrene in human fetal tissue homogenates. *Biochem. Pharmacol.* 21:2269-2272.
18. Kapitulnik, J., Levin, W., Conney, A. H., Yagi, H., and Jerina, D. M. (1977): Benzo(a)pyrene-7,8-dihydrodiol is more carcinogenic than benzo(a)pyrene in newborn mice. *Nature* 266:378-380.

19. Kouri, R. E. (1976): Relationship between levels of aryl hydrocarbon hydroxylase activity and susceptibility to 3-methylcholanthrene and benzo(a)pyrene-induced cancers in inbred strains of mice. In: *Carcinogenesis, A Comprehensive Survey,* Vol. 1., R. Freudenthal and P. Jones, Eds., pp. 139-151, Raven Press, New York.
20. Kouri, R. E., Rude, T. H., Joglekar, R., Dansette, P. M., Jerina, D. M., Atlas, S. A., Owens, I. S., and Nebert, D. W. (1978): 2,3,7,8-Tetrachlorodibenzo-p-dioxin as a co-carcinogen causing 3-methylcholanthrene-initiated subcutaneous tumors in mice genetically "nonresponsive" at Ah locus. *Cancer Res.* 38:2777-2783.
21. Levin, W., Wood, A. W., Chang, R. L., Slaga, T. J., Yagi, H., Jerina, D. M., and Conney, A. H. (1977): Marked differences in the tumor-initiating activity of optically pure (+)- and (-)-trans-7,8,-dihydroxy-7,8-dihydrobenzo(a)pyrene on mouse skin. *Cancer Res.* 37:2721-2725.
22. Lowry, O. H., Rosebrough, N. J., Farr, A. L., and Randall, R. J. (1951): Protein measurement with the folin phenol reagent. *J. Biol. Chem.* 193:265-275.
23. Lucier, G. W., McDaniel, O. S., and Hook, G.E.R. (1975): Nature of the enhancement of hepatic uridine diphosphate glucuronyltransferase activity by 2,3,7,8-tetrachlorodibenzo-p-dioxin in rats. *Biochem. Pharmacol.* 24:325-334.
24. Malaveille, C., Bartsch, H., Grover, P. L., and Sims, P. (1975): Mutagenicity of non-K-region diols and diol-epoxides of benzo(a)-anthracene and benzo(a)pyrene in *S. typhimurium* TA100. *Biochem. Biophys. Res. Commun.* 66:639-700.
25. Moschel, R. C., Baird, W. M., and Dipple, A. (1977): Metabolic activation of the carcinogen 7,12-dimethylbenz(a)anthracene for DNA binding. *Biochem. Biophys. Res. Commun.* 76:1092-1098.
26. Owens, I. S. (1977): Genetic regulation of UDP-glucuronyltransferase induction by polycyclic aromatic compounds in mice. *J. Biol. Chem.* 252:2827-2833.
27. Pohl, R. J., Philpot, R. M., and Fouts, J. R. (1976): Cytochrome P-450 content and mixed-function oxidase activity in microsomes isolated from mouse skin. *Drug Metab. Dispos.* 4:442-450.
28. Poland, A., and Glover, E. (1974): Comparison of 2,3,7,8-tetrachlorodibenzo-p-dioxin, a potent inducer of aryl hydrocarbon hydroxylase, with methylcholanthrene. *Molec. Pharmacol.* 10:349-359.
29. Poland, A., Glover, E., Robinson, J. R., and Nebert, D. W. (1974): Genetic expression of aryl hydrocarbon hydroxylase activity: Induction of monooxygenase activities and cytochrome P-450 formation by 2,3,7,8-tetrachlorodibenzo-p-dioxin in mice genetically

"non-responsive" to other aromatic hydrocarbons. *J. Biol. Chem.* 249:5599-5606.
30. Poland, A., and Kende, A. (1976): 2,3,7,8-Tetrachlorodibenzo-p-dioxin: environmental contaminant and molecular probe. *Fed. Proc.* 35:2404-2411.
31. Sims, P., Grover, P. L., Swaisland, A., Pal, K., and Hewer, A. (1974): Metabolic activation of benzo(a)pyrene proceeds by a diol-epoxide. *Nature* 252:326-328.
32. Slaga, T. J., Bracken, W. M., Gleason, G, Levin, W., Jerina, D. M., and Conney, A. H. (1979): Marked differences in the skin tumor-initiating activities of the optical enantiomers of the diastereomeric benzo(a)pyrene-7,8-diol-9,10-epoxides. *Cancer Res.* 39:67-71.
33. Slaga, T. J., Huberman, E., Selkirk, J. K., Harvey, R. G., and Bracken, W. M. (1977): Carcinogenicity and mutagenicity of benz(a)anthracene diols and diol-epoxides. *Cancer Res.* 38:1699-1704.
34. Slaga, T. J., Viaje, A., Berry, D. L., Bracken, W. M., Buty, S. G., and Scribner, J. D. (1976): Skin tumor-initiating ability of benzo(a)pyrene-4,5,-7,8 and 7,8-diol-9,10-epoxides and 7,8-diol. *Cancer Lett.* 2:115-122.
35. Vigny, P., Duquesne, M., Coulomb, H., Tierney, B., Grover, P. L., and Sims, P. (1977): Fluorescence spectral studies on the metabolic activation of 3-methylcholanthrene and 7,12-dimethylbenz(a)anthracene in mouse skin. *FEBS Lett.* 82:278-282.
36. Weinstein, I. B., Jeffrey, A. M., Jenette, K. W., Blobstein, S. H., Harvey, R. G., Harris, C., Autrup, H., Kasai, H., and Nakanishi, K. (1976): Benzo(a)pyrene diol-epoxides as intermediates in nucleic acid binding in vitro and in vivo. *Science* 193:592-595.
37. Winsnes, A. (1969): Studies on the activation in vitro of glucuronyl-transferase. *Biochem. Biophys. Acta* 191:279-291.
38. Wood, A. W., Chang, R. L., Levin, W., Lehr, R. E., Schaefer-Ridder, M., Karle, J. M., Jerina, D. M., and Conney, A. H. (1977): Mutagenicity and cytotoxicity of benzo(a)anthracene diol-epoxides and tetrahydroepoxides: exceptional activity of the bay region 1,2-epoxides. *Proc. Nat. Acad. Sci.* 74:2746-2750.
39. Wood, A. W., Levin, W., Chang, R. L., Lehr, R. E., Schaefer-Ridder, M., Karle, J. M., Jerina, D. M., and Conney, A. H. (1977): Tumorigenicity of five dihydrodiols of benz(a)anthracene on mouse skin: exceptional activity of benz(a)anthracene-3,4-diol. *Proc. Nat. Acad. Sci.* 74:3176-3179.
40. Wood, A. W., Levin, W., Lu, A.Y.H., Ryan, D., West, S. B., Lehr, R. B., Schaefer-Ridder, M., Jerina, D. M., and Conney, A. H. (1976): Mutagenicity of metabolically activated benz(a)anthracene-3,4-

dihydrodiol. Evidence for bay region activation of carcinogenic polycyclic hydrocarbons. *Biochem. Biophys. Res. Commun.* 72:680-686.
41. Wood, A. W., Levin, W., Ryan, D., Thomas, P. E., Yagi, H., Moh, H. D., Thakker, D. R., Jerina, D. M., and Conney, A. H. (1977): High mutagenicity of metabolically activated chrysene-1,2-dihydrodiol. Evidence for bay region activation of chrysene. *Biochem. Biophys. Res. Commun.* 78:847-854.
42. Wood, A. W., Levin, W., Thomas, P. E., Ryan, D., Karle, J. M., Yagi, H., Jerina, D. M., and Coney, A. H. (1978): Metabolic activation of dibenzo(a,h)anthracene and its dihydrodiols to bacterial mutagens. *Cancer Res.* 38:1967-1973.

# MEMBRANE CHANGES ASSOCIATED WITH AQUEOUS EXTRACTS OF FOSSIL-FUEL GENERATED RESPIRABLE PARTICULATES

T. J. Facklam, J. P. Crowley, M. A. Drum, and A. J. Dennis

    Battelle Columbus Laboratories
    505 King Avenue
    Columbus, Ohio 43201

Membrane proteins of cultured C3H 10T1/2 fibroblasts were examined during and after exposure to subtoxic doses of $H_2O$ extracts of fly-ash samples from fossil-fuel-generated power plants. Two fly-ash samples were tested, one which produced distinct genetic toxicity and a second which has no appreciable biological activity. After treatment with extracts for varying time periods ranging from 18 hours to 2 weeks, the cell membrane proteins were radiolabeled with $^{125}I$ and separated by polyacrylamide electrophoresis. The biologically active fly ash induced several significant permanent alterations in membrane proteins, including the loss of high-molecular-weight proteins of 210K to 280K daltons. However, there was an induction of several proteins in the range of 130K to 140K daltons. Additionally, the C3H cells were treated simultaneously with fly ash extracts and a variety of protease inhibitors. Membrane profiles of the untreated cells were not significantly altered by the protease inhibitors alone. However, fly-ash-extract-treated cells coincubated with phenylmethylsulfonylfluoride exhibited cell-surface membrane protein profiles strikingly similar to those of the control-cell membrane profiles. These data were substantiated by the binding kinetics of iodinated concanavalian A to treated and untreated cells. To determine the active component of $H_2O$ extracts of the fly ash, several purified polycyclic aromatic hydrocarbons (PAH), metals, and fractions of the fly-ash extracts were tested to duplicate the membrane profiles and lectin binding.

## INTRODUCTION

There are a number of cellular events that are directly associated with transformation and the neoplastic state. An important event apparently associated with contact inhibition is a change in cell surface and associated membrane functions. These changes may include the loss of surface proteins such as the large external transformation-sensitive protein (LETS) (1,2), alterations in the carbohydrate content of glycoproteins (3), and changes in lectin agglutinability (4-8). Additionally, transformed cells also have been shown to have increased proteolytic activity (12-14). Proteolytic activity in transformed cells appears to lead to a series of cellular modifications, including altered growth rates (11), loss of contact inhibition (9,15), and loss of LETS protein (10,14).

The alterations of the membranes of transformed cells may be responsible for a number of the changes in cellular behavior, such as loss of contact inhibition, changes in the cytoskeleton, and varying transport rates. There are few data to show whether any of these surface alterations occur during initial exposure to chemical carcinogens or mutagens and prior to actual morphologic transformation. One such potentially mutagenic complex, particulate fly ash from fossil-fuel combustion, is found in the environment and may induce the "transformation-associated" cellular changes.

Studies of the effects of fly ash on biological processes generally have been limited to the detection of potential mutagenic and cytotoxic properties (16,17). We have examined the characteristics of the surface membrane and associated proteases of C3H 10T1/2 fibroblasts exposed to nontoxic but mutagenic doses of extracts of fly-ash particulates.

To determine the biochemical effects of short-term exposure to fly ash, membrane protein profiles were examined by surface iodination with lactoperoxidase and polyacrylamide gel electrophoresis. Alterations in the binding kinetics of lectin to fly-ash-treated cells were also examined. Both of these parameters of surface membrane character were found to be altered by the addition of several protease inhibitors. It was also possible to measure the protease activity that appeared to be induced by the fly-ash extracts.

## MATERIALS AND METHODS

### Cells

C3H 10T1/2 fibroblasts were obtained from Dr. C. Heidelberger and were cultured in Eagle's basal medium with 10 percent fetal calf serum and antibiotics (GIBCO, Grand Island, New York). C3H cells at 90 percent confluence were exposed to 300 $\mu$g/ml of fly-ash extract for 18 hours. Protease inhibitors, tosyl arginine methylester (TAME), and tosyl-lysyl-chloromethylketone (TLCK) were added to the cultures at a final concentration of 50 µg/ml. Phenylmethylsulonylfluoride (PMSF) was dissolved in acetone and then added to the cultures to a final concentration of 0.1 mm. The inhibitors and fly ash were incubated simultaneously with the appropriate cells for 18 hours.

**Fly-Ash Extracts**

Aqueous extracts of fly ash were prepared by mixing 10-g samples of fly ash and 80 ml of $H_2O$ for 2 hours at room temperature. The suspension was centrifuged and filtered through a Nalgene 0.45-µ filter. Dry weights were measured to determine the concentration of samples per ml of extract. Benzene extracts were prepared by adding 25 ml of benzene to 3 g of fly ash and sonicating (Bronson Sonifier) for 60 seconds at one-half maximum setting. The fly ash was centrifuged and the extraction procedure repeated. The two benzene extracts were evaporated to dryness with nitrogen at room temperature. The remaining residue was dissolved into 3 ml of dimethylsulfoxide (DMSO).

**Iodination and Electrophoresis**

The cells first were washed three times with phosphate-buffered saline (PBS) and PBS containing 5 mM glucose, and 1µCi/ml of carrier-free $^{125}I$ (420 mCi/ml, NEN) was added. The iodination reaction was initiated by adding 20 µg/ml lactoperoxidase (Sigma) and 0.1 unit/ml glucose oxidase (Sigma). The reaction was allowed to proceed for 10 minutes at room temperature and then was stopped by the addition of phosphate-buffered iodine (PBI) followed by three washings in PBS. The washed cells were then scraped into PBS containing 0.1 mM PMSF, 1 percent 2-mercaptoethanol, and 1 percent SDS for electrophoresis. Polyacrylamide disc gel electrophoresis was run according to Laemmli (18) with a 5 percent resolving gel and 3 percent stacking gel. The samples were applied to the stacking gel with bromphenol blue as a dye marker. When the dye front had reached the bottom of the gels, the gels were removed, sliced, and counted in a Searle Gamma Counter. The markers for molecular weight determination were cross-linked bovine serum albumin (Sigma) which had molecular weights of 66,000, 132,000, 198,000, and 264,000 daltons in SDS gels. The proteins were iodinated by the Chloramine T-method (19). After labelling and extensive dialysis against PBS, the standards were resolved in the 5 percent gels.

**Conconavalan A Binding Studies**

Con A was radiolabelled with $^{125}I$ using the Hunter and Greenwood Chloramine T method (19). Lectin binding studies were conducted in 10 x

25-mm plastic petri dishes seeded with 1.7 x $10^5$ cells. After an 18-hour treatment with fly ash extract in the presence or absence of protease inhibitor, each plate was washed three times with PBS, and 3 ml of PBS containing 10 $\mu$g/ml $^{125}$I-Con A was added. The plates were incubated for varying time periods on a slowly shaking table. At the end of each incubation, the cells were washed three times with PBS and then with 10 percent TCA. The TCA precipitable fractions then were filtered through Whatman GF/A filters and radioactivity was determined in a Searle Gamma Counter.

**Protease Assays**

C3H cells grown to 90 percent confluency were washed three times with PBS and resuspended in hypotonic butter (10 mM Tris-HCl, pH 7.4, and 1 mM EDTA). The cells were homogenized with a Dounce homogenizer at 0°C and then centrifuged at 10,000 x g for 10 minutes. The resulting supernatant then was assayed for protease and glycosidase activities. N-benzoyl-tyrosine ethylester, BTEE (Sigma), was utilized as a substrate for serine protease activity (20). Glycosidase activities of the homogenates were measured according to the method of Bosmann (21). The following substrates were used: p-nitrophenyl-N-acetyl-$\beta$-D-glucosamide for N-acetyl-$\beta$-D-glucosaminidase, p-nitrophenyl-$\beta$-D-galactoside for $\beta$-D-galactoside, p-nitrophenyl-$\alpha$-D-mannoside for $\alpha$-D-mannosidase, p-nitrophenyl-$\beta$-D-xylopyranoside for $\beta$-D-xylosidase, p-nitrophenyl-$\beta$-D-glucuronide for $\beta$-D--D-glucuronidase, p-nitrophenyl-$\beta$-D-glucopyranoside for $\beta$-D-glucosidase, and p-nitrophenyl-$\alpha$-D-galactopyranoside for $\alpha$-D-galactosidase (Sigma). Enzyme activity was reported as percentage of activity in control cells.

## RESULTS

The gel profile of an untreated C3H lactoperoxidase surface labelled membrane is shown in Figure 1. Solvent-treated control cells contained several lactoperoxidase-accessible proteins with molecular weights of 300K, 280K, and 270 to 265K, two proteins in the 210 to 190K range; and a 160K protein. Addition of a nontoxic dose (300 $\mu$g/ml) of water extract of mutagenic fly ash, as determined in the Ames, E. coli polymerase and V79 transformation assays (22), altered the membrane profile radically (Figure 2a). There appeared to be a quantitative decrease in the 280 to 250K proteins, as well as an increase in a cluster of proteins ranging from 125 to 145K daltons. These surface alterations appeared to be a stable characteristic of fly-ash-treated cells.

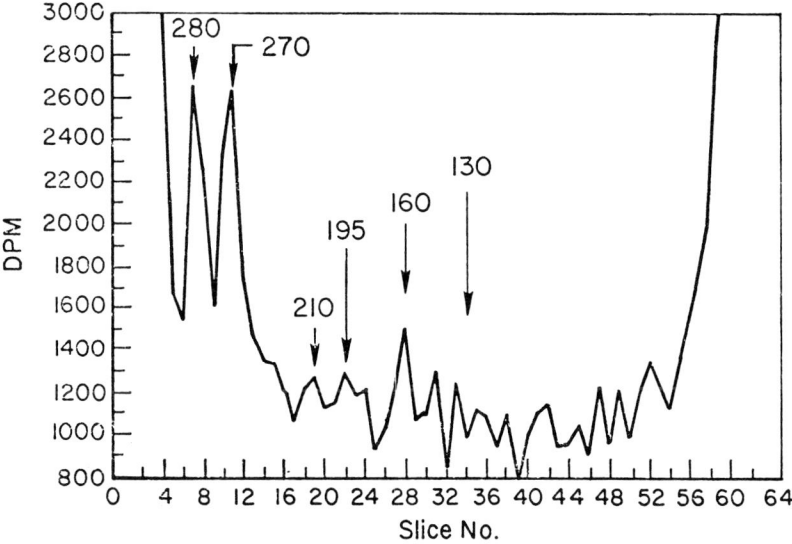

**FIGURE 1.** Polyacrylanide gel profile of C3H fibroblast lactoperoxidase iodinated membrane.

After the C3H fibroblasts were exposed to the water extract for varying time periods, the cells were washed and replaced with fresh media (Figure 2). The data indicate that the characteristic 125 to 145K cluster of proteins appeared rapidly after treatment and was retained 96 hours after the removal of the fly-ash material. This change in the membrane protein profile did not appear to be a transient event, because the cluster of proteins remained long after the extract had been removed.

To examine whether this altered membrane profile could be the result of proteolytic activity, the protease inhibitors TAME, PMSF, and TLCK were added and coincubated with the mutagenic or nonmutagenic fly-ash-extract-treated and untreated cells. At the concentrations of inhibitors used, the cells were not visibly affected, nor were their membrane protein profiles significantly altered. However, the gel profiles of the treated cells were altered from cultures treated with fly ash alone. Figure 3 shows the iodinated membrane profiles for TLCK-treated cells. The mutagenic fly-ash-treated cells now has a suppressed profile similar to that of both the nonmutagenic fly-ash-treated cells and the control. TAME (Figure 4) when added to the fly-ash-treated C3H cultures induced an altered profile similar to that of the TAME-treated control cells. Additions of PMSF to both treated and untreated cultures resulted in very similar membrane profiles (Figure 5). All three inhibitors appeared to alter the fly-ash-treated cell membrane profile so that it resembled the membrane profile for either control cells or cells treated with nonmutagenic fly-ash extracts.

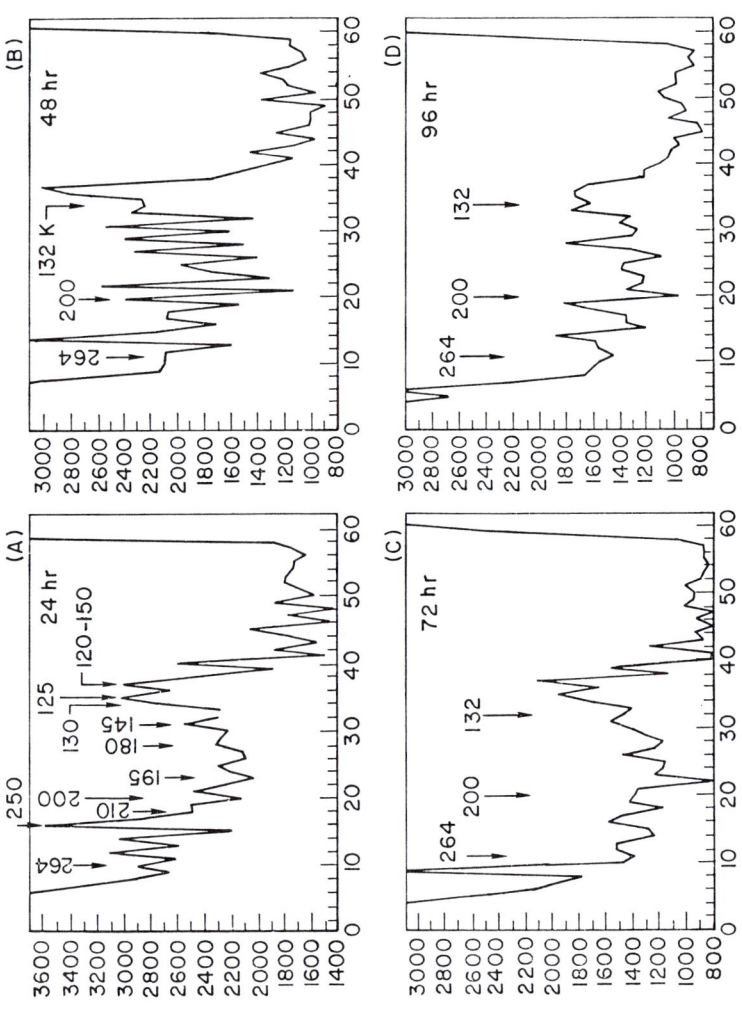

FIGURE 2. Gel profile of iodinated membrane of C3Hs treated with biologically active fly-ash extracts. Cells were exposed to fly-ash extracts for 18 hours (A), 48 hours (B), 72 hours (C), 96 hours (D).

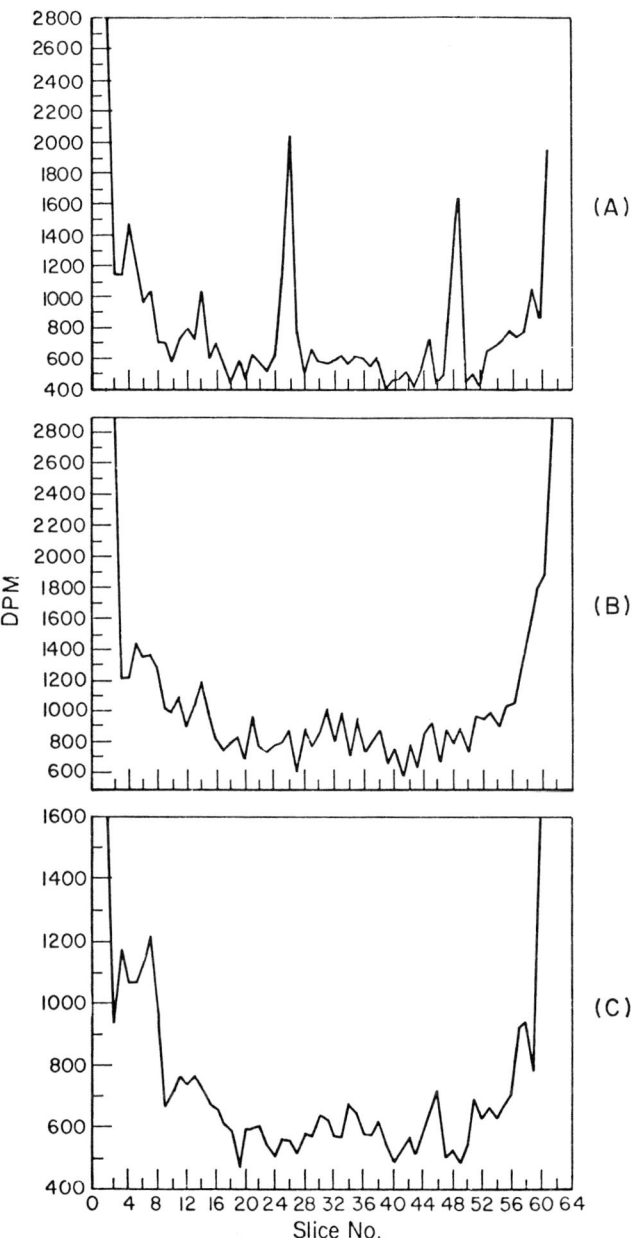

**FIGURE 3.** Gel profiles of radiolabelled membranes of fibroblasts treated with TLCK and fly ash. Cells were incubated with 50μg/ml of TLCK in the presence or absence of fly ash. (A) control cells, (B) biologically active fly-ash-treated cells, (C) nonbiologically active fly-ash-treated cells.

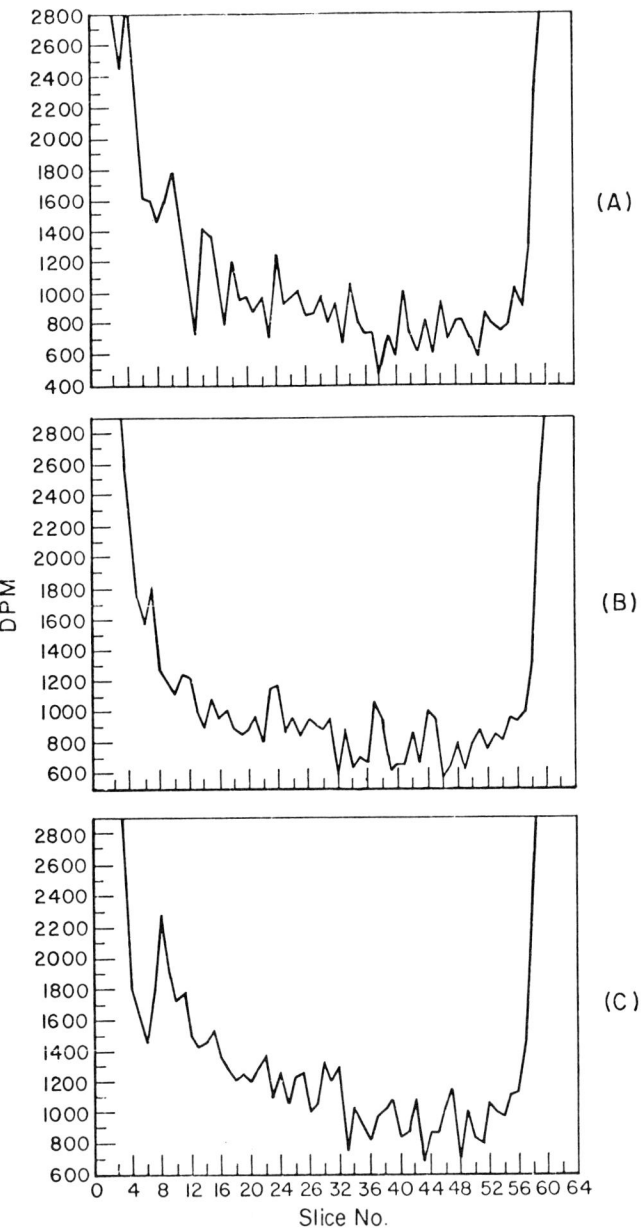

FIGURE 4. Gel profiles of iodinated membranes of C3H cells treated with TAME and cells were incubated with 50µg/ml of TAME in the presence or absence of fly ash. (A) control cells, (B) biologically active fly-ash-treated cells, (C) nonactive fly-ash-treated cells.

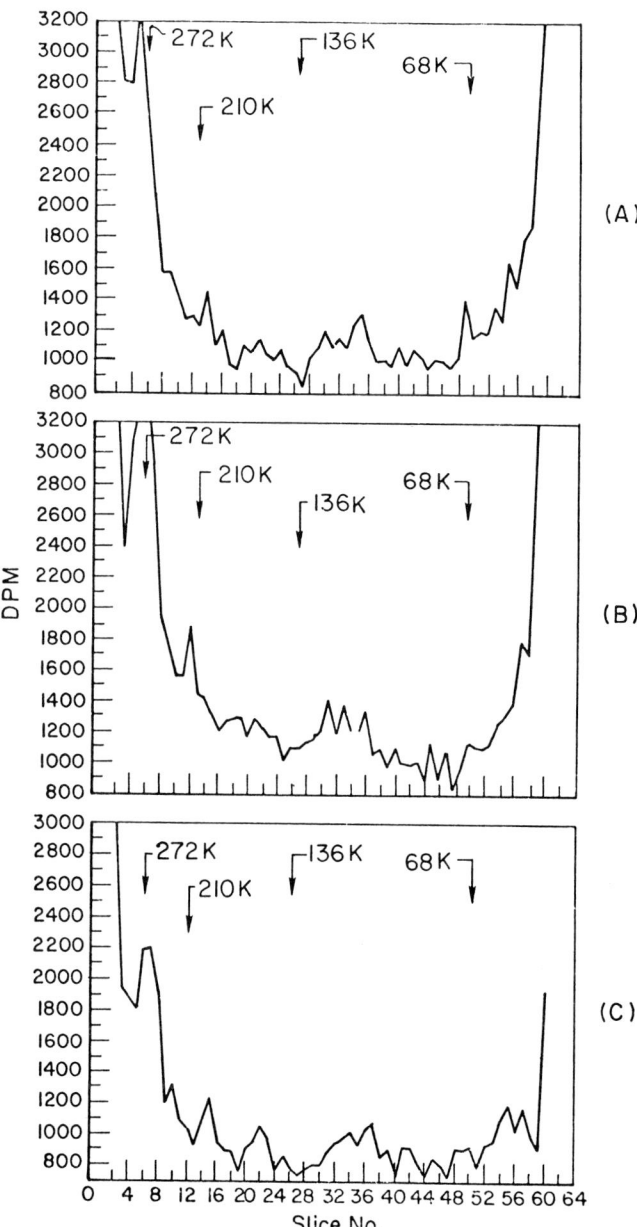

FIGURE 5. Gel profiles of iodinated membranes of C3H cells treated with PMSF and fly ash. Cells were incubated with 0.1 mM of PMSF in the presence or absence of fly ash. (A) control cells, (B) biologically active fly-ash-treated cells, (C) nonactive fly-ash-treated cells.

To examine additional cell surface alterations induced by fly ash, the binding kinetics of a radiolabelled lectin was studied. C3H cells were incubated with $^{125}$I-Con A both with and without prior incubation with fly ash extract. Figure 6a illustrates the differences between the lectin binding efficiency of these treated cells after 18 hours' incubation with fly ash. When methymannoside was added to each plate, the Con A binding was effectively blocked, indicating the specificity of binding.

Addition of the protease inhibitors to these cultures produced a range of Con A binding responses. TAME (Figure 6b) and TLCK (Figure 6c) both altered the binding kinetics of Con A to the treated cultures; however, these alterations in binding did not resemble those of the control cells. PMSF-treated cultures (Figure 6d) appear to have similar lectin binding patterns for both fly-ash-treated and untreated cells. These results suggest that the lectin binding differences attributed to the fly ash could be the result of a proteolytic activity which is sensitive to PMSF.

The membrane profile data and lectin binding data indicate that the addition of one of more protease inhibitors can result in the inhibition of mutagen-induced alterations. To further examine the role of proteases as a possible cause for these surface alterations, cell-free homogenates of both fly-ash-treated and untreated cells were assayed for protease and glycosidase activities. Table 1 shows that the general protease activity increased two-fold in cells treated with the mutagenic fly ash as compared with nonmutagenic or nonactive fly-ash-treated cells. In general, the biologically active fly ash extract also induced the glycosidase activities, whereas the nonactive extract did not. Because of the higher sensitivity of

TABLE 1. Protease and Glycosidase Activity of C3H Fibroblasts Treated with WaterExtracts of Biologically Active and Nonactive Fly Ash[a]

| Enzyme | Activity (% control) | |
|---|---|---|
| | Active F.A. Extract | Nonactive F.A. Extract |
| N-acetyl-$\beta$-D-glucosaminidase | 155 | 93 |
| $\beta$-D-glucuronidase | 143 | 97 |
| $\beta$-D-glucosidase | 121 | — |
| $\beta$-B-xylosidase | 180 | 133 |
| $\beta$-D-galactosidase | 133 | 92 |
| $\alpha$-D-galactosidase | 122 | 92 |
| $\alpha$-D-mannosidase | 94 | 115 |
| Protease | 221 | 98 |

[a]Fibroblasts were treated with 300 µg/ml fly-ash extract for 18 hours.
All activities are reported as percentage of control-cell enzyme activities.

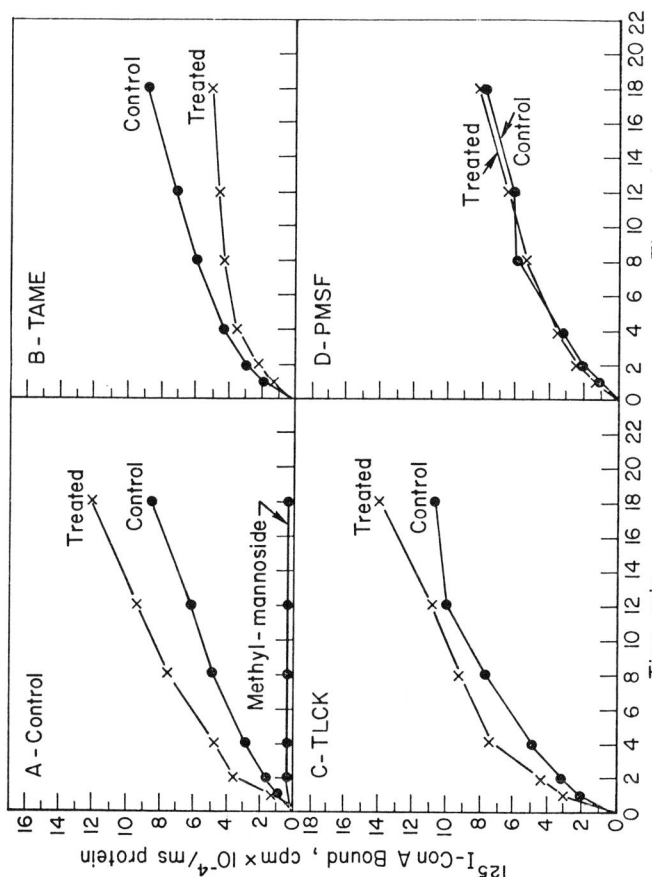

FIGURE 6. Binding kinetics of $^{125}$I-Concanavalin A to C3H cells treated with protease inhibitors and mutagenic fly-ash extracts. (A) $^{125}$I Con A binding kinetics to cells treated with biologically active fly ash, methylmannoside, or to untreated cells. (B) $^{125}$I-Con A binding in the presence 50 μg/ml TAME to cells treated or untreated with fly ash. (C) $^{125}$I-Con A binding in the presence of 50 μg/ml TLCK to cells treated or untreated with fly ash. (D) $^{125}$I-Con A binding in the presence of 0.1 mM PMSF to cells treated or untreated with fly ash.

several of the glycosidase assays, the remaining experiments were restricted to assaying N-acetyl-$\beta$-D-glucosamidase, $\beta$-D-galactosidase, $\alpha$-D-mannosidase, and $\alpha$-D-galactosidase activities.

The data from Table 1 suggest that the biologically active, mutagenic fly-ash extracts contain components that induce proteolytic and glycosidase activity not present in the nonactive fly-ash extracts. To explore this possibility, two of the components of fly ash, metals and polynuclear aromatic hydrocarbons, were incubated with the C3H cells and the glycosidase activity assayed.

A large number of metals are present in fly ashes. Among these are nickel, zinc, cadmium, chromium, and lead. Each of these metals was added as a chloride salt to C3H cells at 1 $\mu$g/ml for 18 hours. Table 2 shows that only zinc induced increases in glycosidase activity. Both cadmium and chromium were inhibitory of the glycosidases assayed.

To examine whether an organic-soluble component of the fly ash could be responsible for the observed enzymatic and associated membrane changes, cells were treated with benzene extracts from the active and three nonactive fly ashes. As Table 3 illustrates, none of the extracts induced

TABLE 2. Glycosidase Activity of C3H Fibroblasts Treated with Primary Metal Constituents of Fly Ash[a]

| Enzyme | Activity (% control) | | | | |
|---|---|---|---|---|---|
| | Ni | Zn | Cd | Cr | Pb |
| N-acetyl-$\beta$-D-glucosaminidase | 106 | 110 | 81 | 44 | 98 |
| $\beta$-D-galactosidase | 104 | 110 | 84 | 57 | 92 |
| $\alpha$-D-mannosidase | 96 | 128 | 71 | 76 | 82 |

[a]C3H fibroblasts were incubated with 1 $\mu$g/ml of metal salts for 18 hours. Enzyme activities are reported as percentage of control-cell activities.

TABLE 3. Glycosidase Activity of C3H Fibroblasts Treated with Benzene Extracts of Biologically Active and Nonactive Fly Ash[a]

| Enzyme | Activity (% control) | | | |
|---|---|---|---|---|
| | Active | Nonactive-1 | Nonactive-2 | Nonactive-3 |
| N-acetyl-$\beta$-D-glucosaminidase | 76.5 | 76.4 | 100 | 102.8 |
| $\beta$-D-galactosidase | 78.3 | 80.3 | 95.2 | 98.1 |
| $\alpha$-D-galactosidase | 75.0 | 68.7 | 89.0 | 98.4 |
| $\alpha$-D-mannosidase | 85.0 | 84.9 | 103.2 | 109.7 |

[a]Fibroblasts were treated with benzene extracts for 18 hours. Enzyme activities are reported as percentage of control-cell activities.

glycosidase activities. Different purified PAH were added to the C3H cells and incubated for 18 hours as an additional test of organic compounds to induce activity, before glycosidase activity was determined. Table 4 shows that the activities N-acetyl-$\beta$-D-glucosaminidase and $\beta$-D-galactosidase were significantly increased by phenanthrene, fluoranthene, anthracene, and benzo(a)pyrene. Also, $\alpha$-D-mannosidase activity was increased by both phenanthrene and fluoranthene. Although the benzene extracts were unable to induce any significant effect on the C3H cells, PAH by themselves were capable of inducing several glycosidase activities.

## DISCUSSION

This study examined the effects of extracts of a mutagenic, biologically active fly ash on the cell surface of C3H fibroblasts. Membrane alterations were measured by lactoperoxidase iodination of the cell surface and lectin binding rates. Additionally, proteolytic activities associated with membrane changes also were determined.

Lactoperoxidase labelling experiments demonstrated that a variety of surface alterations occurred in the fly-ash-treated cells. One fly-ash-induced change, the decrease in a high-molecular-weight protein, is similar to the effects seen with LET which is well documented, as are the increases in other proteins (1). For example, in transformed murine erythrocytes, there are a series of membrane changes, including the loss of a 13,000 dalton protein, the appearance of a 8,000 dalton protein, and an increase in the amount of a 140,000-molecular-weight protein (23). Membrane alterations also occur in viral-transformed cells; these consist of a series of decreases in those proteins associated with normal cells and the appearance of new proteins (24,25,26). Alterations such as the decrease in high-molecular-weight proteins and the appearance of new sets of proteins occur in C3H cells as a result of treatment with fly ash. Such alterations could be the result of fly-ash-induced mutations in gene regulation that redirect the synthesis of new proteins and the turnover of old proteins. An alternative mechanism involves proteolytic cleavages of surface proteins and the resulting change in the membrane profile. The increase in the 125 to 145K protein could be caused by an accumulation of either a high-molecular-weight protein, or a set of proteins which are newly accessible to lactoperoxidase iodination as a consequence of previous surface alterations.

We initially examined the involvement of proteolysis in fly-ash-induced surface changes by using three protease inhibitors. The results indicated that the cells treated with biologically active fly ash exhibited membrane changes that could be altered by protease inhibitors and that this profile could resemble those of control cells and cells treated with nonactive fly

TABLE 4. Glycosidase Activity of C3H Fibroblasts Treated with Various Polynuclear Aromatic Hydrocarbons[a]

| Enzyme | Activity (% control) | | | | |
|---|---|---|---|---|---|
| | Phenanthrene | Fluoranthrene | Anthracene | Benz(a)anthracene | Benzo(a)pyrene |
| N-acetyl-$\beta$-D-glucosaminidase | 187.6 | 190.1 | 154.0 | 92.0 | 143.8 |
| $\beta$-D-galactosidase | 170.8 | 160.5 | 155.7 | 102.3 | 132.1 |
| $\alpha$-D-galactosidase | 85.4 | 81.2 | 104.1 | 83.3 | 100.1 |
| $\alpha$-D-mannosidase | 131.2 | 144.5 | 116.0 | 81.2 | 89.5 |

[a] Fibroblasts were treated with 10 $\mu$g/ml APH in DMSO for 18 hours. Enzyme activities are reported as percentage of control cell activities.

ash. Protease inhibitors in other systems have been shown to suppress various characteristics of transformation. Inhibitors antipain and leupepti were able to suppress radiation-induced transformation of 10T1/2 cells (27). The inhibitors TLCK, TPCK, and TAME can inhibit tumorigenesis induced by DMBA and promotion by croton oil in vivo on mouse skin (28). Growth of transformed 3T3 fibroblasts was inhibited by TAME, TPCK, and TLCK, but the growth of untransformed cells was not affected (9).

From these data, protease inhibitors appear to act on protease-mediated events which are related to carcinogenesis. Despite the deficiencies of using protease inhibitors (29) for these studies, the data indicate that the protease inhibitors exerted a moderating effect on the cells treated with biologically active fly ash. A similar effect of the protease inhibitors was also seen in the alterations in lectin binding.

Cells transformed by chemicals and viruses have alterations in the membrane glycoproteins. This characteristic is reflected in differing agglutinability of normal and transformed cells to lectins and in the changes in the oligosaccharide composition of glycoprotein. We used the binding of the lectin Conconavalan A to probe the membrane glycoproteins of fly-ash-treated cells, which appeared to have a greater ability to bind the lectin than control cells. This binding pattern could be altered by the presence of protease inhibitors. The data indicate that a serine protease (PMSF sensitive) may be active in altering or maintaining the fly-ash-induced lectin binding pattern.

The lectin binding kinetics of fly-ash-induced cells could be the result of a membrane rearrangement or of the same proteolytic action. The fly ash is possibly inducing a protease to act on the cell surface in the same manner that mild trypsin treatments of normal cells cause them to agglutinate with lectins at the same rate as transformed cells.

Additional evidence for further fly-ash-induced cell alterations is the increased protease and glycosidase activities detected in cell-free homogenates. The increased protease activity is similar to increases found in virus-transformed cells (13) in which trypsin and cathepsin-like activities were elevated. Fibrolytic activity (1) and a diisprapylfluorophosphate-sensitive protease activity are also increased in transformed cells (12). Increased glycosidase activities were induced by both PAH and the mutagenic fly ash. These findings are similar to both malignant human cells and viral-transformed cells. Cell-free homogenates from malignant human heart and colon tissue have higher levels of $\beta$-galactosidase, $\alpha$-mannosidase, and protease activity than normal tissue (30). Viral-transformed mouse and hamster fibroblasts have enhanced glycosidase and protease activities (13,31). The glycosidases $\beta$-D-galactosidase, N-acetyl-$\beta$-D-glucosaminidase, and $\alpha$-D-mannosidase showed significant increases ranging from 20 to over 200 percent. These levels of glycosidase activity appear to be correlated with the oncogenically transformed cells. These observed cellular changes induced by this active fly ash may represent events which characterize a pretransformed cell or the first stages of transformation. An example of this type of early change in membranes before morphologic changes occur is the increase in fluorescence labelling of surface proteins similar to fully transformed cells in a temperature-sensitive virus-transformed cell line at the permissive temperature (32). The role of the mutagen in this fly ash, whether a PAH or some other molecule, and the possible connection between a mutational event and altered cell surfaces are unknown. It may be that the observed changes in enzyme activities and membrane proteins are only a cellular response to a noxious agent. However, some component of the mutagen fly ash not found in nonmutagen fly ashes is capable of inducing a number of cellular changes which are characteristic of transformed cells.

## REFERENCES

1. Vaheri, A., and Mosher, D. F. (1978): High molecular weight, cell surface associated glycoprotein (fibronectin) lost in malignant transformation. *Biochem Biophys. Acta* 516:1-25.
2. Hynes, R. O. (1976): Cell surface proteins and malignant transformation. *Biochem. Biophys. Acta* 458:73-107.
3. Warren, L., Burck, C. A., and Tuszynski, G. P. (1978): Glycopeptide changes and malignant transformation, a possible role for carbohydrate in malignant behavior. *Biochem. Biophys. Acta* 516:97-127.
4. Burger, M. M., and Goldberg, A. R. (1967): Identification of a tumor-specific determinant on neoplastic cell surfaces. *Proc. Natl. Acad. Sci. U.S.A.* 57:359-366.
5. Burger, M. M. (1969): A difference in the architecture of the surface membrane of the normal and virally transformed cells. *Proc. Natl. Acad. Sci. U.S.A.* 62:994-1001.
6. Ozanne, B., and Sambrook, J. (1971): Binding of radioactively-labelled concanavalin A and wheat germ agglutinin to normal and virus-transformed cells. *Nature New Biology* 232:156-160.
7. Cline, H. J., and Livingston, D. C. (1971): Binding of $^3$H-concanavalin A by normal and transformed cells. *Nature New Biology* 233:155-157.
8. Shoham, J., and Sachs, L. (1972): Differences in the binding of fluorescent concanavalin A to the surface membrane of normal and transformed cells. *Proc. Natl. Acad. Sci. U.S.A.* 69:2479-2482.
9. Schnebli, H. P., and Burger, M. M. (1972): Selective inhibition of growth of transformed cells by protease inhibitors. *Proc. Natl. Acad. Sci. U.S.A.* 69:3825-3827.

10. Hynes, R. O. (1973): Alteration of cell surface proteins by viral transformation and by proteolysis. *Proc. Natl. Acad. Sci. U.S.A.* 70:3170-3174.
11. Blumberg, P. M., and Robbins, P. W. (1975): Effect of proteases on activation of resting chick embryo fibroblasts and on cell surface proteins. *Cell* 6:137-147.
12. Mahdavi, V., and Hynes, R. O. (1979): Proteolytic enzymes in normal and transformed cells. *Biochem. Biophys. Acta* 583:167-178.
13. Bosmann, G. H. (1972): Elevated glycosidases and proteolytic enzymes in cells transformed by RNA tumor virus. *Biochem. Biophys. Acta* 264:339-343.
14. Unkeless, J. C., Tobia, A., Ossowski, L., Quigley, J. P., Rifkin, D. B., and Reich, E. (1973): An enzymatic function associated with transformation of fibroblasts by oncogenic viruses. *J. Exp. Med.* 137:85-111.
15. Chou, I. N., Black, P. H., and Roblin, R. O. (1974): Non-selective inhibition of transformed cell growth by a protease inhibitor. *Proc. Natl. Acad. Sci. U.S.A.* 71:1748-1752.
16. Chrisp, C. E., Fischer, G. L., and Lammert, J. E. (1978): Mutagenicity of filtrates from respirable coal fly ash. *Science* 199:73-75.
17. Symposium paper.
18. Laemmli, U. K. (1970): Cleavage of structural proteins during the assembly of head bacteriophage T4. *Nature* 277:680-685.
19. Hunter, W. M., and Greenwood, F. C. (1962): Preparation of $^{131}$I labelled human growth hormone of high specific activity. *Nature London*, 194:485-496.
20. Walsh, K. A., and Wilcox, P. E. (1976): Serine proteases. In: *Methods in Enzymology*, G. E. Perlman and L. Lorand, Eds., pp. 31-41, Academic Press, New York.
21. Bosmann, H. B., and Pike, G. Z. (1970): Glycoprotein synthesis and degradation; glycoprotein; N-acetyl glucosamine transferase, proteolytic and glycosidase activity in normal and polyoma virus transformed BHK cells. *Life Sci. U.S.A.* 9:1433-1440.
22. Crowley, J. P., Dennis, A. J., Facklam, T. J., and Margard, W. (1979): Manuscript in preparation.
23. Glass, J., Fischer, S., Lavidor, L. M., and Nunez, T. (1977): External surface membrane proteins in normal and neoplastic murine erythroid cells. *Cancer Res.* 37:1497-1501.
24. Stone, K. R., Smith, R. E., and Joklik, W. K. (1974): Changes in membrane polypeptides that occur when chick embryo fibroblasts and NRK cells are transformed with avian sarcoma viruses. *Virology* 58:86-100.

25. Isaka, T., Yoshida, M., Owada, M., and Toyoshima, K. (1975): Alterations in membrane polypeptides of chick embryo fibroblasts induced by transformation with avian sarcoma viruses. *Virology* 65:226-237.
26. Hynes, R. O., and Wyke, J. A. (1975): Alterations in surface proteins in chicken cells transformed by temperature-sensitive mutants of rous sarcoma virus. *Virology* 64:492-504.
27. Kennedy, A. R., and Little, J. B. (1978): Protease inhibitors suppress radiation-induced malignant transformation in vitro. *Nature, London* 276:825-826.
28. Troll, W., Klassen, A., and Janoff, A. (1970): Tumorigenesis in mouse skin; inhibition by synthetic inhibitors of proteases. *Science* 169:1211-1213.
29. McIlhinney, A., and Hogan, B.L.M. (1974): Effect of inhibitors of proteolytic enzymes on the growth of normal and polyoma transformed BHK cells. *Biochem. Biophys. Res. Comm.* 60:348-354.
30. Bosmann, H. B., and Hall, T. C. (1974): Enzyme activity in invasive tumors of human breast and colon. *Proc. Natl. Acad. Sci. U.S.A.* 71:1833-1837.
31. Thompson, J. E., Gruber, M. Y., and Kruuv, J. (1978): Changes in glycosidase enzyme activity during growth of normal and transformed cells. *Exp. Cell Res.* 111:47-53.
32. Parry, G., and Hawkes, S. P. (1978): Detection of an early surface change during oncogenic transformation. *Proc. Natl. Acad. Sci. U.S.A.* 75:3703-3707.

# SENSITIZED FLUORESCENCE DETECTION OF PAH

### E. M. Smith and P. L. Levins
Arthur D. Little, Inc.
Acorn Park
Cambridge, Massachusetts 02140

## INTRODUCTION

Polynuclear aromatic hydrocarbons (PAH) are among the many polycyclic organic materials (POM) that are commonly encountered as trace-level environmental contaminants in effluents associated with combustion, pyrolysis, and other thermal degradation processes. The PAH category, which may be defined as containing hydrocarbon species with three or more fused aromatic rings, includes some compounds suspected to be potent carcinogens, as well as many isomeric and other noncarcinogenic compounds. Determination of emission levels of PAH is, therefore, important in environmental assessment.

Several procedures, such as gas chromatography/mass spectrometry (GC/MS), have demonstrated applicability for obtaining compound specific information for evaluation of potential health hazards associated with an effluent containing PAH. However, these procedures are necessarily sophisticated because of the large number of possible PAH species requiring state-of-the-art equipment and extensive investment of expert analyst's time. It is not cost effective to apply them routinely to samples that may not, in fact, contain any detectable levels of PAH.

This paper reports the results of a study initiated to determine whether the phenomenon of sensitized fluorescence could be utilized in the analysis of PAH as a class. A major objective was to develop a simple procedure for detection of PAH at much lower levels than is possible with current methods based on fluorescence analysis. This procedure, requiring only instrumentation readily available to most laboratories, would provide a low-cost screening technique to determine whether environmental assessment samples contained levels of PAH such that more detailed analyses should be undertaken.

## THEORY

PAH are inherently fluorescent materials and are known to exhibit sensitized fluorescence. The two processes of directly excited fluorescence and sensitized fluorescence are shown in a simplified energy-level diagram in Figure 1.

Compound A, as depicted, absorbs energy (↑) and is raised to various excited singlet energy levels. An energy release is made vibrationally until the lowest excited singlet state is achieved (↓). The energy is released from this state to the various vibrational levels of the ground state in the form of fluorescent emissions (↓). When compound B is present, a vibrational coupling interaction can occur between the excited states of Compound A and Compound B (6), which results in resonant energy transfer, and fluorescent emissions of Compound B will be observed. As can be seen in the diagram, Compound B must have a common vibrational frequency with Compound A and the lowest vibrational level of B's excited singlet state must be at lower energy than the corresponding level for Compound A. The transfer of energy is most efficient when the acceptor (B) is present in an extremely low molar ratio to the donar (A). Notable examples of sensitized fluorescence are naphthacene in benz(a)anthracene (10) and in anthracene (9) at molar ratios of $10^{-4}$ and $10^{-6}$, respectively.

In most studies involving directly excited fluorescence, the limit of detection of PAH has been on the order of 10 ng/ml in solution or, in the case of thin-layer chromatography, 10 ng/spot. With sensitized fluorescence it was considered likely that by using the analyte as the "minor" constituent of an appropriate mixture, the limit of detection could

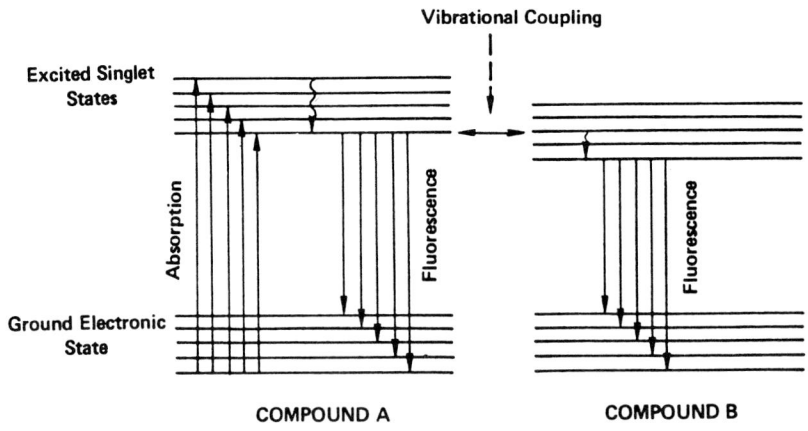

FIGURE 1. Singlet—Singlet Energy Transfer.

be reduced on the order of $10^4$- to $10^6$-fold as noted in the above-mentioned sensitized fluorescence systems.

Among aromatic hydrocarbons, both the absorption and fluorescence shift to longer wavelengths (lower energies) with increasing conjugation; both are also at longer wavelengths for linear multiple ring compounds than for corresponding nonlinear isomers. From the energy considerations, then, lower-molecular-weight aromatic compounds should be sensitizers for PAH of higher molecular weight (anthracene-naphthacene system), and in the case of isomers, a nonlinearly conjugated aromatic compound should sensitize a linearly conjugated compound (benz(a)anthracene-naphthacene system).

Some lower molecular weight aromatic compounds might sensitize the fluorescence of many PAH and thus be of general use for PAH detection in a screening type of test.

## DEVELOPMENT OF SPOT TEST PROCEDURE

### Selection of Analytes

The PAH of interest in environmental assessment are those containing three or more fused rings. Most known carcinogenic PAH contain at least four fused rings. In the benzenoid series, the low-molecular-weight analytes would be the isomers phenanthrene and anthracene of which only anthracene fluoresces in the visible. Higher molecular weight PAH, e.g., pyrene and chrysene, show fluorescent peaks in the ultraviolet region but still have bands in the visible region. PAH containing even more condensed rings all exhibit fluorescence in the visible region of the spectrum.

Similar spectral-emission qualities are found in the methylene-bridged PAH, e.g., fluorene derivatives. Fluorene and the benzofluorenes fluoresce in the ultraviolet region and compounds with more fused rings emit in the visible region.

To meet the objective of keeping the method simple, only materials known to fluoresce in the visible spectral region were studied as analytes. The compounds selected as representative analytes, their corrected fluorescence emission peaks (9), and their carcinogenicity ratings (8) are shown in Table 1. These compounds were chosen because of their range of molecular weights and their availability.

Although only nonheterocyclic compounds were tested as analytes in this study, the heteroatom-bridged aromatic ring structures are considered as analytes that can also be detected by sensitized fluorescence. The lower molecular weight members that fluoresce only in the ultraviolet region should be efficient sensitizers for the higher molecular weight compounds, e.g., carbazole for the benzocarbazoles.

TABLE 1. Model Analytes for Sensitized Fluorescence of PAH

| Compound | Molecular Weight | Emission Peaks[a] (nm) | Carcinogenicity[b] |
|---|---|---|---|
| Anthracene | 178 | 378,400,422 | - |
| Pyrene | 202 | 370,382,392 | - |
| Fluoranthene | 202 | 465 | - |
| Perylene | 252 |  | - |
| Benzo(a)pyrene | 252 | 392,416 | +++ |
| Coronene | 300 | 425,442,540 | - |
| Dibenzo(a,i)pyrene | 302 | 420,450 | +++ |

[a]Reference 9.
[b]-Not carcinogenic, +++ highly carcinogenic; see reference 8.

## Fluorescence Sensitizers

In accord with the energy considerations mentioned earlier, lower molecular weight aromatic hydrocarbons were selected as potential fluorescence sensitizers. These included benzene, naphthalene, fluorene, and phenanthrene, all of which absorb and and fluoresce in the ultraviolet region of the spectrum. Their fluorescence emission bands are given below.

Fluorescence of Sensitizers[5]

| Compound | Emission Wavelength (nm) |
|---|---|
| Benzene | 255 – 300 |
| Naphthalene | 300 – 365 |
| Fluorene | 302 – 370 |
| Phenanthrene | 348 – 407 |

Although sensitized fluorescence was observed with frozen benzene systems, its use was eliminated because of the practical limitations due to its volatility. Theoretical considerations suggest that the energy transfer required for sensitized fluorescence occurs best when the analyte and sensitizer are of similar crystalline structure. For this reason, several attempts were made to use fluorene and phenanthrene as sensitizers. The results with these latter materials were, however, complicated by trace impurities of higher PAH present in every sample available for study, even after several extensive attempts at purification. For these practical reasons, the final development was done using naphthalene as the sensitizer.

## Spot Test Procedure

The full details of the development of the spot-test procedure are available (11). The supplies required for the spot test, in addition to the solution (normally in methylene chloride) of the sample extract, are listed below:

Naphthalene solution (60 $\mu g/\mu l$)
Pipets (Drummond Microcaps, $1\mu l$)
Whatman No. 42 Filter Paper
Ultraviolet Source - 254 nm.

A pencil is used to mark three circles on filter paper, each approximately 1 cm in diameter. With the pipet, a 1 $\mu l$ portion of the sample is applied to each of two of the marked spots and allowed to dry. Similarly, 1 $\mu l$ of the naphthalene sensitizer solution is applied to the remaining blank circle and one sample spot. After the spots have dried they are observed under 254-nm illumination. Differences in intensity and color between the sample/sensitizer spot and either spot alone indicate sensitized fluorescence. The filter paper will appear approximately as shown in Figure 2.

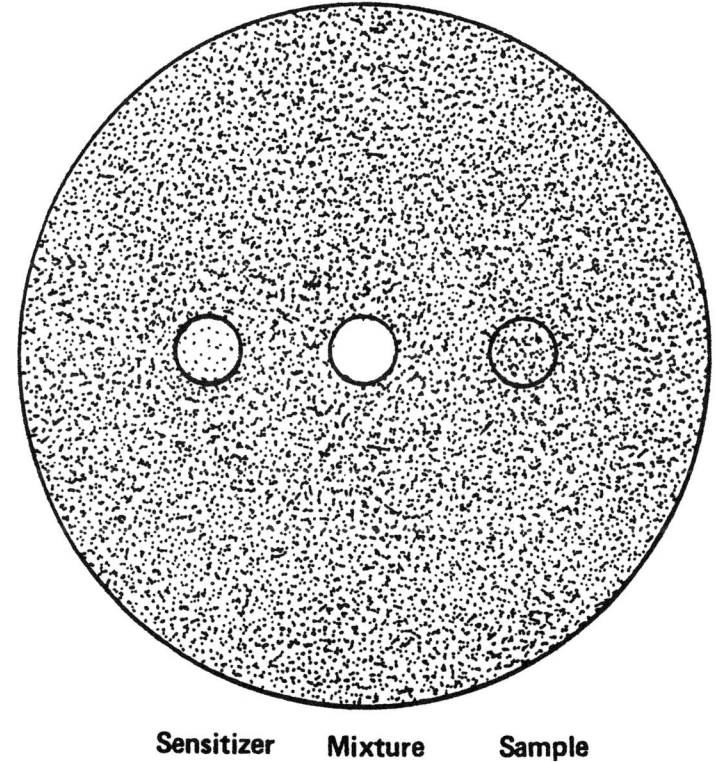

**FIGURE 2.** Representation of 254 nm Illuminated Spot-Test Series.

The detection limits have been found to be between 1 and 10 pg PAH/spot for sensitized fluorescence and approximately 10 ng/spot for non-sensitized (normal) fluorescence. Approximate PAH quantitites using this procedure can be estimated as follows:

| | |
|---|---|
| No Sensitized Fluorescence | <1 pg |
| Weak Sensitized Fluorescence | 1-10 pg |
| Strong Sensitized Fluorescence (but not Self-Fluorescence) | <100 pg |
| Self-Fluorescence | >10,000 pg |

## Detection Limits

Sensitivity-test results obtained from specific PAH are listed below:

| | pg/Spot |
|---|---|
| Anthracene | 10 |
| Pyrene | 10 |
| Fluoranthene | 10 |
| Perylene | 10 |
| Benzo(a)pyrene | 1 |
| Coronene | 10 |
| Dibenzo(a,i)pyrene | <10 |

The detection limits for PAH are thus seen to be 10 pg/spot in most cases and lower in some cases, as for benzo(a)pyrene at 1 pg/spot.

Since the sample is applied from 1 $\mu$l of solution, the detection limit corresponds to 10 pg/$\mu$l or 10 ng/ml. These levels are lower than those typically achieved by GC/MS and are consistent with the needs of environmental assessment studies.

The spot-test detection levels may be compared with those of typical samples collected for environmental surveys by reference to Table 2. In this table, the sample volumes normally collected with three particulate collection devices (EPA Method 5 and SASS trains for source emissions and Hi Vol for ambient) and water are indicated. The spot-test limits have been calculated on the basis of the 10 pg/spot ($\mu$l) limit and the assumption that the sample extract has been concentrated to 1 ml. The sample concentration limits thus calculated may be compared with the health-effects multimedia environmental goals (MEG) limits set for BaP for source and ambient conditions. In each case it is seen that the spot test is more sensitive than required to detect BaP (a worst-case example) at its MEG levels.

TABLE 2. Sensitivity Toxicity Comparisons

|  | Sample Vol. | Spot-Test Limit | BaP MEG Value |
|---|---|---|---|
| Method 5 | 7 m$^3$ | 1 ng/m$^3$ | 20 ng/m$^{3a}$ |
| SASS | 30 m$^3$ | 0.3 ng/m$^3$ | 20 ng/m$^{3a}$ |
| Hi Vol | 2000 m$^3$ | 0.005 ng/m$^3$ | 0.05 ng/m$^{3b}$ |
| Water | 1 l | 10 ng/l | 300 ng/l$^a$ |

[a]Source emission level.
[b]Ambient level.

## Positive Interference Tests

Limited tests were carried out to determine whether other species could give a positive sensitized fluorescence in the absence of PAH. The two mixtures listed below, containing equal amounts of each component, were tested separately and found to give no sensitized fluorescence.

| 1 | 2 |
|---|---|
| tetrachloroethane | n-pentane |
| cumene | n-octane |
| benzaldehyde | n-nonane |
| 2-ethylhexanol | n-decane |
| o-nitrotoluene | n-undecane |
| quinoline | n-tridecane |
| dihexylether | n-pentadecane |
|  | n-heptadecane |

# EVALUATION WITH ENVIRONMENTAL SAMPLES

The sensitized fluorescence spot test has been evaluated on a variety of environmental samples during other studies at the Arthur D. Little laboratories and in the laboratories of other contractors.

Several samples were prepared in conjunction with a study on the incineration of industrial waste (2-4). The extracts of these samples were analyzed by direct-probe low-resolution mass spectrometry (LRMS) for PAH levels and were also studied with the spot test. The comparison in Table 3 shows that the PAH were always indicated by the spot test when present, as shown by the LRMS results, and the relative levels were in the same order of abundance. Both procedures showed no PAH in the F waste sample.

TABLE 3. Comparison of LRMS Data and Sensitized Fluorescence Results on Chemical Waste Incineration Samples

| Waste | MW ≧178 PAH by LRMS | Sensitized Fluorescence | |
|---|---|---|---|
| | | 10-ng Spot | 100-pg Spot |
| A | Low | Moderate | 0 |
| B | Moderate-high | High | Moderate |
| C | Moderate | High | Weak |
| D | High | High | Weak |
| F | None | 0 | 0 |

Smith (12) has prepared a series of utility-industry particulate samples for comparative analysis by GC/MS and the spot test and has obtained the results given in Table 4. The four cases where PAH were found to be present in the GC/MS analysis were positive in the spot test. There were no false negatives in the spot-test results. Two positive indications with the spot test where PAH were not found by GC/MS could be due to the fact that the spot test is 10 to 100 times more sensitive than the GC/MS under the test conditions used for the study.

Benning (1) has used the spot-test procedure as an aid to determine sample collection requirements during field studies of PAH emissions from wood stoves. In cases with low levels of PAH indicated, larger sample volumes were collected. The PAH levels indicated by the spot test in the field were again checked after the samples had been returned to the laboratory. The results (Table 5) show excellent agreement and further indicate the value of this simple test in enhancing environmental studies.

TABLE 4. Comparison of Particulate Extract PAH by GC/MS and Spot Test[a]

| | Spot | GC/MS |
|---|---|---|
| Coke Oven | + | + |
| | + | + |
| Oil Burner | − | − |
| | + | + |
| | + | + |
| Coal Power Plant | − | − |
| | − | − |
| | + | − |
| | − | − |
| | − | − |
| | − | − |
| | + | − |

[a]Data of Smith (12).

TABLE 5. Field/Laboratory Comparison of PAH from Wood Stoves

| Sample | Field | Laboratory |
|--------|-------|------------|
| A | 40-430 | 40-430 |
| B | 44,000-440,000 | 44,000-440,000 |
| C | 20-200 | 20-200 |

[a]Data of Benning (1).

The following example demonstrates the potential of the PAH spot test as a quantitative tool. A coal-tar-shampoo sample was extracted and diluted successively tenfold until sensitized fluorescence was no longer detected with a 1-$\mu$l sample - a total of $10^6$ dilution. From an initial sample weight of 600 mg and the 1 to 10 pg detection limit, the sample was estimated to contain between 0.2 and 2 percent PAH - the concentration on the bottle label was 1 percent.

## CONCLUSIONS

The laboratory and field results obtained to date with the sensitized fluorescence spot test show it to be a simple, sensitive, and reliable method for PAH detection. Use of the spot test as a prescreen can eliminate the need for costly GC/MS analysis when no PAH are present. The spot test is also a convenient aid to the field sampler. PAH of higher molecular weight than frequently detected by GC/MS may also be indicated by the spot test.

## ACKNOWLEDGMENTS

This research was supported by EPA Contract No. 68-02-2150, Dr. Larry D. Johnson, Project Officer.

## REFERENCES

1. Benning, R. (1979): Results of PAH screening by fluorescence, Private Communication from Monsanto Research Corporation, June.
2. Facility Report No. 2, EPA Contract No. 68-01-2966, Subcontract No. A82870DNB-L, November 1976.
3. Facility Report No. 5, EPA Contract No. 68-01-2966, Subcontract No. A82870DNB-L, July 1977.
4. Facility Report No. 6, EPA Contract No. 68-01-2966, Subcontract No. A82870DNB-L, June 1977.

5. Guilbault, G. C. (1973): *Practical Fluorescence, Theory, Methods and Techniques,* Marcel Dekker, Inc., New York, N. Y.
6. Hercules, D., Ed. (1966): *Fluorescence and Phosphorescence Analysis, Principles and Applications,* Interscience, New York, N. Y.
7. Lippsett, F. R., and Dekker, A 1951): Can. J. Phys. 30:165.
8. National Academy of Science (1972): *Particulate Polycyclic Organic Matter.*
9. Porro, T. J., Anacreon, R. E., Flandreau, P. S., and Fagerson, I. S. (1973): *J. Assoc. Off. Agr. Chem.* 56:607.
10. Sawicki, E. (1969): *Talanta* 16:1231.
11. Smith, E. M., and Levins, P. L. Sensitized fluorescence for the detection of polycyclic aromatic hydrocarbons. Report No. EPA-600/7-78-182 on EPA Contract No. 68-02-2150.
12. Smith, T. R. (1979). Evaluation of sensitized fluorescence for polynuclear aromatic hydrocarbon detection. Draft report to EPA from TRW, June 1.

# FLUORESCENCE SPECTROSCOPIC PROPERTIES OF CARCINOGENIC AND AIRBORNE POLYNUCLEAR AROMATIC HYDROCARBONS

G. Heinrich and H. Güsten

Kernforschungszentrum Karlsruhe, Institut für Radiochemie,
7500 Karlsruhe, Postfach 3640, Federal Republic of Germany.

## INTRODUCTION

Among the various analytical techniques for the quantitative determination of airborne polynuclear aromatic hydrocarbons (PAH) in the sub-ppb range, fluorescence spectroscopy is still the most frequently used method (17,25,27). For the analysis of PAH in environmental samples such as smoke condensates (23), automobile exhaust gases (7), tars (14), and water (5), the separation and identification are now generally carried out by more sophisticated techniques, in particular, capillary column gas chromatography (1,13) together with mass spectrometry (6), or high-performance liquid chromatography (18) in combination with UV- or fluorescence spectroscopic detection (20). To analyze PAH in airborne particulate matter, however, less raw material is available for separation and identification. Since fluorescence spectroscopy is generally more sensitive, fluorescence techniques require smaller samples and less sampling time (15). Various new fluorescence spectroscopic techniques, either methodically, such as quencho-fluorimetry (25,28) and the method utilizing the Shpol'skii effect (8,9), or electronically, by selective modulation (21), have achieved a high degree of selectivity and hence they allow analysis of poorly resolved PAH compounds. Some PAH isomers, e. g., benzo(a)pyrene, perylene and benzo(e)pyrene, are difficult to separate by the chromatographic techniques mentioned above. These PAH isomers are unfortunately always associated with each other in environmental samples. Since, however, it is well established that, for example, of the two isomeric benzopyrenes only the benzo(a)pyrene shows a carcinogenic activity, the extensive characterization of PAH mixtures is of importance.

This paper reports the basic photophysical data (absolute fluorescence spectrum, fluorescence quantum yield, and decay time) of a sizable number of airborne and carcinogenic PAH. The knowledge of these data is a prerequisite for the optimal analysis of PAH by fluorescence spectroscopy. It is further shown that the differences in oxygen quenching of the fluorescent state of the various PAH can sometimes be used more conveniently to determine PAH mixtures by fluorescence spectroscopy which are difficult to separate by chromatographic techniques. We measured, in particular, those airborne PAH which constitute the main components in atmospheric aerosols and which have been shown to be groups of $C_{18}H_{12}$ and $C_{20}H_{12}$ isomeric PAH. Among the 14 important PAH we report here the basic photophysical data of six highly purified PAH which have been selected as PAH reference materials by the Community Bureau of Reference of the Commission of the European Communities in Brussels.

## EXPERIMENTAL

The absorption and fluorescence spectra, the fluorescence quantum yields and decay times of some important carcinogenic hydrocarbons (e. g., benzo(a)pyrene, benzo(e)pyrene, indeno (1,2,3-cd)pyrene, benzo(b)fluoranthene, benzo(j)fluoranthene, benzo(b)chrysene and cyclopenta (cd) pyrene), and of other airborne polycyclic aromatic hydrocarbons (PAH) always associated with each other in the environment, were measured with and without oxygen in n-heptane at room temperature.

### Compounds

Benzo(b)fluoranthene (BbF), benzo(j)fluoranthene (BjF), benzo(k)-fluoranthene (BkF), benzo(e)pyrene (BeP), benzo(b)chrysene (BbC), and indeno (1,2,3-cd) pyrene (IcdP) were obtained with 99.5 percent purity as PAH reference materials from the Commission of the European Communities (Brussels, Belgium). The other PAH are of commercial origin and were purified by column chromatography on $Al_2O_3$ in benzene, followed by two-dimensional thin layer chromatography. Purification was continued until the fluorescence excitation spectrum in very dilute solution did not change any longer. Chrysene (C), fluoranthene (F) and triphenylene (T) were purified by zone melting (36 zones on a Desaga zone melting apparatus) and final recrystallization from benzene or ethanol. A sample of cyclopenta(cd)pyrene (CcdP) was kindly supplied by Dr. A. Gold (Harvard University, School of Public Health).

## Spectroscopy

The absorption measurements were carried out with a recording Cary-15 spectrophotometer. During each scan the spectral data (wavelength and optical density) were encoded and punched on paper tape.

The fluorescence spectra were measured using a recording fluorimeter built at our laboratory (4). The fluorescence is excited with monochromatic radiation coming either from a xenon high-pressure lamp in combination with a monochromator, or a low-pressure mercury lamp and an interference filter which is only transparent for the intense light at the wavelength of 254 nm. The fluorescence radiation of the PAH in the cell emitted at an angle of 90 degrees with respect to the excitation light path is focussed into the entrance slit of a monochromator. The light divided into its spectral parts impinges on the cathode of a photomultiplier. The resulting electrical signal is amplified and further handled by means of electronic components which enhance the signal-to-noise ratio and allow for the automatic correction of the spectral sensitivity of the total apparatus. The final spectrum as recorded is a correct representation of the relative number of photons emitted per unit time and unit wave number as a function of the wave number. This spectrum can be stored in a multichannel analyzer and finally printed out on paper tape.

The fluorescence spectrum recorded at the short wavelength side is sometimes dependent on the wavelength of the exciting light and on the concentration of the solution. Such a distortion arises from self-absorption of the fluorescent radiation in the solution. It can be minimized by exciting the solution with light of a wavelength coincident with one of the more intense absorptions of the compound to be measured, or by reducing the concentration of the solution.

The fluorescence quantum yields were evaluated according to the method of Parker and Rees (22) from the areas under the corrected emission spectra with respect to the area of the absolute fluorescence spectra of quinine bisulfate in 1 N $H_2SO_4$, taking into account the different refractive indices of the solvents. The quantum yield was 0.55 at 298° K. The concentration of a sample was carefully adjusted to give the same absorbance as quinine bisulfate at the exciting wavelength of 254 nm and the temperature chosen. Care was taken to avoid errors in the calculation of the quantum yields due to self-absorption effects. The area under the spectrum was then measured using a very dilute solution ($c < 10^{-6} M$) and recalculated for the area of the more concentrated solution by comparison of both intensities at a longer wavelength where re-absorption does not exist.

The oxygen quenching constant $L_q$ was measured by comparison of the fluorescence intensity of a solution in the absence of air, with the fluorescence intensity of the same solution saturated with air.

The device for measuring the decay time of fluorescence was built at our laboratory (4). Pulsed ultraviolet radiation from a triggered hydrogen lamp is absorbed by the fluorescent sample after having passed a broad-band filter which cuts off the long wavelength range of the light. The fluorescence pulses emitted are detected with a fast and high-gain photomultiplier whose resulting electrical pulses are collected by a sampling oscilloscope. The information about the pulse shape is stored in the memory of a multichannel analyzer so that each sample of the pulse is registered in its assigned channel of the 400-channel analyzer. Data from usually up to 400 traces are summed up, stored in the memory, and finally punched on paper tape. Knowing the pulse shape of the exciting pulse, it is possible to determine the fluorescent decay time by means of a computer program according to the phase-plane method (10). By this technique we are able to measure fluorescence lifetimes down to one nanosecond.

All measurements were performed in a quartz cell with a light path of 1 cm. Prior to use in fluorescence measurements, the solution in the cell was degassed to remove oxygen which quenches the fluorescence. Degassing was performed by at least five freeze-evacuation-thaw cycles to a pressure of less than $10^{-7}$ bar.

## RESULTS

The absorption and corrected fluorescence spectra of the benzofluoran-thenes, benzopyrenes, benz(a)anthracene (BaA), BbC, IcdP and CcdP are shown in Figures 1-9. Since n-heptane is both inert and transparent to ultraviolet radiation, it is used as the standard solvent. All the other spectra measured have been published elsewhere (2).

The absorption and fluorescence curves of a compound are presented in the same figure. The abscissa of each figure is the wave number. The reason is that the energy of light is directly proportional to the wave number and, therefore, the energy absorbed or emitted is also proportional to the difference between two wave numbers. A wavelength scale is shown at the top of each graph only for orientation purposes. The absorption spectrum at the left-hand side represents the extinction coefficient, and the fluorescence spectrum at the right side represents the relative photon flux per unit wave number versus the wave number. Other photophysical data belonging to the fluorescence are added to the figures: the fluorescence quantum yield in the absence of air, $Q_o$, the fluorescence lifetime without air, $\tau_o$, and the oxygen quenching constant, $L_q$. $L_q = Q_o/Q$ is the quotient of the fluorescence quantum yield without and with oxygen. The exciting wavelength for the measurements of the quantum efficiencies is 254 nm in each case.

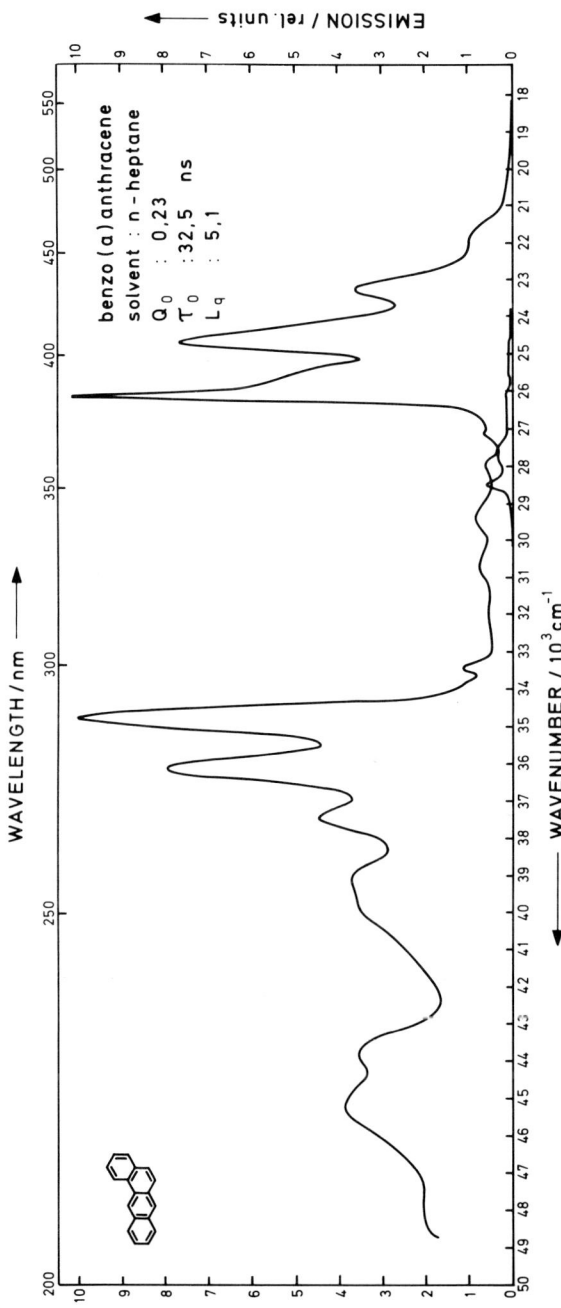

FIGURE 1. Absorption and fluorescence spectra of benzo(a)anthracene in n-heptane at room temperature.

## 988  FLUORESCENCE SPECTROSCOPY OF PAH

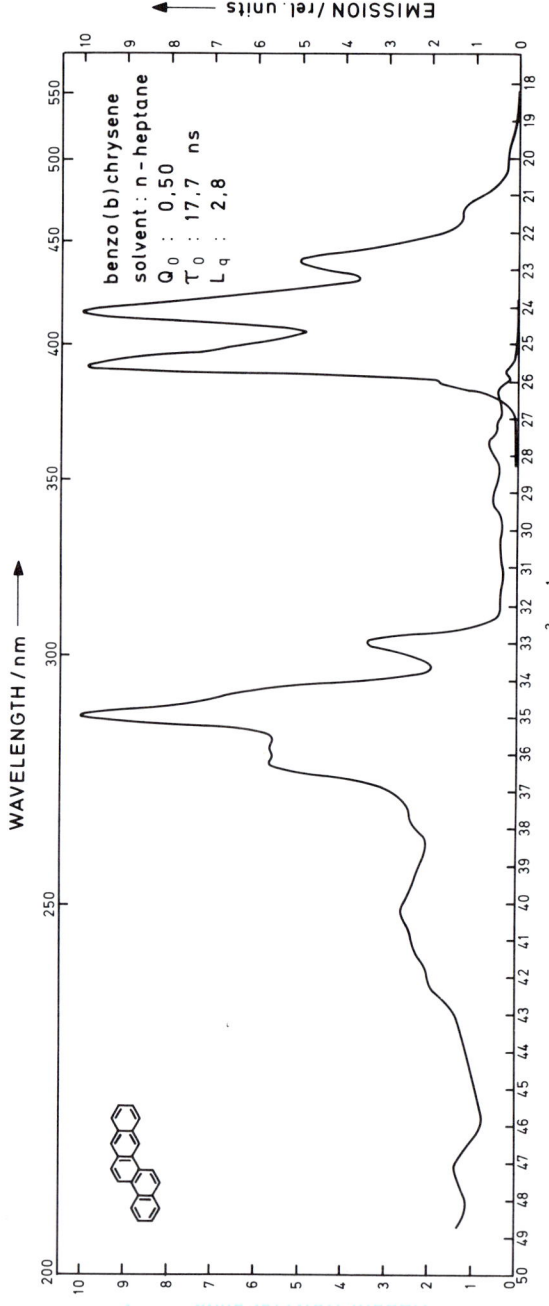

FIGURE 2. Absorption and fluorescence spectra of benzo(b)chrysene in n-heptane at room temperature.

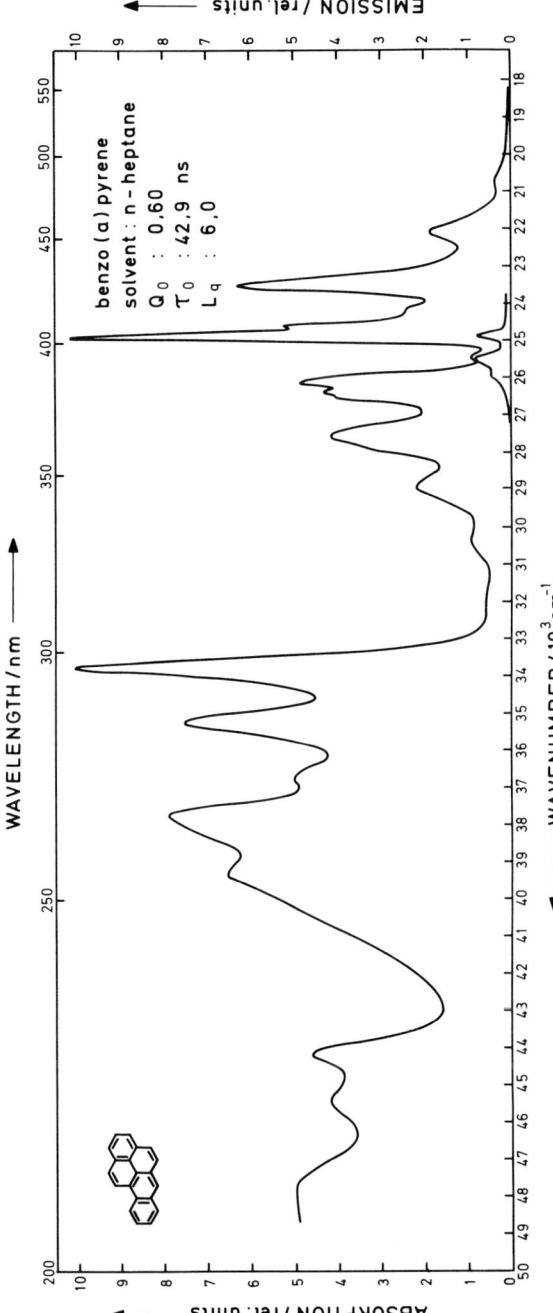

FIGURE 3. Absorption and fluorescence spectra of benzo(a)pyrene in n-heptane at room temperature.

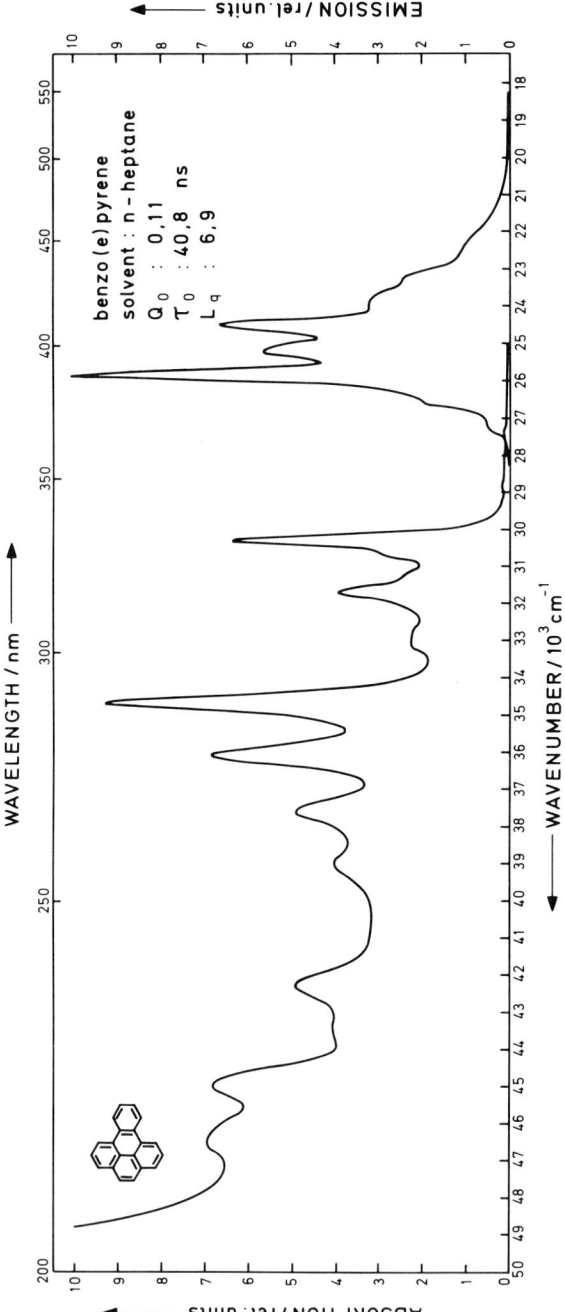

FIGURE 4. Absorption and fluorescence spectra of benzo(e)pyrene in n-heptane at room temperature.

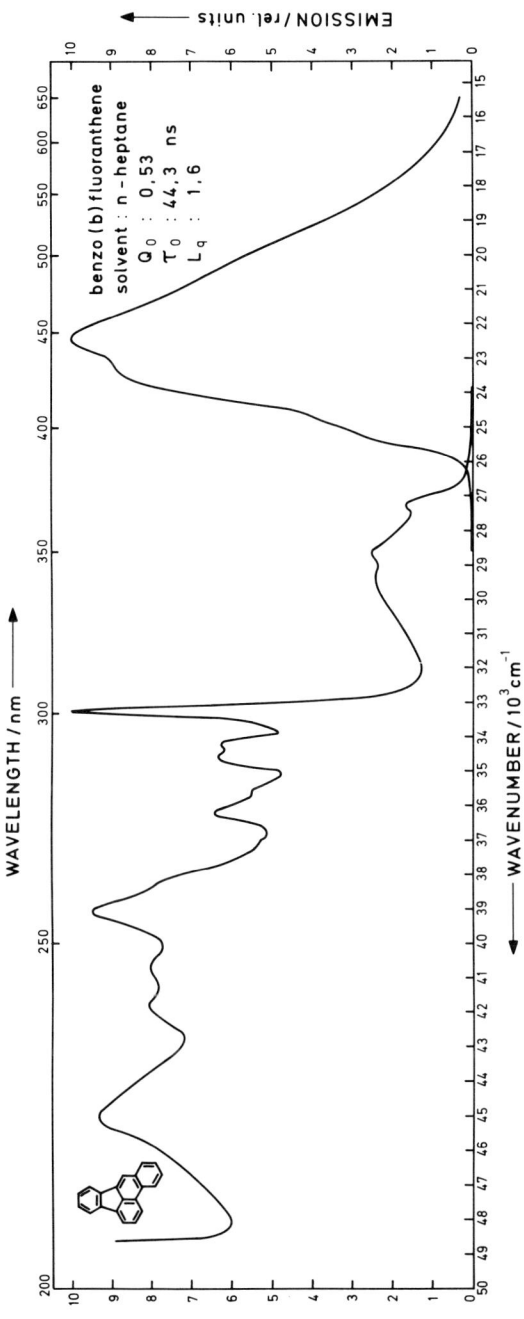

FIGURE 5. Absorption and fluorescence spectra of benzo(b)fluoranthene in n-heptane at room temperature.

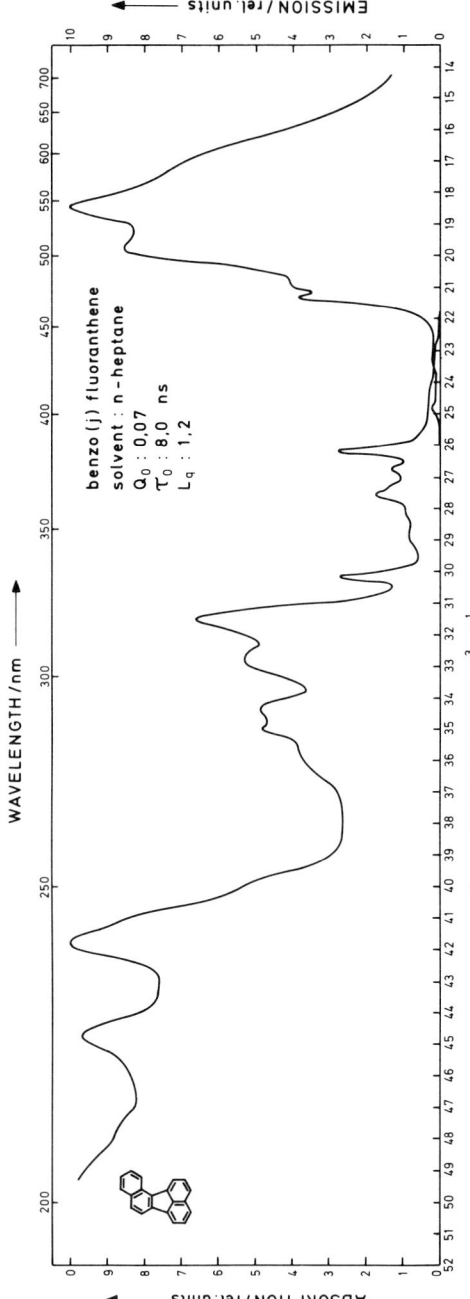

FIGURE 6. Absorption and fluorescence spectra of benzo(j)fluoranthene in n-heptane at room temperature.

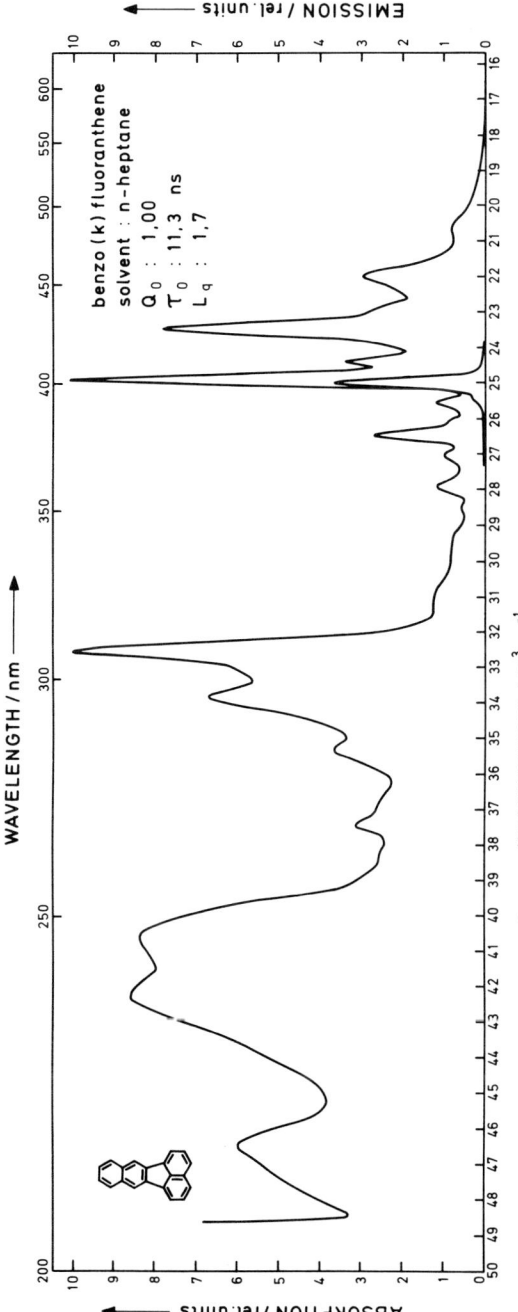

FIGURE 7. Absorption and fluorescence spectra of benzo(k)fluoranthene in n-heptane at room temperature.

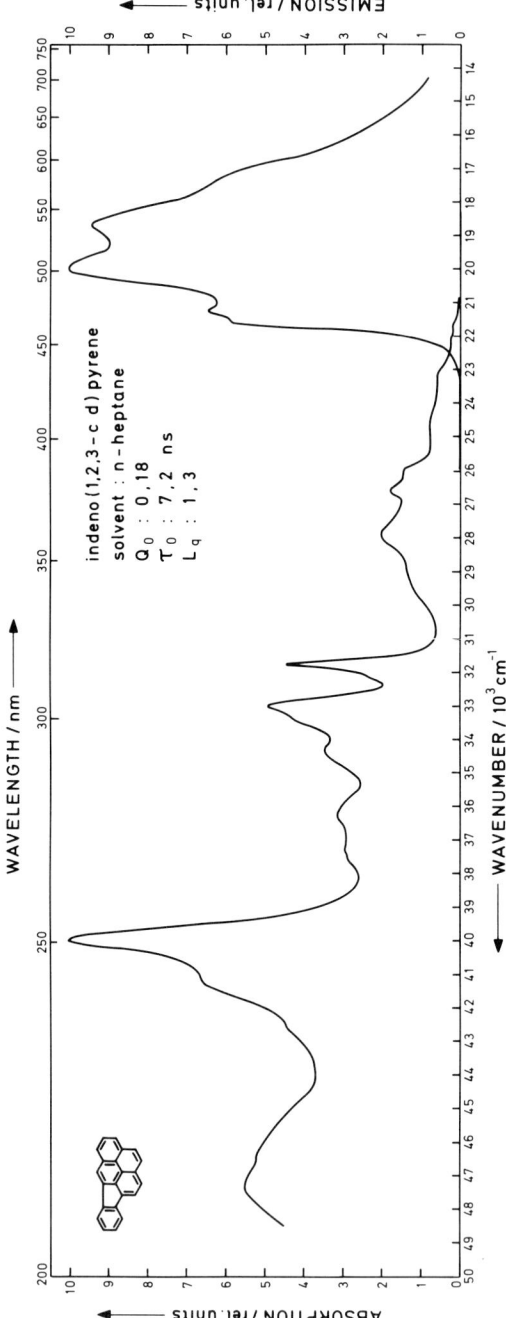

FIGURE 8. Absorption and fluorescence spectra of indeno(1,2,3-cd)pyrene in n-heptane at room temperature.

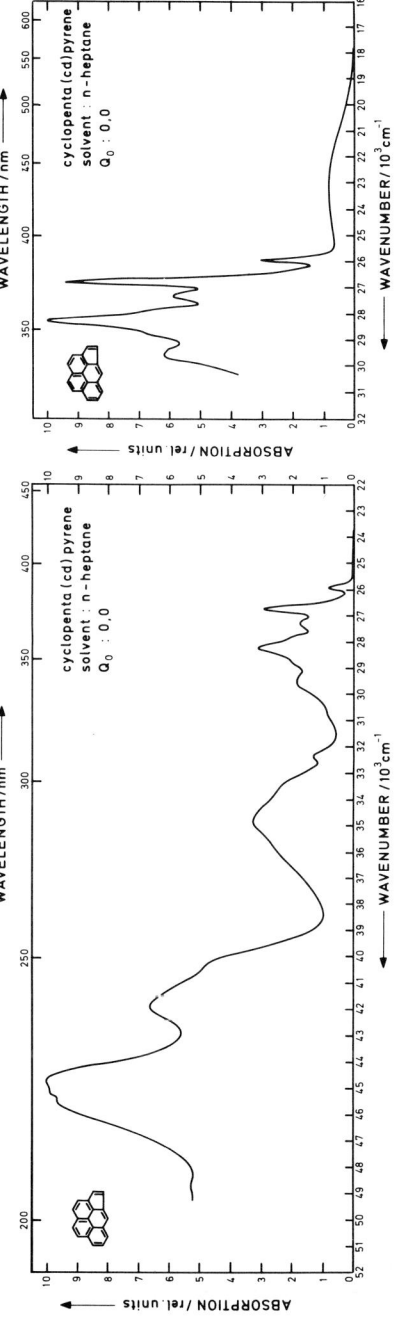

FIGURE 9. Absorption spectrum of cyclopenta(cd)pyrene in dilute (a) and in concentrated (b) solution of n-heptane at room temperature.

All the compounds measured show vibrational structures in both the absorption and the fluorescence spectra. Spectra with well-resolved vibrational bands refer to planar molecules with a high degree of symmetry. Usually, the absorption and emission spectra are characterized by mirror symmetry or mirror similarity. This can be very well seen in Figure 7 for BkF.

The spectral shift of the fluorescence spectrum with respect to the absorption spectrum of a molecule (Stokes shift) can also be correlated with the degree of planarity. All compounds whose spectra are shown in Figures 2-4 and 6-8 have no or only little anti-Stokes shift with the exception of BbF, showing a large Stokes shift, and BaA. In BaA there is a relatively strong anti-Stokes shift (Figure 1), indicating that the emission of fluorescence originates from either vibrationally excited states of $S_1$ or from the second excited singlet state $S_2$ (3) (12).

For all compounds measured, the values of $Q_o$, $\tau_o$, and $L_q$ are presented together with the values of the radiative and nonradiative rate constants $k_F$ and $k_N$ in Table 1. The radiative rate constant $k_F$ is evaluated by means of the equation

$$k_F = Q_o/\tau_o, \qquad [1]$$

and the nonradiative rate constant $k_N$, which represents the radiationless deactivation paths for both internal conversion to the singlet ground state and intersystem crossing to the triplet manifold, is calculated with the equation

$$k_N = 1/\tau_o - k_F. \qquad [2]$$

The quantum yields, the lifetimes, the radiative and nonradiative rate constants, and the oxygen quenching constants of the compounds with four rings, BaA, C, and T, are very similar. This finding is neither true for the compounds with five rings, BaP, BeP, and BbC; the six-ring systems, anthanthrene (A) and benzo (ghi) perylene (BP); nor for the nonalternant hydrocarbons, such as compounds with even and odd numbers of carbon atoms in the ring systems, e. g., fluoranthene (F), BbF, BjF, BkF and IcdP. An extreme difference can be seen between A and BP. The quantum yield 1.0 of A is higher by a factor of more than 3, whereas the lifetime is about 10 times shorter than in BP. BP has the largest oxygen quenching constant of all compounds measured. In BeP the radiationless path of deactivation is favoured more than the emission of fluorescence, just in contrast with BaP. The fluorescent properties of F change with the addition of a further aromatic ring in the benzofluoranthenes. This change is less obvious in BbF but strong in BjF and BkF. In BjF the fluorescence emission is reduced drastically and the radiationless deactivation is increased. In BkF, however, there is no evidence for any nonradiative transition; all the photons absorbed by the molecule are reemitted as fluorescent light.

The effect of oxygen quenching on the first excited singlet state of a compound leads to a reduction in its fluorescence yield. All nonalternant hydrocarbons are relatively immune to oxygen quenching. The highest value is 1.7 for BkF. The alternant hydrocarbons with only even numbers of carbon atoms in the rings show much higher $L_q$ values, except for A. The radiative process which alone takes place in this molecule is so fast that quenching with oxygen cannot compete with it.

The sensitivity of a molecule to oxygen quenching is related to its fluorescence lifetime. When oxygen is the only quenching molecule, it can be shown that the quenching constant $L_q$ is related to the lifetime $\tau_o$ by the Stern-Volmer equation

$$L_q = 1 + \tau_o k_q (O_2), \qquad [3]$$

where $k_q$ is the rate constant for quenching by oxygen, and $(O_2)$ is the concentration of dissolved oxygen in the solvent. Since all our measurements of $L_q$ were made with n-heptane as the solvent, the concentration of oxygen and the value of $k_q$ were the same for each solution. Figure 10 is a plot of $L_q$ versus $\tau_o$. The relationship between $L_q$

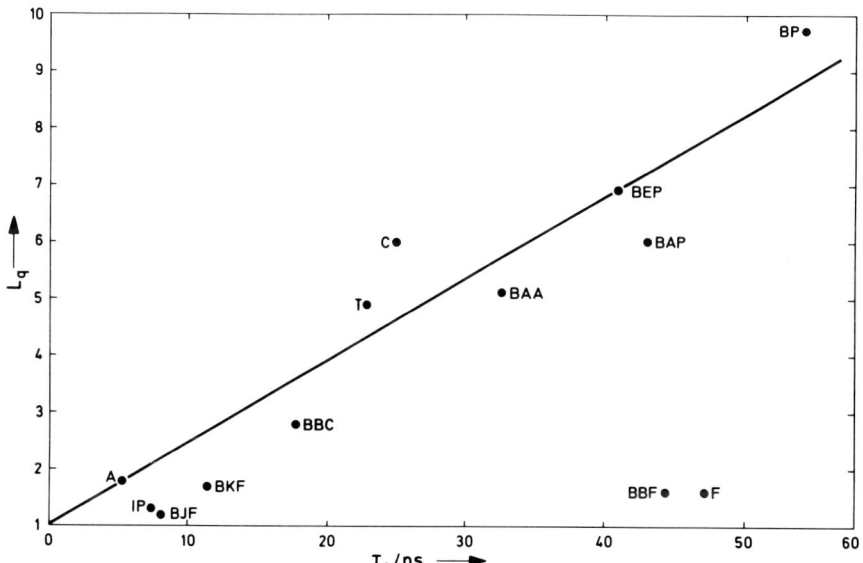

FIGURE 10. Relationship between oxygen quenching constant $L_q$ and fluorescence lifetime $\tau_o$; BaA = benz(a)anthracene, C = chrysene, T = triphenylene, BaP = benzo(a)pyrene, BeP = benzo(e)pyrene, BbC = benzo(b)chrysene, A = anthanthrene, BP = benzo(ghi)perylene, F = fluoranthene, BbF = benzo(b)fluoranthene, BjF = benzo(j)fluoranthene, BkF = benzo(k)fluoranthene, IP = indeno(1,2,3-cd)pyrene.

and $\tau_o$ is quite good. Exceptions are found for F and BbF. Although having relatively long decay times, they are not very sensitive to oxygen quenching. This effect may be explained by the assumption that a conformational change takes place on excitation and the basic chromophore is partially shielded from a direct encounter with an oxygen molecule. The evidence supporting this possibility is that the fluorescence spectra of these two compounds are relatively broad with little structure, and that the Stokes shifts are large. In the other nonalternant hydrocarbons like BjF, BkF, and IcdP the radiative or the nonradiative processes are so fast that oxygen quenching is not very effective.

CcdP is a ubiquitous environmental carcinogen (24) that does not emit any fluorescence. Fluorescence excitation at 440 nm with a xenon high-pressure lamp gave a very weak fluorescence emission ($Q_o < 10^{-3}$) with a maximum about 650 nm.

## DISCUSSION

Fluorescence spectroscopy is a relatively simple, specific, and very sensitive method for the detection and quantitative determination of PAH in our environment. The knowledge of quantitative photophysical data, such as absorption and fluorescence spectra, fluorescence quantum yield, and oxygen quenching constant, is therefore of great value in the analysis of PAH which are isolated from airborne samples, tobacco smoke, tar, etc., especially if some of these PAH which are similar in chemical structure, cannot be completely separated by chromatographic methods.

The analysis by Lao et al (19) of the composition of airborne PAH adsorbed on aerosols and dusts shows that about 60 percent of the PAH emitted to the atmosphere by incomplete combustion of fossil fuels remains limited to only few, but relatively strong carcinogenic compounds. These most frequently emitted PAH are the $C_{18}H_{12}$-isomers BaA, C, and T; the $C_{20}H_{12}$-isomers BkF, BbF, BaP, BeP, and perylene; and the $C_{22}H_{12}$- isomers A and BP. All these groups of isomeric PAH are difficult to separate chromatographically (10,15,16).

The direct analysis of a PAH in a mixture is possible if its fluorescence spectrum is well separated from the corresponding emission spectra of the other components. If the emission spectra overlap, the inspection of the absorption spectrum will indicate the optimum wavelength for excitation of the particular PAH in comparison with the absorption spectra of the other PAH. By excitation at appropriate absorption maxima it is thus frequently possible to obtain reasonably pure emission spectra although the other components may be excited to some extent (16). In a mixture, only the emission of the PAH absorbing at the longest wavelength can be obtained undistorted by the others. Thus it is possible to detect perylene in a mixture by its fluorescence spectrum through excitation at a wavelength above 400 nm, while the emission of other components is suppressed. For example, Sawicki et al (26) have measured the emission spectrum of perylene excited at 430 nm in the presence of 25 other aromatic hydrocarbons. By careful selection of the excitation wavelength, the spectrum of a PAH absorbing only at wavelengths shorter than the other PAH in the mixture can also be obtained nearly undistorted. An example is the mixture of BaP, BeP and perylene, which is difficult to separate chromatographically. The absorption of perylene at a wavelength near 290 nm is low relative to those of the benzopyrenes. On the other side, the quantum efficiency of BeP is low compared with perylene and BaP (see Table 1). By exciting at 290 nm an almost pure emission spectrum of the carcinogenic BaP is obtained in the presence of equal concentrations of perylene and BeP. It is, however, very difficult or impossible to measure the noncarcinogenic BeP in such a mixture.

**TABLE 1.** Fluorescence Spectroscopic Data of Some Polynuclear Aromatic Hydrocarbons in n-Heptane at Room Temperature

| PAH | $Q_o$ | $\tau_o$ (NS) | $K_F$ ($10^6$ sec$^{-1}$) | $K_n$ ($10^6$ sec$^{-1}$) | $L_Q$ |
|---|---|---|---|---|---|
| Benzo(a)anthracene | 0.23 | 32.5 | 7.1 | 23.7 | 5.1 |
| Chrysene | 0.18 | 24.9 | 7.2 | 32.9 | 6.0 |
| Triphenylene | 0.09 | 22.8 | 3.9 | 39.9 | 4.9 |
| Benzo(a)pyrene | 0.60 | 42.9 | 14.0 | 9.3 | 6.0 |
| Benzo(e)pyrene | 0.11 | 40.8 | 2.7 | 21.8 | 6.9 |
| Benzo(b)chrysene | 0.50 | 17.7 | 28.2 | 28.2 | 2.8 |
| Anthanthrene | 1.00 | 5.2 | 192.3 | 0.0 | 1.8 |
| Benzo(ghi)perylene | 0.29 | 54.3 | 5.3 | 13.1 | 9.7 |
| Fluoranthene | 0.35 | 47.0 | 7.4 | 14.1 | 1.6 |
| Benzo(b)fluoranthene | 0.53 | 44.3 | 12.0 | 10.6 | 1.6 |
| Benzo(j)fluoranthene | 0.07 | 8.0 | 8.8 | 116.3 | 1.2 |
| Benzo(k)fluoranthene | 1.00 | 11.3 | 88.5 | 0.0 | 1.7 |
| Indeno(1,2,3-cd)pyrene | 0.18 | 7.2 | 25.0 | 113.9 | 1.3 |

Note: The excitation wavelength for the determination of the quantum yields was 254 nm.

Less successful is the measurement of the carcinogenic compounds BaA and C, which still occur together with T after chromatographic separation, and which can reach nearly 20 percent of all fractions of PAH in an airborne sample (19). All of them have similar absorption and fluorescence properties (2). The quantitative analysis of one of the two carcinogenic compounds is only possible in a mixture if the concentration of the component to be measured is much higher than those of the other components. Because of the high quantum yield and the well-structured emission spectrum, A and BkF can be determined easily in mixtures with other PAH (16). There is no difficulty in measuring A in the presence of BP even if the concentration of BP is much higher than that of A. On the other side, the analysis of BP is not possible in the presence of A at similar concentrations.

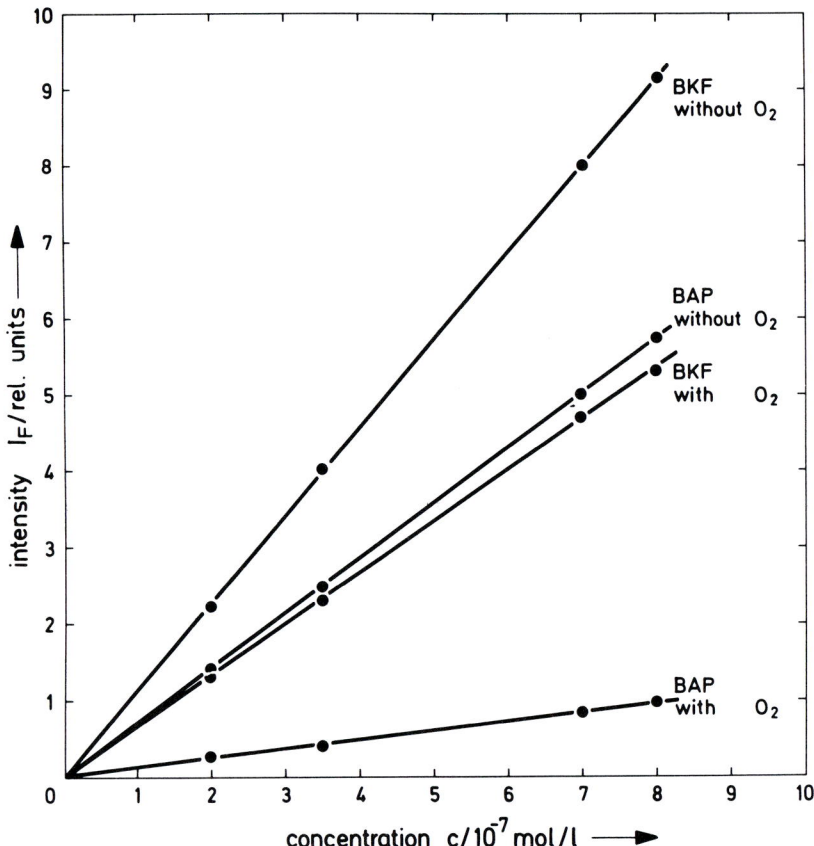

FIGURE 11. Concentration dependency of fluorescence intensity of benzo(a)-pyrene (BaP) are benzo(k)fluoranthene(BkF) with and without oxygen in n-heptane at room temperature.

It was suspected that the noncarcinogenic BkF, which exhibits a nearly identical fluorescence spectrum as BaP (see Figure 3 and 7), was measured in the early investigations on airborne PAH erroneously as the carcinogenic BaP (11,15). In order to obtain the concentration of carcinogenic BaP in such a case it is recommended to increase the selectivity by measuring in the absence of dissolved air, because in a degassed solution the fluorescence quantum yield of BaP is more enhanced than that of BkF. This effect can be seen when comparing the calibration curves of the fluorescence intensity of BaP and BkF with and without oxygen in the solution (Figure 11). The fluorescence of BaP increases by a factor of 6, whereas that of BkF increases only by a factor of 1.7. The measurement with a deaerated solution minimizes the strong interference between the two components in the mixture. Likewise, to some extent the oxygen effect can be utilized in the determination of BP with a $L_q$ value of 9.7 in a mixture with anthanthrene with a low $L_q$ value of 1.8. The described oxygen quenching method may be used for the quantitative determination of other mixtures of PAH isomers in Table 1.

## REFERENCES

1. Beernaert, H. (1979): Gas chromatographic analysis of polycyclic aromatic hydrocarbons. *J. Chromatogr.* 173: 109-118.
2. Berlman, I.B. (1971): *Handbook of Fluorescence Spectra of Aromatic Molecules,* 2nd Ed., Academic Press, New York and London.
3. Birks, J.B., Easterly, C.E., and Christophorou, L.G. (1971): Stokes and anti-Stokes fluorescence of 1,12-benzoperylene in solution. *J. Chem. Phys.* 66: 4231-4236.
4. Blume, H., and Güsten, H. (1977): Spektroskopische Verfahren. In: *Ultraviolette Strahlen,* J. Kiefer, Ed., pp. 311-348, Walter de Gruyter Verlag, Berlin.
5. Borneff, J., and Kunte, H. (1969): Kanzerogene Substanzen in Wasser und Boden. XXVI. Routinemethode zur Bestimmung von polyzyklischen Aromaten im Wasser. *Arch. Hyg.* 153, 220-29.
6. Borwitzky, H., and Schomburg, G. (1979): Separation and identification of polynuclear aromatic compounds in coal tar by using glass capillary chromatography including combined gas chromatography-mass spectrometry. *J. Chromatogr.* 170: 99-124.
7. Candeli, A., Mastrandrea, V., Morozzi, G., and Toccaceli, S. (1974): Carcinogenic air pollutants in the exhaust from a European car operating on various fuels. *Atmos. Environ.* 8: 693-705.
8. Causey, B.S., Kirkbright, G.F., and de Lima, C.G. (1976): Detection and determination of polynuclear aromatic hydrocarbons by luminescence spectrometry utilizing the Shpol'skii effect at 77 K. *Analyst* 101: 367-78
9. Colmsjö, A., and Sternberg, U. (1979): Identification of polynuclear aromatic hydrocarbons by Shpol'skii low temperature fluorescence. *Anal. Chem.* 51: 145-150.

10. Demas, J. N., and Adamson, A. W. (1971): Evaluation of photoluminescence lifetimes. *J. Phys. Chem* 75: 2463-66.
11. Dubois, L., Zdrojewski, A., and Monkman, J. L. (1972): Eine verbesserte Methode zur Bestimmung von Benzo(a)pyren in Luftproben. *Staub-Reinh. Luft* 32: 487-490.
12. Easterly, C. E., and Christophorou, L. G. (1974): Fluorescence emission from the first- and the second-excited singlet states of aromatic hydrocarbons in solution, and their temperature dependences. *J. Chem. Soc., Faraday Frans.* II, 70: 267-273.
13. Giger, W., and Schaffner, C. (1978): Determination of polycyclic aromatic hydrocarbons in the environment by glass capillary gas chromatography. *Anal. Chem.* 50: 243-49.
14. Grant, D. W., and Meiris, R. B. (1977): Application of thin-layer and high-performance liquid chromatography to the separation of polynuclear aromatic hydrocarbons in bituminous materials. *J. Chromatogr.* 142: 339-51.
15. Güsten, H., and Heinrich, G. (1979): Polycyclic aromatic hydrocarbons in the lower atmosphere of Karlsruhe. In :*Polynuclear aromatic hydrocarbons,* P. W. Jones and P. Leber, Eds., pp. 357-370, Ann Arbor Science Publishers, Inc., Ann Arbor.
16. Heinrich, G., and Güsten, H. (1976): Fluoreszenzspektroskopische Bestimmung in der Atmosphäre vorkommender polycyclischer aromatischer Kohlenwasserstoffe. *Z. Anal. Chem.* 278: 257-262.
17. Heinrich, G., and Güsten, H. (1978): Belastung der Atmosphäre durch polycyclische aromatische Kohlenwasserstoffe und Blei im Raume Karlsruhe. *Staub-Reinh. Luft* 38: 94-100.
18. Lankmayr, E. P., and Müller, K. (1979): Polycyclic aromatic hydrocarbons in the environment: high-performance liquid chromatography using chemically modified columns. *J. Chromatogr.* 170: 139-146.
19. Lao, R. C., Thomas, R. S., and Dubois, L. (1973): Application of a gas chromatograph-mass spectrometer-data processor combination to the analysis of the polycyclic aromatic hydrocarbon content of airborne pollutants. *Anal. Chem.* 45: 908-915.
20. Nielsen, T. (1979): Determination of polynuclear aromatic hydrocarbons in automobile exhaust by means of high-performance liquid chromatography with fluorescence detection. *J. Chromatogr.* 170: 147-56.
21. O'Haver, T. C., and Parks, W. M. (1974): Selective modulation: a new instrumental approach to the fluorimetric analysis of mixtures without separation. *Anal. Chem.* 46: 1886-1894.
22. Parker, C. T., and Rees, W. T. (1960): Correction of fluorescence spectra and measurement of fluorescence quantum efficiency. *Analyst* 85: 587-600.

23. Radecki, A., Lamparczyk, H., Grybowski, J., and Halkiewicz, J. (1978): Separation of polycyclic aromatic hydrocarbons and determination of benzo(a)pyrene in liquid smoke preparations, *J. Chromatogr. 150: 527-32.*
24. Ruehle, P. H., Fischer, D. L., and Wiley, J. C. (1979): Synthesis of cyclopenta(cd)pyrene, a ubiquitous environmental carcinogen, *J. Chem. Soc., Chem. Comm.* 302-303.
25. Sawicki, E. (1969): Fluorescence analysis in air pollution research. *Talanta* 16: 1231-1266.
26. Sawicki, E., Hauser, T. R., and Stanley, T. W. (1960): Ultraviolet, visible and fluorescence spectral analysis of polynuclear hydrocarbons. *Int. J. Air Poll.* 2: 253-272.
27. Sawicki, E., Stanley, T. W., Elbert, W. C., Meeker, J., and McPherson, S. (1967): Comparison of methods for the determination of benzo(a)pyrene in particulates from urban and other atmospheres. *Atmos. Environ.* 1: 131-145.
28. Zander, M. (1973): Zur Anwendung des äu βeren Schweratom-Effektes in der Spektrofluorimetrie. *Z. Anal. Chem* 263: 19-23.

# RECENT DEVELOPMENTS IN MATRIX ISOLATION SPECTROSCOPIC ANALYSIS OF POLYNUCLEAR AROMATIC HYDROCARBONS

E. L. Wehry, G. Mamantov, D. M. Hembree, and J. R. Maple

Department of Chemistry
University of Tennessee
Knoxville, Tennessee 37916

## INTRODUCTION

The identification and quantitative determination of individual polycyclic aromatic hydrocarbons (PAH) in complex samples, such as "synthetic liquid fuels", poses an extraordinarily challenging problem to the analytical chemist. In this laboratory, extensive effort has been devoted toward development of matrix isolation spectroscopy (27) as an analytical procedure applicable to qualitative and quantitative analyses in very complex samples. We have thus far concentrated on the use of matrix isolation Fourier transform infrared spectroscopy (14, 15, 23) and molecular fluorescence spectroscopy (6, 15, 21, 23, 26) for identification and quantitation of PAH.

In matrix isolation (MI) spectroscopy, samples are vaporized and then mixed thoroughly with a large excess of a diluent ("matrix") gas; the resulting gaseous mixture then is deposited as a solid at cryogenic temperatures for subsequent spectroscopic measurements. The purpose of the MI operation is implicit in the name of the technique: one wishes to isolate solute molecules from each other in the solid matrix, such that the spectroscopic behavior of any one compound is not influenced by the identities and amounts of other substances present in a complex sample. The use of a cryogenic solvent as the spectroscopic "solvent" offers the very important additional advantage of high spectral resolution, providing spectra which can serve as "fingerprints" for identification of specific sample constituents and minimizing the extent to which the spectra of different compounds present in a complex sample overlap with each other. Thus, the MI sampling method offers both "qualitative" and "quantitative" analytical advantages.

At previous Battelle symposia, we have discussed the fundamentals and experimental procedures used in MI (28, 29); in addition, we have recently given a general review of analytical MI spectroscopy elsewhere (27). In the present report, we describe three recent innovations which promise to enhance significantly the analytical capabilities of MI spectroscopy.

## SAMPLE PREPARATION BY FRACTIONAL SUBLIMATION

In any MI spectroscopic analysis of PAH, the sample must first be vaporized; this operation usually is achieved by straightforward vacuum sublimation (27, 28). It seems evident that this vacuum sublimation step also should be capable of serving as a fractionation procedure; indeed, the analytical utility of microsublimation methods in PAH analysis was demonstrated by Monkman et al (17) nearly 10 years ago. We have recently examined the use of fractional sublimation as an ancillary "separation" procedure in the analysis of PAH mixtures by MI Fourier transform infrared (FTIR) spectroscopy.

The fundamental factor underlying use of fractional sublimation as a separation method is that different PAH exhibit different volatilities, with volatility tending to decrease with increasing number of aromatic rings. Thus, if a sample containing a number of PAH is placed into a conventional Knudsen cell (27, 28) for MI sample preparation, and the temperature of that cell is increased slowly, the composition of the solid spectroscopic sample will change as a function of time. A rapid spectroscopic technique, such as FTIR, can be used to detect the changing composition of the deposited solid sample. A simple example of the use of this technique is shown in Figure 1, wherein the MI FTIR spectrum of a synthetic five-component PAH mixture is shown as a function of sublimation temperature. Only very volatile compounds sublime at an appreciable rate at low Knudsen cell temperatures; thus, only phenanthrene (PH), fluoranthene (F), and pyrene (P) are observed in the fraction which sublimes between 25 and 75°C. Use of a second surface for MI sample deposition, coupled with a higher sublimation temperature range (75 to 200°C) produces a MI FTIR spectrum containing absorption bands due only to the less volatile compounds perylene (PE) and chrysene (C). This very simple example demonstrates how sample fractionation can be achieved as an integral part of sample preparation by MI techniques.

An application of fractional sublimation to the MI FTIR spectrometry of a complex real sample is depicted in Figure 2. This sample, an adsorption chromatography fraction from a wastewater stream in a coking plant, has been demonstrated using gas and liquid chroma-

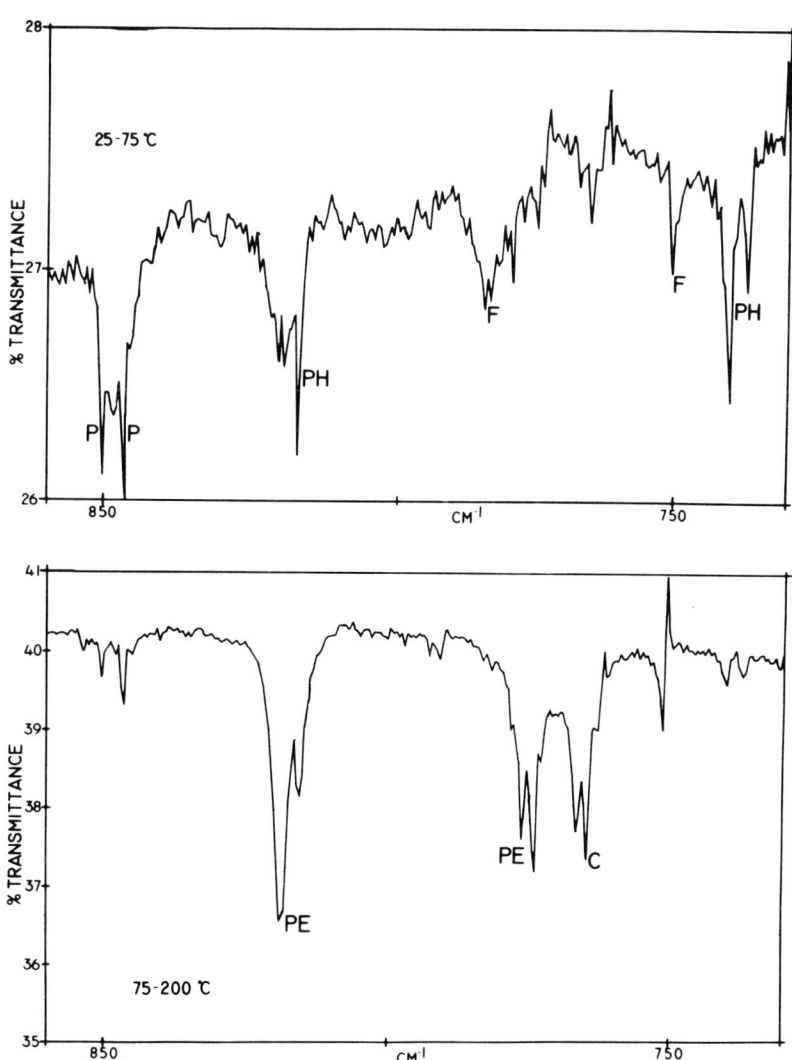

**FIGURE 1.** Fractional sublimation MI FTIR spectra of five-component synthetic PAH mixture. Top: low-temperature region (25-75 C); compounds: *P*, pyrene; *PH*, phenanthrene; *F*, fluoranthene. Bottom: high-temperature region (75-200 C); compounds: *PE*, perylene; *C*, chrysene.

tography and molecular fluorescence spectrometry to contain at least 25 aromatic compounds. Spectra of this sample obtained over two ranges of sublimation temperature are shown in Figure 2. While the comparison is naturally not as spectacular as that for the synthetic mixture shown in

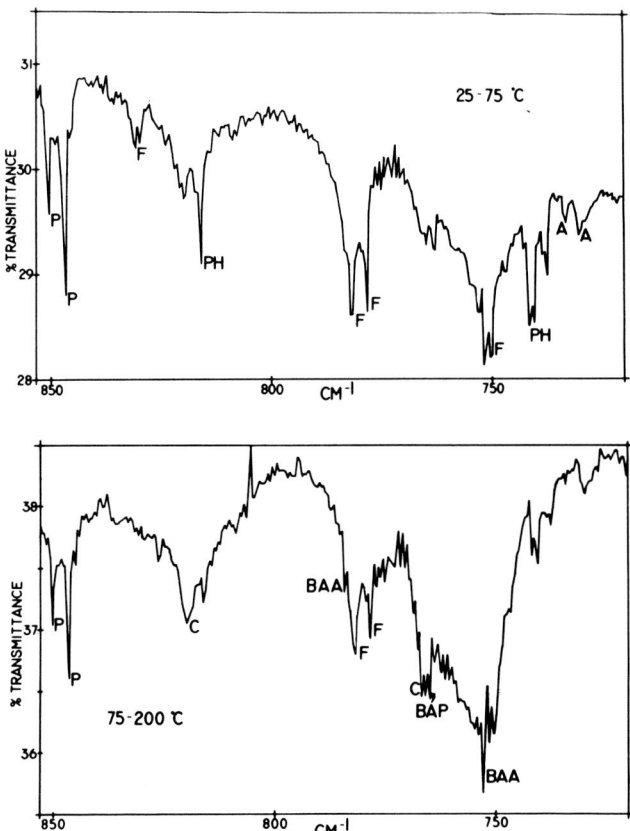

**FIGURE 2.** Fractional sublimation MI FTIR spectra of adsorption chromatography fraction from coking plant wastewater stream. Top: low-temperature region (25-75 C); compounds: *P*, pyrene; *PH*, phenanthrene; *F*, fluoranthene; *A*, anthracene. Bottom: high-temperature region (75-200 C); compounds: *P*, pyrene; *C*, chrysene; *BaA*, benz(a)anthracene; *F*, fluoranthene; *BaP*, benzo(a)pyrene.

Figure 1, it demonstrates clearly that even samples which already have been subjected to chromatographic cleanup can benefit from fractional sublimation in the course of sample preparation in MI spectroscopy. Hence, the vaporization step in MI sampling, which might initially be viewed as something of a nuisance, actually can have significant value in facilitating spectroscopic characterization of complex samples. Obviously, such a simple separation procedure can only augment, rather than supplant, more elaborate chromotographic fractionation methods in the analysis of complex PAH mixtures. Nevertheless, the fact that the sample preparation step also can serve as an additional stage of fractionation is a very attractive, albeit heretofore ignored, feature of MI spectroscopic analysis. Further applications of fractional sublimation in the MI FTIR spectroscopic characterization of complex PAH mixtures will be described in detail elsewhere.

## USE OF UNCONVENTIONAL MATRICES AND LASER EXCITATION IN MI FLUORESCENCE SPECTROMETRY

MI as a sampling technique for the fluorometric analysis of PAH in mixtures has two very important advantages over fluorometry in liquid solution. First, highly structured fluorescence spectra are obtained in cryogenic solid media, which can be used for identification of specific PAH in complex mixtures (15). Second, excellent quantitative precision, wide quantitative linear dynamic range, and freedom from quenching or inner-filter effects all are characteristics of MI fluorescence spectrometry (21). Together with the high sensitivity characteristic of fluorescence spectrometry, these facts all signify that MI fluorescence spectrometry can be employed for reliable quantitative analyses in mixtures at subnanogram levels of individual PAH.

The matrices usually employed (27) in MI spectroscopy are "inert" gases at room temperature (e.g., $N_2$, Ar, Xe); MI fluorescence spectra in such matrices, though much more highly resolved than those obtained in liquid solution (21, 27), tend to exhibit greater spectral bandwidths than fluorescence spectra of PAH obtained in frozen organic solvents (21). The variant of frozen-solution fluorescence spectrometry known as the "Shpol'skii effect" (20), wherein a PAH is dissolved in a liquid solvent (usually an aliphatic hydrocarbon) of comparable molecular dimensions to those of the PAH, with the solution then being frozen at 77 K or a lower temperature for spectroscopic examination as a solid, has been shown to produce fluorescence spectra of extremely high resolution; for this reason, the Shpol'skii frozen solution fluorescence technique recently has received considerable study in terms of its analytical potentialities (3, 7, 13).

The Shpol'skii frozen-solution method indeed is extremely useful for qualitative analysis; however, as discussed previously (21), difficulties often can be encountered in the performance of quantitative analyses when using the technique. In an attempt to ascertain if the high spectral resolution characteristic of Shpol'skii fluorescence spectra can be combined with the high quantitative precision, broad linear dynamic range, and freedom from the quenching characteristic of MI, we have undertaken a detailed investigation of MI fluorescence spectra of PAH in vapor-deposited organic matrices. Organic solvents such as n-heptane, although liquids at room temperature, are sufficiently volatile to be used as matrix gases in MI spectroscopy without difficulty. In previous studies, we demonstrated that highly resolved ("Shpol'skii-quality") fluorescence spectra of PAH indeed can be obtained through MI by using organic matrices, provided that the deposited sample is heated ("annealed") for a very short period of time following deposition (24).

As an example of the improvement in spectral resolution afforded by use of an organic matrix (n-heptane) instead of a "conventional" matrix ($N_2$) in MI fluorescence spectroscopy, Figure 3 shows fluorescence spectra, at one specific wavelength of excitation, of the coking plant water fraction whose FTIR spectra are shown in Figure 2. Two facts are immediately apparent from comparison of these spectra. First, the spectral bandwidths obtained in heptane are substantially smaller than those observed in the $N_2$ matrix; this result is important both for "fingerprinting" and quantitation of specific PAH in multicomponent samples. Second, whereas all the bands in the fluorescence spectrum of this sample in heptane can be assigned to one compound (benz(a)-anthracene), fluorescence from at least two different sample constituents, one of which cannot presently be identified, is observed at the same excitation wavelength in nitrogen. The reason for the greater analytical selectivity observed in the heptane matrix is because the absorption

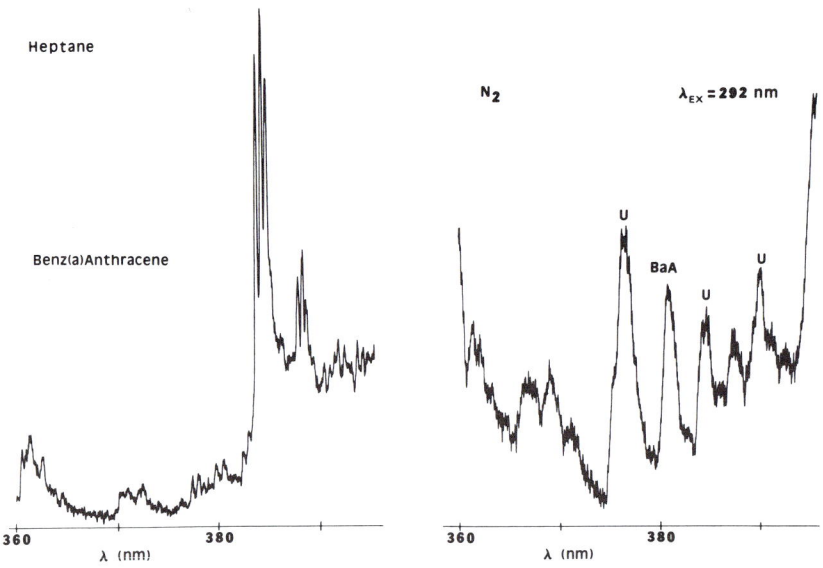

FIGURE 3. MI fluorescence spectra in heptane (left) and nitrogen (right) matrices at 15 K of adsorption chromatography fraction from coking plant wastewater sample. Only benz(a)anthracene (BaA) bands are observed in heptane; in $N_2$, at least one additional unknown component (U) also is observed. Excitation wavelength = 292 nm.

spectra, as well as the emission spectra, of PAH are much sharper in heptane than in nitrogen matrices. Consequently, it is easier to execute "selective excitation" of fluorescence from individual compounds in complex mixtures when an organic matrix, such as heptane, is employed in MI fluorometry of PAH.

In a study of quantitation of benz(a)anthracene (BaA) by MI in heptane matrices, it was determined that the analytical calibration curve was linear from the detection limit (30 pg) to a maximum quantity in excess of 3 $\mu$g; that is, a linear dynamic range exceeding five decades in BaA concentration was achieved. Hence, it can be concluded that our objective of combining the high selectivity and spectral resolution of the Shpol'skii effect with the wide linear quantitative dynamic range of MI fluorescence spectrometry has indeed been achieved by use of organic matrix materials.

In order to take full advantage of the analytical capabilities offered by MI fluorometry in organic matrices, the excitation source should exhibit a narrow bandwidth (i.e., be highly monochromatic) while at the same time exhibiting a variable wavelength of output (i.e., be "tunable"). The combination of a continuum source, such as a xenon lamp, with an excitation monochromator satisfies the latter criterion but not the former. Fortunately, the advent of the commercial dye laser signifies that both criteria can now be satisfied by a single source. As noted above, the highly resolved absorption spectra of PAH in organic matrices present many opportunities for selective excitation, and one therefore observes that the emission spectrum observed for a complex sample can be drasticaly altered by a small change in the wavelength of excitation.

Comparison of Figures 3 and 4 shows how a relatively small alteration in excitation wavelength can produce a dramatic change in the MI fluorescence spectrum of a PAH mixture (the aforementioned coking plant water fraction) in both heptane and nitrogen matrices. In both sets of spectra, a tunable dye laser (pumped by a nitrogen laser) was used as the excitation source; the inherent selectivity of MI fluorometry, especially in organic matrices, is difficult to exploit fully unless a tunable laser is used for excitation (a lamp-monochromator combination operating at a realistic photon flux usually exhibits too large a spectral bandwidth for optimally selective excitation).

The selectivity of MI fluorescence spectrometry can be enhanced in other ways by laser excitation of fluorescence (26). The electronically excited states responsible for fluorescence have finite decay times, typically in the 10 to 100 ns time domain. It is conceivable that two PAH which cannot be distinguished from each other in the spectral domain (due to overlap of their principal fluorescence bands) can be resolved in the time domain if their fluorescence rate constants differ. Lasers which

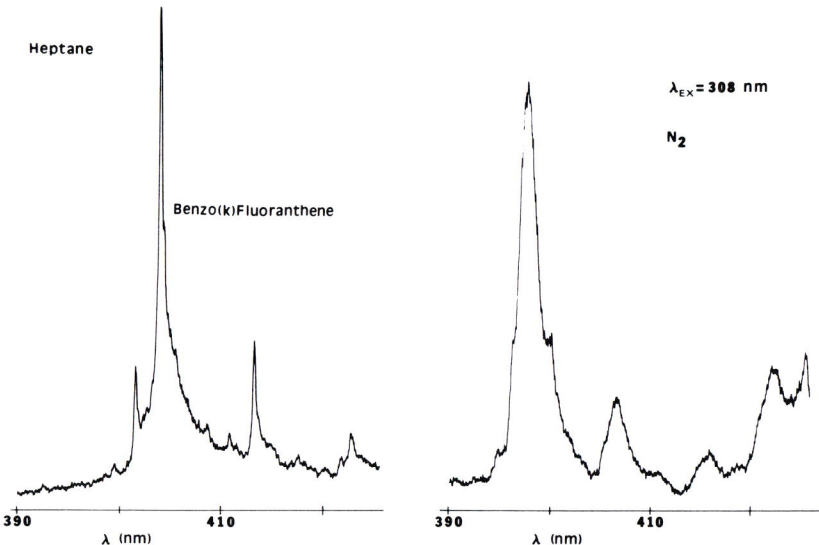

**FIGURE 4.** MI fluorescence spectra in heptane (left) and N₂ (right) matrices of adsorption chromatography fraction from coking plant wastewater sample; excitation wavelength = 308 nm. Only benzo(k)fluoranthene fluorescence bands are observed at this excitation wavelength in either matrix. Both this and the spectrum shown in Figure 3 were excited with a tunable dye laser.

emit pulses of subnanosecond duration now are commercially available. Use of laser-induced, time-resolved fluorescence to distinguish between PAH whose fluorescence spectra in $N_2$ matrices overlap has recently been discussed (6); example applications include resolution of the isometric compounds benzo(a)pyrene and benzo(k)fluoranthene, which are relatively difficult to resolve in the spectral domain but which are easily distinguished from each other by the time-resolution procedure.

It is also worthwhile to inquire whether laser excitation of fluorescence can be used to achieve increased spectral resolution in "conventional" low-temperature matrices. In a Shpol'skii frozen-solution matrix, the solute molecules occupy virtually identical "guest sites" in the crystal structure of the frozen organic solvent (20); the same situation appears to occur when PAH are matrix-isolated in an annealed alkane solvent (24). In a more conventional MI medium, such as nitrogen, wherein the molecular dimensions of the "solvent" (matrix) are much smaller than those of the solutes, the various molecules of a given PAH occupy different types of "lattice sites". The electronic transition energies for those different molecules thus are not precisely the same, and the resulting "inhomogeneous broadening" (1, 16) is responsible for the larger spectral bandwidths for PAH fluorescence observed in $N_2$ than in, for example, n-heptane.

The inhomogeneously broadened width of fluorescence spectral lines in cryogenic solids can be reduced by laser excitation. In a procedure known variously as "site-selection spectroscopy" (SSS) or "fluorescence line narrowing spectroscopy" (FLNS), one uses as an excitation source a laser, the output of which has a bandwidth smaller than the inhomogeneously broadened width of the electronic transition of the PAH in question. The effect of so doing is to ensure that only a certain fraction of the molecules of a given PAH (those which occupy nearly identical sites in the solid sample) is excited, and the resulting "site-selected" fluorescence spectrum thus is much narrower than that obtained for the same compound in the same matrix when a broadband exciting light is used (1, 5, 8, 16, 18, 25). This site-selection fluorescence method recently has been applied to PAH analyses involving frozen solution matrices (2), and we are currently exploring its applicability to the MI fluorometric analysis of PAH (26). The objective of this laser-based spectroscopic methodology is to secure the high spectral resolution obtainable in organic matrices (see Figures 3 and 4) in $N_2$, or indeed in virtually any low-temperature sampling matrix, thereby enhancing further the "fingerprinting" and quantitative capabilities of PAH analysis MI fluorescence spectrometry.

## COMBINED GAS CHROMATOGRAPHY-MATRIX ISOLATION SPECTROSCOPY

The fundamental premise underlying our investigations in analytical matrix isolation spectroscopy is that high spectral resolution is essential to achieve qualitative or quantitative analyses of individual PAH in mixtures without extensive prior sample fractionation. However, a realistic view must be that many samples of current importance (such as shale oils and coal liquids) are so complex that it is virtually inconceivable that any spectroscopic technique, no matter how selective, could be applied profitably without some prior fractionation of the sample. We have already considered the use of the MI sample preparation procedure itself as a separation method via fractional sublimation. For complex samples, more sophisticated separation techniques must be used. It is desirable, when possible, that the separation and spectroscopic analysis proceed as a "one-step" operation using a single instrument. The most important contemporary example of such a "hyphenated" analytical methodology is, of course, combined gas chromatography-mass spectrometry (GC-MS).

Because sampling by MI requires that the sample be vaporized, it seems obvious that MI spectroscopy is suited to monitoring of effluents from GC columns. Probably the most significant advantage of this "GC-MI" technique over GC-MS is the fact that MI FTIR or fluorescence spectroscopy generally can readily distinguish between isometric PAH (23), whereas the mass spectra of most isomeric aromatic compounds are very similar and frequently virtually indistinguishable from each other (12). Thus, we are presently examining both MI fluorescence and FTIR spectroscopy as detection principles for "on-the-fly" identification and quantitation of PAH separated by GC.

Several requirements must be fulfilled for MI spectroscopy to succeed in GC effluent detection. First, the spectroscopic measurement itself must be rapid. Particularly for capillary GC columns, narrow chromatographic peaks (30 sec or less) usually are obtained, and acquisition of a full spectrum within that time interval is required. FTIR is well suited for such an application, inasmuch as the multiplex FTIR technique intrinsically is capable of rapidly producing IR spectra; indeed, the use of FTIR spectrometry as a GC detection method (via gas-phase IR spectra) is already reasonalby well established (10, 11). For fluorescence spectrometry, the "time" problem is more serious. Most fluorescence spectrometers obtain a spectrum by the classical mechanical scanning of a diffraction grating; thus, especially if high spectral resolution is desired, long scan times (5 to 20 min) are common. Obviously, scans of such length are useless in GC column effluent monitoring. Fortunately, two classes of electronic "array detectors" (TV camera tubes and solid-state diode arrays) recently have been developed and now are commercially available. These devices serve essentially as electronic analogues to a photographic plate detector in a classical spectrograph and, as such, do not require mechanical scanning operations to produce spectra. The use of such detectors in fluorescence spectroscopy has recently been discussed (22), and a fluorometric GC detector (using vapor-phase fluorescence spectra) which uses an array detector has been developed (4). Our instrumentation for GC-MI fluorometry uses a TV camera tube ("SIT" vidicon tube) as a detector.

Beyond strictly spectroscopic problems, other experimental aspects of GC-MI experimental design also merit comment. The temperature of a GC effluent usually is relatively high, while MI is a cryogenic technique. Moreover, GC separation of a complex sample may require several hours. These difficulties can be solved by proper cyrostat and interface design and by use of a movable surface upon which the GC effluent is deposited continuously. Under these conditions, the sample thickness never becomes prohibitively large. Moreover, it is possible to examine spectroscopically one portion of the sample (previously deposited) while the sample, as it continues to elute from the column, is deposited at another region of the movable optical surface. Indeed, Reedy and co-workers (19) recently have reported a detailed description of an apparatus for MI FTIR spectroscopic analysis of GC effluents. Our own apparatus for this purpose, which will be described elsewhere, differs somewhat in detail from that of Reedy, Bourne, and Cunningham (19), though the basic principles underlying the two designs are similar.

In both fluorometric and FTIR GC monitoring, the reason for using MI, rather than gas-phase spectroscopic measurements, is the much greater spectral resolution afforded by MI. Comparisons of vapor-phase and MI FTIR spectra (9), or of vapor-phase fluorescence spectra (4) with MI fluorescence spectra (21) (see also Figures 3 and 4 above) are extremely instructive in that regard.

A very simple example of MI FTIR detection of a GC effluent is the separation of a three-component mixture of naphthalenes. As shown in Figure 5, the three compounds eluted are in the order naphthalene ("N"), 2-methylnaphthalene ("2"), and 2,3-dimethylnaphthalene ("2,3"). Figure 6 shows a comparison of the MI FTIR spectra of naphthalene obtained with a Knudsen cell (28) conventional MI sampling system (top) with that of naphthalene obtained as the compound eluted from the GC column (bottom). Though a slight loss of resolution is observed in the "GC-MI spectrum", the spectral detail is sufficient to permit unambiguous identification of naphthalene as the compound emerging from the GC column. Figures 7 and 8 show the MI FTIR spectra of the other two components obtained as they eluted from the column. This very simple example is sufficient to indicate the potential power of GC-MI to provide spectra which are true "fingerprints" of individual PAH.

In a mixture as simple as the three-component PAH sample discussed above, one might regard the use of high-resolution spectroscopic detection as "overkill", though the use of the spectra for identification purposes could be useful. The most intriguing aspect of GC-MI, however, would be its use for very complex samples in which chromatographic separation into individual pure compound zones is infeasible in a realistic time. The ability of both MI FTIR (15, 23) and fluorescence (6, 15, 23) to identify and quantitate individual PAH in mixtures means that "complete" GC separation of difficultly separable PAH (e.g., isomers) need not be achieved in order for qualitative and quantitative analysis of the sample constituents to be accomplished. One may therefore envision a "trade-off" of chromatographic for spectroscopic resolution, so that, for example, GC columns could conceivably be operated at carrier gas flow rates far in excess of optimum, with the saving in analysis time thereby incurred not offset by loss of analytical information.

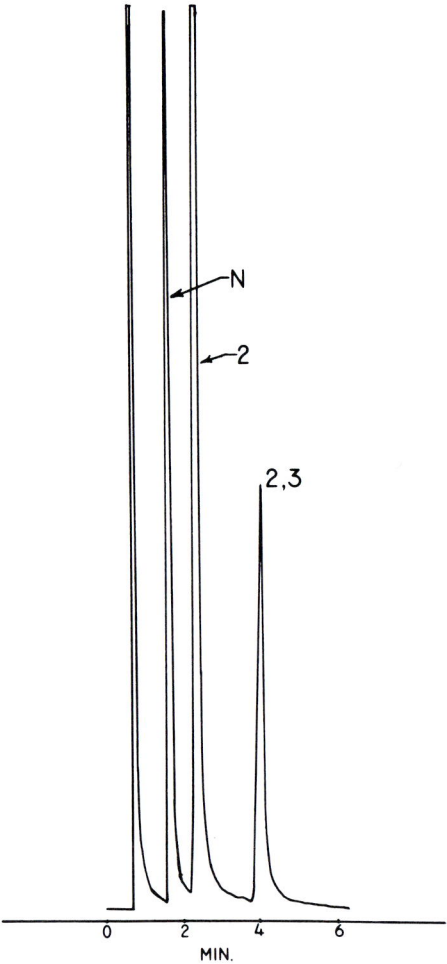

**FIGURE 5.** Gas chromatogram (SE 30 glass capillary SCOT column, 39 m length) of synthetic PAH mixture. Compounds: *N*, naphthalene; *2*, 2-methylnaphthalene; *2, 3*, 2, 3-dimethylnaphthalene.

The use of MI fluorescence and FTIR for "off-line" characterization of the PAH content of samples derived from the liquid chromatography (LC) of synfuel samples has already been described (15). A long-range objective of this work is the use of MI spectroscopy for direct "on-line" characterization of LC column effluents. An obvious requirement for such a method to succeed is that the solvent must be eliminated before the sample can be deposited (unless it is possible to use the chromatographic solvent as the spectroscopic matrix, which may be feasible in MI fluorescence spectroscopy). This problem is encountered in any attempt

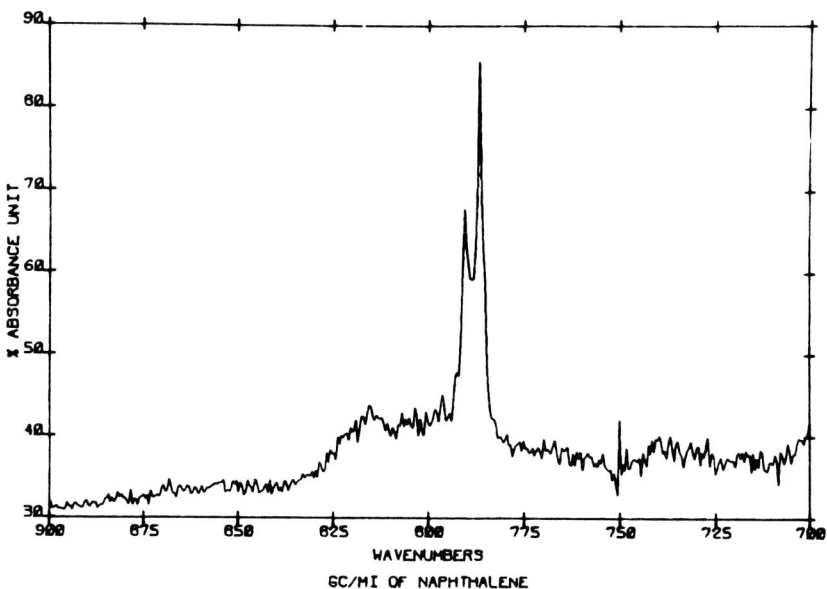

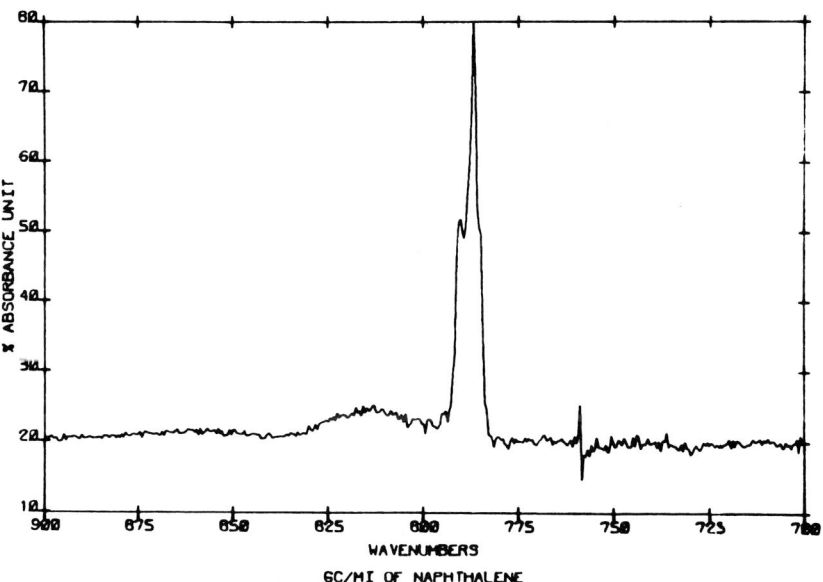

FIGURE 6. MI FTIR spectra of naphthalene. Top: obtained by Knudsen cell vacuum sublimation technique. Bottom: obtained by GC MI of naphthalene eluting from GC (chromatogram shown in Figure 5).

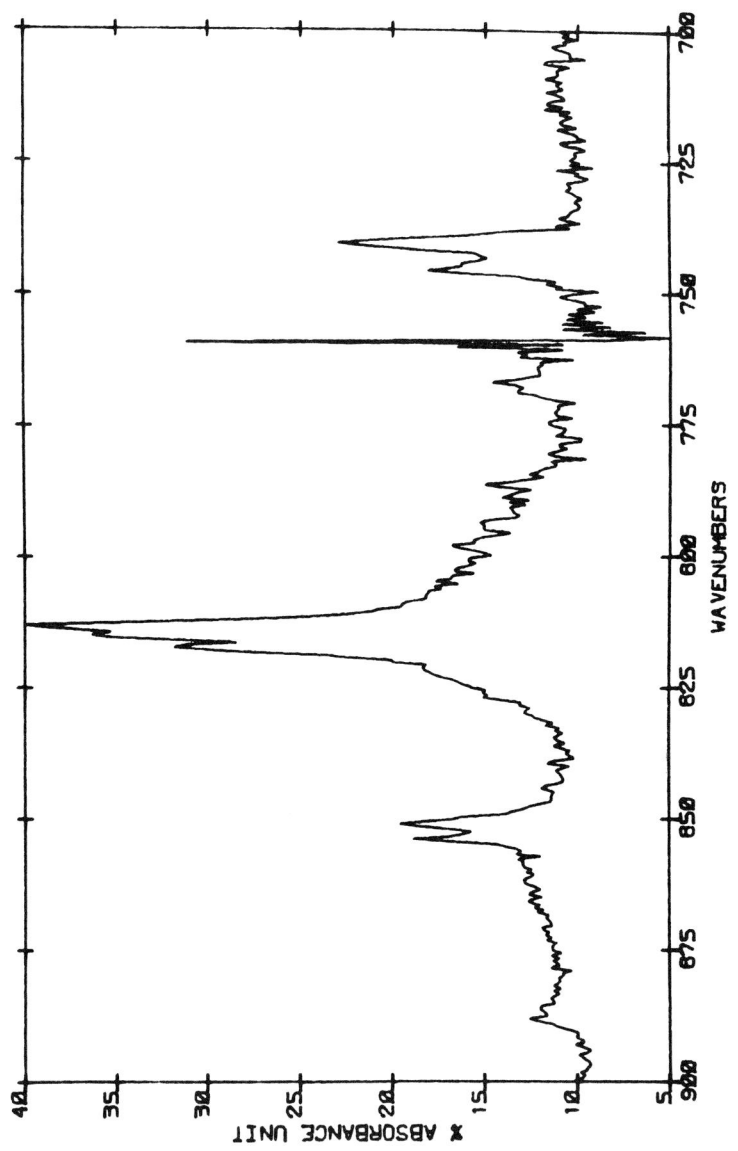

FIGURE 7. MI FTIR spectrum of 2-methylnaphthalene eluant from GC column.

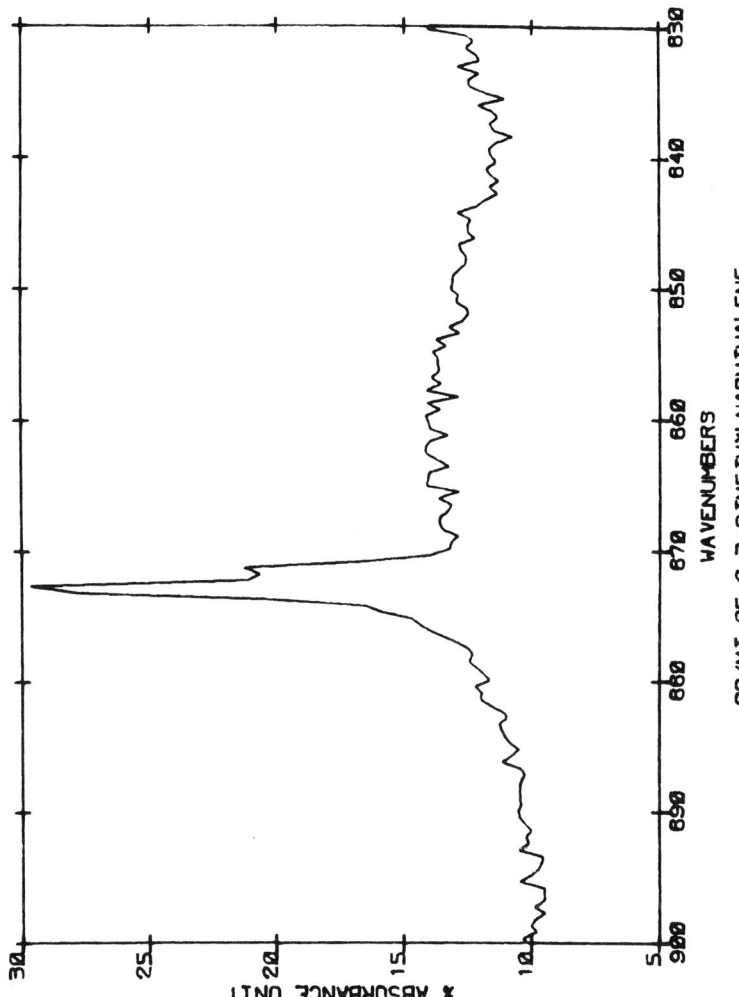

FIGURE 8. MI FTIR spectrum of 2, 3-dimethylnaphthalene eluant from GC column.

to use FTIR spectroscopy (matrix isolation or otherwise) as an LC detection method. As demonstrated by Griffiths (10), it is possible to eliminate the LC solvent simply by properly heating the eluant from the column, provided that the solvent is more volatile than any sample constituent of interest. In a sense, such a procedure represents an extreme case of "separation by fractional volatilization", which we have discussed above as a useful technique in MI spectroscopy.

## CONCLUSION

Several modifications and refinements of the techniques of matrix isolation Fourier transform infrared and fluorescence spectroscopic analysis of PAH have been considered. It should not be inferred that the utility of MI sampling procedures is limited to fluorescence and IR spectroscopy; examples of applications of MI to other forms of spectroscopy have been itemized elsewhere (27). It is our expectation that, with continued development, MI spectroscopy will assume a place as one of several powerful new high-resolution analytical techniques for characterization of complex samples containing PAH and their derivatives.

## ACKNOWLEDGMENTS

Financial support for this research has been derived from the Electric Power Research Institute (Contracts RP-1307 and RP-332) and the National Science Foundation (Grants CHE77-12542 and MPS75-05364).

## REFERENCES

1. Abram, I., Auerbach, R. A., Birge, R. R., Kohler, B. E., and Stevenson, J. M. (1974): Line narrowing and stimulated emission of perylene in n-octane at liquid helium temperatures. *J. Chem. Phys.* 61:3857.
2. Brown, J. C., Edelson, M. C., and Small, G. J (1978): Fluorescence line narrowing spectrometry in organic glasses containing parts-per-billion levels of polycyclic aromatic hydrocarbons. *Anal. Chem.* 50:1394.
3. Colmsjö, A. and Stenberg, U. (1979): Identification of polynuclear aromatic hydrocarbons by Shpol'skii low-temperature fluorescence. *Anal. Chem.* 51:145.

4. Cooney, R. P., Vo-Dinh, T., and Winefordner, J. D. (1977): The SIT image vidicon as a gas-phase fluorescence detector for gas chromatography. *Anal. Chim. Acta* 89:9.
5. Cunningham, K., Morris, J. M., Fünfschilling, J., and Williams, D. F. (1975): Site selection spectroscopy luminescence of solutions with laser excitation. *Chem. Phys. Lett.* 32:581.
6. Dickinson, R. B., Jr. and Wehry, E. L. (1979): Time-resolved matrix-isolation fluorescence spectrometry of mixtures of polycyclic aromatic hydrocarbons. *Anal. Chem.* 51:778.
7. D'Silva, A. P., Oestreich, G. J., and Fassel, V. A. (1976): X-ray excited optical luminescence of polynuclear aromatic hydrocarbons. *Anal. Chem.* 48:915.
8. Flatscher, G., Fritz, K., and Friedrich, J. (1976): Site selection spectroscopy in the temperature range 2K and 300K. *Z. Naturforsch.* 31a:1220.
9. Green, D. W. and Reedy, G. T. (1978): Matrix-isolation studies with Fourier transform infrared. In: *Fourier Transform Infrared Spectroscopy*, Vol. 1. J. R. Ferraro and L. J. Basile, Eds., Academic Press, New York.
10. Griffiths, P. R. (1977): Recent applications of Fourier transform infrared spectrometry in chemical and environmental analysis. *Appl. Spectrosc.* 31:497.
11. Griffiths, P. R. (1978): Gas chromatography and Fourier transform infrared. In: *Fourier Transform Infrared Spectroscopy*, Vol. 1. J. R. Ferraro and L. J. Basile, Eds., p. 143, Academic Press, New York.
12. Hase, A., Lin, P. H., and Hites, R. A. (1976): Analysis of complex polycyclic aromatic hydrocarbon mixtures by computerized GC-MS. In: *Polynuclear Aromatic Hydrocarbons* Vol. 1. R. I. Freudenthal and P. W. Jones, Eds., p. 435, Raven Press, New York.
13. Kirkbright, G. F. and deLima, C. G. (1974): The detection and determination of polynuclear aromatic hydrocarbons by luminescence spectrometry utilizing the Shpol'skii effect at 77 K. *Analyst* 99:338.
14. Mamantov, G., Wehry, E. L., Kemmerer, R. R., and Hinton, E. R. (1977): Matrix isolation Fourier transform infrared spectrometry of polycyclic aromatic hydrocarbons. *Anal. Chem.* 49:86.
15. Mamantov, G., Wehry, E. L., Kemmerer, R. R., Stroupe, R. C., Hinton, E. R., and Goldstein, G. (1978): Characterization of mixtures of polycyclic aromatic hydrocarbons by liquid chromatography and matrix isolation spectroscopy. *Advan. Chem. Ser.* 170:99.
16. McColgin, W. C., Marchetti, A. P., and Eberly, J. H. (1978): The nature of solution spectra. Inhomogeneous broadening and phonon effects in frozen solutions. *J. Am. Chem. Soc.* 100:5622.

17. Monkman, J. L., Dubois, L., and Baker, C. J. (1971): The rapid measurement of polycyclic hydrocarbons in air by microsublimation. *Pure Appl. Chem.* 24:731.
18. Personov, R. I. (1975): The sharp narrowing of spectral bands of organic molecules under laser excitation. *Sov. Phys. Usp.* 18:645.
19. Reedy, G. T., Bourne, S., and Cunningham, P. T. (1979): Gas chromatography/infrared matrix isolation spectrometry. *Anal. Chem.* 51:1535.
20. Shpol'skii, E. V. and Bolotnikova, T. N. (1974): Modern trends in quasi-linear spectral studies. *Pure Appl. Chem.* 37:183.
21. Stroupe, R. C., Tokousbalides, P., Dickinson, R. B. Jr., Wehry, E. L., and Mamantov, G. (1977): Low-temperature fluorescence spectrometric determination of polycyclic aromatic hydrocarbons by matrix isolation. *Anal. Chem.* 49:701.
22. Talmi, Y., Baker, D. C., Jadamec, J. R., and Saner, W. A. (1978): Fluorescence spectrometry with optoelectronic image detectors. *Anal. Chem.* 50:936A.
23. Tokousbalides, P., Hinton, E. R., Jr., Dickinson, R. B., Jr., Bilotta, P. V., Wehry, E. L., and Mamantov, G. (1978): Analysis of the isomeric methylchrysenes by matrix isolation fluorescence and Fourier transform infrared spectrometry. *Anal. Chem.* 50:1189.
24. Tokousbalides, P., Wehry, E. L., and Mamantov, G. (1977): Observation of quasilinear fluorescence spectra (the "Shpol'skii effect") in matrix-isolated polycyclic aromatic hydrocarbons. *J. Phys. Chem.* 81:1769.
25. Vo-Dinh, T., Kreibich, U. T., and Wild, U. P. (1974): Phosphorescence spectra from selected sites of N-ethylcarbazole in n-alkanes. *Chem. Phys. Lett.* 24:352.
26. Wehry, E. L., Gore, R. R., and Dickinson, R. B., Jr. (1980): Laser-excited matrix isolation molecular fluorescence spectrometry. In: *Lasers and Chemical Analysis.* G. M. Hieftje, F. E. Lytle, and J. C. Travis, Eds. (in press). Humana Press, Clifton, N. J.
27. Wehry, E. L. and Mamantov, G. (1979): Matrix isolation spectroscopy. *Anal. Chem.* 51:643A.
28. Wehry, E. L., Mamantov, G., Kemmerer, R. R., Brotherton, H. O., and Stroupe, R. C. (1976): Low-temperature Fourier transform infrared spectroscopy of polynuclear aromatic hydrocarbons. In: *Polynuclear Aromatic Hydrocarbons,* Vol. 1. R. I. Freudenthal and P. W. Jones, Eds., p. 299, Raven Press, New York.
29. Wehry, E. L., Mamantov, G., Kemmerer, R. R., Stroupe, R. C., Tokousbalides, P. T., Hinton, E. R., Hembree, D. M., Dickinson, R. B., Jr., Garrison, A. A., Bilotta, P. V., and Gore, R. R. (1979): Analysis of polycyclic aromatic hydrocarbons by matrix isolation

fluorescence and Fourier transform infrared spectroscopy. In: *Polynuclear Aromatic Hydrocarbons,* Vol. 3, P. W. Jones and R. I. Freudenthal, Eds., p. 133, Raven Press, New York.

# POLYNUCLEAR AROMATIC HYDROCARBONS AND THE MUTAGENICITY OF USED CRANKCASE OILS

E. Peake* and K. Parker**

*Kananaskis Centre for Environmental Research
University of Calgary
Calgary, Alberta
Canada, T2N 1N4
**Biology Department
University of Calgary
Calgary, Alberta
Canada, T2N 1N4

## INTRODUCTION

Of the polynuclear aromatic hydrocarbons (PAH) produced by the automobile engine, about 85 percent is retained in the crankcase oil, with the remaining 15 percent being emitted to the atmosphere as part of the automobile exhaust (12). Attention has focused on the emitted PAH and on the possible health effects of PAH, many of which are mutagenic and some carcinogenic. Relatively little attention has been paid to the 85 percent of PAH which is contained in the used oils (15). Despite intensified efforts during the energy shortage of the past few years to recover and reclaim used oils, much oil is still disposed of haphazardly by the general public. About 500 million gallons was discharged to the environment in the United States during 1976 (9), most of which was poured into sewers or onto waste land. In the province of Alberta where several studies into the fate of PAH are being conducted, some 2.8 million gallons was released to the environment in 1973. As a first step in determining the rate of degradation of PAH in spilled crankcase oils and their environmental effects, motor oil samples were analyzed for PAH by combined gas chromatography-mass spectrometry and examined for mutagenic activity by the Ames test. The results of this first step are reported here.

## METHODOLOGY

### Sample Preparation

Used motor oils A and B were obtained from the oil sumps of two different automobiles with 4-stroke gasoline engines after 5 to 6 months of winter use. Used oil C is a pooled supply of oil obtained from a local service station in Calgary during the winter season. Six unused oils were also examined, including one re-refined oil and one synthetic oil. PAH and other materials of similar polarity were separated from the bulk of the oil by extraction with dimethyl sulfoxide (DMSO) (13). Each oil was extracted four times successively by shaking 4 ml of oil vigorously with an equal volume of DMSO and centrifuging. The four extracts were then combined. Extraction in this manner removed approximately 70 percent of the mutagenic components from the oil into the DMSO, as determined by extrapolation of the mutagenic activity of a fifth extract. A portion of the combined DMSO extract was set aside for the Ames test. The remaining DMSO was shaken vigorously with 10 volumes of water and 10 volumes of cyclohexane, thereby transferring the PAH and other organic material into cyclohexane. The volume of the cyclohexane was reduced to 400 $\mu$l in preparation for analysis by gas chromatography-mass spectrometry (GC-MS).

### Ames Test

The DMSO extract of each oil was assayed for mutagenic activity by the Ames methods (2, 10), with some modifications. All samples were tested with *Salmonella typhimurium* strain TA98; used oils A and B and one unused oil were also tested with strains TA1535, TA1537, TA1538 and TA100. A variation of the Ames method entailed the addition of 0.1 ml of aqueous 0.2 percent ampicillin per 20 ml Penassay broth to strains TA98 and TA100 prior to incubation to prevent loss of the R factor in these strains (14). Each sample was tested with eight doses of DMSO extract, the equivalent of 0.2 to 200 $\mu$l of oil per plate, with and without the addition of S-9 liver homogenate from rats injected with Aroclor 1254 (1). Benzo(a)pyrene was used as a positive control and DMSO as a negative control.

### GC-MS Analysis

The GC-MS system was a Finnigan model 4000 equipped with an INCOS data system. The GC operating conditions were as follows:

column type, 30-m glass capillary with SP2100 liquid phase providing 74,000 effective plates (J and W Scientific); Grob-type injection of 1 μl of sample; injection temperature, 300°C; carrier gas flow, helium 29 cm/sec; initial column temperature, 40°C held for 1 min; temperature programmed at 40°C/min from 40 to 100°C and at 2°C/min from 100 to 280°C; final temperature held for 60 min. The MS operating conditions were as follows: transfer line temperature, 320°C; manifold temperature, 100°C; scan range 45 to 350 amu; scan time 1.0 sec; ionizing voltage, 50 electron volts.

Identification of individual PAH was based upon computer comparisons with standard spectra of the National Bureau of Standards library, with spectra of PAH standards obtained in this laboratory, and upon published spectra and GC retention times (6,7,8). Quantitation was based upon an internal standard of deuterated anthracene and upon response factors obtained with standard PAH analyzed under the same GC-MS conditions as the sample.

## RESULTS AND DISCUSSION

**Ames Test**

The three used crankcase oil samples showed mutagenic activity, both in the presence of S-9 mammalian enzymes and in the absence of microsomal activation. As illustrated in Figure 1, which shows a dose-response curve for used oil A with strain TA98 in the presence of S-9 enzymes, a nonlinear relationship was observed between the number of revertants and the amount of oil extract applied per plate in the range of 0.2 to 200 μl. This effect was found both with and without S-9. When more than 100 μl of extract was applied per plate, a decline in the number of revertants was observed, probably due to toxicity.

With doses of less than 20 μl per plate, a linear dose-response curve was found, as illustrated in Figure 2, for used oil B with strain TA98 and S-9. The number of spontaneous revertants occurring with the addition of DMSO and S-9 but without oil extract is given by the y intercept, in this case 215. A dose-response activity was measured for each oil on the basis of the linear portion of the curve below 20 μl per plate. This is designated as the mutagenic potency of the oil and is expressed as the number of histidine-positive revertants per 100 μl of extracted oil. The mutagenic potency was reproducible within a given test, but varied up to 30 percent from test to test using different cultures of the same *Salmonella* strain. The mutagenic potencies of three used motor oils on the five standard strains of the Ames test are summarized in Table 1.

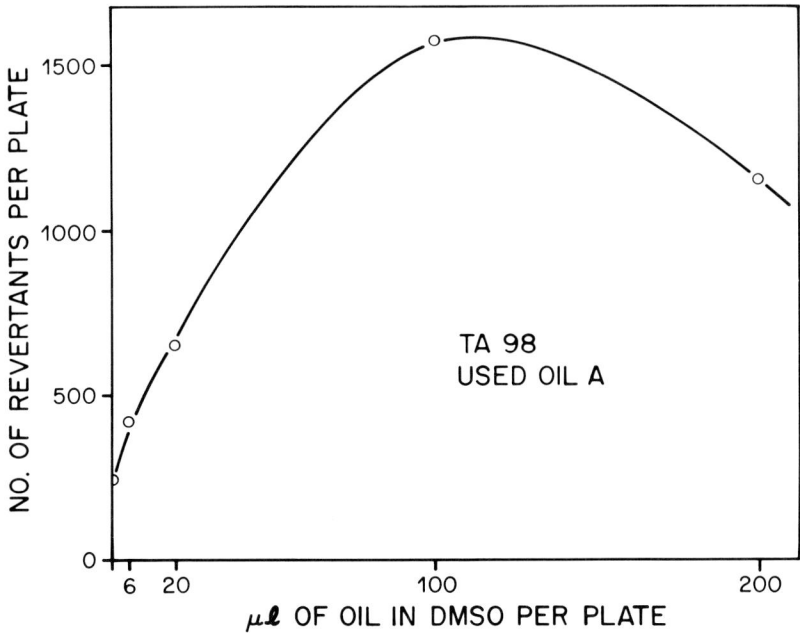

**FIGURE 1.** The mutagenic activity of the DMSO extract of used crankcase oil A as determined by the Ames test with strain TA98 in the presence of S-9. Volume of extract applied per plate (abscissa) represents the equivalent volume of oil extracted in DMSO, in this case 0.2 to 200 µl. The number of histidine-positive revertant colonies per plate is given by the ordinate.

All three used motor oils tested were mutagenic on each of the five tester strains. The mutagenic components of used motor oils induced base-pair substitutions in TA1535 and TA100 and frameshift mutations in TA1537, TA1538, and TA98. Oils A and B, which were collected from the oil sumps of two automobiles, were more active than the pooled supply of oil C, obtained from a local service station in Calgary.

TA100 was clearly the most responsive strain with S-9, having shown potencies of up to 18,000 rev./100 µl for oil B. The frameshift strains, TA1538 and TA98, produced 5,000 to 7,500 rev./100 µl for these same oils with S-9; and strains TA1537 and TA1535 each indicated potencies of 1,000 to 3,000 rev./100 ul with S-9.

There was no significant change in the activity with S-9 of used motor oil B when tested with either TA98 or TA100 after a 2-year period during which the sample was dissolved in DMSO and left in the dark at room temperature. This suggests that the mutagenic components of used oil may be stable indefinitely when stored under these conditions.

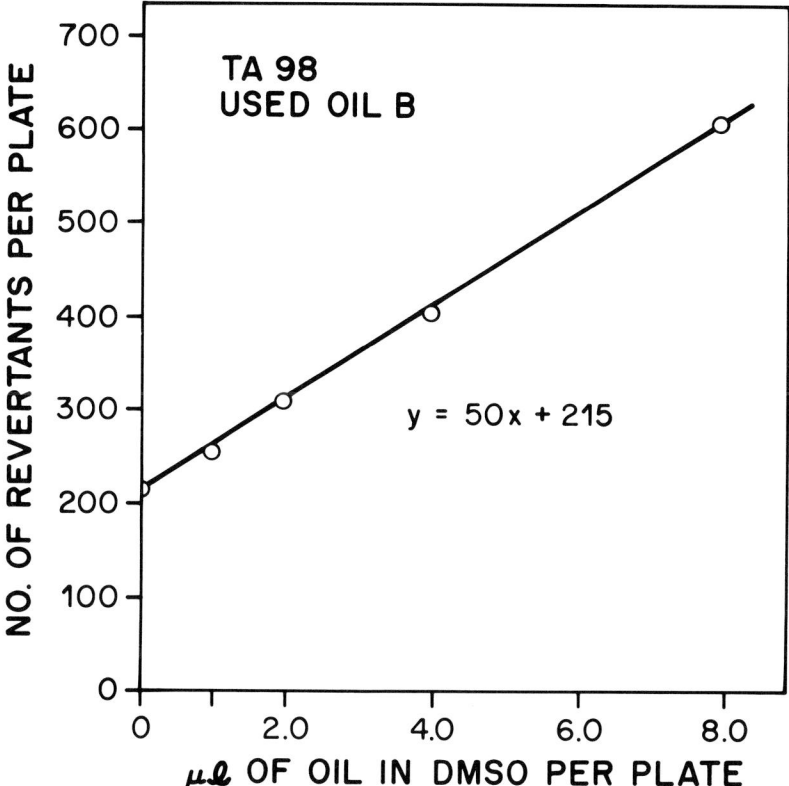

FIGURE 2. The mutagenic activity of the DMSO extract of used crankcase oil B as determined by the Ames test with strain TA98 in the presence of S-9. Volume of extract applied per plate (abscissa) represents the equivalent volume of oil extract in DMSO, in this case 0.2 to 8 µl. The number of histidine-positive revertant colonies per plate is given by the ordinate.

The high mutagenicity of used motor oils prompted the question of whether unused oils were mutagenic. Tests of seven different oils indicated that most unused motor oils show some mutagenic activity which is detected in the absence of S-9 and is usually lost with the addition of S-9. Of the seven unused oils tested on strain TA98, the oil with the highest and most consistent activity was the synthesized engine lubricant, Mobil-1, with a potency of 725 to 830 rev./100 µl extract without S-9 (Table 1). Two other oils, Shell X-100 and Castrol GTX, showed potencies greater than 300 rev./100 µl, but extracts became progressively less active when stored over a 2-week period. The other four oils, including two oils re-refined from used oil, had potencies less than

### TABLE 1. Mutagenicity of Used and Unused Motor Oils

| Motor Oil | S-9 (+/-) | Average Mutagenic Potency[a] (rev. per 100 µl extract) | | | | |
|---|---|---|---|---|---|---|
| | | TA1537 | TA1535 | TA100 | TA1538 | TA98 |
| Used oil A | + | (2,000) | (650) | 17,400 | (5,000) | (6,200) |
| Used oil B | − | (575) | 0 | 2,100 | 1,600 | 1,300 |
|  | + | 2,100 | 2,100 | 8,000 | 5,740 | (6,100) |
| Used oil C | − |  |  |  |  | 530 |
|  | + |  |  |  |  | 1,800 |
| Castrol GTX | − |  |  | (490) |  | 400 |
|  | + |  |  |  |  | 0 |
| Mobil-1 | − |  | 0 | (2,100) | 0 | 770 |
|  | + |  |  |  |  | 0 |
| Shell X-100 | − | 0 | (30) | (360) | (66) | 240 |
|  | + |  |  |  |  | 0 |

[a]Average mutagenic potencies (rev./100µl extract) for used and unused motor oils based on 2 to 4 estimates per oil. Values enclosed by parentheses represent single estimates only. Addition of S-9 is indicated by +.

150 rev./100 µl. Used oils are comparatively more mutagenic than unused oils, especially with S-9. The components of unused oils which appear to be deactivated with the addition of S-9 may also be present in used oils but are not detectable in the latter. Presumably this is because used oils have acquired substances which are more mutagenic in the presence of S-9.

The ratio of activity with S-9 to activity without S-9 was relatively constant for two different used motor oils and with four different strains. Ratios determined from the potencies in Table 1 were as follows: 3.7 (used oil B, TA1537), 3.8 (B, TA100), 3.6 (B, TA1538), 4.7 (B, TA98), and 3.4 (C, TA98). Consistency in this ratio among many different used oils might suggest that their compositions were similar; however, there is no apparent reason to expect these ratios to be similar for the different strains. Thus, used oils appear to be about fourfold more active with the addition of S-9 than the same oils without S-9, whereas unused oils are active only in the absence of S-9.

### Chemical Composition

PAH extracted from used oil with DMSO and subsequently transferred into cyclohexane were analyzed by GC-MS. A reconstructed ion chromatogram of used oil C is shown in Figure 3. The mixture is com-

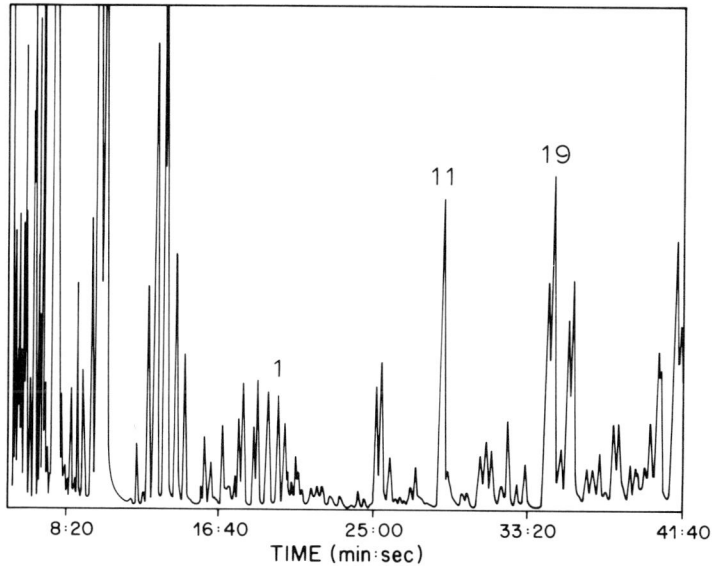
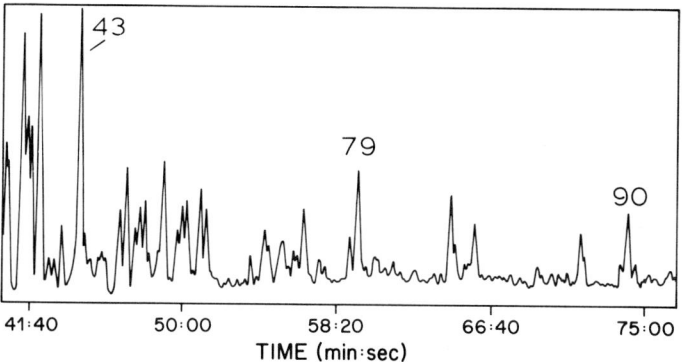
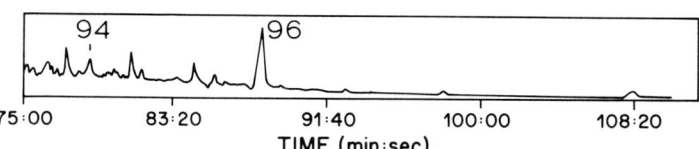

**FIGURE 3.** The reconstructed ion chromatogram (gas chromatogram) of the DMSO extract of used crankcase oil C. Peak numbers refer to PAH identified in Table 3. See text for GC-MS conditions.

plex, with many partially resolved components. The mass spectra of PAH are distinctive, having a strong molecular ion and a weaker doubly charged ion. In the case of alkyl-substituted aromatics the molecular ion may not dominate the spectra but is nevertheless strong and easily recognized. By the use of computer processing of the spectra contained in the reconstructed ion chromatogram of Figure 3, several hundred components were detected as illustrated in Figure 4. Response factors for several groups of compounds were obtained by the analysis of PAH standards under the same GC-MS conditions. Quantitation was then based upon the intensity of the molecular ion and the appropriate response factor. The overall composition of the DMSO extract of used oil C is given in Table 2.

The majority of the 1,760 µg/ml aromatic hydrocarbons found in the DMSO extract were alkylated benzenes (900 µg/ml) and alkyl naphthalenes (440 µg/ml). The quantity of alkyl benzenes in the oil is probably greater than the 900 µg/ml reported since peaks eluted from the GC column shortly after the cyclohexane solvent were not recorded. Substituted benzenes and naphthalenes are not likely mutagenic but may be

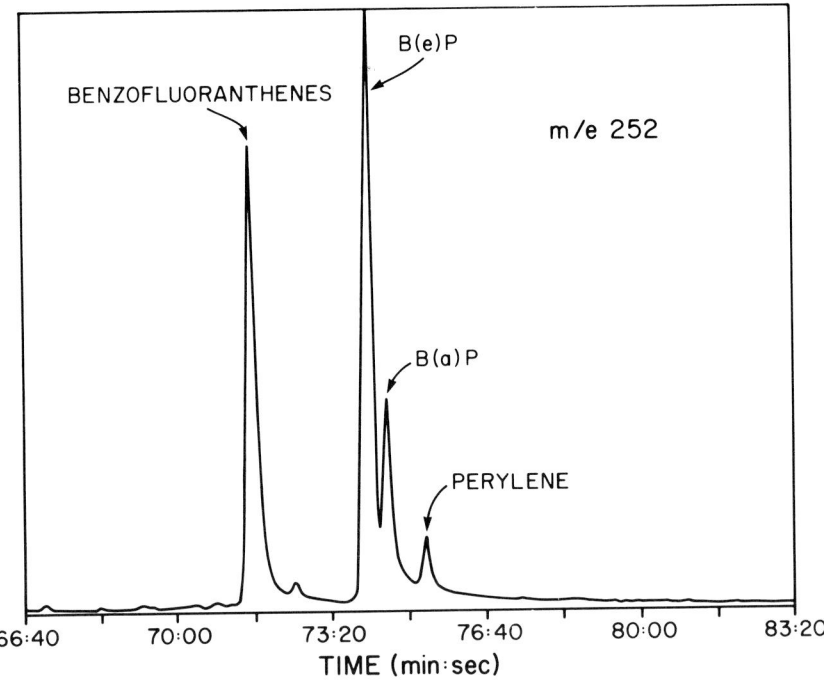

**FIGURE 4.** The reconstructed ion chromatogram of mass 252 of the DMSO extract of used crankcase oil C. See text for GC-MS conditions.

**TABLE 2. Composition of the DMSO Extract of Oil C**

|  | µg/ml of oil |
|---|---|
| Alkyl benzenes | 900 |
| Naphthalene and alkyl naphthalene | 440 |
| Fluorene and alkyl fluorenes | 30 |
| Phenanthrene and alkyl phenanthrenes | 220 |
| Fluoranthene and alkyl fluoranthenes | 60 |
| Benzanthracene, chrysene, triphenylene and their alkyl derivatives | 30 |
| Benzpyrenes and benzfluoranthenes | 22 |
| Benzperylenes | 24 |
| Sulfur containing heterocyclics | 20 |
| Total aromatics | 1,760 |

toxic to *Salmonella* in the Ames test. Toxicity is evident at upper doses by a nonlinear dose-response curve as shown in Figure 1. Of 400 µg/ml PAH with molecular weight 166 or greater, phenanthrene and its alkyl derivatives constitute 220 µg/ml; fluoranthene and its alkyl derivatives, 60 µg/ml; and fluorene and its alkyl derivatives, 30 µg/ml. PAH with mass 226 or greater represent only 80 µg/ml of the total 1,760 µg/ml. It is the latter compounds which have proven to be both mutagenic (3, 4, 11) and carcinogenic (12, 17).

PAH components in used oil C of mass 166 (fluorene) and greater identified by GC-MS are listed in Table 3; sulfur-containing heterocyclic aromatic compounds are listed in Table 4. No attempt was made to identify nitrogen-bearing compounds.

The data given in Tables 2 and 3 is semiquantitative. PAH standards were not available for many of the components identified and therefore response factors were not applied to the data in these tables. Quantitation was based on the intensity of the molecular ion as compared with that of the deuterated anthracene internal standard, peak 12.

Many of the compounds identified in the used oil are among those found previously in atmospheric aerosols (6,7,8). As automobile exhaust is a major contributor of PAH to the atmosphere (12), this similarity is to be expected. However, some differences were found in the relative abundances of PAH in the used oil as compared with atmospheric PAH. Most striking was the predominance of alkyl-substituted aromatic compounds in the oil. For example, eleven alkyl fluorene isomers were found, two in quantities almost equivalent to that of

**TABLE 3. Polycyclic Aromatic Hydrocarbons in Used Motor Oil Identified by Gas Chromatography—Mass Spectrometry**

| Peak | Mol. Wt. | Compound | μg/ml oil[a] |
|---|---|---|---|
| 1 | 168 | Methylbiphenyl | 0.74 |
| 2 | 168 | Methylbiphenyl | 0.36 |
| 3 | 168 | Methylbiphenyl | 0.26 |
| 4 | 166 | Fluorene | 1.47 |
| 5 | 168 | Methylbiphenyl | 0.42 |
| 6 | 168 | Methylbiphenyl | 0.18 |
| 7 | 168 | Methylbiphenyl | 0.09 |
| 8 | 180 | Methylfluorene | 0.10 |
| 9 | 180 | Methylfluorene | 1.19 |
| 10 | 180 | Methylfluorene | 1.08 |
| 11 | 178 | Phenanthrene | 7.80 |
| 12 | 188 | Deuterated Anthracene[i] | 0.50 |
| 13 | 180 | Methylfluorene | 0.08 |
| 14 | 194 | Dimethylfluorene[b] | 0.58 |
| 15 | 218 | | 0.10 |
| 16 | 178 | Anthracene | 0.33 |
| 17 | 194 | Dimethylfluorene[b] | 0.61 |
| 18 | 192 | Methylphenanthrene | 2.63 |
| 19 | 192 | Methylphenanthrene | 3.62 |
| 20 | 192 | Methylphenanthrene | 2.95 |
| 21 | 208 | Trimethylfluorene[c] | 0.12 |
| 22 | 192 | Methylphenanthrene | 2.44 |
| 23 | 208 | Trimethylfluorene[c] | 0.29 |
| 24 | 208 | Trimethylfluorene[c] | 0.36 |
| 25 | 204 | Phenylnaphthalene | 0.90 |
| 26 | 208 | Trimethylfluorene[c] | 0.15 |
| 27 | 208 | Trimethylfluorene[c] | 0.18 |
| 28 | 206 | Dimethylphenanthrene[b] | 0.22 |
| 29 | 206 | Dimethylphenanthrene[b] | 0.16 |
| 30 | 206 | Dimethylphenanthrene[b] | 0.75 |
| 31 | 192 | Methylanthracene | 0.09 |
| 32 | 206 | Dimethylphenanthrene[b] | 2.45 |
| 33 | 206 | Dimethylphenanthrene[b] | 4.21 |
| 34 | 192 | Methylanthracene | 0.28 |
| 35 | 206 | Dimethylphenanthrene[b] | 2.80 |
| 36 | 202 | Fluoranthene | 4.36 |
| 37 | 192 | Methylanthracene | 0.21 |
| 38 | 222 | | 0.08 |
| 39 | 218 | Ethylcyclopenta(def)phenanthrene[b] | 0.79 |
| 40 | 218 | Ethylcyclopenta(def)phenanthrene[b] | 0.46 |
| 41 | 222 | | 0.10 |

## TABLE 3. (Continued)

| Peak | Mol. Wt. | Compound | μg/ml oil[a] |
|---|---|---|---|
| 42 | 220 | Trimethylphenanthrene[c] | 0.10 |
| 43 | 202 | Pyrene | 6.69 |
| 44 | 218 | Ethylcyclopenta(def)phenanthrene[b] | 0.17 |
| 45 | 204 | | 0.39 |
| 46 | 220 | Trimethylphenanthrene[c] | 0.08 |
| 47 | 220 | Trimethylphenanthrene[c] | 0.29 |
| 48 | 230 | Terphenyl | 0.12 |
| 49 | 220 | Trimethylphenanthrene[c] | 1.32 |
| 50 | 206 | Dimethylanthracene[b] | 0.08 |
| 51 | 206 | Dimethylanthracene[b] | 0.10 |
| 52 | 220 | Trimethylphenanthrene[c] | 2.72 |
| 53 | 220 | Trimethylphenanthrene[c] | 0.48 |
| 54 | 218 | Dihydromethylpyrene[d] | 0.13 |
| 55 | 206 | Dimethylanthracene[b] | 0.08 |
| 56 | 220 | Trimethylphenanthrene[c] | 1.16 |
| 57 | 216 | Benzo(a)fluorene | 0.93 |
| 58 | 216 | Benzo(b)fluorene | 1.38 |
| 59 | 216 | Benzo(c)fluorene | 0.44 |
| 60 | 216 | Methylpyrene[d] | 1.19 |
| 61 | 220 | Trimethylanthracene[c] | 0.51 |
| 62 | 218 | Dihydromethylpyrene[d] | 0.32 |
| 63 | 216 | Methylpyrene[d] | 1.14 |
| 64 | 216 | Methylpyrene[d] | 1.14 |
| 65 | 216 | Methylpyrene[d] | 0.78 |
| 66 | 234 | Diethylphenanthrene[e] | 0.18 |
| 67 | 230 | Dimethylpyrene[b,d] | 0.13 |
| 68 | 234 | Diethylphenanthrene[e] | 0.53 |
| 69 | 234 | Diethylphenanthrene[e] | 0.14 |
| 70 | 230 | Dimethylpyrene[b,d] | 0.70 |
| 71 | 234 | Diethylphenanthrene[e] | 0.34 |
| 72 | 230 | Dimethylpyrene[b,d] | 0.28 |
| 73 | 230 | Dimethylpyrene[b,d] | 0.33 |
| 74 | 228 | Benzo(c)phenanthrene | 0.12 |
| 75 | 230 | Dimethylpyrene[b,d] | 0.27 |
| 76 | 244 | Ethylmethylpyrene[b,d] | 0.14 |
| 77 | 228 | Benzo(a)anthracene | 0.87 |
| 78 | 226 | | 0.22 |
| 79 | 228 | Chrysene + triphenylene | 2.48 |
| 80 | 226 | Cyclopenta(cd)pyrene | 0.78 |
| 81 | 242 | Methylbenzo(a)anthracene[f] | 1.68 |
| 82 | 240 | Methylbenzo(mno)fluoranthene[g] | 0.15 |

## TABLE 3. (Continued)

| Peak | Mol. Wt. | Compound | µg/ml oil[a] |
|---|---|---|---|
| 83 | 242 | Methylbenzo(a)anthracene[f] | 0.26 |
| 84 | 242 | Methylbenzo(a)anthracene[f] | 0.23 |
| 85 | 240 | Methylbenzo(mno)fluoranthene[g] | 0.15 |
| 86 | 242 | Methylbenzo(a)anthracene[f] | 0.28 |
| 87 | 256 | Ethylbenzo(a)anthracene[b,f] | 0.44 |
| 88 | 256 | Ethylbenzo(a)anthracene[b,f] | 0.21 |
| 89 | 252 | Benzo(k)fluoranthene | 1.44 |
| 90 | 252 | Benzo(e)pyrene | 1.74 |
| 91 | 252 | Benzo(a)pyrene | 0.36 |
| 92 | 252 | Perylene | 0.13 |
| 93 | 266 | Methylbenzofluoranthene | 0.18 |
| 94 | 266 | Methylbenzopyrene | 0.41 |
| 95 | 276 | h | 0.32 |
| 96 | 276 | Benzo(ghi)perylene | 1.67 |

[a] µg/ml of oil based on the response of deuterated anthracene.
[b] Could be ethyl- or dimethyl.
[c] Could be ethylmethyl-, trimethyl-, or propyl-.
[d] Could be a pyrene or fluoranthene.
[e] Could be diethyl-, ethyldimethyl-, tetramethyl-, methylpropyl- or butyl-.
[f] Could be a derivative of chrysene, triphenylene, benzo(c)phenanthrene or benzo(a)anthracene.
[g] Could be a derivative of benzo(mno)fluoranthene or cyclopenta(cd)pyrene.
[h] Compounds with molecular weight 276 can be any of the following: indeno(1,2,3-cd)pyrene, indeno(1,2,3-cd)fluoranthene, cyclopenta(cd)perylene, phenanthro(10,-1,2,3-cdef)fluorene, acenaphth(1,2-a)acenaphthylene, dibenzo(b,mno)-fluoranthene, dibenzo(e,mno)fluoranthene, dibenzo(f,mno)fluoranthene. Further possibilities are the benzo derivatives of cyclopenta(cd)pyrene and cyclopenta(cd)fluoranthene.
[i] Internal standard.

fluorene. Similarly methyl phenanthrene, methyl anthracene, and methyl pyrene isomers were abundant. Alkyl-substituted aromatics were also found in Calgary atmospheric samples, but the diversity of isomers was not as great (16).

Among the known mutagens or carcinogens found in the oil extract were benzo(a)pyrene, chrysene, benzo(c)phenanthrene, benz(a)anthracene, cyclopenta(c,d)pyrene, and benzo(g,h,i)perylene (3,4,11,12). These components totalled less than 10 µg/ml of oil. Assuming the mutagenic effect of PAH in a mixture to be additive, these compounds alone could not account for much of the observed activity of 18,000 revertants per ml of oil found with strain TA98 in the presence of S-9 for oil C. The mutagenic potency of this oil was the equivalent of 120 µg of benzo(a)pyrene/ml. The many alkyl substituted isomers observed but not specifically identified in the extract of oil C may have contributed to

**TABLE 4. Sulfur-Containing Heterocyclic Aromatic Compounds Identified in Used Motor Oil by Gas Chromatography—Mass Spectrometry**

| Peak | Mol. Wt. | Compound | µg/ml[a] |
|---|---|---|---|
| 1 | 184 | Dibenzothiophene or naphthothiophene | 0.79 |
| 2 | 198 | Methyldibenzothiophene | 1.37 |
| 3 | 198 | Methyldibenzothiophene | 0.67 |
| 4 | 198 | Methyldibenzothiophene | 0.22 |
| 5 | 212 | Dimethyldibenzothiophene | 0.05 |
| 6 | 212 | Dimethyldibenzothiophene | 0.83 |
| 7 | 212 | Dimethyldibenzothiophene | 0.34 |
| 8 | 212 | Dimethyldibenzothiophene | 1.49 |
| 9 | 212 | Dimethyldibenzothiophene | 0.80 |
| 10 | 212 | Dimethyldibenzothiophene | 0.35 |
| 11 | 226 | Trimethyldibenzothiophene | 0.09 |
| 12 | 226 | Trimethyldibenzothiophene | 0.47 |
| 13 | 226 | Trimethyldibenzothiophene | 0.23 |
| 14 | 226 | Trimethyldibenzothiophene | 0.29 |
| 15 | 226 | Trimethyldibenzothiophene | 0.47 |
| 16 | 226 | Trimethyldibenzothiophene | 0.29 |
| 17 | 226 | Trimethyldibenzothiophene | 0.05 |
| 18 | 226 | Trimethyldibenzothiophene | 0.05 |
| 19 | 234 | Benzonaphthothiophene | 0.34 |
| 20 | 248 | Methylbenzonaphthothiophene | 0.18 |
| 21 | 248 | Methylbenzonaphthothiophene | 0.20 |
| 22 | 248 | Methylbenzonaphthothiophene | 0.16 |

[a] µg/ml of oil based on the response of deuterated anthracene.

the mutagenic activity. Several dimethylbenz(a)anthracenes, trimethylbenz(a)anthracenes, and methylethylbenz(a)anthracenes are highly carcinogenic (17) and therefore probably mutagenic; some methyl chrysenes (5) and methylbenz(a)anthracenes (17) are moderately active. However, these components totalled less than 5 µg/ml of oil. Thus the majority of the observed mutagenic activity of the used crankcase oil was caused by compounds other than PAH in the 166 to 302 molecular-weight range. Possibilities include heterocyclic aromatic compounds, higher molecular weight aromatic hydrocarbons, and polar derivatives of PAH (e.g., keto, hydroxy, and nitro derivatives).

## ACKNOWLEDGMENTS

The authors thank Dr. B. N. Ames for providing tester strains of *Salmonella* and Dr. K. Sanderson for advice and guidance in conducting the Ames test.

This work was supported in part by a research grant from the Alberta Environment Research Trust.

## REFERENCES

1. Ames, B. N., Durston, W. E., Yamasaki, E., and Lee, F. D. (1973): Carcinogens are mutagens: a simple test system combining liver homogenates for activation and bacteria for detection. *Proc. Natl. Acad. Sci. U.S.A.* 70:2281-2285.
2. Ames, B. N., McCann, J., and Yamasaki, E. (1975): Methods for detecting carcinogens and mutagens with the *Salmonella*/mammalian-microsome mutagenicity test. *Mutat. Res.* 31:347-363.
3. Andrews, A. W., Thibault, L. H., and Lijinksky, W. (1978): The relationship between carcinogenicity and mutagenicity of some polynuclear hydrocarbons. *Mutat. Res.* 51:311-318.
4. Eisenstadt, E., and Gold, A. (1978): Cyclopenta(c,d)pyrene: a highly mutagenic polycyclic aromatic hydrocarbon. *Proc. Natl. Acad. Sci. U.S.A.* 75:1667-1669.
5. Hoffman, D., Bondinell, W. E., and Wynder, E. L. (1974): Carcinogenicity of methylchrysenes. *Science* 183:215-216.
6. Lao, R. C., Thomas, R. S., Oja, H., and Dubois, L. (1973): Application of a gas chromatograph-mass spectrometer-data processor combination to the analysis of the polycyclic aromatic hydrocarbon content of airborne pollutants. *Anal. Chem.* 45:908-915.
7. Lee, M. L., Novotny, M., and Bartle, K. D. (1976): Gas chromtography/mass spectrometric and nuclear magnetic resonance determination of polynuclear aromatic hydrocarbons in airborne particulates. *Anal. Chem.* 48:1566-1572.
8. Lee. M. L., Vassilaros, D. L., White, C. M., and Novotny, M. (1979): Retention indices for programmed-temperature capillary-column gas chromatography of polycyclic aromatic hydrocarbons. *Anal. Chem.* 51:768-774.
9. Maugh II, T. H. (1976): Rerefined oil: an option that saves oil, minimizes pollution. *Science* 193:1108-1110.
10. McCann, J., and Ames, B. N. (1977): The *Salmonella*/microsome mutagenicity test: predictive value for animal carcinogenicity. In: *Origins of Human Cancer,* H. H. Hiatt, J. D. Watson, and J. A. Winsten, Eds., pp. 1431-1450, Cold Spring Harbor Laboratory, New York.
11. McCann, J., Choi, E., Yamasaki, E., and Ames, B. N. (1975): Detection of carcinogens as mutagens in the *Salmonella*/microsome test: assay of 300 chemicals. *Proc. Natl. Acad. Sci. U.S.A.* 72: 5135-5139.
12. National Academy of Sciences (1972): *Particulate polycyclic organic matter.* Committee on Biologic Effects of Atmospheric Pollutants,

Division of Medical Sciences, National Research Council, Washington, D.C.
13. Natusch, D.F.S., and Tomkins, B. A. (1978): Isolation of Polycyclic organic compounds by solvent extraction with dimethyl sulfoxide. *Anal. Chem.* 50:1429-1434.
14. Parker, K. R. (1979): *Polycyclic aromatic hydrocarbons in automotive lubricants: chemical analysis and mutagenic effects in bacteria.* M.Sc. Thesis, University of Calgary.
15. Payne, J. F., Martins, I., and Rahimtula, A. (1978): Crankcase oils: are they a major mutagenic burden in the aquatic environment? *Science* 200:329-330.
16. Peake, E., and Parker, K. R. (1979): Carcinogens in the atmosphere of the city of Calgary. *Presented at the Pacific Northwest International Section of the Air Pollution Control Association, Edmonton, Alberta,* November 7 to 9, 1979.
17. Searle, C. E., Ed. (1976): *Chemical Carcinogens.* American Chemical Society, Washington, D.C.

# MUTAGENICITY, TUMOR INITIATING ACTIVITY, AND METABOLISM OF TRICYCLIC POLYNUCLEAR AROMATIC HYDROCARBONS

E. LaVoie, L. Tulley, V. Bedenko, and D. Hoffmann

    Naylor Dana Institute for Disease Prevention
    American Health Foundation
    Valhalla, New York 10595

The detection of mutagenic activity within the lower molecular weight fractions of polynuclear aromatic hydrocarbons (PAH) from coal-derived liquids and tars has led to increased concern regarding their potential as environmental carcinogens. While the exact structure of these components has not, in all instances, been completely elucidated, the major group of compounds found within these mutagenic fractions appears to be alkylated tricyclic PAH (7-9).

Fluorene, phenanthrene, and anthracene have been found to be inactive as mutagens in the Ames assay (10-12). Methylated derivatives of tricyclic PAH, however, have been shown to be mutagenic toward *S. typhimurium* TA 100 (10). In this study all the positional isomers of methylfluorene, methylphenanthrene, and methylanthracene were assayed for mutagenic activity. Compounds which were found to be mutagenic were further bioassayed for tumor-initiating activity on mouse skin (HA/ICR). This paper discusses (1) the effects of positional isomers of methylated tricyclic PAH on their metabolic activation to mutagens and (2) the potential of mutagenic activity to suggest the carcinogenic activity of tricyclic PAH.

## MATERIALS AND METHODS

### Mutagenicity Assays

Mutagenicity studies were performed as previously described (1,10) using *S. typhimurium* TA 100 (TA 1535/pKm 101) and TA 98 (TA 1538/pKm 101) provided by Dr. Bruce Ames of the University of

California, Berkeley. The S-9 fraction employed in both mutagenicity assays and in metabolic studies was obtained from the livers of male Fisher-344 rats weighing 300 to 350 g which had been treated 5 days prior to sacrifice with Aroclor$^R$ 1254 (Analabs, Inc.). All compounds assayed for mutagenicity were analyzed for purity using a Waters associates Model ALC/GPC-204 high-speed liquid chromatograph equipped with a Model 400 UV detector and a 10-$\mu$ LiChrosorb RP-18 column (EM Reagents) 4.6 mm x 250 mm. Only test samples at least 99 percent pure under these analytical conditions were employed in this study.

## Metabolism Studies

The in vitro metabolism studies were performed using the identical S-9 mix used in the mutagenicity assays. The S-9 mix contained, per ml, 100 $\mu$moles of potassium phosphate buffer at pH 7.4, 8.0 $\mu$moles of MgCl$_2$, 1.65 $\mu$moles of KCl, 5.0 $\mu$moles of glucose-6-phosphate, 4.0 $\mu$moles of NADP$^+$, and 0.5 ml of S-9 fraction. Incubations were performed at 37°C for 20 minutes using a 25-ml Erlenmeyer flask to which was added 200 $\mu$g of compound in 20 $\mu$l of methanol or dimethylsulfoxide and 2 ml of S-9 mix. The effect of epoxide hydrase inhibition in these studies was examined by employing a 2.0 x 10$^{-3}$ M concentration of TCPO in these incubation mixtures. The incubations were terminated by the addition of 2 ml of ice-cold acetone. The mixture was then extracted (5x) with 10-ml aliquots of ethylacetate. The ethylacetate solution was concentrated in vacuo below 40°C prior to analysis or purification of the metabolites by reverse-phase HPLC. Metabolites of phenanthrene and methylphenanthrenes were profiled by HPLC using a 30 to 100 percent-methanol in water gradient elution system (flow = 2 cc/min) with a 4.5 mm x 250 mm LiChrosorb 10-$\mu$m RP-18 column (EM Reagents). Incubations were performed on a preparative scale (5.0 mg) for mutagenicity studies on individual metabolites. Metabolites were separated into six individual bands by preparative and then thin-layer chromatography (TLC) using 0.5 mm Silica 60 plates (EM Reagents) and 5 percent ethanol in benzene as eluent. On each of two plates was applied 30 percent of the metabolites isolated from TLC. The identification of metabolites that were mutagenic was determined by UV spectra and mass spectra after isolation from HPLC.

## Tumor Initiation

Compounds tested as tumor initiators were applied to the skin of Swiss Albino female mice (Ha/ICR). When the animals had entered the second telogen phase of the hair cycle, the hydrocarbon to be tested was applied as a 0.1 percent solution in acetone (100 $\mu$l) ten times on alternate days to the shaved backs of the mice. Ten days after the last initiator dose, promotion was begun by application thrice weekly of 2.5 µg of tetradecanoyl phorbol acetate in acetone (100 µl) for 20 weeks.

## RESULTS AND DISCUSSION

A series of methylated fluorenes was assayed for mutagenic activity toward *S. typhimurium* TA 98 and TA 100. Among all the positional isomers of methylfluorene, only 9-methylfluorene was found to be mutagenic toward *S. typhimurium* TA 100 in the presence of rat liver homogenate (Table 1). This requirement for a single methyl group at the 9-position for mutagenic activity among methylfluorenes was found to persist within a series of dimethylfluorenes. When 1,9-, 2,3-, and 9,9-dimethylfluorene were assayed for mutagenicity toward TA 100, 1,9-dimethylfluorene was the only isomer found to be active (Table 1). None of the methylated fluorenes in this study were found to be mutagenic in the absence of metabolite activation by rat liver homogenate.

Further studies were undertaken to determine whether this structural requirement for mutagenicity among methylated fluorenes could be extended to other compounds. The absence of mutagenic activity observed when 1,1-diphenylethane was assayed under identical conditions as 9-methylfluorene indicated that the "biphenyl-character" or planarity of the fluorene-ring systems was essential. The lack of mutagenic activity observed for unsubstituted benzofluorenes provided an additional series of PAH for determining the effect of similar methylation at the benzylic position on their mutagenic potential. The synthesis of 11-methylbenzo-(a)fluorene and 7-methylbenzo(c)fluorene was performed via their corresponding benzofluorenones by coupling with methylmagnesium bromide followed by catalytic hydrogenolysis with 5 percent Pd/C (3,14). The preparation of 11-methylbenzo(b)fluorene was accomplished by coupling of the sodium salt of benzo(b)fluorene with methyliodine in dimethylsulfoxide. Each of these three methylated benzofluorenes, unlike their parent hydrocarbons (Table 1), was found to be mutagenic toward TA 100 in the presence of rat liver homogenate (Figures 1 and 2).

While metabolic activation is required, the exact mechanism by which 9-methylfluorene and related compounds elicit their mutagenic activity is not known. Both 9-hydroxymethylfluorene and 9-methyl-9-fluorenol have been assayed for mutagenicity (Table 1). The absence of mutagenic activity toward TA 100 with or without metabolic activation indicates that these hydroxylated derivatives are not proximate mutagens of 9-methylfluorene.

TABLE 1. Mutagenic Activity of Fluorene and Benzofluorene Derivatives Toward S. Typhimurium TA 100

| Compound | Mutagenic Response | Revertants/ nmol[a] | Revertants/ Plate[a] |
|---|---|---|---|
| Fluorene | — | 0.10 | 125 |
| 1-Methyfluorene | — | 0.08 | 95 |
| 2-Methylfluorene | — | 0.11 | 120 |
| 3-Methylfluorene | — | 0.09 | 100 |
| 4-Methylfluorene | — | 0.09 | 100 |
| 9-Methylfluorene | ++ | 0.50 | 550 |
| 1,9-Dimethylfluorene | +++ | 1.37 | 1420 |
| 2,3-Dimethylfluorene | — | 0.11 | 125 |
| 9,9-Dimethylfluorene | — | 0.08 | 110 |
| 1,1-Diphenylethane | — | 0.10 | 115 |
| Benzo(a)fluorene | — | 0.11 | 100 |
| Benzo(b)fluorene | — | 0.16 | 150 |
| Benzo(c)fluorene | — | 0.14 | 125 |
| 11-Methylbenzo(a)fluorene | ++ | 0.29 | 250 |
| 11-Methylbenzo(b)fluorene | +++ | 1.06 | 920 |
| 7-Methylbenzo(c)fluorene | ++ | 0.46 | 400 |
| 9-Hydroxymethylfluorene | — | 0.10 | 110 |
| 9-OH-9-Methylfluorenol | — | 0.11 | 105 |
| Dimethylsulfoxide (50 µl) | — | — | 100-130 |

[a]Based upon their mutagenic activity at a dose of 200 µg.

Phenathrene and all possible isomers of methylphenanthrene were assayed for mutagenicity in the Ames assay. Both 1-methylphenanthrene and 9-methylphenanthrene were found to be active as mutagens toward S. typhimurium TA 100 in the presence of rat liver homogenate (Figure 3). In addition to these assays on monosubstituted phenanthrenes, 2,7-dimethylphenanthrene, 3,6-dimethylphenanthrene, and 4,5-methylenephenanthrene were found to be inactive as mutagens (Table 2). None of the phenanthrenes assayed were mutagenic toward S. typhimurium TA 98 or were active in the absence of rat liver homogenate.

Several studies have been performed on the metabolism of phenanthrene and the mutagenicity of its various metabolites (4,5,13,15). In view of the variation in activity observed for the methylphenanthrene, the influence of methylation on metabolic activation to mutagen or detoxification among the various positional isomers was investigated. The metabolites of 1-methylphenanthrene, formed using the identical liver homogenate preparation employed in the mutagenicity assays, were profiled by HPLC (Figure 4). Structural assignments for these metabolites were made on the basis of their UV spectra and the mass spectra of the isolates obtained from HPLC. The presence of dihydrodiols of 1-hydroxymethylphenanthrene (peaks A and B) and

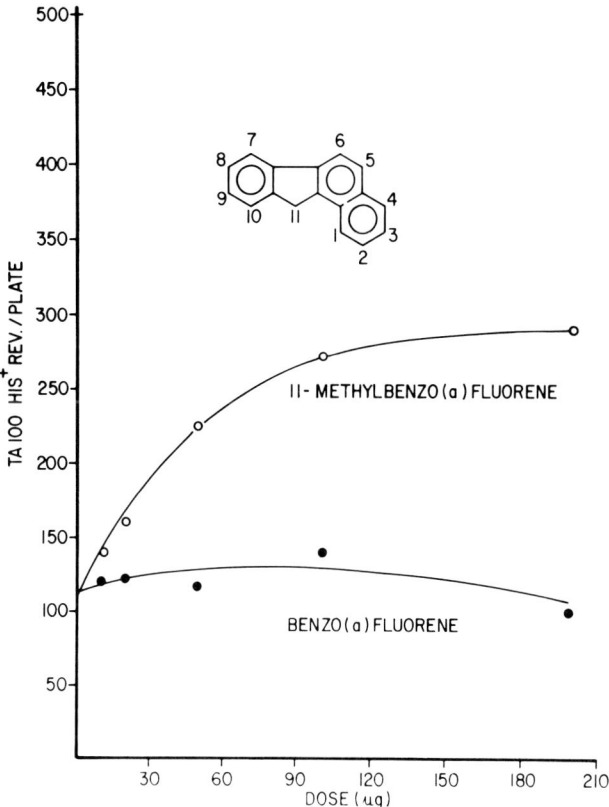

**FIGURE 1.** Mutagenicity of benzo(a)fluorene and 11-methylbenzo(a)fluorene toward *S. typhimurium* **TA 100** in the presence of rat liver homogenate.

dihydrodiols (peaks C and D) was confirmed by their absence in incubations to which trichlorpropylene oxide was added to inhibit epoxide hydrase (Figure 5). Peak E was identified as 1-hydroxymethylphenanthrene. Peaks F, G, and H were 1-methylphenanthrenols. Mutagenicity assays on these metabolites after separation by thin-layer chromatography indicated that peak D was the major proximate mutagen of 1-methylphenanthrene (725 His$^+$ revertants/plate). On the basis of its UV spectrum, this diol was either the 3,4 or 5,6-dihydrodiol of 1-methylphenanthrene. Contrasted against the metabolites obtained from 4-methylphenanthrene, the major differences in metabolism appear in the types of diols formed. In the case of 4-methylphenanthrene, peak B is the 9,10-diol and peak C is either the 7,8 or the 1,2-diol (Figure 6). Similar results were obtained with the nonmutagenic 2- and 3-methylphenanthrenes (Figures 7 and 8). The HPLC Profile of metabolites obtained from incubations with 9-methylphenanthrene is

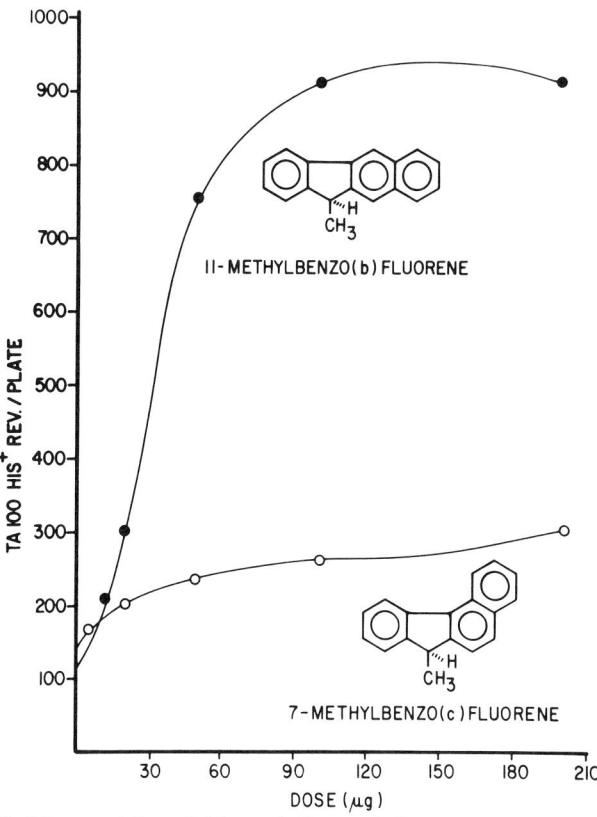

**FIGURE 2.** Mutagenicity of 11-methylbenzo(b)fluorene and 7-methylbenzo(c)-fluorene toward *S. typhimurium* **TA 100** in the presence of rat liver homogenate.

shown in Figure 9. Mutagenicity assays on these metabolites after separation by thin-film layer chromatography demonstrated that peak C contained the proximate mutagens of 9-methylphenanthrene (370 His$^+$ revertants/plate). By UV and mass spectra, this peak was shown to be either (or both) the 3,4- or 5,6-dihydrodiol of 9-methylphenanthrene. The metabolites of 3-methylphenanthrene obtained under identical conditions failed to indicate the presence of a proximate mutagen. Table 3 summarizes the results of the studies on the metabolism of methylphenanthrenes, which indicate that a primary requirement for mutagenic activity within this series of compounds is the inhibition of 9,10-diol formation. In the case of 1-methylphenanthrene and 9-methylphenanthrene, the proximate mutagen was the 3,4- or 5,6-dihydrodiol. The 1,2 or 7,8-dihydrodiol which would be the requisite intermediate for formation of a "bay region" diolepoxide was not implicated as the major proximate mutagen for these compounds.

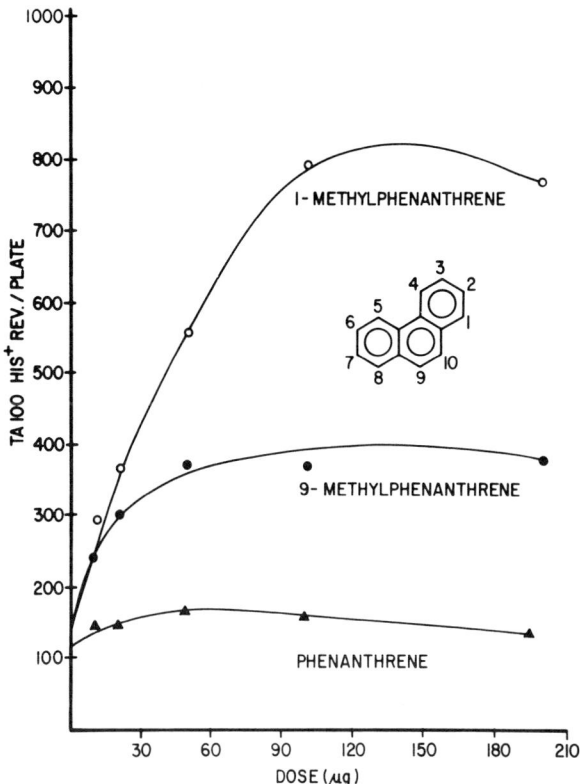

**FIGURE 3.** Mutagenicity of phenanthrene, 1-methylphenanthrene, and 9-methylphenanthrene toward S. typhimurium **TA 100** in the presence of rat liver homogenate.

No mutagenic activity was exhibited by anthracene or any of the three possible isomers of methylanthracene in either TA 98 or TA 100.

Recent studies have evaluated the effectiveness of the *Salmonella*/mammalian microsomal mutagenicity assay system in suggesting the carcinogenic potential of PAH (2,10,11). Because of the limited number of tricyclic PAH known to be mutagenic, the potential of mutagenic activity toward *S. typhimurium* to be an indicator of tumorigenic activity among tricyclic PAH has not been determined. Since tumor-initiating activity on mouse skin has been shown to correlate more closely than complete carcinogenicity with mutagenicity toward *S. typhimurium* (10), the methylphenanthrenes were assayed for tumor-initiating activity. Despite the fact that 1-methylphenanthrene and 9-methylphenanthrene were mutagenic in the *Salmonella*/mammalian microsomal assay system, none of the five possible isomers of methylphenanthrene were found to be

TABLE 2. Mutagenic Activity of Phenanthrene, Anthracene, and Their Alkylated Derivatives Toward S. TYPHIMURIUM TA 100

| Compound | Mutagenic Response | Revertants/ nmol[a] | Revertants/ Plate[a] |
|---|---|---|---|
| Phenanthrene | — | .12 | 150 |
| 1-Methylphenanthrene | +++ | 0.74 | 775 |
| 2-Methylphenanthrene | — | 0.12 | 125 |
| 3-Methylphenanthrene | — | 0.11 | 115 |
| 4-Methylphenanthrene | — | 0.07 | 73 |
| 9-Methylphenanthrene | ++ | 0.37 | 385 |
| 3,6-Dimethylphenanthrene | — | 0.12 | 125 |
| 2,7-Dimethylphenanthrene | — | 0.13 | 130 |
| 4,5-Methylenephenanthrene | — | 0.12 | 130 |
| Anthracene | — | 0.09 | 120 |
| 1-Methylanthracene | — | 0.10 | 105 |
| 2-Methylanthracene | — | 0.12 | 120 |
| 9-Methylanthracene | — | 0.12 | 125 |
| Dimethylsulfoxide (50 µl) | — | — | 100-130 |

[a] Based upon their mutagenic activity at a dose of 200 µg.

active as tumor initiators. Similarly, 9-methylfluorene and 1,9-dimethylfluorene, which are both active as bacterial mutagens, were assayed as tumor initiators. The results, shown in Table 4, demonstrate that 9-methylfluorene was not active as a tumor initiator. Because of the possibility of weak tumor initiating activity for 1,9-dimethylfluorene, further bioassays employing larger experimental groups are in progress.

The inability of the *Salmonella* mutagenicity assay to effectively discern which PAH are noncarcinogenic appears to persist even in the case of the less complex tricyclic PAH. While the bioassays employed for determining carcinogenic activity in this study are not exhaustive, these data suggest cautious interpretation of the significance of the mutagenic activity of PAH in the Ames assay as it may relate to environmental carcinogenesis.

## ACKNOWLEDGMENTS

This study was supported by Grant No. CA 12376.

We thank Dr. Hans D. Sauerland of Ruetgerswerke A.G. Duisburg, Federal Republic of Germany for samples of several methylphenanthrenes and methylfluorenes. We also express our appreciation to Professor Melvin S. Newman of The Ohio State University for providing 2,7-dimethylphenanthrene.

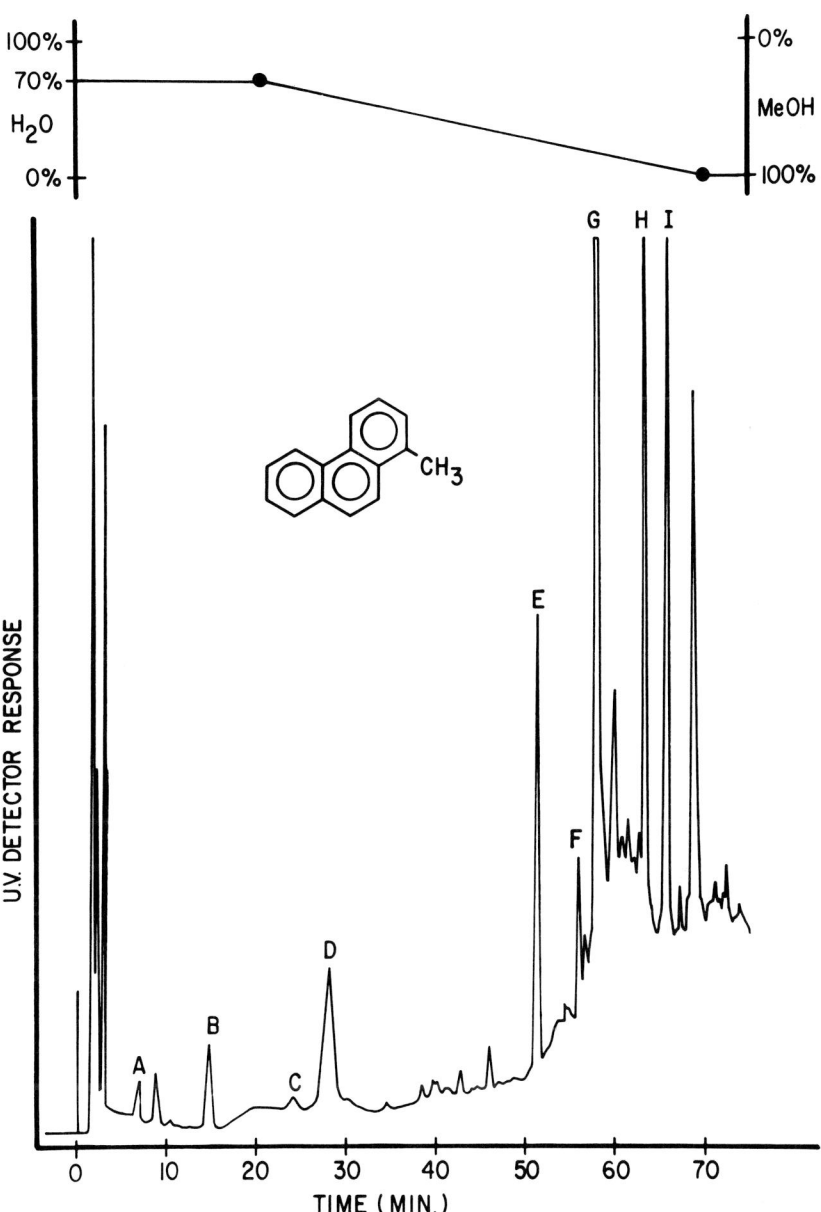

**FIGURE 4.** Metabolites of 1-methylphenanthrene formed upon incubation with rat liver 9000 x g supernatant. The identity of peaks A through I is indicated in Table 3.

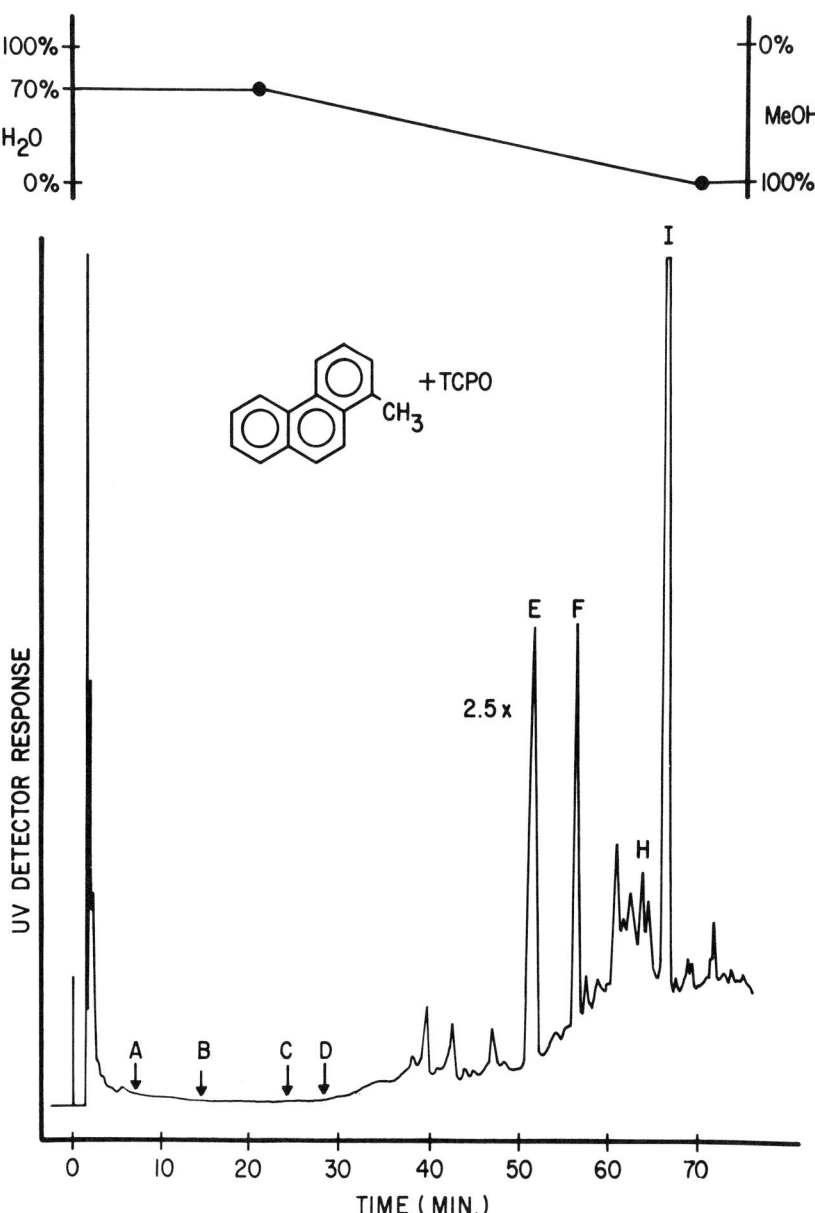

FIGURE 5. Metabolites of 1-methylphenanthrene formed upon incubation with rat liver 9000 x g supernatant and trichloropropylene oxide (TCPO). The identity of peaks E through I is indicated in Table 3. Arrows indicate retention time of diols formed in the absence of TCPO (see Figure 4.)

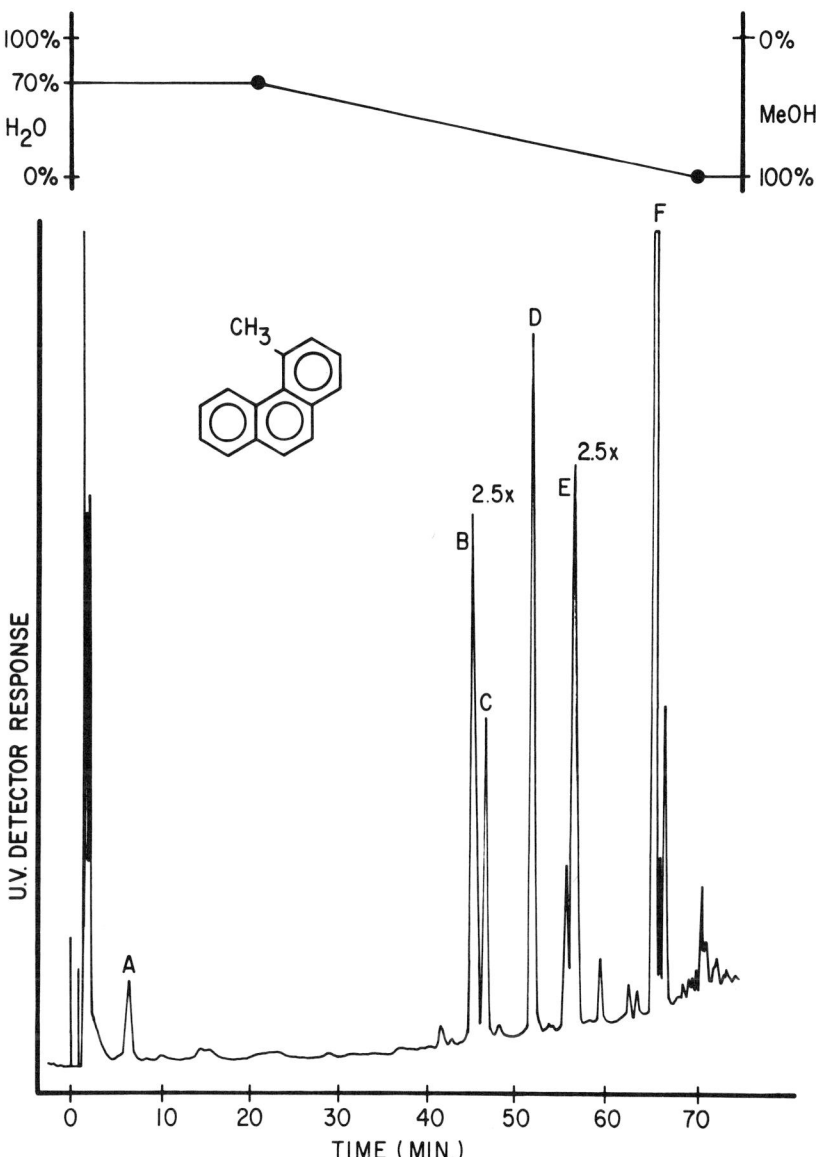

FIGURE 6. Metabolites of 4-methylphenanthrene formed upon incubation with rat liver 9000 x g supernatant. The identity of peaks A through F is indicated in Table 3.

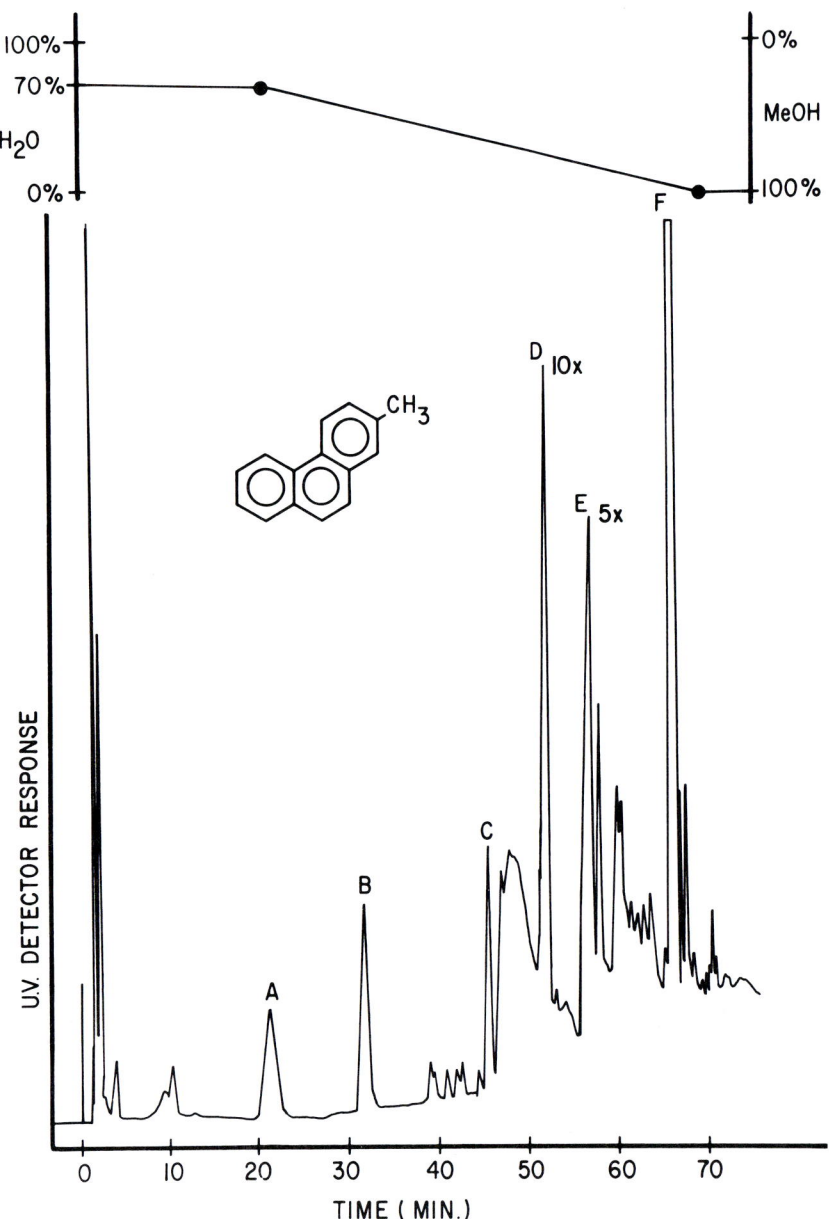

FIGURE 7. Metabolites of 2-methylphenanthrene formed upon incubation with rat liver 9000 x g supernatant. The identity of peaks A-F is indicated in Table 3.

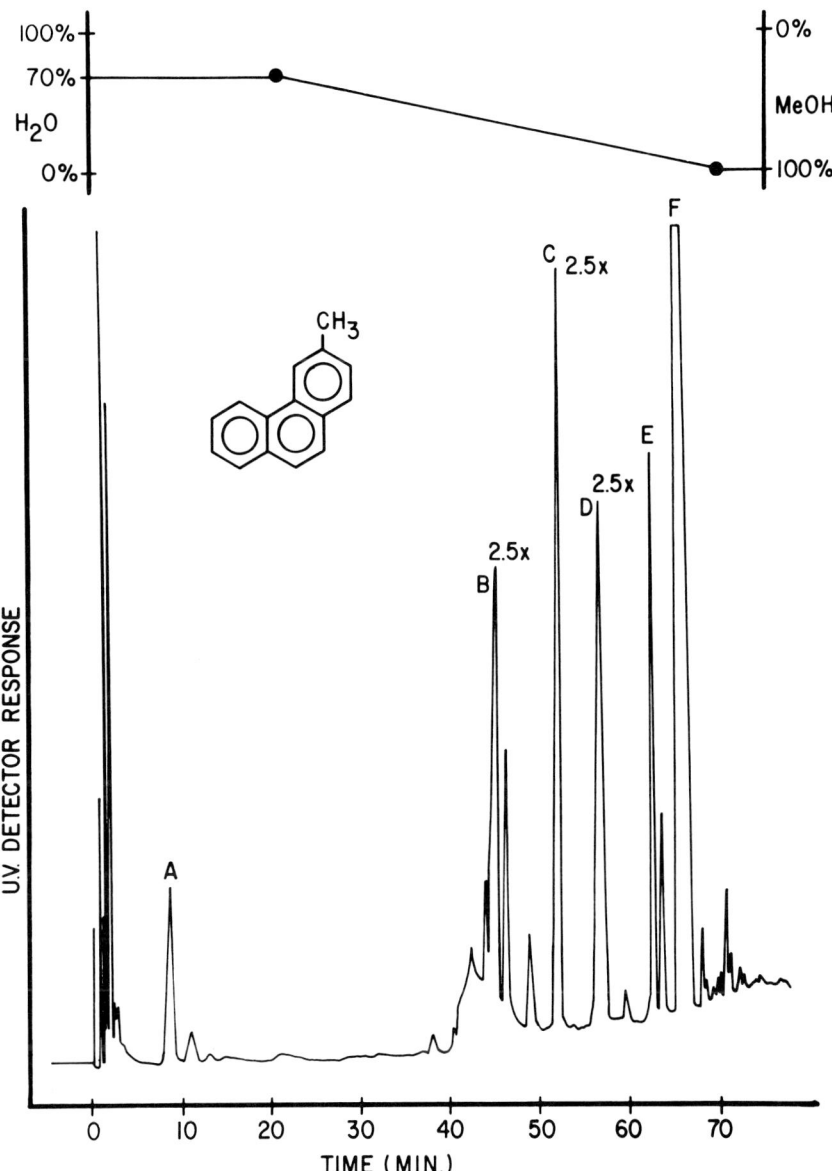

FIGURE 8. Metabolites of 3-methylphenanthrene formed upon incubation with rat liver 9000 x g supernatant. The identity of peaks A-F is indicated in Table 3.

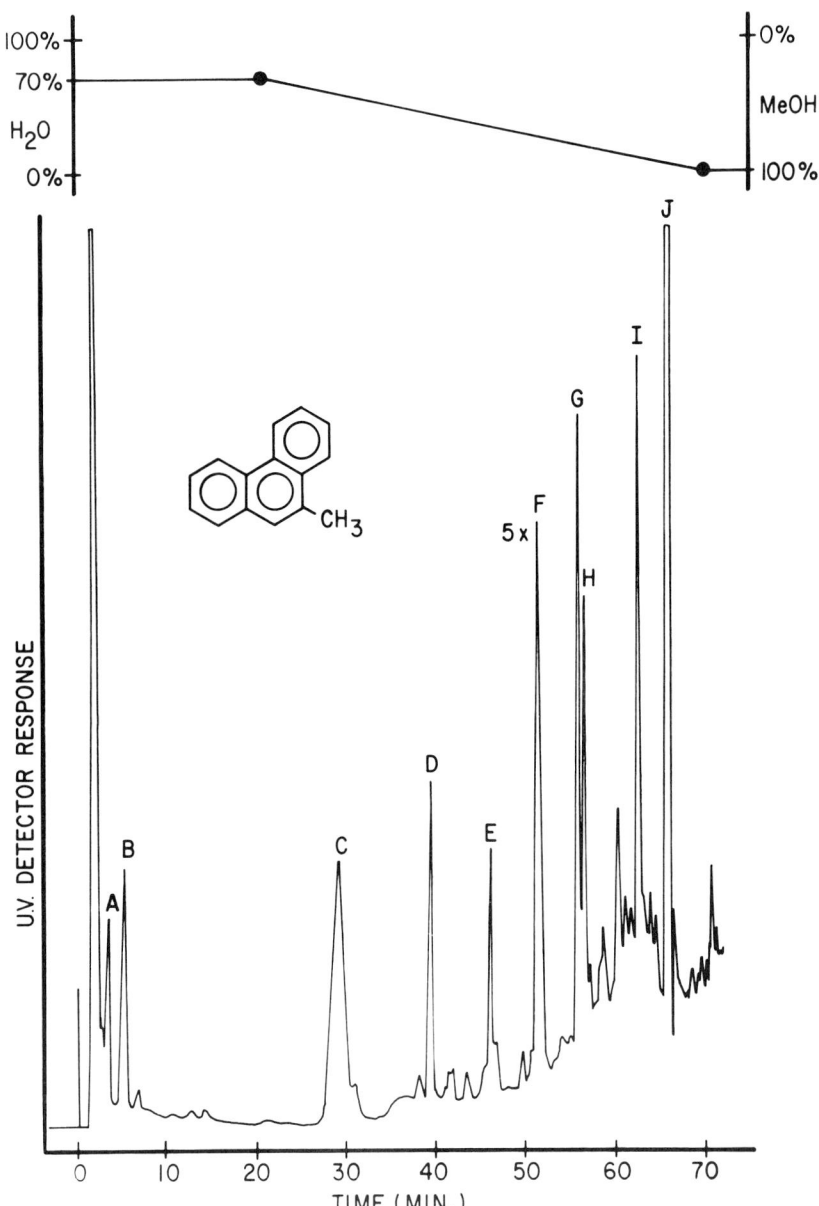

FIGURE 9. Metabolites of 9-methylphenanthrene formed upon incubation with rat liver 9000 x g supernatant. The identity of peaks A-J is indicated in Table 3.

TABLE 3. Metabolites Identified from HPLC Profiles in Figures 4 Through 9

| Metabolites | Methylphenanthrenes | | | | |
|---|---|---|---|---|---|
| | $1\text{-}CH_3$ Peak | $2\text{-}CH_3$ Peak | $3\text{-}CH_3$ Peak | $4\text{-}CH_3$ Peak | $9\text{-}CH_3$ Peak |
| Hydroxymethyldihydrodiols | | | | | |
| $CH_2OH$-3,4 or 5,6-Diol | A | | | | A |
| $CH_2OH$-1,2 or 7,8-Diol | B | | | | B |
| $CH_2OH$-9,10-Diol | | | A | A | |
| Dihydrodiols | | | | | |
| 3,4 or 5,6 | C,D | A,B | | | C |
| 1,2 or 7,8 | | | | C | E |
| 9,10 | | C | B | B | |
| Hydroxymethyl Phenols | | | | | D |
| Hydroxymethyl Phenols | E | D | C | D | F |
| | F,G,H | E | D,E | E | G,H,I |
| Methylphenanthrene | I | F | F | F | J |

TABLE 4. Tumor Initiating Activity of Fluorene and Methylated Fluorenes

| Compound | Total Dose[a] (mg/animal) | Percent Tumor-Bearing Animals | Tumors Per Animal |
|---|---|---|---|
| Fluorene | 1.0 | 5 | .05 |
| 9-Methylfluorene | 1.0 | 5 | .05 |
| 1,9-Dimethylfluorene | 1.0 | 20 | 0.2 |
| Control | 0 | 0 | 0 |

[a] Applied as 10 subdoses on mouse skin; promotion was performed as outlined in the Materials and Methods section with tetradecanoyl acetate.

# REFERENCES

1. Ames, B. N., McCann, J., and Yamasaki, E. (1975): Methods for detecting carcinogens and mutagens with the *Salmonella*/mammalian microsome mutagenicity test. *Mutat. Res.* 31:347-364.

2. Andrews, A. W., Thibault, L. H., and Lijinsky, W. (1978): The relationship between carcinogenicity and mutagenicity of some polynuclear hydrocarbons. *Mutat. Res.* 51:311-318.
3. Badger, G. M. (1941): The synthesis of growth-inhibitory polycyclic compounds. *J. Chem. Soc.* (London), 535-538.
4. Boyland, E., and Sims, P. (1962): The metabolism of phenanthrene in rabbits and rats: dihydrodihydroxy compounds and related glucosiduronic acids. *Biochem. J.* 84:571-582.
5. Bucker, M., Glatt, H. R., Platt, K. L., Avnir, D., Ittak, Y., Blum, J., and Oesch, F. (1979): Mutagenicity of phenanthrene and phenanthrene K-region derivatives. *Mutat. Res.* 66:337-348.
6. Coombs, M. M., Dixon, C., and Kissonerghis, A. (1976): Evaluation of the mutagenicity of compounds of known carcinogenicity, belonging to the benz(a)anthracene, chrysene, and cyclopenta(a)phenanthrene series, using Ames' test. *Cancer Res.* 36:4525-4529
7. Epler, J. L., Young, J. A., Hardegree, A. A., Rao, T. K., Guerin, M. R., Rubin, I. B., Ho, C.-H., and Clark, B. R. (1978): Analytical and biological analysis of test materials from synthetic fuel technologies. Mutagenicity of crude oils determined by the *Salmonella typhimurium*/microsomal activation system. *Mutat. Res.* 57:265-276.
8. Griest, W. H., Tomkins, B. A., Epler, J. L. and Rao, T. K. (1979): Characterization of multialkylated polycyclic aromatic hydrocarbons in energy-related materials. In: *Carcinogenesis,* Vol. 4, P. W. Jones and P. Leber, Eds., pp. 395-409, Raven Press, New York.
9. Guerin, M. R., Epler, J. L., Griest, W. H., Clark, B. R., and Rao, T. K. (1978): Polycyclic aromatic hydrocarbons from fossil fuel conversion processes. In: *Carcinogenesis,* Vol 3, P. W. Jones and R. I. Freudenthal, Eds., pp. 21-33, Raven Press, New York.
10. LaVoie, E. J., Bedenko, V., Hirota, N., Hecht, S. S. and Hoffmann, D. (1979): A comparison of the mutagenicity, tumor initiating activity and complete carcinogenicity of polynuclear aromatic hydrocarbons. In: *Carcinogenesis,* Vol 4, P. W. Jones and P. Leber, Eds., pp. 705-721, Raven Press, New York.
11. McCann, J., Choi, E., Yamasaki, E., and Ames, B. N. (1975): Detection of carcinogens as mutagens in the *Salmonella*/Microsome test: assay of 300 chemicals. *Proc. Nat. Acad. Sci., U.S.* 72:5135-5139.
12. Purchase, J.F.H., and Longstaff, E., Ashly, J., Slytes, J. A., Anderson, D., Lefevre, P. A., and Westwood, F. R. (1978): An evaluation of 6-short-term tests for detecting organic chemical carcinogens. *Brit. J. Cancer* 37:873-959.
13. Sim, P. (1970): Qualitative and quantitative studies on the metabolism of a series of aromatic hydrocarbons by rat-liver preparations. *Biochem. Pharmacol.* 19:795-818.

14. Sprinzak, Y. (1958): Reactions of active methylene compounds in pyridine solution. The ionic autoxidation of fluorene and its derivatives. *J. Amer. Chem. Soc.* 80:5449-5455.
15. Wood, A. W., Chang, R. L., Levin, W., Yagi, H., Jerina, D. M., and Conney, A. H. (1979): Biological activity of the bay region diol epoxides of chrysene and phenanthrene. *Proc. Amer. Assoc. Cancer Res.* p. 222.

# MIXED FUNCTION OXIDASE INDUCIBILITY AND POLYAROMATIC HYDROCARBON METABOLISM IN THE MULLET, SEA CATFISH, AND GULF KILLIFISH

P. Melius*, D. Elam*, M. Kilgore*, B. Tan*, and W. P. Schoor**

*Chemistry Department, Auburn University, Auburn, AL 36830
**U. S. Environmental Protection Agency
Environmental Research Laboratory
Sabine Island, Gulf Breeze, Florida 22561

## INTRODUCTION

The hydroxylation of corticosteroids, the duration and intensity of activity mediated by certain drugs, and the metabolism of polynuclear aromatic hydrocarbons (PAH) are activities attributed to certain inducible microsomal enzymes and hemoproteins that constitute the mixed-function oxidase (MFO) system (7,15,16). In the past two decades significant advances have been made with regard to characterization of mammalian MFO systems (21,23,25,29,20,38,40,18,28). Certain components, particularly cytochrome $P_{450}$ and the associated reductase, are ubiquitous; these constituents have been detected in bacteria, insects, plants, and several animal tissues (47,36,46,11,33). Brodie and Maikel (8) conducted studies which indicated that fish were incapable of microsomal drug oxidation. Later studies of Buhler (9) and Lech (30) demonstrated that some fish species detoxified xenobiotics by hepatic metabolic activity. Recent research efforts in several laboratories have indicated that aquatic organisms possess microsomal oxidation, reduction, and conjugation systems. Generally, the temperature optima and activities of aquatic organism MFO systems are lower than mammalian MFO systems (45,2,39,13,24,6,1,12). Several workers have now detected BaP metabolites from freshwater and marine fish species with both in vivo and in vitro systems (3,39,31,41,27). PAH are stable, increasingly common environmental pollutants that are likely to accumulate in and pose a threat to the marine environment. Our research has therefore been directed toward PAH metabolism and MFO induction by PAH in marine organisms.

## MATERIALS AND METHODS

### Reagents

Bacteriological media and related materials were purchased from Difco Laboratories or KC biologicals. Biochemicals were obtained from Sigma Chemical Company. Organic solvents were obtained from Fischer Scientific. Benzo(a)pyrene (BaP), BaP metabolites, 3-methylcholanthrene (3-MC), and Aroclor® 1254 were provided by Dr. James N. Keith at the ITT Research Institute, Chicago, Illinois.

### Induction of Experimental Animals

Mullet *(Mugil cephalus)* weighing 250 to 350 g and sea catfish *(Arius felis)* weighing 100 to 500 g were collected from coastal waters in northwestern Florida during April and June, respectively. Fish were transferred to flow-through tanks (approximately 250 ℓ seawater/hr, 18 to 22°C) located at the EPA Research Laboratory, Sabine Island in Gulf Breeze, Florida. Gulf killifish (*Fundulus grandis*) were obtained from the staff at the Claude Peteet Mariculture Center located in Gulf Shores, Alabama. Fish were transferred to the EPA Research Laboratory at Gulf Breeze, Florida, and placed in 55-gallon flow-through aquaria (approximately 50 ℓ seawater/hr, 18 to 22°C). Mullet, sea catfish, and gulf killifish were acclimated for 7, 14, and 1 day(s), respectively. Following acclimation, fish were administered a single intraperitoneal (IP) injection of the desired inducing agent. All inducing agents were prepared in sterile corn oil such that 1 kg body mass received the inducing agent in 1 ml corn oil. Each experimental group was maintained in a separate tank. During acclimation and induction periods, sea catfish were fed frozen shrimp and it was assumed that mullet and gulf killifish subsisted on the plankton present in the seawater flowing through the tanks. Information regarding inducing agents, dose levels, and exposure times is summarized in Table 1.

### Preparation of Fish Liver Homogenates

Following the induction period, liver homogenates (S-9 fractions) were prepared in accordance with methods established for the rat (4,17,22). The livers were removed aseptically from exsanguinated animals killed by cervical dislocation and rinsed twice in 10-ml volumes of iced, sterile 0.5 M KCl, transferred to beakers containing 3 ml of 0.15 M

**TABLE 1. Summary of Induction Methods**

| Animal | Designation | Treatment | Duration | No. of Subjects |
|---|---|---|---|---|
| Mullet (Mugel cephalus) | | | | |
| Control #1 | MCC | None | 11D | 6 |
| Control #2 | MC | 1.0 ml sterile corn oil/kg body mass | 11D | 6 |
| Aroclor® treated[a] | M50 A | 50 mg Aroclor® 1254/kg body mass | 11D | 2 |
| Aroclor® treated | M100 A | 100 mg Aroclor® 1254/kg body mass | 11D | 5 |
| Aroclor® treated | M200 A | 200 mg Aroclor® 1254/kg body mass | 11D | 6 |
| Sea Catfish (Arius felis) | | | | |
| Control #1 | CCC | None | 7D | 5 |
| Control #2 | CC | 1.0 sterile corn oil/kg body mass | 7D | 5 |
| 3-MC treated[a] | C10 | 10 mg 3mc/kg body mass | 7D | 6 |
| 3-MC treated | C20 MC | 20 mg 2mc/kg body mass | 7D | 7 |
| 3-MC treated | C40 MC | 40 mg 3mc/kg body mass | 7D | 7 |
| Gulf Killifish (Fundulis grandis) | | | | |
| Control #1 | KCC | None | 7D | 14 |
| Control #2 | KC | 1.0 ml sterile corn oil/kg body mass | 7D | 12 |
| BaP treated[c] | KBP | 30 mg BaP/kg body mass | 7D | 9 |
| 3-MC treated[b] | KMC | 30 mg/kg body mass | 7D | 11 |

[a] Aroclor® 1254 is a registered trademark of Monsanto Chemical Company and is a mixture of various polychlorinated biphenyls.
[b] 3-MC, 3-Methylcholanthrene.
[c] BaP, Benzo(a)pryene.

KCl/g liver mass, minced with sterile scissors, and homogenized on ice in a Potter-Elvehjem homogenizer (10 to 12 strokes) or in a Brinkman Polytron (setting 5 to 6, 30 seconds). The homogenate was centrifuged at 9000 g and 0 to 4°C for 25 minutes. The supernatant was collected aseptically and stored in 0.5 to 1-ml volumes in sterile screw-cap tubes at -20°C and used within 6 weeks of preparation.

## Spectrophotometric Determinations

Protein concentrations were determined by the method of Lowry, et al (32) using bovine serum albumin as the standard. Protein concentrations averaged 25 mg protein/ml S-9 fraction. NADH-Cytochrome $b_5$ reductase (E.C. 1.6.2.2) and NADPH-cytochrome $P_{450}$ reductase (E.C. 1.6.2.4) were determined spectrophotometrically as described by Comai and Gaylor (14). Cytochrome $P_{450}$ content was determined by the method of Omura and Sato (38) and when hemoglobin interference was suspected, by the method of Miyake et al (37). Spectrophotometric determinations were performed using either a Gilford Model 250 spectrophotometer or a Cary 17 spectrophotometer.

## In Vitro Metabolism of BaP

BaP metabolites were formed in vitro and analyzed by high-pressure liquid chromatography (HPLC), employing modified procedures of Selkirk et al (42,43). Metabolism was initiated by adding 100 nmoles BaP in 0.06 ml methanol to a solution containing 50 $\mu$moles tris-hydroxyaminomethane hydrochloride (tris-HCl), pH 7.2 or 7.5, 5 $\mu$moles $MgCl_2$, 1 $\mu$mole nicotinamide adenine dinucleotide phosphate (NADP), 10 $\mu$moles glucose-6-phosphate (G-6-P), 1 unit G-6-P dehydrogenase (E.C. 1.1.1.49), and 1 to 2 mg S-9 protein to yield a final volume of 1.0 ml. Experiments were performed in duplicate and tubes were allowed to equilibriate 5 minutes at the desired temperature before the addition of the substrate. Incubations were conducted at 25, 30, or 37°C with constant shaking (setting 6 on a Precision Scientific Dubnoff metabolic incubator) for 15, 20, or 30 minutes. The reactions were terminated by the addition of 1.0 ml acetone and BaP metabolites were subsequently extracted five times with 2.0-ml volumes of ethyl acetate. Ethyl acetate extracts were dried over 1.0 g anhydrous $MgSO_4$, duplicate samples pooled, and then evaporated to dryness under a stream of nitrogen at room temperature. Metabolites were dissolved in 0.5 ml nitrogen-purged methanol and stored at $-20°C$ under nitrogen until HPLC analyses could be performed.

HPLC analyses were performed using a Lichrosorb® RP-18 (Merck) packed stainless steel column (0.4 cm x 250 cm) in either a Micromeritics Model 7000 B liquid chromatograph equipped with a Schoffel Model 970 spectrofluorometer (excitation: 263 nm; emission 370 nm) or a Waters Model 660 liquid chromatograph in conjunction with a Waters Model 120 filter fluorometer (excitation: 360 nm; emission: 440 nm). Solvent compositions, flow rates, column temperatures, and column pressures may be found in the legends of Figures 1, 2, and 3.

## Salmonella Microsome Mutagenicity Assay

The *Salmonella typhimurium* tester strains were obtained from Dr. B. N. Ames, Biochemistry Department, University of California, Berkeley, California. Organisms were checked routinely for biotin and histidine requirements, crystal violet sensitivity, ampicillin resistance (TA 98 and TA 100, only), and response to appropriate mutagens (4,35,34,5,19). TA 98 proved most useful for our purposes since it responded more favorably to BaP than did TA 1535 or TA 1538, yet possessed a moderate reversion rate in the presence of S-9 protein. TA 98 was grown overnight in Oxoid nutrient broth #II (KC Biologicals) at 37 °C with shaking (setting 6 on a Precision Scientific Dubnoff metabolic incubator). The turbidity of the

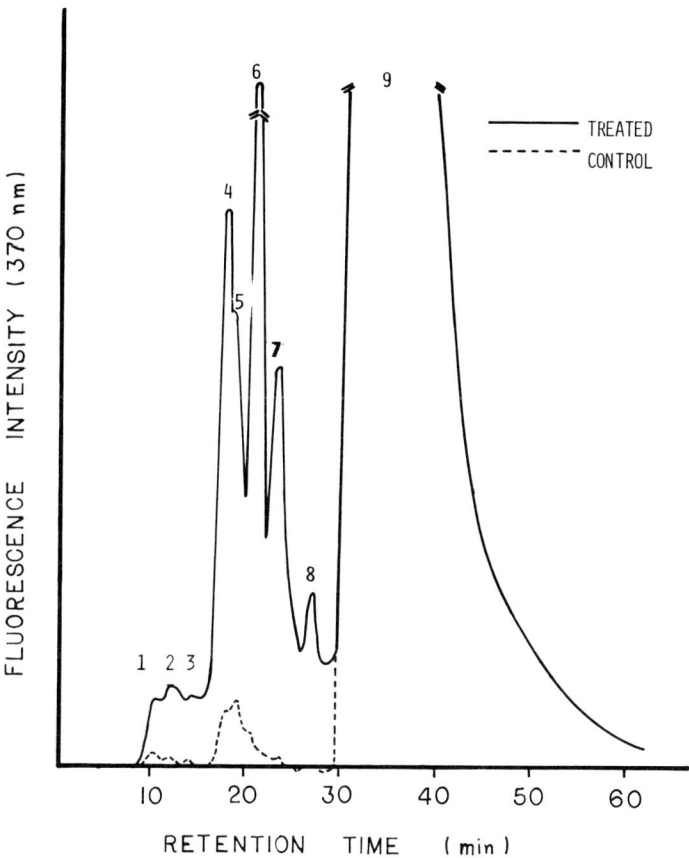

**FIGURE 1.** BaP metabolite profile of control(MC)- and Aroclor® (M200a)-treated mullet. Metabolites were formed in 30 minutes at pH 7.5 and 37°C using 2 mg S-9 protein. Five µl of sample was injected. Metabolites were eluted from the column with 100 percent methanol at 0.3 ml/min, 50°C and 100 psi, using the Micromeritics LC. Metabolites [benzo(a)pyrenes]: (1) trans-9, 10-dihydrodiol; (2) cis-4,5-dihydrodiol; (3) cis-7,8-dihydrodiol; (4) 1,6- or 3,6-quinone; (5) 6,12-quinone; (6) 9-hydroxy; (7) 3-hydroxy; (8) not identified, possibly 6-hydroxy; and (9) BaP.

culture was adjusted to a McFarland 4 (1.2 x $10^9$ total cells/ml) and 0.1 ml was then inoculated into polystyrene tubes containing 2 mg S-9 protein, 0.1 ml of 10 mM $NADP^+$; 0.5 ml of 100 mM tris-HCl, pH 7.4; 0.1 ml of 100 mM G-6-P; and 1 unit of G-6-P dehydrogenase. (The buffer and NADPH generating system were prepared in batches of 100 ml, filter sterilized by passage through Nalge ® 0.45-µm disposable filtration units, and distributed into tubes as 0.8-ml aliquots.) Following the addition

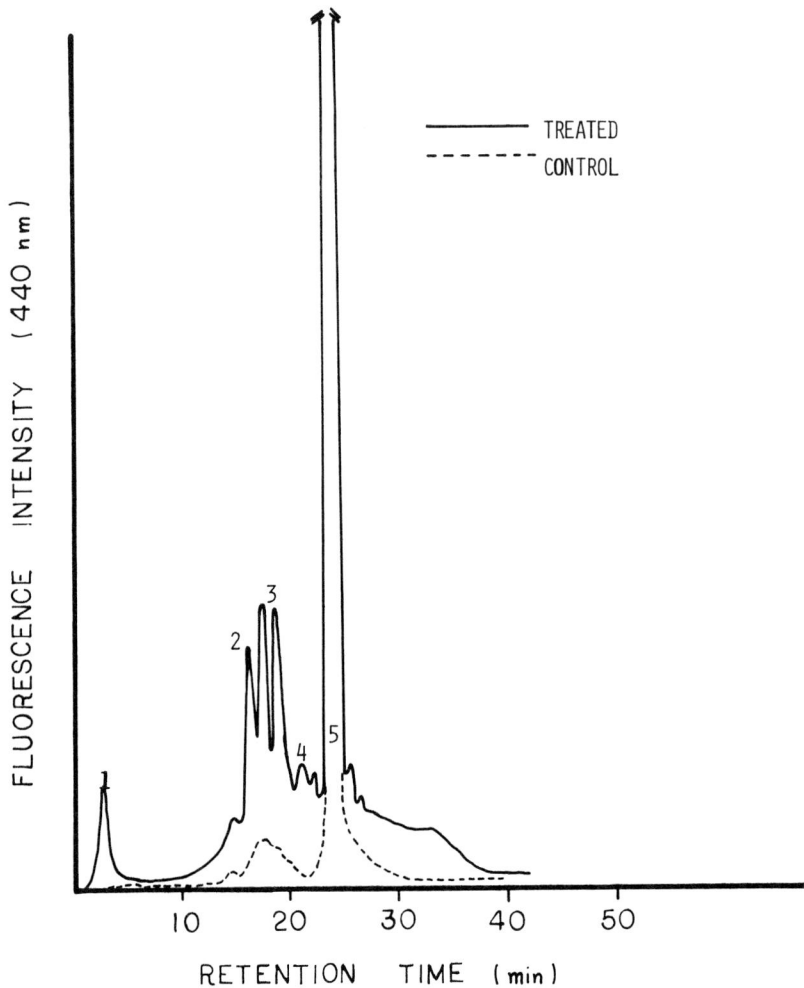

FIGURE 2. BaP metabolite profile of control(CC)- and 3-MC-treated (C20 MC) sea catfish. Metabolites were formed in 15 minutes at pH 7.2 and 30°C using 2 mg S-9 protein. Twenty µl of sample was injected. Metabolites were eluted from the column with a linear gradient of methanol (65 percent $CH_3OH$; 35 percent $H_2O$ to 100 percent $CH_3OH$, $\Delta$ 2.5 percent/min) at 0.5 ml/min, room temperature, and 400 psi using the Waters LC. Metabolites [benzo(a)pyrenes]: (1) not identified suspected tetrol; (2) dihydrodiols; (3) quinones; (4) phenols; and (5) BaP.

of 20 nmoles BaP in 0.1 ml dimethylsulfoxide (DMSO), tubes were incubated for 30 minutes at 25°C with shaking. Two ml of soft agar (0.5 percent NaCl, 0.6 percent Difco Agar) containing biotin and histidine (10 ml of 0.5 mM histidine-biotin/100 ml soft agar) was added to the tubes at the end of the incubation period. The tubes were agitated

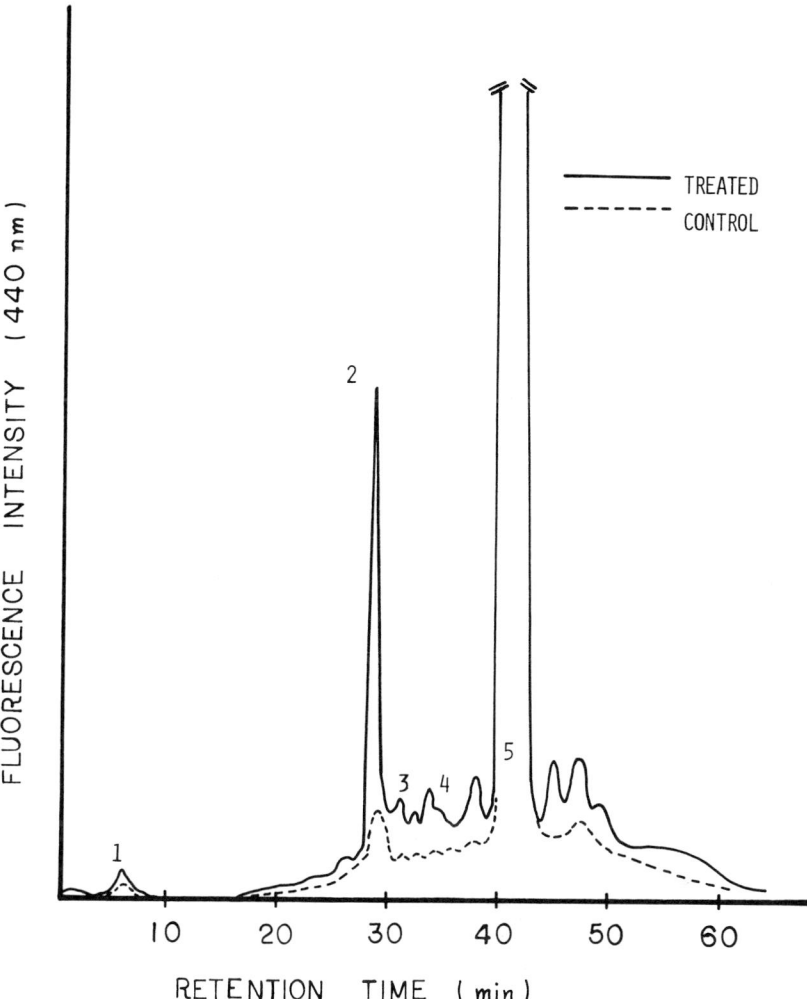

**FIGURE 3.** BaP metabolite profile of control(KC)- and 3-MC-treated (K3OMC) gulf killifish. Metabolites were formed in 20 minutes at pH 7.5 and 25°C using 1 mg S-9 protein. Forty $\mu$l of sample was injected. Metabolites were eluted with a linear gradient of methanol (65 percent $CH_3OH$; 35 percent $H_2O$ to 100 percent $CH_3OH$, $\Delta$ 1.75 percent/min) at 0.5 ml/min, room temperature and 350 psi using the Waters LC. Metabolites [benzo(a)pyrenes]: (1) not identified, suspected tetrol; (2) dihydrodiols; (3) quinones; (4) phenols; and (5) BaP.

gently and their contents poured onto the surface of Davis minimal agar plates (Difco, supplemented with 4 g glucose/l). Revertants were counted after incubation of plates for 48 hours in the dark. Values reported represent the mean ($\bar{Y}$) and standard error of the mean ($s\bar{y}$) for 3 to 5 plate counts.

## RESULTS AND DISCUSSION

### Physical Parameters

Several physical parameters have been monitored and compared in both control and treated fish (Table 2). The liver-mass to body-mass ratios in neither the mullet nor the sea catfish increase significantly upon treatment with Aroclor® 1254 or 3-MC, respectively. Similar results were obtained by Addison et al (1) using hepatic preparations from DDT-treated trout. The liver-mass to body-mass ratios obtained from the BaP treated gulf killifish were approximatley 18 percent higher than those obtained from either the control or the 3-MC treated gulf killifish.

Trends in the liver-protein to liver-mass ratios also vary among the species studied. Of the three species, only the mullet liver-protein to liver-mass ratios approach those observed in the rat, but the mullet ratio does not seem to significantly increase upon treatment as is noted in the rat *(Rattus norvegicus)* (unpublished data). Although the liver-protein to liver-mass ratios of the sea catfish and gulf killifish are 50 to 80 percent lower than those of the mullet, both sea catfish and gulf killifish liver-protein to liver-mass ratios increase significantly in the treated animals.

TABLE 2. Liver Mass/Body Mass Ratios and Liver (S-9) Protein/Liver Mass Ratios of Control and Experimental Animals[a]

| Sample | Liver Mass/Body Mass (g/kg) | S-9 Protein/Liver Mass (mg/g) |
|---|---|---|
| M-CC | 10.7 ± 1.4 | 124 ± 6 |
| M-C | 12.4 ± 0.9 | 125 ± 16 |
| M-50A | 12.0 ± 0.4 | 150 ± 7 |
| M-100A | 11.8 ± 0.5 | 120 ± 7 |
| M-200A | 13.8 ± 1.0 | 133 ± 21 |
| C—CC | 13.7 ± 2.1 | 35.1 ± 1.9 |
| C-C | 10.8 ± 1.0 | 39.3 ± 2.6 |
| C-10MC | 13.3 ± 1.2 | 52.8 ± 4.3 |
| C-20MC | 12.2 ± 1.4 | 49.8 ± 2.2 |
| C-40MC | 15.5 ± 1.2 | 53.6 ± 4.4 |
| K-CC | 20.4 ± 1.2 | 23.7 ± 3.0 |
| K-C | 18.0 ± 2.6 | 27.2 ± 1.8 |
| K-BP | 28.4 ± 3.6 | 37.1 ± 2.7 |
| K-MC | 21.9 ± 1.7 | 40.3 ± 2.1 |

[a] The values reported represent the means of three replicates (±) the standard errors of the means ($\bar{Y} \pm s\bar{y}$).

The ratios increase 40 and 45 percent over the corresponding controls in the 3-MC-treated sea catfish and in the 3-MC- or BaP-treated gulf killifish, respectively. Burns (10) has also noted an increase in liver protein from phenylbutazone-treated killifish *(Fundulus heteroclitus)* and from killifish collected from polluted environments, but Hill et al (26) detected no increase in liver protein in Aroclor® 1254 treated channel catfish *(Ictalurus punctatus)*.

## Enzyme Determinations

NADH-cytochrome $b_5$ reductase and NADPH-cytochrome $P_{450}$ reductase activities have been monitored in both control and treated fish (Table 3). The specific activity of NADH-cytochrome $b_5$ reductase ($\mu$moles ferricyanide reduced/mg protein-min) remains unchanged in treated mullet and decreases slightly in treated sea catfish and gulf killifish, suggesting the NADH-cytochrome $b_5$ reductase is not induced by these treatments. NADPH-cytochrome $P_{450}$ reductase specific activity (nmoles cytochrome c reduced/mg protein-min) increases marginally in

TABLE 3. NADP-Cytochrome $B_5$ Reductase and NADPH-Cytochrome $P_{450}$ Reductase Activities in Control and Experimental Animals[a]

| Sample | Cyt-$b_5$ Reductase Activity[b] ($\mu$moles substrate reduced/mg-min) | Cyt-$P_{450}$ Reductase Activity[c] (nmoles substrate reduced/mg-min) |
|---|---|---|
| M-CC | 3.4 ± 0.4 | 35 ± 7 |
| M-C | 2.1 ± 0.1 | 34 ± 4 |
| M-50A | 2.6 ± 0.3 | 35 ± 5 |
| M-100A | 2.8 ± 0.1 | 38 ± 2 |
| M-200S | 2.6 ± 0.2 | 38 ± 5 |
| C-CC | 4.3 ± 0.1 | ND[d] |
| C-C | 6.5 ± 0.1 | ND |
| C-10MC | 3.6 ± 0.1 | ND |
| C-20MC | 5.5 ± 0.1 | ND |
| C-40MC | 4.5 ± 0.1 | 4.5 ± 0.1 |
| K-CC | 7.1 ± 0.1 | ND |
| K-C | 6.8 ± 0.1 | ND |
| K-BP | 5.0 ± 0.1 | ND |
| K-MC | 4.6 ± 0.1 | 3.3 ± 0.6 |

[a] The values reported represent the means of three replicate (±) the standard errors of the means ($\bar{Y} \pm s\bar{y}$).
[b] Specific activity: $\mu$moles Ferricyanide reduced/mg S-9 protein-min.
[c] Specific activity: nmoles cytochrome C reduced/mg S-9 protein-min.
[d] ND—not detectable.

all treated fish, indicating possible induction of the NADPH dependent cytochrome $P_{450}$ associated reductase. As illustrated in Table 3, the specific activity of NADPH cytochrome $P_{450}$ reductase in the mullet is approximately 10-fold higher than that of the sea catfish and gulf killifish. In both the sea catfish and gulf killifish, the only detectable NADPH-cytochrome $P_{450}$ reductase activity was found in subjects treated with 40 mg 3MC/kg, respectively. Although NADH-cytochrome $b_5$ reductase levels in all fish studied were comparable to those detected in the rat, the levels of NADPH-cytochrome $P_{450}$ reductase in the fish were lower than in similar rat preparations (unpublished data).

All control and treated S-9 fractions were subjected to carbon monoxide difference spectroscopy to determine cytochrome $P_{450}$ content. An approximate twofold increase in cytochrome $P_{450}$ concentration was noted in S-9 fractions from Aroclor® 1254-treated mullet (MC: 0.28 nmoles cytochrome $P_{450}$/mg S-9 protein; M100A; 0.57 nmoles cytochrome $P_{450}$/mg S-9 protein). Cytochrome $P_{450}$ could not be detected in sea catfish and gulf killifish S-9 fractions. Hemoglobin interference was suspected in sea catfish and gulf killifish preparations since the livers were not perfused; however, the selective reduction techniques of Miyake et al (37) produced no change in the difference spectra. Although cytochrome $P_{450}$ is not detectable by spectroscopic methods in our sea catfish and gulf killifish hepatic preparations, BaP metabolism studies and the *Salmonella/ microsome* assay data indicate that cytochrome $P_{450}$ is present in control and present at elevated levels in 3-MC-treated sea catfish and gulf killifish. Cytochrome $P_{450}$ may have eluded spectroscopic detection since it has been shown to be present in low levels in most fish species (6). Burns (10) has also reported that cytochrome $P_{450}$ from killifish *(Fundulus heteroclitus)* is rapidly denatured once the liver is excised from the animal.

Overall, these findings are consistent with data reported for mammalian systems which indicate that cytochrome $P_{450}$ and NADPH-cytochrome $P_{450}$ reductase are the only inducible oxidation-linked components (7,15,16). The increase in protein over that attributed to cytochrome $P_{450}$ and the associated NADPH-dependent reductase may result from secondary physiological alterations, from increased half-life of other protein components, or, in the case of other hemoprotein components, from heme saturation (7).

### High-Pressure Liquid Chromatography (HPLC)

Thin-layer chromatography (TLC), employing benzene-methanol solvent systems and silica gel plates with a fluorescent indicator, was initially used to separate BaP metabolites. These methods indicated that low levels of only 3 to 4 metabolites were being formed. HPLC analyses were performed to achieve better resolution and more sensitive detection of BaP metabolites. The Micromeritics liquid chromatograph (LC) enabled more reproducible retention times and more sensitive metabolite detection than the Waters LC (although reproducible retention times were not achieved with the Waters LC, BaP metabolite elution order appeared constant). The highly reproducible retention times achieved with the Micromeritics LC probably resulted from the increased column temperature which permitted resolution of BaP metabolites isocratically. The Waters LC could operate only at ambient column temperatures, which mandated resolution of BaP metabolites by solvent gradients. We have therefore identified specific metabolites on samples analyzed with the Micromeritics LC and restricted identification to metabolite groups on samples analyzed with the Waters LC.

Figures 1, 2, and 3 indicate that all treated fish are producing higher levels of BaP metabolites than are untreated fish. Graphical integration of peaks to the left of BaP indicates that induced mullet, sea catfish, and gulf killifish produce 15.3, 3.8, and 1.3 times more BaP metabolites than do the corresponding control animals, respectively. Two and three peaks appear to the right of BaP in catfish and gulf killifish, respectively. These nonpolar compounds probably represent induction artifacts.

Eight BaP metabolites were produced by mullet S-9 fraction. The seven identified metabolites have also been detected using rat hepatic preparation (42,43). The first metabolite eluted using sea catfish and gulf killifish S-9 fractions did not chromatograph with any of the dihydrodiol, quinone, or phenol standards. This highly polar compound may be a tetrol or an induction artifact. BaP dihydrodiols, quinones, and phenols have also been formed by trout liver (3,39), human liver (44), and rat liver (42,43). The similarity in metabolite production suggests that fish possess a MFO system that is analogous in mechanism and function to mammalian MFO systems.

## *Salmonella*/Microsome Mutagenicity Assay

Figure 4 indicates that most S-9 fractions from treated animals metabolize BaP to products mutagenic to *Salmonella typhimurium* TA 98. Two mg S-9 protein and 20 nmoles BaP were applied to all plates to facilitate comparisons within and among groups. When only S-9 fractions and the tester strain were incubated together, a fairly constant number (22.9 ± 1.7) of revertants was induced. This indicates that the

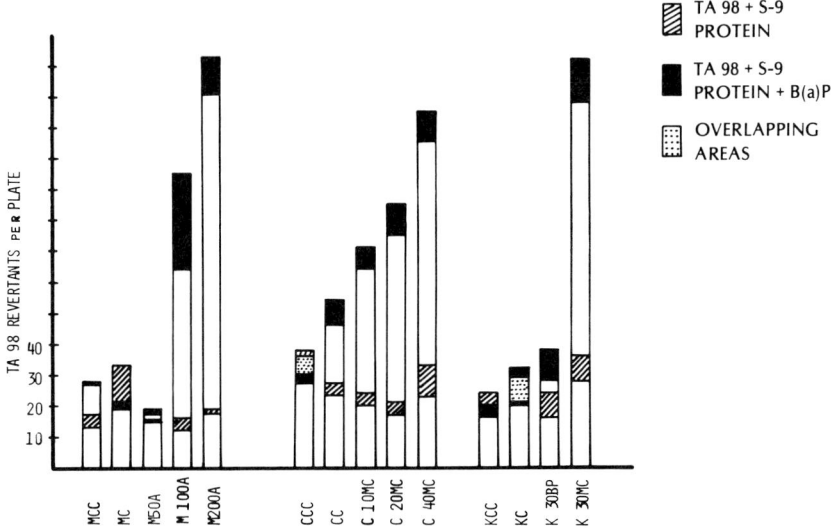

FIGURE 4. Salmonella typhimurium TA 98 revertants induced by BaP in the presence of S-9 protein. All plates contained 2 mg of the indicated S-9 protein and 20 nmoles BaP. The midlines of the shaded regions represent mean values of 3 plate counts for mullet data and 5 plate counts for sea catfish and gulf killifish data. The shaded areas above and below the midline represent the standard errors of the means.

inducing agent contributes little, if any, to the reversion rate. S-9 fractions from mullet treated with 100 and 200 mg Aroclor®/kg body mass in the presence of BaP produced 4 and 6.5 times more revertants than control mullet S-9 fractions, respectively. If it is assumed that the tester-strain reversion rate reflects the relative amounts of MFO system components, the catfish data suggest a relationship between inducing agent level and MFO system component concentrations (correlation coefficient = 0.996). S-9 fractions from catfish induced with 10, 20, and 40 mg 3-MC/kg body mass produced 1.3, 1.6, and 2.2 times more revertants than control catfish S-9 fraction, respectively. S-9 fractions from gulf killifish treated with 30 mg BaP/kg body mass produced the same number of revertants as control S-9 fractions, while 30 mg 3-MC/kg body mass S-9 fractions produced 4.6 times more revertants than the control S-9 fraction. These results clearly indicate that Aroclor® 1254- and 3-MC-treated fish are capable of producing higher levels of mutagenic BaP metabolites than are control or untreated fish.

## CONCLUSION

Our studies have shown that the mullet *(Mugil cephalus)*, the sea catfish *(Arius felis)* and the gulf killifish *(Fundulus grandis)* possess MFO systems which are inducible by Aroclor® 1254 (mullet) and by 3-MC (sea catfish and gulf killifish). BaP treated gulf killifish did not appear to metabolize BaP as efficiently and seemed to produce lower levels of BaP metabolites than did 3-MC-treated gulf killifish. This may have resulted because (1) BaP is a less effective inducer than 30-MC or (2) in vivo BaP metabolite may have induced the conjugation systems, which would result in ethyl acetate insoluble and nonmutagenic in vitro BaP metabolites.

These studies show that certain similarities exist in the mechanics of fish and mammalian MFO systems even though differences exist in the activities of these systems. Chambers and Yarbrough (12) speculate that the less active MFO systems of fish reflect their evolutionary adaptation to an environment in which xenobiotics are naturally diluted.

## ACKNOWLEDGMENTS

This work was supported by EPA Funds, Project Number R806213010, and an Auburn University Grant-in-Aid. We would like to thank Bill Trimble and Vernon Menton of the Claude Peteet Mariculture Center, Alabama Department of Conservation and Natural Resources, Marine Resources Division, Gulf Shores, Alabama, for providing the gulf killifish. We are grateful for the technical assistance provided by Susan Coker and for the assistance in manuscript preparation provided by Kay Hendricks. We appreciate use of the Waters HPLC belonging to Dr. N. Davis, Botany and Microbiology Department, Auburn University.

## REFERENCES

1. Addison, R. F., Zinck, M. E., and Willis, D. E. (1977): Mixed function oxidase enzymes in trout *(Salvelinus fontinalis)* liver: Absence of induction following feeding of p,p' DDT or p,p' DDE. *Comp. Biochem. Physiol.* 57C:39–43.
2. Ahokas, J., Paakonen, R., Ronnholm, K., Raunio, V., Karki, N., and Pelkonen, O. (1977): Oxidative metabolism of carcinogens by trout liver resulting in protein binding and mutagenicity. In: *Microsomes and Drug Oxidations, Proc. 3rd. Int. Symp.,* Berlin, 1976, V. Ullrich, A. Hildebrandt, I. Roots, R. W. Estabrook, and A. H. Conney, Eds., pp. 435–441, Pergamon Press, New York.

3. Ahokas, J. T., Pelkonen, O., and Karki, N. T. (1975): Metabolism of polycyclic aromatic hydrocarbons by a highly active aryl hydrocarbon hydroxylase system in the liver of a trout species. *Biochem. and Biophys. Res. Comm.* 63(3):635–641.
4. Ames, B. N., McCann, J., and Yanasaki, E. (1975): Methods for detecting carcinogens and mutagens with the *Salmonella* microsome mutagenicity test. *Mut. Res.* 31:347–364.
5. Ames, B. N. (1979): Supplement to methods paper (Reference #4). Ames Laboratory Publication, Berkeley.
6. Bend, J. R., and James, M. O. (1979): Xenobiotic metabolism in marine and freshwater species. In: *Biochemical Biophys. Perspectives in Marine Biol.* Vol. 4, D. C. Malius and J. R. Sargent, Eds., pp. 125–188, Academic Press, New York.
7. Bock, K .W., and Remmer, H. (1978): Induction of hepatic hemoproteins. *Handb. Exp. Pharmacol.* 44, (Heme and Hemoproteins), F. DeMatteis and W. N. Aldridge, Eds., pp. 49–80, Springer-Verlag, New York.
8. Brodie, B. B., and Maickel, R. P. (1962): Comparative biochemistry of drug metabolism. In: *Proc. 1st Inter. Pharmacol. Meetg.* 6, B. B. Brodie and E. G. Erdos, Eds., pp. 229–324, Macmillan, New York.
9. Buhler, D. R. (1966): Hepatic drug metabolism in fishes. *Fedn. Proc. Fedn. Am. Soc. Exp. Biol.* 25:343.
10. Burns, K. A. (1976): Microsomal mixed function oxidases in an estuarine fish, *Fundulus heteroclitus,* and their induction as a result of environmental contamination. *Comp. Biochem. Physiol.* 53B: 443–446.
11. Capdevila, J., and Argosin, M. (1977): Multiple forms of housefly cytochrome $P_{450}$. In: *Microsomes and Drug Oxidations, Proc. 3rd. Int. Symp.*, Berlin, 1976, V. Ullrich, A. Hildebrandt, I. Roots, R. W. Estabrook, and A. H. Conney, Eds., pp. 144–151, Pergamon Press, New York.
12. Chambers, J. E., and Yarbrough, J. D. (1976): Xenobiotic biotransformation in fishes. *Comp. Biochem. Physiol.* 55C:77–84.
13. Chan, T. M., Gillett, J. W., and Terriere, L. C. (1967): Interaction between microsomal electron transport systems of trout and male rat in cyclodiene epoxidation. *Comp. Biochem. Physiol.* 20:731–742.
14. Comai, K., and Gaylor, J. L. (1973): Existence and separation of three forms of cytochrome $P_{450}$ from rat liver microsomes. *J. Biol. Chem.* 248 (14):4947–4955.
15. Conney, A. H. (1967): Pharmacological implications of microsomal enzyme induction. *Pharmacol. Reviews* 19 (3):317–366.
16. Conney, A. H., and Burns, J. J. (1972): Metabolic interactions among environmental chemicals and drugs. *Science* 178:576–586.

17. Czygan, P., Greim, H., Garro, A. J., Hutherer, F., Schaffner, F., Popper, H., Rosenthal, P., and Cooper, D. V. (1973): Microsomal metabolism of dimethylnitrosomine and the cytochrome $P_{450}$ dependency of its activation to a mutagen. *Cancer Res.* 33:2983–2986.
18. Dean, W. L., and Coon, M. J. (1977): Immunochemical studies on two electrophoretically homogeneous forms of rabbit liver microsomal cytochrome $P_{450}$: $P-450_{LM_2}$ and $P-450_{LM_4}$. *J. Biol. Chem.* 252(10):3255–3261.
19. DeSerres, F. J., and Shelby, M. D. (1979): The Salmonella mutagenicity assay: Recommendations. *Science* 203:563–565.
20. Estabrook, R. W., Patrizi, V. W., and Prough, R. (1976): The activation of polycyclic hydrocarbons: Cytochromes $P_{450}$, oxygen and electrons. In: *Cancer Enzymology*, J.Schultz and F. Ahmad, Eds., pp. 103–117, Academic Press, New York.
21. Estabrook, R. W., and Werringloer, J. (1977): Cytochrome $P_{450}$—Its role in oxygen activation for drug metabolism. In: *Drug Metabolism Concepts, ACS Symposium,* Vol. 4, D. M. Jerina, Ed., pp. 1–26, American Chemical Society, Washington D.C.
22. Garner, R. C., Miller, E. C., and Miller, J. A. (1972): Liver microsomal metabolism of aflatoxin $B_1$ to a reactive derivative toxic to *Salmonella typhimurium* TA 1530. *Cancer Res.* 32:2058–2066.
23. Gelboin, H. V. (1969): Microsome-dependent binding of benzo(a)pyrene to DNA. *Cancer Res.* 29:1272–1276.
24. Gruger, E. H., Wekell, M. M., Numoto, P. T., and Craddock, D. R. (1977): Induction of hepatic aryl hydrocarbon hydroxylase in salmon exposed to petroleum dissolved in seawater to petroleum and polychlorinated biphenyls, separate and together, in food. *Bull. Environ. Contam. Toxicol.* 17(15):512–520.
25. Heidelberger, C. (1976): Studies on the mechanism of carcinogenesis by polycyclic aromatic hydrocarbons and their derivatives. In: *Carcinogenesis—A Comparative Survey,* 1. R. I. Freudenthal and P. Jones, Eds., pp. 1–8, Raven Press, New York.
26. Hill, D. W., Hejmancik, E., and Camp, B. J. (1976): Induction of hepatic microsomal enzymes by Aroclor® 1254 in *Ictalurus punctatus* (channel catfish). *Bull. Environ. Contam. Toxicol.* 16(4):495–502.
27. Hinton, D. E., and Trump, B. F. (1979): Detection of mutagenic/carcinogenic alteration in fish and its application to human health. Department of Pathology, University of Maryland, School of Medicine, Baltimore, Maryland (Personal communication).
28. Imai, Y., and Sato, R. (1966): Evidence for two forms of $P_{450}$ hemoprotein in microsomal membranes. *Biochem. Biophys. Res. Comm.* 23:5–11.

29. Jerina, D. M., and Daly, J. (1974): Arene oxides: A new aspect of drug metabolism. *Science* 185:573–582.
30. Lech, J. J. (1974): Glucuronide formation in rainbow trout: Effect of salcylamide on the acute toxicity, conjugation and excretion of 3-trifluoromethyl-4-nitrophenol. *Biochem. Pharmacol.* 23:2403–2410.
31. Lee, R. F., Sauerheber, R., and Dobbs, G. H. (1972): Uptake, metabolism and discharge of polycyclic aromatic hydrocarbons by marine fish. *Marine Biology* 17:201–208.
32. Lowry, O. H., Rosebrough, N. J., Farr, A. L., and Randall, R. J. (1951): Protein measurement with Folin phenol reagent. *J. Biol. Chem.* 193:265–275.
33. Madyastha, K. M., and Coscia, C. J. (1979): Detergent solubilized NADPH-cytochrome c ($P_{450}$) reductase from the higher plant, *Catharonthus roseus*. *J. Biol. Chem.* 254 (7):2419–2427.
34. McCann, J., and Ames, B. N. (1979): The *Salmonella*/microsome mutagenicity test: Predictive value for animal carcinogenicity. In: *Origins of Human Cancer,* Book C. H. H. Wiatt, J. D. Watson and J. A. Winsten, Eds., pp. 1421–1450, Cold Spring Harbor Laboratory, Cold Spring Harbor, N.Y.
35. McCann, J., Spingarn, N. E., Kobori, J., and Ames, B. N. (1975): Detection of carcinogens as mutagens: bacterial tester strains with R-factor plasmids. *Proc. Natl. Acad. Sci.* 72:979–983, U.S.
36. Meyer, R. T., and Durrant, J. L. (1979): Preparation of homogenous NADPH Cytochrome c ($P_{450}$) reductase from house flies using affinity chromatography techniques. *J. Biol. Chem.* 254 (3):756–761.
37. Miyake, Y., Gaylor, J. L., and Morris, H. P. (1974): Abnormal microsomal cytochromes and electron transport in Morris hepatomas. *J. Biol. Chem.* 249 (6):1980–1987.
38. Omura, T., and Sato, R. (1964): The carbon-monoxide binding pigment of liver microsomes. *J. Biol. Chem.* 254 (7):2370–2385.
39. Pederson, M. G., Hershberger, W. K., and Jachau, M. R. (1974): Metabolism of 3, 4-Benzypyrene in rainbow trout *(Salmo gairdneri)*. *Bull. Environ. Contam. Toxicol.* 12 (4):481–486.
40. Ryan, D., Lu, A.Y.H., Kawalek, J., West, S. B., and Lewin, W. (1975): Highly purified cytochrome P-448 and P-450 from rat liver microsomes. *Biochem. Biophys. Res. Comm.* 64 (4):1134–1141.
41. Schoor, W. P., and Couch, J. (1979): EPA publication #382. Environmental Protection Agency Laboratory, Sabine Island, Gulf Breeze, Florida (Personal Communication).
42. Selkirk, J. K., Croy, R. G., and Gelboin, H. V. (1974): Benzo(a)pyrene metabolites: efficient and rapid separation by high-pressure liquid chromatography. *Science* 184:169–171.

43. Selkirk, J. K., Croy, R. G., Roller, P. P., and Gelboin, H. V. (1974): High pressure liquid chromatographic analysis of benzo(a)pyrene metabolism and covalent binding and the mechanism of action of 7, 8-benzoflavone and 1,2-epoxy-3,3,3 trichloropropane. *Cancer Res.* 34:3474–3480.
44. Selkirk, J. K., Croy, R. G., Whitlock, J. P., Jr., and Gelboin, H. V. (1975): In vitro metabolism of benzo(a)pyrene by human liver microsomes and lymphocytes. *Cancer Res.* 35:3651–3655.
45. Statham, C. N., Szyjka, S. P., Manahan, L. A., and Lech, J. J. (1977): Fraction and subcellular localization of marker enzymes in rainbow trout liver. *Biochem. Pharmacol.* 26:1395–1400.
46. Tyson, C. A., Lipscomb, J. D., and Gunsalus, I. C. (1972): The roles of Putidaredoxin and $P_{450}$ cam in methylene hydroxylation. *J. Biol. Chem.* 247 (18):577–5784.
47. Ullrich. V., Heldebrondt, A., Roots, I., Estabrook, R. W., and Conney, A. H. (1977): In: *Microsomes* and *Drug Oxidations, Proceedings of 3rd International Symposium.* Berlin (1976), Pergamon Press, New York.

# SUBMICRON SIZE DISTRIBUTIONS OF PARTICULATE POLYCYCLIC AROMATIC HYDROCARBONS IN COMBUSTION SOURCE EMISSIONS

A. H. Miguel and L.M.S. Rübenich

Instituto de Quimica, Universidade Federal do Rio de Janeiro,
21941 Ilha do Fundão, Rio de Janeiro, Brasil

## INTRODUCTION

Recent reports on the size distribution of polynuclear aromatic hydrocarbons (PAH) with respect to particle size have shown that some toxic and potentially toxic species predominate in small particles in urban aerosols (1,3,7,8,9,11,13) and in small particles emitted from combustion sources (4). The distribution of PAH with respect to a range of particle sizes is crucial in the evaluation of the penetration and distribution of these compounds in the repository system. Transport and lifetime of PAH in the atmosphere are also dependent on their particle size distribution. In addition, this type of information may help explain the mechanism of formation of PAH in combusion processes.

A recent study (10) estimated that motor vehicles account for roughly 90 percent of the total benzo(a)pyrene (BaP) emissions into the Los Angeles atmosphere. Possible emissions from steel mill coke ovens were not included in the estimates because suitable emissions factors were not available for this source. In the following sections we report on the distribution of BaP and benzo(ghi)perylene (BghiP) in a coke oven sample, and of BaP in a sample taken inside a tunnel with heavy automobile traffic by means of an eight-state low-pressure impactor with size resolution down to 0.05 $\mu$m in aerodynamic diameter. The PAH distribution of these two sources is compared with that of ambient air and discussed in terms of the origin of PAH in urban atmospheres. Finally, the data were used to estimate the extent of penetration and distribution of these compounds in the respiratory system.

## EXPERIMENTAL

### Reagents

BaP (Eastman Organic Chemicals), BghiP (Aldrich Chemical Co.), and spectrograde solvents were used as supplied.

### Sampling and Analysis

All samples were collected with an eight-stage low-pressure impactor with 50 percent efficiency cutoffs of 4.0, 2.0, 1.0, 0.5, 0.26, 0.12, 0.075, and 0.05 $\mu$m in aerodynamic diameter described elsewhere (5,6). East stage of the impactor contained a glass collection disc 25.4 mm in diameter by 2 mm thick. To the center of each disc, 0.5 $\mu$l of an 8 percent solution of Vaseline® in benzene was applied. The benzene evaporated, leaving a thin film of Vaseline about 5 mm in diameter to prevent particle rebound. A Sargent Welch 1403 vacuum pump was used. The impactor sampled at a rate of 1/min, which was verified before and after sampling. Sampling was carried out for 2 hours on the bench (push side) of a steel mill coke oven and for 6 hours inside an urban automobile traffic tunnel (100,000 vehicles/day). Before entering the impactor, the coke oven sample passed through a 10-mm Dorr-Oliver polyethylene cyclone to remove particles greater than about 8 $\mu$m in aerodyanamic diameter.

Details of the extraction and analysis of the PAH are reported elsewhere (8).

## RESULTS AND DISCUSSION

Data on the concentration of BaP and BghiP in the emission sources studied are presented in Table 1, along with BaP data of ambient air for comparison. In the coke oven sample, the concentrations of BaP and BghiP as a function of particle size are similar. Less than 39 percent of the mass of PAH is associated with particles of geometric mean diameter, $\bar{d}_P$, smaller than 0.71 $\mu$m. In contrast, in the ambient air and tunnel samples, respectively, 82 percent and 100 percent of the BaP mass is found in particles with $\bar{d}_P$ smaller than 0.17 $\mu$M.

The PAH distributions as a function of aerodynamic diameter shown in Figure 1 were obtained using data from Table 1. The coke oven distributions of BaP and BghiP show some important characteristics.

**TABLE 1. BaP and BghiP Concentration, ng/m³ as a Function of Particle Size.**

| Size Range[a] [$d_p(\mu m)$] | Geometric Mean Size [$d_p(\mu m)$] | Tunnel BaP | Coke Oven BaP | Coke Oven BghiP | Ambient Air[b] BaP |
|---|---|---|---|---|---|
| 4.0–8.0[c] | 5.7 | N.D.[d] | 0.17(0.29) | 2.58(5.8) | 0.028(2.8) |
| 2.0–4.0 | 2.8 | N.D. | 12.7(21.7) | 9.83(22.3) | 0.032(3.2) |
| 1.0–2.0 | 1.4 | N.D. | 13.2(22.6) | 7.33(16.6) | 0.039(3.8) |
| 0.5–1.0 | 0.71 | N.D. | 13.9(23.8) | 7.58(17.2) | 0.083(8.2) |
| 0.26–0.5 | 0.36 | 3.50(8.2) | 6.00(10.3) | 5.75(13.0) | 0.090(8.9) |
| 0.12–0.26 | 0.18 | 5.08(11.9) | 6.17(10.6) | 3.92(8.9) | 0.238(23.5) |
| 0.075–0.12 | 0.095 | 30.5(71.2) | 4.92(8.4) | 4.17(9.4) | 0.431(42.5) |
| 0.05–0.075 | 0.061 | 3.75(8.8) | 1.42(2.4) | 3.00(6.8) | 0.072(7.1) |

[a]The aerodynamic diameter, $\bar{d}_p$, refers to particles of unit density.
[b]Data from Reference 8.
[c]Upper limit assumed to be 8.0 $\mu m$.
[d]N.D. = not detected.
The numbers in parenthesis represent the percentage of mass of PAH in the stated size range.

Both are bimodal and similar in shape, despite large differences in the vapor pressures of these two compounds.

In the tunnel and ambient air samples (Figure 1) the distribution of BaP is unimodal and similar in shape. The peak in the concentration is found in particles with aerodynamic diameters between 0.075 and 0.12 $\mu m$ for both samples.

## Deposition in the Respiratory System

The Task Group on Lung Dynamics (12) recommended that the model respiratory tract be composed of three compartments: the nasopharynx (NP), which extends from the nose and mouth down to the larynx; the trachea and bronchial tree (TB), which includes the terminal bronchioles; and the pulmonary compartment (P), consisting of the respiratory bronchioles, alveolar ducts, atria, alveoli, and alveolar sacs. The distributions of BaP and BghiP in the respiratory system were compared for the two source and ambient air distributions (Table 2). These distributions were estimated using published lung deposition data (12) for a tidal volume of 1450 ml and a geometric standard deviation of 1.2. The coke oven sample showed a PAH distribution which is bimodal

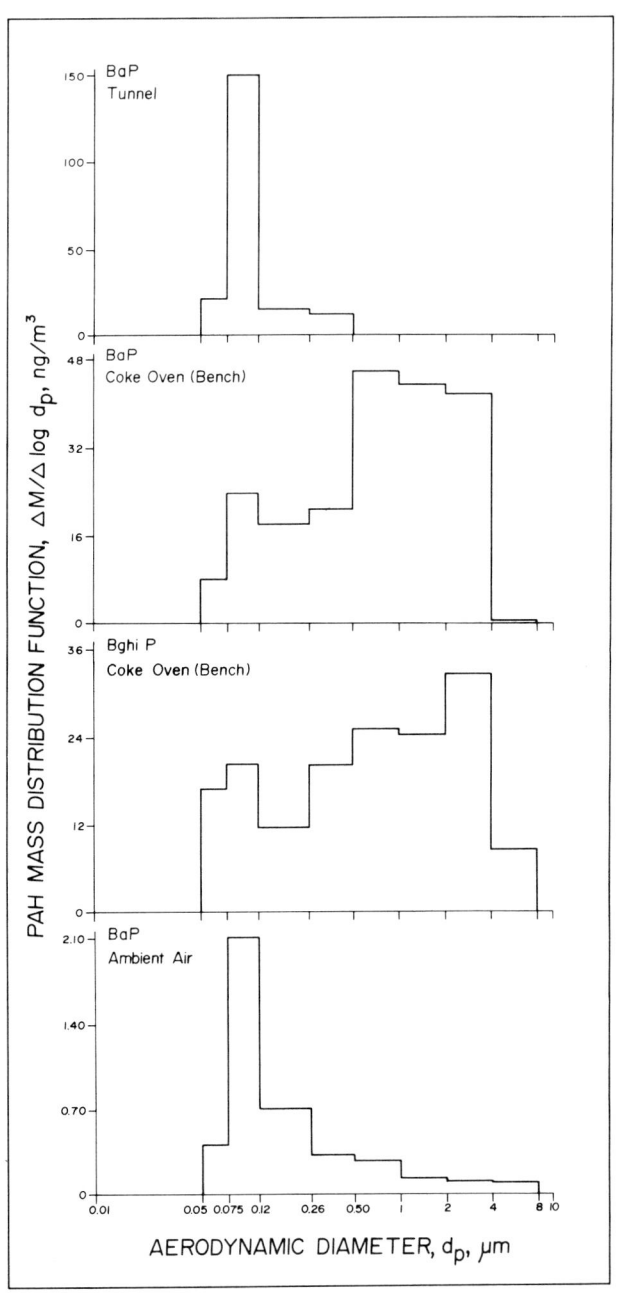

**FIGURE 1.** Distributions of BaP and BghiP as a function of aerodynamic diameter. Total PAH concentrations, in ng/m$^3$, were: tunnel, 42.8 BaP; coke oven, 58.5 BaP and 44.2 BghiP; ambient air, 1.01 BaP.

**TABLE 2. Distribution of BaP and BghiP in Respiratory System.**

| Deposition Compartment | Tunnel Distribution BaP (%) | Coke Oven Distribution BaP (%) | Coke Oven Distribution BghiP (%) | Ambient Air Distribution BaP (%) |
|---|---|---|---|---|
| Nasopharynx (NP) | 0 | 8 | 12 | 3 |
| Trachea and Bronchial Tree (TB) | 5 | 3 | 3 | 4 |
| Pulmonary (P) | 39 | 25 | 26 | 34 |

(Figure 1). For this reason, estimates of respiratory deposition were made by considering each impactor stage separately, using the geometric mean diameter of a given range to obtain the fraction of particles deposited in each respiratory compartment. From this fraction, the percent of PAH deposited in each compartment was calculated (Table 2).

Because of the predominantly small size of the PAH-containing particles, most of the deposition in the respiratory system occurred in the pulmonary compartment. For the PAH-containing coke oven particles, a large fraction of the total deposition occurred in the nose and pharynx. The fraction deposited in the trachea and bronchial tree was similar for both source and ambient air samples.

Although only a few percent of ambient and source PAH deposit in the TB regions (Table 2), Bell (2) has shown that greatly enhanced local doses may occur at bifurcations of the bronchial passages because of preferential impaction of the larger particles there.

## SUMMARY AND CONCLUSIONS

The mass distribution of BaP and BghiP as a function of particle size in a steel mill coke oven sample taken at the push side of the bench is bimodal and similar in shape. More than 60 percent of the mass of both PAH in this source is associated with particles of geometric mean diameter larger than 0.71 $\mu$m.

The distributions of BaP in a tunnel sample was unimodal and similar in shape to that of BaP in ambient air. The peak in the concentration lies in the range of 0.075 to 0.12 $\mu$m in aerodynamic diameter. The similarities of these two distributions indicate that a substantial fraction of the PAH-containing particles in ambient air originates from automobile emissions. In cities where large amounts of coal are burned for generating power and for space heating, the situation may be different.

The data obtained in this study did not follow a log-normal distribution, indicating that the mass median diameter (MMD) of the aerosols associated with the PAH does not coincide with the median diameter of the PAH mode.

Most of the deposition of PAH-containing particles from source and ambient air occurs in the pulmonary compartment of the respiratory system.

The mass distribution of PAH as a function of particle size presented in this study provides the information necessary to calculate their lifetime in air.

## ACKNOWLEDGMENTS

This work was partially supported by the Brazilian Council for Scientific and Technologic Development (CNPq), and FINEP.

The authors thank Professors Wolfgang C. Pfeiffer and Helena A. Trindade for helpful discussions.

## REFERENCES

1. Albagli, A., Oja, H., and Dubois, L. (1974): Size distribution pattern of polycyclic aromatic hydrocarbons in airborne particulates. *Environ. Lett.* 6(4):241-251.
2. Bell, K. A. (1974): Aerosol deposition models of a human lung bifurcation. Ph.D. Thesis. California Institute of Technology, Pasadena, CA.
3. DeMaio, L., and Corn, M. (1966): Polynuclear aromatic hydrocarbons associated with particulates in Pittsburgh air *J. Air Poll. Control Assoc.* 16:67-71.
4. Broddin, G., Van Vaeck, L., and Van Cauwenberghe, K (1977): On the size distribution of polycyclic aromatic hydrocarbon containing particles from a coke oven emission source. *Atmos. Environ.* 11:1061-1064.
5. Hering, S.V., Flagan, R. C., and Friedlander, S. K. (1978): Design and evaluation of a new low pressure impactor, I. *Environ. Sci. Technol.* 12:667-673.

6. Hering, S. V., Friedlander, S. K., Collins, J. G., and Richards, L. W. (1979): Design and evaluation of a new low pressure impactor, II. *Environ. Sci. Technol.* 13:183-188.
7. Kertész-Saringer, M., Mészaros, E., and Varkonyi, T. (1971): On the size distribution of benzo(a)pyrene-containing particles in urban air. *Atmos. Environ.* 5:429-431.
8. Miguel, A. H., and Friedlander, S. K. (1978): Distribution of benzo(a) pyrene and coronene with respect to particle size in Pasadena aerosols in the submicron range. *Atmos. Environ.* 12:2407-2413.
9. Pierce, R. C., and Katz, M. (1975): Dependency of polynuclear aromatic hydrocarbon content on size distribution of atmospheric aerosols. *Environ. Sci. Technol.* 9:347-353.
10. Special Report to the U.S. Environmental Protection Agency on Contract No. 68-03-0434 (1978), White, W. H., Friedlander, S. K., Eds. California Institute of Technology.
11. Starkey, R., and Warpinski, J. (1974): Size distribution of particulate benzo(a)pyrene. *J. Environ. Health* 36(5):503-505.
12. Task Group on Lung Dynamics (1966): Deposition and retention models for internal dosimetry of the human respiratory tract. *Health Physics* 12:173-207.
13. Van Vaeck, L., and Van Cauwenberghe, K. (1978): Cascade impactor measurements of the size distribution of the major classes of organic pollutants in atmospheric particulate matter. *Atmos. Environ.* 12:2229-2239.

# MEASUREMENT AND ENVIRONMENTAL IMPACT OF PAH—SOME CLOSING REMARKS

**P. W. Jones**

Electric Power Research Institute
Palo Alto, California 94303 USA

Concern regarding environmental contamination by polycyclic aromatic hydrocarbons (PAH) arises because many of these species have been demonstrated to be carcinogenic to animals and are most probably carcinogenic to man. Furthermore, the occupational-health risk associated with high-PAH-exposure industries, such as coking and asphalt production, has been well established. A large number of PAH were included in both the Federal Water Pollution Control Act and the U. S. EPA list of Priority Pollutants, and the World Health Organization has recommended maximum levels of PAH for drinking water, further underlying the significance attached to this class of compounds.

PAH occur widely throughout the environment, both as a result of the technological activities of man and as a result of natural production. The PAH load resulting from natural production is difficult to assess, but worldwide PAH production may be estimated as follows, in tons per year(1):

| | |
|---|---:|
| Heating and Power Generation | 260,000 |
| Industrial Processes | 105,000 |
| Incineration and Open Burning | 135,000 |
| Vehicular Transportation | 4,500 |
| Total PAH Emissions | 504,500 |

As a result, occasions for PAH measurement arise from all environmental media, and under a variety of industrial and other technological conditions.

This symposium series was established in the hope that it would serve as a focal point, and effective medium for infomation exchange, for all scientists who were actively concerned with the environmental emission and ultimate impact of PAH. While the format will doubtless continue to be refined with input from participants, the dynamic debate and highly informative presentations which we have all been a part of at this meeting are a testimony to its success. Such success would not have been possible without the participation and support of all of those present.

While it is neither possible nor perhaps desirable for me to attempt to summarize the discussions at this symposium in the short time that remains, there are nevertheless two questions that I would like to raise which directly impact the regulatory aspects of PAH measurements:

1. Why are we concerned with measurement of all individual PAH species?
2. Do we have any assurance that present-day PAH measurements on real-world systems are sufficiently reliable?

In regard to the first point, I would quickly agree that we have seen some fine examples of isomer separation by both HPLC and GC at this symposium. However, I would submit that even these separations have been made on relatively simple PAH mixtures. Those who have attempted separations of alkyl isomers on coal-derived oils are painfully aware of the near impossibility of achieving individual PAH compound resolution; it should be remembered that coal-derived material will probably become increasingly abundant during the next few decades. I would suggest, however, that the effort required to separate such complex PAH mixtures may not be cost- and information-effective.

Much of the effort being directed towards PAH isomer separation is, not surprisingly, being motivated by anticipated regulatory requirements. Any future regulation of PAH emissions will presumably be based upon a health-effects rationale, data on which are totally lacking for most PAH. Since the number of PAH isomers is very large, the potential for synergism both between isomers and with other compounds would appear to be virtually unlimited, which seriously complicates a facile link between PAH levels and observable health effects.

There has been much discussion regarding PAH that could be used as surrogates, or indicators, of total PAH concentration. While this may not be an unrealistic concept for a specific type of emission source, it is clearly unreasonable to propose a universally applicable PAH surrogate for emissions, since PAH emissions from different types of sources may exhibit highly significant qualitative differences in specific compounds emitted. Thus it may be reasonable to select a single PAH isomer as an indicator of source emissions from a specific industry, but a generally applicable measure of emissions from all sources would appear to demand recognition of the total loading of all PAH compounds. In either case, the need for exhaustive separation of all PAH isomers would appear to be an unnecessary academic exercise. On one hand, a relatively straightforward analytical procedure for the selected surrogate PAH could be used; on the other, a procedure for measuring total PAH irrespective of isomers is required. The simple yet effective measurement technique for total PAH described earlier by Arthur D. Little Inc. has clear potential for fulfilling the latter role. In either scenario, it must be remembered in any regulatory posture that correlation of the selected PAH measure with demonstrated health effects must be established, together with an evaluation of threshold levels for onset of such health effects.

Turning to the second point, "Do we have any assurance that present day PAH measurements on real-world systems are sufficiently reliable?" There was a time, maybe 2 or 3 years ago, when any suggestion that PAH data could be in error by more than ±25 percent would have met with scorn. Indeed, many of us were becoming so confident in our methods that we were tempted to design fairly sophisticated field validation experiments, just to show how good our methods really were. This, of course, was tempting providence, and roughly marks the beginning of the present uncertainty surrounding the measurement of PAH.

A recent state-of-the-art report prepared for EPRI by the GCA Corporation (2) on the measurement of emissions from fossil fuel power plants shows some serious defects in work carried out over the past 5 years:

Reports of PAH Measurements

| Activity | Percentage of Reports with Adequate Coverage |
|---|---|
| Source Description | 60 |
| Sampling | 70 |
| Sample Recovery and Preservation | 20 |
| Analysis | 75 |

It is quite clear that one of our least favorite activities is knowing what a sample represents, and ensuring that no physical or chemical changes occur to it before analysis. It would appear that 80 percent of PAH studies in the past 5 years have largely ignored sample recovery and preservation.

The early results from several studies indicate that we have major problems in at least three areas:
  (1) PAH recovery from fly ash
  (2) PAH reactivity on filters
  (3) Reactivity of organics on sorbents.

Work at Oak Ridge National Laboratory, described earlier (3), has shown that ultrasonic extraction of 3 g of fly ash with 10 ml of benzene leads to incomplete PAH recovery, especially for larger molecules:

| PAH Compound | Recovery (%) |
|---|---|
| Naphthalene | 70 |
| Phenanthrene | 62 |
| Benz(a)anthracene | 15 |
| Benz(a)pyrene | 2 |

Even five further extractions of the same fly ash do not offer significant improvement; studies currently under way on this continuing project are investigating this phenomenon further.

Studies at the Ford Motor Company, described earlier (4), indicated that there are serious problems associated with the collection of particulate PAH on filters. Even though this work concerns diesel exhaust, there appears to be no reason why it should not be applicble to collection of the other fossil-fuel combustion effluent. Important findings from the Ford work are:

(1) Degradation of PAH on a filter is dependent upon the nature of the filter. For example ~1 percent recovery of labelled PAH was obtained from quartz or fiber glass, whereas ~60 percent recovery of labelled PAH was obtained from Teflon or Fluoropore.
(2) PAH loss is dependent upon PAH concentration, which suggests that PAH degradation is surface catalyzed by the filter.
(3) PAH loss is proportional to air flow across the filter, which suggests a second-order oxidation (or other) reaction with the gas stream.
(4) Reaction products are mainly oxygenated and consist of large numbers of quinones and hydroxy species.

It may be true that the experimental conditions in the Ford work were not entirely representative of conditions to which PAH are subject during normal exhaust conditions, and some other diesel studies discussed earlier appear to be in contradiction. Nevertheless, I believe that the Ford data are truly symptomatic of problems facing PAH measurement, and its implications should not be ignored.

Recent work has suggested that sorbent resins may catalyze the oxidation of some organic compounds, though PAH do not yet appear to have been studied. It has long been known that sorbents such as Chromosorb, XAD-2 and Tenax give rise to artifacts such as quinones, benzaldehyde, styrene, acetophenone, and phenol. However, it has recently been shown in studies at Research Triangle Institute (5) that deuterium-labelled styrene is oxidized to benzaldehyde during sampling on Tenax:

$$C_6D_5\text{-}CD=CD_2 \xrightarrow[\text{Tenax}]{O_2} C_6D_5\text{-}CD=O$$

It is clear that such oxidation could also take place with PAH, especially the potentially important multialkylated PAH that result from coal conversion processes.

Finally, limited experiments have been carried out in an attempt to evaluate the efficiency of PAH collection using stack gas sampling trains. An early experiment (6) in which effluent was sampled with a single probe and then equally divided between two identical filtration and sorption sampling trains suggested that measurement uncertainty was no worse than 20 percent. This result appeared encouraging, but in fact addressed only precision and not accuracy. While the result may have been reproducible, there was no means of determining the absolute accuracy of the experiments. However, more recent studies, discussed earlier by Sounnichsen et al (7), in which labelled standards were injected into the probe tip during stack-gas sampling have suggested that measurements may be uncertain between a factor of ten high and a factor of one hundred low, if judged by the recoveries of the standards.

## CONCLUSION

The only conclusion that can be made with any assurance, is that we still have a lot to learn with regard to PAH measurements from stationary sources. True, we appear to be fairly successful at analyzing PAH in given samples, and we also appear capable of taking the PAH out of a gas stream entering a probe. But what we do not appear capable of doing is maintaining the integrity of PAH samples both during sampling and during the many steps of workup leading up to analysis. Finally, it is clear that effective correlation of health-effects data and measurement techniques must be achieved before any meaningful regulatory posture can be established.

## REFERENCES

1. Suess, M. J. (1976): *Sci. Total Environ.* 6:239.
2. Zelenski, S. G., and Pangaro, N. (1979): Inventory of emissions from fossil fuel combustion for power generation. Final report on EPRI project No. TPS-78-820.
3. Griest, W. H., Guerin, M. R., Yeatts, L. S., Jr., Reagan, R. R., Rao, T. K., Epler, J. L., Maskarinec, M. P., Buchanan, M. V., and Tomkins, B. A. (1979): Identification and quantification of polynuclear organic matter on particulates from a coal fired power plant. EPRI report No. EA-1092, June.

4. Lee, F., Harvey, T. M., Prater, T. J., and Paputa, M. C. (1979): Chemical analysis of diesel particulate matter and an evaluation of artifact formation. Presented at the ASTM Conference on Sampling and Analysis of Toxic Organics in the Atmosphere, August 6 to 9, 1979, Boulder, Colorado.
5. Pellizzari, E. D., Research Triangle Institute, North Carolina, USA, Private Communication (1979).
6. Knapp, K. T., Bennett, R. L., Jones, P. W., Wilkinson, J. E., and Strup, P. E. (1979): Measurement of polycyclic organic materials and other hazardous organic compounds in stack gases. In: *Polynuclear Aromatic Hydrocarbons—Third International Symposium on Chemistry Biology, Carcinogenesis, and Mutagenesis*, P. W. Jones and P. Leber, Eds., Ann Arbor Science, Ann Arbor.
7. Sonnichsen, T. W., McElroy, M., and Bjørseth, A. (1979): Variability and uncertainty of PAH measurements from coal fired utility boilers. Presented at the Fourth International Symposium on Polynuclear Aromatic Hydrocarbons, October 2-4, 1979, Battelle's Columbus Laboratories, Ohio, USA.

## Author Index

Ali, S.   395
Amin, S.   417
Andren, A. W.   127
Arnott, M. S.   299
Ashurst, S. W.   503
Autrup, H.   89

Baird, W. M.   471
Banwart, W. L.   395
Bassett, D. O.   75
Basu, D.   435
Bayer, U.   153
Bedenko, V.   417, 1041
Benditt, E. P.   489
Bevan, D. R.   879
Bigger, C.A.H.   663
Bjørseth, A.   405, 565, 617
Bond, J. A.   489
Bonnett, W. J.   791
Boutwell, R. K.   935
Brash, D.   523
Bruce, C.   379
Buening, M. K.   675
Busbee, D. L.   299, 917

Calle, L. M.   163
Cannova, F.   193
Cantrell, E. T.   299, 917
Carpenter, R. L.   599
Carver, J. H.   177
Castle, R. N.   59
Caton, J. E.   819
Cavalieri, E.   215, 259
Cazer, F. D.   689
Charpentier, J. M.   899
Chang, R. L.   675
Chaude, O.   899
Chou, M. W.   645, 733

Cohen, G. M.   503
Conney, A. H.   267, 675
Cronen, M. C.   633
Crowley, J. P.   955

D'Ambrosio, S. M.   523
Daniel, F. B.   523, 689
Daub, G. H.   753
Dennis, A. J.   955
DiGiovanni, J.   935
Dipple, A.   663
Drum, M. A.   955
Dudney, C. S.   287
Dumaswala, R. U.   471
Dunn, B. P.   367
Durand, J. P.   899

Edlund, U.   345
Elam, D.   1059
Ellis, L. E.   163
Ezike, J.   543

Facklam, T. J.   955
Felton, J. S.   177
Fu, P. P.   645
Fujimori, E.   319
Furcinitti, P. S.   287

Gammage, R. B.   139, 565
Gibson, E. S.   579
Gibson, R. E.   523
Giorgio, P.   193
Gmur, D. J.   455
Greenberg, A.   193
Griest, W. H.   819
Griffin, A. C.   299
Griffin, G. D.   287

Grimmer, G. 107, 807
Guerin, M. R. 819
Güsten, H. 983

Hanson, R. L. 599
Hardy, R. 379
Harris, C. C. 89
Hart, R. W. 523, 689
Harvey, R. G. 753
Hassett, J. J. 395
Hecht, S. S. 417
Heinrich, G. 983
Hembree, D. M. 1005
Hermann, M. 899
Higgins, C. E. 819
Hoffmann, D. 417, 1041
Hofnung, M. 899
Homola, M. 153
Howard, P. H. 435
Hudgins, W. R. 663
Hughes, M. M. 1
Hunt, G. T. 589

Inbasekaran, M. N. 689
Isaac, R. S. 243
Iyer, R. P. 753

Jacob, J. 807
Jeffrey, A. M. 89
Jerina, D. M. 267, 675
Jones, P. W. 1085
Jones, T. D. 287
Joyce, N. J. 523
Juchau, M. R. 233, 489

Kahn, A. 395
Kaiser, C. 579
Kerr, A. 579
Kieda, C. A. 75
Kilgore, M. 1059
Kim, W. 523
Kishore, G. S. 935
Kittle, J. D., Jr. 163
Knize, M. G. 177
Kocan, R. M. 489
Kootstra, A. 633
Krahn, M. M. 455
Kumar, S. 675

Lakowicz, J. R. 879
Lane, D. A. 199
Lao, R. C. 829
Laub, R. J. 25
LaVoie, E. 417, 1041
Lee, F.S.C. 543
Lee, M. L. 59
Lehr, R. E. 675
Levin, W. 267, 675
Levins, P. L. 973
Lockington, J. N. 579
Lu, A.Y.H. 733
Lubawy, W. C. 243
Lyga, W. 753

Mackie, P. R. 379
MacLeod, M. C. 9
Mamantov, G. 1005
Maple, J. R. 1005
Marshall, M. H. 299
Marshall, M. V. 299, 917
Martin, R. R. 299, 917
May, W. E. 791
McCalla, D. R. 579
McElroy, M. W. 617
McGill, A. S. 379
McLemore, T. L. 299, 917
Means, J. C. 395
Melius, P. 1059
Meuser, J. M. 405
Mhaskar, D. 523
Miguel, A. H. 1077
Moore, C. J. 9
Moore, F. R. 405
Moschel, R. C. 663

Natusch, D.F.S. 1
Naujack, K.-W. 107
Newton, G. J. 599
Norden, B. 345
Nowicki, H. G. 75

Ocasio, I. J. 163
Olufsen, B. 333
Omiecinski, C. J. 233
Ortman, J. 523
Orwig, D. S. 177

Pangaro, N. 589
Parker, K. 1025
Peake, E. 1025
Petroff, N. 899
Pierson, W. R. 543

Quan, E.S.K. 199

Robbins, W. K. 841
Roberts, W. L. 25
Rogan, E. G. 215, 259
Roth, R. 259
Rubenich, L.M.S. 1077

Sakuma, T. 199
Salazar, E. P. 177
Santodonato, J. 435
Schmidt, K. 523
Schmoldt, A. 807
Schneider, D. 107
Schoor, W. P. 1059
Secrist, A., III 753
Selkirk, J. K. 9, 503
Sheikh, Y. M. 689
Siebert, D. 153
Sinha, D. 215
Slaga, T. J. 503, 633, 753, 935
Smillie, R. D. 863
Smith, E. M. 973
Sonnichsen, T. W. 617
Srinivasan, B. N. 319
Strand, J. W. 127
Strup, P. E. 405
Sullivan, P. D. 163
Swarin, S. J. 771

Tada, M. 267
Tan, B. 1059
Taylor, D. R. 1
Thakker, D. R. 267, 675
Thomas, R. S. 829
Tulley, L. 1041

Vandecasteele, J.-P 899
Varanasi, U. 455
Vo-Dinh, T. 139

Walsh, P. J. 287
Wang, D. T. 863
Wani, A. 523
Wehry, E. L. 1005
Weill, N. 899
Whalen, D. L. 675
White, C. M. 59
Whittle, K. J. 379
Wilkinson, J. E. 405
Willey, C. 59
Williams, R. L. 771
Wise, S. A. 791
Wislocki, P. G. 733
Witiak, D. T. 523
Wold, S. 345
Wood, A. W. 675
Wood, S. G. 395
Wray, N. P. 299, 917

Yagi, H. 267, 675
Yang, S. K. 645, 733
Yeatts, L. B., Jr. 819
Yokoyama, R. 193

Zelenski, S. G. 589
Zeller, M. V. 1

# Key Word Index

Alumina, 1
Ames assay, 163, 417, 579, 675, 899, 1025, 1041, 1059
Anthanthrene, 215
Anthanthrene (methyl substituted), 215
Anthracene, 199, 379
Anthracene metabolism, 1041
Anthracene (methyl substituted), 1041
Anthracenes (substituted), 379
Anticarcinogenesis, 935
Antioxidants, 163
Antipyrine clearance, 917
Aorta, chicken, 489
Aorta, rabbit, 489
Aryl hydrocarbon hydroxylase inhibitors, 689
Asbestos, 879
Atherosclerosis, 489
Atmospheric pressure chemical ionization MS system (mobile), 199

BaP, 153, 163, 233, 259, 319, 367, 435, 503, 543, 579, 711, 829, 1077
BaP-bay region diol epoxides, 267
BaP-metabolism, 89, 177, 215, 243, 299, 455, 503, 879, 935, 1059
BaP metabolite mixtures, 287
BaP metabolites, 471, 633, 753
BaP metabolites (dihydrodiols), 267
BaP (methyl substituted), 215, 259
BaP-$NO_2$ interaction, 1
BaP-$SO_3$ interaction, 1
BaP (substituted derivatives), 163
BaP tracers, 819
Bay region substitution, 689
(BBBT) = N,N'-Bis[p-butoxybenzylidene]- $\alpha$- $\alpha$- bi-p-toluidine, 25
Benzanthracene, 215
Benzanthracene, metabolism, 807
Benzanthracene, metabolites, 267, 753
Benzanthracene, methyl substituted, 215
Benzanthracene, methyl substituted metabolism, 645
Benzo(e)pyrene, dihydrodiols and diol epoxides, 675
Benzo(e)pyrene metabolites, 753
Benzofluoranthene metabolism, 417
Bivalves, mytilus edulis, 367, 379
Bronchus, human, 89

C3H10T ½ cells, 955
Carbon black, 879
Carcinogenicity, structure analysis, 345
Carcinogenesis, comparative, 245
Cells, Chinese hamster ovary (CHO), 177
Cells, hamster embryo, 9, 471
Cells, hamster embryo V-79, 471, 675
Cells, lung, 807
Cells, lymphocytes, 299, 917
Cells, pulmonary alveolar macrophages, (PAM), 299, 917
Cells, Syrian hamster embryo (SHE), 177, 523
Cell, survival, 287
Cholanthrene (methyl substituted), 359
Chromosomal histone protein, 633
Chrysene, dihydrodiol, 267
Chrysene, metabolites, 753
Chrysene, methyl substituted, 215
Colon, human, 89
Complex mixtures, 59, 107, 127, 139, 193, 243, 333, 405, 543, 579, 589, 599, 617, 771, 791, 819, 841, 863, 955, 973, 983, 1025
Complex mixtures, standard, 75
Complex mixture, synthetic, 1005
Conjugates BaP, 299
Conjugates, glucuronic acid (BaP), 9
Conjugates, glutathione (BaP), 9
Cysteine-BaP, 319
Cytochrome P-450 and P-448, purified, 733

Distribution coefficients $K_D$, extraction of complex mixtures of PAH, 841
7,12-Dimethyl benzanthracene (DMBA), 523, 663
DMBA, fluorene substituted, 689
DMBA metabolism, 489, 773, 935
DMBA metabolites, 689
DNA adducts, 503
DNA binding, 89, 455, 523, 663
DNA binding, adduct formation, 471
DNA binding, mouse skin, 503
DNA repair, 523
DNA synthesis, 287
Degradation of PAH, storage, 771, 829, 543

Electron spin resonance (ESR), 163, 319
Endonuclease digestion, 471, 523
Environmental impact, 1085

Epoxide hydratase, 489
Epoxide hydratase, inhibition, 153
Esophagus, human, 89
Exhaust gas, diesel, 771

Fetal smooth-muscle cells, human, 489
Filtration media, 543
Fish, gulf killfish, 1059
Fish, mullet, 1059
Fish, plueronectid, 455
Fish, salmonid, 455
Fish, sea catfish, 1059
Flue gas, vapor phase PAH, 599, 617
Fluidized bed combustion, 599
Fluoranthene, 199
$\beta$-Fluorenone, 199
Fluorene, 199
Fluorescent sensitization, 973
Fluorescent spectroscopy, 879, 1059
Food, 435
Fourier transform infrared spectroscopy (FTIR), 1005
Fly ash, 1, 819, 955
Fly ash, coal, 589, 599, 617
Fly ash, extracts, 405

Gas chromatography mixed phase, 25
Gas phase PAH, 199
Gene conversion, in yeast, 153
Glass capillary (GC), 25, 59, 107, 405, 589, 807, 829, 1005
Glass capillary analysis (GC), 333
Glass capillary liquid crystal (GC), 25
Gas chromatography/mass spectroscopy (GC/MS), 127, 379, 589, 617, 973, 1025
GC/MS analysis, 75
Gene loci, multiple, 177
Glutathione transferase, 489

Heavy atom effects on PAH, 139
Hematite, 879
Hemoglobin, 233
Hepatocyte, 503
Heterocycles, sulfur containing, 59
High performance liquid chromatography (HPLC), 1, 9, 177, 193, 243, 259, 267, 299, 417, 471, 489, 503, 543, 589, 645, 689, 733, 771, 791, 863, 899, 1041, 1059
Horseradish peroxidase/hydrogen peroxide, 259
Human exposure, multimedia, 435

Industrial effluents, 863, 1085

Laser excitation MI, 1005
Lectin binding, 955
Lung cancer, human, 917
Lung cells, fetal hamster, 807
Lung perfusion, rabbit, 243
Lymphocytes, human, 299, 917

Macrophages-pulmonary alveolar, human, 299, 917
Mammary gland, rat, 523
Manganic acetate, 259
Methylsilicone (OV-101), 25
Models-theoretical, 345
Monitoring, work place, 565
Monooxygenation, hematin-mediated, 233
Methylcholanthrene, metabolism, 935, 1059
Mutagenesis, 153, 163, 177, 417, 579, 675, 899, 1025, 1041, 1059

Naphthalene, 199, 379
Naphthalene (methyl substituted), 379
(NMP) N-methyl pyrrolidone, extraction, 841

Octanol-water partition coefficients (Kow), 395
Oil, crank case, 899, 1025
Oil, hydrotreated, 899
Organ homogenates, mouse liver, 177
Oxidation, one electron, 259

PAH distribution (by ring size), 379
Particulate extracts, 405
Particulates, 879
Particulates-airborne, 193, 543, 579, 617, 771, 791, 819, 955, 983, 1077
Particulates, automobile, 1077
Particulates, coke oven, 1077
Particulates, diesel, 543, 771
Particulates, lung penetration of, 1077
Petroleum distillate, 899
Phenanthrene, 199
Phenanthrene dihydrodiol, 267
Phenanthrene, metabolism, 1041
Phenanthrene, (methyl substituted), 1041
Phosphorescence, room temperature, 565
Phosphorimetry, room temperature, 139
Polyacrylamide gel electrophoresis (PAGE), 633, 955
Polyphenylsulfone (PS-176), 25
Priority pollutant protocol, EPA recommended, 75
Protease inhibitors, 955
Protoporphorin, 233
Proxy compounds, 565
Pyrene, 199, 379

Risk assessment, 435, 917
RNA synthesis, 287

Sampling, ambient air, 107, 127, 139, 193, 435, 543, 579, 791, 829, 983, 1077
Sampling, drinking water, 333
Sampling, filter, 771
SASS train, 617
Seaweed, 367
Sediments, aquatic, 395
Sediments, marine, 367

Selected ion mass spectrometry (SIMS), 589
Sencar mouse, 503, 753
Serum albumin-BaP, 319
Silica, 979
Silica gel, 1
Skin contamination, 565
Skin painting, CD-1 mice, 675, 935
Skin painting, Ha/ICR mice, 417
Skin painting, Swiss albino mice, 1041
Skin painting, sencar mice, 503, 753
Smoker/nonsmoker comparisons, 299
Soil content, 395
Spectroscopy, fluorescent, 9, 319, 663, 689, 771, 863, 879, 983, 1059
Spectroscopy, matrix isolation (MI), 1005
Structure-activity relationship, 345
Synthoil, 139

Tar, coal derived gasification, 59
TCDD-2,3,7,8-tetrachlorodibenzo-p-dioxin, 503, 935
Tobacco smoke, 243
TPA (12-O-tetra-decanoylphorbol-13-acetate), 753
Trace atmospheric gas analyzer (TAGA™), 199
Tracers, PAH, 617
Trachea, human, 89
Tumorigenicity, 675
Tumor induction, 215, 503, 753, 935, 1041

Waste incineration, 973
Water, drinking, 435
Water, natural, 863
Water, waste, 333, 367, 863
Wood stoves, 973

**Notes**

**Notes**

**Notes**

**Notes**

**Notes**

**Notes**

**Notes**

**Notes**

**Notes**

**Notes**

## DATE DUE

| | | | |
|---|---|---|---|
| OCT 0 2 1982 | | | |
| NOV 0 5 1987 | | | |
| DEC 0 9 1983 | | | |
| DEC 2 1 1982 | | | |
| | | | |
| | | | |
| | | | |
| | | | |
| | | | |
| | | | |
| | | | |
| | | | |
| | | | |
| | | | |

DEMCO 38-297